McGraw-Hill's NEW MCAT

MEDICAL COLLEGE ADMISSION TEST

George J. Hademenos, Ph.D.

Candice J. McCloskey, Ph.D.

Shaun Murphree, Ph.D.

Jennifer M. Warner, M.S.

Kathy A. Zahler, M.S.

Thomas A. Evangelist, M.A. Contributor

McGraw-Hill

New York | Chicago | San Francisco | Lisbon | London | Madrid | Mexico City
Milan | New Delhi | San Juan | Seoul | Singapore | Sydney | Toronto

The McGraw-Hill Companies

Library of Congress Cataloging-in-Publication Data

McGraw-Hill's new MCAT / George Hademenos ... [et al.].
p. cm.
ISBN-13: 978-0-07-147076-6 (alk. paper)
ISBN-10: 0-07-147076-X (alk. paper)
1. Medical colleges—United States—Entrance examinations—Study guides.
I. Hademenos, George J. II. Title: New MCAT.
[DNLM: 1. Medicine—Examination Questions. W 18.2 M4788 2007]
R838.5.M36 2007
610.76—dc22 2006035082

1 2 3 4 5 6 7 8 9 0 QPD/QPD 0 1 2 1 0 9 8 7 6

ISBN-13: P/N 978-0-07-147078-0 of set
978-0-07-147076-6

ISBN-10: P/N 0-07-147078-6 of set
0-07-147076-X

Printed and bound by Quebecor/Dubuque.

Contents

PART III: REVIEWING MCAT GENERAL CHEMISTRY

PART IV: MCAT VERBAL REASONING AND WRITING

PART V: REVIEWING MCAT BIOLOGY

PART VI: REVIEWING MCAT ORGANIC CHEMISTRY

SECTION I: STRUCTURE

SECTION V: CHARACTERIZATION

How to Use This Book

Welcome to *McGraw-Hill's New MCAT*. You've made the decision to pursue a medical career, you've studied hard, you've taken and passed the most difficult science courses, and now you must succeed on this very tough exam. We're here to help you.

This book has been created by a dedicated team of scientists, teachers, and test-prep experts. Together, they have helped thousands of students score high on all kinds of exams, from rigorous science tests to difficult essay-writing assignments. They have pooled their knowledge, experience, and test-taking expertise to make this the most effective MCAT preparation program available.

McGraw-Hill's New MCAT contains a wealth of features to help you do your best. In the days, weeks, or months before you take the test, you can substantially improve your chances of scoring high by using the book and CD-ROM as follows:

- **In Part I, All About the MCAT, you'll learn basic facts about the test, be familiarized with the computerized testing format, and learn about the kinds of questions you're going to encounter.** You'll also find important tips about pacing and guessing. In addition, you can also review some basic test-taking strategies to keep in mind throughout all phases of the exam.
- **In Parts II through VI, you can review every subject you must know for the MCAT.** Parts II, III, V, and VI present detailed coverage of all tested topics in physics, general chemistry, biology, and organic chemistry. Special "Cram Sessions" at the end of each chapter reiterate key terms and concepts for quick and effective review. Many also offer valuable advice for tackling MCAT questions in the particular topic area. Each science review ends with "On Your Own" practice quizzes that you can use as a fast, efficient way to test your mastery of the subject. Part IV shows you how to maximize your scores on the Verbal Reasoning section. It also covers the Writing Sample—the MCAT section most dreaded by many pre-med students. Here you'll find numerous sample essays illustrating exactly what the graders are looking for. Read the samples carefully—and learn how you can score high on this difficult portion of the MCAT.
- **On the CD-ROM, you'll find two complete sample MCATs for practice.** These tests provide you with questions spanning the entire range of subjects and difficulty levels you're likely to find on the MCAT. They also give you an authentic test-taking experience much like what you'll encounter on the actual MCAT. You should take

each test under actual test conditions: set aside the time you'll need to take the entire test at one sitting, screen out distractions, and concentrate on doing your best. Of course, these practice tests can provide only an approximation of how well you will do on the actual MCAT. However, if you approach them as you would the real test, they should give you a very good idea of how well you are prepared. After you take each test, read through the explanations for each question, paying special attention to those you answered incorrectly or had to guess on. If necessary, go back and reread the corresponding sections in the subject reviews in this book.

Different people have different ways of preparing for a test like the MCAT. You must find a preparation method that suits your schedule and your learning style. We have tried to make *McGraw-Hill's New MCAT* flexible enough for you to use in a way that works best for you, but to succeed on this extremely rigorous exam, there's no substitute for serious, intensive review and study. The more time and effort you devote to preparing, the better your chances of achieving your MCAT goals.

Be sure to visit the MCAT*easy* website. Here you'll find hundreds of additional practice questions in all the MCAT subject areas. Use these questions to test your mastery of each subject and to build test-taking speed and accuracy. Site access is free to all purchasers of *McGraw-Hill's New MCAT*. To visit the site, go to

http://www.mcateasy.com

PART I

ALL ABOUT THE MCAT

CHAPTER 1

Introducing the MCAT

Read This Chapter to Learn About:

MCAT BASICS

The Medical College Admission Test (MCAT) is a standardized exam that is used to assess applicants to medical schools. The test is created by the Association of American Medical Colleges (AAMC) in cooperation with its member schools. It is required as part of the admissions process by most U.S. medical schools. The test is administered by Thomson Prometric, a private firm that is a leading provider of technology-based testing and assessment services.

The questions on the MCAT are basically designed to measure your problem-solving and critical-thinking skills. Two test sections assess your mastery of basic concepts in biology, general chemistry, organic chemistry, and physics. For most questions, choosing the correct answer requires more than just a rote response; you must calculate a solution, interpret and evaluate given data, or apply a particular scientific principle to a given situation. There is also a third test section that requires you to read passages on general topics and answer questions by applying your reasoning skills to what you have read. Finally, a fourth test section requires you to write two essays that describe and explain your thinking on given general topics.

According to the AAMC, the skills tested on the MCAT are those identified by medical professionals and educators as essential for success in medical school and in a career as a physician. The importance of the sciences is self-evident; the inclusion of verbal reasoning and writing skills is intended, according to the AAMC, to "encourage undergraduates with broad educational backgrounds to consider careers in the health professions and [to] stimulate premedical students to investigate a wide variety of course offerings outside the natural sciences."

THE NEW COMPUTERIZED MCAT

The MCAT was formerly given in two formats: as a paper-and-pencil exam and as a computer-based test (CBT). However, since January 2007 the test has been offered only on computer. When you take the computerized MCAT, you view the questions on a computer screen and indicate your answers by clicking on on-screen answer ovals. You type your essays using the computer keyboard.

As you work through the on-screen questions, you are able to highlight relevant portions of the reading passages for easy reference. You are also able to strike out answer choices that you know are incorrect. This helps you use the process of elimination to pick the correct answer. You are also allowed to make notes on scratch paper (although all your notes are collected at the end of the test). Within each test section, you are able to go back, review questions you have already answered, and change your answer if you decide to do so. However, once you have finished a test section, you cannot go back to it and make any changes. When you type your essays for the Writing Sample, you are able to use the computer's "Cut," "Paste," and "Copy" functions. However, you are not able to spell-check your work.

Don't be concerned if you're not a whiz with computers; the skills required are minimal, and in any case, on test day you have the opportunity to access a computer tutorial that shows you exactly what you need to do.

WHERE AND WHEN TO TAKE THE MCAT

The computerized MCAT is offered at approximately 275 sites in the United States (including the U.S. territories of Puerto Rico and the Virgin Islands) and at 12 sites in Canada. All of these sites are testing labs operated by Thomson Prometric. The test is also offered at numerous locations outside North America, including sites in Europe, Great Britain, the Middle East, Africa, Asia, and Australia.

There are 22 test dates every year. Two of the dates are in January, and the rest are in the period from April through early September. Most test dates are weekdays, but a few are Saturdays. On some dates, the test is given only in the morning; on others, it is given only in the afternoon. On a few dates, the test is given in both morning and afternoon sessions.

It is a good idea to take the MCAT in the spring or summer of the year before the fall in which you plan to enroll in medical school. That way, you have enough time to submit your scores to meet the schools' application deadlines.

For up-to-date lists of testing sites and also for upcoming test dates, make sure to check the official MCAT website at www.aamc.org/mcat.

HOW TO REGISTER FOR THE MCAT

You can register for the MCAT online at www.aamc.org/mcat. Online registration for each test date begins six months prior to that date. Registration is available until two weeks before the test date. It's a good idea to register early, because seating at the testing centers may be limited and you want to make sure you get a seat at the center of your choice. When you register, you are charged a fee, which you can pay by credit card. If you wish to change your test date, you can do so online.

TAKING THE MCAT MORE THAN ONCE

If your MCAT score is lower than expected, you may want to take the test again. You can take the MCAT up to three times in the same year. However, the AAMC recommends retesting only if you have a good reason to think that you will do better the next time. For example, you might do better if, when you first took the test, you were ill, or you made mistakes in keying in your answers, or your academic background in one or more of the test subjects was inadequate.

If you are considering retesting, you should also find out how your chosen medical schools evaluate multiple scores. Some schools give equal weight to all MCAT scores; others average scores together, and still others look only at the highest scores. Check with admissions officers before making a decision.

YOUR MCAT SCORES

When you take the MCAT, your work on each of the four test sections first receives a "raw score." On the three multiple-choice sections (Physical Sciences, Biological Sciences, and Verbal Reasoning), the raw score is calculated based on the number of questions you answer correctly. No points are deducted for questions answered incorrectly. On the Writing Sample, the raw score is determined by adding the scores that readers assign to each of the two essays that you write. Because two readers read each essay, the raw score represents the sum of four readers' scores. For more information on how the Writing Sample is scored, see Part IV of this book, "MCAT Verbal Reasoning and Writing."

Each raw score is then converted into a scaled score. Using scaled scores helps to make test-takers' scores comparable from one version of the MCAT to another. For the three multiple-choice sections, scaled scores range from 1 (lowest) to 15 (highest). For the Writing Sample, scaled scores are reported on an alphabetic scale that ranges from J (lowest) to T (highest).

Your score report will be mailed to you approximately 30 days after you take the MCAT. You will also be able to view your scores on the online MCAT Testing History (THx) System as soon as they become available. (For details on the THx system, see the MCAT website.) MCAT score reports also include percentile rankings that show how well you did in comparison to others who took the same test.

HOW MEDICAL SCHOOLS USE MCAT SCORES

Medical college admission committees emphasize that MCAT scores are only one of several criteria that they consider when evaluating applicants. When making their decisions, they also consider students' college and university grades, recommendations, interviews, and involvement and participation in extracurricular or health care–related activities that, in the opinion of the admission committee, illustrate maturity, motivation, dedication, and other positive personality traits that are of value to a physician. If the committee is unfamiliar with the college you attend, they may pay more attention than usual to your MCAT scores.

There is no hard-and-fast rule about what schools consider to be an acceptable MCAT score. A few schools accept a score as low as 4, but many require scores of 10 or higher. To get into a top-ranked school, you generally need scores of 12 or higher. A high score on the Writing Sample may compensate for any weaknesses in communication skills noted on the application or at an interview.

Note that many medical schools do not accept MCAT scores that are more than three years old.

REPORTING SCORES TO MEDICAL SCHOOLS

Your MCAT scores are automatically reported to the American Medical College Application Service (AMCAS), the nonprofit application processing service used by nearly all U.S. medical schools. When you use this service, you complete and submit a single application, rather than separate applications to each of your chosen schools. Your scores are submitted to your designated schools along with your application. There is a fee for using AMCAS. If you wish to submit your scores to other application services or to programs that do not participate in AMCAS, you can do so through the online MCAT Testing History (THx) System.

FOR FURTHER INFORMATION

For further information about the MCAT, visit the official MCAT website at

www.aamc.org/mcat

For questions about registering for the test, reporting and interpreting scores, and similar issues, you may also contact:

MCAT CBT Program Office
1000 Lancaster Street
Baltimore, MD 21202

CHAPTER 2

Test Format and Structure

Read This Chapter to Learn About:

- ➤ The Format of the Test
- ➤ The Physical Sciences and Biological Sciences Sections
- ➤ The Verbal Reasoning Section
- ➤ The Writing Sample

THE FORMAT OF THE TEST

The MCAT consists of four separately timed sections. Add in three 10-minute breaks between the sections, and the total testing time comes to 4 hours 50 minutes. The actual testing session takes even longer because it also includes a pre-test optional computer tutorial (5 minutes) and a post-test survey (5 minutes). In all, the AAMC advises you to be prepared to spend about 5½ hours at the testing center.

The four test sections are always given in the same order. The following chart shows the sections in order, with the number of questions and time allowed for each section.

THE PHYSICAL SCIENCES AND BIOLOGICAL SCIENCES SECTIONS

The Physical Sciences and Biological Sciences sections of the MCAT test your mastery of the concepts and principles of physics, general chemistry, biology, and organic chemistry as taught in typical undergraduate college courses with laboratory sessions. The test-makers say that the depth of knowledge required is "limited" and that you do not need to have taken advanced science courses to be prepared for the test. Nevertheless, you must be familiar with all the basic concepts of the subjects tested, and that includes a very large amount of material. And just as important, to answer most MCAT

MCAT: Format of the Test		
Section	**Number of Questions**	**Time Allowed (minutes)**
Computer Tutorial (optional)		
Physical Sciences	52 multiple-choice questions (7–9 passages with 4–8 questions each, plus 10 stand-alone questions)	70
	Break: 10 minutes	
Verbal Reasoning	40 multiple-choice questions (5–6 passages with 5–8 questions each)	60
	Break: 10 minutes	
Writing Sample	2 essay questions (30 minutes for each)	60
	Break: 10 minutes	
Biological Sciences	52 multiple-choice questions (7–9 passages with 4–8 questions each, plus 10 stand-alone questions)	70
Survey (no test content)		
Totals	**146**	**290 (= 4 hours, 50 minutes)**

science questions, you must be able to use that knowledge and your own reasoning skills to analyze scientific information, interpret scientific data, and calculate the solutions to scientific problems. Your reasoning skills are being tested just as much as your knowledge, because in the opinion of the test-makers, those skills are essential for a physician.

Each of the two sections contains the following:

- 7–9 reading passages with 4–8 multiple-choice questions based on each passage (52 passage-based questions in all)
- 10 stand-alone multiple-choice questions

Passage-Based Questions

The passages in the Physical Sciences and Biological Sciences sections are each about 250 words long. Each one describes a situation or problem having to do with a scientific topic. Many also include data presented in the form of a diagram, a chart, or a graph. The questions based on the information in the passage may ask you to do any one of the following:

- Use the information in the passage to calculate the solution to a problem.
- Apply the concept described in the passage to a new situation.
- Determine the probable cause of the situations or phenomena described.
- Identify, interpret, or compare the data presented in a diagram, chart, or graph.

- Select an appropriate method for solving a scientific problem.
- Analyze or interpret the results of a research project.
- Evaluate the validity of an argument for or against a given point of view.

EXAMPLE

The following is an example of an MCAT Biological Sciences passage and related question.

Human Immunodeficiency Virus (HIV) is known to infect T cells and macrophages possessing the CD4 receptor. In addition to the CD4 receptor, there are a variety of co-receptors that are also needed for certain strains of HIV to infect cells. One of these co-receptors is called CCR5 and is needed for the most common strains of HIV (called R5) to infect their host cells. The more CCR5 receptors a cell has, the greater the rate of infection for the cell. Some individuals have the delta32 allele of CCR5, which leads to a decreased risk of HIV infection.

Women are at a much higher risk for contracting HIV from a male partner than a male is for contracting HIV from a female partner. Some of the increased risk for women is dependent on sex hormones. It has been hypothesized that women have different risks of contracting HIV at different points in their reproductive cycles. For example, women may be at a greater risk of contracting HIV following ovulation during the last two weeks of their cycle as opposed to the first two weeks of their cycle prior to ovulation. The difference between risk of infection before and after ovulation relates to sex hormones produced at various points in the reproductive cycle. It seems that the estrogen produced prior to ovulation provides a somewhat protective, although certainly not absolute, role against HIV infection. For women who do become infected with HIV, studies have shown that women in the first 3–5 years after HIV infection carry lower HIV viral loads than men perhaps due to the influence of estrogen. This suggests a major role for sex hormones in the infection and progression of HIV.

1. According to the hypothesis presented concerning estrogen's role in HIV infection, which of these individuals would be at the least risk for HIV infection?

A. A female producing elevated levels of testosterone.

B. A female with an ovarian tumor that causes increased secretion of estrogen.

C. A female who has had her ovaries removed.

D. A male with a reduced testosterone level.

Stand-Alone Questions

The 10 stand-alone questions in each of the two sections are not based on any passage. These questions typically ask you to use your knowledge of a scientific concept or principle to calculate a solution or choose the correct answer from several alternatives.

TOPICS TESTED: PHYSICS

The test-makers list the following physics concepts as topics for questions in the MCAT Physical Sciences section:

- Translational motion
- Force and motion, gravitation
- Equilibrium and momentum
- Work and energy
- Waves and periodic motion
- Sound
- Fluids and solids
- Electrostatics and electromagnetism
- Electronic circuit elements
- Light and geometric optics
- Atomic and nuclear structure

See Part II of this book for a complete review of these topics.

TOPICS TESTED: GENERAL CHEMISTRY

The test-makers list the following general chemistry concepts as topics for questions in the MCAT Physical Sciences section:

- Electronic structure and periodic table
- Bonding
- Phases and phase equilibria
- Stoichiometry
- Thermodynamics and thermochemistry
- Rate processes in chemical reactions—kinetics and equilibrium
- Solution chemistry
- Acids and bases
- Electrochemistry

See Part III of this book for a complete review of these topics.

TOPICS TESTED: BIOLOGY

The test-makers list the following biology concepts as topics for questions in the MCAT Biological Sciences section:

- Molecular biology: Enzymes and metabolism
- Molecular biology: DNA and protein synthesis
- Molecular biology: Eukaryotes
- Genetics
- Microbiology
- Generalized eukaryotic cell

- Specialized eukaryotic cells and tissues
- Nervous and endocrine systems
- Circulatory, lymphatic, and immune systems
- Respiration system
- Skin system
- Digestive and excretory systems
- Muscle and skeletal systems
- Reproductive system and development
- Evolution

See Part V of this book for a complete review of these topics.

TOPICS TESTED: ORGANIC CHEMISTRY

The test-makers list the following organic chemistry concepts as topics for questions in the MCAT Biological Sciences section:

- The covalent bond
- Molecular structure and spectra
- Separations and purifications
- Hydrocarbons
- Oxygen-containing compounds
- Amines
- Biological molecules

See Part VI of this book for a complete review of these topics.

The Math You Need to Know for the MCAT

Many MCAT Physical Sciences and Biological Sciences questions require you to use mathematical knowledge to solve problems. Following is a list of the math concepts you must know for the MCAT:

- Basic arithmetic
- Ratio and proportion
- Arithmetic mean (average) and range of a data set
- Estimating square roots
- Basic algebra
- Exponents
- Logarithms (natural and base 10)
- Scientific notation
- Solving systems of equations
- Quadratic equations

- Graphing functions
- Calculating slope or rate of change
- Reciprocals
- Basic trigonometric functions: sine, cosine, and tangent
- Right triangles
- Metric units of measure
- Probability
- Vector addition, vector subtraction, and the right-hand rule
- Significant digits of measurement
- Relative magnitude of experimental error
- Understanding of basic statistical concepts: standard deviation, association, and correlation (calculating standard deviation is not required)

THE VERBAL REASONING SECTION

The Verbal Reasoning section of the MCAT is intended to measure how well you understand, analyze, and apply information presented in prose texts. This test section contains the following:

- 5 or 6 reading passages, each of which is 500–600 words long. The passages are on topics in the humanities, the social sciences, and areas of the natural sciences not tested elsewhere on the MCAT.
- 6–9 multiple-choice questions per passage, for a total of 40 questions.

The passages are taken from texts that you most likely have never seen before and may be on topics with which you are unfamiliar. However, do not be alarmed; you are not asked questions that require any prior knowledge of the passage topic. Everything you need to know to answer each question is included in the passage.

The questions in the Verbal Reasoning section may ask you to do any of the following:

- Identify the main idea of the passage.
- Analyze an argument presented in the passage and judge its validity.
- Use information from the passage to solve a given problem.
- Determine cause-and-effect relationships for events or conditions described in the passage.
- Evaluate a claim made in the passage based on the strength of the evidence or argument provided to support that claim.
- Identify the reasons or evidence offered in support of a particular viewpoint.
- Recognize stated or unstated assumptions that underlie a viewpoint presented in the passage.
- Identify new facts or results that might undermine a conclusion presented in the passage.

- Apply information from the passage to a new situation.
- Determine the meaning of an unfamiliar word based on its context.

The test-makers have said that the question sets in this section are presented in order from easiest to most difficult. Be aware that this may be the case with the Verbal Reasoning questions you see on test day.

Example

Following is an example of an MCAT Verbal Reasoning question set.

According to Professor John MacKinnon, certain places are more likely than others to house new or rediscovered species. First, scientists should look in areas that are geologically stable. Areas with regular earthquakes or volcanic eruptions are less likely to contain species that go back thousands of years because, obviously, the frequent upheavals are not conducive to steady growth or a comfortable way of life. Second, scientists are most likely to find undiscovered or rediscovered species in remote, isolated areas. Cultivated areas are inhospitable to many animals. Often, cultivation eliminates the trees or shrubs that house and protect animal life. A stable climate is a third thing to look for. Stability of climate ensures that the animals that live in the region have had no reason to leave for warmer or wetter environments.

Although most of the recent discoveries and rediscoveries of animal species have taken place in tropical, humid regions, some have occurred in rarely explored mountain habitats. A fourth key thing scientists should look for, according to MacKinnon, is an area with a variety of unusual species that are specific to that area. Of course, this requirement refers back to requirement number two, isolation. An area that is very isolated or difficult to access will naturally have species that cannot be found anywhere else.

In 1994 American biologist Peter Zahler took his second trip to the isolated valleys of the Diamer region in northern Pakistan. He was searching for the woolly flying squirrel, a dog-sized squirrel last seen in 1924. As he describes it, he had narrowed down his search to two animals, using the criteria *mammal, not too small, relatively unknown,* and *in need of conservation intervention.* Because he did not wish to spend time in the steamy environment of the Congo basin, he eliminated the aquatic genet from his list and decided to concentrate on the woolly flying squirrel, which lived in the high Himalayas where there were fewer bugs and no tropical diseases. Working with a guide, Zahler moved from valley to valley. He quizzed the local residents about the squirrel. Many recalled hearing tales about the strange animal, but all insisted that it was extinct.

Zahler continued to set traps, listen to squirrel legends, and ask questions. He dodged avalanches, skirted valleys patrolled by armed warlords, and crossed mountains on narrow goat paths. Policemen told him that the squirrels indeed existed in caves above the valley, and that they excreted a substance, *salajit,* that was used as an aphrodisiac. One day, two large, machine-gun-toting men entered his camp. After some small talk, one asked whether he would be willing to pay for a live squirrel. The pair claimed to be *salajit* collectors. Zahler agreed to pay top dollar, never envisioning that two hours later, he would be presented with a woolly flying squirrel in a bag. Following that surprise, Zahler located evidence of live squirrels throughout the region. The squirrel certainly obeyed all of MacKinnon's regulations. Despite frequent avalanches, the Himalayas are a geologically stable area with a harsh but stable climate and a number of rare species. Not only were the squirrels' caves remote and isolated, it was often necessary to rappel down or climb up sheer cliffs to reach them.

Each rediscovery gives scientists a vital second chance to protect and preserve our rarest, most fragile species. With scientists and students from Pakistan, Zahler began work on a program to protect the squirrel's habitat. Only the advent of war in Afghanistan forced him to abandon the area, and he remains connected to scientists in Pakistan who are continuing the effort.

1. According to the passage, all of the following statements are true EXCEPT

A. Woolly flying squirrels excrete an unusual substance.

B. The Himalayas contain a variety of unique species.

C. Animals in a drought may leave for a wetter habitat.

D. The woolly flying squirrel vanished for 70 years.

2. The discussion of the aquatic genet shows primarily that

A. some mammals are unlikely ever to be rediscovered.

B. large mammals are more easily tracked and rediscovered.

C. endangered mammals are found in many different biomes.

D. tropical diseases affect a variety of endangered species.

3. Which of MacKinnon's requirements does the home of the woolly flying squirrel meet?

I. Isolation

II. Geographic stability

III. Climatic stability

A. I only

B. II only

C. I and III only

D. I, II, and III

4. The author suggests that Zahler decided to search for the woolly flying squirrel because it

A. lived in an area that was easily accessible.

B. was neither too small nor too tropical.

C. had been spotted recently by *salajit* collectors.

D. was a legendary creature of great renown.

5. According to the passage, which of these locations would be likely to house a new species?

I. The jungles of Borneo

II. The mountains of western Iran

III. The edges of the eastern Sahara

A. I only

B. II only

C. I and II only

D. II and III only

6. According to the author, rediscovery gives scientists a second chance to protect and preserve rare species. Which of the following statements, if true, would most WEAKEN this argument?

A. Ornithologists hoping for a second sighting of the long-lost ivory-billed woodpecker have triggered a stampede of cameras and news media to the fragile forests where it was recently spotted.

B. The Audubon Society has set aside part of a forest on Oahu as a potential breeding ground for an endangered warbler once common on several of the Hawaiian Islands.

C. The northern hairy-nosed wombat lives only in one known area, a protected section of Epping Forest National Park in central Queensland, Australia.

D. Rodent control programs on the Galapagos Islands have managed to stem a threat to endangered petrels, and now 90 percent of their nests result in a successful hatching out of eggs.

See Part IV of this book to review and practice the skills you'll need to succeed on the MCAT Verbal Reasoning section.

THE WRITING SAMPLE

According to the test-makers, the purpose of the Writing Sample is to assess your ability to communicate clearly, logically, and persuasively—a skill that is essential for health care providers who interact with patients, colleagues, and the public at large. The Writing Sample was added to the MCAT at the request of medical school deans and faculty specifically to measure students' abilities in this area.

For the Writing Sample, you are required to write two short essays. The time limit for each one is 30 minutes. For each one, you are given a **prompt**, a short statement that typically presents a particular viewpoint or discusses a policy. Your job is to write a response that accomplishes the following three tasks:

- Explains the meaning of the statement
- Describes a situation that refutes or contradicts the statement, or in which the given principle does not apply
- Proposes a way in which the original statement and your counterexample might be resolved, perhaps by referring to a more general idea or principle

Within the 30-minute time limit, you must organize your thoughts, plan your writing (perhaps by jotting down a quick outline on scratch paper), and type your essay using the computer keyboard. You do not necessarily have to produce a lengthy piece of writing, but you must write enough to accomplish all three of the tasks listed in a way that is cogent, logical, and effective.

MCAT Writing Samples are scored holistically; that is, the readers who score them take into account the depth of your ideas, the strength of your argument, and the clarity of your expression and assign a single rating that assesses the essay as a whole. The scorers recognize that you are writing under time pressure and that your essay is really little more than a first draft, so they do not necessarily give your work a lower rating if you make some errors of grammar or punctuation.

Example

Following is an example of an MCAT Writing Sample prompt.

> Consider this statement:
>
> **There is nothing you can learn from television today that you could not learn better and more thoroughly from the books at the downtown library.**
>
> Write a unified essay in which you perform the following tasks. Explain what you think the above statement means. Describe a specific situation in which you might learn better and more thoroughly from television than from books. Discuss the features that you think determine whether books or television provide a better overall learning experience.

See Part IV of this book to review and practice the skills you need to succeed on the MCAT Writing Sample.

CHAPTER 3

General Test-Taking Strategies

Read This Chapter to Learn About:

- General Strategies for Answering MCAT Questions
- Coping with Exam Pressure

This chapter presents some general test-taking strategies that apply to all the multiple-choice questions on the MCAT. These strategies can help you gain valuable points when you take the actual test.

GENERAL STRATEGIES FOR ANSWERING MCAT QUESTIONS

Take Advantage of the Multiple-Choice Format

All of the questions on the MCAT (except for the Writing Sample essay questions) are in the multiple-choice format, which you have undoubtedly seen many times before. That means that for every question, the correct answer is right in front of you. All you have to do is pick it out from among three incorrect choices, called "distracters." Consequently, you can use the process of elimination to rule out incorrect answer choices. The more answers you rule out, the easier it is to make the right choice.

Answer Every Question

Recall that on the MCAT, there is no penalty for choosing a wrong answer. Therefore, if you do not know the answer to a question, you have nothing to lose by guessing. So make sure that you answer every question. If time is running out and you still have not answered some questions, make sure to enter an answer for the questions that you have not tackled. With luck, you may be able to pick up a few extra points, even if your guesses are totally random.

Make Educated Guesses

What differentiates great test takers from merely good ones is the ability to guess in such a way as to maximize the chance of guessing correctly. The way to do this is to use the process of elimination. Before you guess, try to eliminate one or more of the answer choices. That way, you can make an educated guess, and you have a better chance of picking the correct answer. Odds of one out of two or one out of three are better than one out of four!

Go with Your Gut

In those cases where you're not 100% sure of the answer you are choosing, it is often best to go with your gut feeling and stick with your first answer. If you decide to change that answer and pick another one, you may well pick the wrong answer because you have over-thought the problem. More often than not, if you know something about the subject, your first answer is likely to be the correct one.

Take Advantage of Helpful Computer Functions

On the computerized MCAT, you have access to certain computer functions that can make your work easier. As you work through the on-screen questions, you are able to highlight relevant portions of the reading passages. This helps you save time when you need to find facts or details to support your answer choices. You are also able to cross out answer choices that you know are incorrect. This helps you use the process of elimination to pick the correct answer.

Take Advantage of Helpful Word Processing Functions

When you type your essays for the Writing Sample, you are able to use the computer's "Cut," "Paste," and "Copy" functions. Using these functions can help you to arrange your ideas in logical order and to do so quickly and efficiently. Note, however, that you are not able to use the computer's spell-check function. You must rely on your own spelling and grammar skills.

Use the Scratch Paper Provided

The MCAT is an all-computerized test, so there is no test booklet for you to write in. However, you are given scratch paper, so use it to your advantage. Jot down notes, make calculations, and write out an outline for each of your essays. Be aware, however, that you cannot remove the scratch paper from the test site. All papers are collected from you before you leave the room.

Because you cannot write on the actual MCAT, don't get into the habit of writing notes to yourself on the test pages of this book. Use separate scratch paper instead. Consider it an opportunity to learn to use scratch paper effectively.

COPING WITH EXAM PRESSURE

Keep Track of the Time

Make sure that you're on track to answer all of the questions within the time allowed. With so many questions to answer in a short time period, you're not going to have a lot of time to spare. Keep an eye on your watch or on the computerized timer provided.

Do not spend too much time on any one question. In the Physical Sciences and Biological Sciences sections, you have only 70 minutes to read 7 to 9 passages and answer 52 questions. In the Verbal Reasoning section, you have only 60 minutes to read 5 or 6 passages and answer 40 questions. If you find yourself stuck for more than a minute or two on a question, then you should make your best guess and move on. If you have time left over at the end of the section, you can return to the question and review your answer. However, if time runs out, don't give the question another thought. You need to save your focus for the rest of the test.

Don't Panic If Time Runs Out

If you pace yourself and keep track of your progress, you should not run out of time. If you do, however, don't panic. Because there is no guessing penalty and you have nothing to lose by doing so, enter answers to all the remaining questions. If you are able to make educated guesses, you will probably be able to improve your score. However, even random guesses may help you to pick up a few points. In order to know how to handle this situation if it happens to you on the test, make sure you observe the time limits when you take the practice tests on the CD. Guessing well is a skill that comes with practice, so incorporate it into your preparation program.

If Time Permits, Review Questions You Were Unsure of

Within each test section, the computer allows you to return to questions you have already answered and change your answer if you decide to do so. (However, once you have completed an entire section, you cannot go back to it and make changes.) If time permits, you may want to take advantage of this function to review questions you were unsure of, or to check for careless mistakes.

PART II

REVIEWING MCAT PHYSICS

CHAPTER 1

Mathematics Fundamentals

Read This Chapter to Learn About:

- ➤ Ratio, Proportion, and Percentage
- ➤ Functions: Polynomial, Power, Trigonometric, Exponential, and Logarithmic
- ➤ Graphical Representation of Data
- ➤ Simultaneous Solution of Equations
- ➤ Accuracy, Precision, and Propagation of Errors
- ➤ Significant Figures
- ➤ Probability and Statistics

This physics review begins with a review of the fundamental concepts of college-level mathematics, which serve as an important tool in qualitatively describing relationships and quantitatively calculating values found in all the physics topics covered in this review.

RATIO, PROPORTION, AND PERCENTAGE

Ratio and proportion are mathematical concepts used in physics not only to express the units of a particular quantity (such as the ratio of a speed of 50 miles per hour = 50 miles to 1 hour), but also to convert between different units (such as the proportion of the two expressions of speed: $\frac{50 \text{ miles}}{1 \text{ hour}} = \frac{73 \text{ feet}}{1 \text{ second}}$). In simple terms, a **ratio** is defined as the comparison of two quantities and can be stated as $\frac{x}{y}$ or $x{:}y$. A **proportion** is typically stated as the equality of two ratios, $\frac{a}{b} = \frac{c}{d}$.

A **percentage** defines a quantity as a fraction of a whole or total amount (100%). If a perfect score on an exam is graded as a percentage and thus equivalent to 100%, a student who scores 80% on an exam correctly answered 8 out of every 10 questions posed on the exam. This percentage can be expressed as a ratio of $\frac{80}{100}$ or a proportion of 80:100 = 8:10.

FUNCTIONS: POLYNOMIAL, POWER, TRIGONOMETRIC, EXPONENTIAL, AND LOGARITHMIC

A **function** is defined as a mathematical one-to-one relationship between elements of a given set $X = \{x\}$, referred to as the **domain**, and another set $Y = \{y\}$, referred to as the **range**. The function describes a mathematical relationship between a dependent variable y and an independent variable x, which can represent any unique number within a specified interval. A function, denoted typically as $y = f(x)$, allows you to characterize mathematically a physical system in terms of variables that directly influence the system's behavior and stability. Types of functions demonstrated throughout this review of physics include polynomial functions, power functions, trigonometric functions, exponential functions, and logarithmic functions.

Polynomial Functions

Polynomial functions are the most common of the five types of functions and of the form

$$y = f(x) = a_0 + a_1x + a_2x^2 + a_3x^3 + \cdots + a_nx^n$$

where a_0, a_1, a_2, a_3, a_n are coefficients (constants) and n is an integer that describes the degree or order of the polynomial. Special cases of the polynomial function include the first-order polynomial function (linear function) given by

$$y = f(x) = a_0 + a_1x \qquad a_1 \neq 0$$

Another special case of the polynomial function is the second-order polynomial function (quadratic equation) given by

$$y = f(x) = a_0 + a_1x + a_2x^2 \qquad a_1, a_2 \neq 0$$

Consider the quadratic equation expressed in the form $Ax^2 + Bx + C$ and equal to zero, or

$$Ax^2 + Bx + C = 0$$

The solution x can be determined for the quadratic equation from the following formula:

$$x = \frac{-B \pm \sqrt{B^2 - 4AC}}{2A}$$

Two items to note:

1. It is possible that the solution from the above formula may result in complex roots (i.e., the square root of a negative quantity).
2. The calculation from the formula reveals two possible solutions. When applied to a physics problem, it becomes readily clear by either the sign or the magnitude of the values that one solution is not physically realistic or possible, with the other solution being the correct solution. In some instances (e.g., projectile motion), both solutions may be realistic.

The graphical representation of polynomial functions is illustrated in Figure 1-1. In physics, applications of polynomial functions include the following:

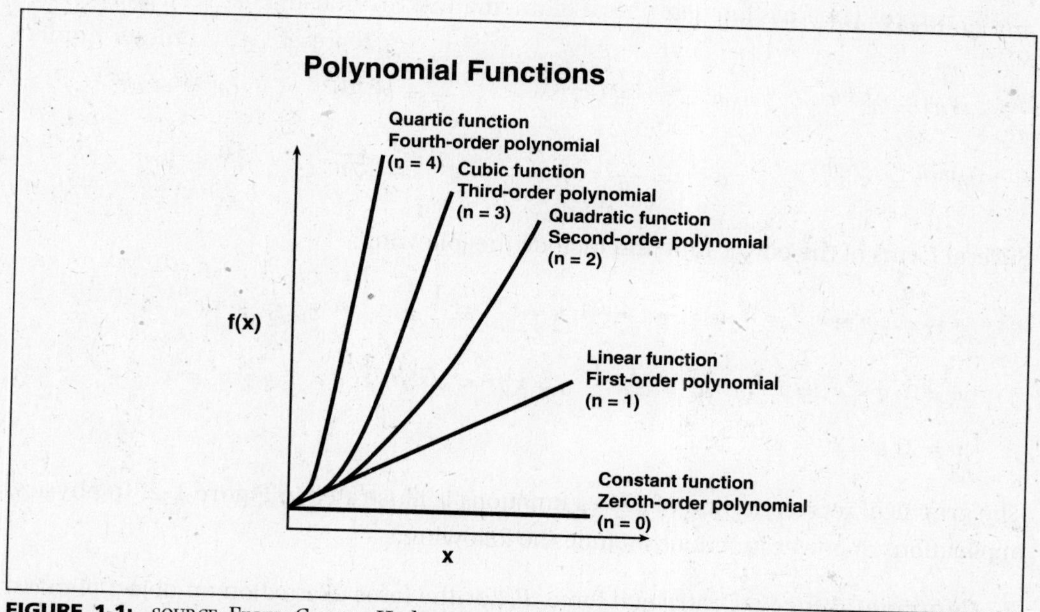

FIGURE 1-1: SOURCE: From George Hademenos, *Schaum's Outline of Physics for Pre-Med, Biology, and Allied Health Students*, McGraw-Hill, 1998; reproduced with permission of The McGraw-Hill Companies.

1. **Linear momentum**. The linear momentum $\boldsymbol{p}$ of an object of mass m moving at a velocity $\mathbf{v}$ is given by the product of the object's mass and velocity, or

   $$\boldsymbol{p} = m\mathbf{v}$$

   Linear momentum is an example of a first-order polynomial function.
2. **Displacement with constant uniform acceleration**. The final position, x_f, of an object moving with constant uniform acceleration is related to its initial position, x_i, by the following formula:

   $$x_f = x_i + v_i\Delta t + \frac{1}{2}a(\Delta t)^2$$

where v_i is the object's initial velocity, a is the object's acceleration, and Δt is the time interval over which motion occurs. This equation of motion is an example of a second-order polynomial function.

Power Functions

Power functions are defined by

$$y = ax^n$$

where a is a constant and n is a real number, representing the mathematical power of the function. Mathematical rules involving powers are summarized next. If p and q are any numbers and $a > 0$ but $a \neq 1$,

$$a^p a^q = a^{p+q} \qquad a^p + a^q \neq a^{p+q} \qquad \frac{a^p}{a^q} = a^{p-q} \qquad (a^p)^q = a^{pq}$$

$$a^{m/n} = \sqrt[n]{a^m} \qquad a^{-n} = \frac{1}{a^n} \qquad a^{1/n} = \sqrt[n]{a} \qquad a^0 = 1$$

Several forms of the power function include the following:

$$n = -2: y = x^{-2} = \frac{1}{x^2} \qquad n = -1: y = x^{-1} = \frac{1}{x} \qquad n = -\frac{1}{2}: y = x^{-1/2} = \frac{1}{\sqrt{x}}$$

$$n = 0: y = x^0 = 1 \qquad n = \frac{1}{2}: y = x^{1/2} = \sqrt{x} \qquad n = 1: y = x^1 = x$$

$$n = 2: y = x^2 = x^2$$

The graphical representation of power functions is illustrated in Figure 1-2. In physics, applications of power functions include the following:

1. **Centripetal force**. Centripetal force, $\mathbf{F}_c$, or the force exerted on an object of mass, m, moving with uniform velocity, $\mathbf{v}$, in a circular path of radius R is given by

$$\mathbf{F}_c = \frac{m\mathbf{v}^2}{R}$$

2. **Power in an electric circuit**. The power, P, or the rate at which the electric energy in an electric circuit dissipates as a result of resistance, R, within the circuit depends on the current, I, and voltage, V, by the relations

$$P = I^2 R \quad \text{or} \quad P = \frac{V^2}{R}$$

Trigonometric Functions

Trigonometric functions are typically used in the description of periodic or oscillatory phenomena, but are also used in the characterization of vector quantities. The family

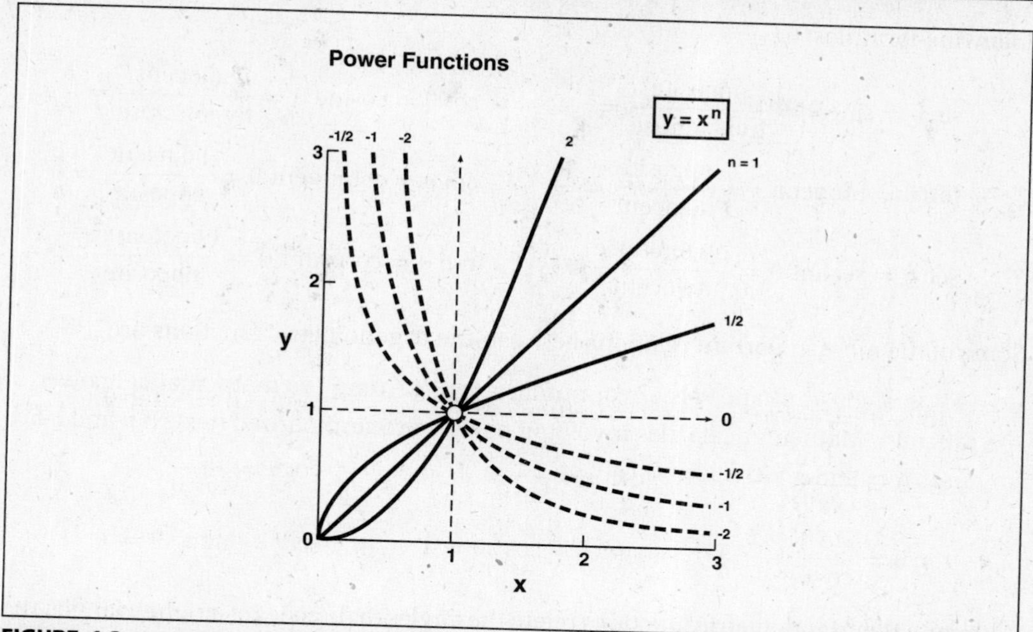

FIGURE 1-2: SOURCE: From George Hademenos, *Schaum's Outline of Physics for Pre-Med, Biology, and Allied Health Students*, McGraw-Hill, 1998; reproduced with permission of The McGraw-Hill Companies.

of trigonometric functions, which includes sin θ, cos θ, tan θ, sec θ, csc θ, and cot θ, can best be explained by drawing reference to a right triangle.

Trigonometric functions are mathematical quantities that relate angles to ratios of sides of a right triangle (a triangle with one angle equal to 90°). In the right triangle illustrated in Figure 1-3, the angle of interest, θ, is related to the following labeled sides: the opposite side (because it is the side opposite the angle of interest and labeled **a**), the adjacent side (because it is the side of the triangle adjacent or next to the angle of interest and labeled **b**), and the hypotenuse (labeled **c**) according to the

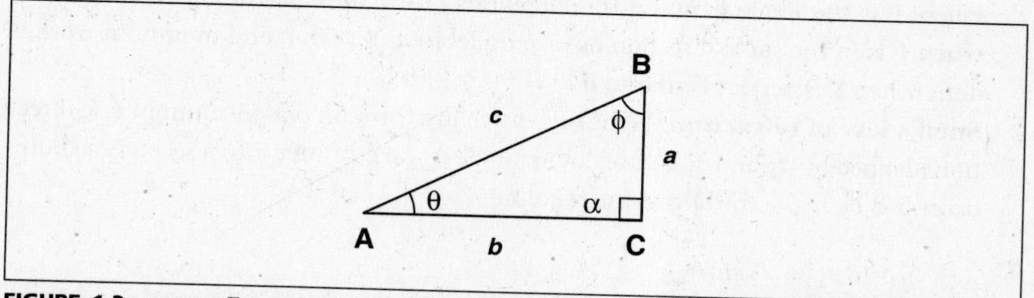

FIGURE 1-3: SOURCE: From George Hademenos, *Schaum's Outline of Physics for Pre-Med, Biology, and Allied Health Students*, McGraw-Hill, 1998; reproduced with permission of The McGraw-Hill Companies.

following formulas:

$$\sin\theta = \text{sine}\ \theta = \frac{\text{opposite}}{\text{hypotenuse}} = \frac{a}{c} \qquad \cos\theta = \text{cosine}\ \theta = \frac{\text{adjacent}}{\text{hypotenuse}} = \frac{b}{c}$$

$$\tan\theta = \text{tangent}\ \theta = \frac{\text{opposite}}{\text{adjacent}} = \frac{a}{b} \qquad \cot\theta = \text{cotangent}\ \theta = \frac{\text{adjacent}}{\text{opposite}} = \frac{b}{a}$$

$$\sec\theta = \text{secant}\ \theta = \frac{\text{hypotenuse}}{\text{adjacent}} = \frac{c}{b} \qquad \csc\theta = \text{cosecant}\ \theta = \frac{\text{hypotenuse}}{\text{opposite}} = \frac{c}{a}$$

Some of the more important relations between the trigonometric functions are

$$\sin(-\theta) = -\sin\theta \qquad \cos(-\theta) = \cos\theta \qquad \tan(-\theta) = -\tan\theta$$

$$\sec\theta = \frac{1}{\cos\theta} \qquad \csc\theta = \frac{1}{\sin\theta} \qquad \cot\theta = \frac{1}{\tan\theta}$$

$$\tan\theta = \frac{\sin\theta}{\cos\theta} \qquad \sin^2\theta + \cos^2\theta = 1 \qquad \sec^2\theta - \tan^2\theta = 1$$

Although the trigonometric functions relate the angles to the sides of a right triangle, the sides of a right triangle are related among themselves by the **Pythagorean Theorem:**

$$c = \sqrt{a^2 + b^2}$$

or

$$\text{Hypotenuse} = \sqrt{(\text{opposite})^2 + (\text{adjacent})^2}$$

The graphical representation of trigonometric functions is illustrated in Figure 1-4.

Applications of trigonometric functions in physics include:

1. **Work**. The work, W, done by a constant force, $\boldsymbol{F}$, exerted on an object over a given displacement, $\boldsymbol{d}$, is given by the relation

   $$W = Fd\cos\theta$$

   where θ is the angle between the directions of $\boldsymbol{F}$ and $\boldsymbol{d}$. Maximum work is done when $\boldsymbol{F}$ is in the same direction as or parallel to $\boldsymbol{d}$ ($\theta = 0^\circ$), and minimum work is done when $\boldsymbol{F}$ is perpendicular to $\boldsymbol{d}$ ($\theta = 90^\circ$).
2. **Snell's law of refraction**. Light rays traveling through one medium in one direction (defined by n_1 and θ_1) refract at an angle θ_2 on entrance into a second medium, described by n_2, according to the relation

   $$n_1 \sin\theta_1 = n_2 \sin\theta_2$$

   where the angles θ_1 and θ_2 are defined with respect to a line perpendicular to the interface of the two media.

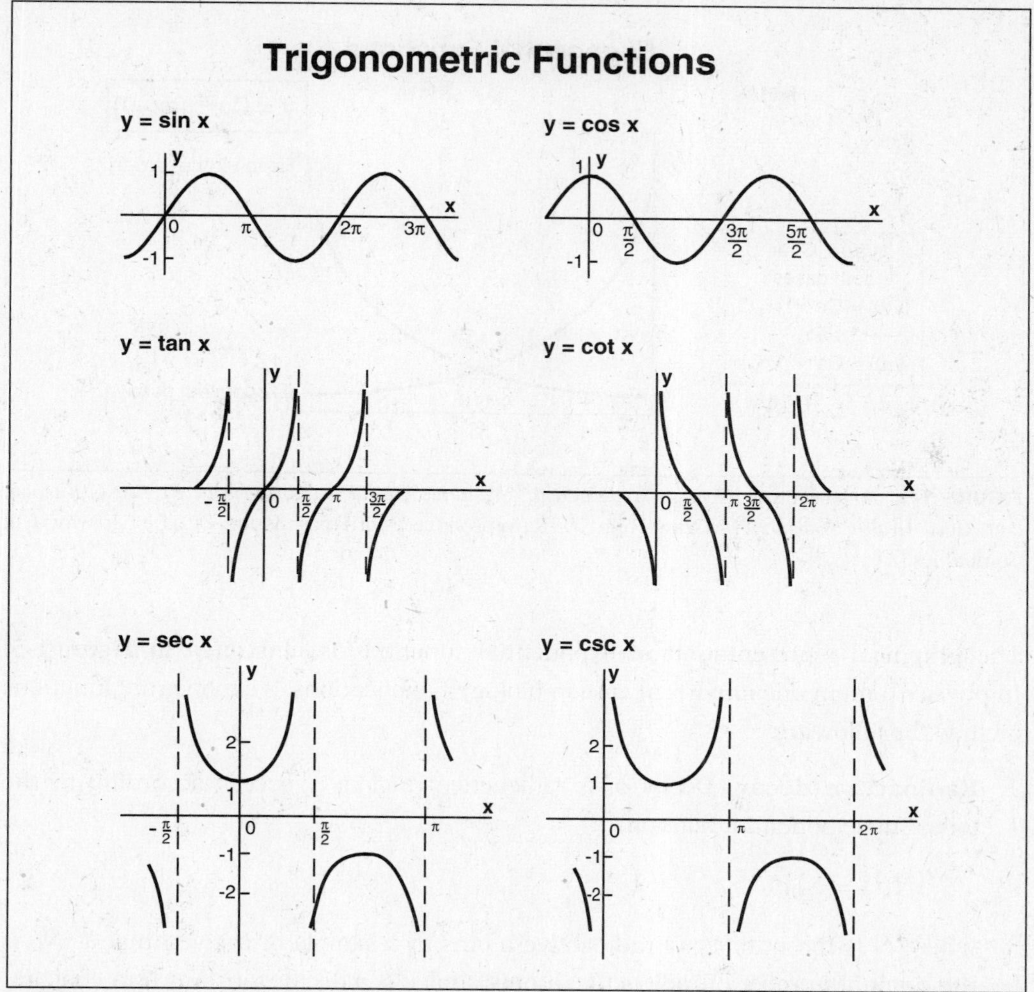

FIGURE 1-4: SOURCE: From George Hademenos, *Schaum's Outline of Physics for Pre-Med, Biology, and Allied Health Students*, McGraw-Hill, 1998; reproduced with permission of The McGraw-Hill Companies.

Exponential Functions

Exponential functions are defined mathematically as

$$f(x) = Ca^{bx}$$

for $a > 0$, where a, b, and C are constants. Exponential functions describe many scientific phenomena, particularly when $b > 0$ (growth processes) or $b < 0$ (decay processes). The y-intercept or $f(x = 0)$ is equal to C for both types of processes. Exponential functions are especially useful when $a = e$ ($e = 2.718$). In this case, exponential functions are expressed mathematically as

$$f(x) = Ce^{bx}$$

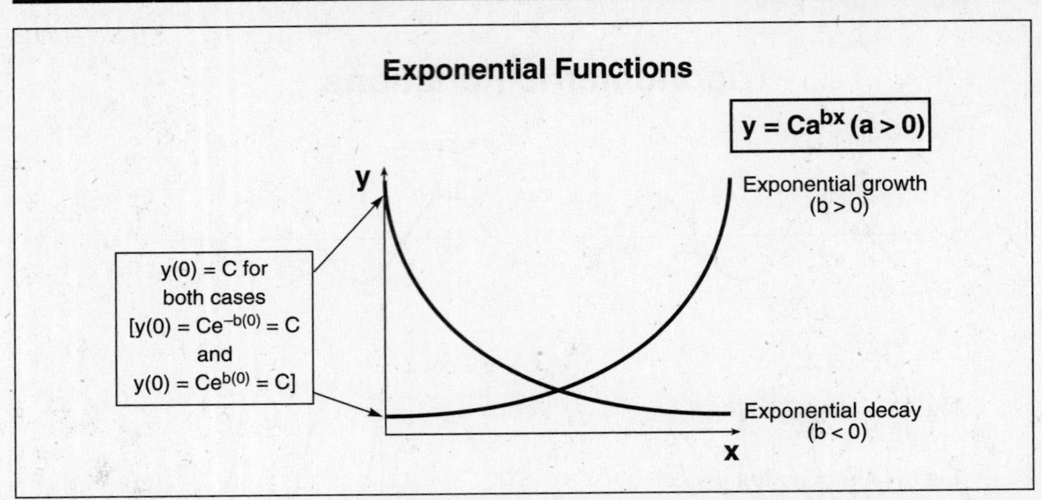

FIGURE 1-5: SOURCE: From George Hademenos, *Schaum's Outline of Physics for Pre-Med, Biology, and Allied Health Students*, McGraw-Hill, 1998; reproduced with permission of The McGraw-Hill Companies.

The graphical representation of exponential functions is illustrated in Figure 1-5. In physics (and in one important case in biology), applications of exponential functions include the following:

1. **Radioactive decay**. Decay of a radioactive nucleus proceeds according to the following exponential relation:

   $$N(t) = N_0 e^{-\lambda t}$$

   where N is the number of radioactive atoms in a sample at a given time t, N_0 is the original number of radioactive atoms, and λ is a decay constant or a constant describing the rate of decay unique to the radioactive nucleus.
2. **Bacterial growth**. The growth of bacteria in a sufficient nutrient medium can be described mathematically by

   $$N(t) = N_0 2^{\frac{t}{T_0}}$$

   where N is the number of bacteria at a given time t, N_0 is the original number of bacteria, and T_0 is the generation or doubling time. It should be noted that the exponential growth of a bacterial population represents one phase of the growth curve. Exponential growth is followed by a stationary phase and eventually a death phase in which the bacterial population declines.

Logarithmic Functions

Logarithmic functions are defined by

$$y = \log_a x$$

for $a > 0$, $a \neq 1$. The argument of a logarithmic function, that is, x, can be determined easily by

$$x = a^{\log_a x} = a^y$$

Given the exponential function $f(x) = e^x$, the logarithm of $f(x) = \log_e x$ can be simplified further to $f(x) = \ln x$.

Several mathematical relations involving logarithmic functions are expressed next:

$$\log y_1 y_2 = \log y_1 + \log y_2 \qquad y_1 y_2 > 0$$

$$\log \frac{y_1}{y_2} = \log y_1 - \log y_2 \qquad y_2 \neq 0, \frac{y_1}{y_2} > 0$$

$$\log (y_1)^{y_2} = y_2 \log y_1 \qquad y_1 > 0$$

The graphical representation of logarithmic functions is illustrated in Figure 1-6. In physics (and in one important case in chemistry), applications of logarithmic functions include the following:

1. **Decibel scale of sound intensity**. The intensity level or loudness of a sound is defined in terms of decibels (dB) and expressed mathematically as

$$\text{dB} = 10 \log_{10} \frac{I}{I_0}$$

where I is the intensity of a given sound and I_0 is a reference sound intensity equal to 1×10^{-12} W/m^2 (watts/meter^2) corresponding to the lower limit of sound intensity detectable by the human ear.

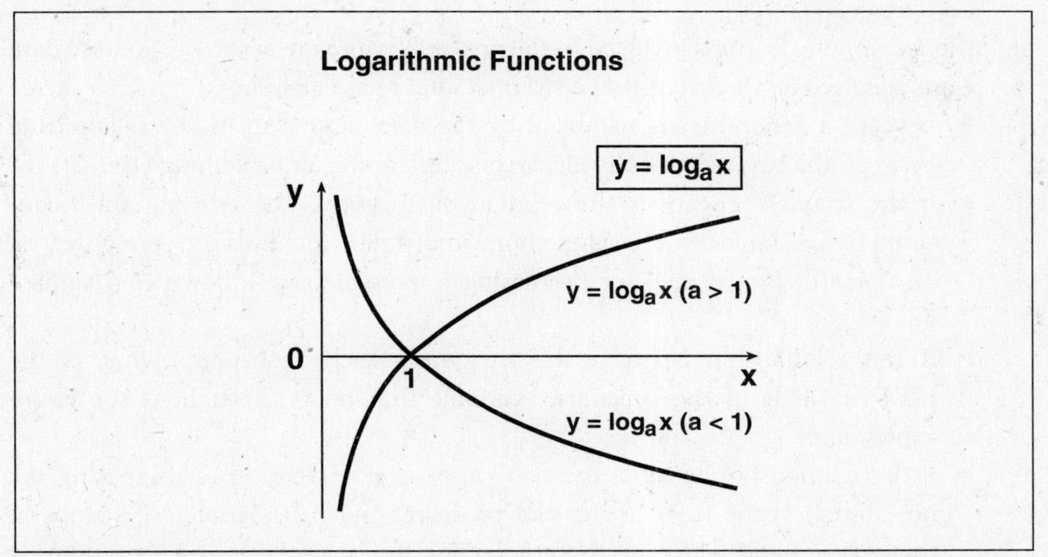

FIGURE 1-6: SOURCE: From George Hademenos, *Schaum's Outline of Physics for Pre-Med, Biology, and Allied Health Students*, McGraw-Hill, 1998; reproduced with permission of The McGraw-Hill Companies.

2. **The pH of a solution**. The pH of a solution represents the hydrogen ion activity H^+ of a solution and is defined by

$$pH = -\log[H^+]$$

GRAPHICAL REPRESENTATION OF DATA

An important aspect of physics as well as in all facets of science is the ability to evaluate and interpret data presented in graphs. Graphs represent an important means for illustrating information obtained from a scientific experiment or observations. They can reveal relationships and trends among the graphed variables, and allow the scientist to draw relevant conclusions about the experimental variables and the underlying principles involved in the scientific question being answered.

When you are asked to evaluate and interpret graphs, keep in mind the following points:

1. All graphs should have a short title describing the experiment that was conducted.
2. The variable represented on each axis should be labeled clearly on its relevant axis with appropriate units noted. The **independent variable** (the variable that is free to move) is displayed on the x-axis (abscissa), whereas the **dependent variable** (the variable that depends on the value of the independent variable) is displayed on the y-axis (ordinate).
3. The scale and range for each axis should be evident from the graph. On the graph where the data is presented, a legend should be noted (e.g., 1 square = 1 meter, 5 squares = 10 m/s).
4. After noting the features involved in the presentation of the graph, look at the data points themselves. Questions that a scientist might pose include:
    - Is there a general trend exhibited by the data, and if so, is the relationship between the two variables linear, inverse, quadratic, or something else?
    - If the trend is linear, is the relationship between the two variables constant (represented by a straight horizontal line), increasing (represented by an upward slanted line), or decreasing (represented by a downward slanted line)?
    - If the relationship between the two variables is constant, what is the numeric value of the dependent variable that holds throughout the entire experiment?
    - If the relationship between the two variables is increasing or decreasing, by how much is the trend increasing or decreasing (i.e., What is the slope of the curve?), and where do these functions begin on the y-axis (i.e., What is the y-intercept of the curve?). This information can be quantified for a linear function by calculating its slope and y-intercept.

Calculating the Slope and y-Intercept

Consider the function describing a line:

$$y = mx + b$$

Two important features of a function presented by the above equation are the **slope**, m, and the **y-intercept**, b, of the function. These features are illustrated in Figure 1-7. Some relationships between variables may originate at the origin (0,0) of the coordinate system, whereas others may originate from or pass through a point along the y-axis other than the origin (0,0). This value, the point at which the curve intercepts or passes through the y-axis, is known as the **y-intercept**.

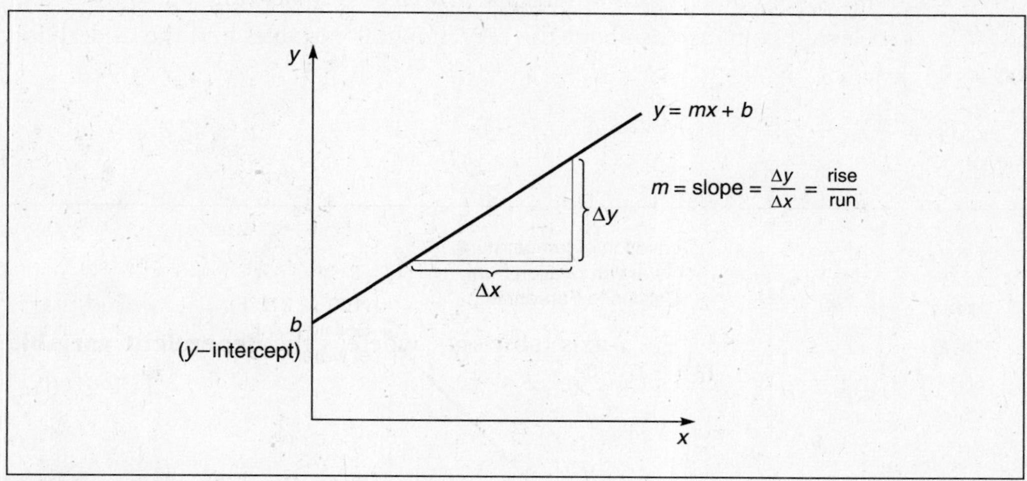

FIGURE 1-7

The slope defines the rate of change of the curve in the y-axis with respect to the corresponding change in the x-axis (also known as **the rise over the run**), and is given mathematically as

$$\text{Slope} = m = \frac{\Delta y}{\Delta x} = \frac{y_2 - y_1}{x_2 - x_1}$$

EXAMPLE: Applying these concepts with given information can help you derive equations that might not be easy to remember. For example, consider the equation that relates the temperature of an object expressed in units of the Fahrenheit scale, T_F, to the temperature of an object expressed in units of the Celsius scale, T_C:

$$T_F = \frac{9}{5}T_C + 32^\circ$$

If you cannot remember the relation, you can derive it by knowing the freezing point and the boiling point of water in both temperature scales. The freezing point of water is 0 °C or 32 °F, whereas the boiling point of water is 100 °C or 212 °F. These are essentially two data points [(0 °C, 32 °F) and (100 °C, 212 °F)] that can be plotted on a graph, with T_C being the independent variable (represented along the x-axis) and T_F being the dependent variable (represented along the y-axis). Plotting the two points on the graph, as shown in Figure 1-8, reveals a straight line that originates at 32 °F along the y-axis. This is the y-intercept. The slope of the line can be calculated as follows:

$$m = \frac{\Delta y}{\Delta x} = \frac{y_2 - y_1}{x_2 - x_1} = \frac{212\ ^\circ\mathrm{F} - 32\ ^\circ\mathrm{F}}{100\ ^\circ\mathrm{C} - 0\ ^\circ\mathrm{C}} = \frac{180\ ^\circ\mathrm{F}}{100\ ^\circ\mathrm{C}} = 1.8\frac{^\circ\mathrm{F}}{^\circ\mathrm{C}}$$

The slope can be expressed in fraction form as

$$\frac{9}{5}\frac{^\circ\mathrm{F}}{^\circ\mathrm{C}}$$

Thus, if you substitute all of this information into the standard equation for a line:

$$y = mx + b$$

or

$$T_F = \frac{9}{5}T_C + 32^\circ$$

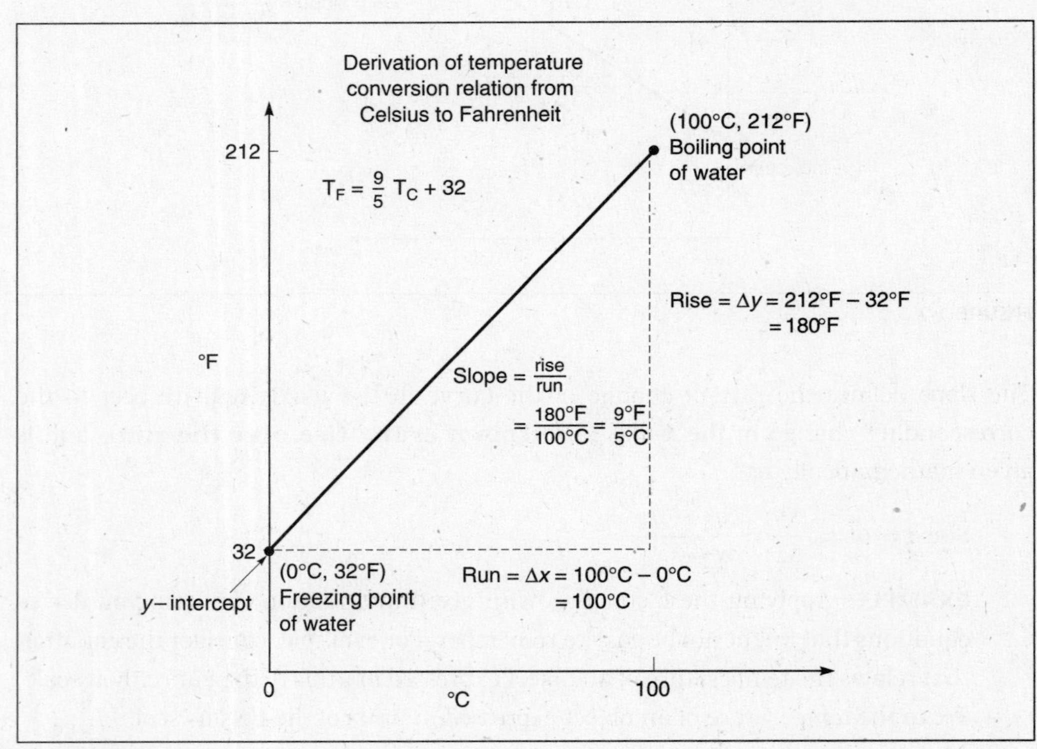

FIGURE 1-8

SIMULTANEOUS SOLUTION OF EQUATIONS

Most problems encountered in physics involve a scenario in which you are asked to solve for an unknown quantity by using an equation relating the given information in the problem to the unknown quantity. Simple substitution of numeric values and possible elementary algebra operations give you the desired answer. In some instances, however, a problem in physics involves a scenario with two unknown quantities. Solving a problem with two unknown quantities requires you to use two equations that each describe the same system, but in terms of different variables. You can obtain the desired answers by solving the system of equations simultaneously. The best way to explain the simultaneous solution of a system of equations is through an example.

EXAMPLE: Consider the object of mass, $M = 25$ kg (kilograms), suspended by two strings of different lengths oriented at different angles, as shown in Figure 1-9. String 1 is of length $L = 0.5$ m (meter), oriented at angle 45° with respect to the horizontal, and is subjecting a force on the object equal to T_1. String 2 is of length $L = 0.80$ m, oriented at an angle 20° with respect to the horizontal, and is subjecting a force on the object equal to T_2. You must find the magnitude of the tension forces T_1 and T_2.

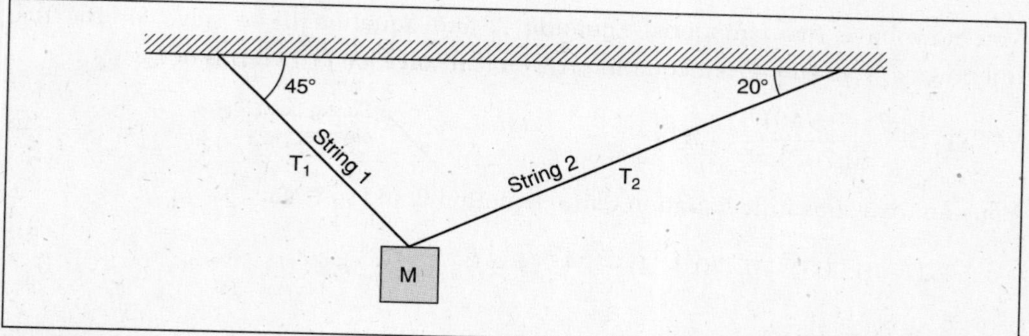

FIGURE 1-9

To solve the problem, begin by noting all given information:

Object	String 1	String 2
$M = 25$ kg	$L_1 = 0.5$ m	$L_2 = 0.8$ m
	$\theta_1 = 45°$	$\theta_2 = 20°$
$W = (25\text{ kg})\left(9.8\ \frac{\text{m}}{\text{s}^2}\right) = 245$ N (newtons)	$T_1 = ?$	$T_2 = ?$

Newton's Second Law of Motion, namely, $\Sigma \boldsymbol{F} = m\boldsymbol{a}$, applies to this problem. In words, Newton's Second Law (covered in more detail in Chapter 4 of this review) states, "The sum of the forces ($\Sigma \boldsymbol{F}$) acting on an object is equal to the mass (m) of the object times its acceleration ($\boldsymbol{a}$)." Applying Newton's Second Law to the sum of the forces in the x direction and the y direction yields the two equations from which the two unknowns can be determined: $\Sigma F_x = ma_x$ and $\Sigma F_y = ma_y$. Because the object is not accelerating in either direction, both a_x and $a_y = 0$. Identifying all of the forces acting only in the x direction yields

$$\Sigma F_y = ma_y$$

$$-T_1 \cos\theta_1 + T_2 \cos\theta_2 = 0$$

$$-T_1 \cos(45°) + T_2 \cos(20°) = 0$$

$$-T_1 (0.707) + T_2 (0.940) = 0 \qquad \text{(A)}$$

Similarly, identifying the forces acting only in the y direction yields

$$\Sigma F_x = ma_x$$

$$T_1 \sin\theta_1 + T_2 \sin\theta_2 - W = 0$$

$$T_1 \sin(45°) + T_2 \sin(20°) - 245\text{ N} = 0$$

$$T_1 (0.707) + T_2 (0.342) - 245\text{ N} = 0 \qquad \text{(B)}$$

You now have two equations (Equation A and Equation B) to solve for the two unknowns T_1 and T_2. From Equation A, you can solve for T_1 in terms of T_2:

$$T_1 = \frac{T_2 (0.940)}{(0.707)} = 1.3T_2 \qquad \text{(C)}$$

You can then substitute Equation C into Equation B and solve for T_2:

$$1.3T_2 (0.707) + T_2 (0.342) - 245\text{ N} = 0$$

$$1.26T_2 = 245\text{ N}$$

$$T_2 = \frac{245\text{ N}}{1.26} = 194\text{ N}$$

Now substitute this value into Equation A to determine the second unknown, T_1:

$$-T_1 (0.707) + (194\text{ N}) (0.940) = 0$$

$$T_1 = \frac{182.36\text{ N}}{0.707} = 258\text{ N}$$

ACCURACY, PRECISION, AND PROPAGATION OF ERRORS

Accuracy refers to how close a measurement is to the standard or accepted value, and depends on the quality of the measuring device. For example, if you place a 200.0-g (gram) object on a balance, you should observe a reading of 200.0 g. A reading of 200.1 g is a more accurate measurement than 210 g. The **error** of a measurement refers to the variance between the experimental measurement and the standard or accepted value and can be calculated by

$$\text{Percentage error} = \frac{\text{observed value} - \text{accepted value}}{\text{accepted value}} \times 100\%$$

Precision is the ability of a measurement to be reproduced consistently. If you place the 200.0-g object or sample on a balance five times and each time record a measurement of 200.1 g, the measurement would be both accurate and precise. If you place the 200.0-g object or sample on a balance five times and each time record a measurement of 210.0 g, the measurement would be inaccurate but would still be precise. The goal of a scientist is to obtain both accurate and precise measurements in an experiment.

To understand the terms **accuracy** and **precision**, think of a person throwing darts. The person in Figure 1-10 who is throwing a group of 5 darts at a bullseye target can obtain the following possible outcomes:

High accuracy, high precision	All 5 darts strike the center of the bullseye and are grouped together.
High accuracy, low precision	All 5 darts strike the center of the bullseye, but are not grouped together.
Low accuracy, high precision	All 5 darts strike a target in the outer ring of the target, but are all grouped together.
Low accuracy, low precision	All 5 darts are thrown and strike different locations within the target area.

For a scientist conducting an experiment, the ideal goal is to have high accuracy and high precision.

SIGNIFICANT FIGURES

Significant figures are a reflection of the experimental accuracy of a measurement or quantitative value. All nonzero digits in a numeric result are always significant. When zeroes appear in a numeric result, do not be misled and automatically interpret them as significant figures. The purpose of zeroes is (1) to convey magnitude and (2) to identify the location of a decimal point. It is the presence and position of the zero that determines its significance.

Rule 1: Leading zeroes, such as those zeroes in the measurements 0.0126 or 0.000579, are not significant. The number of significant figures in these measurements is 3.

Rule 2: Embedded zeroes, such as those in the measurements 8.046 or 3.901, are always significant. The number of significant figures in these measurements is 4.

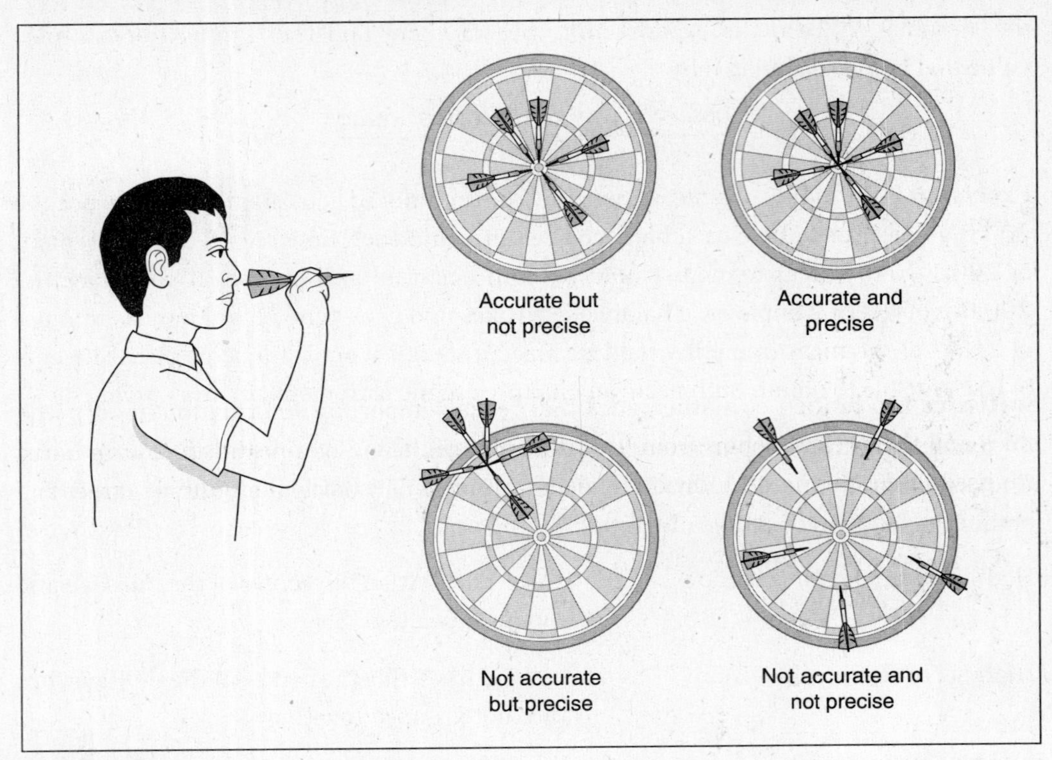

FIGURE 1-10

Rule 3: Trailing zeroes are significant only if a decimal point is present. The number of significant figures in the measurement 600 is 1, whereas the number of significant figures in 0.600 is 3.

Rule 4: In the addition or subtraction of measurements, the number of significant figures in the answer cannot be greater than that found in the least accurate measurement.

Rule 5: In the multiplication or division of measurements, the number of significant figures in the answer cannot be greater than that found in the least accurate measurement.

PROBABILITY AND STATISTICS

The **probability** of an event is a numeric measure of the likelihood that the given event may or may not occur. If you assume that there are n different but equally possible outcomes to an event, and that a number s of these outcomes are considered successes and a number $f = n - s$ are considered failures, then the probability of success p for a given event is $p = \frac{s}{n}$. The probability of failure q for a given event is $q = \frac{f}{n}$.

Consider a single playing card drawn from a deck of 52 cards. What is the probability that the selected card is (1) a black card and (2) not the queen of clubs? A single card can be drawn from a deck in $n = 52$ different ways. Because there are 2 black suits with 13 cards per suit, a black card can be drawn from the deck in $s = 26$ different ways. Thus, the probability that the selected card is a black card is

$$p = \frac{s}{n} = \frac{26}{52} = \frac{1}{2}$$

Because there is only one way to draw the queen of clubs, the probability of selecting the queen of clubs is $p = \frac{s}{n} = \frac{1}{52}$. The probability of not selecting the queen of clubs, q, is

$$q = 1 - p = \frac{52}{52} - \frac{1}{52} = \frac{51}{52}$$

Statistics is a branch of mathematics that reveals important information about variables and their relationships from groups or sets of data. Several statistical terms that are used in describing an individual variable from a data set include **mean, median, mode,** and **range.** Consider the following data set of masses measured of a diverse group of seven classroom objects: 2.0 g, 3.5 g, 2.0 g, 5.5 g, 8.4 g, 9.1 g, and 12.3 g. The **mean** is the average value of the measurements, or

$$\frac{2.0\text{ g} + 3.5\text{ g} + 2.0\text{ g} + 5.5\text{ g} + 8.4\text{ g} + 9.1\text{ g} + 12.3\text{ g}}{7} = \frac{42.8\text{ g}}{7} = 6.1\text{ g}$$

The **median** is the middle or center value of the data set when the values are arranged in ascending or descending order. In the case of the data set described earlier, the median is 5.5 g. If there are an even number of measurements in the data set, the median is found by calculating the average of the two middle measurements (i.e., adding the two middle numbers and dividing by 2). The **mode** is the measurement that is the most frequent or common value in the data set. In this example, the mode is 2.0 g, because it occurs twice, more than any of the other measurements, each of which occurs only once. The **range,** or the difference between the largest and smallest values in this data set, is 12.3 g − 2.0 g = 10.3 g.

CRAM SESSION

Mathematics Fundamentals

Ratio: The comparison of two numbers or quantities; can be stated as $\frac{x}{y}$ or $x{:}y$.

Proportion: Expresses the relationship of the relative size of two quantities and can be stated as the equality of two ratios, $\frac{a}{b} = \frac{c}{d}$.

Percentage: A quantity as a fraction of a whole or total amount (100%).

Function: A mathematical relationship between a dependent variable y and an independent variable x, which can represent any number within a specified interval. Five general types of functions are described in the following table:

Function Type	Definition	Function Graph	Physics Examples
Polynomial	$y = f(x) = a_0 = a_1x + a_2x^2 + a_3x^3 + \cdots + a_nx^n$ where a_0, a_1, a_2, a_3, a_n are coefficients (constants) and n is an integer, describing the degree or order of the polynomial	Polynomial Functions Quartic function Fourth-order polynomial (n = 4) Cubic function Third-order polynomial (n = 3) Quadratic function Second-order polynomial (n = 2) Linear function First-order polynomial (n = 1) Constant function Zeroth-order polynomial (n = 0) f(x) x	Linear momentum; displacement with constant uniform acceleration
Power	$y = ax^n$ where a is a constant and n is a real number, the power of the function	Power Functions y = x^n -1/2 -1 -2 2 n = 1 1/2 0 -1/2 -1 -2 y 0 1 2 3 x 1 2 3	Centripetal force; power in an electric circuit
Trigonometric	$y = \cos\theta$; $\sin\theta$; $\tan\theta$; $\sec\theta$; $\csc\theta$; $\cot\theta$	See Figure 1-4	Work; Snell's law

Function Type	Definition	Function Graph	Physics Examples
Exponential	$f(x) = Ca^{bx}$ for $a > 0$, where a, b, and C are constants. When $b > 0$, growth processes; when $b < 0$, decay processes	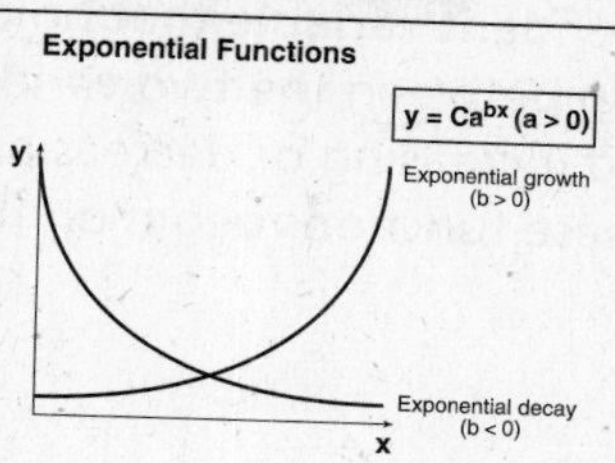 Exponential Functions $y = Ca^{bx}$ $(a > 0)$ Exponential growth $(b > 0)$ Exponential decay $(b < 0)$	Radioactive decay; bacterial growth
Logarithmic	$y = \log_a x$ for $a > 0$, $a \neq 1$ where $x = a^{\log_a x} = a^y$	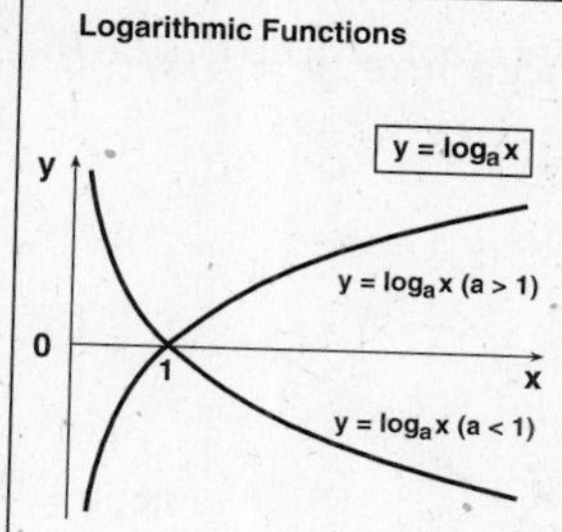 Logarithmic Functions $y = \log_a x$ $y = \log_a x$ $(a > 1)$ $y = \log_a x$ $(a < 1)$	Decibel scale of sound intensity; pH of a solution

Art in this table from George Hademenos, *Schaum's Outline of Physics for Pre-Med, Biology, and Allied Health Students,* McGraw-Hill, 1998; reproduced with permission of The McGraw-Hill Companies.

Evaluating Graphed Data

Graphs represent a means for illustrating information obtained from a scientific experiment or observations. Graphs can reveal relationships and trends between the graphed variables and allow the scientist to draw relevant conclusions about the experimental variables and the underlying principles involved in the scientific question being answered. Important features to consider in evaluating a graph include the following:

1. All graphs should have a short title describing the experiment that was conducted.
2. The variable represented on each axis should be clearly labeled on its relevant axis with appropriate units noted. The **independent variable** (the variable that is free to move) is displayed on the *x*-axis (abscissa). The **dependent variable** (the variable that depends on the independent variable) is displayed on the *y*-axis (ordinate).
3. The scale and range for each axis should be evident from the graph. On the graph, a legend should be noted (e.g., 1 square = 1 meter, 5 squares = 10 m/s).
4. Analysis of the features of a graph might involve the following questions:
   - Is there a general trend exhibited by the data, and if so, is the relationship between the two variables linear, inverse, quadratic, or something else?
   - If the trend is linear, is the relationship between the two variables constant (a straight horizontal line), increasing (an upward slanted line), or decreasing (a downward slanted line)?

- If the relationship between the two variables is constant, what is the numeric value of the dependent variable that holds throughout the entire experiment?
- If the relationship between the two variables is increasing or decreasing, by how much is the trend increasing or decreasing (i.e., What is the slope of the curve?) and where do these functions begin on the *y*-axis (i.e., What is the *y*-intercept of the curve?)?

Calculating the Slope and *y*-Intercept

From the function describing a line: $y = mx + b$

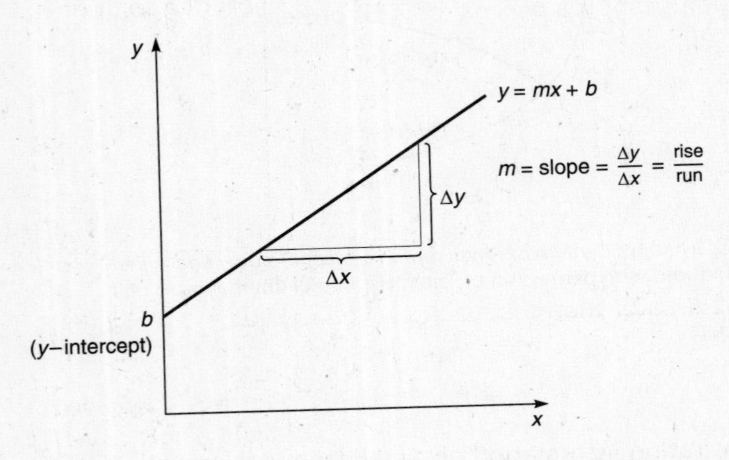

m is the slope of the function, a quantity that defines the rate of change of the curve in the *y*-axis with respect to the corresponding change in the *x*-axis (also known as **the rise over the run**) and is given mathematically as

$$\text{Slope} = m = \frac{\Delta y}{\Delta x} = \frac{y_2 - y_1}{x_2 - x_1}$$

b is the *y*-intercept of the function or the point at which the function intercepts or passes through the *y*-axis.

Accuracy refers to how close a measurement is to the standard or accepted value, and depends on the quality of the measuring device.

Error of a measurement refers to the variance between the experimental measurement and the standard or accepted value and can be calculated by

$$\text{Percentage error} = \frac{\text{observed value} - \text{accepted value}}{\text{accepted value}} \times 100\%$$

Precision: The ability of a measurement to be reproduced consistently.

Significant figures are a reflection of the experimental accuracy of a measurement or quantitative value.

- All nonzero digits in a numeric result are always significant.
- The purpose of zeroes is to (1) convey magnitude and (2) identify the location of a decimal point.

The presence and position of a zero determines its significance according to the following rules:

Rule 1: Leading zeroes, such as those zeroes in the measurements 0.0126 or 0.000579, are not significant. The number of significant figures in these measurements is 3.
Rule 2: Embedded zeroes, such as those zeroes in the measurements 8.046 or 3.971, are always significant. The number of significant figures in these measurements is 4.
Rule 3: Trailing zeroes are only significant if a decimal point is present. The number of significant figures in the measurement 600 is 1, whereas the number of significant figures in 0.600 is 3.

Other rules for significant figures applied to the addition, subtraction, multiplication, and division of measurements include:

Rule 4: In the addition or subtraction of measurements, the number of significant figures in the answer cannot be greater than that found in the least accurate measurement.
Rule 5: In the multiplication or division of measurements, the number of significant figures in the answer cannot be greater than that found in the least accurate measurement.

The **probability** of an event is a numeric measure of the likelihood that the given event may or may not occur. If there are n different but equally possible outcomes to an event, with an s number of these outcomes considered successes and an amount, $f = n - s$, considered failures, then the probability of success, p, for a given event is $p = \frac{s}{n}$ and the probability of failure, q, for a given event is $q = \frac{f}{n}$.

Statistics: A branch of mathematics that reveals important information about variables and their relationships from groups or sets of data. Several statistical terms that are used in describing an individual variable from a data set include **mean, median, mode,** and **range**.

Mean: The average value of measurements of a data set.

Median: The middle or center value of a data set.

Mode: The measurement that is the most frequent or common value in a data set.

Range: The difference between the largest and smallest values in a data set.

CHAPTER 2

Physics Fundamentals

Read This Chapter to Learn About:

- ➤ Introduction to Physics
- ➤ Orders of Magnitude/Scientific Notation
- ➤ Dimensions
- ➤ Units of Measurement
- ➤ Unit Conversions
- ➤ Estimating Quantities Using Unit Conversions
- ➤ Problem-Solving Strategies in Physics

Physics is a branch of science named for the Greek word meaning **nature.** As you begin studying the various topics contained within physics, it is important to first see the big picture of physics followed by the appropriate tools needed to understand concepts and solve problems within each of these topics. In this chapter, we introduce the fundamental properties and tools of physics addressed in all of the topics covered in this review.

INTRODUCTION TO PHYSICS

Physics is a branch of science that attempts to describe and explain phenomena and experiences that we consistently observe and depend upon on a daily basis. Doors swinging open, water coming out of a faucet, electricity powering a lamp, coffee brewing in a pot, cars moving down the freeway, people walking up stairs—all of these rather routine, almost mundane events are explored and explained in physics. In your early childhood years, you probably enjoyed the fruits of physics by sliding down a slide, swinging on a swing, riding on a merry-go-round, kicking a soccer ball, hitting a baseball, or throwing a football. All of these activities are deeply rooted in physics.

Throughout this review, there are many examples that show how the principles of physics apply to the real world, as well as many sample problems solved in detail.

ORDERS OF MAGNITUDE/ SCIENTIFIC NOTATION

The spectrum of phenomena commonly observed in physics ranges from the extremely small [e.g., the diameter of an atom, ≈0.0000000001 meter (m)] to the extremely large (e.g., the diameter of Earth, ≈12,756,000 m). In order to avoid the rather cumbersome task of writing such values with many numeric places, you can express these quantities by using either scientific notation or scientific prefixes.

Scientific notation simplifies the presentation of numeric values by expressing the value of a number between 1 and 10, multiplied by 10 raised to a power determined by the number of zeroes in the numeric result. For example, a value of 5000 can be expressed in scientific notation as 5.0×10^3, which is 5.0 multiplied by 10 raised to the power equal to the number of zeroes that follow the 5, that is, 3.

Scientific prefixes are defined symbols that represent a numeric quantity in powers of 10. For example, consider the value of 1 kilometer. The base unit of the quantity is the meter, and the prefix *kilo* represents 1000. Taken together, 1 kilometer is equivalent to 1000 meters. Examples of other common scientific prefixes and their corresponding numeric values are shown in the following table.

Scientific Notation/Prefixes for Values of Various Magnitudes

Expanded	Scientific Notation	Prefix	Symbol	Magnitude	Example
0.000000001	1×10^{-9}	nano	n	one-billionth	nanometer
0.000001	1×10^{-6}	micro	μ	one-millionth	micrometer (micron)
0.001	1×10^{-3}	milli	m	one-thousandth	millisecond
0.01	1×10^{-2}	centi	c	one-hundredth	centimeter
1	1×10^{0}			one	meter
1,000	1×10^{3}	kilo	k	one thousand	kilogram
1,000,000	1×10^{6}	mega	M	one million	megawatt
1,000,000,000	1×10^{9}	giga	G	one billion	gigabyte

DIMENSIONS

In physics, typical quantities that might be measured include the length of a lab table, the mass of a textbook, or the time required for an object to strike the ground when dropped from a known height. These quantities are described by **dimensions** or physical descriptions of a quantity. In physics, three basic dimensions are used, corresponding to the three examples of quantities noted previously: length (L), mass (M), and time (T). These are not the only basic dimensions in physics; in fact, in physics there are seven basic dimensions: length, mass, time, temperature, amount of a substance, electric current, and luminous intensity.

A quantity may also be described by some combination of these three dimensions. For example, consider the quantity of **force**, defined by Newton's Second Law of Motion as the product of mass and acceleration. The dimension of mass is M, and acceleration (defined as the change in velocity over the change in time) is L/T^2. So, the dimension of force is $(M)(L/T^2)$, or ML/T^2.

UNITS OF MEASUREMENT

Although the dimension indicates the type of physical quantity expressed by a physical measurement, units indicate the amount of the physical quantity. Each of the dimensions described in the previous section (i.e., length, mass, and time) is measured in terms of a **unit**, which indicates the amount of a physical quantity. An appropriate unit for a specific quantity depends on the dimension of the quantity. For example, let's say you want to know the length of a pencil. Because the dimension of interest is length, the pencil can be measured in terms of various units, such as centimeters, meters, inches, or feet—all of which describe length. The choice of unit used to describe the length of the pencil depends on the size of the pencil; you probably would not measure the length of a pencil in terms of miles or kilometers.

Although there are several systems of units known in science, the system of units that has been adopted by the science community and that will be used throughout this review is the SI system of units (Système International d'Unités), also known as the metric system. In the **SI system of units**, mass is measured in kilograms, length is measured in meters, and time is measured in seconds, as shown in the following table.

Physical Quantities and SI Units of Measurement

Physical Quantity	Dimension	SI Unit
Length	L	meter, m
Mass	M	kilogram, kg
Time	T	second, s

UNIT CONVERSIONS

All physics equations must be equal in both dimensions and units. For example, consider the quantity speed. The units for speed must be consistent in terms of dimension $\left(\frac{L}{T}\right)$ and in units $\left(\frac{\text{km}}{\text{h}} \text{ or } \frac{\text{m}}{\text{s}}\right)$ in order for the final result to be physically logical. **Conversion factors** convert units through ratios of equivalent quantities from one system of measure to another. Following are some examples of the process of unit conversions:

1. Determine the number of grams (g) in 5 kilograms (kg).

$$5 \text{ kg} \times \frac{1000 \text{ g}}{1 \text{ kg}} = 5000 \text{ g}$$

In this problem, 5 kg is multiplied by the conversion factor $\frac{1000 \text{ g}}{1 \text{ kg}}$. The conversion factor is equal to 1 because 1 kg = 1000 g, and any value (e.g., 5 kg) multiplied by 1 retains its value, even if expressed in different units. Although multiplication of a quantity by a conversion factor changes the numeric value of the quantity, it does not change the measurement of the physical quantity because objects of mass 5 kg and 5000 g are identical. The arrangement of the conversion factor is important, because the final result must have units of grams. In order for this to happen, the conversion factor must have the kilogram unit on the bottom so that the unit cancels with the kilogram unit in 5 kg, yielding a quantity with a unit in grams.

2. Determine the number of seconds in 1 year. To find the number of seconds in 1 year, begin with the quantity of 1 year and multiply that quantity by appropriate conversion factors such that the last unit standing is seconds.

$$1 \text{ year} \times \frac{365 \text{ days}}{1 \text{ year}} \times \frac{24 \text{ hours}}{1 \text{ day}} \times \frac{60 \text{ minutes}}{1 \text{ hour}} \times \frac{60 \text{ seconds}}{1 \text{ minute}}$$

$$= \left(\frac{1 \times 365 \times 24 \times 60 \times 60}{1 \times 1 \times 1 \times 1}\right) \text{seconds} = 3.15 \times 10^7 \text{ s}$$

ESTIMATING QUANTITIES USING UNIT CONVERSIONS

You may use this same conversion technique to estimate quantities. However, in making such estimations, you need to make several assumptions—and the more sound and robust your assumptions are, the more realistic your estimations will be.

EXAMPLE: Calculate the number of heartbeats that occur in an individual over a lifetime.

SOLUTION: Before you begin to solve this problem, you must make several assumptions including: (1) the average heartbeat rate of the individual over the course of a lifetime = 1.1 beats/s, and (2) the number of years in a lifetime = 80 years. The problem requires you to convert from 1.1 beats per second to the total number of beats occurring over 80 years.

$$\frac{\text{Number of beats}}{\text{Lifetime (80 years)}} = \frac{1.1 \text{ beats}}{1 \text{ second}} \times \frac{60 \text{ seconds}}{1 \text{ minute}} \times \frac{60 \text{ minutes}}{1 \text{ hour}} \times \frac{24 \text{ hours}}{1 \text{ day}}$$

$$\times \frac{365 \text{ days}}{1 \text{ year}} \times \frac{80 \text{ years}}{1 \text{ lifetime}}$$

$$= \left(\frac{1.1 \times 60 \times 60 \times 24 \times 365 \times 80}{1 \times 1 \times 1 \times 1 \times 1 \times 1}\right) \frac{\text{beats}}{\text{lifetime}}$$

$$= 2.78 \times 10^9 \; \frac{\text{beats}}{\text{lifetime}}$$

By this estimate, the number of heartbeats in a person's lifetime of 80 years is 2.78 billion beats. However, this calculation is an estimate of a quantity, and not an exact answer, primarily because of factors that were not addressed, such as: (1) the extra days not factored in during leap years; (2) the change in heart rate that occurs between night and day; (3) the effect of different emotions, medications, and illnesses; (4) family history, genetic disorders, and illnesses that affect the circulatory system; and (5) the moment that the heart begins to beat includes the time during fetal development in which the defined circulatory system begins circulating blood within the fetus (typically in the second trimester).

PROBLEM-SOLVING STRATEGIES IN PHYSICS

In each chapter of this review, there are helpful hints for solving problems specific to the topic being discussed. Although each physics problem is based on unique topics and given scenarios, there are several effective strategies that you can apply to each problem as you attempt to solve it. The problem-solving strategies in physics include:

- Read the problem carefully, underlining key words, terms, and values so that you understand the processes reflected in any accompanying figures or diagrams.
  - What is the problem about?
  - What scenario is being described?
  - What information is given to solve the problem?

- Draw a rough sketch of the processes described in the problem.
- Write down all pertinent descriptive information.
  - Which physical quantities are given in the problem?
  - What is/are the unknown quantities?
- Write down all relevant equations that relate the unknown quantity to the given quantities.
- As you substitute all given quantities into the equation to solve for the unknown, be sure to do the following:
  - Perform all mathematical/algebra operations correctly (e.g., when substituting into an equation that contains a squared quantity, make sure that the quantity is indeed squared. When rearranging the equation to solve for an unknown, make sure that the expression for the unknown is obtained using fundamental algebraic operations).
  - Treat units of each quantity consistently according to the equations into which they are substituted.
- Finally, as you solve the problem and inspect the answer, make sure the answer is reasonable. Ask yourself: Does the answer make sense? Remember, just because an answer might match one of the possible choices in a particular problem does not mean the answer is the correct choice. On the MCAT, most questions are in the multiple-choice format with four answer choices, one of which is correct. To assess not only whether you understand the concept being tested, but also whether you comprehend the underlying principles and can perform the necessary calculations, the test-writers often include wrong answer choices that represent answers you might reach if you misapplied an equation or made an error in your figuring. So do not assume that your answer is correct just because it matches one of the answer choices. When you finish your calculations, recheck your information, review your calculations, and, most important, check your answer to make sure that it is reasonable. That is a good way to determine whether it is correct.

CRAM SESSION

Physics Fundamentals

Physics: Describes and explains all of the phenomena and experiences that we consistently observe and depend upon on a daily basis.

Scientific notation: Simplifies the presentation of numeric values by expressing the value of a number between 1 and 10, multiplied by 10 raised to a power determined by the number of zeroes in the numeric result. For example, a value of 5000 can be expressed in scientific notation as 5.0×10^3, which is 5 multiplied by 10 raised to the power equal to the number of zeroes that follow the 5; that is, 3.

Scientific prefixes: Defined symbols that represent a numeric quantity in powers of 10.

Scientific Notation/Prefixes for Values of Various Magnitudes					
Expanded	**Scientific Notation**	**Prefix**	**Symbol**	**Magnitude**	**Example**
0.000000001	1×10^{-9}	nano	n	one-billionth	nanometer
0.000001	1×10^{-6}	micro	μ	one-millionth	micrometer (micron)
0.001	1×10^{-3}	milli	m	one-thousandth	millisecond
0.01	1×10^{-2}	centi	c	one-hundredth	centimeter
1	1×10^{0}			one	meter
1,000	1×10^{3}	kilo	k	one thousand	kilogram
1,000,000	1×10^{6}	mega	M	one million	megawatt
1,000,000,000	1×10^{9}	giga	G	one billion	gigabyte

Dimensions: Physical descriptions of a quantity.

Units: Measure the amount of the physical quantity.

SI system of units (Système International d'Unités): Also referred to as the **metric system,** the system of units in which mass is measured in kilograms, length is measured in meters, and time is measured in seconds.

Physical Quantities and SI Units of Measurement		
Physical Quantity	**Dimension**	**SI Unit**
Length	L	meter, m
Mass	M	kilogram, kg
Time	T	second, s

Unit conversion: A process of using conversion factors to change units through ratios of equivalent quantities expressed in different units. For example, the number of seconds in a year can be determined by

$$1\text{ year} \times \frac{365\text{ days}}{1\text{ year}} \times \frac{24\text{ hours}}{1\text{ day}} \times \frac{60\text{ minutes}}{1\text{ hour}} \times \frac{60\text{ seconds}}{1\text{ minute}}$$

$$= \left(\frac{1 \times 365 \times 24 \times 60 \times 60}{1 \times 1 \times 1 \times 1}\right)\text{seconds} = 3.15 \times 10^{7}\text{ s}$$

Problem-Solving Strategies in Physics

1. Read the problem carefully, underlining key words, terms, and values, as well as understanding the processes reflected in any accompanying figures or diagrams.
2. Draw a rough sketch of the processes described in the problem.
3. Write down all pertinent descriptive information.
4. Write down all relevant equations that relate the unknown quantity to the given quantities.
5. As you substitute all given quantities into the equation to solve for the unknown, be sure that
   - ➤ All mathematical/algebraic operations are performed correctly.
   - ➤ Units of each quantity are treated consistently according to the equations into which they are substituted.
6. As you solve the problem and inspect the answer, make sure the answer is reasonable.

CHAPTER 3

Kinematics

Read This Chapter to Learn About:

- ➤ Displacement, Velocity, and Acceleration
- ➤ Graphical Representation of Motion
- ➤ Uniformly Accelerated Motion
- ➤ Free-Fall Motion
- ➤ Vectors and Scalars
- ➤ Vector Addition and Subtraction
- ➤ Projectile Motion

On a scale as small as an atom or as large as a planet, motion is an important constant critical to all living things. It is easy to imagine that if all motion stopped, from the atom to the planets and everything in between, then life would cease to exist. The science of motion is referred to as **kinematics**. This chapter focuses on kinematics and the concepts and equations that describe the motion of objects.

DISPLACEMENT, VELOCITY, AND ACCELERATION

Displacement (Δx) is defined as a distance in a given direction. Consider a typical trip to work in which you travel from point A (home, starting position) to point B (work, final position). Expressed in units of length (e.g., meters, kilometers, yards, or miles), displacement is a measure of the **length** [difference between the final position (x_f) and the initial position (x_i)] required to get from point A to point B, regardless of the path taken.

$$\text{Displacement: } \Delta x = x_f - x_i$$

Because displacement is a directed distance, it is possible that the displacement can be either positive or negative. A negative displacement, just as is the case for any negative quantity in physics, does not imply that the distance is a negative quantity; rather, it implies that the direction of displacement is opposite to that direction considered positive.

A similar quantity related to length is distance. Although displacement is the length between the final point and starting point, **distance** is the length required to get from the starting point to the ending point, dependent on the path taken. If you leave home, drop by the post office to mail some letters, take your children to school, and stop by the grocery store before reaching work, the path taken is much longer than the straight path between home and work. In this case, the distance (home → post office → school → grocery store → work) is much larger than the displacement (home → work).

Velocity (v) is the rate of change of displacement or the rate at which a directed distance between point A and point B is covered over a quantity of time. Velocity, expressed in dimensions of length per unit time, is the speed of an object in a given direction.

$$\text{Average velocity: } v = \frac{\Delta x}{\Delta t} = \frac{x_f - x_i}{t_f - t_i}$$

Average velocity represents velocity over a time interval. **Instantaneous velocity** represents velocity at a given instant of time, similar to the velocity of a car indicated by its speedometer.

Acceleration (a) is the rate at which velocity changes. Acceleration, expressed in dimensions of length per unit time squared, is the change of velocity in a given direction over a defined time interval.

$$\text{Average acceleration: } a = \frac{\Delta v}{\Delta t} = \frac{v_f - v_i}{t_f - t_i}$$

Average acceleration represents average velocity over a time interval. **Instantaneous acceleration** represents acceleration at a given instant of time.

GRAPHICAL REPRESENTATION OF MOTION

In a typical motion graph, time [typically in seconds (s)] is always represented by the x-axis (horizontal), whereas the variable along the y-axis could be displacement [in meters (m), for example] or velocity [in meters per second (m/s), for example]. If there is an accompanying table with data points that are represented in a graph, the left column contains values representative of the independent variable (typically time) and is plotted along the horizontal or x-axis, whereas the right column contains values of the dependent variable (either displacement or velocity, respectively) and is plotted along the vertical or y-axis.

If you graph displacement versus time for an object, the velocity of the object is the slope of the line on the graph:

$$\text{Slope} = \frac{\text{rise}}{\text{run}} = \frac{y_2 - y_1}{x_2 - x_1} = \frac{\text{displacement (meters)}}{\text{time (seconds)}} = \text{velocity}\left(\frac{\text{meters}}{\text{second}}\right)$$

If you graph velocity versus time for an object, the acceleration of the object can be represented by the slope of the line on the graph:

$$\text{Slope} = \frac{\text{rise}}{\text{run}} = \frac{y_2 - y_1}{x_2 - x_1} = \frac{\text{velocity (meters/second)}}{\text{time (seconds)}}$$

$$= \text{acceleration}\left(\frac{\text{meters}}{\text{second}^2}\right)$$

Velocity and acceleration represent a change in a variable (displacement or velocity, respectively) as a function of time. Figure 3-1 shows examples of graphs of displacement, velocity, and acceleration. Note that the slope and the y-intercept of each graphed quantity depend on the location of the object and the rate of change of motion of the object.

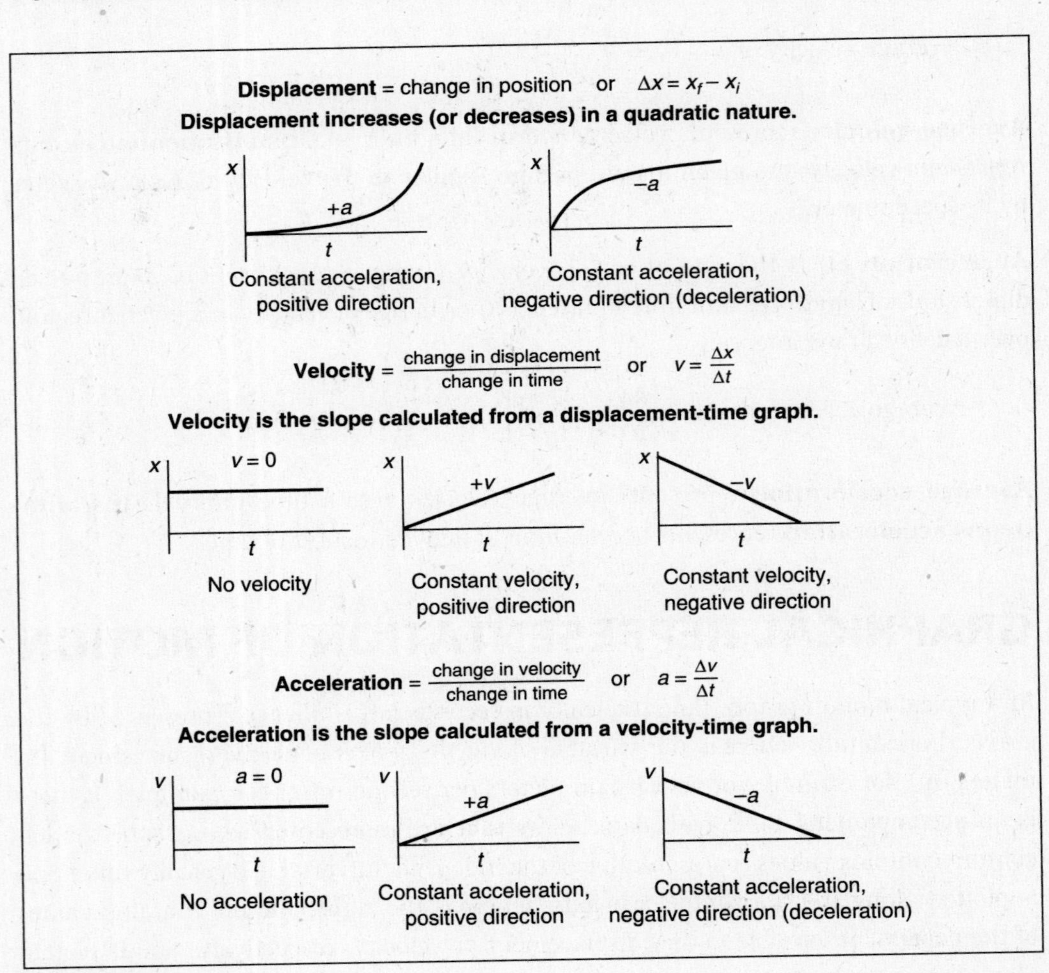

FIGURE 3-1

As you evaluate and interpret the graphs of motion data, you can obtain important information by considering the following questions:

- What is the physical significance of the data?
  - What do the graphed data tell me about the nature and behavior of an object?
  - What are the quantities presented in the columns of data found in a table and/or the quantities represented along the x- and y-axes of a graph?
  - If the x-axis is time and the y-axis is displacement (distance in a given direction), then the slope of the graph gives velocity.
  - If the x-axis is time and the y-axis is velocity (speed in a given direction), then the slope of the graph gives acceleration.
- Does y change as x is varied? If so, how?
  - If there is a straight horizontal displacement–time graph, there is no change in y, and thus the velocity equals zero.
  - If the graph is linear, the slope is constant; therefore, the velocity is constant.
  - If the graph is nonlinear (increasing/decreasing in a quadratic nature), the slope is changing, indicating either a positive acceleration or negative acceleration (deceleration).
- If there is motion, what is its quantitative value (in magnitude and units)?
  - The quantitative value of motion implies the numeric value of the slope of the graph.
  - If the x-axis is time and the y-axis is displacement, then the slope of the graph yields velocity, as calculated next.

$$\text{Slope} = \frac{\text{rise}}{\text{run}} = \frac{y_2 - y_1}{x_2 - x_1} = \frac{\text{displacement (meters)}}{\text{time (seconds)}}$$

$$= \text{velocity}\left(\frac{\text{meters}}{\text{second}}\right)$$

  - If the x-axis is time and the y-axis is velocity, then the slope of the graph yields the acceleration, as calculated next.

$$\text{Slope} = \frac{\text{rise}}{\text{run}} = \frac{y_2 - y_1}{x_2 - x_1} = \frac{\text{velocity (meters/second)}}{\text{time (seconds)}}$$

$$= \text{acceleration}\left(\frac{\text{meters}}{\text{second}^2}\right)$$

UNIFORMLY ACCELERATED MOTION

Up to this point, we have defined and discussed motion in terms of displacement, velocity, and acceleration. In order for you to be able to use these quantities and apply them to physics problems, you must be able to determine relations between them. For motion of an object with constant uniform acceleration along the x-axis, you can apply the following four equations:

1. Displacement with constant uniform acceleration

$$\Delta x = \frac{1}{2}\left(v_i + v_f\right)\Delta t$$

$$\text{Displacement} = \frac{1}{2}\,(\text{initial velocity} + \text{final velocity})\,(\text{time interval})$$

2. Final velocity with constant uniform acceleration

$$v_f = v_i + a\Delta t$$

$$\text{Final velocity} = \text{initial velocity} + (\text{acceleration})\,(\text{time interval})$$

3. Displacement with constant uniform acceleration

$$\Delta x = v_i\Delta t + \frac{1}{2}a(\Delta t)^2$$

$$\text{Displacement} = (\text{initial velocity})\,(\text{time interval}) + \frac{1}{2}\,(\text{acceleration})\,(\text{time interval})^2$$

4. Final velocity after any displacement

$$v_f^2 = v_i^2 + 2a\Delta x$$

$$(\text{Final velocity})^2 = (\text{initial velocity})^2 + 2(\text{acceleration})\,(\text{displacement})$$

You might ask why there are two equations used to calculate final velocity and two equations used to calculate displacement. Careful inspection of the equations indicates that each equation depends on four different variables. This gives you maximum flexibility to determine any related quantity of motion about an object given a specific scenario, depending on the given information.

EXAMPLE: A car initially moving at 20 m/s uniformly accelerates at 2.5 m/s^2. Find the final speed and displacement of the car after 8.0 s.

SOLUTION: The given information from this problem includes:

$v_i = 20$ m/s $a = 2.5$ m/s^2 $\Delta t = 8.0$ s

To determine the final speed of the car, use the following equation:

$$v_f = v_i + a\Delta t = 20\,\frac{\text{m}}{\text{s}} + \left(2.5\,\frac{\text{m}}{\text{s}^2}\right)(8.0\ \text{s}) = 40\,\frac{\text{m}}{\text{s}}$$

To determine the displacement of the car, use the following equation:

$$\Delta x = v_i\Delta t + \frac{1}{2}a\,(\Delta t)^2 = \left(20\,\frac{\text{m}}{\text{s}}\right)(8.0\ \text{s}) + \frac{1}{2}\left(2.5\,\frac{\text{m}}{\text{s}^2}\right)(8.0\ \text{s})^2 = 240\ \text{m}$$

FREE-FALL MOTION

The equations for uniformly accelerated motion presented previously dealt with motion primarily moving along the x-axis (from left to right, and vice versa). However, the same concepts, definitions, variables, and equations also apply to motion along the y-axis (from up to down, and vice versa). In the case of motion along the y-axis, a constant acceleration is caused by gravity. Any and all types of motion involving the y-axis, such as throwing a ball in the air or dropping a pencil, occurs under the influence of gravity. But what is it about gravity that factors into the motion of such objects? Gravity is a force that pulls objects to Earth. That pull causes the objects to accelerate in a constant manner. The acceleration due to gravity, g, is given numerically by $g = -9.8\ \text{m/s}^2$ and always acts downward. Thus, the four equations of motion given above that described motion of an object in the x-direction can also be applied to an object moving in the y-direction by making two minor changes: (1) displacement which was originally noted in the equations of motion as Δx now becomes Δy; and (2) the acceleration, a, now is replaced by g, the acceleration due to gravity.

1. Displacement with constant uniform acceleration

$$\Delta y = \frac{1}{2}\left(v_i + v_f\right)\Delta t$$

2. Final velocity with constant uniform acceleration

$$v_f = v_i + g\Delta t$$

3. Displacement with constant uniform acceleration

$$\Delta y = v_i\Delta t + \frac{1}{2}g\left(\Delta t\right)^2$$

4. Final velocity after any displacement

$$v_f^2 = v_i^2 + 2g\Delta y$$

A special case of motion along the y-axis is **free-fall motion**, which refers to motion of an object that is dropped from a certain height or y-direction and allowed to fall toward the ground. The first equation indicates how far an object has fallen per given time:

$$\Delta y = \frac{1}{2}g(\Delta t)^2$$

(Distance in the y-direction) = ½(acceleration due to gravity)(time in flight)^2

Note that this is the same equation as No. 3, displacement with constant uniform acceleration, with the initial velocity $v_i = 0$.

The other equation indicates how fast an object falls per given time:

$$v_f = g\Delta t$$

(Velocity in the y-direction) = (acceleration due to gravity)(time in flight)

Note that this is the same equation as No. 2, final velocity with constant uniform acceleration, with the initial velocity $v_i = 0$.

EXAMPLE: A stone is dropped from a bridge that is 15 m in height. Determine the stone's velocity as it strikes the water below.

SOLUTION: The given information from this problem includes:

- $v_i = 0$ m/s, because it is dropped or released from rest
- $a = -9.8$ m/s^2, because it is accelerating downward as a result of gravity
- $\Delta y = -15.0$ m (This height is negative, because the origin is the object at the top of the bridge, and as it moves downward, it is moving in the negative direction.)

To determine the final speed of the stone, use the following equation:

$$v_f^2 = v_i^2 + 2g\Delta y = \left(0\ \frac{\text{m}}{\text{s}}\right)^2 + 2\left(-9.8\ \frac{\text{m}}{\text{s}^2}\right)(-15.0\ \text{m}) = 294\ \frac{\text{m}^2}{\text{s}^2}$$

$$v_f = \sqrt{294\ \frac{\text{m}^2}{\text{s}^2}} = 17.1\ \frac{\text{m}}{\text{s}}$$

VECTORS AND SCALARS

In physics, measurements of physical quantities, processes, and interactions can be classified according to two types. Some measurements, such as displacement, velocity, and acceleration, require both a magnitude (size) as well as a direction, while some measurements such as distance and speed are presented only as a magnitude or size. Those quantities that require both a magnitude and direction are known as **vector quantities**, whereas those quantities that require only magnitude are referred to as **scalar quantities**. Examples of scalar and vector quantities follow.

Scalar Quantities	Vector Quantities
Measurement	
Time	
Mass	
Area	
Volume	
Kinematics	
Distance	Displacement
Speed	Velocity
	Acceleration

Scalar Quantities	Vector Quantities
Dynamics	
Work	Force
Energy	Momentum
	Torque

A vector quantity is generally noted by a bold-faced letter, sometimes with an arrow drawn over the top of the letter. For example, a vector quantity S can be represented by the symbols, **S** or $\vec{S}$. The magnitude of the vector quantity S is typically indicated by the vector quantity symbol encased within the absolute value bars, |**S**| or $|\vec{S}|$. In addition, a vector quantity is represented graphically by an arrow with the size of the arrow corresponding to the size or magnitude of the vector quantity and the orientation of the arrow corresponding to the direction of the vector quantity, as shown in Figure 3-2.

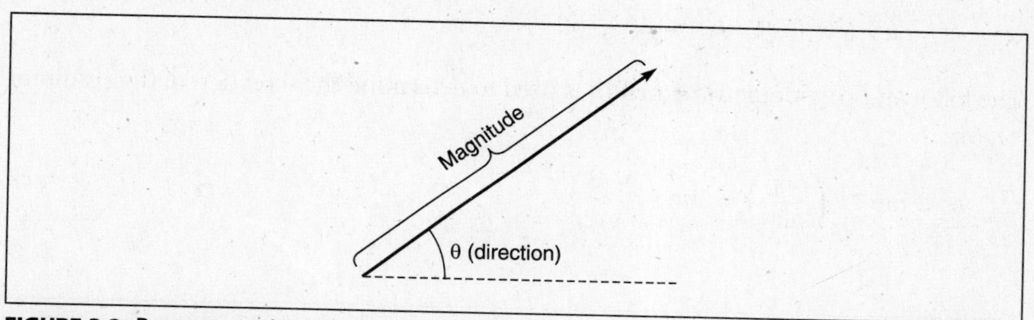

FIGURE 3-2: Representation of a vector quantity.

VECTOR ADDITION AND SUBTRACTION

Although scalar and vector quantities are similar in that they both can be added, they are different in that scalar quantities can be added arithmetically, but vector quantities must be added in ways that take into account their direction as well as their magnitude. One method of vector addition is the **head-to-tail method**. A second method is called the **component method**.

Head-to-Tail Method of Vector Addition

Let's say you have two vectors, **A** and **B**. In this method of vector addition, one of the vectors, **A**, is drawn to scale with its tail positioned at the origin. At the head of this vector, the tail of the second vector is drawn according to its scale. The vector sum or resultant, indicated by **R**, is the magnitude and direction of the arrow drawn from the tail of the first vector to the head of the second vector. This method can be extended to include more than two vectors.

If the two vectors **A** and **B** are one-dimensional (i.e., lie along the same dimension), then they may be added as shown in Figure 3-3.

FIGURE 3-3

If the vectors are perpendicular to one another (i.e., vector **A** lies along the x-axis and vector **B** lies along the y-axis, as shown in Figure 3-4), then the resultant vector, **R**, represents the hypotenuse of the right triangle formed by the two perpendicular vectors, **A** and **B**. The magnitude of **R** can be found using the Pythagorean theorem:

$$c^2 = a^2 + b^2 \quad \text{or} \quad R^2 = A^2 + B^2$$

The following trigonometric function is used to determine the direction of the resultant vector:

$$\theta = \tan^{-1}\left(\frac{\text{opp}}{\text{adj}}\right) = \tan^{-1}\left(\frac{B}{A}\right)$$

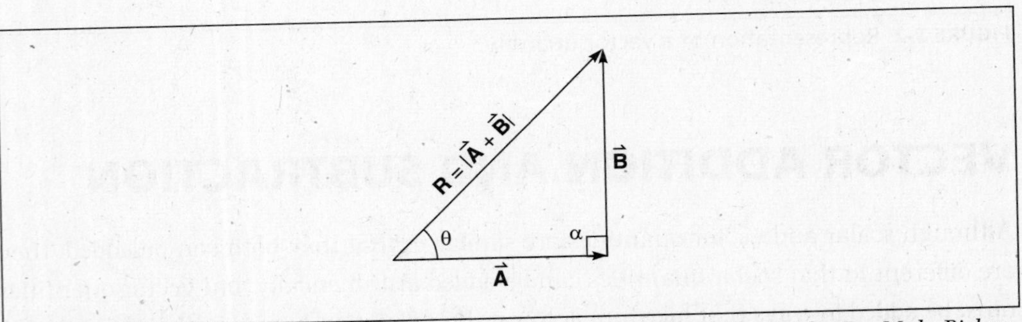

FIGURE 3-4: SOURCE: From George Hademenos, *Schaum's Outline of Physics for Pre-Med, Biology, and Allied Health Students*, McGraw-Hill, 1998; reproduced with permission of The McGraw-Hill Companies.

Component Method of Vector Addition

The component method of vector addition requires that the x and y components of each vector be determined. In a two-dimensional coordinate system, a vector can be positioned solely along the x direction, solely along the y direction, or can have components both in the x and y directions. A vector quantity oriented at an angle implies that a component of the quantity is in the x direction and a component of the quantity is in the y direction, as shown in Figure 3-5. To determine exactly how much of a

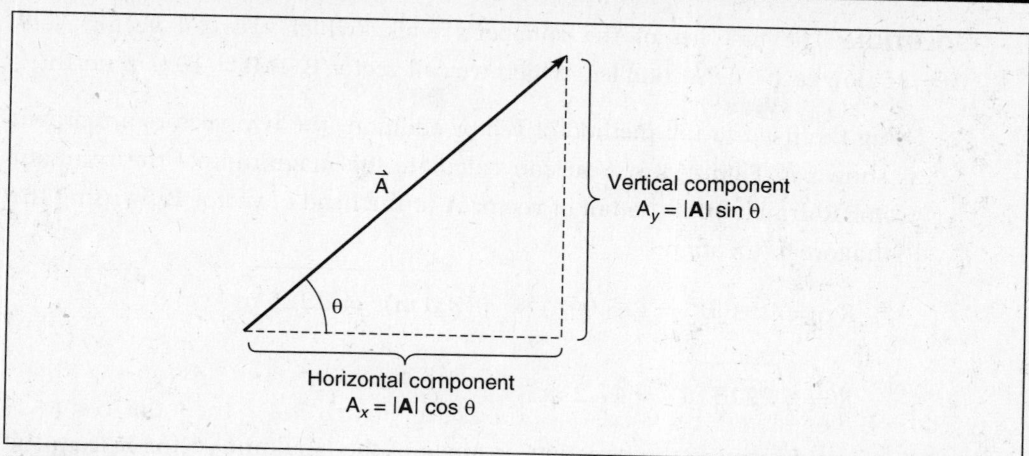

FIGURE 3-5

vector is in the x or the y directions, you use the following trigonometric functions to resolve the vector (**A**) in terms of its horizontal or x components (A_x) and its vertical or y components (A_y):

Horizontal component: $\cos\theta = \dfrac{\text{adj}}{\text{hyp}} = \dfrac{A_x}{A}$

Vertical component: $\sin\theta = \dfrac{\text{opp}}{\text{hyp}} = \dfrac{A_y}{A}$

The angles $\theta = 90^\circ$, 180°, 270°, and 360° are special cases. Assuming that $\theta = 0^\circ$ corresponds to the $+x$ axis,

- A vector **A** at 90° has the components: $A_x = 0$; $A_y = +|\mathbf{A}|$
- A vector **A** at 180° has the components: $A_x = -|\mathbf{A}|$; $A_y = 0$
- A vector **A** at 270° has the components: $A_x = 0$; $A_y = -|\mathbf{A}|$
- A vector **A** at 360° has the components: $A_x = +|\mathbf{A}|$; $A_y = 0$

Nevertheless, each component of the vectors is added, with the sum of the x components now representing the x component of the resultant vector, R_x, and the sum of the y components representing the y component of the resultant vector, R_y. Once you know the x and y components of the resultant vector, you can determine the magnitude of the resultant vector by using the Pythagorean theorem:

$$R^2 = R_x^2 + R_y^2$$

You can also determine the direction by using the trigonometric function:

$$\theta = \tan^{-1}\left(\frac{\text{opp}}{\text{adj}}\right) = \tan^{-1}\left(\frac{R_y}{R_x}\right)$$

EXAMPLE: On a walk across a large parking lot, a shopper walks 25.0 m to the east and then turns and walks 40.0 m to the north. Find the resultant of the two displacement vectors representing the shopper's walk using: (1) the head-to-tail method of vector addition and (2) the component method of vector addition.

SOLUTION: The first leg of the shopper's walk, which we call vector **A**, is **A** = 25.0 m east. The second leg, which we call vector **B**, is **B** = 40.0 m north.

1. Using the head-to-tail method of vector addition, the two vectors are drawn as shown in Figure 3-6. You can calculate the magnitude of the resultant vector **R** drawn from the tail of vector **A** to the head of vector **B** by using the Pythagorean theorem:

$$R^2 = A^2 + B^2 = (25.0\text{ m})^2 + (40.0\text{ m})^2 = 2225\text{ m}^2$$

$$R = \sqrt{2225\text{ m}^2} = 47.2\text{ m}$$

You can determine the direction or angle of the resultant vector **R** from the trigonometric relationship:

$$\theta = \tan^{-1}\left(\frac{\text{opp}}{\text{adj}}\right) = \tan^{-1}\left(\frac{B}{A}\right) = \tan^{-1}\left(\frac{40.0\text{ m}}{25.0\text{ m}}\right)$$

$$= 57.9^\circ \text{ (as measured from the positive } x\text{-axis)}$$

2. Use the component method of vector addition to resolve vectors **A** and **B** into their x and y components:

Vector **A**:	$A_x = 25.0\text{ m}$	$A_y = 0\text{ m}$
Vector **B**:	$B_x = 0\text{ m}$	$B_y = 40.0\text{ m}$
Vector **R**:	$R_x = A_x + B_x$	$R_y = A_y + B_y$
	$= 25.0\text{ m} + 0\text{ m}$	$= 0\text{ m} + 40.0\text{ m}$
	$= 25.0\text{ m}$	$= 40.0\text{ m}$

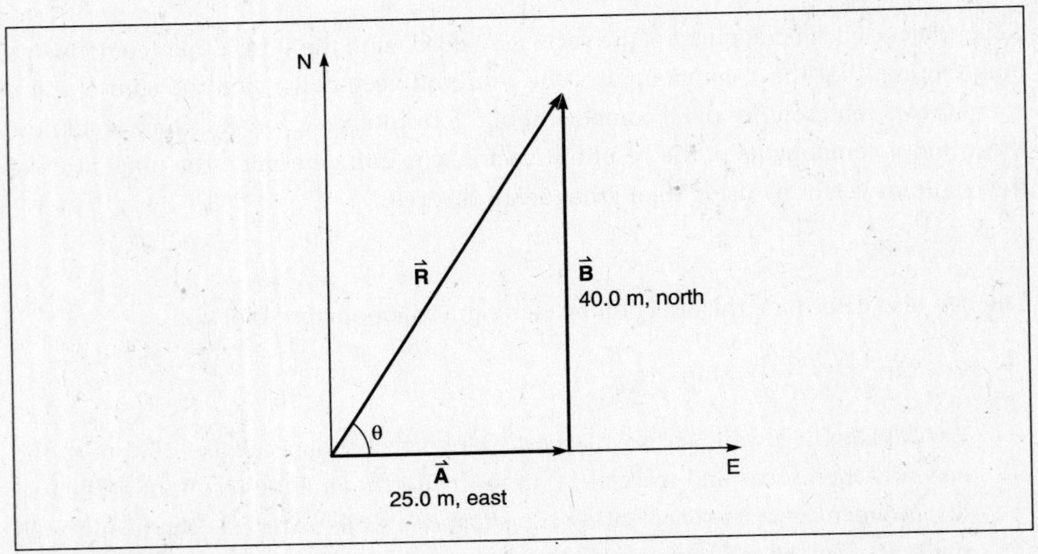

FIGURE 3-6

You can find the magnitude of the resultant vector **R** by using the Pythagorean theorem:

$$R^2 = R_x^2 + R_y^2 = (25.0\ \text{m})^2 + (40.0\ \text{m})^2 = 2225\ \text{m}^2$$

$$R = \sqrt{2225\ \text{m}^2} = 47.2\ \text{m}$$

The direction or angle of the resultant vector **R** can be determined from the trigonometric relationship

$$\theta = \tan^{-1}\left(\frac{\text{opp}}{\text{adj}}\right) = \tan^{-1}\left(\frac{R_y}{R_x}\right) = \tan^{-1}\left(\frac{40.0\ \text{m}}{25.0\ \text{m}}\right)$$

$$= 57.9^\circ \text{ (as measured from the positive } x\text{-axis)}$$

Vector Subtraction

Subtraction of a vector can be done by reversing the direction of the vector and adding it to the remaining vectors, as shown in Figure 3-7. Reversing the sign of the vector simply changes the direction of the vector without affecting its magnitude.

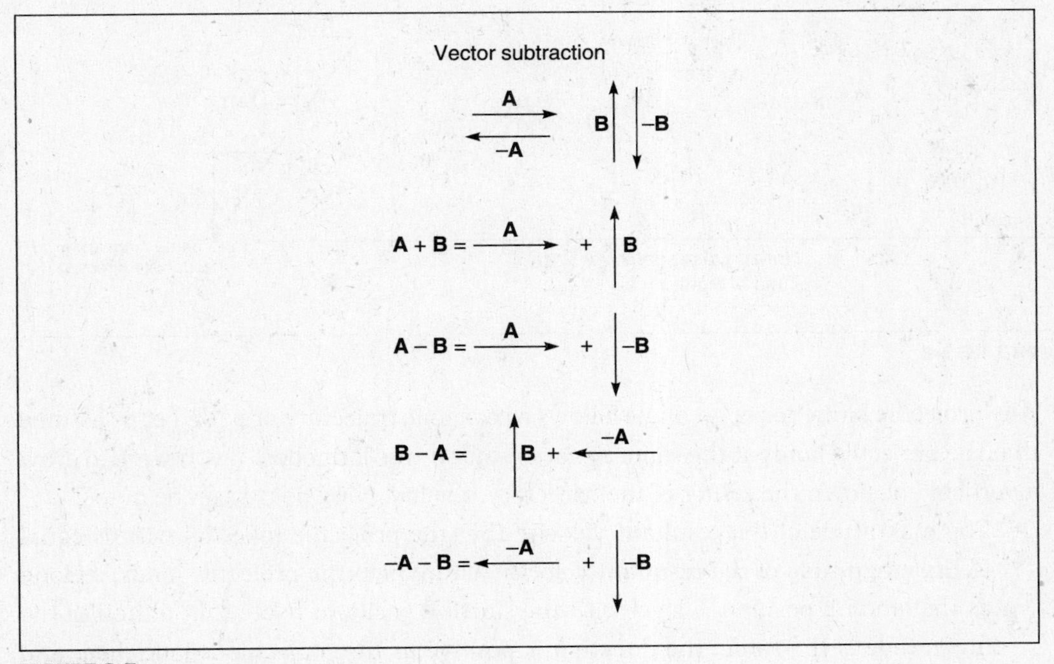

FIGURE 3-7

PROJECTILE MOTION

Motion is a physical concept that involves the movement or change in position of an object over time. The previous section dealt with one-dimensional motion, or the motion of an object moving **only** along the x-axis (from left to right, and vice versa) or **only** along the y-axis (from up to down, and vice versa). Next we discuss the concept of projectile motion, or motion of an object moving along **both** the x- and y-axes.

Although projectile motion involves motion in both the x- and y-axes, you do not need additional equations to solve such problems. All of the equations previously discussed, those equations pertaining to motion in the x direction and those equations pertaining to motion in the y direction, can be applied to solving projectile motion problems—as long as you consider the x and y motions separately. In addition to the equations for use in solving projectile motion problems, you can gain a tremendous amount of information by looking at the symmetry of projectile motion, as shown in Figure 3-8.

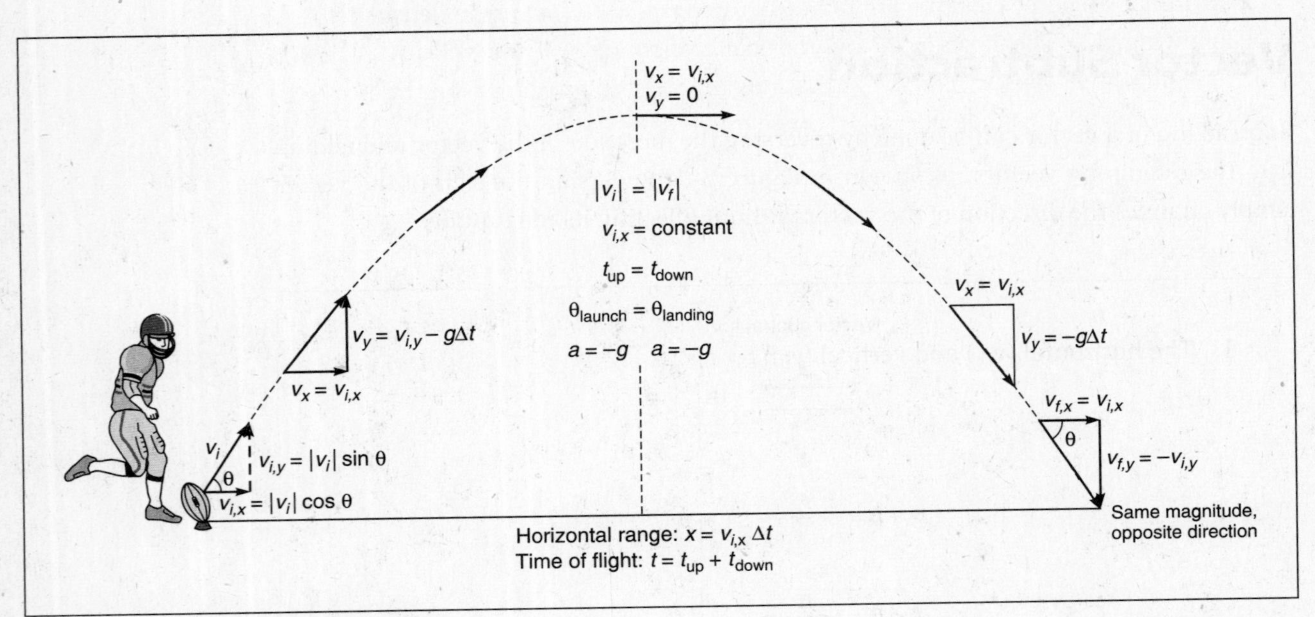

FIGURE 3-8

Any projectile launched at an angle follows a parabolic trajectory or path. Let us assume that the projectile lands at the same elevation that it was launched. If you were to draw a vertical line down the center of the trajectory, the following points can be noted:

- The magnitude of the resultant velocity that the projectile takes off with is equal to the magnitude of the resultant velocity with which the projectile lands, as long as the landing position is level with the starting position. (Note that although the magnitude is the same, the direction is positive as the projectile is launched, and the direction is negative as the projectile lands.)
- The horizontal component of the initial velocity in the first half of the trajectory is equal to the horizontal component of the initial velocity in the second half of the trajectory. (In other words, the horizontal component of the velocity is constant and the same throughout the entire trajectory.)
- The magnitude of the launch angle of the projectile is equal to the magnitude of the landing angle of the projectile.

- The distance traveled in the first half of the trajectory is equal to the distance traveled in the second half of the trajectory.
- The time required for the projectile to travel in the first half of the trajectory is equal to the time required for the projectile to travel in the second half of the trajectory.
- Because gravity is always acting downward, the vertical component of the initial velocity decreases in the first half of the trajectory at a rate of 9.8 m/s for every second in flight until it reaches zero at its maximum elevation. The vertical component of the velocity then increases at the same rate as the projectile accelerates in the second half of the trajectory until it lands.

EXAMPLE: A football is kicked from a tee with an initial velocity of 20 m/s and an initial angle of elevation of 40°. Assuming $g = -9.8\ \text{m/s}^2$, determine the following: (1) The horizontal and vertical components of the velocity; (2) the football's total time in flight; (3) the maximum height attained by the football; (4) the horizontal range or total distance traveled by the football; and (5) the final horizontal and vertical components of the velocity of the football prior to striking the ground.

SOLUTION: In this problem, let y represent the vertical height and choose up as the positive direction. Also, $g = -9.8\ \text{m/s}^2$.

1. The horizontal (v_{ix}) and vertical (v_{iy}) components of the velocity are

$$v_{ix} = v_i \cos 40° = \left(20\ \frac{\text{m}}{\text{s}}\right)(0.766) = 15.3\ \frac{\text{m}}{\text{s}}$$

$$v_{iy} = v_i \sin 40° = \left(20\ \frac{\text{m}}{\text{s}}\right)(0.643) = 12.9\ \frac{\text{m}}{\text{s}}$$

2. The equation of motion needed to solve this problem is

$$\Delta y = v_{iy}\Delta t + \frac{1}{2}g\Delta t^2$$

Because the height of the projectile at kickoff and at landing is the same, $\Delta y = 0$. Substituting into the previous equation and solving for t yields

$$0 = \left(12.9\ \frac{\text{m}}{\text{s}}\right)\Delta t + \left(-4.9\ \frac{\text{m}}{\text{s}^2}\right)\Delta t^2$$

The previous equation is a quadratic equation, and thus the solution can be found using

$$x = \frac{-B \pm \sqrt{B^2 - 4AC}}{2A}$$

where $x = \Delta t$, $A = -4.9\ \text{m/s}^2$, $B = 12.9$ m/s; and $C = 0$. On substitution, the solution yields two roots: $t = 0$ s and 2.63 s. The root $t = 0$ s is the time at kickoff and the root $t = 2.63$ s is the time at landing.

3. The maximum height attained by the football can be determined from the equation of motion

$$v_{fy}^2 = v_{iy}^2 + 2g\Delta y$$

where $v_{fy} = 0$ m/s (because at its maximum height, the football stops in the vertical direction), $v_{iy} = 12.9$ m/s, and $g = -9.8$ m/s^2. Substituting the appropriate values yields

$$\Delta y = \frac{-v_{iy}^2}{-2g} = 8.5 \text{ m}$$

4. The horizontal distance traveled by the football is given by the equation of motion $\Delta x = \bar{v}\Delta t$, where $\bar{v}$ is the average velocity along the horizontal axis, determined by the horizontal component of the initial velocity, which remains constant throughout the trajectory. The horizontal component of the velocity can be calculated from $v_{ix} = v_i \cos\theta = (20 \text{ m/s}) \cos 40^\circ = 15.3$ m/s, and Δt is the total time in flight given by the solution in part 2 of this problem ($\Delta t = 2.63$ s). Therefore, the horizontal distance, or range of the football, is

$$\Delta x = \bar{v}\Delta t = \left(15.3 \frac{\text{m}}{\text{s}}\right)(2.63 \text{ s}) = 40.2 \text{ m}$$

5. The horizontal velocity of the football remains constant, so $v_x = v_{ix} = 15.3$ m/s. The vertical velocity is given by the equation of motion

$$v_{fy} = v_{iy} + g\Delta t = \left(12.9 \frac{\text{m}}{\text{s}}\right) + \left(-9.8 \frac{\text{m}}{\text{s}^2}\right)(2.63 \text{ s}) = -12.9 \frac{\text{m}}{\text{s}}$$

You should note that $v_{iy} = -v_{fy}$. The two velocity quantities are equal in magnitude, with the negative sign implying that the velocity is directed downward.

As stated earlier, this discussion and application of equations involving projectile motion assumes that the projectile lands at the same elevation that it was launched. How would these equations apply to projectile motion if the projectile lands at a higher or lower elevation than the one from which it was launched? In these cases, the displacement along the vertical axis (Δy) is different, as noted in the drawings in Figure 3-9.

EXAMPLE: An amateur golfer strikes a golf ball at an angle of 50° with respect to the horizontal axis and strikes a tree 45 m away at a point 2 m above ground level. Determine the initial speed of the golf ball.

SOLUTION: First calculate the time the golf ball is in flight by using the equations:

$$\Delta x = v_{i,x}\Delta t = v_i \cos\theta_i \Delta t$$

$$\Delta t = \frac{\Delta x}{v_i \cos\theta_i} = \frac{45 \text{ m}}{v_i \cos 50^\circ}$$

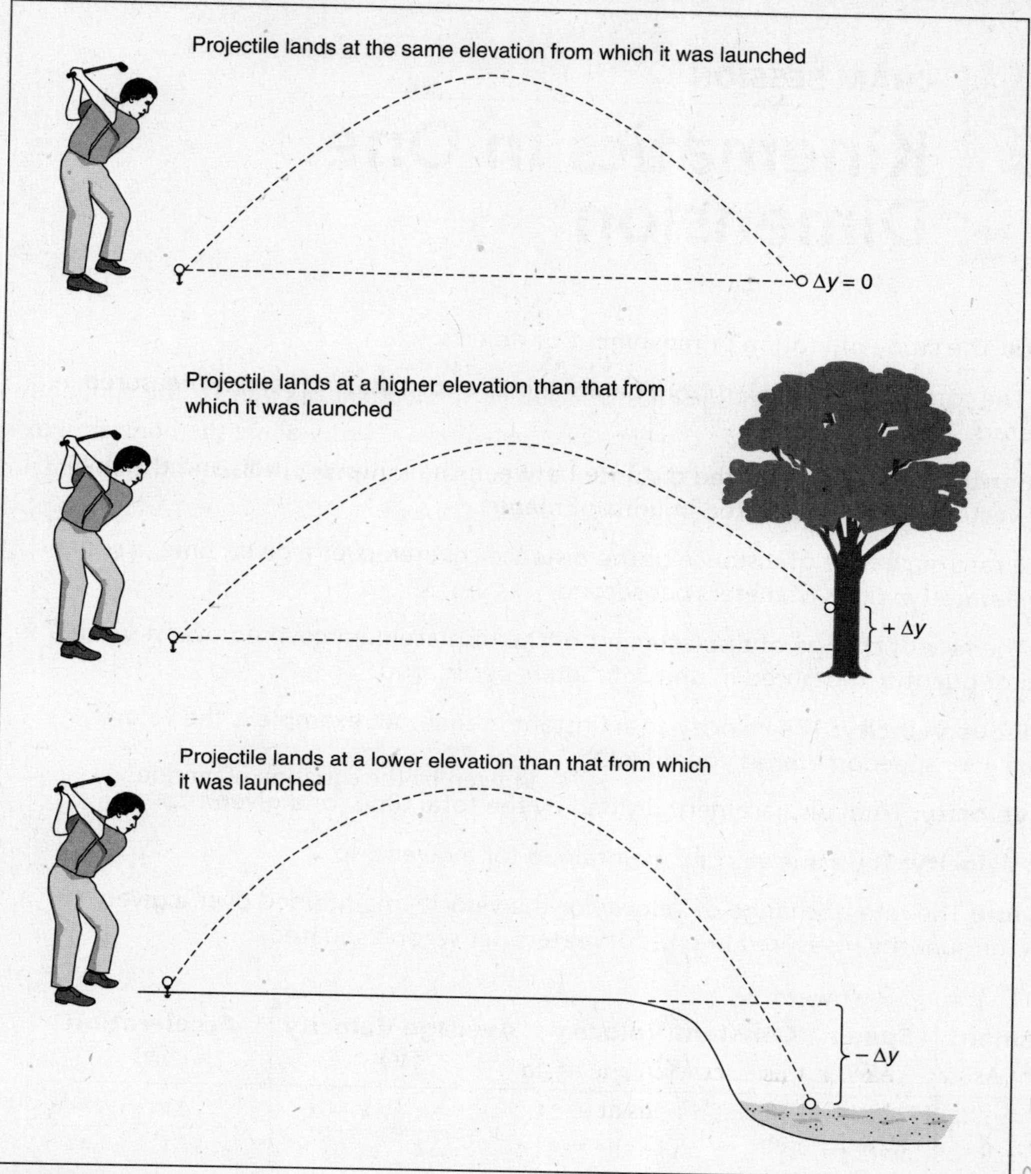

FIGURE 3-9

Substituting this expression of Δt into

$$\Delta y = v_i \sin \theta_i \Delta t + \frac{1}{2} g\,(\Delta t)^2$$

$$2 \text{ m} = v_i \sin 50° \left(\frac{45 \text{ m}}{v_i \cos 50°}\right) + \frac{1}{2}\left(-9.8 \ \frac{\text{m}}{\text{s}^2}\right)\left(\frac{45 \text{ m}}{v_i \cos 50°}\right)^2$$

Solving for v_i yields

$$v_i = 21.6 \text{ m/s}$$

CRAM SESSION

Kinematics in One Dimension

Kinematics: The study of motion or movement of objects.

Distance: The total length traveled from beginning to end; a scalar quantity measured in units of meters.

Displacement: Straight-line directed distance between the initial position and the final position; a vector quantity measured in units of meters.

Speed: The rate of change of distance or the distance covered over a given time; a scalar quantity measured in units of meters per second.

Velocity: The rate of change of displacement or the displacement covered over a given time; a vector quantity measured in units of meters per second.

Instantaneous velocity: The velocity at an instant in time. An example is the velocity recorded by your speedometer as you drive.

Average velocity: Total displacement divided by the total time for a given trip.

Constant velocity: The same velocity maintained for a given trip.

Acceleration: The rate of change of velocity or the velocity maintained over a given time; a vector quantity measured in units of meters per second squared.

Displacement (Δx)	Speed (v)	Constant Velocity (v)	Average Velocity ($\bar{v}$)	Acceleration (a)
$\Delta x = x_f - x_i$	$v = \dfrac{d}{t}$	$v = \dfrac{x}{t}$	$v = \dfrac{\Delta x}{\Delta t} = \dfrac{x_f - x_i}{\Delta t}$	$a = \dfrac{\Delta v}{\Delta t} = \dfrac{v_f - v_i}{\Delta t}$

Equations of motion: Four equations of motion can be derived in terms of all of the variables based on these relationships. Each equation consists of four unique variables; thus, the equation to use for a given problem depends on which four variables are described in the scenario. These equations apply to one-dimensional motion, either along the x-axis or along the y-axis. The only two differences are that (1) $\Delta x \rightarrow \Delta y$ and (2) $a = -g$ (the acceleration due to gravity).

Involved Variables	Equation of Motion in *x* Direction	Equation of Motion in *y* Direction	Involved Variables
$\Delta x, v_i, v_f, \Delta t$	$\Delta x = \frac{1}{2}(v_i + v_f)\,\Delta t$	$\Delta y = \frac{1}{2}(v_i + v_f)\,\Delta t$	$\Delta y, v_i, v_f, \Delta t$
$v_i, v_f, \Delta t, a$	$v_f = v_i + a\Delta t$	$v_f = v_i + g\Delta t$	$v_i, v_f, \Delta t, g$
$\Delta x, v_i, \Delta t, a$	$\Delta x = v_i\Delta t + \frac{1}{2}a\,(\Delta t)^2$	$\Delta y = v_i\Delta t + \frac{1}{2}g\,(\Delta t)^2$	$\Delta y, v_i, \Delta t, g$
$v_i, v_f, \Delta x, a$	$v_f^2 = v_i^2 + 2a\Delta x$	$v_f^2 = v_i^2 + 2g\Delta y$	$v_i, v_f, \Delta y, g$

Strategies for Solving One-Dimensional Kinematics Problems

1. Identify and list all given information (known and unknown variables).
   - ➤ Do not assume that all information given in a problem is required to solve the problem.
   - ➤ Look for key words that might be just as important as values: **rest, drop** implies $v_i = 0$, and **stop** implies $v_f = 0$. Also, words like **constant, speed up, increase, slow down,** and **decrease** are important in describing the acceleration of the object.
   - ➤ Problems commonly involve the sign of g. Because gravity acts downward, g should be a negative quantity. However, care should be taken to ensure the proper sign convention of all other quantities as well. For example, an object dropped from a cliff will have fallen −5 m after 1 second. The negative value implies a distance in the $-y$ direction.
2. Make sure all units are consistent (in SI system of units), and if not, perform required conversions.
3. Choose an equation that can be solved with the known variables. Each problem should include information about four variables. Match these four variables with the outer columns to find the correct equation.
4. Substitute all variables in proper units in the chosen equation, perform the necessary algebraic operations, arrive at the solution, and then ask yourself: Does the answer make sense?

CRAM SESSION

Vectors and Scalars

Vector: A physical quantity that possesses both magnitude and direction. Examples of vector quantities include displacement, velocity, acceleration, and force.

Scalar: A physical quantity that possesses magnitude only. Examples of scalar quantities include mass, distance, speed, and temperature.

Vector Representation

A vector is generally noted by a bold-faced letter or a letter with an arrow drawn over the top of the letter. For example, a vector quantity S can be represented by the symbols **S** or $\vec{S}$. The magnitude of the vector quantity S is typically indicated by the vector quantity symbol encased within absolute value bars, |**S**| or $|\vec{S}|$. In addition, a vector quantity is represented graphically by an arrow, with the size of the arrow corresponding to the size or magnitude of the vector quantity and the orientation of the arrow corresponding to the direction of the vector quantity.

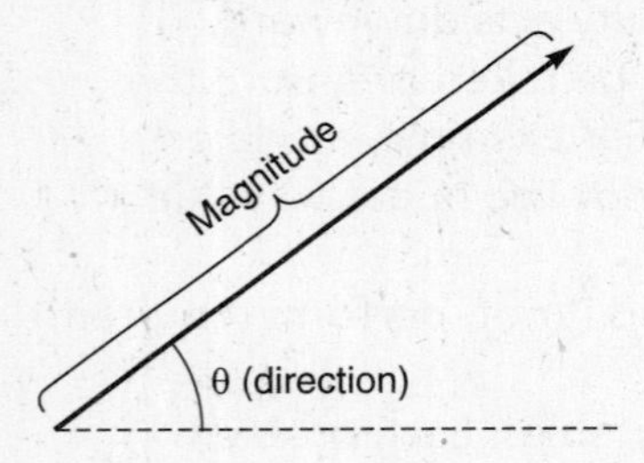

Vector Addition

Although scalar quantities and vector quantities are similar in that they both can be added, they are different in that scalar quantities can be added arithmetically whereas vector quantities must be added in ways that take into account their direction as well as their magnitude. One method of vector addition is the head-to-tail method; another method is the component method.

Head-to-tail method of vector addition: In this method of vector addition, one of the vectors, **A**, is drawn to scale with its tail positioned at the origin. At the head of this vector, the tail of the second vector, **B**, is drawn according to its scale. The vector sum or resultant, indicated by **R**, is the magnitude and direction of the arrow drawn from the tail of the first vector to the head of the second vector. The magnitude of **R** can be found using Pythagorean's theorem: $C^2 = A^2 + B^2$ or $R^2 = A^2 + B^2$.

The following trigonometric function is used to determine the direction of the resultant vector:

$$\theta = \tan^{-1}\left(\frac{\text{opp}}{\text{adj}}\right) = \tan^{-1}\left(\frac{B}{A}\right)$$

Component method of vector addition: The component method of vector addition requires that the *x* and *y* components of each vector be determined. To determine exactly how much of a vector is in the *x* or the *y* directions, you use the following trigonometric functions to resolve the vector (**A**) in terms of its horizontal or *x* components (A_x) and its vertical or *y* components (A_y):

Horizontal component: $\cos\theta = \dfrac{\text{adj}}{\text{hyp}} = \dfrac{A_x}{A}$ Vertical component: $\sin\theta = \dfrac{\text{opp}}{\text{hyp}} = \dfrac{A_y}{A}$

Each component of the vectors is added, with the sum of the *x* components now representing the *x* component of the resultant vector, R_x, and the sum of the *y* components representing the *y* component of the resultant vector, R_y. Once the *x* and *y* components of the resultant vector are known, the magnitude of the resultant vector can be determined using Pythagorean's theorem, $R^2 = R_x^2 + R_y^2$, and the direction obtained from using the trigonometric function:

$$\theta = \tan^{-1}\left(\frac{\text{opp}}{\text{adj}}\right) = \tan^{-1}\left(\frac{R_y}{R_x}\right)$$

CRAM SESSION

Kinematics in Two Dimensions

Projectile motion: Motion of objects moving in two dimensions simultaneously. Examples of projectile motion include a kicked soccer ball, a passed football, and a thrown baseball.

Trajectory: Parabolic path of a projectile in motion.

Projectile Motion Problems

In projectile motion problems, an object is moving in the x and y directions simultaneously. This means that all of the equations discussed under Kinematics in One Dimension also apply here. There are two general types of projectile motion problems: horizontally launched projectile motion and projectile motion launched at an angle.

Horizontally launched projectile motion

In this type of projectile motion, an object moving at constant velocity rolls off of a horizontal surface and drops a certain height to the ground below.

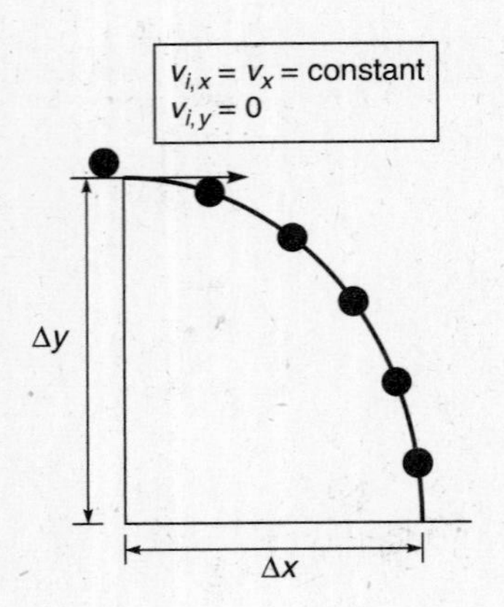

If given Δy and Δx, you can find v_x by:

(a) finding the time required for the object to strike the ground using $\Delta y = \frac{1}{2}g(\Delta t)^2$ and

(b) using $v_x = \frac{\Delta x}{\Delta t}$

If given Δy and v_x, you can find Δx by:

(a) finding the time required for the object to strike the ground using $\Delta y = \frac{1}{2}g(\Delta t)^2$ and

(b) using $\Delta x = v_x \Delta t$

Projectile motion launched at an angle

In this type of projectile motion, an object is launched from a horizontal surface at an angle measured with respect to the $+x$-axis and follows a parabolic trajectory. Problems can involve the projectile landing at an elevation higher or lower than the plane from which it was launched, as well as the projectile landing on a plane at the same elevation as the starting plane.

***x* direction:**

Initial velocity component, $v_{ix} = v_i \cos\theta$

Velocity component, $v_{fx} = v_{ix} =$ constant

Acceleration component, $a_x = 0$ m/s^2

***y* direction:**

Initial velocity component, $v_{iy} = v_i \sin\theta$

Velocity component, $v_{fy} = v_{iy} + g\Delta t$

Acceleration component, $a_y = -g = -9.8$ m/s^2

Another important point of consideration in analyzing projectile motion problems is symmetry.

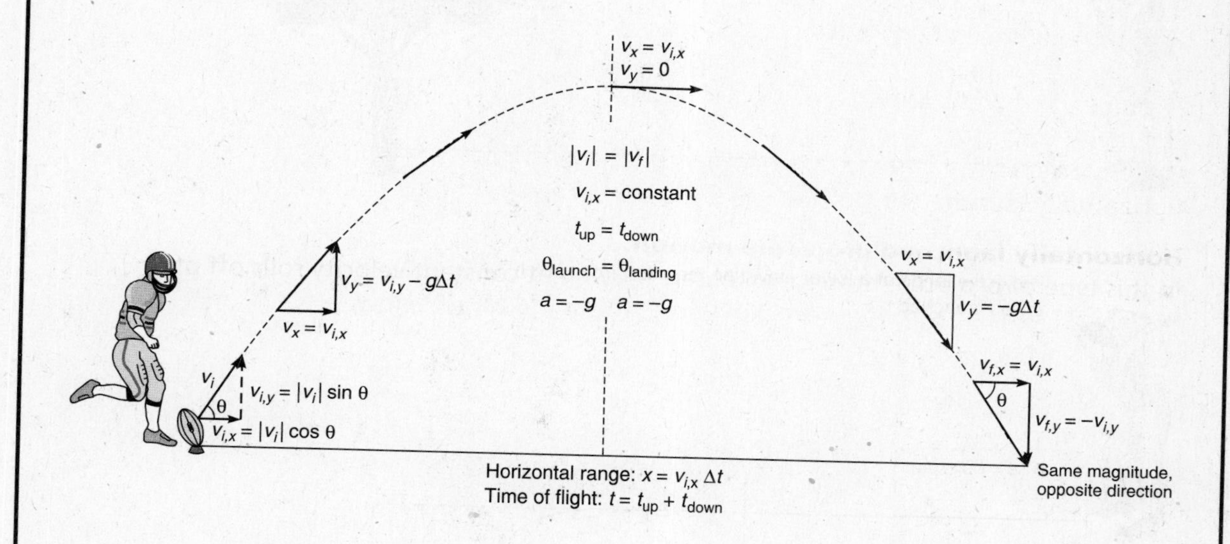

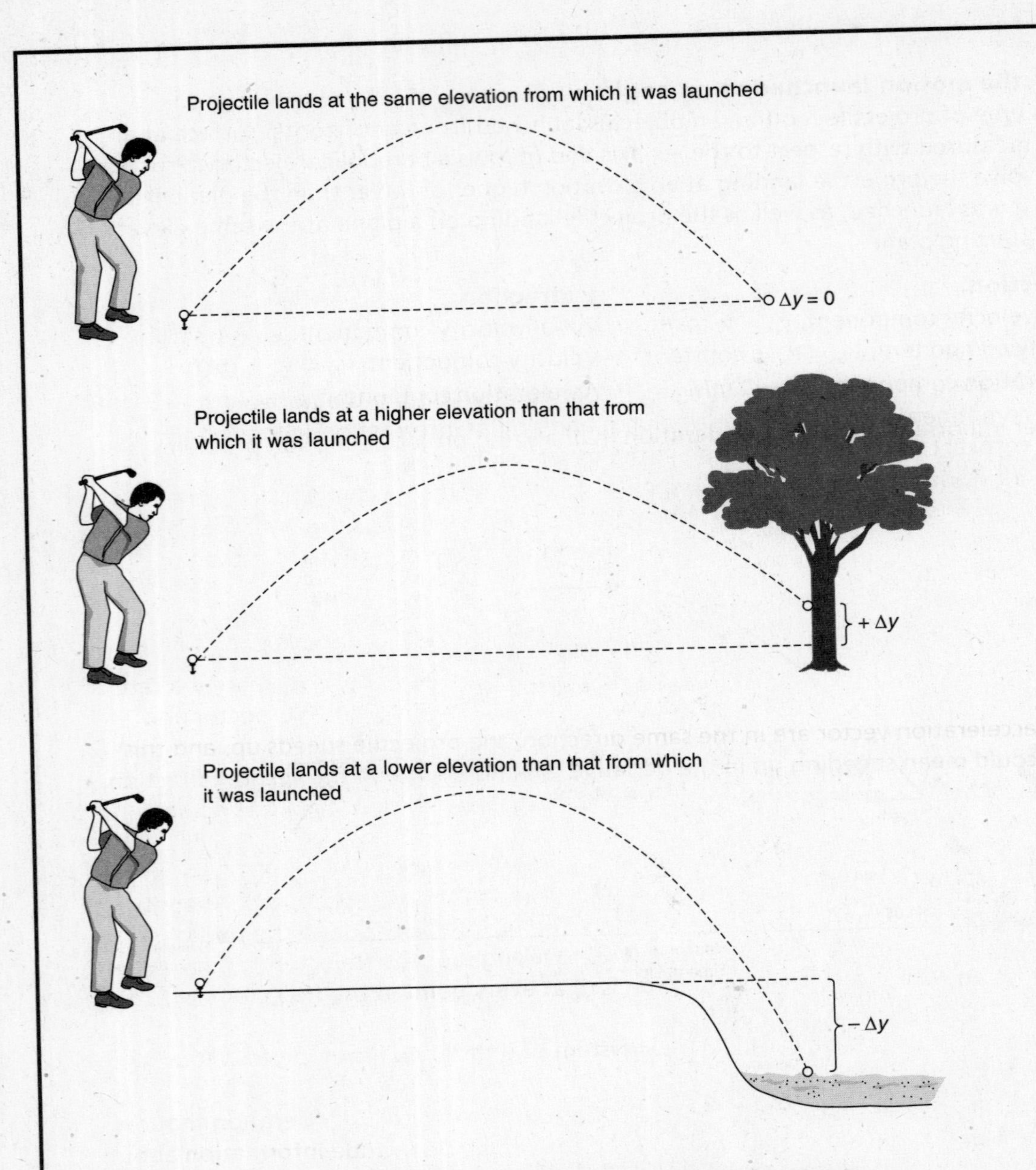

Strategies for Solving Two-Dimensional Kinematics Problems

1. Identify the type of projectile motion scenario (horizontally launched projectile motion or projectile motion launched at an angle).
2. Identify and list all given information (known and unknown variables).
   a. If the problem involves a horizontally launched projectile:
   - What is the unknown variable?
   - Write down all given information. (Remember, the mass of the object has not been discussed, is not present in any of the equations of motion, and therefore is extraneous information not needed to solve these problems.)

- ➤ The time required for the projectile to reach the ground after leaving the horizontal surface can be found from the free-fall equation $\Delta y = \frac{1}{2}g(\Delta t)^2$.
- ➤ Problems involve the sign of g. Because gravity always acts downward, g should be a negative quantity ($g = -9.8$ m/s^2). However, care should be taken to ensure the proper sign convention of all other quantities as well. For example, an object dropped from a cliff will have fallen -5 m after 1 second. The negative value implies a distance in the $-y$ direction.

b. If the problem involves a projectile launched at an angle, survey the information in the problem by considering the symmetry in the problem.

- ➤ Create a 2-column table listing all information related to the horizontal component of motion in one column and similar information related to the vertical component of motion in the other column.
- ➤ For horizontal motion, the horizontal initial velocity, v_{ix}, can be calculated from $v_{ix} = v_i \cos\theta$. The horizontal velocity is constant throughout the entire trajectory of the projectile. Because the horizontal velocity is constant, the projectile is not accelerating in the x-direction.
- ➤ For vertical motion, the vertical initial velocity, v_{iy}, can be calculated from $v_{iy} = v_i \sin\theta$. As the projectile is launched, it is slowed down at a rate of 9.8 m/s every second ($=g$) until it is totally stopped. It then begins to speed up at a rate of 9.8 m/s every second until it strikes the ground. When the velocity vector and acceleration vector are in the same direction, the projectile speeds up, and this could mean speeding up in the negative direction. Thus, in the vertical direction, whether it is moving upward or downward, the projectile is accelerating at $g = -9.8$ m/s^2.
- ➤ NOTE: Once it is launched, the projectile is subject to gravity at every point in the trajectory, including at the very top. At the top of the trajectory, even though the vertical component of velocity is 0 m/s, it still has velocity due to the horizontal component of velocity, which is constant throughout the trajectory, and acceleration in the vertical direction is g, at every point in the trajectory, including at the point of maximum height.

3. Make sure all units are consistent (in SI system of units), and if not, perform the required conversions.
4. Choose an equation specific to direction or component under consideration that can be solved with the known variables. Each problem should include information about four variables. Match these four variables with the outer columns in the Equations of Motion table in the Cram Session on Kinematics in One Dimension to find the correct equation.
5. Substitute all variables in proper units in the chosen equation, perform the necessary algebraic operations, arrive at the solution, and then ask yourself: Does the answer make sense?

CHAPTER 4

Forces and Newton's Laws

Read This Chapter to Learn About:

- ➤ Forces in Nature
- ➤ Mass, Inertia, and Weight
- ➤ Center of Gravity/Center of Mass
- ➤ Newton's Three Laws of Motion
- ➤ Free-Body Diagrams
- ➤ Law of Gravitation
- ➤ Translational and Rotational Equilibrium
- ➤ Translational and Rotational Motion
- ➤ Uniform Circular Motion and Centripetal Force

Chapter 3 discussed kinematics, or the concepts and equations that describe the motion of an object being uniformly accelerated. The variables and equations that describe an object's motion in one dimension (along the *x*- and *y*-axes) and two dimensions (projectile motion) were defined and applied to specific problems and examples. This chapter explains the causes and the factors involved in motion.

FORCES IN NATURE

Given any object, whether an electron, a person, or a planet, what causes the object to move? The answer is simple—a force. **Force** can be defined simply as a push or a pull. Forces cause objects to move. Forces occur everywhere in nature on every object—but the mere presence of a force does not mean that an object will necessarily move. It is electric forces that are responsible for the movement of current through an electric circuit, mechanical forces that can cause a person to move, and gravitational forces that cause planets to move. The reasons are discussed later as part of Newton's Laws of Motion.

A force is a vector quantity expressed in a unit called the newton (N):

$$1 \text{ newton} = 1 \frac{\text{kilogram} \cdot \text{meter}}{\text{second}^2} \quad \text{or} \quad 1 \frac{\text{kg} \cdot \text{m}}{\text{s}^2}$$

All forces, whether they are electric, mechanical, or gravitational, are expressed in units of newtons, regardless of their source. Gravitational forces are discussed later in this chapter and electric forces are discussed in Chapter 12. Weight, normal force, friction, and tension are four specific types of mechanical forces that exist in nature and are defined as follows:

1. **Weight** is the force exerted on an object by gravity. It is also referred to as the force due to gravity. Any object that has mass has weight. Weight can be calculated by the equation

   Weight = (mass) (acceleration due to gravity)

   $$\boldsymbol{W} = m\boldsymbol{g}$$

2. **Normal force** is the force exerted on an object by a surface. It is also referred to as the support force. Normal force always acts perpendicular to the surface that is supporting the object.
3. **Friction** is a force generated by the properties of the interface between a moving object and a surface. Friction acts in a direction opposite to an object's motion. The force due to friction, $\mathbf{F}_f$, is defined as

   Frictional force = (coefficient of friction) (normal force)

   $$\mathbf{F}_f = \mu\mathbf{N}$$

   There are two types of frictional forces corresponding to the state of motion of the object. If the object is stationary, then a frictional force (static friction) is acting on the object to prevent motion, described by

   $$\mathbf{F}_{f,s} \leq \mu_s\mathbf{N}$$

   where μ_s is the coefficient of static friction. If the object is set in motion, then the object is subject to kinetic (sliding) friction, defined as

   $$\mathbf{F}_{f,k} = \mu_k\mathbf{N}$$

   Static friction is generally greater in magnitude than kinetic friction because an object requires a larger force to start an object in motion than it does to be kept in motion.
4. **Tension** is the force exerted by a string, rope, or cable on a suspended object.

MASS, INERTIA, AND WEIGHT

Before you can understand Newton's Laws of Motion, you must make a distinction between three physical terms that are often confused: mass, inertia, and weight.

- **Mass**, measured in SI units of kilograms, is the amount of substance that an object has.
- **Inertia** is an object's resistance to motion. What makes an object resistant to motion is not an object's size, but its mass. The more mass an object has, the more inertia it has, making the object more resistant to motion. Inertia does not have a physical unit, but is indicated by the object's mass.
- **Weight**, a force exerted on an object due to gravity, is often confused with mass. These are two very different quantities. Weight is a vector quantity expressed in units of newtons, whereas mass is a scalar quantity expressed in units of kilograms. The weight of an object is calculated by multiplying the mass of the object times the acceleration due to gravity, or

  $$\text{Weight} = (\text{mass})\,(\text{acceleration due to gravity})$$

  The mass of an object does not change unless material is added or taken away. The weight of an object can change, for example, on a different planet, where the acceleration due to gravity is different. The mass of a person on Earth is the same as the mass of that person on Mars, as the mass always remains the same. However, the weight of the person differs because each planet has a different gravitational field and hence a different value of g (the acceleration due to gravity). Because the g on Mars is $g = -3.8\ \text{m/s}^2$, there is less of a pull on the person due to gravity, and hence the weight of the person on Mars is less.

CENTER OF GRAVITY/CENTER OF MASS

Center of Gravity

Although each and every atom and molecule composing a given object is subject to Earth's gravitational force, it becomes convenient to consider a single point or location within the object where the gravitational force is concentrated. The **center of gravity** of an object is that point where the force of gravity is considered to act.

Center of Mass

Up to now, our discussion of physical principles has been applied to a particle or point source where the mass is concentrated at a single point. However, the mass of many objects ranging from a molecule to the human body is distributed over the object. For example, the human body is composed of a number of extended objects (i.e., arms, legs, head, chest, and abdomen), all of which possess mass. To simplify problems involving the statics of rigid bodies, it becomes convenient to identify a single point where the total mass can be concentrated, and it can be determined from knowledge of the mass of the extended object and its distance from a central point defined as the origin. In two-dimensional coordinates, the **center of mass** is given by

$$x_{CM} = \frac{\sum xm}{\sum m} = \frac{x_1m_1 + x_2m_2 + x_3m_3 + \cdots + x_nm_n}{m_1 + m_2 + m_3 + \cdots + m_n}$$

$$y_{CM} = \frac{\sum ym}{\sum m} = \frac{y_1m_1 + y_2m_2 + y_3m_3 + \cdots + y_nm_n}{m_1 + m_2 + m_3 + \cdots + m_n}$$

Usually, the center of mass and center of gravity are located at the same point and vary only for extremely large objects, where the acceleration due to gravity may vary from point to point within the object.

NEWTON'S THREE LAWS OF MOTION

Newton's First Law of Motion

An object at rest remains at rest, and an object in motion remains in motion unless acted upon by an unbalanced force.

Newton's First Law is also known as the "Law of Inertia." Inertia is an object's resistance to motion. The more inertia an object has, the harder it is to move the object. Thus, the more mass an object has, the more inertia it has, and the harder it is for you to move that object.

EXAMPLE: When you are in a car driving down the highway going 65 mph, not only is the car going 65 mph, but you and everyone else in the car are also moving at a velocity of 65 mph. The car continues to move at a velocity of 65 mph. Now let's say that the driver runs into a telephone pole, bringing the car to a complete stop. The telephone pole stops the car, but what about the driver and passengers? The pole does not stop the driver and passengers, so they continue moving at 65 mph until something does stop them—usually the steering wheel and the windshield.

Newton's Second Law of Motion

The acceleration of an object is directly proportional to the net external force acting on the object, and inversely proportional to the object's mass.

In equation form, Newton's Second Law of Motion can be expressed as

$$\Sigma \mathbf{F} = m\mathbf{a}$$

Force = (mass) (acceleration)

Force is measured in units of newtons (N), where $1\ N = 1\ kg \cdot m/s^2$. The Σ is the Greek capital letter sigma and signifies a sum. In this context, the sum of the forces equals an object's mass times its acceleration. Also, because force is a vector, this equation applies to forces acting along the x and y directions, or

$$\Sigma F_x = ma_x$$

$$\Sigma F_y = ma_y$$

EXAMPLE: Golf ball versus bowling ball: Which hits the ground first? When released from rest from the same height at the same time, which object will strike the ground first—the golf ball or the bowling ball? Any object released from rest and dropped is accelerated at a constant rate by gravity ($g = -9.8\ m/s^2$). Because g is constant, the force due to gravity is smaller for the golf ball because of the smaller mass of the golf ball in comparison to the bowling ball. In other words,

$$g = \frac{F_{\text{golf ball}}}{m_{\text{golf ball}}} = \frac{F_{\text{bowling ball}}}{m_{\text{bowling ball}}} = \text{constant}$$

Gravity exerts a greater pulling force on the greater mass such that its value remains constant. Therefore, neglecting air resistance, the golf ball and the bowling ball hit the ground at the same time.

Newton's Third Law of Motion

For every action force exerted on an object, there is an equal yet opposite reaction force exerted by the object.

This law is also known as the Law of Action and Reaction.

EXAMPLE: A person standing exerts a force on the ground equal to his/her weight. The ground, in turn, exerts an equal yet opposite force on the person, supporting the weight of the person. The force exerted by the ground on the person is the normal force, as defined previously.

FREE-BODY DIAGRAMS

An important technique used to solve problems involving forces acting on an object is free-body diagrams. A **free-body diagram** is a diagram in which an object is isolated, and all forces acting on the object are identified and represented on the diagram.

In creating free-body diagrams, you should

- Isolate the object in an imaginary coordinate system in which the object represents the origin.
- Identify and represent all forces (with appropriate magnitude and direction) acting on the object.

- Resolve all forces in terms of their x and y components.
- Substitute into the appropriate Newton's Second Law of Motion, either $\Sigma F_x = ma_x$ or $\Sigma F_y = ma_y$, setting the sum of all forces in each direction equal to ma if the object is accelerating and equal to 0 if the object is not accelerating.
- Solve for the unknown variable.

An example of using free-body diagrams to solve a problem is in the case where an object of mass m is accelerating down a ramp with a coefficient of kinetic friction μ_k and inclined at an angle θ, as shown in Figure 4-1.

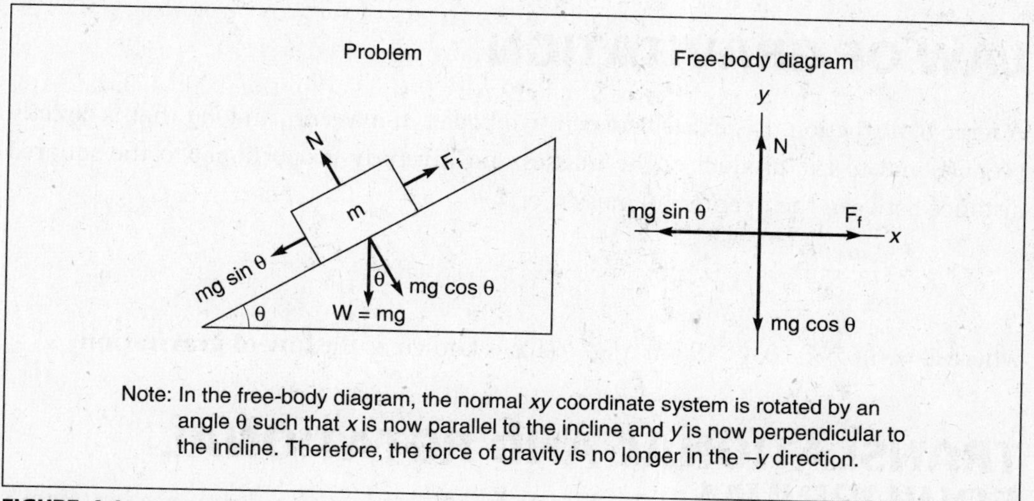

FIGURE 4-1: SOURCE: From George Hademenos, *Schaum's Outline of Physics for Pre-Med, Biology, and Allied Health Students*, McGraw-Hill, 1998; reproduced with permission of The McGraw-Hill Companies.

The object is at rest on the inclined plane of $\theta = 30^\circ$, and you are asked to determine the coefficient of static friction, μ_s, between the object and the inclined plane. The coordinate system is rotated such that the x-axis is now along the incline. As noted in the free-body diagram, there are three forces identified from the scenario: the weight of the object acting down (for which there is a component acting in the x direction and a component in the y direction), the normal force acting perpendicular to the inclined plane, and the frictional force that acts opposite to the direction of motion. Substituting these forces into Newton's Second Law,

$$\Sigma F_x = ma_x \qquad \Sigma F_y = ma_y$$

$$\mathbf{F}_{f,s} - mg\sin\theta = 0 \qquad \mathbf{N} - mg\cos\theta = 0$$

By definition, the static frictional force is $\mathbf{F}_{f,s} = \mu_s\mathbf{N}$, where $\mathbf{N}$ can be found from the forces acting in the y direction: $\mathbf{N} = mg\cos\theta$. So, the frictional force is

$$\mathbf{F}_{f,s} = \mu_s\mathbf{N} = \mu_s mg\cos\theta$$

Substituting this expression back into the equation for the forces acting in the x direction:

$$\mu_s mg \cos\theta - mg \sin\theta = 0$$

Eliminating mg and solving for μ_s:

$$\mu_s \cos\theta - \sin\theta = 0$$

$$\mu_s = \frac{\sin\theta}{\cos\theta} = \tan\theta = \tan 30^\circ = 0.58$$

LAW OF GRAVITATION

A force of attraction, F_G, exists between two bodies of mass m_1 and m_2 that is directly proportional to the product of the masses and inversely proportional to the squared distance between their centers of mass r, or

$$F_G = G\frac{m_1 m_2}{r^2}$$

where $G = 6.67 \times 10^{-11}\ \text{N}\cdot\text{m}^2/\text{kg}^2$. This is known as the **law of gravitation**.

TRANSLATIONAL AND ROTATIONAL EQUILIBRIUM

Equilibrium is the condition in which the sum of all forces acting on an object is equal to zero and the object is balanced. An object can be in either translational and/or rotational equilibrium, depending on the forces acting on the object.

Forces cause acceleration of an object in the direction of the force. An object is in **translational equilibrium** when the sum of forces is equal to zero, or

$$\Sigma \mathbf{F} = 0$$

or

$$\Sigma F_x = 0, \qquad \Sigma F_y = 0$$

Another type of force, when exerted on an object, causes rotational motion. This type of force is referred to as **torque**, characterized by the symbol τ. It occurs when an external force acts at a given distance from a fixed pivot point (also known as a lever arm, r), as shown in Figure 4-2. Torque is defined as

$$\text{Torque} = (\text{force})(\text{lever arm}) \sin\theta$$

$$\tau = \mathbf{F}\mathbf{r} \sin\theta$$

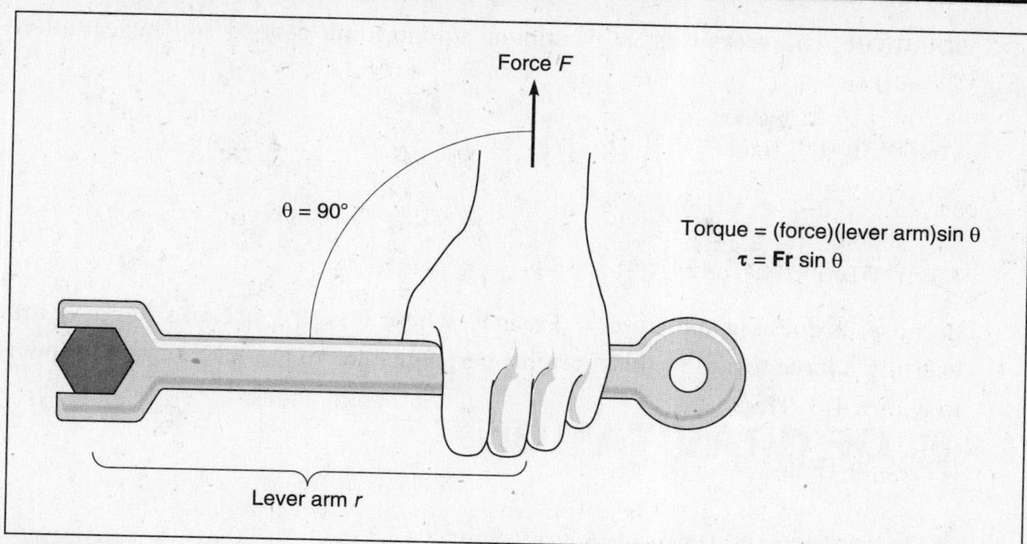

FIGURE 4-2

where θ is the angle between the direction of the force and the direction of the lever arm. Because it is a rotational force, torque can also be expressed analogous to the translational force ($\mathbf{F} = m\boldsymbol{a}$) by

$$\tau = I\alpha$$

where I is the moment of inertia or an object's resistance to rotational motion and $\boldsymbol{\alpha}$ is the angular acceleration of the rotating object. Torque is a vector quantity with SI units of newton · meter (N · m).

Torque occurs as the result of the rotation of an object, and an object can rotate in either a clockwise or a counterclockwise direction, resulting in τ_{cw} and τ_{ccw}, respectively. To determine the direction of the torque, you must first identify an axis of rotation and the point where each external force acts on the object. Once each torque about the axis of rotation has been calculated, torque is positive if the force causes a counterclockwise rotation; torque is negative if the force causes a clockwise rotation. An object is in rotational equilibrium when the sum of torques or rotational forces about any point is equal to zero, or

$$\Sigma\tau = 0$$

or

$$\Sigma\tau_{cw} = \Sigma\tau_{ccw}$$

EXAMPLE: Suppose Tori of mass 42 kg and Cori of mass 70 kg join their sister Lori of mass 35 kg on a seesaw. If Tori and Lori are seated 2.0 and 3.5 m from the pivot, respectively, on the same side of the seesaw, where must Cori sit on the other side of the seesaw in order to balance it?

SOLUTION: The seesaw is in rotational equilibrium, defined by the equation $\Sigma\tau = 0$ or

$$\tau_{Tori} + \tau_{Lori} + \tau_{Cori} = 0$$

or

$$\tau_{Tori} + \tau_{Lori} = -\tau_{Cori}$$

Because torque is given as $\tau = \mathbf{Fr} \sin\theta$, where θ is 90° because the girls are exerting a force (equal to their weight) perpendicular to the level arm, as shown in Figure 4-3. Thus,

$$(\mathbf{Fr})_{Tori} + (\mathbf{Fr})_{Lori} = -(\mathbf{Fr})_{Cori}$$

$$(42 \text{ kg})(9.8 \text{ m/s}^2)(2.0 \text{ m}) + (35 \text{ kg})(9.8 \text{ m/s}^2)(3.5 \text{ m})$$

$$= -(70 \text{ kg})(9.8 \text{ m/s}^2)(\mathbf{r}_{Cori})$$

Solving for $\mathbf{r}_{Cori}$

$$\mathbf{r}_{Cori} = -2.95 \text{ m}$$

with the minus sign indicating a position on the opposite side of the seesaw.

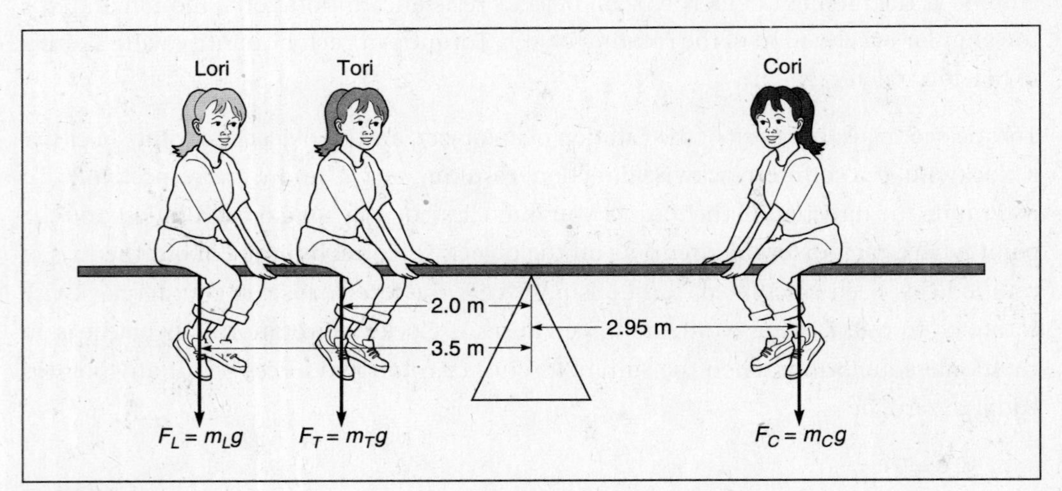

FIGURE 4-3

TRANSLATIONAL AND ROTATIONAL MOTION

Translational motion, or the motion of an object caused by forces exerted in the direction of motion, has already been described through Newton's Three Laws of Motion.

These laws of motion can also be extended to describe the **rotational motion** of an object, as noted next:

	Translational Motion	Rotational Motion
Newton's First Law	An object at rest or in motion remains at rest or in motion unless acted upon by an unbalanced force.	A body in rotational motion will continue its rotation until acted upon by an unbalanced torque.
Newton's Second Law	The magnitude of force is equal to the product of an object's mass and its acceleration or $\mathbf{F} = m\boldsymbol{a}$.	The magnitude of a torque is equal to the product of an object's moment of inertia and its angular acceleration or $\tau = I\alpha$.
Newton's Third Law	For every action force acting on an object, there is a reaction force exerted by the object equal in magnitude and opposite in direction to the original force.	For every torque acting on an object, there is another torque exerted by the object equal in magnitude and opposite in direction to the original torque.

UNIFORM CIRCULAR MOTION AND CENTRIPETAL FORCE

An object moving with constant speed in a circular path undergoes uniform circular motion. Because the object is moving at constant speed does not mean the object is not accelerating—in fact, it is. Although the magnitude of velocity (speed) is constant for an object in uniform circular motion, it is constantly changing direction. If the object is changing direction, it has a changing velocity and thus is accelerating. This acceleration, called **centripetal** (or center-seeking) **acceleration**, is perpendicular to the tangential velocity and is directed toward the center of rotation. The magnitude of centripetal acceleration, a_c, is defined as

$$a_c = \frac{v^2}{r}$$

and has typical units of acceleration (m/s^2).

Because an object in uniform circular motion has mass and is accelerating, there is a force directed inward toward the center. It is referred to as the **centripetal force** (F_c) defined by

$$F_c = ma_c = m\frac{v^2}{r}$$

A centripetal force is a vector quantity and has units of newtons.

CRAM SESSION

Forces and Newton's Laws

Force: A push or a pull.

Newton's Laws of Motion: Scientific laws that describe how and why motion occurs.

Newton's First Law of Motion: An object at rest remains at rest and an object in motion remains in motion unless acted on by an unbalanced force (also known as the Law of Inertia).

Newton's Second Law of Motion: $\Sigma \mathbf{F} = ma$, which applies to all motion in any direction (e.g., $\Sigma F_x = ma_x$ for motion in the x direction; $\Sigma F_y = ma_y$ for motion in the y direction).

Newton's Third Law of Motion: For every action force, there is an equal yet opposite reaction force (also known as the Law of Action and Reaction).

Law of Gravitation: States that a force of attraction, F_G, exists between two bodies of mass m_1 and m_2 that is directly proportional to the product of the masses and inversely proportional to the squared distance between their centers of mass r, or $F_G = \frac{Gm_1m_2}{r^2}$, where $G = 6.67 \times 10^{-11}$ N · m²/kg².

Torque: A force that, when exerted on an object, causes rotational motion; Torque = (force)(lever arm) sin θ can also be expressed as $\tau = I\alpha$, where I is the object's moment of inertia and α is angular acceleration. Torque is a vector quantity with SI units of newton · meter (N · m).

Equilibrium: The condition in which the sum of all forces acting on an object is equal to zero and the object is balanced.

Translational equilibrium: A condition that occurs when the sum of forces is equal to zero, or $\Sigma \mathbf{F} = 0$ or $\Sigma F_x = 0$, $\Sigma F_y = 0$.

Rotational equilibrium: An object is in rotational equilibrium when the sum of torques or rotational forces about any point is equal to zero, or $\Sigma\tau = 0$ or $\Sigma\tau_{cw} = \Sigma\tau_{ccw}$.

	Translational Motion	Rotational Motion
Newton's First Law	An object at rest or in motion remains at rest or in motion unless acted upon by an unbalanced force.	A body in rotational motion will continue its rotation until acted upon by an unbalanced torque.

	Translational Motion	Rotational Motion
Newton's Second Law	The magnitude of force is equal to the product of an object's mass and its acceleration, or $\mathbf{F} = m\mathbf{a}$.	The magnitude of a torque is equal to the product of an object's moment of inertia and its angular acceleration, or $\tau = I\alpha$.
Newton's Third Law	For every action force acting on an object, there is a reaction force exerted by the object equal in magnitude and opposite in direction to the original force.	For every torque acting on an object, there is another torque exerted by the object equal in magnitude and opposite in direction to the original torque.

An object in uniform circular motion is accelerating with **centripetal** (or center-seeking) **acceleration**, and is perpendicular to the tangential velocity and directed toward the center of rotation. The magnitude of centripetal acceleration, a_c, is defined as

$$a_c = \frac{v^2}{r}$$

and has typical units of acceleration (m/s^2). An object of mass m in uniform circular motion exerts a force toward the center and is referred to as the **centripetal force** (F_c), defined by

$$F_c = ma_c = m\frac{v^2}{r}$$

A centripetal force is a vector quantity and has units of newtons.

Free-Body Diagrams

A free-body diagram is a diagram in which an object is isolated and all forces acting on the object are identified and represented on the diagram. In performing free-body diagrams, one should

- Isolate the object in an imaginary coordinate system where the object represents the origin.
- Identify and represent all forces (with appropriate magnitude and direction) acting on the object in the diagram.
- Resolve all forces in terms of their x and y components.
- Substitute into the appropriate Newton's Second Law of Motion, either $\Sigma F_x = ma_x$ or $\Sigma F_y = ma_y$, setting the equation equal to ma if the object is accelerating and equal to 0 if the object is not accelerating.
- Solve for the unknown variable.

Examples of several common scenarios are presented in the following table:

Scenario	Free-Body Diagram	Application of Newton's Second Law
An object in equilibrium lying on a level surface 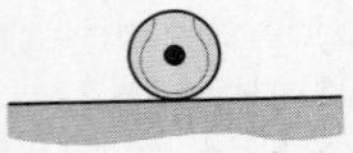	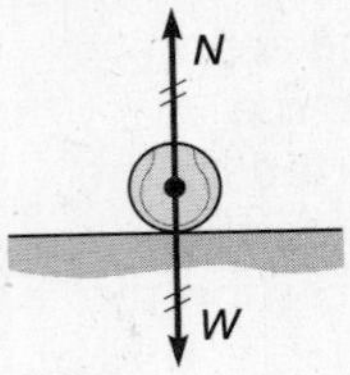 N, W	Object is in equilibrium, thus $\Sigma \mathbf{F} = m\mathbf{a} = 0$ x component: None y component: $\Sigma F_y = ma_y = 0$ $N - W = 0$ $N = W = mg$
An object in equilibrium lying on an inclined surface with friction preventing motion 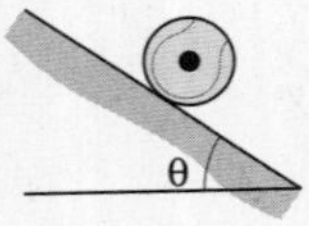	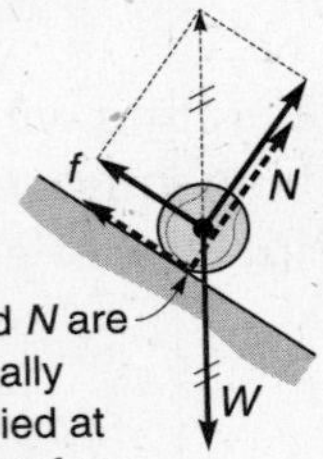 f, N, W f and N are actually applied at the surface.	Object is in equilibrium, thus $\Sigma \mathbf{F} = m\mathbf{a} = 0$ x component: $\Sigma F_x = ma_x = 0$ $-f + W \sin\theta = 0$ $f = W \sin\theta$ y component: $\Sigma F_y = ma_y = 0$ $N - W \cos\theta = 0$ $N = W \cos\theta$
An object in equilibrium suspended by a string 	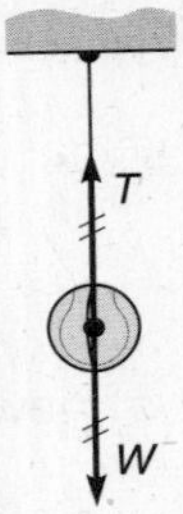 T, W	Object is in equilibrium thus $\Sigma \mathbf{F} = m\mathbf{a} = 0$ x component: None y component: $\Sigma F_y = ma_y = 0$ $T - W = 0$ $T = W = mg$
An object in equilibrium suspended by two strings of unequal lengths 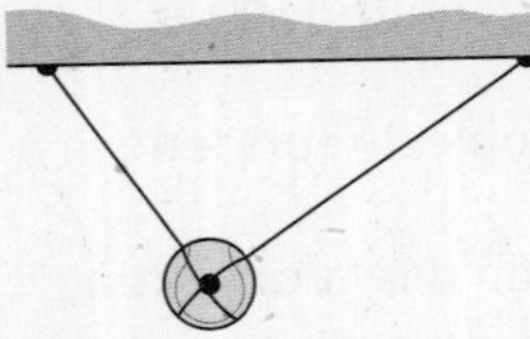	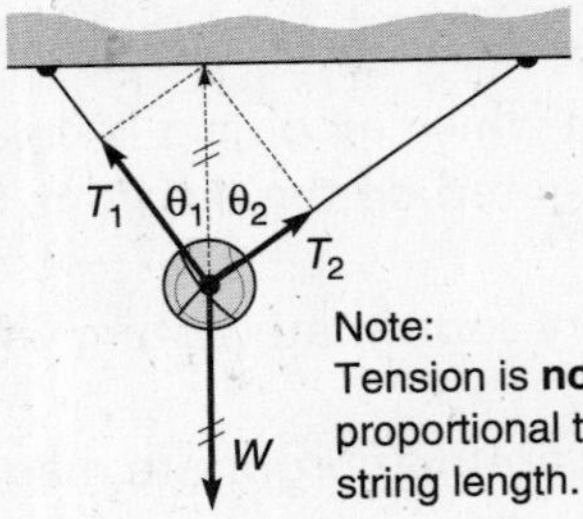 T_1, θ_1, θ_2, T_2, W Note: Tension is **not** proportional to string length.	Object is in equilibrium, thus $\Sigma \mathbf{F} = m\mathbf{a} = 0$ x component: $\Sigma F_x = ma_x = 0$ $-T_1 \sin\theta_1 + T_2 \sin\theta_2 = 0$ $T_1 \sin\theta_1 = T_2 \sin\theta_2$ y component: $\Sigma F_y = ma_y = 0$ $T_1 \cos\theta_1 + T_2 \cos\theta_2 - W = 0$ $T_1 \cos\theta_1 + T_2 \cos\theta_2 = W$
An object is falling and subject to no friction 	 W	Object is in motion, thus $\Sigma \mathbf{F} = m\mathbf{a}$ x component: None y component: $\Sigma F_y = ma_y$ $-W\ (= mg) = ma_y$

Scenario	Free-Body Diagram	Application of Newton's Second Law
An object is falling at constant (terminal) velocity 	 f W	Object is in motion, thus $\Sigma \mathbf{F} = m\mathbf{a}$ x component: None y component: $\Sigma F_y = ma_y$ $f - W = ma_y$ $f - mg = ma_y$

CHAPTER 5

Particle Dynamics: Work, Energy, and Power

Read This Chapter to Learn About:

- ➤ Work
- ➤ Pulley Systems
- ➤ Mechanical Energy: Kinetic and Potential
- ➤ Work–Kinetic Energy Theorem
- ➤ Conservation of Energy
- ➤ Power

A force applied to an object results in an acceleration of the object in the direction of the force. This is explained by Newton's Second Law and, in effect, explains why the object is set into motion. How the object responds to this force, ultimately resulting in motion, is the basis for this chapter.

WORK

Work represents the physical effects of an external force applied to an object that results in a net displacement in the direction of the force. Work is a scalar quantity and is expressed in SI units of joules (J) or newton meters (N · m).

By definition, the work done by a constant force acting on a body is equal to the product of the force **F** and the displacement **d** that occurs as a direct result of the force, provided that **F** and **d** are in the same direction. Thus

$$W = Fd$$

$$\text{Work} = (\text{force})(\text{displacement})$$

If **F** and **d** are not parallel but **F** is at some angle θ with respect to **d**, then

$$W = Fd \cos \theta$$

because it is only the portion of the force acting in the direction of motion that causes the object to move its displacement **d**. When the entire force is exerted on an object parallel to its direction of motion, then $\theta = 0^\circ$, and thus $\cos 0^\circ = 1$. This in turn results in the formula $\mathbf{W} = \mathbf{Fd}$ (when **F** is parallel to **d**). When **F** is perpendicular to **d**, $\theta = 90^\circ$ and $\cos 90^\circ = 0$. No work is done in this case.

EXAMPLE: Calculate the work done by a mother who exerts a force of 75 N at an angle of 25° below the horizontal (*x*-axis) in pushing a baby in a carriage a distance of 20 meters (m).

SOLUTION: The given information in this problem includes

$$\mathbf{F} = 75\text{ N} \quad \theta = -25^\circ \quad \mathbf{d} = 20\text{ m}$$

Substituting this given information into the equation $W = Fd \cos \theta$ yields

$$W = Fd \cos \theta = (75\text{ N})(20\text{ m})\left[\cos(-25^\circ)\right] = 1360\text{ J}$$

PULLEY SYSTEMS

Simple machines are devices that allow people to make work easier by changing the direction of the force applied to the load. One type of simple machine is the pulley, which consists of a rope threaded through a groove within the rim of a wheel. As the user applies a force (effort) to one end of a rope threaded over a pulley moving the string a certain distance (effort distance), the other end of the pulley containing the attached load is moved a certain distance (load distance). The pulley reduces the amount of force required to move a load over a given distance. This effect is magnified with the number of pulleys in a pulley system. As the number of pulleys and supporting ropes increases for a pulley system, a smaller amount of force is required, although the user must pull the rope over a greater distance.

Two terms that describe the capability of a pulley are its mechanical advantage and its efficiency. The **mechanical advantage**, MA, of a pulley is the amount by which the force required to move a load is reduced. It is described by

$$\text{MA} = \frac{F_{out}}{F_{in}}$$

where F_{out} is the output force (force exerted by the pulley system on the load) and F_{in} is the input force (force exerted on the pulley system by the person/machine).

The **efficiency** of a pulley is the ratio comparing the magnitude of the work output, W_{out}, relative to the work input, W_{in}, or

$$\text{Efficiency} = \frac{W_{out}}{W_{in}} = \frac{F_{out}d_{out}}{F_{in}d_{in}}$$

EXAMPLE: Two movers who are using a pulley system to lift a dining table of mass 230 kilograms (kg) by a distance of 3.5 m exert a force of 750 N. If the pulley system has an efficiency of 72%, determine the length of rope that must be pulled to accomplish the move.

SOLUTION: The equation needed to solve this problem is

$$\text{Efficiency} = \frac{W_{out}}{W_{in}} = \frac{F_{out}d_{out}}{F_{in}d_{in}}$$

where d_{in} represents the unknown quantity. The force exerted by the pulley system on the dining table is equal to the weight of the dining room table, or

$$F_{out} = \text{weight of the table} = m_{table}g$$

Therefore,

$$d_{in} = \frac{F_{out}d_{out}}{F_{in}(\text{efficiency})} = \frac{(230\text{ kg})\left(9.81\ \frac{\text{m}}{\text{s}^2}\right)(3.5\text{ m})}{(750\text{ N})(0.72)} = 14.6\text{ m}$$

MECHANICAL ENERGY: KINETIC AND POTENTIAL

Work describes the effect of a force exerted on an object, causing the object to move a given displacement. Work is performed on a continual basis by athletes when they throw, hit, or kick objects; by children when they slide down a slide, swing on a swing, or jump off a jungle gym; and by cars when they drive down the road. What is it that allows athletes, children, or cars to perform work? The answer is energy. Energy exists in many different forms, such as mechanical energy, solar energy (energy provided by the sun), chemical energy (energy provided by chemical reactions), elastic potential energy (energy provided by elastic objects such as a spring or rubber band), thermal energy (energy provided as a result of a temperature difference), sound energy (energy transmitted by propagating sound waves), radiant energy (energy transmitted by light waves), electric energy, magnetic energy, atomic energy, and nuclear energy.

Energy is a property that enables an object to do work. In fact, **energy** is defined as the ability of an object to do work. The more energy an object has, the more work the object can perform. The unit of measure for energy is the same as that for work, namely the joule. The next form of energy we discuss is mechanical energy, which exists in two types: kinetic energy and potential energy.

Kinetic energy is the energy of a body that is in motion. If the body's mass is m and its velocity is v, its kinetic energy is

$$\text{Kinetic energy (KE)} = \frac{1}{2}mv^2$$

EXAMPLE: Determine the kinetic energy of a 1500-kg car that is moving with a velocity of 30 m/s.

SOLUTION:

$$\text{KE} = \frac{1}{2}mv^2 = \frac{1}{2}(1500\text{ kg})(30\text{ m/s})^2 = 6.75 \times 10^5\text{ J}$$

The energy that is stored within a body because of its location above the Earth's surface (or reference level) is called **gravitational potential energy**. The gravitational potential energy of a body of mass m at a height h above a given reference level is

$$\text{Gravitational potential energy (PE)} = mgh$$

where g is the acceleration due to gravity. Gravitational potential energy depends on the object's mass and location (height above a level surface), but does not depend on how it reaches this height. In other words, gravitational potential energy is path independent. Any force, such as the force due to gravity, that performs work on an object independent of the object's path of motion is termed a **conservative force**.

EXAMPLE: A 4.5-kg book is held 80 centimeters (cm) above a table whose top is 1.2 m above the floor. Determine the gravitational potential energy of the book (1) with respect to the table and (2) with respect to the floor.

SOLUTION:

1. Here, $h = 80\text{ cm} = 0.8\text{ m}$, so

$$\text{PE} = mgh = (4.5\text{ kg})(9.8\text{ m/s}^2)(0.8\text{ m}) = 35.3\text{ J}$$

2. The book is $h = 0.8\text{ m} + 1.2\text{ m} = 2.0\text{ m}$ above the floor, so its PE with respect to the floor is

$$\text{PE} = mgh = (4.5\text{ kg})(9.8\text{ m/s}^2)(2.0\text{ m}) = 88.2\text{ J}$$

WORK–KINETIC ENERGY THEOREM

Work and energy are interrelated quantities. Energy is the ability to do work, and work can only be done if energy is present. A moving object has kinetic energy and maintains the same kinetic energy if it proceeds at the same speed. However, if the object speeds up or slows down, work is required. The net work required to change the kinetic energy of an object is given by the **work–kinetic energy theorem**:

Net work = change in kinetic energy

$W_{net} = \Delta KE$

CONSERVATION OF ENERGY

Energy is a conserved quantity. According to the **law of conservation of energy**, energy can neither be created nor destroyed, only transformed from one kind to another. Examples of this law can be found in many instances in nature, including the physiology of the human body and the working principles of a car. Consider a child on a swing. At the highest point of the arc, the child's energy is exclusively potential energy, because the child is not moving. As the child then begins moving downward toward the bottom of the arc, the potential energy of the child is transformed into kinetic energy. At the bottom point of the arc, the child's energy is exclusively kinetic energy.

In general, the initial mechanical energy (sum of potential energy and kinetic energy before an interaction) is equal to the final mechanical energy (sum of potential energy and kinetic energy after an interaction), or

Initial mechanical energy = final mechanical energy

$$(PE_{initial} + KE_{initial}) = (PE_{final} + KE_{final})$$

EXAMPLE: A child of mass 25.0 kg is at the top of a slide that is 3.8 m high. What is the child's velocity at the bottom of the slide?

SOLUTION: The child's velocity v at the bottom of the slide can be determined from the law of conservation of energy. The potential energy of the child at the top of the slide is transformed into kinetic energy at the bottom of the slide.

$$PE_{initial} = KE_{final}$$

or, simplifying,

$$mgh = \frac{1}{2}mv^2$$

Solving for v

$$v = \sqrt{2gh} = \sqrt{(2)\left(9.8\text{ m/s}^2\right)(3.8\text{ m})} = \sqrt{74.5}\text{ m/s} = 8.6\text{ m/s}$$

POWER

Power is the rate at which work is done by a force. Thus,

$$P = \frac{W}{t}$$

$$\text{Power} = \frac{\text{work}}{\text{time}}$$

The more power something has, the more work it can perform in a given time. The SI unit for power is the watt, where 1 watt (W) = 1 joule/second (J/s).

When a constant force **F** does work on a body that is moving at the constant velocity **v**, if **F** is parallel to **v**, the power involved is

$$P = \frac{W}{t} = \frac{Fd}{t} = F\left(\frac{d}{t}\right) = Fv$$

because $d/t = v$, that is,

$$P = Fv$$

$$\text{Power} = (\text{force})(\text{velocity})$$

EXAMPLE: As part of an exercise activity, a 60-kg student climbs up a 7.5-m rope in 11.6 s. What is the power output of the student?

SOLUTION: The problem can be solved with the following equation:

$$P = \frac{W}{t} = \frac{Fd}{t} = \frac{mgh}{t} = \frac{(60\text{ kg})\left(9.8\text{ m/s}^2\right)(7.5\text{ m})}{11.6\text{ s}} = 380\text{ W}$$

CRAM SESSION

Particle Dynamics: Work, Energy, and Power

Work: Represents the physical effects of an external force applied to an object that results in a net displacement in the direction of the force. Work is a scalar quantity and is expressed in SI units of joules (J) or newton meters (N · m). The work done by a constant force acting on a body is equal to the product of the force **F** and the displacement **d** that occurs as a direct result of the force. If **F** and **d** are in the same direction, then $W = Fd$. If **F** and **d** are not parallel but **F** is at some angle θ with respect to **d**, then $W = Fd \cos \theta$ because it is that is that portion of the force acting in the direction of motion that is object to move its displacement, **d**.

Pulley: A type of simple machine that consists of a rope threaded through a groove within the rim of a wheel.

Mechanical advantage (MA) of a pulley: The amount by which the required force to move a load is reduced given by

$$\text{MA} = \frac{F_{out}}{F_{in}}$$

where F_{out} is the output force and F_{in} is the input force. Output force is the force exerted by the pulley system on the load. Input force is the force exerted on the pulley system by the person/machine.

Efficiency (of a pulley): The ratio comparing the magnitude of the work output, W_{out}, relative to the work input, W_{in}, or

$$\text{Efficiency} = \frac{W_{out}}{W_{in}} = \frac{F_{out}d_{out}}{F_{in}d_{in}}$$

Energy: The ability to do work. It is a scalar quantity measured in SI units of joules. Energy exists in several forms including mechanical, solar, chemical, thermal, sound, radiant, magnetic, electric, nuclear, and atomic. Two types of mechanical energy are kinetic energy and gravitational potential energy.

Kinetic energy: Energy of an object in motion given by the equation

$$\text{KE} = \frac{1}{2}mv^2$$

Gravitational potential energy: Energy that is stored within a body because of its location, and is given by the equation

$$\text{PE} = mgh$$

Work–kinetic energy theorem: The net work required to change the kinetic energy of an object is given by:

Net work = change in kinetic energy or $W_{net} = \Delta KE$

Conservative force: Any force, such as the force due to gravity, that performs work on an object independent of the object's path of motion.

Law of conservation of energy: Energy can neither be created nor destroyed, only transformed from one kind to another.

Initial mechanical energy = final mechanical energy

$(PE_{initial} + KE_{initial}) = (PE_{final} + KE_{final})$

Power: The rate at which work is done by a force or power is Power $= \frac{\text{work}}{\text{time}}$. Power is a scalar quantity expressed in SI units of watts, where 1 watt (W) = 1 joule/second.

Strategies for Solving Work, Energy, and Power Problems

Work

- Identify the magnitude and direction of the net force exerted on an object and the displacement of the object that occurs as a result of the force.
  - How much of the force is exerted in the direction of the displacement, i.e., what is the angle that the force exerts with respect to the direction of motion?
  - Is work positive or negative? If the force is opposite to the displacement, then work is negative.
  - If the force is in the same direction as the displacement, then work is positive.
  - If the force is exerted along the $+y$-axis, then the force exerted on the object required to displace it is the weight of the object and thus work is being done against gravity.
- Make sure all quantities involved in the work calculation are expressed in SI units, i.e., force in newtons, displacement in meters leading to work in units of joules.

Energy

- Remember that energy is the ability to do work. In order for work to be done on an object, energy must be involved.
- What type(s) of energy are involved in the problem?
  - If an object of mass m is suspended a certain height (h) from a surface, then gravitational potential energy (PE) is involved and can be calculated from $PE = mgh$.
  - If an object of mass m is in motion, then the object has kinetic energy (KE) which can be calculated from

$$KE = \frac{1}{2}mv^2.$$

  - If the object is in motion at a certain height above the reference point, then the object has both GPE and KE.

- Energy is a conserved quantity meaning that the total energy before an event or interaction is equal to the total energy after the event or interaction. In conservation of energy problems involving gravitational potential energy and kinetic energy, mass is irrelevant in the calculations as it is common to both relations and can thus cancel out.
- Make sure all quantities involved in the energy calculations are expressed in SI units and that all forms of energy, regardless of their source or type, are expressed in units of joules.

Power

- Power is the rate at which work is done. Problems will provide information regarding work done on the object as well as a time interval over which the work was done. You can then solve for power using Power = work/time.
- Is time expressed in units of seconds? All other units should be expressed in SI units.
- Another way to calculate power is through the product of force and velocity. If a velocity is given in the problem, make sure the units are meters/second.

CHAPTER 6

Momentum and Impulse

Read This Chapter to Learn About:

- ➤ Linear Momentum
- ➤ Impulse
- ➤ Conservation of Linear Momentum
- ➤ Elastic Collisions
- ➤ Inelastic Collisions

What is the proper way for a skydiver to land? How should a boxer best respond to a punch? What should a person do during an accidental fall? The answers to all of these questions lie in the topics of momentum and impulse.

LINEAR MOMENTUM

Linear momentum, **p**, is defined as the product of an object's mass and its velocity, or

$$\mathbf{p} = m\mathbf{v}$$

Linear momentum is a vector quantity whose direction is that of the object's velocity and is given in units of kilogram meter per second (kg m/s).

EXAMPLE: What is the momentum of a 1500-kg car traveling at a velocity of 35 kilometers per hour (km/h)?

SOLUTION: The momentum can be calculated by the equation $\mathbf{p} = m\mathbf{v}$, where $\mathbf{v}$ is the velocity in m/s. Because the velocity in the problem is given in units of km/h, one must convert this value into the appropriate units:

$$\mathbf{v} = \left(35\ \frac{\text{km}}{\text{h}}\right)\left(\frac{1000\ \text{m}}{1\ \text{km}}\right)\left(\frac{1\ \text{h}}{3600\ \text{s}}\right) = 9.7\ \frac{\text{m}}{\text{s}}$$

Substituting this value for **v** into the expression for momentum yields

$$\mathbf{p} = m\mathbf{v} = (1500 \text{ kg})\left(9.7 \frac{\text{m}}{\text{s}}\right) = 14{,}550 \text{ kg m/s}$$

IMPULSE

A moving object, which originally has a momentum $\mathbf{p}_i$, may eventually undergo a change in momentum through either an increase or decrease in velocity, resulting in a final momentum, $\mathbf{p}_f$. The change in momentum, $\Delta\mathbf{p}$, is defined as

$$\Delta\mathbf{p} = \mathbf{p}_f - \mathbf{p}_i = m(\mathbf{v}_f - \mathbf{v}_i)$$

The change in momentum, also referred to as the physical quantity **impulse**, I, is related to the impact force acting to speed up or slow down the object multiplied by the time interval during which this force acts. This relation can be determined if one begins with the primary definition of a force

$$F = ma = m\left(\frac{\Delta v}{\Delta t}\right)$$

Multiplying both sides of the equation by Δt yields

$$m\Delta v = F\Delta t$$

Change in momentum = impulse

The units of impulse are the same units as momentum: kg m/s or newton second ($\text{N} \cdot \text{s}$). The relation between the impulse and change in momentum is known as the **impulse–momentum theorem**.

EXAMPLE: A 0.45-kg baseball is thrown to a batter at 30 m/s and is hit with a velocity of 20 m/s in the direction of the pitcher. Assuming that the ball is in contact with the bat for 0.02 s, determine (1) the impulse of the batter and (2) the average force exerted on the ball by the bat.

SOLUTION: Begin with the assumption that the pitcher pitches the ball along the positive x direction, and because the batter hits in the direction of the pitcher, the ball leaves the bat in the negative x direction.

1. The impulse can be determined by

$$I = \Delta\mathbf{p} = m(v_f - v_i) = (0.45 \text{ kg})(-20 \text{ m/s} - 30 \text{ m/s})$$
$$= (0.45 \text{ kg})(-50 \text{ m/s}) = -22.5 \text{ kg m/s}$$

2. The average force exerted on the ball by the bat is

$$F = \frac{m(v_f - v_i)}{\Delta t} = \frac{-22.5 \text{ kg m/s}}{0.02 \text{ s}} = -1125 \text{ N}$$

CONSERVATION OF LINEAR MOMENTUM

Just like energy, linear momentum is a conserved quantity. If the sum of all external forces acting on a system is zero, then the total linear momentum of the system remains unchanged or constant. A common example of the **conservation of linear momentum** is the collision of two bodies. Two types of collisions are possible: elastic collisions and inelastic collisions.

ELASTIC COLLISIONS

An **elastic collision** is a collision in which two bodies approach each other, interact with each other, and part on impact with a different momentum. In an elastic collision, both linear momentum and kinetic energy are conserved. For a system of two bodies that undergoes an elastic collision, the following can be stated:

Total linear momentum before collision = total linear momentum after collision

$$m_{1b}\mathbf{v}_{1b} + m_{2b}\mathbf{v}_{2b} = m_{1a}\mathbf{v}_{1a} + m_{2a}\mathbf{v}_{2a}$$

Total kinetic energy before collision = total kinetic energy after collision

$$\frac{1}{2}m_{1b}v_{1b}^2 + \frac{1}{2}m_{2b}v_{2b}^2 = \frac{1}{2}m_{1a}v_{1a}^2 + \frac{1}{2}m_{2a}v_{2a}^2$$

where b represents the state of the object *before* the collision and a represents the state of the object *after* the collision. Figure 6-1 illustrates an elastic collision.

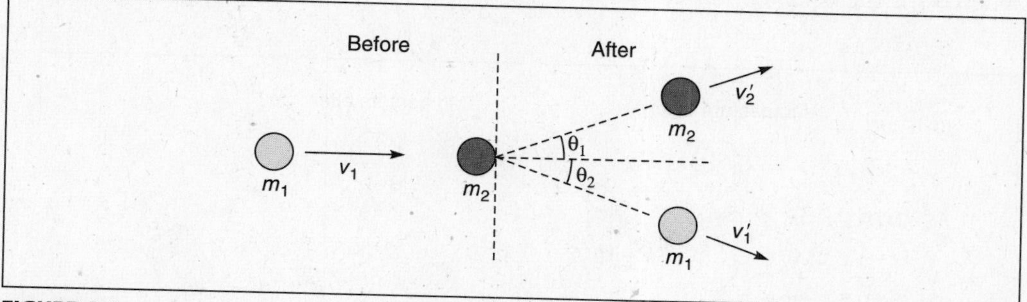

FIGURE 6-1

INELASTIC COLLISIONS

An **inelastic collision** can be one in which two bodies approach each other, collide, stick together, and part as a combined single body with a different velocity in a different direction, as shown in Figure 6-2. Yet another scenario is one in which a single body breaks up into two or more smaller bodies, each moving with different velocities in

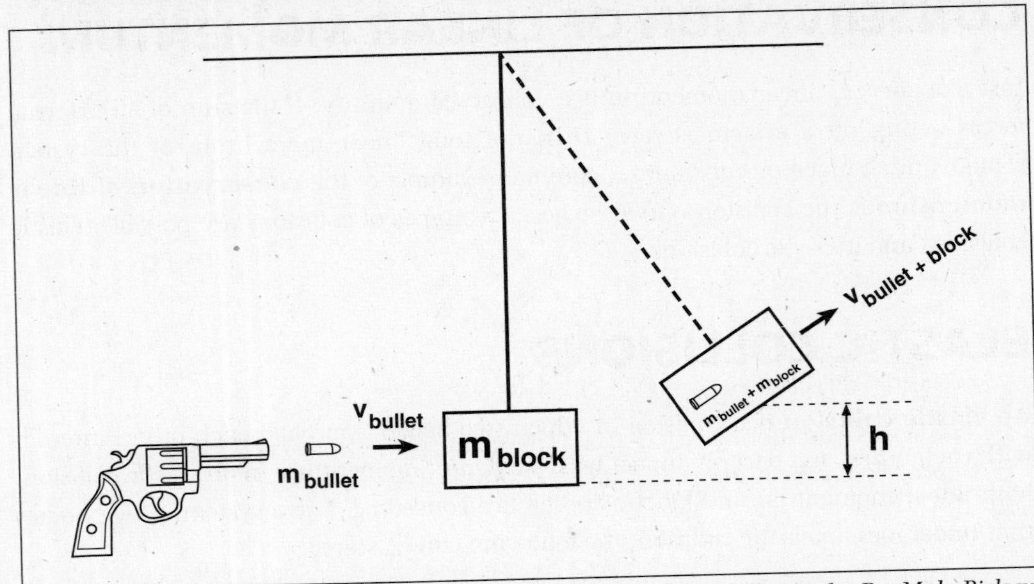

FIGURE 6-2: SOURCE: From George Hademenos, *Schaum's Outline of Physics for Pre-Med, Biology, and Allied Health Students*, McGraw-Hill, 1998; reproduced with permission of The McGraw-Hill Companies.

different directions, as shown in Figure 6-3. For a system of two bodies that undergoes a perfectly inelastic collision, only linear momentum is conserved and the following can be stated:

Total linear momentum before collision = total linear momentum after collision

$$m_{1b}\mathbf{v}_{1b} + m_{2b}\mathbf{v}_{2b} = (m_{1a} + m_{2a})\mathbf{v}_f$$

where $\mathbf{v}_f$ is the velocity of the combined objects.

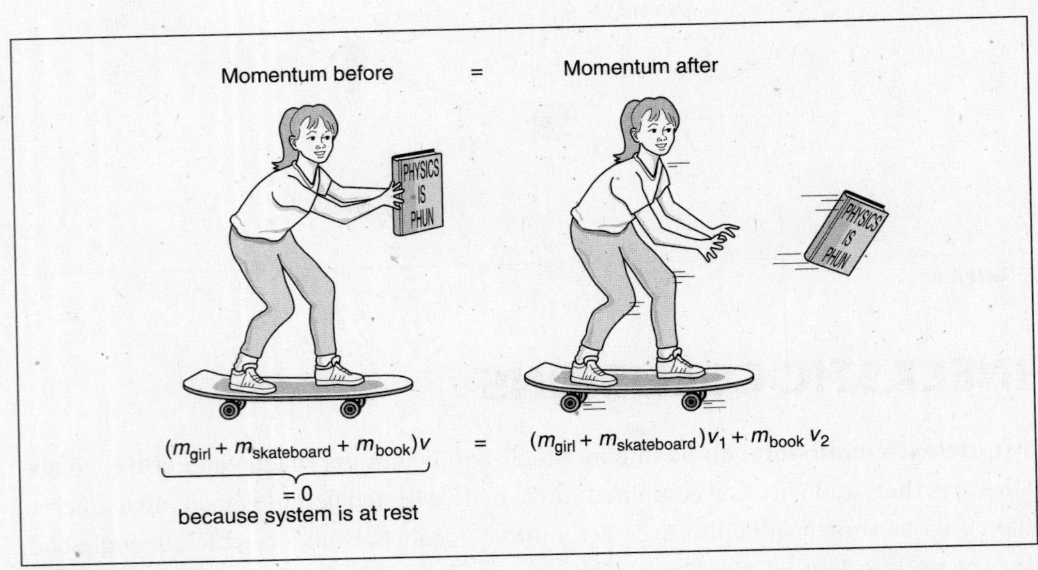

FIGURE 6-3

EXAMPLE: An example of an inelastic collision is the ballistic pendulum, which consists of a wooden block suspended by a cord secured to a level surface, as shown in Figure 6-2. A projectile is fired horizontally into the wooden block and becomes embedded in the block. Assume that a 25-gram (g) bullet is fired horizontally at a 3-kg block of wood suspended by a cord. If the initial velocity of the bullet on firing is 110 m/s, determine the height of the swing.

SOLUTION: The ballistic pendulum is an example of an inelastic collision and, by conservation of momentum, it can be stated that

$$m_{\text{bullet},b}\mathbf{v}_{\text{bullet},b} + m_{\text{block},b}\mathbf{v}_{\text{block},b} = (m_{\text{bullet},a} + m_{\text{block},a})\mathbf{v}_f$$

From this equation, every variable is given with the exception of $\mathbf{v}_f$. Making appropriate substitutions and solving for $\mathbf{v}_f$ yields

$$(0.025 \text{ kg})(110 \text{ m/s}) + (3 \text{ kg})(0 \text{ m/s}) = (0.025 \text{ kg} + 3 \text{ kg})\mathbf{v}_f$$

$$\mathbf{v}_f = 0.91 \text{ m/s}$$

On impact with the block, the bullet possesses kinetic energy, which is converted to potential energy during the movement of the pendulum to its highest point. The height of this swing can be determined through the conservation of energy:

$$\text{KE}_{\text{before}} = \text{PE}_{\text{after}}$$

$$\frac{1}{2}(m_{\text{bullet}} + m_{\text{block}})v_f^2 = (m_{\text{bullet}} + m_{\text{block}})gh$$

$$\frac{1}{2}v_f^2 = gh$$

$$h = \frac{v_f^2}{2g} = \frac{\left(0.91 \frac{\text{m}}{\text{s}}\right)^2}{2\left(9.8 \frac{\text{m}}{\text{s}^2}\right)} = 0.042 \text{ m} = 4.2 \text{ cm}$$

CRAM SESSION

Momentum and Impulse

Linear momentum: Defined as the product of an object's mass, m, and its velocity, $\mathbf{v}$, or $\mathbf{p} = m\mathbf{v}$. Linear momentum is a vector quantity whose direction is that of the object's velocity and is expressed in SI units of kg m/s.

Impulse: Also known as the change in momentum, a physical quantity that describes the impact force, $\boldsymbol{F}$, acting to speed up or slow down the object multiplied by the time interval, Δt, during which this force acts, or $m\Delta v = F\Delta t$.

The units of impulse are the same units as momentum: kg m/s or N · s.

The relationship between the impulse and change in momentum is known as the **impulse–momentum theorem.**

Law of conservation of linear momentum: A conserved quantity; thus, if the sum of all external forces acting on a system is zero, then the total linear momentum of the system remains unchanged or constant.

A common example of the conservation of linear momentum is found in collisions, of which there are two types: elastic collisions and inelastic collisions.

Elastic collision: A collision in which two bodies approach each other, interact with each other, and part on impact with a different velocity in a different direction. In an elastic collision, both linear momentum and kinetic energy are conserved. For a system of two bodies that undergo an elastic collision:

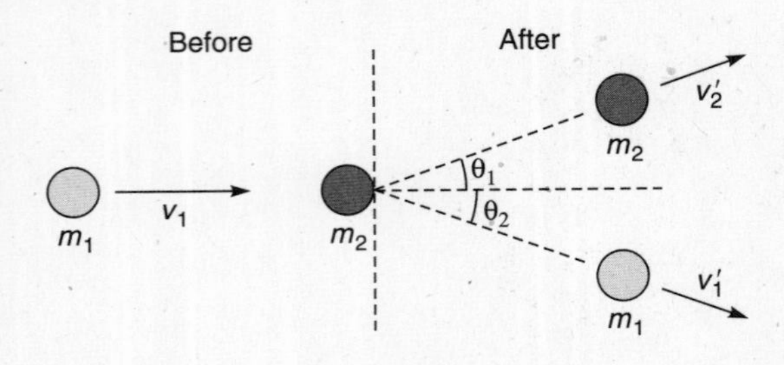

Total linear momentum before collision = total linear momentum after collision

$$m_{1b}\mathbf{v}_{1b} + m_{2b}\mathbf{v}_{2b} = m_{1a}\mathbf{v}_{1a} + m_{2a}\mathbf{v}_{2a}$$

Total kinetic energy before collision = total kinetic energy after collision

$$\frac{1}{2}m_{1b}v_{1b}^2 + \frac{1}{2}m_{2b}v_{2b}^2 = \frac{1}{2}m_{1a}v_{1a}^2 + \frac{1}{2}m_{2a}v_{2a}^2$$

where *b* represents the state of the object *before* the collision and *a* represents the state of the object *after* the collision.

Inelastic collision: A collision in which two bodies approach each other, collide, stick together, and part as a combined single body with a different velocity in a different direction. A second example of an inelastic collision is one in which a single body/entity breaks up into two or more smaller bodies, each moving with different velocities in different directions.

Example: Inelastic collision

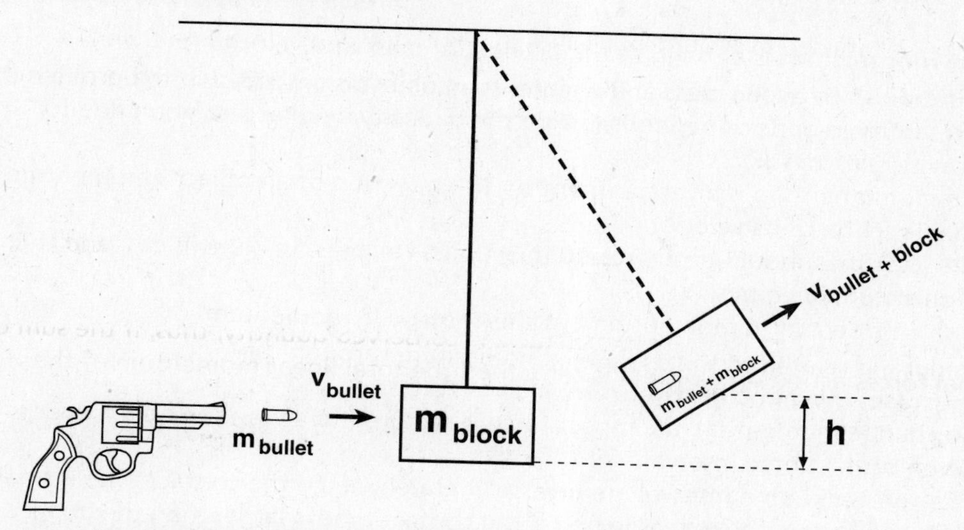

SOURCE: From George Hademenos, *Schaum's Outline of Physics for Pre-Med, Biology, and Allied Health Students*, McGraw-Hill, 1998; reproduced with permission of The McGraw-Hill Companies.

Example: Inelastic collision

$$\underbrace{(m_{girl} + m_{skateboard} + m_{book})v}_{=0} = (m_{girl} + m_{skateboard})v_1 + m_{book}\, v_2$$

because system is at rest

In an inelastic collision, only linear momentum is conserved. For a system of two bodies that undergo an inelastic collision:

Total linear momentum before collision = total linear momentum after collision

$$m_{1b}\mathbf{v}_{1b} + m_{2b}\mathbf{v}_{2b} = (m_{1a} + m_{2a})\mathbf{v}_f$$

where b represents the state of the object *before* the collision and a represents the state of the object *after* the collision.

Strategies for Solving Momentum and Impulse Problems

1. **Linear momentum** is defined as the product of mass and velocity ($\mathbf{p} = m\mathbf{v}$).
   - Distinguish between mass and weight. If an object is described in the problem by a quantity in units of newtons, divide the quantity by g (acceleration due to gravity; 9.81 m/s^2).
   - Linear momentum is a vector quantity. The direction of an object's momentum is the direction of its velocity.
   - All quantities should be expressed in SI units, i.e., mass in kg, $\mathbf{v}$ in m/s, and thus momentum in kg m/s.
2. Impulse is not equal to momentum but the **change** in momentum.
   - Impulse is related to the change in momentum through the **impulse–momentum theorem:** $m\Delta v = F\Delta t$.
   - To change momentum, an impact force is required to act on the object over a given contact time interval.
   - The force and time interval are inversely related; i.e., if the contact time is small, then the impact force is large and if the contact time is large, then the impact force is small.
   - In some problems, information might be given that describes the motion of an object before it changes momentum, e.g., the time an object moves a certain distance so that its velocity can be calculated. This time is different from the contact time required to change momentum. Make sure you read the problem carefully to distinguish between all of the information given in a problem, especially the time quantities.
   - In problems involving an object moving at some initial velocity, striking a surface and causing it to rebound with a velocity in an opposite direction, remember the change in velocity is not the difference between the two velocities but rather the sum of the two velocities because of their opposite directions. Once you define the direction and hence the original velocity you consider positive, then the other velocity becomes negative and when subtracted from the original velocity, the velocities are added.
3. Linear momentum is a conserved quantity and applies to the physics of collisions.
   - **Elastic collisions**
     a. Draw a picture of the two objects before their collision and then after the collision.

b. Write all information (mass, velocity, direction) given in the problem about the two objects before and after the collision. The mass of the two objects does not change as a result of the collision; only the velocity changes.

c. Make sure that all quantities are expressed in SI units.

d. Both linear momentum and kinetic energy are conserved and should be used to determine the unknown quantities.

➤ **Inelastic collisions**

a. Draw a picture of the two objects before their collision and then after the collision. There are two types of inelastic collisions: (1) a collision in which two bodies approach each other, collide, stick together, and part as a combined single body with a different velocity in a different direction; and (2) a single body/entity breaks up into two or
more smaller bodies, each moving with different velocities in different directions. Make sure the proper representation of the two objects is noted as you begin to solve the problem.

b. Write all information (mass, velocity, direction) given in the problem about the two objects before and after the collision. The mass of the two objects does not change as a result of the collision, but rather combines as the two objects move after the collision as a single entity.

c. Make sure that all quantities are expressed in SI units.

d. Only linear momentum is conserved and should be used to determine the unknown quantity.

CHAPTER 7

Solids and Fluids

Read This Chapter to Learn About:

SOLIDS

- Elastic Properties of Solids

FLUIDS AT REST

- Density and Specific Gravity
- Pressure: Atmospheric and Hydrostatic
- Pascal's Principle
- Buoyant Force and Archimedes' Principle
- Surface Tension

FLUIDS IN MOTION

- Viscosity
- Continuity Equation
- Bernoulli's Equation
- Laminar and Turbulent Flow

This chapter reviews two separate but equally important topics of physics: solids and fluids. An object acted upon by an external force such as pushing, pulling, twisting, bending, or stretching responds to these forces through a deformation or change in the size and/or shape of the object. How the object responds to these forces—a property referred to as **elasticity,** which is the physical basis for the structure of solids—is important. Although fluids differ from solids in terms of structure and composition, fluids possess inertia, as defined by their density, and are thus subject to the same physical interactions as solids. All of these interactions as they pertain to fluids at rest and fluids in motion are discussed in this chapter.

SOLIDS

ELASTIC PROPERTIES OF SOLIDS

Elasticity is an inherent property of all materials describing their response to an external force. It is characterized by the presence and degree of deformation of the object and depends primarily on the size and internal structure of the object and on the magnitude of the force. The elasticity of a substance can be characterized by two physical parameters: stress and strain.

Stress S is defined as the magnitude of an external force F required to induce the given deformation of an object divided by the surface area A over which the force is exerted perpendicularly, or

$$\text{Stress} = \frac{\text{force}}{\text{area}}$$

$$S = \frac{F}{A}$$

and is expressed in units of newton meters squared ($N \cdot m^2$).

Strain ε is a quantitative measure of the fractional extent of deformation of an object produced as a result of applied stress, and it is a unitless quantity. **Strain** is defined as the ratio of the increase or decrease in a particular dimension in the deformed state to that dimension in its initial undeformed state, and can be expressed mathematically in general form as

$$\text{Strain} = \frac{\text{extent of deformation}}{\text{undeformed state}}$$

$$\varepsilon = \frac{x_d - x_n}{x_n}$$

where x_d and x_n represent the dimensional quantities of the elastic object in the deformed and undeformed (relaxed) states, respectively. These quantities are calculated differently, depending on the type of elasticity demonstrated. Tensile strain and shear strain can be determined from the relation $\frac{\Delta L}{L}$ (ΔL is the change in length of the object and L is the original length) and volumetric strain can be found from $-\frac{\Delta V}{V}$ (ΔV is the change in volume of the object and V is the original volume).

The **elastic modulus** of a solid, a physical constant that describes the elastic properties of a solid, is defined as

$$\text{Elastic modulus} = \frac{\text{stress}}{\text{strain}}$$

Elastic modulus is similar to stress in that their units are similar: $N \cdot m^2$. There are three types of elastic moduli that are used to characterize the deformation of solids, described later and depicted in Figure 7-1.

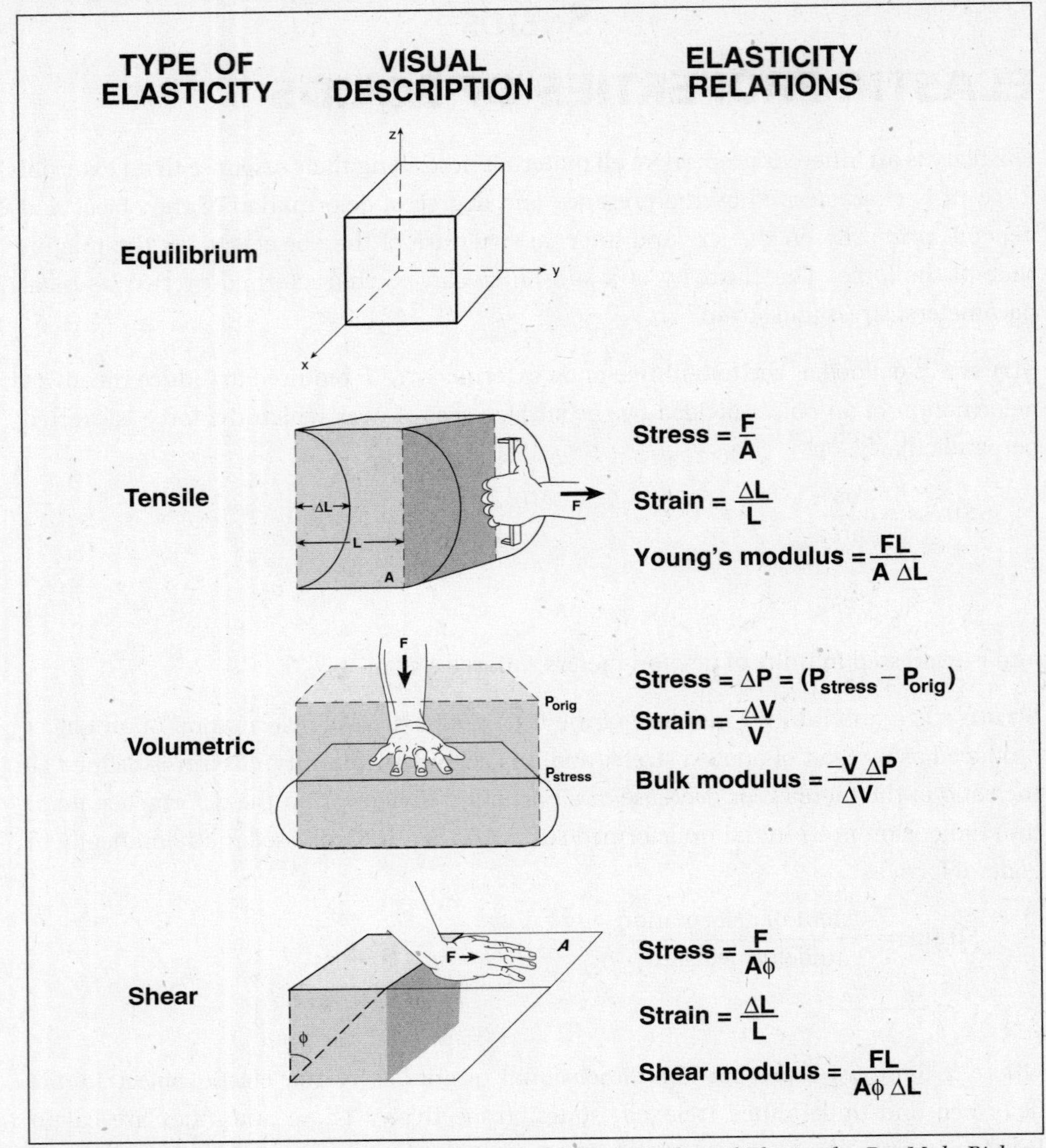

FIGURE 7-1: SOURCE: From George Hademenos, *Schaum's Outline of Physics for Pre-Med, Biology, and Allied Health Students*, McGraw-Hill, 1998; reproduced with permission of The McGraw-Hill Companies.

Young's modulus Y is given as the ratio of tensile stress to tensile strain. **Tensile stress** and **tensile strain** refer to the stress and strain induced by a pulling force exerted on the object. Young's modulus is defined as

$$\text{Young's modulus} = \frac{\text{tensile stress}}{\text{tensile strain}} = \frac{FL}{A\Delta L}$$

where F is the force applied to the solid, A is the cross-sectional area of the solid, L is the original length of the solid, and ΔL is the fractional amount of deformation as a result of the applied force.

Bulk modulus B is the elastic modulus due to volumetric compression of an elastic solid and is defined as

$$\text{Bulk modulus} = -\frac{V\Delta P}{\Delta V}$$

where V is volume and P is pressure. The minus sign is used to negate the negative effects of volumetric change, making the bulk modulus positive.

Shear modulus G relates the shear stress to the shear or angular strain. **Shear stress** and **shear strain** refer to the stress and strain induced by a sliding force exerted on the object. Shear modulus is defined as

$$\text{Shear modulus} = \frac{\text{shear stress}}{\text{shear strain}} = \frac{FL}{A\phi\Delta L}$$

where F is the force applied to the solid, A is the cross-sectional area of the solid, L is the original length of the solid, ϕ is the angle by which the elastic solid is displaced by the shearing stress, and ΔL is the fractional amount of deformation as a result of the applied force.

EXAMPLE: A 2.5-kilogram (kg) mass is attached to one end of a sample of artery with 0.05-centimeter (cm) radius and 15-millimeter (mm) length. The other end of the artery sample is secured to a rigid support. Determine the elongation of the artery sample if $Y = 1 \times 10^8\ \text{N} \cdot \text{m}^2$.

SOLUTION: The mass attached to the artery sample exerts a force $F = mg$. The strain of the artery is given by

$$\varepsilon = \frac{\Delta L}{L} = \frac{S}{Y} = \frac{F}{AY}$$

and it can be used to determine the change in length, ΔL, of the artery sample

$$\Delta L = \frac{FL}{AY} = \frac{(\text{mg})L}{(\pi r^2)Y} = \frac{(2.5\ \text{kg})\left(9.8\ \frac{\text{m}}{\text{s}^2}\right)(15 \times 10^{-3}\ \text{m})}{(3.14)(0.5 \times 10^{-3}\ \text{m})^2\left(1 \times 10^8\ \frac{\text{N}}{\text{m}^2}\right)} = 4.7\ \text{mm}$$

Fluids at Rest

DENSITY AND SPECIFIC GRAVITY

Density, ρ, is a physical property of a fluid, given as mass per unit volume, or

$$\rho = \frac{\text{mass}}{\text{unit volume}} = \frac{m}{V}$$

Density represents the fluid equivalent of mass and is given in units of kilograms per cubic meter $\left(\frac{kg}{m^3}\right)$, grams per cubic centimeter $\left(\frac{g}{cm^3}\right)$, or grams per milliliter $\left(\frac{g}{mL}\right)$. Density, a property unique to each substance, is independent of shape or quantity but is dependent on temperature and pressure.

Specific gravity (Sp. gr.) of a given substance is the ratio of the density of the substance ρ_{sub} to the density of water ρ_w, or

$$\text{Sp. gr.} = \frac{\text{density of substance}}{\text{density of water}} = \frac{\rho_{sub}}{\rho_w}$$

where the density of water ρ_w is 1.0 g/cm^3 or 1.0×10^3 kg/m^3. Assuming that equal volumes are chosen, the specific gravity can also be expressed in terms of weight:

$$\text{Sp. gr.} = \frac{\text{weight of substance}}{\text{weight of water}} = \frac{w_{sub}}{w_w}$$

Specific gravity is a pure number that is unitless.

EXAMPLE: Determine the size of container needed to hold 0.7 g of ether, which has a density of 0.62 g/cm^3.

SOLUTION: The volume of a fluid can be found from the relation for density:

$$V = \frac{m}{\rho} = \frac{0.7\text{ g}}{0.62\ \frac{\text{g}}{\text{cm}^3}} = 1.129\text{ cm}^3 = 1.129\text{ mL}$$

PRESSURE: ATMOSPHERIC AND HYDROSTATIC

Pressure P is defined as a force F acting perpendicular to a surface area, A, of an object and is given by

$$\text{Pressure} = \frac{\text{force}}{\text{area}} = \frac{F}{A}$$

Pressure is a scalar quantity and is expressed in units of N · m^2. Two specific types of pressure particularly applicable to fluids include atmospheric pressure and hydrostatic pressure.

Atmospheric pressure P_{atm} represents the average pressure exerted by Earth's atmosphere, and is defined numerically as 1 atm but can also be expressed as 760 millimeters of mercury (mmHg) or 1.01×10^5 pascals (Pa).

Hydrostatic pressure P_{hyd} is the fluid pressure exerted at a depth h in a fluid of density ρ and is given by

$$P_{hyd} = \rho g h$$

The total pressure, P, exerted on a contained fluid is the sum of the atmospheric pressure and the hydrostatic pressure:

$$P = P_{atm} + P_{hyd} = P_{atm} + \rho g h$$

PASCAL'S PRINCIPLE

Pascal's principle states: "An external pressure applied to a confined fluid will be transmitted equally to all points within the fluid."

EXAMPLE: An example of Pascal's principle is the hydraulic jack, shown in Figure 7-2. If a force of 300 N is applied to a piston of 1-cm^2 cross-sectional area, determine the lifting force transmitted to a piston of cross-sectional area of 100 cm^2.

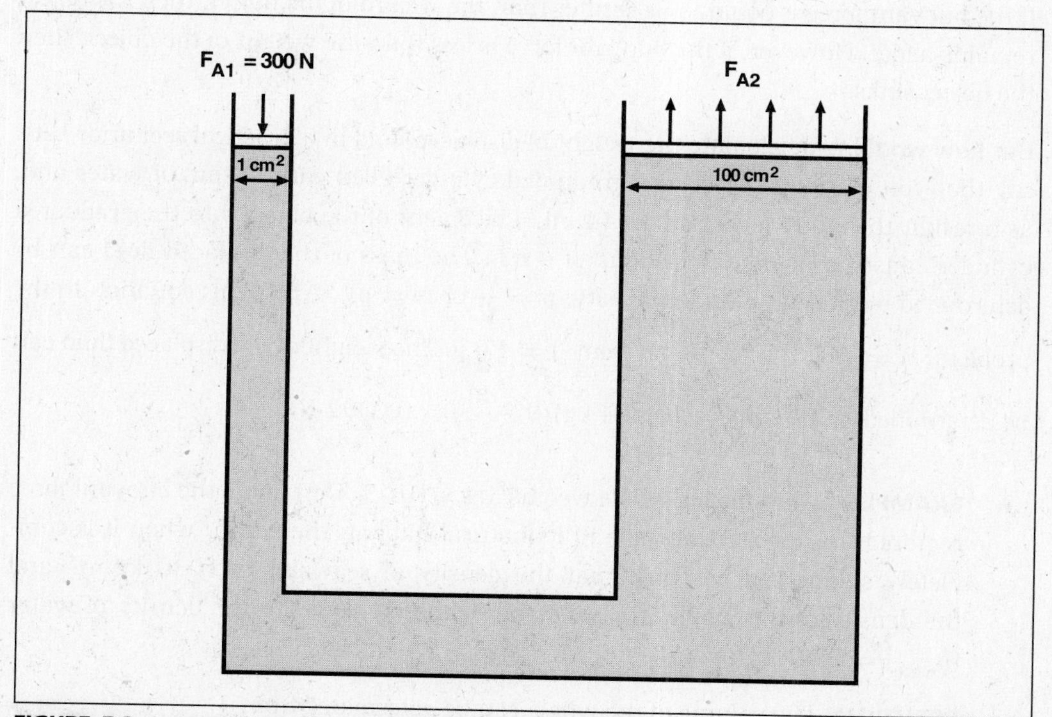

FIGURE 7-2: SOURCE: From George Hademenos, *Schaum's Outline of Physics for Pre-Med, Biology, and Allied Health Students*, McGraw-Hill, 1998; reproduced with permission of The McGraw-Hill Companies.

SOLUTION: According to Pascal's principle,

$$P_{1\ \text{cm}^2} = P_{100\ \text{cm}^2}$$

$$\left(\frac{F}{A}\right)_{1\ \text{cm}^2} = \left(\frac{F}{A}\right)_{100\ \text{cm}^2}$$

Making the appropriate substitutions yields

$$\frac{300\ \text{N}}{1\ \text{cm}^2} = \frac{F}{100\ \text{cm}^2}$$

Solving for F yields

$$F = 30{,}000\ \text{N}$$

BUOYANT FORCE AND ARCHIMEDES' PRINCIPLE

Archimedes' principle states: "A body immersed wholly or partially in a fluid is subjected to a buoyant force that is equal in magnitude to the weight of the fluid displaced by the body," or

Buoyant force = weight of displaced fluid

If the buoyant force is equal to or greater than the weight of the object, then the object remains afloat. However, if the buoyant force is less than the weight of the object, then the object sinks.

But how would you calculate the weight of displaced fluid in a practical scenario? Let's say that you place an object in a graduated cylinder filled with 40 mL of water and, as a result, the water level rises to 44 mL. Placement of the object into the graduated cylinder caused a change in volume of 4 mL. The mass of the displaced fluid can be determined by the equation for density: $\rho = \frac{m}{V}$ or $m = \rho V$. Apply this equation to the problem, $m = \rho V = \left(1.00\ \frac{\text{g}}{\text{cm}^3}\right)\left(4\ \text{cm}^3\right) = 4.0\ \text{g}$. The weight of the displaced fluid can be determined by $W = mg = (0.004\ \text{kg})\left(9.8\ \frac{\text{m}}{\text{s}^2}\right) = 0.0392\ \text{N}$.

EXAMPLE: A humpback whale weighs 5.4×10^5 N. Determine the buoyant force required to support the whale in its natural habitat, the ocean, when it is completely submerged. Assume that the density of seawater is 1030 kg/m^3 and the density of the whale ρ_{whale} is approximately equal to the density of water ($\rho_{\text{water}} = 1000\ \text{kg/m}^3$).

SOLUTION: The volume of the whale can be determined from

$$m = \rho V = \frac{W}{g}$$

Solving for V yields

$$V_{whale} = \frac{W_{whale}}{\rho_{whale} g} = \frac{5.4 \times 10^5 \text{ N}}{(1000 \frac{\text{kg}}{\text{m}^3})(9.8 \frac{\text{m}}{\text{s}^2})} = 55.1 \text{ m}^3$$

The whale displaces 55.1 m^3 of water when submerged. Therefore, the buoyant force BF which is equal to the weight of displaced water, is given by

$$\text{BF} = W_{seawater} = \rho_{seawater} g V_{whale} = \left(1030 \frac{\text{kg}}{\text{m}^3}\right)\left(9.8 \frac{\text{m}}{\text{s}^2}\right)\left(55.1 \text{ m}^3\right) = 5.6 \times 10^5 \text{ N}$$

SURFACE TENSION

Surface tension γ is the tension, or force per unit length, generated by cohesive forces of molecules on the surface of a liquid acting toward the interior. Surface tension is given as force per unit length and defined as the ratio of the surface force F to the length d along which the force acts, or

$$\gamma = \frac{F}{d}$$

Surface tension is given in units of $\frac{\text{N}}{\text{m}}$.

FLUIDS IN MOTION

VISCOSITY

Viscosity η is a measurement of the resistance or frictional force exerted by a fluid in motion. The SI unit of viscosity is newton seconds per square meter $\left(\frac{\text{N} \cdot \text{s}}{\text{m}^2}\right)$. Viscosity is typically expressed in units of poise (P), where

$$1 \text{ poise} = 0.1 \frac{\text{N} \cdot \text{s}}{\text{ms}^2}$$

CONTINUITY EQUATION

The **equation of continuity** is, in essence, an expression of the conservation of mass for a moving fluid. Specifically, the equation of continuity states that: (1) a fluid maintains constant density regardless of changes in pressure and temperature, and (2) flow measured at one point along a vessel is equal to the flow at another point along the vessel, regardless of the cross-sectional area of the vessel. Fluid flow Q, expressed in terms of the cross-sectional area of the vessel A, is given as

$$Q = Av$$

where v is the velocity of the fluid. Thus, the equation of continuity at any two points in a vessel is given as

$$Q = A_1 v_1 = A_2 v_2 = \text{constant}$$

BERNOULLI'S EQUATION

Bernoulli's equation, the fluid equivalent of conservation of energy, states that the energy of fluid flow through a rigid vessel by a pressure gradient is equal to the sum of the pressure energy, kinetic energy, and the gravitational potential energy, or

$$E_{tot} = P + \frac{1}{2}\rho v^2 + \rho g h = \text{constant}$$

An important application of Bernoulli's equation involves fluid flow through a vessel with a region of expansion or contraction. Bernoulli's equation describing fluid flow through a vessel with sudden changes in geometry can be expressed as

$$\left(P + \frac{1}{2}\rho v^2 + \rho g h\right)_1 = \left(P + \frac{1}{2}\rho v^2 + \rho g h\right)_2$$

where 1 describes the energy of fluid flow in the normal region of the vessel and 2 describes the energy of fluid flow in the obstructed or enlarged region. An illustration of these two different scenarios of Bernoulli's equation as applied to vessel disease is seen in Figures 7-3 and 7-4. Note that Bernoulli's equation assumes a smooth transition between the two regions within a vessel with no turbulence in fluid flow. However, in the practical applications depicted in Figures 7-3 and 7-4, there is turbulence noted in Figure 7-3 about the expansion of the blood vessel, and in Figure 7-4 about the surface of the atherosclerotic plaque. Nevertheless, in many cases the

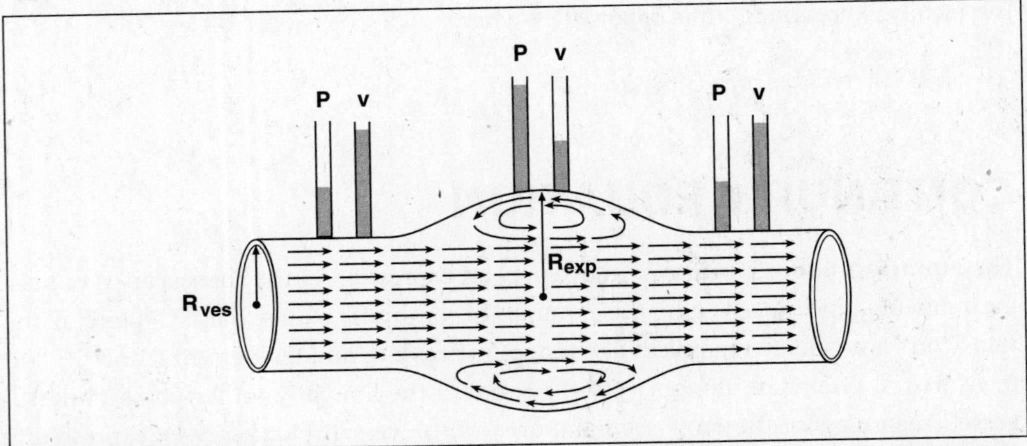

FIGURE 7-3: SOURCE: From George Hademenos, *Schaum's Outline of Physics for Pre-Med, Biology, and Allied Health Students*, McGraw-Hill, 1998; reproduced with permission of The McGraw-Hill Companies.

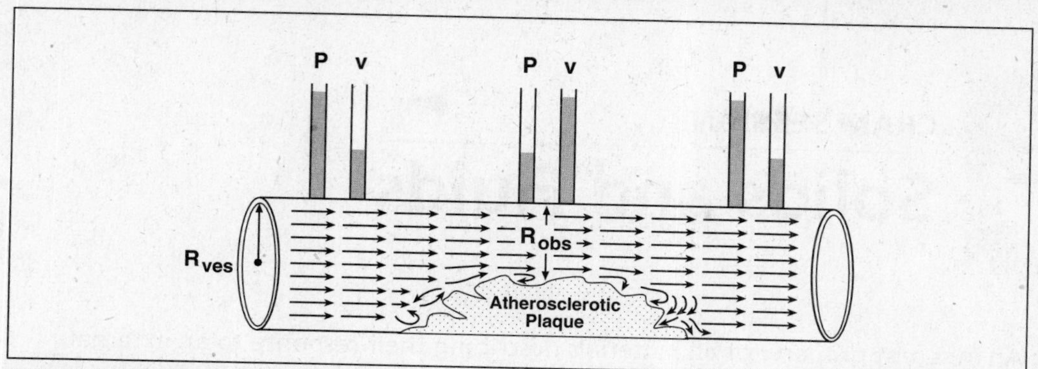

FIGURE 7-4: SOURCE: From George Hademenos, *Schaum's Outline of Physics for Pre-Med, Biology, and Allied Health Students,* McGraw-Hill, 1998; reproduced with permission of The McGraw-Hill Companies.

vessel expansion in Figure 7-3 and the atherosclerotic plaque in Figure 7-4 are not significantly pronounced, so Bernoulli's principle can still be applied to these scenarios.

LAMINAR AND TURBULENT FLOW

Fluid flow can be characterized according to two different types of flow: laminar flow and turbulent flow. Consider a fluid with density ρ and viscosity η flowing through a tube of diameter d. In **laminar flow,** the fluid flows as continuous layers stacked on one another within a smooth tube and moving past one another with some velocity. However, as the velocity of the fluid increases, fluid particles fluctuate between these ordered layers, causing random motion. When the fluid's velocity exceeds a threshold known as the critical velocity, it now is described as **turbulent flow.** The critical velocity depends on the parameters of the fluid and the vessel according to the relation

$$v = \frac{\eta \mathrm{Re}}{\rho d}$$

where Re is a dimensionless quantity known as the Reynold's number.

CRAM SESSION

Solids and Fluids

Elasticity: An inherent property of all materials describing their response to an external force, characterized by the presence and degree of deformation of the object. Two quantities describe the elasticity of an object: stress and strain.

Stress: The magnitude of an external force F required to induce the given deformation of an object divided by the surface area A over which the force is exerted perpendicularly, or $S = \frac{F}{A}$; expressed in units of $N \cdot m^2$.

Strain: The ratio of the increase or decrease in a particular dimension in the deformed state to that dimension in its initial, undeformed state; can be expressed mathematically as $\varepsilon = \frac{x_d - x_n}{x_n}$, where x_d and x_n represent the dimensional quantities of the elastic object in the deformed and undeformed (relaxed) states, respectively.

Elastic modulus: A physical constant that describes the elastic properties of a solid, defined by: Elastic modulus $= \frac{\text{stress}}{\text{strain}}$; expressed in units of $N \cdot m^2$. The three types of elastic moduli that are used to characterize the elasticity of solids are described below.

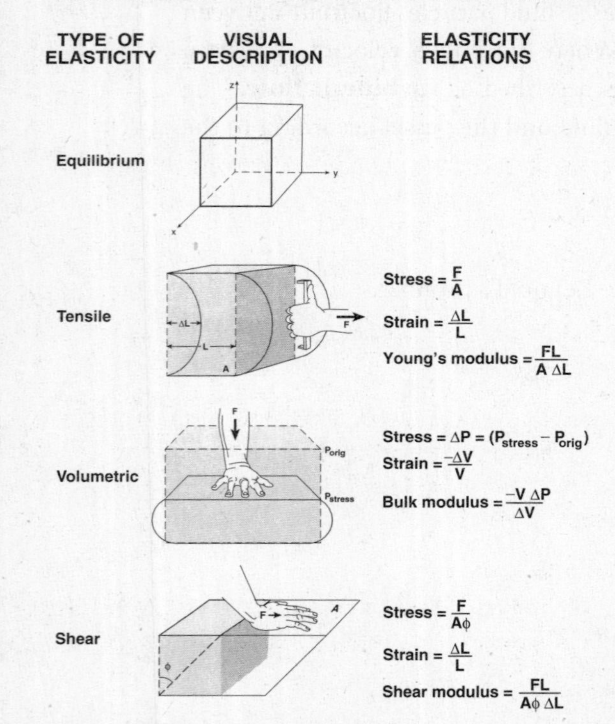

SOURCE: From George Hademenos, *Schaum's Outline of Physics for Pre-Med, Biology, and Allied Health Students*, McGraw-Hill, 1998; reproduced with permission of The McGraw-Hill Companies.

Density: Density ρ is a physical property of a fluid, given as mass per unit volume, or $\rho = \frac{\text{mass}}{\text{unit volume}} = \frac{m}{V}$, density represents the fluid equivalent of mass and is given in units of $\frac{\text{kg}}{\text{m}^3}$, $\frac{\text{g}}{\text{cm}^3}$, or $\frac{\text{g}}{\text{mL}}$. Density, a property unique to each substance, is independent of shape or quantity but is dependent on temperature and pressure.

Specific gravity (Sp. gr.): The ratio of the density of a given substance ρ_{sub} to the density of water ρ_w, or

$$\text{Sp. gr.} = \frac{\text{density of substance}}{\text{density of water}} = \frac{\rho_{\text{sub}}}{\rho_w}$$

where the density of water ρ_w is 1.0 g/cm^3 or 1.0×10^3 kg/m^3. Assuming that equal volumes are chosen, the specific gravity can also be expressed in terms of weight:

$$\text{Sp. gr.} = \frac{\text{weight of substance}}{\text{weight of water}} = \frac{w_{\text{sub}}}{w_w}$$

Specific gravity is a pure number that is unitless.

Pressure: Pressure P is a force F acting perpendicular to a surface area A of an object and given by

$$\text{Pressure} = \frac{\text{force}}{\text{area}} = \frac{F}{A}$$

Pressure is a scalar quantity and is expressed in units of N $\cdot$ m^2. Two specific types of pressure applicable to fluids are atmospheric pressure and hydrostatic pressure. Atmospheric pressure P_{atm} represents the average pressure exerted by Earth's atmosphere and is defined numerically as 1 atm, but can also be expressed as 760 mmHg or 1.01×10^5 Pa. Hydrostatic pressure P_{hyd} is fluid pressure exerted at a depth h in a fluid of density ρ and is given by $P_{\text{hyd}} = \rho g h$. The total pressure P exerted on a contained fluid is the sum of the P_{atm} and P_{hyd}:

$$P = P_{\text{atm}} + P_{\text{hyd}} = P_{\text{atm}} + \rho g h$$

Pascal's principle: An external pressure applied to a confined fluid is transmitted equally to all points within the fluid.

Archimedes' principle: A body immersed wholly or partially in a fluid is subjected to a buoyant force that is equal in magnitude to the weight of the fluid displaced by the body; or

$$\text{Buoyant force} = \text{weight of displaced fluid}$$

If the buoyant force is greater than or equal to the weight of the displaced fluid, then the object floats. If the buoyant force is less than the weight of the displaced fluid, then the object sinks.

Viscosity: Viscosity η is a measurement of the resistance or frictional force exerted by a fluid in motion. The SI unit of viscosity is $\frac{\text{N} \cdot \text{s}}{\text{m}^2}$. Viscosity is typically expressed in units of poise (P), where 1 poise $= 0.1 \frac{\text{N} \cdot \text{s}}{\text{ms}^2}$.

Equation of continuity: An expression of the conservation of mass for a moving fluid, it states that

1. A fluid maintains constant density regardless of changes in pressure and temperature
2. Flow measured at one point along a vessel is equal to the flow at another point along the vessel, regardless of the cross-sectional area of the vessel. If fluid flow Q is expressed in terms of the cross-sectional area of the vessel A as

   $$Q = Av$$

   where v is the fluid's velocity, then the equation of continuity at any two points in a vessel is

   $$Q = A_1v_1 = A_2v_2 = \text{constant}$$

Bernoulli's equation: The fluid equivalent of conservation of energy, it states that the energy of fluid flow through a rigid vessel by a pressure gradient is equal to the sum of the pressure energy, kinetic energy, and the gravitational potential energy, or

$$E_{\text{tot}} = P + \frac{1}{2}\rho v^2 + \rho gh = \text{constant}$$

Laminar flow: A condition in which a fluid flows as continuous layers stacked on one another within a smooth tube and moving past one another with some velocity.

Turbulent flow: A condition that occurs when the fluid particles fluctuate between these ordered layers as the velocity of the fluid increases. The random, chaotic motion of the fluid that results is described as turbulent flow.

Strategies for Solving Solids and Fluids Problems

1. Regardless of problems dealing with solids or fluids, density is either given or must be determined. Density is defined as $\rho = \frac{\text{mass}}{\text{unit volume}} = \frac{m}{V}$.
   - ➤ For a solid with uniform geometry, volume is calculated using a ruler to identify length in all three dimensions.
   - ➤ For a solid with nonuniform geometry, volume is calculated by dropping the object into a graduated cylinder with water up to some level and then noting the increase in volume by the difference between the two water levels.
   - ➤ For a fluid, volume is calculated by using a graduated cylinder.
2. For elastic moduli, note the type and magnitude of the force applied to the elastic object.

3. Pressure is defined as Pressure $= \frac{\text{force}}{\text{area}} = \frac{F}{A}$, where F is in SI units of newtons and A is in units of m^2.

- If the mass of an object is given, then the force it exerts is its weight which is determined by: $W = mg$.
- If the area is given in units of cm, then to convert to m^2: $cm^2 = [(10^{-2})\ m]^2 = 10^{-4}\ m^2$.
- Maintain organization if atmospheric pressure and hydrostatic pressures are involved and keep their calculations separate.

CHAPTER 8

Temperature and Heat

Read This Chapter to Learn About:

- ➤ Temperature
- ➤ Thermal Expansion
- ➤ Heat
- ➤ Heat Transfer
- ➤ Laws of Thermodynamics

Temperature and heat are often closely linked as two terms describing the same concept: whether an object feels hot or cold to the touch. From the feverish symptoms you feel when ill or the baking temperature of a casserole to the shiver you have when out in an ice storm or the cold you feel when touching icy metal, you have a qualitative sense of temperature. Heat, however, is a type of energy, often referred to as **thermal energy.** Thermal energy depends on temperature difference. This chapter discusses the relationship between temperature and heat and the basic laws explaining their interaction.

TEMPERATURE

Temperature is a physical property of a body that reflects its warmth or coldness. Temperature is a scalar quantity that is measured with a thermometer and is expressed in units that are dependent on the temperature measurement scale used. Three common scales of temperature measurement are Fahrenheit (°F), Celsius (°C), and Kelvin (K).

(The SI unit for temperature is degrees Celsius.) They are defined according to absolute zero as well as the freezing point and boiling point of water:

	Fahrenheit (°F)	Celsius (°C)	Kelvin (K)
Absolute Zero	−460	−273	0
Freezing Point of Water	32	0	273
Boiling Point of Water	212	100	373

Absolute zero is the lowest possible threshold of temperature and is defined as 0 K.

The temperature scales described in this table are related according to the following equations:

Fahrenheit ⇔ Celsius: $T_F = \frac{9}{5}T_C + 32°$ $\quad T_C = \frac{5}{9}(T_F - 32°)$

Celsius ⇔ Kelvin: $T_C = T_K - 273°$ $\quad T_K = T_C + 273°$

EXAMPLE: Normal body temperature is 98.6°F. Convert this temperature to degrees Celsius.

SOLUTION: To convert degrees Fahrenheit to degrees Celsius, begin with the equation

$$T_C = \frac{5}{9}(T_F - 32°)$$

Solving for T_C yields

$$T_C = \frac{5}{9}(98.6° - 32°) = 37°\text{C}$$

EXAMPLE: Derive a relationship between the Fahrenheit and Kelvin temperature scales.

SOLUTION: The Fahrenheit and Celsius temperature scales and the Celsius and Kelvin temperature scales are related by the equations given previously.

Substituting the first Celsius ⇔ Kelvin equation into the first Fahrenheit ⇔ Celsius equation yields

$$T_F = \frac{9}{5}(T_K - 273°) + 32° = \frac{9}{5}T_K - \frac{9}{5}(273°) + 32°$$

$$= \frac{9}{5}T_K - 491.4° + 32° = \frac{9}{5}T_K - 460°$$

So the desired equation is

$$T_F = \frac{9}{5}T_K - 460°$$

THERMAL EXPANSION

Thermal expansion is a physical phenomenon in which increases in temperature can cause substances in the solid, liquid, and gaseous states to expand.

Linear Expansion of Solids

A solid subjected to an increase in temperature ΔT experiences an increase in length ΔL that is proportional to the original length L_0 of the solid. The relation for the linear expansion of solids is

$$\Delta L = \alpha L_0 \Delta T$$

where the proportionality constant α is the coefficient of linear expansion and is expressed in units of inverse temperature.

Volumetric Expansion of Liquids

A liquid subjected to an increase in temperature ΔT experiences an increase in volume ΔV that is proportional to the original volume V_0 of the liquid. The relation for the volumetric expansion of liquids is

$$\Delta V = \beta V_0 \Delta T$$

where the proportionality constant β is the coefficient of volumetric expansion and is equal to 3α. As is the case for α, the units of β are inverse temperature.

Volumetric Expansion of Gases

The volumetric expansion of gases can be summarized by two important gas laws: Charles's law and Boyle's law, which can be combined in the Ideal Gas Law.

Charles's law states that at constant pressure P, the volume V occupied by a given mass of gas is directly proportional to the absolute temperature T by

$$V = kT \quad \text{or} \quad \frac{V_1}{T_1} = \frac{V_2}{T_2}$$

$P =$ constant where k is a proportionality constant.

Boyle's law states that at constant temperature, the volume V occupied by a given mass of gas is inversely proportional to the pressure P exerted on it:

$$PV = k \quad \text{or} \quad P_1 V_1 = P_2 V_2$$

$T =$ constant where k is a proportionality constant.

The **Ideal Gas Law** combines the relations from Charles's law and Boyle's law to yield

$$PV = nRT \quad \text{or} \quad \frac{P_1V_1}{T_1} = \frac{P_2V_2}{T_2}$$

where R is the universal gas constant ($R = 0.082$ L atm/mol K) and n is the number of moles.

EXAMPLE: A metal cylindrical rod of length $L = 4.0$ meters (m) is heated from 25°C to 225°C. Given a coefficient of linear expansion of 11×10^{-6} °C^{-1}, determine the change in length following expansion.

SOLUTION: The equation needed to solve this problem is

$$\Delta L = \alpha L_o \Delta T = \left(11 \times 10^{-6}\ {}^\circ\text{C}^{-1}\right)(4.0\ \text{m})\left(225^\circ\text{C} - 25^\circ\text{C}\right) = 8.8\ \text{mm}$$

HEAT

Heat is closely associated with temperature, but they are very different quantities. Heat, or thermal energy, is a form of energy that depends on a change of temperature and can be converted to work and other forms of energy. The quantity of heat Q gained or lost by a body is related to the temperature difference ΔT by the following equation:

$$Q = mc\Delta T$$

where m is the mass of the body and c is a proportionality constant termed the **specific heat capacity.** Specific heat capacity, a constant dependent on the material of the object, is the quantity of heat required to raise the temperature of 1 kilogram (kg) of the substance by 1°C. Because heat is a form of energy, the SI unit of energy, the joule (J), is used to measure heat. The specific heat capacity has units of $\frac{\text{J}}{\text{kg}\ {}^\circ\text{C}}$.

Heat of transformation is similar to the specific heat capacity, but it accounts for changes in the phase of the body. The specific heat of a body assumes that no change in phase occurs during a temperature change. In order for a substance to change states of matter—that is, from solid to liquid or from liquid to gas—heat energy must be added to or removed from the substance. The amount of heat required to change the phase of 1 kg of a substance is the heat of transformation L. Thus, the total amount of heat Q gained or lost by a substance of mass m during a change between phases is

$$Q = mL$$

where L is the heat of transformation unique to the substance. The heat of transformation exists in two forms, according to the particular phase transformation:

- **Heat of fusion** L_f is the amount of heat energy required to change 1 kg of solid matter to liquid or the amount of energy released when changing 1 kg of liquid matter to solid.

- **Heat of vaporization** L_V is the amount of heat energy required to change 1 kg of liquid matter to gas or the amount of energy released when changing 1 kg of gas matter to liquid.

This is summarized in the following table.

Phase Transformation	Descriptive Term	Temperature Point	Heat of Transformation
Solid state → Liquid state	Melting	Melting point	Heat of fusion
Liquid state → Solid state	Freezing	Melting point	Heat of fusion
Liquid state → Gas state	Boiling	Boiling point	Heat of vaporization
Gas state → Liquid state	Condensation	Boiling point	Heat of vaporization

HEAT TRANSFER

There are three mechanisms of heat transfer: conduction, convection, and radiation, each dependent on the state of matter of the object.

Conduction is the method of heat energy transfer that occurs in solids—for example, the warming of a spoon when cream is stirred in coffee. In conduction, heat energy is transferred by collisions between the rapidly moving molecules of the hot region and the slower-moving molecules of the cooler region. A portion of the kinetic energy from the rapidly moving molecules is transferred to the slower-moving molecules, causing an increase in heat energy at the cooler end and a subsequent increase in the flow of heat.

Convection is the method of heat energy transfer that occurs in liquids and gasses—for example, the cooling of coffee after cream is poured into it. Convection represents the transfer of heat energy due to the physical motion or flow of the heated substance, carrying heat energy with it to cooler regions of the substance. In contrast to conduction, convection is the primary mechanism of heat transfer in fluids.

Radiation is the method of heat energy transfer that occurs in space, for example, Earth's surface being warmed by the sun. Radiation represents the transfer of heat energy by electromagnetic waves that are emitted by rapidly vibrating, electrically charged particles. The electromagnetic waves propagate from the heated body or source at the speed of light.

LAWS OF THERMODYNAMICS

First Law of Thermodynamics

The total energy of a body—the sum of the potential energy and kinetic energy—is represented by its internal energy U. The internal energy of a body can be altered either by performing work on it or by adding heat to it. Thus, the change in internal energy ΔU is related to the heat energy Q transferred to the body and the work W done by the system according to

$$\Delta U = Q - W$$

This relation describes the principle of **conservation of energy.** The proper sign convention for each of the quantities represented in the above equation can be summarized as follows:

W: + for work done *by* system
 – for work done *on* system

Q: + for heat flow *into* system
 – for heat flow *out of* system

Second Law of Thermodynamics

A system subjected to a spontaneous change responds such that its disorder or entropy S increases or at least remains the same. A spontaneous change means that no work is done. For example, heat energy from a hot object always flows toward a cold object, but not vice versa. For a given system in equilibrium, the change in entropy ΔS is related to the added heat ΔQ by

$$\Delta S = \frac{\Delta Q}{T}$$

Entropy is expressed in units of J/K. The temperature is in kelvins.

EXAMPLE: Thirty-five grams of ice initially at 0°C begins to melt. Determine the entropy of the melting process, given the heat of fusion for water is 80 calories per gram (cal/g).

SOLUTION: The heat needed to melt the ice is

$$\Delta Q = mL_f = (35\text{ g})\left(80\ \frac{\text{cal}}{\text{g}}\right) = 2800\text{ cal}$$

The entropy can now be calculated as

$$\Delta S = \frac{\Delta Q}{T} = \frac{2800\text{ cal}}{273\text{ K}} = 10.25\ \frac{\text{cal}}{\text{K}}$$

Converting to joules gives

$$\left(10.25\ \frac{\text{cal}}{\text{K}}\right)\left(4.184\ \frac{\text{J}}{\text{cal}}\right) = 42.9\ \frac{\text{J}}{\text{K}}$$

CRAM SESSION

Temperature and Heat

Temperature: A physical property of a body that reflects its warmth or coldness. Temperature is a scalar quantity that is measured with a thermometer and can be expressed in units of Fahrenheit (°F), Celsius (°C), and Kelvin (K), although the SI unit for temperature is degrees Celsius. The scales are defined according to absolute zero as well as the freezing point and boiling point of water:

	Fahrenheit (°F)	Celsius (°C)	Kelvin (K)
Absolute Zero	−460	−273	0
Freezing Point of Water	32	0	273
Boiling Point of Water	212	100	373

Temperature scale conversions: You can convert temperature measurements between temperature scales using the following relations:

Fahrenheit ⇔ Celsius: $T_F = \frac{9}{5}T_C + 32°$ $\quad T_C = \frac{5}{9}(T_F - 32°)$

Celsius ⇔ Kelvin: $T_C = T_K - 273°$ $\quad T_K = T_C + 273°$

Thermal expansion: A physical phenomenon in which increases in temperature can cause substances in the solid, liquid, and gaseous state to expand.

➤ **Linear expansion of solids**

$$\Delta L = \alpha L_o \Delta T$$

where ΔL = change in length, ΔT = change in temperature, L_o = original length, α = the coefficient of linear expansion.

➤ **Volumetric expansion of liquids**

$$\Delta V = \beta V_o \Delta T$$

where ΔV = change in volume, ΔT = change in temperature, V_o = original volume, β = the coefficient of volumetric expansion = 3α.

- **Volumetric expansion of gases:** Described by the gas laws.
  - **Charles's law:** At constant pressure, the volume V occupied by a given mass of gas is directly proportional to the absolute temperature T by:

    $$V = kT \quad \text{or} \quad \frac{V_1}{T_1} = \frac{V_2}{T_2}$$

    P = constant where k is a proportionality constant.
  - **Boyle's law:** At constant temperature, the volume V occupied by a given mass of gas is inversely proportional to the pressure P exerted on it:

    $$PV = k \quad \text{or} \quad P_1V_1 = P_2V_2$$

    T = constant where k is a proportionality constant.
  - **Ideal Gas Law:** This law combines Charles's law and Boyle's law to yield

    $$PV = nRT \quad \text{or} \quad \frac{P_1V_1}{T_1} = \frac{P_2V_2}{T_2} \quad \text{where } R \text{ is the universal gas constant}$$

    (R = 0.082 L atm/mol K) and n is the number of moles of the gas.

Heat: A form of energy (thermal energy) that depends on the change of temperature, and can be converted to work and other forms of energy. The quantity of heat Q gained or lost by a body is related to the temperature difference ΔT by

$$Q = mc\Delta T$$

where m is the mass of the body and c is the specific heat capacity. Because heat is a form of energy, it is measured according to the SI unit of measurement of energy, Joules.

Specific heat capacity is the quantity of heat required to raise the temperature of 1 kg of substance by 1°C.

In order for a substance to change states of matter—that is, from solid to liquid or from liquid to gas—heat energy must be added to or removed from the substance:

$$Q = mL$$

where L is the heat of transformation unique to the substance.

Heat of transformation is the quantity of heat required to change the phase/state of 1 kg of a substance and exists in two forms:

- **Heat of fusion:** L_f—the amount of heat energy required to change 1 kg of solid matter to liquid or the amount of energy released when changing 1 kg of a liquid matter to solid.
- **Heat of vaporization:** L_v—the amount of heat energy required to change 1 kg of liquid matter to gas or the amount of energy released when changing 1 kg of gas matter to liquid.

Phase Transformation	Descriptive Term	Temperature Point	Heat of Transformation
Solid state → Liquid state	Melting	Melting point	Heat of fusion
Liquid state → Solid state	Freezing	Melting point	Heat of fusion
Liquid state → Gas state	Boiling	Boiling point	Heat of vaporization
Gas state → Liquid state	Condensation	Boiling point	Heat of vaporization

Heat transfer: There are three mechanisms of heat transfer, each depending on the state of matter of the object:

- **Conduction** is the method of heat energy transfer that occurs in solids, e.g., warming of a spoon when cream is stirred in the coffee.
- **Convection** is the method of heat energy transfer that occurs in liquids and gasses, e.g., cooling of coffee after cream is poured in it.
- **Radiation** is the method of heat energy transfer that occurs in space, e.g., Earth's surface being warmed by the Sun.

Laws of Thermodynamics

- **First Law of Thermodynamics:** Describes the principle of the conservation of energy. It states that the change in internal energy, ΔU, is related to the heat energy, Q, transferred to the body and the work done by the system W according to

$$\Delta U = Q - W$$

W: + for work done *by* system — Q: + for heat flow *into* system

− for work done *on* system; — − for heat flow *out of* system

- **Second Law of Thermodynamics:** A system subjected to a spontaneous change responds such that its disorder or entropy increases or at least remains the same. For a given system in equilibrium, the change in entropy ΔS is related to the added heat ΔQ by

$$\Delta S = \frac{\Delta Q}{T}.$$

Entropy is expressed in units of J/K.

Strategies for Solving Temperature and Heat Problems

1. **Temperature:** Use the appropriate conversion relation to determine the temperature expressed in the correct units. Often, it is simple arithmetic that results in the wrong temperature quantity, and hence the wrong answer to the problem.

2. **Heat:** Heat is very different from temperature—like apples and oranges. Although heat does depend on temperature in the form of change in temperature, heat is an energy and temperature is simply a number.
   - ➤ Heat flows from hot to cold.
   - ➤ Each object/material has its own specific heat capacity.

   Make sure the value used for c conforms to the units used for the other quantities; c can be expressed in units of cal/(g °C) or J/(kg K). The units for thermal energy (cal, J) and mass (g, kg) can easily be converted. The units for the temperature difference need no conversion because you are concerned with the change in temperature and not specific temperature quantities. A change in Celsius is equivalent to a change in Kelvin, with the only exception that the two scales are shifted by 273.
3. **Heat of transformation:** When making calculations involving an object/substance that changes phases/states of matter, the appropriate phase change and its corresponding heat of transformation must be addressed and included in the calculation of total heat energy.

CHAPTER 9

Vibrations and Waves

Read This Chapter to Learn About:

- Simple Harmonic Motion
- Amplitude, Frequency, Period, and Phase Angle
- Mass–Spring System
- Simple Pendulum
- Wave Types: Transverse Waves and Longitudinal Waves
- Wave Speed
- Wave Interactions

Although vibrations and waves are two separate physical entities, they have one thing in common: they both describe oscillatory motion, or motion that moves in a cyclic or periodic manner. The simplest type of oscillatory motion is **simple harmonic motion**, in which an object or body moves about an equilibrium point as a result of a restoring force proportional to the object's distance from equilibrium. The characteristics and physical behavior of vibrations and waves play an important role in two areas of physics—sound and light—each of which is discussed in upcoming chapters of this review. This chapter focuses on the basics of vibrations and waves.

SIMPLE HARMONIC MOTION

Simple harmonic motion is the simplest type of oscillatory motion in which an object or body moves about an equilibrium point as a result of a restoring force. A **restoring force** is a force that satisfies the following two criteria: (1) It is always directed toward the object's equilibrium position, and (2) it is linearly proportional to the object's displacement from the equilibrium position.

An example of simple harmonic motion, shown in Figure 9-1, is the motion of a mass attached to a spring. When the object of mass M is pulled or displaced a distance $-A$ from the point of equilibrium and released, the restoring force of the spring pulls the object a distance $+A$ above the point of equilibrium, causing the spring to be compressed. The object then oscillates between the limits of $-A$ and $+A$, resembling a sinusoidal curve.

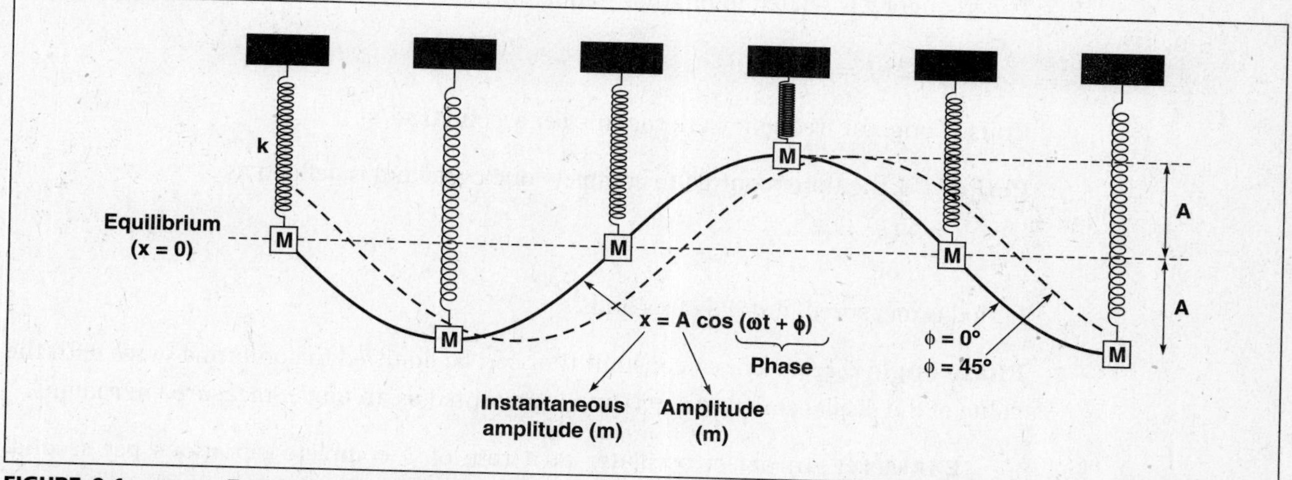

FIGURE 9-1: SOURCE: From George Hademenos, *Schaum's Outline of Physics for Pre-Med, Biology, and Allied Health Students*, McGraw-Hill, 1998; reproduced with permission of The McGraw-Hill Companies.

AMPLITUDE, FREQUENCY, PERIOD, AND PHASE ANGLE

The motion of an object in simple harmonic motion, shown graphically in Figure 9-1, consists of a continuous cycle of peaks (crests) and valleys (troughs). Four physical characteristics that describe the motion of such an object are amplitude, frequency, period, and phase. The influence of each of these characteristics on the object in simple harmonic motion can be seen from the equation describing the position of the oscillating object from its point of equilibrium:

$$x = A\cos(\omega t + \phi)$$

where x is the position as a function of time (in meters), A is the amplitude of the oscillatory motion (in meters), ω is the angular velocity (in radians per second), t is time (in seconds), and ϕ is the phase angle (in radians).

Amplitude A is the maximum magnitude of displacement from equilibrium. Units of amplitude are typically assigned according to the dimension of length but can be represented by any physical quantity whose magnitude varies in an oscillatory manner.

Frequency f is the number of cycles in a time interval and is given by

$$f = \frac{1}{T}$$

where T is the period of the oscillation or the time required to complete one cycle. Units of frequency are hertz (Hz) or cycles/second. Cycles are, in essence, unitless, and therefore frequency is actually given in units of inverse time (1/s).

The frequency is related to angular frequency by

$$\omega = 2\pi f = \frac{2\pi}{T}$$

Units of angular frequency are radians per second (rad/s).

Period T is the time required to complete one cycle and is defined as

$$T = \frac{2\pi}{\omega}$$

Period is measured in units of seconds.

Phase angle ϕ represents a constant that can be adjusted to conform a wave with the value of the displacement at $t = 0$ and is presented as an angle, measured in radians.

EXAMPLE: An object oscillates at a rate of 3 complete vibrations per second. Determine (1) the angular frequency and (2) the period of the motion.

SOLUTION:

(1) The angular frequency of the vibrating object can be determined from

$$\omega = 2\pi f = (2)\,(3.14)\left(3\,\frac{\text{cycles}}{\text{s}}\right) = 18.8\,\frac{\text{cycles}}{\text{s}} = 18.8\text{ Hz}$$

(2) The period of motion of the vibrating object is

$$T = \frac{2\pi}{\omega} = \frac{(2)\,(3.14)}{18.8\,\frac{\text{cycles}}{\text{s}}} = 0.33\text{ s}$$

MASS–SPRING SYSTEM

One example of simple harmonic motion is an object of mass M attached to a spring and is referred to as the **mass–spring system**. This type of motion is illustrated in Figure 9-1. The restoring force F of the spring that guarantees simple harmonic motion is proportional to the object's displacement x, as given by the relationship known as **Hooke's law**:

$$F = -kx$$

where k is the spring constant. The spring constant, a value unique to each spring, indicates the amount of force required to either stretch or compress a spring a certain distance and is expressed in units of N/m. Because $\boldsymbol{F} = m\boldsymbol{a}$, then the acceleration of an object in simple harmonic motion can be determined:

$$ma = -kx$$

$$a = -\left(\frac{k}{m}\right)x$$

The period of an object moving in a mass–spring system is

$$T = 2\pi\sqrt{\frac{m}{k}}$$

The frequency and angular frequency of this object can be found from the following expressions:

$$\text{Frequency: } f = \frac{1}{T} = \frac{1}{2\pi\sqrt{\frac{m}{k}}} = \frac{1}{2\pi}\sqrt{\frac{k}{m}}$$

$$\text{Angular frequency: } \omega = 2\pi f = \frac{2\pi}{T} = \frac{2\pi}{2\pi\sqrt{\frac{m}{k}}} = \sqrt{\frac{k}{m}}$$

EXAMPLE: A marble attached to a spring is displaced 4.0 centimeters (cm) from its point of equilibrium and is released. If the marble reaches 4.0 cm on the opposite side of the point of equilibrium in 0.85 s, determine the following concerning the motion of the marble:

1. Amplitude
2. Period
3. Frequency
4. Equation of motion

SOLUTION:

1. The amplitude is the maximum distance from the point of equilibrium attained by the oscillating marble, or $A = 4.0$ cm.
2. The period is the time required for the marble to complete one entire cycle. The time taken to traverse from -4.0 cm below the equilibrium to 4.0 cm above the equilibrium point was 0.85 s and constituted one-half of a cycle. The time taken to complete one cycle is the period, or $T = 2 \times 0.85\text{ s} = 1.7\text{ s}$.
3. The frequency of the oscillating marble is $f = \frac{1}{T}$, where T is the period of motion. Therefore,

   $$f = \frac{1}{T} = \frac{1}{1.7\text{ s}} = 0.59\text{ Hz}$$

4. In order to derive an equation of motion of the marble, the characteristics of motion of the marble are displayed in the form of

   $$x = A\cos\omega t$$

The amplitude of the oscillating marble is 4.0 cm. The angular frequency can be determined from the period by

$$\omega = \frac{2\pi}{T} = \frac{2\pi}{1.7\ \text{s}} = 3.7\ \frac{\text{rad}}{\text{s}}$$

Thus, the equation of motion for the oscillating marble is

$$x = 4.0 \cos 3.7t\ (\text{cm})$$

SIMPLE PENDULUM

Another example of simple harmonic motion is the simple pendulum. A simple pendulum consists of a mass or bob attached to one end of a massless cord, as shown in Figure 9-2.

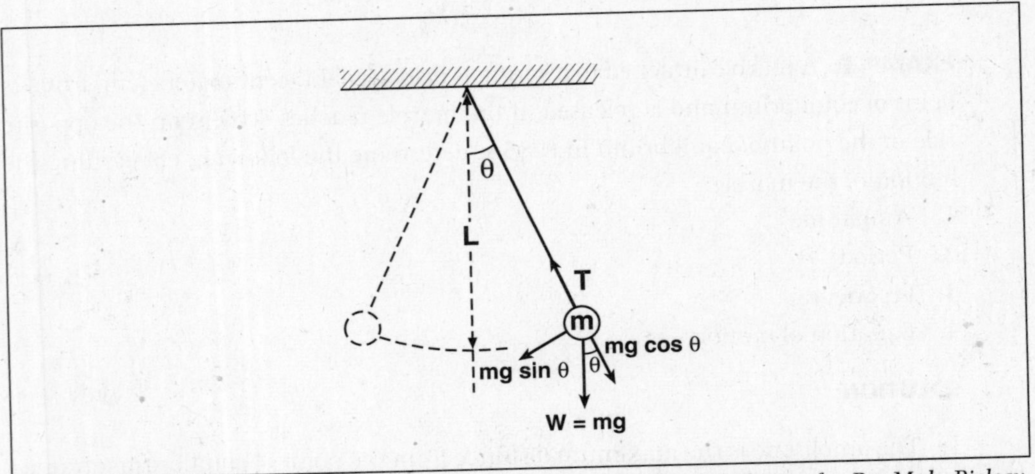

FIGURE 9-2: SOURCE: From George Hademenos, *Schaum's Outline of Physics for Pre-Med, Biology, and Allied Health Students*, McGraw-Hill, 1998; reproduced with permission of The McGraw-Hill Companies.

The pendulum bob is pulled to one side and released, causing the bob to move back and forth in periodic motion under the force of gravity. The period of a simple pendulum is given by

$$T = 2\pi\sqrt{\frac{L}{g}}$$

where L is the length of the cord and g is the gravitational acceleration constant.

WAVE TYPES: TRANSVERSE WAVES AND LONGITUDINAL WAVES

Waves are vibrations that are created by a disturbance and move in space with respect to time. There are two types of waves classified according to the orientation between the direction of the disturbance and the direction of motion. **Transverse waves** are waves generated by a disturbance that acts perpendicular to the direction of wave motion. Examples of transverse waves include light waves and waves in a plucked string. **Longitudinal waves** are waves generated by a disturbance that acts parallel to the direction of wave motion. Examples of longitudinal waves include sound waves and waves moving down the length of a spring due to pushing/pulling one end. Both types of waves are illustrated in Figure 9-3.

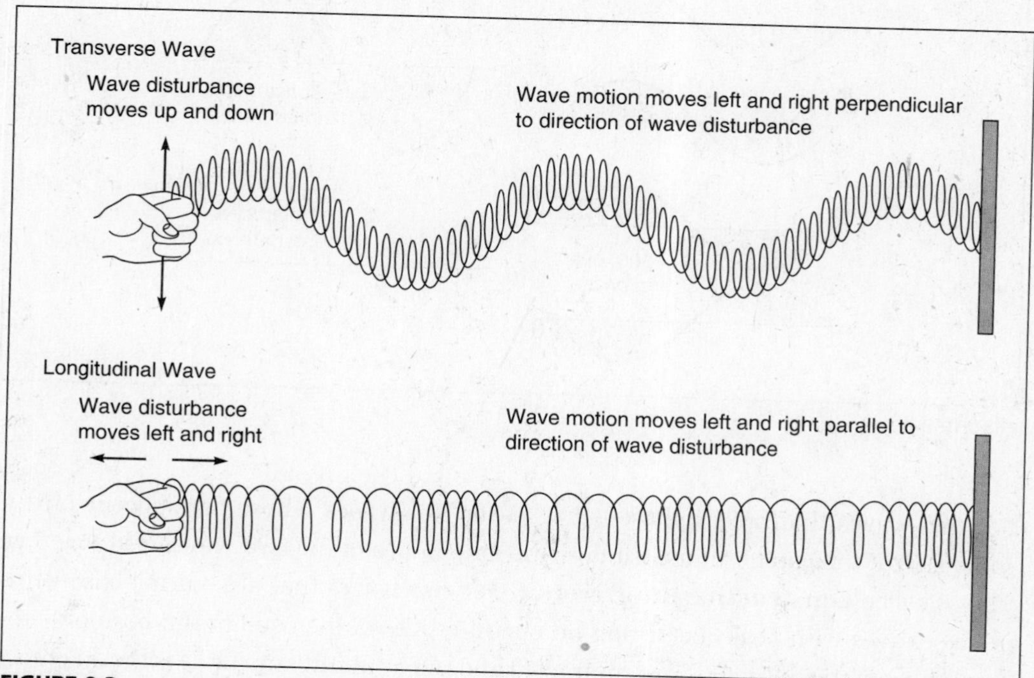

FIGURE 9-3

WAVE SPEED

The physical characteristics describing simple harmonic motion (amplitude, frequency, and period) still apply in defining a wave, describing it, and discussing its units. Because a wave moves, it exhibits speed. The speed of a wave, v, is defined by

$$v = \lambda f$$

where λ is the wavelength or distance measured between adjacent peaks and f is the frequency of the wave. Speed is expressed in terms of meters per second (m/s).

WAVE INTERACTIONS

We have focused on the nature and behavior of a single wave moving in space. This section describes the interactions that can occur between two waves. The **principle of superposition** states that the wave that results from the interaction between two or more waves is the algebraic sum or overlapping of the individual waves, also referred to as superposition. These resultant waves undergo interference, the type of which depends on the phase of the two interacting waves, as shown in Figure 9-4.

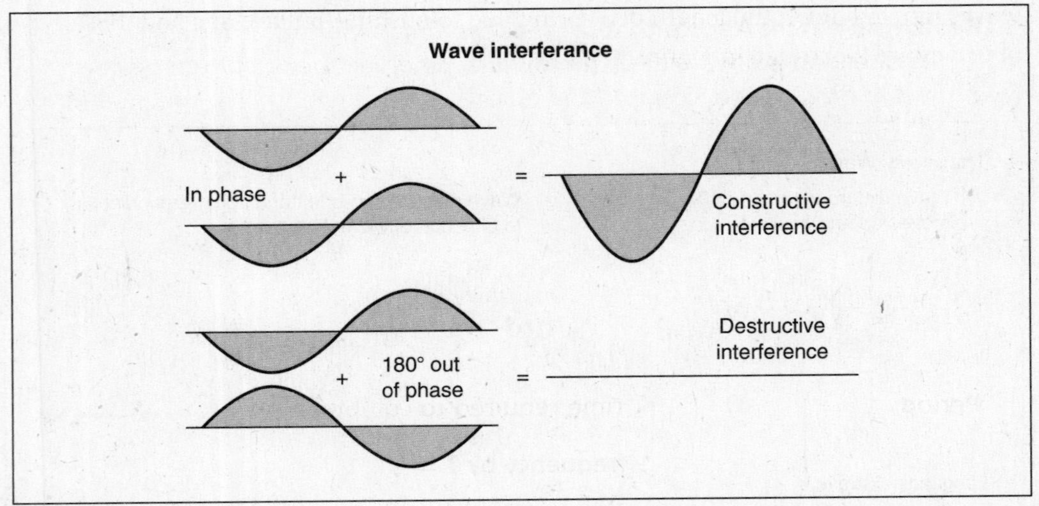

FIGURE 9-4

For two waves that are in phase—that is, for two waves whose peaks occur on the same side of the equilibrium position—the amplitudes of the two waves add together. This is called **constructive interference**. For two waves that are out of phase—that is, two waves with peaks occurring on opposite sides of the equilibrium position—the amplitudes of the two waves are subtracted and the resultant amplitude is the difference between the two amplitudes. This is called **destructive interference**.

CRAM SESSION

Vibrations and Waves

Simple harmonic motion: Motion in which an object or body moves about an equilibrium point as a result of a restoring force.

Restoring force: A force that satisfies the following two criteria:

1. Its direction is always directed toward the object's equilibrium position.
2. It is linearly proportional to the object's displacement from the equilibrium position.

Characteristics of simple harmonic motion: Shown in the following table:

Quantity	Symbol	Definition	Unit
Amplitude	A	Maximum displacement from equilibrium	Meters
Wavelength	λ	Distance between two adjacent, successive points on a wave	Meters
Period	T	Time required to complete one cycle; related to frequency by $T = \frac{1}{f}$	Seconds
Frequency	f	Number of cycles per unit time; related to period by $f = \frac{1}{T}$	Hertz
Phase angle	ϕ	Constant that can be adjusted to conform a wave with the value of the displacement at $t = 0$	Radians

Examples of simple harmonic motion: Shown in the following table:

Example	Diagram	Restoring Force	Period
Mass–spring system	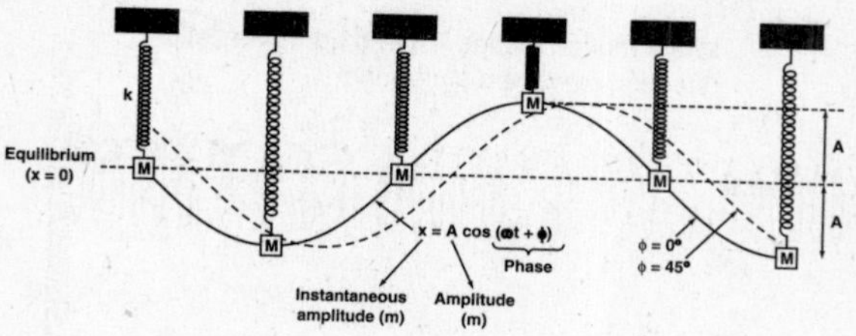	$F = -kx$	$T = 2\pi\sqrt{\frac{m}{k}}$

Example	Diagram	Restoring Force	Period
Simple pendulum	θ; L; T; m; mg cos θ; mg sin θ; W = mg	$F = mg \sin\theta$	$T = 2\pi\sqrt{\frac{L}{g}}$

Art in this table from George Hademenos, *Schaum's Outline of Physics for Pre-Med, Biology, and Allied Health Students,* McGraw-Hill, 1998; reproduced with permission of The McGraw-Hill Companies.

Waves: Vibrations that are created by a disturbance and move in space with respect to time. There are two types of waves classified according to the orientation between the direction of the disturbance and the direction of motion.

- **Transverse waves** are waves generated by a disturbance that acts perpendicular to the direction of wave motion. Examples of transverse waves include light waves and waves in a plucked string.
- **Longitudinal waves** are waves generated by a disturbance that acts parallel to the direction of wave motion. Examples of longitudinal waves include sound waves and waves moving down the length of a spring due to pushing and pulling one end.

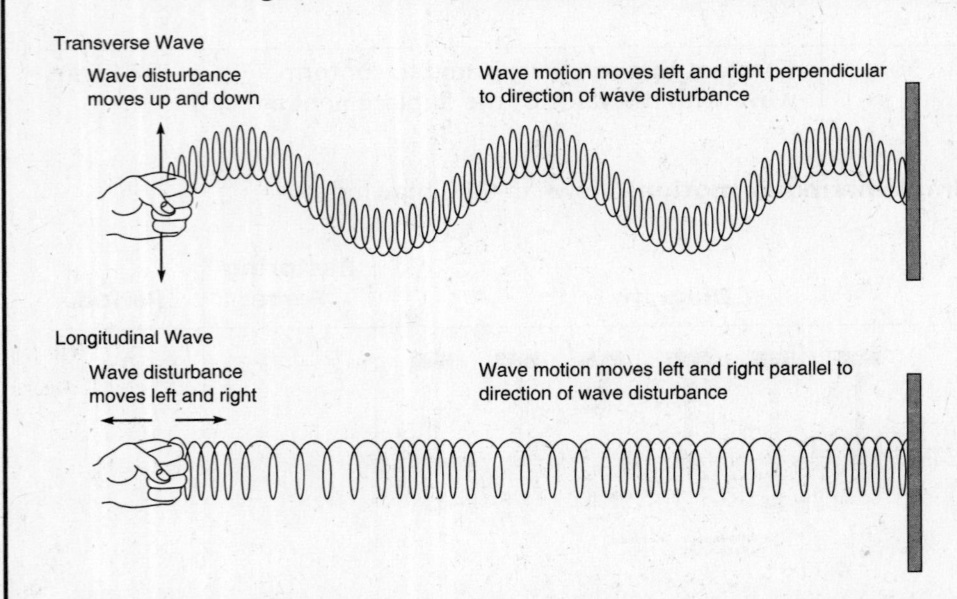

Wave speed: The speed of a wave, v, is defined by: $v = \lambda f$, where λ = wavelength and f = frequency. Speed is expressed in terms of m/s.

Wave interactions

- **Principle of superposition** States that the resultant wave that occurs from the interaction between two or more waves is the algebraic sum or overlapping of the individual waves; also referred to as superposition.
- **Constructive interference** States that for two waves that are in phase (i.e., the peaks of the two waves occur on the same side of the equilibrium position), the amplitudes of the two waves are added together.
- **Destructive interference** States that for two waves that are out of phase (i.e., the peaks of the two waves occur on opposite sides of the equilibrium position), the amplitudes of the two waves are subtracted, and the resultant amplitude is the difference between the two amplitudes.

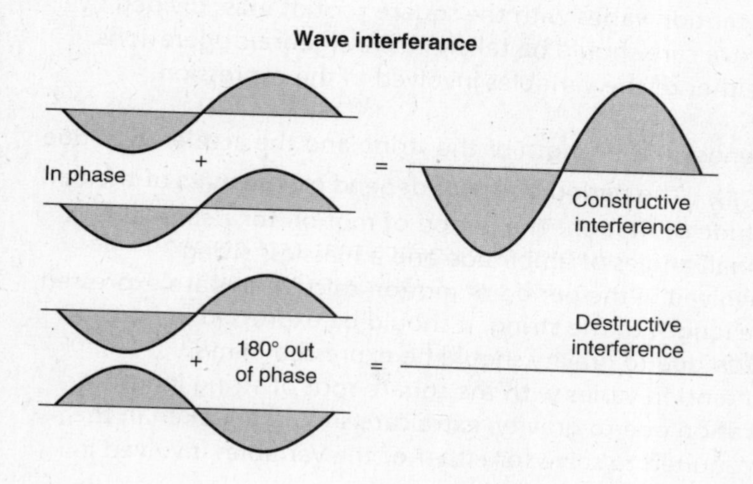

Strategies for Solving Vibration and Wave Problems

1. **Characteristics of simple harmonic motion**
   - **Amplitude** is measured from the equilibrium point to the maximum point of displacement—on either side of the equilibrium point.
   - **Period** is the time to complete 1 cycle or one round trip of the object in simple harmonic motion. If the time it takes for the object to move from one maximum point to the opposite maximum point is 1 s, then the period is 2 s because the object passes the point twice during its round trip.
   - **Frequency** is the number of cycles per unit time. Therefore, if an object vibrates 100 cycles in 400 s, then its frequency is 100 cycles/400 s or 0.25 Hz. Once the frequency is known, the period can be determined by taking the inverse of the frequency or 4 s.

2. **Examples of simple harmonic motion** Problems involving these examples most likely include period calculations and each of the two systems discussed in this chapter have different expressions for the period.

➤ **Mass–spring system**

a. Period of motion depends on the mass of the object and the spring constant, i.e., $T = 2\pi\sqrt{\frac{m}{k}}$. If given the weight of an object attached to a spring, convert to mass by dividing the quantity by g ($= 9.8$ m/s^2). The spring constant is determined by the force exerted on the spring causing the displacement divided by the amount of displacement. If the mass is hanging vertically from the spring, the restoring force (force required to stretch or compress the spring) is given by Hooke's law ($F = -kx$).

b. Ensure all variables involved in the period of motion calculations are expressed in SI units. If given the mass of the object, it should be expressed in units of kilograms. The spring constant should be expressed in N/m.

c. Because the period of motion varies with the square root of mass divided by the spring constant, extra care should be taken in the algebraic operations required to solve for either of the variables involved in the expression.

➤ **Simple pendulum**

a. Period of motion depends on the length of the string and the acceleration due to gravity, i.e., $T = 2\pi\sqrt{\frac{L}{g}}$. The period *does not* depend on the mass of the bob or the angle or amplitude of motion. The period of motion for a simple pendulum holds for small angles of amplitude and a massless string.

b. Ensure all variables involved in the period of motion calculations are expressed in SI units. If given the length of the string, it should be expressed in units of meters. The acceleration due to gravity should be expressed in m/s^2.

c. Because the period of motion varies with the square root of string length divided by the acceleration due to gravity, extra care should be taken in the algebraic operations required to solve for either of the variables involved in the expression.

CHAPTER 10

Sound

Read This Chapter to Learn About:

- Characteristics of Sound Waves
- Doppler Effect
- Sound Intensity and Sound Level
- Resonance
- Standing Waves
- Frequencies in a Stretched String
- Frequencies in a Pipe
- Beats

Any sound that you hear—whether it be the whisper of a librarian, a lawnmower your neighbor is using, a car moving down the road, or a jet preparing for takeoff—begins with a vibration, a vibration that moves in space, or a wave. In Chapter 9, you learned that sound is an example of longitudinal waves, or waves that are generated by a disturbance that moves parallel to the direction of motion of the wave. In this chapter, the characteristics and behavior of sound waves are discussed.

CHARACTERISTICS OF SOUND WAVES

Sound waves are longitudinal waves that can propagate through all forms of matter—solids, liquids, and gases. Sound waves are generated by the motion of molecules or particles of a medium vibrating back and forth in a direction parallel to the direction of wave motion. As an object such as a tuning fork is struck, the vibration of the prong causes the air molecules to vibrate. As the air molecules vibrate, they alternate between being forced together (**compressions**) and then being separated at distances greater than the normal spacing (**rarefactions**). This example of a longitudinal wave is depicted in Figure 10-1.

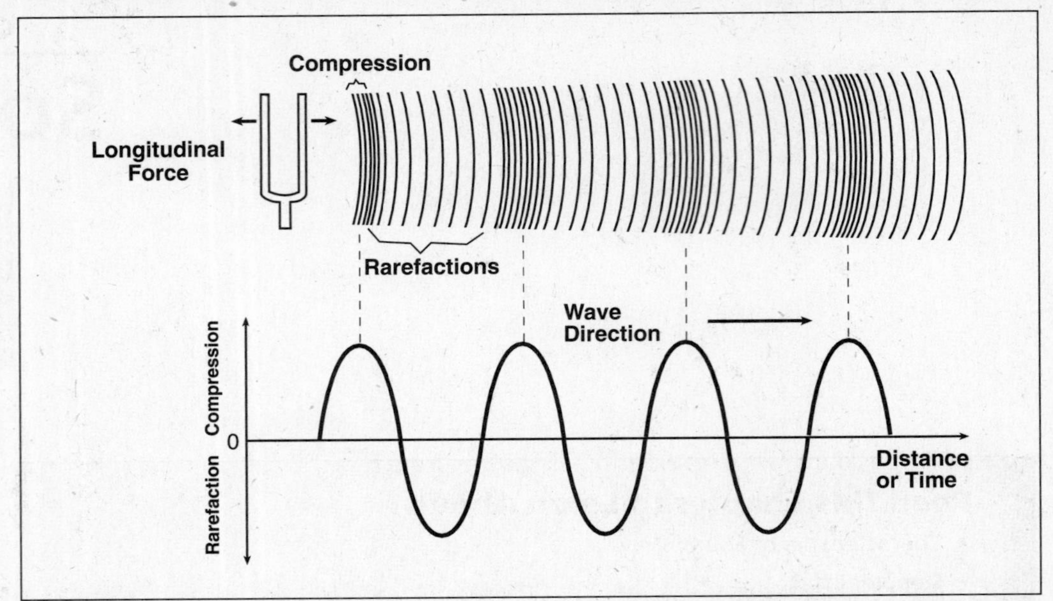

FIGURE 10-1: SOURCE: From George Hademenos, *Schaum's Outline of Physics for Pre-Med, Biology, and Allied Health Students*, McGraw-Hill, 1998; reproduced with permission of The McGraw-Hill Companies.

The movement of sound requires molecules in a medium. The farther apart the molecules in the medium are spaced, the slower the speed of sound through the medium. Conversely, when molecules are tightly spaced, speed travels the fastest. Therefore, the speed of sound is the fastest in solids (where molecules are tightly spaced) and faster in liquids than in gases (where molecules are widely spaced).

The speed of sound through air is approximately 344 meters/second (m/s; 760 miles/hour). In contrast, the speed of sound in water is 1480 m/s. The range of frequencies of sound audible or detectable by the human ear is 20 hertz (Hz) to 20 kilohertz (kHz). Frequencies above 20 kHz are termed **ultrasonic waves**.

DOPPLER EFFECT

The **Doppler effect** refers to the shift in frequency of a transmitted sound that is caused by a change in distance between the source of the sound and the observer. Consider a source emitting a sound of frequency f_S detected by an observer as a frequency f_O. If the source is moving toward the observer, the observer perceives an increase in the sound frequency. If the source is moving away from the observer, the observer perceives a decrease in frequency. The altered frequency detected by the observer is known as the **Doppler-shifted frequency**. These qualitative relationships can be expressed by

$$f_O = f_S\left(\frac{v \pm v_O}{v \pm v_S}\right)$$

where v is the speed of sound in the given medium.

There are several specific cases of the Doppler effect:

- Moving sound source toward a fixed observer

$$f_O = f_S\left(\frac{v}{v - v_S}\right)$$

- Moving sound source away from a fixed observer

$$f_O = f_S\left(\frac{v}{v + v_S}\right)$$

- Fixed sound source with observer moving toward source

$$f_O = f_S\left(\frac{v + v_O}{v}\right)$$

- Fixed sound source with observer moving away from source

$$f_O = f_S\left(\frac{v - v_O}{v}\right)$$

EXAMPLE: A police car, in pursuit of a driver suspected of speeding, is traveling at 30 m/s when the siren is turned on, operating at a frequency of 1.2 kHz. Given that the speed of sound in air is 344 m/s, determine the frequency heard by a stationary witness as

1. The police car approaches the observer
2. The police car passes the observer

SOLUTION:

1. For the case of the moving source approaching a stationary observer, the Doppler-shifted frequency detected by the observer is given by

$$f_O = f_S\left(\frac{v}{v - v_S}\right) = 1200\ \text{Hz}\left(\frac{344\ \dfrac{\text{m}}{\text{s}}}{344\ \dfrac{\text{m}}{\text{s}} - 30\ \dfrac{\text{m}}{\text{s}}}\right) = 1315\ \text{Hz}$$

2. For the case of the moving source passing a stationary observer, the Doppler-shifted frequency detected by the observer is given by

$$f_O = f_S\left(\frac{v}{v + v_S}\right) = 1200\ \text{Hz}\left(\frac{344\ \dfrac{\text{m}}{\text{s}}}{344\ \dfrac{\text{m}}{\text{s}} + 30\ \dfrac{\text{m}}{\text{s}}}\right) = 1104\ \text{Hz}$$

SOUND INTENSITY AND SOUND LEVEL

Sound intensity represents the rate of energy transported by the sound wave per unit area perpendicular to its direction of motion. The sound level β, expressed in units of decibels (dB), is defined in terms of sound intensity I as

$$\beta = 10\log\left(\frac{I}{I_O}\right)$$

where I_o is the threshold for human hearing [$= 1 \times 10^{-12}$ watts per square meter (W/m^2)] and represents the intensity of the weakest sound detectable by the human ear. The following table lists representative values of sound intensity I and their corresponding sound level β.

Sound Intensity and Sound Level of Representative Sounds

Sound Intensity (W/m^2)	Sound Level (dB)	Representative Sounds
1×10^{-12}	0	Threshold of hearing
1×10^{-10}	20	Distant whisper
1×10^{-8}	40	Normal outdoor sounds
1×10^{-6}	60	Normal conversation
1×10^{-4}	80	Busy traffic
1×10^{-2}	100	Siren at 30 m
1	120	Loud indoor rock concert (threshold of pain)
1×10^{2}	140	Jet airplane
1×10^{4}	160	Bursting of eardrums

EXAMPLE: A particular sound level was measured at 75 dB. Determine its sound intensity.

SOLUTION: Using the expression for the sound level intensity, you have

$$\beta = 10 \log \left(\frac{I}{I_o} \right)$$

$$75 \text{ dB} = 10 \log \left(\frac{I}{1 \times 10^{-12} \frac{W}{m^2}} \right)$$

Solving for I yields

$$I = \left(1 \times 10^{-12} \frac{W}{m^2} \right) \left(\log^{-1} \frac{75 \text{ dB}}{10} \right) = 3.16 \times 10^{-5} \frac{W}{m^2}$$

RESONANCE

Many objects, such as a tuning fork or plucked string, vibrate at a specific frequency. This frequency is referred to as the object's **natural frequency**. When a periodic, external force strikes the object with a frequency equal to the natural frequency of the object, the amplitude of the object's motion increases and hence energy is absorbed by the object, a condition referred to as **resonance**. Examples of resonance can be found in engineering in the analysis of structure failure in response to severe weather/natural disasters, in medicine with magnetic resonance imaging and lithotripsy of kidney stones, and in music with the tonal qualities of certain string and brass instruments.

STANDING WAVES

Consider a wave generated on a stretched string with one side connected to a rigid surface. As the wave propagates toward the fixed end, it reflects, inverts, and continues back and forth across the string, causing the string to vibrate. In such instances, two waves (original and reflected waves) of equal frequencies and amplitudes are moving in opposite directions along the string, creating a **standing wave**, as shown in Figure 10-2.

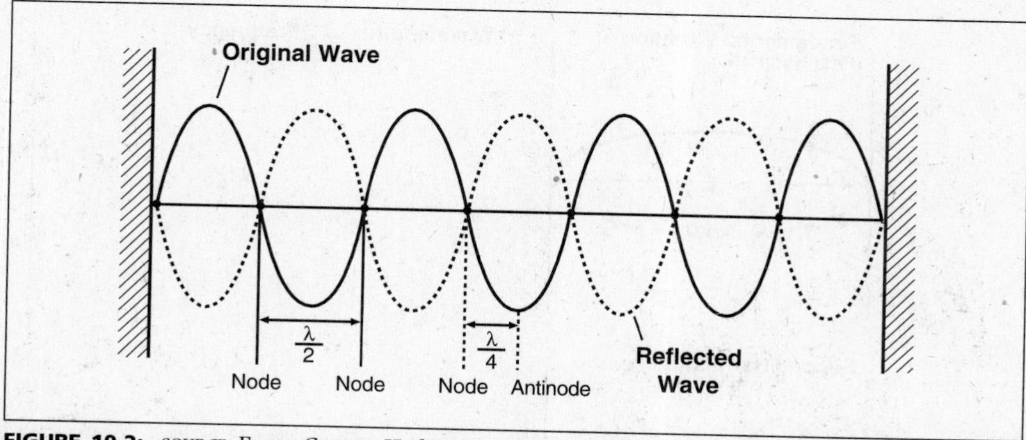

FIGURE 10-2: SOURCE: From George Hademenos, *Schaum's Outline of Physics for Pre-Med, Biology, and Allied Health Students*, McGraw-Hill, 1998; reproduced with permission of The McGraw-Hill Companies.

The points with no displacement from the horizontal axis are termed **nodes**, and the points that are maximally displaced from the horizontal axis—that is, points at the peaks and valleys of each wave—are termed **antinodes**. The distance between two adjacent displacement nodes is one-half wavelength and the distance from node to antinode is one-quarter wavelength. The frequency, f, of a standing wave is given by

$$f = \frac{v}{\lambda}$$

where v is the velocity of the wave.

FREQUENCIES IN A STRETCHED STRING

A stretched string of length L with both ends secured to a rigid surface possesses nodes at both ends. Because the distance between two adjacent nodes is $\lambda/2$, then

$\frac{\lambda}{2} = L$ or $\lambda = 2L$. Substituting this value into the expression for the frequency yields

$$f = \frac{v}{\lambda} = \frac{v}{2L}$$

This is the lowest frequency that the string can accommodate and is termed the **fundamental frequency**, or first harmonic. **Harmonic frequencies**, second and greater, are integer multiples of the first harmonic and can be determined from the generalized relation

$$f = \frac{nv}{2L} \qquad n = 1, 2, 3, \ldots$$

The first, second, and third harmonics are illustrated in Figure 10-3.

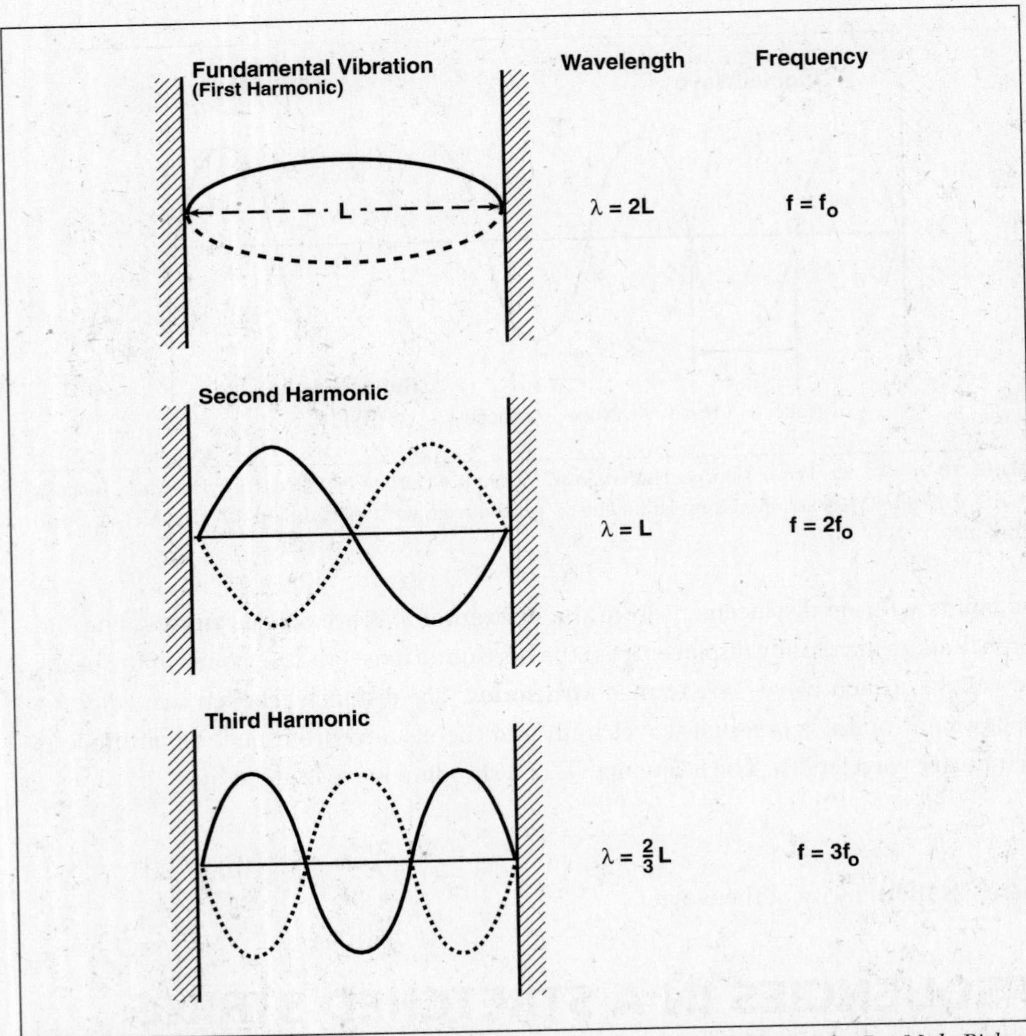

FIGURE 10-3: SOURCE: From George Hademenos, *Schaum's Outline of Physics for Pre-Med, Biology, and Allied Health Students*, McGraw-Hill, 1998; reproduced with permission of The McGraw-Hill Companies.

FREQUENCIES IN A PIPE

The frequencies in a pipe can be derived in a fashion similar to that for the stretched string. However, the pipe offers more conditions to consider, such as both ends open, both ends closed, and one end open. A node can be found at the end of a closed pipe whereas an antinode is present toward the open end of the pipe. Given that the distance from node to node is $\lambda/2$ and the distance from node to antinode is $\lambda/4$, the frequencies for a pipe of length L are as follows:

PIPE WITH BOTH ENDS OPEN:

$$L = 2\frac{\lambda}{4} = \frac{\lambda}{2}$$

Fundamental frequency: $f_1 = \dfrac{v}{2L}$

Harmonics: $f_n = \dfrac{nv}{2L} \quad (n = 1, 2, 3, \ldots)$

PIPE WITH BOTH ENDS CLOSED:

$$L = \frac{\lambda}{2}$$

Fundamental frequency: $f_1 = \dfrac{v}{2L}$

Harmonics: $f_n = \dfrac{nv}{2L} \quad (n = 1, 2, 3, \ldots)$

PIPE WITH ONE END OPEN:

$$L = \frac{\lambda}{4}$$

Fundamental frequency: $f_1 = \dfrac{v}{4L}$

Harmonics: $f_n = \dfrac{nv}{4L} \quad (n = 1, 2, 3, \ldots)$

EXAMPLE: Determine the velocity of waves in an open pipe of 1.3 m length if the fundamental frequency is 225 Hz.

SOLUTION: For an open pipe, the fundamental frequency is given by

$$f_1 = \frac{v}{2L}$$

Solving for v gives

$$v = f_1\,(2L) = (225\text{ Hz})\,(2)\,(1.3\text{ m}) = 585\ \frac{\text{m}}{\text{s}}$$

BEATS

Beats are produced during interference by sound waves at slightly different frequencies. The beat frequency f_{beat} is equal to the difference in the frequencies of the individual sound waves f_1 and f_2 or

$$f_{beat} = f_1 - f_2$$

EXAMPLE: Determine the number of beats per second that are heard when two tuning forks of frequencies 256 Hz and 264 Hz are struck simultaneously.

SOLUTION: The beat frequency or the difference between frequencies of the two tuning forks determines the number of beats per second, or

$$f_{beat} = f_1 - f_2 = 264\,\text{Hz} - 256\,\text{Hz} = 8\,\text{Hz}$$

CRAM SESSION

Sound

Sound waves: Longitudinal waves that can propagate through all forms of matter—solids, liquids, and gases. Sound waves are generated by the motion of molecules or particles of a medium vibrating back and forth in a direction parallel to the direction of wave motion. A longitudinal wave, shown below, consists of a series of **compressions** (air molecules being forced together) and **rarefactions** (air molecules separated at distances greater than their normal spacing).

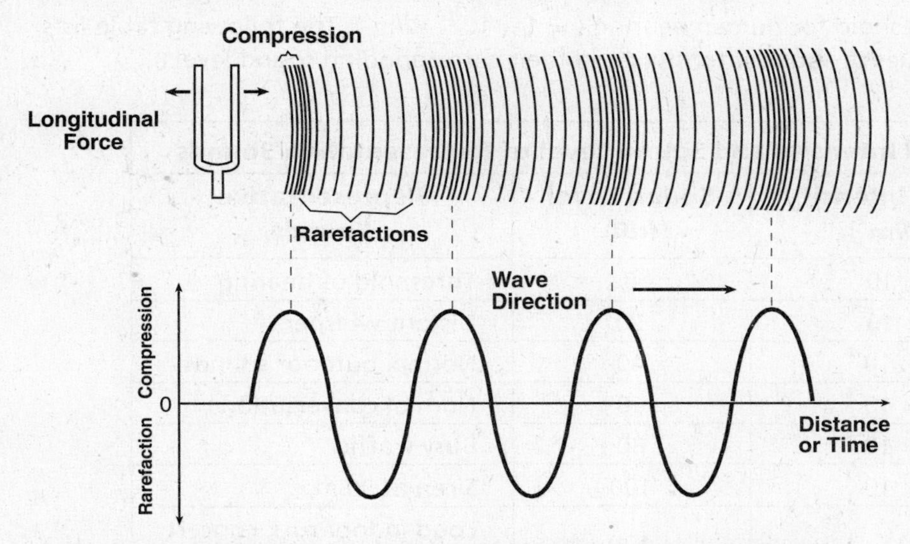

SOURCE: From George Hademenos, *Schaum's Outline of Physics for Pre-Med, Biology, and Allied Health Students*, McGraw-Hill, 1998; reproduced with permission of The McGraw-Hill Companies.

Doppler effect: The shift in frequency of a transmitted sound caused by a change in path length between source and observer. Consider a source emitting a sound of a frequency f_s detected by an observer as a frequency f_o, referred to as the Doppler-shifted frequency. Assuming v is the speed of sound in the given medium, the specific cases of the Doppler effect are as follows:

- Moving sound source toward a fixed observer

$$f_o = f_s\left(\frac{v}{v - v_s}\right)$$

- Moving sound source away from a fixed observer

$$f_o = f_s\left(\frac{v}{v + v_s}\right)$$

- Fixed sound source with observer moving toward source

$$f_o = f_s\left(\frac{v + v_o}{v}\right)$$

- Fixed sound source with observer moving away from source

$$f_o = f_s\left(\frac{v - v_o}{v}\right)$$

Sound intensity: The rate of energy transported by the wave per unit area perpendicular to its direction of motion. The level of sound intensity β, expressed in units of decibels (dB), is defined in terms of intensity I as

$$\beta = 10 \log\left(\frac{I}{I_o}\right)$$

where I_o is the threshold for human hearing ($= 1 \times 10^{-12}$ W/m^2). The following table lists representative values of sound intensity I and their corresponding sound level β.

Sound Intensity and Sound Level of Representative Sounds		
Sound Intensity (W/m^2)	**Sound Level (dB)**	**Representative Sounds**
1×10^{-12}	0	Threshold of hearing
1×10^{-10}	20	Distant whisper
1×10^{-8}	40	Normal outdoor sounds
1×10^{-6}	60	Normal conversation
1×10^{-4}	80	Busy traffic
1×10^{-2}	100	Siren at 30 m
1	120	Loud indoor rock concert (threshold of pain)
1×10^{2}	140	Jet airplane
1×10^{4}	160	Bursting of eardrums

Resonance: A condition that occurs when a periodic, external force strikes the object with a frequency equal to the natural frequency of the object, resulting in an increase in the amplitude and hence energy absorbed by the object.

Standing wave: A wave generated when two waves (original and reflected waves) of equal frequencies and amplitudes are moving in opposite directions along the string, as shown in the figure on the following page.

Nodes: Points with no displacement from the horizontal axis.

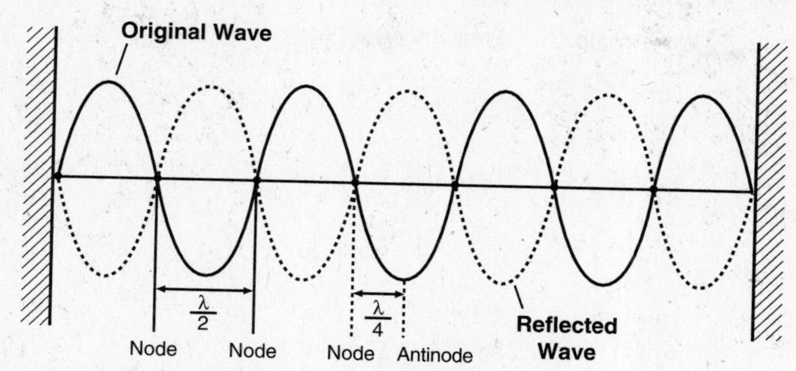

SOURCE: From George Hademenos, *Schaum's Outline of Physics for Pre-Med, Biology, and Allied Health Students,* McGraw-Hill, 1998; reproduced with permission of The McGraw-Hill Companies.

Antinodes: Points that are maximally displaced from the horizontal axis, that is, points at the peaks and valleys of each wave. The distance between two adjacent displacement nodes is one-half wavelength. The frequency, f, of a standing wave is given by

$$f = \frac{v}{\lambda}$$

where v is the velocity of the wave.

Frequencies in a Stretched String

A stretched string of length L with both ends secured to a rigid surface possesses nodes at both ends. The frequencies in a stretched string are:

- Fundamental frequency (first harmonic)

$$f = \frac{v}{\lambda} = \frac{v}{2L}$$

- Harmonic frequencies

$$f = \frac{nv}{2L} \quad n = 1, 2, 3, \ldots$$

The first, second, and third harmonics are illustrated in the figure on the following page.

Frequencies in a Pipe

The frequencies in a pipe offer more conditions to consider, such as both ends open, both ends closed, and one end open. Given that the distance from node to node is $\lambda/2$ and the distance from node to antinode is $\lambda/4$, the frequencies for a pipe of length L are as follows:

- **Pipe with both ends open:** $L = 2\frac{\lambda}{4} = \frac{\lambda}{2}$
  - Fundamental frequency: $f_1 = \frac{v}{2L}$
  - Harmonics: $f_n = \frac{nv}{2L}$ $(n = 1, 2, 3, \ldots)$

Harmonic	Wavelength	Frequency
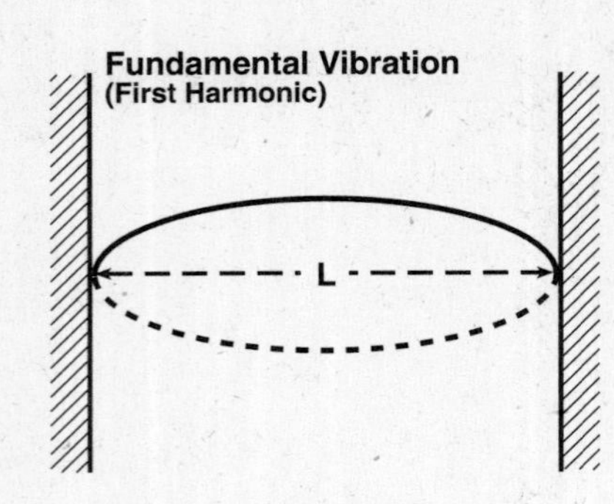	$\lambda = 2L$	$f = f_o$
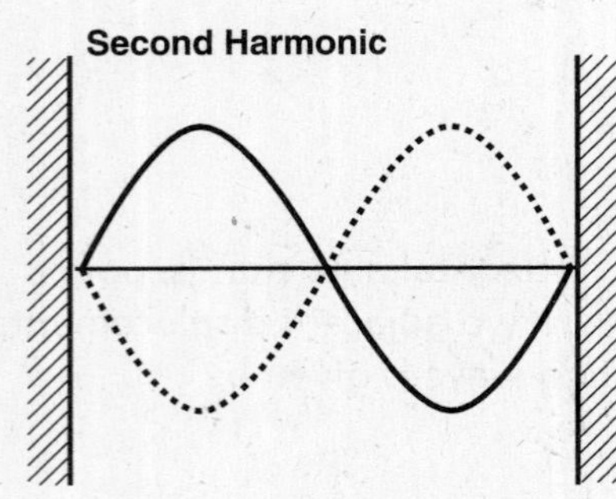	$\lambda = L$	$f = 2f_o$
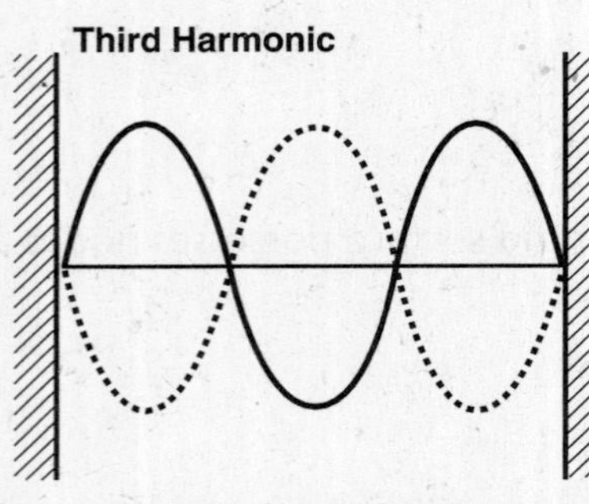	$\lambda = \frac{2}{3}L$	$f = 3f_o$

SOURCE: From George Hademenos, *Schaum's Outline of Physics for Pre-Med, Biology, and Allied Health Students,* McGraw-Hill, 1998; reproduced with permission of The McGraw-Hill Companies.

- **Pipe with both ends closed:** $L = \frac{\lambda}{2}$
  - Fundamental frequency: $f_1 = \frac{v}{2L}$
  - Harmonics: $f_n = \frac{nv}{2L} \quad (n = 1, 2, 3, \ldots)$
- **Pipe with one end open:** $L = \frac{\lambda}{4}$
  - Fundamental frequency: $f_1 = \frac{v}{4L}$
  - Harmonics: $f_n = \frac{nv}{4L} \quad (n = 1, 2, 3, \ldots)$

Beats: Produced during interference by sound waves at slightly different frequencies. The beat frequency f_{beat} is equal to the difference in the frequencies of the individual sound waves f_1 and f_2 or $f_{\text{beat}} = f_1 - f_2$.

CHAPTER 11

Light and Geometric Optics

Read This Chapter to Learn About:

LIGHT

- Electromagnetic Spectrum
- Reflection and Refraction of Light
- Total Internal Reflection
- Dispersion of Light
- Polarization
- Interference
- Diffraction

GEOMETRIC OPTICS

- Mirrors
- Thin Lenses

We are accustomed to thinking about light in terms of the output of a lamp or a fluorescent lightbulb illuminating a room. This type of light—visible light—represents only a very small subset of the physics definition of light waves, formally known as **electromagnetic waves**. Electromagnetic waves are an example of transverse waves, meaning that the direction of the disturbance is perpendicular to or at right angles to the direction of motion. This chapter introduces the concept of electromagnetic waves, defines them, and explains their characteristics and behavior as they interact with various surfaces or media, such as mirrors and lenses, to form images (geometric optics).

LIGHT

ELECTROMAGNETIC SPECTRUM

Light is a form of energy that travels as waves in space. But as described in this context, light is much more than just the visible light that one sees emitted from a lightbulb. Visible light is a small part of the **electromagnetic spectrum**, a collection of waves that travel at the same speed [the speed of light in a vacuum, denoted by $c = 3 \times 10^8$ meters/second (m/s)] but have different wavelengths. A **wavelength** of a wave is defined as the distance between successive peaks of the wave.

Electromagnetic waves are transverse waves, meaning they occur as a result of a disturbance acting in a direction at right angles or perpendicular to the direction of wave motion. The disturbance that generates electromagnetic waves is coupled oscillating electric and magnetic fields, as shown in Figure 11-1.

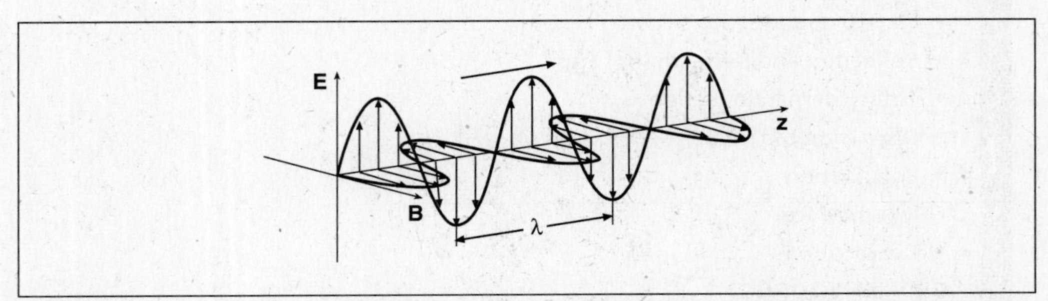

FIGURE 11-1: SOURCE: From George Hademenos, *Schaum's Outline of Physics for Pre-Med, Biology, and Allied Health Students*, McGraw-Hill, 1998; reproduced with permission of The McGraw-Hill Companies.

The **electromagnetic spectrum** is a compilation of electromagnetic waves, arranged with the larger wavelengths (radio waves) to the left and the smaller wavelengths (gamma rays) to the right, as represented in Figure 11-2 and described in the

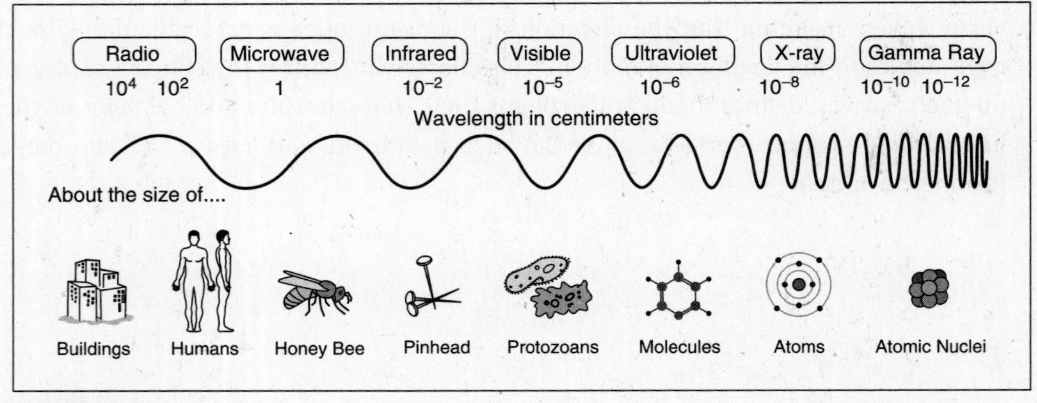

FIGURE 11-2

following table. Notice as the waves move from radio waves to gamma rays, the frequencies become larger and the wavelengths become smaller. The wavelength is inversely related to the frequency by

$$v = f\lambda$$

where $v = c$, the speed of light.

Arrangement of Waves in the Electromagnetic Spectrum		
Type of Electromagnetic Wave	**Range of Frequencies**	**Range of Wavelengths**
Radio waves	27 Hz – 1.0×10^9 Hz	1.0×10^7 m – 0.30 m
Microwaves	1.0×10^9 Hz – 3.0×10^{11} Hz	0.30 m – 0.001 m
Infrared (IR) waves	3.0×10^{11} Hz – 4.3×10^{14} Hz	0.001 m – 7.0×10^{-7} m
Visible light	4.3×10^{14} Hz – 7.5×10^{14} Hz	7.0×10^{-7} m – 4.0×10^{-7} m
Ultraviolet (UV) light	7.5×10^{14} Hz – 5.0×10^{15} Hz	4.0×10^{-7} m – 6.0×10^{-8} m
X-rays	5.0×10^{15} Hz – 5.0×10^{19} Hz	6.0×10^{-8} m – 1.0×10^{-13} m
Gamma rays	5.0×10^{19} Hz – 4.0×10^{25} Hz	1.0×10^{-13} m – 6.0×10^{-18} m

Of particular note is that visible light, that region of the electromagnetic spectrum that is visible to the human eye, is the smallest of all of the types of waves. Taken altogether, visible light is seen as white but can be separated into six colors (i.e., violet, blue, green, yellow, orange, and red), with each color defined by its characteristic wavelength. The wavelengths for visible light range from 4.0×10^{-7} m [400 nanometers (nm); 1 nm = 1×10^{-9} m], which represents the color violet, to 7.0×10^{-7} m, which represents the color red.

EXAMPLE: Radio waves have a typical wavelength of 10^2 m. Determine their frequency.

SOLUTION:

$$f = \frac{c}{\lambda} = \frac{3 \times 10^8 \frac{\text{m}}{\text{s}}}{1 \times 10^2 \text{ m}} = 3 \times 10^6 \text{ Hz}$$

EXAMPLE: What is the speed of waves from radio station KQRS that broadcasts at a frequency of 710 kilohertz (kHz)?

SOLUTION: All electromagnetic waves, including radio waves, move with the same speed, the speed of light. Thus, the speed of these waves is 3×10^8 m/s.

REFLECTION AND REFRACTION OF LIGHT

As light waves propagate through one medium and approach a second medium, they can interact with the boundary between the two media in one of two separate ways: reflection and refraction. Consider a light ray that propagates through one medium and strikes the boundary at an incidence angle of θ_i with the normal (axis perpendicular to the plane surface), as shown in Figure 11-3.

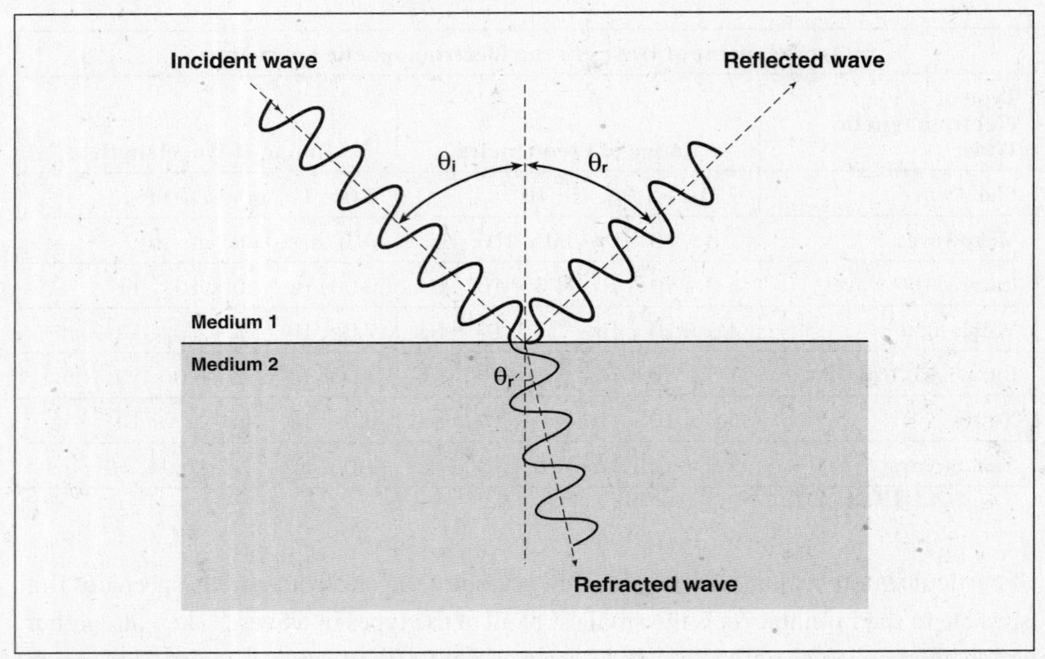

FIGURE 11-3: SOURCE: From George Hademenos, *Schaum's Outline of Physics for Pre-Med, Biology, and Allied Health Students*, McGraw-Hill, 1998; reproduced with permission of The McGraw-Hill Companies.

The light ray can reflect or bounce off the surface boundary at an angle of reflection θ_r, also measured with respect to the normal, which is equal in magnitude to the angle of incidence. This is known as the **law of reflection**, represented by

$$\theta_i = \theta_r$$

Angle of incidence = angle of reflection

In addition to reflection, the light ray can bend or refract as it enters the second medium. The angle at which the light ray refracts, $\theta_{r'}$, is dependent on the ratio of the indices of refraction between the two media. The index of refraction n for a given material is defined as

$$n = \frac{\text{speed of light in vacuum}}{\text{speed of light in material}} = \frac{c}{v}$$

The index of refraction for common substances is as follows:

Air	$n = 1.00$	Ethyl alcohol	$n = 1.36$
Glass	$n = 1.52$	Polystyrene	$n = 1.55$
Water	$n = 1.33$	Sodium chloride	$n = 1.54$
Diamond	$n = 2.42$	Acetone	$n = 1.36$

The angle of refraction $\theta_{r'}$ with respect to the normal is related to the incident angle by the **law of refraction**, also known as **Snell's law**:

$$n_1 \sin \theta_i = n_2 \sin \theta_{r'}$$

where n_1 and n_2 are the indices of refraction of the first medium and second medium, respectively.

EXAMPLE: A ray of light strikes the surface of a swimming pool at an angle of 45° from the normal. Determine the angle of reflection.

SOLUTION: From the law of reflection, the angle of incidence is equal to the angle of reflection. Thus, $\theta_i = \theta_r = 45^\circ$ from the normal.

EXAMPLE: A light ray in air ($n = 1.00$) strikes the surface of a glass pane ($n = 1.52$) at an angle of incidence of 35° from the normal. Determine the angle of refraction.

SOLUTION: From Snell's law:

$$n_1 \sin \theta_i = n_2 \sin \theta_{r'}$$

$$1.00 \sin 35^\circ = 1.52 \sin \theta_{r'}$$

$$\sin \theta_{r'} = \frac{1.00 \sin 35^\circ}{1.52} = 0.377$$

$$\theta_{r'} = \sin^{-1} 0.377 = 22.1^\circ$$

TOTAL INTERNAL REFLECTION

Total internal reflection is a phenomenon that occurs when light passes from a medium or material with a high index of refraction to a medium with a low index of refraction. For example, consider a light ray passing from glass ($n = 1.52$) to air ($n = 1.00$) at an angle of incidence θ_i, as shown in Figure 11-4. In this case, the angle of refraction is greater than the angle of incidence. In fact, as one increases the angle of incidence, the light ray has an angle of refraction of 90° and is subsequently refracted along the interface between the two media. The angle of incidence that causes light to be refracted at 90° from the normal is called the **critical angle** θ_c, given by

$$\sin \theta_c = \frac{n_2}{n_1}$$

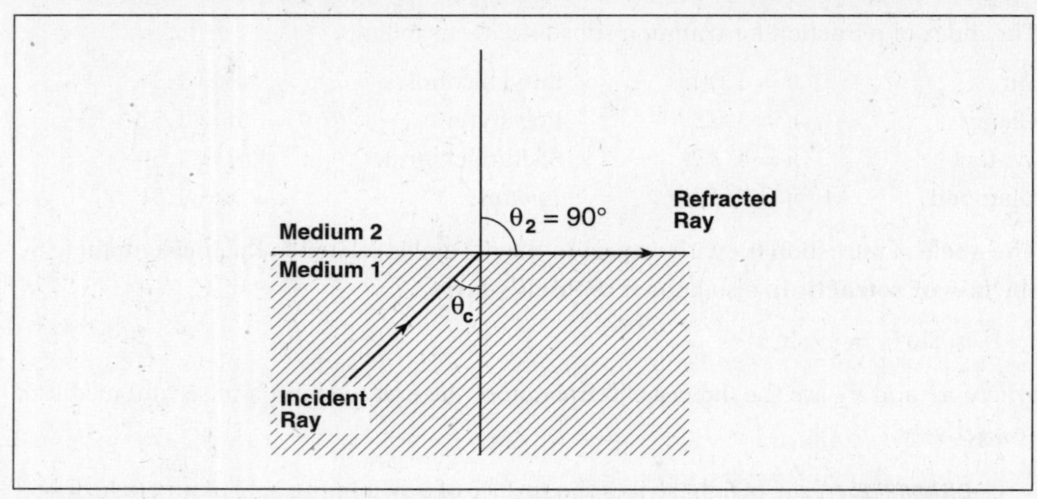

FIGURE 11-4: SOURCE: From George Hademenos, *Schaum's Outline of Physics for Pre-Med, Biology, and Allied Health Students*, McGraw-Hill, 1998; reproduced with permission of The McGraw-Hill Companies.

EXAMPLE: Assuming air as the external medium, determine the critical angle for total internal reflection for diamond ($n = 2.42$).

SOLUTION: Using the expression for the critical angle

$$\sin\theta_c = \frac{n_2}{n_1} = \frac{n_{\text{air}}}{n_{\text{diamond}}} = \frac{1.00}{2.42} = 0.413$$

$$\theta_c = \sin^{-1} 0.413 = 24.4°$$

DISPERSION OF LIGHT

Dispersion is the separation of any traveling wave into separate frequency components. An example of dispersion is the separation of white light into its wavelength components and its corresponding elementary colors as it passes through a prism, as shown in Figure 11-5. As it passes through a prism, white light refracts at different

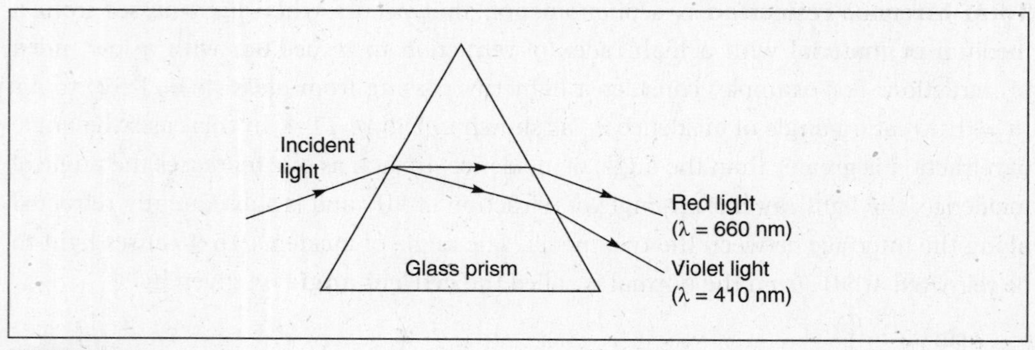

FIGURE 11-5

angles with the smallest component wavelength (violet) being bent the farthest and the largest component wavelength (red) being bent the least.

POLARIZATION

The disturbance that generates electromagnetic (light) waves is coupled electric and magnetic fields that act in a direction perpendicular to the direction of wave motion. In typical light sources, the electric field vectors are oriented in random directions but still in a direction perpendicular to wave motion. (The same is true for magnetic field vectors, but for the purpose of our discussion on polarization, only the electric field vectors are significant.) A light wave with the electric field vectors oscillating in more than one plane is termed **unpolarized light**.

Polarization is a process by which the electric field vectors of the light waves are all aligned along the same direction; in other words, they are all parallel. Polarization can be accomplished in several ways, with a common one being filters constructed of special materials (such as those found in sunglasses) that are able to block an oscillating plane of electric field vectors.

INTERFERENCE

Interference refers to the superposition of two waves traveling within the same medium when they interact. When applied to light waves, interference occurs only when the waves have the same wavelength and a fixed phase difference (i.e., the difference in which the peaks of one wave lead or lag the peaks of the other wave remains constant with time). These types of waves are referred to as **coherent waves**. When coherent waves with the same amplitude are combined, **constructive interference** occurs when the two waves are in phase and **destructive interference** occurs when the two waves are out of phase (have a phase difference of 180°).

Consider the experimental setup depicted in Figure 11-6, where two coherent light waves of wavelength λ impinge on two narrow slits separated by a distance d. The result is a fringe pattern of alternating bright and dark fringes where the bright fringes represent constructive interference and the dark fringes represent destructive interference. The position of the bright fringes (referred to as **maxima**) and the dark fringes (referred to as **minima**) can be determined from the relations

Bright fringes (maxima) $\quad d(\sin\theta) = m\lambda \qquad m = 0, 1, 2, \ldots$

Dark fringes (minima) $\quad d(\sin\theta) = \left(m + \frac{1}{2}\right)\lambda \qquad m = 0, 1, 2, \ldots$

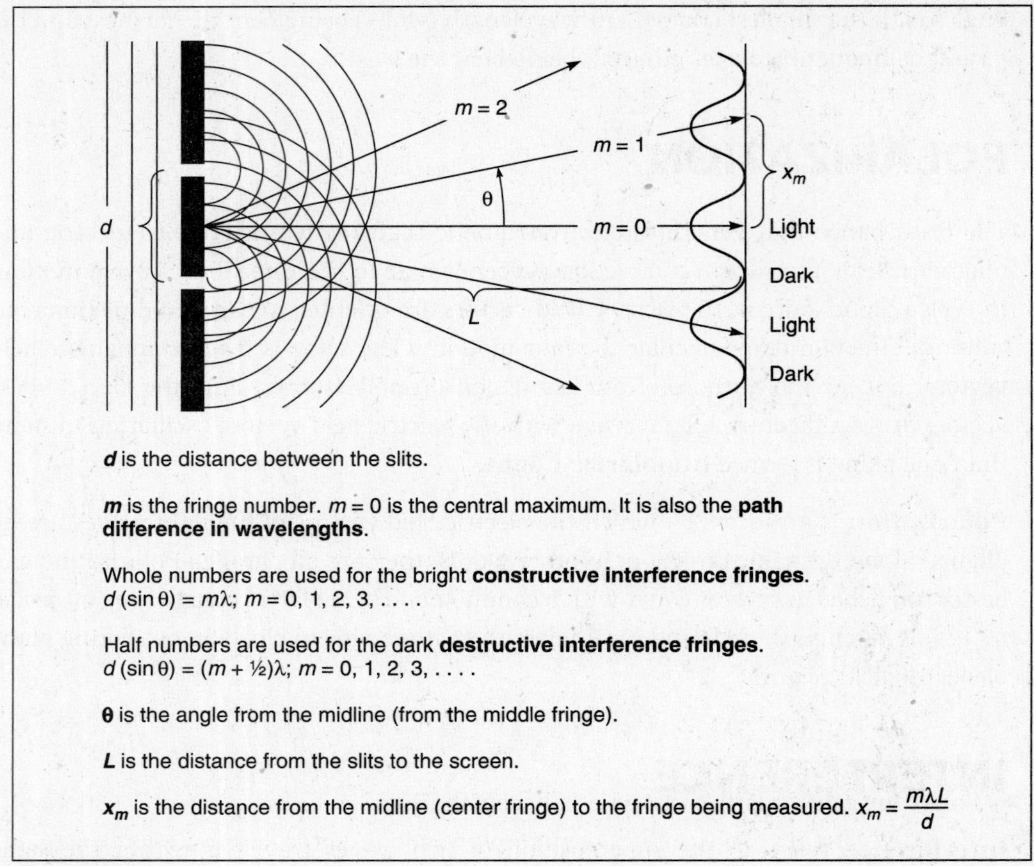

FIGURE 11-6

where $d \sin \theta$ represents the path difference, or the difference between distances that the waves travel to reach the fringe position on the screen.

DIFFRACTION

Diffraction describes the ability of light waves to bend or spread through an aperture (opening) or around an obstacle. Consider a light wave with a wavelength λ that impinges on a narrow slit of width d. The light spreads out or diverges (a process known as **diffraction**) onto a viewing screen and results in a diffraction pattern consisting of alternating bright fringes and dark fringes. The position of the dark fringes observed at angles θ with respect to the normal or perpendicular of the screen is

$$n\lambda = d \sin \theta$$

where $n = 1, 2, 3, \ldots$ is the order number of the dark fringe with respect to the central fringe.

GEOMETRIC OPTICS

Geometric optics is a branch of physics concerned with the propagation of light used to form images. In typical problems of geometric optics, a light source is allowed to strike one of two basic types of optical components: mirrors and thin lenses.

Mirrors are optical components that reflect light rays from a light source to form an image. Two types of mirrors are **plane mirrors** and **spherical mirrors**. Spherical mirrors can be further subdivided into **concave** (curved inward) mirrors and **convex** (curved outward) mirrors.

Thin lenses are optical components that refract light rays from a light source to form an image. Thin lenses are generally curved and referred to as spherical thin lenses. Two general forms of spherical thin lenses are concave (curved inward) and convex (curved outward). These two components are discussed in greater detail later.

Terms describing parameters critical to problems in geometric optics include the following:

- **Object distance**, o, is the distance between the object and the optical component.
- **Image distance**, i, is the distance between the image and the optical component.
- **Focal length**, f, is the distance between the image and the optical component observed when the source is imaged from an infinite distance. The point at which the incoming light rays converge to form the image is known as the **focal point**, F.
- **Radius of curvature**, R, the radius of a circle that most closely approximates the curvature of the spherical optical component. The point representing the center of this circle is termed the **center of curvature**, C.

Other terms that describe the image formed as a result of these optical components include the following:

- A **real image** is an image formed by the convergence of light rays and is characterized by a **positive image distance**, i.
- A **virtual image** is an image formed by the divergence of light rays and is characterized by a **negative image distance**, i.
- **Magnification**, m, of an optical component (thin lenses or mirror) is defined as:

$$m = -\frac{i}{o} = \frac{h_i}{h_o}$$

$$\text{Magnification} = -\frac{\text{image distance}}{\text{object distance}} = \frac{\text{image height}}{\text{object height}}$$

A negative magnification value indicates that the image is inverted, whereas a positive magnification value indicates that the image is upright or erect.

In solving problems in geometric optics, you are presented with an optical component and an object (light source) positioned a given distance from the component and asked to determine the image distance from the component where the image is formed and the type of image formed. This objective can be accomplished using two different techniques: an analytical technique and a graphical technique. The analytical technique consists of a mathematical relation between the image distance i, object distance o, focal length f, and radius of curvature R. The graphical technique involves ray tracing for the location of the image formed by the optical component. Ray tracing graphically follows the path of three principal light rays emitted from the object and interacting with the optical component.

MIRRORS

Plane Mirrors

Light rays emanating from an object toward a plane mirror appear to converge at a point within the mirror, as shown in Figure 11-7. The image created by a plane mirror is always virtual and erect—that is, the image appears to be positioned the same distance from the mirror as the distance from the object to the mirror, but on the other side of the mirror.

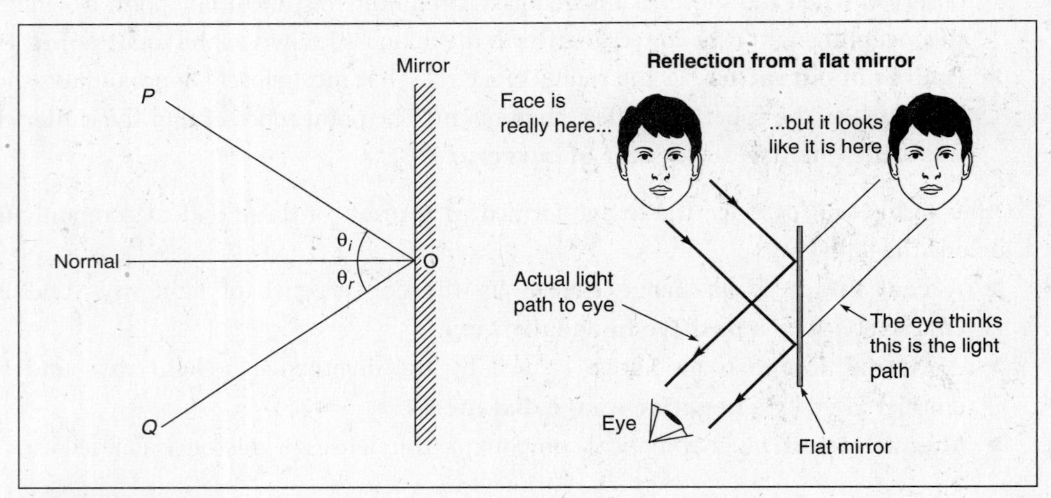

FIGURE 11-7

Spherical Mirrors

Spherical mirrors are mirrors that are curved and exist primarily in two forms, depending on their curvature. Concave mirrors are curved inward, and convex mirrors are curved outward. The spherical mirrors are represented by their radius of curvature R, which is related to their focal length f by

Concave mirror $R = +2f$

Convex mirror $R = -2f$

In Figure 11-8a, parallel light rays emanating from an object toward a concave mirror reflect and ultimately converge at the real focal point F. Similarly, in Figure 11-8b, parallel light rays emanating from an object toward a convex mirror reflect and ultimately

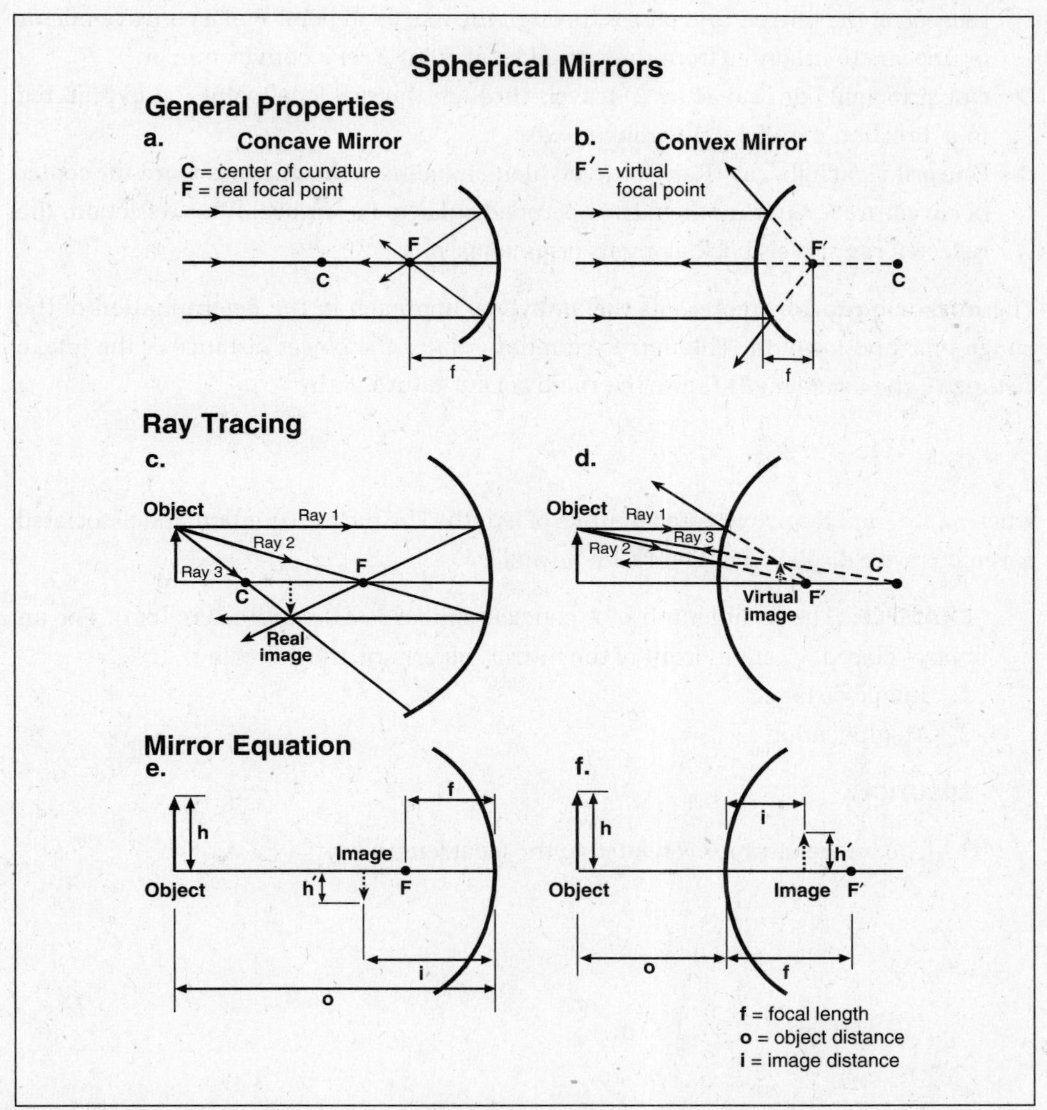

FIGURE 11-8: SOURCE: From George Hademenos, *Schaum's Outline of Physics for Pre-Med, Biology, and Allied Health Students*, McGraw-Hill, 1998; reproduced with permission of The McGraw-Hill Companies.

diverge, such that the reflected rays appear to converge at a virtual focal point F' behind the mirror.

Ray tracing represents a graphical approach by which the image position and image size formed by a mirror can be determined. Ray tracing follows the path of three different principal light rays emanating from an object (typically represented by an arrow), interacting with the mirror, and either converging or diverging at the real or virtual focal point, depending on whether the lens is convex or concave, as shown in Figure 11-8c and d.

➤ One principal light ray (Ray 1) travels parallel to the optical axis of the mirror, reflects off the mirror, and passes through the real focal point F of a concave mirror or appears to originate from the virtual focal point F' of a convex mirror.
➤ One principal light ray (Ray 2) travels through the real focal point F and reflected in a direction parallel to the mirror axis.
➤ One principal light ray (Ray 3) travels along a radius of the mirror toward the center of curvature C, striking the mirror perpendicular to its surface. After reflection, the reflected ray travels back along its original path.

The **mirror equation** represents the analytical approach in the determination of the image type and location. The mirror equation relates the object distance o, the image distance i, the focal length f, and the radius of curvature, R, by

$$\frac{1}{o} + \frac{1}{i} = \frac{1}{f} = \pm\frac{2}{R}$$

where o, i, f, and R are expressed in units of length. The mirror equation and associated parameters are displayed in Figure 11-8e and f.

EXAMPLE: The focal length of a concave mirror is 10 centimeters (cm). For an object placed 30 cm in front of the mirror, determine the following:

1. Image distance
2. Magnification

SOLUTION:

1. The image distance is related to the focal length by

$$\frac{1}{o} + \frac{1}{i} = \frac{1}{f}$$

$$\frac{1}{30\text{ cm}} + \frac{1}{i} = \frac{1}{10\text{ cm}}$$

$$i = 15\text{ cm}$$

2. The magnification is given by

$$m = -\frac{i}{o} = -\frac{15\text{ cm}}{30\text{ cm}} = -0.5$$

Referring to the following table of sign conventions for spherical mirrors, the image is real, inverted, and located between the focal point and the center of curvature.

Sign Conventions for Spherical Mirrors		
Parameter	**Positive**	**Negative**
Object distance, o	Real object	Virtual object
Image distance, i	Real image	Virtual image
Focal length, f	Concave mirror	Convex mirror
Radius of curvature, R	Concave mirror	Convex mirror
Magnification, m	Erect image	Inverted image

THIN LENSES

Thin lenses, like mirrors, exist as convex or concave lenses. However, lenses form images by refracting light rays, as opposed to mirrors that form images by reflecting light rays. Thin lenses have two surfaces and are thus characterized by two radii of curvature R_1 (side of lens closest to the object) and R_2 (side of lens opposite to the object). A convex lens has a positive R_1 and a negative R_2, whereas a concave lens has a negative R_1 and a positive R_2. For a convex (also referred to as converging) lens, as shown in Figure 11-9a, all light rays emanating from an object refract through the lens and converge to generate an image at the real focal point F on the side of the lens opposing the object. For a concave (also referred to as diverging) lens, as shown in Figure 11-9b, all light rays emanating from an object refract through the lens and diverge, but generate an image at the virtual focal point F' on the same side of the lens as the object.

Ray tracing can also be used to determine image position and image size formed by a thin lens. As in the case for mirrors, three particular principal rays of light are followed from the object and ultimately interacting with the lens, as shown in Figure 11-9c and d.

- One principal light ray (Ray 1) travels parallel to the optical axis of the lens, refracts through the lens, and passes through a second real focal point F_2 for a convex lens or appears to originate from a second virtual focal point F_2' for a concave lens.
- One principal light ray (Ray 2) travels diagonally, directed toward and penetrating through the center of the lens without bending.
- One principal light ray (Ray 3) travels through the first real focal point F_1 for a converging lens before it refracts and travels parallel to the optical axis. For a diverging lens, Ray 3 moves through the center of the lens in a straight line.

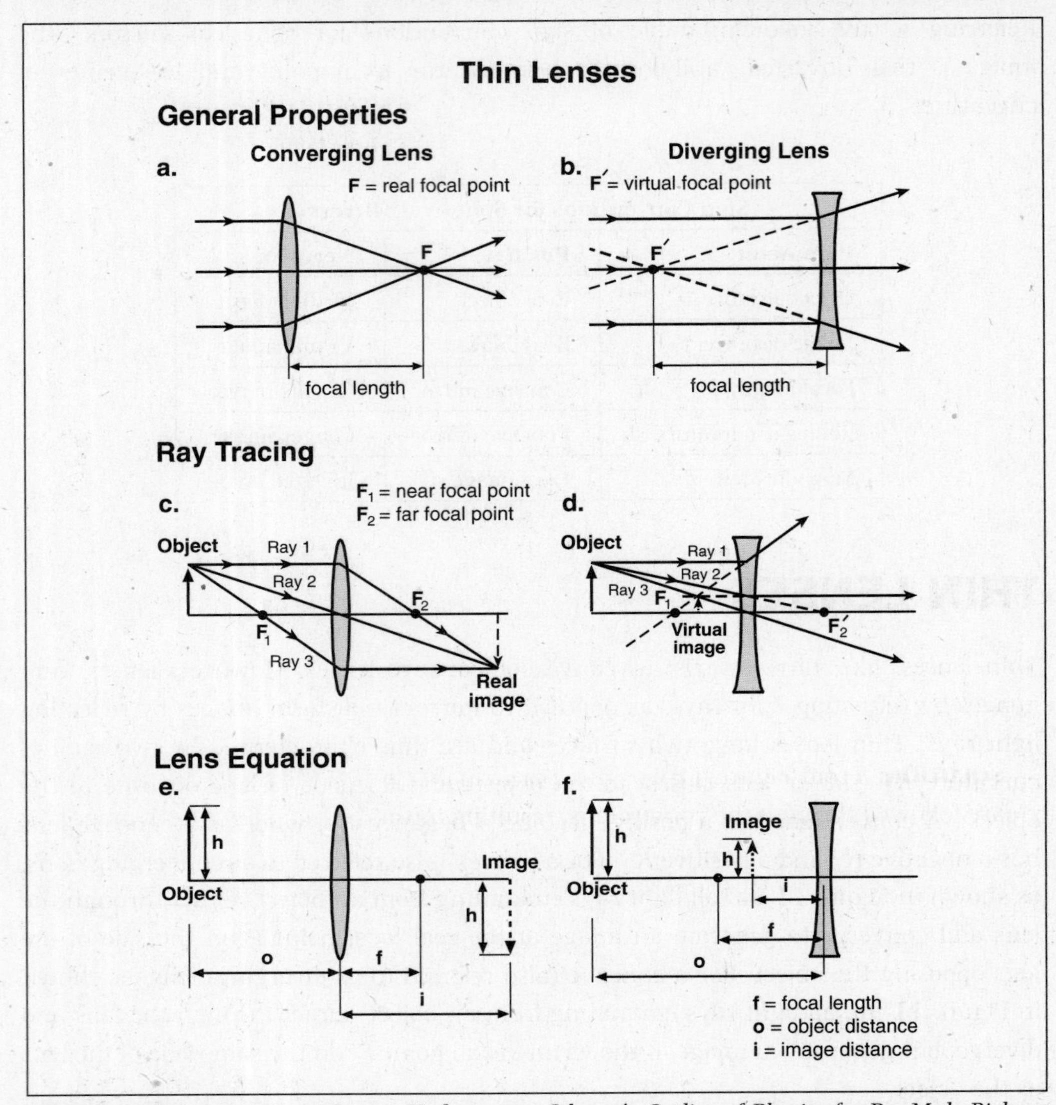

FIGURE 11-9: SOURCE: From George Hademenos, *Schaum's Outline of Physics for Pre-Med, Biology, and Allied Health Students*, McGraw-Hill, 1998; reproduced with permission of The McGraw-Hill Companies.

The **lens equation** relates the object distance o, image distance i, and the focal length f of a lens according to

$$\frac{1}{f} = \frac{1}{i} + \frac{1}{o}$$

The lens equation and associated parameters are illustrated in Figure 11-9e and f.

Also of importance in creating the lens is the **lens maker's equation**, which can be expressed as

$$\frac{1}{f} = (n - 1)\left(\frac{1}{R_1} - \frac{1}{R_2}\right)$$

where f is the focal length, n is the index of refraction for the lens material, R_1 is the radius of curvature of the lens closest to the object, and R_2 is the radius of curvature of the lens farthest from the object.

Sign Conventions for Spherical Lenses		
Parameter	**Positive**	**Negative**
Object distance, o	Real object	Virtual object
Image distance, i	Real image	Virtual image
Focal length, f	Converging lens	Diverging lens
Magnification, m	Erect image	Inverted image

EXAMPLE: An object of height 5 cm is positioned 24 cm in front of a convex lens with a focal length of 15 cm. Determine the location and height of the produced image, using the following:

1. Ray tracing
2. The lens equation

SOLUTION: From Figure 11-10, three principal rays emanating from the object are followed through the convex lens, resulting in the formation of a real image.

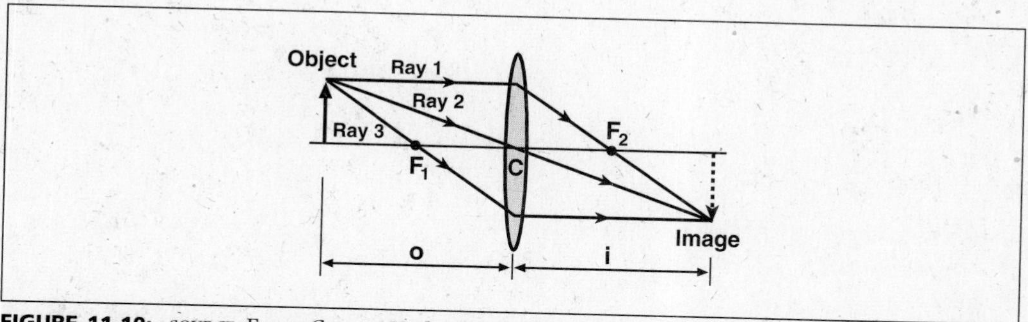

FIGURE 11-10: SOURCE: From George Hademenos, *Schaum's Outline of Physics for Pre-Med, Biology, and Allied Health Students*, McGraw-Hill, 1998; reproduced with permission of The McGraw-Hill Companies.

- Ray 1 strikes the lens toward the top and refracts through it. Following refraction, Ray 1 passes through the principal focal point F.
- Ray 2 passes through the center of the convex lens and continues until it converges with Ray 1.
- Ray 3 strikes the lens toward the bottom and refracts through it. Following refraction, Ray 3 continues parallel to the optical axis until it converges with Rays 1 and 2.

The point where all three rays converge represents the size and location of the image. From the ray tracing in Figure 11-10, it can be concluded that the image is real, inverted, magnified, and located farther from the lens than the object.

The lens equation can be used to find the image distance:

$$\frac{1}{i} = \frac{1}{f} - \frac{1}{o} = \frac{1}{15\text{ cm}} - \frac{1}{24\text{ cm}} = 0.025\frac{1}{\text{cm}}$$

so

$$i = \frac{1}{0.025\frac{1}{\text{cm}}} = 40\text{ cm}$$

A positive value of i indicates the image is real and on the side of the lens opposite the object.

The height of the object h_i can be determined from the magnification m

$$m = -\frac{i}{o} = -\frac{40\text{ cm}}{24\text{ cm}} = -1.67$$

The magnification m is related to the image height by

$$m = \frac{h_i}{h_o}$$

$$h_i = mh_o = (-1.67)\,(5\text{ cm}) = -8.35\text{ cm}$$

The negative value of h_i indicates that the image is inverted.

CRAM SESSION

Light

Light: A form of energy that travels as waves in space at a speed $c = 3 \times 10^8$ m/s (in a vacuum).

Electromagnetic waves: Transverse waves generated by a disturbance of coupled oscillating electric (**E**) and magnetic (**B**) fields acting in a direction at right angles or perpendicular to the direction of wave motion, as shown in the figure below.

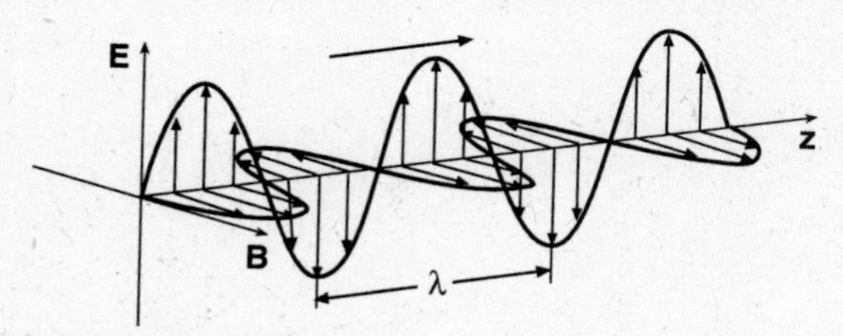

SOURCE: From George Hademenos, *Schaum's Outline of Physics for Pre-Med, Biology, and Allied Health Students*, McGraw-Hill, 1998; reproduced with permission of The McGraw-Hill Companies.

Electromagnetic spectrum: A compilation of electromagnetic waves, arranged with the larger wavelengths (radio waves) to the left and the smaller wavelengths (gamma rays) to the right, as represented in the figure below.

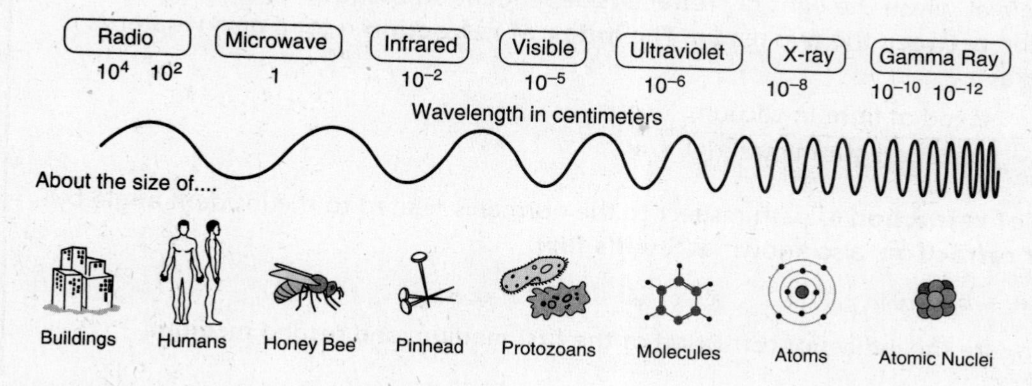

Reflection and Refraction of Light

As light waves propagate through one medium and approach a second medium, they can interact with the boundary between the two media in one of two ways: reflection and refraction. Consider a light ray that propagates through one medium and strikes the boundary at an incidence angle of θ_i with the normal (axis perpendicular to the plane surface), as shown in the figure below.

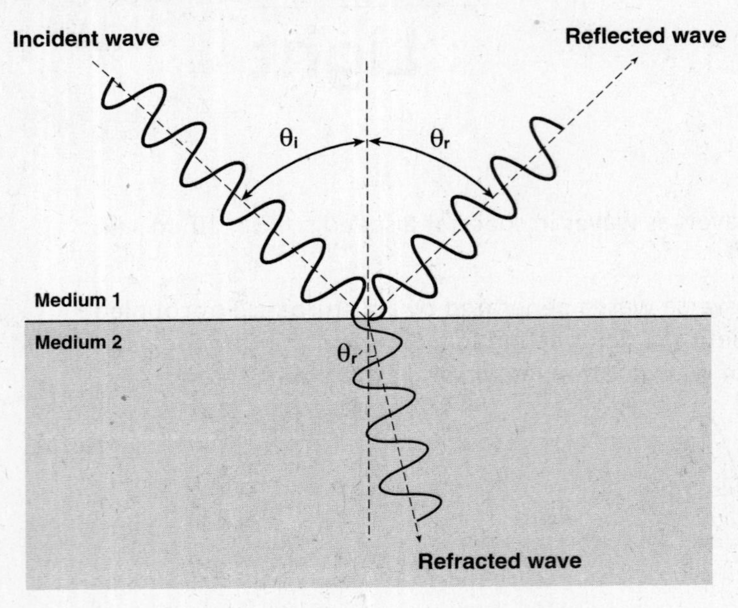

SOURCE: From George Hademenos, *Schaum's Outline of Physics for Pre-Med, Biology, and Allied Health Students*, McGraw-Hill, 1998; reproduced with permission of The McGraw-Hill Companies.

- **Reflection** The light ray can reflect or bounce off the surface boundary at an angle of reflection θ_r, also measured with respect to the normal, which is equal in magnitude to the angle of incidence, known as the **law of reflection** or $\theta_i = \theta_r$.
- **Refraction** The light ray can also bend or refract as it enters the second medium. The angle at which the light ray refracts is dependent on the ratio of the indices of refraction between the two media. The **index of refraction** n for a given material is defined as

$$n = \frac{\text{speed of light in vacuum}}{\text{speed of light in material}} = \frac{c}{v}$$

The **angle of refraction** $\theta_{r'}$ with respect to the normal is related to the incident angle by the **law of refraction**, also known as **Snell's law**:

$$n_1 \sin\theta_i = n_2 \sin\theta_{r'}$$

where n_1, n_2 are the indices of refraction of the first medium and second medium, respectively.

Total internal reflection: A phenomenon that occurs when light passes from a medium or material with a high index of refraction to a medium with a low index of refraction. The angle of incidence that causes total internal reflection is called the **critical angle** θ_c, given by $\sin\theta_c = \frac{n_2}{n_1}$.

Dispersion: The separation of any traveling wave into separate frequency components. An example of dispersion is the separation of white light into its wavelength components and its corresponding elementary colors as it passes through a glass prism.

Polarization: A process by which the electric field vectors of the light waves are all aligned along the same direction; in other words, they are all parallel.

Interference: The superposition of two waves traveling within the same medium when they interact. Consider the experimental setup depicted in the figure below where two coherent light waves of wavelength λ impinge on two narrow slits separated by a distance d. The result will be a fringe pattern of alternating bright and dark fringes where the bright fringes represent constructive interference and the dark fringes represent destructive interference.

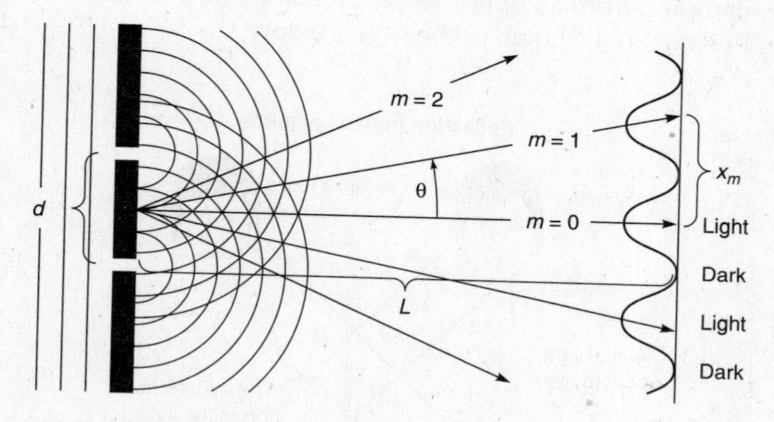

d is the distance between the slits.

m is the fringe number. $m = 0$ is the central maximum. It is also the **path difference in wavelengths**.

Whole numbers are used for the bright **constructive interference fringes.**
$d(\sin\theta) = m\lambda$; $m = 0, 1, 2, 3, \ldots$

Half numbers are used for the dark **destructive interference fringes.**
$d(\sin\theta) = (m + \frac{1}{2})\lambda$; $m = 0, 1, 2, 3, \ldots$

θ is the angle from the midline (from the middle fringe).

L is the distance from the slits to the screen.

x_m is the distance from the midline (center fringe) to the fringe being measured. $x_m = \frac{m\lambda L}{d}$

Diffraction: Describes the ability of light waves to bend or spread through an aperture (opening) or around an obstacle.

CRAM SESSION

Geometric Optics

Geometric optics: A branch of physics concerned with the propagation of light used to form images. In typical problems, a light source is allowed to strike one of two basic types of optical components: mirrors and thin lenses.

Mirrors

- **Plane mirror:** Light rays emanating from an object toward a plane mirror appear to converge at a point within the mirror, as shown in the figure below.

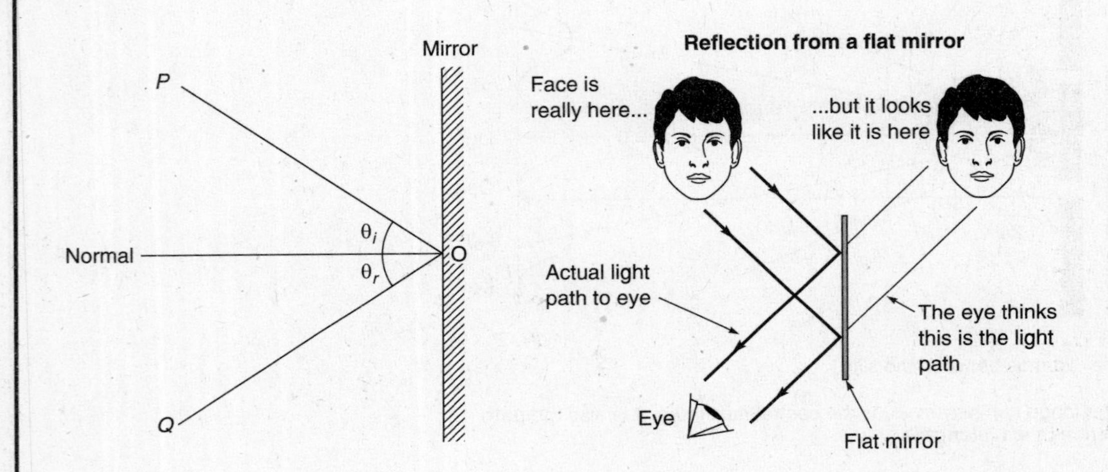

The image created by a plane mirror is always virtual and erect; that is, the image appears to be located the same distance from the mirror as the distance from the object to the mirror.

- **Spherical mirror:** Spherical mirrors are mirrors that are curved and exist primarily in two forms, depending on their curvature. **Concave mirrors** are curved inward, and **convex mirrors** are curved outward. The spherical mirrors are represented by their radius of curvature R, which is related to their focal length f by

Concave mirror: $R = +2f$ Convex mirror: $R = -2f$

The formation of images using spherical mirrors is described in the figure below.

Spherical Mirrors

General Properties

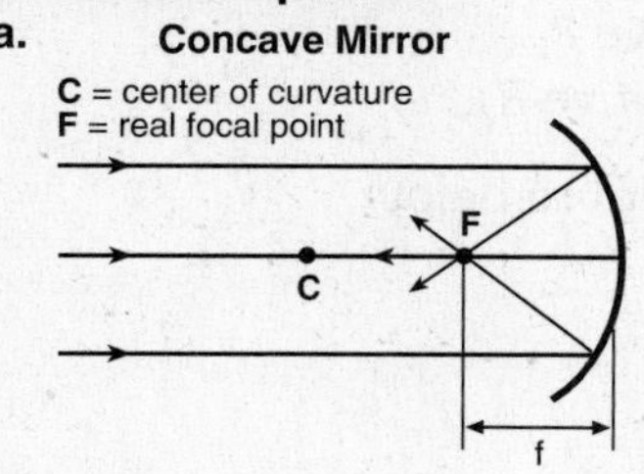

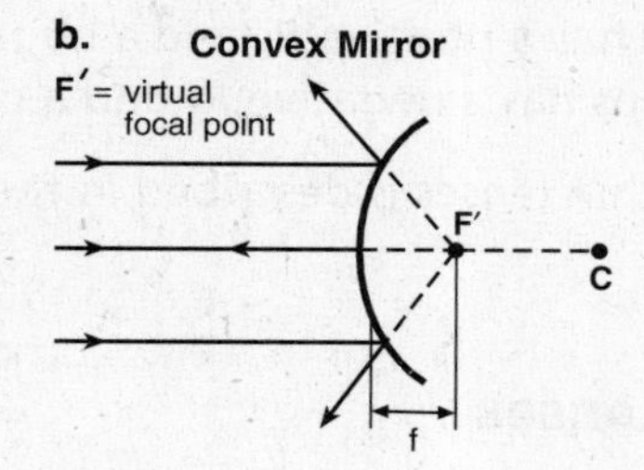

Ray Tracing

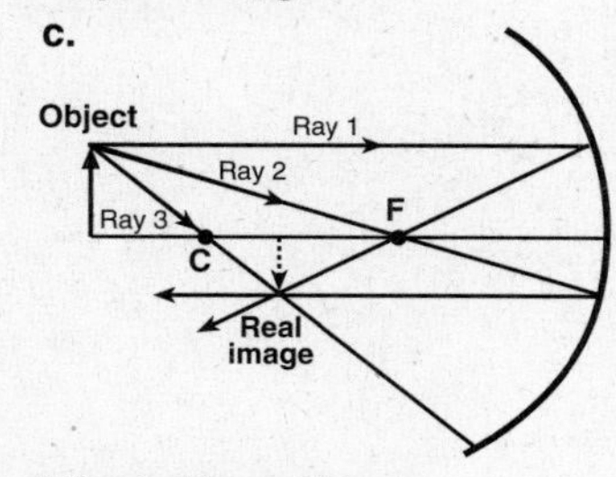

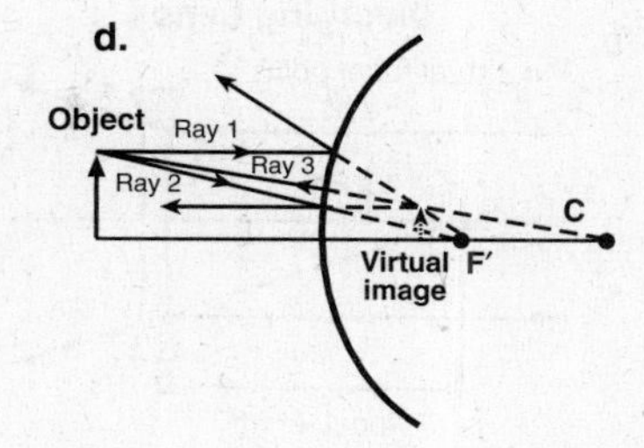

Mirror Equation

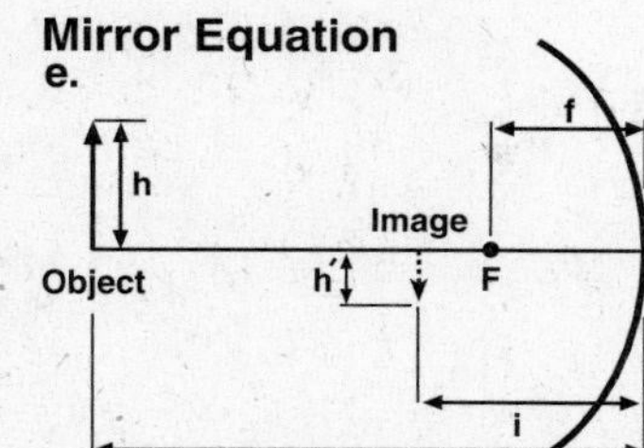

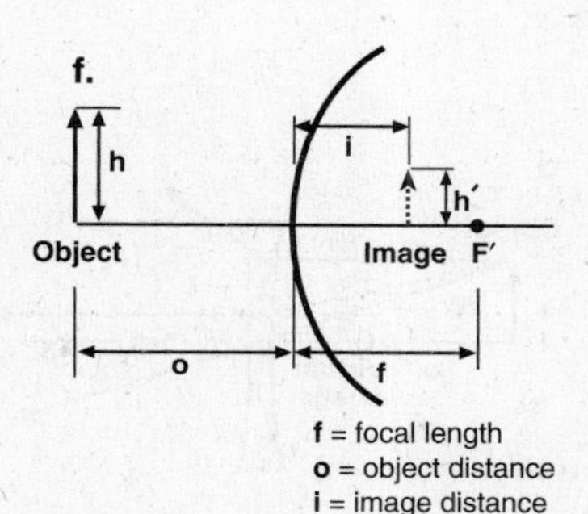

f = focal length
o = object distance
i = image distance

SOURCE: From George Hademenos, *Schaum's Outline of Physics for Pre-Med, Biology, and Allied Health Students*, McGraw-Hill, 1998; reproduced with permission of The McGraw-Hill Companies.

Sign Conventions for Spherical Mirrors

Parameter	Positive	Negative
Object distance, *o*	Real object	Virtual object
Image distance, *i*	Real image	Virtual image
Focal length, *f*	Concave mirror	Convex mirror
Radius of curvature, *R*	Concave mirror	Convex mirror
Magnification, *m*	Erect image	Inverted image

Thin lenses: Thin lenses, like mirrors, exist as convex or concave lenses. However, lenses form images by refracting light rays, as opposed to mirrors that form images by reflecting light rays. Thin lenses have two surfaces and are thus characterized by two radii of curvature R_1 (side of lens closest to the object) and R_2 (side of lens opposite to the object).

- **Convex lens:** A convex lens has a positive R_1 and a negative R_2.
- **Concave lens:** A concave lens has a negative R_1 and a positive R_2.

The formation of images using thin lenses is described in the figure below.

Thin Lenses

General Properties

a. Converging Lens

F = real focal point

F

focal length

b. Diverging Lens

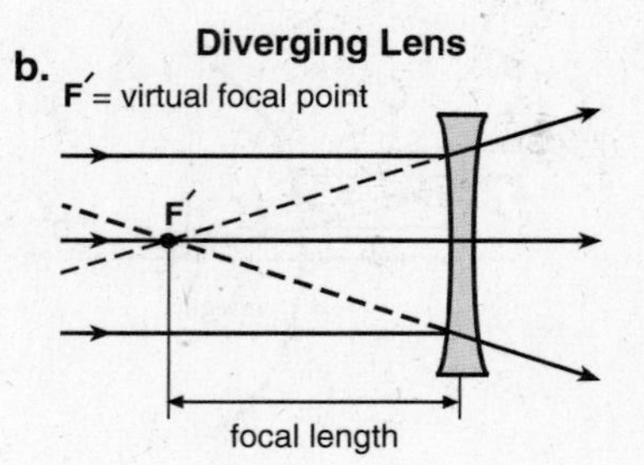

Ray Tracing

c.

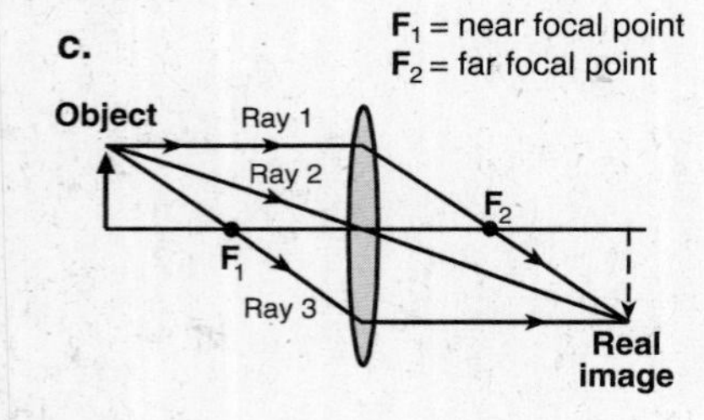

d.

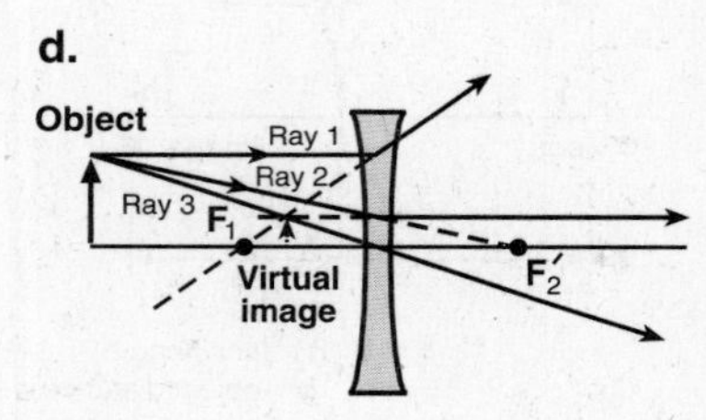

Lens Equation

e.

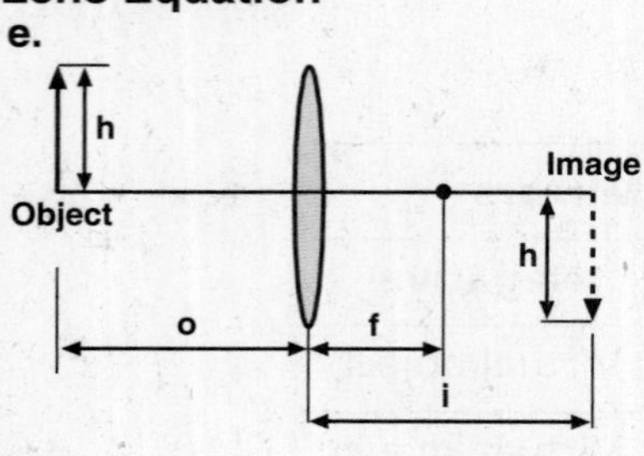

f.

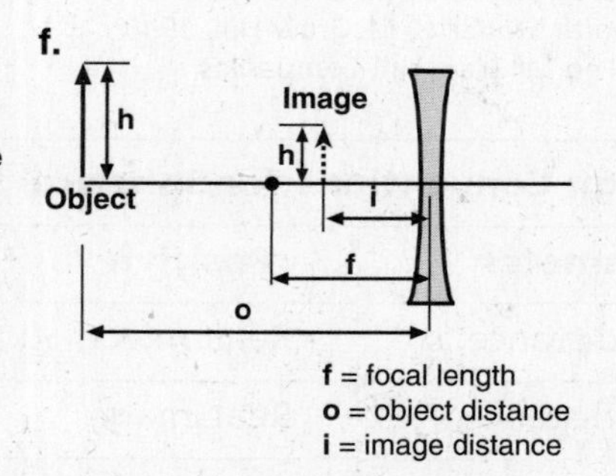

f = focal length
o = object distance
i = image distance

SOURCE: From George Hademenos, *Schaum's Outline of Physics for Pre-Med, Biology, and Allied Health Students*, McGraw-Hill, 1998; reproduced with permission of The McGraw-Hill Companies.

Sign Conventions for Spherical Lenses		
Parameter	**Positive**	**Negative**
Object distance, *o*	Real object	Virtual object
Image distance, *i*	Real image	Virtual image
Focal length, *f*	Converging lens	Diverging lens
Magnification, *m*	Erect image	Inverted image

CHAPTER 12

Electrostatics

Read This Chapter to Learn About:

- Electric Charge
- Electric Forces: Coulomb's Law
- Electric Fields
- Electric Potential
- Electric Dipole

A simple static shock, the beating of the heart, the operation of household appliances, and the devastating damage inflicted by a lightning bolt—all of these examples in nature involve applications of electrostatics. **Electrostatics** is the study of electrically charged particles—their properties such as mass and charge, their behavior such as conservation of charge, and their interactions such as the repulsive or attractive forces that occur and the calculation of the magnitude of such forces through Coulomb's law. This chapter reviews the fundamental concepts of electrostatics.

ELECTRIC CHARGE

Electric charge q is a physical property of the basic building blocks of the atom, a fundamental property of all matter. The SI unit of charge is the **coulomb,** abbreviated C. Although charge can be positive or negative, the magnitude of charge is $e = 1.6 \times 10^{-19}$ C. Considering the particles of the atom, the charge of the positively charged proton is $+1.6 \times 10^{-19}$ C, and the charge of the negatively charged electron is -1.6×10^{-19} C. Two like charges (either two positive charges or two negative charges) repel each other. Positive and negative charges attract each other.

Electric charge is a conserved quantity and thus follows **conservation of charge:**

> Electric charge can neither be created nor destroyed, only transferred. The net charge of a system remains constant.

ELECTRIC FORCES: COULOMB'S LAW

Coulomb's law describes the electrostatic force $\mathbf{F}_{el}$ between two charged particles q_1 and q_2, separated by a distance r:

$$F_{el} = \frac{1}{4\pi\varepsilon_o}\frac{q_1 q_2}{r^2} = k\frac{q_1 q_2}{r^2}$$

where ε_o is the permittivity constant, defined as $\varepsilon_o = 8.85 \times 10^{-12}\ \mathrm{C^2/N \cdot m^2}$. Values for k are

$$k = \frac{1}{4\pi\varepsilon_o} = 9.0 \times 10^9\ \frac{\mathrm{N \cdot m^2}}{\mathrm{C^2}}$$

Electrostatic force, as is the case for all types of forces, is a vector quantity and is expressed in units of newtons (N). The direction of the electrostatic force is based on the charges involved. Unlike charges generate an attractive (negative) force and the direction is toward the other charge; like charges generate a repulsive (positive) force, and the direction is away from the other charge.

EXAMPLE: Determine the electrostatic force between two alpha particles of charge $+2e$ (3.2×10^{-19} C) separated by 10^{-13} m.

SOLUTION: The electrostatic force can be determined using Coulomb's law,

$$F_{el} = \left(9 \times 10^9\ \frac{\mathrm{N \cdot m^2}}{\mathrm{C^2}}\right)\frac{(3.2 \times 10^{-19}\ \mathrm{C})(3.2 \times 10^{-19}\ \mathrm{C})}{(1 \times 10^{-13}\ \mathrm{m})^2}$$

$$= 9.22 \times 10^{-2}\ \mathrm{N, repulsive}$$

ELECTRIC FIELDS

Electric field *E* defines the electric force exerted on a positive test charge positioned at any given point in space. A positive test charge, q_o, is similar in most respects to a true charge except that it does not exert an electrostatic force on any adjacent or nearby charges. Thus, the electric field of a positive test charge provides an idealized distribution of electrostatic force generated by the test charge and is given by

$$\mathbf{E} = \frac{\mathbf{F}_{el}}{q_o}$$

Because ***E*** is a vector quantity, the direction is dependent on the identity of the charge. Because the test charge is positive, if the other charge is negative, an attractive force is generated and the direction of ***E*** is toward the negative charge. Likewise, if the other charge is positive, a repulsive force is generated and the direction of ***E*** is away from the positive charge.

Electric field *E* is expressed in units of newtons per coulomb. If ***E*** is known, it is possible to determine the electrostatic force exerted on any charge q placed at the same position as the test charge using

$$\boldsymbol{F}_{\mathrm{el}} = q\boldsymbol{E}$$

An electric field can be produced by one or more electric charges. The electric field of a point charge, which always points away from a positive charge toward a negative charge, can be calculated by direct substitution of Coulomb's law into the expression for ***E***:

$$\mathbf{E} = \frac{\boldsymbol{F}_{\mathrm{el}}}{q_o} = \frac{k\frac{qq_o}{r^2}}{q_o} = k\frac{q}{r^2}$$

For more than one charge in a defined region of space, the total electric field $\boldsymbol{E}_{\mathrm{tot}}$, because it is a vector quantity, is the vector sum of the electric field generated by each charge $\boldsymbol{E}_q$ in the distribution, or

$$\boldsymbol{E}_{\mathrm{tot}} = \boldsymbol{E}_{q1} + \boldsymbol{E}_{q2} + \boldsymbol{E}_{q3} + \boldsymbol{E}_{q4} + \cdots$$

Electric field lines represent a visual display of the electric field that uses imaginary lines to represent the magnitude and direction of the electric field or the distribution of the electrostatic force over a region in space. The lines of force from a positive charge are directed away from the positive charge, whereas the lines of force of a negative charge are directed toward the negative charge, as depicted in Figure 12-1. The magnitude of the force is greater in the region closer to the charge and becomes weaker farther from the charge.

ELECTRIC POTENTIAL

The concept of potential in this regard is similar to the potential that was discussed in Chapter 5. We discussed the fact that potential energy becomes stored by an object as a result of work done to raise the object against a gravitational field. The **electric potential** V at some point B becomes stored as a result of work done, W, against an electric field to move a positive test charge from infinity (point A) to that point (point B), or

$$V = \frac{-W}{q_o}$$

$$\text{Electric potential} = \frac{\text{work}}{\text{charge}}$$

The electric potential is a scalar quantity that can be positive, negative, or zero, depending on the sign and magnitude of the point charge as well as the work done. Electric potential is expressed in units of joules per coulomb (J/C) = volts (V).

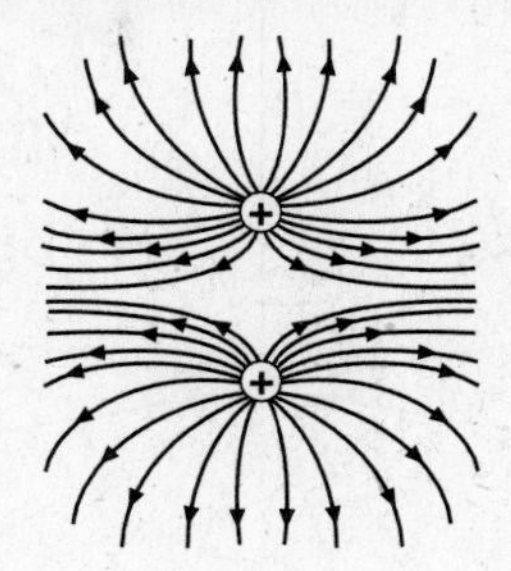

Attractive Electrostatic Force of Unlike Charges

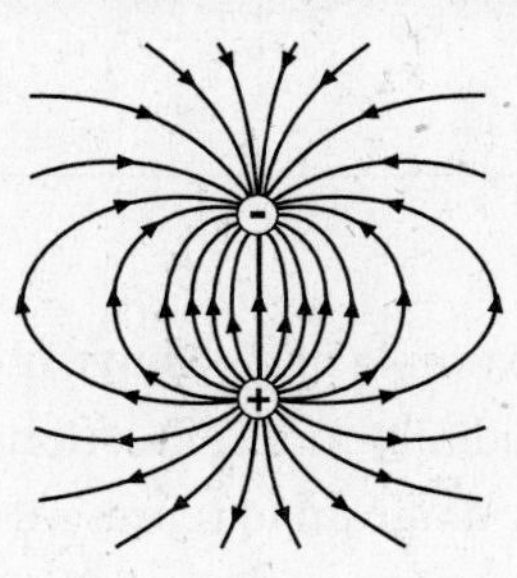

FIGURE 12-1: SOURCE: From George Hademenos, *Schaum's Outline of Physics for Pre-Med, Biology, and Allied Health Students,* McGraw-Hill, 1998; reproduced with permission of The McGraw-Hill Companies.

The **electric potential difference,** ΔV, between any two points A and B in an electric field is related to the work done by the electrostatic force to move the charge from point A to point B as:

$$\Delta V = V_B - V_A = \frac{-W_{AB}}{q_o}$$

The absolute electric potential at a point A that exists at a distance r from a charged particle at point B depends on the magnitude of charge at point B as well as the distance according to the following formula:

$$V = k\frac{q}{r}$$

This is only true if one assumes that $V \to 0$ at a point infinitely far away. Because the electric potential is a scalar quantity, the electric potential for n charges can be determined by adding the electric potential values calculated for each of the n charges:

$$V = \sum_{i=1}^{n} V_i = V_{q1} + V_{q2} + V_{q3} + V_{q4} + \cdots + V_{qn}$$

Equipotential surfaces are a graphical method of representing the electric potential of any charge distribution as concentric circles that are normal or perpendicular to

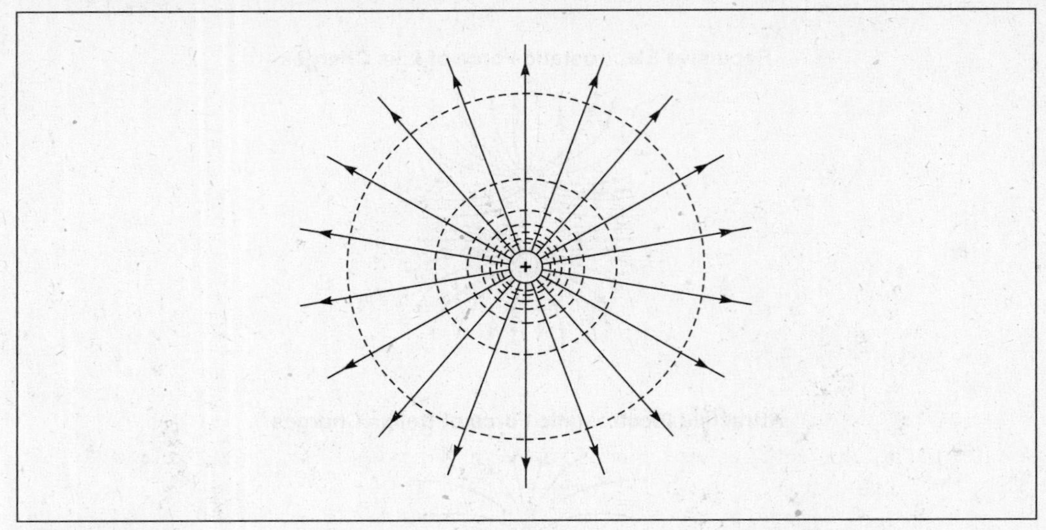

FIGURE 12-2

electric field lines. Consider the example of a positive charge in Figure 12-2. The electric field vectors are pointed away radially in all directions from the charge. Equipotential surfaces can also be drawn on the diagram to represent the electric potential of a positive charge at any distance r from the charge. Recall that the electric potential of a point charge is given by $V = k\frac{q}{r}$. On the surface of the charge where r is at its minimum, the electric potential is at its greatest, and thus a solid circle is drawn about the point charge to represent the largest magnitude of the potential at the surface of the point charge. As the distance r increases, the electric potential is represented as concentric circles that become larger in circumference.

EXAMPLE: The Bohr model of the hydrogen atom describes electron motion in a circular orbit of radius 0.53 Å (angstrom, where 1 angstrom $= 1 \times 10^{-10}$ m) about the nuclear proton. Determine the following for the orbiting electron:

1. The electric field
2. The electric potential

SOLUTION:

1. The electric field $\boldsymbol{E}$ experienced by the orbiting electron is

$$E = k\frac{q}{r^2}$$

where $q = -1.6 \times 10^{-19}$ C and r is the distance between the electron and the proton. Substituting values yields

$$E = \left(9 \times 10^9 \frac{\text{N} \cdot \text{m}^2}{\text{C}^2}\right)\frac{-1.6 \times 10^{-19}\ \text{C}}{(0.53 \times 10^{-10}\ \text{m})^2} = -5.14 \times 10^{11}\ \frac{\text{N}}{\text{C}}$$

where the minus sign indicates that the electric field is directed toward the electron.

2. The electric potential V of the electron is

$$V = k\frac{q}{r}$$

Substituting values yields

$$V = \left(9 \times 10^9 \frac{\text{N} \cdot \text{m}^2}{\text{C}^2}\right) \frac{1.6 \times 10^{-19}\ \text{C}}{0.53 \times 10^{-10}\ \text{m}} = 27.2 \frac{\text{N} \cdot \text{m}}{\text{C}} = 27.2\ \text{V}$$

ELECTRIC DIPOLE

An **electric dipole** consists of two point charges, $+q$ and $-q$, typically equal in magnitude but opposite in sign separated by a small distance d, as shown in Figure 12-3.

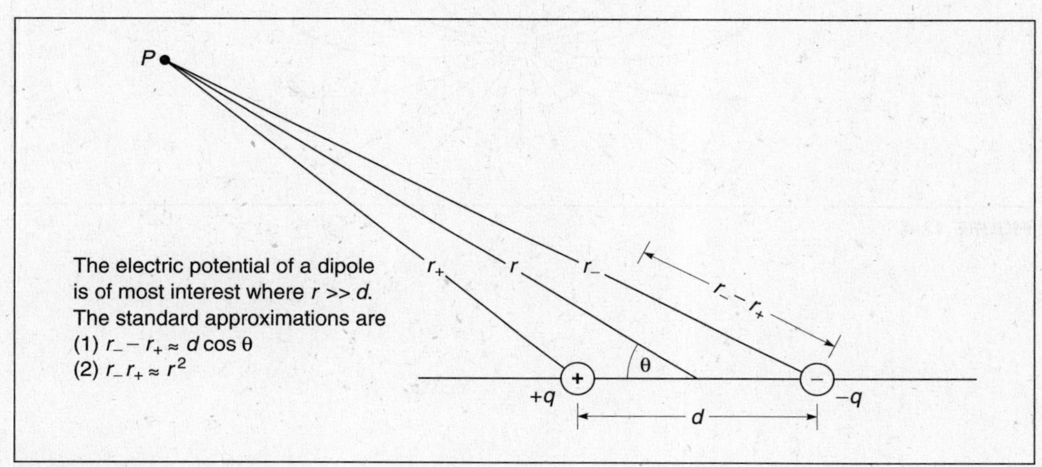

FIGURE 12-3

The electric potential, V, can be found for the electric dipole through the superposition of the two potentials:

$$V = V_{+q} + V_{-q} = k\frac{+q}{r_+} + k\frac{-q}{r_-} = kq\left(\frac{1}{r_+} - \frac{1}{r_-}\right) = kq\left(\frac{r_- - r_+}{r_+ r_-}\right)$$

The objective of this discussion is to calculate the potential of the dipole at a point much farther from the dipole than the distance of separation between the two charges. With such conditions, the following approximations can be made:

1. $r_- r_+ \approx r^2$
2. $r_- - r_+ \approx d \cos\theta$

Substituting these approximations into the equation for electric potential given previously yields an expression for the potential of an electric dipole:

$$V = k\frac{qd\cos\theta}{r^2}$$

where the product qd is referred to as the physical quantity, dipole moment $\mathbf{p}$. The dipole moment is a vector quantity, pointing in a direction from the negative charge to the positive charge, with units of coulomb meters (Cm). The equipotential surfaces of an electric dipole are shown in Figure 12-4.

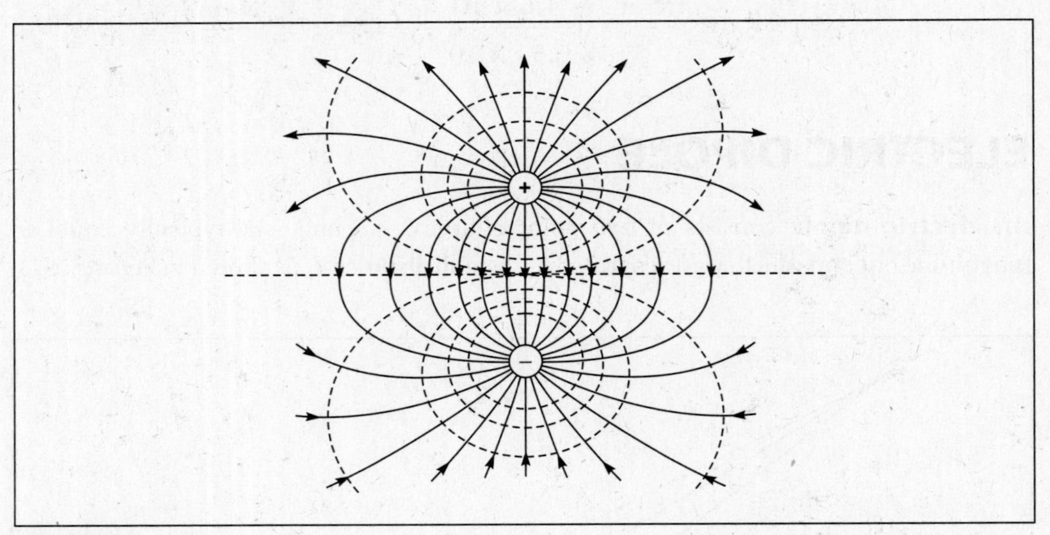

FIGURE 12-4

CRAM SESSION

Electrostatics

Electric charge *q*: A physical property of the basic building blocks of the atom, a fundamental property of all matter.

- The SI unit of charge is the coulomb (C).
- Although charge can be positive or negative, the magnitude of charge is $e = 1.6 \times 10^{-19}$ C.
- The charge of the positively charged proton is $+1.6 \times 10^{-19}$ C; the charge of the negatively charged electron is -1.6×10^{-19} C.
- Like charges (either two + charges or two − charges) repel each other; unlike charges attract.

Conservation of charge: Electric charge can neither be created nor destroyed, only transferred. The net charge of a system remains constant.

Electric Forces

Coulomb's law: Describes the electrostatic force $\boldsymbol{F}_{el}$ between two charged particles q_1 and q_2 separated by a distance r:

$$F_{el} = \frac{1}{4\pi\varepsilon_o}\frac{q_1 q_2}{r^2} = k\frac{q_1 q_2}{r^2} \text{ where } k = \frac{1}{4\pi\varepsilon_o} = 9.0 \times 10^9 \frac{\text{N} \cdot \text{m}^2}{\text{C}^2}$$

Electrostatic force: A vector quantity and expressed in units of newtons, as for all types of forces. The direction of the electrostatic force is based on the charges involved.

- Unlike charges generate an attractive (negative) force and the direction is toward the other charge.
- Like charges generate a repulsive (positive) force, and the direction is away from the other charge.

Electric field: Defines the electric force exerted on a positive test charge positioned at any given point in space. A positive test charge, q_o, is similar in most respects to a true charge except that it does not exert an electrostatic force on any adjacent or nearby charges. Thus, the electric field of a positive test charge provides an idealized distribution of electrostatic force generated by the test charge and is given by $\boldsymbol{E} = \frac{\boldsymbol{F}_{el}}{q_o}$. Because $\boldsymbol{E}$ is a vector quantity, the direction is dependent on the identity of the charge.

- Because the test charge is positive, if the other charge is negative, an attractive force is generated and the direction of $\boldsymbol{E}$ is toward the negative charge. Likewise, if the other charge is positive, a repulsive force is generated and the direction of $\boldsymbol{E}$ is away from the positive charge.

- Electric field can also be determined from Coulomb's law:

$$\boldsymbol{E} = \frac{\boldsymbol{F}_{\text{el}}}{q_o} = \frac{k\dfrac{qq_o}{r^2}}{q_o} = k\frac{q}{r^2}$$

- The total electric field from a distribution of charges is the vector sum of the electric field generated by each charge: $\boldsymbol{E}_{\text{tot}} = \boldsymbol{E}_{q1} + \boldsymbol{E}_{q2} + \boldsymbol{E}_{q3} + \boldsymbol{E}_{q4} + \cdots$
- Electric field $\boldsymbol{E}$ is expressed in units of newtons per coulomb.

Electric field lines: Represent a visual display of the electric field that uses imaginary lines to represent the magnitude and direction of the electric field or the distribution of the electrostatic force over a region in space.

- The lines of force from a positive charge are directed away from the positive charge, whereas the lines of force of a negative charge are directed toward the negative charge, as depicted in the figure below.
- The magnitude of the force is greater in the region closer to the charge and becomes weaker farther from the charge.

Repulsive Electrostatic Force of Like Charges

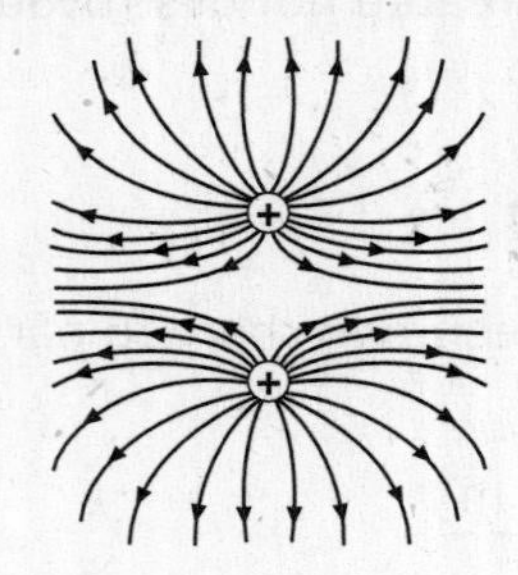

Attractive Electrostatic Force of Unlike Charges

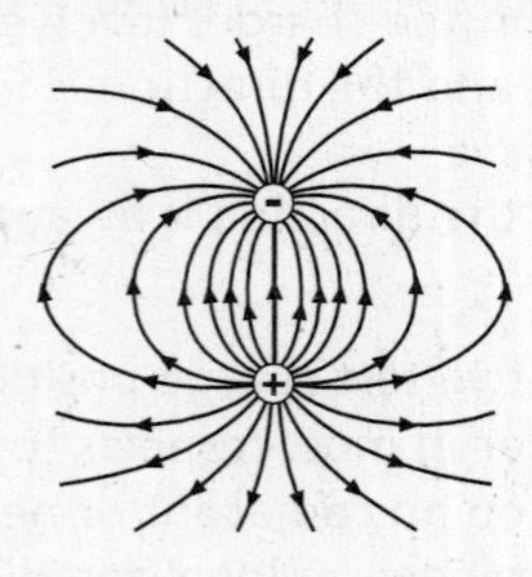

SOURCE: From George Hademenos, *Schaum's Outline of Physics for Pre-Med, Biology, and Allied Health Students*, McGraw-Hill, 1998; reproduced with permission of The McGraw-Hill Companies.

Electric potential: The **electric potential** V at some point B becomes stored as a result of work done, W, against an electric field to move a positive test charge q, from infinity (point A) to that point (point B), or $V = \frac{-W}{q_o}$. The electric potential is a scalar quantity that can be either positive, negative, or zero, depending on the sign and magnitude of the point charge as well as the work done. Electric potential is expressed in units of joules per coulomb = volts.

Electric potential difference: The **electric potential difference** ΔV between any two points A and B in an electric field is related to the work done by the electrostatic force to move the charge from point A to point B as:

$$\Delta V = V_B - V_A = \frac{-W_{AB}}{q_o}$$

The absolute electric potential at a point A that exists at a distance r from a charged particle at point B depends on the magnitude of charge at point B as well as the distance according to the following formula:

$$V = k\frac{q}{r}$$

Because the electric potential is a scalar quantity, the electric potential for n charges can be determined by adding the electric potential values calculated for each of the n charges

$$V = \sum_{i=1}^{n} V_i = Vq_1 + Vq_2 + Vq_3 + Vq_4 + \cdots + Vq_n$$

Equipotential surfaces: A graphical method of representing the electric potential of any charge distribution as concentric circles that are normal or perpendicular to electric field lines, as shown for a positive charge in the figure below.

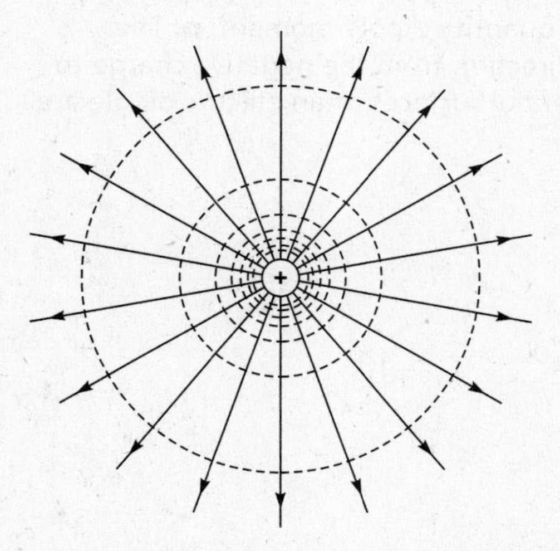

The electric field lines are represented as vectors pointed away radially in all directions from the charge. Equipotential surfaces can also be drawn on the diagram to represent the electric potential of a positive charge at any distance r from the charge. Recall that the electric potential of a point charge is given by $V = k\frac{q}{r}$. On the surface of the charge where r is at its minimum, the electric potential is at its greatest, and thus a solid circle is drawn about the point charge to represent the largest magnitude of the potential at the surface of the point charge. As the distance, r, increases, the electric potential is represented as concentric circles that become larger in circumference.

Electric dipole: Two point charges $+q$ and $-q$ typically equal in magnitude but opposite in sign separated by a small distance d, as shown in the figure below.

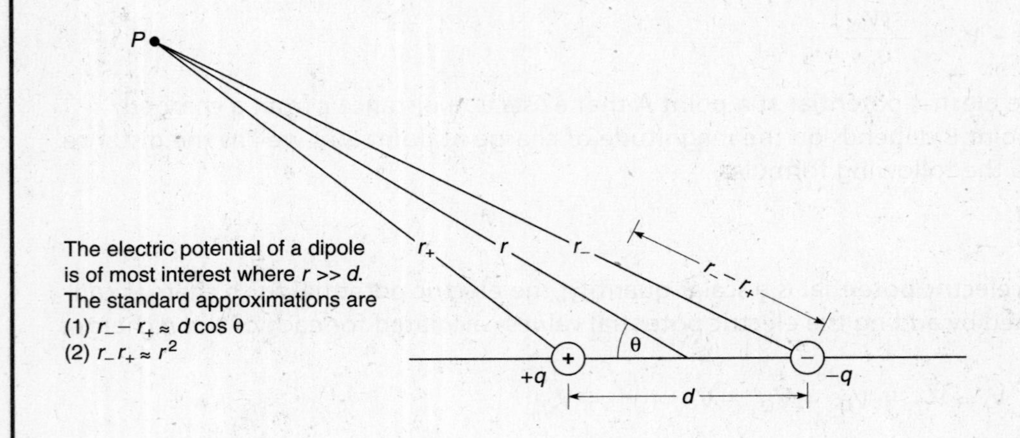

The electric potential, V, for the electric dipole is

$$V = k\frac{qd \cos\theta}{r^2}$$

where the product qd is referred to as the physical quantity dipole moment, **p.** The dipole moment is a vector quantity, pointing in a direction from the negative charge to the positive charge, with units of Cm. The equipotential surfaces of an electric dipole are shown in the figure below.

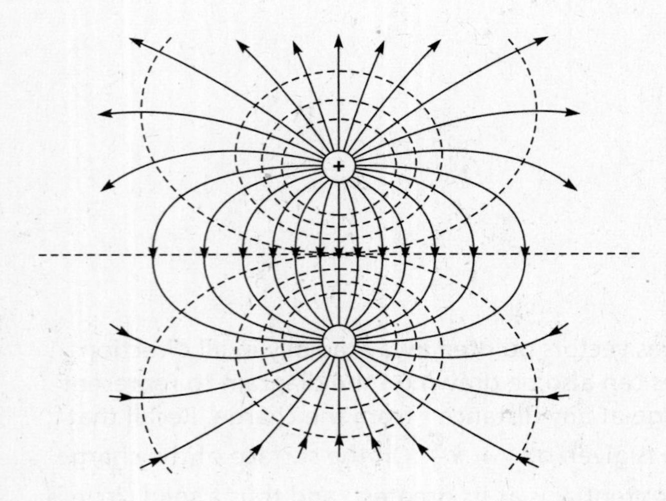

CHAPTER 13

Electric Circuits

Read This Chapter to Learn About:

DIRECT CURRENT (DC) CIRCUITS

- Fundamental Quantities of DC Circuits: Current, Voltage, and Resistance
- Ohm's Law
- Capacitors and Dielectrics
- Electric Power
- Kirchhoff's Rules of Circuit Analysis

ALTERNATING CURRENT (AC) CIRCUITS

- Fundamental Quantities of AC Circuits: Voltage and Current
- Ohm's Law for AC Circuits

You do not have to look far to realize the significance of electric circuits: they are involved in providing light to a room and power to an appliance. There are two types of electric circuits: direct current (DC) circuits and alternating current (AC) circuits. In a DC circuit, electricity or electric charge flows at a constant rate from a voltage source (such as a battery) through a series of connected conductors (typically copper wires) to a device, such as a lightbulb, siren, or motor, that uses the electric charge to operate. AC circuits, the type of electric circuits used in residential structures, use voltage that oscillates with time. This chapter provides a detailed review of both direct current and alternating current electric circuits.

DIRECT CURRENT (DC) CIRCUITS

In a **direct current (DC) circuit**, electricity in the form of an electric charge generated by a voltage source (such as a battery) flows as current through an arrangement of circuit elements or devices (e.g., a lightbulb, an alarm, a motor) that are all connected by a **conductor**, or a material that allows electric charge to flow easily through it.

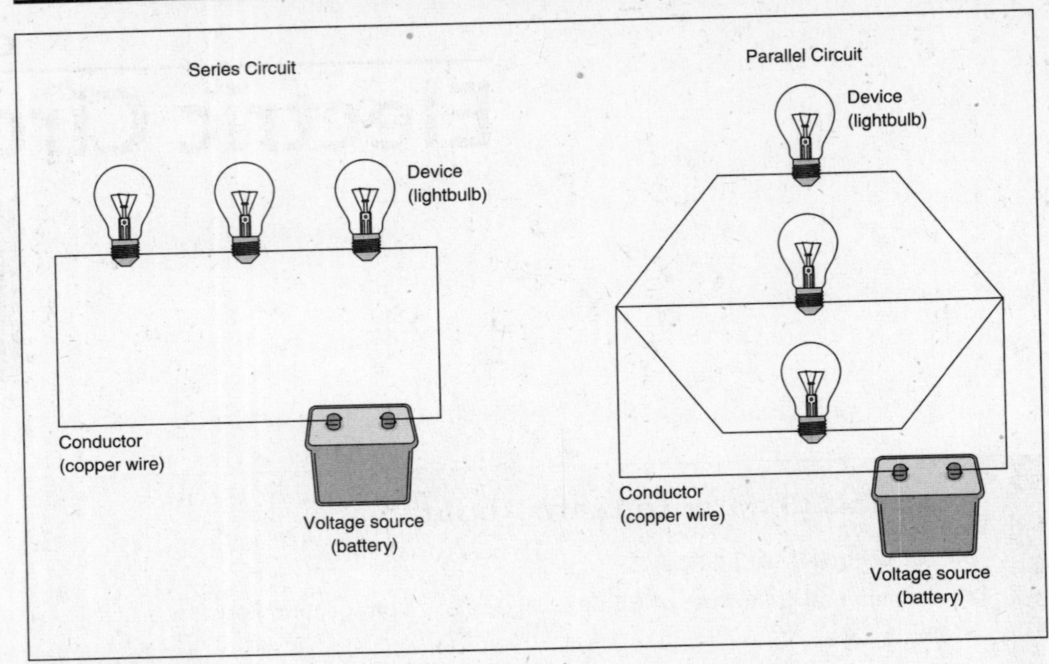

FIGURE 13-1

These circuit elements may be connected in series or parallel, as shown in Figure 13-1. In a **series circuit**, two or more circuit elements are connected in sequence, one after the other. In a **parallel circuit**, two or more circuit elements are connected in a branching arrangement, such that a charge may pass through one element or the other. Regardless of the type of circuit, these elements interact with and can directly influence the flow of charge through the circuit.

FUNDAMENTAL QUANTITIES OF DC CIRCUITS: CURRENT, VOLTAGE, AND RESISTANCE

The three fundamental quantities that describe the properties of an electric circuit are current, voltage, and resistance.

Current

Current, I, is the rate of motion of electric charge and is expressed as:

$$\text{Current} = \frac{\text{electric charge moving through a region}}{\text{time required to move the charge}}$$

$$I = \frac{\Delta q}{\Delta t}$$

where current is measured in SI units of **amperes**, A.

- **Current in a series circuit:** Current that flows through one circuit element connected in series must also flow through the remaining elements connected in the series circuit. Therefore, the net or total current that flows through a series circuit remains the same through each of the individual elements or

$$I_{net} = I_1 = I_2 = I_3 = \cdots = I_n$$

where n refers to the nth element in the series circuit.

- **Current in a parallel circuit:** In a parallel circuit, one circuit element branches into two or more connected elements. The current that flows through the single element, separates, and in the process divides its current among the branch elements. The amount of current that enters each of the branch elements depends on the resistance of that element. In any event, the net or total current in a parallel circuit is equal to the sum of the currents that flow through all connected branching elements, or

$$I_{net} = I_1 + I_2 + I_3 + \cdots + I_n$$

The flow of current in a series and parallel circuit is illustrated in Figure 13-2.

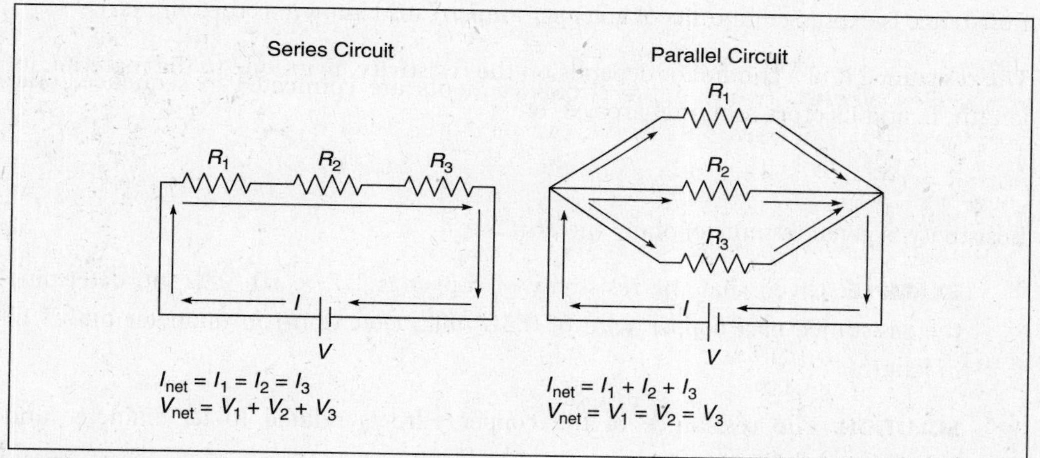

FIGURE 13-2

Voltage

Electromotive force, $\mathscr{E}$ (often abbreviated emf), is the voltage or potential difference generated by a battery or power source when no current is flowing. When current is flowing through a conductor, the conductor generates an internal resistance, thereby yielding a voltage drop of IR_{int} unique to the conductor. Therefore, the voltage V generated by an emf is

$$V = \mathscr{E} - IR$$

Electromotive force is expressed in units of volts (V).

- **Voltage in series** The net voltage of electromotive forces connected in series is the sum of the individual voltages of each electromotive force, or

$$V_{net} = V_1 + V_2 + V_3 + \cdots + V_n$$

- **Voltage in parallel** The net voltage of electromotive forces connected in parallel is identical to that of the voltage from an individual electromotive source, or

$$V_{net} = V_1 = V_2 = V_3 = \cdots = V_n$$

Resistance

Resistance, R, is the inherent property of a conductor by which it resists the flow of electric current and represents a measure of the potential difference V that must be supplied to a circuit to drive current, I, through the circuit. Resistance is defined as

$$\text{Resistance} = \frac{\text{potential difference}}{\text{current}}$$

$$R = \frac{V}{I}$$

Resistance is expressed in units of volts per ampere, also known as the ohm (Ω).

The resistance R of a conductor depends on the resistivity, ρ, unique to the material, its length, L, and its cross-sectional area, A, or

$$R = \rho\frac{L}{A}$$

Resistivity is given in units of ohm · meter (Ω · m).

EXAMPLE: Given that the resistivity of copper is 1.7×10^{-8} Ω · m, determine the resistance of a copper wire of 0.30 millimeter (mm) in diameter and 5 m in length.

SOLUTION: The resistance of the copper wire is related to its diameter and length by

$$R = \rho\frac{L}{A} = \rho\frac{L}{\pi r^2} = \left(1.7 \times 10^{-8}\ \Omega \cdot \text{m}\right)\frac{5.0\ \text{m}}{3.14\left(0.15 \times 10^{-3}\ \text{m}\right)^2} = 1.2\ \Omega$$

- **Resistors in Series:** In a circuit consisting of three resistors connected in series, as in Figure 13-3, the current must flow through the path presented by the resistors in series. To simplify circuit calculations, the equivalent resistance R_{eq} can be calculated in terms of the resistance of the individual components:

$$R_{eq} = R_1 + R_2 + R_3 + \cdots + R_n$$

EXAMPLE: Two resistances $R_1 = 8\ \Omega$ and $R_2 = 6\ \Omega$ are connected in series. Determine the equivalent resistance.

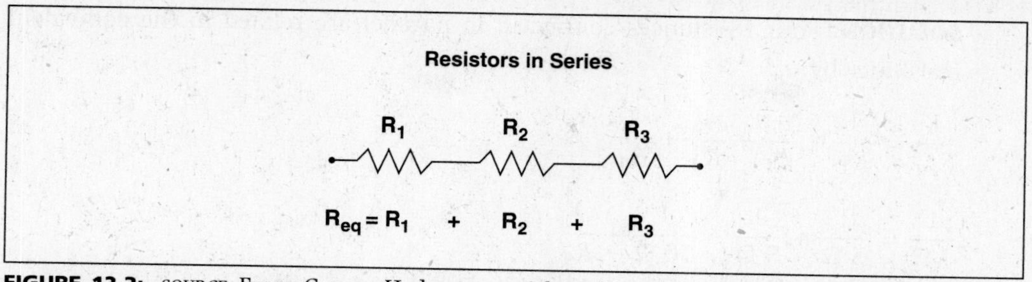

Resistors in Series

R_1 R_2 R_3

$R_{eq} = R_1 + R_2 + R_3$

FIGURE 13-3: SOURCE: From George Hademenos, *Schaum's Outline of Physics for Pre-Med, Biology, and Allied Health Students*, McGraw-Hill, 1998; reproduced with permission of The McGraw-Hill Companies.

SOLUTION:

$R_{eq} = R_1 + R_2 = 8\ \Omega + 6\ \Omega = 14\ \Omega$

➤ **Resistors in parallel:** In a circuit consisting of three resistors connected in parallel, as in Figure 13-4, the current must flow through the path presented by the resistors in parallel. To simplify circuit calculations, the equivalent resistance R_{eq} can be calculated in terms of the resistance of the individual components:

$$\frac{1}{R_{eq}} = \frac{1}{R_1} + \frac{1}{R_2} + \frac{1}{R_3} + \cdots + \frac{1}{R_n}$$

where the equivalent resistance R_{eq} is always less than the smallest value of resistance of the individual components.

EXAMPLE: A circuit with four resistances connected in parallel yields an equivalent resistance of 1 Ω. If $R_1 = 5\ \Omega$, $R_2 = 5\ \Omega$, and $R_3 = 10\ \Omega$, determine R_4.

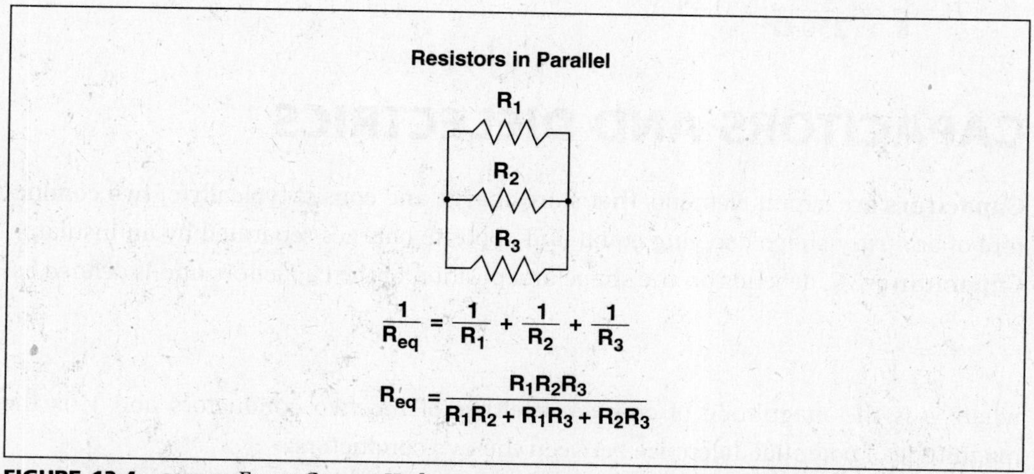

Resistors in Parallel

R_1 R_2 R_3

$$\frac{1}{R_{eq}} = \frac{1}{R_1} + \frac{1}{R_2} + \frac{1}{R_3}$$

$$R_{eq} = \frac{R_1R_2R_3}{R_1R_2 + R_1R_3 + R_2R_3}$$

FIGURE 13-4: SOURCE: From George Hademenos, *Schaum's Outline of Physics for Pre-Med, Biology, and Allied Health Students*, McGraw-Hill, 1998; reproduced with permission of The McGraw-Hill Companies.

SOLUTION: Four resistances connected in parallel are related to the equivalent resistance by

$$\frac{1}{R_{eq}} = \frac{1}{R_1} + \frac{1}{R_2} + \frac{1}{R_3} + \frac{1}{R_4}$$

$$\frac{1}{1\ \Omega} = \frac{1}{5\ \Omega} + \frac{1}{5\ \Omega} + \frac{1}{10\ \Omega} + \frac{1}{R_4}$$

Solving for R_4 yields $R_4 = 2\ \Omega$.

OHM'S LAW

Ohm's law states that the voltage V across a resistor R is proportional to the current I through it, and can be written in equation form as

$$V = IR$$

An increase in voltage drives more electrons through the wire (conductor) at a greater rate, thus the current increases. If the voltage remains the same and the wire is a poor conductor, the resistance against the flow of electrons increases, thereby decreasing the current. It is possible to change I by manipulating V and R, but it is not possible to change V by manipulating I because current flow is due to the difference in voltage, not vice versa. The current is a completely dependent variable.

EXAMPLE: The filament of a lightbulb has a resistance of 250 Ω. Determine the current through the filament if 120 V is applied to the lamp.

SOLUTION: The current is related to the voltage by Ohm's law

$$I = \frac{V}{R} = \frac{120\ \text{V}}{250\ \Omega} = 0.48\ \text{A}$$

CAPACITORS AND DIELECTRICS

Capacitors are circuit elements that store charge and consist typically of two conductors of arbitrary shape carrying equal and opposite charges separated by an insulator. **Capacitance**, C, depends on the shape and position of the capacitors and is defined as

$$C = \frac{q}{V}$$

where q is the magnitude of charge on either of the two conductors and V is the magnitude of potential difference between the two conductors.

The SI unit of capacitance is coulomb/volt, collectively known as the farad (F). Because capacitors store positive and negative charge, work is done in separating the two types of charge, which is stored as electric potential energy W in the capacitor, given by

$$W = \frac{1}{2}qV = \frac{1}{2}CV^2 = \frac{1}{2}\frac{q^2}{C}$$

where V is the potential difference and q is charge.

EXAMPLE: To restore cardiac function to a heart attack victim, a cardiac defibrillator is applied to the chest in an attempt to stimulate electrical activity of the heart and restore the heartbeat. A cardiac defibrillator consists of a capacitor charged to approximately 7.5×10^3 V with stored energy of 500 watt seconds ($W \cdot s$). Determine the charge on the capacitor in the cardiac defibrillator.

SOLUTION: The energy stored in a capacitor is

$$E = \frac{1}{2}CV^2$$

and the charge on the capacitor is

$$q = CV$$

Solving for C gives

$$C = \frac{q}{V}$$

Substituting into the expression for E, you get

$$E = \frac{1}{2}\left(\frac{q}{V}\right)V^2 = \frac{1}{2}qV$$

Solving for q results in

$$q = \frac{2E}{V} = \frac{2 \cdot 500 \text{ W} \cdot \text{s}}{7.5 \times 10^3 \text{ V}} = 0.13 \text{ C}$$

The most common type of capacitor is the parallel-plate capacitor consisting of two large conducting plates of area A and separated by a distance d. The capacitance of a parallel-plate capacitor is

$$C = \kappa\varepsilon_o\frac{A}{d}$$

where κ is a dielectric constant (dimensionless) and $\varepsilon_o = 8.85 \times 10^{-12}\ \text{C}^2/\text{N} \cdot \text{m}^2 = 8.85 \times 10^{-12}$ F/m. For vacuum, $\kappa = 1$.

A **dielectric** is an insulator that is usually used to fill the gap between the plates of a capacitor because it increases the capacitance. Dielectric materials are characterized by a dielectric constant, κ, which relates the new capacitance of the capacitor to its original capacitance by the relation:

$$C = \kappa C_o$$

where C_o is the capacitance of a capacitor with an empty gap and C is the capacitance of a capacitor filled with a dielectric. The dielectric constant is equal to 1 for vacuum and slightly greater than 1 for air ($\kappa = 1.00054$).

- **Capacitors in series:** The effective capacitance C_{eff} of capacitors connected in series, as shown in Figure 13-5, is given by

$$\frac{1}{C_{\text{eff}}} = \frac{1}{C_1} + \frac{1}{C_2} + \frac{1}{C_3} + \cdots + \frac{1}{C_n}$$

where C_n is the nth capacitor connected in series.

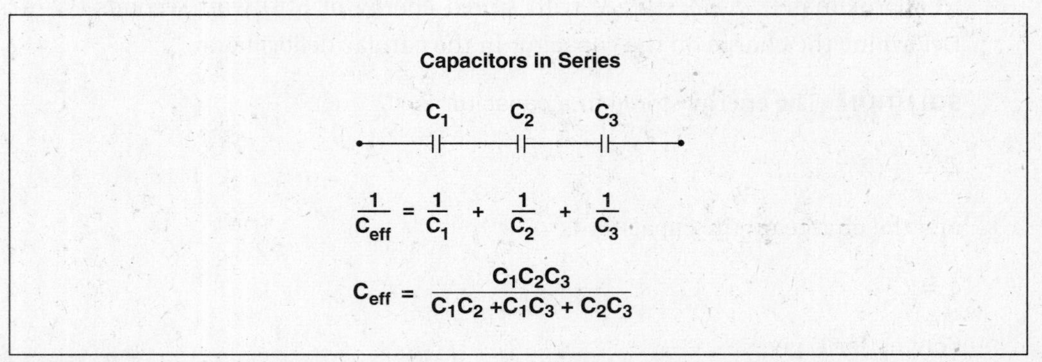

FIGURE 13-5: SOURCE: From George Hademenos, *Schaum's Outline of Physics for Pre-Med, Biology, and Allied Health Students*, McGraw-Hill, 1998; reproduced with permission of The McGraw-Hill Companies.

- **Capacitors in parallel:** The effective capacitance C_{eff} of capacitors connected in parallel, as shown in Figure 13-6, is given by

$$C_{\text{eff}} = C_1 + C_2 + C_3 + \cdots + C_n$$

where C_n is the nth capacitor connected in parallel.

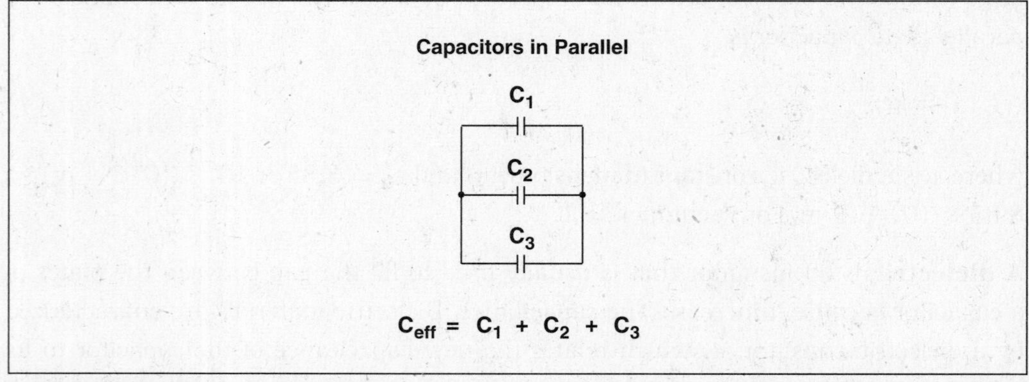

FIGURE 13-6: SOURCE: From George Hademenos, *Schaum's Outline of Physics for Pre-Med, Biology, and Allied Health Students*, McGraw-Hill, 1998; reproduced with permission of The McGraw-Hill Companies.

Summary of Circuit Element Quantities of DC Circuits		
Circuit Element	**Series**	**Parallel**
Voltage	$V_{net} = V_1 + V_2 + V_3 + \cdots + V_n$	$V_{net} = V_1 = V_2 = V_3 = \cdots = V_n$
Current	$I_{net} = I_1 = I_2 = I_3 = \cdots = I_n$	$I_{net} = I_1 + I_2 + I_3 + \cdots + I_n$
Resistance	$R_{eq} = R_1 + R_2 + R_3 + \cdots + R_n$	$\frac{1}{R_{eq}} = \frac{1}{R_1} + \frac{1}{R_2} + \frac{1}{R_3} + \cdots + \frac{1}{R_n}$
Capacitance	$\frac{1}{C_{eff}} = \frac{1}{C_1} + \frac{1}{C_2} + \frac{1}{C_3} + \cdots + \frac{1}{C_n}$	$C_{eff} = C_1 + C_2 + C_3 + \cdots + C_n$

ELECTRIC POWER

As current flows through a resistor, electric power P is dissipated into the resistor according to the following equivalent expressions:

$$P = IV = I^2R = \frac{V^2}{R}$$

The unit for electric power is the watt, abbreviated W.

KIRCHHOFF'S RULES OF CIRCUIT ANALYSIS

In a simple circuit (e.g., a voltage source connected to a single device via a single resistance), one can easily determine the current through the single resistance using Ohm's law. However, as the elements in a given circuit increase in number and complexity, it becomes a much more difficult task to calculate current and voltage though all circuit elements. Analysis of complex electric circuits is typically performed according to Kirchhoff's rules of circuit analysis:

- **Node or junction rule:** The algebraic sum of all currents entering a junction or node must be equal to the sum of all currents exiting the node.
- **Loop or circuit rule**: The algebraic sum of potential drops encountered while traversing the circuit in either a counterclockwise or a clockwise direction must be equal to zero. Within a current loop, if a resistor is traversed in the direction of current, the change in voltage is $-IR$; in the opposite direction of current, the change in voltage is $+IR$. If a seat of voltage (emf) is traversed in the direction of voltage, the change in emf is $+V$; in the opposite direction of voltage, it is $-V$.

EXAMPLE: For the electric circuit in Figure 13-7, determine I_1, I_2, I_3 given $R_1 = 5\ \Omega$, $R_2 = 8\ \Omega$, and $R_3 = 11\ \Omega$, $\mathcal{E}_1 = 1$ V, and $\mathcal{E}_2 = 5$ V.

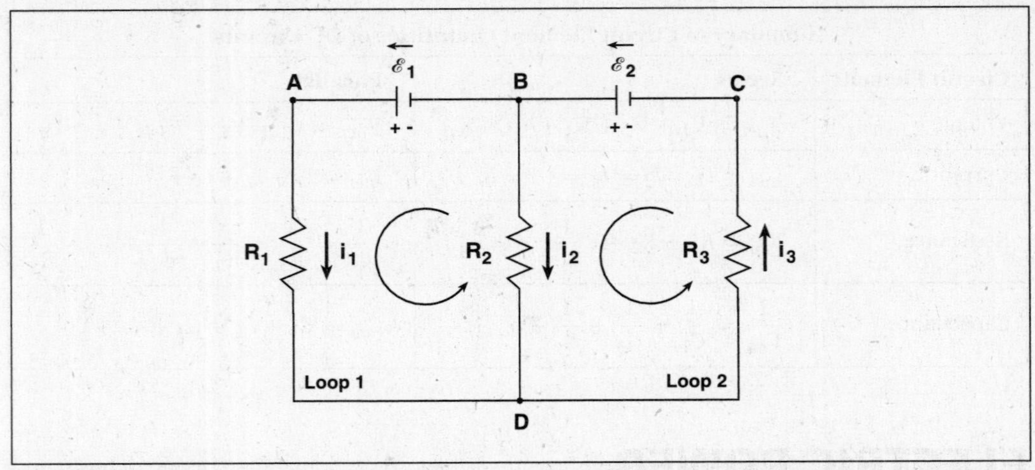

FIGURE 13-7: SOURCE: From George Hademenos, *Schaum's Outline of Physics for Pre-Med, Biology, and Allied Health Students*, McGraw-Hill, 1998; reproduced with permission of The McGraw-Hill Companies.

SOLUTION: The first step is to identify all nodes and loops within the circuit. Of the four nodes in the circuit, consider node *B*. For node *B*, nodal analysis reveals the following equation:

$$I_1 + I_2 - I_3 = 0$$

Applying the loop theorem to the circuit yields

Loop 1: $\mathscr{E}_1 - I_1R_1 + I_2R_2 = 0$

Loop 2: $\mathscr{E}_2 - I_2R_2 - I_3R_3 = 0$

We now have three equations that can be used to solve for the three unknowns: I_1, I_2, and I_3. Making the appropriate substitutions gives

$$I_1 + I_2 - I_3 = 0 \quad (1)$$

$$1\text{ V} - (5\ \Omega)I_1 + (8\ \Omega)I_2 = 0 \quad (2)$$

$$5\text{ V} - (8\ \Omega)I_2 - (11\ \Omega)I_3 = 0 \quad (3)$$

From Equation 1, you can write

$$I_1 = -I_2 + I_3$$

Then, substituting in Equation 2 and solving for I_2 yields

$$I_2 = (0.38)I_3 - 0.077 \quad (4)$$

Substituting into Equation 3, you can obtain the value for I_3:

$$I_3 = 0.39\text{ A}$$

Substituting into Equation 4 yields

$I_2 = 0.07$ A

Finally, substituting values for I_2 and I_3 into Equation 1 yields I_1:

$I_1 = 0.32$ A

ALTERNATING CURRENT (AC) CIRCUITS

Alternating current (AC) circuits are similar to direct current (DC) circuits with one exception: the type of voltage source driving current through the circuit. In DC circuits, the voltage source is continuous and constant, resulting in current flow in only one direction, whereas in AC circuits, the voltage is oscillatory or sinusoidal, resulting in periodic changes in direction of current flow. AC electricity is used in most households (typically at an oscillatory frequency of 60 Hz) because its voltage can easily be changed by a transformer.

FUNDAMENTAL QUANTITIES OF AC CIRCUITS: VOLTAGE AND CURRENT

Voltage in AC circuits is produced by an electric power generator, which generates an oscillating potential across its terminals. As a result, the voltage potential is sinusoidal, as shown in Figure 13-8, changing direction and hence magnitude over time, and can be expressed mathematically as

$$V = V_o \sin \omega t = V_o \sin 2\pi f t$$

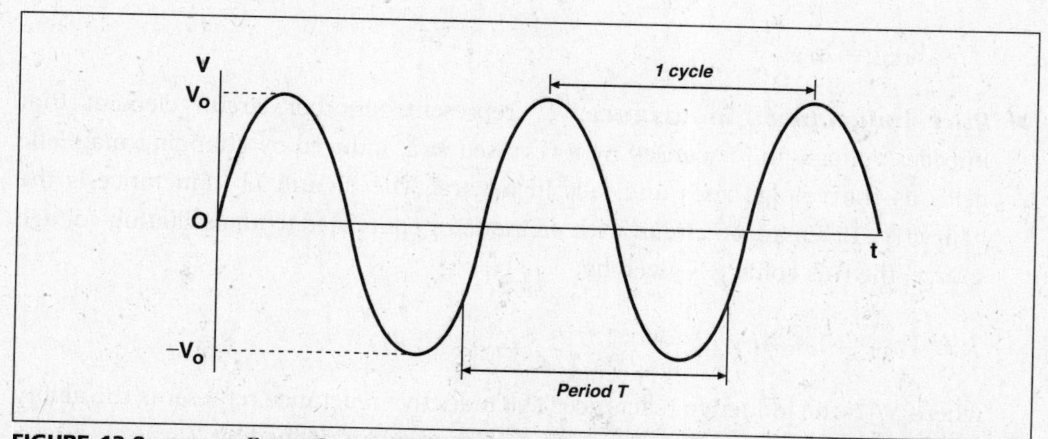

FIGURE 13-8: SOURCE: From George Hademenos, *Schaum's Outline of Physics for Pre-Med, Biology, and Allied Health Students*, McGraw-Hill, 1998; reproduced with permission of The McGraw-Hill Companies.

where V is the instantaneous voltage, in volts; V_o is the amplitude or maximum value of the voltage, in volts; ω is the angular velocity, in radians per second (rad/s); f is the frequency, in hertz (Hz); and t is time, in seconds (s). Because it continuously changes over time, AC voltage can be expressed as an effective or root mean square (rms) voltage, given by

$$V = V_{\text{rms}} = \frac{V_{\text{max}}}{\sqrt{2}}$$

Current in AC circuits, similar to voltage, is also sinusoidal and is defined as

$$I = I_o \sin \omega t = I_o \sin 2\pi f t$$

Because it also continuously changes over time, AC current can be expressed as an effective or root mean square (rms) current, given by

$$I = I_{\text{rms}} = \frac{I_{\text{max}}}{\sqrt{2}}$$

EXAMPLE: Determine the maximum or peak voltage from an AC rms voltage of 120 V.

SOLUTION: The maximum voltage is related to the rms voltage by

$$V_{\text{max}} = \sqrt{2}V_{\text{rms}} = \sqrt{2} \times 120 \text{ V} = 169.7 \text{ V}$$

OHM'S LAW FOR AC CIRCUITS

Ohm's law for an AC circuit exists in four distinct forms, corresponding to the type and arrangement of circuit elements:

- **Pure resistance:** For an AC circuit with a resistor connected to an oscillating power source, the rms voltage V_{rms} is related to the rms current I_{rms} according to

$$V_{\text{rms}} = I_{\text{rms}}R$$

- **Pure inductance: Inductance**, L, represents another circuit element that impedes voltage and is caused by a reversed emf, induced by changing magnetic fields as the voltage rises and falls in a wire. The SI unit of inductance is the henry (H). Given an AC circuit with an inductor connected to an oscillating voltage source, the rms voltage is given by

$$V_{\text{rms}} = I_{\text{rms}}X_L$$

where X_L is the inductive reactance. The inductive reactance represents the ability of the inductor to resist the flow of an AC circuit and is defined by

$$X_L = 2\pi f L = \omega L$$

Here, X_L is expressed in units of ohms, given that L is in henrys, ω is in rad/s, and f is in hertz.

➤ **Pure capacitance:** Capacitance represents the ability to store charge. Given an AC circuit with a capacitor connected to an oscillating voltage source, the rms voltage is given by

$$V_{\text{rms}} = I_{\text{rms}} X_C$$

where X_C is the capacitive reactance, defined by

$$X_C = \frac{1}{2\pi f C}$$

Here, X_C is expressed in units of ohms, given that C is in farads and f is in hertz.

➤ **Resistance, inductance, and capacitance in combination:** Given an AC circuit with a resistor, capacitor, and inductor connected in series to an oscillating voltage source, as shown in Figure 13-9, the rms voltage is given by

$$V_{\text{rms}} = I_{\text{rms}} Z$$

where Z is the impedance, in ohms, defined by

$$Z = \sqrt{R^2 + (X_L - X_C)^2}$$

Impedance represents the total resistance or opposition to current flow in AC circuits.

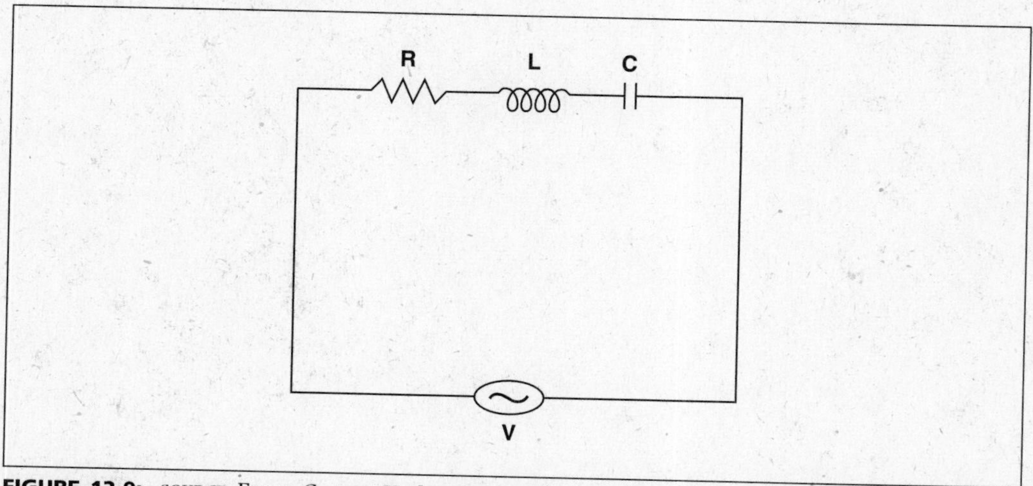

FIGURE 13-9: SOURCE: From George Hademenos, *Schaum's Outline of Physics for Pre-Med, Biology, and Allied Health Students*, McGraw-Hill, 1998; reproduced with permission of The McGraw-Hill Companies.

Continuously Changing Voltage and Current Values

Because the voltage and current are oscillatory in alternating current circuits, their values continuously change with respect to phase and are dependent on the circuit element:

Pure resistance	Voltage and current are in phase
Pure inductance	Voltage leads the current by 1/4 cycle, or 90°
Pure capacitance	Voltage lags the current by 1/4 cycle, or 90°

The phase angle ϕ which describes the phase between the voltage, V, and current, I, is

$$\tan \phi = \frac{X_L - X_C}{R}$$

Power Dissipated in an AC Circuit

Voltage from an AC generator through an AC circuit with any type of impedance gives rise to a current I through the impedance, distinguished by the phase angle ϕ between V and I. The average power P_{avg} dissipated by the impedance is given by

$$P_{avg} = I_{rms} V_{rms} \cos \phi$$

For an AC circuit with pure resistance, $\cos \phi = 1$, and $\cos \phi = 0$ for an AC circuit with either pure capacitance or pure inductance.

CRAM SESSION

Electric Circuits

Direct current (DC) circuits: Electricity in the form of electric charge that is generated by a voltage source (e.g., a battery) and flows as current through an arrangement of circuit elements or devices (e.g., a lightbulb, an alarm, a motor) that are all connected by a conductor, or a material that allows electric charge to flow easily through it. These circuit elements may be connected in series or parallel.

- In a series circuit, two or more circuit elements are connected in sequence.
- In a parallel circuit, two or more circuit elements are connected in a branching arrangement.

Current: The rate of motion of electric charge and is expressed as

$$\text{Current} = \frac{\text{electric charge moving through a region}}{\text{time required to move the charge}} \quad \text{or} \quad I = \frac{\Delta q}{\Delta t}$$

where current is measured in SI units of amperes, abbreviated A.

Voltage: Potential difference generated between a positive terminal *a* and a negative terminal *b* of a battery or power source when no current is flowing. Voltage, also referred to as electromotive force, is expressed in units of volts (V).

Resistance: An inherent property of a conductor that resists flow of electric current (I) and represents a measure of the potential difference V that must be supplied to a circuit to drive current through the circuit. Resistance, R, is defined as

$$\text{Resistance} = \frac{\text{potential difference}}{\text{current}} \quad \text{or} \quad R = \frac{V}{I}$$

Resistance is expressed in units of volts per ampere, also known as the ohm (Ω).

Ohm's law: States that the voltage V across a resistor R is proportional to the current I through it and can be written in equation form as $V = IR$.

Capacitors: Circuit elements that store charge and consist typically of two conductors of arbitrary shape carrying equal and opposite charges separated by an insulator. Capacitance C depends on the shape and position of the capacitors and is defined as $C = \frac{q}{V}$ where q is the magnitude of charge on either of the two conductors and V is the magnitude of potential difference between the two conductors. The SI unit of capacitance is coulomb/volt, collectively known as the farad (F).

Summary of Circuit Element Quantities of DC Circuits		
Circuit Element	**Series**	**Parallel**
Voltage	$V_{net} = V_1 + V_2 + V_3 + \cdots + V_n$	$V_{net} = V_1 = V_2 = V_3 = \cdots = V_n$
Current	$I_{net} = I_1 = I_2 = I_3 = \cdots = I_n$	$I_{net} = I_1 + I_2 + I_3 + \cdots + I_n$
Resistance	$R_{eq} = R_1 + R_2 + R_3 + \cdots + R_n$	$\frac{1}{R_{eq}} = \frac{1}{R_1} + \frac{1}{R_2} + \frac{1}{R_3} + \cdots + \frac{1}{R_n}$
Capacitance	$\frac{1}{C_{eff}} = \frac{1}{C_1} + \frac{1}{C_2} + \frac{1}{C_3} + \cdots + \frac{1}{C_n}$	$C_{eff} = C_1 + C_2 + C_3 + \cdots + C_n$

Electric power: Power dissipated by current flow into the resistor is given by

$$P = IV = I^2R = \frac{V^2}{R}$$

The unit for electric power is the watt, abbreviated W.

Kirchhoff's rules of circuit analysis: Two rules for calculating the current and voltage through circuit elements in complex circuits are

- **Node or junction rule:** The algebraic sum of all currents entering a junction or node must be equal to the sum of all currents exiting the node.
- **Loop or circuit rule:** The algebraic sum of potential drops encountered while traversing the circuit in either a counterclockwise or a clockwise direction must be equal to zero.

Alternating current (AC) circuits: Similar to direct current (DC) circuits with one exception: the type of voltage source driving current through the circuit. In DC circuits, the voltage source is continuous and constant, resulting in current flow in only one direction, while in AC circuits, the voltage is oscillatory or sinusoidal, resulting in periodic changes in direction of current flow.

Fundamental quantities of AC circuits

- **AC voltage** is sinusoidal and can be expressed mathematically as

  $$V = V_o \sin \omega t = V_o \sin 2\pi f t$$

  where V is the instantaneous voltage (V); V_o is the maximum value of the voltage (V); ω is the angular velocity (rad/s); f is the frequency (Hz); and t is time (s).

  AC voltage (root mean square): Given by

  $$V = V_{rms} = \frac{V_{max}}{\sqrt{2}}$$

- **AC current** is sinusoidal and can be expressed mathematically as

$$I = I_o \sin \omega t = I_o \sin 2\pi f t$$

where I is the instantaneous current (A); I_o is the maximum value of the current (A); ω is the angular velocity (rad/s); f is the frequency (Hz); and t is time (s).

AC current (root mean square): Given by

$$I = I_{rms} = \frac{I_{max}}{\sqrt{2}}$$

Ohm's law for AC circuits: Ohm's law for AC circuits exists in four distinct forms, corresponding to the type and arrangement of circuit elements:

Pure resistance: $V_{rms} = I_{rms} R$

Pure inductance: $V_{rms} = I_{rms} X_L$ where X_L is the inductive reactance given by $X_L = 2\pi f L = \omega L$

Pure capacitance: $V_{rms} = I_{rms} X_C$ where X_C is the capacitive reactance, defined by $X_C = \dfrac{1}{2\pi f C}$

Resistance, inductance, and capacitance in combination: $V_{rms} = I_{rms} Z$ where Z is the impedance, defined by $Z = \sqrt{R^2 + (X_L - X_C)^2}$

CHAPTER 14

Magnetism

Read This Chapter to Learn About:

- Magnetic Field
- Magnetic Force on a Charged Particle in Motion
- Magnetic Force on a Current-Carrying Wire

The working principles of a television monitor, the physical properties of Earth and the basis for the most effective navigational tool in history (the compass) all have a common thread—**magnetism**. A property of matter discovered by the ancient Chinese in the mineral lodestone, magnetism is involved in a wide spectrum of applications in science and technology and represents an important subject in physics. This chapter covers the basics of magnetism.

MAGNETIC FIELD

Magnetic field B is an attractive or repulsive force field generated by a moving charged particle. For a standard bar magnet with a north pole (where the lines of force begin) and a south pole (where the lines of force end), the magnetic field is attractive for opposite poles and repulsive for like poles.

Magnetic field is a vector quantity. The unit of magnetic field is the tesla (T), where

$$1 \text{ tesla} = 1 \frac{\text{newton}}{\text{ampere} \cdot \text{meter}} = 1 \frac{\text{weber}}{(\text{meter})^2}$$

Magnetic fields can also be measured in units of gauss (G), a unit used for measuring smaller magnetic fields. These two units are related through the conversion ratio $1 \text{ T} = 1 \times 10^4 \text{ G}$.

Magnetic fields are created in the presence of moving charges or current. You can calculate the magnitude of the magnetic field and determine its direction based on the behavior or configuration of the current. Current can flow through a long, straight wire; through a wire loop or coil; through a long solenoid (or a long, straight cylinder consisting of many loops of wire wrapped around the cylinder); or through a toroid (or a long, cylinder of wire coils that is bent in the form of a doughnut). If current is flowing through each of these four configurations, the magnitude of the magnetic field can be calculated according to the formulas noted in Figure 14-1, where the constant in each of the equations, $\mu_o = 4\pi \times 10^{-7}$ T · m/A, is termed the **permeability of free space**. The direction of the magnetic field can be determined using the **right-hand rule**, as depicted in the figure.

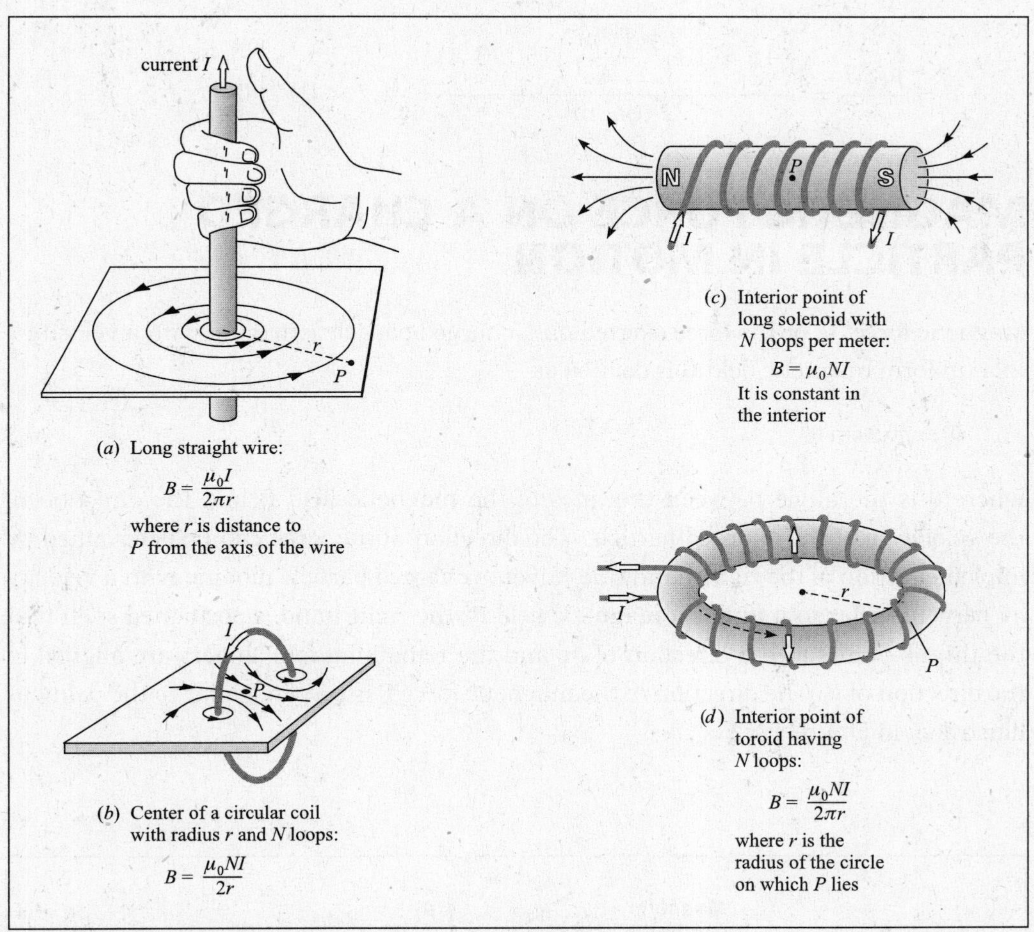

FIGURE 14-1: SOURCE: From Frederick J. Bueche and Eugene Hecht, *Schaum's Outline of College Physics*, 10th ed., McGraw-Hill, 2006; reproduced with permission of The McGraw-Hill Companies.

EXAMPLE: Earth can produce magnetic fields as high as 600 milligauss (mG). Express this value of magnetic field in terms of teslas.

SOLUTION: The units of tesla and gauss are related according to

$$1\ \text{T} = 10^4\ \text{G}$$

Therefore,

$$600 \text{ mG} = \left(600 \times 10^{-3} \text{ G}\right)\left(\frac{1 \text{ T}}{10^4 \text{ G}}\right) = 6 \times 10^{-5} \text{ T}$$

EXAMPLE: A circular conducting coil of diameter 0.4 m has 50 loops of wire and a current of 3 A flowing through it. Determine the magnetic field generated by the coil.

SOLUTION: The equation for the magnetic field in the center of a circular loop or coil is

$$B = \frac{\mu_0 NI}{2r} = \frac{\left(4\pi \times 10^{-7} \frac{\text{T} \cdot \text{m}}{\text{A}}\right)(50)(3 \text{ A})}{2\,(0.2 \text{ m})} = 4.7 \times 10^{-4} \text{ T}$$

MAGNETIC FORCE ON A CHARGED PARTICLE IN MOTION

Magnetic force, $\vec{\mathbf{F}}$, or the force exerted on a charged particle q moving with a velocity $\vec{\mathbf{v}}$ in a uniform magnetic field **B** is defined as

$$F = qvB \sin \theta$$

where θ is the angle between the lines of the magnetic field **B** and the direction of the velocity v of the charged particle. The direction of the force can be determined by implementation of the right-hand rule. Given a charged particle moving with a velocity $q\mathbf{v}$ perpendicular to a uniform magnetic field **B**, the right hand is positioned such that the thumb points in the direction of $q\mathbf{v}$ and the remaining four fingers are aligned in the direction of **B**. The direction of the magnetic force **F** is perpendicular to the palm, as illustrated in Figure 14-2.

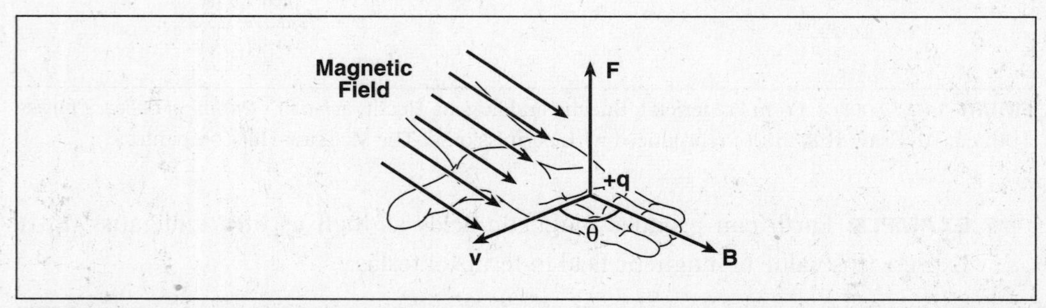

FIGURE 14-2: SOURCE: From George Hademenos, *Schaum's Outline of Physics for Pre-Med, Biology, and Allied Health Students*, McGraw-Hill, 1998; reproduced with permission of The McGraw-Hill Companies.

MAGNETIC FORCE ON A CURRENT-CARRYING WIRE

The magnetic force exerted on a current-carrying wire of current I of length L placed in a uniform magnetic field **B** is defined as

$$F = ILB \sin \theta$$

where L is the length of the wire (conductor) and θ is the angle between the current and the magnetic field. The direction of the magnetic force on the wire can be found by orienting the thumb of the right hand along the axis of the wire with the remaining fingers in the direction of the magnetic field. The magnetic force is directed upward from the aligned palm, as shown in Figure 14-3.

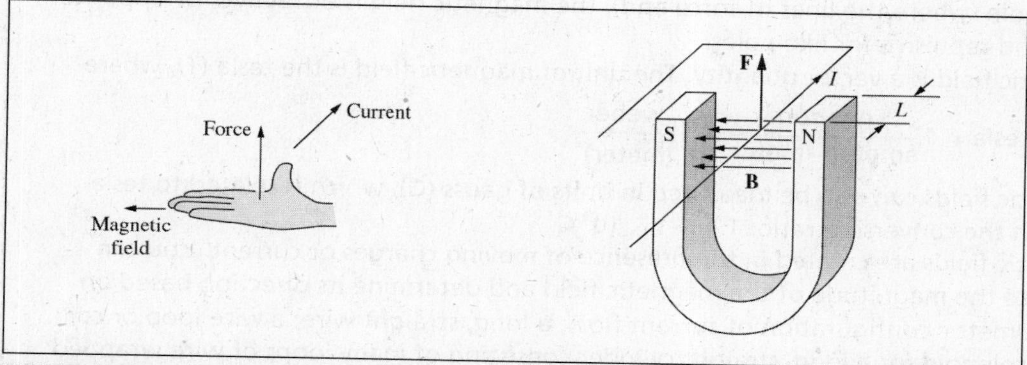

FIGURE 14-3: SOURCE: From Arthur Beiser, *Schaum's Outline of Applied Physics*, 4th ed., McGraw-Hill, 2004; reproduced with permission of The McGraw-Hill Companies.

EXAMPLE: A wire of length 40 centimeters (cm) carrying a current of 30 A is positioned at an angle of 50° to a uniform magnetic field of 10.0×10^{-4} watts per square meter (W/m^2). Determine the magnitude and direction of the force exerted on this wire.

SOLUTION: The magnetic force **F** exerted on the wire of length L with current I placed in a magnetic field **B** is given by:

$$F = ILB \sin \theta$$

$$= (30 \text{ A})(40 \times 10^{-2} \text{ m})(10.0 \times 10^{-4} \text{ W/m}^2)(\sin 50°) = 9.2 \times 10^{-3} \text{ N}$$

Using the right-hand rule where the fingers are aligned with the magnetic field lines and the thumb is aligned with the current direction, the force is directed perpendicularly into the page.

CRAM SESSION

Magnetism

Magnetism: A property of matter of some materials that result in an attractive or repulsive force; it was discovered by the ancient Chinese in the mineral lodestone.

Magnetic field B: An attractive or repulsive force field generated by a moving charged particle.

- For a standard bar magnet with a north pole (where the lines of force begin) and a south pole (where the lines of force end), the magnetic field is attractive for opposite poles and repulsive for like poles.
- Magnetic field is a vector quantity. The unit of magnetic field is the tesla (T), where

$$1 \text{ tesla} = 1\frac{\text{newton}}{\text{ampere} \cdot \text{meter}} = 1\frac{\text{weber}}{(\text{meter})^2}$$

- Magnetic fields can also be measured in units of gauss (G), which is related to tesla through the conversion ratio: $1\text{ T} = 1 \times 10^4\text{ G}$.
- Magnetic fields are created in the presence of moving charges or current. You can calculate the magnitude of the magnetic field and determine its direction based on the geometric configuration of current flow: a long, straight wire; a wire loop or coil; a long solenoid (or a long, straight cylinder consisting of many loops of wire wrapped around the cylinder); or through a toroid (or a long cylinder of wire coils that is bent in the form of a doughnut), as shown in the figure below.

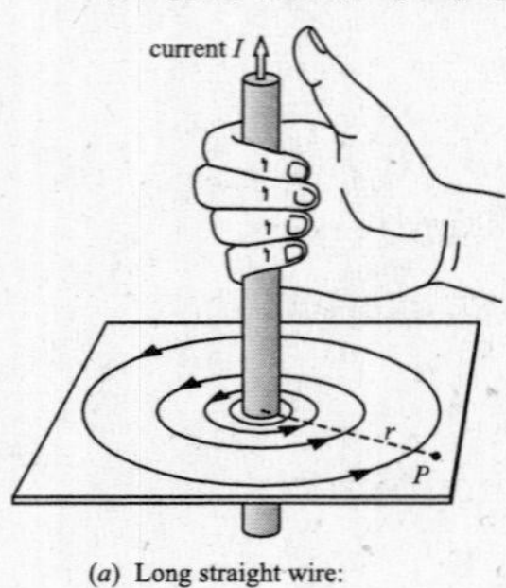

(*a*) Long straight wire:

$$B = \frac{\mu_0 I}{2\pi r}$$

where r is distance to P from the axis of the wire

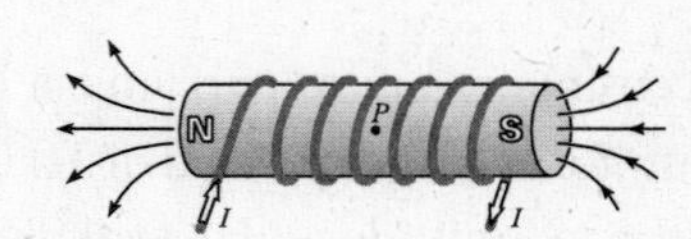

(*c*) Interior point of long solenoid with N loops per meter:

$$B = \mu_0 NI$$

It is constant in the interior

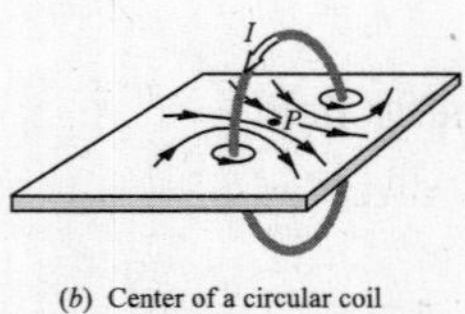

(*b*) Center of a circular coil with radius r and N loops:

$$B = \frac{\mu_0 NI}{2r}$$

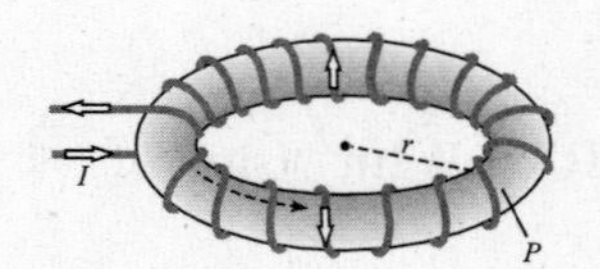

(*d*) Interior point of toroid having N loops:

$$B = \frac{\mu_0 NI}{2\pi r}$$

where r is the radius of the circle on which P lies

SOURCE: From Frederick J. Bueche and Eugene Hecht, *Schaum's Outline of College Physics*, 10th ed., McGraw-Hill, 2006; reproduced with permission of The McGraw-Hill Companies.

Magnetic force F: Force exerted on a charged particle q moving with a velocity **v** in a uniform magnetic field **B** is defined as

$$F = qvB \sin \theta$$

where θ is the angle between the lines of the magnetic field **B** and the direction of the velocity **v** of the charged particle.

➤ The direction of the force can be determined by implementation of the **right-hand rule**. Given a charged particle moving with a velocity q**v** perpendicular to a uniform magnetic field **B**, the right hand is positioned such that the thumb points in the direction of q**v** and the remaining four fingers are aligned in the direction of **B**. The direction of the magnetic force **F** is perpendicular to the palm, as shown below.

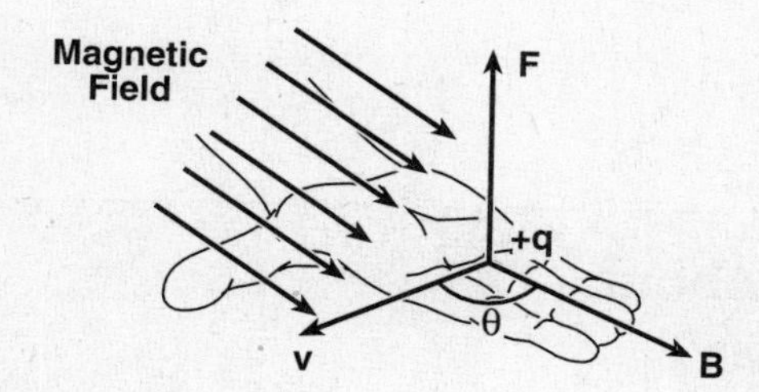

SOURCE: From George Hademenos, *Schaum's Outline of Physics for Pre-Med, Biology, and Allied Health Students*, McGraw-Hill, 1998; reproduced with permission of The McGraw-Hill Companies.

Magnetic force on a current-carrying wire: The magnetic force exerted on a current-carrying wire of current I of length L placed in a uniform magnetic field **B** is defined as

$$F = ILB \sin \theta$$

where L is the length of the wire (conductor) and θ is the angle between the current and the magnetic field.

➤ The direction of the magnetic force on the wire can be found by orienting the thumb of the right hand along the axis of the wire with the remaining fingers in the direction of the magnetic field. The magnetic force is directed upward from the aligned palm, as shown in the figure below.

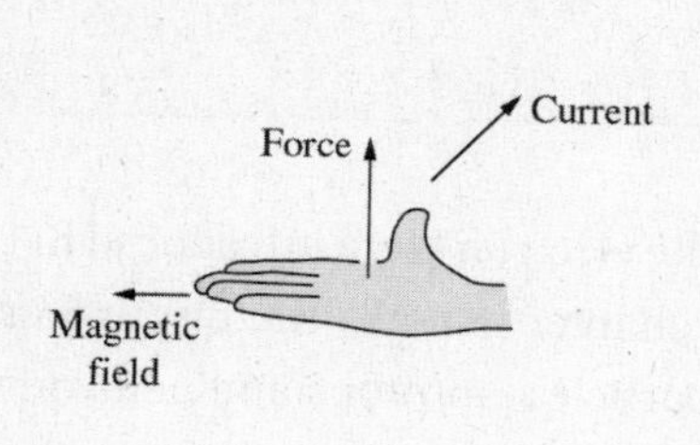

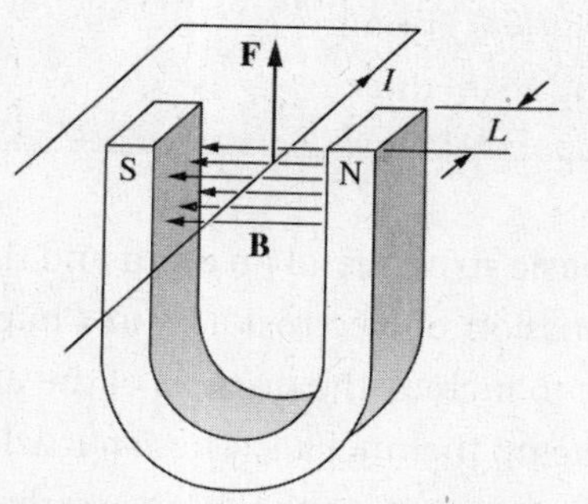

SOURCE: From Arthur Beiser, *Schaum's Outline of Applied Physics*, 4th ed., McGraw-Hill, 2004; reproduced with permission of The McGraw-Hill Companies.

CHAPTER 15

Atomic and Nuclear Physics

Read This Chapter to Learn About:

ATOMIC STRUCTURE AND PHENOMENA

- ➤ Wave-Particle Duality of Light
- ➤ Blackbody Radiation
- ➤ Photoelectric Effect
- ➤ Bohr Model of the Hydrogen Atom
- ➤ Fluorescence

NUCLEAR STRUCTURE AND PHENOMENA

- ➤ Nuclear Structure
- ➤ Categorization of Elements
- ➤ Isotopes
- ➤ Radioactive Decay
- ➤ Nuclear Binding Energy
- ➤ Nuclear Fission
- ➤ Nuclear Fusion

The basic structure of an atom and the physics of the electron were introduced in the discussion of electrostatics in Chapter 12. This chapter extends the discussion of the atom to include the nucleus of the atom and the particles—protons and neutrons—that make up the nucleus. These particles are responsible for the energy and stability of the atom and whether the atom is radioactive.

ATOMIC STRUCTURE AND PHENOMENA

WAVE-PARTICLE DUALITY OF LIGHT

It seems reasonable to believe that light, or electromagnetic waves, should behave like waves and only like waves. Wavelike behavior refers to experimental effects seen in interference and diffraction (discussed in Chapter 11). It turns out that, as evidenced from a series of landmark experiments and observations, light can also exhibit particle-like behavior. This is the basis for the **wave-particle duality of light.** When electromagnetic waves interact with matter such as atoms, light behaves as a beam of small packets or bundles of light referred to as **photons,** or **quanta of light.** Photon energy, E, is related to its frequency, ν, or its wavelength, λ, by the following relation:

$$E = h\nu = h\frac{c}{\lambda}$$

where h is known as Planck's constant ($h = 6.626 \times 10^{-34}$ J · s) and c is the speed of light, $c = 3 \times 10^{8}$ m/s.

Two of the landmark experiments and observations referred to earlier in defining the wave-particle duality of light—blackbody radiation and the photoelectric effect—were fundamental in describing the processes that occur at the atomic level and are described next.

BLACKBODY RADIATION

Blackbody radiation involves the electromagnetic radiation emitted by a blackbody. Suppose light is allowed to enter a hollow object through a small opening. Part of the light is absorbed by the interior walls of the object, and part of the light is reflected between the walls. The reflected light continues to be reflected among the inner walls until it too is absorbed by the walls. However, in addition to absorbing the energy from the electromagnetic radiation, the hollow object, or blackbody, also emits energy. A **blackbody** is, by definition, a perfect absorber and emitter of radiation energy. In fact, at thermal equilibrium, the blackbody emits energy equal to the amount of energy that it absorbs.

It was the emission of radiation that prompted questions and further investigations: How much radiation energy (intensity) is emitted by a blackbody, and what are the wavelengths of the emitted radiation? It was found that the intensity and the spectrum of wavelengths of the emitted radiation depend on temperature, above 700 K, as represented in Figure 15-1. The experimental observations of a blackbody radiator—namely, the spectrum of wavelengths of emitted radiation—could be

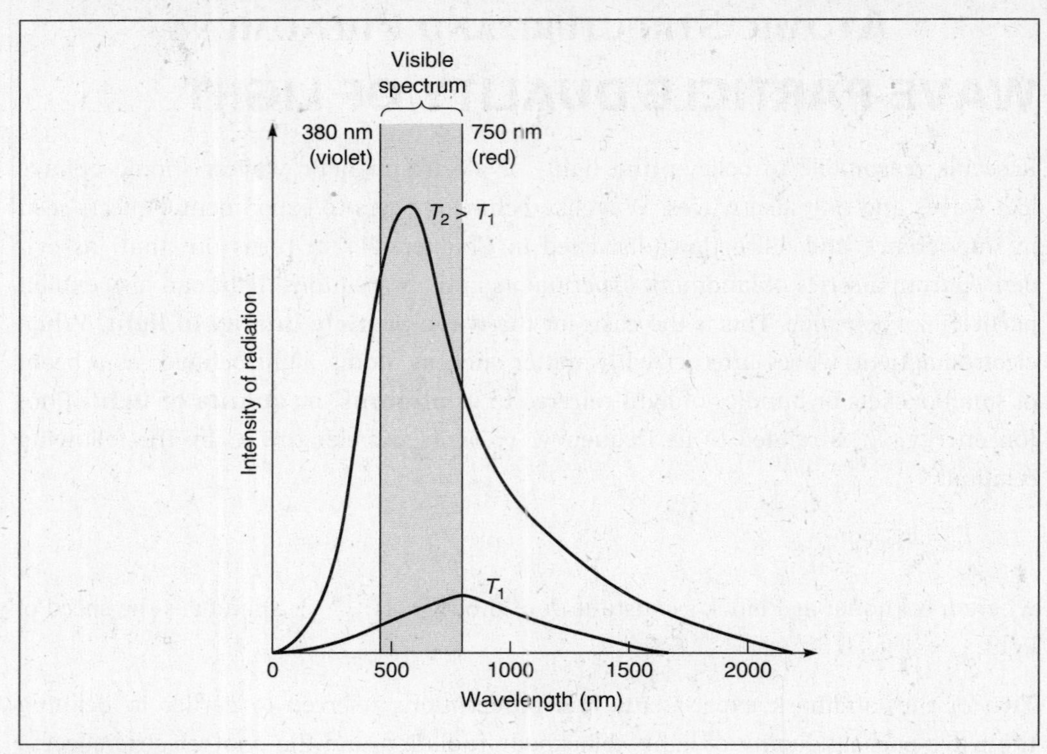

FIGURE 15-1

explained only if the radiation energy were absorbed by and emitted from the blackbody in discrete amounts of energy. This observation led to the derivation of the equation

$$E = h\nu = h\frac{c}{\lambda}$$

PHOTOELECTRIC EFFECT

Electrons are bound to the surface of a metal with an energy known as the **work function.** The work function, W_{min}, represents the minimum amount of work required to liberate the electron from the surface. When a beam of light strikes the surface of the metal, the energy of the photon is transferred to the electron. If the photon energy ($E = h\nu$) is greater than the work function of the bound electron, electrons (or photoelectrons) are ejected from the surface with a maximum kinetic energy, KE_{max}, given by

$$\text{KE}_{\text{max}} = h\nu - W_{\text{min}}$$

This is referred to as the **photoelectric effect.**

BOHR MODEL OF THE HYDROGEN ATOM

In 1913, physicist Niels Bohr advanced a model of the hydrogen atom in which an electron moves in a circular orbit around the proton in the nucleus. It was explained that the attractive electric force between the positively charged proton and the negatively charged electron kept the electron in circular orbit. In Bohr's model of the atom, the electron could only be found in stable orbits or discrete energy states where no electromagnetic radiation was emitted by the atom. According to Bohr, radiation energy was emitted by the atom when the electron jumped between energy states. This model provided an explanation of atomic spectra.

In the Bohr atom, the lowest energy level, or **ground state** (characterized by the integer $n = 1$), required for the electron to maintain a circular orbit closest to the nucleus is -13.6 electron volts (eV). In order for an electron to orbit in excited energy states about the nucleus, energy must be given to the electron. Energy is provided in the form of electromagnetic radiation or light. When an electron absorbs electromagnetic radiation of a certain frequency, it jumps to the corresponding excited ground state with an energy equivalent to the frequency. When the electron returns to its ground state, the electron emits electromagnetic radiation of frequency equal to the energy difference between the two energy states. The relation for energy levels in a Bohr atom is given by

$$E = -(13.6\text{ eV})\frac{Z^2}{n^2} \qquad n = 1, 2, 3, \ldots$$

and is depicted for the hydrogen atom ($Z = 1$) in Figure 15-2.

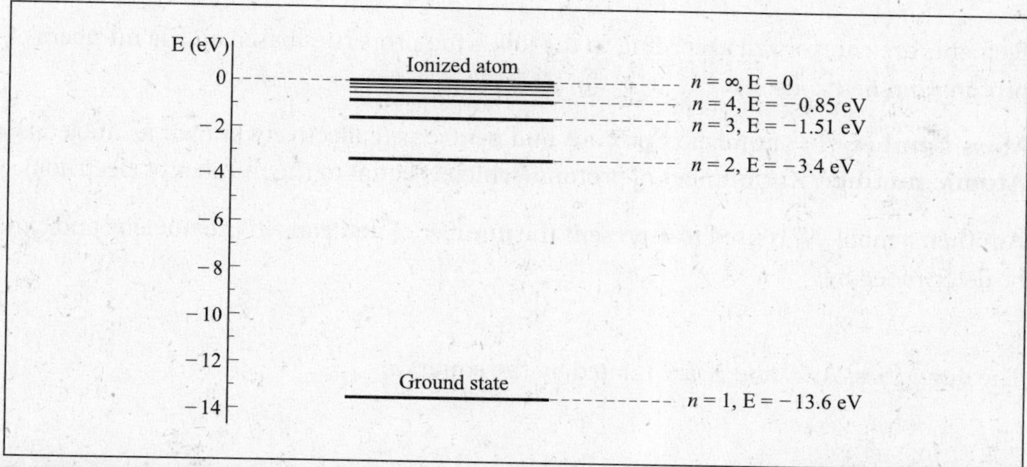

FIGURE 15-2: SOURCE: From Frederick J. Bueche and Eugene Hecht, *Schaum's Outline of College Physics*, 10th ed., McGraw-Hill, 2006; reproduced with permission of The McGraw-Hill Companies.

FLUORESCENCE

Fluorescence occurs when certain atoms or molecules absorb light at a particular wavelength and, after a very brief time interval, emit light at longer wavelengths. The light that is absorbed is typically ultraviolet light (high frequency/smaller wavelength), whereas the emitted light is at wavelengths longer than those of ultraviolet light and may possibly be within the visible region of the electromagnetic spectrum.

NUCLEAR STRUCTURE AND PHENOMENA

NUCLEAR STRUCTURE

The smallest building block, or unit, of all matter is the **atom.** All atoms consist of three basic particles: **protons, neutrons,** and **electrons.** The protons and neutrons are tightly packed in a positively charged nucleus situated at the center of the atom, and the electrons are in continual orbit around the nucleus. The protons and neutrons, collectively known as nucleons, account for the majority of the mass of the atom. The proton charge is identical in magnitude to that of the electron but is positive, whereas the electron is negatively charged. The neutron is neutrally charged (uncharged) and is slightly heavier than the proton.

Mass of the proton	$m_p = 1.67 \times 10^{-27}$ kg
Mass of the neutron	$m_n = 1.68 \times 10^{-27}$ kg
Mass of the electron	$m_e = 9.11 \times 10^{-31}$ kg

CATEGORIZATION OF ELEMENTS

Elements are categorized according to the following properties based on the numbers of protons and neutrons:

Mass number A = number of protons and neutrons (collectively known as nucleons)
Atomic number Z = number of protons (which is equal to the number of electrons)

Another symbol, N, is used to represent the number of neutrons in the nucleus and can be determined by

$$N = A - Z$$

The quantities, A, Z, and N are related by the equation

$$A = N + Z$$

In symbolic form, an atom is represented by

$$^{A}_{Z}X$$

For example, consider the elements carbon $^{12}_{6}C$, sulfur $^{32}_{16}S$, and potassium $^{39}_{19}K$.

Carbon has 12 protons and neutrons ($A = 12$), 6 protons ($Z = 6$), and 6 neutrons ($N = 12 - 6 = 6$). Sulfur has 32 protons and neutrons ($A = 32$), 16 protons ($Z = 6$), and 16 neutrons ($N = 32 - 16 = 16$). Potassium has 39 protons and neutrons ($A = 39$), 19 protons ($Z = 19$), and 20 neutrons ($N = 39 - 19 = 20$).

ISOTOPES

Isotopes represent the class of nuclei with the same number of protons (Z) but different number of neutrons (N) and thus of nucleons (A). They usually do not decay into different nuclei. As an example, the nuclides $^{1}_{1}H$, $^{2}_{1}H$ (deuterium), and $^{3}_{1}H$ (tritium) are all isotopes of hydrogen. The nuclides $^{10}_{6}C$, $^{11}_{6}C$, $^{12}_{6}C$, $^{13}_{6}C$, $^{14}_{6}C$, and $^{15}_{6}C$ are all isotopes of carbon.

RADIOACTIVE DECAY

Radioactive decay is a nuclear phenomenon exhibited by radioactive isotopes or elements with an atomic number Z greater than that of lead ($Z = 82$). In these elements, which contain generally more neutrons (N) than protons (Z), the repulsive electric forces in the nucleus become greater than the attractive nuclear forces, making the nuclei unstable. In nature, the radioactive element strives toward a stabilized state of existence and, in the process, spontaneously emits particles (photons and charged and uncharged particles), and in so doing transforms to a different nucleus and hence a different element. This process is referred to as radioactive decay and is dependent on the amount and identity of the radioactive element.

Given a radioactive element originally with N_o number of atoms, the number of atoms present at any time t is

$$N(t) = N_o e^{-\lambda t}$$

where λ is a decay constant, defined by

$$\lambda = \frac{0.693}{T_{1/2}}$$

Here, $T_{1/2}$ is the **half-life** of the radioactive element and represents the time required for one-half of the radioactive atoms to remain unchanged. Half-lives for radioactive elements range from fractions of seconds (e.g., polonium-212 [^{212}Po], $T_{1/2} = 3 \times 10^{-7}$ s) to billions of years (e.g., uranium-238 [^{238}U], $T_{1/2} = 4.5 \times 10^{9}$ yr). Radioactive decay is an exponential curve and is illustrated in Figure 15-3.

Radionuclides typically undergo radioactive decay of three common types.

- **Alpha decay**, caused by the repulsive electric forces between the protons, involves the emission of an alpha particle (or a helium nucleus that consists of two protons

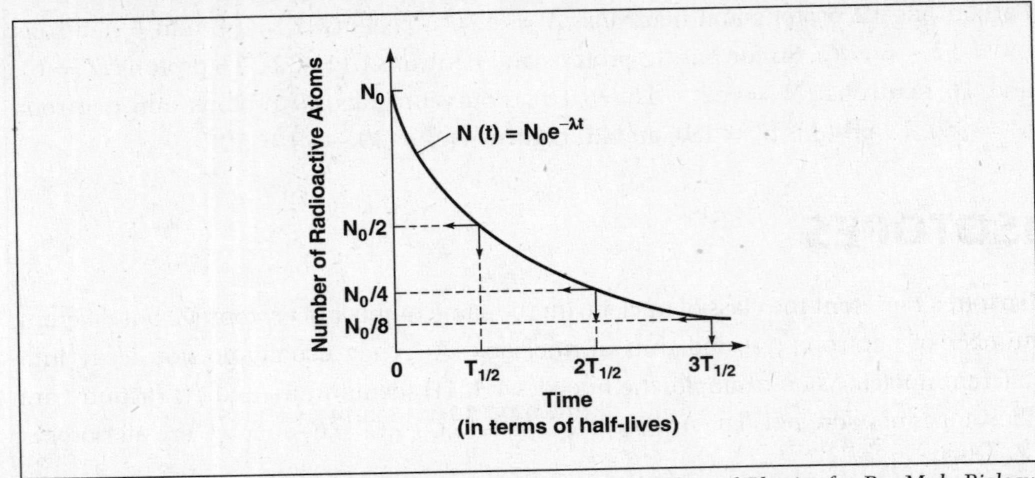

FIGURE 15-3: SOURCE: From George Hademenos, *Schaum's Outline of Physics for Pre-Med, Biology, and Allied Health Students*, McGraw-Hill, 1998; reproduced with permission of The McGraw-Hill Companies.

and two neutrons) by nuclei with many protons. In alpha decay, the radioactive nucleus **decreases** in *A* (mass number) by 4 and **decreases** in Z (number of protons) by 2.

- **Beta decay**, which occurs in nuclei that have too many neutrons, can occur by emission of a β particle. A β^- particle is an electron, and a β^+ particle is a positively charged electron, known as a **positron.** In β^- decay, the radioactive nucleus **remains unchanged** in *A* and **increases** in Z by 1. In β^+ decay, the radioactive nucleus **remains unchanged** in *A* and **decreases** in Z by 1.
- **Gamma decay** occurs by the emission of highly energetic photons. In gamma decay, the radioactive nucleus **remains unchanged** in *A* and **remains unchanged** in Z.

NUCLEAR BINDING ENERGY

Applied to the atomic nucleus, the conservation of mass implies that the mass of the nucleus is equal to the sum of the masses of protons and neutrons, the particles that are housed in the nucleus. However, this is not the case with the rest mass of a nucleus, which is less than the sum of the rest masses of the protons and neutrons. The reason for this difference in mass, referred to as the **mass defect,** is that negative energy is required to bind the individual proton and neutron particles within the nucleus. This energy, referred to as the **nuclear binding energy,** is given by Einstein's equation that describes the conversion between mass and energy, or

$$\text{Nuclear binding energy} = \left\{(Zm_p)c^2 + (Nm_n)c^2\right\} - M_{\text{nuc}}c^2$$

where $M_{\text{nuc}}c^2$ is the mass defect.

NUCLEAR FISSION

Nuclear fission is a process in which a large nucleus (generally with an atomic mass number, $A > 230$) is bombarded with a neutron and, as a result, splits into two or three medium-size nuclei as well as additional neutrons. An example of a nuclear fission reaction involves uranium:

$$^{235}_{92}\mathrm{U} + {}^{1}_{0}n \rightarrow \left[{}^{236}_{92}\mathrm{U}\right] \rightarrow {}^{141}_{56}\mathrm{Ba} + {}^{92}_{36}\mathrm{Kr} + \kappa\,{}^{1}_{0}n$$

Uranium + neutron → Compound nucleus → Barium + Krypton + Additional neutrons

where κ is an integer, representing an integer number of neutrons that could result from a fission reaction.

NUCLEAR FUSION

Nuclear fusion is a process in which two small nuclei (generally those with $A < 20$) combine or fuse to form a single, larger nucleus. Examples of fusion reactions are

- Fusion of a proton and a neutron to form a deuteron or ${}^{1}_{1}\mathrm{H} + {}^{1}_{0}n \rightarrow {}^{2}_{1}\mathrm{H}$
- Fusion of two deuterons to form an α-particle or ${}^{2}_{1}\mathrm{H} + {}^{2}_{1}\mathrm{H} \rightarrow {}^{4}_{2}\mathrm{He}$

CRAM SESSION

Atomic and Nuclear Physics

Atomic Structure and Phenomena

Wave-particle duality of light

Physical behavior of light as a wave and as a particle.

Experimental evidence of light as waves: Interference and diffraction.

Experimental evidence of light as particles: Blackbody radiation and photoelectric effect. Light behaves as a beam of small packets or bundles of light referred to as **photons** or **quanta of light.**

Photon energy *E*: Related to its frequency, ν, or its wavelength, λ, by the following relation:

$$E = h\nu = h\frac{c}{\lambda}$$

where h is known as Planck's constant ($h = 6.626 \times 10^{-34}$ J · s) and c is the speed of light, $c = 3 \times 10^{8}$ m/s.

Blackbody radiation

Phenomenon involving the electromagnetic radiation emitted by a blackbody, which, by definition, is a perfect absorber and emitter of radiation energy. Suppose light is allowed to enter a hollow object (blackbody) through a small opening. Part of the light is absorbed by the interior walls, and part of the light is reflected between the walls. The reflected light continues to be reflected among the inner walls until it too is absorbed by the walls. However, in addition to absorbing the energy from the electromagnetic radiation, the blackbody also emits energy. In fact, at thermal equilibrium, the blackbody emits energy equal to the amount of energy that the object absorbed. It was the radiation emission that prompted questions and further investigations:

How much radiation energy (intensity) was emitted by the blackbody, and what were the wavelengths of the emitted radiation?

It was shown that the intensity and the spectrum of wavelengths of the emitted radiation depended on temperature, above 700 K, as shown in the figure on the next page. The experimental observations of a blackbody radiator (i.e., the spectrum of wavelengths of emitted radiation) could be explained only if the radiation energy was absorbed by and emitted from the blackbody in discrete amounts of energy—an observation that led to the derivation of the equation

$$E = h\nu = h\frac{c}{\lambda}$$

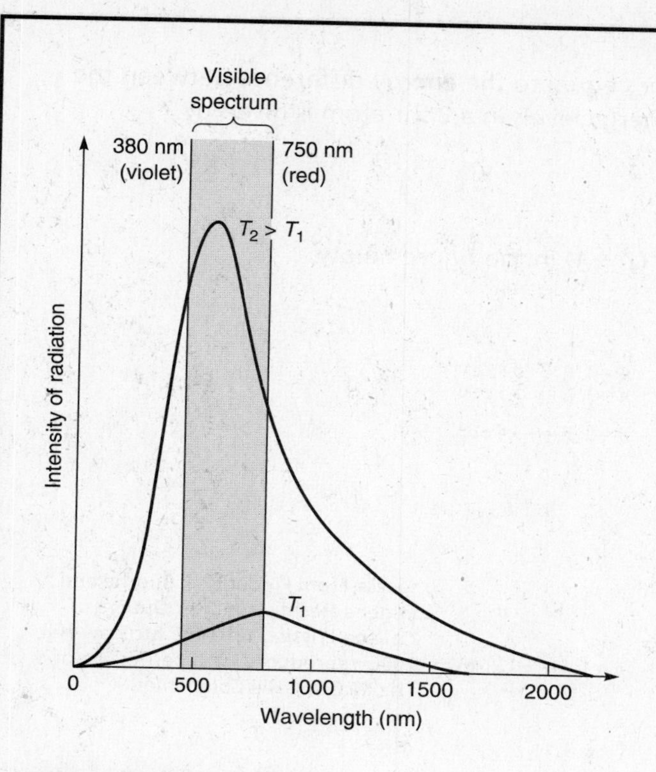

Photoelectric effect

On the surface of a metal, electrons are bound to the surface with an energy known as the **work function,** W_{min}, which represents the minimum amount of work required to liberate the electron from the surface. When a beam of light strikes the surface of the metal surface, the energy of the photon is transferred to the electron. If the photon energy ($E = h\nu$) is greater than the work function of the bound electron, electrons (or photoelectrons) are ejected from the surface with a maximum kinetic energy given by

$$KE_{max} = h\nu - W_{min}$$

This phenomenon is referred to as the photoelectric effect.

Bohr model of the hydrogen atom

In 1913, physicist Niels Bohr advanced a model of the hydrogen atom in which

- An electron moves around the proton in the nucleus in a circular orbit maintained by an attractive electric force between the positively charged proton and the negatively charged electron.
- The electron could only be found in stable orbits or discrete energy states where no electromagnetic radiation was emitted by the atom. Radiation energy was emitted by the atom when the electron jumped between energy states. This model provided an explanation of atomic spectra. The lowest energy level or ground state (characterized by the integer $n = 1$) required for the electron to maintain a circular orbit closest to the nucleus is -13.6 eV. In order for an electron to orbit in excited energy states about the nucleus, energy in the form of electromagnetic radiation must be given to the electron. When the electron returns to its ground state, the electron emits

electromagnetic radiation of frequency equal to the energy difference between the two energy states. The relation for energy levels in a Bohr atom is given by

$$E = -(13.6\ \text{eV})\frac{Z^2}{n^2} \qquad n = 1, 2, 3, \ldots$$

and is shown for the hydrogen atom ($Z = 1$) in the figure below.

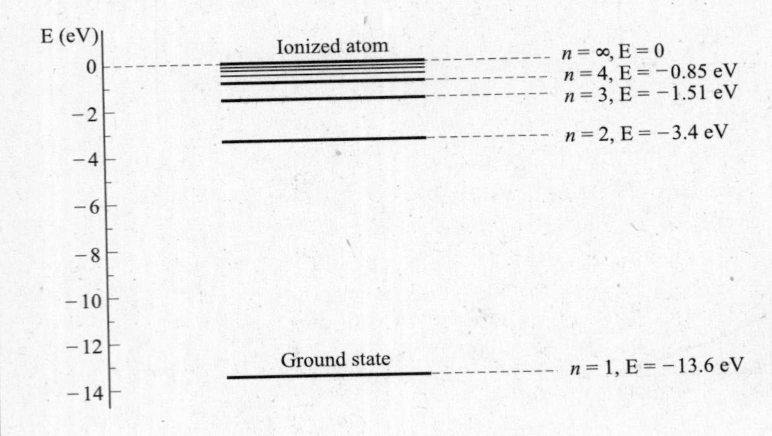

SOURCE: From Frederick J. Bueche and Eugene Hecht, *Schaum's Outline of College Physics*, 10th ed., McGraw-Hill, 2006; reproduced with permission of The McGraw-Hill Companies.

Nuclear Structure and Phenomena

Nuclear structure

The smallest building block or unit of all matter is the atom. All atoms consist of three basic particles: **protons, neutrons, and electrons.**

- **Protons** located in the nucleus.

  Mass of the proton: $m_p = 1.67 \times 10^{-27}$ kg

  Charge of the proton: $q_p = +1.6 \times 10^{-19}$ C

- **Neutrons** located in the nucleus.

  Mass of the neutron: $m_n = 1.68 \times 10^{-27}$ kg

  Charge of the neutron: uncharged

- **Electrons** in continual orbit about the nucleus.

  Mass of the electron: $m_e = 9.11 \times 10^{-31}$ kg

  Charge of the electron: $q_e = -1.6 \times 10^{-19}$ C

Elements are categorized according to the following properties based on the numbers of protons and neutrons:

- **Mass number** A = number of protons and neutrons (collectively known as nucleons)
- **Atomic number** Z = number of protons (which is equal to the number of electrons)
- **Neutron number** N = number of neutrons in the nucleus and can be determined by $N = A - Z$

In symbolic form, an atom is represented by:

$$^{A}_{Z}X$$

For example, carbon $^{12}_{6}C$ has 12 protons and neutrons ($A = 12$), 6 protons ($Z = 6$), and 6 neutrons ($N = 12 - 6 = 6$).

Isotopes

Represent the class of nuclei with the same number of protons (Z) but different number of nucleons (A). They usually do not decay into different nuclei. As an example, the nuclides $^{10}_{6}C$, $^{11}_{6}C$, $^{12}_{6}C$, $^{13}_{6}C$, $^{14}_{6}C$, and $^{15}_{6}C$ are all isotopes of carbon.

Radioactive decay

A nuclear phenomenon exhibited by radioisotopes or elements with an atomic number Z greater than that of lead ($Z = 82$). In these elements, which contain generally more N than Z, the repulsive electric forces in the nucleus become greater than the attractive nuclear forces, making the nuclei unstable. The radioactive element, in nature, strives toward a stabilized state of existence and, in the process, spontaneously emits particles (photons and charged and uncharged particles) in the transformation to a different nucleus and hence a different element. This process is referred to as radioactive decay and is dependent on the amount and identity of the radioactive element.

Given a radioactive element originally with N_o number of atoms, the number of atoms present at any time t is

$$N(t) = N_o e^{-\lambda t}$$

where λ is a decay constant, defined by

$$\lambda = \frac{0.693}{T_{1/2}}$$

Here, $T_{1/2}$ is the **half-life** of the radioactive element and represents the time required for one-half of the radioactive atoms to remain unchanged.

Radioactive decay is an exponential curve and is illustrated in the figure below.

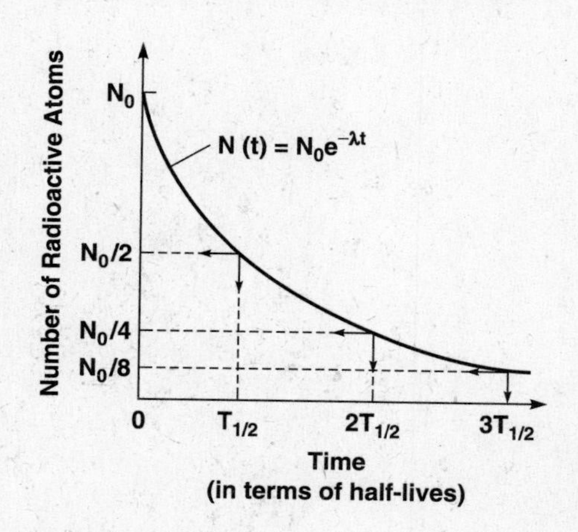

SOURCE: From George Hademenos, *Schaum's Outline of Physics for Pre-Med, Biology, and Allied Health Students*, McGraw-Hill, 1998; reproduced with permission of The McGraw-Hill Companies.

Radionuclides typically undergo radioactive decay of three common types.

- **Alpha decay,** caused by the repulsive electric forces between the protons, involves the emission of an alpha particle (or a helium nucleus that consists of two protons and two neutrons) by nuclei with many protons. In alpha decay, the radioactive nucleus **decreases** in A (mass number) by 4 and **decreases** in Z (number of protons) by 2.
- **Beta decay,** which occurs in nuclei that have too many neutrons, can occur by emission of a β particle. A β^- particle is an electron, and a β^+ particle is a positively charged electron, known as a **positron.** In β^- decay, the radioactive nucleus **remains unchanged** in A and **increases** in Z by 1. In β^+ decay, the radioactive nucleus **remains unchanged** in A and **decreases** in Z by 1.
- **Gamma decay** occurs by the emission of highly energetic photons. In gamma decay, the radioactive nucleus **remains unchanged** in A and **remains unchanged** in Z.

Nuclear fission is a process in which a large nucleus (generally with an atomic mass number, $A > 230$) is bombarded with a neutron and, as a result, splits into two or three medium-size nuclei as well as additional neutrons.

Nuclear fusion is a process in which two small nuclei (generally those with $A < 20$) combine or fuse to form a single, larger nucleus.

Glossary of Physics Terms

Acceleration (a) The rate at which velocity changes. Acceleration, expressed in units of length per unit time squared, is the change of velocity in a given direction over a defined time interval. It is measured in SI units of meters per second squared.

Accuracy How close a measurement is to the standard or accepted value; it depends on the quality of the measuring device.

Alpha decay A mode of radioactive decay caused by the repulsive electric forces between the protons, involving the emission of an alpha particle (or a helium nucleus that consists of 2 protons and 2 neutrons) by nuclei with many protons. In alpha decay, the radioactive nucleus decreases in A by 4 and decreases in Z by 2.

Alternating current (AC) circuits Similar to direct current (DC) circuits with one exception: the type of voltage source driving current through the circuit. In DC circuits, the voltage source is continuous and constant, resulting in current flow in only one direction, whereas in AC circuits, the voltage is oscillatory or sinusoidal, resulting in periodic changes in direction of current flow.

Amplitude The maximum magnitude of displacement from equilibrium. Units of amplitude are typically assigned according to the dimension of length, but can be represented by any physical quantity whose magnitude varies in an oscillatory manner.

Antinodes The points on a standing wave that are maximally displaced from the horizontal axis; that is, points at the peaks and valleys of each wave.

Archimedes' principle "A body immersed wholly or partially in a fluid is subjected to a buoyant force that is equal in magnitude to the weight of the fluid displaced by the body."

Atmospheric pressure (P_{atm}) Represents the average pressure exerted by Earth's atmosphere and is defined numerically as 1 atm, but can also be expressed as 760 millimeters of mercury (mmHg) or 1.01×10^5 pascals (Pa).

Average acceleration Represents average velocity over a time interval.

Bernoulli's equation The fluid equivalent of conservation of energy; states that the energy of fluid flow through a rigid vessel by a pressure gradient is equal to the sum of the pressure energy, kinetic energy, and the gravitational potential energy, or

$$E_{tot} = P + \frac{1}{2}\rho v^2 + \rho gh = \text{constant}$$

Beta decay A mode of radioactive decay that occurs in nuclei that have too many neutrons and can occur by emission of a β particle. A β^- particle is an electron, and a β^+ particle is a positively charged electron, known as a positron. In β^- decay, the radioactive nucleus *remains unchanged* in A and *increases* in Z by 1. In β^+ decay, the radioactive nucleus *remains unchanged* in A and *decreases* in Z by 1.

Boyle's law States that, at constant temperature, the volume V occupied by a given mass of gas is inversely proportional to the pressure P exerted on it: $PV = k$ or $P_1V_1 = P_2V_2$, $T = \text{constant}$, where k is a proportionality constant.

Bulk modulus (*B*) The elastic modulus due to volumetric compression of a solid; defined as: Bulk modulus $= -\frac{V\Delta P}{\Delta V}$ where V is volume and P is pressure.

Capacitance (C) Depends on the shape and position of the capacitors and is defined as $C = \frac{q}{V}$, where q is the magnitude of charge on either of the two conductors and V is the magnitude of potential difference between the two conductors. The SI unit of capacitance is coulomb/volt, collectively known as the farad (F).

Capacitors Circuit elements that store charge and consist typically of two conductors of arbitrary shape carrying equal and opposite charges separated by an insulator.

Center of gravity That point within an object where the force of gravity is considered to act.

Center of mass That point within an object where the total mass is concentrated.

Centripetal acceleration Acceleration of an object moving in uniform circular motion.

Centripetal force A "center-seeking" force exerted on an object in uniform circular motion directed inward toward the center of the circular motion; defined by

$$F_c = ma_c = m\frac{v^2}{r}$$

where a_c is centripetal acceleration.

Charles's law States that, at constant pressure, the volume V occupied by a given mass of gas is directly proportional to the absolute temperature T by $V = kT$ or $\frac{V_1}{T_1} = \frac{V_2}{T_2}$, $P = \text{constant}$, where k is a proportionality constant.

Coherent waves Light waves that have the same wavelength and a fixed phase difference.

Conduction The method of heat energy transfer that occurs in solids.

Conservation of charge States that electric charge can neither be created nor destroyed, only transferred. The net charge of a system remains constant.

Conservation of energy law States that energy can neither be created nor destroyed, only transformed from one kind to another. The total energy of a system remains constant.

Conservative force Any force, such as the force due to gravity, that performs work on an object independent of the object's path of motion.

Constructive interference For two waves that are in phase (i.e., the peaks of the two waves occur on the same side of the equilibrium position), the amplitudes of the two waves add together.

Convection The method of heat energy transfer that occurs in liquids.

Coulomb's law Describes the electrostatic force $\boldsymbol{F}_{el}$ between two charged particles q_1 and q_2 separated by a distance r:

$$F_{el} = \frac{1}{4\pi\varepsilon_o}\frac{q_1 q_2}{r^2} = k\frac{q_1 q_2}{r^2}$$

where ε_o is the permittivity constant, defined as $\varepsilon_o = 8.85 \times 10^{-12}\ \mathrm{C^2/N \cdot m^2}$.

Current The rate of motion of electric charge, expressed as

$$\text{Current} = \frac{\text{electric charge moving through a region}}{\text{time required to move the charge}}$$

where current is measured in SI units of amperes.

Density (ρ) A physical property of a fluid, given as mass per unit volume, or

$$\rho = \frac{\text{mass}}{\text{unit volume}} = \frac{m}{V}$$

expressed in SI units of $\frac{\mathrm{kg}}{\mathrm{m^3}}$, $\frac{\mathrm{g}}{\mathrm{cm^3}}$, or $\frac{\mathrm{g}}{\mathrm{mL}}$.

Dependent variable The variable that depends on the value of the independent variable and is displayed on the y-axis (ordinate) of a scientific graph.

Destructive interference For two waves that are out of phase (i.e., the peaks of the two waves occur on opposite sides of the equilibrium position), the amplitudes of the two waves are subtracted and the resultant amplitude is the difference between the two amplitudes.

Dielectric An insulator that is usually used to fill the gap between the plates of a capacitor because it increases the capacitance.

Diffraction The ability of the light waves to bend or spread through an aperture (opening) or around an obstacle.

Dimensions Physical descriptions of a quantity. In physics, the three basic dimensions frequently used are length (L), mass (M), and time (T).

Direct current (DC) circuit Electricity in the form of electric charge generated by a voltage source (such as a battery) flowing as current through an arrangement of circuit elements or devices (e.g., a lightbulb, alarm, motor) all connected by a conductor or a material that allows electric charge to flow easily through it.

Displacement (Δx) Defined as a distance in a given direction between the starting point and the ending point, independent of the path taken; measured in SI units of meters.

Distance The length required to get from the starting point to the ending point, dependent on the path taken; measured in SI units of meters.

Domain Elements of a given data set representing an independent variable of a function.

Doppler effect Refers to the shift in frequency of a transmitted sound caused by a change in path length between source and observer.

Efficiency A descriptive quantity of a pulley, describes the ratio comparing the magnitude of the work output, W_{out}, relative to the work input, W_{in}, or

$$\text{Efficiency} = \frac{W_{out}}{W_{in}}$$

Elastic collision A collision in which two bodies approach each other, interact with each other, and part on impact with a different velocity in a different direction. In an elastic collision, both linear momentum and kinetic energy are conserved.

Elastic modulus A physical constant that describes the elastic properties of a solid, is defined as

$$\text{Elastic modulus} = \frac{\text{stress}}{\text{strain}}$$

Electric charge A physical property of the basic building blocks of the atom and a fundamental property of all matter. The SI unit of charge is the coulomb, abbreviated C. Although charge can be positive or negative, the magnitude of charge is $e = 1.6 \times 10^{-19}$ C.

Electric dipole Consists of two point charges $+q$ and $-q$ typically equal in magnitude but opposite in sign separated by a small distance.

Electric field (E) Defines the electric force exerted on a positive test charge positioned at any given point in space. Electric field, $\mathbf{E}$, is a vector quantity and expressed in units of newtons/coulomb.

Electric potential (V) At some point B, electric potential becomes stored as a result of work done, W, against an electric field to move a positive test charge from infinity (point A) to that point (point B), or

$$\text{Electric potential} = \frac{-\text{work}}{\text{charge}}$$

$$V = \frac{-W}{q_o}$$

The electric potential is a scalar quantity that can be either positive, negative, or zero, depending on the sign and magnitude of the point charge as well as the work done. Electric potential is expressed in units of $\frac{\text{joules}}{\text{coulomb}} = \text{volts}$.

Electric potential difference (ΔV) Describes the work done by the electrostatic force to move the charge from point A to point B in an electric field:

$$\Delta V = V_B - V_A = \frac{-W}{q_o}$$

Electromagnetic spectrum A compilation of electromagnetic waves, arranged with the larger wavelengths (radio waves) to the left and the smaller wavelengths (gamma rays) to the right.

Electromagnetic waves An example of transverse waves, meaning they occur as a result of a disturbance acting in a direction at right angles or perpendicular to the direction of wave motion. The disturbance that generates electromagnetic waves is coupled oscillating electric and magnetic fields.

Electromotive force The voltage or potential difference generated between a positive terminal a and a negative terminal b of a battery or power source when no current is flowing. Electromotive force is expressed in units of volts.

Electron One of three basic particles of the atom; it orbits the nucleus of an atom and is negatively charged.

Energy The ability of an object to do work. The SI unit of energy is the joule.

Equation of continuity An expression of the conservation of mass for a moving fluid; it states that: (1) a fluid maintains constant density regardless of changes in pressure and temperature, and (2) flow measured at one point along a vessel is equal to the flow at another point along the vessel, regardless of the cross-sectional area of the vessel.

Equilibrium The condition in which the sum of all forces acting on an object is equal to zero and the object is balanced.

Equipotential surfaces A graphical method of representing the electric potential of any charge distribution as concentric circles that are normal or perpendicular to electric field lines.

Exponential function A type of function described by the form

$$f(x) = Ca^{bx} \qquad \text{for } a > 0$$

where a, b, and C are constants. Exponential functions describe many scientific phenomena, particularly when $b > 0$ (growth processes), or $b < 0$ (decay processes) and when $a = e$ ($e = 2.718$).

First Law of Thermodynamics The change in internal energy ΔU is related to the heat energy transferred to the body Q and the work done by the system W according to

$$\Delta U = Q - W$$

Fluorescence Phenomenon that occurs when certain atoms or molecules absorb light at a particular wavelength and, after a very brief time interval, emit light at longer wavelengths.

Force A push or a pull, measured in SI units of newtons.

Free-body diagram A diagram in which an object is isolated and all forces acting on the object are identified and represented on the diagram.

Free-fall motion Motion of an object that is dropped from a certain height or y-direction and allowed to fall toward the ground.

Frequency (ν) The number of cycles in a time interval; given by $\nu = \frac{1}{T}$, where T is the period of the oscillation or the time required to complete one cycle. The unit of frequency is the hertz (Hz) or cycles/s. Cycles are unitless, and therefore frequency is given in units of inverse time (1/s).

Friction A force generated by the properties of the interface between a moving object and the surface it is moving on.

Function A mathematical one-to-one relationship between a dependent variable y and an independent variable x.

Fundamental frequency Also known as the first harmonic, it is the lowest frequency produced by either a stretched string or pipe.

Gamma decay A mode of radioactive decay that occurs by the emission of highly energetic photons. In gamma decay, the radioactive nucleus remains unchanged in A and remains unchanged in Z.

Geometric optics A branch of physics concerned with the propagation of light used to form images.

Gravitation, law of A force of attraction, F_G, exists between two bodies of mass m_1 and m_2 that is directly proportional to the product of the masses and inversely proportional to the squared distance between their centers of mass r, or $F_G = G\frac{m_1 m_2}{r^2}$ where $G = 6.67 \times 10^{-11}$ N·m^2/kg^2.

Gravitational potential energy The energy that is stored within a body because of its location. The gravitational potential energy (GPE) of a body of mass m at a height h above a given reference level is GPE $= mgh$, where g is the acceleration due to gravity.

Harmonic frequencies Are integer multiples of the fundamental frequency or first harmonic attained by either a stretched string or pipe.

Heat Or thermal energy, a form of energy that depends on the change of temperature. The quantity of heat, Q, gained or lost by a body is related to the temperature difference, ΔT, by the following equation $Q = mc\Delta T$, where m is the mass of the body and c is a proportionality constant termed the specific heat capacity. Because heat is a form of energy, the SI unit of measurement is the joule.

Heat of fusion The amount of heat energy required to change 1 kg of solid matter to liquid or released when changing 1 kg of liquid matter to solid.

Heat of transformation The amount of heat required to change the phase of 1 kg of a substance. Thus, the total amount of heat Q gained or lost by a substance of mass m during a change between phases is: $Q = mL$, where L is the heat of transformation unique to the substance.

Heat of vaporization The amount of heat energy required to change 1 kg of liquid matter to gas or released when changing 1 kg of gas matter to liquid.

Hydrostatic pressure (P_{hyd}) The fluid pressure exerted at a depth h in a fluid of density ρ; given by $P_{hyd} = \rho gh$.

Ideal Gas Law Combines the relations from Charles's law and Boyle's law to yield

$$PV = nRT \quad \text{or} \quad \frac{P_1 V_1}{T_1} = \frac{P_2 V_2}{T_2}$$

where R is the universal gas constant ($R = 0.082$ L atm/mol K) and n is the number of moles.

Impulse Defined as the change in momentum and is related to the impact force acting to speed up or slow down the object multiplied by the time interval during which this force acts. Units of impulse are kg m/s or N · s.

Impulse-momentum theorem States that the impulse is related to the change in momentum of an object according to the relation $m\Delta v = F\Delta t$.

Independent variable The variable that is free to move in an experiment and is displayed on the x-axis (abscissa) of a scientific graph.

Inelastic collision A collision in which two bodies approach each other, collide, stick together, and part as a combined single body with a different velocity in a different direction. In an inelastic collision, only linear momentum is conserved.

Inertia An object's resistance to motion.

Instantaneous acceleration Represents acceleration at a given instant of time.

Instantaneous velocity Represents velocity at a given instant of time, similar to the velocity of a car indicated by its speedometer.

Interference Refers to the superposition of two waves traveling within the same medium when they interact.

Isotopes A class of nuclei with the same number of protons (Z) but different number of nucleons (A).

Kinetic energy The energy of a body that is in motion. If the body's mass is m and its velocity is v, then kinetic energy $= \frac{1}{2}mv^2$.

Kirchhoff's rules of circuit analysis Two rules used to analyze complex electric circuits:

Node or Junction Rule The algebraic sum of all currents entering a junction or node must be equal to the sum of all currents exiting the node.

Loop or Circuit Rule The algebraic sum of potential drops encountered while traversing the circuit in either a counterclockwise or clockwise direction must be equal to zero.

Laminar flow A state of fluid flow in which the fluid flows as continuous layers stacked one on another within a smooth tube and moving past one another with some velocity.

Law of reflection States that a light ray that reflects or bounces off the surface boundary at an angle of reflection θ_r, also measured with respect to the normal, is equal in magnitude to the angle of incidence θ_i, or $\theta_i = \theta_r$.

Lens equation Relates the object distance o, image distance i, and the focal length f of a lens according to $\frac{1}{f} = \frac{1}{i} + \frac{1}{o}$.

Lens maker's equation Used in creating the lens, it is given by

$$\frac{1}{f} = (n-1)\left(\frac{1}{R_1} - \frac{1}{R_2}\right)$$

where f is the focal length, n is the index of refraction for the lens material, R_1 is the radius of curvature of the lens closest to the object, and R_2 is the radius of curvature of the lens farthest from the object.

Linear momentum (p) Defined as the product of an object's mass (m) and its velocity ($\mathbf{v}$), or $\mathbf{p} = m\mathbf{v}$. Linear momentum is a vector quantity whose direction is that of the object's velocity and is given in units of kg m/s.

Logarithmic function A type of function described by the form $y = \log_a x$ for $a > 0$, $a \neq 1$.

Longitudinal waves Waves generated by a disturbance that acts parallel to the direction of wave motion. An example of longitudinal waves is sound waves.

Magnetic field (B) An attractive or repulsive force field generated by a moving charged particle.

Magnetic force (F) A force exerted on a charged particle q moving with a velocity $\mathbf{v}$ in a uniform magnetic field $\mathbf{B}$ is defined as $F = qvB \sin\theta$ where θ is the angle between the lines of the magnetic field $\mathbf{B}$ and the direction of the velocity of the charged particle $\mathbf{v}$.

Mass The amount of substance that an object has, measured in SI units of kilograms.

Mass defect The difference between the sum of the rest masses of the protons and neutrons and the rest mass of a nucleus.

Mean The average value of measurements in a data set.

Mechanical advantage A descriptive quantity of a pulley, denotes the amount by which the required force to move a load is reduced, defined by $MA = \frac{F_{out}}{F_{in}}$, where F_{out} is the output force and F_{in} is the input force.

Median The middle or center value of the data set.

Mirror equation Represents the analytical approach in the determination of the image type and location formed by a mirror. The mirror equation relates the object distance o, the image distance i, the focal length f, and the radius of curvature, R, by $\frac{1}{o} + \frac{1}{i} = \frac{1}{f} = \pm\frac{2}{R}$, where o, i, f, and R are expressed in units of length.

Mirrors Optical components that reflect light rays from a light source to form an image.

Mode The measurement that is the most frequent or common value in a data set.

Neutron One of three basic particles of the atom; it is found in the nucleus of an atom and is neutrally charged.

Newton's Laws of Motion Three laws that describe the motion of an object:

Newton's First Law of Motion: Law of Inertia

Newton's Second Law of Motion: $\Sigma\mathbf{F} = m\mathbf{a}$

Newton's Third Law of Motion: Law of Action and Reaction

Nodes The points on a standing wave that lie along the horizontal axis.

Normal force The force exerted on an object by a surface.

Nuclear binding energy The negative energy required to hold the nucleons (protons and neutrons) together within the nucleus and is defined as the difference between the rest energies of the constituent nucleons and the rest energy of the final (stable) nucleus.

Nuclear fission A process in which a large nucleus (generally with an atomic number, $A > 230$) is bombarded with a neutron and, as a result, splits into two or three medium-size nuclei as well as additional neutrons.

Nuclear fusion A process in which two small nuclei (generally those with $A < 20$) combine or fuse to form a single, larger nucleus.

Ohm's law States that the voltage V across a resistor R is proportional to the current I through it, and can be written in equation form as $V = IR$.

Pascal's principle States that "An external pressure applied to a confined fluid will be transmitted equally to all points within the fluid."

Percentage A mathematical quantity described as a fraction of a whole or total amount (100%).

Period (*T*) The time required to complete one cycle, defined as $T = \frac{2\pi}{\omega}$. Period is measured in units of seconds.

Phase angle Represents a constant that can be adjusted to conform with the value of the displacement of a wave at $t = 0$ and is presented as an angle, measured in radians.

Photons Characterization of light as a beam of small packets or bundles of light. Photon energy, E, is related to its frequency, ν, or its wavelength, λ, by the following relation: $E = h\nu = h\frac{c}{\lambda}$, where h is known as Planck's constant ($h = 6.626 \times 10^{-34}$ J · s) and c is the speed of light, $c = 3 \times 10^8$ m/s.

Polarization A process by which the electric field vectors of the light waves are all aligned along the same direction.

Polynomial function A type of function described by the form

$$y = f(x) = a_0 + a_1x + a_2x^2 + a_3x^3 + \cdots + a_nx^n$$

where a_0, a_1, a_2, a_3, a_n are coefficients (constants) and n is an integer that describes the degree or order of the polynomial.

Power The rate at which work is done by a force; defined by Power $= \frac{\text{work}}{\text{time}}$; the SI unit for power is watt (W), where 1 W = 1 joule/second.

Power function A type of function described by the form $y = ax^n$, where a is a constant and n is a real number, representing the mathematical power of the function.

Precision The ability of a measurement to be consistently reproduced.

Pressure Defined as a force F acting perpendicular to a surface area A of an object and is given by Pressure $= \frac{\text{force}}{\text{area}} = \frac{F}{A}$. Pressure is a scalar quantity and is expressed in units of N/m^2.

Principle of superposition States that the resultant wave that occurs from the interaction between two or more waves is the algebraic sum or overlapping of the individual waves also referred to as superposition.

Probability A numeric measure of the likelihood that a given event may or may not occur.

Projectile motion Motion of an object along both x- and y-axes.

Proportion A mathematical relationship of the relative size of two quantities, typically stated as the equality of two ratios, $\frac{a}{b} = \frac{c}{d}$.

Proton One of three basic particles of the atom; it is found in the nucleus of an atom and is positively charged.

Quanta of light See *Photons.*

Radiation The method of heat energy transfer that occurs in space.

Radioactive decay A nuclear phenomenon exhibited by radioisotopes or elements with an atomic number Z greater than that of lead ($Z = 82$). In nature, the radioactive element is unstable, and in the process of returning toward a stabilized state of existence, spontaneously emits particles (photons and charged and uncharged particles) in the transformation to a different nucleus, and hence a different element.

Range

Functions: Elements of a given data set representing a dependent variable of a function.

Statistics: Difference between the largest and smallest values in a data set.

Ratio A comparison of two numbers or quantities that can be stated as $\frac{x}{y}$ or $x{:}y$.

Ray tracing Represents a graphical approach by which the image position and image size formed by a mirror or a lens can be determined.

Resistance (*R*) An inherent property of a conductor that resists flow of electric current and represents a measure of the potential difference V that must be supplied to a circuit to drive current I through the circuit. Resistance is defined as

$$R = \frac{V}{I}$$

Resistance is expressed in SI units of ohms (Ω).

Resonance A condition that occurs when a periodic external force strikes the object with a frequency equal to the natural frequency of the object, resulting in an increase in the amplitude, and hence energy is absorbed by the object.

Rotational motion The rotational motion of an object caused by forces exerted in the direction of motion.

Scalar A physical quantity that has magnitude only.

Scientific notation Presentation of numeric values by expressing the value of a number between 1 and 10, multiplied by 10 raised to a power determined by the number of zeroes in the numeric result.

Scientific prefixes Defined symbols that represent a numeric quantity in powers of 10.

Second Law of Thermodynamics A system subjected to a spontaneous change responds such that its disorder or entropy increases or at least remains the same. For a given system in equilibrium, the change in entropy ΔS is related to the added heat ΔQ by

$$\Delta S = \frac{\Delta Q}{T}$$

Entropy is expressed in units of J/K.

Shear modulus Relates the shear stress to the shear or angular strain and is defined as

$$\text{Shear modulus} = \frac{\text{shear stress}}{\text{shear strain}} = \frac{FL}{A\phi\Delta L}$$

where F is the force applied to the solid, A is the cross-sectional area of the solid, L is the original length of the solid, ϕ is the angle by which the elastic solid is displaced by the shearing stress, and ΔL is the fractional amount of deformation as a result of the applied force.

Significant figures A reflection of the experimental accuracy of a measurement or quantitative value.

Simple harmonic motion The simplest type of oscillatory motion in which an object or body moves about an equilibrium point as a result of a restoring force.

Slope The rate of increase or decrease of a graphed function; defined as

$$\text{Slope} = \frac{\text{rise}}{\text{run}} = \frac{\text{change in } y\text{-coordinates}}{\text{change in } x\text{-coordinates}} = \frac{y_2 - y_1}{x_2 - x_1}$$

Snell's law Also known as the law of refraction, states that the angle of refraction θ_r' with respect to the normal is related to the incident angle θ_i by the law of refraction, or $n_1 \sin\theta_i = n_2 \sin\theta_r'$, where n_1 and n_2 are the indices of refraction of the first medium and second medium, respectively.

Sound waves Longitudinal waves that can propagate through all forms of matter—solids, liquids, and gases; they are generated by the motion of molecules or particles of a medium vibrating back and forth in a direction parallel to the direction of wave motion.

Specific gravity (sp. gr.) For a given substance, it is the ratio of the density of the substance ρ_{sub} to the density of water ρ_w, or

$$\text{Sp. gr.} = \frac{\text{density of substance}}{\text{density of water}} = \frac{\rho_{sub}}{\rho_w}$$

where the density of water ρ_w is 1.0 g/cm^3, or 1.0×10^3 kg/m^3. Assuming that equal volumes are chosen, the specific gravity can also be expressed in terms of weight:

$$\text{Sp. gr.} = \frac{\text{weight of substance}}{\text{weight of water}} = \frac{w_{sub}}{w_w}$$

Specific gravity is a pure number that is unitless.

Specific heat capacity A constant dependent upon the material of the object, it is the quantity of heat required to raise the temperature of one kilogram of a substance by 1°C.

Statistics A branch of mathematics that reveals important information about variables and their relationships from groups or sets of data.

Strain The ratio of the increase in a particular dimension in the deformed state to that dimension in its initial, undeformed state, and can be expressed mathematically as

$$\text{Strain} = \frac{\text{extent of deformation}}{\text{undeformed state}}$$

Stress Defined as the magnitude of an external force required to induce the given deformation of an object divided by the surface area over which the force is exerted perpendicularly, or

$$\text{Stress} = \frac{\text{force}}{\text{area}}$$

and is expressed in units of N/m^2.

Surface tension The tension, or force per unit length, generated by cohesive forces of molecules on the surface of a liquid acting toward the interior.

SI system of units (SI = Système International d'Unités) Commonly known as the metric system of measurement, a system of units where mass is measured in kilograms, length is measured in meters, and time is measured in seconds.

Temperature A physical property of a body that reflects its warmth or coldness.

Tension The force exerted by a string, rope, or cable on a suspended object.

Thermal expansion A physical phenomenon in which increases in temperature can cause substances in the solid, liquid, and gaseous states to expand.

Thin lenses Optical components that refract light rays from a light source to form an image.

Torque Rotational equivalent of force; occurs when an external force acts at an angle to a given distance from a fixed pivot point (also known as a lever arm, r). Torque is defined as Torque = (force)(lever arm) $\sin\theta$, where θ is the angle between the external force and the lever arm.

Total internal reflection A phenomenon that occurs when light passes from a medium with a high index of refraction to a medium with a low index of refraction.

Translational equilibrium The condition that occurs when the sum of forces acting on an object is equal to zero, or

$$\Sigma \mathbf{F} = 0 \quad \text{or} \quad \Sigma F_x = 0, \quad \Sigma F_y = 0$$

Translational motion The motion of an object caused by forces exerted in the direction of motion.

Transverse waves Waves generated by a disturbance that acts perpendicular to the direction of wave motion. Examples of transverse waves include light waves and waves in a plucked string.

Trigonometric function A type of function that involves the mathematical quantities $\sin\theta$, $\cos\theta$, $\tan\theta$, $\sec\theta$, $\csc\theta$, $\cot\theta$. Trigonometric functions are typically used to describe periodic or oscillatory phenomena.

Turbulent flow State of fluid flow characterized by its velocity exceeding a threshold known as the critical velocity, a value determined by the Reynold's number; the flow becomes unsteady and chaotic.

Units Indicate the amount of a physical quantity.

Vector A physical quantity that has both magnitude and direction.

Velocity (v) The rate of change of displacement or the rate at which a directed distance between point A and point B is covered over a required time. It is measured in SI units of meters per second.

Viscosity A measurement of the resistance or frictional force exerted by a fluid in motion. The SI unit of viscosity is $\frac{\text{N} \cdot \text{s}}{\text{m}^2}$. Viscosity is typically expressed in units of poise (P), where 1 poise $= 0.1\frac{\text{N} \cdot \text{s}}{\text{m}^2}$.

Weight The force exerted on an object due to gravity.

Work Represents the physical effects of an external force applied to an object that results in a net displacement in the direction of the force. Work is a scalar quantity and is expressed in SI units of joules (J) or newton-meters ($\text{N} \cdot \text{m}$).

Work–kinetic energy theorem States that the kinetic energy of an object undergoes a change when a net force does work on an object, or

$$\text{Net work} = \text{change in kinetic energy}$$

***y*-Intercept** The point of a scientific graph where the function intercepts the y-axis or the point whose x-coordinate is 0.

Young's modulus The ratio of tensile stress to tensile strain and is defined as

$$\text{Young's modulus} = \frac{\text{tensile stress}}{\text{tensile strain}} = \frac{FL}{A\Delta L}$$

where F is the force applied to the solid, A is the cross-sectional area of the solid, L is the original length of the solid, and ΔL is the fractional amount of deformation as a result of the applied force.

ON YOUR OWN

MCAT Physics Practice

This section contains fifteen quizzes, one for each of the fifteen chapters of the physics review section that you have just completed. Use these quizzes to test your mastery of the concepts and principles covered in the physics review. Detailed solutions for every question are provided following the last quiz. If you need more help, go back and reread the corresponding review chapter.

CHAPTER 1 MATHEMATICS FUNDAMENTALS

Quiz

1. Given a temperature in units of Fahrenheit, it can be converted to degrees Celsius according to the relation $T_{°C} = \frac{5}{9}(T_{°F} - 32)$, which is a linear equation. The y-intercept is

   A. 32.0°C **B.** 17.8°C **C.** −17.8°C **D.** 57.6°C

2. Temperature can be expressed in units of degrees Fahrenheit, degrees Celsius, and kelvin. Given a temperature expressed in degrees Fahrenheit, which equation represents the correct conversion to kelvin?

   A. $T_K = \frac{5}{9}T_{°F} + 255$ **B.** $T_K = \frac{5}{9}T_{°F} - 273$

   C. $T_K = \frac{5}{9}T_{°F} - 255$ **D.** $T_K = \frac{5}{9}T_{°F} + 273$

3. The Lineweaver-Burk equation of enzyme kinetics is $\frac{1}{v} = \frac{K_m}{V_{max}}\frac{1}{[S]} + \frac{1}{V_{max}}$, where v is the enzyme velocity and $[S]$ is substrate concentration. The y-intercept of this equation is

   A. $-\frac{1}{V_{max}}$ **B.** $\frac{1}{V_{max}}$ **C.** V_{max} **D.** $\frac{K_m}{V_{max}}$

4. An example of a power function is Poiseuille's law, which describes fluid flow through a rigid tube. Poiseuille's law states that fluid flow Q depends on the tube radius R, tube length L, fluid viscosity η and pressure gradient ΔP according to the relation $Q = \frac{\pi \Delta P}{8\eta L}R^4$. By how much will blood flow be reduced through a vessel reduced by one-fourth of its original radius?

   A. 0.32 **B.** 0.63 **C.** 0.75 **D.** 0.86

5. Coulomb's law describes the electric force that exists between two charged objects and is defined by $F = k_C\frac{q_1 q_2}{r^2}$, where F is the force between the two charged objects, measured in newtons (N); q_1, q_2 are the charges of the two objects, measured in coulombs (C); and r is the distance of separation of the two charged objects, measured in meters (m). Given this information, the units for the constant k_C is

   A. $\frac{NC}{m}$ **B.** $\frac{Nm}{C}$ **C.** $\frac{NC^2}{m^2}$ **D.** $\frac{Nm^2}{C^2}$

6. A teacher recorded the following exam scores from a select group of students: 93, 75, 92, 76, 86, 56, 88, 96. The median of this data set is

   A. 85 **B.** 86 **C.** 87 **D.** 88

7. A student is making observations on the average rainfall and made the following measurements over a 7-day period:

Day	1	2	3	4	5	6	7
Rainfall (inches)	1	0	1.5	0.5	1.3	3.4	2.3

The average daily rainfall based on these measurements is

A. 1.4 inches **B.** 1.7 inches **C.** 2.3 inches **D.** 2.6 inches

8. During a 7-week grading period, a student earned the following grades on weekly quizzes: 77, 88, 72, 86, 94, 87, 96. The range of these quiz scores over the 7-week period is

A. 14 **B.** 21 **C.** 24 **D.** 26

9. How many significant figures are in the result 204.0?

A. 1 **B.** 2 **C.** 3 **D.** 4

10. Of the numeric quantities (a) 0.0889, (b) 60.06, (c) 12,400,000, (d) 578,000, and (e) 0.02586, which one(s) are expressed to 4 significant figures?

A. a, b, e **B.** b, e **C.** a, b **D.** b

Quiz

1. Density, ρ, is defined as $\rho = \frac{m}{V}$, where m is mass and V is volume. The dimension of density is

   A. $\frac{M}{L}$ **B.** ML^3 **C.** $\frac{M}{L^2}$ **D.** $\frac{M}{L^3}$

2. Of the following units, which one is not an SI unit?

   A. kg **B.** ft **C.** s **D.** °C

3. If the displacement of an object is described by $x = 4+9t$, the units of the coefficient 9 are

   A. m **B.** ms **C.** $\frac{\text{m}}{\text{s}}$ **D.** $\frac{\text{m}}{\text{s}^2}$

4. If one were to express the surface area (SA) of a spherical cell in terms of its volume, the resultant expression would be

   A. $4.8V$ **B.** $4.8V^{\frac{1}{3}}$ **C.** $4.8V^{\frac{2}{3}}$ **D.** 4.8

5. A light-year, the distance light travels in a year, is 9.5×10^{12} km. If 1 inch = 2.54 centimeters, how many inches are in a light-year?

   A. 3.7×10^{15} in. **B.** 3.7×10^{17} in. **C.** 9.5×10^{15} in. **D.** 9.5×10^{17} in.

6. Under normal conditions, blood is pumped from the heart into the circulatory system under a systolic pressure of 120 mmHg. This quantity of pressure in terms of newtons per meter squared is approximately (NOTE: 1 mmHg = 1333 N/m^2):

   A. 1.3×10^4 N/m^2 **B.** 1.3×10^5 N/m^2

   C. 1.6×10^4 N/m^2 **D.** 1.6×10^5 N/m^2

7. The angular speed of centrifuge rotors is often expressed in revolutions per minute. The dimensions of angular speed are

   A. T **B.** T^{-1} **C.** LT **D.** T^2

8. The dimensions of a dollar bill are 2.6 in. × 6.1 in., giving an area of 15.9 in.^{2}. How many dollar bills would it take to completely cover the surface of a standard football field of dimensions 120 yd × 55 yd?

   A. 350,000 bills **B.** 420,000 bills **C.** 550,000 bills **D.** 750,000 bills

9. Keeping in mind that acceleration is change in velocity over change in time, the dimensions of acceleration are

A. $\frac{L}{T^2}$ B. $\frac{M}{T^2}$ C. $\frac{L}{T}$ D. $\frac{M}{L^2}$

10. The number 589,000,000 expressed in scientific notation is

A. 58.9×10^8 B. 5.89×10^8 C. 5.89×10^9 D. 5.89×10^{10}

CHAPTER 3 KINEMATICS
Quiz

1. During a particular trip, a car traveled 800 meters north in 200 seconds, and then traveled 600 meters east in 150 seconds. The car's average velocity (magnitude and direction with respect to the positive x-axis) was

   A. 2.8 m/s, 37° **B.** 2.8 m/s, 53° **C.** 4.6 m/s, 37° **D.** 4.6 m/s, 53°

2. A golf ball is struck from a tee with an initial velocity v_i at an angle θ_i with the horizontal and follows the trajectory as shown in the following figure. Which of the following pairs of graphs best represents the horizontal and vertical components of the velocity, v, of the golf ball as functions of time t?

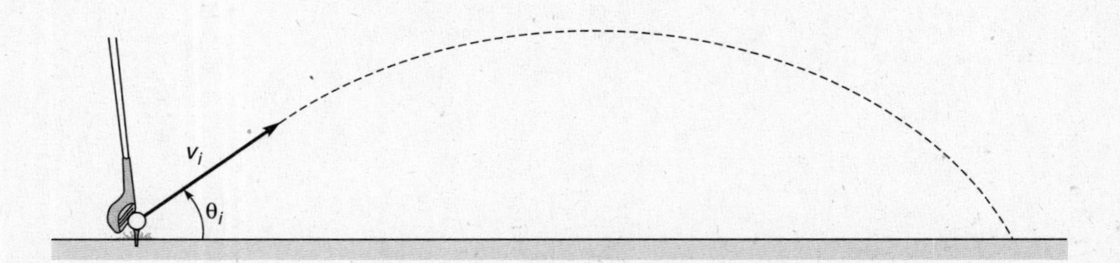

Horizontal component of velocity | Vertical component of velocity

A.
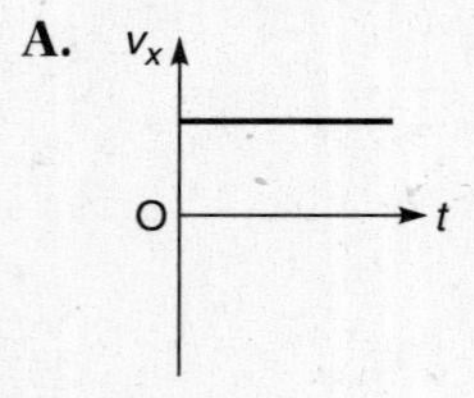

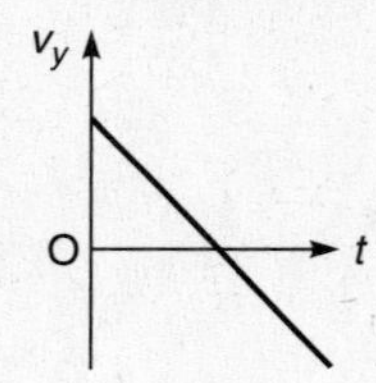

B.
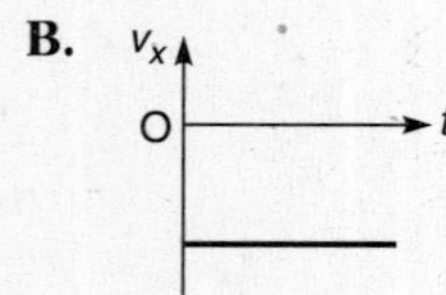

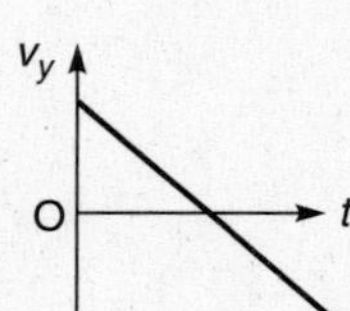

C.
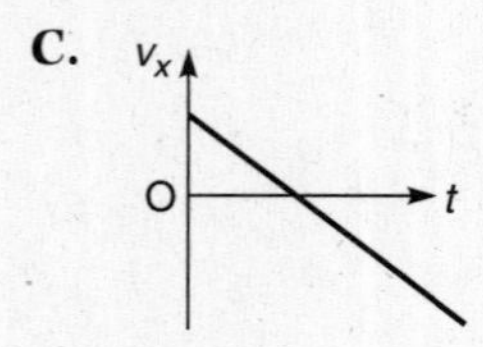

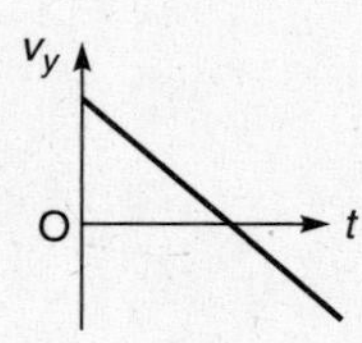

D.
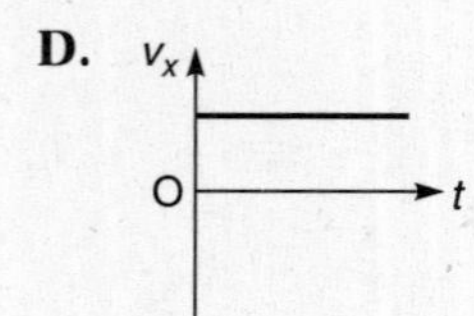

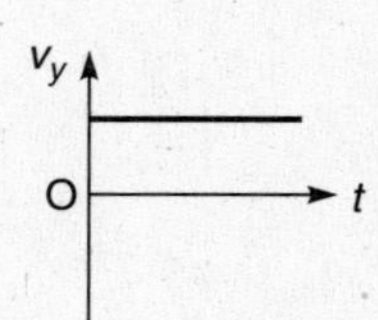

3. For the golf ball trajectory described in Problem 2, which of the following pairs of graphs best represents the horizontal and vertical components of the acceleration, a, of the golf ball as functions of time t?

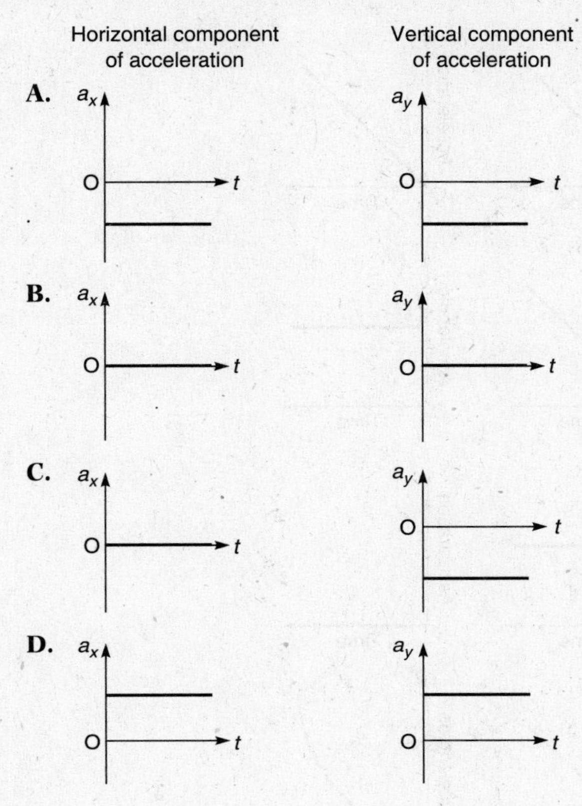

4. A car is uniformly accelerated from rest. Which of the following pairs of graphs shows the distance traveled versus time, speed versus time, and acceleration versus time for an object uniformly accelerated from rest?

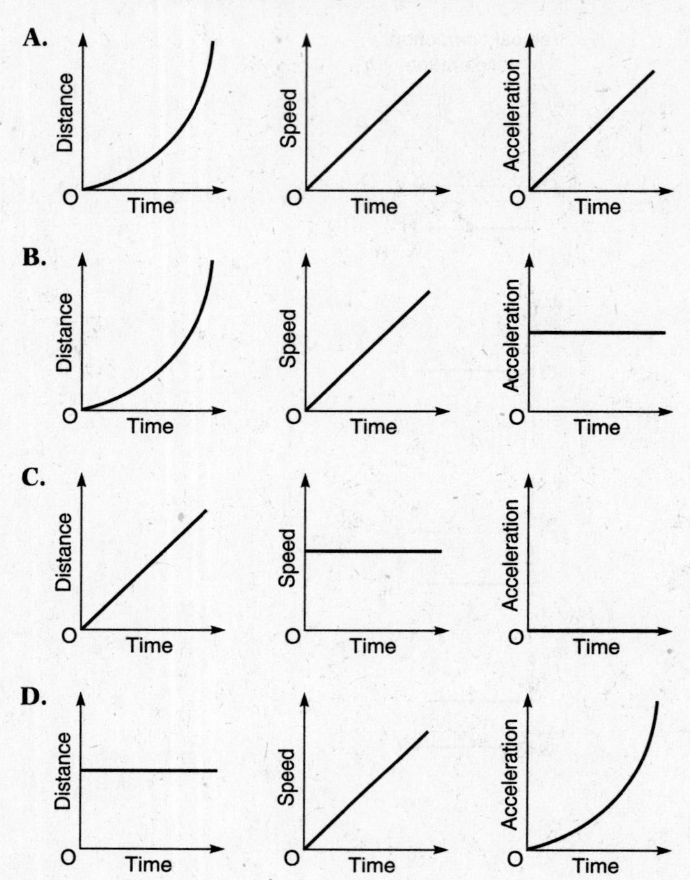

5. A physics teacher, standing on a chair, holds a tennis ball in each hand at the same point above the floor. She then drops one of the balls and throws down the other ball at the same time. Which of the following is true regarding the final velocity and the acceleration of the motion of the two tennis balls?

A. $v_{f,\text{drop}} < v_{f,\text{thrown}}$; $a_{\text{drop}} = a_{\text{thrown}}$

B. $v_{f,\text{drop}} < v_{f,\text{thrown}}$; $a_{\text{drop}} < a_{\text{thrown}}$

C. $v_{f,\text{drop}} = v_{f,\text{thrown}}$; $a_{\text{drop}} < a_{\text{thrown}}$

D. $v_{f,\text{drop}} > v_{f,\text{thrown}}$; $a_{\text{drop}} = a_{\text{thrown}}$

6. A projectile is shot vertically downward with an initial velocity of 20 m/s from the top of a 150-m-high cliff. The final velocity of the projectile as it strikes the ground is

A. 36.9 m/s **B.** 41.5 m/s **C.** 57.2 m/s **D.** 65.7 m/s

7. An outfielder can throw a baseball straight up to a height, h. If the baseball player now throws the baseball toward home plate at an angle of $45°$ with the horizontal, the maximum height that the baseball can now attain with respect to height h is

A. $\frac{h}{2\sqrt{2}}$ B. $h/2$ C. h D. $\frac{h}{\sqrt{2}}$

8. During its morning route, a school bus first uniformly accelerates on its way to its first stop, maintains a constant speed, and then uniformly decelerates as it approaches its first stop. The graph that best describes the velocity versus time of the bus is

A.
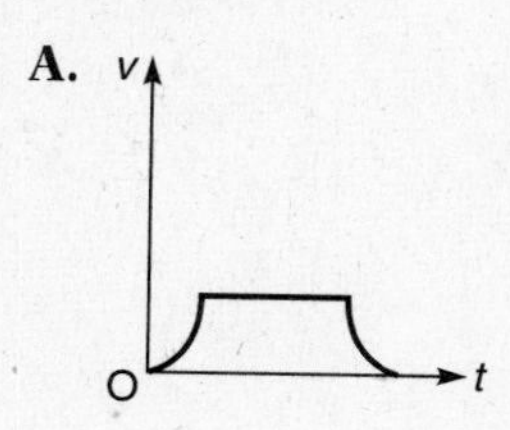

B.
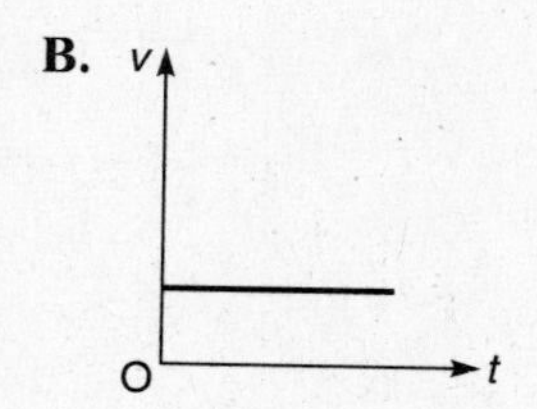

C.
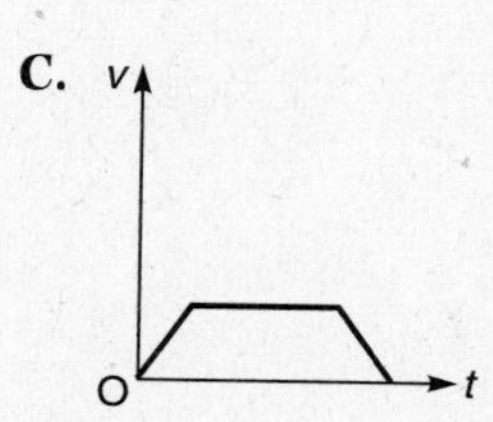

D.
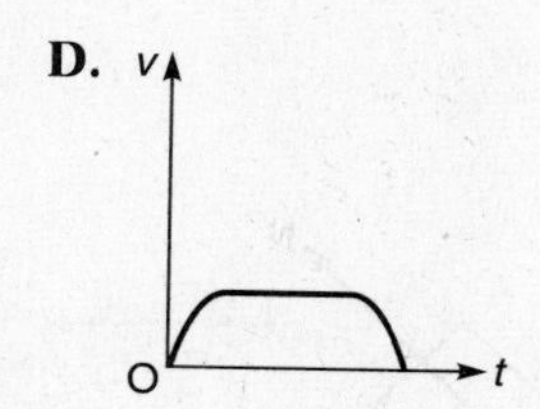

9. An object is dropped from a height of 75 m. The time taken for the object to travel the first meter is

A. 0.13 s B. 0.26 s C. 0.39 s D. 0.45 s

10. Using the information from the previous problem, the distance that the object travels during the last second of motion is

A. 33.8 m B. 41.2 m C. 56.4 m D. 63.6 m

Quiz

1. A 3-kg wooden block slides down a ramp inclined at 40°, as shown in the following figure, with an acceleration of $3\frac{m}{s^2}$. Which of the following free-body diagrams best represents the normal force **N**, the frictional force **f**, and the gravitational force **W** that act on the block?

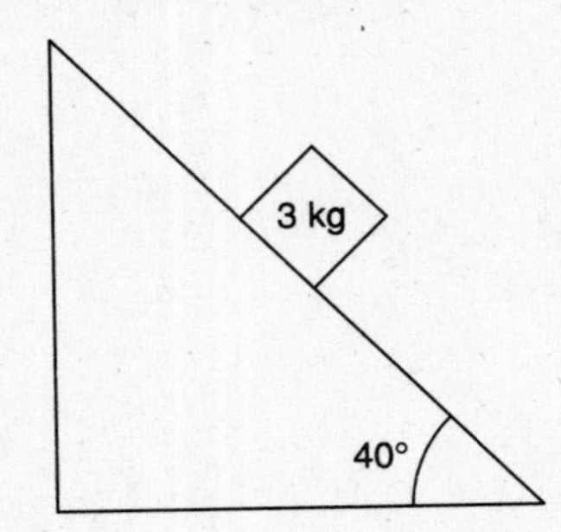

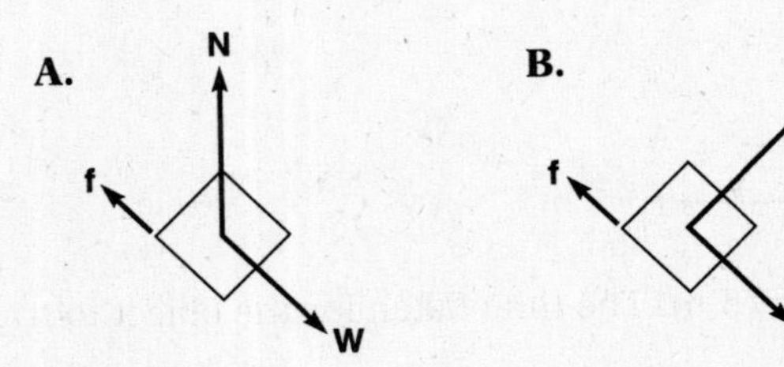

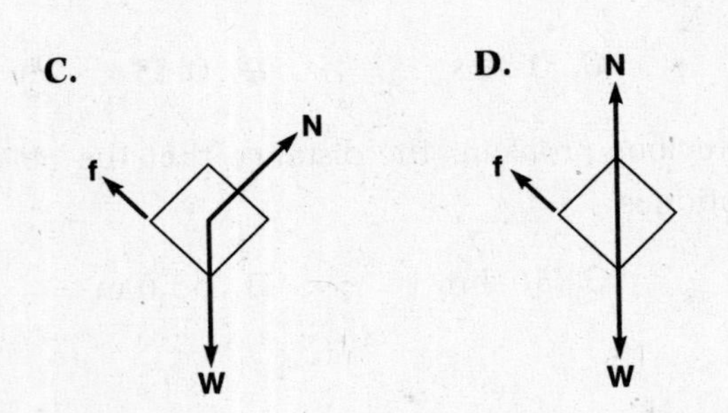

2. For the block described in Problem 1, the magnitude of the frictional force is approximately

   A. 5 N **B.** 10 N **C.** 16 N **D.** 20 N

3. Referring to the scenario in Problem 1, the coefficient of friction between the block and the surface of the inclined plane is

   A. 0.21 **B.** 0.27 **C.** 0.35 **D.** 0.44

4. A football of mass m has just been kicked from the tee and is in its trajectory as seen in the following figure. Assuming no air resistance, the free-body diagram that best describes the force(s) acting on the football is

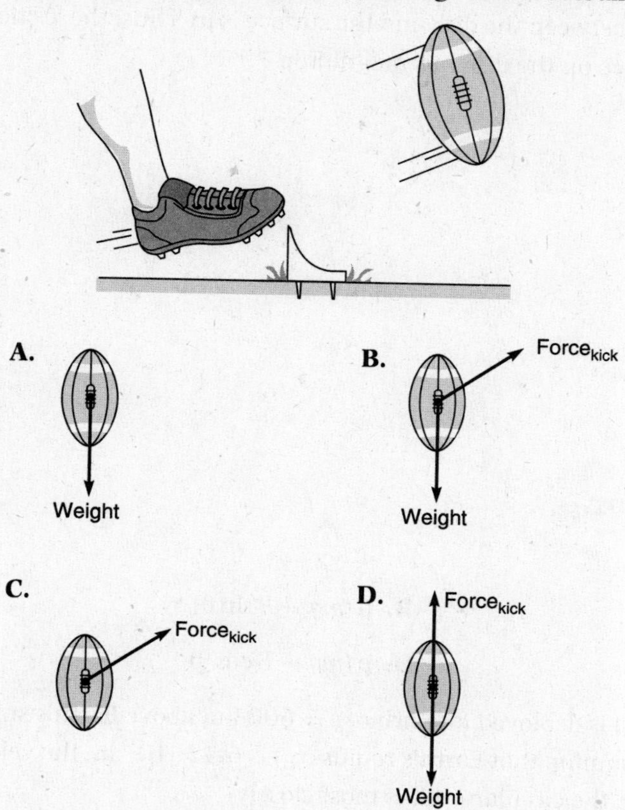

5. An object is subjected to an external force, F, acting at an angle of θ with the horizontal, causing it to move along a horizontal surface at constant speed v. The normal force is determined to be $W - F \sin \theta$. Which of the following scenarios is being described?

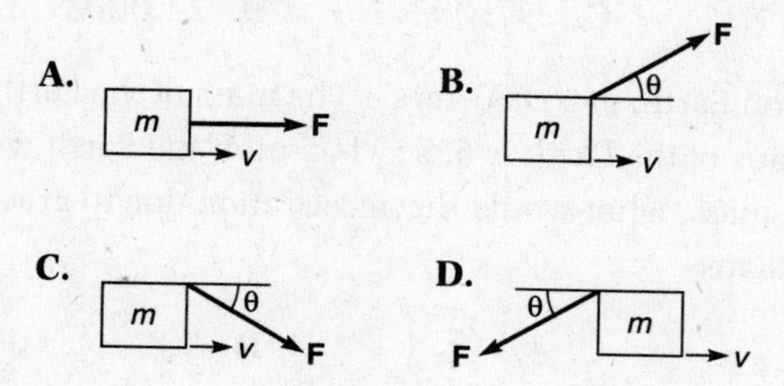

6. In shuffleboard, a player uses a long stick to push a weighted disk across a smooth surface. In the figure, a player uses a stick to apply a force of magnitude, F, directed at an angle θ to the disk of mass m, causing it to move along a smooth surface. The coefficient of friction between the disk and the surface is μ. Thus, the frictional force exerted by the surface on the disk has magnitude

A. $\mu(mg + F \sin \theta)$ B. $\mu(mg - F \sin \theta)$

C. $\mu(mg + F \cos \theta)$ D. $\mu(mg - F \cos \theta)$

7. An artificial satellite (m_s) is deployed into orbit $r_s = 600$ km above Earth's surface ($m_E = 6 \times 10^{24}$ kg). Assuming that Earth's radius (r_E) is 6.4×10^6 m, the velocity of the satellite moving in the circular orbit is most closely

A. $6500 \frac{m}{s}$ B. $7000 \frac{m}{s}$ C. $7500 \frac{m}{s}$ D. $7900 \frac{m}{s}$

8. The weight of a space scientist of mass $m_{ss} = 68$ kg on Pluto, which has a mass (m_{Pl}) of 5×10^{23} kg and a radius (r_{Pl}) of 4×10^5 m, would be approximately

A. 14,000 N B. 17,000 N C. 21,500 N D. 24,000 N

9. The acceleration due to gravity on Earth, g, is 9.81 m/s^2. The mass of the Earth is $m_E = 6 \times 10^{24}$ kg, and the radius of the Earth is 6.4×10^6 m. If the Earth were doubled in radius and halved in mass, what would the acceleration due to gravity be in terms of its original dimensions?

A. $0.6 \frac{m}{s^2}$ B. $1.2 \frac{m}{s^2}$ C. $4.9 \frac{m}{s^2}$ D. $9.8 \frac{m}{s^2}$

10. A molecule of H_2O consists of one oxygen atom bonded to two hydrogen atoms separated by an angle of 105°, as shown in the following figure. The mass of the oxygen atom is 26.6×10^{-27} kg, and the mass of the hydrogen atom is 1.67×10^{-27} kg, positioned 0.96 angstrom (Å) from the oxygen atom (1 angstrom = 1×10^{-10} m). The x- and y-coordinates, respectively, of the center of mass of a water molecule are

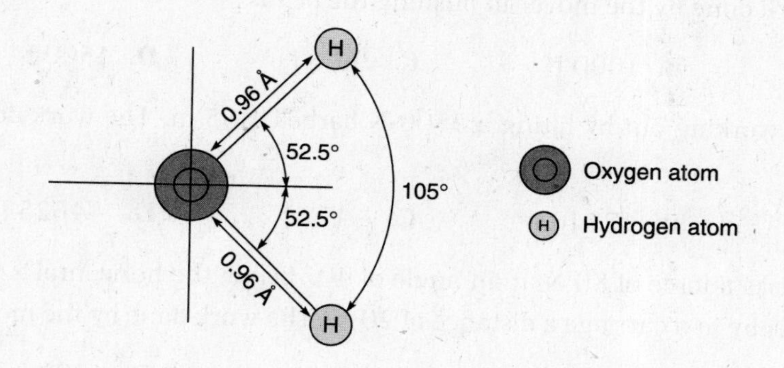

A. 6.5×10^{-12} m; 6.5×10^{-12} m

B. -6.5×10^{-12} m; -6.5×10^{-12} m

C. 6.5×10^{-12} m; 0.0 m

D. 0.0 m; 6.5×10^{-12} m

CHAPTER 5 PARTICLE DYNAMICS: WORK, ENERGY, AND POWER

Quiz

1. A mover exerts a constant horizontal force of 150 N to move a box 30 m across a floor. The work done by the mover in pushing the box is

   A. 150 J B. 1000 J C. 2600 J D. 4500 J

2. An athlete is working out by lifting a 2500-N barbell 1.85 m. The work done by the athlete is

   A. 4625 J B. 472 J C. −472 J D. −4625 J

3. A mother exerts a force of 80 N at an angle of 40° below the horizontal x-axis in pushing her baby in a carriage a distance of 20 m. The work done by the mother is

   A. 8040 J B. 1226 J C. 2485 J D. 4800 J

4. A 6-kg block slides down a ramp that is 2.5 m in height and inclined at an angle of 35°. The coefficient of friction of the ramp is 0.42. The work done by friction as the block slides down the ramp is

   A. 23 J B. 45 J C. 89 J D. 180 J

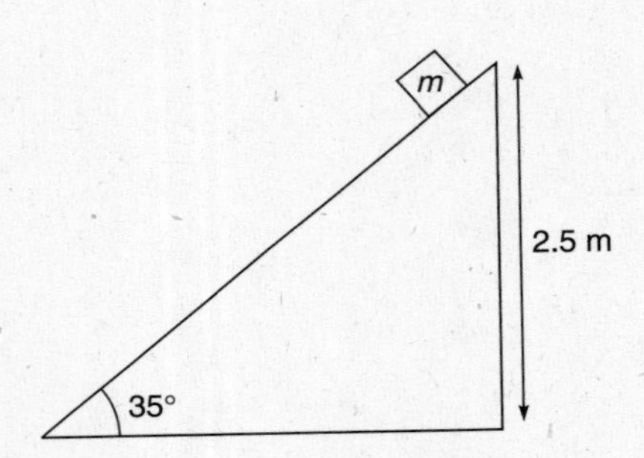

QUESTIONS 5–7

A 200-N child slides down a frictionless slide of height 3.5 m inclined at 55° from the horizontal axis. At the point of 1.75 m above the ground, the child has a kinetic energy of 350 J.

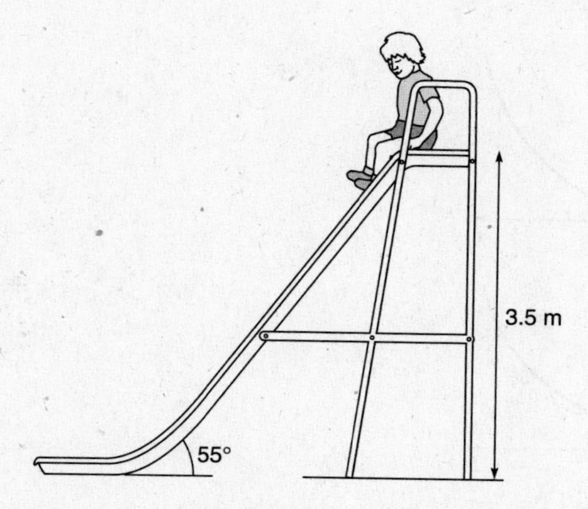

5. The gravitational potential energy of the child is

A. 350 J B. 700 J C. 1000 J D. 1800 J

6. The length of the slide is

A. 3.5 m B. 4.3 m C. 7.0 m D. 8.4 m

7. The speed of the child at the bottom of the slide is

A. 2.3 m/s B. 4.2 m/s C. 6.9 m/s D. 8.3 m/s

8. A 2.5-kg ball is set into motion with a horizontal speed on a tabletop and rolls off the table edge until it strikes the floor below a certain distance from the base of the table. The pair of graphs that best represents the potential energy and kinetic energy of the ball as a function of time as the ball rolls off the table until it strikes the ground as a function of time is

A. PE O t

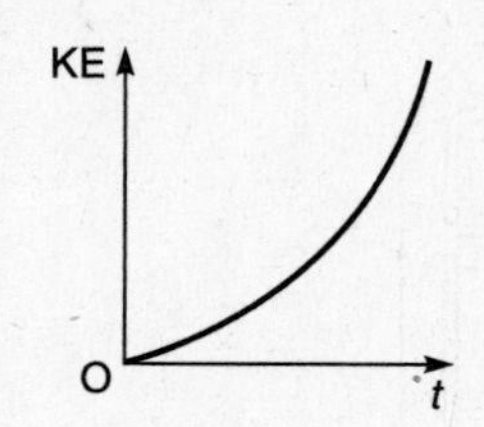

B. PE O t

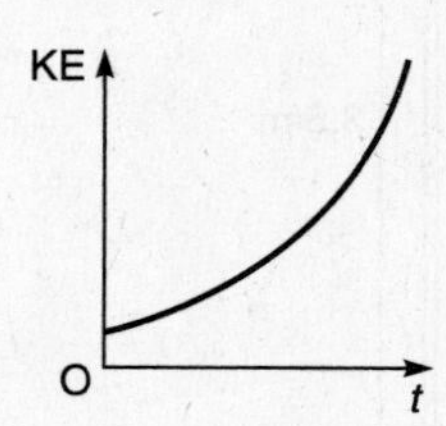

C. PE O t

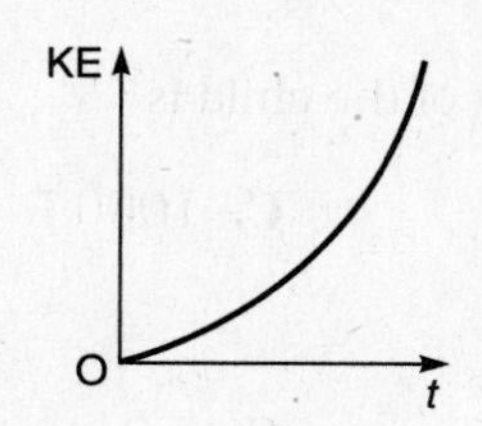

D. PE O t

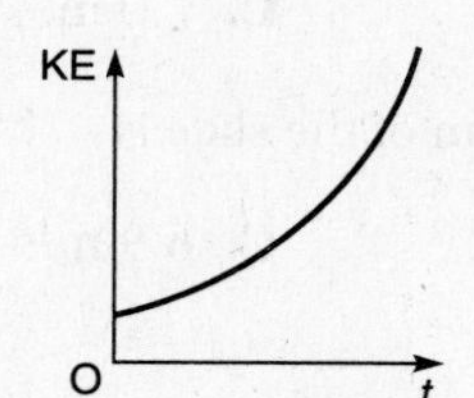

9. A 35-kg stone is dropped from a height of 50 m. The kinetic energy of the stone as it strikes the ground is

A. 1.7×10^4 J **B.** 3.4×10^4 J **C.** 1.7×10^5 J **D.** 3.4×10^5 J

10. In a particular experiment, a student dropped a marble from a known height, h_i, and graphed the potential energy of the marble at various heights as it reached the floor. The graph is displayed next. The slope of the curve represents which of the following:

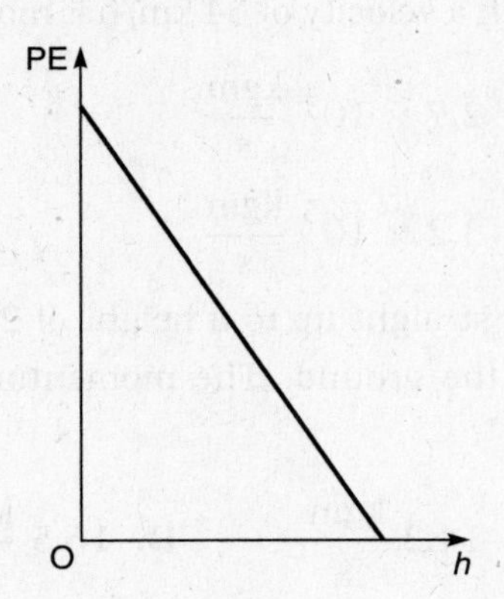

A. weight of the marble

B. mass of the marble

C. total energy of the marble

D. potential energy of the marble

CHAPTER 6 **MOMENTUM AND IMPULSE**

Quiz

1. The momentum of a 1500-kg car moving with a velocity of 64 km/h is most nearly

   A. $2.5 \times 10^4 \frac{kgm}{s}$ B. $2.7 \times 10^4 \frac{kgm}{s}$

   C. $2.7 \times 10^5 \frac{kgm}{s}$ D. $3.2 \times 10^5 \frac{kgm}{s}$

2. A baseball player throws a 0.45-kg baseball straight up to a height of 20 m and catches the ball in his glove 1.2 m above the ground. The momentum of the baseball as it strikes his glove is

   A. $8.6 \frac{kgm}{s}$ B. $9.8 \frac{kgm}{s}$ C. $14.2 \frac{kgm}{s}$ D. $16.5 \frac{kgm}{s}$

QUESTIONS 3 AND 4

3. A 0.40-kg baseball is thrown horizontally to a batter at 30 m/s and is hit with a velocity of 20 m/s in the direction of the pitcher. Assuming that the ball is in contact with the bat for 0.02 s, the impulse exerted by the batter is

   A. $20 \frac{kgm}{s}$ B. $4 \frac{kgm}{s}$ C. $-4 \frac{kgm}{s}$ D. $-20 \frac{kgm}{s}$

4. From Problem 1, the average force exerted on the ball by the bat is

   A. 500 N   B. 750 N   C. 1000 N   D. 1250 N

5. A 2000-kg car is moving at a speed of 25 m/s. The force required to stop the car in 7.0 s is

   A. 5.9×10^3 N   B. 7.1×10^3 N   C. 8.6×10^3 N   D. 9.2×10^3 N

6. A 20-kg child standing on a 4-kg skateboard jumps off and, in the process, kicks the skateboard, imparting a velocity of 4.7 m/s to the skateboard. The velocity of the child is

   A. $0.41 \frac{m}{s}$ B. $0.52 \frac{m}{s}$ C. $0.76 \frac{m}{s}$ D. $0.94 \frac{m}{s}$

7. Kevin (mass = 24 kg) is running at constant speed of 4.6 m/s before jumping into a 8-kg wagon initially at rest. The resultant speed of the wagon with Kevin in it is most nearly

   A. $3.5 \frac{m}{s}$ B. $4.7 \frac{m}{s}$ C. $6.2 \frac{m}{s}$ D. $8.1 \frac{m}{s}$

8. An example of an inelastic collision is the ballistic pendulum, which consists of a wooden block suspended by a cord secured to a level surface, as shown in the following figure. A projectile or bullet is fired horizontally into the wooden block and becomes embedded in the block. In one such pendulum, a 6-g bullet is fired horizontally at a 3-kg block of wood suspended by a cord. If the initial velocity of the bullet on firing is 620 m/s, the height of the swing of the pendulum is

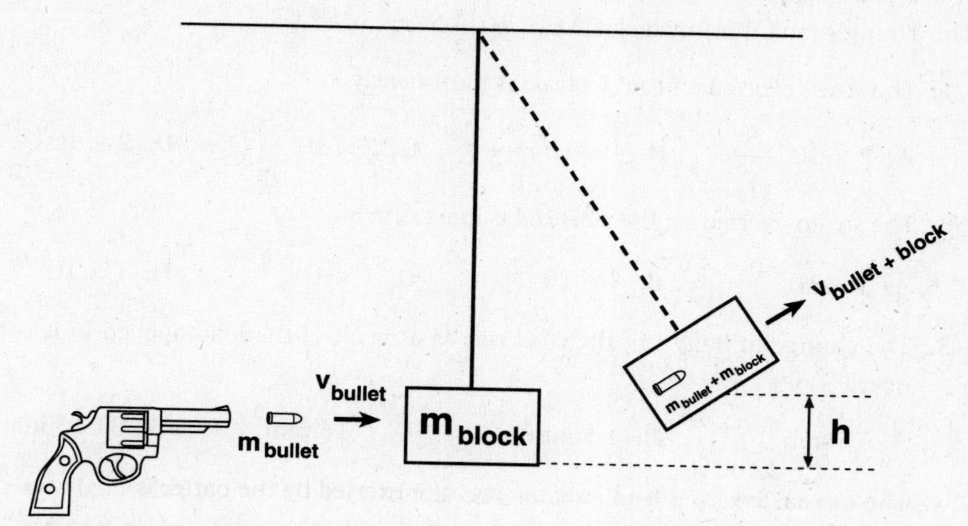

SOURCE: From George Hademenos, *Schaum's Outline of Physics for Pre-Med, Biology, and Allied Health Students*, McGraw-Hill, 1998; reproduced with permission of The McGraw-Hill Companies.

A. 4.2 cm **B.** 5.6 cm **C.** 7.8 cm **D.** 8.9 cm

QUESTIONS 9 AND 10

A 4.2-kg bowling ball moving in the positive x direction with a speed of 4.7 m/s collides with a 7.0-kg bowling ball initially at rest. The two bowling balls engage in a head-on collision.

9. The magnitude of the final velocity of the smaller bowling ball is most nearly

A. $1.2 \frac{m}{s}$ **B.** $-1.2 \frac{m}{s}$ **C.** $4.7 \frac{m}{s}$ **D.** $-4.7 \frac{m}{s}$

10. The magnitude of the final velocity of the larger bowling ball is

A. $1.2 \frac{m}{s}$ **B.** $-1.2 \frac{m}{s}$ **C.** $3.5 \frac{m}{s}$ **D.** $-3.5 \frac{m}{s}$

Quiz

QUESTIONS 1–3

A steel rod 0.25 m in diameter and 1.5 m in length can support a total load of 10,000 kg. The Young's modulus for steel is 2.0×10^{11} N/m^2.

1. The stress exerted on the steel rod is most nearly

A. $2 \times 10^6 \frac{N}{m^2}$ **B.** $4 \times 10^6 \frac{N}{m^2}$ **C.** $7 \times 10^6 \frac{N}{m^2}$ **D.** $9 \times 10^6 \frac{N}{m^2}$

2. The strain exerted on the steel rod is most nearly

A. 1×10^{-4} **B.** 5×10^{-4} **C.** 1×10^{-5} **D.** 1×10^{-6}

3. The change in length of the steel rod as a result of the load applied to it is most nearly

A. 15 mm **B.** 1.5 mm **C.** 0.15 mm **D.** 0.015 mm

4. The top surface of a lead cube of side length $L = 0.05$ m is subjected to a shear force of 500 N. Assuming that the shear modulus, G, for lead is $5.5 \times 10^9 \frac{N}{m^2}$, the shear stress exerted on the cube is

A. $2 \times 10^4 \frac{N}{m^2}$ **B.** $4 \times 10^4 \frac{N}{m^2}$ **C.** $2 \times 10^5 \frac{N}{m^2}$ **D.** $4 \times 10^6 \frac{N}{m^2}$

5. Thirty milliliters of a solution drawn into a 5-g syringe has a mass of 80 g. The density of the solution in the syringe is

A. $1.8 \frac{g}{cm^3}$ **B.** $2.5 \frac{g}{cm^3}$ **C.** $4.2 \frac{g}{cm^3}$ **D.** $5.7 \frac{g}{cm^3}$

6. Assuming that the surface area covered by the bottom of a foot is 150 cm^2, the pressure exerted by a standing 80-kg person is

A. $1.3 \times 10^4 \frac{N}{m^2}$ **B.** $1.3 \times 10^5 \frac{N}{m^2}$ **C.** $2.6 \times 10^4 \frac{N}{m^2}$ **D.** $2.6 \times 10^5 \frac{N}{m^2}$

7. Given that the density of water is $1.0 \frac{g}{cm^3}$, the pressure in a swimming pool at a depth of 180 cm is most nearly (atmospheric pressure $= 1.013 \times 10^5 \frac{N}{m}$)

A. $1.2 \times 10^4 \frac{N}{m^2}$ **B.** $1.2 \times 10^5 \frac{N}{m^2}$ **C.** $2.4 \times 10^4 \frac{N}{m^2}$ **D.** $2.4 \times 10^5 \frac{N}{m^2}$

8. A force of 2000 N is applied to a piston of 0.5 m^2 of cross-sectional area in a hydraulic jack. The lifting force transmitted to a piston of cross-sectional area of 2 m^2 is

A. 1000 N **B.** 2000 N **C.** 4000 N **D.** 8000 N

9. Blood is pumped from the heart at a rate of 5 L/min into the aorta of typical radius 1.2 cm. The velocity of blood as it enters the aorta is most nearly

A. 9 $\frac{\text{cm}}{\text{s}}$ **B.** 12 $\frac{\text{cm}}{\text{s}}$ **C.** 18 $\frac{\text{cm}}{\text{s}}$ **D.** 32 $\frac{\text{cm}}{\text{s}}$

10. The average length of a human aorta is 30 cm. If blood flows with a velocity $v = 18.4 \frac{\text{cm}}{\text{s}}$, the time required for the blood to travel the length of the aorta is most nearly

A. 0.5 s **B.** 0.9 s **C.** 1.2 s **D.** 1.6 s

CHAPTER 8 TEMPERATURE AND HEAT

Quiz

1. If one wanted to determine a temperature measurement expressed in degrees Fahrenheit from a temperature expressed in kelvins, the equation that would correctly yield the desired temperature is most nearly

   A. $T_F = \frac{9}{5}T_K - 460^\circ$ **B.** $T_F = \frac{9}{5}\left(T_K - 460^\circ\right)$

   C. $T_F = \frac{9}{5}T_K + 460^\circ$ **D.** $T_F = \frac{9}{5}\left(T_K + 460^\circ\right)$

2. Normal body temperature is 98.6°F. This temperature expressed in kelvins is

   A. −310 K **B.** −273 K **C.** 273 K **D.** 310 K

3. The temperature that is equivalent in both the Celsius and Fahrenheit temperature scales is

   A. −40° **B.** 0° **C.** 40° **D.** −273°

4. A copper rod 1.0 m in length is maintained at 25°C. Assuming the coefficient of linear expansion for copper is 1.7×10^{-5} °C^{-1}, the increase in length when heated to 40°C is

   A. 25 cm **B.** 25 mm **C.** 2.5 mm **D.** 0.25 mm

5. The volume of a brass sphere maintained at 110°C is 500 cm^3. The change in volume of the sphere at −10°C, given the coefficient of volumetric expansion $\beta = 57 \times 10^{-6}$ °C^{-1}, is

   A. 2.5 cm^3 **B.** 3.4 cm^3 **C.** 4.7 cm^3 **D.** 6.2 cm^3

6. The amount of heat required to raise the temperature of 5 kg of water from 20°C to 42°C is most nearly (NOTE: The specific heat capacity of water is c_{water} = 1.0 kcal kg^{-1} °C^{-1})

   A. 110 kcal **B.** 240 kcal **C.** 310 kcal **D.** 500 kcal

QUESTIONS 7–9

7. An insulated container contains 4.5 kg of ice at 0°C. The amount of heat required to change the amount of ice to water at 0°C is (NOTE: The latent heat of fusion for water is L_f = 80 cal g^{-1})

   A. 160 kcal **B.** 220 kcal **C.** 300 kcal **D.** 360 kcal

8. A 600-W immersion heater is placed in the insulated container. The amount of time required for the heater to change the 4.5-kg container of ice to water is (NOTE: 1 kcal = 4184 J)

A. 42 min **B.** 48 min **C.** 52 min **D.** 58 min

9. From Problem 8, let us assume that we would like the phase transfer to occur within 30 minutes. The power of an immersion heater required to change the state from ice to water within this time period is

A. 750 W **B.** 840 W **C.** 920 W **D.** 1000 W

10. In brewing tea, 0.5 kg of water is heated to 38°C and then poured into a 600-g pot (c_{pot} = 0.25 kcal kg^{-1} $°C^{-1}$) initially at 20°C. Given the specific heat capacity of water is c_{water} = 1.0 kcal kg^{-1} $°C^{-1}$, the resultant temperature of the water in the pot at thermal equilibrium is most nearly

A. 34°C **B.** 41°C **C.** 58°C **D.** 72°C

CHAPTER 9 VIBRATIONS AND WAVES

Quiz

1. The frequency of a wave with a period of 1 millisecond is

A. 100 Hz **B.** 1 kHz **C.** 10 kHz **D.** 1 MHz

QUESTIONS 2–4

A simple pendulum consists of a bob of mass m attached to a string of length l that is suspended from a level, secure surface, as shown in the figure below. The bob is pulled to the side at an angle of release, θ, and released where the bob is now in simple harmonic motion and swings with a period T.

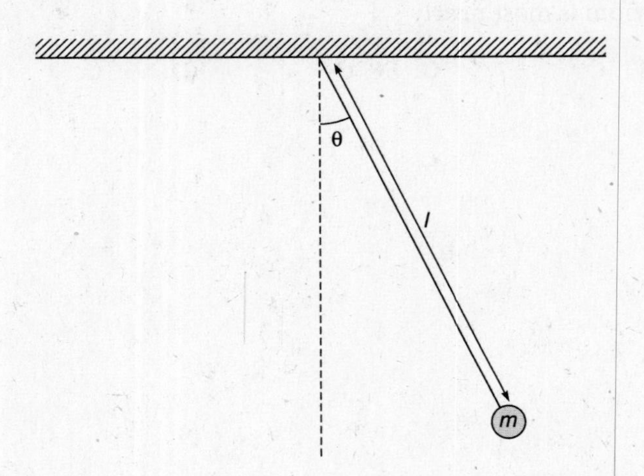

2. The mass of the pendulum bob is now doubled. Doubling the mass of the pendulum bob changes the period to

A. $2T$ **B.** $\sqrt{T}$ **C.** T **D.** $T/2$

3. The length of the pendulum is now doubled. Doubling the length of the pendulum changes the period to

A. $2T$ **B.** $\sqrt{2}T$ **C.** T **D.** $T/2$

4. The angle of release of the pendulum bob is now doubled. Doubling the angle of release of the pendulum bob changes the period to

A. $2T$ **B.** $\sqrt{T}$ **C.** T **D.** $T/2$

QUESTIONS 5 AND 6

A simple pendulum consists of an object of mass 2 kg attached to a string of length of 25 cm.

5. The period of the pendulum is most nearly

   A. 0.25 s B. 0.5 s C. 1.0 s D. 1.25 s

6. The frequency of the pendulum is most nearly

   A. 1 Hz B. 2 Hz C. 5 Hz D. 10 Hz

QUESTIONS 7 AND 8

An object of mass *m* is attached to a spring of constant *k* suspended from a level, secure surface, as shown in the following figure. The object is pulled down a distance Δx then released where the object is now in simple harmonic motion and moves up and down with a period *T*.

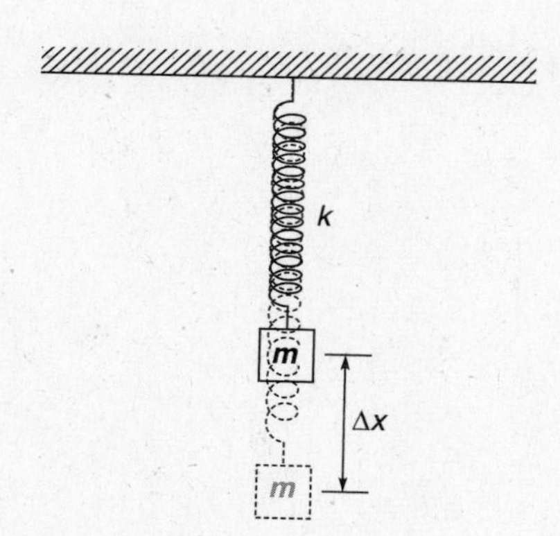

7. The mass of the object is now quadrupled. Quadrupling the mass of the object changes the period to

   A. $2T$ B. $\sqrt{2}T$ C. T D. $T/2$

8. The length of the spring is now quadrupled. Quadrupling the length of the spring changes the period to

   A. $2T$ B. $\sqrt{2}T$ C. T D. $T/2$

9. The frequency of a simple pendulum can be determined from

A. $2\pi\sqrt{\frac{l}{g}}$ B. $2\pi\sqrt{\frac{g}{l}}$ C. $\frac{1}{2\pi}\sqrt{\frac{l}{g}}$ D. $\frac{1}{2\pi}\sqrt{\frac{g}{l}}$

10. A simple pendulum consists of a mass $m = 2.5$ kg released from rest at an angle of 15°. The restoring force exerted on the pendulum bob is most nearly

A. 1.3 N B. 2.5 N C. 6.3 N D. 23.7 N

CHAPTER 10 SOUND

Quiz

1. Middle C of the musical scale has a frequency of 256 Hz. The period of this sound is

A. 2.4 ms **B.** 3.3 ms **C.** 3.9 ms **D.** 4.5 ms

2. Assuming the speed of sound v in air at 25°C is 344 $\frac{m}{s}$, the wavelength for a wave of frequency 104 MHz is

A. 3.3 μm **B.** 4.6 μm **C.** 7.2 μm **D.** 9.1 μm

3. Jim shouts his wish down a wishing well, only to hear his echo 0.50 s later. Assuming the speed of sound in air at 25°C is 344 $\frac{m}{s}$, the depth of the well is

A. 43 m **B.** 86 m **C.** 172 m **D.** 344 m

4. A stretched string of length L with both ends attached to rigid surfaces is plucked and allowed to vibrate. The first harmonic or fundamental frequency f_1 can be determined by

A. vL **B.** $\frac{v}{L}$ **C.** $\frac{v}{2L}$ **D.** $\frac{v}{4L}$

5. A stretched string of length L is forced to vibrate. The wavelength of the vibrating stretched string in its third harmonic λ_3 is

A. $3L$ **B.** L **C.** $\frac{1}{3}L$ **D.** $\frac{2}{3}L$

6. The fourth harmonic frequency of a 1.5-m organ pipe open at both ends is 459 Hz. The fundamental frequency of the pipe is most nearly

A. 98 Hz **B.** 115 Hz **C.** 268 Hz **D.** 459 Hz

7. From the previous question, the speed of sound through the 1.5-m pipe is

A. 115 $\frac{m}{s}$ **B.** 230 $\frac{m}{s}$ **C.** 345 $\frac{m}{s}$ **D.** 459 $\frac{m}{s}$

8. A band at a rock concert played their music at a sound intensity of 5 $\frac{W}{m^2}$. The sound level of the music in decibels is

A. 127 dB **B.** 135 dB **C.** 142 dB **D.** 160 dB

QUESTIONS 9 AND 10

A police car in pursuit of a driver suspected of speeding is traveling at 24 $\frac{m}{s}$ when its siren is turned on, operating at a frequency of 1.2 kHz. The speed of sound in air is 344 $\frac{m}{s}$.

9. The frequency heard by a stationary witness as the police car approaches the observer is

A. 1175 Hz **B.** 1290 Hz **C.** 1320 Hz **D.** 1420 Hz

10. The frequency heard by a stationary witness as the police car passes the observer is

A. 1000 Hz **B.** 1113 Hz **C.** 1122 Hz **D.** 1235 Hz

CHAPTER 11 LIGHT AND GEOMETRIC OPTICS Quiz

1. Radio station KABC broadcasts at a frequency of 960 kHz. The period of the radio wave transmissions are

A. 1 ms **B.** 10 ms **C.** 100 ms **D.** 1 μs

2. A frequency of ultraviolet waves is 2×10^{16} Hz. The wavelength of these waves is

A. 1×10^{-8} m **B.** 1.5×10^{-8} m **C.** 2×10^{-8} m **D.** 2.5×10^{-8} m

3. Of radio waves, ultraviolet waves, and gamma rays, the type of electromagnetic waves that travels the fastest is

A. radio waves

B. ultraviolet waves

C. gamma rays

D. They all travel at the same speed.

4. The speed of a light ray traveling through a pane of glass ($n = 1.52$) is most nearly

A. $2.0 \times 10^8 \frac{m}{s}$ **B.** $2.4 \times 10^8 \frac{m}{s}$ **C.** $2.7 \times 10^8 \frac{m}{s}$ **D.** $3.0 \times 10^8 \frac{m}{s}$

5. A light ray in air ($n_{air} = 1.00$) strikes the surface of a tank of water ($n_{water} = 1.33$) at an angle of incidence of 35° with respect to the normal. The angle of refraction of the light ray as it enters the tank of water is most nearly

A. 15° **B.** 21° **C.** 26° **D.** 34°

QUESTIONS 6–9

A pencil of height 4.5 cm is placed 20 cm in front of a concave mirror with a focal length of 15 cm.

6. The image distance of the pencil is

A. 15 cm **B.** 30 cm **C.** 40 cm **D.** 60 cm

7. The magnification of the mirror is

A. −4.0 **B.** −3.0 **C.** 3.0 **D.** 4.0

8. The height of the image of the pencil is

A. −13.5 cm **B.** 13.5 cm **C.** −7.5 cm **D.** 7.5 cm

9. The type of image formed by the concave mirror in this problem is

A. real and inverted

B. real and upright

C. virtual and inverted

D. virtual and upright

10. A figurine is placed 10 cm in front of a converging lens with a focal length of 40 cm. The image distance is most nearly

A. −10 cm　　B. −13 cm　　C. 13 cm　　D. 40 cm

CHAPTER 12 ELECTROSTATICS Quiz

QUESTIONS 1–4

1. The electric force exerted between two electric charges $q_1 = +6.0\ \mu C$ and $q_2 = -5.0\ \mu C$ separated by 10 cm is

   A. −2700 N **B.** −3000 N **C.** −6500 N **D.** −9000 N

2. The new electric force that is generated by doubling q_1 and q_2 is

   A. −2700 N **B.** −5400 N **C.** −10,800 N **D.** −21,600 N

3. The new electric force that is generated by halving the distance of separation between q_1 and q_2 is

   A. −2700 N **B.** −5400 N **C.** −10,800 N **D.** −21,600 N

4. The new electric force that is generated by changing the sign of q_1 and q_2 is

   A. −2700 N **B.** −5400 N **C.** −10,800 N **D.** −21,600 N

QUESTIONS 5 AND 6

5. Four point charges are assembled at the four corners of a square, as shown in the following figure, with $q_1 = +8.0\ \mu C$, $q_2 = +4.0\ \mu C$, $q_3 = -4.0\ \mu C$, $q_4 = -2.0\ \mu C$. The length of one side of the square is equal to 10.0 cm. The magnitude of the net electric force acting on q_4 is most closely

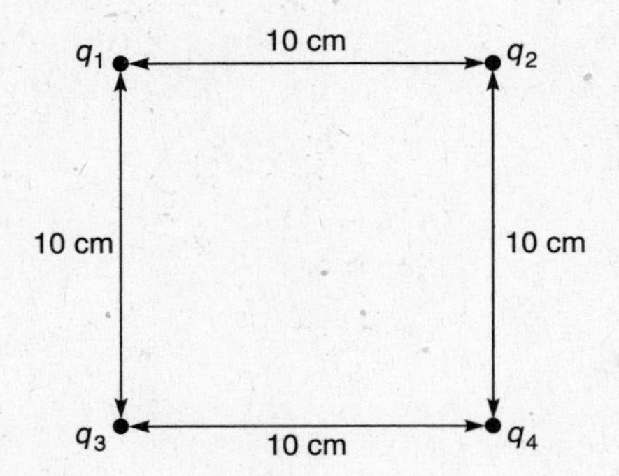

   A. 8 N **B.** 12 N **C.** 16 N **D.** 24 N

6. For the problem described in Question 5, the direction of the net electric force acting on q_4 is most nearly

   A. 45° **B.** 67° **C.** 81° **D.** 86°

QUESTIONS 7 AND 8

7. Two conducting spheres of mass $m = 4$ kg and equal charge $q = +7$ μC are each suspended by strings of length $l = 1.0$ m, as shown in the following figure, and held at an angle of θ by a repelling electric force of 10 N. The angle θ that each sphere makes with respect to the vertical is most nearly

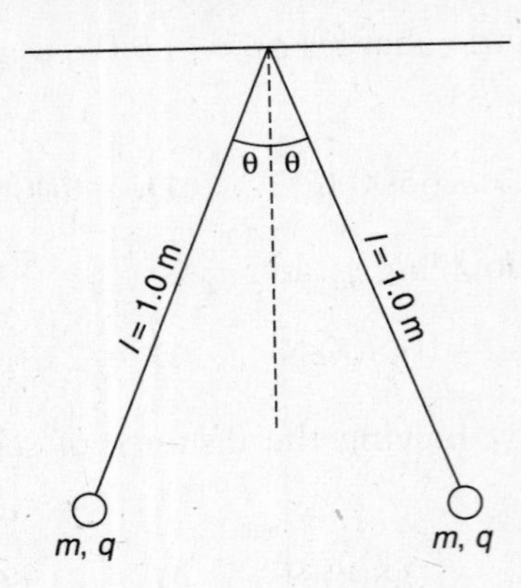

A. 6° B. 8° C. 12° D. 15°

8. For the conducting spheres described in Problem 7, the tension in the string suspending the conducting sphere is

A. 10 N B. 18 N C. 40 N D. 55 N

9. Three point charges of $q_1 = +5.0$ μC, $q_2 = -10.0$ μC, $q_3 = +5.0$ μC are spaced 0.10 m equidistant along the base of an isosceles triangle as shown in the following figure. Each side of the triangle is 0.20 m in length. The magnitude of the net electric field acting on point A is

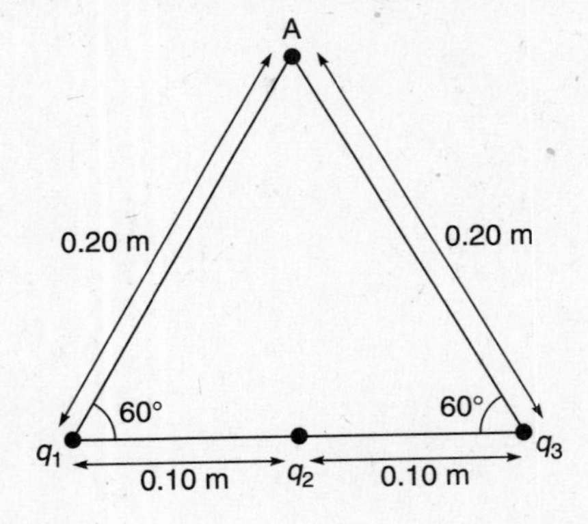

A. $1.6 \times 10^5 \frac{N}{C}$ B. $3.2 \times 10^5 \frac{N}{C}$ C. $1.6 \times 10^6 \frac{N}{C}$ D. $3.2 \times 10^6 \frac{N}{C}$

10. The magnitude of the electric potential of a point charge ($q = 8.0 \times 10^{-6}$ C) at a distance 0.4 m from the point charge is

A. 1.8×10^5 V B. 3.6×10^5 V C. 1.8×10^6 V D. 3.6×10^6 V

CHAPTER 13 ELECTRIC CIRCUITS
Quiz

QUESTIONS 1–3

Three resistors $R_1 = 5\ \Omega$, $R_2 = 10\ \Omega$, and $R_3 = 15\ \Omega$ are used in a particular electric circuit.

1. The equivalent resistance of the circuit assuming the resistors are all connected in series is

A. 2.7 Ω **B.** 11 Ω **C.** 18.3 Ω **D.** 30.0 Ω

2. The equivalent resistance of the circuit assuming the resistors are all connected in parallel is

A. 2.7 Ω **B.** 11 Ω **C.** 18.3 Ω **D.** 30.0 Ω

3. The equivalent resistance of the circuit assuming that R_1 and R_2 are connected in parallel, which in turn is connected in series with R_3, is

A. 2.7 Ω **B.** 11 Ω **C.** 18.3 Ω **D.** 30.0 Ω

4. An electric circuit has an arrangement of resistor as shown in the following figure. The equivalent resistance of the entire circuit is

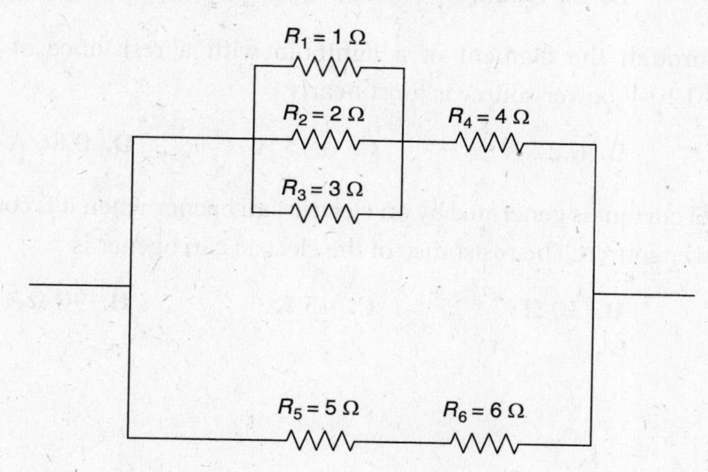

A. 3.2 Ω **B.** 5.5 Ω **C.** 7.3 Ω **D.** 9.4 Ω

QUESTIONS 5 AND 6

5. During a heart attack, the normal heart rhythm is restored by a cardiac defibrillator, which transmits approximately 15 A of current through the chest during a time interval of 7 ms. The amount of charge that flows during this time is

   A. 0.1 C B. 1.0 C C. 10.0 C D. 100.0 C

6. The number of electrons that flow during a single discharge of the cardiac defibrillator is

   A. 1.6×10^{17} electrons

   B. 3.2×10^{17} electrons

   C. 6.6×10^{17} electrons

   D. 9.5×10^{17} electrons

7. Given that the resistivity of copper is $1.7 \times 10^{-8}\ \Omega \cdot m$, the resistance of a copper wire 0.30 mm in diameter and 0.75 m in length is most nearly

   A. 0.2 Ω B. 0.6 Ω C. 1.2 Ω D. 2.4 Ω

8. Given that the resistivity of aluminum is $2.6 \times 10^{-8}\ \Omega \cdot m$, the diameter of an aluminum wire of 2 m in length and with a resistance of 3 Ω is

   A. 0.006 mm B. 0.06 mm C. 0.12 mm D. 1.2 mm

9. The current through the filament of a lightbulb with a resistance of 450 Ω connected to a 120-V power source is most nearly

   A. 0.14 A B. 0.27 A C. 0.45 A D. 0.86 A

10. Four amperes of current is generated by an electric can opener when it is connected to a 120-V power source. The resistance of the electric can opener is

    A. 16 Ω B. 30 Ω C. 45 Ω D. 90 Ω

CHAPTER 14 MAGNETISM

Quiz

1. Magnetic force is a quantity that is measured in units of

   A. Tesla   B. Gauss   C. Newton   D. Ampere

QUESTIONS 2–7

A beam of electrons (mass $m_e = 9.11 \times 10^{31}$ kg) moving with a velocity $v = 8.0 \times 10^{6}$ m/s is projected into a uniform magnetic field **B** of 5×10^{-4} T acting perpendicular to the direction of electron motion. As a result, the electrons move in a circular path or arc.

2. The general expression for the radius r of the circular arc of the beam of electrons of charge q moving in a perpendicular uniform magnetic field **B** is

   A. $r = \dfrac{m_e v}{qB}$ B. $r = \dfrac{m_e q}{vB}$ C. $r = \dfrac{qB}{m_e v}$ D. $r = \dfrac{Bv}{q m_e}$

3. The expression for the radius of the circular arc that describes the motion of the beam of electrons in the previous problem is based on

   A. Newton's First Law of Motion

   B. Newton's Second Law of Motion

   C. Newton's Third Law of Motion

   D. Ampere's Law

4. From the information provided in the passage, the radius of the circular arc that the electrons are moving is

   A. 1 cm   B. 3 cm   C. 6 cm   D. 9 cm

5. The period of the circulating electrons or the time required for the electrons to complete one circular path is most nearly

   A. 4.7×10^{-8} s   B. 5.9×10^{-8} s   C. 7.2×10^{-8} s   D. 9.1×10^{-8} s

6. The frequency of the circulating electrons or the number of complete circular paths per unit time is most nearly

   A. 1.0×10^{7} Hz   B. 1.4×10^{7} Hz   C. 2.2×10^{7} Hz   D. 3.7×10^{7} Hz

7. The angular frequency of the circulating electrons is most nearly

   A. $2 \times 10^7 \frac{rad}{s}$ B. $5 \times 10^7 \frac{rad}{s}$ C. $7 \times 10^7 \frac{rad}{s}$ D. $9 \times 10^7 \frac{rad}{s}$

8. A straight wire of length 5 m carries a current of 10 A. The distance of the point where the magnetic field is equivalent to the earth's magnetic field of $\mathbf{B} = 5 \times 10^{-5}$ T is (NOTE: The permeability of free space $\mu_o = 4\pi \times 10^{-7}$ T · m/A)

   A. 4 cm B. 40 cm C. 4 m D. 5 m

9. A circular conducting coil of diameter 0.4 m has 50 loops of wire and a current of 3 A flowing through it. The magnitude of the magnetic field at the center of the coil is

   A. 2.1×10^{-4} T B. 4.7×10^{-4} T C. 6.3×10^{-4} T D. 8.6×10^{-4} T

10. A solenoid of diameter 2.0 cm, 10.0 cm in length, and constructed from 400 turns of wire has a current of 5 A flowing through it. The magnitude of the magnetic field in the interior of the solenoid is

   A. 1.3 mT B. 1.9 mT C. 2.1 mT D. 2.5 mT

CHAPTER 15 ATOMIC AND NUCLEAR PHYSICS Quiz

QUESTIONS 1–3

Uranium-235 is an element that has a mass number of 235 and an atomic number of 92.

1. The number of protons of uranium-235 is

A. 92 **B.** 143 **C.** 235 **D.** 327

2. The number of neutrons of uranium-235 is

A. 92 **B.** 143 **C.** 235 **D.** 327

3. The number of electrons of uranium-235 is

A. 92 **B.** 143 **C.** 235 **D.** 327

QUESTIONS 4–6

A certain element has 10 neutrons and 9 electrons.

4. The atomic number, Z, of this element is

A. 1 **B.** 9 **C.** 10 **D.** 19

5. The mass number, A, of this element is

A. 1 **B.** 9 **C.** 10 **D.** 19

6. The identity of the chemical element with 10 neutrons and 9 electrons is

A. Fluorine **B.** Potassium **C.** Oxygen **D.** Calcium

7. The frequency of a 140-keV gamma ray is (NOTE: Planck's constant is $h = 4.135 \times 10^{-15}$ eV·s)

A. 2.1×10^{19} Hz **B.** 3.4×10^{19} Hz **C.** 5.6×10^{19} Hz **D.** 6.8×10^{19} Hz

8. Oxygen-15 is a radioisotope with a half-life of 2.1 min. The decay constant λ of oxygen-15 is

A. $1.2\times10^{-3}\ s^{-1}$ **B.** $3.6\times10^{-3}\ s^{-1}$ **C.** $5.5\times10^{-3}\ s^{-1}$ **D.** $7.9\times10^{-3}\ s^{-1}$

QUESTIONS 9 AND 10

9. The isotope radium-226 decays according to the reaction

$$^{226}_{88}\text{Ra} \rightarrow\, ^{222}_{86}\text{Rn} +\, ^{A}_{Z}\text{X}$$

The identity of the unknown element $^{A}_{Z}X$ is

A. $^{1}_{1}\text{H}$ **B.** $^{2}_{1}\text{H}$ **C.** $^{2}_{1}\text{H}$ **D.** $^{4}_{2}\text{He}$

10. The type of radioactive decay illustrated by the reaction in Problem 9 is

A. alpha decay

B. beta decay

C. gamma decay

D. nuclear fission

CHAPTER 1 MATHEMATICS FUNDAMENTALS Answers and Explanations

1. **C** The y-intercept can be found from the general equation of a line: $y = mx + b$, where x and y represent the independent and dependent variables, m is the slope of the function, and b is the y-intercept. The equation given in the problem can be simplified as

$$T_{^\circ C} = \frac{5}{9}(T_{^\circ F} - 32) = \frac{5}{9}T_{^\circ F} - \frac{5}{9}(32) = \frac{5}{9}T_{^\circ F} - 17.8$$

According to the equation of a line, the y-intercept is -17.8.

2. **A** The temperature conversion equation between degrees Celsius and degrees Fahrenheit is

$$T_{^\circ C} = \frac{5}{9}(T_{^\circ F} - 32) \qquad (1)$$

The conversion between degree Celsius and kelvin is

$$T_{^\circ C} = T_K - 273 \qquad (2)$$

Substituting Equation 2 into Equation 1 yields

$$T_K - 273 = \frac{5}{9}(T_{^\circ F} - 32)$$

$$T_K - 273 = \frac{5}{9}T_{^\circ F} - \left(\frac{5}{9}\right)32 = \frac{5}{9}T_{^\circ F} - 17.8$$

Solving for T_K yields

$$T_K = \frac{5}{9}T_{^\circ F} - 17.8 + 273$$

Simplifying the equation,

$$T_K = \frac{5}{9}T_{^\circ F} + 255$$

3. **B** Although there are inverse quantities in the Lineweaver-Burk equation, one can still draw similarities to the equation of a line ($y = mx + b$). In doing so, the y-intercept, b, is $\frac{1}{V_{max}}$.

4. **A** Poiseuille's law states that $Q = \frac{\pi \Delta P}{8\eta L}R^4$ where, for this problem, the vessel radius is the only parameter of concern and all other flow parameters (e.g., $\frac{\pi \Delta P}{8\eta L}$ are considered constant). In other words, Poiseuille's law can be restated as $Q_{orig} = (\text{constant})\,R^4$. A vessel reduced by one-fourth of its original radius implies that the new radius is $\frac{3}{4}R$ and the new flow is $Q_{new} = (\text{constant})\left(\frac{3}{4}R\right)^4 = 0.32\,(\text{constant})\,(R)^4 = 0.32Q_{orig}$.

5. **D** In any equation, the units on the left side must equal the units on the right side. In Coulomb's law, the variable on the left side is force, expressed in units of newtons, which means the combined units on the right side must also equal newtons. Because charge squared is in the numerator and distance squared is in the denominator, the constant must involve units such that the units for distance squared cancels with the quantity in the denominator and the units for charge squared cancels with quantity in the numerator, but also includes the unit of force (newtons). The answer that includes the correct placement of the units in the necessary arrangement is choice D.

6. **C** The median is the middle value of a data set. However, because there is an even number of values in this data set, the median is calculated by the average of the middle two values, once assembled in increasing order. The test scores in increasing order are 56, 75, 76, 86, 88, 92, 93, 96. The median is then determined by

$$\text{Median} = \frac{86 + 88}{2} = \frac{174}{2} = 87$$

7. **A** The average value of a data set is the sum of all the measurements divided by the number of measurements, or

$$\text{Average rainfall (inches)} = \frac{1.0\text{ in} + 0\text{ in} + 1.5\text{ in} + 0.5\text{ in} + 1.3\text{ in} + 3.4\text{ in} + 2.3\text{ in}}{7} = \frac{10\text{ in}}{7} = 1.4\text{ in}$$

8. **C** The range of a data set is defined as the difference between the highest and lowest values in the data set. In the data set of weekly quizzes, the highest score is 96 and the lowest score is 72. Thus, the range of this data set is $96 - 72 = 24$.

9. **C** Referring to the rules involving significant figures, the quantity 204.0 has 3 significant figures.

10. **B** Referring to the rules involving significant figures,

(a) 0.0889	3 significant figures
(b) 60.06	4 significant figures
(c) 12,400,000	3 significant figures
(d) 578,000	3 significant figures
(e) 0.02586	4 significant figures

Only (b) and (e) have 4 significant figures.

CHAPTER 2 PHYSICS FUNDAMENTALS Answers and Explanations

1. **D** Density is defined as mass divided by volume. The dimension of mass is simply mass or M and the dimension of volume is L^3 ($L \times L \times L$). So, the dimension of density is $\frac{M}{L^3}$.

2. **B** The only unit that is not an SI unit is ft. The correct expression of length as an SI unit is meter (m).

3. **C** The coefficient 9 is multiplied by t (time) that has units of seconds. Because x is displacement and measured in units of meters, then the units of 9 when multiplied by seconds yields meters. So, the units of 9 are $\frac{\text{m}}{\text{s}}$: $x(\text{m}) = 4(\text{m}) + 9\left(\frac{\text{m}}{\text{s}}\right) t(\text{s})$.

4. **C** The cell is assumed to be spherical, and in order to solve this problem you need to know the surface area (SA) and volume (V) of a sphere:

$$\text{SA} = 4\pi R^2 \text{ and } V = \frac{4}{3}\pi R^3$$

Solving the expression for volume in terms of R $\left(R = \sqrt[3]{\frac{3V}{4\pi}}\right)$ and then substituting into the expression for surface area yields

$$\text{SA} = 4\pi\left(\sqrt[3]{\frac{3V}{4\pi}}\right)^2 = 4\pi\left(\frac{3V}{4\pi}\right)^{\frac{2}{3}} = 4\pi\left(\frac{3}{4\pi}\right)^{\frac{2}{3}} \cdot (V)^{\frac{2}{3}} = 4.8V^{\frac{2}{3}}$$

5. **B** Converting light-years from kilometers to inches can be done using the appropriate conversion factors:

$$9.5 \times 10^{12} \text{ km} \times \left(\frac{1000 \text{ m}}{1 \text{ km}}\right) \times \left(\frac{100 \text{ cm}}{1 \text{ m}}\right) \times \left(\frac{1 \text{ in}}{2.54 \text{ cm}}\right)$$

$$= \frac{9.5 \times 10^{12} \times 1000 \times 100}{2.54} \text{ in} = 3.7 \times 10^{17} \text{ in}$$

6. **D** One can convert blood pressure from units of mmHg to units of N/m^2 using the given conversion factor:

$$120 \text{ mmHg} \times \left(\frac{1333 \frac{\text{N}}{\text{m}^2}}{1 \text{ mmHg}}\right) = 1.6 \times 10^5 \frac{\text{N}}{\text{m}^2}$$

7. **B** Revolution is not a recognized unit and thus is dimensionless. Time is a recognized unit, and because it is in the denominator, the dimensions of angular speed is T^{-1}.

8. **C** This problem is asking how many "areas" of dollar bills would equal the area of a football field, or $\text{Area}_{\text{foot}} = \text{Area}_{\text{dol}} \times$ no. of dollar bills. Therefore, no. of dollar

bills $= \dfrac{\text{Area}_{\text{foot}}}{\text{Area}_{\text{dol}}}$. However, before one can directly substitute into the equation for the no. of dollar bills, the units between the two quantities must be consistent (i.e., one must convert 15.9 in^2 to square yards).

$$15.9\ \text{in}^2 \times \left(\frac{1\ \text{yd}}{36\ \text{in}}\right) \times \left(\frac{1\ \text{yd}}{36\ \text{in}}\right) = \frac{15.9}{36 \times 36}\ \text{yd}^2 = 1.2 \times 10^{-2}\ \text{yd}^2$$

Thus,

$$\text{No. of dollar bills} = \frac{\text{Area}_{\text{foot}}}{\text{Area}_{\text{dol}}} = \frac{6.6 \times 10^3\ \text{yd}^2}{1.2 \times 10^{-2}\ \text{yd}^2}$$

$= 5.5 \times 10^5$ dollar bills, or 550,000 dollar bills.

9. A The dimensions of acceleration can be determined by

$$a = \frac{\Delta v}{\Delta t} \Rightarrow \frac{\frac{L}{T}}{T} = \frac{L}{T^2}$$

10. B Scientific notation simplifies the presentation of numeric values by expressing the value as a number between 1 and 10 raised to a power determined by the number of zeros in the value. Thus, 589,000,000 $\Rightarrow 5.89 \times 10^8$.

CHAPTER 3 KINEMATICS
Answers and Explanations

1. B The average velocity $\overline{v}$ is defined as

$$\overline{v} = \frac{\Delta x}{\Delta t}$$

where Δx is the change in displacement and Δt is the time interval over which the change in displacement occurs. Because the displacement varies between both legs of the trip in magnitude and direction, the change in displacement can be found by determining the resultant vector of the two displacement vectors using vector addition

$$\Delta x = \sqrt{(600\text{ m})^2 + (800\text{ m})^2} = 1000\text{ m}$$

The change in time can be found by summing the time of each of the two legs of the trip, or

$$\Delta t = 200\text{ s} + 150\text{ s} = 350\text{ s}$$

Thus,

$$\overline{v} = \frac{\Delta x}{\Delta t} = \frac{1000\text{ m}}{350\text{ s}} = 2.8\ \frac{\text{m}}{\text{s}}$$

The direction can be found by determining the angle that the net displacement makes with the $+x$-axis using the relation

$$\theta = \tan^{-1}\left(\frac{y}{x}\right) = \tan^{-1}\left(\frac{800\text{ m}}{600\text{ m}}\right) = 53°$$

2. A As the golf ball follows the parabolic trajectory, the horizontal component of its velocity, which can be calculated by $v_{i,x} = v_i \cos\theta_i$, is positive and remains constant throughout the entire trajectory. Based on this information, one can readily exclude option C as a possible answer. The vertical component of velocity begins with its maximum (positive) value at the origin, calculated by $v_{i,y} = v_i \sin\theta_i$, and then follows according to the equation $v_{f,y} = v_{i,y} + g\Delta t$, where $g = -9.8\text{ m/s}^2$. In other words, the equation that describes the vertical component of velocity throughout the trajectory is

$$v_{f,y} = v_i\,(\sin\theta) - g\Delta t$$

Thus, at $\Delta t = 0$, the vertical component of velocity is at its positive maximum value, which decreases until it reaches zero at its maximum altitude (represented by the point where the graph intersects the x-axis). After the golf ball stops temporarily, the vertical component of velocity begins to increase, but, because the golf ball is moving downward, it is negative. The negative values are indicated by the graph moving linearly under the positive x-axis. When the golf ball lands, the magnitude of the vertical component of velocity is equal to its value at launch but its direction is opposite to that of its launch. The pair of graphs that best represents both components of velocity in projectile motion is choice A.

3. **C** The horizontal component of velocity of the golf ball moving in projectile motion is constant. Thus, the horizontal component of acceleration (the change in velocity over change in time) is zero, represented by a line drawn over the $+x$-axis. Based on this information, choices A and D can readily be excluded. The vertical component of acceleration is constant in projectile motion and is equal to g ($= -9.8$ m/s^2). The graph of the vertical component of acceleration is a straight line drawn from -9.8 m/s^2.

4. **B** This problem can best be solved by identifying the equations that describe the relationships depicted in the series of graphs.

Distance versus time:	$\Delta x = v_i \Delta t + \frac{1}{2} a (\Delta t)^2$
Speed versus time:	$v_f = v_i + at$
Acceleration versus time:	$a =$ constant

The equation describing the relationship of distance versus time is quadratic with respect to time, eliminating choices C and D as possible answers. The velocity increases linearly with respect to time, shown in both choices A and B. However, acceleration is constant and positive because the car is accelerated or speeding up, and choice B is the correct response.

5. **A** The final velocity of the two tennis balls can be calculated from the equation $v_{f,y} = v_{i,y} + at$, where $a = -g$. The difference in the two cases is with the initial velocity. The thrown ball has an initial velocity, whereas the initial velocity of the dropped ball is zero because it was dropped from rest. Thus, $v_{f,\text{drop}} < v_{f,\text{thrown}}$. Because both tennis balls are moving under the influence of gravity, they both have the same acceleration due to gravity, and $a_{\text{drop}} = a_{\text{thrown}}$.

6. **D** This problem requires the use of the free-fall equation: $v_f^2 = v_i^2 + 2g\Delta y$, where $v_i = 20$ m/s, $g = -9.8$ m/s^2, and $\Delta y = -200$ m. (NOTE: The minus sign represents the downward direction of both the acceleration due to gravity as well as the height of the cliff measured from the point of origin and extending downward the length of the cliff.) Substituting all known values into the equation:

$$\left(v_f^2\right) = \left(20\ \frac{\text{m}}{\text{s}}\right)^2 + 2\left(-9.8\ \frac{\text{m}}{\text{s}^2}\right)(-200\ \text{m})$$

$$= 400\ \frac{\text{m}^2}{\text{s}^2} + 3920\ \frac{\text{m}^2}{\text{s}^2} = 4320\ \frac{\text{m}^2}{\text{s}^2}$$

$$v_f = 65.7\ \frac{\text{m}}{\text{s}}$$

7. **B** This problem can be solved using the equation $v_f^2 = v_i^2 + 2g\Delta y$, where $\Delta y = h$ and $v_f = 0$. Throwing the ball straight up yields a height of $h = \frac{v_i^2}{2g}$. For the baseball thrown at an angle of 45°,

$$h = \frac{(v_i \sin 45°)^2}{2g} = \frac{\left(v_i \left[\frac{\sqrt{2}}{2}\right]\right)^2}{2g} = \frac{v_i^2 \left(\frac{2}{4}\right)}{2g} = \frac{v_i^2 \left(\frac{1}{2}\right)}{2g}$$

which is one-half the height attained by the baseball thrown straight up.

8. **C** The bus route involved three different stages: (1) uniform acceleration; (2) constant speed; and (3) uniform deceleration. An object in uniform acceleration must have a linearly increasing speed, and consequently an object in uniform deceleration must have a linearly decreasing speed. The middle leg of the trip is constant speed and is represented on a graph of velocity versus time as a straight line. The answer that correctly addresses each of these three stages is choice C.

9. **D** The distance traveled by the object can be determined from the equation of motion

$$\Delta y = v_i \Delta t + \frac{1}{2} a (\Delta t)^2$$

However, because the object is dropped from rest, Δy is -1.0 m, $v_i = 0$, $a = -g$. The time can be found by

$$\Delta t = \sqrt{\frac{2(-1.0 \text{ m})}{-9.8 \frac{\text{m}}{\text{s}^2}}} = 0.45 \text{ s}$$

10. **A** The distance traveled by the object can be determined from the equation of motion: $\Delta y = v_i \Delta t + \frac{1}{2} a (\Delta t)^2$, where $a = -g$. To determine the distance the object has fallen during its last second of motion, the total time for the projectile to strike the ground is needed. Thus, because the object is dropped from rest, $\Delta y = -75$ m, $v_i = 0$, $a = -g$, the total time can be found by $-75 \text{ m} = -\frac{1}{2} g (\Delta t)^2$, and Δt is

$$\Delta t = \sqrt{\frac{2(-75 \text{ m})}{-9.8 \frac{\text{m}}{\text{s}^2}}} = 3.9 \text{ s}$$

As the object reaches its last second of motion, it has traveled a distance during the time interval 3.9 s – 1.0 s = 2.9 s, equal to

$$-\Delta y = -\frac{1}{2} g (\Delta t)^2 \text{ or } \Delta y = \frac{1}{2} \left(9.8 \frac{\text{m}}{\text{s}^2}\right) (2.9 \text{ s})^2 = 41.2 \text{ m}$$

Thus, the total distance (Δy_{fs}) traveled by the object during the final second of motion is $\Delta y_{fs} = 75 \text{ m} - 41.2 \text{ m} = 33.8 \text{ m}$.

CHAPTER 4 FORCES AND NEWTON'S LAWS Answers and Explanations

1. **C** The normal force is the support force provided by the surface in contact with the object and always acts perpendicular to the surface. By this fact alone, choices A and D can be eliminated from further consideration. Frictional force always opposes motion and thus, because the object is sliding down the ramp, the frictional force is represented by an upward arrow. The gravitational force or weight of the object always acts downward. Although the gravitational force in both choices B and C are depicted by downward arrows, the correct representation of the gravitational force is straight down, and not downward along the ramp. Thus, the correct answer is choice C.

2. **B** The frictional force can be determined by applying Newton's Second Law to the object's motion along the surface of the inclined plane and calculating the sum of forces occurring in the x direction.

$$\Sigma \mathbf{F}_x = m\mathbf{g} \sin\theta - \mathbf{f} = m\mathbf{a}$$

Solving for $\mathbf{f}$ yields

$$\mathbf{f} = m\mathbf{g} \sin\theta - m\mathbf{a}$$

$$\mathbf{f} = (3 \text{ kg})(9.8 \text{ m/s}^2)(\sin 40°) - (3 \text{ kg})(3 \text{ m/s}^2) = 9.89 \text{ N}$$

and the answer closest to this value is choice B.

3. **D** The frictional force is defined as

$$\mathbf{f} = \mu\mathbf{N}$$

where $\mathbf{f}$ is the frictional force, μ is the coefficient of friction, and $\mathbf{N}$ is the normal force. The frictional force was calculated in the previous problem to be

$$\mathbf{f} = 9.89 \text{ N}$$

The normal force can be found by applying Newton's Second Law to the forces acting in the y direction, that is,

$$\Sigma F_y = \text{N} - mg\cos\theta = 0$$

$\Sigma \mathbf{F}_y = 0$ because, although the object is moving, motion is along the surface of the ramp and not in the y direction. Therefore,

$$\text{N} - mg\cos\theta = 0$$

or

$$\text{N} = mg\cos\theta = (3 \text{ kg})\left(9.8 \frac{\text{m}}{\text{s}^2}\right)\cos(40°) = 22.5 \text{ N}$$

Substituting into the expression for the frictional force yields the coefficient of friction,

$$f = \mu N \Rightarrow \mu = \frac{f}{N} = \frac{9.89N}{22.5N} = 0.44$$

4. A A free-body diagram is a representation of all forces acting on an object at a particular instant. In this case, the particular instant is just after the football has been kicked. At that point, the only force acting on the football is that due to gravity (weight), causing the ball to slow down on its ascent at a rate of 9.8 m/s every second. Thus, the correct answer is choice A.

5. B In order to determine the normal force, one must apply Newton's Second Law in the y-direction to each of the four scenarios shown. For the scenario in option A, the only forces in the y direction are the normal force **N** (acting upward) and the weight, **W** (acting downward). In addition, because the object is not moving in the y direction, Newton's Second Law in the y direction is equal to zero. So, choice A, $\Sigma F_y = N - W = 0$, or $N = W$, which is not the correct answer.

For the scenario in choice B, the forces in the y direction are the normal force **N** (acting upward), the y component of the force **F** (also acting upward) and the weight, **W** (acting downward). So, choice B, $\Sigma F_y = N + F \sin\theta - W = 0$, or $N = W - F \sin\theta$, which is the correct answer. For the scenarios in choices C and D, the forces in the y direction are the normal force **N** (acting upward), the y component of the force **F** (which in this case now acts downward), and the weight, **W** (acting downward). So, choices C and D, $\Sigma F_y = N - F \sin\theta - W = 0$, or $N = W + F \sin\theta$, which are not the correct answers.

6. A Frictional force can be determined by the equation

$$\boldsymbol{f} = \mu \boldsymbol{N}$$

where **N** is the normal force. The normal force can be found by applying Newton's Second Law in the y direction to the scenario described in the problem. A free-body diagram of the forces acting on the weighted disk is given:

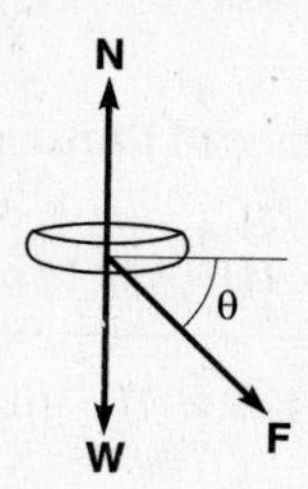

$$\Sigma F_y = N + F \sin(-\theta) - W = N - F \sin\theta - W = 0$$

Solving for the normal force **N**

$$N = W + F \sin\theta = mg + F \sin\theta$$

and the frictional force, $\boldsymbol{f}$, is

$$\boldsymbol{f} = \mu \boldsymbol{N} = \mu\,(mg + F \sin\theta)$$

and the correct answer is choice A.

7. C The satellite that is moving in a circular orbit is acted on by a centripetal force

$$F_c = \frac{m_s v^2}{r}$$

The centripetal force occurs as a result of the attractive gravitational force, $\boldsymbol{F_g}$

$$F_g = G\frac{M_E m_s}{r^2}$$

and

$$F_c = F_g$$

$$\frac{m_s v^2}{r} = G\frac{M_E m_s}{r^2}$$

$$v^2 = G\frac{M_E}{r}$$

where $G = 6.67 \times 10^{-11}\,\frac{\text{Nm}^2}{\text{kg}^2}$, $m_E = 6 \times 10^{24}$ kg, and r_o, the radius of the orbit measured from the center of the Earth, is $r_o = r_E + r_s = 6.4 \times 10^6\ \text{m} + 0.6 \times 10^6\ \text{m} = 7.0 \times 10^6\ \text{m}$.

$$v^2 = G\frac{M_E}{r} = \left(6.67 \times 10^{-11}\,\frac{\text{Nm}^2}{\text{kg}^2}\right)\frac{\left(6 \times 10^{24}\ \text{kg}\right)}{\left(7.0 \times 10^6\ \text{m}\right)} = 5.7 \times 10^7\,\frac{\text{m}^2}{\text{s}^2}$$

$$v = 7549\,\frac{\text{m}}{\text{s}}$$

8. A The gravitational force exerted by the scientist on the surface of Pluto is given by

$$W = mg = G\frac{M_{Pl} m_{ss}}{r_{Pl}^2} = \left(6.67 \times 10^{-11}\,\frac{\text{Nm}^2}{\text{kg}^2}\right)\frac{\left(5 \times 10^{23}\ \text{kg}\right)(68\ \text{kg})}{\left(4.0 \times 10^5\ \text{m}\right)^2}$$

$$= 1.42 \times 10^4\ \text{N}$$

9. B Begin with

$$mg = G\frac{M_E m}{r_E^2}$$

where m is the mass of any object on the surface of the Earth. The acceleration due to gravity can be calculated by

$$g = G\frac{M_E}{r_E^2}$$

If the mass of the Earth were halved and the radius doubled, the acceleration due to gravity is

$$g = G\frac{M_E}{r_E^2} = G\frac{\left(\frac{M_E}{2}\right)}{(2r_E)^2}$$

$$= \left(6.67 \times 10^{-11}\frac{\text{Nm}^2}{\text{kg}^2}\right)\frac{\left(\frac{6 \times 10^{24}\text{ kg}}{2}\right)}{\left[2\left(6.4 \times 10^6\text{ m}\right)\right]^2}$$

$$= \left(6.67 \times 10^{-11}\frac{\text{Nm}^2}{\text{kg}^2}\right)\frac{3 \times 10^{24}\text{ kg}}{163.8 \times 10^{12}\text{ m}}$$

$$= 1.2\frac{\text{m}}{\text{s}^2}$$

10. C The x- and y-coordinates of the center of mass can be found by

$$x_{cm} = \frac{\Sigma m_i x_i}{\Sigma m_i}, \quad y_{cm} = \frac{\Sigma m_i y_i}{\Sigma m_i}$$

The water molecule has three particles (2 hydrogen atoms and 1 oxygen atom). Assuming that the origin coincides with the oxygen atom,

$$x_{cm} = \frac{\Sigma m_i x_i}{\Sigma m_i} = \frac{m_O x_O + m_H x_{H,top} + m_H x_{H,bottom}}{m_O + m_H + m_H}$$

$$y_{cm} = \frac{\Sigma m_i y_i}{\Sigma m_i} = \frac{m_O y_O + m_H y_{H,top} + m_H y_{H,bottom}}{m_O + m_H + m_H}$$

where $m_O = 26.6 \times 10^{-27}$ kg, $m_H = 1.67 \times 10^{-27}$ kg. The x- and y-coordinates of the oxygen atom is 0.0, because the oxygen atom is considered to be the origin. The x- and y-coordinates of the hydrogen atom must be determined using trigonometry.

$$x_{H,top} = (0.96 \times 10^{-10}\text{ m})\cos(52.5°) = 5.85 \times 10^{-11}\text{ m}$$

$$y_{H,top} = (0.96 \times 10^{-10}\text{ m})\sin(52.5°) = 7.58 \times 10^{-11}\text{ m}$$

$$x_{H,bottom} = (0.96 \times 10^{-10}\text{ m})\cos(-52.5°) = 5.85 \times 10^{-11}\text{ m}$$

$$y_{H,bottom} = (0.96 \times 10^{-10}\text{ m})\sin(-52.5°) = -7.58 \times 10^{-11}\text{ m}$$

Thus,

$$x_{cm} = \frac{\left(26.6 \times 10^{-27}\ \text{kg}\right)(0.0) + \left(1.67 \times 10^{-27}\ \text{kg}\right)\left(5.85 \times 10^{-11}\ \text{m}\right) + \left(1.67 \times 10^{-27}\ \text{kg}\right)\left(5.85 \times 10^{-11}\ \text{m}\right)}{26.6 \times 10^{-27}\ \text{kg} + 1.67 \times 10^{-27}\ \text{kg} + 1.67 \times 10^{-27}\ \text{kg}}$$

$$= \frac{0.0 + 9.76 \times 10^{-38}\ \text{kg} \cdot \text{m} + 9.76 \times 10^{-38}\ \text{kg} \cdot \text{m}}{29.9 \times 10^{-27}\ \text{kg}}$$

$$= \frac{19.52 \times 10^{-38}\ \text{kg} \cdot \text{m}}{29.9 \times 10^{-27}\ \text{kg}}$$

$$= 0.65 \times 10^{-11}\ \text{m}$$

$$= 6.5 \times 10^{-12}\ \text{m}$$

$$y_{cm} = \frac{\left(26.6 \times 10^{-27}\ \text{kg}\right)(0.0) + \left(1.67 \times 10^{-27}\ \text{kg}\right)\left(7.58 \times 10^{-11}\ \text{m}\right) + \left(1.67 \times 10^{-27}\ \text{kg}\right)\left(-7.58 \times 10^{-11}\ \text{m}\right)}{26.6 \times 10^{-27}\ \text{kg} + 1.67 \times 10^{-27}\ \text{kg} + 1.67 \times 10^{-27}\ \text{kg}}$$

$$= \frac{0.0 + 12.68 \times 10^{-38}\ \text{kg} \cdot \text{m} - 12.68 \times 10^{-38}\ \text{kg} \cdot \text{m}}{29.9 \times 10^{-27}\ \text{kg}}$$

$$= 0.0\ \text{m}$$

CHAPTER 5 PARTICLE DYNAMICS: WORK, ENERGY, AND POWER
Answers and Explanations

1. D The work done by the worker in moving the box can be determined from the equation

$$\text{Work} = (\text{force}) \cdot (\text{distance}) \cdot \cos\theta$$

where $\theta = 0^\circ$ because the force is exerted on the box horizontally. In other words, the angle between the force and the direction of motion is 0° or are parallel. Therefore,

$$\text{Work} = (150\text{ N}) \cdot (30\text{ m}) \cdot \cos 0^\circ = 4500\text{ J}$$

2. D The work done by the athlete in lifting the barbell can be determined from the equation

$$\text{Work} = (\text{force}) \cdot (\text{distance}) \cdot \cos\theta$$

where $\theta = 180^\circ$ because the force due to gravity is acting in a direction opposite to the direction of motion (gravity acts downward while the barbell is lifted upward). Therefore,

$$\text{Work} = (2500\text{ N}) \cdot (1.85\text{ m}) \cdot \cos 180^\circ = -4625\text{ J}$$

3. B The work done by the mother in pushing the carriage can be determined from the equation

$$\text{Work} = (\text{force}) \cdot (\text{distance}) \cdot \cos\theta$$

where $\theta = -40^\circ$ because the angle is 40° below the horizontal axis. Therefore,

$$\text{Work} = (80\text{ N}) \cdot (20\text{ m}) \cdot \cos(-40^\circ) = 1225.6\text{ J}$$

4. C The work done by friction as the block slides down the ramp is given by

$$\text{Work} = (\text{force}) \cdot (\text{distance})$$

where the force is the frictional force and distance is the length of the ramp that the block covers. The frictional force exerted on the block is

$$\boldsymbol{f} = \mu\mathbf{N}$$

where $\mathbf{N} = \mathbf{W}\cos\theta = m\mathbf{g}\cos\theta$. Thus,

$$\boldsymbol{f} = \mu\mathbf{N} = (0.42)(6\text{ kg})(9.8\text{ m/s}^2)(\cos 35^\circ) = 20.2\text{ N}$$

The distance can be found by visualizing the ramp as a right triangle, where the base of the triangle is the height of the ramp, given by 2.5 m, and the variable of interest (ramp distance) is the hypotenuse. These sides of the right triangle are related through the sine of the slope angle by

$$\sin\theta = \frac{\text{opposite}}{\text{hypotenuse}} \Rightarrow \text{hypotenuse} = \frac{\text{opposite}}{\sin 35^\circ} = \frac{2.5\text{ m}}{0.574} = 4.4\text{ m}$$

The energy lost due to friction, E_f, is

$$E_f = (\text{frictional force}) \cdot (\text{slope distance}) = 20.2\text{ N} \cdot 4.4\text{ m} = 88.8\text{ J}$$

5. B The conservation of energy states

$$(\text{Total mechanical energy})_{\text{top}} = (\text{total mechanical energy})_{\text{bottom}}$$

Because the slide is frictionless, the child has gravitational potential energy (GPE) at the top of the slide that is transformed to kinetic energy (KE) at the bottom of the slide. At a point halfway above the ground, the kinetic energy is 350 J, which means that starting at the top of the slide, the child had 0 J of kinetic energy. As the child slides down the slide, the child gains kinetic energy while losing potential energy. At a point halfway above the ground, the child has lost 50% of his potential energy while gaining 50% of his kinetic energy, meaning at the bottom of the slide, the child attains a total kinetic energy of 700 J. If the child has 700 J of kinetic energy at the bottom of the slide, the child must have had 700 J of potential energy at the top of the slide, based on the assumption that the slide was frictionless. Thus, the correct answer is choice B.

6. B The length of the slide can be determined using trigonometry. It is known that

$$\sin\theta = \frac{\text{opposite side}}{\text{hypotenuse}}$$

where $\theta = 55^\circ$, the opposite side is 3.5 m, and the hypotenuse is the desired quantity.

$$\text{hypotenuse} = \frac{\text{opposite side}}{\sin\theta} = \frac{3.5\text{ m}}{\sin 55^\circ} = \frac{3.5\text{ m}}{0.819} = 4.27\text{ m}$$

7. D From Problem 5, the kinetic energy of the child at the bottom of the slide was 700 J. Substituting into the expression for kinetic energy

$$\text{KE} = \frac{1}{2}mv^2 = \frac{1}{2}\left(\frac{200\text{ N}}{9.8\,\frac{\text{m}}{\text{s}^2}}\right) \cdot (v)^2 = 700\text{ J}$$

$$700\text{ J} = (10.2\text{ kg})\, v^2$$

Solving for $\mathbf{v}$,

$$\mathbf{v} = 8.28\text{ m/s}$$

8. B As the ball rolls off the tabletop, it has its maximum potential energy that steadily decreases as the height of the ball with respect to the ground reduces, until it eventually becomes zero once it hits the ground. Also, as the ball rolls off the table, the ball has some kinetic energy because it does have a net velocity due to its horizontal velocity. The net velocity increases as the vertical velocity of the ball increases for each second that gravity is acting on the ball. The kinetic energy depends on the square of the velocity. Thus, the pair of graphs that best represents the potential energy and kinetic energy of the falling marble is given by choice B.

9. A In order to calculate the kinetic energy of the stone as it strikes the ground, one must calculate the final velocity of the stone using the equation of motion:

$$v_f^2 = v_i^2 + 2g\Delta y$$

where v_f = final velocity, v_i = initial velocity (= 0 m/s because the stone was dropped from rest), $g = -9.8\ \text{m/s}^2$, and $\Delta y = -50$ m (because the distance is measured from its starting point to its final point 50 m downward). Substituting into the equation yields

$$v_f^2 = 2\left(-9.8\ \frac{\text{m}}{\text{s}^2}\right)(-50\ \text{m}) = 980\ \frac{\text{m}^2}{\text{s}^2}$$

$$v_f = \sqrt{980\ \frac{\text{m}^2}{\text{s}^2}} = 31.3\ \frac{\text{m}}{\text{s}}$$

The kinetic energy is found from the equation

$$\text{KE} = \frac{1}{2}mv^2 = \frac{1}{2}(35\ \text{kg})\left(31.3\ \frac{\text{m}}{\text{s}}\right)^2 = 1.7 \times 10^4\ \text{J}$$

given by choice A.

10. A The potential energy is a maximum at the height h_i and becomes zero once it hits the floor. The potential energy as a function of height is given by the equation PE = mgh, which is linear ($y = mx + b$, where the y-intercept $b = 0$). If potential energy is the y variable and h is the x variable, the slope of the equation is mg, or the weight of the marble, given by choice A.

CHAPTER 6 MOMENTUM AND IMPULSE

Answers and Explanations

1. B The momentum, p, can be determined using the equation

$$p = mv$$

where m = mass of the object and v is the object's velocity. The mass of the car is in units of kilograms, but the velocity, which is in units of km/h, must be converted to meters per second.

$$64\frac{\text{km}}{\text{h}} \cdot \left(\frac{1000\text{ m}}{1\text{ km}}\right) \cdot \left(\frac{1\text{ h}}{3600\text{ s}}\right) = 17.8\ \frac{\text{m}}{\text{s}}$$

The momentum can now be calculated

$$p = mv = (1500\text{ kg}) \cdot \left(17.8\ \frac{\text{m}}{\text{s}}\right) = 26{,}700\ \frac{\text{kgm}}{\text{s}}$$

2. A Before one can determine the momentum, one must first calculate the velocity of the baseball as it strikes the glove. At the highest point, the initial speed, v_i, of the baseball is 0 m/s. The distance that the baseball travels, $\Delta y = 20\text{ m} - 1.2\text{ m} = 18.8\text{ m}$. (The height is actually -18.8 m because it is measured downward from its initial position at its maximum height.) Because the baseball is in free-fall from its maximum height, acceleration is g ($= -9.8\text{ m/s}^2$). The final velocity, v_f, can be determined by

$$v_f^2 = v_i^2 + 2g\Delta y = \left(0\ \frac{\text{m}}{\text{s}}\right)^2 + 2\left(-9.8\ \frac{\text{m}}{\text{s}^2}\right)(-18.8\text{ m}) = 368.5\ \frac{\text{m}^2}{\text{s}^2}$$

$$v_f = 19.2\ \frac{\text{m}}{\text{s}}$$

The momentum of the baseball as it strikes the glove is

$$p = mv = (0.45\text{ kg}) \cdot \left(19.2\ \frac{\text{m}}{\text{s}}\right) = 8.64\ \frac{\text{kgm}}{\text{s}}$$

3. D One can assume that the ball moves toward the batter along the positive x direction and, once hit by the bat, moves toward the pitcher in the negative x direction. The impulse, which is the change in momentum, can be calculated from

$$I = \Delta p = m\Delta v = m\left(v_f - v_i\right) = (0.40\text{ kg})\left(-20\ \frac{\text{m}}{\text{s}} - 30\ \frac{\text{m}}{\text{s}}\right) = -20\ \frac{\text{kgm}}{\text{s}}$$

4. C The average force exerted on the baseball by the bat is related to the impulse by

$$I = F\Delta t = m\Delta v$$

From the previous problem, the impulse I is $-20\ \frac{\text{kgm}}{\text{s}}$. The contact time, $\Delta t = 0.02$ s and the impact force is

$$F = \frac{m\Delta v}{\Delta t} = \frac{-20\ \frac{\text{kgm}}{\text{s}}}{0.02\ \text{s}} = -1000\ \text{N}$$

where the negative sign implies a force exerted in a direction opposite to the motion of the baseball.

5. B The force required to stop the car is related to the change in momentum by

$$F\Delta t = m\Delta v = m\left(v_f - v_i\right)$$

$$F = \frac{m\left(v_f - v_i\right)}{\Delta t} = \frac{2000\ \text{kg}\left(0\ \frac{\text{m}}{\text{s}} - 25\ \frac{\text{m}}{\text{s}}\right)}{7.0\ \text{s}} = 7.1 \times 10^3\ \text{N}$$

6. D This problem illustrates an example of an inelastic collision, described by the conservation of momentum equation

(Total momentum before collision) = (total momentum after collision)

In this case, the "before collision" is the combined momenta of the child and the skateboard, whereas the "after collision" is the individual momenta of the child and the skateboard as they part in different directions.

$$\left(m_{\text{child}} + m_{\text{skateboard}}\right) v_f = \left(m_{\text{child}} v_{\text{child}} + m_{\text{skateboard}} v_{\text{skateboard}}\right)$$

$$(20\ \text{kg} + 4\ \text{kg})\left(0\ \frac{\text{m}}{\text{s}}\right) = \left(20\ \text{kg} \cdot v_{\text{child}} + 4\ \text{kg} \cdot 4.7\ \frac{\text{m}}{\text{s}}\right)$$

$$20\ \text{kg} \cdot v_{\text{child}} = -\left(4\ \text{kg} \cdot 4.7\ \frac{\text{m}}{\text{s}}\right) = -18.8\ \frac{\text{kgm}}{\text{s}}$$

$$v_{\text{child}} = \frac{-18.8\frac{\text{kgm}}{\text{s}}}{20\ \text{kg}} = -0.94\ \frac{\text{m}}{\text{s}}$$

where the negative sign implies a direction opposite to the skateboard.

7. A This problem illustrates an example of an inelastic collision, described by the conservation of momentum equation

(Total momentum before collision) = (total momentum after collision)

In this case, the "before collision" is the individual momenta of the child and the wagon as they approach one another, whereas the "after collision" is the combined momenta of the child and the wagon as they move in the same direction.

$$(m_{\text{child}} v_{\text{child}} + m_{\text{wagon}} v_{\text{wagon}}) = (m_{\text{child}} + m_{\text{wagon}})\, v_f$$

$$\left(24\ \text{kg} \cdot 4.6\ \frac{\text{m}}{\text{s}} + 8\ \text{kg} \cdot 0\ \frac{\text{m}}{\text{s}}\right) = (24\ \text{kg} + 8\ \text{kg})\,(v_f)$$

$$110.4\ \frac{\text{kgm}}{\text{s}} = (32\ \text{kg})\,(v_f)$$

$$v_f = \frac{110.4\ \frac{\text{kgm}}{\text{s}}}{32\ \text{kg}} = 3.45\ \frac{\text{m}}{\text{s}}$$

8. C In order to calculate the height attained by the ballistic pendulum, one must know the velocity with which the bullet and wooden block begin their swing upward. This velocity, $v_{\text{bullet+block}}$, can be found using the conservation of momentum and the equation that describes an inelastic collision

(Total momentum before collision) = (total momentum after collision)

In this case, the "before collision" is the individual momenta of the bullet and the wooden block as they approach one another, whereas the "after collision" is the combined momenta of the bullet and the wooden block as they move by swinging upward.

$$(m_{\text{bullet}} v_{\text{bullet}} + m_{\text{block}} v_{\text{block}}) = (m_{\text{bullet}} + m_{\text{block}})\, v_{\text{bullet+block}}$$

$$\left(0.006\ \text{kg} \cdot 620\ \frac{\text{m}}{\text{s}} + 3\ \text{kg} \cdot 0\ \frac{\text{m}}{\text{s}}\right) = (0.006\ \text{kg} + 3\ \text{kg})\,(v_{\text{bullet+block}})$$

$$3.72\ \frac{\text{kgm}}{\text{s}} = (3.006\ \text{kg})\,(v_{\text{bullet+block}})$$

$$v_{\text{bullet+block}} = \frac{3.72\ \frac{\text{kgm}}{\text{s}}}{3.006\ \text{kg}} = 1.24\ \frac{\text{m}}{\text{s}}$$

The wooden block with the embedded bullet now has kinetic energy that is transferred to potential energy as it swings upward. Using the conservation of energy

$$(\text{Kinetic energy})_{\text{bottom}} = (\text{potential energy})_{\text{top}}$$

$$\frac{1}{2}\,(m_{\text{bullet+block}})\,(v_{\text{bullet+block}})^2 = (m_{\text{bullet+block}})\,(g)\,(h)$$

The mass is present on both sides of the equation and thus cancels out, yielding

$$\frac{1}{2}\,(v_{\text{bullet+block}})^2 = (g)\,(h)$$

or, solving for h and substituting,

$$h = \frac{(v_{\text{bullet+block}})^2}{2g} = \frac{\left(1.24\ \frac{\text{m}}{\text{s}}\right)^2}{2 \cdot 9.8\ \frac{\text{m}}{\text{s}^2}} = 0.078\ \text{m} = 7.8\ \text{cm}$$

9. B Solution of this problem requires the conservation of momentum in two dimensions or

$$(\text{Total momentum before collision})_x = (\text{total momentum after collision})_x$$

$$(m_{\text{sbb}}v_{\text{sbb}} + m_{\text{lbb}}v_{\text{lbb}})_{x,\text{before}} = (m_{\text{sbb}}v_{\text{sbb}} + m_{\text{lbb}}v_{\text{lbb}})_{x,\text{after}}$$

$$\left(4.2\ \text{kg} \cdot 4.7\ \frac{\text{m}}{\text{s}} + 7.0\ \text{kg} \cdot 0\ \frac{\text{m}}{\text{s}}\right)_{x,\text{before}} = (4.2\ \text{kg} \cdot v_{\text{sbb}} + 7.0\ \text{kg} \cdot v_{\text{lbb}})_{x,\text{after}}$$

$$\left(19.7\ \frac{\text{kgm}}{\text{s}}\right)_{x,\text{before}} = (4.2\ \text{kg} \cdot v_{\text{sbb}} + 7.0\ \text{kg} \cdot v_{\text{lbb}})_{x,\text{after}}$$

This equation has two unknown variables and requires a second equation in order to determine both unknowns. The second equation that can be used in elastic collisions is the conservation of kinetic energy

$$\left(\left(\frac{1}{2}\right)m_{\text{sbb}}v_{\text{sbb}}^2 + \left(\frac{1}{2}\right)m_{\text{lbb}}v_{\text{lbb}}^2\right)_{\text{before}} = \left(\left(\frac{1}{2}\right)m_{\text{sbb}}v_{\text{sbb}}^2 + \left(\frac{1}{2}\right)m_{\text{lbb}}v_{\text{lbb}}^2\right)_{\text{after}}$$

$$\left(\left(\frac{1}{2}\right)(4.2\ \text{kg})\left(4.7\ \frac{\text{m}}{\text{s}}\right)^2 + \left(\frac{1}{2}\right)(7.0\ \text{kg})\left(0\ \frac{\text{m}}{\text{s}}\right)^2\right)_{\text{before}}$$

$$= \left(\left(\frac{1}{2}\right)(4.2\ \text{kg})v_{\text{sbb}}^2 + \left(\frac{1}{2}\right)(7.0\ \text{kg})v_{\text{lbb}}^2\right)_{\text{after}}$$

$$46.4\ \text{J} = (2.1\ \text{kg})\ v_{\text{sbb}}^2 + (3.5\ \text{kg})\ v_{\text{lbb}}^2$$

Rearranging and solving the previous equation for v_{lbb} in terms of v_{sbb} yields

$$v_{\text{lbb}} = \sqrt{\frac{46.4\ \text{J} - (2.1\ \text{kg})\ v_{\text{sbb}}^2}{3.5\ \text{kg}}}$$

Substituting into the equation derived from the conservation of momentum

$$19.7\ \frac{\text{kgm}}{\text{s}} = 4.2\ \text{kg} \cdot v_{\text{sbb}} + 7.0\ \text{kg} \cdot \sqrt{\frac{46.4\ \text{J} - (2.1\ \text{kg})\ v_{\text{sbb}}^2}{3.5\ \text{kg}}}$$

or

$$19.7\ \frac{\text{kgm}}{\text{s}} - 4.2\ \text{kg} \cdot v_{\text{sbb}} = 7.0\ \text{kg} \cdot \sqrt{\frac{46.4\ \text{J} - (2.1\ \text{kg})\ v_{\text{sbb}}^2}{3.5\ \text{kg}}}$$

Squaring both sides yields

$$388.1\ \frac{\text{kg}^2\text{m}^2}{\text{s}^2} - 165.4\ \frac{\text{kg}^2\text{m}}{\text{s}} v_{\text{sbb}} + 17.6\ \text{kg}^2 v_{\text{sbb}}^2$$

$$= 49\ \text{kg}^2 \cdot \frac{46.4\ \text{J} - (2.1\ \text{kg})\ v_{\text{sbb}}^2}{3.5\ \text{kg}}$$

Multiplying through and combining like terms yields the quadratic equation

$$-261.5\ \frac{\text{kg}^2\text{m}^2}{\text{s}^2} - 165.4\ \frac{\text{kg}^2\text{m}}{\text{s}} v_{\text{sbb}} + 47.6\ \text{kg}^2 v_{\text{sbb}}^2 = 0$$

which can be solved using the solution of the quadratic equation

$$x = \frac{-b \pm \sqrt{b^2 - 4ac}}{2a}$$

where

$$a = 47.6\ \text{kg}^2 v_{\text{sbb}}^2$$

$$b = -165.4\ \frac{\text{kg}^2\text{m}}{\text{s}} v_{\text{sbb}}$$

$$c = -261.5\ \frac{\text{kg}^2\text{m}^2}{\text{s}^2}$$

The solutions from the quadratic equation, representing the velocity of the smaller bowling ball, are $x = +4.70\ \frac{\text{m}}{\text{s}}$ and $-1.18\ \frac{\text{m}}{\text{s}}$. The solution $x = +4.70\ \frac{\text{m}}{\text{s}}$ cannot be correct, because this would imply that the smaller bowling ball still retains its initial velocity after a head-on collision with a larger bowling ball and is physically impossible. Thus, the correct solution, and the answer to this problem, is $x = -1.18\ \frac{\text{m}}{\text{s}}$.

10. C The final velocity of the larger bowling ball can be solved using either the conservation of momentum equation or the conservation of kinetic energy equation. Opting for the conservation of momentum equation

$$\left(19.7\ \frac{\text{kgm}}{\text{s}}\right)_{x,\text{before}} = \left(4.2\ \text{kg} \cdot \left(-1.2\ \frac{\text{m}}{\text{s}}\right) + 7.0\ \text{kg} \cdot v_{\text{lbb}}\right)_{x,\text{after}}$$

Simplifying and solving for v_{lbb} yields

$$v_{\text{lbb}} = 3.53\ \frac{\text{m}}{\text{s}}$$

CHAPTER 7 SOLIDS AND FLUIDS
Answers and Explanations

1. A From the problem, the load applied to the steel rod exerts a tensile stress on the rod. In order to calculate the stress on the rod, you must first calculate the cross-sectional area of the rod, or

$$A = \pi r^2 = (3.14)\left(\frac{0.25 \text{ m}}{2}\right)^2 = 0.05 \text{ m}^2$$

The tensile stress can now be calculated

$$\text{Stress} = \frac{\text{force}}{\text{area}} = \frac{(10{,}000 \text{ kg})\left(9.8 \frac{\text{m}}{\text{s}^2}\right)}{0.05 \text{ m}^2} = 1.96 \times 10^6 \frac{\text{N}}{\text{m}^2}$$

2. C The tensile strain exerted on the steel rod can be determined from the definition for Young's modulus

$$\text{Young's modulus} = \frac{\text{tensile stress}}{\text{tensile strain}}$$

$$\text{Tensile strain} = \frac{\text{tensile stress}}{\text{Young's modulus}} = \frac{1.96 \times 10^6 \frac{\text{N}}{\text{m}^2}}{2 \times 10^{11} \frac{\text{N}}{\text{m}^2}}$$

$$= 9.8 \times 10^{-6} \approx 1 \times 10^{-5}$$

3. D The change in length of the steel rod, ΔL, can be determined using the expression for the tensile strain

$$\text{Tensile strain} = \frac{\Delta L}{L}$$

$$\Delta L = \text{tensile strain} \cdot L = \left(9.8 \times 10^{-6}\right)(1.5 \text{ m}) = 1.48 \times 10^{-5} \text{ m} \approx 0.015 \text{ mm}$$

4. C In order to determine the shear stress, one must first calculate the surface area of the cube

$$A = L^2 = (0.05 \text{ m})^2 = 0.0025 \text{ m}^2$$

The shear stress can be calculated by

$$\text{Shear stress} = \frac{\text{tangential force}}{\text{surface area}} = \frac{500 \text{ N}}{2.5 \times 10^{-3} \text{ m}^2} = 2.0 \times 10^5 \frac{\text{N}}{\text{m}^2}$$

5. B The density of a solution can be determined from

$$\rho = \frac{m}{V}$$

where ρ is the density of the solution (in units of $\frac{g}{cm^3}$), m is the mass of the solution (in units of g), and V is volume (in units of cm^3). The mass of the solution, m_{sol}, is the combined mass of the solution and syringe, m_C, minus the mass of the syringe, m_{syr}, or

$$m_C = m_{sol} + m_{syr}$$

$$m_{sol} = m_C - m_{syr} = 80 \text{ g} - 5 \text{ g} = 75 \text{ g}$$

The volume of the solution in the syringe is

$$V_{syr} = 30 \text{ mL} = 30 \text{ cm}^3$$

The density of the solution is

$$\rho = \frac{m}{V} = \frac{75 \text{ g}}{30 \text{ cm}^3} = 2.5 \frac{\text{g}}{\text{cm}^3}$$

6. C Pressure is defined by the relationship

$$\text{Pressure} = \frac{\text{force}}{\text{area}}$$

where the force in this problem is the weight of the person, or

$$\text{Weight} = (\text{mass})(\text{acceleration due to gravity}) = (80 \text{ kg})(9.8 \frac{\text{m}}{\text{s}^2}) = 784 \text{ N}$$

Because the SI unit of length is meters, the surface area of 150 cm^2 must be converted to m^2

$$150 \text{ cm}^2 \cdot \left(\frac{1 \text{ m}}{100 \text{ cm}}\right) \cdot \left(\frac{1 \text{ m}}{100 \text{ cm}}\right) = \frac{150}{10{,}000} \text{m}^2 = 0.015 \text{ m}^2$$

For a person, the total surface area for 2 feet is

$$\text{Area} = 2 \cdot 0.015 \text{ m}^2 = 0.03 \text{ m}^2$$

$$\text{Pressure} = \frac{\text{force}}{\text{area}} = \frac{784 \text{ N}}{0.03 \text{ m}^2} = 26{,}133 \frac{\text{N}}{\text{m}^2} = 2.6 \times 10^4 \frac{\text{N}}{\text{m}^2}$$

7. B The pressure exerted at a certain depth within a swimming pool depends on the atmospheric pressure as well as the hydrostatic pressure due to the water above the depth in question. The total pressure exerted at a depth of 180 cm, or 1.8 m, is

$$P_{total} = P_{atm} + P_{hyd} = P_{atm} + \rho g h$$

where ρ is the density of water, g is the acceleration due to gravity, and h is the height (or depth) of the column of water.

$$P_{total} = P_{atm} + P_{hyd} = \left(1.013 \times 10^5 \frac{\text{N}}{\text{m}^2}\right) + \left(1000 \frac{\text{kg}}{\text{m}^3}\right)\left(9.8 \frac{\text{m}}{\text{s}^2}\right)(1.8 \text{ m})$$

$$= 1.19 \times 10^5 \frac{\text{N}}{\text{m}^2}$$

8. D This problem is an illustration of Pascal's principle, which states: "An external pressure applied to a confined fluid will be transmitted equally to all points within the fluid." According to Pascal's principle,

$$P_{0.5\text{m}^2} = P_{2\text{m}^2}$$

$$\left(\frac{F}{A}\right)_{0.5\text{m}^2} = \left(\frac{F}{A}\right)_{2\text{m}^2}$$

$$\left(\frac{2000\ \text{N}}{0.5\ \text{m}^2}\right) = \left(\frac{F}{2\ \text{m}^2}\right)$$

$$F = 8000\ \text{N}$$

9. C The volumetric flow rate, Q, is related to flow velocity, v, by the following relation:

$$Q = vA$$

The volumetric flow rate must first be converted to units of $\frac{\text{cm}^3}{\text{s}}$:

$$Q = 5\frac{\text{L}}{\text{min}} \times \frac{1000\ \text{cm}^3}{1\ \text{L}} \times \frac{1\ \text{min}}{60\ \text{s}} = 83.3\frac{\text{cm}^3}{\text{s}}$$

The cross-sectional area, A, is

$$A = \pi r^2 = (3.14)\,(1.2\ \text{cm})^2 = 4.52\ \text{cm}^2$$

Thus, the blood flow velocity is

$$v = \frac{83.3\ \frac{\text{cm}^3}{\text{s}}}{4.52\ \text{cm}^2} = 18.4\frac{\text{cm}}{\text{s}}$$

10. D The time, t, required to cover a distance or length, x, is given by

$$\text{Velocity} = \frac{\text{length}}{\text{time}} \Rightarrow \text{Time} = \frac{\text{length}}{\text{velocity}} = \frac{30\ \text{cm}}{18.4\ \frac{\text{cm}}{\text{s}}} = 1.63\ \text{s}$$

CHAPTER 8 TEMPERATURE AND HEAT
Answers and Explanations

1. A The Fahrenheit and Celsius temperature scales are related by the following:

$$T_F = \frac{9}{5}T_C + 32° \quad (1)$$

The Celsius and kelvin temperature scales are related by the following:

$$T_C = T_K - 273° \quad (2)$$

Substituting Equation 2 into Equation 1 yields

$$T_F = \frac{9}{5}(T_K - 273°) + 32° = \frac{9}{5}T_K - 491.4° + 32° = \frac{9}{5}T_K - 459.4°$$

2. D In Problem 1, the correct relation converting a temperature measurement from kelvin to degrees Fahrenheit was determined to be

$$T_F = \frac{9}{5}T_K - 460°$$

Rearranging this equation such that temperature expressed in kelvin is the dependent variable yields

$$T_K = \frac{5}{9}(T_F + 460°) = \frac{5}{9}(98.6°\text{ F} + 460°) = 310\text{ K}$$

3. A Defining the unknown temperature as T_x, the relation between temperature in degrees Fahrenheit and degrees Celsius is

$$T_x = \frac{9}{5}T_x + 32°$$

Solving for T_x,

$$-32° = \frac{9}{5}T_x - T_x = \frac{4}{5}T_x$$

$$T_x = \frac{-32° \cdot 5}{4} = \frac{-160°}{4} = -40°$$

4. D The increase in length can be determined from

$$\Delta L = \alpha L_o \Delta T = \left(1.7 \times 10^{-5}°\text{C}^{-1}\right)(1.0\text{ m})\left(40°\text{C} - 25°\text{C}\right)$$

$$= 2.55 \times 10^{-4}\text{ m} = 0.25\text{ mm}$$

5. B The increase in volume of the brass sphere due to temperature can be determined from

$$\Delta V = \beta V_o \Delta T = \left(57 \times 10^{-6}°\text{C}^{-1}\right)\left(500\text{ cm}^3\right)(-10°\text{C} - 110°\text{C}) = 3.42\text{ cm}^3$$

6. A The amount of heat, Q, required to raise the temperature of a substance can be determined from the equation

$$Q = mc\Delta T$$

where m = mass of the substance; c is the specific heat capacity of the substance; and ΔT is the change in temperature. Making the appropriate substitutions yields

$$Q = (5 \text{ kg}) \left(1.0 \frac{\text{kcal}}{\text{kg} \cdot {}^\circ\text{C}}\right) (42^\circ\text{C} - 20^\circ\text{C}) = 110 \text{ kcal}$$

7. D The amount of heat required to change the state from solid to liquid without changing the temperature can be determined from the phase change equation

$$Q = mL_f$$

where m is the amount of substance and L_f is the latent heat of fusion of the substance. Thus, the amount of heat required for the phase change is

$$Q = (4500 \text{ g}) \left(80 \frac{\text{cal}}{\text{g}}\right) = 360 \text{ kcal}$$

8. A From Problem 7, it was determined that the amount of heat required to change the state of the amount of ice in the insulated container was 360 kcal. However, the power output of the immersion heater, in watts, is defined as joules/s. So the amount of heat in kcal must be converted to joules using the conversion factor given in the problem:

$$360 \text{ kcal} \cdot \left(\frac{4184 \text{ J}}{\text{kcal}}\right) = 1.51 \times 10^6 \text{ J}$$

The time, t, required to change the state from solid to liquid is

$$t = \frac{1.51 \times 10^6 \text{ J}}{600 \frac{\text{J}}{\text{s}}} = 2516.7 \text{ s} \cdot \left(\frac{1 \text{ min}}{60 \text{ s}}\right) = 41.9 \text{ min} \approx 42 \text{ min}$$

9. B From the previous problem, it was determined that the amount of heat in joules required to change the state from solid to liquid was 1.51×10^6 J. The power of an immersion heater needed to change the state within 30 minutes (1800 seconds) is

$$P = \frac{1.51 \times 10^6 \text{ J}}{1800 \text{ s}} = 839 \text{ W} \approx 840 \text{ W}$$

10. **A** This problem is an extension of the conservation of energy in that (assuming that no heat energy is lost to its surroundings), the heat of the hotter object is transferred to the cooler object until thermal equilibrium is reached.

(Heat lost by water) = (heat gained by the pot)

$$(mc\Delta T)_{\text{water}} = (mc\Delta T)_{\text{pot}}$$

$$\left((0.5\text{ kg})\left(1.0\frac{\text{kcal}}{\text{kg}\cdot{}^{\circ}\text{C}}\right)(38^{\circ}\text{C} - T_e)\right)_{\text{water}}$$

$$= \left((0.6\text{ kg})\left(0.25\frac{\text{kcal}}{\text{kg}\cdot{}^{\circ}\text{C}}\right)(T_e - 20^{\circ}\text{C})\right)_{\text{pot}}$$

$$\left(\left(0.5\frac{\text{kcal}}{\text{kg}\cdot{}^{\circ}\text{C}}\right)(38^{\circ}\text{C} - T_e)\right)_{\text{water}} = \left(\left(0.15\frac{\text{kcal}}{\text{kg}\cdot{}^{\circ}\text{C}}\right)(T_e - 20^{\circ}\text{C})\right)_{\text{pot}}$$

Multiplying through, collecting like terms, and solving for T_e yields $T_e = 33.8^{\circ}\text{C}$.

CHAPTER 9 VIBRATIONS AND WAVES Answers and Explanations

1. B The frequency, f, of a wave is related to its period, T:

$$f = \frac{1}{T}$$

For a period of 1 millisecond, or 1×10^{-3} s, the frequency is

$$f = \frac{1}{T} = \frac{1}{1 \times 10^{-3}\ \text{s}} = 1000\ \frac{\text{cycles}}{\text{s}} = 1000\ \text{Hz} = 1\ \text{kHz}$$

2. C The period of a pendulum is defined by the relation

$$T = 2\pi\sqrt{\frac{l}{g}}$$

where l is the length of the string of the pendulum and g is the acceleration due to gravity. The period of the pendulum is not dependent on the mass of the pendulum bob. Thus, the period of a pendulum with a mass $2m$ is the same as the period of a pendulum with a mass m.

3. B The period of a pendulum is defined by the relation

$$T = 2\pi\sqrt{\frac{l}{g}}$$

where l is the length of the string of the pendulum and g is the acceleration due to gravity. Doubling the length of the pendulum changes the period of the pendulum by

$$T = 2\pi\sqrt{\frac{2l}{g}} = \sqrt{2} \cdot 2\pi\sqrt{\frac{l}{g}} = \sqrt{2}T$$

4. C The period of a pendulum is defined by the relation

$$T = 2\pi\sqrt{\frac{l}{g}}$$

where l is the length of the string of the pendulum and g is the acceleration due to gravity. The period of the pendulum is not dependent on the angle of release of the pendulum bob. Thus, the period of a pendulum with an angle of release 2θ is the same as the period of a pendulum with an angle of release θ.

5. C The period of a pendulum is defined by the relation

$$T = 2\pi\sqrt{\frac{l}{g}}$$

where l is the length of the string of the pendulum and g is the acceleration due to gravity. Converting 25 cm to 0.25 m and substituting into the previous relation

$$T = 2\pi\sqrt{\frac{0.25\text{ m}}{9.8\,\frac{\text{m}}{\text{s}^2}}} = 1.0\text{ s}$$

6. A The frequency of a pendulum is related to its period by the relation

$$T = \frac{1}{f} = \frac{1}{1.0\text{ s}} = 1\text{ Hz}$$

7. A The period of a mass-spring system is defined by the relation

$$T = 2\pi\sqrt{\frac{m}{k}}$$

where m is the mass of the oscillating object and k is the spring constant of the spring. The effect of quadrupling the mass of the object on the period of the oscillating object is

$$T = 2\pi\sqrt{\frac{4m}{k}} = \sqrt{4}\cdot 2\pi\sqrt{\frac{m}{k}} = 2T$$

8. C The period of a mass-spring system is defined by the relation

$$T = 2\pi\sqrt{\frac{m}{k}}$$

where m is the mass of the oscillating object and k is the spring constant of the spring. The length of the spring is not a factor in the period of the oscillating mass, only the spring constant, which is dependent on a particular spring, not the length of a spring. Thus, the period of the oscillating object with a quadrupled spring length is the same as the period of the oscillating object with the original spring length.

9. D The frequency, f, is inversely related to the period, T, which for a simple pendulum is

$$T = 2\pi\sqrt{\frac{l}{g}}$$

$$f = \frac{1}{T} = \frac{1}{2\pi\sqrt{\frac{l}{g}}} = \frac{1}{2\pi}\sqrt{\frac{g}{l}}$$

10. C The restoring force is that force that acts on the oscillating pendulum bob to restore it to its equilibrium point. For a pendulum, the restoring force, F_r, is

$$F_r = mg\sin\theta = (2.5\text{ kg})\left(9.8\,\frac{\text{m}}{\text{s}^2}\right)\sin(15^\circ) = 6.34\text{ N}$$

CHAPTER 10 SOUND
Answers and Explanations

1. **C** The period is related to the frequency by

$$T = \frac{1}{f} = \frac{1}{256 \text{ Hz}} = 3.9 \times 10^{-3} \text{ s} = 3.9 \text{ ms}$$

2. **A** The wavelength is related to the frequency by

$$\lambda = \frac{v}{f} = \frac{344 \frac{\text{m}}{\text{s}}}{104 \times 10^6 \text{ Hz}} = 3.31 \times 10^{-6} \text{ m} = 3.31 \text{ μm}$$

3. **B** The speed of sound is related to the distance traveled by the sound wave by the standard kinematic equation

$$v = \frac{\Delta x}{\Delta t}$$

where Δx is the distance traveled and Δt is the time required to cover that distance. The time referred to in the question, however, is the time that Jim heard an echo of his voice, that is, the time required for the sound wave to make a round trip. In order to find the time required for the sound wave to reach the depths of the wishing well, this time is simply divided by 2, yielding $\Delta t = 0.25$ s.

$$v = \frac{\Delta x}{\Delta t} \Rightarrow \Delta x = v \cdot \Delta t = \left(344 \frac{\text{m}}{\text{s}}\right)(0.25 \text{ s}) = 86 \text{ m}$$

4. **C** By definition, the first harmonic or fundamental frequency, f, of a stretched string is given by

$$f = \frac{v}{2L}$$

5. **D** The wavelength of a stretched string varies within the harmonic series as

$$\lambda = \frac{2L}{n}$$

where n is the harmonic number. Because the string is in its third harmonic, $n = 3$ and the wavelength is

$$\lambda = \frac{2L}{n} = \frac{2L}{3}$$

6. **B** The fundamental frequency, f_1, is related to the harmonic series of frequencies, f_n, by

$$f_n = nf_1$$

where n is the harmonic number. For this problem, $f_4 = 4f_1 = 458.7$ Hz. Solving for f_1 yields

$$f_1 = \frac{458.7 \text{ Hz}}{4} = 114.7 \text{ Hz} \approx 115 \text{ Hz}$$

7. C The fundamental frequency of a pipe open at both ends is

$$f_1 = \frac{v}{2L}$$

From the previous problem, the fundamental frequency was determined to be 115 Hz. Also, the length of the pipe was given as 1.5 m. Thus, the speed of sound through the pipe can be determined by

$$v = 2Lf_1 = (2)\,(1.5\text{ m})\,(115\text{ Hz}) = 345\,\frac{\text{m}}{\text{s}}$$

8. A The sound level β is related to sound intensity, I, by

$$\beta = 10\log\frac{I}{I_o}$$

where I_o is the threshold of hearing ($= 1 \times 10^{-12}\,\frac{\text{W}}{\text{m}}$). Thus,

$$\beta = 10\log\frac{I}{I_o} = 10\log\frac{5\,\frac{\text{W}}{\text{m}^2}}{1 \times 10^{-12}\,\frac{\text{W}}{\text{m}^2}} = 127\text{ dB}$$

9. B This problem requires the use of the Doppler effect and the equation that describes the Doppler-shifted frequency, f_o, of a source moving toward a stationary observer

$$f_o = f_s\left(\frac{v}{v - v_s}\right)$$

where f_s = frequency of the moving source, v = speed of sound, and v_s = speed of the moving source. Making the appropriate substitutions yields

$$f_o = 1200\text{ Hz}\left(\frac{344\,\frac{\text{m}}{\text{s}}}{344\,\frac{\text{m}}{\text{s}} - 24\,\frac{\text{m}}{\text{s}}}\right) = 1290\text{ Hz}$$

10. C This problem requires the use of the Doppler effect and the equation that describes the Doppler-shifted frequency, f_o, of a source moving passing a stationary observer

$$f_o = f_s\left(\frac{v}{v + v_s}\right)$$

where f_s = frequency of the moving source, v = speed of sound, and v_s = speed of the moving source. Making the appropriate substitutions yields

$$f_o = 1200\text{ Hz}\left(\frac{344\,\frac{\text{m}}{\text{s}}}{344\,\frac{\text{m}}{\text{s}} + 24\,\frac{\text{m}}{\text{s}}}\right) = 1122\text{ Hz}$$

CHAPTER 11 LIGHT AND GEOMETRIC OPTICS Answers and Explanations

1. D The frequency f of a wave is related to its period, T:

$$f = \frac{1}{T}$$

For a frequency of 960,000 Hz, the period is

$$T = \frac{1}{f} = \frac{1}{960{,}000 \text{ Hz}} = 1 \ \mu\text{s}$$

2. B The wavelength of electromagnetic waves is related to frequency by

$$\lambda = \frac{v}{f}$$

where v is the speed of light in a vacuum ($c = 3.0 \times 10^8 \ \frac{\text{m}}{\text{s}}$). Substituting into the previous equation yields

$$\lambda = \frac{v}{f} = \frac{3.0 \times 10^8 \frac{\text{m}}{\text{s}}}{2 \times 10^{16} \text{ Hz}} = 1.5 \times 10^{-8} \text{ m}$$

3. D All types of electromagnetic radiation travel at the same speed—the speed of light ($c = 3.0 \times 10^8 \ \frac{\text{m}}{\text{s}}$).

4. A By definition, the index of refraction is given by the ratio

$$n = \frac{\text{speed of light in vacuum}}{\text{speed of light in material}} = \frac{c}{v}$$

Thus, the speed of light through a pane of glass is

$$v = \frac{c}{n} = \frac{3.0 \times 10^8 \ \frac{\text{m}}{\text{s}}}{1.52} = 1.97 \times 10^8 \ \frac{\text{m}}{\text{s}} \approx 2.0 \times 10^8 \ \frac{\text{m}}{\text{s}}$$

5. C Using Snell's law of refraction

$$n_{\text{air}} \sin \theta_i = n_{\text{water}} \sin \theta_{r'}$$

$$(1.00) \sin (35°) = (1.33) \sin \theta_{r'}$$

$$\sin \theta_{r'} = \frac{(1.00) \sin (35°)}{(1.33)} = 0.43$$

$$\theta_{r'} = \sin^{-1} (0.43) = 25.5° \approx 26°$$

6. D The image distance is related to the focal length by the mirror equation

$$\frac{1}{o} + \frac{1}{i} = \frac{1}{f}$$

where o is the object distance, i is the image distance, and f is the focal length.

$$\frac{1}{i} = \frac{1}{f} - \frac{1}{o} = \frac{1}{15\text{ cm}} - \frac{1}{20\text{ cm}} = \frac{1}{60\text{ cm}}$$

$$i = 60\text{ cm}$$

7. B The magnification, M, of the mirror is defined as

$$M = -\frac{i}{o} = -\frac{60\text{ cm}}{20\text{ cm}} = -3.0$$

8. A In addition to the negative ratio of image distance to object distance, the magnification, M, of the mirror can also be expressed as

$$M = \frac{\text{image height}}{\text{object height}} = \frac{h_i}{h_o}$$

$$h_i = Mh_o = (-3.0)\,(4.5\text{ cm}) = -13.5\text{ cm}$$

The negative image height indicates that the image is inverted.

9. A The positive value for the image distance indicates that the image of the pencil is a real image. The negative value of the magnification indicates that the image is inverted with respect to the object.

10. B This problem can be solved using the thin lens equation

$$\frac{1}{o} + \frac{1}{i} = \frac{1}{f}$$

where o is the object distance, i is the image distance, and f is the focal length.

$$\frac{1}{i} = \frac{1}{f} - \frac{1}{o} = \frac{1}{40\text{ cm}} - \frac{1}{10\text{ cm}} = -\frac{3}{40\text{ cm}}$$

$$i = -\frac{40\text{ cm}}{3} = -13.3\text{ cm}$$

CHAPTER 12 ELECTROSTATICS
Answers and Explanations

1. **A** The electric force can be determined using Coulomb's law:

$$F = \left(9 \times 10^9 \frac{\text{N} \cdot \text{m}^2}{\text{C}^2}\right) \frac{\left(+6.0 \times 10^{-6}\ \text{C}\right)\left(-5.0 \times 10^{-6}\ \text{C}\right)}{\left(1 \times 10^{-2}\ \text{m}\right)^2}$$

$$= -2.7 \times 10^3\ \text{N} = -2700\ \text{N}$$

Because the two charges are unlike in sign, the electric force is attractive.

2. **C** From Problem 1, the electric force that exists between the two charges was found to be −2700 N. Doubling the two charges increases the original force by 4 since the two charges are multiplied together in Coulomb's law. The new electric force = $4 \cdot -2700\ \text{N} = -10{,}800\ \text{N}$.

3. **C** Coulomb's law is given by

$$F = k\frac{q_1 q_2}{r^2}$$

where r is the distance of separation of the two charges. Halving the variable r yields

$$F = k\frac{q_1 q_2}{\left(\frac{r}{2}\right)^2} = 4 \cdot k\frac{q_1 q_2}{(r)^2} = 4 \cdot (-2700\ \text{N}) = -10{,}800\ \text{N}$$

4. **A** Changing the sign of the two electric charges with q_1 going from positive to negative and q_2 going from negative to positive does not change the magnitude of the electric force, nor does it change the overall sign, as the product of the two charges is still negative. Thus, the new electric force is the same as the original electric force.

5. **B** The charge of interest, q_4, is the charge in the lower left corner, and the task is to determine the net effect of the force exerted by each of the three remaining charges on the charge of interest. This is done by using Coulomb's law to determine the force exerted by q_1 on q_4 ($\mathbf{F}_{4,1}$), the force exerted by q_2 on q_4 ($\mathbf{F}_{4,2}$), and the force exerted by q_3 on q_4 ($\mathbf{F}_{4,3}$) or

$$F_{4,1} = k\frac{q_1 q_4}{r_{4,1}^2} = \left(9 \times 10^9\ \frac{\text{N} \cdot \text{m}^2}{\text{C}^2}\right) \frac{\left(+8.0 \times 10^{-6}\ \text{C}\right)\left(-2.0 \times 10^{-6}\ \text{C}\right)}{(0.14\ \text{m})^2}$$

$$= -7.4\ \text{N, attractive}$$

$$F_{4,2} = k\frac{q_2 q_4}{r_{4,2}^2} = \left(9 \times 10^9 \frac{\text{N} \cdot \text{m}^2}{\text{C}^2}\right)\frac{\left(+4.0 \times 10^{-6}\ \text{C}\right)\left(-2.0 \times 10^{-6}\ \text{C}\right)}{(0.10\ \text{m})^2}$$
$$= -7.2\ \text{N, attractive}$$

$$F_{4,3} = k\frac{q_3 q_4}{r_{4,3}^2} = \left(9 \times 10^9 \frac{\text{N} \cdot \text{m}^2}{\text{C}^2}\right)\frac{\left(-4.0 \times 10^{-6}\ \text{C}\right)\left(-2.0 \times 10^{-6}\ \text{C}\right)}{(0.10\ \text{m})^2}$$
$$= +7.2\ \text{N, repulsive}$$

The minus or plus signs associated with the electric force quantities are an indication as to whether the force is attractive or positive, respectively, and not an indication of position. For example, $\mathbf{F}_{4,2}$ is a negative quantity, but due to the orientation of the two charges, the force acting on q_4 by q_2 is attractive, and is thus being pulled in the $+y$ direction. All three electric force vectors are drawn as a free-body diagram on q_4, as shown next:

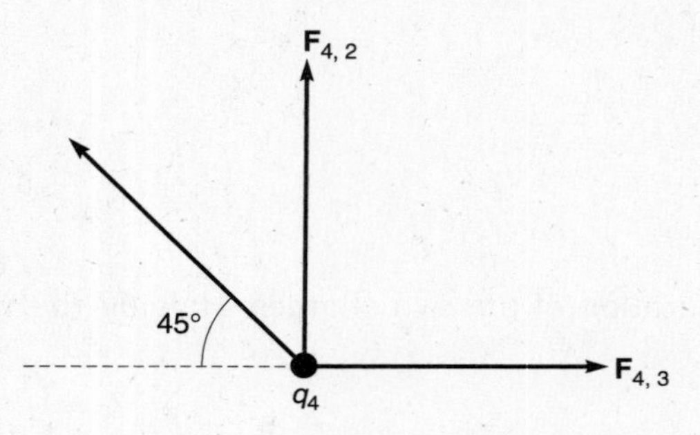

Each electric force vector is now resolved into its corresponding x and y components and arranged in a table as shown next:

Force	*x* Component	*y* Component
$\mathbf{F}_{4,1}$	$7.4\ \text{N} \cos(135^\circ) = -5.2\ \text{N}$	$7.4\ \text{N} \sin(135^\circ) = 5.2\ \text{N}$
$\mathbf{F}_{4,2}$	0	+7.2 N
$\mathbf{F}_{4,3}$	+7.2 N	0
$\mathbf{F}_{\text{net}}$	$\mathbf{F}_{\text{net},x} = +7.2\ \text{N} - 5.2\ \text{N} = +2.0\ \text{N}$	$\mathbf{F}_{\text{net},y} = +5.2\ \text{N} + 7.2\ \text{N} = +12.2\ \text{N}$

The magnitude of the net force can be calculated by

$$F_{\text{net}} = \sqrt{\left(F_{\text{net},x}\right)^2 + \left(F_{\text{net},y}\right)^2} = \sqrt{(2.0\ \text{N})^2 + (12.2\ \text{N})^2} = 12.4\ \text{N}$$

6. C The direction with respect to the x-axis of the net force can be calculated by

$$\theta = \tan^{-1}\left(\frac{F_{\text{net},y}}{F_{\text{net},x}}\right) = \tan^{-1}\left(\frac{12.2\ \text{N}}{2.0\ \text{N}}\right) = \tan^{-1}(6.1) = 80.7^\circ$$

7. **D** Each sphere with the same mass and charge is subjected to the same forces and thus the angle is the same for both charges. Applying Newton's Second Law in the x and y direction to the left sphere yields

$$\Sigma F_x = ma_x \text{ and } \Sigma F_y = ma_y$$

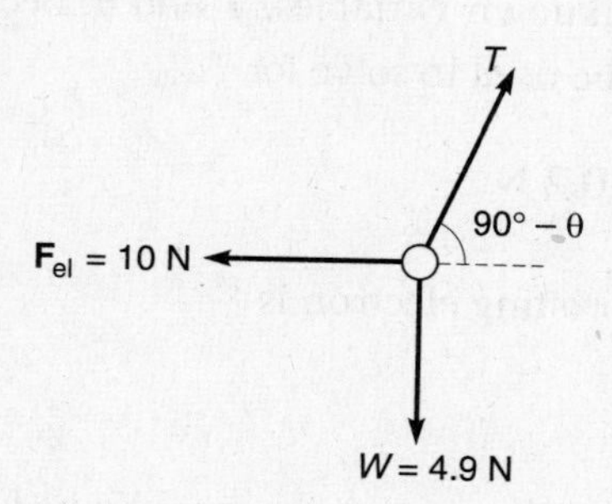

$$\Sigma F_x = T\cos(90° - \theta) - 10\text{ N} = 0$$

and

$$\Sigma F_y = T\sin(90° - \theta) - 4.9\text{ N} = 0$$

By definition,

$$\cos(90° - \theta) = \sin\theta$$

and

$$\sin(90° - \theta) = \cos\theta$$

Thus,

$$\Sigma F_x = T\sin\theta - 10\text{ N} = 0 \qquad (1)$$

and

$$\Sigma F_y = T\cos\theta - 39\text{ N} = 0 \qquad (2)$$

Solving Equation 1 for T yields

$$T = \frac{10\text{ N}}{\sin\theta}$$

Substituting into Equation 2 yields

$$\left(\frac{10\text{ N}}{\sin\theta}\right)\cos\theta - 39\text{ N} = 0$$

$$\cot\theta = \frac{39\text{ N}}{10\text{ N}} = 3.9$$

$$\cot\theta = \frac{1}{\tan\theta}$$

$$\tan\theta = \frac{1}{3.9} = 0.26$$

$$\theta = \tan^{-1}(0.26) = 14.6°$$

8. C In the solution for Problem 7, two equations were derived from Newton's second law. The two equations contained two unknown variables, T and θ. Because θ is know known, either of the equations can be used to solve for T:

$$T = \frac{39\text{ N}}{\cos\theta} = \frac{39\text{ N}}{\cos(15°)} = \frac{39\text{ N}}{0.97} = 40.2\text{ N}$$

9. C The electric field E experienced by the orbiting electron is

$$E = k\frac{q}{r^2}$$

The electric field at point A from each of the three point charges can be found by

$$E_1 = \left(9 \times 10^9 \frac{\text{N}\cdot\text{m}^2}{\text{C}^2}\right)\frac{+5.0 \times 10^{-6}\text{ C}}{(0.20\text{ m})^2} = 1.1 \times 10^6 \frac{\text{N}}{\text{C}}$$

$$E_2 = \left(9 \times 10^9 \frac{\text{N}\cdot\text{m}^2}{\text{C}^2}\right)\frac{-10.0 \times 10^{-6}\text{ C}}{(0.17\text{ m})^2} = -3.1 \times 10^6 \frac{\text{N}}{\text{C}}$$

$$E_3 = \left(9 \times 10^9 \frac{\text{N}\cdot\text{m}^2}{\text{C}^2}\right)\frac{+5.0 \times 10^{-6}\text{ C}}{(0.20\text{ m})^2} = 1.1 \times 10^6 \frac{\text{N}}{\text{C}}$$

The net electric field can be determined by first conceptualizing the electric field vectors on a free-body diagram of point A, as shown next:

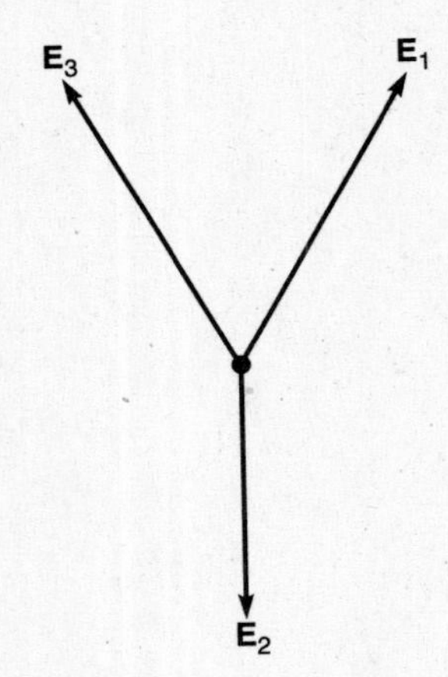

The net electric field can be determined by calculating the components of each of the electric field vectors

Field	x Component	y Component
E_1	1.1×10^6 N/C · cos (60°) = 5.5×10^5 N/C	1.1×10^6 N/C · sin (60°) = 9.5×10^5 N/C
E_2	0	-3.1×10^6 N/C
E_3	1.1×10^6 N/C · cos (120°) = -5.5×10^5 N/C	1.1×10^6 N/C · sin (120°) = 5.5×10^5 N/C
E_{net}	$E_{net,x} = 5.5 \times 10^5$ N/C $- 5.5 \times 10^5$ N/C = 0 N	$E_{net,y} = 9.5 \times 10^5$ N/C -3.1×10^6 N/C $+ 5.5 \times 10^5$ N/C = -1.6×10^6 N/C

The magnitude of the net electric field can be calculated by

$$E_{net} = \sqrt{(E_{net,x})^2 + (E_{net,y})^2}$$

$$= \sqrt{\left(0\ \frac{N}{C}\right)^2 + \left(-1.6 \times 10^6\ \frac{N}{C}\right)^2} = 1.6 \times 10^6\ \frac{N}{C}$$

10. A The electric potential V of the point charge is

$$V = k\frac{q}{r}$$

Substituting values given in the problem yields

$$V = \left(9 \times 10^9\ \frac{N \cdot m^2}{C^2}\right) \frac{8.0 \times 10^{-6}\ C}{0.4\ m} = 1.8 \times 10^5\ V$$

CHAPTER 13 ELECTRIC CIRCUITS
Answers and Explanations

1. D The equivalent resistance of resistors connected in series can be determined from

$$R_{eq} = R_1 + R_2 + R_3 = 5\ \Omega + 10\ \Omega + 15\ \Omega = 30\ \Omega$$

2. A The equivalent resistance of resistors connected in parallel can be determined from

$$\frac{1}{R_{eq}} = \frac{1}{R_1} + \frac{1}{R_2} + \frac{1}{R_3} = \frac{1}{5\ \Omega} + \frac{1}{10\ \Omega} + \frac{1}{15\ \Omega} = \frac{11}{30\ \Omega}$$

$$R_{eq} = \frac{30\ \Omega}{11} = 2.7\ \Omega$$

3. C One must first find the equivalent resistance of the two resistors R_1 and R_2 connected in parallel resistance of resistors connected in parallel by

$$\frac{1}{R_{eq}} = \frac{1}{R_1} + \frac{1}{R_2} = \frac{1}{5\ \Omega} + \frac{1}{10\ \Omega} = \frac{3}{10\ \Omega}$$

$$R_{eq,p} = \frac{10\ \Omega}{3} = 3.3\ \Omega$$

This resistance is connected with R_3 in series and thus the equivalent resistance of the resistors connected in the circuit is

$$R_{eq} = R_{eq,p} + R_3 = 3.3\ \Omega + 15\ \Omega = 18.3\ \Omega$$

4. A The arrangement of resistances can be simplified by first determining the equivalent resistance of the parallel arrangement of R_1, R_2, and R_3:

$$\frac{1}{R_{eq,1,2,3}} = \frac{1}{R_1} + \frac{1}{R_2} + \frac{1}{R_3} = \frac{1}{1\ \Omega} + \frac{1}{2\ \Omega} + \frac{1}{3\ \Omega} = \frac{11}{6\ \Omega}$$

$$R_{eq,1,2,3} = \frac{6\ \Omega}{11} = 0.55\ \Omega$$

$R_{eq,1,2,3}$ is in series with R_4, so the equivalent resistance of the upper strand is

$$R_{eq,upper} = R_{eq,1,2,3} + R_4 = 0.55\ \Omega + 4\ \Omega = 4.55\ \Omega$$

R_5 and R_6 are connected in series in the lower strand, so the equivalent resistance is thus

$$R_{eq,lower} = R_5 + R_6 = 5\ \Omega + 6\ \Omega = 11\ \Omega$$

The upper and lower strands of the circuit are connected in parallel, and the equivalent resistance of the entire circuit is

$$\frac{1}{R_{eq}} = \frac{1}{R_{eq,upper}} + \frac{1}{R_{eq,lower}} = \frac{1}{4.55\ \Omega} + \frac{1}{11\ \Omega} = \frac{15.55}{50.05\ \Omega}$$

$$R_{eq} = \frac{50.05\ \Omega}{15.55} = 3.2\ \Omega$$

5. A The current, I, is related to the amount of charge, Δq, that flows during a time interval, Δt, by

$$I = \frac{\Delta q}{\Delta t} \Rightarrow \Delta q = I \cdot \Delta t = (15\ \text{A})\ (0.007\ \text{s}) = 0.105\ \text{C}$$

6. C One electron has 1.6×10^{-19} C of charge. Thus, the number of electrons in 0.105 C of charge is

$$\text{Number of electrons} = (0.105\ \text{C}) \cdot \left(\frac{1\ \text{electron}}{1.6 \times 10^{-19}\ \text{C}}\right) = 6.6 \times 10^{17}\ \text{electrons}$$

7. A The resistance of the copper wire is related to its diameter and length by

$$R = \frac{\rho L}{A} = \frac{\rho L}{\pi r^2} = \frac{\left(1.7 \times 10^{-8}\ \Omega \cdot \text{m}\right)(0.75\ \text{m})}{(3.14)\left(0.15 \times 10^{-3}\ \text{m}\right)^2} = 0.18\ \Omega$$

8. C The resistance of the copper wire is related to its diameter and length by

$$R = \frac{\rho L}{A} = \frac{\rho L}{\pi r^2} \Rightarrow r = \sqrt{\frac{\rho L}{\pi R}} = \sqrt{\frac{\left(1.7 \times 10^{-8}\ \Omega \cdot \text{m}\right)(2\ \text{m})}{(3.14)(3\ \Omega)}} = 6.0 \times 10^{-5}\ \text{m}$$

$$\text{Diameter} = 2 \cdot \text{radius} = 2 \cdot 6.0 \times 10^{-5}\ \text{m} = 1.2 \times 10^{-4}\ \text{m} = 0.12\ \text{mm}$$

9. B The current is related to the resistance and voltage by Ohm's law:

$$I = \frac{V}{R} = \frac{120\ \text{V}}{450\ \Omega} = 0.27\ \text{A}$$

10. B The current is related to the resistance and voltage by Ohm's law:

$$I = \frac{V}{R} \Rightarrow R = \frac{V}{I} = \frac{120\ \text{V}}{4\ \text{A}} = 30\ \Omega$$

CHAPTER 14 MAGNETISM
Answers and Explanations

1. **C** All quantities of force (e.g., weight, tension, friction, electrostatic) are expressed in units of newtons. Both tesla and gauss are units of magnetic field, and ampere represents units of electric current.

2. **A** As the beam of electrons moves in a circular arc, a magnetic force $F = qvB$ acts on the electrons, keeping them in a circular path. As a result of their circular motion, the beam of electrons are accelerating with a centripetal acceleration, $a_c = \frac{v^2}{r}$. Substituting into Newton's Second Law of Motion,

$$\Sigma \mathbf{F} = m\boldsymbol{a}$$

$$qvB = m_e \frac{v^2}{r}$$

Solving for r yields

$$r = \frac{m_e v}{qB}$$

3. **B** As seen from the derivation presented in the solution of Problem 2, the expression for the radius of the circular arc was based on Newton's Second Law of Motion.

4. **D** From Problem 2, the expression for the radius of the circular arc is

$$r = \frac{m_e v}{qB}$$

Substituting the corresponding values provided in the passage,

$$r = \frac{\left(9.11 \times 10^{-31}\ \text{kg}\right)\left(8 \times 10^{6}\ \frac{\text{m}}{\text{s}}\right)}{\left(1.6 \times 10^{-19}\ \text{C}\right)\left(5 \times 10^{-4}\ \text{T}\right)} = 0.09\ \text{m} = 9\ \text{cm}$$

5. **C** The period, T, of a circulating object is related to the radius, r, of the circular path and the speed, v, of the circulating electrons by

$$T = \frac{2\pi r}{v}$$

where r is the radius of circular motion, which is given by

$$r = \frac{mv}{qB}$$

This, the period can be rewritten as

$$T = \frac{2\pi}{v} \cdot \frac{mv}{qB} = \frac{2\pi m}{qB}$$

Substituting given information yields

$$T = \frac{2\pi m}{qB} = \frac{2 \cdot (3.14)\left(9.11 \times 10^{-31}\ \text{kg}\right)}{\left(1.6 \times 10^{-19}\ \text{C}\right)\left(5 \times 10^{-4}\ \text{T}\right)} = 7.15 \times 10^{-8}\ \text{s}$$

6. B The frequency, f, is related to the period, T, of a circulating object by

$$f = \frac{1}{T} = \frac{1}{7.2 \times 10^{-8}\ \text{s}} = 1.39 \times 10^{7}\ \text{Hz}$$

7. D The angular frequency is related to the frequency, f, of a circulating object by

$$\omega = 2\pi f = 2 \cdot (3.14) \cdot 1.39 \times 10^{7}\ \text{Hz} = 8.7 \times 10^{7}\ \frac{\text{rad}}{\text{s}}$$

8. A The magnitude of the magnetic field, **B**, generated by a long, straight wire is given by

$$B = \frac{\mu_o I}{2\pi r} \Rightarrow r = \frac{\mu_o I}{2\pi B} = \frac{\left(4\pi \times 10^{-7}\ \frac{\text{Tm}}{\text{A}}\right)(10\ \text{A})}{2 \cdot (3.14)\left(5 \times 10^{-5}\ \text{T}\right)} = 0.04\ \text{m} = 4\ \text{cm}$$

9. B The magnitude of the magnetic field, **B**, at the center of a wire loop of radius r is given by

$$B = \frac{N\mu_o I}{2r} = \frac{(50)\left(4\pi \times 10^{-7}\ \frac{\text{Tm}}{\text{A}}\right)(3\ \text{A})}{2\ (0.2\ \text{m})} = 4.7 \times 10^{-4}\ \text{T}$$

10. D The magnitude of the magnetic field, **B**, in the interior of the solenoid is given by

$$B = N\mu_o I = (400)\left(4\pi \times 10^{-7}\ \frac{\text{Tm}}{\text{A}}\right)(5\ \text{A}) = 2.5 \times 10^{-3}\ \text{T} = 2.5\ \text{mT}$$

CHAPTER 15 ATOMIC AND NUCLEAR PHYSICS Answers and Explanations

1. **A** Uranium has an atomic number $Z = 92$, which indicates the number of protons. Thus, uranium-235 has 92 protons.

2. **B** The number of neutrons is the mass number A minus the atomic number Z. Thus, the number of neutrons of uranium-235 is

$$N = A - Z = 235 - 92 = 143 \text{ neutrons}$$

3. **A** Because the normal atom is electrically neutral, the atom has Z electrons. Thus, uranium has 92 electrons.

4. **B** The atomic number, Z, is the number of electrons, which is equal to the number of protons. From the information in the problem, the element has 9 electrons. Thus, $Z = 9$.

5. **D** The mass number, A, is equal to the number of neutrons and protons. The element has 9 electrons, which means it also has 9 protons. In addition, the number of neutrons of this element is given as 10. Thus,

$$A = 10 \text{ neutrons} + 9 \text{ protons} = 19$$

6. **A** The chemical element with an atomic mass, $A = 19$, and an atomic number, $Z = 9$, is fluorine.

7. **B** The energy of a gamma ray is related to its frequency by

$$E = h\nu \Rightarrow \nu = \frac{E}{h} = \frac{1.4 \times 10^5 \text{ eV}}{4.135 \times 10^{-15} \text{ eV} \cdot \text{s}} = 3.4 \times 10^{19} \text{ Hz}$$

8. **C** The decay constant is related to the half-life of a radioactive compound by

$$\lambda = \frac{0.693}{t_{1/2}} = \frac{0.693}{(2.1 \text{ min}) \cdot \left(\frac{60 \text{ s}}{1 \text{ min}}\right)} = 5.5 \times 10^{-3} \text{ s}^{-1}$$

9. **D** By the conservation of mass, the mass number, A, and the atomic number, Z, must be equal on either side of the reaction. Thus,

$$226 = 222 + A_{unk} \quad \text{or} \quad A_{unk} = 226 - 222 = 4$$

$$88 = 86 + Z_{unk} \quad \text{or} \quad Z_{unk} = 88 - 86 = 2$$

The unknown element has a mass number of $A = 4$ and an atomic number of $Z = 2$ which is a helium atom or ${}^4_2He$.

10. **A** A radioactive decay with a byproduct of a helium atom is, by definition, alpha decay.

PART III

REVIEWING MCAT GENERAL CHEMISTRY

CHAPTER 1

Atoms and Molecules

Read This Chapter to Learn About:

- ➤ The Structure of an Atom
- ➤ Isotopes
- ➤ Chemical Formulas
- ➤ The SI Unit System
- ➤ The Mole
- ➤ The Nomenclature of Compounds

STRUCTURE OF THE ATOM

An atom is made up of a **nucleus** surrounded by a cloud of **electrons.** The nucleus contains **protons** and **neutrons** and has a positive charge because each proton carries a +1 charge and each neutron is neutral. Every electron has a −1 charge.

In an atom of an **element,** the sum of the positive and negative charges must be zero. If there is a net charge, the species is called an **ion.** Ions have more or fewer electrons than the atoms from which they are derived. The charge of an ion is written to the upper right of the symbol. A **cation** is positively charged. An **anion** is negatively charged.

The number of protons in the nucleus of an atom or ion is the **atomic number.** It often appears on the periodic table in red. Hydrogen has an atomic number of 1, helium, 2, and so on. The neutron number cannot be determined from the periodic table.

There can be different forms of an element, called **isotopes,** that differ in the number of neutrons in the nucleus. Isotopes exist in nature, and the percentage of each isotope in a normal sample of an element is called its **natural abundance.** For example, neon has three isotopes, neon-20, neon-21, and neon-22, as shown in the following table.

Natural Abundances of the Isotopes of Neon		
Isotope	Exact Mass	Natural Abundance (%)
^{20}Ne	19.992	90.51
^{21}Ne	20.994	0.27
^{22}Ne	21.991	9.22

Each isotope of an element has an exact mass that corresponds to its **mass number** when rounded. The **mass number** is the sum of the protons and neutrons in the nucleus. Thus, neon-20 has 10 protons and 10 neutrons; neon-21 has 10 protons and 11 neutrons; and neon-22 has 10 protons and 12 neutrons. The mass number is written in the upper left corner of the symbol.

The following table shows the number of protons, neutrons, and electrons for some specific isotopes, some of which have a net charge.

Composition and Net Charges for Some Isotopes and Their Ions				
Symbol with Mass Number	Number of Protons	Number of Neutrons	Number of Electrons	Charge
$^{58}Ni^{+2}$	28	30	26	+2
^{60}Ni	28	32	28	0
^{2}H	1	1	1	0
$^{2}H^{+}$	1	1	0	+1

AVERAGE MOLAR MASS

The average **molar mass** is the weighted average of the masses of all the known isotopes for an element; this number is found on the periodic table. A natural sample of carbon, for instance, has an average molar mass of 12.01. This means that 1 **mole** of carbon atoms has the mass 12.01 grams (g). One mole is defined as an **Avogadro's number** of atoms, or 6.022×10^{23} atoms.

The molar mass of a molecule is calculated by adding up the molar masses of the individual elements in the molecule.

EXAMPLE: Determine the molar mass of potassium permanganate, $KMnO_4$.

K 39.10 g/mole
Mn 54.94 g/mole
O 16.00 g/mole × 4 atoms

Summing these, the molar mass of potassium permanganate is 158.04 g/mole.

CHEMICAL FORMULAS

Formulas are written as $A_xB_yC_z$, where A, B, and C are symbols of elements and x, y, and z represent the number of each type of element in the formula.

The **molecular formula** of a compound is the actual formula of the fundamental unit of the substance. For instance, the molecular formula of sucrose is $C_6H_{12}O_6$. The **empirical formula** is the lowest whole-number ratio of the different elements in the formula. For most ionic substances, the empirical formula and the molecular formula are the same. But for sucrose, the empirical formula is CH_2O, showing a ratio of 1:2:1 for the elements carbon, hydrogen, and oxygen.

IONIC VERSUS COVALENT COMPOUNDS

An **ionic compound** is the combination of a metal and a nonmetal. Metals are located left of the stairline in the periodic table. A **covalent compound** is a combination of nonmetals. Nonmetals are located to the right of the periodic table, and include hydrogen.

Examples of ionic compounds: Fe_2O_3 LiOH CaH_2
Examples of covalent compounds: HCl CH_4 NH_3

THE SI UNIT SYSTEM

The units of measurement used in chemistry are those of the Système Internationale d'Unités, or SI, and are listed in the following table.

Base Units of the SI Unit System

Base Unit	Name	Abbreviation	Equivalency
Length	Meter	m	100 cm
Mass	Kilogram	kg	1000 g
Time	Second	s	
Temperature	Kelvin	K	
Amount	Mole	mol	
Volume	Cubic meter	m^3	1000 L
Density	Mass/volume	kg/m^3	
Energy	Joule	J	$kg\ m^2/s^2$

Some of the units used in chemistry are complex units. These include the unit for volume, the liter (L), and the unit for energy, the joule (J). The definition of the liter is that 1 L = 1000 cm^3, where 1 cm^3 = 1 mL. The definition of the joule is 1 J = kg m^2/s^2.

The prefixes are from the metric system and are listed in the following table.

SI and Metric Prefixes

Prefix Name	Abbreviation	Equivalency
giga	G	1×10^{9}
mega	M	1×10^{6}
kilo	k	1×10^{3}
deci	d	1×10^{-1}
centi	c	1×10^{-2}
milli	m	1×10^{-3}
micro	μ	1×10^{-6}
nano	n	1×10^{-9}

FACTOR ANALYSIS FOR CONVERSIONS

The process of **factor analysis,** which is also called **dimensional analysis,** is used to convert from one unit system to another. To set up a problem using factor analysis, the value and the unit that you are converting is written as a factor (in parentheses). Then it is multiplied by the conversion equation, which has also been turned into a factor so that the original unit cancels.

EXAMPLE: Convert 411 nm to m.

$$(411\ \cancel{\text{nm}})\left(\frac{1 \times 10^{-9}\ \text{m}}{1\ \cancel{\text{nm}}}\right) = 4.11 \times 10^{-7}\ \text{m}$$

EXAMPLE: Convert 12.6 h to ms.

$$(12.6\ \cancel{\text{h}})\left(\frac{3600\ \cancel{\text{s}}}{1\ \cancel{\text{h}}}\right)\left(\frac{1\ \text{ms}}{10^{-3}\ \cancel{\text{s}}}\right) = 4.54 \times 10^{7}\ \text{ms}$$

MASS RELATIONSHIPS

Mass relationships include the percentage composition by mass of a compound, which is used to determine the empirical formula of that compound.

Percentage Composition by Mass

The percentage composition by mass of each element in a compound is found by determining the total mass of each element and then dividing each element's mass by the total mass of the compound.

EXAMPLE: Calculate the percentage composition by mass of each element in potassium dichromate, $K_2Cr_2O_7$.

To solve this problem, find the mass of each element:

K 39.10 g/mole × 2 mole = 78.20 g

Cr 52.00 g/mole × 2 mole = 104.0 g

O 16.00 g/mole × 7 mole = 112.0 g

On summing these values, the total mass is 294.2 g. The percentage composition is

K = 78.20 g/294.2 g × 100 = 26.58%

Cr = 104.0 g/294.2 g × 100 = 35.35%

O = 112.0 g/294.2 g × 100 = 38.07%

Determining an Empirical Formula

The percentage composition by mass data can be used to determine the empirical formula of a compound. Assume you have 100 g of the substance, and then determine the molar ratio of the elements in the substance. To do this, divide the percentage (converted to a decimal) of each element by the molar mass of the element, and then divide each of the results by the largest result.

EXAMPLE: Calculate the empirical formula of a compound that is 34.47% cobalt, 28.10% carbon, and 37.43% oxygen.

Co 34.47 g ÷ 58.93 g/mole = 0.585 moles Co

C 28.10 g ÷ 12.01 g/mole = 2.34 moles C

O 37.43 g ÷ 16.00 g/mole = 2.34 moles O

Divide each value by 0.585 to obtain a ratio

Co 0.585 mol ÷ 0.585 mol = 1

C 2.34 mol ÷ 0.585 mol = 4

O 2.34 mol ÷ 0.585 mol = 4

This is a whole number ratio, so the empirical formula is CoC_4O_4.

THE CONCEPT OF THE MOLE

The mole is a fundamental idea in chemistry. It relates amounts of substances to the number of particles in the amount, regardless of mass. In other words, 1 mole of feathers is the same number of particles as *q* mole of marbles, namely, 6.022×10^{23} particles. This is called **Avogadro's number.**

One can use Avogadro's number to calculate the number of particles (atoms, ions, or molecules) in any quantity of a substance.

EXAMPLE: Calculate the number of atoms in 72.1 g gold (Au).

$$(72.1\ \cancel{g}\ \text{Au})\left(\frac{1\ \cancel{\text{mole}}}{197\ \cancel{g}}\right)\left(\frac{6.022 \times 10^{23}\ \text{atoms}}{1\ \cancel{\text{mole}}}\right) = 2.20 \times 10^{23}\ \text{atoms Au}$$

EXAMPLE: Calculate the mass of $Mg(OH)_2$ that corresponds to 8.903×10^{16} molecules.

$$(8.903 \times 10^{16}\ \cancel{\text{molecules}})\left(\frac{1\ \cancel{\text{mole Mg(OH)}_2}}{6.022 \times 10^{23}\ \cancel{\text{molecules}}}\right)\left(\frac{58.33\ \text{g Mg(OH)}_2}{1\ \cancel{\text{mole Mg(OH)}_2}}\right)$$

$$= 8.62 \times 10^{-6}\ \text{g}$$

NOMENCLATURE

Nomenclature is the naming of compounds, whether by a systematic method or by use of common names. It is a system of rules used to name ionic and covalent compounds.

Nomenclature of Ionic Compounds

Ionic compounds consist of cations and anions. The name of the cation comes first, and it is named as the element. There are a few complex cations, but the most important one is ammonium ion, which has the formula NH_4^{+1}.

The main group elements form cations in a systematic manner. Elements from group I always form +1 cations. Elements from group II always form +2 cations. The metals Al and Ga from group III always form +3 cations. In general, the metals always form cations.

Anions are generally formed from nonmetals, or groups of nonmetals that are covalently bonded. The simple cations and anions and many of the complex anions are listed in the following table.

To name a compound such as Na_3N, the cation Na^{+1} is called **sodium** and the anion N^{-3} is called **nitride.** Therefore, the compound is sodium nitride.

Simple and Complex Anions and Cations

Main group cations and simple anions

Group IA	IIA	IIIA	IVA	VA	VIA	VIIA
H^+ Hydrogen						H^- Hydride
Li^+ Lithium	Be^{2+} Beryllium	Al^{3+} Aluminum		N^{3-} Nitride	O^{2-} Oxide	F^- Fluoride
Na^+ Sodium	Mg^{2+} Magnesium	Ga^{3+} Gallium			S^{2-} Sulfide	Cl^- Chloride
K^+ Potassium	Ca^{2+} Calcium	In^{3+} Indium			Se^{2-} Selenide	Br^- Bromide
Rb^+ Rubidium	Sr^{2+} Strontium				Te^{2-} Teleride	I^- Iodide
Cs^+ Cesium	Ba^{2+} Barium					

Complex anions

Cyanide	CN^-				
Hydroxide	OH^-				
Peroxide	O_2^{-2}				
Permanganate	MnO_4^-				
Chromate	CrO_4^{-2}				
Dichromate	$Cr_2O_7^{-2}$				
Carbonate	CO_3^{-2}	Hydrogen carbonate	HCO_3^-		
Phosphate	PO_4^{-3}	Hydrogen phosphate	HPO_4^{-2}	Dihydrogen phosphate	$H_2PO_4^-$
Phosphite	PO_3^{-3}	Hydrogen phosphite	HPO_3^{-2}	Dihydrogen phosphite	$H_2PO_3^-$
Sulfate	SO_4^{-2}	Hydrogen sulfate	HSO_4^-		
Sulfite	SO_3^{-2}	Hydrogen sulfite	HSO_3^-		
Nitrate	NO_3^-				
Nitrite	NO_2^-				
Perchlorate	ClO_4^-	Perbromate	BrO_4^-	Periodate	IO_4^-
Chlorate	ClO_3^-	Bromate	BrO_3^-	Iodate	IO_3^-
Chlorite	ClO_2^-	Bromite	BrO_2^-	Iodite	IO_2^-
Hypochlorite	ClO^-	Hypobromite	BrO^-	Hypoiodite	IO^-

Complex cation

Ammonium	NH_4^+				

AlN is aluminum nitride; Mg_3N_2 is magnesium nitride; MgO is magnesium oxide.

To write the formula of calcium hydride, the calcium cation is Ca^{+2} and the anion hydride is H^{-1}. Two hydrides are required to balance the +2 charge on the calcium ion, in order to net out to zero overall. So the formula is CaH_2.

A good method to determine how many cations and how many anions are in a formula is the lowest common denominator method. The lowest common denominator of the charges is determined (ignoring the signs). This gives the total + charges, and the total – charges necessary.

EXAMPLE: Determine the formula of sodium oxide.

Sodium is Na^{+1} and oxide is O^{-2}. The lowest common denominator of 1 and 2 (the charges) is 2. Thus, the positive charges must total +2 and the negative charges must total −2. So there must be 2 sodium cations and 1 oxide anion, and the formula is Na_2O.

Transition metal cations vary in their possible charges; the charge of the metal must be written in the name as a roman numeral. Thus, copper (II) oxide means that the copper cation is Cu^{+2}. Because oxide is always O^{-2}, the charges balance, and the formula is CuO.

Copper (I) oxide means that the copper cation is Cu^{+1}. Oxide is O^{-2}. The lowest common denominator is 2; thus, there must be two Cu^{+1} cations for every oxide anion. The formula is Cu_2O.

The complex anions must be memorized. Their names, formulas, and net charge are given in the previous table.

EXAMPLE: Write the formula of sodium phosphate.

The sodium cation is Na^{+1} and the phosphate anion is PO_4^{-3}. The lowest common denominator of the charges is 3. Thus, there must be 3 sodium cations and 1 phosphate anion to result in a net zero charge for the molecule. So, the formula is Na_3PO_4.

Nomenclature of Binary Covalent Compounds

Binary covalent compounds contain two elements in varying proportions. The first element is named as the element. The second is named as if it was an anion. The elements do not have charges (there are no ions here), so there is no way to determine how many of each there must be. The actual number of each element in the formula must be stated in the name by using a prefix. If there is more than one of the first element, a prefix goes in front of the name. The prefixes are as follows: mono, di, tri, tetra, penta, hexa, hepta, octa, nona, and deca. These correspond to 1 through 10. Some examples are

NO	Nitrogen monoxide
P_4O_{10}	Tetraphosphorus decaoxide
N_2H_4	Dinitrogen tetrahydride
PCl_5	Phosphorus pentachloride

Some of the binary covalent compounds have common names, such as water, which is H_2O. Others are ammonia, NH_4; hydrogen peroxide, H_2O_2; phosphine, PH_3; and arsine AsH_3.

Nomenclature of Covalent Acids

Acids are compounds that are derived by adding enough protons to a simple or complex anion to make it a neutral compound. Some examples of acids formed from simple and complex ions are as follows:

Simple Anion	Acid
F^{-1}	HF
Br^{-1}	HBr
S^{-2}	H_2S

Complex Anion	Acid
CN^{-1}	HCN
CO_3^{-2}	H_2CO_3
PO_4^{-3}	H_3PO_4

Acid names are based on the anion name. If the anion name ends with *-ide*, the acid name starts with *hydro-*, and the *-ide* is changed to *-ic* acid.

EXAMPLES:

HF	hydrofluoric acid	from the anion fluoride
HCN	hydrocyanic acid	from the anion cyanide
H_2S	hydrosulfuric acid	from the anion sulfide

If the anion name ends with *-ate*, the *-ate* ending is changed to *-ic* acid.

EXAMPLES:

H_2CO_3	carbonic acid	from the anion carbonate
H_2SO_4	sulfuric acid	from the anion sulfate
H_3PO_4	phosphoric acid	from the anion phosphate
$HClO_4$	perchloric acid	from the anion perchlorate

If the anion name ends with *-ite*, the *-ite* ending is changed to *-ous* acid.

EXAMPLES:

H_2SO_3	sulfurous acid	from the anion sulfite
$HClO$	hypochlorous acid	from the anion hypochlorite
HNO_2	nitrous acid	from the anion nitrite

CRAM SESSION

Atoms and Molecules

1. The **nucleus** consists of the positive protons and the neutral neutrons and contains all the mass. The electron cloud that surrounds the nucleus consists of electrons, each with a −1 charge.
2. The **average molar mass of an element** is a weighted average of the exact masses of all the individual isotopes of that element. The **molar mass of a compound** is the sum of all the molar masses of all the elements in the compound's molecular formula.
3. The **molecular formula** shows the number and type of each element in a molecule.
4. The SI units for length, mass, time, volume, temperature, amount, and energy are used, along with metric prefixes, to express values in the SI system.
5. The **percentage composition by mass** expresses the percentage by mass of each element in a compound. It can be used to determine the empirical formula of a compound.
6. The **empirical formula** shows the lowest whole-number ratio of the elements in a molecule of a substance.
7. One mole of a substance contains **Avogadro's number** of particles, or 6.022×10^{23} particles.
8. Nomenclature is the naming of chemical compounds. The rules are different for ionic compounds, binary covalent compounds, and covalent acids.
   A. **Ionic compounds** are named by naming the cation followed by the anion. Any transition metal cation must have its charge included in the name as a roman numeral.
   B. **Binary covalent compounds** consist of two elements. They require prefixes to indicate exactly how many of each element type there is in the formula of the compound. Acids are named using the name of the anion they contain.

CHAPTER 2

Electronic Structure and the Periodic Table

Read This Chapter to Learn About:

- ➤ The Hydrogen Atom
- ➤ Quantum Mechanics
- ➤ Electron Shells
- ➤ Electron Configurations

The electron of the hydrogen atom is known to have a wavelike nature. When hydrogen atoms are heated to glowing, they emit light in a quantized series of discrete lines. The wavelike nature of the hydrogen atom is described fully by the **Schrödinger wave equation**, which has four solutions called **wave functions**, represented by ψ.

When ψ is squared, a three-dimensional probability map for finding the electron around the nucleus results. This map shows where the likelihood of finding the electron is 95% or greater. This area is called an **orbital**.

There are four orbitals, one for each of the ψ^2 areas. The orbitals are called *s*, *p*, *d*, and *f*. These orbitals are described by four variables, which are called **quantum numbers**, contained in each wave function. A quantum number indicates the energy of the orbital.

THE QUANTUM NUMBER, *n*

The quantum number, *n*, is a positive integer that describes the size and energy level of the orbital. Orbitals are grouped according to their *n* value. All orbitals with the same *n* value are said to be in the *n*-shell.

Each orbital has a specific energy level associated with it. The energy levels of the orbitals are quantized—that is, only certain energy levels exist. If an electron that is in a low-energy orbital absorbs heat energy, it can use the energy to reach a higher-energy orbital, which is called an **excited state**. Which orbital it can reach depends on how much energy is absorbed.

Once in the excited state, the electron drops back to its original ground state. When it does this, it releases the energy as light energy. Figure 2-1 shows an electron in the 2-shell absorbing enough energy to reach the 5-shell, then dropping back down to the 2-shell. This process is called the **5 → 2 transition.**

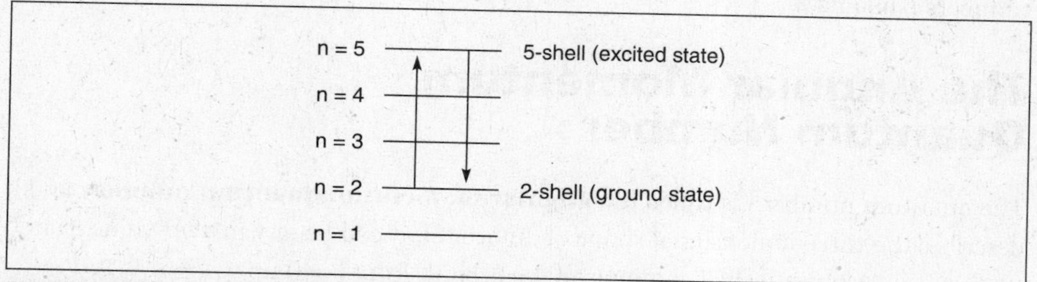

FIGURE 2-1: The 5 → 2 transition of the hydrogen atom.

When the 5 → 2 transition occurs, the electron emits light. It is a narrow band of blue light that can be seen in the visible spectrum of the light emitted by a glowing sample of hydrogen.

THE HYDROGEN EMISSION SPECTRUM

There are four lines in the visible spectrum of the hydrogen atom. All of them have $n = 2$ as the ground state. This group of four lines is called the **Balmer series**. The energy of each of these transitions was calculated by Niels Bohr to be

$$E = R_H(1/n_i^2 - 1/n_f^2)$$

where $R_H = 2.18 \times 10^{-18}$ J, n_i = initial state, and n_f = final state. Bohr determined that the energies emitted by electrons were quantized, but he thought that the states represented orbits of energy around the nucleus. The Bohr model states that energy absorption leads to an electron occupying a higher energy orbit. When the electron decays to its original orbit, light is emitted.

Einstein determined that the energy of one photon of light would be equivalent to $h\nu$, where h is Planck's constant 6.63×10^{-34} J s and ν is the frequency of the light, with the unit s^{-1}. From wave physics, the frequency $\nu = c/\lambda$ where λ is the wavelength and c is the speed of light, 3.00×10^8 m/s.

$$E = h\nu = hc/\lambda$$

Quantum mechanics describes the electron of the hydrogen atom as having wave nature. The quantum numbers that come out of the wave function solutions to the Schrodinger equation describe the different orbitals, their sizes, shapes, and energy levels. The Bohr model explains the discrete lines found in the hydrogen emission spectrum, but its value ends there.

OTHER QUANTUM NUMBERS

In addition to the quantum number n, which describes the shell, there are the quantum numbers l, m_l and m_s.

The Angular Momentum Quantum Number

The quantum number l is called the **angular momentum quantum number**, and it describes the three-dimensional shape of an orbital. It can be any integer value from 0 up to $n - 1$. A given n shell contains all the orbitals from $l = 0$ up to $l = n - 1$.

The $l = 0$ orbital is called the **s orbital**. It is spherical with the nucleus at the center of the sphere. The s orbital can hold up to 2 electrons.

The $l = 1$ orbital is called the ***p* orbital**. It is dumbbell shaped with 2 lobes and a node (0% probability) at the nucleus. The p orbital can hold up to 6 electrons.

The $l = 2$ orbital is called the ***d* orbital** and it has various shapes, including 4 lobes. The d orbital can hold up to 10 electrons.

The $l = 3$ orbital is called the ***f* orbital** and it also has various shapes, including 8 lobes. The f orbital can hold up to 14 electrons.

The Magnetic Quantum Number

The quantum number m_l is called the **magnetic quantum number**. It describes the orientation of the orbital about an x, y, z coordinate system. Each possible orientation can hold up to 2 electrons maximum. For each orbital there are $2l + 1$ different orientations. The quantum number m_l is all integer values from $-l$ up to $+l$.

For the s orbital $l = 0$ and $m_l = 0$. There is one orientation; it is labeled $m_l = 0$ and it can hold up to 2 electrons maximum.

For the p orbital, $l = 1$ and $m_l = -1, 0, +1$. There are three orientations, one along the x axis, one along the y axis, and the third along the z axis. Each orientation has an m_l label, and each orientation can hold 2 electrons, for a total of 6 electrons.

For the d orbital $l = 2$ and $m_l = -2, -1, 0, +1, +2$. There are 5 orientations and each has an m_l label. The d orbital can hold a total of 10 electrons.

For the f orbital, $l = 3$ and $m_l = -3, -2, -10, +1, +2, +3$. There are 7 orientations and each has an m_l label. The f orbital can hold a total of 14 electrons.

The Spin Quantum Number

Two electrons can occupy the individual orientations of each orbital. Both electrons in an orientation have a -1 charge. They do not repel each other, as might be expected, because of the spin quantum number.

When a charged particle spins, it acts like a bar magnet. When an electron spins in a clockwise manner, it acts like a magnet with—say, North up. If the other electrons spins in a counterclockwise manner, it has the opposite orientation—say, North down. So, the 2 electrons pair up in a very stable manner. One of the electrons has spin $+1/2$, and the other has spin $-1/2$.

Electrons are often designated by an arrow, using $\uparrow$ for spin $+1/2$ and $\downarrow$ for spin $-1/2$ (although this is arbitrary).

Thus, for 2 electrons to occupy the same orbital orientation, they must have opposite spins.

PAULI EXCLUSION PRINCIPLE

Every electron can be described by a unique set of quantum numbers n, l, m_l, and m_s.

EXAMPLE: Write all the possible quantum numbers for a 5p electron.

$n = 5 \quad l = 1 \quad m_l = -1, 0, +1 \quad m_s = \pm 1/2$

A 5p electron can have one of six possible sets of quantum numbers, as shown in Figure 2-2.

n	5	5	5	5	5	5
ℓ	1	1	1	1	1	1
m_ℓ	−1	0	+1	−1	0	+1
m_s	+1/2	+1/2	+1/2	−1/2	−1/2	−1/2

FIGURE 2-2: The six possible sets of quantum numbers for a 5p electron.

If the $5p$ orbital is full, it will contain 6 electrons. Each electron has a different set of quantum numbers. This is called the **Pauli exclusion principle**—no two electrons in an atom can have the same set of quantum numbers.

THE SHELLS

The 7 shells contain orbitals based on the quantum numbers. The 1-shell contains the $1s$ orbital for a total of 2 electrons. The 2-shell contains the $2s$ and the $2p$ orbitals for a total of 8 electrons. The 3-shell contains the $3s$, $3p$, and $3d$ orbitals, for a total of 18 electrons.

The 4-shell contains the $4s$, $4p$, $4d$, and $4f$ orbitals. The 5-shell contains the $5s$, $5p$, $5d$, and $5f$ orbitals. There is room for a $5g$ orbital, but an element with this many electrons has not yet been discovered.

The 6-shell contains the $6s$, $6p$, and $6d$ orbitals. The 7-shell contains the $7s$ and $7p$ orbitals. Higher orbitals are also possible for these two shells, but elements with that many electrons are unknown.

The order of the orbitals from lowest to highest energy can be determined by using a mnemonic device made by listing the orbitals in each shell, then following the arrows as shown in Figure 2-3. This is called the **Aufbau principle**.

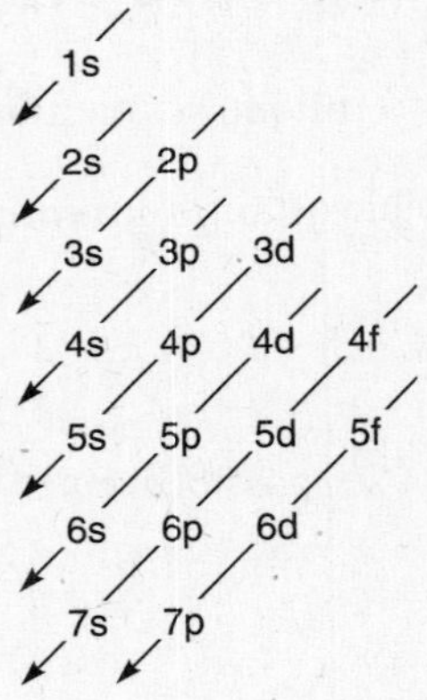

1s
2s 2p
3s 3p 3d
4s 4p 4d 4f
5s 5p 5d 5f
6s 6p 6d
7s 7p

FIGURE 2-3: The order of the orbitals, in terms of increasing energy level.

ORBITAL DIAGRAMS OF MULTIELECTRON ATOMS

Going beyond hydrogen in the periodic table, an electron must be added for each subsequent element. Thus helium has two electrons, lithium has three, etc.

The first orbital is the lowest energy orbital, and the first electron always occupies this orbital. Because the first orbital is 1*s*, it holds 2 electrons. The second electron goes into the 1*s* orbital as well, but with an opposite spin to the first electron. The 1*s* orbital is now full. Figure 2-4 illustrates the filling of the 1*s* orbital.

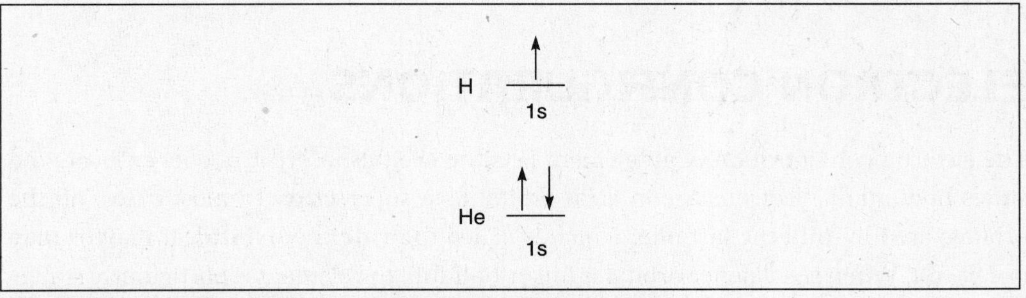

FIGURE 2-4: The orbital diagrams of hydrogen and helium.

The next higher energy orbital is the 2*s* orbital. Lithium's third electron must go into this orbital. Going on to boron, which has 5 electrons, the next orbital, the 2*p*, is utilized. Figure 2-5 illustrates the orbital diagram of lithium and of boron.

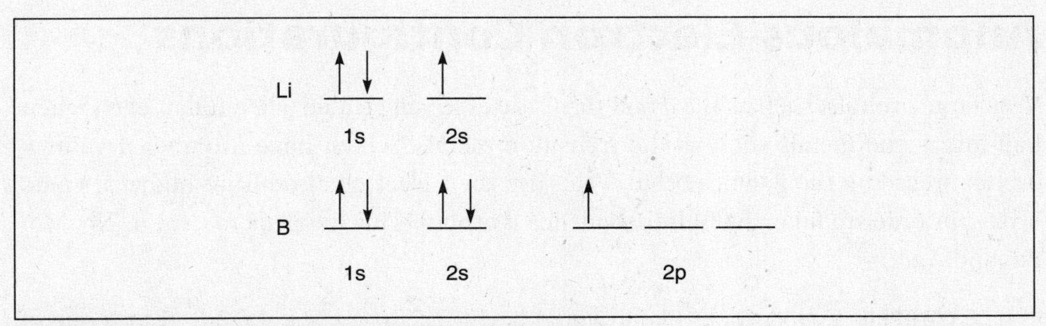

FIGURE 2-5: The orbital diagrams of lithium and boron.

With carbon, there are 2 electrons in the 2*p* orbital. The second electron goes into the second orientation, and it has the same spin as the first electron, as per **Hund's rule**, which states that each orientation must get 1 electron before any is filled. This maximizes the number of parallel spins, as shown in Figure 2-6.

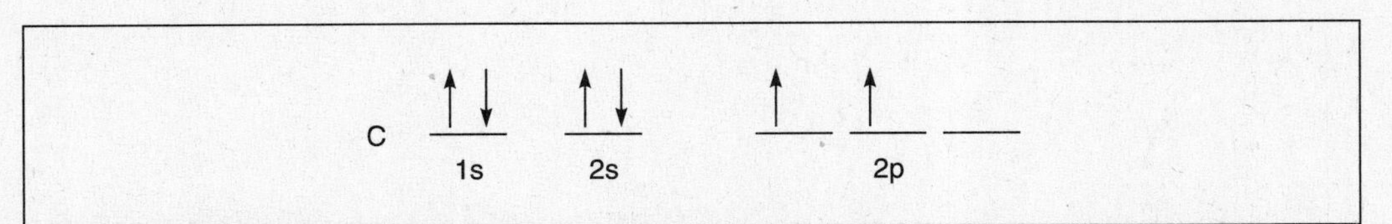

FIGURE 2-6: The orbital diagram of carbon.

Oxygen has 4 electrons in the 2*p* orbital. Each orientation gets a single electron, then the first orientation gets the fourth, as shown in Figure 2-7.

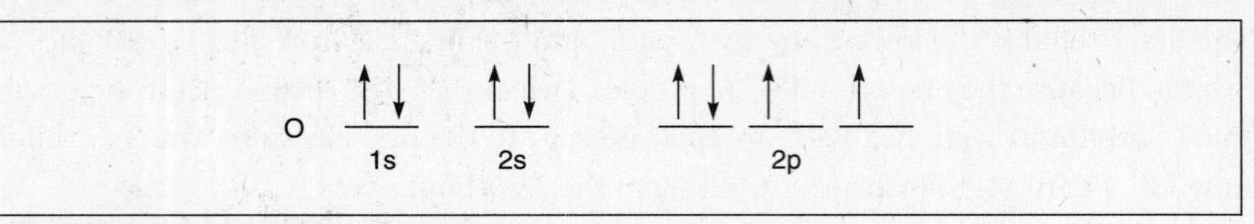

FIGURE 2-7: The orbital diagram of oxygen.

ELECTRON CONFIGURATIONS

The electron configuration of an element lists the orbitals in order of energy level, and states how many electrons are in each orbital as a superscript. In most cases, all the orbitals are full until the last one, which is called the **valence orbital**. It may or may not be full. When the valence orbital is full or half-full, the element is particularly stable.

EXAMPLES

H is $1s^1$
He is $1s^2$
O is $1s^2 2s^2 2p^4$

Anomalous Electron Configurations

Very large orbitals, such as the *d* and the *f*, are especially stable when full or even when half-full. Some metals such as the transition metals, which have a higher *n* value *s* orbital preceding the valence orbital, can use the *s* electrons from that higher *n* value orbital in order to fill or half-fill the valence *d* orbital. This happens to Cr, Cu, Nb, Mo, Pd, and Ag.

EXAMPLE: Copper looks like it would be $1s^2 2s^2 2p^6 3s^2 3p^6 4s^2 3d^9$. But if one of the 4*s* electrons goes into the 3*d* orbital, the valence orbital will be full. So copper is actually $1s^2 2s^2 2p^6 3s^2 3p^6 4s^1 3d^{10}$.

CRAM SESSION

Electronic Structure and the Periodic Table

1. The **Bohr model** shows quantized energy levels for the light emitted by the hydrogen electron, which can be calculated using the formula $\Delta E = R_H (1/n_i^2 - 1/n_f^2)$, where R_H is the Rydberg constant 2.18×10^{-18} J.
2. The **energy of one photon of light** is equivalent to $h\nu$, where h is Planck's constant 6.63×10^{-34} J s and ν is the frequency of the light wave. The frequency is equivalent to c/λ, where c is the speed of light 3.00×10^{8} m/s and λ is the wavelength in meters.
3. The **quantum numbers** describe the shell that an electron is in, the orbital in that shell, the orientation of the orbital in the set, and the spin of the electron in its orbital. According to the **Pauli exclusion principle**, there can be no more than 2 electrons in a given orbital. These must have opposite spin, thus giving each electron a unique set of quantum numbers.
4. Orbitals increase in energy as they get farther from the nucleus, and they follow a specific order. In filling the orbitals, electrons fill each orbital following **Hund's rule**, which states that each orientation must get 1 electron before any orientation is filled.
5. The **electron configuration** is a shorthand notation that lists each orbital, in order, with its complement of electrons. The last orbital, which may or may not be full, is called the **valence orbital**. Atoms with a full valence orbital are especially stable.

CHAPTER 3

Trends in the Periodic Table

Read This Chapter to Learn About:

- The Periodic Table
- Metals, Nonmetals, and Metalloids
- Main Groups of Elements
- Ionization Energy, Electron Affinity, and Electronegativity
- The Size of Atoms and Ions
- Bonding Between Ions

The chemical properties of elements are closely associated with the electron configuration of their outermost shells. The elements are arranged into groups (the columns) and periods (the rows). Within a group, every atom has the same number of electrons in its valence orbital, and they share similar chemical properties. Within a period, electrons are added sequentially from left to right to fill the orbitals within the shells. The period number (1–7) corresponds exactly to the shell number for the *s* and *p* orbitals within that period (Figure 3-1).

GROUPS OF ELEMENTS

Groups IA through VIIIA are called the **representative**, or **main group, elements**. They have either *s* or *p* orbitals for their valence orbital. The total number of electrons in the valence shell of each A group is equivalent to the group number. For instance, carbon in group IVA has the electron configuration $2s^2 2p^2$ and has 4 electrons in its 2-shell. It is the number of electrons in the valence orbital that gives a group its characteristic chemical properties.

1 IA	2 IIA	3	4	5	6	7	8	9	10	11	12	13 IIIA	14 IVA	15 VA	16 VIA	17 VIIA	18 VIIIA
1 H 1.0079																	2 He 4.0026
3 Li 6.941	4 Be 9.0122											5 B 10.81	6 C 12.011	7 N 14.007	8 O 15.999	9 F 18.998	10 Ne 20.179
11 Na 22.989	12 Mg 24.305											13 Al 26.981	14 Si 28.086	15 P 30.974	16 S 32.06	17 Cl 35.453	18 Ar 39.948
19 K 39.098	20 Ca 40.08	21 Sc 44.956	22 Ti 47.88	23 V 50.941	24 Cr 51.996	25 Mn 54.938	26 Fe 55.847	27 Co 58.933	28 Ni 58.69	29 Cu 63.546	30 Zn 65.38	31 Ga 59.72	32 Ge 72.59	33 As 74.922	34 Se 78.96	35 Br 79.904	36 Kr 83.80
37 Rb 85.468	38 Sr 87.62	39 Y 88.906	40 Zr 91.22	41 Nb 92.905	42 Mo 95.94	43 Tc (98)	44 Ru 101.07	45 Rh 102.91	46 Pd 106.42	47 Ag 107.87	48 Cd 112.41	49 In 114.82	50 Sn 118.69	51 Sb 121.75	52 Te 127.60	53 I 126.90	54 Xe 131.29
55 Cs 132.91	56 Ba 137.33	57 *La 138.90	72 Hf 178.49	73 Ta 180.95	74 W 183.85	75 Re 186.21	76 Os 190.2	77 Ir 192.22	78 Pt 195.08	79 Au 196.97	80 Hg 200.59	81 Tl 204.38	82 Pb 207.2	83 Bi 208.98	84 Po (209)	85 At (210)	86 Rn (222)
87 Fr (223)	88 Ra 226.0	89 #Ac 227.03	104 Rf (261)	105 Db (262)	106 Sg (263)	107 Bh (262)	108 Hs (265)	109 Mt (266)	110 Uun (269)	111 Uuu (272)	112 Uub (277)						

*Lanthanides	58 Ce 140.12	59 Pr 140.91	60 Nd 144.24	61 Pm (145)	62 Sm 150.36	63 Eu 151.96	64 Gd 157.25	65 Tb 158.92	66 Dy 162.50	67 Ho 164.93	68 Er 167.26	69 Tm 168.93	70 Yb 173.04	71 Lu 174.97
#Actinides	90 Th 232.03	91 Pa 231.03	92 U 238.03	93 Np 237.05	94 Pu (244)	95 Am (243)	96 Cm (247)	97 Bk (247)	98 Cf (251)	99 Es (254)	100 Fm (257)	101 Md (257)	102 No (255)	103 Lr (256)

FIGURE 3-1: The periodic table.

The B groups are called the **nonrepresentative elements**, or **the transition metals**. They are found in the middle of the periodic table. They have a *d* orbital for their valence orbital.

The **lanthanides** and the **actinides** are found at the bottom of the table. They have an *f* orbital for their valence orbital.

METALS AND NONMETALS

The elements can be divided into two major categories, the metals and the nonmetals. The dividing line between them is a stairline that starts at boron and goes down to astatine. The elements to the left of the stairline are metals; those to the right of it are the nonmetals.

Metals have certain properties in common. They have luster, are malleable, and can conduct electricity and heat. They lose electrons easily to form cations, and most of their bonding is ionic in nature.

The **nonmetals** have certain properties in common as well. The ones with lower molar masses tend to be gases in the elemental state. The elements from groups V, VI, and VII can gain electrons to form anions, although most of the bonding of nonmetals, including hydrogen, is covalent.

The **metalloids** are the compounds along the stairline that share properties of both metals and nonmetals. Silicon, for instance, has luster and malleability. Others in this group include germanium, arsenic, antimony, tellurium, polonium, and astatine. As one goes down a column, the main group elements of groups III through VII increase their metallic character. These elements are often used as **semiconductors**.

THE MAIN GROUPS OF METALS AND NONMETALS

Group IA is called the **alkali metal group** and the valence orbital is ns^1. Hydrogen is in this group, but it has properties rather different from those of the other elements in the group. The others, lithium, sodium, potassium, rubidium, and cesium, are soft, low-melting, lustrous metals. They react violently with water, forming a +1 ion as a hydroxide salt, and hydrogen gas. They become more reactive with increasing atomic number. Almost all of the compounds made with these metals are soluble in water.

The group IIA metals are called the **alkaline earth metals**. Their valence shell electron configuration is ns^2. Beryllium is in this group, but it has properties rather different than the others because it is a nonmetal. The others, magnesium, calcium, strontium, barium, and radium, are fairly soft with low density. They readily form oxides and hydroxides; they usually give up their valence electrons to form +2 ions. The oxides and hydroxides of this group are insoluble in water for the most part, and do not decompose when heated. Barium hydroxide, of this group, is soluble and is also a strong base.

The group VIA elements are in the **oxygen group**. This group contains oxygen, sulfur, selenium, and tellurium. Their valence shell electron configuration is $ns^2 np^4$. They tend to form −2 ions, and they often have a −2 oxidation state in covalent compounds. The number of oxidation states increases with atomic number. Oxygen reacts readily with most metals; this reactivity decreases going down the column. Selenium and tellurium are semiconductors, whereas sulfur is an electrical insulator.

The group VIIA elements are called the **halogens**. This group includes fluorine, chlorine, bromine, iodine, and astatine. Their valence shell electron configuration is $ns^2 np^5$. They have a tendency to gain 1 electron to form a −1 ion. They can make ionic bonds as a −1 ion, or they can share electrons in covalent bonds with other nonmetals. The elemental halogens exist as diatomic molecules, except for astatine. Most of the chemistry of the halogens involves oxidation–reduction reactions in water solution. Fluorine is the strongest oxidizer in the group, and it is the most readily reduced.

The elements in group VIIIA are called the **noble gases**. This group includes helium, neon, argon, krypton, xenon, and radon. Their valence shell electron configuration is $ns^2 np^6$; in other words, their valence orbital is full. This makes them very stable elements and, for the most part, they are nonreactive. They do not form ions. Xenon, krypton, and radon react with the very electronegative elements oxygen and fluorine to make a few covalent compounds. The noble gases are monatomic in the elemental form.

THE TRANSITION METALS

The transition metals have a *d* orbital as their valence orbital. Some of them, as seen in Chapter 2, use higher *n* value *s* orbital electrons in order to fill or half-fill their *d* orbital, thus stabilizing it.

The transition metals lose electrons easily and readily form ionic compounds. With the transition metals, it is the *d* orbital that gives the element its physical and chemical properties. They can form various different ionic states, ranging from +1 to +8. They tend to lose electrons from a higher *n* value shell first. Cations that have a half-full or full *d* orbital are especially stable.

Many transition metals form more than one stable oxide. Usually, the oxide that has a higher percentage of oxygen forms most easily.

IONIZATION ENERGY

The **ionization energy, *I*,** of an atom is the energy required to remove an electron from the valence shell, making it an ion.

The ionization energy for the hydrogen atom is 1312 kJ/mole. Elements that have fewer electrons in their valence orbital have lower ionization energies. The elements in group I have the lowest first ionization energies in their respective periods; those in group VIII have the highest. It is more difficult to remove an electron from a full orbital than from one that is not full. If the removal of an electron results in a full valence orbital, the ionization energy is lower.

The ionization energy decreases as one goes down a group, because electrons that are held in higher *n* value shells are farther from the nucleus, and held less tightly (Figure 3-2).

The Ionization Energies (kJ/mol) of the First 20 Elements

Z	Element	First	Second	Third	Fourth	Fifth	Sixth
1	H	1,312					
2	He	2,373	5,251				
3	Li	520	7,300	11,815			
4	Be	899	1,757	14,850	21,005		
5	B	801	2,430	3,660	25,000	32,820	
6	C	1,086	2,350	4,620	6,220	38,000	47,261
7	N	1,400	2,860	4,580	7,500	9,400	53,000
8	O	1,314	3,390	5,300	7,470	11,000	13,000
9	F	1,680	3,370	6,050	8,400	11,000	15,200
10	Ne	2,080	3,950	6,120	9,370	12,200	15,000
11	Na	495.9	4,560	6,900	9,540	13,400	16,600
12	Mg	738.1	1,450	7,730	10,500	13,600	18,000
13	Al	577.9	1,820	2,750	11,600	14,800	18,400
14	Si	786.3	1,580	3,230	4,360	16,000	20,000
15	P	1,012	1,904	2,910	4,960	6,240	21,000
16	S	999.5	2,250	3,360	4,660	6,990	8,500
17	Cl	1,251	2,297	3,820	5,160	6,540	9,300
18	Ar	1,521	2,666	3,900	5,770	7,240	8,800
19	K	418.7	3,052	4,410	5,900	8,000	9,600
20	Ca	589.5	1,145	4,900	6,500	8,100	11,000

FIGURE 3-2: Variation of the first ionization energy with atomic number.

The first ionization energy is the lowest because subsequent electrons are more difficult to remove because of the positive charge produced. It is easiest to remove an electron from a partially filled orbital, and more difficult from a filled valence orbital.

ELECTRON AFFINITY

Ionization energy is the energy required to form a cation. **Electron affinity** is the energy change that occurs when electrons are added to the valence orbital, producing an anion. Energy is given off when anions are formed; thus, a negative sign accompanies the energy difference to indicate its direction of flow. The greater the electron affinity of an atom, the more stable the anion that is formed. It would be expected that group VII elements would have the largest (most negative) electron affinity, because only 1 electron is required to fill the valence orbital.

Generally, the electron affinity increases going across a period to group VII. It then drops, and then increases again going across the next period. Within a group, the electron affinities are approximately equal. Electron affinity is lower to produce a half-full or a full valence orbital, and it is higher if one is adding an electron to an already half-full or full orbital.

ELECTRONEGATIVITY

Electronegativity is the ability of an atom in a molecule to pull electron density of a bond toward itself. It is used most often with covalently bonded atoms.

Electronegativity generally increases going across a period and decreases going down a group. Fluorine has the highest, at 4.0 on the **Pauling scale,** followed by oxygen at 3.5 then chlorine and nitrogen at 3.0. These numbers are averages of the absolute values for the ionization energy and the electron affinity for an atom.

EFFECTIVE NUCLEAR CHARGE

The **nuclear charge** of an atom would appear to be the same as its atomic number, or number of protons. However, the electrons in the innermost orbitals can have a shielding effect, so that the effective nuclear charge, Z_{eff}, is less than the atomic number Z, by the electron shielding effect. If this shielding were perfect, the maximum number of electrons in the innermost two shells would be 10. So, any electrons in the third shell (such as a 3*s* electron) would be attracted to the nucleus by a charge equal to $Z - 10$ for an atom of atomic number Z.

Z_{eff} increases going across a row and up a column; the attractive force on the electrons increases and the atomic radius decreases.

ATOMIC RADIUS

The size of atoms is related to the number of electrons and shells that it has. Generally, the size decreases going across a period, because as the electron number increases, the attraction to the nucleus increases (Z_{eff} increases), thus the atomic radius decreases.

Going down a group, the shell number increases and Z_{eff} decreases, thus the atomic radius increases (Figure 3-3).

Increasing atomic radius

Increasing atomic radius

1A	2A	3A	4A	5A	6A	7A	8A
H 37							He 31
Li 152	Be 112	B 85	C 77	N 70	O 73	F 72	Ne 70
Na 186	Mg 160	Al 143	Si 118	P 110	S 103	Cl 99	Ar 98
K 227	Ca 197	Ga 135	Ge 123	As 120	Se 117	Br 114	Kr 112
Rb 248	Sr 215	In 166	Sn 140	Sb 141	Te 143	I 133	Xe 131
Cs 265	Ba 222	Tl 171	Pb 175	Bi 155	Po 164	At 142	Rn 140

FIGURE 3-3: Atomic radii (in picometers) of the main group elements.

IONIC RADIUS

The sizes of ions depend on whether they have lost or gained electrons. The more electrons an ion has, the larger its radius. Thus, Fe^{+2} is larger than Fe^{+3}, but both are smaller than an Fe atom. If two ions are **isoelectronic** (have the same number of electrons), then the radius decreases across a row and increases down a column, as do neutral atoms.

BONDING BETWEEN IONS

Bonds between atoms that have an electronegativity difference of more than 2 are ionic. An **ionic bond** consists of an electrostatic attraction between a positive ion and a negative ion. It occurs when an element with a low ionization energy encounters an element with a high electron affinity, such as when sodium encounters chlorine:

$$2Na + Cl_2 \rightarrow 2Na^{+1}Cl^{-1}$$

Ions arrange themselves into a lattice network, where no two like ions are neighbors. The energy required to break up the lattice into individual ions is called the **lattice energy, *U***. The energy given off when a lattice is created is $-U$, where $U = kz_1z_2/d$.

In this equation, k is a proportionality constant, z_1 is the charge on the cation, z_2 is the charge on the anion, and d is the average distance between their nuclei. The lattice energy is greatest when the charges are large and the diameters are small. Thus, LiF has a greater lattice energy than LiI, because F is smaller than I. By the same reasoning, AlI_3 has a greater lattice energy than NaI because the charges are greater.

CRAM SESSION

Trends in the Periodic Table

1. Elements within the same group share chemical properties because they have the same number of electrons in their valence orbital.
2. The main group elements include the **alkali metal group**, the **alkaline earth metals**, the **oxygen group**, the **halogens**, and the **noble gases**. These all have *s* or *p* valence orbitals. The transition metals have a *d* orbital for their valence orbital.
3. Trends in the periodic table include **ionization energy**, the energy required to remove an electron from the valence orbital, and **electron affinity**, the energy required to add an electron to the valence orbital. Ionization energies and electron affinities increase with group number.
4. **Electronegativity** is the ability of an atom to pull bonding electrons toward itself. It increases going across a period and decreases going down a group. Fluorine is the most electronegative atom.
5. **Effective nuclear charge** is the charge of a nucleus minus the shielding effect of the innermost electrons. It increases going across a row, thus decreasing the atomic radius of the atoms going across the row.
6. The radius of ions depends on whether electrons have been gained or lost to form the ion. Ions with more electrons are larger. Ions that have the same number of electrons are compared according to their location in the row. Isoelectronic ions get smaller from left to right in a row.

CHAPTER 4

Lewis Dot Structures, Hybridization, and VSEPR Theory

Read This Chapter to Learn About:

- Lewis Structures
- Resonance Structures
- The Covalent Bond
- Hybridized Orbitals
- Valence Shell Electron Pair Repulsion (VSEPR) Theory
- Polar Covalent Bonds and Dipole Moment

Covalent molecules can be represented graphically by Lewis structures or resonance structures. Hybridization of a covalently bonded compound and valence shell electron pair repulsion (VSEPR) theory, which allows one to predict the shapes of molecules, further clarify the nature of covalent bonding and covalent compounds.

LEWIS STRUCTURES

Lewis structures are a means of representing the bonds and lone pairs in a covalent molecule, using all of the valence electrons of each atom so that every atom has a full shell.

For the second-row atoms carbon, nitrogen, oxygen, and fluorine, 8 electrons are needed to fill the shell. Boron is an exception and has only 6 electrons in its shell when it is bonding normally.

For hydrogen, 2 electrons fill its shell.

The third-row atoms phosphorus and sulfur have *d* orbitals, so their shells can hold up to 12 electrons.

Figure 4-1 shows some examples of good Lewis structures.

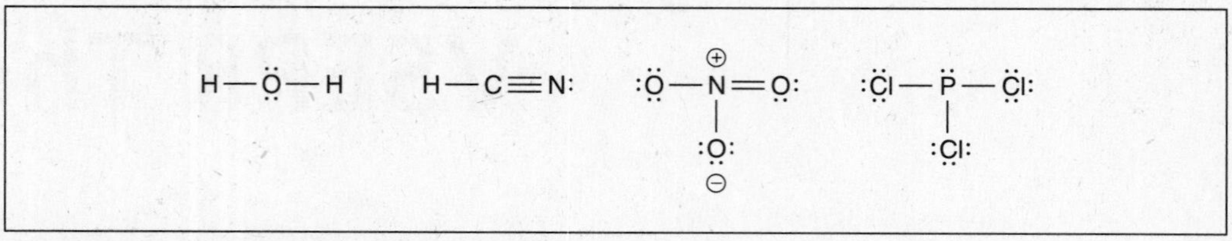

FIGURE 4-1: Examples of good Lewis structures.

Drawing Lewis Structures

To start drawing Lewis structures, the number of valence electrons in the molecule must be determined. This is found by summing the number of valence electrons brought into the molecule by each atom. This value is the same as the atom's group number. For example, hydrogen in group IA has one valence electron. Carbon in group IVA has 4 valence electrons.

To place the atoms properly, the least electronegative atom most often goes in the center, with the more electronegative atoms around it. Molecules prefer to be as spherical as possible, with the most positive atom in the center, much like atoms.

The bonding that each atom is capable of comes from a study of the Lewis dot symbols for atoms in groups IA through VIIA. The dot symbol shows the valence electrons placed around the atom one by one onto four sides. The electrons are placed singly up until 4, and then they begin to double up into what are called **lone pairs.** The Lewis dot symbols are shown in Figure 4-2.

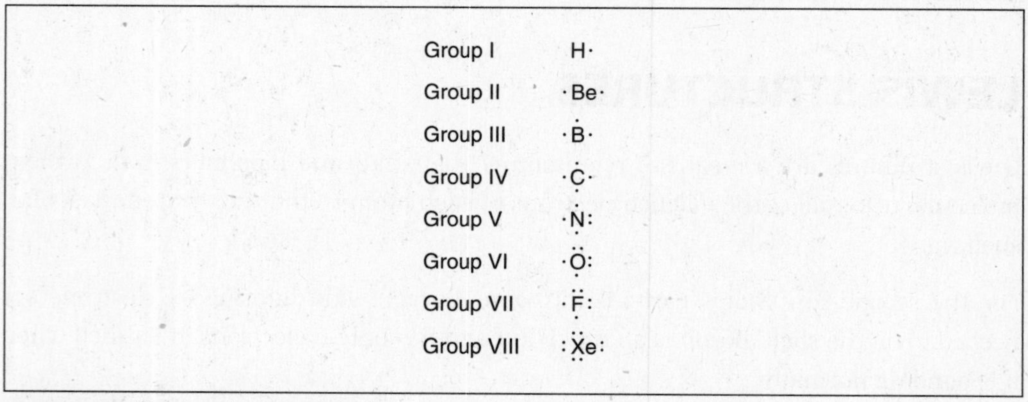

Group I H
Group II Be
Group III B
Group IV C
Group V N
Group VI O
Group VII F
Group VIII Xe

FIGURE 4-2: The Lewis dot symbols.

The dot symbol indicates the number of bonds and lone pairs that a second-row atom has that is most representative of that group. When this bonding prevails, the atom is neutral.

Group IA	Has 1 single dot	Makes 1 single bond
Group IIA	Has 2 single dots	Makes 2 single bonds
Group IIIA	Has 3 single dots	Makes 3 single bonds
Group IVA	Has 4 single dots	Makes 4 bonds: 4 single bonds 2 single bonds and 1 double bond 2 double bonds 1 single bond and 1 triple bond
Group VA	Has 3 single dots and 1 lone pair	Makes 3 bonds and 1 lone pair: 3 single bonds 1 single bond and 1 double bond 1 triple bond
Group VIA	Has 2 single dots and 2 lone pairs	Makes 2 bonds and 2 lone pairs: 2 single bonds 1 double bond
Group VIIA	Has 1 single dot and 3 lone pairs	Makes 1 single bond with 3 lone pairs

There are some exceptions to these rules—namely,

P can make 3 bonds with 1 lone pair, or 5 bonds
S can make 2 bonds with 2 lone pairs, 4 bonds with 1 lone pair, or 6 bonds
Cl, Br, and I can make up to 4 bonds, but single bonds only
N can have an odd number of electrons, such as in NO or NO_2

Formal Charge

After the bonding has been determined, any atom that is not doing its neutral, representative bonding has a formal charge. Formal charges occur when the number of electrons that an atom "owns" in a molecule is different than the number of electrons it brought into the molecule.

For second-row atoms, this difference cannot be more than 1. In other words, a second-row atom can "act like" its immediate neighbors in the periodic table. If it is acting like the neighbor to the left, it carries a +1 formal charge. If it is acting like the neighbor to the right, it carries a −1 formal charge. An example is the ammonium ion NH_4^+. The nitrogen is making 4 bonds, like carbon. It carries a +1 charge.

Any easy way to calculate the formal charge on an atom is

$$\text{Formal charge} = \text{group \#} - (\text{\# bonds on the atom} - \text{\# unshared electrons on the atom})$$

Any second-row atom not bonding according to its Lewis dot symbol has a formal charge. The charge must be written next to the atom to complete the Lewis structure.

Summary of Rules for Drawing Lewis Structures

- Count the total valence electrons. For ions, add electrons for a negative charge, and subtract electrons for positive charges.
- Place the least electronegative atom in the center.
- Place the other atoms around the central atom.
- Try for neutral, representative bonding; if this is not possible, a second-row atom can bond like its immediate neighbor in the periodic table.
- Make sure that all shells are full. Eight electrons in a shell is best, if possible.
- Any atom not bonding "normally" has a formal charge. Compute the formal charge and write it near the atom.

EXAMPLE: NO_2^+

- There are 16 valence electrons. N brings in 5; each O brings in 6; then subtract 1 for the positive charge.
- The nitrogen atom goes in the center.
- Normal bonding does not work; this would require more than 16 electrons.
- Make the oxygens bond normally.
- Make sure all the shells are full.
- Nitrogen is acting like carbon, making 4 bonds, so it has a +1 formal charge.
- The structure is

$$\ddot{\underset{\cdot\cdot}{O}}{=}\underset{\oplus}{N}{=}\ddot{O}:$$

RESONANCE STRUCTURES

Sometimes there is more than one good Lewis structure for the same molecule. These are called **resonance structures.** The atoms are in the same locations; only the electron distribution differs. The molecule SO_2 has two resonance structures, shown in Figure 4-3.

$$:\ddot{O}=\ddot{S}=\ddot{O}: \longleftrightarrow :\ddot{O}-\ddot{S}=\ddot{O}:$$

FIGURE 4-3: The resonance structures of SO_2.

Resonance structures are indicated using the arrow notation $\longleftrightarrow$, which signifies that both structures exist simultaneously. In effect, neither exists. The actual molecule has bonding that is somewhere in between that of the individual resonance structures. This makes it difficult or even impossible to draw the actual bonding. Each resonance structure is a contributor to the actual molecule, which is a hybrid. The hybrid most resembles the most stable resonance structure.

THE COVALENT BOND

Lewis structures show the shared electron bonds that allow each atom to have a full shell. This is called a **covalent bond,** and it is the strongest type of bond between nonmetals. The average distance between the two nuclei is called the **average bond length.**

There are two types of covalent bonds, sigma (σ) and pi (π). The **sigma** (σ) **bond** is made from overlapping hybridized orbitals. The **pi** (π) **bond** is made from overlapping *p* orbitals. There is one and only one sigma bond between any two atoms.

A—B	A=B	A≡B
Single bond	Double bond	Triple bond
1 σ	1 σ	1 σ
	1 π	2 π

HYBRIDIZED ORBITALS

Hybridization is a model for bonding that equates the *s* and *p* orbitals in the 2-shell and combines them to become a new set of equivalent orbitals.

There are 4 total orbitals in the 2-shell, the 2*s* orbital and the 3 orbitals of the 2*p*. These are mathematically combined to produce 4 new hybridized orbitals.

There are three possible ways to combine the 2*s* and 2*p* orbitals:

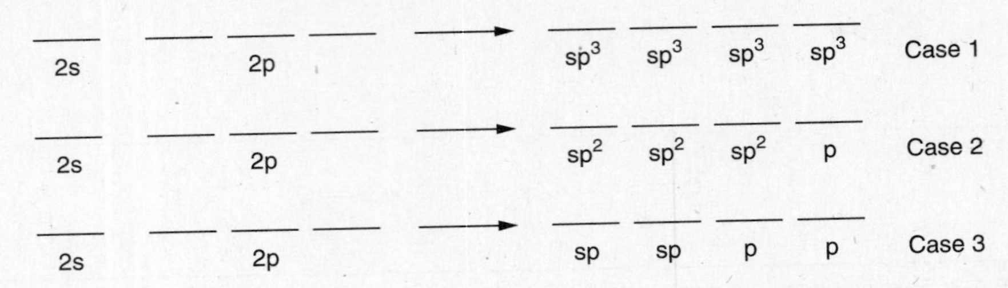

All electrons in the 2-shell are distributed among the new orbitals using the **Aufbau principle.** When there are 2 electrons in a hybridized orbital, it is filled, and this is the lone pair. Any *p* orbital is not filled. Each *p* orbital has 1 electron and gets another electron by sharing with another atom to form a pi bond.

NITROGEN CASE 1: Nitrogen has five electrons in the 2-shell so they will be distributed as follows:

2s 2p → sp^3 sp^3 sp^3 sp^3 Case 1

The four sp^3 orbitals have a tetrahedral geometry, with 109.5° bond angles between them.

The ammonia molecule, NH_3, is sp^3-hybridized. Each sigma bond to a hydrogen is formed by overlapping an sp^3 orbital of N with the 1*s* orbital of H. The last sp^3 orbital holds the lone pair.

NITROGEN CASE 2: The three sp^2 orbitals have trigonal planar geometry, with 120° bond angles between them.

2s 2p → sp^2 sp^2 sp^2 p Case 2

The nitrate ion NO_2^- is sp^2-hybridized. Two sp^2 orbitals are used to make the sigma bonds to the oxygens. The full sp^2 orbital holds a lone pair. The *p* orbital is used to make the pi bond to one of the oxygens.

NITROGEN CASE 3: The two *sp* orbitals have linear geometry, with 180° bond angles between them.

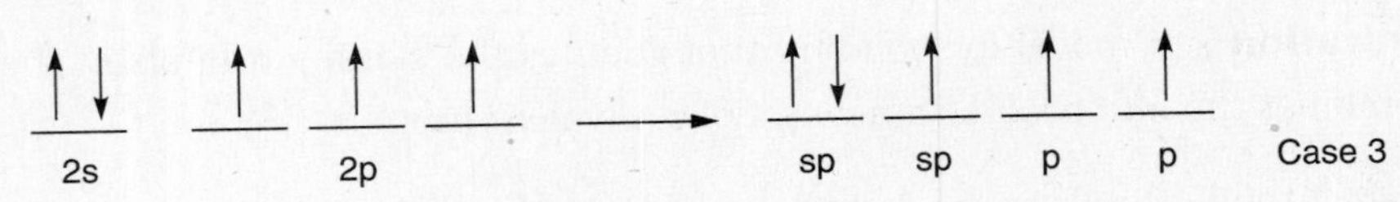

The nitrogens in N_2 are both *sp*-hybridized. Each nitrogen uses an *sp* orbital to make a sigma bond between them. The full *sp* orbital is holding the lone pair. Then there are two *p* orbitals that are used to make two pi bonds, resulting in a triple bond between the nitrogens.

Summary of Hybridization of Second-Row Atoms

To determine the hybridization of a covalently bonded atom, count the number of sigma bonds and lone pairs on the atom.

Total (σ) Bonds and Lone Pairs	Hybridization	Geometry	Bond Angle
4	sp^3	Tetrahedral	109.5°
3	sp^2	Trigonal planar	120°
2	sp	Linear	180°

Other Types of Hybridization

For elements beyond the second row, *d* orbitals are available to hybridize with the *s* and *p* orbitals. Up to two *d* orbitals can be used to make five sp^3d orbitals or six sp^3d^2 orbitals.

Total (σ) Bonds and Lone Pairs	Hybridization	Geometry	Bond Angle
6	sp^3d^2	Octahedral	90°
5	sp^3d	Trigonal bipyramid	120° and 90°

VALENCE SHELL ELECTRON PAIR REPULSION THEORY

Valence shell electron pair repulsion (VSEPR) theory allows one to predict the shapes of molecules. It states that areas of electron density around an atom will get as far from each other as possible. Thus if there are two such areas, such as in O=C=O, the shape of the molecule is linear, with 180° bond angles.

If there are three areas of electron density, the shape is trigonal planar, with 120° bond angles. This occurs in the SO_2 molecule (there is a lone pair on the S).

Lone pairs are larger areas of electron density than the density of bonds. They take up more space. So the bonds of SO_2 are forced together into a 119° bond angle because of the lone pair on the sulfur.

The shape of the bonds, ignoring the lone pair, is bent, as shown in Figure 4-4.

FIGURE 4-4: The bent shape of SO_2.

If there are four areas of electron density around an atom, the geometry is tetrahedral, with 109.5° bond angles. If any of these areas is a lone pair, the bonds are forced closer together, as shown in Figure 4-5 for NH_3 and H_2O, which have bond angles of 107° and 104.5°, respectively.

FIGURE 4-5: Illustration of the decreasing bond angle with increasing number of lone pairs on a central tetrahedral atom.

If there are four bonds, the shape is tetrahedral. If there are three bonds (and a lone pair), the shape of the bonds is trigonal pyramidal. If there are two bonds (and two lone pairs), the shape of the bonds is bent.

For molecules that have five areas of electron density around the central atom, the geometry is trigonal bipyramidal. Three are in the same plane, 120° apart. The remaining two bisect the plane and are normal to it (90°), as shown in Figure 4-6.

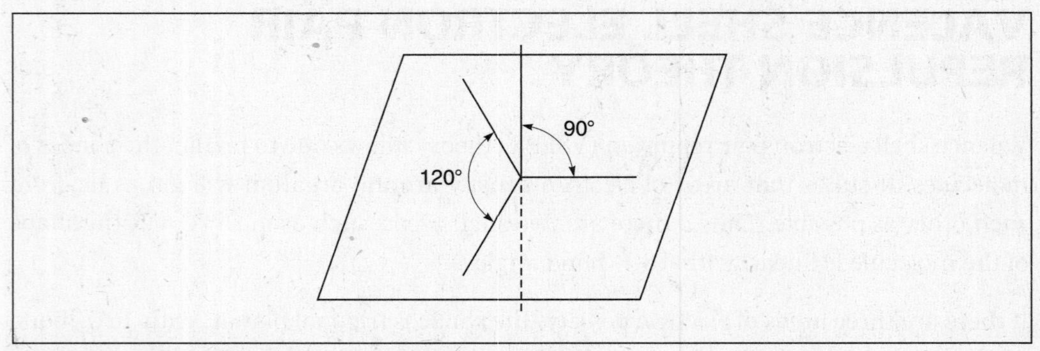

FIGURE 4-6: The trigonal bipyramidal bond angles.

If the central atom has five bonds, the bonds are trigonal bipyramidal. If the central atom has four bonds (and one lone pair), the bonds are in a "seesaw" shape, with the lone pair in the plane.

If the central atom has three bonds (and two lone pairs), the bonds are in a T shape, with the lone pairs in the plane. If the central atom has two bonds (and three lone pairs), the bonds are linear and all the lone pairs are in the plane.

For molecules that have six areas of electron density around the central atom, all six are 90° apart to form an octahedral array, as shown in Figure 4-7.

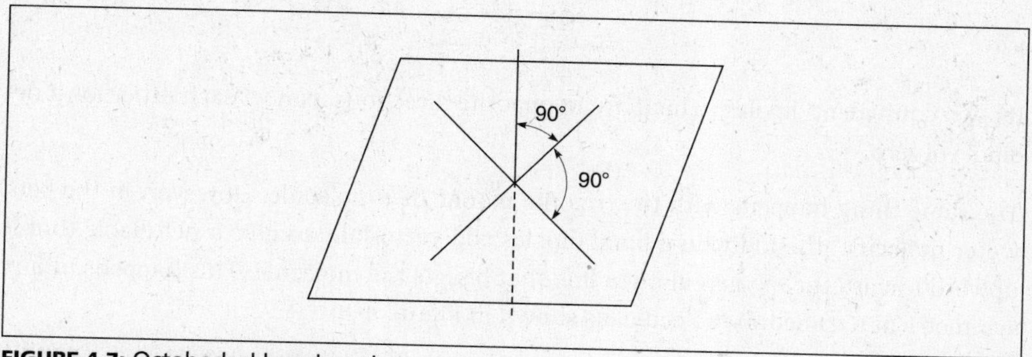

FIGURE 4-7: Octahedral bond angles.

If the central atom has five bonds (and one lone pair), the bonds are in a square pyramidal shape.

If the central atom has four bonds (and two lone pairs) the bonds are in the plane for a square planar shape, and the lone pairs oppose each other.

If the central atom has three bonds (and three lone pairs), the bonds are in a T shape.

If the central atom has two bonds (and four lone pairs), the bonds oppose each other in a linear array.

POLAR COVALENT BONDS AND DIPOLE MOMENT

The Lewis structure can be inspected for the presence of polar covalent bonds. Any bond between two atoms that differ in electronegativity by at least 0.5 unit is polarized, with a partially negative charge (δ^-) at the more electronegative atom, and a partially positive charge (δ^+) at the less electronegative atom.

Each polar bond in a molecule has a dipole moment, which is a measure of its polarity. Dipole moments are vector quantities and have direction as well as magnitude. They are represented by the arrow notation

The dipole moment of each bond in a molecule can add vectorially to give the overall dipole value of the molecule. To do this requires knowledge of the molecular geometry of the molecule.

Because CO_2 is the linear molecule

Its two equivalent dipoles, which are in opposite directions, cancel each other for a net dipole of zero.

The same thing happens with the trigonal planar BF_3 molecule. However, in the bent water molecule, the individual bond dipoles add vectorially to give a net dipole that is upward toward the oxygen along a line that bisects the molecule. This happens in any symmetrical tetrahedral molecule, as shown in Figure 4-8.

FIGURE 4-8: Symmetrical tetrahedral molecules that have a net dipole.

CRAM SESSION

Lewis Dot Structures, Hybridization, and VSEPR Theory

Lewis structures represent the bonding electrons and lone pairs of every atom in a covalently bonded molecule.

1. The Lewis dot symbol for each atom shows the same number of electrons as the group number of the atom, and represents the electrons of the shell as equivalent dots.
2. The octet rule requires that C, N, O, and F atoms each have 8 electrons surrounding them in any Lewis structure.
3. **Resonance structures** are different Lewis structures for the same molecule. They differ only in the placement of electrons.

Hybridization is a model for covalent bonding that hybridizes the atomic orbitals.

1. Hybridization combines the *s* and *p* orbitals of the 2-shell into a new set of equivalent orbitals. The four new orbitals may be sp^3, or sp^2 plus a *p* orbital, or *sp* plus two *p* orbitals.
2. Sigma bonds and lone pairs use hybridized orbitals, whereas pi bonds use *p* orbitals.
3. For atoms with an expanded octet, one *d* orbital may be used to form five sp^3d orbitals, or two *d* orbitals may be combined with the *s* and *p* orbitals to form six sp^3d^2 orbitals.

Valence shell electron pair repulsion (VSEPR) theory predicts the shape of a molecule based on the repulsion of areas of electron density. Areas of electron density include bonded atoms and lone pairs.

1. If there are four areas of electron density around an atom, the orbital geometry of the atom is **tetrahedral.**
2. If there are three areas of electron density around an atom, the orbital geometry of the atom is **trigonal planar.**
3. If there are two areas of electron density around an atom, the orbital geometry of the atom is **linear.**

For atoms with expanded shells, five areas of electron density result in an orbital geometry called **trigonal bipyramid.** Six areas of electron density result in an orbital geometry called **octahedral.** Molecular geometries indicate only the shape of the bonds.

Covalent bonds can be **polar** or **nonpolar.** They are polar if the two atoms of the bond differ sufficiently in electronegativity. Individual bond dipoles can be added vectorially, using the shape of the molecule, to give the overall dipole (direction and magnitude) of the molecule.

CHAPTER 5

Gases

Read This Chapter to Learn About:

- Pressure
- Ideal Gases
- Using the Ideal Gas Law
- Partial Pressure of Gases—Dalton's Law
- The Kinetic Molecular Theory of Gases
- Real Gases

The volume, pressure, and temperature of gases are all related. These relationships for ideal gases, in which there is no intermolecular interaction, have been described through a series of laws—Boyle's law, Charles' law, and Avogadro's law—that make up the Ideal Gas Law. That law and Dalton's law concerning partial pressures in a mixture of gases provide a model for the behavior of ideal gases—the kinetic molecular theory of gases. The van der Waals constants allow for the correction for real gas behavior from that of ideal gases.

PRESSURE

A gas exerts pressure on the walls of its container. The pressure within a container is measured using a **manometer**. The atmosphere exerts pressure onto the Earth due to gravity. Atmospheric pressure is measured using a mercury **barometer**.

A barometer consists of an evacuated tube in a dish of mercury. The atmospheric pressure pushes the mercury into the tube. At sea level, the mercury rises 760 millimeters (mm). Thus, the conversion equation is 1 atmosphere (atm) = 760 mm Hg.

Pressure is a force per unit area. It has various units, including pounds per square inch (psi), Pascal units (Pa), or torr (mm Hg). Torr units allow pressure to be in terms of length.

IDEAL GASES

Ideal gases are those in which there is no intermolecular interaction between the molecules. The molecules are many times farther apart than their diameter. Boyle's law, Charles' law, and Avogadro's law describe the relationship between pressure, volume, and temperature in these gases, respectively.

Boyle's Law

Boyle's law derives from experiments done in the 1660s. Boyle determined that the pressure is inversely related to the volume of a gas at constant temperature. In other words, the pressure times the volume equals a constant for a given amount of gas at a constant temperature:

$$PV = \text{constant}$$

Charles' Law

Charles' law derives from experiments done in the early 1800s. Charles determined that the volume of a gas is directly related to temperature. However, he needed a temperature scale that consisted of positive numbers only. The Kelvin scale, with 0 K as the lowest possible temperature, must be used with Charles' law. To obtain a temperature in K, simply add 273.15 to the temperature in degrees Celsius (°C). Charles' law states that the volume divided by the temperature equals a constant for a given amount of gas at a constant pressure:

$$V/T = \text{constant}$$

Avogadro's Law

Avogadro determined that 1 mole (6.022×10^{23} particles) of any gas occupies the same volume at a given temperature and pressure. The use of moles (mol) allows for the counting of particles, thus ignoring the masses of the particles. Moles, with the symbol n, are part of the constant in Boyle's law and Charles' law.

Ideal Gas Constant

Combining the three laws, one gets

$$PV = n \text{ constant } T$$

The constant is called the **ideal gas constant**. It is given the symbol R, and has the value 0.08206 L atm/mol K.

Avogadro used this ideal gas law to determine that 1 mole of any ideal gas would occupy 22.4 L at 0°C and 1 atm pressure.

USING THE IDEAL GAS LAW

There are two major ways to use the Ideal Gas Law. The first method involves changing conditions. If there are initial conditions that are changed, one can solve for any unknown final condition. The mole amount, n, and the gas constant, R, are constant, thus

$$\frac{P_i V_i}{T_i} = \frac{P_f V_f}{T_f}$$

If five of the six variables are given, the sixth variable can be solved for. If any value remains constant, it falls out of the equation.

EXAMPLE: A sample of gas has volume 3.14 L at 512 mm Hg and 45.6°C. Calculate the volume at 675 mm Hg and 18.2°C.

1. Change all temperatures to Kelvin

$$45.6°\text{C} = 318.8\ \text{K} \qquad 18.2°\text{C} = 291.4\ \text{K}$$

2. Rearrange the equation and solve for V_f

$$V_f = \frac{P_i V_i T_f}{T_i P_f} = \frac{(512\ \text{mm Hg})\ (3.14\ \text{L})\ (291.4\ \text{K})}{(318.8\ \text{K})\ (675\ \text{mm Hg})} = 2.18\ \text{L}$$

The second way to use the Ideal Gas Law is under a set of conditions. There are four variables (P, V, T, and n). If three of them are given, the fourth can be solved by using $PV = nRT$.

EXAMPLE: Calculate the molar mass of a gas if a 12.8-gram (g) sample occupies 9.73 L at 21.0°C and 754 mm Hg.

1. Change the temperature to Kelvin and the pressure to atm

$$21.0°\text{C} = 294.2\ \text{K}$$

$$P = 752\ \text{mm Hg}/760\ \text{mm Hg/atm} = 0.989\ \text{atm}$$

2. Solve for n

$$n = \frac{PV}{RT} = \frac{(0.989\ \text{atm})\ (9.73\ \text{L})}{(0.08206\ \text{L atm/mol K})\ (294.2\ \text{K})} = 0.399\ \text{mole}$$

3. Calculate the molar mass

$$\text{mass/mole} = 12.8\ \text{g}/0.399\ \text{mole} = 32.0\ \text{g/mole}$$

PARTIAL PRESSURE OF GASES—DALTON'S LAW

Dalton's law of partial pressures concerns mixtures of gases. It states that each gas in a mixture exerts its own pressure, and the total of each gas's partial pressure equals the total pressure in the container.

At constant V and T, $P_A + P_B + P_C = P_{total}$ for gases A, B, and C in the mixture. Thus, $P_{total} = n_{total}RT/V$. Each gas consists of a fraction of the entire amount, n_{total}. Each mole fraction is calculated

$$X_A = n_A/n_{total} \qquad X_B = n_B/n_{total} \qquad X_C = n_C/n_{total}$$

And each partial pressure is

$$P_A = X_A P_{total} \qquad P_B = X_B P_{total} \qquad P_C = X_C P_{total}$$

EXAMPLE: Calculate the partial pressure of each gas in a balloon that contains 50.97 g nitrogen, 23.8 g helium, and 19.5 g argon at a total pressure of 2.67 atm.

1. Calculate the moles of each gas

$$(50.97 \text{ g } N_2)\ (1 \text{ mole } N_2/28.02 \text{ g}) = 1.819 \text{ mole } N_2$$

$$(23.8 \text{ g He})\ (1 \text{ mole He}/4.003 \text{ g}) = 5.95 \text{ mole He}$$

$$(19.5 \text{ g Ar})\ (1 \text{ mole Ar}/39.95 \text{ g}) = 0.488 \text{ mole Ar}$$

2. Sum the moles

$$\text{Sum} = 8.26 \text{ mole total}$$

3. Determine the mole fraction X of each gas

$$X_{N_2} = 1.819 \text{ mole}/8.26 \text{ mole} = 0.220$$

$$X_{He} = 5.95 \text{ mole}/8.26 \text{ mole} = 0.720$$

$$X_{Ar} = 0.488 \text{ mole}/8.26 \text{ mole} = 0.0591$$

4. Multiply each mole fraction by the total pressure in the balloon

$$P_{N_2} = 0.220 \times 2.67 \text{ atm} = 0.587 \text{ atm } N_2$$

$$P_{He} = 0.720 \times 2.67 \text{ atm} = 1.92 \text{ atm He}$$

$$P_{Ar} = 0.0591 \times 2.67 \text{ atm} = 0.158 \text{ atm Ar}$$

THE KINETIC MOLECULAR THEORY OF GASES

The **kinetic molecular theory of gases** is a model for gas behavior. It consists of five assumptions concerning ideal gases and explains Boyle's, Charles', Avogadro's, and Dalton's laws. The five assumptions are

- A gas consists of very small particles that move randomly.
- The volume of each gas particle is negligible compared to the spaces between particles.
- There are no intermolecular attractive forces between the gas particles.
- When gas particles collide with each other or with the walls of the container, there is no net gain or loss of kinetic energy.
- The average kinetic energy of each gas particle is proportional to the temperature.

These assumptions are related to Boyles' law, $P \propto 1/V$. Because the pressure is related to the collisions each particle has, the more crowded the particles, the more collisions there are, and the higher the pressure. So, as the volume is decreased, crowding and collisions increase, and the pressure increases.

These assumptions are also related to Charles' law, $V \propto T$. Because the particles move faster when the temperature is increased, it follows that there are more collisions when the particles are moving faster. Thus, the higher the temperature, the greater the volume required to keep the pressure constant.

The assumptions also relate to Avogadro's law, $V \propto n$. As the number of particles increases, the number of collisions increase (pressure and temperature are constant). Thus, the volume must increase to contain the particles.

As for Dalton's law, $P_{\text{total}} = P_A + P_B + P_C$, all gas particles have effectively the same size if their size is negligible compared to the volume between them. Thus, the type of gas is immaterial, and only the total number of gas particles matters.

The kinetic energy of a gas particle is called E_k

$$E_k = \tfrac{1}{2}mv^2$$

where v is the velocity and m is the mass, and v can be shown to be equal to

$$v = (3RT/M)^{1/2}$$

where R is the gas constant 8.314 J/mole K, M is the molar mass, and T is the temperature in K.

The heavier the gas particle, the slower it moves.

REAL GASES

A real gas deviates somewhat in its behavior from the ideal gas, because the basic assumptions about an ideal gas are not always strictly true.

For example, under conditions of high pressure and/or small volume, gas particles can get close enough to exhibit intermolecular attractive forces. Thus, at certain pressures, the actual volume is somewhat smaller than predicted by the Ideal Gas Law.

When pressures get even higher, repulsive forces between particles become important, and the actual volume becomes somewhat larger than predicted by the Ideal Gas Law.

The **van der Waals constants** a and b, which are characteristic for each type of gas, allow for the correction for real gas behavior as follows.

The correction for pressure is $(P + an^2/V^2)$, and the correction for volume is $(V - nb)$. So, the van der Waals equation for real gases becomes

$$(P + an^2/V^2)(V - nb) = nRT$$

CRAM SESSION

Gases

1. The pressure of gases in a container can be in the units atm, mm Hg, psi, or torr, to name a few. The units most often used in chemistry are mm Hg and atm.
2. The **Ideal Gas Law** $PV = nRT$ is derived from work done on gases by Boyle, Charles, and Avogadro. **Boyle** determined that pressure is inversely related to volume at constant temperature. **Charles** determined that temperature and volume are directly related at constant pressure. The Kelvin scale, with absolute zero being the coldest possible temperature, is necessary for all temperature work with gases. In this gas law, n is the number of moles of gas, and is directly related to volume. The **gas constant** R is 0.08206 L atm/mole K. The gas law can be used to calculate the density of a gas at a certain temperature and pressure. It can also be used to determine the molar mass of a gas.
3. **Dalton's law of partial pressures** states that in a mixture of gases, each gas exerts its own pressure, and that each gas's partial pressure is directly related to the mole fraction of that gas in the mixture.
4. The **kinetic theory of gases** is a model for gas behavior that explains Charles', Boyle's, Avogadro's, and Dalton's laws. It consists of five assumptions. These assumptions are not strictly true under extreme conditions. There are correction factors that can be applied.
5. The equation for real gas behavior, also called the **van der Waals equation**, is $(P + an^2/V^2)(V - nb) = nRT$, where a and b are constants for each type of gas.

CHAPTER 6

Intermolecular Forces and Phase Equilibria

Read This Chapter to Learn About:

- ➤ Intermolecular Forces
- ➤ Phase Equilibria
- ➤ Vapor Pressure
- ➤ Properties of Solutions
- ➤ Colloids

Forces between molecules in a liquid or solid determine the density of the substance as well as its boiling and freezing points. In addition, these intermolecular forces determine the crystal lattice of a solid and various properties of a liquid, including vapor pressure. These forces also affect the properties of solutions and colloids.

INTERMOLECULAR FORCES

Intermolecular forces are the attractive forces between molecules in a liquid or solid sample. The intermolecular forces determine how far apart the molecules are, which, in turn, determines the density of the sample. The strength of the intermolecular forces in liquids and solids also determines their melting and boiling temperatures.

Other properties of liquids that are the result of the strength of intermolecular force are surface tension, viscosity, capillary action, and vapor pressure. The stronger the intermolecular forces, the stronger each of these properties will be. In solids, intermolecular forces determine the structure of the crystal lattice.

There are several types of intermolecular forces. The intermolecular interactions that are found in liquids include dipole–dipole interactions, hydrogen bonding, and dispersion forces. The intermolecular interaction that is found in ionic solids is ion–ion forces.

Dipole–dipole interactions are found in liquid covalently bonded compounds that contain a polar bond, called a **dipole**. Dipole–dipole interactions are also electrostatic attractive forces. They occur between the negative end of one dipole and the positive end of another dipole in another molecule. Figure 6-1 gives an example of dipole–dipole interactions between PCl_3 molecules in the liquid phase.

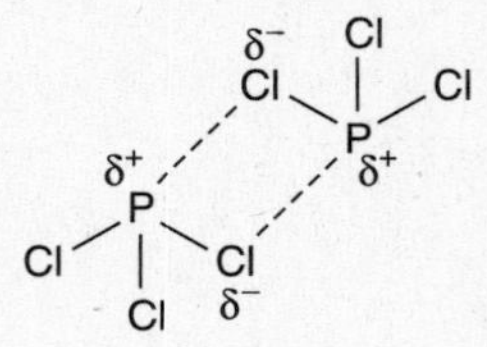

FIGURE 6-1: Dipole–dipole interactions in PCl_3.

Hydrogen bonding is found in molecules that contain an O—H, N—H, and/or F—H bond. These bonds have fairly large dipole moments. The electrostatic attraction of the negative end of one polar bond (an O, F, or N) with the positive end (H) of another polar bond in another molecule is a stronger type of dipole–dipole interaction. Hydrogen bonding is the major attractive force that holds large biomolecules, such as proteins and DNA, in their specialized shapes. Figure 6-2 shows hydrogen bonding in the water molecule.

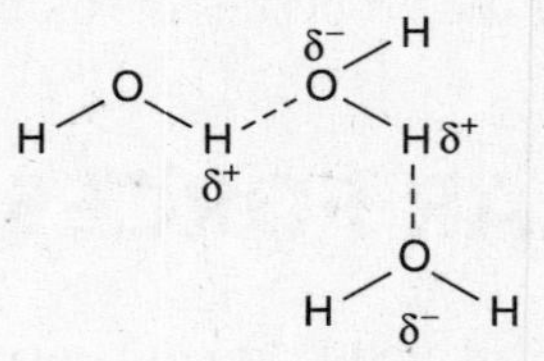

FIGURE 6-2: Hydrogen bonding in water.

Dispersion forces are found in all compounds, but they are the major attractive force between nonpolar covalent molecules. In the nonpolar bond, a short-lived dipole is produced when the electron density in the bond fluctuates normally. At a certain instant, the electron density may be greater at one end of the bond, resulting in a dipole. This short-lived dipole induces the formation of another dipole in a nearby molecule, which, in turn, induces another to form, and so on, dispersing dipoles throughout the sample.

Ion–ion forces are simply the electrostatic attractive forces between cations and anions. In the crystal lattice, all the negatively charged ions alternate with positively charged ions.

PHASE EQUILIBRIA

Phase changes occur due to a change in temperature, pressure, or both. The most common phases are solid, liquid, and gas. The names of the phase changes are as follows:

Phase-to-Phase Change	
Solid to liquid	Melting
Solid to gas	Sublimation
Liquid to solid	Freezing
Liquid to gas	Evaporation
Gas to liquid	Condensation
Gas to solid	Deposition

A phase diagram shows the different phases that a substance is in over a range of temperatures and pressures. A simple phase diagram is shown in Figure 6-3.

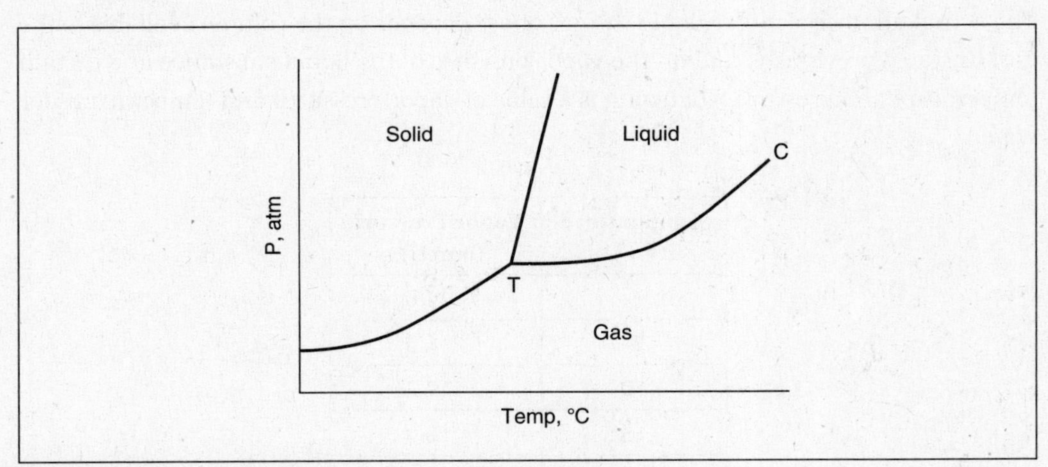

FIGURE 6-3: A typical pressure–temperature diagram.

The lines in Figure 6-3 represent the boundaries between the phases. Each point on a line represents a certain temperature and pressure for that phase transition. The solid–liquid boundary line extends indefinitely, but the liquid–gas boundary ends at the point C, which is called the **critical point**, and it represents the beginning of the supercritical fluid phase. At all temperatures and pressures beyond those at point C, the substance is in a phase that is in between liquid and gas called **supercritical fluid**.

Point *T* is called the **triple point**. At this pressure and temperature, all three phases coexist in an equilibrium mixture. This means that the actual number of molecules in each phase remains the same, but the individual molecules are changing their phase.

The boiling and freezing points at 1 atmosphere (atm) pressure are called the **normal boiling point** and the **normal freezing point**. To determine these values, find the temperature that corresponds to 1 atm pressure on the liquid–gas boundary and on the liquid–solid boundary.

The **density** of a substance increases with pressure. Therefore, the relative densities of the phases can be determined by following the phases as the pressure is increased. The phase that exists at the highest pressures is the densest phase. For water, the densest phase is the liquid phase. For most other substances, the solid phase is the densest phase.

VAPOR PRESSURE

If a liquid is in a closed container at a constant temperature and pressure, a certain number of molecules leave the liquid phase and enter the gas phase. As long as the volume available to the gas phase remains constant, the number of molecules in the gas phase remains constant. There is equilibrium between the gas and the liquid phases. Specific individual molecules may enter or leave the gas phase, but the actual number of molecules in the gas phase remains constant under these conditions.

The actual number of molecules in the gas phase depends on the volume available to the gas, as stated previously, and on the vapor pressure of the liquid substance at a certain temperature and pressure. Following is a table of vapor pressures and temperatures for water.

Temperature (°C)	Vapor Pressure (mm Hg)
0	4.6
5	6.5
10	9.2
15	12.8
20	17.5
21	18.6
22	19.8
23	21.1
25	22.4
26	23.8

Temperature (°C)	Vapor Pressure (mm Hg)
27	26.7
28	28.3
29	30.0
30	31.8
35	42.2
40	55.3
45	71.9
50	92.5
55	118.0
60	149.4
65	187.8
70	233.7
80	355.1
90	525.8
100	760.0

Calculations Involving Vapor Pressure

Substituting vapor pressure for P, the number of moles of gas in the gas phase can be calculated from the gas law $PV = nRT$.

EXAMPLE: Calculate the percentage of water in the gas phase when 235 grams (g) water is placed in a 6.44-liter (L) container at 65.0°C.

➤ Find the vapor pressure of water at 65.0°C (187.5 mm Hg)

➤ Convert the vapor pressure from mm Hg to atm

(187.5 mm Hg)(1 atm/760 mm Hg) = 0.247 atm

➤ Calculate $n = PV/RT$:

(0.247 atm)(6.44 L)/(0.08206)(338.2 K) = 0.0573 mole water

➤ Convert n to mass

(0.0573 mole)(18.02 g/mole) = 1.03 g water in the gas phase

➤ Calculate the percentage of water in the gas phase

(1.03 g/235 g) × 100 = 0.438% in the gas phase

Change of Vapor Pressure with Temperature

The vapor pressure at any temperature can be calculated if the boiling point is known, using the **Clausius–Clapyron equation.** If T_1 is the normal boiling point, P_1 is 1 atm.

$$\ln(P_2/P_1) = (\Delta H_{vap}/R)(1/T_1 - 1/T_2)$$

PROPERTIES OF SOLUTIONS

The properties of solutions are known as **colligative properties**. When a **solute** is dissolved in a **solvent**, the boiling point of the solution increases, the freezing point decreases, and the vapor pressure decreases compared to those of the pure solvent. The magnitude of these changes depends on the amount of solute present; in other words, the concentration of the solution affects the actual change.

Units of Concentration

The unit of concentration used with the equations of colligative properties is **molality, *m***. It is used in the equations for boiling point elevation and for freezing point depression. Molality is determined by dividing the mole solute by the kilogram (kg) solvent.

$$m = \text{mole solute/kg solvent}$$

Another type of concentration unit, used with vapor pressure lowering, is **mole fraction, *X***. The mole fraction of a solute is determined by dividing the mole solute by the total moles of solute plus solvent.

$$X_{solute} = \text{mole solute/mole total}$$

Vapor Pressure Lowering in a Solution (Raoult's Law)

A solution always has a vapor pressure that is lower than that of the pure solvent. The amount that the vapor pressures change is calculated using the equation

$$\Delta P = X_{solute}P^\circ$$

where P° is the vapor pressure of the pure solvent.

Once the change is calculated, the actual vapor pressure of the solution is found by subtracting the change from the vapor pressure of the pure solvent

$$P = P^\circ - \Delta P$$

EXAMPLE: Calculate the vapor pressure of a solution that contains 18.5 g glucose ($C_6H_{12}O_6$) in 82.5 g water at 24.0°C.

1. Look up the vapor pressure of water at 24.0°C (22.4 mm Hg)
2. Calculate the mole fraction of glucose in the solution

$$(18.5 \text{ g glucose})(1 \text{ mole}/180.2 \text{ g}) = 0.1027 \text{ mole glucose}$$

$$(82.5 \text{ g } H_2O)(1 \text{ mole}/18.02 \text{ g}) = 4.58 \text{ mole water}$$

$$X_{\text{glucose}} = 0.1027 \text{ mole}/(4.58 + 0.1027 \text{ mole}) = 0.0219$$

3. Calculate $\Delta P = X_{\text{glucose}}P^{\circ} = (0.0219)\,(22.4 \text{ mm Hg}) = 0.491 \text{ mm Hg}$
4. Calculate the actual vapor pressure of the solution

$$P^{\circ} - \Delta P = 22.4 - 0.491 = 21.9 \text{ mm Hg}$$

Boiling Point Elevation in a Solution

A solution always has a boiling point higher than that of the pure solvent. The magnitude of the change is calculated using the equation

$$\Delta T_b = iK_b m$$

where i = number of particles per molecule when dissolved, called the **Van't Hoff factor** and K_b = molal boiling point depression constant for the solvent.

The actual boiling point of the solution is found using the equation

$$T_b = \text{BP}_{\text{normal}} - \Delta T_b$$

where $\text{BP}_{\text{normal}}$ = the normal boiling point of the solvent.

Freezing Point Depression in a Solution

A solution always has a freezing point that is lower than that of the pure solvent. The magnitude of the change is calculated using the equation

$$\Delta T_f = iK_f m$$

where i = number of particles per molecule when dissolved (the Van't Hoff factor) and K_f = molal freezing point depression constant for the solvent. The actual freezing point of the solution is found using the equation

$$T_f = \text{FP}_{\text{normal}} - \Delta T_f$$

where $\text{FP}_{\text{normal}}$ = the normal freezing point of the solvent.

EXAMPLE: Calculate the boiling point and the freezing point of a solution of antifreeze that contains 455 g ethylene glycol ($C_2H_6O_2$) (EG) in 675 g water.

K_f for water is 1.86°C/m

K_b for water is 0.512°C/m

1. Determine the molality of the solution

   m = mole solute/kg solvent

   (455 g ethylene glycol)(1 mole/62.07 g) = 7.33 mole EG

   M = 7.33 mole ethylene glycol/0.675 kg water = 10.9 m EG

2. Determine the i value.
   Because ethylene glycol is a covalent molecule, it does not break up into separate particles when dissolving, so $i = 1$.
3. Calculate ΔT_f

   $\Delta T_f = (1.86°\text{C/m})(1)(10.9\text{ m}) = 20.3°\text{C}$

4. Calculate T_f

   $T_f = 0°\text{C} - 20.3°\text{C} = -20.3°\text{C}$

5. Calculate ΔT_b

   $\Delta T_b = (0.512°\text{C/m})(1)(10.9\text{ m}) = 5.6°\text{C}$

6. Calculate T_b

   $T_b = 100.0°\text{C} + 5.6°\text{C} = 105.6°\text{C}$

Osmotic Pressure

If a solution is held within a semipermeable membrane, such as a dialysis bag, and the bag is placed in pure water, water molecules will enter the bag to dilute the solution. The water within the bag will at some point attain a pressure equilibrium with the water outside the bag. This pressure is defined as the osmotic pressure Π of the solution, and it can be calculated using the formula

$$\Pi V = nRT$$

where V is the volume of the bag, or container, n is the moles of solute, R is the gas constant 0.08206 L atm/mole K, and T is the temperature in K.

Another equation that is often used is

$$\Pi V = cRTi$$

where c is the molar concentration.

Osmotic pressure is typically used to estimate the molar mass of a large biomolecule. Its osmotic pressure is determined separately, and then n is calculated.

EXAMPLE: Calculate the molar mass of a small protein if a solution that contains 5.6 milligrams (mg) in 35.0 milliliters (mL) water shows an osmotic pressure of 1.2 mm Hg at 37.0°C.

1. Convert 1.2 mm Hg to atm

$$(1.2 \text{ mm Hg})(1 \text{ atm}/760 \text{ mm Hg}) = 0.00158 \text{ atm}$$

2. Convert 37.0°C to K

$$273.2 + 37.0 = 310.2 \text{ K}$$

3. Calculate n

$$n = \Pi V/RT = (0.00158 \text{ atm})(0.0350 \text{ L})/(0.08206 \text{ L atm/mol K})(310.2 \text{ K}) = 2.2 \times 10^{-6} \text{ mole}$$

4. Convert 5.6 mg to grams

$$(5.6 \text{ mg})(1 \text{ g}/1000 \text{ mg}) = 0.0056 \text{ g}$$

5. Calculate the molar mass

$$0.0056 \text{ g}/2.2 \times 10^{-6} \text{ mole} = 2545 \text{ g/mole}$$

Dissolved Gases—Henry's Law

Henry's law for dissolved gases states that the amount of gas dissolved is directly proportional to the pressure of the gas above the solution. It is given by the equation

$$P = k_H X$$

where X is the mole fraction of the gas, P is the pressure of the gas above the solution, and k_H is Henry's law constant for the solution.

Henry's law works best for gas molecules that remain intact in the solvent and that do not react with the solvent in any way.

Henry's law is also given sometimes as $P = k_H c$, where c is the molarity of the gas.

EXAMPLE: Calculate the solubility of oxygen in water at 25°C if the partial pressure is 125 mm Hg.

1. The k_H for oxygen is 7.8×10^2 L atm/mol
2. Convert 125 mm Hg to atm

$$(125 \text{ mm Hg})(1 \text{ atm}/760 \text{ mm Hg}) = 0.164 \text{ atm}$$

3. Calculate $c = P/k_H = 0.164 \text{ atm}/7.8 \times 10^2 \text{ L atm/mole} = 2.1 \times 10^{-4}$ mole/L

COLLOIDS

Colloids consist of a suspension of small particles in a medium. They are distinguished from true solutions by the **Tyndall effect**, which shows the scattering of light by the suspended particles. A true solution allows light to pass through unseen.

Examples of colloidal suspensions include milk, butter, smoke, gelatin, paint, and fog.

CRAM SESSION

Intermolecular Forces and Phase Equilibria

1. Intermolecular forces hold molecules together in the liquid and solid states. They include **dispersion forces,** which are found in all liquid and solid samples, but which predominate in nonpolar molecules. **Hydrogen bonding** is the predominant force between molecules that contain an O—H or N—H bond. **Dipole–dipole interactions** are found in other polar covalent molecules.
2. A **phase diagram** shows all phases that a substance is in at various temperatures and pressures. The normal boiling point and melting point can be determined from a phase diagram, as well as the relative densities of the phases.
3. The **vapor pressure** of a liquid allows the calculation of the amount that will go into the gas phase in a certain volume container. Vapor pressure is temperature-dependent and rises until the boiling point is reached, where the vapor pressure is equal to the external pressure, 1 atm. The vapor pressure at any temperature can be calculated, if the boiling point and the ΔH_{vap} of the substance are known.
4. **Colligative properties of solutions** show how the vapor pressure decreases, the boiling point increases, and the melting point decreases as solute is added.
5. **Osmotic pressure** can be used to estimate the molar mass of a biomolecule in solution.
6. **Colloids** are not true solutions. They show the Tyndall effect by scattering light that is passed through the colloid.

CHAPTER 7

Chemical Equations and Stoichiometry

Read This Chapter to Learn About:

- Stoichiometry
- Balancing Chemical Equations
- Limiting Reagents
- Determining Theoretical Yield

Stoichiometry is the process of using a balanced chemical equation to calculate quantities of substances used or produced in the reaction. The amount of product formed from a given amount of starting material can be calculated. Or the amount of starting material needed to form a given amount of product can be calculated. In other words, starting with an amount of any substance in the chemical equation, one can calculate the amount of any other substance.

BALANCING CHEMICAL EQUATIONS

A balanced chemical equation is necessary to do any calculations using the equation. A chemical equation has starting material, or **reactant**, on the left, and **product** on the right. There is an arrow in between that signifies that the reactant is turning into product. An example of a chemical equation is

$$KClO_3 \rightarrow KCl + O_2$$

This equation states that the substance $KClO_3$, potassium chlorate, is converted into KCl, potassium chloride, and O_2, oxygen gas.

This equation is not balanced, however. There are three oxygen atoms on the left side of the equation and only two on the right side. In order for an equation to be balanced, there must be the same number of each type of atom on both sides of the arrow.

One method that can be useful in balancing equations is as follows:

1. Find an atom that has a subscript on one side of the equation, which shows in only one compound on the other side.
2. Turn this subscript into a coefficient in front of the formula containing that atom on the other side.
3. Then follow through until all the atoms are balanced.

EXAMPLE:

$KClO_3 \rightarrow KCl + O_2$

1. Start with the subscript 3 on the oxygen on the left side.
2. Turn it into a coefficient in front of the O_2 on the right side.

   $KClO_3 \rightarrow KCl + 3O_2$

3. This results in six oxygens on the right side. So a 2 is placed in front of $KClO_3$

   $2KClO_3 \rightarrow KCl + 3O_2$

4. This results in two potassiums and two chlorines on the left side; so a 2 is placed in front of the KCl on the right side

   $2KClO_3 \rightarrow 2KCl + 3O_2$

5. The equation is now balanced.

As stated previously, stoichiometry requires a balanced equation. The significance of the coefficients in the balanced equation is that they provide the lowest whole-number **molar ratio** of the substances in the reaction.

One way that different substances can be related is by their molar ratio. It is by the use of moles that quantities of one substance can be converted into quantities of another substance in the equation. Volume or molecules may also be used, but not mass.

EXAMPLE: Calculate the mass of oxygen gas that will be produced when 48.5 grams (g) $KClO_3$ reacts according to the equation

$2KClO_3 \rightarrow 2KCl + 3O_2$

1. First, the mass of the $KClO_3$ must be converted into moles:

   $(48.5\text{ g } KClO_3)\ (1\text{ mole}/122.6\text{ g}) = 0.396\text{ mole } KClO_3$

2. Next, the moles $KClO_3$ are converted into mole O_2 using their molar ratio from the balanced equation

   $(0.396\text{ mole } KClO_3)\ (3\text{ mole } O_2/2\text{ mole } KClO_3) = 0.594\text{ mole } O_2$

3. Then the moles O_2 are converted into mass

   $(0.594 \text{ mole } O_2)(32.00 \text{ g/mole}) = 19.0 \text{ g } O_2$

4. 19.0 g O_2 is the theoretical yield of oxygen formed when 48.5 g $KClO_3$ reacts according to the previous equation.

To summarize, the process is always

Mass A → Mole A	By using the molar mass of A
Mole A → Mole B	By using the coefficients of each
Mole B → Mass B	By using the molar mass of B

LIMITING REAGENTS

When the amount of two starting materials is given, one of them may be in excess. This means that it is not all used up in the reaction; there is some left over. This occurs because there is not enough of the other starting material to react with it all. The other starting material is said to be the **limiting reagent**. The limiting reagent does get all used up, and it limits the amount of product that can be produced.

To work a limiting reagent problem, the product theoretical yield is calculated from each starting material separately. Whichever one gives the smaller amount of product is the limiting reagent, and the smaller amount is the correct theoretical yield.

EXAMPLE: Calculate the yield of ammonia produced when 59.7 g of hydrogen reacts with 75.2 g of nitrogen according to the equation

$N_2 + H_2 \rightarrow NH_3$

1. First, the equation must be balanced. A 2 must go in front of the NH_3 and then a 3 must go in front of the H_2 to give

   $N_2 + 3H_2 \rightarrow 2NH_3$

2. Next, the product yield is calculated from the given amount of H_2

   $(59.7 \text{ g } H_2)(1 \text{ mole } H_2/2.016 \text{ g } H_2)(2 \text{ mole } NH_3/3 \text{ mole } H_2)$

   $(17.03 \text{ g } NH_3/1 \text{ mole } NH_3) = 335 \text{ g } NH_3$

3. Then the product yield is calculated from the given amount of N_2

   $(75.2 \text{ g } N_2)(1 \text{ mole } N_2/28.02 \text{ g } N_2)(2 \text{ mole } NH_3/1 \text{ mole } N_2)$

   $(17.03 \text{ g } NH_3/1 \text{ mole } NH_3) = 91.4 \text{ g } NH_3$

4. Therefore, the limiting reagent is N_2 and the theoretical yield is 91.4 g NH_3.

BALANCING OXIDATION–REDUCTION REACTIONS

Oxidation–reduction reactions often cannot be balanced in the usual way. They involve the transfer of electrons from one species to another, and the electrons must be balanced in addition to balancing the atoms. This type of balancing is explained in Chapter 15 on electrochemistry.

CRAM SESSION

Chemical Equations and Stoichiometry

1. **Stoichiometry** is a process that uses a balanced chemical equation to convert moles of one substance into moles of another substance in the equation.
2. A balanced chemical equation has the same number of atoms of each type on both sides of the equation.
3. Because substances are weighed in gram quantities, the mass of the given substance must be converted into moles, using the molar mass of the substance. Then the coefficients of the given substance and the target substance are used to convert the moles of the given substance into the moles of the target substance. Finally, the moles of the target substance are converted into mass units, using its molar mass.
4. If there are two reactants, and one of the reactants is in excess, it is not all used up in the reaction. The other reactant, which is completely converted into product, is called the **limiting reagent**. It limits the amount of product that can be formed.

CHAPTER 8

Reactions in Solution

Read This Chapter to Learn About:

- ➤ Concentration Units in Solutions
- ➤ Precipitation Reactions
- ➤ Acid–Base Reactions
- ➤ Oxidation–Reduction Reactions

A solution consists of a solute dissolved in a solvent. This chapter deals with the different types of reactions that occur in solutions.

CONCENTRATION UNITS

There are various concentration units that can be used to describe how much solute is dissolved in a certain amount of solution. In Chapter 5, the concentration units of molality and mole fraction are discussed and are used with the colligative properties of solutions. This chapter deals with reactions that take place in solution. The concentration unit that is used to make solutions for reaction is **molarity**, which has the symbol **M**, and is defined as

$$M = \text{mole solute/liter (L) solution}$$

This definition is also a working equation. There are three variables; if two are known, the third can be calculated.

EXAMPLE: Calculate the molarity of a solution that is prepared by dissolving 1.192 grams (g) of oxalic acid ($C_2H_2O_4$) in water to a total of 100.0 milliliters (mL) of solution.

1. Convert the mass of oxalic acid to moles

$$(1.192 \text{ g}) (1 \text{ mole}/90.04 \text{ g}) = 0.01324 \text{ mole oxalic acid}$$

2. Convert the milliliters to liters

$$(100.0 \text{ mL})\ (1 \text{ L}/1000 \text{ mL}) = 0.1000 \text{ L}$$

3. Calculate the molarity

$$M = 0.01324 \text{ mole}/0.1000 \text{ L} = 0.1324 \text{ M}$$

REACTIONS IN SOLUTION

There are at least three major categories of reactions that take place using solutions of reactants. They include precipitation reactions, where a solid product is formed; acid–base reactions, which produce an ionic compound and water; and oxidation–reduction reactions, which involve the transfer of electrons from one species to another.

Precipitation Reactions

Sometimes, when two solutions are mixed together, an insoluble compound, called a **precipitate**, is formed. This occurs whenever a cation from one of the solutions and an anion from the other solution together form an insoluble compound. Whether or not a particular pair-up of cation and anion result in an insoluble compound requires knowledge of the solubility rules for ionic compounds in water.

SOLUBILITY RULES

- All compounds with a group IA or ammonium cation are soluble.
- All compounds with a nitrate anion are soluble.
- All compounds with a chloride anion are soluble, unless the cation is lead, silver, or mercury.
- All compounds with a sulfate anion are soluble, unless the cation is barium, strontium, calcium, lead, silver, or mercury.
- All carbonates, phosphates, and sulfides are insoluble, unless the cation is ammonium or from group IA.
- All hydroxides are insoluble, unless the cation is ammonium, barium, or from group IA.

NET IONIC EQUATIONS FOR PRECIPITATION REACTIONS

To write the net ionic equation means writing only the cation and the anion that are doing the reacting on the left side of the equation arrow and the insoluble product that they form on the right side of the arrow. It is written as

$$\text{cation} + \text{anion} \rightarrow \text{insoluble compound}$$

EXAMPLE: Write the net ionic equation for the precipitation reaction that occurs when a solution of silver (I) nitrate is mixed with a solution of magnesium chloride.

1. List the two possible products.
   - Silver (I) chloride
   - Magnesium nitrate
2. Determine, using the solubility rules, which of these is insoluble in water.
   - A silver cation with chloride anion is insoluble, so silver (I) chloride is insoluble.
3. Write the two ions that produce the compound.

   $Ag^{+1} + Cl^{-1}$

4. Complete the equation by writing the formula of the product on the right.

   $Ag^{+1} + Cl^{-1} \rightarrow AgCl$

5. Make sure the equation is balanced; this one is already balanced.

Write the net ionic equation that occurs when a solution of potassium phosphate is mixed with a solution of calcium chloride.

1. Write the two possible products.
   - Potassium chloride
   - Calcium phosphate
2. Determine which of these is insoluble.
   - Calcium phosphate is insoluble because all phosphates are insoluble, except for group IA cations and ammonium.
3. Write the ions that react.

   $Ca^{+2} + PO_4^{-3}$

4. Complete the equation by writing the formula of the product that is formed.

   $Ca^{+2} + PO_4^{-3} \rightarrow Ca_3(PO_4)_2$

5. Balance the equation.

   $3Ca^{+2} + 2PO_4^{-3} \rightarrow Ca_3(PO_4)_2$

Acid–Base Reactions

An acid reacts with a base to form an ionic compound, often called a **salt**, plus water.

ACIDS

An **acid** is a compound that produces **H^+ (hydronium ion)** in water solution. There are strong acids and weak acids. A strong acid is defined as an acid that dissociates fully into two ions in water solution.

The six strong acids are HCl, HBr, HI, HNO_3, H_2SO_4, and $HClO_4$.

All other acids are **weak acids**, defined as an acid that dissociates only to a very small extent in water solution. Some examples are HF, HCN, H_2CO_3, H_3PO_4, acetic acid, and oxalic acid.

BASES

A **base** is a compound that produces **OH^- (hydroxide ion)** in water solution. There are strong bases and weak bases. A strong base fully dissociates into its ions in water solution. There are four strong bases, hydroxides of groups IA and IIA, that are fully soluble in water. They are LiOH, NaOH, KOH, and $Ba(OH)_2$.

There are many weak bases, but most of them are organic analogs of ammonia, NH_3. Ammonia dissolves in water, but only a very small percentage of NH_3 molecules produce NH_4OH, which then dissociates into NH_4^+ and OH^-.

TYPES OF ACID–BASE REACTIONS

When a strong acid reacts with a strong base, the ionic compound that forms is neutral in water solution. The ionic compound is formed by taking the cation from the base and the anion from the acid.

$HCl + KOH \rightarrow KCl + H_2O$ KCl is neutral

$H_2SO_4 + NaOH \rightarrow NaHSO_4 + H_2O$ $NaHSO_4$ is neutral

When a strong acid reacts with a weak base, the ionic compound that forms is acidic in water solution.

$HCl + NH_4OH \rightarrow NH_4Cl + H_2O$ NH_4Cl is acidic

$HClO_4 + NH_4OH \rightarrow NH_4ClO_4 + H_2O$ NH_4ClO_4 is acidic

When a weak acid reacts with a strong base, the ionic compound that forms is basic in water solution.

$HF + NaOH \rightarrow NaF + H_2O$ NaF is basic

$HCN + KOH \rightarrow KCN + H_2O$ KCN is basic

ACID–BASE TITRATIONS

An **acid–base titration** is the addition of a quantity of base of known concentration, using a burette, to a known volume of acid of unknown concentration, until the endpoint is reached. An indicator compound, such as phenolphthalein, is added so that a color change occurs at the endpoint. At that point, an equivalent amount of base has been added, and the volume that contains that amount is determined from the initial and final volumes on the burette. The concentration of the acid solution can then be calculated.

EXAMPLE: A 25.0-mL sample of HCl is titrated with 0.0750 M NaOH. The titration requires 32.7 mL of base to reach the endpoint. Calculate the molarity of the HCl.

1. Write the equation for the reaction.

$$HCl + NaOH \rightarrow NaCl + H_2O$$

2. The stoichiometry of HCl to NaOH is 1:1.
3. Calculate the moles of base added in the titration.

$$(0.0750 \text{ M})\ (0.0327 \text{ L}) = 0.00245 \text{ mole NaOH added}$$

4. Calculate the moles of acid in the solution.

$$(0.00245 \text{ mole base})\ (1 \text{ mole acid}/1 \text{ mole base}) = 0.00245 \text{ mole acid}$$

5. Calculate the molarity of the acid.

$$M = \text{mole/L} = 0.00245 \text{ mole}/0.0250 \text{ L} = 0.098 \text{ M acid}$$

Oxidation–Reduction Reactions

Oxidation–reduction reactions, the third type of reaction in solution, involve the transfer of one or more electrons from one species in the reaction to another.

Any species that loses electrons is **oxidized**. Any species that gains electrons is **reduced**. Oxidation and reduction can be tracked using the oxidation number of each atom. Each atom in a formula has an oxidation number. It is similar to a charge; for simple ions, the charge and the oxidation number are the same. For atoms of an element, the oxidation number is always zero.

RULES FOR OXIDATION NUMBERS

- The oxidation number of an atom in a pure element is zero.
- The oxidation number of a simple ion is the same as its charge.
- The oxidation number of oxygen is usually −2, except in peroxides, where it is −1.
- The oxidation number of hydrogen in a covalent compound is +1.
- The oxidation number of hydrogen in an ionic compound is −1.
- The oxidation number of fluorine is always −1.
- The sum of all the oxidation numbers in a neutral compound is always zero.

When the oxidation number of an atom decreases from one side of the equation to the other side, that atom is reduced. If an oxidation number increases, that atom is oxidized.

EXAMPLE: State which atom is oxidized and which is reduced in the following equation:

$Fe + Cu^{+2} \rightarrow Fe^{+2} + Cu$

1. Because Fe is an element, its oxidation number is 0.
2. Because Cu^{+2} is a simple ion, its oxidation number is +2.
3. Fe^{+2} is a simple ion; its oxidation number is +2.
4. Cu is an element; its oxidation number is 0.
5. Therefore Fe is going from 0 to +2; Fe is being oxidized.
6. Cu is going from +2 to 0; Cu is reduced.

Try another problem. Which atom is oxidized and which is reduced?

$2Sb + 3Cl_2 \rightarrow 2\ SbCl_3$

1. Sb is an element; oxidation number is 0.
2. Cl_2 is an element; oxidation number of each Cl is 0.
3. Sb in $SbCl_3$ is the ion Sb^{+3}; oxidation number +3.
4. Cl in $SbCl_3$ is the ion Cl^{-1}; oxidation number −1.
5. Therefore,

$Sb^0 \rightarrow Sb^{+3}$ oxidation

$Cl^0 \rightarrow Cl^{-1}$ reduction

COMMON TYPES OF OXIDATION–REDUCTION REACTIONS

A few of the more common classifications of oxidation–reduction reactions are combination reactions, decomposition reactions, single displacement reactions, and combustion reactions.

A **combination reaction** occurs when two substances combine to form a third substance.

$CaO\ (s) + SO_2\ (g) \rightarrow CaSO_3\ (s)$

A **decomposition reaction** occurs when a single substance decomposes to two or more substances.

$2KClO_3 \rightarrow 2KCl + 3O_2$

A single **displacement reaction** occurs when an element reacts with an ionic compound and replaces the cation in the compound. The original cation becomes an element.

$Cu + 2AgNO_3 \rightarrow 2Ag + Cu(NO_3)_2$

A **combustion reaction** occurs when a substance reacts with oxygen to produce one or more oxides.

$$4Fe + 3O_2 \rightarrow 2Fe_2O_3$$

OXIDATION–REDUCTION TITRATIONS

Oxidation–reduction titration often does not require an indicator compound to be added because there is often an automatic color change on reaction.

EXAMPLE: A 25.0 mL sample of MnO_4^- is titrated with 0.0485 M I^- solution. To reach the endpoint, 18.7 mL of the I^- solution is required. Calculate the molarity of the MnO_4^- solution. The balanced equation is

$$10I^- + 16H^+ + 2MnO_4^- \rightarrow 5I_2 + 2Mn^{+2} + 8H_2O$$

1. Calculate the moles of I^- added.

$$(0.0485 \text{ mole/L})(0.0187 \text{ L}) = 9.07 \times 10^{-4} \text{ mole } I^- \text{ added}$$

2. Calculate the moles of MnO_4^-.

$$(9.07 \times 10^{-4} \text{ mole } I^-)(2 \text{ mole } MnO_4^-/10 \text{ mole } I^-)$$
$$= 1.81 \times 10^{-4} \text{ mole } MnO_4^-$$

3. Calculate the molarity of MnO_4^-.

$$M = 1.81 \times 10^{-4} \text{ mole}/0.0250 \text{ L} = 0.00726 \text{ M } MnO_4^-$$

CRAM SESSION

Reactions in Solution

1. **Molarity** (M) is the most common concentration unit used for solutions. Molarity equals the moles of solute per liter of solution.
2. A **precipitation reaction** is a reaction that can take place when two solutions are mixed together. If the cation from one solution can react with the anion from the other solution to form an insoluble compound, a solid precipitate forms. The net ionic equation for the reaction shows the cation and anion that react to form the insoluble compound that is the product.
3. An **acid–base reaction** takes place between an acid and a base, and forms an ionic compound and water as the products.
4. There are two kinds of acids, strong and weak. A **strong acid** dissociates fully into two ions in water solution, forming H^+ and the conjugate base A^-. There are six strong acids: HCl, HBr, HI, HNO_3, H_2SO_4, and $HClO_4$. **Weak acids** do not dissociate fully, and produce only a small amount of H^+ in water solution.
5. There are two kinds of bases: strong and weak. A **strong base** dissociates fully into hydroxide ion and a cation. The most common strong bases are LiOH, NaOH, KOH, and $Ba(OH)_2$. A **weak base** produces a small amount of hydroxide ion by reacting with water to a small extent. The most common weak base is ammonia NH_3.
6. The ionic compound formed from the reaction of an acid with a base can be neutral, acidic, or basic, depending on the acid and base used to form it.
   - **Strong acid and strong base** react to form a neutral salt.
   - **Strong acid and weak base** react to form an acidic salt.
   - **Weak acid and strong base** react to form a basic salt.
7. An **acid–base titration** can be done with any combination of acid and base, but it often is the reaction of a weak acid with a strong base. The base solution, of known concentration, is added to a known volume of acid solution until the endpoint is reached. This is seen as a color change due to the addition of an indicator compound prior to titration. The volume of base that was added is determined, and the moles of base added are calculated. The corresponding moles of acid are then determined. The concentration of the acid is then calculated.
8. An **oxidation–reduction reaction** involves the transfer of electrons. Whether an atom is oxidized or reduced in a reaction is tracked by change in oxidation number.
   - If an atom's oxidation number increases, the atom is oxidized.
   - If an atom's oxidation number decreases, that atom is reduced.
9. There are numerous classifications of oxidation–reduction reactions, including decomposition, single displacement, combustion, and combination reactions.

CHAPTER 9

Thermochemistry

Read This Chapter to Learn About:

- ➤ Heat
- ➤ Enthalpy
- ➤ Heat/Mass Relationships
- ➤ Heat Capacity
- ➤ Specific Heat
- ➤ Heats of Vaporization and Fusion
- ➤ Constant Pressure Calorimetry
- ➤ Hess' Law

Thermochemistry is the study of thermal changes that occur in a chemical reaction. Thermal energy is convertible with kinetic energy and potential energy and must obey the law of conservation of energy.

HEAT

Heat is an energy that flows into or out of a system due to a difference between the temperature of the system and its surroundings when they are in thermal contact. Heat flows from hotter to cooler areas and has the symbol q. If heat is added to a system, q is positive. If heat is removed from a system, q is negative.

The units of heat are joules (J) or calories. One calorie equals 4.184 J. The definition of **calorie** is the heat needed to raise the temperature of 1 gram (g) of water by 1 degree Celsius ($^{\circ}$C).

A reaction from which heat is given off and has a negative q is called an **exothermic reaction.** A reaction that requires heat to be added and has a positive q is called an **endothermic reaction**.

ENTHALPY

Enthalpy is a state function and it is used to denote heat changes in a chemical reaction. It is given the symbol H. Because it depends only on the initial and final states of the system, the difference between these states is ΔH. At standard pressure (1 atm), $q = \Delta H$.

ΔH_f° is called the **enthalpy of formation** of a substance. It denotes the heat that is absorbed or given off when a substance is produced from its elements at standard temperature (25°C) and pressure (1 atm).

ΔH_{rxn}° is called the **enthalpy of reaction**. It denotes the amount of heat that is given off ($-\Delta H$) or absorbed ($+\Delta H$) by a reaction at standard temperature and pressure.

The ΔH_{rxn}° can be calculated from ΔH_f° values. For the reaction

$$aA + bB \rightarrow cC + dD$$

the ΔH_{rxn}° is

$$\Delta H_{rxn}^\circ = [\sum c\,(\Delta H_f(C)) + d\,(\Delta H_f\,(D))] - [\sum a\,(\Delta H_f\,(A)) + b\,(\Delta H_f\,(B))]$$

If the equation is doubled, the ΔH_{rxn}° doubles as well.

The ΔH_f° of a compound can be calculated using bond dissociation energies, found in the following table. The formula is

$$\Delta H_f^\circ = \sum \text{BE of bonds broken} - \sum \text{BE of the bonds formed}$$

where BE is the bond energy per mole of bonds.

Dissociation Energies of Some Single Bonds in kJ/mole

	H	C	N	O	S	F	Cl	Br	I
H	432								
C	411	346							
N	386	305	167						
O	459	358	201	142					
S	363	272			226				
F	565	485	283	190	284	155			
Cl	428	327	313	218	255	249	240		
Br	362	285		201	217	249	216	190	
I	295	213		201		278	208	175	149

Data from Huheey, J. E.; Keiter, E. A.; and Keiter, R. L., *Inorganic Chemistry*, 4th ed. New York: Harper Collins, 1993; pp. A21–A34.

EXAMPLE: Determine the ΔH_f° of HF using bond dissociation energies.

1. First write the equation for the formation of HF from its elements:

$$H_2\ (g) + F_2\ (g) \rightarrow 2HF\ (g)$$

2. For H_2, the BE = 432 kJ/mole.
3. For F_2, the BE = 155 kJ/mole.
4. For HF, the BE = 565 kJ/mole.
5. Fill the values into the equation and solve for ΔH_f°:

$$\Delta H_f^\circ = [(1\text{ mole }H_2)(432\text{ kJ/mole}) + (1\text{ mole }F_2)(155\text{ kJ/mole})] - [(2\text{ mole HF})\ (565\text{ kJ/mole})] = -544\text{ kJ}$$

HEAT/MASS RELATIONSHIPS

The ΔH_{rxn}° can be used to determine the heat formed from any amount of a starting material. If you inspect the equation

$$2Na + 2H_2O \rightarrow 2NaOH + H_2 \qquad \Delta H_{rxn}^\circ = -368\text{ kJ}$$

which says that when these amounts of starting materials react (2 moles Na; 2 moles H_2O), 368 kJ of heat are given off. The equation also states that when the amounts of product shown here are formed (2 moles NaOH; 1 mole H_2), 368 kJ of heat are given off.

Knowing this, you can calculate the heat given off when any amount of starting material reacts or when any amount of product is formed.

EXAMPLE: Calculate the heat produced when 147 g hydrogen gas reacts according to the equation

$$N_2\ (g) + 3H_2\ (g) \rightarrow 2NH_3\ (g) \qquad \Delta H = -92\text{ kJ}$$

1. The equation states that 3 moles H_2 produce 92 kJ heat.
2. Calculate the number of moles that 147 g H_2 represents.

$$(147\text{ g }H_2)\ (1\text{ mole }H_2/2.016\text{ g}) = 72.9\text{ mole }H_2$$

3. Use 3 mole H_2 = 92 kJ heat as the conversion factor

$$(72.9\text{ mole }H_2)\ (92\text{ kJ}/3\text{ mole }H_2) = 2.2 \times 10^3\text{ kJ heat}$$

HEAT CAPACITY

Heat capacity is the heat needed to raise the temperature of a sample of substance by 1°C. It has the symbol C and the units J/°C.

Thus, $q = C\Delta T$, where ΔT is the difference in the initial and final temperatures.

SPECIFIC HEAT

Specific heat is the heat needed to raise the temperature of 1 g of a substance by 1°C. It has the symbol c and the units J/g °C. Thus, $q = cm\Delta T$, where ΔT is the difference in the initial and final temperatures. The following table lists the specific heats of some common substances.

The Specific Heat of Some Common Substances

Substance	Specific Heat in J/g °C
AL	0.900
Au	0.129
C (diamond)	0.502
Cu	0.385
Fe	0.444
Hg	0.139
H_2O	4.184

EXAMPLE: Calculate the heat needed to raise the temperature of 85.3 g of iron from 35.0°C to 275.0°C.

1. The ΔT is $275.0 - 35.0 = 240.0°C$
2. The c for iron is 0.444 J/g °C, given from the table.
3. Plug the values into the equation and solve for q:

$$q = cm\Delta T = (0.444 \text{ J/g °C})(85.3 \text{ g})(240.0°\text{C}) = 9.09 \times 10^3 \text{ J}$$

HEATS OF VAPORIZATION AND FUSION

The equation for heat given previously cannot be used for a phase change, because there is no temperature difference during a phase change. The temperature remains the same during the freezing/melting process or during the boiling process. So, for an equation that denotes a phase change, such as

$$H_2O\ (s) \rightarrow H_2O\ (l) \qquad \Delta H_{fus} = 6.01 \text{ kJ/mole}$$

The heat change is the heat needed to convert 1 mole of solid water to 1 mole of liquid water at 0°C, and is called the **heat of fusion**, ΔH_{fus}.

For the boiling process,

$$H_2O\ (l) \rightarrow H_2O\ (g) \qquad \Delta H_{vap} = 44.6 \text{ kJ/mole}$$

The heat change is the heat needed to convert 1 mole of liquid water to 1 mole of water vapor at 100.0°C, and is called the **heat of vaporization**, ΔH_{vap}.

CONSTANT PRESSURE CALORIMETRY

This type of calorimetry uses an insulated container at room pressure to measure the heat changes that occur during a physical or chemical process. Because the calorimeter is insulated, no heat can leave or enter the system, and all of the heat changes that occur within the calorimeter net to zero. In other words, the heat that is given off equals the heat that is gained.

Heat released = heat absorbed and ΔT = higher temperature − lower temperature

The calorimeter itself is ignored in the following calculations.

EXAMPLE: Calculate the final temperature of the system when an 18.4-g sample of aluminum at 215°C is added to a calorimeter that contains 120.0 g water at 24.5°C.

1. The c for water is 4.184 J/g °C.
2. The c for aluminum is 0.900 J/g °C, given from the table.
3. The ΔT for the water is $x - 24.5$°C, where x is the final temperature. For water, x is the higher temperature.
4. The ΔT for the aluminum is 215°C − x, where x is the final temperature. For the aluminum, 215°C is the higher temperature.
5. Plugging into the equation $mc\Delta T_{water} = mc\Delta T_{Al}$ and solving for x

$$(120.0\text{ g})(4.184\text{ J/g °C})(x - 24.5\text{°C}) = (18.4\text{ g})(0.900\text{ J/g °C}) \times (215\text{°C} - x)$$

$$x = 30.6\text{°C}$$

HESS' LAW

Some reactions are the sum of individual reaction steps. To sum reactions, any compound that shows up on the left side of one equation and on the right side of another equation cancels out. **Hess' law** states that the sum of the ΔHs for each individual equation sum to the ΔH_{rxn} for the net equation.

	$2C + \cancel{2}O_2$	$\rightarrow$	$\cancel{2CO_2}$	ΔH_1
	$\cancel{2CO_2}$	$\rightarrow$	$2CO + \cancel{O_2}$	ΔH_2
sum	$2C + O_2$	$\rightarrow$	$2CO$	$\Delta H_{sum} = \Delta H_1 + \Delta H_2$

EXAMPLE: Determine the ΔH of the following reaction:

$2S + 3O_2 \rightarrow 2SO_3$

given these reactions and their ΔHs

$SO_2 \rightarrow S + O_2$ $\quad \Delta H_1 = 297$ kJ

$2SO_3 \rightarrow 2SO_2 + O_2$ $\quad \Delta H_2 = 198$ kJ

1. First, reverse the first equation and reverse the sign of ΔH_1.
2. Next, double the first equation and double the ΔH_1.
3. Next, reverse the second equation and reverse the sign of ΔH_2.

$$\begin{array}{lllll} & 2S + 2O_2 & \rightarrow & \cancel{2SO_2} & \Delta H_1 = -594\text{ kJ} \\ & \cancel{2SO_2} + O_2 & \rightarrow & 2SO_3 & \Delta H_2 = -198\text{ kJ} \\ \hline \text{sum} & 2S + 3O_2 & \rightarrow & 2SO_3 & \Delta H_{sum} = -792\text{ kJ} \end{array}$$

CRAM SESSION

Thermochemistry

1. Heat has the symbol q, which can be positive or negative, depending on the direction of the heat flow. A positive q means that heat is added to the system; a negative q means that heat is given off by the system.
2. Heat is related to the state function called **enthalpy**, which is the difference in heat between the reactant and the product of a chemical reaction. The **standard change in enthalpy of formation of a substance**, ΔH_f°, can be calculated from bond dissociation energies.
3. The **standard change in enthalpy of a reaction**, ΔH_{rxn}°, can be used to convert a specific amount of a reactant or product to heat either given off or consumed in the reaction.
4. Heat capacity and specific heat are properties of substances.
   A. **Heat capacity** is the heat required to raise the temperature of a substance by 1°C. Heat capacity has the symbol C.
   B. **Specific heat** is the heat required to raise the temperature of 1 g of a substance by 1°C. Specific heat has the symbol c. Specific heat is directly related to heat by the equation $q = mc\Delta T$, where m is the mass of substance and ΔT is the difference in temperature.
5. **Heat of vaporization** and **heat of fusion** give the amount of heat that is required to either vaporize or melt a sample completely, respectively. During the phase change, there is no temperature difference.
6. **Calorimetry** is the study of the heat changes that occur within an insulated system. If no heat is gained from or lost to the surroundings, the net heat change within is zero. The total heat gained within the system equals the total heat lost within the system.
7. **Hess' law** can be used to calculate the ΔH_{rxn} of an equation that is the sum of a series of other equations, by summing the individual ΔH_{rxn} of each equation in the series.

CHAPTER 10

Kinetics

Read This Chapter to Learn About:

- Reaction Rates
- Reaction Law
- Reaction Orders
- The Rate Constant *k* and Its Dependence on Temperature
- Half-life of a First-Order Reaction
- Catalysts

Kinetics is the study of how fast reactions occur; this is called the **reaction rate**. Reaction rates are dependent on the temperature at which the reaction is taking place, on the concentrations of the reactants, and on whether a catalyst is present.

REACTION RATES

Reaction rates are measured experimentally. If the reactant is colored, its absorbance can be followed. At the beginning of the reaction, the absorbance is at a maximum, and it decreases as the reactant disappears (turns into product; see Figure 10-1).

The rate equals the change in concentration with time. The rate is fastest at the beginning of the reaction. An **instantaneous rate** is the change in the concentration of the reactant divided by the change in time, or

$$\Delta[\text{reactant}]/\Delta t$$

when the change in time, or Δt, is a very small increment of time. This is seen as the tangent to the curve, at any point. The tangent has the steepest slope at the beginning of the reaction (fastest rate), and the flattest slope at the end of the reaction (slowest rate). So, the reaction slows with time.

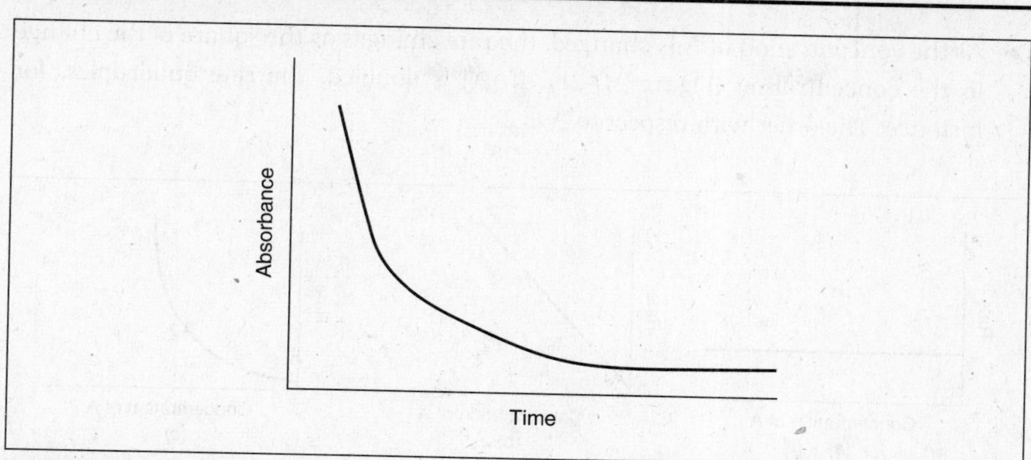

FIGURE 10-1

Rate Law

Kinetics studies how the reaction rate changes when the concentrations of the reactants change. This rate data is used to determine the rate law for the reaction. The **rate law** expresses how the rate is affected by the concentrations and by the rate constant.

For a reaction

$$A + 2B \rightarrow \text{products}$$

the rate law is written

$$\text{Rate} = k[A]^x [B]^y$$

where [A] is the molarity of A, [B] is the molarity of B, x is the order with respect to A, and y is the order with respect to B.

Ways Rates Relate to Changes in the Concentration of the Reactant

Rates usually relate to the concentration of a reactant in one of three ways: by not changing, by changing proportionately to the change in concentration, or by changing as the square of the change of the concentration.

- As the concentration of A is changed, the rate does not change (Figure 10-2*a*). The rate is unaffected; the order with respect to A is 0.
- As the concentration of A is changed, the rate changes proportionally (Figure 10-2*b*). If [A] is doubled, the rate doubles, and so on. The order with respect to A is 1.

- As the concentration of A is changed, the rate changes as the square of the change in the concentration (Figure 10-2*c*). If [A] is doubled, the rate quadruples, for instance. The order with respect to A is 2.

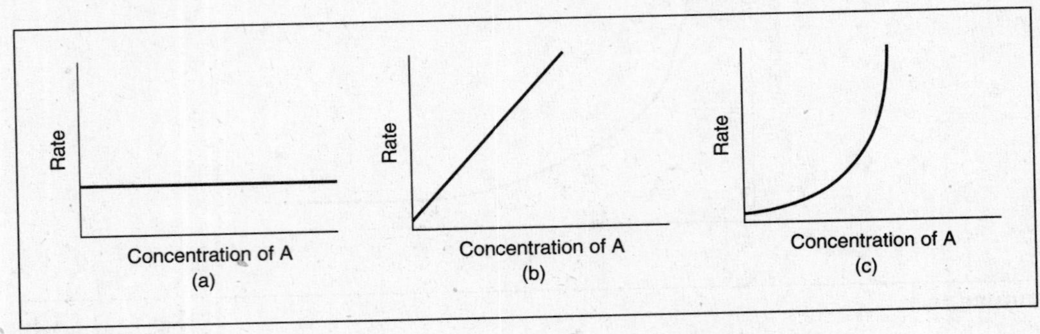

FIGURE 10-2

REACTION ORDERS

The orders of each reactant are thus determined by observing how the rate is affected by the change in the concentration of each reactant. Only one order can be determined at a time. All other concentrations must remain constant; only the concentration of the reactant being studied is changed. The rates are shown under various conditions; these are used to determine the rate law and the rate constant, *k*, is then calculated.

EXAMPLE: For the reaction

$A + B + C \rightarrow$ products

determine the rate law and calculate the rate constant using the following data:

Run	[A]	[B]	[C]	Rate (M/s)
1	0.003	0.001	1.0	1.8×10^4
2	0.003	0.002	1.0	3.6×10^4
3	0.006	0.002	1.0	7.2×10^4
4	0.003	0.001	2.0	7.2×10^4

- From run 1 to 2, the [B] is doubled. The rate doubles, so the order of B is 1. (The rate is directly proportional to the concentration of B.)
- From run 2 to 3, the [A] is doubled. The rate doubles, so the order of A is 1. (The rate is directly proportional to the concentration of A.)
- From run 1 to 4, the [C] is doubled. The rate quadruples, so the order of C is 2. (The rate is proportional to the square of the concentration of A.)

➤ So the rate law is

$$\text{Rate} = k[\text{A}]\,[\text{B}]\,[\text{C}]^2$$

➤ Using the data from run 1, the rate constant k is calculated

$$1.8 \times 10^{-4} = k[0.003]\,[0.001]\,[1]^2$$

$$k = 60\ 1/\text{M}^3\text{s}$$

USE OF THE RATE CONSTANT *k*

The rate constant can be used to determine the concentration of the reactant at any point. For a first-order reaction (overall) the equation is:

$$\ln [\text{A}]_0 = kt\ [\text{A}]_t$$

where ln means natural log, $[\text{A}]_0$ is the concentration of A at the beginning of the reaction ($t = 0$), and $[\text{A}]_t$ is the concentration of A at time t; k is the rate constant, and t is the time.

EXAMPLE: For a first-order reaction

A → product

the rate constant is $3.02 \times 10^{-3}\ \text{s}^{-1}$. The concentration of A initially is 0.025 M.

1. Calculate the [A] after 172 s.
2. Calculate the time for the [A] to reach 1.13×10^{-3} M.
3. Calculate the time needed for 98% of the A to react.
4. Plug in $[\text{A}]_0 = 0.0250$ M, $t = 172$ s, $k = 3.02 \times 10^{-3}\ \text{s}^{-1}$

$$\ln 0.0250 - \ln [\text{A}]_t = (3.02 \times 10^{-3}\ \text{s}^{-1})\,(172\ \text{s})$$

$$-3.69 - \ln [\text{A}]_t = 0.519$$

$$-3.69 - 0.52 = \ln [\text{A}]_t$$

$$-4.21 = \ln [\text{A}]_t$$

$$0.015\ \text{M} = [\text{A}]_t$$

5. $$\ln \frac{0.0250}{0.00113} = (3.02 \times 10^{-3}\ \text{s}^{-1})\,t$$

$$t = 1.03 \times 10^3\ \text{s}$$

6. When 98% of A has reacted, 2% is left, so

$$\ln \frac{100\%}{2\%} = (3.02 \times 10^{-3}\ \text{s}^{-1})\,t$$

$$t = 1.30 \times 10^3\ \text{s}$$

HALF-LIFE OF A FIRST-ORDER REACTION

The half-life of a reaction is the time needed for 50% of the reactant to react. It is given the symbol $t_{1/2}$ and has time units.

$$\ln \frac{100}{50} = kt_{1/2} \text{ is the same as } \ln 2 = kt_{1/2}$$

So

$$\ln 2 = 0.693$$

and the half-life equation is

$$0.693 = kt_{1/2}$$

for a first-order reaction.

DEPENDENCE OF *k* ON TEMPERATURE

If the **energy of activation**, E_a, for a reaction is known (this is the minimum energy needed for reaction to occur), and the rate constant for one temperature is known, the rate constant at any other temperature can be calculated using the **Arrhenius equation**

$$\ln \frac{k_2}{k_1} = \frac{E_a}{R}(1/T_1 - 1/T_2)$$

where k_1 is the rate constant at temperature T_1, k_2 is the rate constant at temperature T_2. T_1 and T_2 are in Kelvin, $R = 8.314$ joules (J)/mole K, and E_a is the activation energy.

EXAMPLE: For a certain first-order reaction, the rate constant at 190°C is 2.61×10^{-5} and the rate constant at 250°C is 3.02×10^{-3}. Calculate the energy of activation of this reaction.

$$k_1 = 2.61 \times 10^{-3} \quad \text{at } T_1 = 463.2 \text{ K}$$

$$k_2 = 3.02 \times 10^{-3} \quad \text{at } T_2 = 523.2 \text{ K}$$

So

$$\ln \frac{3.02 \times 10^{-3}}{2.61 \times 10^{-3}} = \frac{E_a}{8.314}\left(\frac{1}{463.2} - \frac{1}{523.2}\right)$$

$$\ln 1.16 = \frac{E_a(2.48 \times 10^{-4})}{8.314}$$

$$\frac{0.148(8.314)}{2.48 \times 10^{-4}} = E_a$$

$$5.0 \times 10^3 \text{ J/mole} = E_a$$

CATALYSTS

Catalysts speed a reaction rate by lowering the **activation energy** of the reaction. The catalyst may be involved in the formation of the activated complex, the high-energy compound that is partly starting material and partly product, but it is regenerated at the end of the reaction.

Catalysts can be in the same phase as the reactants (homogeneous) or in a different phase (heterogeneous). They can be inorganic or organic. Biological catalysts are called **enzymes**, and they catalyze every biochemical reaction in the body. Enzymes are very specific for their starting material, which is called its **substrate**. They often interact with the substrate like a key fitting into a lock. A three-dimensional site on the substrate acts as the lock, and the enzyme fits into the site, called the **active site**, like a key. Enzymes work optimally within a very narrow range of temperature and pH.

CRAM SESSION

Kinetics

1. **Kinetics** is the study of reaction rates. Reaction rates depend on the temperature, the concentrations of the reactants, the orders of the reactants, the rate constant of the reaction, and whether a catalyst is present.
2. The **reaction rate** is fastest at the start of a reaction. Then it slows with time in a logarithmic fashion.
3. The **rate law of the reaction** expresses how the rate is affected by the concentrations of the starting materials, their orders, and the rate constant. Changing the concentration of one reactant at a time and measuring the rate determine the rate law. The order of each reactant is then determined by seeing how the rate changes on changing the concentration.
   A. If the rate change and the concentration change are directly proportional, the order of that substance is 1.
   B. If the rate changes as the square of the concentration changes, the order of the substance is 2.
   C. If the rate does not change with concentration change, the order of the substance is 0.
   D. Once the rate law is determined, the rate constant is calculated.
4. If the rate constant for a first-order reaction is known, the equation $\ln \{[A]_0/[A]_t\} = kt$ can be used to determine the concentration at any time t (if $[A]_0$ is given), or to calculate the time needed to reach a target concentration.
5. The time required for 50% of the reactant to react is called the **half-life**, and it has the symbol $t_{1/2}$.
6. The equation

$$\ln (k_2/k_1) = (E_a/R)\,(1/T_1 - 1/T_2)$$

   can be used to calculate the rate constant at any temperature, if the rate constant at another temperature is known and the E_a is known.
7. **Catalysts** affect reaction rate by lowering the activation energy barrier. Catalysts are regenerated at the end of the reaction.

CHAPTER 11

Equilibrium

Read This Chapter to Learn About:

- Equilibrium Point
- Equilibrium Expression
- Equilibrium Constant K
- Reaction Quotient
- Le Chatelier's Principle

Equilibrium occurs with reactions that are reversible when the forward reaction rate equals the reverse reaction rate. Using various formulas, the rate at which a reaction occurs before reaching equilibrium and the concentrations of the products can be determined.

EQUILIBRIUM POINT

For a reversible reaction, the starting material (on the left) is converted into the product (on the right) at a rate equal to $k[A]^x$

$$A \rightleftarrows B$$

At a certain point in the progress of the reaction, some of the B that has formed begins to convert to A at a rate equal to $k_{-1}[B]^y$. There comes a point when the forward rate and the reverse rate are equal, and the concentrations of A and B no longer change, although each reaction is still proceeding. This is called the **equilibrium point**. The net numbers of A and B present in the reaction flask remain the same. It appears that the reaction has stopped.

EXAMPLE: If one begins with 100 molecules of A, A begins immediately to turn into B. When enough B's are present, they start to turn back into A. At the equilibrium point, there are, for example, 60 A's and 40 B's. The reactions, both forward and reverse, are still proceeding, but the reaction rates are equal, so there remain at all times 60 A's and 40 B's.

EQUILIBRIUM CONSTANT *K*

The **equilibrium constant *K*** describes the extent to which the forward reaction proceeds before reaching the equilibrium point. Is there a lot of A left over, or just a little? K is a constant value at a constant temperature; it does change with temperature.

If K is large (>1), there is mostly product and very little starting material left at the equilibrium point. This is a **product-favored reaction**.

If K is small ($<10^{-4}$), there is mostly starting material left over, and very little product formed at the equilibrium point. This is a **reactant-favored reaction**.

With a medium K (between 1 and 10^{-4}), a lot of product is formed, but there are still substantial quantities of starting material left.

Factors Affecting Equilibrium Constant—Equilibrium Expression

The equilibrium constant K depends on temperature and is related to the amounts of starting material and product in the following manner:

$$aA + bB \rightleftarrows cC \qquad K_c = \frac{[C]^c}{[A]^a[B]^b}$$

where a, b, and c are the molar coefficients and [ ] = mole per liter (mol/L)

EXAMPLE:

$$N_2 + 3\,H_2 \rightleftarrows 2NH_3$$

$$K_p = \frac{(P_{NH_3})^2}{(P_{N_2})(P_{H_2})^3} \text{ in terms of pressure}$$

Only gases (with pressures) and species in solution (with concentrations) affect the equilibrium. Solids and liquids do not appear in the equilibrium expression.

EXAMPLE:

$$C\,(s) + 2Cl_2\,(g) \rightleftarrows CCl_4\,(g)$$

$$K = \frac{(P_{CCl_4})}{(P_{Cl_2})^2}$$

Problems Involving a Small *K*

Given the K and the initial conditions (concentrations or pressures), you can calculate the equilibrium conditions of the products.

If the K is $\leq 10^{-4}$, the reaction does not proceed far before equilibrium. Very little product is formed, and the amount of starting material has changed very little. You can neglect the change in the starting material because it is such a small amount.

EXAMPLE: If you start with 0.240 mole SO_3 in a 3.00-L container, calculate the equilibrium concentrations of SO_2 and O_2 in the reaction O_2.

$2SO_3 \rightleftarrows 2SO_2 + O_2 \quad K = 9.6 \times 10^{-8}$

➤ The balanced chemical equation is used to set up an ICE (Initial, Change, Equilibrium) table

$$2SO_3 \leftrightarrows 2SO_2 + O_2$$

Initial	0.080 M	0	0	Initial amounts in M or pressure.
Δ	$-2x$	$+2x$	$+x$	For every 2 moles of SO_3 that react, 2 moles of SO_2 and 1 mole of O_2 are formed.
Equilibrium	0.080 M	$2x$	x	Add the values in the column to get the equilibrium value: you can neglect the change to the SO_2; these values go into the K expression.

➤ Write the K expression:

$$K = \frac{[SO_2]^2\,[O_2]}{[SO_3]^2}$$

➤ Plug in the equilibrium values and solve for x:

$$9.6 \times 10^{-8} = \frac{[2x]^2\,[x]}{[0.080]^2}$$

➤ $x = 5.4 \times 10^{-4}$

➤ So the

$[SO_2] = 2x = 1.1 \times 10^{-3}$ M

$[O_2] = x = 5.4 \times 10^{-4}$ M

Problems Involving a Large or Medium *K*

You can calculate the K if given the initial conditions (concentrations or pressures) and either the percentage of starting material that reacts or one of the actual equilibrium values.

EXAMPLE: You have 0.40 atm of CH_4 and 0.54 atm of H_2S initially. Calculate the K of the following reaction if 25% of the CH_4 remains at the equilibrium point.

$$CH_4\ (g) + 2H_2S\ (g) \rightleftarrows CS_2\ (g) + 4H_2\ (g)$$

Initial	0.40	0.65	0	0
Δ	$-x$	$-2x$	$+x$	$+4x$
At equilibrium	0.10	0.05	0.30	1.20

- The numbers on the bottom line were determined using the information given in the problem. If you start with 0.40 atm of CH_4 and 25% remains at the equilibrium point, then (0.40 atm) (0.25) = 0.10 atm is what remains of the CH_4 at equilibrium.
- Then using the equation $0.40 - x = 0.10$ you get $x = 0.30$ atm.
- Plugging in this value for x on the change line and adding the columns, the other equilibrium values are obtained.
- Now you write the K expression

$$K = \frac{(P_{CS_2})\,(P_{H_2})^4}{(P_{CH_4})\,(P_{H_2S})^2}$$

- Plug in the bottom line values to calculate K

$$K = \frac{(0.30)\,(1.20)^4}{(0.10)\,(0.05)^2} = 2.5 \times 10^3$$

REACTION QUOTIENT

At any point other than the equilibrium point, the concentrations or pressures are different than the equilibrium values. If you are given a set of values that represent the composition at some moment in the progression of the reaction, you can plug these values into the K expression. The result is not K, of course. It is called Q, the **reaction quotient**.

Q is a number larger or smaller than K. Because the reaction is always striving for equilibrium, comparison of Q to K allows you to determine the direction that the reaction is proceeding at the moment.

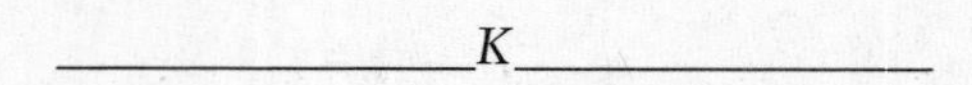

Shown here is K on a number line. If Q is smaller than K, it is left of K. To get to K, the reaction must go to the right ($\longrightarrow$).

If Q is greater than K, it is to the right of K. To get to K, the reaction must go to the left ($\longleftarrow$).

LE CHATELIER'S PRINCIPLE

Le Chatelier's principle states that if you subject a system that is at equilibrium, to some change in conditions, the equilibrium shifts so as to counteract the change. For example, note the following reaction:

$$2FeCl_3\ (s) + 3H_2O\ (g) \rightleftarrows Fe_2O_3\ (s) + 6HCl\ (g) \quad \text{endothermic}$$

Change	Counteraction	Direction of Equilibrium Shift
Add some water vapor	Remove the water vapor	$\longrightarrow$
Remove some HCl	Make more HCl	$\longrightarrow$
Remove some Fe_2O_3	No counteraction	No change
Add some HCl	Remove some HCl	$\longleftarrow$
Increase the volume (decrease the pressure)	Increase the pressure (make more gas molecules)	$\longrightarrow$
Decrease the volume (increase the pressure)	Decrease the pressure (remove gas molecules)	$\longleftarrow$
Heat the reaction	Remove the heat	$\longrightarrow$
Cool the reaction	Increase the heat content	$\longleftarrow$

CRAM SESSION

Equilibrium

1. **Equilibrium** occurs with reactions that are reversible. When the forward reaction rate equals the reverse reaction rate, the reaction is at equilibrium. At this point, the net number of moles of starting material and product does not change.
2. The **equilibrium constant *K*** describes the extent to which a reaction occurs before reaching the equilibrium point.
   A. If *K* is very small, the amount of starting material does not change significantly by the time the reaction reaches equilibrium.
   B. If *K* is larger than approximately 10^{-4}, the initial amount of starting material changes significantly by the time the reaction reaches equilibrium.
   C. For a reaction with a very large *K,* there is very little starting material at the equilibrium point.
3. The *K* expression is used to calculate the concentration (or pressure) of products at the equilibrium point. If at least one of the product concentrations is given, *K* itself can be calculated. Only gases and species in solution appear in the *K* expression.
4. If nonequilibrium concentrations (or pressures) are provided, a **reaction quotient** can be calculated. By comparing the reaction quotient to the equilibrium constant, the direction at which the reaction is proceeding at that point can be determined.
   A. If the reaction quotient is larger than *K,* the reaction goes to the left to reestablish equilibrium.
   B. If the reaction quotient is smaller than *K,* the reaction goes to the right to reestablish equilibrium.
5. **Le Chatelier's principle** states that a reaction at equilibrium shifts in order to undo a change that is made to the system at equilibrium. The reaction always shifts to counteract the change, unless the change is made to a solid or liquid component of the reaction.

CHAPTER 12

Solubility Equilibria

Read This Chapter to Learn About:

- ➤ Equilibria of Solubility of "Insoluble" Ionic Compounds
- ➤ Effect of Adding Ions to a Solution at Equilibrium—Common Ion Effect
- ➤ Precipitation Calculations
- ➤ The Effect of pH on the Solubility of Basic Compounds

EQUILIBRIA OF SOLUBILITY OF "INSOLUBLE" IONIC COMPOUNDS

Insoluble ionic compounds are actually soluble to a very small extent. These solubilities are equilibrium-based processes, with a very small equilibrium constant K.

The solubility equation is written with the solid on the left and the ions in solution on the right, as shown in the following examples:

$$CaC_2O_4\ (s) \leftrightarrows Ca^{+2}\ (aq) + C_2O_4{}^{-2}\ (aq)$$
$$PbI_2\ (s) \leftrightarrows Pb^{+2}\ (aq) + 2I^{-}\ (aq)$$
$$Ca_3(PO_4)_2\ (s) \leftrightarrows 3Ca^{+2}\ (aq) + 2PO_4{}^{-3}\ (aq)$$

The K expression for solubility product constant—called K_{sp}—is written

$$K_{sp} = [Ca^{+2}]\,[C_2O_4{}^{-2}]$$
$$K_{sp} = [Pb^{+2}]\,[I^{-}]^2$$
$$K_{sp} = [Ca^{+2}]^3\,[PO_4{}^{-3}]^2$$

The solubility of the compound (in M units) can be calculated from the K_{sp} data

➤ Write the equation

$$CaF_2\ (s) \leftrightarrows Ca^{+2}\ (aq) + 2F^{-}\ (aq)$$

➤ Fill in the values

	$\rightarrow Ca^{+2}$	$+\ 2F^-$
Initial	0	0
Δ	$+x$	$+2x$
At equilibrium	x	$2x$

➤ Write the K_{sp} expression

$$K_{sp} = [Ca^{+2}]\,[F^-]^2$$

➤ Plug in value for K_{sp} (found in the following table) and solve for x

$$4.0 \times 10^{-11} = x(2x)^2 = 4x^3$$

$$x = [Ca^{+2}] = 2.2 \times 10^{-4}\ M$$

Solubility Products of Some Slightly Soluble Ionic Compounds at 25°C

Compound	K_{sp}
Aluminum hydroxide $[Al(OH)_3]$	1.8×10^{-33}
Barium carbonate $(BaCO_3)$	8.1×10^{-9}
Barium fluoride (BaF_2)	1.7×10^{-6}
Barium sulfate $(BaSO_4)$	1.1×10^{-10}
Bismuth sulfide (Bi_2S_3)	1.6×10^{-72}
Cadmium sulfide (CdS)	8.0×10^{-28}
Calcium carbonate $(CaCO_3)$	8.7×10^{-9}
Calcium fluoride (CaF_2)	4.0×10^{-11}
Calcium hydroxide $[Ca(OH)_2]$	8.0×10^{-6}
Calcium phosphate $[Ca_3(PO_4)_2]$	1.2×10^{-26}
Chromium (III) hydroxide $[Cr(OH)_3]$	3.0×10^{-29}
Cobalt (II) sulfide (CoS)	4.0×10^{-21}
Copper (I) bromide (CuBr)	4.2×10^{-8}
Copper (I) iodide (CuI)	5.1×10^{-12}
Copper (II) hydroxide $[Cu(OH)_2]$	2.2×10^{-20}
Copper (II) sulfide (CuS)	6.0×10^{-37}
Iron (II) hydroxide $[Fe(OH)_2]$	1.6×10^{-14}
Iron (III) hydroxide $[Fe(OH)_3]$	1.1×10^{-36}
Iron (II) sulfide (FeS)	6.0×10^{-19}

Solubility Products of Some Slightly Soluble Ionic Compounds at 25°C (*Cont.*)

Compound	K_{sp}
Lead (II) carbonate ($PbCO_3$)	3.3×10^{-14}
Lead (II) chloride ($PbCl_2$)	2.4×10^{-4}
Lead (II) chromate ($PbCrO_4$)	2.0×10^{-14}
Lead (II) fluoride (PbF_2)	4.1×10^{-8}
Lead (II) iodide (PbI_2)	1.4×10^{-8}
Lead (II) sulfide (PbS)	3.4×10^{-8}
Magnesium carbonate ($MgCO_3$)	4.0×10^{-5}
Magnesium hydroxide [$Mg(OH)_2$]	1.2×10^{-11}
Manganese (II) sulfide (MnS)	3.0×10^{-14}
Mercury (I) chloride (Hg_2Cl_2)	3.5×10^{-18}
Mercury (II) sulfide (HgS)	4.0×10^{-54}
Nickel (II) sulfide (NiS)	1.4×10^{-24}
Silver bromide (AgBr)	7.7×10^{-13}
Silver carbonate (Ag_2CO_3)	8.1×10^{-12}
Silver chloride (AgCl)	1.6×10^{-10}
Silver iodide (AgI)	8.3×10^{-17}
Silver sulfate (Ag_2SO_4)	1.4×10^{-5}
Silver sulfide (Ag_2S)	6.0×10^{-51}
Strontium carbonate ($SrCO_3$)	1.6×10^{-9}
Strontium sulfate ($SrSO_4$)	1.6×10^{-9}
Tin (II) sulfide (SnS)	1.0×10^{-26}
Zinc hydroxide [$Zn(OH)_2$]	1.8×10^{-14}
Zinc sulfide (ZnS)	3.0×10^{-23}

Conversely, the K_{sp} can be calculated using solubility data.

EXAMPLE: Calculate the K_{sp} of lead (II) iodide if the solubility is 1.2×10^{-3} M.

The solubility equals x; therefore, $x = 1.2 \times 10^{-3}$.

➤ Write the equation.

$$PbI_2\ (s) \leftrightarrows Pb^{+2}\ (aq) + 2I^{-}\ (aq)$$

➤ Fill in the values.

	Pb^{+2}	+	$2I^-$
Initial	0		0
Δ	$+1.2 \times 10^{-3}$ (x)		$+2.4 \times 10^{-3}$ $(2x)$
At equilibrium	$+1.2 \times 10^{-3}$		$+2.4 \times 10^{-3}$

➤ Write K_{sp} expression.

$$K_{sp} = [Pb^{+2}]\,[I^-]^2$$

➤ Plug in values and solve for K_{sp}.

$$K_{sp} = (1.2 \times 10^{-3})\,(2.4 \times 10^{-3})^2$$

$$K_{sp} = 6.9 \times 10^{-9}$$

EFFECT OF ADDING IONS TO A SOLUTION AT EQUILIBRIUM—COMMON ION EFFECT

When one of the ions is added to a solution at equilibrium, the equilibrium shifts to remove the added ion; thus, the solubility lessens and more precipitate forms.

EXAMPLE: Calculate the solubility of calcium oxalate in 0.15 M calcium chloride.

$K_{sp} = 2.3 \times 10^{-9}$

➤ Write the equation.

$$CaC_2O_4\ (s) \leftrightarrows Ca^{+2}\ (aq) + C_2O_4^{-2}\ (aq)$$

➤ Fill in the values.

	Ca^{+2}	+	$C_2O_4^{-2}$
Initial	0.15		0
Δ	$+x$		$+x$
At equilibrium	$0.15 + x$		x

➤ Assume the change is negligible.

➤ Write the K_{sp} expression.

$$K_{sp} = [Ca^{+2}]\,[C_2O_4^{-2}]$$

➤ Plug in values and solve for x.

$$2.3 \times 10^{-9} = (0.15)x$$

$$x = [C_2O_4^{-2}] = 1.5 \times 10^{-8}\ M$$

PRECIPITATION CALCULATIONS

This is similar to a reaction quotient problem. Instead of calculating K_{sp}, the Q_{sp} is calculated using nonequilibrium values for the concentrations. If Q_{sp} is greater than K_{sp}, a precipitate forms (the reaction goes to the left).

EXAMPLE: A solution is 0.050 M in Pb^{+2} and 0.10 M in Cl^-. Will a precipitate form?

$K_{sp} = 1.6 \times 10^{-5}$

➤ Write the equation.

$$PbCl_2\ (s) \leftrightarrows Pb^{+2}\ (aq) + 2Cl^-\ (aq)$$

➤ Fill in the values.

	Pb^{+2}	+ $2Cl^-$
Initial	0.050	0.10
Δ	$+x$	$+2x$
At equilibrium	0.050	0.10

➤ Assume the change is negligible.

➤ Write the Q_{sp} expression.

$$Q_{sp} = [Pb^{+2}]\,[Cl^-]^2$$

➤ Fill in the values and solve for Q_{sp}

$$Q_{sp} = (0.050)(0.10)^2$$

$$Q_{sp} = 5.0 \times 10^{-4}$$

➤ Compare with K_{sp}. Q_{sp} is larger, so yes, a precipitate will form.

EXAMPLE: Will a precipitate form when 50.0 milliliters (mL) of 0.0010 M barium chloride is added to 50.0 mL of 0.00010 M sodium sulfate?

$K_{sp} = 1.1 \times 10^{-10}$

➤ Recognize that the precipitate would be barium sulfate, so you must calculate the concentrations of barium ion and sulfate ion.

➤ Calculate $[Ba^{+2}]$.

$$(0.0010\ M)\ (0.0500\ L) = 5.0 \times 10^{-5}\ \text{mole}\ Ba^{+2}$$

$$[Ba^{+2}] = 5.0 \times 10^{-5}\ \text{mole}/0.100\ L = 5.0 \times 10^{-4}\ M$$

➤ Calculate $[SO_4^{-2}]$.

$$(0.00010\ M)\ (0.050\ L) = 5.0 \times 10^{-6}\ \text{mole}\ SO_4^{-2}$$

$$[SO_4^{-2}] = 5.0 \times 10^{-6}\ \text{mole}/0.100\ L = 5.0 \times 10^{-5}\ M$$

➤ Write the equation.

$$BaSO_4\ (s) \leftrightarrows Ba^{+2}\ (aq) + SO_4^{\ -2}\ (aq)$$

➤ Write the Q_{sp} expression.

$$Q_{sp} = [Ba^{+2}]\,[SO_4^{\ -2}]$$

➤ Plug in values and solve for Q_{sp}.

$$Q_{sp} = (5.0 \times 10^{-4})\,(5.0 \times 10^{-5})$$

$$Q_{sp} = 2.5 \times 10^{-8}$$

➤ Compare with K_{sp}. Q_{sp} is larger, so yes, a precipitate will form.

THE EFFECT OF pH ON THE SOLUBILITY OF BASIC COMPOUNDS

Lowering the pH increases the solubility of basic ionic compounds. If you calculate the solubility of a basic compound in pure water as a reference, you can compare the solubility at a lower pH.

EXAMPLE (Part I—Reference): Calculate the solubility of nickel (II) hydroxide in pure water.

$K_{sp} = 2.0 \times 10^{-15}$

➤ Write the equation.

$$Ni(OH)_2\ (s) \leftrightarrows Ni^{+2}\ (aq) + 2\ OH^-\ (aq)$$

➤ Fill in the values.

	Ni^{+2}	+ $2OH^-$
Initial	0	0
Δ	$+x$	$+2x$
At equilibrium	x	$2x$

➤ Write the K_{sp} expression.

$$K_{sp} = [Ni^{+2}]\,[OH^-]^2$$

➤ Fill in the values and solve for x.

$$2.0 \times 10^{-15} = x(2x)^2$$

$$x = [Ni^{+2}] = 7.9 \times 10^{-6}\ M$$

EXAMPLE (Part II): Calculate the solubility of nickel (II) hydroxide at pH 8.

➤ The pOH = 6; therefore, $[OH^-] = 1 \times 10^{-6}$ M. (See Chapter 13.)

➤ Plug in the values using this concentration of OH^-.

$$2.0 \times 10^{-15} = x(1 \times 10^{-6})^2$$

$$x = [Ni^{+2}] = 2 \times 10^{-3}\ M$$

EXAMPLE (Part III): What is the pH of a saturated solution of nickel (II) hydroxide in pure water?

$[OH^-] = 2x = 2(7.9 \times 10^{-6}) = 1.6 \times 10^{-5}$ M

pOH = 4.8

pH = 9.2

This says that at any pH less than 9.2, the solubility is $>7.9 \times 10^{-6}$ M, and as shown previously, at pH 8, the solubility is 2×10^{-3} M.

CRAM SESSION

Solubility Equilibria

1. Insoluble ionic compounds are actually soluble to a very small extent, which is expressed by the ***K_{sp}*, the solubility product constant** of the compound.
2. Using the K_{sp} the solubility of a compound (in molarity) can be calculated. And conversely, given the solubility of the compound, the K_{sp} can be calculated.
3. The **common ion effect** causes the equilibrium to shift to the left when one of the ions is added to the solution. Thus, solubility decreases whenever a common ion is present. The new solubility can be calculated.
4. When two solutions are mixed together:
   A. A precipitate forms if the Q_{sp} for the precipitation reaction is greater than the K_{sp} of the product compound.
   B. If the Q_{sp} is less than the K_{sp}, no precipitate forms, and more solid dissolves if possible.
5. If an insoluble ionic compound is basic, its solubility increases as pH is lowered. The solubility of basic compounds at various pHs can be calculated using the same procedure as that for a common ion effect.

Acid–Base Chemistry

Read This Chapter to Learn About:

- Acids
- Bases
- The pH Scale
- Polyprotic Acids
- Salts
- Buffers
- Titration

Acidity is a measure of the concentration of H^+ in a dilute water solution. It is measured using the **pH scale**, which is based on a process that water undergoes called **autoionization**:

$$H_2O\,(l) \rightleftarrows H^+(aq) + OH^-\,(aq) \qquad K_w = 1 \times 10^{-14}$$

This is an equilibrium process that lies very far to the left. The K for this, in other words, is very small. It is called K_w (for water) and $K_w = 1 \times 10^{-14}$ at 25°C. This means that approximately 1 out of every 10^{14} molecules dissociates into ions in this manner.

An acid produces H^+ in water solution. A base produces OH^- (hydroxide) in water solution. When an acid reacts with a base, a salt is formed. This chapter discusses these compounds and their interactions.

THE pH SCALE

The pH scale is from 0 to 14. It is a logarithmic scale (powers of 10). Each pH is 10-fold less acidic than the next lower pH value.

At pH 7, a solution is neutral.

$$[H^+] = [OH^-]$$

Writing the K expression,

$$K_w = [H^+][OH^-] = 1 \times 10^{-14}$$

At pH 7,

$$[H^+] = [OH^-] = 1 \times 10^{-7}\text{ M}$$

From this, one can see that the pH = – log $[H^+]$ and that pOH = – log $[OH^-]$.

When the pH is 7, the pOH is 7 as well.

$$\text{pH} + \text{pOH} = 14 \text{ always}$$

When $[H^+] = 1 \times 10^{-3}$ M, the pH = 3; and when $[OH^-] = 1 \times 10^{-11}$ M, the pOH = 11.

EXAMPLE: Calculate the pH if $[H^+]$ is 3.5×10^{-5} M.

- Because the exponent is –5, the pH is near 5.
- $\text{pH} = -\log(3.5 \times 10^{-5}) = 4.5$

EXAMPLE: Calculate $[H^+]$ when pH = 8.4.

- $8.4 = -\log[H^+]$
- $[H^+] = 10^{-8.4} = 4 \times 10^{-9}$ M

EXAMPLE: Calculate $[OH^-]$ when pH = 10.6.

- $\text{pOH} = 14 - 10.6 = 3.4$
- $[OH^-] = 10^{-3.4} = 4 \times 10^{-4}$ M

ACIDS

A practical definition of an **acid** is a species that produces H^+ in water solution.

Strong Acids

A strong acid is an acid where every molecule dissociates into H^+ and the counter ion (the conjugate base). There are six strong acids—HCl, HBr, HI, HNO_3, H_2SO_4, and $HClO_4$.

The equation for the dissociation of HCl is written

$$HCl\,(aq) \rightarrow H^+\,(aq) + Cl^-\,(aq)$$

Because there is no reverse reaction taking place, strong acids do not involve any equilibrium process. Simple stoichiometry suffices. The concentration of the proton and the concentration of the conjugate base Cl^- are equal to the initial concentration of the acid.

If 1000 HCls are dissolved in water, they dissociate into 1000 H^+ and 1000 Cl^- ions.

Weak Acids

Weak acids have the word "acid" as part of their name, and they are not one of the six strong acids.

For weak acids, only 1 out of every 10,000–100,000 molecules undergoes dissociation to proton and conjugate base

$$HA\ (aq) \rightleftarrows H^+\ (aq) + A^-\ (aq)$$

Most of the molecules are still in the HA form. This is an equilibrium process, with a very small K_a (for acid). The K_as are in the range of 10^{-4} to 10^{-6}, typically. The K_a expression is written

$$K_a = \frac{[H^+][A^-]}{[HA]}$$

EXAMPLE: Calculate the pH of 0.15 M HNO_3.

➤ This is a strong acid, so the proton concentration is equal to the acid concentration $[H^+] = [HNO_3] = 0.15$ M.

➤ $pH = -\log 0.15 = 0.8$

EXAMPLE: Calculate the pH of 0.15 M HF.

$K_a = 6.7 \times 10^{-4}$

This is a weak acid. The steps for solving this problem are

➤ Write the equation.

$$HF\ (aq) \rightleftarrows H^+\ (aq) + F^-\ (aq)$$

➤ Fill in the table.

At equilibrium 0.15 x x

(Small K so the initial HF concentration equals the HF concentration at equilibrium.)

➤ Write the K expression.

$$K_a = \frac{[H^+][F^-]}{[HF]}$$

➤ Fill in the equilibrium values.

$$6.7 \times 10^{-4} = \frac{x^2}{0.15}$$

- Solve the $[H^+]$ for x.

$$x = [H^+] = 1 \times 10^{-2}\ M$$

- Calculate the pH.

$$pH = -\log (1 \times 10^{-2}) = 2$$

BASES

A base produces OH^- (hydroxide) in water solution.

Strong Bases

If the base is a strong base, every molecule dissociates into ions in water solution. The most common strong bases are LiOH, NaOH, KOH, and $Ba(OH)_2$.

$$KOH\ (aq) \rightarrow K^+\ (aq) + OH^-\ (aq)$$

$$Ba(OH)_2\ (aq) \rightarrow Ba^{2+}\ (aq) + 2\ OH^-\ (aq)$$

There is no equilibrium here; stoichiometry suffices to determine $[OH^-]$ and, thus, pH. For LiOH, KOH, and NaOH, the base concentration equals the hydroxide concentration.

$$[KOH] = [OH^-]$$

For $Ba(OH)_2$, the base concentration is half of the hydroxide concentration.

$$[Ba(OH)_2] = \tfrac{1}{2}\,[OH^-]$$

EXAMPLE: Calculate the pH of 0.065 M KOH. This is a strong base, with 1:1 stoichiometry.

- The $[KOH] = [OH^-] = 0.065$ M.
- $pOH = -\log 0.065 = 1.2$
- $pH = 14 - 1.2 = 12.8$

EXAMPLE: Calculate the pH of 0.065 M $Ba(OH)_2$. This is a strong base, with 1:2 stoichiometry.

- $[Ba(OH)_2] = \tfrac{1}{2}\,[OH^-] = 0.065$
- Therefore, $[OH^-] = 0.13$ M
- $pOH = -\log 0.13 = 0.9$
- $pH = 14 - 0.9 = 13.1$

Weak Bases

We use ammonia for the weak base problems because almost all weak bases are organic derivatives of ammonia. When ammonia is dissolved in water, approximately 1 out of every 100,000 molecules produces an OH^-.

The equation is

$$NH_3\ (aq) + H_2O\ (aq) \rightleftarrows NH_4^+\ (aq) + OH^-\ (aq)$$

This is an equilibrium process and the K_b (for base) is 1.8×10^{-5}. The K_b expression is written

$$K_b = \frac{[NH_4^+][OH^-]}{[NH_3]}$$

EXAMPLE: Calculate the pH of 0.065 M NH_3. $K_b = 1.8 \times 10^{-5}$. This is a weak base; there are certain steps to follow.

➤ Write the equation.

$$NH_3\ (aq) + H_2O\ (l) \leftrightarrows NH_4^+\ (aq) + OH^-\ (aq)$$

➤ Fill in the table.

	NH_3	$\leftrightarrows$	NH_4^+	+	OH^-
At equilibrium	0.065		x		x

➤ Write the K expression.

$$K_b = \frac{[NH_4^+][OH^-]}{[NH_3]}$$

➤ Fill in the equilibrium values.

➤ Solve for x, the $[OH^-]$.

$$1.8 \times 10^{-5} = \frac{x^2}{0.065}$$

$$x = [OH^-] = 1.1 \times 10^{-3}\ M$$

➤ Calculate the pOH.

$$pOH = 3$$

➤ Calculate the pH.

$$pH = 11$$

POLYPROTIC ACIDS

Polyprotic acids are acids that have more than one proton; therefore, they have more than one K_a. There is just one K_a per proton removed. In the following equations, all substances are in aqueous solution.

➤ Carbonic acid has 2 protons (both weak)

$$H_2CO_3$$

➤ For the removal of the first proton,

$$H_2CO_3 \rightleftarrows H^+ + HCO_3^{-1}$$

- For the removal of the second proton,

$$HCO_3^{-1} \rightleftarrows H^+ + CO_3^{-2}$$

- K_{a1} for the first proton = 4.3×10^{-7} Weak acid
- K_{a2} for the second proton = 4.8×10^{-11} Weaker acid
- Phosphoric acid has 3 protons (all weak)

H_3PO_4

- Removal of first proton,

$$H_3PO_4 \rightleftarrows H^+ + H_2PO_4^{-1} \quad K_{a1} = 6.9 \times 10^{-3}$$

3H → 2H

- Removal of second proton,

$$H_2PO_4^{-1} \rightleftarrows H^+ + HPO_4^{-2} \quad K_{a2} = 6.2 \times 10^{-8}$$

2H → 1H

- Removal of third proton,

$$HPO_4^{-2} \rightleftarrows H^+ + PO_4^{-3} \quad K_{a3} = 4.8 \times 10^{-13}$$

1H → 0H

pH of a Polyprotic Acid Solution

If one is calculating the pH of a solution of carbonic acid or phosphoric acid, only the first (and largest) K_{a1} is used. The amount of proton produced by the subsequent K_as is negligible compared to the first. In all of the equations that follow, every substance is in aqueous solution except the water, which is liquid.

Salts of Polyprotic Acids

$NaHCO_3$ is the salt of NaOH and H_2CO_3. It is a basic salt (K_b); thus, it produces OH^- and a conjugate acid

$$HCO_3^{-1} + H_2O \rightleftarrows OH^- + H_2CO_3 \quad K_{b1} = 2.3 \times 10^{-8}$$

Na_2CO_3 is also a salt of NaOH and H_2CO_3. It is a basic salt (K_b); thus, it produces OH^- and a conjugate acid

$$CO_3^{-2} + H_2O \rightleftarrows OH^- + HCO_3^{-1} \quad K_{b2} = 2.1 \times 10^{-4}$$

There are three salts of phosphoric acid

NaH_2PO_4
Na_2HPO_4
Na_3PO_4

All of these are basic salts; they produce OH^- and a conjugate acid

$H_2PO_4^{-1} + H_2O \rightleftarrows OH^- + H_3PO_4$ $K_{b1} = 1.4 \times 10^{-12}$

$HPO_4^{-2} + H_2O \rightleftarrows OH^- + H_2PO_4^{-1}$ $K_{b2} = 1.6 \times 10^{-7}$

$PO_4^{-3} + H_2O \rightleftarrows OH^- + HPO_4^{-2}$ $K_{b3} = 2.1 \times 10^{-2}$

EXAMPLE: Calculate the pH of 0.1185 M phosphoric acid solution.

➤ For the pH of a polyprotic acid, we use only the K_{a1}.

$$K_{a1} = 6.9 \times 10^{-3}$$

➤ Write the equation.

$$H_3PO_4 \rightleftarrows H^+ + H_2PO_4^{-1}$$

➤ Fill in table.

0.1185 M x x

➤ Write K_a expression.

$$K_a = \frac{[H^+][H_2PO_4^{-1}]}{[H_3PO_4]}$$

➤ Fill in values and solve for x.

$$6.9 \times 10^{-3} = \frac{x^2}{0.1185}$$

$$x = [H^+] = 8.2 \times 10^{-4} \text{ M}$$

➤ Calculate the pH.

$$\text{pH} = 3.1$$

EXAMPLE: Calculate the pH of 0.0165 M sodium hydrogen phosphate.

➤ Write the formula.

$$Na_2HPO_4$$

➤ Recognize that this is a basic salt, so it is a base, so you need a K_b.

➤ Write the equation.

$$HPO_4^{-2} + H_2O \rightleftarrows OH^- + H_2PO_4^{-1}$$

➤ Fill in the table.

0.0165 M x x

➤ Determine which K_b is needed.

K_{b1} involves 2H → 3H

K_{b2} involves 1H → 2H

K_{b3} involves 0H → 1H

- Because our salt has 1 H and is going to the conjugate acid with 2 H, we need K_{b2}.

$$K_{b2} = 1.6 \times 10^{-7}$$

- Write the K_b expression.

$$K_{b2} = \frac{[OH^-][H_2PO_4^-]}{[HPO_4^{-2}]}$$

- Fill in values and solve for x.

$$1.6 \times 10^{-7} = \frac{x^2}{0.0165}$$

$$x = [OH^-] = 5.1 \times 10^{-5} \text{ M}$$

- Calculate pOH and pH.

$$pOH = 4.3$$

$$pH = 9.7$$

SOLUTIONS OF SALTS

When an acid reacts with a base, in stoichiometric quantities, an ionic compound (a salt) is formed, along with water

$$\text{acid} + \text{base} \rightarrow \text{salt} + H_2O$$

The ionic compound is neutral, acidic, or basic (when dissolved in water), depending on which acid and base were used to make it.

CASE 1:

Strong acid + Strong base $\longrightarrow$ Neutral salt $\quad pH = 7$

CASE 2:

Strong acid + Weak base $\longrightarrow$ Acidic salt $\quad pH < 7$

CASE 3:

Weak acid + Strong base $\longrightarrow$ Basic salt $\quad pH > 7$

We look next at salts from cases 2 and 3 only. In the following equations, all substances are in aqueous solution except water, which is liquid.

CASE 2 EXAMPLES:

$$HCl + NH_4OH \longrightarrow NH_4Cl + H_2O$$

$$HNO_3 + NH_4OH \longrightarrow NH_4NO_3 + H_2O$$

CASE 3 EXAMPLES:

$$HF + LiOH \longrightarrow LiF + H_2O$$

$$HCN + KOH \longrightarrow KCN + H_2O$$

$$CH_3COOH + KOH \longrightarrow CH_3COOK + H_2O$$

Acidic Salts

Acidic salts are formed when a strong acid reacts with a weak base, as in case 2.

EXAMPLE: Calculate the pH of 0.078 M ammonium nitrate.

$K_a = 5.6 \times 10^{-10}$

- Recognize that NH_4NO_3 is an acidic salt.
- Therefore, it is an acid (produces H^+).
- Write the equation.

$$NH_4^+ \text{ (aq)} \rightleftharpoons NH_3 \text{ (aq)} + H^+ \text{ (aq)}$$

- Fill in the table.

At equilibrium 0.078 x x

- Write the K expression.

$$K_a = \frac{[NH_3][H^+]}{[NH_4^+]}$$

- Fill in the values and solve for x, the $[H^+]$.

$$5.6 \times 10^{-10} = \frac{x^2}{0.078}$$

$$x = [H^+] = 6.6 \times 10^{-6} \text{ M}$$

- Calculate pH.

$$\text{pH} = 5.2$$

Basic Salts

Basic salts are formed when a strong base reacts with a weak acid, as in case 3.

EXAMPLE: Calculate the pH of 0.035 M KCN.

$K_b = 1.6 \times 10^{-5}$

- Recognize that KCN is a basic salt.
- Therefore, it is a base (produces OH^-).

➤ Write the equation. The only ion here that can produce OH^- from water is the cyanide. You must have a negative ion to remove the H^+ from H_2O, leaving OH^-.

$$CN^- (aq) + H_2O (l) \rightleftarrows HCN (aq) + OH^- (aq)$$

➤ Fill in the table,

At equilibrium 0.035 x x

➤ Write the K expression.

$$K_b = \frac{[HCN][OH^-]}{[CN^-]}$$

➤ Fill in the values and solve for x, the $[OH^-]$.

$$1.6 \times 10^{-5} = \frac{x^2}{0.035}$$

$$x = [OH^-] = 7.5 \times 10^{-4} \text{ M}$$

➤ Calculate pOH.

pOH = 3.1

➤ Calculate pH.

pH = 10.9

COMMON ION EFFECT

The common ion effect occurs when you add strong acid or strong base to a solution of weak acid or weak base at equilibrium. The added base or acid reacts to form more of one of the components in equilibrium.

EXAMPLE: Calculate the concentration of acetate ion in a solution that is 0.10 M HAc and 0.010 M HCl.

$K_a = 1.8 \times 10^{-5}$

➤ Write the equation.

$$HAc (aq) \rightleftarrows H^+(aq) + Ac^- (aq)$$

➤ Fill in the table.

	HAc	⇄	H^+	+	Ac^-
Initial	0.10		0.010		0
Δ	$-x$		$+x$		$+x$
At equilibrium	$0.10 - x$		$0.010 + x$		x

- Assume the change is negligible.
- Write the K_a expression.

$$K_a = \frac{[H^+][Ac^-]}{[HAc]}$$

- Plug in values.

$$1.8 \times 10^{-5} = \frac{(0.010)\,x}{0.10}$$

- Solve for x.

$$x = 1.8 \times 10^{-4}\ M = [Ac^-]$$

EXAMPLE: Calculate the pH of a solution that is 0.10 M HAc and 0.20 M Ac^-.

- Write the equation.

$$HAc\ (aq) \rightleftarrows H^+\ (aq) + Ac^-\ (aq)$$

- Fill in the values.

	HAc	⇄	H^+	+	Ac^-
Initial	0.10		0.0		0.20
Δ	$-x$		$+x$		$+x$
	$0.10 - x$		x		$0.20 + x$

- Assume the change is negligible.
- Write the K_a expression.

$$K_a = \frac{[H^+][Ac^-]}{[HAc]}$$

- Plug in the values.

$$1.8 \times 10^{-5} = \frac{x\,(0.20)}{0.10}$$

- Solve for x.

$$x = [H^+] = 9 \times 10^{-6}\ M$$

- Calculate the pH.

$$pH = 5.0$$

BUFFERS

A **buffer** is a solution that resists change in pH when small amounts of acid or base are added. It is usually a solution that is made of approximately equal concentrations of an acid and its conjugate base. The calculation of the pH of a buffer uses exactly the same method as for a common ion.

EXAMPLE: Calculate the pH of a buffer made by mixing 60.0 milliliters (mL) of 0.100 M NH_3 with 40.0 mL of 0.100 M NH_4Cl.

$K_a = 5.6 \times 10^{-10}$

- Calculate the molarity of each component in the mixture.

$$(0.0600\ L)(0.100\ M) = 0.006\ mole/0.100\ L = 0.060\ M\ NH_3$$

$$(0.0400\ L)(0.100\ M) = 0.004\ mole/0.100\ L = 0.040\ M\ NH_4Cl$$

- Write the equation and fill in the values.

	NH_4	$\rightleftarrows$	H^+	+	NH_3
Initial	0.040		0		0.060
Δ	$-x$		$+x$		$+x$
At equilibrium	$0.040 - x$		x		$0.060 + x$

- Assume the change is negligible.
- Write the K_a expression.

$$K_a = \frac{[H^+][NH_3]}{[NH_4^+]}$$

- Plug in the values and solve for x.

$$5.6 \times 10^{-10} = \frac{x(0.060)}{0.040}$$

$$x = [H^+] = 3.7 \times 10^{-10}\ M$$

- Calculate the pH.

$$pH = 9.4$$

TITRATION

There are two major types of acid–base titrations. One is the addition of strong base to strong acid; the other is the addition of strong base to weak acid. The pH of the endpoint solution can be calculated for the second case.

Addition of Strong Base to Strong Acid

- The initial pH is the pH of the strong acid solution.

$$[H^+] = [HA]$$

- At the midpoint, calculate the moles of acid left.
- Then calculate the M of the acid solution.
- Calculate the pH at the midpoint.
- At the endpoint, the pH is 7, because the salt is neutral.

Addition of Strong Base to Weak Acid

- The initial pH is the pH of the weak acid solution.

  $K_a = x^2/[HA]$
  $x = [H^+]$
  $pH = -\log [H^+]$

- At the midpoint, use the **Henderson–Hasselbach equation** after calculating the molarity of each component.

  $$pH = pK_a + \log \frac{[A^-]}{[HA]} \quad \text{where } pK_a = -\log K_a$$

- At the endpoint, it's the pH of a basic salt solution.
  - Calculate the molarity of the salt.
  - Calculate the K_b.

    $K_b = x^2/[A^-]$
    $x = [OH^-]$

  - Calculate pOH.
  - Calculate pH.

EXAMPLE: A 25.0-mL sample of 0.0875 M acetic acid is titrated with 0.150 M NaOH.

$K_a = 1.8 \times 10^{-5}$ $\quad pK_a = 4.74$

- Calculate the pH of the initial acid solution.

  $HA\ (aq) \rightleftarrows H^+\ (aq) + A^-\ (aq)$

  $$K_a = \frac{[H^+][A^-]}{[HA]}$$

  $1.8 \times 10^{-5} = x^2/0.0875$

  $x = [H^+] = 1.25 \times 10^{-3}$ M

  $pH = 2.9$

- Calculate the pH at the point when 5.0 mL of base has been added.

  (0.005 L) (0.150 M NaOH) = 7.5×10^{-4} mole base added

  - Reacts with 7.5×10^{-4} mole acid.

    (0.0250 L) (0.0875 M acid) = 2.188×10^{-3} mole acid initially

  - Subtract 7.5×10^{-4} mole acid that is now gone to get

    1.438×10^{-3} mole acid in 30 mL = 0.0479 M acid and

    7.5×10^{-4} mole conj base in 30 mL = 0.0250 M base

- Plug into Henderson–Hasselbach equation

$$pH = 4.74 + \log\frac{0.0250}{0.0479}$$

$$pH = 4.5$$

- Calculate the pH at the endpoint of the titration:
  Here, you have added 2.188×10^{-3} mole NaOH to completely neutralize the initial amount of acid. This requires 2.188×10^{-3} mole/0.150 M = 0.0146 L NaOH solution

  $$[A^-] = 2.188 \times 10^{-3} \text{ mole}/0.0396 \text{ L total} = 0.055 \text{ M conj base}$$

  - Calculate $K_b = K_w/K_a = 5.6 \times 10^{-10}$

  $$K_b = x^2/0.055 \text{ M} = 5.6 \times 10^{-10}$$

  $$x = [OH^-] = 5.5 \times 10^{-6} \text{ M}$$

  $$pOH = 5.3$$

  $$pH = 8.7$$

CRAM SESSION

Acid–Base Chemistry

1. The **pH scale** is a scale that is based on the concentration of H^+ ion in dilute water solutions. Pure water undergoes an **autoionization** process, with K_w equal to 1.0×10^{-14} at 25°C. The **pH scale** is from 0 to 14, with pH 7 neutral, <7 acidic, and >7 basic.
   A. The pH of a solution can be calculated using the equation $pH = -\log[H^+]$.
2. **Weak acid solutions** have a small K_a and the pH is calculated using the same process as for any small K problem.
3. For a **weak base**, the K_b is used. The weak base equation must show the weak base reacting with water to form hydroxide ion and the conjugate acid of the weak base. The equation $K_a K_b = K_w$ is used to calculate K_b from K_a.
4. **Polyprotic acids** have more than one proton, and there is a K_a for each proton removed sequentially. The conjugate base formed when a proton is removed is an anion, and thus is a basic salt. The pH of a solution of basic salt is a K_b problem, where the basic salt produces hydroxide ion and the conjugate acid of the base.
5. The pH of an acidic salt solution is calculated in the same manner as a weak acid, using a K_a.
6. The addition of a common ion shifts the equilibrium to the left. **Buffer solutions,** which resist change in pH on small additions of acid or base, consist of solutions that contain both the acid and its conjugate base in approximately equal amounts. The pH of a buffer solution is usually near the pK_a of the acid.
7. The **Henderson–Hasselbach equation** can be used to calculate the pH of a buffer solution.
8. In a **titration**, the buffer region is approximately midway in the titration, in the vicinity of pK_a. The Henderson–Hasselbach equation can be used to calculate the pH at this point in a titration.

CHAPTER 14

Thermodynamics

Read This Chapter to Learn About:

- The Laws of Thermodynamics
- Spontaneity and Entropy
- Predicting the Sign of Entropy, *S*
- Free Energy, *G*
- The Relationship between Free Energy and Work, the Equilibrium Constant, and Temperature

Thermodynamics is the study of the relationships of enthalpy, entropy (disorder of a system), and free energy. The laws of thermodynamics and certain formulas explain the relationships between free energy, entropy, work, temperature, and equilibrium.

FIRST LAW OF THERMODYNAMICS

The first law of thermodynamics states that the change in the internal energy of a system $\Delta U = q + w$, where q = heat and w = work. U is the sum of the kinetic and potential energies. It is a state function; that is, it depends only on the initial and final states, and not on the path between. The **heat *q*** is the energy into or out of a system because of a temperature difference between the system and its surroundings. **Work** is the energy that results when a force moves an object some **distance, *d***.

The enthalpy H is equal to the quantity $U + PV$, where H is also a state function. ΔH is the difference between the initial and final states; at atmospheric pressure, $\Delta H = q$.

SPONTANEITY AND ENTROPY

A **spontaneous process** occurs in the direction a chemical reaction is written; in other words, it proceeds to the right. The entropy, with symbol S, is a measure of the **disorder of a system**. S is also a state function, and ΔS is the difference between the initial and final states.

If ΔS is positive, the process results in more disorder. If ΔS is negative, the process results in more order.

SECOND LAW OF THERMODYNAMICS

The second law of thermodynamics states that the total **entropy** of a system and its surroundings always increases for a spontaneous process. This law relates spontaneity to entropy. The effect on the surroundings must be taken into account.

For a spontaneous process, ΔS must be $> q/T$, and at equilibrium, $\Delta S = q/T = \Delta H/T$.

THIRD LAW OF THERMODYNAMICS

The third law of thermodynamics states that a substance that is perfectly crystalline at 0 K has an entropy of zero. The change in entropy can be negative or positive, however.

S° is the standard entropy at 25°C and 1 atm. The ΔS° for a reaction is the difference between the products' entropies and the reactants' entropies.

PREDICTING THE SIGN OF ΔS_{rxn}

If disorder is increasing, ΔS_{rxn} is positive. Some examples of processes with increasing disorder include more gas produced than used up; a solid converted to a liquid; and a solid dissolved in a solvent.

If order is increasing, ΔS_{rxn} is negative. Some examples of processes that have increasing order include less gas produced than used up; a liquid converted to a solid; and a solid precipitated from a solution.

FREE ENERGY AND SPONTANEITY

Free energy has the symbol G and is defined as $H - TS$. One can predict the spontaneity of a process from the sign of ΔG°_{rxn}. Free energy is useful because it eliminates the need to worry about the surroundings.

The equation used is $\Delta G^\circ = \Delta H^\circ - T\Delta S^\circ$. When ΔG° is negative, the reaction is spontaneous as written. When ΔG° is positive, the reaction is spontaneous in the opposite direction.

WORK AND ΔG

The maximum work for a spontaneous reaction is equivalent to ΔG. The free energy change is the maximum energy available to do useful work.

THE RELATIONSHIP BETWEEN THE EQUILIBRIUM CONSTANT *K* AND ΔG

When not at equilibrium, $\Delta G = \Delta G^\circ + RT \ln Q$, where Q is the reaction quotient. Therefore, because $\Delta G = 0$ at equilibrium, $\Delta G^\circ = -RT \ln K$.

EXAMPLE: Calculate K for a reaction for which $\Delta G^\circ = -13.6$ kilojoules (kJ)

$$\ln K = \frac{-13,600 \text{ J}}{(-8.314 \text{ J/mol K})\,(298.2 \text{ K})} = 5.49$$

$K = e^{5.49} = 242$ large K, spontaneous reaction

ΔG AND TEMPERATURE

Because $\Delta H^\circ - T\Delta S^\circ = \Delta G^\circ$, the signs of each variable determine the sign of ΔG°. If both ΔH° and ΔS° are positive, ΔG° is negative at high temperatures. If both ΔH° and ΔS° are negative, ΔG° is negative at low temperatures. If ΔH° is negative and ΔS° is positive, ΔG° is always negative. The last case is if ΔH° is positive and ΔS° is negative; in this case, ΔG° is always positive.

EXAMPLE: Estimate the temperature above which a certain reaction becomes spontaneous, if $\Delta H = 178.3$ kJ and $\Delta S = 159$ J/K.

- At the point between spontaneity and nonspontaneity, at equilibrium, $\Delta G^\circ = 0$
- Plug in the values into $T = \Delta H^\circ / \Delta S^\circ$.

$$T = 178{,}300 \text{ J}/159 \text{ J/K} = 1121 \text{ K}$$

RELATIONSHIP OF *K* TO TEMPERATURE

The equilibrium constant changes with temperature according to the equation

$$\ln \frac{K_2}{K_1} = \frac{\Delta H}{R}\left(\frac{1}{T_1} - \frac{1}{T_2}\right)$$

CRAM SESSION

Thermodynamics

1. **Thermodynamics** is the study of the relationships of enthalpy, entropy, and free energy. A **spontaneous reaction** proceeds to the right as written.
2. Spontaneity is related to **entropy**, which is a measure of the disorder of a system.
   A. If a reaction or process is producing disorder, the ΔS is positive.
   B. If the reaction or process is producing order, the ΔS is negative.
3. The **change in free energy**, ΔG, for a reaction, is directly related to spontaneity.
   A. If the ΔG for a reaction is negative, the reaction is spontaneous.
   B. If the ΔG for a reaction is positive, the reaction is spontaneous in the opposite direction.
   C. Because $\Delta G = \Delta H - T\Delta S$, the signs of ΔH and ΔS contribute to the spontaneity of the reaction.
4. The free energy change is also related to K, the **equilibrium constant**. A spontaneous reaction has a large K and the ΔG is negative. This can be shown by the equation $\Delta G^\circ = -RT \ln K$, where ΔG° is the free energy change at standard conditions of 25°C and 1 atm pressure, and the reaction is at equilibrium.
5. When a reaction is not at equilibrium, the $\Delta G = \Delta G^\circ + RT \ln Q$. Q is the **reaction quotient**, and it is determined by using the nonequilibrium concentrations or pressures of the products and reactants.
6. Equilibrium constants are temperature-dependent. If the ΔH_{rxn} and the equilibrium constant at one temperature are known, the equilibrium constant at another temperature can be calculated using the equation $\ln (K_2/K_1) = (\Delta H/R)(1/T_1 - 1/T_2)$.

CHAPTER 15

Electrochemistry

Read This Chapter to Learn About:

➤ Redox Reactions

➤ Voltaic Cells

➤ Electrolytic Cells

Electrochemistry is the study of oxidation–reduction reactions. These are reactions in which electrons are transferred from one species to another. The species that gains electrons is **reduced**; the species that loses electrons is **oxidized**. A reduction–oxidation, or **redox**, reaction that is spontaneous produces electricity. A nonspontaneous redox reaction requires electricity to run. Batteries (**voltaic** or **galvanic cells**) operate by producing electricity via a spontaneous redox reaction. **Electrolytic cells** require electricity to make the reaction occur; they are used in certain industrial processes, such as the purification of aluminum from its ore. These processes are called **electrolysis reactions**.

BALANCING OXIDATION–REDUCTION REACTIONS

Balancing a redox equation is not the same as balancing a nonredox equation. You have to take the electron change into account.

How to Balance a Redox Equation in an Acidic Solution

1. Divide the equation into two half-reactions, one for oxidation and one for reduction.
2. Balance all the atoms.

   ➤ Start with the main atoms.

- Then use H_2O to balance O.
- Then use H^+ to balance H.

3. Determine the total electron change for each half-reaction, using the oxidation numbers of each species that is changing.
4. Determine the lowest common denominator (LCD) of the total electron change.
5. Multiply through each half-reaction so that the electron change equals the LCD.
6. Add the two half-reactions.
7. Cancel where applicable.

How to Balance a Redox Equation in Basic Solution

1. Balance as for an acidic solution
2. Change all H^+ to H_2O.
3. Put the same number of OH^- on the other side.
4. Add and cancel where applicable.

EXAMPLE IN ACID SOLUTION:

$Cr_2O_7^{-2} + C_2O_4^{-2} \rightarrow Cr^{+3} + CO_2$ (The phase labels have been left out for clarity.)

- Divide into two half-reactions.

$$Cr_2O_7^{-2} \rightarrow Cr^{+3}$$
$$C_2O_4^{-2} \rightarrow CO_2$$

- Balance the atoms, starting with the main atoms.

$$Cr_2O_7^{-2} \rightarrow 2Cr^{+3}$$
$$C_2O_4^{-2} \rightarrow 2CO_2$$

- Use H_2O to balance the oxygens.

$$Cr_2O_7^{-2} \rightarrow 2Cr^{+3} + 7H_2O$$
$$C_2O_4^{-2} \rightarrow 2CO_2$$

- Use H^+ to balance the hydrogens.

$$14H^+ + Cr_2O_7^{-2} \rightarrow 2Cr^{+3} + 7H_2O$$
$$C_2O_4^{-2} \rightarrow 2CO_2$$

- Determine the oxidation number of each atom that is changing and find the total electron change for each half-reaction.
  - Cr on the left is +6; Cr on the right is +3. This is a change of 3 electrons, but there are 2 Crs changing, so the total electron change is 6 for this half-reaction.

- C on the left is +3; C on the right is +4. This is a change of 1 electron, but there are 2 Cs changing, so the total electron change is 2 for this half-reaction.
- Find the LCD of the electron change. The LCD of 6 and 2 is 6.
- Balance the electrons; the electron change for both half-reactions must be 6.
  - The first half-reaction is already a change of 6 electrons, so it stays as is. The second half-reaction is a change of 2, so it must be multiplied by 3.

$$14H^{+} + Cr_2O_7^{-2} \rightarrow 2Cr^{+3} + 7H_2O \quad \Delta e^{-} = 6$$

$$3C_2O_4^{-2} \rightarrow 6CO_2 \quad \Delta e^{-} = 6$$

  - Sum the two half-reactions.

$$14H^{+} + Cr_2O_7^{-2} + 3C_2O_4^{-2} \rightarrow 2Cr^{+3} + 7H_2O + 6CO_2$$

- There is no canceling to do, so this is the final balanced equation.

EXAMPLE IN BASIC SOLUTION:

$Co^{+2} + H_2O_2 \rightarrow Co(OH)_3 + H_2O$ (The phase labels have been left out for clarity.)

- Divide into two half-reactions.

$$Co^{+2} \rightarrow Co(OH)_3$$

$$H_2O_2 \rightarrow H_2O$$

- Balance the atoms, using H_2O to balance oxygen, and H^{+} to balance hydrogen.

$$3H_2O + Co^{+2} \rightarrow Co(OH)_3 + 3H^{+}$$

$$2H^{+} + H_2O_2 \rightarrow 2H_2O$$

- Determine the electron change for each species that is changing.
  - Co is +2 on the left; it is +3 on the right; this is a 1 electron change, and there is 1 atom changing, so the total electron change is 1.
  - O is −1 on the left, it is −2 on the right; this is a 1 electron change, but there are 2 Os changing, so the total electron change is 2.
- The LCD of 1 and 2 is 2. Therefore the *n* value is 2.
  - The first half-reaction must be multiplied by 2.

$$6H_2O + 2Co^{+2} \rightarrow 2Co(OH)_3 + 6H^{+}$$

$$2H^{+} + H_2O_2 \rightarrow 2H_2O$$

- Sum the two half-reactions.

$$6H_2O + 2Co^{+2} + 2H^{+} + H_2O_2 \rightarrow 2Co(OH)_3 + 6H^{+} + 2H_2O$$

- Cancel out where the same species shows up on both sides.

$$4H_2O + 2Co^{+2} + H_2O_2 \rightarrow 2Co(OH)_3 + 4H^{+}$$

➤ Change all H^+ to H_2O's and put the same number of OH^- on the other side.

$$4OH^- + 4H_2O + 2Co^{+2} + H_2O_2 \rightarrow 2Co(OH)_3 + 4H_2O$$

➤ Cancel again.

$$4OH^- + 2Co^{+2} + H_2O_2 \rightarrow 2Co(OH)_3$$

VOLTAIC CELLS

Voltaic cells (also called **galvanic cells**) are constructed so that a redox reaction produces an electric current. A diagram of a voltaic cell is Figure 15-1.

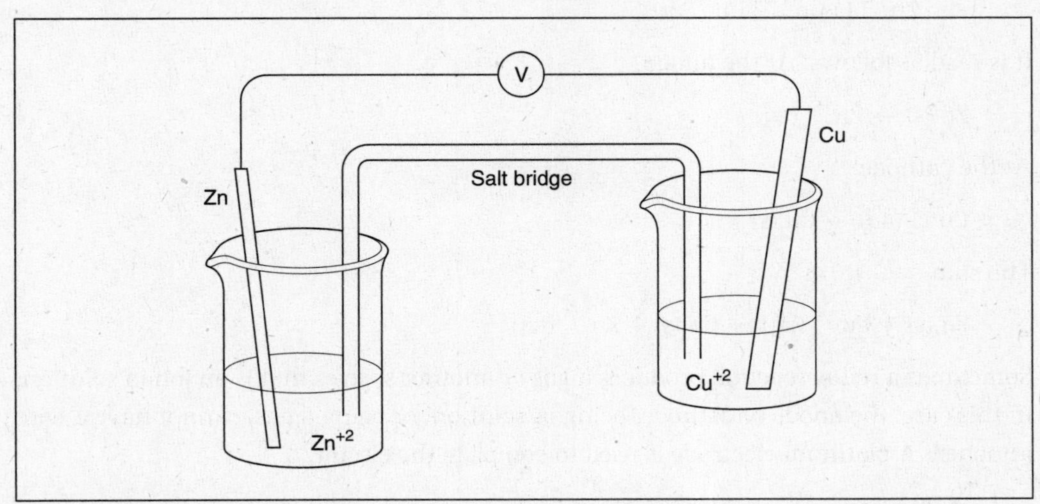

FIGURE 15-1: Voltaic cell with Zn/Zn^{+2} anode and Cu/Cu^{+2} cathode.

One half-cell is the **anode**. This is where the oxidation takes place. The electrode wears away as the reaction proceeds, with the metal electrode becoming an ion in the solution.

The other half-cell is the **cathode**. This is where the reduction is taking place. The electrode builds up as the reaction proceeds, with the metal ions in the solution plating out as pure metal.

The two half-cells are connected by a salt bridge between the solutions and by a wire between the electrodes.

The anode gives up electrons

$$Zn\ (s) \rightarrow Zn^{+2}\ (aq) + 2e^-$$

The cathodes uses the electrons

$$Cu^{+2}\ (aq) + 2e^- \rightarrow Cu\ (s)$$

Cell Notation for Voltaic Cells

Rather than draw the voltaic cell, a notation can be used instead. Cell notation is drawn as follows:

anode | anode's ion | | cathode's ion | cathode

The single line indicates a phase change; the double line indicates the salt bridge between the half-cells. The salt bridge is necessary to complete the electrical circuit and allow the reaction to take place. The concentration of the ions is often written directly after the ion.

The cell notation for the cell drawn above looks like:

$Zn \mid Zn^{+2} \mid\mid Cu^{+2} \mid Cu$

It is read as follows: At the anode

$Zn\ (s) \rightarrow Zn^{+2}\ (aq)$

At the cathode

$Cu^{+2}\ (aq) \rightarrow Cu\ (s)$

The sum

$Zn\ (s) + Cu^{+2}\ (aq) \rightarrow Cu\ (s) + Zn^{+2}\ (aq)$

Sometimes a redox reaction produces a gas or another species that is an ion in solution. In this case, the anode or cathode, being in solution, or being a gas, cannot have a wire attached. A platinum electrode is used to complete the circuit.

EXAMPLE:

$Zn\ (s) + 2Fe^{+3}\ (aq) \rightarrow Zn^{+2}\ (aq) + 2Fe^{+2}\ (aq)$

$Zn \mid Zn^{+2} \mid\mid Fe^{+3}, Fe^{+2} \mid Pt$

Cell Voltage

The **electromotive force, emf**, of a cell, or the cell voltage generated by the reaction, is called E_{cell}. If you know the E_{cell} (by measuring it in the lab), you can calculate the work, w, produced by the cell using the equation

$w_{max} = -nFE_{cell}$

where n = LCD of the mole e^-, $F = 9.65 \times 10^4$ coulombs (C)/mole e^-, and 1 joule (J) = 1 volt-coulomb (V-C).

EXAMPLE: Calculate the work done by the cell as shown next if the E_{cell} = 0.650 V.

$Pt \mid Hg_2^{+2} \mid Hg\ (l) \mid\mid H_2 \mid H^+ \mid Pt$

➤ Calculate the n value.

$n = 2$ because $Hg_2^{+2} \rightarrow Hg$ is a $2e^-$ change
$H_2 \rightarrow 2H^+$ is a $2e^-$ change

➤ $w_{max} = (-2)(9.65 \times 10^4 \text{ C})(0.650 \text{ V}) = -1.25 \times 10^5 \text{ J}$
➤ The sign is negative because work is produced.

Standard Cell Potential

E°_{cell} is the symbol given to the cell potential under the standard conditions of 25°C, 1 M concentrations for all solutions, and 1 atm pressure for all gases. The E°_{cell} is calculated from the table of reduction potentials, as shown next.

Standard Electrode Reduction Potentials in Aqueous Solution at 25°C

Reduction Half-Reaction	**Standard Potential (V)**
$Li^+ (aq) + e^- \rightarrow Li (s)$	−3.04
$Ba^+ (aq) + e^- \rightarrow Ba (s)$	−2.71
$Mg^{+2} (aq) + 2e^- \rightarrow Mg (s)$	−2.38
$Al^{+3} (aq) + 3e^- \rightarrow Al (s)$	−1.66
$Zn^{+2} (aq) + 2e^- \rightarrow Zn (s)$	−0.76
$Cr^{+3} (aq) + 3e^- \rightarrow Cr (s)$	−0.74
$Fe^{+2} (aq) + 2e^- \rightarrow Fe (s)$	−0.41
$Cd^{+2} (aq) + 2e^- \rightarrow Cd (s)$	−0.40
$Ni^{+2} (aq) + 2e^- \rightarrow Ni (s)$	−0.23
$Sn^{+2} (aq) + 2e^- \rightarrow Sn (s)$	−0.14
$Pb^{+2} (aq) + 2e^- \rightarrow Pb (s)$	−0.13
$Fe^{+3} (aq) + 3e^- \rightarrow Fe (s)$	−0.04
$2 H^+ (aq) + 2e^- \rightarrow H_2 (g)$	0.00
$Sn^{+4} (aq) + 2e^- \rightarrow Sn^{+2} (aq)$	0.15
$Cu^{+2} (aq) + e^- \rightarrow Cu^+ (aq)$	0.16
$Cu^{+2} (aq) + 2e^- \rightarrow Cu (s)$	0.34
$I_2 (s) + 2e^- \rightarrow 2I^- (aq)$	0.54
$Fe^{+3} (aq) + e^- \rightarrow Fe^{+2} (aq)$	0.77
$Ag^+ (aq) + e^- \rightarrow Ag (s)$	0.80

Standard Electrode Reduction Potentials in Aqueous Solution at 25°C (*Cont.*)	
Reduction Half-Reaction	**Standard Potential (V)**
Br_2 (l) + $2e^-$ → $2Br^-$ (aq)	1.07
O_2 (g) + $4H^+$ (aq) + $4e^-$ → $2H_2O$ (l)	1.23
$Cr_2O_7^{-2}$ (aq) + $14H^+$ (aq) + $6e^-$ → $2Cr^{+3}$ (aq) + $7H_2O$ (l)	1.33
Cl_2 (g) + $2e^-$ → $2Cl^-$ (aq)	1.36
MnO_4^- (aq) + $8H^+$ (aq) + $5e^-$ → Mn^{+2} (aq) + $4H_2O$ (l)	1.49
H_2O_2 (aq) + $2H^+$ (aq) + $2e^-$ → $2H_2O$ (l)	1.78
$S_2O_8^{-2}$ (aq) + $2e^-$ → $2SO_4^{-2}$ (aq)	2.01
F_2 (g) + $2e^-$ → $2F^-$ (aq)	2.87

Table of Reduction Potentials

The table of reduction potentials is used as follows:

- This is a table of **reductions**.
- For the **oxidation** half-reaction (it is **above** the reduction half-reaction in the table), the reaction is reversed and the sign of the value is reversed.
- For a spontaneous reaction, the anode reaction is **above** the cathode reaction.
- The anode reaction is the **reverse** of the reduction reaction.
- **Never change the values** except for the sign.
- Add the two values to give E°_{cell}.
- The weakest reduction reactions are at the top of the table. (They have the most negative potential.)
- The strongest reduction reactions are at the bottom of the table. (They have the most positive potentials.)
- The E°_{cell} must always be positive for a spontaneous reaction that produces a voltage.

EXAMPLE: Determine E°_{cell} for the following cell:

| Al | Al^{+3} | | I_2 | I^- | Pt

- The anode is an oxidation. The table of reduction potentials gives the value −1.66 V for the reduction half-reaction of Al^{+3} to Al. Therefore, the value for the oxidation half-reaction is +1.66 V.
- The cathode is the reduction. The table of reduction potentials gives the value +0.54 V for this half-reaction.
- Adding the two values gives +2.2 V for the E°_{cell}. It is positive, as it must be for a spontaneous reaction.

The Standard Cell Potential and the Standard Free Energy Change

The ΔG° can be calculated from E°_{cell} using the following equation:

$$\Delta G^\circ = -nFE^\circ_{\text{cell}}$$

For a spontaneous reaction, E°_{cell} is positive, and ΔG° is negative.

EXAMPLE: Calculate the ΔG° for the cell given previously.

- Al to Al^{+3} is a $3e^-$ change; I_2 to I^- is a $2e^-$ change; the $n = 6$
- Plug in the values and calculate ΔG°:

$$\Delta G^\circ = -(6)(9.65 \times 10^4 \text{ J/V})(2.2 \text{ V}) = -1.3 \times 10^6 \text{ J}$$

The Equilibrium Constant and the Standard Cell Potential

The equilibrium constant for a reaction can be calculated from E°_{cell} using the equation

$$E^\circ_{\text{cell}} = (0.0257/n) \ln K$$

EXAMPLE: For the precious cell, with $n = 6$ and $E^\circ_{\text{cell}} = 2.19$ V, the equilibrium constant is

$$2.2 = (0.0257/6) \ln K$$

$$513 = \ln K$$

$$K = e^{513} = \text{too large to calculate}$$

Nonstandard Cell Potentials

The **Nernst equation** is used to calculate the nonstandard cell potential; this occurs when the concentrations are other than 1 M. The Nerst equation is given as

$$E_{\text{cell}} = E^\circ_{\text{cell}} - (0.0257/n) \ln Q$$

The equation requires the following:

- A balanced equation
- The Q expression and the value of Q
- The n value
- The value of E°_{cell}
- The values of E°_{cell}, n, and Q to be plugged in to calculate the nonstandard E_{cell}

EXAMPLE: Calculate the nonstandard cell potential for the cell

$Al \mid Al^{+3}$ (0.150 M) $\mid\mid Zn^{+2}$ (0.075 M) $\mid Zn$

$3e^-$ $2e^-$

- Determine the n value is 6.
- Write the balanced equation

$$2Al\,(s) + 3Zn^{+2}\,(aq) \rightarrow 2Al^{+3}\,(aq) + 3Zn\,(s)$$

- Write the Q expression and calculate Q

$$Q = \frac{[Al^{+3}]^2}{[Zn^{+2}]^3} = \frac{(0.150)^2}{(0.075)^3} = 53.3$$

- Calculate the E°_{cell}

$$\begin{array}{ll} Al \rightarrow Al^{+3} & +1.66 \\ Zn^{+2} \rightarrow Zn & \underline{-0.76} \\ & +0.90 \end{array}$$

- Plug into the Nernst equation.

$$E_{cell} = 0.90 - \{(0.0257/6)\ln 53.3\}$$

$$E_{cell} = 0.88\ V$$

ELECTROLYTIC CELLS

Electrolytic cells are constructed of nonspontaneous redox reactions that require electricity to make them run. To calculate the current (in amperes) required to deposit a certain mass of metal in an electroplating experiment, you must have the following:

- The Faraday constant, F
- n
- The time in seconds
- The molar mass of the metal

EXAMPLE: Calculate the current needed to deposit 365 milligrams (mg) silver in 216 minutes (min) from aqueous silver ion.

- From $Ag^{+1} \rightarrow Ag$ is a $1e^-$ change
- Start with the gram amount, and convert it by steps into amperes
- $A = C/s$

$$(0.365\text{ g Ag})\left(\frac{1\text{ mole Ag}}{107.9\text{ g Ag}}\right)\left(\frac{1\text{ mole }e^-}{1\text{ mole Ag}}\right)\left(\frac{9.65 \times 10^4\text{ C}}{\text{mole }e^-}\right)$$

$$\times\left(\frac{1}{1.30 \times 10^4\text{ s}}\right) = 0.0251\text{ A}$$

EXAMPLE: Calculate the mass of iodine formed when 8.52 milliamperes (mA) flows through a cell containing I^- for 10.0 min.

➤ From I^- to I_2 is a total of $2e^-$ change.

➤ Start with the amp and convert it by steps into g I_2.

➤ A × s = C.

$$(8.52 \times 10^{-3}\ \text{A})\ (600\ \text{s}) \left(\frac{1\ \text{mole e}^-}{9.65 \times 10^4\ \text{C}}\right) \left(\frac{1\ \text{mole I}_2}{2\ \text{mole } e^-}\right) \left(\frac{254\ \text{g I}_2}{\text{mole I}_2}\right)$$

$$= 6.73 \times 10^{-3}\ \text{g I}_2$$

CRAM SESSION

Electrochemistry

1. **Electrochemistry** is the use of spontaneous oxidation–reduction reactions to generate electricity or the use of electricity to run a nonspontaneous redox reaction.
2. To **balance a redox equation**, the electrons must be balanced as well as the atoms.
   A. The equation is divided into two half-reactions, one the oxidation half and the other the reduction half of the equation. Then each half-reaction is balanced. The main atoms are balanced first. The oxygens are balanced using H_2O. Finally, the hydrogens are balanced using H^+.
   B. The electron change for each half-reaction is determined, and both half-reactions must be multiplied through so that they both involve the same electron change, *n*. This balances the electrons.
   C. Then the two half-reactions are added together, canceling where appropriate.
3. A **voltaic cell** involves an electrochemical redox reaction, divided into two half-cells, to generate a cell voltage called the **cell potential**. One of the half-cells is called the **anode**. This is the half-cell where oxidation is taking place. The other half-cell, the **cathode**, is where the reduction is taking place.
   A. The anode electrode wears away as the reaction proceeds; a metal electrode strip wears away as metal atoms become metal ions in solution.
   B. The cathode electrode builds up as the metal ions in the solution become solid metal atoms that plate out onto the electrode surface.
   C. The half-cells are connected by a **salt bridge**, which allows migration of anions toward the anode and cations toward the cathode.
   D. The two electrodes are connected by a wire, with a voltmeter between them.
4. The maximum work that a cell can generate depends on the *n* value, the Faraday constant *F*, and the cell potential E_{cell}. The equation is

$$w_{max} = -nFE_{cell}$$

5. The **standard cell potential** for a redox reaction is calculated using the table of reduction potentials.
   A. The anode reaction is the reverse of the reaction found in the table, and its value's sign must be reversed as well.
   B. The cathode reaction is exactly the same as that found in the table, and its value is then added to the value for the anode reaction. This sum is the standard cell potential for the equation, E°_{cell}.

6. Standard conditions are 1 atm pressure for all gases and 1 M concentrations for all solutions. Using E°_{cell}, the standard free energy change for the reaction can be calculated using the equation

$$\Delta G^\circ = -nFE^\circ_{cell}$$

The equilibrium constant for the reaction can also be calculated from E°_{cell}, using the equation

$$E^\circ_{cell} = (0.0257/n) \ln K.$$

7. The **Nernst equation** is used to determine the nonstandard cell potential, E_{cell}. Nonstandard conditions usually consist of concentrations other than 1 M. Instead of K, the reaction quotient Q is calculated. The nonstandard cell potential is calculated using the equation

$$E_{cell} = E^\circ_{cell} - [(0.0257/n) \ln Q].$$

8. **Electrolytic cells** require electricity to run a nonspontaneous redox reaction. **Electroplating** is an example of the use of an electrolytic cell. The mass of a metal required can be converted into amperage using the molar mass of the metal, the electron change per mole of the metal, the Faraday constant, and the time in seconds.

Glossary of General Chemistry Terms

Acid A compound that produces hydronium ion (simplified as H^+) in water solution. (Chapter 8)

Acid–base reaction A reaction in which an acid reacts with a base and forms an ionic compound and water. (Chapter 8)

Acidic salt An ionic compound that is the product of a strong acid and a weak base and has an acidic pH in water solution. (Chapter 13)

Acidity The measure of the concentration of hydronium ion in a dilute aqueous solution. (Chapter 13)

Alkali metals The metals of group IA. (Chapter 3)

Alkaline earth metals The metals of group IIA. (Chapter 3)

Anion A negatively charged ion. (Chapter 1)

Anode The electrode in an electrochemical cell where oxidation takes place. (Chapter 15)

Angular momentum quantum number (*l*) The quantum number that describes the three-dimensional shape of an orbital; it is all integer values from 0 up to $n - 1$. (Chapter 2)

Arrhenius equation An equation that allows the calculation of the activation energy of a reaction given the rate constants at two different temperatures. (Chapter 10)

Atom The smallest particle of a substance, made up of a nucleus surrounded by an electron cloud. (Chapter 1)

Atomic number (*Z*) The number assigned to an atom on the periodic table that is equivalent to the number of protons in the nucleus of that atom. (Chapter 1)

Atomic radius The radius of an atom, found by dividing by 2 the distance between two nuclei in a diatomic molecule. (Chapter 3)

Atmospheric pressure (atm) The force exerted by the gases of the atmosphere onto the surface of the Earth due to gravity. (Chapter 5)

Autoionization The dissociation that water undergoes into hydronium ions and hydroxide ions, with K_W equal to 1.0×10^{-14} at 25°C. (Chapter 13)

Avogadro's law States: "A mole of any gas occupies the same volume at a given temperature and pressure." (Chapter 5)

Balanced chemical equation Chemical equation that has the same number of each type of atom on both sides of the equation. (Chapter 7)

Balmer series The series of lines seen in the visible part of the emission spectrum of the hydrogen atom. It consists of four lines that arise from the 6 to 2, the 5 to 2, the 4 to 2, and the 3 to 2 transitions. (Chapter 2)

Barometer Instrument used to measure atmospheric pressure. (Chapter 5)

Base Compound that produces hydroxide ion in aqueous solution. (Chapter 8)

Basic salt Ionic compound that is the product of a strong base and a weak acid, and has a basic pH in aqueous solution. (Chapter 13)

Boiling point elevation A colligative property of solutions that increases the boiling point of a solvent due to the presence of solute particles. (Chapter 6)

Boyle's law States: "The pressure of a gas is inversely related to its volume at constant temperature." (Chapter 5)

Buffer A solution that resists change in pH on small additions of acid or base. (Chapter 13)

Calorie A unit of heat energy that is equivalent to 4.184 J. (Chapter 9)

Calorimetry A procedure that uses a calorimeter to measure the heat changes that occur within a system. (Chapter 9)

Catalyst A substance that speeds a reaction rate by lowering the activation energy of the reaction; it is recovered at the end of the reaction. (Chapter 10)

Cathode The electrode in an electrochemical cell where reduction takes place. (Chapter 15)

Cation A positively charged ion. (Chapter 1)

Cell notation A shorthand method for drawing a voltaic cell. (Chapter 15)

Cell voltage (E_{cell}) The voltage generated by an electrochemical cell; it can also be called **electromotive force (emf)** or **cell potential**. (Chapter 15)

Charles' law States: "The volume of a gas is directly related to temperature at constant pressure." (Chapter 5)

Chemical formula The representation of the fundamental unit of a substance that uses symbols and subscripts to state the elements present and their molar ratio. (Chapter 1)

Coefficient The number found in front of a formula in a balanced chemical equation that represents the mole quantity of that substance in the balanced equation. (Chapter 7)

Colligative properties Properties of solutions that result in a change in the boiling point, the melting point, and the vapor pressure of a solvent due to the presence of solute particles. (Chapter 6)

Colloid Consists of small particles suspended in a medium. (Chapter 6)

Combination reaction An oxidation–reduction reaction where two substances combine into one substance. (Chapter 8)

Combustion reaction An oxidation–reduction reaction that occurs when a substance is burned in oxygen and produces one or more oxides. (Chapter 8)

Common ion effect The effect on the solubility of an ionic compound due to the presence of one of its ions in the solution. (Chapter 12)

Covalent compound A compound that is made of atoms held together by covalent bonds; the formula has nonmetals as its elements. (Chapter 1)

Critical point The point on a pressure–temperature phase diagram that marks the end of the liquid–gas boundary line; at all temperatures and pressures beyond this point, the substance exists in the supercritical fluid phase. (Chapter 6)

Dalton's law States: "In a mixture of gases, each gas exerts it own partial pressure that is directly related to the mole fraction of the gas in the mixture." (Chapter 5)

Decomposition reaction An oxidation–reduction reaction where a single substance decomposes into two or more substances. (Chapter 8)

Dipole A covalent bond that has a partially negative end at the more electronegative atom, and a partially positive end at the less electronegative atom. (Chapter 6)

Dipole–dipole interactions The intermolecular force found in liquid or solid samples of covalent molecules that contain a polar bond. (Chapter 6)

Dipole moment A number that is the measure of the polarity of a covalent bond. (Chapter 4)

Dispersion forces The intermolecular force found in liquid or solid samples of covalent molecules that contain nonpolar bonds. (Chapter 6)

Effective nuclear charge The number of protons in the nucleus less the shielding effect of the innermost electrons. (Chapter 3)

Electrochemistry The study of oxidation–reduction reactions that can be used either to generate electricity or that require electricity to proceed. (Chapter 15)

Electrolytic cell An electrochemical cell that requires electricity to run a nonspontaneous oxidation–reduction reaction. (Chapter 15)

Electron The negatively charged particle with negligible mass that is found in a cloud surrounding the nucleus of an atom. (Chapter 1)

Electron affinity The energy change that occurs when an electron is added to a valence orbital, producing an anion. (Chapter 3)

Electron configuration Lists the orbitals in order of their energies and gives the number of electrons in each orbital as a superscript. (Chapter 2)

Electronegativity The ability of an atom in a molecule to pull the electron density of a covalent bond toward itself. (Chapter 3)

Empirical formula The chemical formula of a substance that shows the lowest whole-number ratio of the elements in the substance. (Chapter 1)

Endothermic reaction A reaction that requires heat to proceed; has a positive ΔH_{rxn}. (Chapter 9)

Enthalpy (*H*) The state function that is used to denote heat changes in a chemical process. (Chapters 9, 14)

Enthalpy of formation (ΔH_f°) The heat that is absorbed or given off when a substance is produced from its elements at standard temperature and pressure. (Chapter 9)

Enthalpy of reaction (ΔH_{rxn}°) The heat that is given off or absorbed by a reaction at standard temperature and pressure. (Chapter 9)

Entropy (*S*) The state function that is a measure of the disorder of a system. (Chapter 14)

Equilibrium The point in a reversible reaction where the rate of the forward reaction equals the rate of the reverse reaction. (Chapter 11)

Equilibrium constant (*K*) Describes the extent to which a reversible reaction proceeds before reaching the equilibrium point. (Chapter 11)

Equilibrium expression Expression that relates the concentrations of the products and starting materials to the equilibrium constant. (Chapter 11)

Exact mass The mass of 1 mole of a pure isotope of an element. (Chapter 1)

Excited state A high-energy state attained by an electron on absorbing heat energy. (Chapter 2)

Exothermic reaction A reaction that produces heat; has a negative ΔH°_{rxn}. (Chapter 9)

Factor analysis Process used to convert from one unit system to another by using conversion factors set up so as to cancel out the original unit. (Chapter 1)

First law of thermodynamics states that the change in the internal energy of a system equals the sum of the heat and the work energies of the system. (Chapter 14)

Formal charge The charge on an atom in a Lewis structure that occurs when the number of electrons brought into the molecule by that atom differs from the number of electrons that the atom has in the structure. (Chapter 4)

Free energy (*G*) The measure of the energy of a reaction that is available to do work; the sign of the change in free energy (ΔG) is used to determine whether the reaction is spontaneous. (Chapter 14)

Freezing point depression A colligative property of solutions that decreases the melting point of a solvent due to the presence of solute particles. (Chapter 6)

Gas constant (*R*) Has the value 0.08206 L atm/mole K when used with gases; has the value 8.314 J/mole K when used with thermodynamic functions. (Chapter 5)

Group The elements in the same column of the periodic table. (Chapter 3)

Half-life ($t_{1/2}$) The time it takes for 50% of the reactant to convert to product. (Chapter 10)

Half-reaction Either the oxidation half or the reduction half of an oxidation–reduction reaction. (Chapter 15)

Halogens The elements in group VIIIA. (Chapter 3)

Heat Energy that flows into or out of a system due to a difference in temperature between the system and its surroundings when they are in thermal contact. (Chapter 9)

Heat capacity (*C*) The heat required to raise the temperature of a substance by 1°Celsius; has the units J/°C. (Chapter 9)

Heat of fusion (ΔH_{fus}) The heat required to convert 1 mole of substance in the solid state to the liquid state at the melting point of the substance. (Chapter 9)

Heat of vaporization (ΔH_{vap}) The heat required to convert 1 mole of substance in the liquid state to the gaseous state at the boiling point of the substance. (Chapter 9)

Henry's law States: "The amount of gas that dissolves in a solvent is directly proportional to the pressure of that gas above the solution." (Chapter 6)

Hess' law States: "The sum of the ΔHs for individual reactions equal the ΔH_{rxn} of the sum reaction." (Chapter 9)

Hund's rule When filling an orbital, each orientation gets 1 electron before any are paired up in order to maximize the number of parallel spins. (Chapter 2)

Hybridized orbitals Orbitals that are hybrids of atomic orbitals of the same shell. (Chapter 4)

Hydrogen-bonding The intermolecular force that is found in liquid or solid samples of covalent compounds that contain an O—H, F—H, or an N—H bond. (Chapter 6)

Ideal gas A gas that exhibits no intermolecular interactions. (Chapter 5)

Intermolecular interactions Attractive forces between molecules in the liquid or solid state. (Chapter 6)

Ion A species that has a net charge. (Chapter 1)

Ionic compound A compound that is made of cations and anions so that the net charge is zero. (Chapter 1)

Ionic radius The radius of an ion. (Chapter 3)

Ionization energy The energy required to remove an electron from the valence shell, producing a cation. (Chapter 3)

Isotope Different forms of an element that have different numbers of neutrons, but equivalent protons, in the nucleus. (Chapter 1)

Joule (J) The SI unit of heat energy that is equivalent of kg m^2/s^2. (Chapter 9)

Kinetic energy Energy of motion. (Chapter 5)

Kinetic molecular theory of gases A model for gas behavior that consists of five assumptions that explain the behavior of ideal gases. (Chapter 5)

Kinetics The study of reaction rates. (Chapter 10)

K_w The ion product constant for water; equivalent to 1.0×10^{-14} at 25°C. (Chapter 13)

Lanthanides Elements that have the 4*f* orbital as the valence orbital; also called the **rare earth metals**. (Chapter 3)

Lattice network The orderly array of ions in a solid compound, characterized by alternating positions of positive and negative charge. (Chapter 3)

Le Chatelier's principle States: "If a system that is at equilibrium is subjected to a change in conditions, the equilibrium will shift to counteract the change." (Chapter 11)

Lewis structure A model that shows the positions of all valence electrons in a molecule, either as bonds or as lone pairs. (Chapter 4)

Limiting reagent The reactant in a reaction that is completely used up, and therefore is the compound that is used to calculate the theoretical yield of product formed. (Chapter 7)

Linear The geometry of the orbitals of an *sp*-hybridized atom where all bond angles are 180°. (Chapter 4)

Lone pair A pair of electrons that occupies a hybridized orbital. (Chapter 4)

Magnetic quantum number (m_l) The quantum number that describes the orientations allowed for an orbital; it has all integer values from $-l$ to $+l$. (Chapter 2)

Main group elements The elements of groups IA–VIIIA that have either an *s* orbital or a *p* orbital as valence orbital; also called **representative elements**. (Chapter 3)

Manometer Instrument used to measure the pressure of gases inside a container. (Chapter 5)

Metal Element found to the left of the stairline on the periodic table; metals always lose electrons to form cations in compounds. (Chapter 3)

Metalloid Element found along the stairline of the periodic table that show characteristics of both metals and nonmetals. (Chapter 3)

Molality (m) A concentration unit that is equivalent to mole solute per kg of solvent. (Chapters 6, 8)

Molarity (M) A concentration unit that is equivalent to mole solute per liter of solution. (Chapter 8)

Molar mass The weighted average of the masses of all the known isotopes of an element; it is the mass of 1 mole of a natural sample of the element. (Chapter 1)

Mole Number of particles equivalent to 6.022×10^{23}. (Chapter 1)

Molecular formula The actual chemical formula of the fundamental unit of a substance. (Chapter 1)

Mole fraction (X) Calculated by dividing the mole amount of one substance in a mixture by the total moles. (Chapter 8)

Natural abundance The percentage of an isotope in a natural sample of element. (Chapter 1)

Nernst equation Used to calculate the nonstandard cell potential when concentrations of solutions are other than 1 M. (Chapter 15)

Net ionic equation A balanced equation that includes only the ions that are reacting and the insoluble product that is formed in a precipitation reaction. (Chapter 8)

Neutron A neutral particle with mass that is found in the nucleus of an atom. (Chapter 1)

Nomenclature The system of rules used to name ionic and covalent compounds. (Chapter 1)

Nonmetals The elements found to the right of the stairline in the periodic table, plus hydrogen. (Chapter 3)

Nucleus The core of an atom that consists of the protons and neutrons, and is the location of the mass of the atom. (Chapter 1)

Octahedral The geometry of the orbitals of an sp^3d^2 hybridized atom, where all the bond angles are 90°. (Chapter 4)

Orbital A three-dimensional area about the nucleus of an atom where the probability of finding an electron is greater than 95%. (Chapter 2)

Orbital diagram Pictures the orbitals as sets of lines, one for each orientation of the orbital, and shows the electrons as arrows. (Chapter 2)

Order of reactant The exponent of the reactant's concentration in the rate law of a reaction; the most common orders are zero-order, first-order, and second-order. (Chapter 10)

Osmotic pressure The pressure that water exerts across a semipermeable membrane due to a difference in the concentrations on either side of the membrane. (Chapter 6)

Oxidation Loss of electrons. (Chapter 15)

Oxidation number A value assigned to an atom or ion that is used to track the gain or loss of electrons in an oxidation–reduction reaction. (Chapter 8)

Oxidation–reduction reaction Reaction where electrons are transferred from one species to another. (Chapters 8, 15)

Oxygen group The elements of group VIA. (Chapter 3)

Pauli exclusion principle States: "No two electrons in an atom can have the same set of quantum numbers." (Chapter 2)

Percent composition by mass The percent by mass of each element in a compound; it is determined by dividing the mass of each element by the total mass of the compound. (Chapter 1)

Period The elements in the same row of the periodic table. (Chapter 3)

pH scale A logarithmic scale from 0 to 14 of values that are equivalent to the negative log of the hydronium ion concentration; less than 7 is acidic, 7 is neutral, and greater than 7 is basic. (Chapter 13)

Phase change A change from one state of matter to another; it occurs at constant temperature. (Chapter 6)

Phase diagram A plot of pressure versus temperature that shows the different phases for a substance. (Chapter 6)

Photon The smallest particle of light possible; it has energy equivalent to $h\nu$. (Chapter 2)

Pi bond (π) A covalent bond between two atoms that is in addition to the sigma bond; is made from overlapping *p* orbitals. (Chapter 4)

Polyprotic acid An acid that has more than one acidic proton. (Chapter 13)

Precipitate The insoluble solid product of a reaction. (Chapter 8)

Precipitation reaction Occurs when two solutions are mixed together, and the cation from one solution reacts with the anion from the other solution to form an insoluble product. (Chapter 8)

Pressure (*P*) Force exerted by gases due to collisions with the walls of the container. (Chapter 5)

Principle quantum number (*n*) The quantum number that describes the shell; can be any integer from 1 to 7. (Chapter 2)

Proton A positively charged particle that has mass and is found in the nucleus of an atom. (Chapter 1)

Quantum numbers The variables found in each wave function of the Schrodinger equation; they include n, l, m_l, and m_s. (Chapter 2)

Raoult's law Allows the calculation of the vapor pressure lowering. (Chapter 6)

Rate constant (*k*) Temperature-dependent constant for a reaction. (Chapter 10)

Rate law States the relationship of the rate of a reaction to the rate constant and to the concentrations of the reactants and their orders. (Chapter 10)

Reaction rate How fast a reaction proceeds; it is fastest at the start of a reaction (the initial rate), and can be calculated as the change of concentration of the starting material in a given time period. (Chapter 10)

Reaction quotient (*Q*) A value obtained by using nonequilibrium concentrations or pressures in the equilibrium expression for a reaction. (Chapter 11)

Real gases Gases that exhibit intermolecular forces at high pressure and/or small volume. (Chapter 5)

Reduction The gain of electrons. (Chapter 15)

Resonance structures Different Lewis structures for the same compound; differ only in the placement of valence electrons. (Chapter 4)

Salt bridge Connects the two half-cell solutions of a voltaic cell, and allows the movement of anions toward the anode and cations toward the cathode. (Chapter 15)

Schrodinger equation A wave equation that describes the wave nature of the electron of the hydrogen atom. (Chapter 2)

Second law of thermodynamics States: "The total entropy of a system and its surroundings always increases for a spontaneous process." (Chapter 14)

SI unit system Agreed-on unit system for the sciences worldwide. (Chapter 1)

Sigma bond (σ) The main bond between two atoms in a covalent compound or ion. (Chapter 4)

Single displacement reaction An oxidation–reduction reaction that occurs when an element reacts with an ionic compound; the element becomes a cation and the original cation becomes an element. (Chapter 8)

Solubility The extent to which a compound dissolves in a solvent; it is temperature-dependent; its units can be g/L or mole/L. (Chapter 12)

Solubility product constant (K_{sp}) The equilibrium constant for the dissolving of a sparingly soluble inorganic compound in water. (Chapter 12)

Solute A substance that is dissolved in a solvent. (Chapter 8)

Solution Consists of a solute dissolved in a solvent; is transparent. (Chapter 8)

Solvent Substance that dissolves a solute; is present in the greatest quantity. (Chapter 8)

Specific heat (c) The heat required to raise the temperature of 1 gram of a substance by 1° Celsius; has the units J/g°C. (Chapter 9)

Spin quantum number (m_s) The quantum number that describes the spin of an electron in one individual orientation of an orbital; has value $\pm 1/2$. (Chapter 2)

Spontaneous reaction A reaction that proceeds to the right as written. (Chapter 14)

Standard cell voltage (E°_{cell}) The voltage generated by an electrochemical cell under the standard conditions of 25°C, 1 M concentrations, and 1 atm pressure. (Chapter 15)

Stoichiometry The process that uses a balanced chemical equation to calculate quantities of substances used or produced by the reaction. (Chapter 7)

Strong acid An acid that fully dissociates in water solution. (Chapters 8, 13)

Strong base A base that fully dissociates in water solution. (Chapters 8, 13)

Tetrahedral Geometry of the orbitals of an sp^3 hybridized atom, where all the bond angles are 109.5°. (Chapter 4)

Thermochemistry The study of heat changes that occur in a chemical reaction. (Chapter 9)

Third law of thermodynamics States: "A substance that is perfectly crystalline at 0 K has an entropy of zero." (Chapter 14)

Titration The process of adding a solution of known concentration, by buret, to another solution (the titrant) of known volume, in order to perform a reaction that is complete at the endpoint, where a color change occurs due to the presence of an indicator compound; with the volume added known, the concentration of the titrant solution can be calculated. (Chapter 8)

Transition elements Those metals that have a *d* orbital as their valence orbital; also called the **nonrepresentative elements**. (Chapter 3)

Trigonal bipyramid The geometry of the orbitals of an sp^3d hybridized atom, where the bond angles of the three atoms in the plane are 120°, and those of the two atoms that bisect the plane are 90° from the plane. (Chapter 4)

Trigonal planar The geometry of the orbitals of an sp^2 hybridized atom, where all bond angles are 120°. (Chapter 4)

Triple point (*T*) The point on a pressure–temperature phase diagram where three phases coexist in equilibrium. (Chapter 6)

Valence shell electron pair repulsion (VSEPR) theory Predicts the shape of molecules by assuming that areas of electron density will get as far from each other as possible. (Chapter 4)

Van der Waals equation Contains correction factors for volume and pressure to better approximate the behavior of real gases. (Chapter 5)

Vapor pressure Property of a liquid that results in an equilibrium being established between the liquid state and the vapor state when the liquid is in a closed container. (Chapter 6)

Vapor pressure lowering A colligative property of solutions that lowers the vapor pressure of a solvent due to the presence of solute particles. (Chapter 6)

Voltaic cell Uses a spontaneous oxidation–reduction reaction, divided into two half-cells, to generate electricity. (Chapter 15)

Weak acid An acid that dissociates only to a small extent in water solution. (Chapters 8, 13)

Weak base A base that produces only a very small amount of hydroxide ion in water solution. (Chapters 8, 13)

ON YOUR OWN

MCAT General Chemistry Practice

This section contains fifteen quizzes, one for each of the fifteen chapters of the general chemistry review section that you have just completed. Use these quizzes to test your mastery of the concepts and principles covered in the general chemistry review. Detailed solutions for every question are provided following the last quiz. If you need more help, go back and reread the corresponding review chapter.

CHAPTER 1 ATOMS AND MOLECULES Quiz

1. How many protons, neutrons, and electrons are in the $^{52}Cr^{+3}$ ion?

____

2. There are two isotopes of bromine, ^{79}Br and ^{81}Br. The natural abundance of each is approximately 50%. What is the approximate molar mass of a normal sample of bromine?

____

3. Fill in the following chart:

Symbol (with mass #)	Protons (#)	Neutrons (#)	Electrons (#)	Charge
A. $^{89}Y^{+3}$	____	____	____	+3
B. ____	28	30	____	+2
C. ____	47	62	46	____
D. ^{51}V	____	____	18	____

4. Calculate the number of atoms in a 28.1-g sample of carbon.

5. Calculate the molar mass of $Ca_3(PO_4)_2$.

6. Convert 87.2 ft^3 to m^3.

7. Convert 32.5 m/s to mi/h.

8. Convert 18.4 g gold to mole.

9. State whether each of the following is covalent or ionic:

A. NaH ______________________

B. HCN ______________________

C. N_2O ______________________

D. AlN ______________________

10. Write the formula of each of the following compounds

A. potassium permanganate ______________________

B. sodium sulfide ______________________

C. osmium(VIII) oxide ______________________

D. manganese(III) sulfate ______________________

E. copper(I) carbonate ______________________

F. calcium nitride ______________________

11. Name each of the following:

A. Na_2SO_3 ______________________

B. $Fe(IO_3)_2$ ______________________

C. $Al(NO_3)_3$ ______________________

D. V_2O_3 ______________________

E. CrS_3 ______________________

12. Calculate the average mass of 1 copper atom.

13. Calculate the mass percent of nitrogen in each of the following compounds:

A. ammonium nitrate

B. ammonia (NH_3)

C. ammonium sulfate

CHAPTER 2 ELECTRONIC STRUCTURE AND THE PERIODIC TABLE Quiz

1. How many orbitals are in the 2-shell?

2. Calculate the number of electrons needed to fill the 3-shell.

3. To what transition does the blue line of the hydrogen atom refer?

4. Name all the possible quantum number values for an electron in each of the following orbitals:

   A. 4*p*

   B. 5*f*

   C. 7*s*

5. Calculate the energy of the 6 → 2 transition of the hydrogen electron.

6. Calculate the wavelength of the 6 → 2 transition of the hydrogen electron.

7. The quantum number *l*

   A. can be all integers from 0 up to *n*

   B. describes the shape of the orbital

   C. is called the principal quantum number

   D. describes all the possible orientations of an orbital

8. The quantum number m_l

A. can be all integers from 0 up to n

B. describes the shape of the orbital

C. is called the principal quantum number

D. describes all the possible orientations of an orbital

9. Write the complete electron configuration for

A. copper ______

B. copper (II) ion ______

C. calcium ______

D. ruthenium (III) ion ______

E. sulfur ______

F. selenide ______

10. State all the orbitals in the 4-shell.

______

11. Place the following orbitals in order of energy, with the lowest energy orbital first:

$2p$ $3d$ $3s$ $4s$ $3p$ $4d$

______

12. When an atom absorbs energy, an electron in an outer orbital can undergo transition to a higher energy orbital. Which of the following electron configurations represents an excited state for copper?

A. $1s^2 2s^2 2p^6 3s^2 3p^6 4s^1 3d^{10}$

B. $1s^2 2s^2 2p^6 3s^2 3p^5 4s^2 3d^{10}$

13. Calculate the number of photons in a beam of laser light that has wavelength 510 nm and 1.1 J total energy.

14. Calculate the energy of 1 photon of blue light of wavelength 409 nm.

15. What is the Pauli exclusion principle, and to which quantum number does it refer?

16. If the quantum numbers for an electron are $n = 4$, $l = 0$, $m_l = 0$, and $m_s = 1/2$, what orbital is this electron in?

CHAPTER 3 TRENDS IN THE PERIODIC TABLE

Quiz

1. What characteristic of the representative elements within a group most closely determines their chemical properties?

    ____

2. Elements in group IIA usually make ions with what charge?

    A. +1

    B. +2

    C. −1

    D. −2

3. Group VIIA elements are called the

    A. halogens

    B. halides

    C. noble gases

    D. alkali metals

4. What is the major characteristic of the noble gases?

    ____

    ____

5. What characteristic of the transition metals most closely defines their chemical properties?

    ____

    ____

6. On the periodic table, how do you determine whether an element is a metal or a nonmetal?

    ____

    ____

7. Name four characteristics of metals.

    ____

    ____

    ____

8. Name three properties of nonmetals.

____

____

____

For question 9, circle the letter of your choice.

9. Elements along the stairline can have the property of being a

A. conductor

B. semiconductor

C. cation

D. anion

10. What is the name of the energy required to remove an electron from the valence orbital of an element?

____

11. Which has a lower ionization energy, sodium or magnesium?

____

12. Which has a lower ionization energy, aluminum or gallium?

____

For question 13, circle the letter of your choice.

13. Ionization energy is lowest when a valence *p* orbital is

A. full

B. partially full

14. State the definition of electron affinity.

____

____

15. Which representative group has the highest electron affinity, and why?

____

____

For question 16, circle the letter of your choice.

16. Electronegativity is a property most commonly used with

A. ions

B. elements

C. covalently bonded atoms

D. transition metals

17. Which is the larger atom, zinc or bromine?

____

18. Which is the smaller atom, oxygen or selenium?

____

19. Which is larger, Ni or Ni^{+2}?

____

CHAPTER 4 LEWIS DOT STRUCTURES, HYBRIDIZATION, AND VSEPR THEORY Quiz

1. Write a proper Lewis structure, including all the valence electrons and any formal charges, for each of the following:

A. O_3

B. N_3^-

C. I_3^-

D. SO_2

E. SO_4^{-2}

F. PCl_5

G. XeF_4

H. OF_2

2. State the hybridization of the central atom in SO_2.

____

____

3. What is the molecular geometry of the SO_2 molecule?

____

____

4. What is the hybridization of the central atom in the I_3^- ion?

5. What is the molecular geometry of the central atom in the I_3^- ion?

6. What is the hybridization of the central atom in the ion BrF_4^+?

7. What is the molecular geometry of the BrF_4^+ ion?

8. What is the hybridization of the central atom in XeF_4?

9. What is the molecular geometry of XeF_4?

10. State the electronic geometry of the central atom in SO_4^{-2}.

11. Determine the overall direction of the dipole moment for the PCl_3 molecule.

CHAPTER 5 GASES

Quiz

1. A 4.75-L balloon is at 18.4°C and 1.35 atm. Calculate its volume at 0.0°C and 0.485 atm.

2. Calculate the molar mass of a gas that has a density of 3.15 g/L at 685 mm Hg and 22.5°C.

3. A balloon contains 18.7 g sulfur dioxide, 3.95 g hydrogen chloride gas, and 5.85 g nitrogen at a total pressure of 1.48 atm. Calculate the partial pressure of each gas in the mixture.

4. Explain how the assumptions of the molecular theory of gases relate to Boyle's law.

____

____

____

5. Calculate the kinetic energy of one molecule of xenon gas at 25.0°C.

6. Calculate the real pressure of 55.0 g nitrogen gas in a 2.000-L container at 19.6°C. For nitrogen, $a = 1.39$ L^2 atm/mol^2 and $b = 0.0391$ L/mol.

CHAPTER 6 INTERMOLECULAR FORCES AND PHASE EQUILIBRIA Quiz

1. State the major type of intermolecular force found in

A. liquid bromine ______________________________

B. solid phenol (C_6H_5OH) ______________________________

C. solid iron (III) oxide ______________________________

D. liquid PBr_3 ______________________________

E. liquid butane C_4H_{10} ______________________________

2. A substance has the following phase diagram:

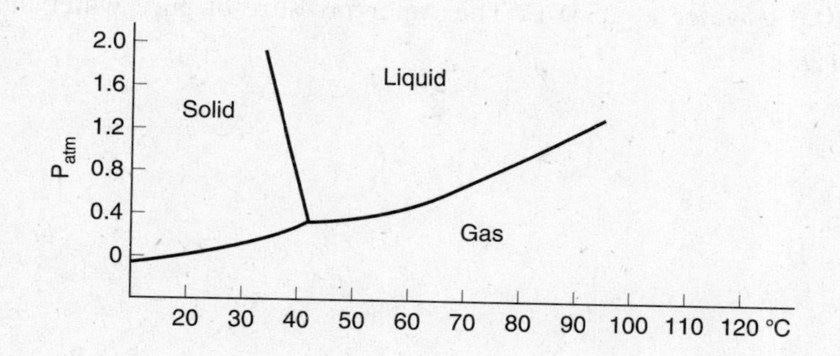

A. What is the normal boiling point of this substance?

____

B. What is the normal melting point of this substance?

____

C. State the pressure and temperature at the critical point.

____

D. Which phase is most dense?

____

E. At room temperature and pressure, in what phase is this substance?

____

F. In what phase is this substance at 2.0 atm and 110°C?

____

3. A 650.0-mL sample of CH_2Cl_2 at 10.4°C is placed in a closed container of volume 5.75 L. Calculate the mass of CH_2Cl_2 that will vaporize if the vapor pressure is 214.5 mm Hg at 10.4°C.

4. Calculate the heat of vaporization of $SiCl_4$ if it has a vapor pressure of 100.0 mm Hg at 5.4°C and a normal boiling point of 56.8°C.

5. Calculate the vapor pressure of a solution that is made by dissolving 26.8 g sucrose ($C_{12}H_{22}O_{11}$) in 320.0 g water at 25.0°C. The vapor pressure of pure water at 25.0°C is 23.8 mm Hg.

6. Calculate the boiling point and the freezing point of a solution that is made by dissolving 5.74 g of sodium hydrogen carbonate in 85.0 g water.

7. A solution of a protein that is 2.5 mg/mL has an osmotic pressure of 8.3 mm Hg at 25.0°C. Calculate the molar mass of this protein.

CHAPTER 7 CHEMICAL EQUATIONS AND STOICHIOMETRY Quiz

1. Balance the following equations:

A. $N_2O_5 \rightarrow N_2O_4 + O_2$

B. $NH_3 + CuO \rightarrow Cu + N_2 + H_2O$

C. $Al + H_2SO_4 \rightarrow Al_2(SO_4)_3 + H_2$

D. $S_8 + O_2 \rightarrow SO_2$

E. $CH_4 + Br_2 \rightarrow CBr_4 + HBr$

2. Calculate the molar mass of each of the following compounds:

A. ammonium sulfate ______________________

B. strontium chlorite ______________________

C. tin (II) fluoride ______________________

D. silver (I) carbonate ______________________

3. Perform the following conversions:

A. 8.752 g copper (II) oxide to moles

B. 97.4 mole ammonia to grams

4. A 2500.0-g sample of coal contains 1.8% FeS_2, an impurity that produces SO_2 when the coal is burned. Calculate the mass of SO_2 that is formed when this sample of coal burns, according to the (unbalanced) equation

$$FeS_2 + O_2 \rightarrow Fe_2O_3 + SO_2$$

5. Calcium phosphate reacts with sulfuric acid to form calcium sulfate and phosphoric acid. Calculate the mass of sulfuric acid needed to react completely with 150.0 g of calcium phosphate according to the (unbalanced) equation

$$Ca_3(PO_4)_2 + H_2SO_4 \rightarrow CaSO_4 + H_3PO_4$$

6. Calculate the mass of carbon dioxide that forms when 587.5 g C_3H_8 reacts with 1.25 kg O_2 according to the (unbalanced) equation

$$C_3H_8 + O_2 \rightarrow CO_2 + H_2O$$

CHAPTER 8 REACTIONS IN SOLUTION Quiz

1. Calculate the molarity of a solution that is prepared by dissolving 56.1 g of sodium phosphate to a total volume of 475.0 mL solution.

2. State whether each of the following is soluble or insoluble in water:

   A. lithium carbonate ____________________

   B. lead (II) nitrate ____________________

   C. calcium sulfide ____________________

   D. silver (I) chloride ____________________

   E. zinc (II) hydroxide ____________________

   F. sodium phosphate ____________________

3. Write the net ionic equation for the reaction that occurs when a solution of ammonium sulfide is mixed with a solution of copper (I) chloride.

4. Write the net ionic equation for the reaction that occurs when a solution of sodium phosphate is mixed with a solution of copper (II) nitrate.

5. State whether the ionic product of the following reactions is neutral, acidic, or basic:

   A. ammonium hydroxide and perchloric acid ____________________

   B. benzoic acid and sodium hydroxide ____________________

   C. hydroiodic acid and lithium hydroxide ____________________

   D. iodic acid and potassium hydroxide ____________________

6. A 25.0-mL sample of hydrobromic acid is titrated with 0.0985 M sodium hydroxide. The titration requires 14.6 mL of base to reach the endpoint. Calculate the molarity of the hydrobromic acid solution.

7. State the oxidation number of each atom in the following compounds:

   A. Mg_3N_2 ____________

   B. P_4 ____________

   C. SO_3 ____________

   D. N_2O_5 ____________

   E. CH_4 ____________

   F. $CaCl_2$ ____________

   G. $NaNO_3$ ____________

   H. NaH ____________

8. State the species that is oxidized and the species that is reduced in the following equations:

   A. $P_4 + 10F_2 \rightarrow 4PF_5$

   ____

   B. $3HNO_2 \rightarrow HNO_3 + 2NO + H_2O$

   ____

9. The following balanced equation can be used to determine the concentration of hydrogen peroxide:

$$2MnO_4^- + 5H_2O_2 + 6H^+ \rightarrow 5O_2 + 2Mn^{+2} + 8H_2O$$

   A 25.0-mL sample of H_2O_2 required 24.84 mL of 0.02311 M MnO_4^- solution to reach the endpoint. Calculate the molarity of the H_2O_2 solution.

CHAPTER 9 THERMOCHEMISTRY Quiz

1. Calculate the ΔH_f° of H_2S using bond dissociation energies.

2. Calculate the heat required to raise the temperature of 6.22 kg of copper from 21.5°C to 323.4°C, using the specific heat of copper, 0.385 J/g°C.

3. A calorimeter contains 100.0 g of water at 30.0°C. A 17.5-g sample of copper that is at −17.0°C is dropped into the water. Calculate the final temperature of the system. The specific heat of copper is 0.385 J/g°C.

4. Calculate the heat that is produced when 2.8×10^3 g NH_3 is combusted according to the following equation:

$$4NH_3\ (g) + 5O_2\ (g) \rightarrow 4NO\ (g) + 6H_2O\ (g) \quad \Delta H_{rxn} = -906\ kJ$$

5. Calculate the mass of CH_4 that is necessary to produce 3.96×10^3 kJ heat from the reaction:

$$CH_4 + 2O_2 \rightarrow CO_2 + 2H_2O \quad \Delta H_{rxn} = -890.3\ kJ$$

6. Calculate the ΔH_{rxn} for the reaction

$$C_2H_4 + H_2 \rightarrow C_2H_6$$

given the following equations:

$$C_2H_4 + 3O_2 \rightarrow 2CO_2 + 2H_2O \qquad \Delta H_1 = -1411 \text{ kJ}$$

$$2C_2H_6 + 7O_2 \rightarrow 4CO_2 + 6H_2O \qquad \Delta H_2 = -3119 \text{ kJ}$$

$$2H_2 + O_2 \rightarrow 2H_2O \qquad \Delta H_3 = -572 \text{ kJ}$$

CHAPTER 10 KINETICS
Quiz

1. Use the following data to determine the rate law and calculate the rate constant for the reaction

$A + 2B \rightarrow$ products

Run	[A]	[B]	Rate (m/s)
1	0.100	0.100	5.30×10^{-6}
2	0.200	0.100	2.20×10^{-5}
3	0.400	0.100	8.80×10^{-5}
4	0.100	0.300	1.65×10^{-5}
5	0.100	0.600	3.30×10^{-5}

2. A certain compound decomposes in a first-order manner with a rate constant of $5.1 \times 10^{-4}\ s^{-1}$ at 45°C.

A. Calculate the concentration after 3.2 min, if the initial concentration is 0.25 M.

B. Calculate how long it will take for the concentration to reach 0.12 M.

C. Calculate how long it will take for 62% of the compound to decompose.

3. The rate constant for a certain first-order reaction is $0.0346\ s^{-1}$ at 25°C. The E_a is 50.2 kJ/mole. Calculate the rate constant at 38.6°C.

4. Calculate the energy of activation for a first-order reaction that has a half-life of 438 min at 70°C and 19.8 min at 100°C.

CHAPTER 11 EQUILIBRIUM

Quiz

1. For the following reaction at 1000°C, $K = 3.80 \times 10^{-5}$. Calculate the concentration of the product at equilibrium, if the reaction is begun with 0.0456 mole I_2 in a 2.30-L flask at 1000°C.

   $I_2 \rightleftarrows 2I$

2. For the following reaction, initial quantities of 0.0655 mole NO and 0.0328 mole Br_2 in a 2.5-L flask produce an equilibrium mixture that contains 0.0389 mole NOBr. Calculate K for this reaction.

   $2NO + Br_2 \rightleftarrows 2NOBr$

3. For the following reaction, $K = 1.25 \times 10^{-4}$. Calculate the percentage of N_2O_4 dissociation when the initial quantity is 0.030 M N_2O_4.

   $N_2O_4 \rightleftarrows 2NO_2$

4. With an initial quantity of 2.00 M NOBr at 77°C, the following reaction proceeds to the extent of 9.4%. Calculate *K* for this reaction

$$2NOBr \rightleftarrows 2NO + Br_2$$

5. For the following reactions at 24°C, $K = 3.1 \times 10^{-4}$. For each set of concentrations, determine the direction the reaction is proceeding:

$$2NOBr \rightleftarrows 2NO + Br_2$$

A. 0.0610 M NOBr 0.0151 M NO 0.0108 M Br_2

B. 0.115 M NOBr 0.0169 M NO 0.0142 M Br_2

C. 0.181 M NOBr 0.0123 M NO 0.02101 M Br_2

6. An 8.0-L vessel is charged with 0.850 mole N_2 and 0.850 mole O_2. Calculate the equilibrium concentrations for the following reaction if $K = 9.22 \times 10^{-5}$.

$$N_2 + O_2 \rightleftarrows 2NO$$

7. In what direction will the equilibrium shift upon a decrease in volume for each of the following systems at equilibrium?

A. A (s) $\rightleftarrows$ 2B (s) ____________________

B. 2A (g) $\rightleftarrows$ B (l) ____________________

C. A (g) $\rightleftarrows$ B (g) ____________________

D. A (g) $\rightleftarrows$ 2B (g) ____________________

8. In what direction will the equilibrium shift upon heating each of the following systems at equilibrium?

A. A $\rightleftarrows$ 2B $\Delta H^\circ = 20.0$ kJ ____________________

B. A + B $\rightleftarrows$ C $\Delta H^\circ = -5.5$ kJ ____________________

C. A $\rightleftarrows$ B $\Delta H^\circ = 0.0$ kJ ____________________

9. In what direction will the equilibrium shift when each of the following changes is made to this reaction at equilibrium?

$$NH_4SH\ (s) \rightleftarrows NH_3\ (g) + H_2S\ (g) \quad \Delta H \text{ is positive}$$

A. Some H_2S is added. ____________________

B. Some NH_3 is removed. ____________________

C. Some NH_4SH is added. ____________________

D. The reaction is cooled. ____________________

E. The volume is increased. ____________________

CHAPTER 12 SOLUBILITY EQUILIBRIA Quiz

1. Calculate the K_{sp} of magnesium oxalate if the solubility is 0.0093 M. Oxalate ion is $C_2O_4{}^{-2}$.

2. Calculate the solubility of lead (II) fluoride.

3. Calculate the solubility of lead (II) chromate in 0.15 M lead chromate.

4. Determine if a precipitate will form when 45.0 mL of 0.015 M calcium chloride is added to 55.0 mL of 0.010 M sodium carbonate.

5. Calculate the solubility of zinc (II) hydroxide at pH 7.

CHAPTER 13 ACID–BASE CHEMISTRY
Quiz

1. Calculate the pH of 0.050 M HCl.

2. Calculate the pH of 0.0850 M barium hydroxide.

3. Calculate the pH if the hydroxide concentration in the solution is 1.5×10^{-9} M.

4. Calculate the pH of a solution that is made by dissolving 6.78 g barium hydroxide to a total volume of 1.00 L.

5. Calculate the $[H^+]$ of a vinegar solution that has pH 4.5.

6. Calculate the hydroxide concentration of a weak base solution that has pH 9.6.

7. Calculate the K_a for a weak acid if a 0.20 M solution has pH 3.2.

8. Calculate the pH of 0.025 M boric acid. $K_a = 5.9 \times 10^{-10}$.

9. Calculate the concentration of an acetic acid solution that has pH 2.7. $K_a = 1.8 \times 10^{-5}$.

10. Calculate the K_b for a weak base if a 0.15 M solution has pH 11.3.

11. Calculate the pH of 0.0811 M sodium hydrogen carbonate.

12. Calculate the pH of 0.025 M sodium propionate. $K_a = 1.3 \times 10^{-5}$.

13. Calculate the pH of 0.010 M sodium cyanide. $K_a = 3.5 \times 10^{-4}$.

14. Calculate the pH of a solution that is 0.10 M KNO_2 and 0.15 M HNO_2. $K_a = 4.5 \times 10^{-4}$.

15. Calculate the pH of a buffer that is made by adding 45.0 mL of 0.15 M NaF to 35.0 mL of 0.10 M HF. $K_a = 6.8 \times 10^{-4}$.

16. Calculate the pH of a solution after 40.0 mL of 0.10 M NaOH is added to 25.0 mL of 0.10 M acetic acid. $K_a = 1.8 \times 10^{-5}$.

17. Calculate the pH of the endpoint when a 50.0-mL sample of 0.1109 M propionic acid is titrated with 0.150 M NaOH. $K_a = 1.3 \times 10^{-5}$.

CHAPTER 14 THERMODYNAMICS

Quiz

1. For each of the following reactions, is the entropy change positive or negative?

   A. $2H_2\ (g) + O_2\ (g) \rightarrow 2H_2O\ (l)$ ____________________

   B. $2O_3\ (g) \rightarrow 3O_2\ (g)$ ____________________

   C. $2C\ (s) + O_2\ (g) \rightarrow 2CO\ (g)$ ____________________

   D. $2H_2O\ (l) + 2SO_2\ (g) \rightarrow 2H_2S\ (g) + 3O_2\ (g)$ ____________________

2. Calculate ΔG° for a reaction that has $K = 1.6 \times 10^{-14}$ at 25°C.

3. Calculate the temperature above which a certain reaction becomes spontaneous if the $\Delta H^\circ = -461$ kJ and the $\Delta S^\circ = -202$ J/K.

4. Calculate the nonstandard free energy for the following reaction at 25°C when $P_{H_2} = 0.19$ atm and the $P_{CH_4} = 0.35$ atm.

   $$C\ (s) + 2H_2\ (g) \rightleftarrows CH_4\ (g) \quad \Delta G^\circ = -75.0\ \text{kJ}$$

5. A certain reaction has an equilibrium constant of 50.0 at 400.0 K. The ΔH_{rxn} is −9.41 kJ/mole. Calculate the equilibrium constant at 357 K.

CHAPTER 15 ELECTROCHEMISTRY

Quiz

1. Balance the following equations in acid. The phase labels have been left out for clarity.

A. $Cu + NO_3^- \rightarrow Cu^{+2} + NO$

B. $MnO_2 + HNO_2 \rightarrow Mn^{+2} + NO_3^-$

C. $PbO_2 + Mn^{+2} + SO_4^{-2} \rightarrow PbSO_4 + MnO_4^-$

D. $HNO_2 + Cr_2O_7^{-2} \rightarrow Cr^{+3} + NO_3^-$

E. $H_2S + NO_3^- \rightarrow NO_2 + S_8$

2. Balance the following equations in base:

A. $Cr(OH)_4^- + H_2O_2 \rightarrow CrO_4^{-2} + H_2O$

B. $MnO_4^- + Br^- \rightarrow MnO_2 + BrO_3^-$

C. $Zn + NO_3^- \rightarrow NH_3 + Zn(OH)_4^{-2}$

D. $S^{-2} + I_2 \rightarrow SO_4{}^{-2} + I^-$

3. Write the balanced chemical equation for each of the following cells:

A. $| Pt | H_2 | | H^+ | Br_2 | Br^- \; Pt$

B. $Zn | Zn^{+2} | | Ag^+ | Ag$

4. Calculate the maximum work that can be obtained from 20.0 g of nickel in the following cell if the $E_{cell} = 0.97$ V.

$Ni | Ni^{+2} | | Ag^+ | Ag$

Standard Electrode Reduction Potentials in Aqueous Solution at 25°C	
Reduction Half-Reaction	**Standard Potential (V)**
Li^{+} (aq) + e^{-} → Li (s)	−3.04
Ba^{+} (aq) + e^{-} → Ba (s)	−2.71
Mg^{+2} (aq) + $2e^{-}$ → Mg (s)	−2.38
Al^{+3} (aq) + $3e^{-}$ → Al (s)	−1.66
Zn^{+2} (aq) + $2e^{-}$ → Zn (s)	−0.76
Cr^{+3} (aq) + $3e^{-}$ → Cr (s)	−0.74
Fe^{+2} (aq) + $2e^{-}$ → Fe (s)	−0.41
Cd^{+2} (aq) + $2e^{-}$ → Cd (s)	−0.40
Ni^{+2} (aq) + $2e^{-}$ → Ni (s)	−0.23
Sn^{+2} (aq) + $2e^{-}$ → Sn (s)	−0.14
Pb^{+2} (aq) + $2e^{-}$ → Pb (s)	−0.13
Fe^{+3} (aq) + $3e^{-}$ → Fe (s)	−0.04
$2H^{+}$ (aq) + $2e^{-}$ → H_2 (g)	0.00
Sn^{+4} (aq) + $2e^{-}$ → Sn^{+2} (aq)	0.15
Cu^{+2} (aq) + e^{-} → Cu^{+} (aq)	0.16
Cu^{+2} (aq) + $2e^{-}$ → Cu (s)	0.34
Cu^{+} (aq) + e^{-} → Cu (s)	0.52
I_2 (s) + $2e^{-}$ → $2I^{-}$ (aq)	0.54
Fe^{+3} (aq) + e^{-} → Fe^{+2} (aq)	0.77
Ag^{+} (aq) + e^{-} → Ag (s)	0.80
Br_2 (l) + $2e^{-}$ → $2Br^{-}$ (aq)	1.07
O_2 (g) + $4H^{+}$ (aq) + $4e^{-}$ → $2H_2O$ (l)	1.23
$Cr_2O_7^{-2}$ (aq) + $14H^{+}$ (aq) + $6e^{-}$ → $2Cr^{+3}$ (aq) + $7H_2O$ (l)	1.33
Cl_2 (g) + $2e^{-}$ → $2Cl^{-}$ (aq)	1.36
MnO_4^{-} (aq) + $8H^{+}$ (aq) + $5e^{-}$ → Mn^{+2} (aq) + $4H_2O$ (l)	1.49
H_2O_2 (aq) + $2H^{+}$ (aq) + $2e^{-}$ → $2H_2O$ (l)	1.78
$S_2O_8^{-2}$ (aq) + $2e^{-}$ → $2SO_4^{-2}$ (aq)	2.01
F_2 (g) + $2e^{-}$ → $2F^{-}$ (aq)	2.87

5. Use the preceding table of reduction potentials to calculate whether

A. Fe^{+3} oxidizes I^- to I_2 under standard conditions

B. Mn^{+2} reduces Zn^{+2} to Zn

C. Ni^{+2} oxidizes I^- to I_2

D. Zn reduces Pb^{+2} to Pb

E. Ni^{+2} oxidizes Zn to Zn^{+2}

6. Calculate the nonstandard cell potential of the following cell at 25°C:

$Al \mid Al^{+3}\ (0.015\ M) \mid\mid Cl_2\ (1\ atm) \mid Cl^-\ (0.025\ M) \mid Pt$

7. Calculate the nonstandard cell potential of the following cell at 25°C:

$$Ag \mid Ag^{+}\ (0.005\ M) \mid\mid Fe^{+3}\ (0.015\ M),\ Fe^{+2}\ (0.025\ M) \mid Pt$$

8. Calculate the equilibrium constant for the following reaction. The phase labels have been left out for clarity.

$$O_2 + 4H^{+} + 4Fe^{+2} \rightleftarrows 4Fe^{+3} + 2H_2O$$

9. Calculate the coulombs required to produce 3.61×10^{3} g of aluminum from Al^{+3} ion.

10. Calculate the time needed to produce 1.18 kg of chlorine gas from chloride ion if the current is 5.00×10^{2} A.

CHAPTER 1 ATOMS AND MOLECULES Answers and Explanations

1. Cr^{+3} has 24 protons, 21 electrons, and 28 neutrons.

2. Because it is a 50/50 mixture, the average molar mass of bromine is approximately 80.

3. A. $^{89}Y^{+3}$ has 39 protons, 50 neutrons, and 36 electrons.

B. $^{58}Ni^{+2}$ has 28 protons, 30 neutrons, and 26 electrons.

C. $^{109}Ag^{+1}$ has 47 protons, 62 neutrons, and 46 electrons.

D. $^{51}V^{+5}$ has 23 protons, 28 neutrons, and 18 electrons.

4. (28.1 $\cancel{g\,C}$) (1 mole C/12.01 $\cancel{g\,C}$) = 2.34 mole C
(2.34 $\cancel{mole\,C}$) (6.022×10^{23} atoms/$\cancel{mole}$) = 1.41×10^{24} atoms C

5. Ca 40.08 g/mole × 3 = 120.2 g/mole

P 30.97 g/mole × 2 = 61.94 g/mole

O 16.00 g/mole × 8 = 128.0 g/mole

Total is 310.14 g/mole

6. $(87.2\ ft^3)\ (12\ in/ft)^3\ (2.54\ cm/in)^3\ (1\ m/100\ cm)^3 = 2.47\ m^3$

7. (32.5 m/s) (1 mi/1609 m) (3600 s/h) = 72.7 mi/h

8. (18.4 g Au) (1 mole Au/196.97 g) = 0.0934 mole Au

9. A. ionic **B.** covalent **C.** covalent **D.** ionic

10. A. $KMnO_4$ **B.** Na_2S **C.** OsO_4

D. $Mn_2(SO_4)_3$ **E.** Cu_2CO_3 **F.** Ca_3N_2

11. A. sodium sulfite **B.** iron (II) iodate **C.** aluminum nitrate

D. vanadium (III) oxide **E.** chromium (VI) sulfide

12. (63.55 g/mole) (1 mole/6.022×10^{23} atom) = 1.055×10^{-22} g/atom

13. A. NH_4NO_3 molar mass is 80.05 g/mole

$$\%N = 28.02/80.05 \times 100\% = 35.00\%N$$

B. NH_3 molar mass is 17.03 g/mole

$$\%N = 14.01/17.03 \times 100\% = 82.27\%N$$

C. $(NH_4)_2SO_4$ molar mass is 132.15 g/mole

$$\%N = 28.02/132.15 \times 100\% = 21.20\%N$$

CHAPTER 2 ELECTRONIC STRUCTURE AND THE PERIODIC TABLE Answers and Explanations

1. There are four orbitals in the 2-shell, one 2*s* orbital and three 2*p* orbitals.

2. The 3*s* holds 2, the 3*p* holds 6, and the 3*d* holds 10 electrons, for a total of 18 electrons.

3. The blue line in the emission spectrum of hydrogen is the $n = 5$ to $n = 2$ transition.

4. **A.** $n = 4$ $l = 1$ $m_l = -1, 0, +1$ $m_s = \pm 1/2$

B. $n = 5$ $l = 3$ $m_l = -3, -2, -1, 0, +1, +2, +3$ $m_s = \pm 1/2$

C. $n = 7$ $l = 0$ $m_l = 0$ $m_s = \pm 1/2$

5. $\Delta E = (2.18 \times 10^{-18})(1/36 - 1/4) = -4.84 \times 10^{-19}$ J

6. $\lambda = hc/E = (6.63 \times 10^{-34}$ J s$)(3.00 \times 10^{8}$ m/s$)/4.84 \times 10^{-19}$ J
$= 4.11 \times 10^{-7}$ m

7. **B** The quantum number l describes the shape of the orbital.

8. **D** The quantum number m_l describes all the possible orientations of an orbital.

9. **A.** Cu $1s^2 2s^2 2p^6 3s^2 3p^6 4s^1 3d^{10}$

B. Cu^{+2} $1s^2 2s^2 2p^6 3s^2 3p^6 3d^9$

C. Ca $1s^2 2s^2 2p^6 3s^2 3p^6 4s^2$

D. Ru^{+3} $1s^2 2s^2 2p^6 3s^2 3p^6 4s^2 3d^{10} 4p^6 4d^5$

E. S $1s^2 2s^2 2p^6 3s^2 3p^4$

F. Se^{-2} $1s^2 2s^2 2p^6 3s^2 3p^6 4s^2 3d^{10} 4p^6$

10. One 4*s* orbital, three 4*p* orbitals, five 4*d* orbitals, and seven 4*f* orbitals.

11. *2p 3s 3p 4s 3d 4d*

12. **B** An electron from the 3*p* orbital is excited into the 4*s* orbital.

13. $E_{photon} = hc/\lambda = (6.63 \times 10^{-34}$ J s$)(3.00 \times 10^{8}$ m/s$)/5.10 \times 10^{-7}$ m
$= 3.90 \times 10^{19}$ J

1.1 J/3.90×10^{-19} J/photon $= 2.8 \times 10^{18}$ photons

14. $E_{photon} = hc/\lambda = (6.63 \times 10^{-34}$ J s$)(3.00 \times 10^{8}$ m/s$)/4.09 \times 10^{-7}$ m
$= 4.86 \times 10^{-19}$ J

15. The Pauli exclusion principle says that no two electrons in an atom can have the same set of quantum numbers. It refers to the spin quantum number, because two electrons in the same orientation can have the same n, l, and m_l values, but their m_s values must be different.

16. 4s

CHAPTER 3 TRENDS IN THE PERIODIC TABLE Answers and Explanations

1. The electron configuration of the outermost shells, particularly the valence orbital, determines chemical properties.

2. **B** +2, because they lose 2 electrons to obtain a noble gas configuration.

3. **A** Halogens (halides are ions).

4. They have a full valence orbital and thus are chemically inactive.

5. The number of electrons in the outermost *d* orbital determines the chemical properties of the transition metals.

6. Any element to the left of the stairline is a metal, except hydrogen and the metalloids; those to the right of the stairline are nonmetals.

7. Metals have luster and are malleable; they conduct electricity and heat; they lose electrons to form cations; metal ions are involved in ionic bonding.

8. Most nonmetals tend to be gases in the elemental state, especially if they are diatomic. Nonmetals from groups VA, VIA, and VIIA gain electrons to become anions. Nonmetal ions are involved in ionic bonding. Nonmetals usually combine to other nonmetals with covalent bonding.

9. **B** Semiconductors.

10. Ionization energy.

11. Sodium has the lower E_i because only 1 electron needs to be removed for the ion to gain a noble gas configuration.

12. Gallium, because its outermost electrons are farther from the nucleus, and thus are held less tightly.

13. **B** Partially full. The E_i is lowest when removal of an electron results in a full or half-full valence orbital.

14. Electron affinity is the energy change that occurs when electrons are added to a valence orbital to produce an anion.

15. Group VIIA has the highest electron affinity because only 1 electron is required to fill the valence orbital.

16. **C** Covalently bonded atoms.

17. Zinc is larger; it is to the left of bromine in the period.

18. Oxygen is smaller; it is above selenium in the group.

19. Nickel is larger; it has more electrons.

CHAPTER 4 LEWIS DOT STRUCTURES, HYBRIDIZATION, AND VSEPR THEORY Answers and Explanations

1. A. O_3 has 18 valence electrons. The minimum bonding places the three oxygens in a row. Place the usual bonding on one of the end oxygens (two bonds, two lone pairs). This gives the central oxygen three bonds, which is its maximum. Fill in all the shells with lone pairs. Place a plus charge on the central oxygen and a negative charge on the end oxygen that has only one bond.

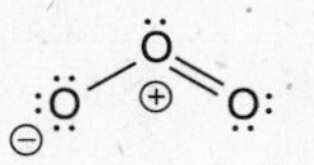

B. N_3^- has 16 valence electrons. Minimum bonding places the three nitrogens in a row. No usual bonding is allowed here. Place a double bond on each end nitrogen. This gives the central nitrogen four bonds, which is its maximum number. Fill in the shells with lone pairs. Place a plus charge on the central nitrogen, and a negative charge on each end nitrogen.

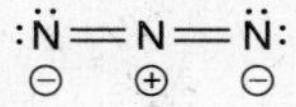

C. I_3^- has 22 valence electrons. Minimum bonding places the three iodines in a row. Fill in the shells of the two outer iodines with lone pairs, so that they have a total of eight e^-s in their shells. Put the rest of the lone pairs (3) on the central iodine. Place a negative charge on the central iodine.

D. SO_2 has 18 valence electrons. Sulfur is in the middle. Place a single bond to one oxygen and a double bond to the other, and put the remaining lone pair on the central sulfur. This gives every atom an octet. The sulfur has a +1 charge, and the singly bonded oxygen has a −1 charge.

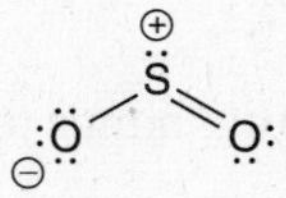

E. SO_4^{-2} has 32 valence electrons. Sulfur is in the middle, with four single bonds to each of the oxygens. Fill in all the oxygens' shells with lone pairs. Place a negative charge on each oxygen and a +2 charge on the sulfur. This gives every atom an octet.

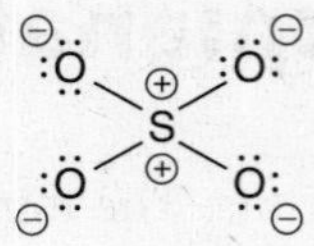

F. PCl_5 has 40 valence electrons. Place the P in the middle, with five single bonds, each going to a chlorine atom. Fill in all the chlorines' shells with lone pairs.

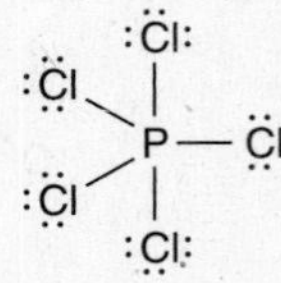

G. XeF_4 has 36 valence electrons. Place the Xe in the middle with four single bonds, each going to a fluorine atom. Fill in all the fluorines' shells with lone pairs. This leaves two lone pairs to place on the Xe atom.

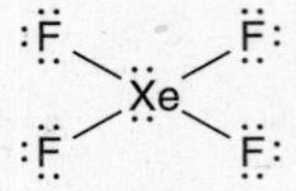

H. OF_2 has 20 valence electrons. The oxygen goes in the middle, with two single bonds, each going to a fluorine atom. Fill in each shell with lone pairs.

2. The sulfur in SO_2 has two sigma bonds and 1 lone pair, so it is sp^2.

3. The sulfur in SO_2 is bonded to two atoms and has one lone pair, so the molecular geometry is bent.

4. The middle iodine in I_3^- has two sigma bonds and three lone pairs, so it is sp^3d.

5. The middle iodine in I_3^- is bonded to two atoms and has three lone pairs, so the molecular geometry is linear.

6. The Br in BrF_4^+ has four sigma bonds and 1 lone pair, so it is sp^3d.

7. The Br in BrF_4^+ is bonded to four atoms with one lone pair, so the molecular geometry is see-saw.

8. The Xe in XeF_4 has four sigma bonds and two lone pairs, so it is sp^3d^2.

9. The Xe in XeF_4 is bonded to four atoms with two lone pairs, so the molecular geometry is square planar.

10. The S in SO_4^{-2} has four sigma bonds and no lone pairs, so it is sp^3. With four atoms bonded, it is tetrahedral in its molecular geometry.

11. The P in PCl_3 is sp^3, so the electronic geometry is tetrahedral. Each P—Cl bond is polarized with the negative end at the Cl. The sum dipole is through the P on a line with the lone pair, directed away from the lone pair.

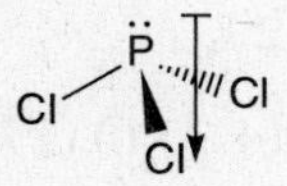

CHAPTER 5 GASES

Answers and Explanations

1. $V_i = 4.75$ L $\quad V_f =$ unknown

$P_i = 1.35$ atm $\quad P_f = 0.485$ atm

$T_i = 291.6$ K $\quad T_f = 273.2$ K

$V_f = P_i V_i T_f / T_i P_f = (1.35 \text{ atm})(4.75 \text{ L})(273.2 \text{ K})/(291.6 \text{ K})(0.485 \text{ atm})$

$V_f = 12.4$ L

2. $V = 1.00$ L $\quad P = 0.901$ atm

$T = 295.7$ K

$n = PV/RT = (0.901 \text{ atm})(1.00 \text{ L})/(0.08206 \text{ L atm/mol K})(295.7 \text{ K})$

$n = 0.03713$ mole

molar mass = g/mole = 3.15 g/0.03713 mole = 84.8 g/mole

3. SO_2 $\quad$ 18.7 g/64.07 g/mole = 0.292 mole $P_{O_2} = 0.479 \times 1.48$ atm $= 0.709$ atm

HCl $\quad$ 3.95 g/36.46 g/mole = 0.108 mole $P_{HCl} = 0.177 \times 1.48$ atm $= 0.262$ atm

N_2 $\quad$ 5.85 g/28.02 g/mole = 0.209 mole $P_{N_2} = 0.343 \times 1.48$ atm $= 0.508$ atm

X is calculated by dividing each mole amount by the total 0.609 moles. Each partial pressure is determined by multiplying each X value by the total pressure.

4. Boyle's law states that the volume is inversely proportional to the pressure. As the volume decreases, crowding and collisions increase, and the pressure increases.

5. $E_k = \frac{1}{2} mv^2$, where $v = (3RT/M)^{1/2}$

v at $25^\circ C = \{(3)(8.314 \text{ J/mole K})(298.2 \text{ K})/131.3 \text{ g/mole}\}^{1/2}$

$v = 7.526$

The mass of 1 molecule of xenon = 131.3 g/mole/6.022×10^{23} particles/mol $= 2.18 \times 10^{-22}$ g.

$E_k = \frac{1}{2}(2.18 \times 10^{-22} \text{ g})(7.526)^2 = 6.17 \times 10^{-21}$ J

6. 55.0 g/28.02 g/mole = 1.96 mole

$$[P+(1.39)(1.96)^2/(2.000)^2][(2.000-(1.96)(0.0391)]=(1.96)(0.08201)\times(292.8\,\text{K})$$

$$[P+1.33][1.92]=47.1$$

$$P+1.33=24.5$$

$$P=23.2\text{ atm}$$

CHAPTER 6 INTERMOLECULAR FORCES AND PHASE EQUILIBRIA
Answers and Explanations

1. **A.** Dispersion, because bromine is a nonpolar covalent compound.

   B. Hydrogen-bonding, because phenol contains an O—H bond.

   C. Ion–ion forces, because Fe_2O_3 is ionic.

   D. Dipole–dipole, because PBr_3 is a polar covalent compound.

   E. Dispersion, because C_4H_{10} is a nonpolar covalent compound.

2. **A.** At 1 atm, the boiling point is 55°C.

   B. At 1 atm, the melting point is 40°C.

   C. At the critical point, the pressure is 1.8 atm and the temperature is 90°C.

   D. The liquid phase is the most dense; it is the predominant phase at high pressures.

   E. Room temperature is 25°C and room pressure is 1 atm; at this T and P, this substance is solid.

   F. At 110°C and 2.0 atm, this substance is in the supercritical fluid phase.

3. Solve for $n = PV/RT = (0.282 \text{ atm})(5.75 \text{ L})/(0.08206 \text{ L atm/mol K})(283.6 \text{ K})$

$$n = 0.0697 \text{ mole}$$

$$\text{mass} = (0.0697 \text{ mole})(84.93 \text{ g/mole}) = 5.92 \text{ g}$$

4. $P_1 = 0.1316$ atm at $T_1 = 278.6$ K

$$P_2 = 1 \text{ atm at } T_2 = 330.0 \text{ K}$$

$$\ln(1 \text{ atm}/0.1316 \text{ atm}) = (\Delta H_{vap}/8.314 \text{ J/mole K}) \times (1/278.6 \text{ K} - 1/330.0 \text{ K})$$

$$2.03 = \Delta H_{vap}(6.72 \times 10^{-5})$$

$$\Delta H_{vap} = 3.02 \times 10^4 \text{ J/mole}$$

5. Calculate the moles of sucrose

$$(26.8 \text{ g})(1 \text{ mole}/342.3 \text{ g}) = 0.0783 \text{ mole sucrose}$$

   Calculate the moles of water

$$(320.0 \text{ g})(1 \text{ mole}/18.02 \text{ g}) = 17.76 \text{ mole } H_2O$$

Calculate the mole fraction of sucrose

$$X_{sucrose} = 0.0783/(0.0783 + 17.76) = 4.39 \times 10^{-3}$$

$$\Delta P = X_{sucrose}P° = (4.39 \times 10^{-3})(23.8 \text{ mm Hg}) = 0.104 \text{ mm Hg}$$

$$P = P° - \Delta P = 23.8 - 0.1 = 23.7 \text{ mm Hg}$$

6. Sodium hydrogen carbonate is $NaHCO_3$. Its i value is 2; its molar mass is 84.02 g/mole.

$$(5.74 \text{ g})\ (1 \text{ mole}/84.01 \text{ g}) = 0.0683 \text{ mole } NaHCO_3$$
$$\text{molality} = 0.0683 \text{ mole}/0.0850 \text{ kg} = 0.804 \text{ m}$$

$$\Delta T_f = imK_f = (2)(0.804 \text{ m})(1.86°\text{C/m}) = 2.99°\text{C}$$

$$T_f = 0°\text{C} - 2.99°\text{C} = -2.99°\text{C}$$

$$\Delta T_b = imK_b = (2)(0.804 \text{ m})(0.512°\text{C/m}) = 0.823°\text{C}$$

$$T_b = 100°\text{C} + 0.8°\text{C} = 100.8°\text{C}$$

7. Calculate $n = \Pi V/RT = (0.0109 \text{ atm})\ (0.001 \text{ L})/(0.08206 \text{ L atm/mol K})\ (298.2 \text{ K})$

$$n = 4.46 \times 10^{-7} \text{ mole}$$

$$\text{molar mass} = \text{mass/mole} = 0.0025 \text{ g}/4.46 \times 10^{-7} \text{ mole}$$

$$= 5.6 \times 10^{3} \text{ g/mole}$$

CHAPTER 7 CHEMICAL EQUATIONS AND STOICHIOMETRY Answers and Explanations

1. **A.** $2N_2O_5 \rightarrow 2N_2O_4 + O_2$

B. $2NH_3 + 3CuO \rightarrow 3Cu + N_2 + 3H_2O$

C. $2Al + 3H_2SO_4 \rightarrow Al_2(SO_4)_3 + 3H_2$

D. $S_8 + 8O_2 \rightarrow 8SO_2$

E. $CH_4 + 4Br_2 \rightarrow CBr_4 + 4HBr$

2. **A.** $(NH_4)_2SO_4$ 132.15 g/mole

B. $Sr(ClO_2)_2$ 222.10 g/mole

C. SnF_2 156.7 g/mole

D. Ag_2CO_3 275.8 g/mole

3. **A.** CuO 79.55 g/mole

$$(8.752\text{ g})(1\text{ mole}/79.55\text{ g}) = 0.1100\text{ mole}$$

B. NH_3 17.03 g/mole

$$(97.4\text{ mole})(17.03\text{ g/mole}) = 1.66 \times 10^3\text{ g}$$

4. $(2500.0\text{ g})(0.018) = 45\text{ g }FeS_2$

The balanced equation is

$$2FeS_2 + 11/2O_2 \rightarrow Fe_2O_3 + 4SO_2$$

$$(45\text{ g }FeS_2)(1\text{ mole}/120.0\text{ g}) = 0.375\text{ mole }FeS_2$$

$$(0.375\text{ mole }FeS_2)(4\text{ mole }SO_2/2\text{ mole }FeS_2) = 0.75\text{ mole }SO_2$$

$$(0.75\text{ mole }SO_2)(64.07\text{ g/mole}) = 48\text{ g }SO_2$$

5. The balanced equation is

$$Ca_3(PO_4)_2 + 3H_2SO_4 \rightarrow 3CaSO_4 + 2H_3PO_4$$

$$[150.0\text{ g }Ca_3(PO_4)_2](1\text{ mole}/310.2\text{ g}) = 0.4836\text{ mole }Ca_3(PO_4)_2$$

$$[0.4836\text{ mole }Ca_3(PO_4)_2]$$
$$\times[3\text{ mole }H_2SO_4/1\text{ mole }Ca_3(PO_4)_2] = 1.451\text{ mole }H_2SO_4$$

$$(1.451\text{ mole }H_2SO_4)(98.09\text{ g/mole}) = 142.3\text{ g }H_2SO_4$$

6. The balanced equation is

$$C_3H_8 + 5O_2 \rightarrow 3CO_2 + 4H_2O$$

From C_3H_8

$$(587.5 \text{ g } C_3H_8)(1 \text{ mole}/44.09 \text{ g } C_3H_8)(3 \text{ mole } CO_2/1 \text{ mole } C_3H_8)$$
$$\times (44.01 \text{ g/mole } CO_2) = 1759 \text{ g } CO_2$$

From O_2

$$(1.25 \times 10^3 \text{ g } O_2)(1 \text{ mole}/32.0 \text{ g } O_2)(3 \text{ mole } CO_2/5 \text{ mole } O_2)$$
$$\times (44.01 \text{ g/mole } CO_2) = 1.03 \times 10^3 \text{ g } CO_2$$

This answer is correct; it is the smaller value.

CHAPTER 8 REACTIONS IN SOLUTION
Answers and Explanations

1. Sodium phosphate is Na_3PO_4. It has molar mass 163.9 g/mole.

   $(56.1\ g)(1\ mole/163.9\ g) = 0.342\ mole$

   $M = mole/L = 0.342\ mole/0.4750\ L = 0.721\ M$

2. **A.** Soluble, because it is a group IA compound.

   B. Soluble, because it is a nitrate.

   C. Insoluble, because it is a sulfide, and not one of the exceptions.

   D. Insoluble, because it is an exception to the chloride rule.

   E. Insoluble, because it is a hydroxide, and not one of the exceptions.

   F. Soluble, because it is a group IA compound.

3. The insoluble pair-up is copper (I) ion with sulfide ion

   $2Cu^{+1} + S^{-2} \rightarrow Cu_2S$

4. The insoluble pair-up is copper (II) ion with phosphate anion

   $3Cu^{+2} + 2PO_4^{-3} \rightarrow Cu_3(PO_4)_2$

5. **A.** The product of a weak base and a strong acid is acidic.

   B. The product of a weak acid and a strong base is basic.

   C. The product of a strong acid and a strong base is neutral.

   D. The product of a weak acid and a strong base is basic.

6. The balanced equation is

   $HBr + NaOH \rightarrow NaBr + H_2O$

   $(0.0146\ L)(0.0985\ mole/L) = 1.438 \times 10^{-3}$ mole NaOH

   Because the stoichiometry is 1:1, there are also 1.438×10^{-3} mole HBr

   $M_{HBr} = 1.438 \times 10^{-3}\ mole/0.0250\ L = 0.0575\ M$ HBr

7. **A.** Mg is +2 N is −3 These are simple ions.

   B. P is 0 This is an element.

   C. S is +6 O is −2

   D. N is +5 O is −2

E. C is −4	H is +1	H is always +1 in a covalent compound.
F. Ca is +2	Cl is −1	These are simple ions.
G. Na is +1	N is +5	O is −2
H. Na is +1	H is −1	These are simple ions.

8. A. P is 0 on the left and +5 on the right; it is oxidized. F is 0 on the left and −1 on the right; it is reduced.

B. N is +3 on the left and +5 on the right; it is oxidized. N is +3 on the left, and +2 on the right; it is reduced.

9. Mole = M × L = (0.02311 mole/L) (0.02484 L) = 5.741×10^{-4} mole MnO_4^-
(5.741×10^{-4} mole MnO_4^-) (5 mole H_2O_2/2 mole MnO_4^-) = 1.435×10^{-3} mole H_2O_2
$M_{H_2O_2} = 1.435 \times 10^{-3}$ mole/0.0250 L = 0.0574 M H_2O_2

CHAPTER 9 THERMOCHEMISTRY
Answers and Explanations

1. The balanced equation for the formation of H_2S from its elements is

$$8H_2 + S_8 \longrightarrow 8H_2S$$

For H_2, the balanced equation is 432 kJ/mole × 1 bond = 432 kJ/mole; for S_8, the balanced equation is 226 kJ/mole × 8 bonds = 1808 kJ/mole; for H_2S, the balanced equation is 363 kJ/mole × 2 bonds = 726 kJ/mole.

$$\Delta H_f = [8 \text{ mole } H_2(432 \text{ kJ/mole}) + 1 \text{ mole } S_8(1808 \text{ kJ/mole})]$$
$$- [8 \text{ mole } H_2S(726)]$$
$$\Delta H_f = (-544 \text{ kJ/mole})/8$$
$$\Delta H_f = -68 \text{ kJ/mole}$$

2. $q = mc\Delta T = (6.22 \text{ kg})\ (0.385 \text{ kJ/kg }^\circ\text{C})\ (301.9^\circ\text{C}) = 723 \text{ kJ}$

3. Total heat lost = total heat gained

$$(100.0 \text{ g } H_2O)\ (4.184 \text{ J/g}^\circ\text{C})\ (T_f - 30^\circ\text{C}) = -(17.5 \text{ g Cu})\ (0.385 \text{ J/g}^\circ\text{C}) \times (T_f - (-17^\circ\text{C}))$$
$$T_f = -\ 29.3^\circ\text{C}$$

4. First, convert mass of NH_3 to moles

$$(2.8 \times 10^3 \text{ g})(1 \text{ mole}/17.03 \text{ g}) = 164.4 \text{ mole } NH_3$$

Next, convert mole NH_3 to kJ

$$(164.4 \text{ mole } NH_3)(906 \text{ kJ produced}/4 \text{ mole } NH_3) = 3.7 \times 10^4 \text{ kJ produced}$$

5. First, convert kJ to moles

$$(3.96 \times 19^3 \text{ kJ})(1 \text{ mole } CH_4/890 \text{ kJ}) = 4.45 \text{ mole } CH_4$$

Next, convert mole CH_4 to mass

$$(4.45 \text{ mole } CH_4)(16.04 \text{ g/mole}) = 71.4 \text{ g } CH_4$$

6. The first equation is fine

$$C_2H_4 + 3O_2 \rightarrow 2CO_2 + 2H_2O \qquad \Delta H = -1411 \text{ kJ}$$

The second equation must be reversed and halved

$$2CO_2 + 3H_2O \rightarrow C_2H_6 + \frac{7}{2}O_2 \qquad \Delta H = 1560 \text{ kJ}$$

The third equation must be halved

$$H_2 + {}^1\!/\!_2O_2 \rightarrow H_2O \qquad \Delta H = -286 \text{ kJ}$$

The equations sum to the target equation, and the sum of the ΔHs is -137 kJ.

CHAPTER 10 KINETICS
Answers and Explanations

1. [A] is doubling from run 1 to run 2, and the rate is quadrupling, so the order of [A] is 2. [B] is doubling from run 4 to run 5, and the rate is doubling, so the order of [B] is 1. Therefore, the rate law is

$$\text{rate} = k[\text{A}]^2[\text{B}]$$

Using the values from run 3 (any run will do)

$$8.80 \times 10^{-5}\ \text{M/s} = k(0.400)^2(0.100)$$

$$k = 5.50 \times 10^{-3}\ 1/\text{M}^2\text{s}$$

2. **A.** $\ln (0.25/x) = (5.1 \times 10^{-4}\ \text{s}^{-1})\ (192\ \text{s})$

$$\ln 0.25 - \ln x = 0.0979$$

$$-1.38 - \ln x = 0.0979$$

$$\ln x = -1.48$$

$$x = 0.23\ \text{M}$$

B. $\ln (0.25/0.12) = (5.1 \times 10^{-4}\ \text{s}^{-1})\ t$

$$0.73/5.1 \times 10^{-4}\ \text{s}^{-1} = t = 1.4 \times 10^3\ \text{s}$$

C. $\ln (100/38) = (5.1 \times 10^{-4}\ \text{s}^{-1})\ t$

$$t = 1.9 \times 10^3\ \text{s}$$

3. $k_1 = 0.0346$ $\quad T_1 = 298.2\ \text{K}$
$k_2 = x$ $\quad T_2 = 311.8\ \text{K}$

$$\ln (x/0.0346) = (50,200\ \text{J/mol}/8.314\ \text{J/mol K})(1/298.2\ \text{K} - 1/311.8\ \text{K})$$

$$\ln x - \ln 0.0346 = 0.883$$

$$\ln x - (-3.36) = 0.883$$

$$\ln x = -2.48$$

$$x = 0.0837\ \text{s}^{-1}$$

4. When $t_{1/2} = 438$ min $\quad k_1 = 1.58 \times 10^{-3}$ min^{-1} $\quad T_1 = 343.2$ K
When $t_{1/2} = 19.8$ min $\quad k_2 = 0.0350$ min^{-1} $\quad T_2 = 373.2$ K

$$\ln(0.0350/1.58 \times 10^{-3}) = (E_a/8.314 \text{ J/mol K})(1/343.2 \text{ K} - 1/373.2 \text{ K})$$

$$3.10 = E_a(2.82 \times 10^{-5})$$

$$E_a = 1.1 \times 10^5 \text{ J/mole}$$

CHAPTER 11 EQUILIBRIUM
Answers and Explanations

1. $[I_2]_{init} = 0.0456$ mole/2.30 L = 0.0199 M
At equilibrium, $2x$ of product is formed, and the $[I_2]_{eq} = 0.0199$ M.

$$K = [I]^2/[I_2]$$

$$3.8 \times 10^{-5} = (2x)^2/0.0199$$

$$x = 4.35 \times 10^{-4} \text{ m}$$

$$[I]_{eq} = 2x = 8.70 \times 10^{-4} \text{ M}$$

2. $[NO]_{init} = 0.0655$ mole/2.5 L = 0.0262 M
$[Br_2]_{init} = 0.0328$ mole/2.5 L = 0.0131 M
$[NOBr]_{eq} = 0.0389$ mole/2.5 L = 0.0156 M

	$2NO$	+	Br_2	$\rightleftarrows$	$2NOBr$
Init	0.0262		0.0131		0
Δ	$-2x$		$-x$		$+2x$
At equilibrium	0.0106		0.0053		0.0156

Because $2x = 0.0156$
$x = 0.0078$

This value is plugged into the Δ line to get the equilibrium concentrations.

$$K = [NOBr]^2/[NO]^2[Br_2] = (0.0156)^2/(0.0106)^2\,(0.0053) = 409$$

3. The equilibrium concentration of NO_2 is $2x$, and $[N_2O_4]_{eq} = 0.030$ M.

$$K = [NO_2]^2/[N_2O_4]$$

$$1.25 \times 10^{-4} = (2x)^2/0.030$$

$$x = 9.7 \times 10^{-4} \text{ M}$$

$$\% \text{ dissociation} = (9.7 \times 10^{-4} \text{ M}/0.030 \text{ M}) \times 100\% = 3.23\%$$

$$[NO_2]_{eq} = 2x = 1.9 \times 10^{-3} \text{ M}$$

4. The amount of NOBr that reacts is $2x$, and the amount of NOBr that reacts is 9.4% of the original 2.00 M. Thus, (2.00 M) (0.094) = 0.19 M = $2x$. So,

$$[NOBr]_{eq} = 2.00 - 0.19 = 1.81 \text{ M}$$

and

$$[NO]_{eq} = 2x = 0.19\ M$$

and

$$[Br_2]_{eq} = x = 0.10\ M$$

$$K = [NO]^2[Br_2]/[NOBr]^2 = (0.19)^2(0.10)/(1.81)^2 = 1.1 \times 10^{-3}$$

5. A. $Q = (0.0151)^2(0.0108)/(0.0619)^2 = 6.6 \times 10^{-4}$ $Q > K$, so the reaction is going left.

B. $Q = (0.0169)^2(0.0142)/(0.115)^2 = 3.1 \times 10^{-4}$ $Q = K$, so the reaction is at equilibrium.

C. $Q = (0.0123)^2(0.0201)/(0.181)^2 = 9.3 \times 10^{-5}$ $Q < K$, so the reaction is going right.

6.

$$[N_2]_{init} = 0.850\ \text{mole}/8.0\ \text{L} = 0.11\ M$$

$$[O_2]_{init} = 0.850\ \text{mole}/8.0\ \text{L} = 0.11\ M$$

The equilibrium concentration of NO is $2x$. The equilibrium concentrations of N_2 and O_2 are the same as their initial concentrations.

$$K = [NO]^2/[N_2][O_2]$$

$$9.22 \times 10^{-5} = (2x)^2/(0.11)^2$$

$$x = 5.1 \times 10^{-4}\ M$$

$$[NO]_{eq} = 2x = 1.0 \times 10^{-3}\ M$$

7. If the volume is decreased, the pressure is increased. The equilibrium shifts to the side that has fewer moles of gas in order to decrease the pressure.

A. No change

B. To the right

C. No change

D. To the left

8. If the reaction is heated, the reaction shifts to get rid of the heat.

A. This reaction is endothermic, so the reaction shifts to the right.

B. This reaction is exothermic, so the reaction shifts to the left.

C. This reaction is thermoneutral, so there is no shift.

9. The equilibrium shifts to undo the change.

   A. Shifts to the left, to get rid of H_2S.

   B. Shifts to the right, to make more NH_3.

   C. No shift, because it is a solid.

   D. Shifts to the left, because the reaction is endothermic.

   E. Shifts to the right, to make more gas molecules.

CHAPTER 12 SOLUBILITY EQUILIBRIA Answers and Explanations

1. $MgC_2O_4\ (s) \rightleftarrows Mg^{+2}\ (aq) + C_2O_4^{-2}\ (aq)$

0.0093 0.0093 M

$K_{sp} = [Mg^{+2}][C_2O_4^{-2}] = (0.0093)^2 = 8.6 \times 10^{-5}$

2. $PbF_2\ (s) \rightleftarrows Pb^{+2}\ (aq) + 2F^-\ (aq)$

x $2x$

$K_{sp} = [Pb^{+2}][F^-]^2$

$4.1 \times 10^{-8} = x(2x)^2 = 4x^3$

$x = \text{solubility} = 2.2 \times 10^{-3}\ M$

3. $PbCrO_4\ (s) \rightleftarrows Pb^{+2}\ (aq) + CrO_4^{-2}\ (aq)$

x 0.15 M

$K_{sp} = [Pb^{+2}][CrO_4^{-2}]$

$2.0 \times 10^{-14} = x(0.15)$

$x = \text{solubility} = 1.3 \times 10^{-13}\ M$

4. $CaCO_3$ would be the precipitate; its K_{sp} is 8.7×10^{-9}.

$Q_{sp} = [Ca^{+2}][CO_3^{-2}]$

$[Ca^{+2}] = (0.015\ M)(45.0\ mL)/1000\ mL = 6.75 \times 10^{-3}\ M$

$[CO_3^{-2}] = (0.010\ M)(55.0\ mL)/1000\ mL = 5.5 \times 10^{-3}\ M$

$Q_{sp} = (6.75 \times 10^{-3})(5.5 \times 10^{-3}) = 3.7 \times 10^{-5}$

$Q_{sp} > K_{sp}$, so yes, a precipitate forms.

5. $Zn(OH)_2\ (s) \rightleftarrows Zn^{+2}\ (aq) + 2OH^-\ (aq)$

$K_{sp} = [Zn^{+2}][OH^-]^2$

$4.5 \times 10^{-17} = x(1 \times 10^{-7})^2$

$x = \text{solubility} = 4.5 \times 10^{-17}\ M$

CHAPTER 13 ACID–BASE CHEMISTRY Answers and Explanations

1. This is a strong acid, so $[H^+] = 0.050$ M pH $= -\log 0.050 = 1.3$

2. This is a strong base

$$Ba(OH)_2\ (aq) \rightarrow Ba^{+2}\ (aq) + 2OH^-\ (aq)$$

So,

$$[OH^-] = 0.0850 \times 2 = 0.170\ M$$

$$pOH = -\log(0.170) = 0.8$$

$$pH = 13.2$$

3. $[OH^-] = 1.5 \times 10^{-9}$ M

$$pOH = -\log(1.5 \times 10^{-9}) = 8.8$$

$$pH = 5.2$$

4. $[Ba(OH)_2]$ = mole/L = 0.0396 M = x

$$6.78\ g\,(1\ mole/171.3\ g) = 0.0396\ g$$

$$[OH^-] = 2x = 0.0792\ M$$

$$pOH = -\log(0.0792) = 1.10$$

$$pH = 12.9$$

5. $[H^+] = 10^{-4.5} = 3.2 \times 10^{-5}$ M

6. pOH = 4.4

$$[OH^-] = 10^{-4.4} = 4.0 \times 10^{-5}\ M$$

7. $HA\ (aq) \rightleftarrows H^+\ (aq) + A^-\ (aq)$

0.20 $\quad x \quad x$

$$x = [H^+] = 10^{-3.2} = 6.3 \times 10^{-4}\ M$$

$$K_a = (6.3 \times 10^{-4})^2/0.20 = 2.0 \times 10^{-6}$$

8. $K_a = [H^+][A^-]/[HA]$

$$5.9 \times 10^{-10} = x^2/0.025$$

$$x = [H^+] = 3.8 \times 10^{-6}\ M$$

$$pH = 5.4$$

9. $[H^+] = 10^{-2.7} = 2.0 \times 10^{-3}$ M

$K_a = (2.0 \times 10^{-3})^2/[HA] = 1.8 \times 10^{-5}$

$[HA] = 0.22$ M

10. pOH = 2.7

$[OH^-] = 10^{-2.7} = 2.0 \times 10^{-3}$ M

$K_b = (2.0 \times 10^{-3})^2/0.15 = 2.7 \times 10^{-5}$

11. HCO_3^- (aq) + H_2O (l) ⇄ OH^- (aq) + H_2CO_3 (aq)

$K_{b1} = x^2/0.0811 = 2.3 \times 10^{-8}$

$x = [OH^-] = 4.3 \times 10^{-5}$ M

pOH = 4.4

pH = 9.6

12. Pr^- (aq) + H_2O (l) ⇄ OH^- (aq) + HPr (aq)

$K_b = x^2/[Pr^-] = 7.7 \times 10^{-10} = K_w/K_a$

$7.7 \times 10^{-10} = x^2/0.025$

$x = [OH^-] = 4.4 \times 10^{-7}$ M

pOH = 5.4

pH = 8.6

13. CN^- (aq) + H_2O (l) ⇄ OH^- (aq) + HCN (aq)

$K_b = x^2/[CN^-] = 2.9 \times 10^{-11} = K_w/K_a$

$2.9 \times 10^{-11} = x^2/0.010$

$x = [OH^-] = 5.3 \times 10^{-7}$ M

pOH = 6.3

pH = 7.7

14. HNO_2 (aq) ⇄ H^+ (aq) + NO_2^- (aq)
0.15 x 0.10

$K_a = [H^+][NO_2^-]/[HNO_2]$

$4.5 \times 10^{-4} = x(0.10)/0.15$

$x = [H^+] = 6.8 \times 10^{-4}$ M

pH = 3.2

15. $pH = pK_a + \log \{[A^-]/[HA]\}$

$$pK_a = -\log (6.8 \times 10^{-4}) = 3.17$$

$$[F^-] = (45.0 \text{ mL}) (0.15 \text{ M})/80.0 \text{ mL} = 0.0844 \text{ M}$$

$$[HF] = (25.0 \text{ mL}) (0.10 \text{ M})/80.0 \text{ mL} = 0.0438 \text{ M}$$

$$pH = 3.17 + \log \{0.0844/0.0438\}$$

$$pH = 3.17 + 0.28$$

$$pH = 3.5$$

16. There were (0.10 M) (0.0250 L) = 2.50×10^{-3} moles acetic acid in the flask. There was added (0.10 M) (0.040 L) = 4.00×10^{-3} mole OH^-. Only 2.50×10^{-3} mole OH^- reacted with the acetic acid. They reacted to form 2.50×10^{-3} mole acetate. There are 1.50×10^{-3} mole OH^- left over in the 65.0 mL of solution. Therefore,

$$[OH^-] = 1.50 \times 10^{-3} \text{ mole}/0.0650 \text{ L} = 2.31 \times 10^{-2} \text{ M}$$

$$pOH = 1.6$$

$$pH = 12.4$$

17. There were (0.0500 L) (0.1109 M) = 5.55×10^{-3} mole acid HA in the flask. This required 5.55×10^{-3} mole of base, and produced 5.55×10^{-3} mole A^-. The volume of base added = 5.55×10^{-3} mole/0.150 M = 0.0370 L. Therefore, the concentration of the A^- at the endpoint is

$$[A^-] = 5.50 \times 10^{-3} \text{ mole}/(0.0500 \text{ L} + 0.0370 \text{ L}) = 0.063 \text{ M}$$

This is a basic salt solution.

$$K_b = x^2/0.063 = 7.7 \times 10^{-10} \text{ M}$$

$$x = [OH^-] = 7 \times 10^{-6}$$

$$pOH = 5.2$$

$$pH = 8.8$$

CHAPTER 14 THERMODYNAMICS Answers and Explanations

1. **A.** Results in more order, so the ΔS is negative.

   B. Results in more disorder, so the ΔS is positive.

   C. Results in more disorder, so the ΔS is positive.

   D. Results in more disorder, so the ΔS is positive.

2. $\Delta G^\circ = -(8.314 \text{ J/mole K})(298.2 \text{ K})(\ln 1.6 \times 10^{-14}) = 7.9 \times 10^4 \text{ J/mole}$

3. Because $\Delta G = 0$, $T = \Delta H^\circ/\Delta S^\circ = -461{,}000 \text{ J}/-202 \text{ J/K} = 2.28 \times 10^3 \text{ K}$

4. $Q = P_{CH_4}/(P_{H_2})^2 = 0.35/(0.19)^2 = 9.7$

   $\Delta G = (-75{,}000 \text{ J/mole}) + (8.314 \text{ J/mole K})(298 \text{ K})(\ln 9.7)$

   $\Delta G = -6.9 \times 10^4 \text{ J/mole}$

5. $\ln K_2 - \ln 50.0 = (-9410 \text{ J/mole}/8.314 \text{ J/mole K})(1/400\ K - 1/357\ K)$

   $\ln K_2 = 0.34 + 3.9 = 4.2$

   $K_2 = 69$

CHAPTER 15 ELECTROCHEMISTRY Answers and Explanations

(The phase labels have been left out for clarity.)

1. **A.** $Cu + NO_3^- \rightarrow Cu^{+2} + NO$

$$3 \mid Cu \rightarrow Cu^{+2} \qquad \Delta e = 2$$
$$2 \mid 4H^+ + NO_3^- \rightarrow NO + 2H_2O \qquad \Delta e = 3$$
$$n = 6$$

$$3Cu \rightarrow 3Cu^{+2}$$
$$8H^+ + 2NO_3^- \rightarrow 2NO + 4H_2O$$

Sum is

$$8H^+ + 3Cu + 2NO_3^- \rightarrow 3Cu^{+2} + 2NO + 4H_2O$$

B. $MnO_2 + HNO_2 \rightarrow Mn^{+2} + NO_3^-$

$$4H^+ + MnO_2 \rightarrow Mn^{+2} + 2H_2O \qquad \Delta e = 2$$
$$H_2O + HNO_2 \rightarrow NO_3^- + 3H^+ \qquad \Delta e = 2$$
$$n = 2$$

Sum is

$$H^+ + MnO_2 + HNO_2 \rightarrow Mn^{+2} + NO_3^- + H_2O$$

C. $PbO_2 + Mn^{+2} + SO_4^{-2} \rightarrow PbSO_4 + MnO_4^-$

$$5 \mid 4H^+ + PbO_2 + SO_4^{-2} \rightarrow PbSO_4 + 2H_2O \qquad \Delta e = 2$$
$$2 \mid 4H_2O + Mn^{+2} \rightarrow MnO_4^- + 8H^+ \qquad \Delta e = 5$$
$$n = 10$$

$$20H^+ + 5PbO_2 + 5SO_4^{-2} \rightarrow 5PbSO_4 + 10H_2O$$
$$8H_2O + 2Mn^{+2} \rightarrow 2MnO_4^- + 16H^+$$

Sum is

$$5PbO_2 + 5SO_4^{-2} + 2Mn^{+2} + 4H^+ \rightarrow 5PbSO_4 + 2MnO_4^- + 2H_2O$$

D. $HNO_2 + Cr_2O_7^{-2} \rightarrow Cr^{+3} + NO_3^-$

$$3 \mid H_2O + HNO_2 \rightarrow NO_3^- + 3H^+ \qquad \Delta e = 2$$
$$14H^+ + Cr_2O_7^{-2} \rightarrow 2Cr^{+3} + 7H_2O \qquad \Delta e = 6$$
$$n = 6$$

$$3H_2O + 3HNO_2 \rightarrow 3NO_3^- + 9H^+$$
$$14H^+ + Cr_2O_7^{-2} \rightarrow 2Cr^{+3} + 7H_2O$$

Sum is

$$5H^+ + 3HNO_2 + Cr_2O_7^{-2} \rightarrow 3NO_3^- + 2Cr^{+3} + 4H_2O$$

E. $H_2S + NO_3^- \rightarrow NO_2 + S_8$

$8H_2S \rightarrow S_8 + 16H^+$ $\quad \Delta e = 16$
$16 \mid 2H^+ + NO_3^- \rightarrow NO_2 + H_2O$ $\quad \Delta e = 1$
$n = 16$

$8H_2S \rightarrow S_8 + 16H^+$
$32H^+ + 16NO_3^- \rightarrow 16NO_2 + 16H_2O$

Sum is

$$16H^+ + 8H_2S + 16NO_3^- \rightarrow S_8 + 16NO_2 + 16H_2O$$

2. A. $Cr(OH)_4^- + H_2O_2 \rightarrow CrO_4^{-2} + 4H^+$ in base solution

$2 \mid Cr(OH)_4^- \rightarrow CrO_4^{-2} + 4H^+$ $\quad \Delta e = 3$
$3 \mid 2H^+ + H_2O_2 \rightarrow H_2O + H_2O$ $\quad \Delta e = 2$
$n = 6$

$2Cr(OH)_4^- \rightarrow 2CrO_4^{-2} + 8H^+$
$6H^+ + 3H_2O_2 \rightarrow 6H_2O$

Sum is

$$2Cr(OH)_4^{-2} + 3H_2O_2 \rightarrow 2CrO_4^{-2} + 6H_2O + 2H^+$$

Change all H^+ to H_2O and add the same number of OH^- to other side

$$2OH^{-1} + 2Cr(OH)_4^- + 3H_2O_2 \rightarrow 2CrO_4^{-2} + 6H_2O + 2H_2O$$

Final sum is

$$2OH^- + 2Cr(OH)_4^- + 3H_2O_2 \rightarrow 2CrO_4^{-2} + 8H_2O$$

B. $MnO_4^- + Br^- \rightarrow MnO_2 + BrO_3^-$ in base solution

$2 \mid 4H^+ + MnO_4^- \rightarrow MnO_2 + 2H_2O$ $\quad \Delta e = 3$
$3H_2O + Br^- \rightarrow BrO_3^- + 6H^+$ $\quad \Delta e = 6$
$n = 6$

$8H^+ + 2MnO_4^- \rightarrow 2MnO_2 + 4H_2O$
$3H_2O + Br^- \rightarrow BrO_3^- + 6H^+$

Sum is

$$2H^+ + 2MnO_4^- + Br^- \rightarrow 2MnO_2 + BrO_3^- + H_2O$$

Change all H^+ to H_2O and add the same number of OH^- to other side

$$2H_2O + 2MnO_4^- + Br^- \rightarrow 2MnO_2 + BrO_3^- + H_2O + 2OH^-$$

Final sum is

$$H_2O + 2MnO_2^- + Br^- \rightarrow 2MnO_2 + BrO_3^- + 2OH^-$$

C. $Zn + NO_3^- \rightarrow NH_3 + Zn(OH)_4^{-2}$ in base solution

$4 \mid 4H_2O + Zn \rightarrow Zn(OH)_4^{-2} + 4H^+$ $\quad \Delta e = 2$

$9H^+ + NO_3^- \rightarrow NH_3 + 3H_2O$ $\quad \Delta e = 8$

$n = 8$

$16H_2O + 4Zn \rightarrow 4Zn(OH)_4^{-2} + 16H^+$

$9H^+ + NO_3^- \rightarrow NH_3 + 3H_2O$

Sum is

$$13H_2O + 4Zn + NO_3^- \rightarrow 4Zn(OH)_4^{-2} + NH_3 + 7H^+$$

Change all H^+ to H_2O and add the same number of OH^- to other side

$$7OH^- + 13H_2O + 4Zn + NO_3^- \rightarrow 4Zn(OH)_4^{-2} + NH_3 + 7H_2O$$

Final sum is

$$7OH^- + 6H_2O + 4Zn + NO_3^- \rightarrow 4Zn(OH)_2^{-2} + NH_3$$

D. $S^{-2} + I_2 \longrightarrow SO_4^{-2} + I^-$ in base solution

$4H_2O + S^{-2} \rightarrow SO_4^{-2} + 8H^+$ $\quad \Delta e = 8$

$4 \mid I_2 \rightarrow 2I^-$ $\quad \Delta e = 2$

$n = 8$

$4H_2O + S^{-2} \rightarrow SO_4^{-2} + 8H^+$

$4I_2 \rightarrow 8I^-$

Sum is

$$4H_2O + S^{-2} + 4I_2 \rightarrow SO_4^{-2} + 8H^+ + 8I^-$$

Change all H^+ to H_2O and add the same number of OH^- to the other side

$$4H_2O + S^{-2} + 4I_2 + 8OH^- \rightarrow SO_4^{-2} + 8H_2O + 8I^-$$

Final sum is

$$S^{-2} + 4I_2 + 8OH^- \rightarrow SO_4^{-2} + 4H_2O + 8I^-$$

3. A. $H_2 \rightarrow 2H^+$ $\quad \Delta e = 2$

$Br_2 \rightarrow 2Br^-$ $\quad \Delta e = 2$

Sum is

$$H_2 + Br_2 \rightarrow 2H^+ + 2Br^-$$

B. $Zn \rightarrow Zn^{+2}$ $\quad \Delta e = 2$

$Ag^+ \rightarrow Ag$ $\quad \Delta e = 1$

The second equation must be multiplied by 2, so the sum is

$$Zn + 2\,Ag^+ \rightarrow Zn^{+2} + 2Ag$$

4. $w_{max} = -nFE_{cell}$

$w_{max} = -(2)(9.65 \times 10^4 \text{ J/V})(0.97 \text{ V}) = -1.9 \times 10^5 \text{ J}$

5. A. $I^- \rightarrow I_2$ +0.54 V Oxidation

$Fe^{+3} \rightarrow Fe^{+2}$ −0.77 V Reduction

$E^\circ_{cell} = -0.23$ V Positive, so Yes

B. $Zn^{+2} \rightarrow Zn$ −0.76 V Reduction

$Mn^{+2} \rightarrow MnO_4^-$ +1.49 V Oxidation

$E^\circ_{cell} = 0.73$ V Positive, so Yes

C. $I^- \rightarrow I_2$ −0.54 V Oxidation

$Ni^{+2} \rightarrow Ni$ −0.23 V Reduction

$E^\circ_{cell} = -0.77$ V Negative, so No

D. $Pb^{+2} \rightarrow Pb$ −0.13 V Reduction

$Zn \rightarrow Zn^{+2}$ +0.76 V Oxidation

$E^\circ_{cell} = +0.63$ V Positive, so Yes

E. $Zn \rightarrow Zn^{+2}$ +0.76 V Oxidation

$Ni^{+2} \rightarrow Ni$ −0.23 V Reduction

$E^\circ_{cell} = +0.53$ V Positive, so Yes

6. $2 \mid Al \rightarrow Al^{+3}$ $\Delta e = 3$

$3 \mid Cl_2 \rightarrow 2Cl^-$ $\Delta e = 2$ $n = 6$

The sum is

$$2Al + 3Cl_2 \rightarrow 2Al^{+3} + 6Cl^-$$

$$Q = [Al^{+3}]^2[Cl^-]^6/(P_{Cl_2})^3 = (0.015)^2(0.025)^6 = 5.5 \times 10^{-14}$$

For the oxidation,

$Al \rightarrow Al^{+3}$ +1.66 V

For the reduction,

$Cl_2 \rightarrow 2Cl^-$ +1.36 V

$E^\circ_{cell} = +3.02$ V

Plugging the values into the Nernst equation

$$E_{cell} = (3.02 \text{ V}) - \{(0.0257/6) \ln 5.5 \times 10^{-14}\}$$

$$E_{cell} = 3.02 \text{ V} - (-0.13 \text{ V})$$

$$E_{cell} = 3.15 \text{ V}$$

7. $Ag \rightarrow Ag^{+}$ $\Delta e = 1$

$Fe^{+3} \rightarrow Fe^{+2}$ $\Delta e = 1$ $n = 1$

The sum equation is

$$Ag + Fe^{+3} \rightarrow Ag^{+} + Fe^{+2}$$

$$Q = [Ag^{+}][Fe^{+2}]/[Fe^{+3}] = (0.005)(0.025)/(0.015) = 8.3 \times 10^{-3}$$

For the oxidation,

$$Ag \rightarrow Ag^{+} \qquad 0.80\ V$$

For the reduction

$$Fe^{+3} \rightarrow Fe^{+2} \qquad +0.77\ V$$

$$E^{\circ}_{cell} = -\ 0.03\ V$$

Plugging the values into the Nernst equation

$$E_{cell} = -0.03\ V - (0.0257 \ln 8.3 \times 10^{-3})$$

$$E_{cell} = -0.03 - (-0.12)$$

$$E_{cell} = 0.09\ V$$

8. $O_2 + 4H^{+} \rightarrow 2H_2O$ $\Delta e = 4$ $E = +1.23\ V$

$4 \mid Fe^{+2} \rightarrow Fe^{+3}$ $\Delta e = 1$ $E = -0.77\ V$

$n = 4$ $E^{\circ}_{cell} = +0.46\ V$

Plugging the values into the equation

$$E^{\circ}_{cell} = 0.0257/n\ \ln K$$

$$0.46 = (0.0257/4)\ \ln K$$

$$71.6 = \ln\ K$$

$$e^{71.6} = K = 1.2 \times 10^{31}$$

9. $(3.61 \times 10^{3}\ g)$ (1 mole Al/26.98 g) (3 mole e^{-}/1 mole Al) $(9.65 \times 10^{4}$ C/ mole $e^{-})$ $= 3.87 \times 10^{7}$ C

10. $(9.65 \times 10^{4}$ amp · s/mole $e^{-})$ (1/500 amp) (2 mole e^{-}/1 mole Cl_2) (1 mole Cl_2/70.90 g) $(1.18 \times 10^{3}\ g) = 6.42 \times 10^{3}\ s = 107\ min = 1.78\ h$

PART IV

MCAT VERBAL REASONING AND WRITING

CHAPTER 1

Verbal Reasoning

Read This Chapter to Learn About:

- ➤ Verbal Reasoning Questions and What They Test
- ➤ How You Can Strengthen Your Reading Comprehension Skills
- ➤ How You Can Improve Your Reading Speed

WHAT VERBAL REASONING TESTS

In contrast to the sections on Physical and Biological Sciences, the Verbal Reasoning section of the MCAT does not test specific knowledge. Instead, it assesses your ability to comprehend, evaluate, apply, and synthesize information from an unfamiliar written text. Its format is familiar to anyone who has attended school in the United States. Most reading comprehension tests look just like it.

The MCAT Verbal Reasoning section consists of 5 or 6 passages of about 500 to 600 words, each of which is followed by a set of multiple-choice questions. There are 40 questions in all. The passages are nonfiction and may be on topics from the humanities, from social sciences, or from those areas of the natural sciences that are not routinely tested elsewhere in the exam. The expectation is that you are not familiar with the content of a given passage, or that if you are familiar with it, you are not an expert.

For this reason, it is not possible to **study** for the Verbal Reasoning section of the MCAT. That being said, however, there are some things you may do to **prepare** for it.

HOW VERBAL REASONING IS SCORED

As with all the multiple-choice sections of the MCAT, you receive 1 point for each correct answer on the Verbal Reasoning section, with no penalty for incorrect answers (which makes it okay to guess!). After that, the numeric score is scaled on a 15-point scale, with extra weight given to the more difficult questions, and less weight given to the easier questions. Your score for this section appears as a number from 0 to 15.

PREPARING FOR VERBAL REASONING

By this stage in your educational career, you should have a pretty good sense of your test-taking skills. If you have achieved solid scores on reading comprehension tests in the past, the MCAT Verbal Reasoning section should be no problem at all. If your comprehension skills are not quite as good as they should be, if you freeze when faced with difficult reading passages, if you read very slowly, or if English is not your first language, you should take the time to work through this section of the book.

Read

The best way to learn to read better is to read more. If you read only materials in your chosen discipline, you are limiting yourself in a way that may show up on your MCAT score. Reading broadly in subject areas that do not, at first glance, hold much appeal for you trains you to focus your attention on what you are reading. Pick up a journal in a field you are not familiar with. Read an article. Summarize the key ideas. Decide whether the author's argument makes sense to you. Think about where the author might go next with his or her argument. Finally, consider how the content of the article relates to your life or to the lives of people you know.

All of this sounds like a chore, but it is the key to making yourself read actively. An active reader interacts with a text rather than bouncing off it. Success on the MCAT Verbal Reasoning section requires active reading.

You can use any of the following strategies to focus your attention on your reading. You may use many of them already, quite automatically. Others may be just what you need to shift your reading comprehension into high gear.

ACTIVE READING STRATEGIES

- **Monitor your understanding.** When faced with a difficult text, it's all too easy to zone out and skip through challenging passages. You do not have that luxury when the text you are reading is only 500 words long and is followed by 8 questions that require your understanding. Pay attention to how you are feeling about a text.

Are you getting the author's main points? Is there something that makes little or no sense? Are there words that you do not know? Figuring out what makes a passage hard for you is the first step toward correcting the problem. Once you figure it out, you can use one of the following strategies to improve your connection to the text.

- **Predict.** Your ability to make predictions is surprisingly important to your ability to read well. If a passage is well organized, you should be able to read the introductory paragraph and have a pretty good sense of where the author is going with the text. Practice this one starting with newspaper articles, where the main ideas are supposed to appear in the first paragraph. Move on to more difficult reading. See whether your expectation of a text holds up through the reading of the text. Making predictions about what you are about to read is an immediate way to engage with the text and keep you engaged throughout your reading.
- **Ask questions.** Keep a running dialog with yourself as you read. You don't have to stop reading; just pause to consider, "What does this mean? Why did the author use this word? Where is he or she going with this argument? Why is this important?" This becomes second nature after a while. When you become acclimated to asking yourself questions as you read a test passage, you may discover that some of the questions you asked appear in different forms on the test itself.
- **Summarize.** You do this when you take notes in class or when you prepare an outline as you study for an exam. Try doing it as you read unfamiliar materials, but do it in your head. At the end of a particularly dense paragraph, try to reduce the author's verbiage to a single, cogent sentence that states the main idea. At the end of a longer passage, see whether you can restate the theme or message in a phrase or two.
- **Connect.** Every piece of writing is a communication between the author and the reader. You connect to a text first by bringing your prior knowledge to that text and last by applying what you learn from the text to some area of your life. Even if you know nothing at all about architecture or archaeology, your lifetime of experience in the world carries a lot of weight as you read an article about those topics. Connecting to a text can lead to "Aha!" moments as you say to yourself, "I knew that!" or even, "I never knew that!" If you barrel through a text passively, you do not give yourself time to connect. You might as well tape the passage and play it under your pillow as you sleep.

Pace Yourself

The Verbal Reasoning section is timed. If you are a slow reader, you are at a decided disadvantage. You have 60 minutes to read 5 or 6 passages and answer 40 questions. That gives you about 10 minutes for each passage and question set. It would be a shame to lose points because you failed to complete a passage or two.

Studies have shown that people read 25 percent slower onscreen than they normally read. Because the MCAT is entirely computer-based, you may benefit from practicing reading longer passages onscreen. Try http://www.bartleby.com/. This website, which bills itself as "Great Books Online," contains a number of long classic works to be read onscreen. You can find fiction and nonfiction of all sorts with which to practice your onscreen reading skills.

You do not need to speed-read to perform well on the Verbal Reasoning section, but you might benefit from some pointers that speed-readers use.

SPEED-READING STRATEGIES

- **Avoid subvocalizing.** It's unlikely that you move your lips while you read, but you may find yourself "saying" the text in your head. This slows you down significantly, because you are slowing down to speech speed instead of revving up to reading speed. You do not need to "say" the words; the connection between your eyes and your brain is perfectly able to bypass that step.
- **Don't regress.** If you don't understand something, you may run your eyes back and forth and back and forth over it. Speed-readers know this as "regression," and it's a big drag on reading speed. It's better to read once all the way through and then reread a section that caused you confusion.
- **Bundle ideas.** Read phrases rather than words. Remember, you are being tested on overall meaning, which is not derived from single words but rather from phrases, sentences, and paragraphs. If you read word by word, your eye stops constantly, and that slows you down. Read bundles of meaning, and your eyes flow over the page, improving both reading speed and comprehension.

Preview

When it comes to taking tests, knowing what to expect is half the battle. The MCAT's Verbal Reasoning section assesses a variety of reading skills, from the most basic comprehension skills to the higher-level application and synthesis skills. Here is a breakdown of skills you should expect to see tested.

FOUR KINDS OF QUESTIONS

- **Comprehension.** These questions look at the author's main idea and support for his or her hypothesis. Expect questions on Finding the Main Idea, Locating Supporting Details or Evidence, Choosing Accurate Summaries or Paraphrases, Comparing and Contrasting, Interpreting Vocabulary, Identifying Hypotheses, and Asking Clarifying Questions.

SAMPLE QUESTION STEMS

In the context of the passage, the word X means . . .
The main argument of the passage is . . .
The central thesis of the passage is . . .
The discussion of X shows primarily that . . .
According to the passage, all of these are true EXCEPT . . .

➤ **Evaluation.** These questions deal with your understanding of the author's assumptions and viewpoints. Expect questions on Analyzing an Argument; Judging Credibility; Assessing Evidence; and Distinguishing Among Fact, Opinion, and Unsupported Assertions.

SAMPLE QUESTION STEMS

Which of the following statements is NOT presented as evidence . . .
The passage suggests that the author would most likely believe . . .
Which of the following assertions does the author support . . .
The author's claim that X is supported by . . .

➤ **Application.** These questions deal with the purpose and structure of the passage and may require you to apply concepts or hypotheses to real-life situations. Expect questions on Making Predictions, Solving Problems, Identifying Cause and Effect, Drawing Conclusions, and Making Generalizations.

SAMPLE QUESTION STEMS

According to the passage, why . . .
The passage implies that . . .
Based on the information in the passage, which of these outcomes . . .
The passage suggests that . . .
According to the passage, X would best be described as . . .

➤ **Synthesis.** These questions work outward from the specific material presented in the passage. Expect questions on Applying New Evidence, Modifying Conclusions, Devising Alternate Solutions, and Combining Information.

SAMPLE QUESTION STEMS

The ideas in this passage would be most useful to . . .
Which of the following claims would weaken the author's argument . . .
Which new information would most challenge the claim . . .
How would this information affect the claim . . .

Because the format of the test is a familiar one, you may not need to preview the format itself. However, you may benefit from these tips on taking reading comprehension tests.

Test-Taking Tips for Verbal Reasoning

1. **Preview** the passage. Read the first paragraph. Skim the passage.
2. **Skim** the question stems (the part of each question that does not include the answer choices). This gives you a quick idea of what to look for as you read.
3. **Read** the passage, using your active reading strategies.
4. **Answer** the questions. Questions on the Verbal Reasoning section of the MCAT move from easiest to hardest within a question set, so answering them in order makes sense. However, if you are stumped on a given question, skip ahead and come back later.

PRACTICING VERBAL REASONING

It is certainly true that the more you practice reading comprehension, the better you are likely to perform on the MCAT. Here are 10 practice passages followed by question sets and explanatory answers. Follow along and see how well your comprehension compares with the answers given. Remember that the easier questions come first in a question set. Notice whether you have trouble with those more basic questions or with the higher-level questions that follow. Try to use your active reading strategies as you read each passage. Can you make it through each passage and question set in 10 minutes or so?

SAMPLE PASSAGE 1: SOCIAL SCIENCES

Fifty years ago, only New Yorkers lived in what is now termed a "megacity," an agglomeration of more than 10 million people living and working in an urban environment. In contrast, today there are more than 40 megacities, most in less developed countries, and more urban centers are expected to explode in population by the year 2015. Demographers and globalization experts are already referring to the 21st century as "the urban century."

Already, more people on our Earth live in cities than live in rural areas. This is an enormous change in population trends, and it skews the entire planet in ways we haven't begun to analyze.

Although some cities have seen immigration expand their borders, for most megacities, it is migration from within the country that has caused the city to grow. An example is China, where some 150 million rural inhabitants have migrated to cities in just the last ten years. In many cases, the cities house the only possibilities of employment in this global economy. That is what has grown Mombai (Bombay), India, from a large city to a megacity of more than 18 million in just a few years. It's the cause of the explosion of population in Lagos, Nigeria; Karachi, Pakistan; Dhaka, Bangladesh; and Jakarta, Indonesia.

Whereas just a few years ago, most large cities were in developed nations, now the largest are suddenly in the less developed countries of South America, Africa, and Asia. Imagine the pressure on the infrastructure of these already poor cities as the influx of workers pushes services to the breaking point. Slums and shantytowns spring up around the outskirts of the cities, and government is powerless to affect the disadvantaged workers, leaving them exposed to corrupt local officials or urban gangs. Imagine, too, what happens in the rural areas that these people have abandoned. China faces a desperate shortage of agricultural labor. So do other areas of Asia and Africa.

According to UN statistics, by the year 2030, more than 60 percent of the world population will be urban, up from 30 percent in 1950. Unlike the population growth in developed nations, the birth rate in less developed nations is high, meaning that the cities continue to grow even as migration slows from the rural areas. Megacities such as New York have populations that have leveled off over time. Despite its location in a less developed nation, even a megacity such as Mexico City has a slow rate of growth compared to Asian and African cities such as Mombai or Lagos.

It is difficult to imagine what the growth of the megacities will mean to the world in the 21st century. Demographers foresee ecological overload, homelessness, uncontrolled traffic, and an infrastructure strained to the breaking point. Despite the notion that industrial jobs improve the lot of the workers, it is already possible to see that megacities are creating a new, even deeper division between rich and poor, as the poor concentrate in the outskirts of town and the rich barricade themselves behind walls and in towers.

1. The main argument of the passage is that

A. megacities are more often found in less developed nations but strain the resources of developed nations.

B. the growth in population and number of megacities means foreseeable changes, many of them negative.

C. the movement of population bases from rural to urban locations decimates the countryside and limits our ability to grow food.

D. we must begin to fight back against the growth of megacities in the less developed nations of the world.

2. The passage suggests that demographers

A. have not been able to keep pace with the growth of cities.

B. focus primarily on population trends in the developing world.

C. are observing the growth of the world's cities with concern.

D. work hand in hand with the UN to plan for the future.

3. The author's use of UN statistics helps

   A. strengthen her argument that urbanization is radically changing the world.

   B. contradict demographers' claims about megacities and their effects.

   C. indicate that the results of urbanization include poverty and crime.

   D. complement her assertion that birth rate is the main reason for urban growth.

4. According to this passage, why might skyscrapers be a sign of divisiveness?

   A. They cost too much to build.

   B. They are found only in developed nations.

   C. They separate rich from poor.

   D. They house businesses, not people.

5. Which new information, if true, might CHALLENGE the author's contention that cities will continue to grow despite a slowing of migration from the countryside?

   A. Scientists are creating new strains of rice and wheat that require far less in the way of hands-on care.

   B. The number of people living below the poverty level will climb in less developed and developed nations.

   C. Inflationary trends in heating oil and gasoline prices will limit most people's discretionary spending.

   D. New methods of birth control will limit the population explosion in the developing world.

SAMPLE PASSAGE 1: ANSWERS AND EXPLANATIONS

1. **B** This is a **Comprehension** question on **Finding the Main Idea**. This kind of question does not require you to go beyond the boundaries of the passage. You should think, "What is the author trying to say?" In fact, the author never says choice A at all. Although the point is made that megacities are more often found in less developed nations, the second half of that statement does not appear in the passage. Nor does the author discuss decimation of the countryside and limitations in our ability to grow food, as choice C would indicate. And although you might infer choice D, the author never makes any such assertion. The best answer, the one that best conforms simply to the words on the page, is choice B.

2. **C** This **Comprehension** question has to do with **Supporting Details and Evidence**. You can answer it easily if you scan the passage for the word **demographer** and then see what information is directly presented. In paragraph 1, demographers "are referring to the 21st century as 'the urban century.'" In paragraph 6,

"demographers foresee ecological overload, homelessness, uncontrolled traffic, infrastructure strained to the breaking point." Based on your quick scan of the passage, there is no evidence to support choices A, B, or D. The answer is clearly choice C.

3. **A** This **Evaluation** question requires you to **Analyze an Argument**. Questions like this ask you to explain why an author included certain information. Scan the essay to locate the reference to UN statistics and you see that those statistics tell us, "by the year 2030, more than 60 percent of the world population will be urban, up from 30 percent in 1950." In other words, there has been a dramatic change in the look of the world, thanks to urbanization. This correlates to choice A. It does not contradict demographers' claims (choice B), nor does it say anything about poverty, crime, or birth rate (choices C and D). Notice that this kind of question asks you to think "outside" the passage a bit more than Comprehension questions 1 and 2 did.

4. **C** This **Application** question looks at **Cause and Effect**, not as directly presented in the essay, but rather as implied by the essay. In fact, you may have been alarmed to see that the word **skyscraper** never appears in the passage at all, so that scanning the passage does not help. Remember that questions get harder within a question set. This question, the fourth one in the set, is harder than the first three.

   The clue to the answer is in the last line of the essay. "Megacities," the author claims, "are creating a new, even deeper division between rich and poor, as the poor concentrate in the outskirts of town and the rich barricade themselves behind walls and in towers." Those "towers" are the high-rise apartment buildings in which the rich dwell, whereas the poor live in shanties on the cities' outskirts. The only answer supported by the text is choice C.

5. **D** This **Synthesis** question asks you to **Apply New Evidence** to an existing argument. This kind of question takes you furthest from the passage itself, in an effort to get you to recognize its real-world applications. You might begin by scanning to locate the part of the passage that deals with the contention mentioned: that cities will continue to grow despite a slowing of migration from the countryside. It appears in paragraph 5, and its causative link is the notion that "the birth rate in less developed nations is high." If you take away this cause, as choice D would do, then the author's supposition that cities will continue to grow has no foundation. None of the other answer choices would collapse her argument as well.

SAMPLE PASSAGE 2: HUMANITIES

Muralism has long been a Mexican tradition, perhaps dating back to the Aztecs, who recorded their history on the walls of their pyramids. The covering of a white wall with political art made the careers of David Alfara Siqueiros, Jose Clemente Orozco, and the best-known of them all, Diego Rivera.

Siqueiros, born in Chihuahua, studied art from an early age. He organized a student strike at the age of 15 and later worked to unseat the Mexican dictator Huerta, attaining the rank of captain during the revolution that was taking place. He later brought his tactical knowledge to the world of organized labor, where his activism led to lengthy jail terms. That is where he created some of his finest artworks on canvas. During the 1930s, he went to Spain to join the anti-fascist forces. His life was that of a soldier–artist, and some considered him a dangerous, subversive gangster.

Orozco, too, studied art as a youth and was inspired by the Mexican Revolution. One of his famous murals depicts the Holy Trinity as a worker, a soldier, and a peasant. Later he turned his focus to the dehumanizing effect of large cities on the people who live there. When he wasn't painting vast murals, Orozco was drawing political cartoons.

Rivera, the third of these Mexican Social Realists, *los tres grandes*, remains the most famous through sheer force of personality. His storytelling, his love affairs, his radicalism, and his love–hate relationship with the land of his birth informed his life and his paintings. He incorporated Mexican folklore and cultural icons into his murals in an effort to educate working people in their own history.

In the Chicano neighborhoods of the southwestern United States, political muralism still explodes onto bare walls in the form of graffiti. Edward Seymour's 1949 invention of canned spray paint provided would-be artists with an easy mode of expression, and the graffiti mural took off as an art form in the 1960s and 1970s. It began as outlaw art, which surely would have appealed to an outlaw such as Siqueiros. Despite the new artists' lack of formal training, some members of the outlaw group managed to create something beautiful while making political statements about poverty, injustice, diversity, and racism.

One extraordinary thing about this kind of public art is that it is truly for everyone. You do not need to enter the halls of a museum to see it; it resides on the walls of your local bodega or school or health clinic; you bounce your ball off of it in the basketball or handball court; you cover it over with posters for your favorite band or flyers about your lost pet.

Certainly, many of the graffiti artists were reviled as nuisances, and their art was erased. For some, however, graffiti would prove a launching point into the world of fine art. Today, modern murals in Austin, San Antonio, Los Angeles, and Tucson, among others, attest to the power of the Mexican tradition of the muralist as purveyor of political thought.

6. In the context of the passage, the word *radicalism* means

A. extremism.

B. intolerance.

C. discrimination.

D. fanaticism.

7. The author probably mentions Orozco's political cartoons as a way of illustrating

A. the lack of seriousness in Orozco's art.

B. how multitalented Orozco was.

C. Orozco's intertwining of politics and art.

D. why Orozco's work fell out of favor.

8. The author's claim that Siqueiros might approve of Chicano graffiti is supported by

A. details about Siqueiros's role in the Spanish Civil War.

B. the description of Siqueiros as an army captain.

C. the fact that Siqueirios moved from Mexico to Spain.

D. information about Siqueiros's gangster past.

9. According to this passage, graffiti would best be described as

A. an attractive way of making a political statement.

B. a pale imitation of the Mexican muralists' work.

C. a nuisance that must be tolerated by urbanites.

D. a way to educate the masses in their own history.

10. In 1933, Diego Rivera was dismissed from his job painting a mural for Rockefeller Center and charged with willful propagandizing for including a portrait of Lenin in the center. How does this anecdote affect the author's contention that Rivera's goal was to educate the workers in their own history?

A. It refutes it.

B. It supports it.

C. It supports the claim only if you believe that propaganda is educational.

D. It refutes the claim only if you believe that Communism is part of Mexican history.

SAMPLE PASSAGE 2: ANSWERS AND EXPLANATIONS

6. A This **Comprehension** question has to do with **Interpreting Vocabulary**. The first thing to do is to scan the passage and search for the word **radicalism**. It appears in paragraph 4, in the context of qualities that informed Rivera's artwork. It is not used in a negative connotation, as choices B and C would indicate.

Choice D is almost right, but again, it has a more negative connotation than the passage implies. Therefore, choice A is the best answer.

7. **C** This **Evaluation** question asks you to **Analyze an Argument**. Authors rarely include information without a reason, and the MCAT often asks you to identify and assess the reasons behind the inclusion of a passage or phrase. Here, Orozco's art is discussed in its relation to politics, from his Trinity of workers to his political cartoons. Although choice B might be correct in a different context, choice C is the better answer.

8. **D** This **Evaluation** question requires you to **Assess Evidence**. Ideally, a writer does not include a statement without adequate support. Here, the statement is that Siqueiros might approve of Chicano graffiti. If you scan to find the exact reference in paragraph 5, you see that the author states, "It began as outlaw art, which surely would have appealed to an outlaw such as Siqueiros." That is a clear giveaway that choice D is the best response; Chicano art was outlaw art, and Siqueiros was considered by many to be an outlaw, or gangster.

9. **A** This **Application** question calls for **Making Generalizations** about a topic—in this case, graffiti. To do this correctly, you must put together the author's statements about the topic and draw a conclusion from the information you are given.

   Paragraphs on graffiti appear at the end of the passage. Although the author refers to it as "outlaw art" and mentions that some people considered it a nuisance, that is clearly not the author's own impression. Phrases such as "managed to create something beautiful while making political statements about poverty, injustice, diversity, and racism" and "public art . . . that . . . is truly for everyone" put a positive spin on the topic. The author does not believe that graffiti is a "pale imitation" (choice B); instead, she indicates that the graffiti artists are following in the muralists' tradition. Choice D appears as a description of Rivera's work and is not relevant to this discussion of graffiti. That makes choice A the most logical answer.

10. **B** This **Synthesis** question asks you to **Apply New Evidence** to an existing argument. It also requires you to have in your repertoire some background information that is not found in the passage itself. The inclusion of Lenin in a mural would tend to support the notion that Rivera's intention was to educate the workers, primarily because a large part of Lenin's philosophy has to do with the aims and needs of the working class. You do not need to believe that propaganda is educational (choice C) to understand the connection of Lenin to workers. Believing that Communism is a part of Mexican history would serve to support the claim, not to refute it (choice D). The best answer is choice B.

SAMPLE PASSAGE 3: NATURAL SCIENCES

For years, anecdotal evidence from around the world indicated that amphibians were under siege. Finally, proof of this hypothesis is available, thanks to the concerted, Internet-based effort of scientists involved with the Global Amphibian Assessment.

Amphibians have a unique vulnerability to environmental changes thanks to their permeable skin and their need of specific habitats to allow their metamorphosis from larva to adult. Studies indicate that they are at risk due to global climactic change, reduction in the ozone layer leading to an increased exposure to ultraviolet rays, interference with migratory pathways, drainage of wetlands, pollution by pesticides, erosion and sedimentation, and exposure to unknown pathogens thanks to the introduction of non-native species. In other words, human progress is responsible for the losses this population is suffering.

The permeable skin of frogs, newts, and other amphibians provides easy entry for a variety of pollutants. Increases in ultraviolet radiation appear to have a deadly effect on eggs, causing mortality and deformities in the amphibians studied. Most amphibians need a clear pathway between land and water to complete their life cycle. Road building and swamp drainage for construction has eliminated many of the dispersal routes for amphibians worldwide. The same sensitivity that allows pollution to damage young amphibians makes them susceptible to fungal and other pathogens transmitted from species released by pet owners. One emerging infectious disease in particular has led to entire populations being wiped out in Australia and the Americas.

Scientists have long considered amphibians a barometer of environmental health. In areas where amphibians are declining precipitously, environmental degradation is thought to be a major cause. Amphibians are not adaptable. They must have clean water in which to lay their eggs. They must have clean air to breathe after they grow to adulthood. Their "double life" as aquatic and land-dwelling animals means that they are at risk of a double dose of pollutants and other hazards.

The Global Amphibian Assessment concluded that nearly one-third of the world's amphibian species are under immediate threat of extinction. Nearly half of all species are declining in population. The largest numbers of threatened species are in Colombia, Mexico, and Ecuador, but the highest percentages of threatened species are in the Caribbean. In Haiti, for example, nine out of ten species of amphibians are threatened. In Jamaica, it's eight out of ten, and in Puerto Rico, seven out of ten. Of all the species studied around the world, only 1 percent saw any kind of increase in population.

Certainly, this is a disaster for amphibians, but scientists rush to point out that it may be equally a disaster for the rest of us on Earth. Even recent pandemics among amphibians may be caused by global changes. True, amphibians are ultrasensitive to such changes, but can reptiles, fish, birds, and mammals be far behind? The threat to equatorial amphibians may be simply the first indication of a global catastrophe to come. The frogs and newts of our world are warning us that continued habitat destruction in the name of progress might ultimately destroy us all.

11. The central thesis of the passage is that

A. the extinction of amphibians is due to global warming.

B. amphibians really are a barometer of environmental health.

C. only equatorial amphibians are currently under siege.

D. amphibians' "double life" on land and in water may end up saving them.

12. The passage implies that the Global Amphibian Assessment has done science a favor by

A. setting forth a hypothesis that connects the environment to species decline.

B. eliminating the need to study the connection between extinction and environment.

C. refuting a contention that had existed purely through anecdotal evidence.

D. collecting data to prove something that was previously just a hypothesis.

13. Which of the following assertions does the author support with an example?

I. The permeable skin of amphibians allows for the entry of pollutants.

II. Amphibians are susceptible to unfamiliar pathogens.

III. Most threatened species are in the Caribbean.

A. I only

B. III only

C. I and II only

D. II and III only

14. Information from this passage might be most valuable to

A. a high school or middle school science teacher.

B. a doctor with a specialty in tropical diseases.

C. a representative to a conference on global warming.

D. a pet store owner with a specialty in rare breeds.

15. According to this passage, the continued degradation of the environment may well lead to

A. global pandemics.

B. floods and droughts.

C. human deformities.

D. decline among insects.

Format DOs and DON'Ts

The MCAT uses a special format for certain questions. It lists three choices headed with Roman numerals I, II, and III and asks you to decide which one—or which combination of two or three—is correct. DON'T assume that the answer will always be a combination. DO use logical thinking to eliminate choices and save time.

Suppose the answers are

A. I only

B. III only

C. I and II only

D. II and III only

If you know that I is NOT possible, then you have eliminated choices A and C as answers. If you know that II IS possible, then you are down to choices C and D.

SAMPLE PASSAGE 3: ANSWERS AND EXPLANATIONS

11. B This **Comprehension** question asks you to **Find the Main Idea**. It may be the most common question you find on the MCAT Verbal Reasoning Test, and it nearly always appears first in a question set. To find the main idea, you must read actively and summarize as you go. There are no titles on MCAT passages to give away the central thesis; it's up to you as a reader to derive it from the text.

Review the answer choices. The author indicates that choice A is a possibility. Just because this detail is included does not make it the central thesis. Although the major threat is to equatorial amphibians (choice C), it is not true that "only" those species are in danger, nor is that the main idea. There is no support for choice D. The main idea, instead, revolves around choice B. The new study indicates that previous anecdotal evidence was correct: Amphibians really **are** a barometer of environmental health.

12. D This **Application** question requires you to **Draw Conclusions** about an assertion that is not directly stated. The answer may be inferred from the opening paragraph, which thanks the scientists involved for offering "proof of this hypothesis" through their "concerted, Internet-based efforts." The Assessment did not set forth the hypothesis (choice A); that had already been done via anecdotal evidence. They certainly did not eliminate the need for more study (choice B); nor did they refute the contention (choice C). They proved it (choice D).

13. **D** This is an **Evaluation** question that asks about **Unsupported Assertions**. It often appears in this particular format on the MCAT (see box). Unless you have a photographic memory, the only way to answer a question like this one is to return to the text and find the assertions listed, which are most likely not worded quite the way they are in the question. Assertion I appears in paragraph 3. The paragraph goes on to discuss other problems with permeability, but it never gives specific examples of pollutants harming amphibians. Because assertion I is unsupported, choices A and C are incorrect. Because choices B and D both include assertion III, you can assume that assertion III is supported in the text (as it is, with a list of countries). The only question you should have is about assertion II. This assertion is made toward the end of paragraph 3, and it is, in fact, supported by an example: "One emerging infectious disease in particular has led to entire populations being wiped out in Australia and the Americas." So, the best answer is choice D.

14. **C** This is one sort of **Synthesis** question that has to do with **Combining Information**. The question is, "What possible practical application might there be for this passage?" The answer is the application that best reflects the information provided in the passage. Although a teacher (choice A) might use the information in class, it would not be of immediate value there. A doctor (choice B) or a pet store owner (choice D) might be interested in the passage, but again, its value would be less than it would for a representative to a conference on global warming (choice C), who might be able to apply the information directly to an argument there.

15. **A** This **Application** question asks you to **Make Predictions**, which you must do based on the text rather than on any external knowledge of the topic you may have. The author never speaks of floods and droughts (choice B), so even though they may be an effect of environmental degradation, they cannot be the correct answer. Likewise, insects (choice D) are never mentioned, although "reptiles, fish, birds, and mammals" are. Choice C is a stretch, although it is possible. The final paragraph deals with an extrapolation of the information about amphibians to other species, and it mentions the idea that even the pandemics affecting amphibians now may be caused by global changes (which are, in turn, caused by environmental degradation). This makes choice A the best answer of the four.

Notice in this sample set that the most difficult question is not the one that requires highest-order thinking, or Synthesis. Although in most cases, Comprehension questions will appear first and Synthesis questions last, the MCAT order really depends on the difficulty of concepts rather than the hierarchy of thinking skills.

SAMPLE PASSAGE 4: HUMANITIES

The best-known poet in the history of Australian literature is a man who never existed. According to the story, Ernest Lalor Malley was born in 1918 and emigrated to

Australia as a child. He worked as an insurance salesman and wrote poetry as a sideline and hobby, a fact discovered only after his death from Grave's disease at age 25. At that time, his sister Ethel, while going through Malley's meager possessions, discovered a sheaf of poems.

She sent the poems to a young editor named Max Harris, who ran a modernist magazine called *Angry Penguins*. The magazine had come under fire from many artistic conservatives, who considered the new modernist poetry humorless and nonsensical. Among these conservatives were the young poets Harold Stewart and James McAuley.

Harris adored Ern Malley's poems, which numbered 17 in all, and promptly produced a special edition of *Angry Penguins* that would contain them all. Immediately, controversy arose. Although many of Harris's colleagues found the poems fascinating, others began to question and even to ridicule the poems, claiming that they represented a hoax perpetrated on the literary scene by Max Harris.

The controversy found its way into the local newspapers, which made it their business to track down Ern Malley and put an end to the story one way or the other. It took only a few weeks for the Sydney *Sun* to determine that poets Harold Stewart and James McAuley had composed all 17 poems in a single afternoon, paraphrasing chunks of text from a dictionary of quotations, the works of Shakespeare, and whatever came into their minds. They had then invented the backstory for the hapless Ern Malley and submitted the whole to their least favorite editor, Max Harris.

The point, of course, was to prove that modernist poetry was indiscriminate and meaningless, and to some degree, that point prevailed. *Angry Penguins* closed up shop, and the modernist movement in Australia was completely derailed. Max Harris was actually tried and convicted for publishing obscene poetry. Perversely, he continued throughout his life to insist that hoax or not, Stewart and McAuley had managed to write real poetry, poetry that had meaning and substance. Surprisingly, others agreed. Although some critics who had leaped onto the Malley bandwagon when the poems first appeared now backpedaled furiously, others insisted that the poems had literary merit.

The poems continued to have a life of their own, and even today they are frequently reissued and anthologized. American poets John Ashbery and other members of the New York School of poetry, who championed a kind of surrealist, transcendental verse, became major devotees of the mythical Malley. According to Michael Heyward in *The Ern Malley Affair*, Ashbery would regularly print a legitimate poem next to a Malley poem for his creative writing class's final exam, and would ask students to identify the hoax. The results were usually about 50–50.

So, what is the moral of this cautionary tale? Certainly an editor should never be so trusting, and the provenance of all written works should be carefully researched. However, because Malley's work today remains better known than that of Stewart or McAuley under their real names, perhaps the lesson has to do with never taking oneself too seriously.

16. According to the passage, at the time of the hoax, McAuley and Stewart were all of these things EXCEPT

A. Australian.

B. conservative.

C. youthful.

D. salesmen.

17. Which of the following statements is NOT presented as evidence that the Ern Malley hoax had a profound influence on the direction of Australian poetry?

A. *Angry Penguins* closed up shop.

B. The New York School championed surrealist verse.

C. The modernist movement was completely derailed.

D. Some critics backpedaled furiously.

18. The passage implies that McAuley and Stewart were motivated by

A. greed.

B. principle.

C. spite.

D. rage.

19. Given the information in this passage, if you were given two unfamiliar poems, one by Ern Malley and the other by a distinguished poet, which of these outcomes would most likely occur?

A. You would be as likely to praise the Ern Malley poem as the real poem.

B. You would identify the Ern Malley poem as the legitimate work.

C. You would be able to discern the true poem from the hoax.

D. You would think both poems had equal merit.

20. Which of the following incidents best supports the author's belief that editors should never be as trusting as Max Harris was?

A. For centuries, some scholars have insisted that either Bacon or Marlowe wrote works attributed to Shakespeare.

B. Reporter Stephen Glass repeatedly falsified quotations in widely read stories for the *New Republic*.

C. Investigative reporter Seymour Hersh first broke the story of the massacre of civilians at My Lai.

D. Editor Harold Ross encouraged writers for the *New Yorker* to publish the truth always tinged with humor.

Format DOs and DON'Ts

Watch out for questions that require you to identify a negative. Questions that are worded in the following ways may trick you:

"All of these are true EXCEPT . . ."

"Which of these statements is NOT . . ."

DO look for the capitalized word in the question stem. The MCAT style is to capitalize the negative word. That's your clue that the answer requires identification of a negative. DON'T fall into the trap of choosing choice A because it's true when the question really asks you to find the statement that's false.

SAMPLE PASSAGE 4: ANSWERS AND EXPLANATIONS

16. D This identifying-a-negative format (see box) occurs from time to time on the MCAT's Verbal section. This **Comprehension** question tests your ability to **Locate Supporting Details or Evidence**. In questions of this kind, everything you need to know is found directly in the text. The last line of paragraph 2 contains much of what you need: "Among these conservatives were the young poets Harold Stewart and James McAuley." The men were conservative (choice B) and youthful (choice C). Because everyone in the story is Australian save for the poets of the New York School mentioned at the end, it is safe to assume that choice A is true of McAuley and Stewart as well. Only Ern Malley is identified as a salesman (choice D); there is no indication that McAuley or Stewart is anything other than a poet.

17. B This **Evaluation** question asks you to **Assess Evidence**. Look at questions like this one as you would mathematical proofs. The hypothesis is given in the question stem: The Ern Malley hoax had a profound influence on the direction of Australian poetry. You must then determine which of the answer choices supports that hypothesis and which one does not. The one that does not is the correct answer.

Here, the fact that the literary magazine closed following the hoax (choice A) does support the hypothesis. The fact that the modernist movement was derailed (choice C) does support the hypothesis. The fact that some critics backpedaled (choice D) does support the hypothesis. The fact that the New York School championed surrealist verse (choice B) does not; the New York School is American, not Australian, and its love of surrealist verse has nothing to do with the effect of the hoax on Australian poetry.

18. C This **Application** question asks you to **Draw Conclusions** about characters' motivations. It is a skill most often applied to fiction, but it works in this case for nonfiction as well. The word *implies* tells you that the answer is not directly stated. It is up to you to return to the passage and make inferences based on your own understanding of human nature. Although the author indicates that the two men wanted to prove the ridiculousness of modernist poetry, which might be a matter of principle (choice B), their use of the innocent Max Harris seems spiteful, making choice C the better answer. There is no support for either choice A or choice D. McAuley and Stewart have nothing to gain monetarily by the hoax, and they seem amused rather than irate.

19. A This **Application** question requires you to **Make Predictions** about a hypothetical situation, based on the information that was laid out for you in the passage. The important clue is in paragraph 6, which mentions that John Ashbery used to perform just this experiment for his classes, with results that approximated 50–50. That information supports choice A, which calls for the same results. Because even professional critics were fooled, there is no reason to believe you would do any better or worse than Ashbery's students did.

20. B This **Synthesis** question requires you to **Combine Information** from the passage with new information to come up with a conclusion. The belief as stated is that "editors should never be as trusting as Max Harris was," and your job is to find the one answer choice that best supports this premise. Of the answers given, choices C and D pretty clearly do not support the premise, because they do not involve a question of trust between editor and writer. Choices A and B may involve trust, but in the case of choice A, it is more between editor and reader than editor and writer. Therefore choice B, in which a writer included false information in published materials, is the choice that best reflects the author's belief.

SAMPLE PASSAGE 5: SOCIAL SCIENCES

". . . for not an orphan in the wide world can
be so deserted as the child who is an outcast
from a living parent's love."

—Charles Dickens, *Dombey & Son*

Of course, in Dickens's day, and indeed, into more modern times, the child outcast from a living parent's love was likely to find him- or herself in the poorhouse or the workhouse or out on the streets. Today, in the United States, we deal with many such children—more than half a million at any one time—through the foster care system and the department of social services.

Certainly most of the children who find themselves in the foster care system are not really "outcast from a living parent's love," but most do have living parents.

The goal of the foster care system is to reunite families that have been torn apart by circumstance.

The circumstances are many. According to the U.S. Department of Health and Human Services, about 60 percent of foster children come into the system for their own protection; in other words, as a result of abuse or neglect or both. Around 16 percent are there because their parents are "absent," perhaps in prison, perhaps simply vanished into the void of drug abuse, alcoholism, or mental illness. Another 14 percent or so are in the system due to their own criminal or delinquent behavior. A small minority, say 4 percent, are in foster care because their birth parent or parents cannot cope with their child's handicap.

Less than 1 percent of foster children are in the system because their parents have relinquished all rights to them. For the remaining 99 percent, then, the sponsoring agencies must try to create a plan by which the child and family can be reconciled. Sometimes this happens quickly; 20 percent of children remain in care for less than 6 months. Sometimes it happens slowly or not at all; 32 percent of children remain in care for more than 3 years.

In April 2005, the Casey Family Programs published a study they had performed in connection with several universities in which they looked at hundreds of former foster children in Washington and Oregon. Although each state differs in the rules for its foster care system, most withdraw care when a child turns 18. The Casey study looked at young adults between the ages of 20 and 33, and what they found was disturbing. Following their release from the system, around 25 percent of the young adults had been homeless for at least some period of time. At the point of the study, about 35 percent were living at or below the poverty level. More than 50 percent suffered from one or more mental health problems, from anxiety to depression to more serious issues.

Of the foster children studied, 56 percent had received a high school diploma, whereas 29 percent had received a GED instead. In the general population, 82 percent graduate from high school, and 5 percent receive a GED.

Other studies tend to corroborate the notion that foster care is not necessarily a positive step on the road to adulthood. According to the National Association of Social Workers, some 80 percent of prison inmates spent part of their childhood in the foster care system. And although that particular statistic might reveal more about the circumstances that led children to foster care than about the system itself, the National Center on Child Abuse and Neglect did find that children were 11 times more likely to be abused in state care than in their own homes. Could the Dickensian alternatives be worse?

21. In the context of the passage, the word *reconciled* means

A. acquiescent.

B. resigned.

C. reunited.

D. resolved.

22. The central thesis of the passage is that

A. foster care has greatly improved since Dickens's day.

B. foster care is not as beneficial as it should be.

C. foster care relies on the kindness of strangers.

D. foster care's goal is the creation of a new, functional family.

23. Based on the information in the passage, which of these outcomes is MOST likely when a young boy is placed in foster care?

A. He will be reunited with his family within 3 years.

B. He will spend part of his adulthood in prison.

C. He will remain in the system due to delinquent behavior.

D. He will be moved between two or three foster homes.

24. The passage implies that

A. withdrawal of care at age 18 may be detrimental to foster children.

B. abuse and neglect are just as prevalent in foster homes as on the street.

C. handicapped children are twice as likely to end up in foster care.

D. a GED diploma is equivalent in value to a high school degree.

25. Suppose a new study of former foster children found that those who were in foster care due to incorrigibility and delinquent behavior were those who later found themselves in prison for some part of their adulthood. This new information would most CHALLENGE the implication that

A. foster care does not succeed in reuniting families.

B. state care alone increases the odds of future incarceration.

C. children are more likely to be abused in foster care.

D. many children enter foster care because of absent parents.

SAMPLE PASSAGE 5: ANSWERS AND EXPLANATIONS

21. C This **Comprehension** question asks you to **Interpret Vocabulary**. The word in question appears in the fourth paragraph of the passage. The child and family are to be reconciled, which means that they are to be brought back together. Although choices A, B, and D are synonyms for *reconciled*, only choice C has this denotation of bringing back together.

22. B To answer this **Comprehension** question on **Finding the Main Idea** correctly, ask yourself, "What point is the author trying to make?" In quoting the largely negative Casey Family Programs study, it seems clear that the point is not choice A. If you have any doubt about that, the final sentence of the passage should clarify the author's stance. Choice C is never mentioned, and choice D is wrong—the goal of foster care, according to the passage, is "to create a plan by which the child and family can be reconciled." The best summary of the central thesis is choice B.

23. A This is an **Application** question that asks you to **Make Predictions.** The key to this question is the phrase "MOST likely." Choice B has some likelihood. We do know that 80 percent of prisoners spent some time in foster care, but without having the numbers of prisoners compared to the number of foster children, we cannot calculate those odds. Choices C and D probably have some likelihood as well, but neither is mentioned in the passage. Statistics in paragraph 4 indicate that for 99 percent of foster children, the goal is reconciliation. Of those, 20 percent are reconciled within 6 months, and 68 percent are reconciled within 3 years. Choice A is correct.

24. A You must **Draw Conclusions** to answer this **Application** question. To draw a conclusion, you must use information that actually exists in the passage. Abuse and neglect are clearly prevalent in foster homes (choice B), but whether they are as prevalent there as on the street is never implied. Only 4 percent of the children in foster care are there because of some handicap, and because we do not know total figures, we cannot draw the conclusion that is given in choice C. The implication in choice D is challenged by the author's statistics in paragraph 6. If a GED diploma were equivalent in value to a high school degree, it would hardly matter that foster children are much more likely to get GED diplomas than are children in the general population. The Casey study looked at young adults who were formerly in foster care and found that only a few years after their release from the system, many were doing poorly. The implication is that without the care, they often fail, so choice A is the best answer.

25. B This **Synthesis** question requires you to **Apply New Evidence**. The new study would indicate that the delinquency itself led to future incarceration, and that the foster care system was only incidental. That in turn would mean that foster care on its own was not the cause of future incarceration (choice B). The new study would not affect choices A, C, or D at all.

SAMPLE PASSAGE 6: NATURAL SCIENCES

A land bridge is land exposed when the sea recedes, connecting one expanse of land to another. One land bridge that still exists today is the Sinai Peninsula, which attaches the Middle East to North Africa. Central America is another land bridge, this one connecting North to South America.

Historical land bridges are many. There was a bridge connecting the British Isles to the European continent, and there was one between Spain and Morocco at what is now the Straits of Gibraltar. There were bridges connecting Japan to China and Korea. One of the most famous land bridges was the Bering Land Bridge, often known as Beringia, which connected Alaska to Siberia across what is now the Bering Strait.

The Bering Land Bridge was not terribly long. If it still existed today, you could drive it in your car in about an hour. It appeared during the Ice Age, when enormous sheets of ice covered much of Europe and America. The ice sheets contained huge amounts of water north of the Equator, and because of this, the sea level dropped precipitously, perhaps as much as 400 feet, revealing landmasses such as the Bering Land Bridge.

At this time, the ecology of the northern hemisphere was that of the Mammoth Steppe. It was a dry, frigid land filled with grasses, sedges, and tundra vegetation. It supported many large, grazing animals including reindeer, bison, and musk oxen, as well as the lions that fed on them. It also contained large camels, giant short-faced bears, and woolly mammoths.

The Bering Land Bridge may have been somewhat wetter than other areas of the Mammoth Steppe, because it was bordered north and south by ocean and fed by ocean breezes. Many of the animals of the Mammoth Steppe used the bridge to cross from east to west and back again. Eventually, their human hunters tracked them from Asia to North America.

Ethnologists and geologists generally believe that humans used the Bering Land Bridge to populate the Americas, which up until about 24,000 years ago had no sign of human life. Ethnologists use evidence such as shared religions, similar houses and tools, and unique methods of cleaning and preserving food to show the link between the people of coastal Siberia and the people of coastal Alaska.

There are those among the Native American population who dispute the land bridge theory. For one thing, it contradicts most native teachings on the origins of the people. For another, it seems to undermine the notion that they are truly "native" to the North American continent.

The waters returned about 12,000 years ago, and today global warming means that the sea level is rising steadily. It is difficult to stand on the shore of the Seward Peninsula today and picture walking across to Russia. It is easier to imagine if you look at a map. The borders are so close that the continents seem to kiss. Just a quick holding back of the waters, à la Moses in *The Ten Commandments*, and you can picture a Pleistocene hunter chasing a woolly mammoth across the land bridge to the untamed continent beyond.

26. According to this passage, the first people in North America lived

A. in what is now Central America.

B. west of the Bering Strait.

C. below sea level.

D. in what is now Alaska.

27. Based on information in the passage, about how long was the Bering Land Bridge?

A. Between 50 and 75 miles long

B. Between 90 and 120 miles long

C. Around 150 miles long

D. Around 200 miles long

28. According to the passage, which of these would be considered a land bridge?

A. The Isthmus of Panama

B. The Chesapeake Bay

C. The Strait of Hormuz

D. The Khyber Pass

29. Which of the following assertions does the author support with an example or examples?

I. Ethnologists use evidence to show links between the people of Siberia and those of coastal Alaska.

II. There are many historical land bridges.

III. The Mammoth Steppe supported many large, grazing animals.

A. I only

B. III only

C. I and II only

D. I, II, and III

30. According to the passage, why does Beringia no longer exist?

A. The Ice Age ended.

B. Ozone depletion raised the sea levels.

C. It was replaced with enormous sheets of ice.

D. The continents drifted apart.

31. The author suggests that some contemporary Native Americans object to the land bridge theory because

A. it equates them with Pleistocene man.

B. it challenges their history and status.

C. it relies on disputed science.

D. it belies the importance of southern tribes.

32. Which of the following findings best supports the author's contention that Siberia was once connected to North America?

A. Native American legends from the American Northwest feature enormous whales and large fish.

B. People of coastal Siberia have features that distinguish them from people in the rest of Russia.

C. Hunters in both Siberia and coastal Alaska continue to hunt seals, walrus, and sea lions.

D. Large animal fossils found in both places prove that identical species once populated both regions.

33. Which new information, if true, would most CHALLENGE the claim that humans first moved to North America across the Bering Land Bridge?

A. A new translation of an Inuit legend about Raven and a giant flood.

B. New fossil records that place the Ice Age 1000 years earlier than believed.

C. The discovery of human fossils in Kansas that predate the Ice Age.

D. DNA proof that the musk ox of Siberia differed from the musk ox of Alaska.

SAMPLE PASSAGE 6: ANSWERS AND EXPLANATIONS

26. D This straightforward **Comprehension** question asks you to **Locate Supporting Details**. The Bering Land Bridge is described as connecting Siberia to what is now Alaska. If humans walked across from Siberia, the first ones in North America would have emerged into Alaska.

27. A This is a very simple **Application** question that requires you to **Solve a Problem** using details from the passage. According to the passage, you could drive across the bridge, if it still existed, in about an hour. Based on that information, you can infer that the bridge was about as long as a typical mileage per hour, which would put it at choice A, between 50 and 75 miles.

28. A Again, this is an **Application** question, this time requiring you to **Make Generalizations**. A land bridge is a piece of land with water on either side that connects two larger pieces of land. That makes it equivalent to an isthmus (choice A). A bay (choice B) is a body of water, a strait (choice C) is a channel of water connecting two pieces of land, and a pass (choice D) is a narrow piece of land between mountains.

29. D Here, you are asked to **Analyze an Argument** by answering an **Evaluation** question. "Ethnologists use evidence such as shared religions, similar houses and tools, and unique methods of cleaning and preserving food to show the link between the people of coastal Siberia and the people of coastal Alaska." That supports assertion I. "There was a bridge connecting the British Isles to the European continent, and there was one between Spain and Morocco at what is now the Straits of Gibraltar. There were bridges connecting Japan to China and Korea." That supports assertion II. "It supported many large, grazing animals including reindeer, bison, and musk oxen, as well as the lions that fed on them." That supports assertion III. Because all three assertions are supported with examples, the correct response is choice D.

30. A You must **Draw Conclusions** to answer this **Application** question. There is no evidence to support choices B, C, or D. The Bering Land Bridge disappeared when the waters came back. The waters went away in the first place because the large ice slabs formed during the Ice Age displaced them. When that age ended, the bridge was covered over. The correct answer is choice A.

31. B This **Application** question involves **Identifying Cause and Effect**. According to the passage, some Native Americans object to the theory because "it contradicts most native teachings on the origins of the people" and "it seems to undermine the notion that they are truly 'native' to the North American continent." In other words, it challenges their history and status.

32. D Read this **Synthesis** question carefully to **Apply New Evidence** to the author's assertions. Your challenge is to find the response that BEST supports the idea that the continents were once connected. Answer A affects only North America. Answer B affects only Siberia. Answer C is possible, but it does not provide the unambiguous evidence that answer D does.

33. C With this **Synthesis** question, you must **Modify Conclusions** based on new information. Again, you must look for the answer that BEST challenges the idea that humans first moved from Siberia to North America. Neither choice A nor choice B would prove or disprove the notion. Choice D would throw some doubt on the theory that the two continents were connected, but only choice C would indicate that humans were already in North America prior to the forming of the land bridge.

SAMPLE PASSAGE 7: HUMANITIES

Pop! goes the weasel . . .

It's a meaningless phrase from a nursery rhyme, but some students of linguistics believe that it derives from Cockney rhyming slang, a form of English slang that originated in the East End of inner London. The Cockneys, traditionally, were those working-class citizens born within earshot of the bells of St. Mary-le-Bow, Cheapside. The word itself was a slap at the ignorant townsfolk by country gentry, who likened their urban brothers and sisters to deformed eggs, known as "cokeney" or "cock's eggs."

The Cockneys developed their own vernacular over a hundred years or so, and during the early part of the 19th century, rhyming slang became an integral part of this argot. It was associated in many Londoners' minds with the underworld, because it could easily be used as a sort of code. For their own self-preservation, Scotland Yard began to publish translations of the slang in police manuals, and thus the strange colloquialisms began to cross out of the East End and into the general population.

The rules behind the rhyming slang are as simple as the result is difficult to comprehend. A speaker puts together words, the last of which rhymes with the word he or she means to denote. For example, *loaf of bread* might mean "head." The difficulty comes as the slang becomes widespread and the original rhyme is discarded as superfluous, so that *loaf* means "head" in a sentence such as "He gave his loaf a thump."

Our own phrase "Put up your dukes" may derive from Cockney rhyming slang. Originally, the theory goes, the rhyme was "Duke of York" = "fork," an old slang term meaning the hand.

But back to "Pop! goes the weasel." Supposedly, *weasel* was from *weasel and stoat*, and referred to a coat. To pop one's weasel was to pawn one's coat, a relatively common practice on the East End as one's paycheck failed to stretch from one week to the next.

Sometimes, clearly, there was a certain deliberate irony in the choices of rhymes. For example, what are we to make of the fact that *trouble*, as in *trouble and strife*, is Cockney slang meaning "wife"? Sometimes, too, there was deliberate obfuscation. Would that bobby on the corner recognize himself as a *bluebottle* (bottle and stopper = "copper")? Would he know you were talking about him if you referred to *ducks and geese* ("police")? And finally, much Cockney rhyming slang was euphemistic, with rhymes substituted for words considered obscene or impolite.

Rhyming slang has moved out of the East End and throughout the English-speaking world. The phrase *eighty-sixed* means "nixed," or "thrown out." It is strictly American rhyming slang from the time of the Great Depression. Australia, too, has its own very elaborate rhyming slang. Many elements of rhyming slang have worked their way into the everyday speech of Londoners of all classes, and often their origins have been entirely forgotten.

Recently, rhyming slang has become popular again, with modern pop culture figures working their way into the mix, from Britney Spears ("beers") to Dame Judy (Dench), ("stench"). As with Pig Latin, the fun is in confounding the listener and creating a code that only the "in" group can decipher. So wipe that puzzled look off your *boat* (boat race = "face") and try to *rabbit* (rabbit and pork = "talk") the way *saucepans* (saucepan lids = "kids") do today!

34. The central thesis of the passage is that

A. Cockney rhyming slang developed as a means of communicating among immigrant populations without a common language.

B. Cockney rhyming slang has seen a comeback among the country gentry in rural England.

C. Cockney rhyming slang was and continues to be a witty way to fashion an exclusive coded language.

D. Cockney rhyming slang has moved from the underworld into the everyday language of pop figures.

35. The discussion of Scotland Yard shows primarily that

A. to the police, rhyming slang was a troublesome barrier.

B. most people who use rhyming slang are criminals.

C. rhyming slang moved from north to south through Britain.

D. code breaking is one critical job of the legal profession.

36. The author of the passage indicates that

A. Cockney rhyming slang is too limited to be useful.

B. certain examples of Cockney rhyming slang are ironic.

C. rhyming slang is restricted to a certain class.

D. the Cockneys of old were aptly named.

37. According to the passage, why might slang-users have called a policeman a "bluebottle"?

A. To offend him

B. To confuse him

C. To irritate him

D. To amuse him

38. The ideas in this passage would be most useful for

A. a tourist.

B. a politician.

C. a police officer.

D. a linguist.

39. The passage suggests that today's rhyming slang is

A. less complex than in the past.

B. limited to the underworld.

C. used by entertainers.

D. popular with teenagers.

40. According to the rules for forming Cockney rhyming slang, which word might be rhyming slang for *house*?

A. *Cat*, from "cat and mouse"

B. *Rat*, from "dirty rat"

C. *Home*, from "house and home"

D. *Louse*, from "dirty louse"

SAMPLE PASSAGE 7: ANSWERS AND EXPLANATIONS

34. C The words *central thesis* tell you that this is a **Comprehension** question that tests your ability to **Find the Main Idea**. Neither choice A nor choice B has any support in the passage. Choice D is close, but the passage suggests that pop figures are a topic of rhyming slang and not the perpetrators of it. The statement that best expresses the main idea of the passage as a whole is choice C.

35. A This **Comprehension** question asks you to **Choose an Accurate Summary**. The word *primarily* is critical here; it indicates that you must think about the author's main reason for including a section of the passage. The discussion of Scotland Yard tells of police publishing translations of rhyming slang in their manuals. The implication is that rhyming slang was causing them problems, and they needed to understand it better to perform their jobs. Choice A is the best paraphrase of this. Criminals did use rhyming slang, but whether most people who use it are criminals (choice B) is open to debate. The word *Scotland* may fool you into selecting choice C, but Scotland Yard is in London, not in Scotland. The author has no reason to emphasize code breaking as an important part of law enforcement, so choice D is not a good answer.

36. B This is an **Application** question involving **Drawing Conclusions**. The author makes no statement, indirect or otherwise, in support of choice A or choice D. Although choice C was once true, the final paragraph of the passage implies that rhyming slang has moved beyond a single class. The best answer is choice B; the author states, "Sometimes, clearly, there was a certain deliberate irony in the choices of rhyme."

37. B Again, this **Application** question asks you to **Draw Conclusions** from what is stated. Skim the passage to find the reference to "bluebottle." The author says, "Sometimes, too, there was deliberate obfuscation. Would that bobby on the corner recognize himself as a *bluebottle* (bottle and stopper = 'copper')?" The mission of the rhymer was to confuse the police.

38. D To answer this **Synthesis** question that asks you to **Combine Information**, think: Who would be most likely to read and use information from a passage of this sort? Because it tells about a particular element of the history of language, the most likely answer is choice D, a linguist.

39. D This kind of **Application** question requires **Making Generalizations**. There is no support for choices A or B, and choice C represents a misunderstanding of the final paragraph of the passage. The best answer is choice D, for the author ends by suggesting that the reader learn to talk the way kids do today.

40. A This is an **Application** question that forces you to apply what you've learned to **Solve a Problem** that is outside the bounds of the passage itself. Cockney rhyming slang begins with a rhyming phrase, but often, as the author states, "the original rhyme is discarded as superfluous." In the passage, the example given is *loaf of bread* for "head" becoming simply *loaf.* The only example from the answer choices that follows this pattern is choice A, with *and mouse* dropping off *cat and mouse* to leave simply *cat.*

SAMPLE PASSAGE 8: HUMANITIES

"He medalled twice in the Empire State Games." "Learn to parent more effectively." "The rise in prices will impact the bottom line."

If your toes curl when you hear nouns used as verbs, as in the underlined words above, you are not alone in your word-snobbishness. Nevertheless, it's worth remembering that the greatest English writer, William Shakespeare, was notorious for coining words in just this way. He changed nouns to verbs and verbs to adjectives with aplomb, and most of his conversions remain with us to this day.

Many of us are familiar with Shakespeare's clichés, those phrases that are now part of our lore of proverbs and expressions. We would not "break the ice" or "give the devil his due" if it weren't for the Bard. No one would have a "heart of gold," lie "dead as a doornail," or "come full circle."

When it comes to word coinage, it is difficult to determine which words are truly original with Shakespeare and which are simply words that first appeared in print with his works. A large percentage of Shakespearean coinages appear to be words that have familiar roots but are used as a new part of speech.

Some of the new words are nouns formed from verbs. For example, *accused* in *Richard II* is used to denote a person accused of a crime. The verb was well known, but there is no earlier example of this noun usage. *Scuffle* in *Antony and Cleopatra* denotes a fight and is taken from a verb already in existence. Shakespeare created other nouns by adding the noun-forming suffix *-ment* to a familiar verb: *amazement* and *excitement* appear for the first time in several of his plays.

More typical was the change of a noun to a verb. He used *to blanket* in *King Lear* and *to champion* in *Macbeth. To petition* shows up in *Antony and Cleopatra,* and *to humor* appears in *Love's Labours Lost.* Other new verbs coined from existing nouns include *to lapse,* meaning "to fail"; *to cake,* meaning "to encrust"; and *to rival,* meaning "to compete with." We have no way of knowing whether these coinages provoked the same spinal shivers that good grammarians experience nowadays at the sound of *to parent*.

Many of Shakespeare's most successful coinages were descriptive in nature. He invented *raw-boned* in *I Henry VI*, and there are few words today that better describe a certain kind of man. Similar hyphenates include *bold-faced* from the same play, *cold-blooded* from *King John*, *hot-blooded* from *The Merry Wives of Windsor*, and *well-bred* from *II Henry IV*.

Occasionally, coinages were stolen from another language. There's *alligator* in *Romeo and Juliet* from the Spanish and *bandit* in *II Henry VI* from the Italian. There's *to negotiate* and *to denote* from the Latin.

Of course, not all of Shakespeare's coinages survived the centuries to come. We don't use the verb *to friend* anymore; nor do we call our friends *co-mates*. Something that is secured is not referred to as *virgined*, and we don't call a messy room *indigest*. Nevertheless, to study Shakespeare is to see how rapidly a language can change and grow and perhaps to recognize that to sneer at new coinages is to ignore the way English has evolved from the beginning.

41. In the context of the passage, the word *notorious* means

A. renowned.

B. disreputable.

C. iniquitous.

D. dishonorable.

42. The central thesis of the passage is that

A. Shakespeare gave us many important maxims and adages.

B. Coined words are often grating to the ears.

C. Shakespeare teaches us a lot about how language grows.

D. Modern coinages are largely stolen from Shakespeare.

43. Based on the information in the passage, which of these is MOST likely to happen to the coinage *to friend*?

A. It will come back into common use.

B. It will remain strictly Shakespearean.

C. It will be changed to an adjective.

D. It will be replaced with *co-mates*.

44. The author apparently believes that

A. Shakespeare took unnecessary license with the Queen's English.

B. Shakespeare's coinages are far more inventive than coinages today.

C. changing nouns to verbs is an appallingly lazy way of coining words.

D. coined words are an important and natural evolution of language.

45. Suppose scholars found a long-lost medieval manuscript that used the word *accused* as a noun. This new information would most CHALLENGE the passage's implication that

A. nouns were frequently coined from verbs.

B. *Richard II* was an original drama.

C. Shakespeare coined the word *accused*.

D. word coinage is a relatively new pastime.

SAMPLE PASSAGE 8: ANSWERS AND EXPLANATIONS

41. A This kind of **Comprehension** question involves **Interpreting Vocabulary**. According to the passage, "the greatest English writer, William Shakespeare, was notorious for coining words in just this way." Although each of the answer choices is a potential synonym for *notorious*, only *renowned* fits the context.

42. C As do many **Comprehension** questions, this one asks you to **Find the Main Idea** of the passage. The "central thesis" of a passage is not just a passing subtheme, as choice A is; nor is it a minor detail, as choice B is. The central thesis of this passage has to do with the way that coined words add to the English language, and the author uses Shakespeare as the primary example of how this works.

43. B This is an **Application** question involving **Making Predictions**. "We don't use the verb *to friend* anymore," says the author, using this as an example of coined words that have not survived the test of time. Because it has not survived, it is unlikely to return (choice A), and there is no support for the notion that it might change form (choice C) or be replaced by another archaic word (choice D). It will probably remain strictly Shakespearean, so the answer is choice B.

44. D In this **Application** question, you are **Drawing Conclusions** about the author's intent. As always, you must choose the best or most likely answer. The author's interest in Shakespeare's coinages makes choice A unlikely, and there is no support for choice C. Choice B is possible, but there is not enough comparison to make it the best choice. The best or most likely answer is choice D; the author certainly implies that language evolves naturally, and that coinages are an important part of that evolution.

45. C This **Synthesis** question provides new information that might require you to **Modify Conclusions**. The discovery would not challenge but might rather support the idea that nouns were frequently coined from verbs (choice A). It would prove nothing about the originality of *Richard II* (choice B), and there is nothing in the passage that indicates that word coinage is new (choice D). An earlier finding of the word's use, however, would clearly contradict the assertion that Shakespeare coined the word. Choice C is correct.

SAMPLE PASSAGE 9: SOCIAL SCIENCES

They firmly practiced celibacy, yet their original leader, "Mother Ann," was a young woman who had birthed and lost four children. They believed in severe simplicity, yet their rituals were filled with the strange, ecstatic dances that led to their name. The American Shaking Quakers, or Shakers, were a radical group whose very contradictions would both empower and destroy them.

The Quakers were well established both in England and in America when James Wardley began to lead his flock down a new path. Inspired by the millennial French Prophets, he instilled Quaker worship with communion with the dead, visions, and shaking. Like all radicals at the time, members of the Shaking Quakers were often harassed and imprisoned for their unusual ways of worship. During one such imprisonment, a young parishioner named Ann Lee had a vision in which she determined that she embodied the Second Coming of Christ, the female version of God as Father and Mother.

Mother Ann then became the leader of the congregation, and when another vision told her that a place had been prepared for her people, she decided to take her flock to safer harbor in America. A small group came with her. They settled on a commune near Albany, New York. Despite the American Revolution roiling around them, the group remained pacifistic, which led to further ostracism and harassment from the colonists nearby.

The tiny Shaker band got a new lease on life in 1779 when Joseph Meacham and his band of followers converted to their religion. It's important to remember what a radical leap this involved. The Shakers rejected the trinity in favor of a belief in the duality of the Holy Spirit, a spirit both male and female. They believed that the new millennium was not in the future but had already begun with Mother Ann's vision. They believed in celibacy as a symbol of a return to pre-Adam purity, and this relinquishing of typical sexual roles would lead to unusual gender equality. Unity of the commune was the goal, and unity was achieved through the suppression of individuality. Shakers lived in common dormitories, shared all worldly goods, and dressed and ate simply.

Equality and community appealed to the people from other cultures who joined the Shakers. They attracted Native Americans, free blacks, and non-Christians, and all were welcomed into the commune. Under Meacham's guidance, the commune began missions to other areas, eventually establishing 17 additional communes.

Of course, a religion with celibacy at its base must rely on missionary work for its survival. A commune, no matter how isolationist, must sometimes look outward.

Many years after Mother Ann's death, Shaker communities used mail-order catalogs to sell the simple, functional furniture we know them for today.

The height of the religion was in the mid-19th century. By the end of the 20th century, only one tiny Shaker commune survived. The commune, near New Gloucester, Maine, has only a handful of elderly, female members. They continue to farm the land, make baskets and woven goods, and maintain a museum and library that are open to the public.

46. The central thesis of the passage is that

A. no religion can survive without a mission.

B. celibacy was the best and worst thing about the Shakers.

C. the Shakers were a paradoxical sect.

D. our culture owes a lot to the Shakers.

47. The discussion of the trinity shows primarily that

A. Shaker beliefs differed radically from prevailing Christian faith.

B. Quakers believed in a duality of spiritual and corporeal being.

C. the Shakers rejected the concept of the new millennium.

D. a return to pre-Adamite purity was part of the Shaker commitment.

48. According to the passage, why might free blacks have joined the Shakers?

A. Because of their antiwar stance

B. Because of their free education

C. Because of their inclusiveness

D. Because of their simplicity

49. The author of the passage indicates that

A. Shaker values are tied to the distant past.

B. the Shakers were doomed to extinction.

C. pacifism was confined to the New World.

D. Native Americans were easily converted.

50. The ideas in this passage would be most useful for

A. a student of British history.

B. a craftsperson.

C. a physical anthropologist.

D. a religious studies major.

51. The passage suggests that the Shakers' use of mail-order catalogs was

A. restrictive.

B. inappropriate.

C. fruitless.

D. practical.

52. Below is a favorite Shaker symbol. Based on the passage, which of these statements does the symbol most likely represent?

A. Better to have your gold in the hand than in the heart.

B. Give your hands to work and your hearts to God.

C. Now join your hands, and with your hands your hearts.

D. A friend reaches for your hand and touches your heart.

SAMPLE PASSAGE 9: ANSWERS AND EXPLANATIONS

46. C Again, the term *central thesis* indicates a **Comprehension** question on **Finding the Main Idea**. The central thesis is stated clearly in paragraph 1: "The American Shaking Quakers, or Shakers, were a radical group whose very contradictions would both empower and destroy them."

47. A This kind of **Comprehension** question asks you to **Choose an Accurate Summary**. The example of the trinity is given specifically to point out how the Shakers differed from traditional doctrine. Although choice D is true of the Shakers, it has nothing to do with their rejection of the notion of the Father, Son, and Holy Spirit.

48. C This is a fairly simple **Application** question involving **Identifying Cause and Effect**. According to the passage, the Shakers were a remarkably inclusive sect whose belief in equality and community "attracted Native Americans, free blacks, and non-Christians." There is no support for choices A or D and no mention at all of choice B.

49. B To answer this **Application** question, you need to **Draw Conclusions**. You must find the answer that best shows what is implied by the passage. Although the Shaker sect is old, the author never implies that their values are outmoded (choice A). There is no support for choices C or D. However, the discussion of celibacy makes it fairly clear that the Shaker sect could not go on forever. The best answer is choice B.

50. D **Synthesis** questions like this ask you to **Combine Information** from the passage with what you know. Think: Who would be most likely to read and use information from a passage of this sort? Although the Shakers started as a sect in England, their relevance is to American history rather than British (choice A). A craftsperson (choice B) might study Shaker crafts, but that is only a minor part of this passage. A physical anthropologist (choice C) would get little out of the passage, but a religious studies major (choice D) might find it useful.

51. D This **Application** question asks you to **Make Generalizations**. It might have been out of character, but the author does not present it as inappropriate (choice B). Nor was it fruitless (choice C); it seems to have been a practical way for the Shakers to sell their crafts.

52. B This is an **Application** question that asks you to **Solve a Problem** by applying what you have learned to something new. Choice B is actually a quotation from Mother Ann. The Shakers were all about work and faith, not about money (choice A) or even friendship (choices C and D).

SAMPLE PASSAGE 10: SOCIAL SCIENCES

The real estate bubble, the dot-com craze—for sheer lunacy and truly disastrous consequences, few economic swings hold a candle to tulipomania.

In 1593, a botanist brought back samples of tulips to Holland from Constantinople. He planted them in a garden with the intent of studying their medicinal value. His garden was ransacked, the bulbs were stolen, and that was the beginning of the tulip trade in Holland.

Rich Dutch homeowners coveted the new, colorful plants, and soon many of the finest houses in Holland had small plots of tulips. A mosaic virus attacked the introduced species, weakening the plant stock while at the same time causing odd streaks of color upon the petals of the flowers. These rather pretty alterations made the plants even more desirable, and shortly thereafter, prices began to rise out of control. By the year 1635, buyers were speculating on tulips, purchasing promissory notes while the bulbs were still in the ground. The trade in tulip futures became frenzied, with the prices of the most popular, damaged plants doubling and tripling overnight. Buyers bought on spec, assuming that they could turn around and sell at a profit once spring came.

Tulips have a built-in rarity, in that it takes years to grow one from seed, and most bulbs produce only one or two "offsets," or bulb clones, annually. That scarcity kept the value up even when new varieties were introduced on the market.

The tulip craze soon found ordinary citizens selling everything they owned for a single bulb, some of which were valued at the equivalent of thousands of dollars in today's currency; quite literally, bulbs were worth their weight in gold. Contemporary documents show that people traded oxen, silver, land, and houses for one tulip bulb. During the winter of 1636, when the craze was at its peak, a single bulb future might change hands half a dozen times in one day. Planting bulbs and growing flowers became wholly beside the point; the object was simply to buy and sell and resell.

As wild as this market was, it was transacted entirely outside of the established Stock Exchange in Amsterdam. It was a people's exchange. Typically, sales took place at auctions, but often they were transacted at pubs or in town squares.

Like any craze in which potential profits seem too good to be true, tulipomania was fated to end badly. In February 1637, at a bulb auction in Haarlem, the bottom fell out when no one agreed to pay the inflated prices. The ensuing panic took a matter of a few weeks; prices fell, bulb dealers refused to honor existing contracts, and the government had to leap in to try to bail the country out, offering 10 cents on the dollar for bulb contracts until even that could not be sustained. Eventually, a panel of judges declared that all investment in tulips was gambling and not recoverable investment. Holland slowly fell into an economic depression that lasted for years and eventually overflowed its borders into the rest of Europe.

53. The central thesis of the passage is that

A. tulipomania was a localized craze with localized effects.

B. tulipomania shows the potential devastating effects of a craze.

C. the Stock Exchange cannot prevent people from unregulated trading.

D. the Stock Exchange developed as an answer to unregulated trading.

54. The passage suggests that the author would most likely believe that

A. trading in futures may be ill-advised.

B. trading in commodities is safer than trading stock.

C. trading stocks is relatively risk-free.

D. trading at auctions is usually unwise.

55. Which of the following assertions does the author support with an example?

I. Tulips have a built-in rarity.

II. Sales were transacted in pubs and town squares.

III. Holland fell into a depression.

A. I only

B. III only

C. I and II only

D. I, II, and III

56. The ideas in this passage would probably be most interesting to readers of

A. *Better Homes and Gardens.*

B. *Consumer Reports.*

C. *Condé Nast Traveler.*

D. *The Economist.*

57. The author refers to the trade in tulips as a "people's exchange." This term is used to indicate

A. fairness.

B. informality.

C. low prices.

D. efficiency.

58. The 17th-century speculation on prospective tulip growth might be compared to today's

A. fluctuation of oil and natural gas prices.

B. exchange trading of options and futures.

C. hedge funds in the foreign exchange market.

D. issuance of risk-free municipal bonds.

59. Auctions in the Netherlands continue to handle between 60 and 70 percent of the world's flower production and export. What question might this information reasonably suggest about tulipomania?

A. Whether it resulted not only in a depression but also in the death of the auction system of commodity trading

B. Whether it led the Dutch government to take over substantial debt for bulb dealers and their patrons

C. Whether it changed the Dutch way of doing business to one that was more formal and controlled

D. Whether it forced the Dutch to look into new, non-agricultural products for export and import

SAMPLE PASSAGE 10: ANSWERS AND EXPLANATIONS

53. B This **Comprehension** question asks you to **Find the Main Idea** of the passage. Although choice C may be inferred from the passage, it is not the main idea. Choice A is belied by the fact that the depression in Holland ended up moving across

Europe, and choice D is incorrect, because the Stock Exchange already existed at the time of tulipomania. The passage primarily shows the damaging effects of a craze (choice B).

54. A To answer this **Application** question, you must **Make Predictions** based on what the author has said in the passage. The author implies that the "people's exchange" was less safe than the Stock Exchange, so choice B is unlikely, but there is still no indication that trading stocks is "risk-free" (choice C). There is no opinion given about the wisdom of trading at auctions (choice D); certainly it does not work well in tulipomania, but generalizing from that to the world at large is not possible. It is more viable to generalize that trading in futures may be ill-advised (choice A), especially because the answer includes the qualifying word *may*.

55. A This **Evaluation** question involves **Analyzing an Argument**. There is no example given of sales transacted in pubs and town squares (II), nor is there an example to show how Holland fell into a depression (III). The assertion that tulips have a built-in rarity (I) is supported by two examples: "It takes years to grow one from seed, and most bulbs produce only one or two 'offsets,' or bulb clones, annually." Because only I is correct, the answer is choice A.

56. D This is a relatively simple **Synthesis** question that requires you to **Combine Information** to determine the practical value of the passage. A gardener (choice A) might find some interesting facts about tulips in the passage, but in general, the passage has to do with economics, so choice D is the best answer.

57. B Although this involves vocabulary, it moves beyond simple comprehension to **Application**, requiring you to **Draw Conclusions** about the author's intent. The so-called "people's exchange" was neither fair (choice A) nor efficient (choice D), and it certainly did not hold prices down (choice C). It was an informal trading structure outside the realm of the traditional Stock Exchange.

58. B This is not a simple compare-and-contrast question; it is a **Synthesis** question that asks you to **Combine Information** from the passage with what you know. "Speculation on prospective tulip growth" equals trade in commodities futures, which today takes place in exchange trading on the mercantile exchange.

59. C Again, the question works outward from the specific material given in the passage, making it a **Synthesis** question, this time on **Applying New Evidence**. This question gives you a fact about the present and asks you to apply it to your understanding of the past. If Holland continues to trade tulips at auction, the auction system cannot be dead (choice A), and agricultural products must still be part of the Dutch economy (choice D). Choice B in no way corresponds to the fact from today. Choice C, on the other hand, is a reasonable question. Now that Holland appears to be successful at the bulb trade, something must have happened to change the way things worked.

CRAM SESSION

Verbal Reasoning

Passages on the Verbal Reasoning section consist of 500–600 words in the nonfiction disciplines of the humanities, social sciences, and natural sciences. There are 5 or 6 passages with 40 multiple-choice questions to be read and completed in 60 minutes.

You can't **study** for Verbal Reasoning, but you can **prepare**. Following are some suggestions of what you can do to prepare:

- **Read**, using Active Reading Strategies
- **Pace yourself**, using Speed-Reading Strategies
- **Preview** the test

Questions on the Verbal Reasoning section progress from easy to hard within each question set. They test the following skills:

- **Comprehension**, including Main Idea, Details, Summaries, Compare and Contrast, and Vocabulary
- **Evaluation**, including Analysis of Argument, Fact and Opinion, and Unsupported Assertions
- **Application**, from Predictions to Cause and Effect to Conclusions and Generalizations
- **Synthesis**, which has to do with the application of new evidence and alternate conclusions

CHAPTER 2

The Writing Sample

Read This Chapter to Learn About:

- ➤ The Writing Sample and What It Tests
- ➤ How Your Essays Are Scored
- ➤ The Three Tasks Your Essay Must Accomplish
- ➤ How to Use the Writing Process to Create a First-Rate Essay

WHAT THE WRITING SAMPLE TESTS

The Writing Sample became part of the MCAT in 1991 in response to complaints from deans and faculty about the communications skills of entry-level medical students. It was also a response to the development of a philosophy that incorporated doctor–patient communication into medical training. As brilliant a scientist as you may be, the philosophy posits, if you cannot speak and write coherently, you cannot be a model physician.

The MCAT Writing Sample consists of two 30-minute essays in response to two given prompts. It measures your ability to construct a central thesis, synthesize ideas, present those ideas in a logical way, and write using the conventions of Standard English.

Unlike most of the writing you may have done in school, the essays for the Writing Sample are not in a single writing form. Instead, they combine expository writing with persuasive writing, requiring you to interpret a statement, state an opposing viewpoint, and support it with factual examples. Like the Verbal Reasoning section of the MCAT, the Writing Sample presumes no particular prior knowledge on your part, but it does expect you to have a well-rounded background of information on which to draw.

HOW THE ESSAY IS SCORED

Instead of assessing spelling, organization, punctuation, and support independently, the MCAT scorers look at your essay holistically. That is, they look at the piece as a whole and rate it on its overall success in responding to the prompt. Two scorers rate your first essay, and a different pair of scorers rates your second. They assign each essay a score between 1 and 6, where 6 is the highest rating possible. If the essay is completely illegible, if you leave it blank, or if you write in a foreign language, the score is X, for "not ratable."

The 1–6 Scoring Rubric	
X	Response blank, illegible, or not in English.
1	Lack of understanding evident. Organization nonexistent or unclear. Ideas undeveloped. Multiple errors in capitalization, punctuation, grammar, and/or usage.
2	Organization unclear or scattershot. Ideas underdeveloped. Multiple errors in capitalization, punctuation, grammar, and/or usage.
3	Organization somewhat more coherent. Ideas developed but lacking complexity. Several errors in capitalization, punctuation, grammar, and/or usage.
4	Organization coherent. Ideas developed but lacking complexity. Some errors in capitalization, punctuation, grammar, and/or usage.
5	Organization coherent and focused. Ideas complex and well developed. Few errors in capitalization, punctuation, grammar, and/or usage.
6	Organization coherent and focused. Ideas complex, sophisticated, and substantially developed. Almost no errors in capitalization, punctuation, grammar, and/or usage.

It is rare that two scorers rate a single essay more than 1 point apart, but in cases where they do, a third scorer may be called in to rate the essay. Notice that the scorers do not rate the length of the essay, and they are not supposed to be influenced by cross-outs or revisions, although complete illegibility obviously counts against you.

Once each pair of scorers assigns numbers to your essays, the numbers are combined and converted to a letter score, with J being the lowest and T the highest. Essays scored J–L are considered below average. Essays scored M–Q are average. Essays scored R–T are above average.

PREPARING FOR THE WRITING SAMPLE

You probably know by now whether you're a good writer, a competent writer, or a person who cannot communicate using the written word. The MCAT Writing Sample requires a very specific type of writing that may seem unfamiliar at first. By reviewing this section and practicing with some prompts, you can improve your ability to submit the kind of writing the MCAT scorers expect.

Write

Practice on letters to the editor, which are a form of writing quite similar to the one required on the MCAT. You don't have to submit your letters, just write them. Pick an issue in your community that is generating some heat, and pick a side. Then write a letter that expresses your opinion and gives solid reasons and examples to support your ideas.

Your typing skills should be excellent. You do not need the added pressure of having to hunt-and-peck. If you need keyboarding practice, now is the time—before you find yourself in the process of taking the test.

Pace Yourself

It is not easy to write under time pressure. You have only 30 minutes per essay. You should spend part of that time preparing to write and part of that time reviewing your writing (see Using the Writing Process later in this chapter). As you complete sample essays in this book, time yourself. Think about breaking each half hour as shown in Figure 2-1.

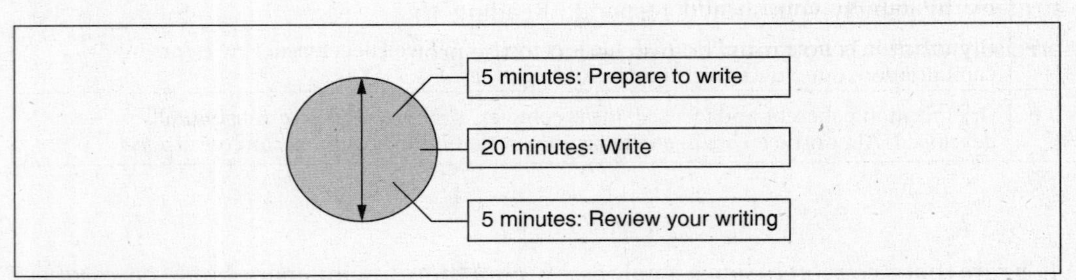

FIGURE 2-1: Budgeting time for MCAT Writing.

Spending a little time organizing your thoughts and a little time cleaning up your mechanics pays off in your score. Remember, you are judged on your organization and on your proper use of Standard English. The more you write and time yourself, the better you will be at judging when your 20-minute writing period is up.

Preview

The MCAT Writing Sample is a very specific piece of writing on a very specific kind of prompt. Knowing what a typical prompt looks like will help you understand how to address the two prompts you will see later on the MCAT.

THE MCAT PROMPT

The MCAT Prompt begins with a statement. (See Figure 2-2.) This statement typically expresses an opinion with which you might or might not agree. Your personal feeling about the statement is not always relevant to the writing of the essay itself, however.

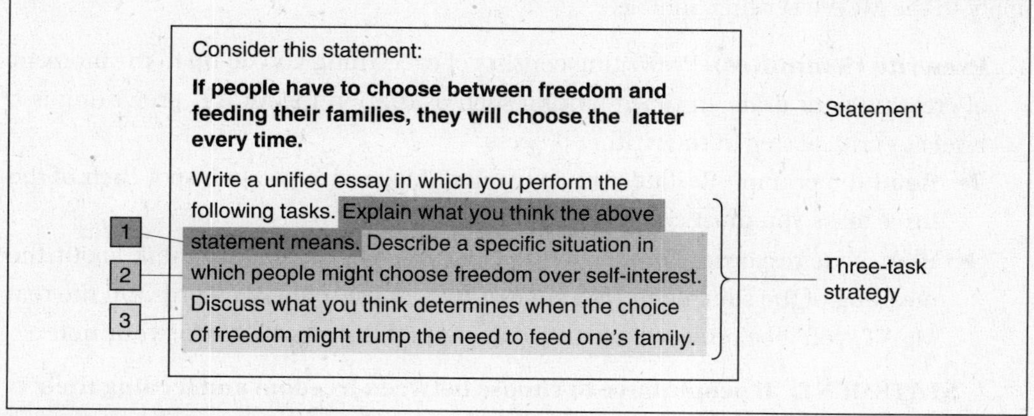

FIGURE 2-2: MCAT Writing Prompt.

The statement is followed by a paragraph in which the MCAT presents you with the strategy by which you should respond. Reading this paragraph carefully tells you precisely what it is you must do to respond to the prompt.

1. **Explain** what the statement means. This is a matter of your own interpretation, based on your understanding of the statement and your background knowledge of the topic. Here you should define terms, use specific examples, and restate the statement in your own words.
2. **Describe** a situation in which the statement might be contradicted or inapplicable. Consider that situation in detail. In other words, you must think up a situation—from real life, from your personal knowledge, from literature, or from your imagination—that contradicts the statement in the way described by the prompt. Always use the prompt as your guide here. Do not go beyond its boundaries and apply the statement to situations that are not relevant.
3. **Discuss** the factors that affect the statement's validity or inaccuracy or that might reconcile the statement with the contradiction you described. Again, use the prompt as your guide. This might mean that you find the fine line that separates the statement from the situation you described. It might mean that you restate the generalization in the statement in a way that resolves the conflict.

This is a lot to do in 30 minutes, which is why your initial preparation is so important. In addition, a response that simply includes three paragraphs that address first task 1, then task 2, and finally task 3 will not receive a high score. The MCAT scorers are looking for an integrated response, one in which the three steps flow logically in whatever order works best.

USING THE WRITING PROCESS

You may have used the writing process in school, in which case you have a leg up on your competition. Here is a quick review of that process, which you can and should apply to the MCAT Writing Sample.

1. **Prewrite (5 minutes).** Prewriting consists of everything you do up to the moment of creating your draft. In timed writing such as that on the MCAT, prewriting is a brief but critical step in the writing process.
   - **Read** the prompt. Restate it in your mind. Make sure you can spot each of the three tasks you must complete.
   - **Plan** your response. For step 1, that means that you should think about the meaning of the statement. Jot a note or two, or list some key words. On the real MCAT, you'll be provided with scratch paper that you can use for your notes.

   STATEMENT: "If people have to choose between freedom and feeding their families, they will choose the latter every time."

   freedom or food? people choose food

   For step 2, you should think of an example to contradict this. More than one example will not necessarily earn you extra points, and it might throw you off track. It's probably better to focus on one.

   PROMPT: "Describe a specific situation in which people might choose freedom over self-interest."

   possible examples–~~slavery~~? ~~American Revolution~~
   Early Americans in search of religious freedom left comforts of Europe for the unknown.

   For step 3, you should generalize to determine when the contradiction might come into play.

   PROMPT: "Discuss what you think determines when the choice of freedom might trump the need to feed one's family."

   when people are desperate
   when beliefs and way of life are threatened

2. **Draft (20 minutes).** Drafting is stating your ideas in writing. Use your prewriting notes to assist you with organization. Separate ideas into paragraphs. Focus on answering the prompt, not on introducing new concepts, expressing personal asides, or using big vocabulary words.

   Notice how this writer fused her prewriting ideas into her draft. By doing that, she could be sure to cover all three required tasks.

   It is cynical to believe that if people are faced with a choice between liberty for themselves and their people and personal self-interest, that they will always choose food and shelter over the more exalted concept of liberty.

A useful example of the triumph of liberty over self-interest might be that of the European
~~original~~ settlers of our own country.

Most of the early colonists came in hope of a better life, it is true, but that "better life" meant both the freedom to own land and to worship as they
Whether they were Puritans or Quakers,
preferred. ~~T~~hey arrived on American shores with next to nothing. There was no guarantee at all that life in the American colonies would be better for them in monetary terms, but it was worth the great leap of faith for them and their families to escape the persecution they faced in their native lands. Freedom, in other words, was worth giving up the known for the unknown, as it would be for immigrants to our shores for the next three centuries.

It seems reasonable to presume that a certain desperation is needed to separate people from their creature comforts. How many of us today would give up
e
our BMWs and Hummers, even to ensure less dependance on foreign oil?. No, our beliefs and ways of life must be seriously and obviously threatened, as they were in the cases of our early colonists, for us to value liberty most of all.

3. **Revise and edit (5 minutes).** In timed writing like that on the MCAT, no one expects a perfect draft. Nevertheless, you should take some time to check your work for errors. As you have seen, punctuation, capitalization, grammar, and usage errors count against you. So do errors involving faulty organization and inadequate support. Keep in mind that on the real test, you are not able to use the computer's "spell-check" function.

As you can see above, this writer corrected a couple of minor errors, clarified, and added a specific example. Because any corrections on the new MCAT are made using "Cut," "Paste," and "Copy" functions on the computer, these editing examples are shown simply to give you the idea of the kinds of things you might change and add during the Revise and Edit process. Below is the corrected draft.

It is cynical to believe that if people are faced with a choice between liberty for themselves and their people and personal self-interest, that they will always choose food and shelter over the more exalted concept of liberty. A useful example of the triumph of liberty over self-interest might be that of the European settlers of our own country.

Most of the early colonists came in hope of a better life, it is true, but that "better life" meant both the freedom to own land and to worship as they preferred. Whether they were Puritans or Quakers, they arrived on American shores with next to nothing. There was no guarantee at all that life in the American colonies would be better for them in monetary terms, but it was worth the great leap of faith for them and their families to escape the persecution they faced in their native lands. Freedom, in other words, was worth giving up the known for the unknown, as it would be for immigrants to our shores for the next three centuries.

It seems reasonable to presume that a certain desperation is needed to separate people from their creature comforts. How many of us today would give up our BMWs and Hummers, even to ensure less dependence on foreign oil? No, our beliefs and ways of life must be seriously and obviously threatened, as they were in the cases of our early colonists, for us to value liberty most of all.

A Mini-Handbook of English Grammar, Usage, and Mechanics

Abbreviations Avoid them in formal writing.

Adjectives/Adverbs Use *good* as an adjective; use *well* as an adverb.

Antecedents Make sure that your pronouns match their preceding nouns.

Apostrophes Use an apostrophe and *s* to show possession in a singular noun. Use an apostrophe alone to show possession in a plural noun. Exceptions are *children's, men's, women's.*

Between/Among Use *between* to refer to two things. Use *among* to refer to three or more.

Capital letters Use capital letters for proper nouns and adjectives. Capitalize the first, last, and all important words in a title or multiple-word proper noun. Do not capitalize the names of seasons.

Commas Use a comma between two independent clauses in a compound sentence. Use a comma to separate all parts of a series. Use a comma after a long introductory prepositional phrase or after a pair of short introductory prepositional phrases. Use a comma to separate city and state and after the state in a sentence such as this one: "He likes Missoula, Montana, better than Naples, Florida." Use a comma to set off an appositive: "My father, a brilliant scientist, won a prize." Use a comma to set off nonessential phrases and clauses: "Her eyes, which were an unusual shade of aqua, held Myron's attention."

Dangling participle Keep modifiers close to the words they modify: "Nailed to the wall, I noticed a special bulletin" as opposed to the correct "Nailed to the wall, a special bulletin caught my eye" or "I noticed a special bulletin nailed to the wall."

Fewer/Less Use *fewer* with plural words: *fewer coins.* Use *less* with singular words: *less money.*

Fragment To avoid fragments, make sure each sentence has a subject and a verb.

Its/It's *Its* is a possessive pronoun. *It's* means "it is."

Lay/Lie You can lie down, or you can lay something down. You cannot lay down or lie something down. *Lie* never takes an object.

Misplaced modifier Keep modifiers close to the words they modify: "At three years of age, my mother taught me to read" as opposed to the correct "My mother taught me to read when I was three years of age."

Negatives Use only one per sentence. Exception: neither/nor.

Paragraph Make sure a reader can tell where your new paragraphs begin. Do not write a one-page response in a single block paragraph.

Passive voice Eschew it. Write actively. (NOT "Your essay should be written in an active way.")

Plural nouns Spell them correctly. Add *s* to most singular nouns. Add *es* to nouns ending in *s, x, sh, ch*, or *z*. Change *y* to *i* and add *es* to nouns ending in a consonant and *y*.

Principal/Principle A *principal* heads a school. Something that is *principal* is critically important. A *principle* is a law or code of conduct.

Quotation marks Use them for direct quotations or for titles of stories or poems. Keep commas and periods inside quotation marks; keep colons and semicolons outside. If question marks and exclamation points are part of the quotation itself, they go inside. If not, they go outside.

Run-ons When you're writing quickly, it's easy to connect too many thoughts in a river of language. As you revise, cut down run-on sentences to a more reasonable length.

Semicolons Use a semicolon to separate two independent clauses; before *for example* followed by a list; or between items in a series when the items themselves include commas: "Jon carried the blanket; the picnic basket; and the red, white, and blue bunting."

Slang Slang is never appropriate in formal writing.

Spelling It counts. If you don't know how to spell a word, use a simpler word.

Verb agreement Use singular verb forms with singular subjects or with compound singular subjects joined with *or* or *nor*. Use plural verb forms with plural subjects or with compound subjects joined with *and*. Use singular verb forms with the pronouns *anybody, anyone, each, either, everybody, everyone, nobody, no one, somebody,* and *someone*. Use plural verb forms with the pronouns *both, few, several,* and *many*. The pronouns *all, any, most, none,* and *some* may take singular or plural verbs depending on their antecedents.

Verb tense Maintain consistency of tenses. When in doubt, read your writing "aloud" in your head to check whether your tenses work.

Who/Whom *Who* is used as a subject or predicate nominative. *Whom* is used as a direct or indirect object. In other words, if you would use *he*, use *who*. If you would use *him*, use *whom*.

PRACTICING WRITING

Try your hand at the following Writing Sample prompts. You might try the first one without timing yourself and then complete each of the others within the 30-minute deadline. If you prefer to input these as you will on the real MCAT, open a file on your computer rather than writing the responses by hand. Use the steps in the Writing Process. Make sure to respond to all three tasks in each prompt. Afterward, compare your response to the rated responses that follow to see how your writing measures up. The final two prompts allow you to grade another test-taker's response to see whether you are able to read and respond the way MCAT graders might.

WRITING SAMPLE 1

Consider this statement:

A democracy without opposition is no democracy at all.

Write a unified essay in which you perform the following tasks. Explain what you think the above statement means. Describe a specific situation in which a democracy could exist without opposition. Discuss what you think determines whether opposition is critical to a thriving democracy.

Essay 1A

Can a valid democracy exist without opposition? If there is no protest movement, no opposition party, no challenging voice, can it really be called a democracy? The question suggests that our notions of free speech and the right to protest are integral to the existence of democracy. That is certainly something that those of us raised in the age of protest rallies and virulent letters to the editor might posit, but in fact, democracies have existed and can exist without opposition.

The democracy of ancient Greece lacked many of the freedoms that we take for granted in our present-day democracy. Although slavery was endemic and protest kept under wraps, the democracy of Athens and the Greek city-states thrived for centuries without being overturned. In fact, some might say that the Golden Age of Greece was the Golden Age of participatory democracy, a democracy that was more representative of the voting population than anything we have had since.

Nevertheless, this really is not a model that appeals to us today. As we traipse around the globe assisting democratic movements, we argue regularly that freedom of the press and freedom of speech must accompany the basic freedom to choose who will govern the state. This goes back to the ideology of our founding fathers, who chose to use protest and dissent to overthrow the oppressive rule of King George III. Such examples as the Boston Tea Party stand clearly in our national consciousness as a proper mode of conduct with which to achieve our desired goals.

So it may be fair to say that although democracies have existed in the past without opposition, our definition of democracy has evolved. Today, opposition is part of the foundation of our two-party system. Opposition is the motivator that gets people to the polls. Opposition is the root cause of many of our best endeavors. We thrive not without, not because of, but <u>despite</u> opposition, and that makes our democracy even stronger than that of the Golden Age.

SCORE: 6

This essay responds clearly and coherently to all three tasks in the prompt. In paragraph 1, the writer restates and explains the statement. In paragraph 2, the writer gives an example that contradicts the statement. In paragraphs 3 and 4, the writer presents opinions on whether opposition is necessary to a democracy. Whether or not one agrees with the writer's premise, the writing itself is strong, reasoned, and easy to follow.

Although only one example of a contradiction is given, the example is supported with specifics, and the analysis of the example is clearly applied to the statement. Each paragraph is well defined, and the paragraphs flow into one another with logical transitions. The writer ends by reapplying the statement to democracies today and concludes that our definition of democracy has evolved. The essay ends on a unifying note, tying the present to the past and the statement to its contradictory example. The use of parallelism and repetition adds to what is a nicely constructed response.

Essay 1B

The statement suggests that it is not possible to have a true democracy without opposition, or challenge to the ruling party. It suggests that we cannot call it "democracy" if it exists without conflict.

Sadly, we might say that our democracy today has altogether too much conflict for a healthy democracy. We may have reached a point where opposition overwhelms useful achievement, where the political parties are so at odds that the logjam cannot be forded. When Congress is at an impasse, laws are not made. When the Administration and Congress are at odds, nothing is accomplished in the name of the people. When the ruling parties are fighting, a spirit of hatred and mistrust overwhelms us. So what does democracy mean when built-in opposition thwarts its achievements?

How much better it might be to have the no-party system of the Inuit in the Northwest Territories of Canada! In that far-off, vast, mostly rural community, people vote for individuals, not parties. That is an example of a democracy where opposition is not inbred into the system. Candidates are chosen on their own merits rather than on how they fit into a party system or how they oppose the party in power. It is a thriving, successful democracy that exists without the opposition we expect here.

It is hard to imagine such a system being imported here (although I know of some villages where it is used in local elections). No, we must enjoy the infighting and partisanship that is part and parcel of our democracy. Otherwise, why in the world would we put up with it?

SCORE: 4

This essay addresses all three tasks in the prompt, but the smoothness of Essay 1A is missing here. In paragraph 1, the writer restates and explains the statement. In paragraph 3, the writer gives an example that contradicts the statement, that of the Inuit in the Northwest Territories. Whereas Essay 1A used an example of a democracy that limited protest, this essay uses an example of a democracy that limits partisanship, a different but equally valid contradiction to the statement.

Support for the example is somewhat limited; the writer simply defines the no-party system. A few of the phrases used are awkward ("the logjam cannot be forded") rather than poetic. Nevertheless, the writer has a good command of English, and the writing is clear and understandable, with a nice variety of sentence types to retain the reader's attention.

Essay 1C

Can democracy be democracy if it is without opposition? Some would say yes, others no. This democracy ideal is the world's most vanted ideal, everybody wants to choose their own governing, and beyond that, to have a say in that governing. Without opposition, is it not totallitarianism? Or can it be still called democracy? That is the question with which we are faced.

Imagine, first of all, a democracy without opposition. Everything goes smoothly, there is no conflict. The President is allowed to govern without anyones complaint. The Senate is allowed to make laws and charge taxes without anyones disagreement. Things get accomplished, and the people have nothing to complain about because everyones wants and needs are met.

This imaginary place still has elections every year. But is it a democracy, or totallitarianism? Suppose everybody agrees about everything, then what is left to vote about? Why bother to vote or to choose between different candidates if everybody thinks the same and there is no conflicting ideas? I would not imagine that many people would bother to vote on election day if there were no differences in the candidates, or would they? It is certainly doubtful that they would bother. It's hard enough now getting people out to vote, if there were no real reason to, even more would stay home.

So I would have to say that opposition is critical to a thriving democracy. Othewise, it might as well be called a dictatership. In a totallitarian dictatership, after all, doesn't everyone always agree? They always agree with the person in charge, there is no freethinking at all, and what is democratic about that.

SCORE: 2

Although it is possible to distill meaning from this essay, it is difficult to do so because of the writer's misspellings, awkward phrasing, and illogical flow of ideas. The writer has attempted to address the three tasks in the writing prompt, which keeps the essay from dropping to a score of 1. The statement is parsed in paragraph 1, and in paragraphs 2 and 3, the writer puts forth a kind of contradiction to the statement. The fact that the example is imaginary does not in itself detract from the response, but the lack of clear support for the writer's argument is problematic.

Some of the response resembles stream-of-consciousness, which indicates that the writer failed to prewrite and simply wrote whatever came to mind. Statements such as "some would say yes, others no" do not add to the response and seem to be the ramblings of someone who has not formulated a cogent argument. Although the response is actually longer than that in Essay 1B, much of its verbiage is repetitive incoherence that does not add to or support the writer's main idea.

WRITING SAMPLE 2

Consider this statement:

Liberty must be defined as the right to do anything that does not interfere with another person's pursuit of happiness.

Write a unified essay in which you perform the following tasks. Explain what you think the above statement means. Describe a specific situation in which genuine liberty might in fact interfere with someone else's pursuit of happiness. Discuss what you think determines whether or not liberty requires restrictions such as that described here.

Essay 2A

It is possible to define liberty as the ability to do whatever one wants as long as one's pleasure does not interfere with that of one's neighbor. That is, in fact, the assumption that provides the underpinning for many of our modern laws. The notion that what goes on "behind closed doors" is permissible, whereas those same things would not be acceptable if done on a street corner or in one's front yard, is one that, after many years, struck down sodomy laws in many states.

The question becomes whether keeping one's liberty locked up makes it truly liberty, or whether true liberty must by definition allow us to do what we want in public. There are always people who test this question. Just last year in a park in my hometown, several women were arrested for sunbathing topless in public. Their defense was that if men could do it, so should they be allowed to take off their tops. The city, not surprisingly, took the position that their state of undress interfered with the ability of others in the park to take their lunches or play with their children in relative peace and quiet.

These same women would be free to march around nude in their homes, but out in the open air, such behavior is considered criminal. The question then becomes: Are they really free if they can only practice their freedom in a constrained area? Doesn't freedom necessarily mean complete, total freedom, free from the crack of the police baton if the laws interfere with that freedom?

In fact, of course, in a civilized country none of us is really that free. Laws are established to protect each of us from our neighbors' freedoms. We post our land to keep people from straying onto it to shoot small animals. We are free to do what we want on our land, but others have their freedom removed at its borders. In a totally free state, our neighbors could walk right into our homes, take chips out of the larder, and make themselves comfortable on our barcaloungers. Where is our freedom in such a case? Even though it seems contradictory, it's probably just as well that we define liberty as having a modicum of constraint.

SCORE: 5

The essay-writer responds thoroughly and thoughtfully to the prompt, using everyday examples from real life to present a cohesive argument. The redefinition appears in the opening sentence: "It is possible to define liberty as the ability to do whatever one wants as long as one's pleasure does not interfere with that of one's neighbor." The writer goes on to provide a contradiction in the form of women protesters in a park, women whose liberty interferes with others and yet who are truly free. The writer then clarifies why freedom requires restrictions—because without them, our neighbors could take advantage of us and therefore impede our own freedom.

The language used is sophisticated and controlled, although there are a few sentences that run on a bit. That plus a somewhat weak or abrupt ending keep this essay from a perfect score.

Essay 2B

Liberty must be defined as the right to do anything that does not interfere with another person's pursuit of happiness. In other words, we can do what we like as long as we do not make anyone else unhappy or step on their toes.

One example of a time when true liberty might in fact interfere with someone's pursuit of happiness might be a situation like that of Edna in *The Awakening* by Kate Chopin. At first, Edna is a person who is truly not free. Like a bird in a cage, she beats herself on the bars. She is trapped in a world of domesticity, her husband views her as a possession.

Edna starts to break out of her cage, and she finds herself in the world of art. She falls in love with another man. She spends less and less time with her husband and family as she explores her own freedom.

Of course, a woman like Edna, a woman in her time and place could never be entirely free, but the point is that her increasing liberty, as much as it is true freedom, causes harm to her family. She may feel herself awakening and finding her own freedom, but it comes at someone elses expense. She no longer thinks of them as she explores her own possibilities.

Like Edna, a lot of adults today try to express themselves and be free. Perhaps they really are free. But their children often suffer the consequences. So is one person's freedom always another person's burden?

It is better to achieve one's own freedom without harming others. That is what the author of this statement must have meant. It is impossible to define our freedom independently of the freedom of others, and others cannot be free if our freedom makes them unhappy. I agree with the original premise: that liberty must be defined as the right to do anything that does not interfere with another person's pursuit of happiness.

SCORE: 3

The essay-writer wastes some time in repeating the exact wording of the prompt, which is not wrong, simply time-consuming. The redefinition lacks precision and thoughtfulness, and the contradiction, drawn from literature, is not particularly strong or convincing.

The essay evinces a reasonable control of language, although there are errors of punctuation and pronoun usage. The flow of language or of ideas from sentence to sentence or paragraph to paragraph is often unclear. Some of the sentences are poorly constructed; for example, "she finds herself in the world of art" could have more than one meaning. Although the writer has addressed all three tasks given in the prompt, the ideas are not thoroughly developed, nor is the argument easy to follow.

Essay 2C

Liberty or freedom, should be defined as something that makes everyone happy. In other words, if it does not make you happy, it is not really freedom.

An example of freedom that does not make everyone happy is freedom of speech. Some people think that protesting the war by splashing their own blood on a recruiting station is freedom of speech. That does not make the Army happy. It does not make the police happy. It does not make law enforcement officials or those who perpetrate our legal system happy, and it does not make a lot of the population of our society happy, either. It is like saying fire in a crowded theater, for certain people.

Another example of freedom that does not make everyone happy is freedom of religion. One persons religion could interfer with the way another person believes. For example, prayer in school. Some people believe in it, but whose prayer is supposed to be used? If one person has one way to pray, another might have offense at that and want to do another way of prayer that does not meet the first persons approval. Other freedoms include freedom of the press and freedom to bear arms. Some people are happy about that, others think it might be a terrible idea to have guns in the home. But there still freedoms, nevertheless.

So there are freedoms that we have write in our own constitution that do not make everyone happy, but they are still freedoms. Making everyone happy should not be a restriction on our freedoms, it is more about making the majority happy or keeping people from interfering with each others happiness.

SCORE: 1

This writer begins right off the bat with a misstatement of the meaning of the prompt. The statement in the prompt does not imply that freedom is something that makes everyone happy, it implies that individual liberty depends on not causing unhappiness to others. Those are related but different positions.

This misunderstanding leads to confusion in the second task. The writer gives examples of freedoms that do not make everyone happy, and despite serious difficulties with language, the examples are reasonable ones. However, they do not provide a response to the writing assignment. That fact, added to the writer's unsophisticated use of English and a significant failure to develop ideas, earns this response the lowest score.

WRITING SAMPLE 3

Consider this statement:

Although one is supposed to be innocent until proven guilty, nowadays, being acquitted of a crime does not imply to the public that one is innocent of the crime.

Write a unified essay in which you perform the following tasks. Explain what you think the above statement means. Describe a specific situation in which an acquittal did indeed indicate innocence as far as the public is concerned. Discuss what you think determines whether or not acquittal should mean innocence.

GRADE IT YOURSELF

Read the following essay. Use the scoring rubric found earlier in this chapter to give it a number score. Then compare your score with the one given by a reader on the facing page.

Essay 3

The writer suggests that despite the fact that our system of justice demands that we believe defendants innocent until they are proved otherwise in a court of law, the public does not always accept that finding of innocence. That was certainly true in a number of recent, highly-publicized crimes. The O.J. Simpson trial and the trial of Robert Blake, both men accused of murdering their wives or ex-wives, are examples of two situations in which the defendants were acquitted, but many doubted their innocence.

In other, less scurrilous trials, however, an acquittal may in fact demonstrate innocence to the people. One recent trial in my own hometown involved a woman accused of poisoning her fiancé. The District Attorney pushed the case forward despite medical records that proved the man in question had a congenital heart defect that seemed more than likely to have caused his untimely death. At trial, certain other facts came out that gave the jury more than reasonable doubt. The woman had a sound allibi that could not be shaken. The DA had relied on the notion that her job with a pharmacy would lead the jury to assume she knew all about poisons, but in fact, her background had to do with sales, and she was not licensed to dispense medicine. Even more important in a small town like this one, scores of people came forward to testify about the couple's close, loving relationship and the woman's devestation at her fiancé's demise. On her acquittal, the headlines read "At Last!" There was no doubt in anyone's mind that she was innocent, and the final indication of people's beliefs came in November, when the DA was voted out of office.

According to our code of law, acquittal must mean innocence. That is not always an easy judgment to make, but it's critical to our sense of fair play.

MY SCORE AND NOTES:

____

____

____

____

____

____

READER'S SCORE: 4

This response is well written and thoughtful. The response to tasks 1 and 2 is complete, and the example that contradicts the statement works well. However, task 3 is given short shrift, perhaps because the writer ran out of time. The prompt asks writers to discuss "what determines whether or not acquittal should mean innocence." Paragraph 3 touches on this, but does not go into enough detail. That, plus a couple of spelling errors (*alibi, devastation*) keep this from achieving a better score.

WRITING SAMPLE 4

Consider this statement:

Because their actions affect local taxpayers to such a degree, school officials and employees should reside in the district in which they work.

Write a unified essay in which you perform the following tasks. Explain what you think the above statement means. Describe a specific situation in which residing in the district might not be a useful prerequisite for school employment. Discuss what you think determines whether or not residency should be a requirement for school officials and employees.

GRADE IT YOURSELF

Read the following essay. Use the scoring rubric found earlier in this chapter to give it a number score. Then turn the page to compare your score with the one given by a reader.

Essay 4

The prompt says that people who work for a school district should live in that district, because the things they do affect the people who pay taxes in that district. In other words, if what you do raises our taxes, you ought to be affected as well. That is a strong argument for living within the district in which you work that might be applied to such other citizens as police and fire officials, town government, mayors, etcetera.

I know of several people who worked at my suburban high school who lived as far away as 50 miles causing them to have a long and difficult commute especially in inclement weather such as the blizzard we had one year. This is another good reason to require people to live within a district. However, their living far away did not make these people bad officials or employees. Far from it, I know of several cases where teachers from far away could be found working longer hours than those who could walk to school.

Still, the implication to taxpayers is that those who pay taxes elsewhere don't worry or have concern about whether or not our taxes are rising, and that may in some cases be the fact. However, my principle at that time benefited from living in a district where there were even more financial difficulties than mine and therefore was able to review his own situation in that particular district and apply newfound knowledge about financial situations to my district with good effects; for example, learning from the district he lived about ways to improve cafeteria finances and applied them to our school saving the taxpayers about $50,000 they were paying to support the financially strapped cafeteria program. Other things he learned from that district where he lived included about better ways of handling special education again with a savings for my district.

In other words, is it really necessary to live in the district where you work, if you are a school employee. Not if the district where you live can give you insides into ways to improve your working district that will make that district more financially viable.

MY SCORE AND NOTES:

___

___

___

___

READER'S SCORE: 2

The response has the run-on quality typical of essays that are drafted without prewriting. At times, the writer's language goes completely off the track as she struggles for something to add. Paragraph 2 is largely extraneous. Spelling errors (*principal*) and misuse of language ("give you insides" instead of "give you insights") add to the difficulties for a reader. The writer was able to restate the prompt reasonably, and the example used as a contradiction seems plausible, but deficiencies in language make the response so hard to follow that it rates only a 2.

CRAM SESSION

The Writing Sample

The MCAT Writing Sample consists of two 30-minute essays in response to two given prompts. It measures your ability to construct a central thesis, synthesize ideas, present those ideas in a logical way, and write using the conventions of Standard English.

You can't **study** for the Writing Sample, but you can **prepare**. Following are some ideas of what you can do to prepare:

- **Write**, practicing especially your persuasive writing.
- **Pace yourself** by dividing your 30-minute writing period into sections.
- **Preview** the test.

Each writing prompt contains three tasks. All three tasks must be completed to generate a reasonable score. You must **explain** a given statement in your own words, **describe** a possible contradiction to that statement, and **discuss** the factors that affect the statement's validity or inaccuracy or that might reconcile the statement with the contradiction you described.

Use the Writing Process as you complete your response.

- **Prewrite** for 5 minutes, reading the prompt and planning your response.
- **Draft** for 20 minutes, using your prewriting notes to organize your ideas.
- **Revise** for 5 minutes, checking for punctuation, capitalization, grammar, and usage errors as well as errors of organization and support.

PART V

REVIEWING MCAT BIOLOGY

CHAPTER 1

The Cell

Read This Chapter to Learn About:

- ➤ Prokaryotic and Eukaryotic Cells
- ➤ Structures and Organelles of Eukaryotic Cells
- ➤ The Cell Membrane
- ➤ Transport Across the Cell Membrane

The cell is the basic unit of life. As a general rule, all cells are small in size in order to maintain a large **surface area-to-volume ratio**. Having a large surface area relative to a small volume allows cells to perform vital functions at a reasonably fast rate, which is necessary for survival. Cells that are too large have small surface area-to-volume ratios and have difficulties getting the nutrients they need and expelling wastes in a timely manner.

MAJOR CATEGORIES OF CELLS

There are two major categories of cells: **prokaryotic** and **eukaryotic**. Bacteria are prokaryotic cells. Other organisms, including amoeba, fungi, plants, and animals, have eukaryotic cells. Both types of cells contain a variety of structures that are used to perform specific functions within the cell. There are some similarities between the two cell types, but there are also some significant differences. The following table depicts the major structural differences between prokaryotic and eukaryotic cells.

One primary factor that differentiates eukaryotic cells from prokaryotic cells is the presence of **organelles**. Organelles are membrane-bound compartments within the cell that have specialized functions. They help with cellular organization and ensure that specific reactions occurring in one organelle do not interfere with those occurring in another organelle.

Major Differences between Eukaryotic and Prokaryotic Cells		
Characteristic	**Eukaryotic Cells**	**Prokaryotic Cells**
Cell size	Relatively larger	Relatively smaller
Presence of membrane-bound organelles	Present	Absent
Organization of genetic material	Linear pieces of DNA organized as chromosomes housed within the nucleus	A single loop of DNA floating in the cytoplasm
Oxygen requirements	Generally need oxygen to produce energy during cellular respiration	May not require oxygen to produce energy during cellular respiration

This chapter focuses on eukaryotic cells; prokaryotic cells will be revisited in Chapter 8.

STRUCTURE OF A TYPICAL EUKARYOTIC CELL

Many different structures and organelles are found in eukaryotic cells, each with a specialized function. Figure 1-1 shows the structure of a typical eukaryotic cell, and the following table summarizes the major cellular components found in eukaryotic cells. The cell membrane, the boundary of the cell that selectively allows the passage of materials into and out of the cell, is discussed separately.

Major Cell Structures and Their Functions	
Organelle or Structure	**Function**
Cytoplasm	Liquid portion of the cell in which organelles are suspended
Nucleus	Stores DNA (hereditary material) of the cell
Nucleolus	Housed within the nucleus; makes rRNA used to produce ribosomes
Ribosomes	Involved in protein synthesis
Smooth endoplasmic reticulum	Produces lipids for the cell and is also involved in detoxification in liver cells
Rough endoplasmic reticulum	Produces and chemically modifies proteins
Golgi complex	Sorts contents from the endoplasmic reticulum and routes them to appropriate locations in the cell or marks them for secretion from the cell
Lysosomes	Vacuoles containing enzymes needed for cellular digestion and recycling

Organelle or Structure	Function
Peroxisomes	Vacuoles that digest fatty acids and amino acids; also break down the metabolic waste product hydrogen peroxide to water and oxygen
Mitochondria	Perform aerobic cellular respiration to produce ATP (energy) for the cell during aerobic cellular respiration
Cytoskeleton	Hollow microtubules used for structural support, organelle movement, and cell division Microfilaments used for cell movement Intermediate filaments that assist in structural support for the cell
Cell wall	A rigid structure composed of cellulose on the outer surface of plant cells; provides structural support and prevents desiccation
Chloroplasts	Used for the process of photosynthesis in plants
Central vacuole	Used to store water, nutrients, and wastes in plants

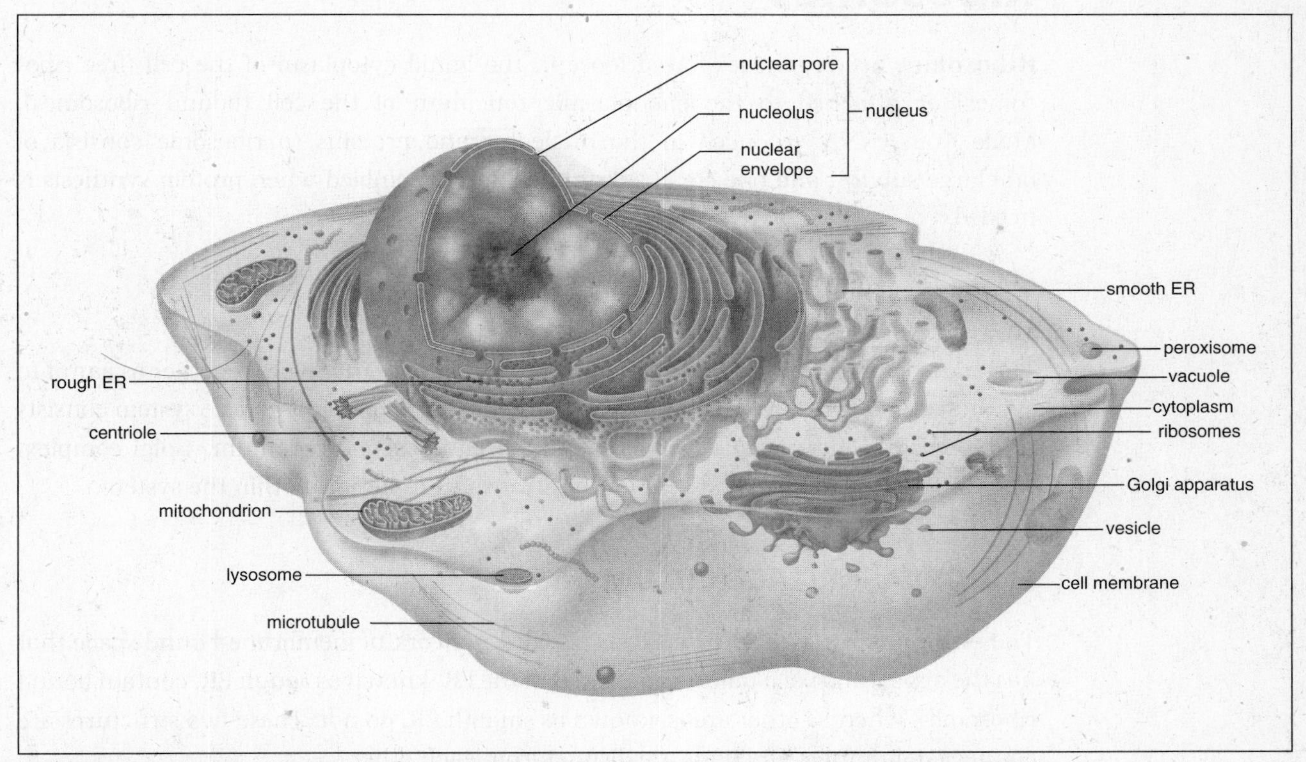

FIGURE 1-1: Animal cell structure. A typical animal cell contains a variety of structures and membrane-bound organelles. SOURCE: From Sylvia S. Mader, *Biology*, 8th ed., McGraw-Hill, 2004; reproduced with permission of The McGraw-Hill Companies.

Cytoplasm

The liquid portion of the cell is the **cytoplasm**, which consists of water, nutrients, ions, and wastes and can be the site of a variety of chemical reactions within the cell. The organelles and other structures of the cell are suspended within the cytoplasm.

Nucleus

All eukaryotic cells contain a membrane-bound **nucleus**. The nucleus houses the genetic material of the cell in the form of chromosomes, which consist of deoxyribonucleic acid (DNA) associated with specialized proteins. The outer boundary of the nucleus is referred to as the **nuclear membrane** (also termed the nuclear envelope); it keeps the contents of the nucleus separate from the rest of the cell. The nuclear membrane has pores that allow certain substances to enter and exit the nucleus. Within the nucleus, there is a **nucleolus**, which makes the ribosomal ribonucleic acid (rRNA) needed to produce ribosomes.

Ribosomes

Ribosomes are organelles found loose in the liquid cytoplasm of the cell (free ribosomes) or attached to the endoplasmic reticulum of the cell (bound ribosomes). Made from rRNA produced in the nucleolus and proteins, a ribosome consists of one large subunit and one small subunit that are assembled when protein synthesis is needed.

The Endomembrane System

The endomembrane system consists of several organelles that work together as a unit to synthesize and transport molecules within the cell. The endomembrane system consists of the smooth endoplasmic reticulum, rough endoplasmic reticulum, Golgi complex, lysosomes, peroxisomes, and vesicles that transport materials within the system.

ENDOPLASMIC RETICULUM

The **endoplasmic reticulum** (ER) is a folded network of membrane-bound space that has the appearance of a maze. Some areas of the ER, known as rough ER, contain bound ribosomes, whereas other areas, known as smooth ER, do not. These two structures are connected, but their functions are distinct from each other.

The primary function of the **smooth ER** is the production of lipids needed by the cell. In certain types of eukaryotic cells (such as those in the liver), the smooth ER also plays a critical role in the production of detoxifying enzymes. The primary function of the **rough ER**, which has bound ribosomes, is related to protein synthesis. The ribosomes produce proteins that enter the rough ER where they are chemically modified and moved to the smooth ER.

The combined contents of the smooth ER and the rough ER are shipped by vesicles to the Golgi complex for sorting. Vesicles are tiny pieces of membrane that will break off and carry the contents of the ER throughout the endomembrane system.

GOLGI COMPLEX

Vesicles from the ER arrive at the **Golgi complex** (also called the **Golgi apparatus**) and deliver their contents, which include proteins and lipids. These molecules are further modified, repackaged, and tagged for their eventual destination. The contents of the Golgi complex leave via vesicles, with many of them moved to the cell membrane for secretion out of the cell.

LYSOSOMES

Lysosomes are membrane-bound **vacuoles** (large sacs) that contain digestive enzymes used to break down substances that enter the lysosomes. Cellular structures that are old, damaged, or unnecessary can be degraded in the lysosomes, as can substances taken into the cell by endocytosis. To function properly, the lysosomes must be acidic (pH lower than 7). If lysosomes rupture, the cell itself is destroyed. In some cases, cells purposefully rupture their lysosomes in an attempt to commit cellular suicide in a process known as **apoptosis**.

PEROXISOMES

Peroxisomes are another type of vacuole found within the endomembrane system. They are capable of digesting fatty acids and amino acids. Enzymes within the peroxisomes break down toxic hydrogen peroxide, a metabolic waste product, to water and oxygen gas. Peroxisomes also assist with the degradation of alcohol in the liver and kidney cells.

Mitochondria

Mitochondria are the organelles responsible for the production of energy in the cell. They perform **aerobic cellular respiration** that ultimately creates **adenosine triphosphate (ATP)**, which is the preferred source of energy for cells. Because all cells require energy to survive, the process of cellular respiration is a vital one for the cell.

Mitochondria have some interesting and unusual features: They are bound by an inner and outer membrane, they contain their own DNA distinct from the nuclear DNA, and they can self-replicate. These unique features have led to the development of the **endosymbiotic theory**, which suggests that mitochondria are the evolutionary remnants of bacteria that were engulfed by other cells long ago in evolutionary time.

The Cytoskeleton

The **cytoskeleton** is composed of three types of fibers that exist within the cytoplasm of the cell. These fibers have a variety of functions including structural support, maintenance of cell shape, and cell division.

Microtubules are responsible for structural support and provide tracks that allow for the movement of organelles within the cell. They are one type of hollow fiber, made of the protein tubulin. A specialized grouping of microtubules is the **centriole**. A pair of centrioles is used within animal cells to assist with cell division.

Microfilaments are made of the protein actin, which assists with cellular movement.

Intermediate filaments, the final type of fiber found in the cytoskeleton, provide structural support for the cell. These fibers vary in composition, depending on the cell type.

Structures That Allow for Movement

Certain types of animal cells contain additional structures, such as cilia and flagella, that allow for movement. **Cilia** are hairlike structures that move in synchronized fashion on the surface of some cells. For example, cilia on the surface of cells lining the respiratory tract constantly move in an attempt to catch and remove bacteria and particles that may enter the respiratory tract. Some animal cells, such as sperm, contain a **flagellum**, which essentially act as a tail to allow for movement. Both cilia and flagella are composed of 9 pairs of microtubules arranged circularly around a pair of microtubules. This is referred to as a 9 + 2 arrangement. The sliding of microfilaments powered by ATP is what allows for the movement of these structures.

Structures Unique to Plant Cells

Plant cells generally have all of the structures and organelles described to this point. However, there are a few additional structures that are unique to plant cells. These include a cell wall, chloroplasts, and a central vacuole.

The **cell wall** is composed of cellulose (fiber) and serves to protect the cell from its environment and desiccation. The **chloroplasts** within a plant cell contain the green pigment **chlorophyll**, which is used in the process of photosynthesis. Chloroplasts are similar to mitochondria in that they have their own DNA and replicate independently. Endosymbiotic theory is also used to explain their existence in plants. Finally, plant cells contain a large **central vacuole** that serves as reserve storage for water, nutrients, and waste products. The central vacuole typically takes up the majority of space within a plant cell. Figure 1-2 shows the typical structure of a plant cell.

CELLULAR ADHESION

Even though a cell is bound by its cell membrane, it must be able to interact with the outside environment and other cells. **Cellular junctions** are connections between the membranes of cells that allow them to adhere to one other and to communicate with one other. These junctions occur in three forms: gap junctions, tight junctions, and adhesion junctions.

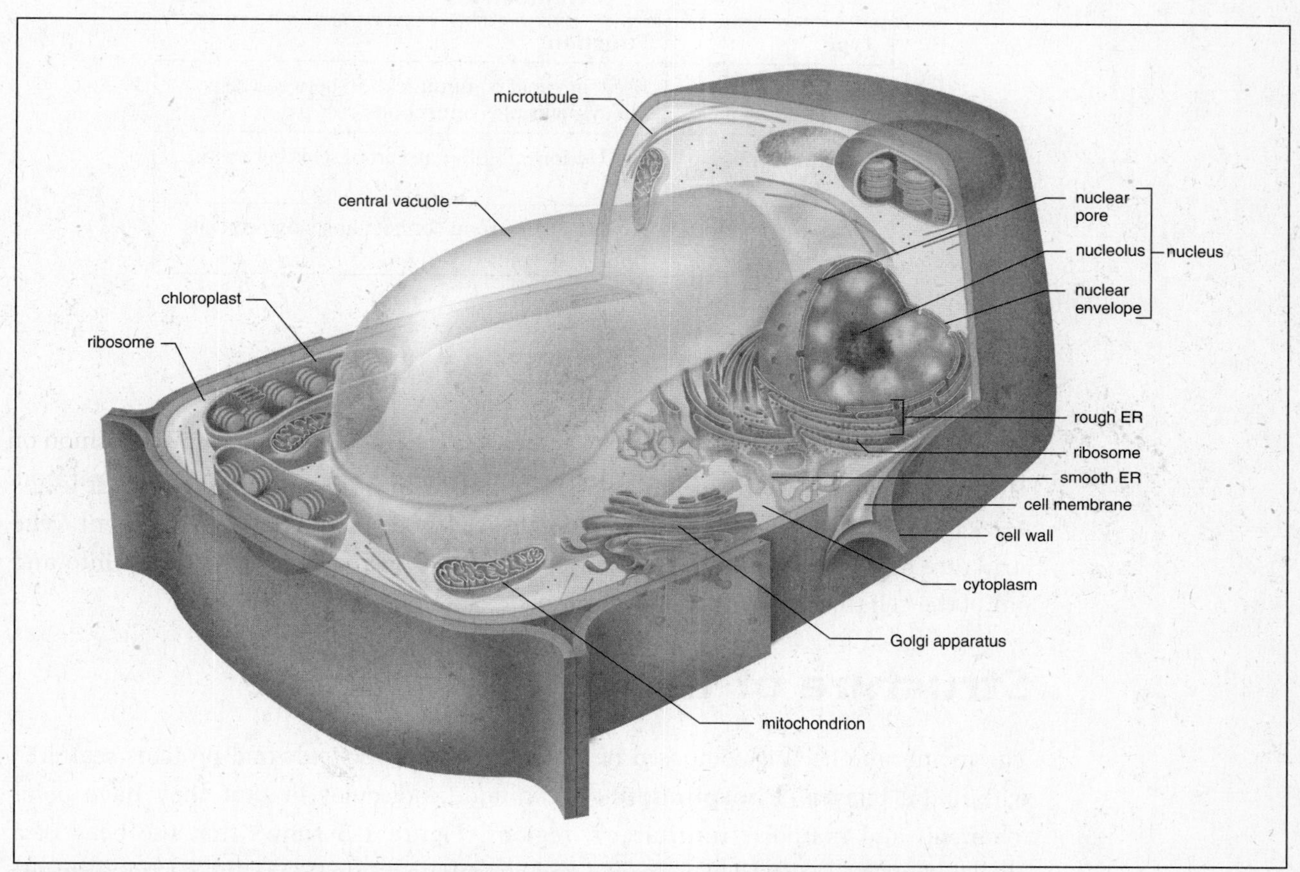

FIGURE 1-2: Plant cell structure. A typical plant cell contains structures and organelles similar to animal cells with the addition of chloroplasts, a cell wall, and a central vacuole. SOURCE: From Sylvia S. Mader, *Biology*, 8th ed., McGraw-Hill, 2004; reproduced with permission of The McGraw-Hill Companies.

In **gap junctions**, the cytoplasm of two or more cells connects directly. These connections serve as channels to allow for the rapid movement of substances between cells. One location where gap junctions are prominent is within the cells of the heart muscle.

Tight junctions are used to attach cells together producing a leak-proof seal. This is critical in areas of the body where it would not be desirable to leak fluids. Tight junctions are common in places such as the stomach lining, internal body cavities, and the outer surfaces of the body.

Adhering junctions are used to attach cells in areas that need to stretch, such as the skin and bladder.

A comparison of junction types is seen in the following table.

Cell Junctions	
Type	**Function**
Gap junctions	Used for rapid communication between cells via cytoplasmic connections
Tight junctions	Used to form tight, waterproof seals between cells
Adhering junctions	Used to form strong connections between cells that need to stretch

THE CELL MEMBRANE AND MOVEMENT ACROSS IT

The outer boundary of the cell is the **cell membrane** (**plasma membrane**). It forms a selectively permeable barrier between the cell and its external environment. The structure of the membrane itself allows for the selective passage of materials into and out of the cell through various mechanisms.

Structure of the Cell Membrane

The membrane itself is composed of a bilayer of phospholipids and proteins scattered within the bilayer. **Phospholipids** are unique molecules in that they have polar (charged) and nonpolar (uncharged) regions. Figure 1-3 shows that the head of a phospholipid is composed of a glycerol and phosphate group (PO_4) that carries a charge and is **hydrophilic** (water-loving). The tails of the phospholipid are fatty acids, which are not charged and are **hydrophobic** (water-repelling). Phospholipids spontaneously arrange themselves in a bilayer in which the heads align themselves toward the inside and outside of the cell where water is located, and the fatty acid tails are sandwiched between the layers. Nonpolar molecules have an easier time crossing the bilayer than other types of molecules.

In addition to the phopholipid bilayer, there are some other substances present in the cell membrane. The **fluid-mosaic model** seen in Figure 1-4 shows the basic membrane structure. **Cholesterol** is found embedded within the interior of the membrane; it regulates the fluidity of the membrane. Proteins are scattered within the bilayer.They function in membrane transport, enzymatic activity, cell adhesion, and

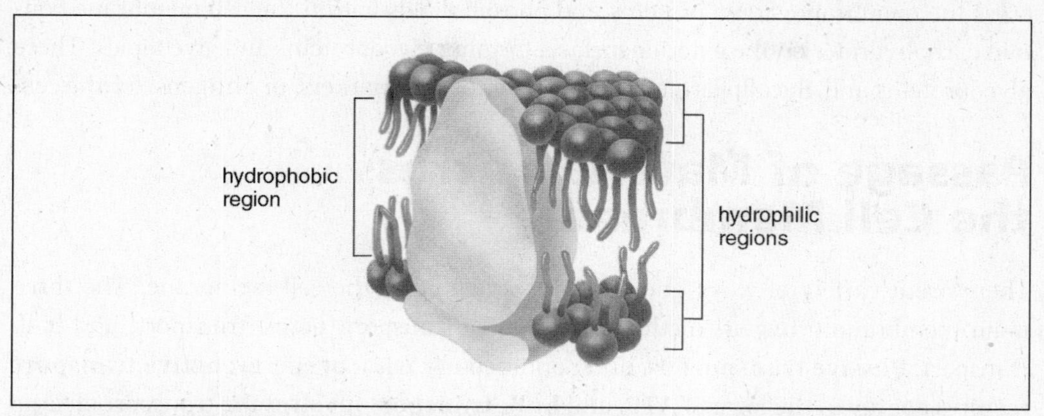

FIGURE 1-3: Phospholipid structure. Phospholipids contain polar regions, which are hydrophilic, and nonpolar regions, which are hydrophobic. SOURCE: From Sylvia S. Mader, *Biology*, 8th ed., McGraw-Hill, 2004; reproduced with permission of The McGraw-Hill Companies.

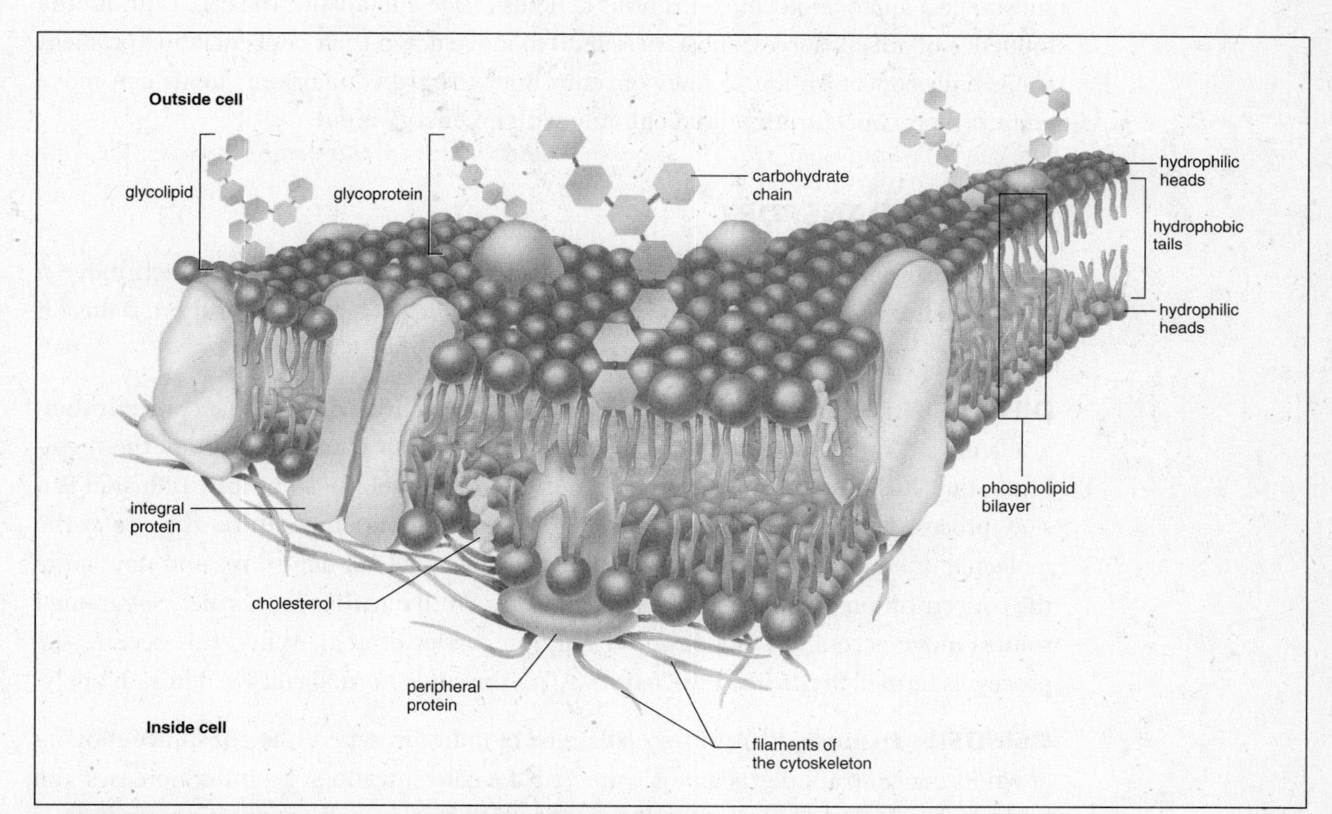

FIGURE 1-4: The fluid-mosaic model. The cell membrane is composed of a phospholipid bilayer in which proteins are embedded. The membrane is anchored to the cell via cytoskeleton fibers. SOURCE: From Sylvia S. Mader, *Biology*, 8th ed., McGraw-Hill, 2004; reproduced with permission of The McGraw-Hill Companies.

cell communication, and also serve as receptors for specific substances that may need to cross the membrane. Some proteins and phospholipids within the cell membrane contain carbohydrates on the exterior surface, forming glycoproteins and glycolipids. These glycoproteins and glycolipids often serve as identifying markers, or antigens, for the cell.

Passage of Materials across the Cell Membrane

There are a variety of ways that substances can cross the cell membrane. The three main membrane transport methods are passive transport, active transport, and bulk transport. **Passive transport** occurs spontaneously without energy; **active transport** requires energy in the form of ATP, and **bulk transport** involves the transport of large items or large quantities of an item using specific mechanisms.

Concentration gradients are a key consideration with movement across the cell membrane. The **concentration gradient** refers to a relative comparison of solutes (dissolved substances) and overall concentrations of fluids inside and outside the cell. Without the influence of outside forces, substances tend to move down their concentration gradient (from high concentration to low concentration) toward equilibrium. Items can move against their concentration gradient only with an energy input.

PASSIVE TRANSPORT

Passive transport mechanisms include diffusion and osmosis, both of which move a substance from an area of high concentration to an area of low concentration. Diffusion and osmosis are spontaneous processes and do not require ATP.

DIFFUSION: Diffusion is the movement of small solutes down their concentration gradients. In other words, dissolved particles move from whichever side of the membrane that has more of them to the side of the membrane that has less. Diffusion is a slow process by nature, but its rate can be influenced by temperature, the size of the molecule attempting to diffuse (large items are incapable of diffusion), and how large the concentration gradient is. Diffusion continues until **equilibrium** is met. Some small solutes move across the membrane through a carrier protein. When this occurs, the process is termed **facilitated diffusion**. All of the same normal rules of diffusion apply.

OSMOSIS: Osmosis is a very specific type of diffusion where the substance moving down its concentration gradient is water. As the concentration of solutes increases, the concentration of water decreases. In an attempt to have equally concentrated solutions both inside and outside the cell, osmosis occurs if the solute itself is unable to cross the cell membrane because it is too large. Simply put, osmosis moves water from the side of the membrane that has more water (and less solute) to the side of the membrane that has less water (and more solute).

When the concentrations of solutes inside and outside the cell are equal, the solutions are termed **isotonic** and there is no net movement of water into or out of the cell. In most situations, isotonic solutions are the goal for cells. A solution that has more water and less solute relative what it is being compared to is termed **hypotonic**, whereas a solution that has less water and more solute relative to what it is being compared to is termed a **hypertonic** solution. When cells are placed in hypertonic solutions, water leaves the cell via osmosis, which can cause cells to shrivel. Cells placed in hypotonic solutions gain water via osmosis and potentially swell and burst. The osmotic effects of each type of solution can be seen in Figure 1-5.

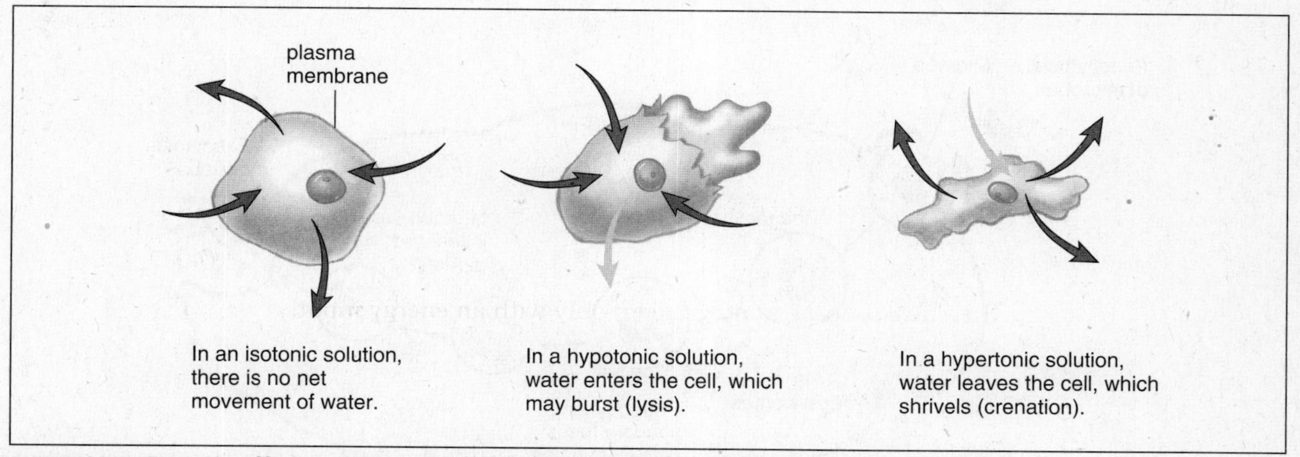

FIGURE 1-5: Osmotic effects on animal cells. The arrows indicate the movement of water. In an isotonic solution, a cell neither gains nor loses water; in a hypertonic solution, the cell loses water; in a hypotonic solution, the cell gains water. SOURCE: From Sylvia S. Mader, *Biology*, 8th ed., McGraw-Hill, 2004; reproduced with permission of The McGraw-Hill Companies.

ACTIVE TRANSPORT

In contrast to passive transport, active transport is used to move solutes against their concentration gradient from the side of the membrane that has less solute to the side that has more. The solutes move via transport proteins in the membrane that act as pumps. Because this is in contrast to the spontaneous nature of passive transport, energy in the form of ATP must be invested to pump solutes against their concentration gradients. Individual items can be actively transported through protein pumps such as the proton pumps used in the electron transport chain of aerobic cellular respiration. There are also co-transporters that move more than one item at a time by active transport. The **sodium–potassium pump** is an example of a co-transporter described in more detail in Chapter 11. Active transport mechanisms are essential to maintaining membrane potentials (charged states) within a variety of specialized cells within the body.

BULK TRANSPORT

The methods for membrane transport described thus far are limited by the size of molecules and do not consider the movement of large items (or large quantities of an item) across the membrane. Endocytosis and exocytosis are used to move large items across the cell membrane and can be seen in Figure 1-6.

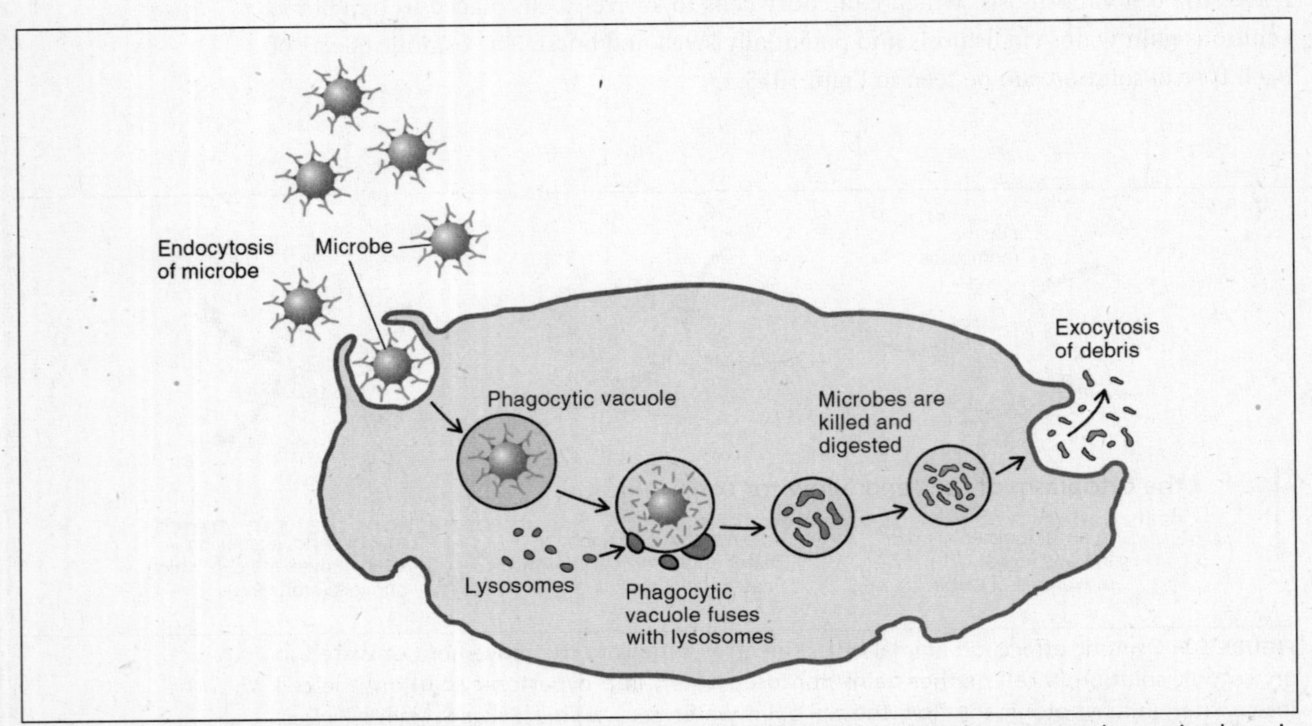

FIGURE 1-6: Endocytosis and exocytosis. Cells engulf large substances such as microbes via endocytosis, and excrete large substances via exocytosis. SOURCE: From Eldon D. Enger, Frederick C. Ross, and David B. Bailey, *Concepts in Biology*, 11th ed., McGraw-Hill, 2005; reproduced with permission of The McGraw-Hill Companies.

ENDOCYTOSIS: Endocytosis is used to bring items into the cell. The membrane surrounds the item to form a vesicle that pinches off and moves into the cell. When liquids are moved into the cell this way, the process is termed **pinocytosis**. When large items, such as other cells, are brought into the cell, the process is termed **phagocytosis**. White blood cells are notorious for performing phagocytosis on items such as bacterial cells. Finally, there is **receptor-mediated endocytosis**, where the molecule to be moved into the cell must first bind to a cell membrane receptor before it can be transported.

EXOCYTOSIS: Exocytosis is used to transport molecules out of the cell. In this case, vesicles containing the substance to be transported move toward the cell membrane and fuse with the membrane. This releases the substance to the outside of the cell.

CRAM SESSION

The Cell

1. Cells must have a small surface area-to-volume ratio to maximize their efficiency.
2. The two major cell types are prokaryotic and eukaryotic.
   a. Eukaryotic cells are larger than prokaryotic cells; have membrane-bound organelles including a nucleus; have linear chromosomes; and usually require oxygen to perform cellular respiration.
3. The cytoplasm, ribosomes, and cytoskeleton are structures always found in eukaryotic cells. The organelles found in eukaryotic cells are the nucleus, smooth endoplasmic retiuclum, rough endoplasmic reticulum, Golgi complex, lysosomes, peroxisomes, and mitochondria. Cell walls, chloroplasts, and the central vacuole are unique to plant cells. Review the functions of the structures and organelles in the table titled "Major Cell Structures and Their Functions."
4. Cells adhere to and communicate with other cells via junctions. Gap junctions connect the cytoplasm of cells and allow for communication. Tight junctions form waterproof seals between cells. Adhering junctions provide strong connections that can stretch.
5. The cell membrane is a phospholipid bilayer that is selectively permeable. The fluid-mosaic model in Figure 1-4 describes the basic structure. Substances cross the cell membrane in several ways.
   a. Passive transport moves a small solute down its concentration gradient from the side of the membrane that has more of the solute to the side of the membrane that has less without any energy requirements.
      (1) Diffusion moves a small solute across the membrane. Facilitated diffusion moves a small solute across the membrane through a carrier protein.
      (2) Osmosis moves water from the side of the membrane that is less concentrated (has more water) to the side that is more concentrated (has less water).
   b. Active transport uses membrane pumps that require ATP to move a small solute against its concentration gradient from the side of the membrane that has less solute to the side that has more solute.
   c. Bulk transport moves large solutes or large quantities of a solute across the membrane.
      (1) Endocytosis moves solutes into the cell. Pinocytosis refers to moving liquid into the cell and phagocytosis refers to moving solids into the cell.
      (2) Exocytosis transports solutes out of the cell.

CHAPTER 2

Enzymes, Energy, and Cellular Metabolism

Read This Chapter to Learn About:

- ➤ Enzyme Structure and Function
- ➤ The Preferred Energy Source for Cells
- ➤ Aerobic and Anaerobic Cellular Respiration

The cell, the basic unit of life, acts as a biochemical factory, using food to produce energy for all the functions of life, including growth, repair, and reproduction. This work of the cell involves many complex chemical processes, including respiration and energy transfer, in which specific enzymes are used to facilitate the reactions.

ENZYMES

Enzymes are a special category of proteins that serve as biological **catalysts**, speeding up chemical reactions. The enzymes, with names often ending in the suffix –ase, are essential to the maintenance of **homeostasis**, or a stable internal environment, within a cell. The maintenance of a stable cellular environment and the functioning of the cell are essential to life.

Enzymes function by lowering the **activation energy** (Figure 2-1) required to initiate a chemical reaction, thereby increasing the rate at which the reaction occurs. Enzymes are involved in **catabolic** reactions that break down molecules, as well as in **anabolic** reactions that are involved in biosynthesis. Most enzymatic reactions are reversible. Enzymes are unchanged during a reaction and are recycled and reused.

Enzyme Structure

As stated earlier, enzymes are proteins and, like all proteins, are made up of amino acids. Interactions between the component amino acids determine the overall shape of an enzyme and it is this shape that is critical to an enzyme's ability to catalyze a reaction.

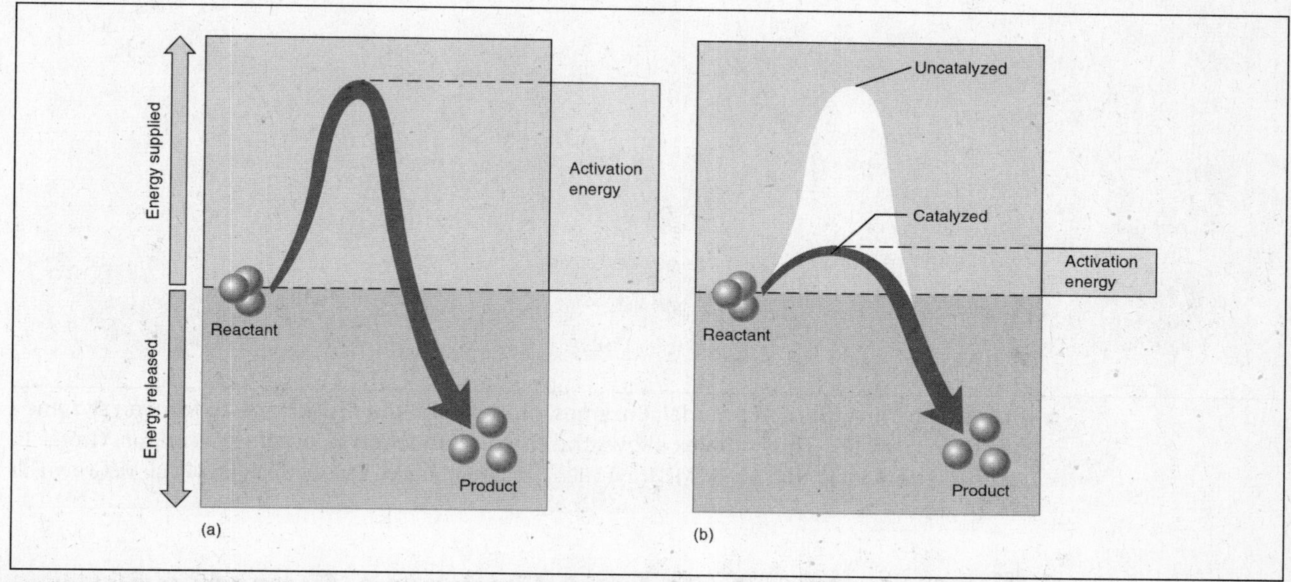

FIGURE 2-1: Lowering activation energy. (*a*) Activation energy is the amount of energy needed to destabilize chemical bonds. (*b*) Enzymes serve as catalysts to lower the amount of activation needed to initiate a chemical reaction. SOURCE: From George B. Johnson, *The Living World,* 3d ed., McGraw-Hill, 2003; reproduced with permission of The McGraw-Hill Companies.

The area on an enzyme where it interacts with another substance, called a **substrate**, is the enzyme's **active site**. Based on its shape, a single enzyme typically only interacts with a single substrate (or single class of substrates); this is known as the enzyme's **specificity**. Any changes to the shape of the active site render the enzyme unable to function.

Enzyme Function

The **induced fit model** is used to explain the mechanism of action for enzyme function seen in Figure 2-2. Once a substrate binds loosely to the active site of an enzyme, a conformational change in shape occurs to cause tight binding between the enzyme and the substrate. This tight binding allows the enzyme to facilitate the reaction. A substrate with the wrong shape cannot initiate the conformational change in the enzyme necessary to catalyze the reaction.

Some enzymes require assistance from other substances to work properly. If assistance is needed, the enzyme has binding sites for cofactors or coenzymes. **Cofactors** are various types of ions such as iron and zinc (Fe^{2+} and Zn^{2+}). **Coenzymes** are organic molecules usually derived from vitamins obtained in the diet. For this reason, mineral and vitamin deficiencies can have serious consequences on enzymatic functions.

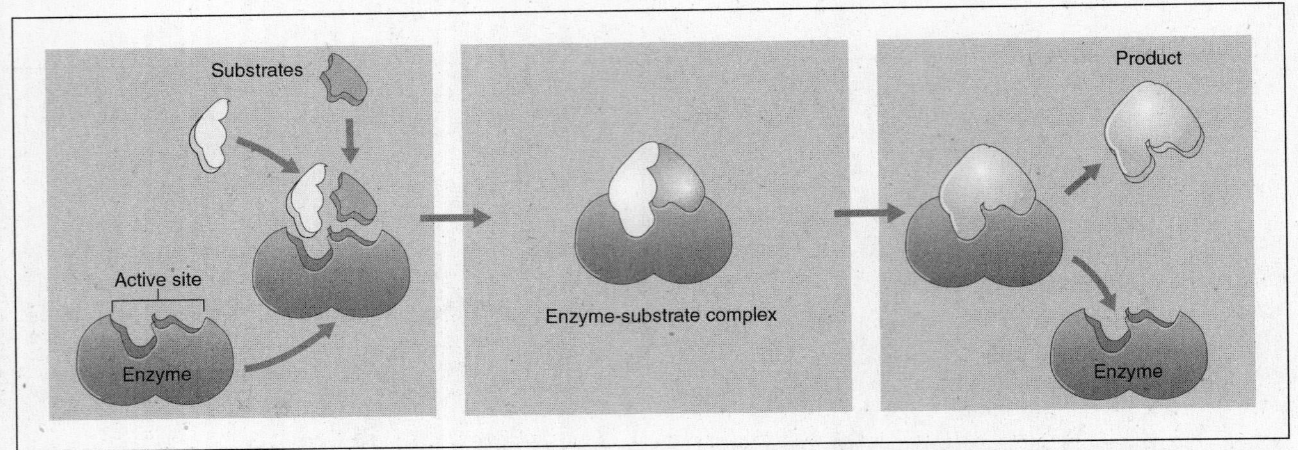

FIGURE 2-2: The induced fit model. Enzymes interact with their substrates to form an enzyme–substrate complex. This complex allows the chemical reaction to occur. SOURCE: From George B. Johnson, *The Living World*, 3d ed., McGraw-Hill, 2003; reproduced with permission of The McGraw-Hill Companies.

Factors That Affect Enzyme Function

There are several factors that can influence the activity of a particular enzyme. The first is the concentration of the substrate and the concentration of the enzyme. Reaction rates stay low when the concentration of the substrate is low, whereas the rates increase when the concentration of the substrate increases. Temperature is also a factor that can alter enzyme activity. Each enzyme has an optimal temperature for functioning. In humans this is typically body temperature (37°C). At lower temperatures, the enzyme is less efficient. Increasing the temperature beyond the optimal point can lead to enzyme **denaturation**, which renders the enzyme useless. Enzymes also have an optimal pH in which they function best, typically around 7 in humans, although there are exceptions. Extreme changes in pH can also lead to enzyme denaturation. The denaturation of an enzyme is not always reversible.

Control of Enzyme Activity

It is critical to be able to regulate the activity of enzymes in cells to maintain efficiency. This regulation can be carried out in several ways. Feedback, or **allosteric inhibition**, illustrated in Figure 2-3, acts somewhat like a thermostat to regulate enzyme activity. Many enzymes contain allosteric binding sites and require signal molecules such as repressors and activators to function. As the product of a reaction builds up, repressor molecules can bind to the **allosteric site** of the enzyme, causing a change in the shape of the active site. The consequence of this binding is that the substrate can no longer interact with the active site of the enzyme and the activity of the enzyme is temporarily slowed or halted. When the product of the reaction declines, the repressor molecule dissociates from the allosteric site. This allows the active site of the enzyme to resume its normal shape and normal activity. Some allosteric enzymes stay inactive unless activator molecules are present to allow the active site to function.

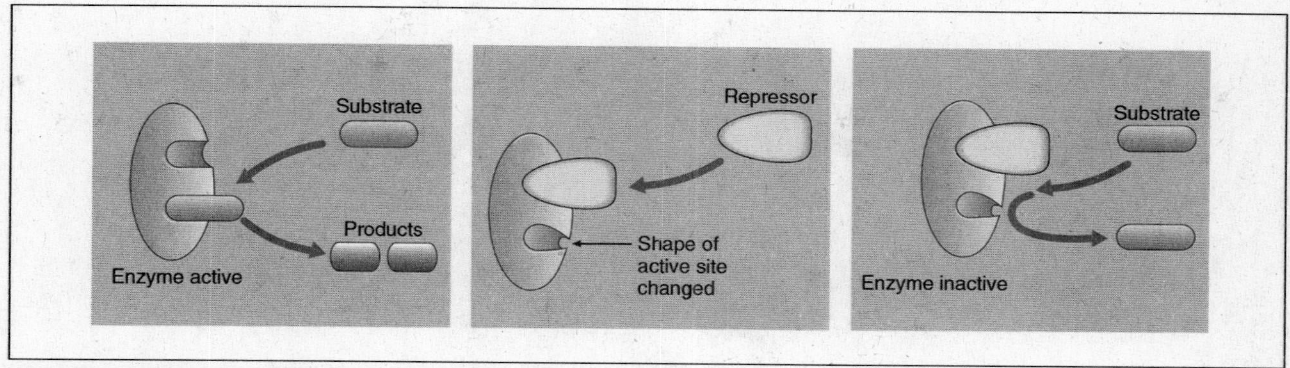

FIGURE 2-3: Allosteric inhibition of an enzyme. Repressors can be used to regulate the activity of an enzyme. SOURCE: From George B. Johnson, *The Living World,* 3d ed., McGraw-Hill, 2003; reproduced with permission of The McGraw-Hill Companies.

Inhibitor molecules also regulate enzyme action. A **competitive inhibitor** is a molecule that resembles the substrate in shape so much that it binds to the active site of the enzyme, thus preventing the substrate from binding. This halts the activity of the enzyme until the competitive inhibitor is removed or is out-competed by an increasing amount of substrate. **Noncompetitive inhibitors** bind to allosteric sites and change the shape of the active site, thereby decreasing the functioning of the enzyme. Increasing levels of substrate have no effect on noncompetitive inhibitors, but the activity of the enzyme can be restored when the noncompetitive inhibitor is removed.

ENERGY

The preferred source of energy for cells is **adenosine triphosphate** (ATP). This molecule is a hybrid consisting of the nitrogenous base adenine (found in nucleotides), the sugar ribose, and three phosphate groups (PO_4). The structure of ATP can be seen in Figure 2-4. The breaking of the bond that attaches the last PO_4 to the molecule results in the release of energy that is used in a variety of cellular processes. The resulting molecule is **adenosine diphoshpate** (ADP). A PO_4 must be reattached to ADP to regenerate ATP. Unfortunately, the process of adding a PO_4 to ADP is not a simple one. The chemical reactions of cellular respiration are used to achieve this goal.

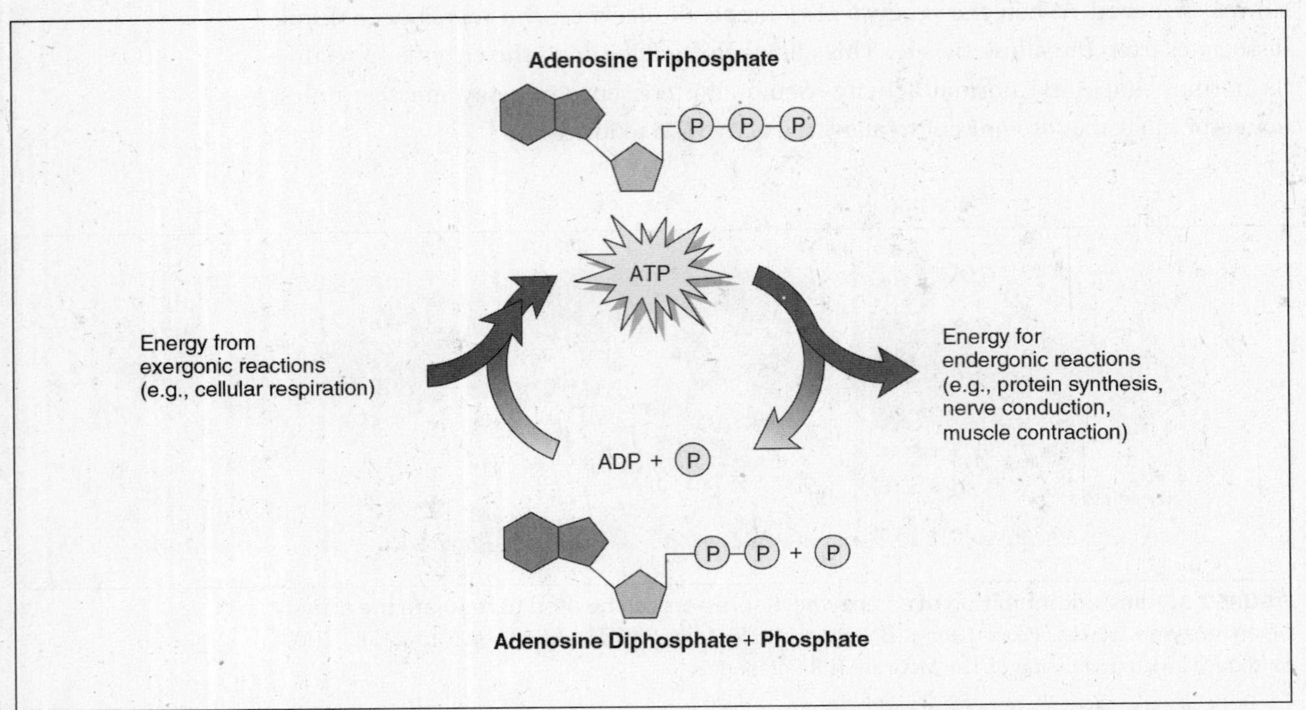

FIGURE 2-4: The ATP cycle. When a PO_4 is removed from ATP, energy is released and ADP is the resulting product. Cellular respiration is a primary means of adding a PO_4 to ADP to generate ATP. SOURCE: From Sylvia S. Mader, *Biology*, 8th ed., McGraw-Hill, 2004; reproduced with permission of The McGraw-Hill Companies.

CELLULAR RESPIRATION AND METABOLISM

Cellular metabolism encompasses the total of all anabolic and catabolic reactions that occur within the cell. These critical reactions rely on a variety of enzymes to increase their rate to an appropriate level. Anabolic reactions require energy whereas catabolic reactions release energy. These reactions are coupled so that the energy released from one catabolic reaction can be used to fuel an anabolic reaction.

A critical catabolic reaction in cells is the breakdown of glucose to release energy in the form of ATP, or more specifically, to convert ADP back to ATP. The glucose is made in the anabolic reactions of photosynthesis that occurs in plants and is usually obtained by animals from their diet. The breakdown of glucose, which must be done by all living organisms to produce energy, occurs through the process of **cellular respiration**.

Cellular respiration can be done aerobically (using oxygen) or anaerobically (without oxygen). The aerobic pathway has a much higher ATP yield than the anaerobic pathway. The production of ATP in either pathway relies on the addition of a PO_4 to ADP. This can be achieved through **substrate-level phosphorylation** when ATP synthesis is directly coupled to the breakdown of glucose, or via **oxidative phosphorylation**, where ATP synthesis involves an intermediate molecule.

Aerobic Respiration

The aerobic pathway can be demonstrated by the following reaction, where glucose ($C_6H_{12}O_6$) and oxygen (O_2) interact to produce carbon dioxide (CO_2), water (H_2O), and ATP:

$$C_6H_{12}O_6 + 6O_2 \rightarrow 6CO_2 + 6H_2O + ATP$$

Aerobic cellular respiration begins with the process of **glycolysis**, is followed by the **Kreb's cycle** (also called the *citric acid cycle*), and concludes with the **electron transport chain**, which is fueled by protons (H^+) and electrons and is the step that produces the majority of ATP.

During glycolysis and the Kreb's cycle, glucose is systematically broken down and small amounts of ATP are generated by substrate-level phosphorylation. Carbon dioxide is released as a waste product. However, the most important part of these steps is the breakdown of glucose that allows for electron carrier molecules to collect protons and electrons needed to run the electron transport chain, where ATP is produced in mass quantities.

Electron carrier molecules are a critical part of aerobic cellular respiration. These electron carrier molecules include **nicotinamide adenine dinucleotide** (NAD^+) and **flavin adenine dinucleotide** (FAD). When these molecules pick up electrons and protons, they are reduced to NADH and $FADH_2$. The reduced forms of the carrier molecules deliver protons and electrons to power the electron transport chain. Once these items are delivered to the electron transport chain, the molecules return to their oxidized forms, NAD^+ and FAD.

An overview of all the reactions of aerobic cellular respiration, including starting and ending points, is shown in the following table.

Summary of Aerobic Cellular Respiration			
Step	**Location in the Cell**	**Starting Products**	**Ending Products**
Glycolysis	Cytoplasm	Glucose, ATP, ADP, NAD^+	Pyruvate, ATP, NADH
Kreb's cycle	Matrix of mitochondria	Acetyl-CoA, ADP, NAD^+, FAD	CO_2, ATP, NADH, $FADH_2$
Electron transport chain	Cristae membrane of mitochondria	NADH, $FADH_2$, O_2, ADP	NAD^+, FAD, H_2O, ATP

GLYCOLYSIS

Overall, the process of glycolysis breaks down glucose into two molecules of **pyruvate**. This happens in the cytoplasm of the cell and is the starting step for both aerobic and anaerobic cellular respiration.

During the process of glycolysis, 2 ATP molecules are invested while 4 ATP molecules are gained via substrate level phosphorylation. This leaves a net gain of 2 ATP for the process. In addition, NAD^+ is reduced to NADH, which is used in a later step (the electron transport chain) for oxidative phosphorylation. Figure 2-5 shows the major chemical reactions of glycolysis.

At the end of glycolysis, the pyruvate molecules can be further broken down either aerobically or anaerobically. In the aerobic pathway, the subsequent reactions—the Krebs cycle and electron transport—occur in the mitochondria.

KREB'S CYCLE

The two pyruvate molecules remaining at the end of glycolysis are actively transported to the mitochondria, specifically to the matrix of the mitochondria. The structure of a mitochondrion, illustrated in Figure 2-6, is important to its function in cellular respiration. A key feature of this organelle is its double membrane. The space between the inner and outer membrane is termed the **intermembrane space**. The space bounded by the inner membrane is a liquid called the **matrix**, and it is here that the Kreb's cycle occurs. The folded inner membrane is called the **cristae membrane** and is the site of the next step in the process, the electron transport chain.

The pyruvate molecules are modified in order to enter into the reactions of the Kreb's cycle. This modification, termed **pyruvate decarboxylation**, involves the oxidation of pyruvate and the release of CO_2, as illustrated in Figure 2-7. The remnants of pyruvate are a two-carbon acetyl group. Coenzyme A (CoA) is added to the acetyl group creating **acetyl-CoA**, which is capable of entering the Kreb's cycle. These modifications to pyruvate also allow for the reduction of NAD^+ to NADH, which is used later in the electron transport chain. Because there are two pyruvate molecules, this step produces 2 CO_2, 2 acetyl-CoA, and 2 NADH molecules.

The acetyl-CoA that has been formed enters the Kreb's cycle by combining with a four-carbon molecule (oxaloacetate) to form the six-carbon molecule citric acid. The remaining reactions of the Kreb's cycle are seen in Figure 2-8. These reactions include the removal of the two carbons that entered as acetyl-CoA and their release as CO_2, the production of one ATP molecule via substrate level phosphorylation, and the rearrangement of the intermediate products to form the starting molecule of oxaloacetate.

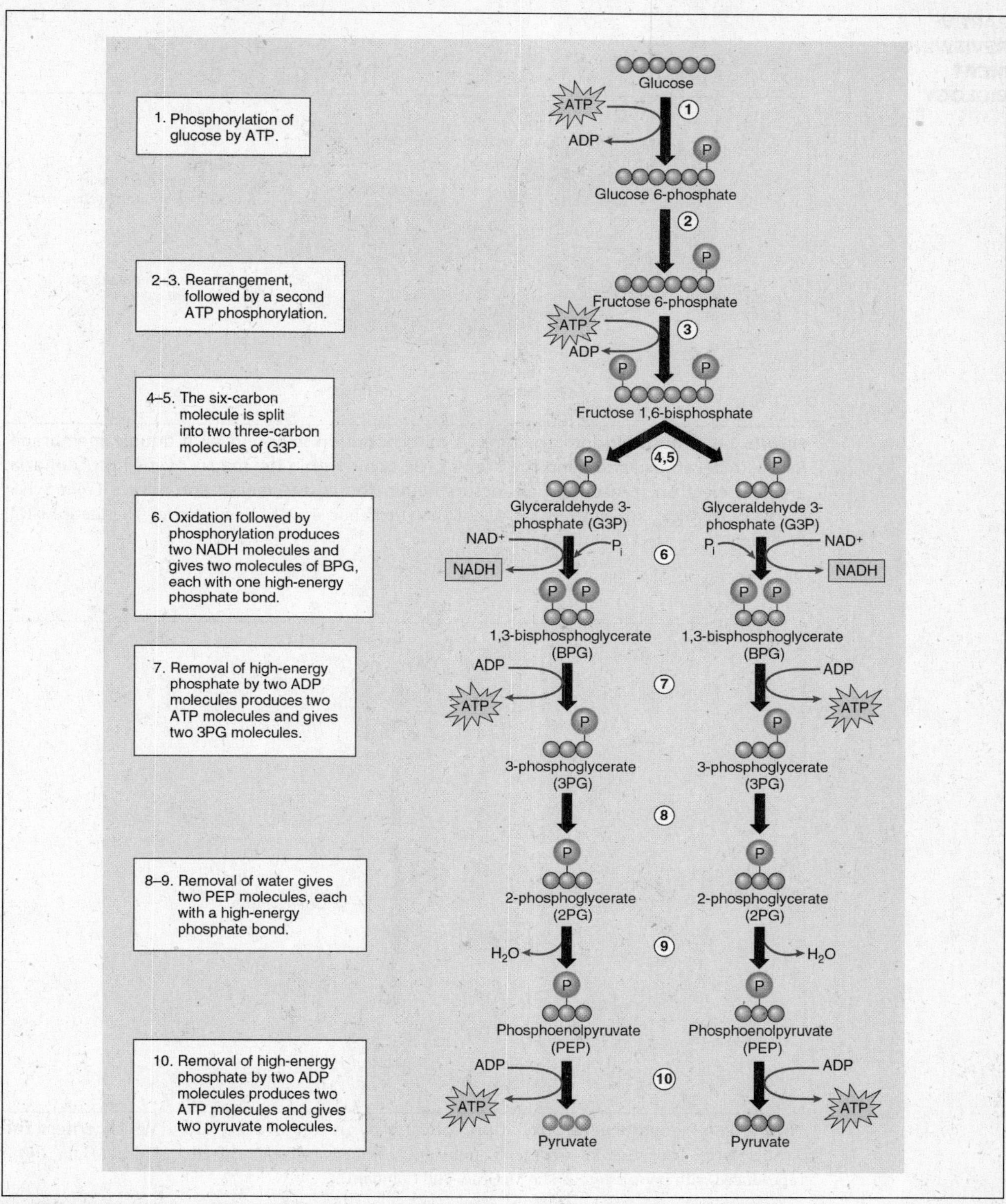

FIGURE 2-5: Glycolysis is a series of chemical reactions that ultimately convert glucose to pyruvate. SOURCE: From George B. Johnson, *The Living World*, 3d ed., McGraw-Hill, 2003; reproduced with permission of The McGraw-Hill Companies.

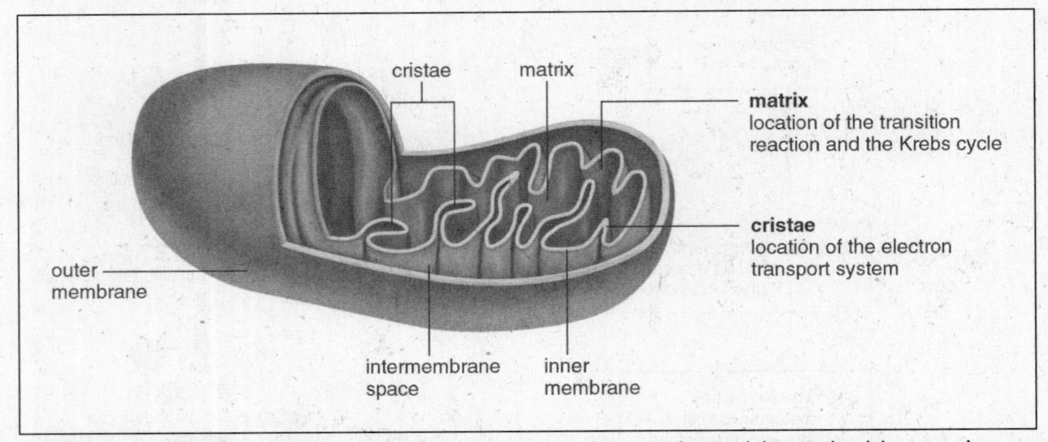

FIGURE 2-6: Mitochondrion structure. A mitochondrion is bound by a double membrane. Pyruvate decarboxylation and the Kreb's cycle occur within the matrix of the mitochondria, and the electron transport chain occurs within the cristae membrane. SOURCE: From Sylvia S. Mader, *Biology*, 8th ed., McGraw-Hill, 2004; reproduced with permission of The McGraw-Hill Companies.

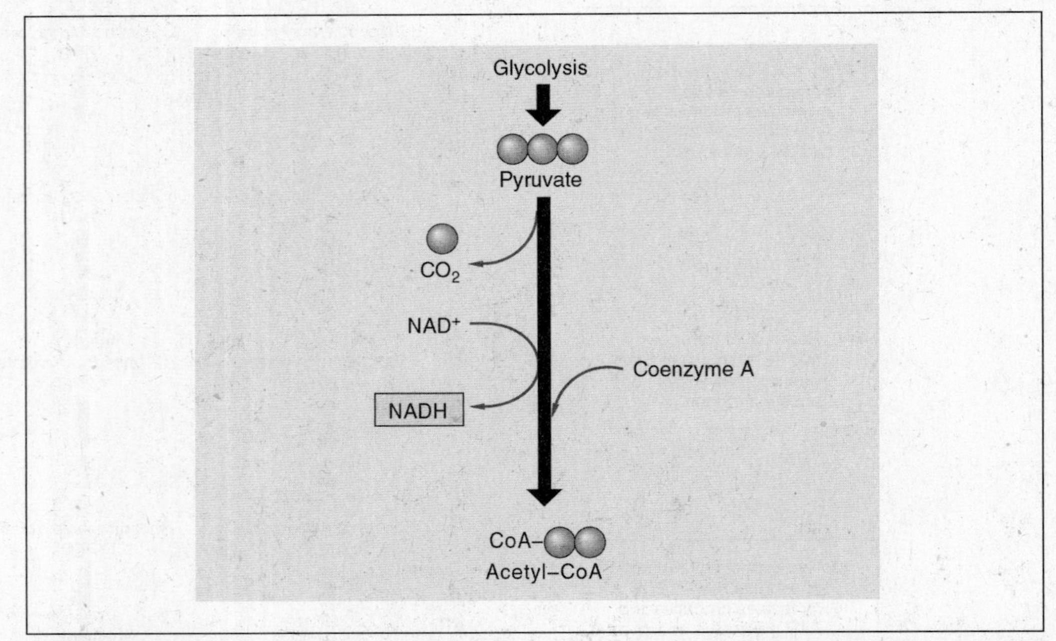

FIGURE 2-7: Pyruvate decarboxylation converts pyruvate to acetyl-CoA, which enters the Kreb's cycle. SOURCE: From George B. Johnson, *The Living World*, 3d ed., McGraw-Hill, 2003; reproduced with permission of The McGraw-Hill Companies.

In this way, the cycle is able to continue. In addition, the rearrangement of intermediates in the process allows for the reduction of NAD^+ to NADH and FAD to $FADH_2$. Because there are two acetyl-CoA molecules, this cycle must turn twice, with the result being the production of 4 CO_2, 2 ATP, 6 NADH, and 2 $FADH_2$ molecules. The glucose has been fully broken down into CO_2, which is released. At this point, the process continues with the electron transport chain.

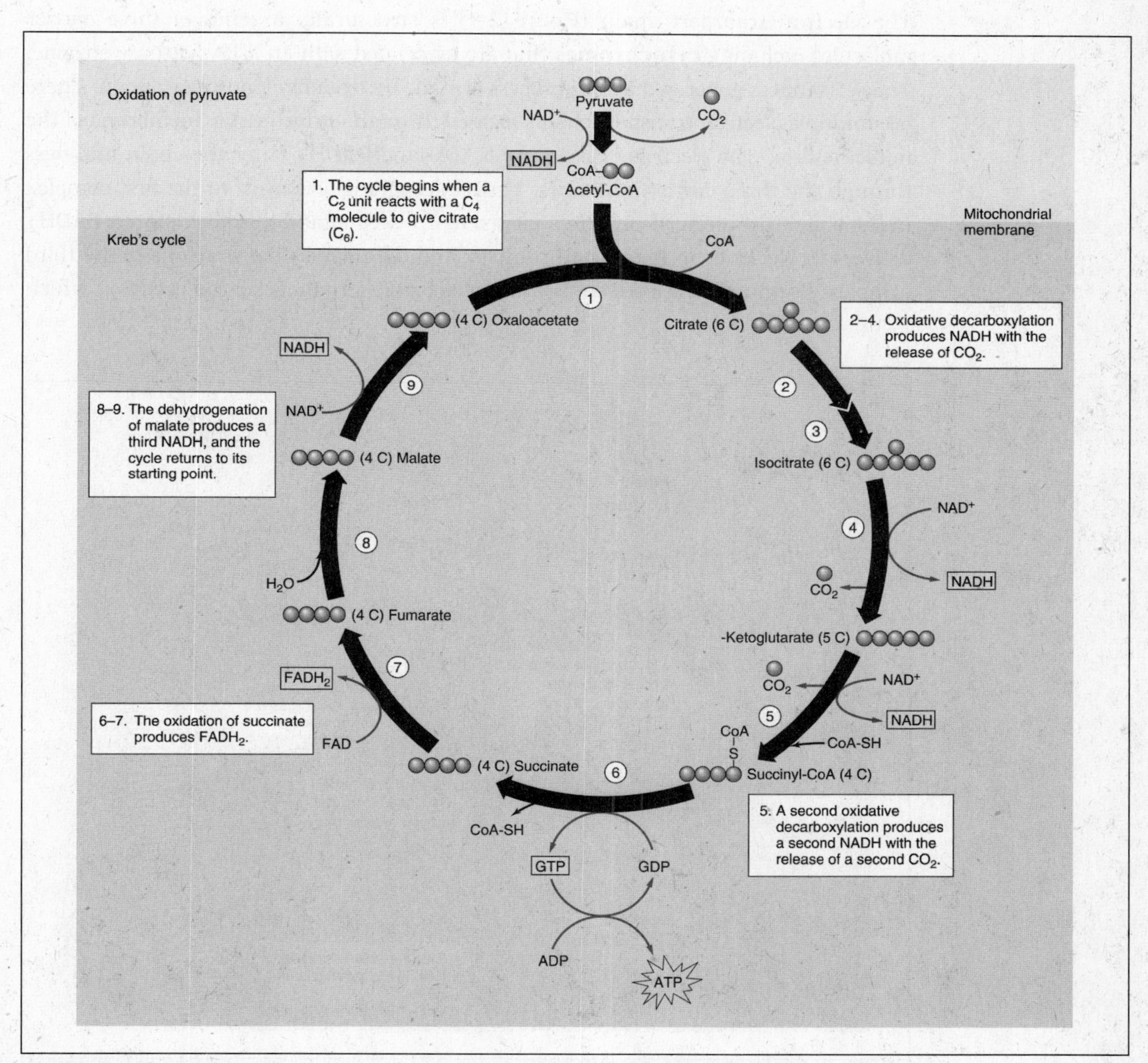

FIGURE 2-8: The Kreb's cycle consists of a series of oxidative and decarboxylation reactions that occur in the matrix of the mitochondria. SOURCE: From George B. Johnson, *The Living World*, 3d ed., McGraw-Hill, 2003; reproduced with permission of The McGraw-Hill Companies.

THE ELECTRON TRANSPORT CHAIN

During glycolysis and the Kreb's cycle, NAD^+ and FAD have been reduced to NADH and $FADH_2$. Once they are reduced, they move toward the electron transport chains located in the cristae membrane of the mitochondria. They are oxidized by releasing the electrons and protons that they carry to the chain. At this point, the oxidized forms of NAD^+ and FAD return to glycolysis and the Kreb's cycle to be used again and again.

The electron transport chain (Figure 2-9) is structurally a series of three carrier molecules including **cytochromes** that are associated with an ATP synthase enzyme. The ATP that is generated in this process is made by oxidative phosphorylation. There are multiple electron transport chains located throughout the cristae membrane of the mitochondria. The electrons carried by NADH and $FADH_2$ enter the chain and pass through the three carrier molecules. NADH delivers its electrons to the first complex in the chain; the electrons are then passed to the second and third complexes. $FADH_2$ delivers its electrons to the second complex and then passes the electrons to the third complex. Eventually the electrons are accepted by the terminal electron acceptor, which

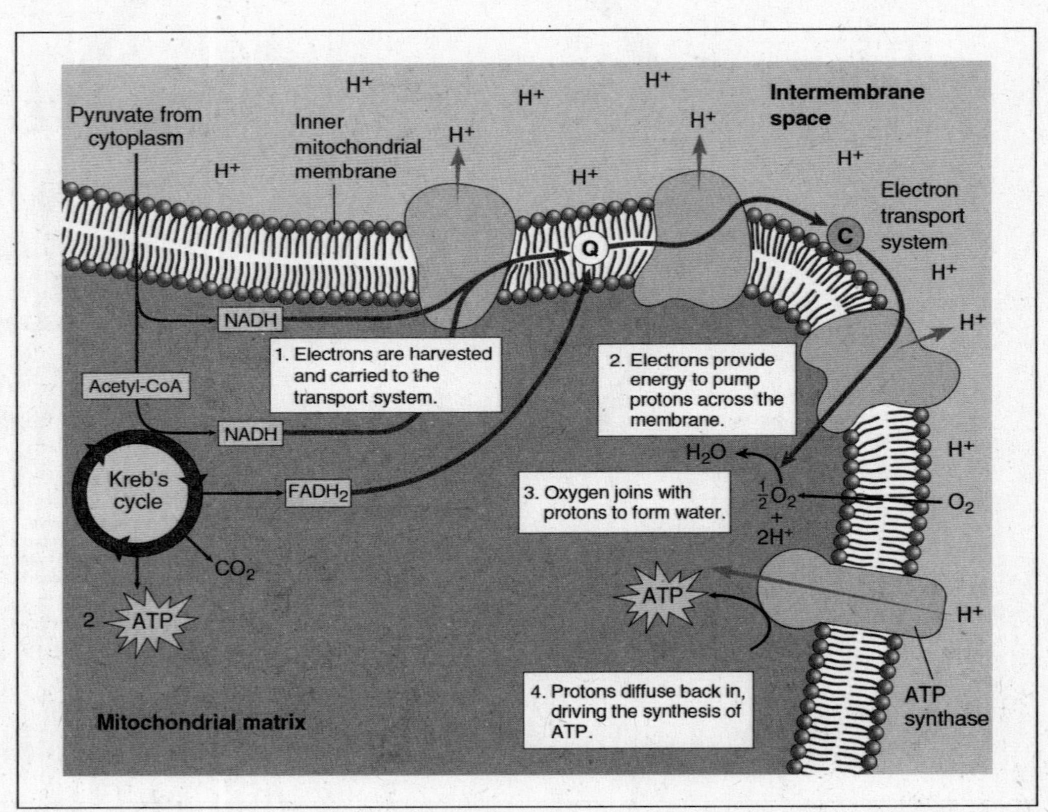

FIGURE 2-9: The electron transport chains exist within the cristae membrane of the mitochondria. ATP is produced via chemiosmosis. SOURCE: From George B. Johnson, *The Living World*, 3d ed., McGraw-Hill, 2003; reproduced with permission of The McGraw-Hill Companies.

is the oxygen inhaled through the respiratory system (hence the aerobic label). As the oxygen picks up the electrons, it also picks up 2 protons (H^+), which, in turn, produce water.

The **chemiosmotic theory** is used to explain how ATP is produced in this process. The energy from the movement of the electrons donated by NADH and $FADH_2$ is used to pump protons into the intermembrane space. These protons build up, creating a proton gradient. The protons move through the **ATP synthase** enzyme via passive transport (diffusion). As each proton moves through the ATP synthase enzyme, ATP is phosphorylated, producing 1 ATP molecule.

For every NADH that donates electrons to the chain, 3 protons are pumped into the intermembrane space, 3 protons re-enter through the ATP synthase, and thus 3 ATP are made. For every $FADH_2$, 2 protons are pumped into the intermembrane space, 2 protons re-enter through the ATP synthase, and 2 ATP are made. A grand total of 32 ATP are produced in the electron transport chain. Combining this number of ATP molecules with the 4 ATP molecules produced by substrate level phosphorylation in glycolysis and the Kreb's cycle leads to a total of 36 ATP molecules made in aerobic cellular respiration for each glucose molecule. A summary of ATP made throughout the process of aerobic respiration is shown in the following table.

Summary of ATP Production in Aerobic Cellular Respiration

	Products of Previous Steps	**ATP Made during Chemiosmotic Phosphorylation in the Electron Transport Chain**
Glycolysis	2 NADH 2 ATP	2 NADH = 6 ATP (However, the net gain is 4 ATP due to the use of ATP to actively transport pyruvate to the mitochondria)
Kreb's cycle (including pyruvate decarboxylation)	8 NADH 2 $FADH_2$ 2 ATP	8 NADH = 24 ATP 2 $FADH_2$ = 4 ATP
Total	4 ATP (substrate level phosphorylation from glycolysis and Kreb's cycle)	32 ATP (oxidative phosphorylation from the electron transport chain)

If oxygen is not available to accept electrons, the electrons in the electron transport chain build up, essentially shutting down the electron transport chain. Not only does ATP production drastically decline, but now that NADH cannot be oxidized to NAD^+ by the electron transport chain, there is not enough NAD^+ to continue glycolysis. In this case, the use of fermentation (a part of the anaerobic pathway) is necessary to complete the oxidation of NADH to NAD^+ in order to continue glycolysis.

Anaerobic Respiration

There are times when oxygen is either not available or not utilized by cells to perform aerobic respiration. For animals, this may occur when the oxygen demands of the cells cannot be met for brief periods of time. Unfortunately, anaerobic respiration produces very little ATP compared to aerobic cellular respiration, and thus it cannot meet the ATP demands of larger organisms for extended periods of time. However, in some smaller organisms, such as certain bacteria and yeasts, anaerobic respiration can be used permanently or for lengthy periods of time.

The first step in the anaerobic pathway is glycolysis. Once glycolysis occurs and pyruvate is generated, the anaerobic pathway continues with a second step, **fermentation**. Depending on the organism, fermentation can occur in one of two ways—lactic acid fermentation or alcoholic fermentation. The primary benefit of either type of fermentation is that it allows for the oxidation of NADH to NAD^+ that is necessary for glycolysis to continue in the absence of a functional electron transport chain. It is important to note that fermentation itself produces no ATP. The only ATP created during anaerobic respiration (glycolysis and fermentation) is from the glycolysis step. Therefore, it is absolutely critical to regenerate the NAD^+ needed to continue with glycolysis. The total net gain of ATP from anaerobic respiration is 2 ATP as compared to 36 from complete aerobic respiration.

LACTIC ACID FERMENTATION

Lactic acid fermentation occurs in some types of bacteria and fungi as well as in the muscle cells of animals when oxygen levels are not sufficient to meet the demands of aerobic respiration. In this step, pyruvate is reduced to lactic acid, thus regenerating the NAD^+ needed to continue glycolysis. In humans, large amounts of lactic acid are responsible for muscle soreness after major exertion.

ALCOHOLIC FERMENTATION

Some organisms such as certain bacteria and yeast use **alcoholic fermentation.** In this step, pyruvate is decarboxylated, which produces CO_2, and then reduced to form ethanol. As with lactic acid fermentation, NAD^+ is recycled so that glycolysis may continue.

Metabolism of Fats and Protein

So far, glucose has been discussed as the only starting material for cellular respiration. However, there are times that glucose may not be available. If cellular respiration stopped completely due to a lack of fuel, death would occur quickly. For this reason, there must be backup sources that can be used to fuel the process when glucose levels are insufficient.

When glucose levels are low, the body uses other forms of carbohydrates, such as **glycogen** stored in the liver and muscles. When those reserves are depleted, fat is used next, and, as a last resort, protein is used. When fats and proteins are used to fuel cellular respiration, they must first be converted to glucose or a glucose derivative. Fats can be broken down to fatty acids and glycerol. Fatty acids can be converted by beta oxidation into acetyl-CoA that can enter the Kreb's cycle. Glycerol can be converted to an intermediate of glycolysis. As proteins are broken down into amino acids, they can be chemically modified into -keto acids and then converted into acetyl-CoA, pyruvate, or various intermediates of the Kreb's cycle.

CRAM SESSION

Enzymes, Energy, and Cellular Metabolism

1. Enzymes are a class of proteins that speed chemical reactions by lowering activation energy.
   a. The induced-fit model explains the loose interaction of a substrate at the active site of an enzyme followed by a conformational change in shape that allows the enzyme to bind tightly and facilitate the reaction.
   b. An enzyme becomes denatured when the shape of the active site is altered.
   c. The allosteric site of an enzyme can be used to regulate enzyme activity. Competitive and noncompetitive inhibitors also regulate enzyme activity.
2. Adenosine triphosphate (ATP) is the preferred energy source for all organisms. The hydrolysis of ATP results in ADP and PO_4.
3. Cellular respiration is used to phosphorylate ADP to generate ATP as glucose is broken down. Substrate level phosphorylation is directly coupled to glucose breakdown, whereas oxidative phosphorylation involves intermediate molecules such as NAD^+ and FAD.
   a. Aerobic cellular respiration breaks down glucose to CO_2. It involves glycolysis, the Kreb's cycle, and the electron transport chain. NAD^+ and FAD collect electrons and protons to be reduced to NADH and $FADH_2$ during glycolysis and the Kreb's cycle. In the electron transport chain, electrons serve as a motor to pump protons across the cristae membrane of the mitochondria to produce a proton gradient. Protons move passively through the ATP synthase, which is coupled to phosphorylation of ADP to create ATP. Oxygen serves as the terminal electron acceptor and ultimately binds to protons to produce water. The net yield of ATP from aerobic cellular respiration is 36.
   b. Anaerobic cellular respiration consists of glycolysis and fermentation. In lactic acid fermentation, pyruvate is converted to lactic acid. In alcoholic fermentation, pyruvate is converted to ethanol and CO_2. Either fermentation step is used to oxidize NADH to NAD^+ so glycolysis can continue. The net gain from anaerobic respiration is 2 ATP.
4. If glucose is absent, fats and proteins can be used for cellular respiration. The molecules must either be converted to glucose or to an intermediate molecule to be used for cellular respiration.

CHAPTER 3

DNA Structure, Replication, and Technology

Read This Chapter to Learn About:

- ➤ DNA Replication and Repair
- ➤ Restriction Enzymes, Recombinant Plasmids, and Gene Cloning
- ➤ DNA Amplification and Sequencing

Deoxyribonucleic acid (DNA) is the genetic, or hereditary, material of the cell. The information encoded in DNA ultimately directs the synthesis of proteins within cells. These proteins determine all biological characteristics. When a cell divides, DNA self-replicates to ensure that progeny cells receive the same DNA instructions as the parent cell.

DNA STRUCTURE

DNA is a nucleic acid polymer consisting of the **nucleotide** monomers seen in Figure 3-1. A nucleotide is a hybrid molecule consisting of a sugar (deoxyribose), a phosphate group (PO_4), and a nitrogenous base. There are four nitrogenous bases used to make a DNA nucleotide: adenine (A), thymine (T), cytosine (C), and guanine (G). Each nucleotide differs only by the nitrogenous base used. In total, there are four possible nucleotides used in DNA.

The nitrogenous bases of each nucleotide are classified as a purine or pyrimidine based on their chemical structure. A **purine** is a double-ringed structure, whereas **pyrimidines** are single-ring structures. The nitrogenous bases adenine and guanine are purines; cytosine and thymine are pyrimidines.

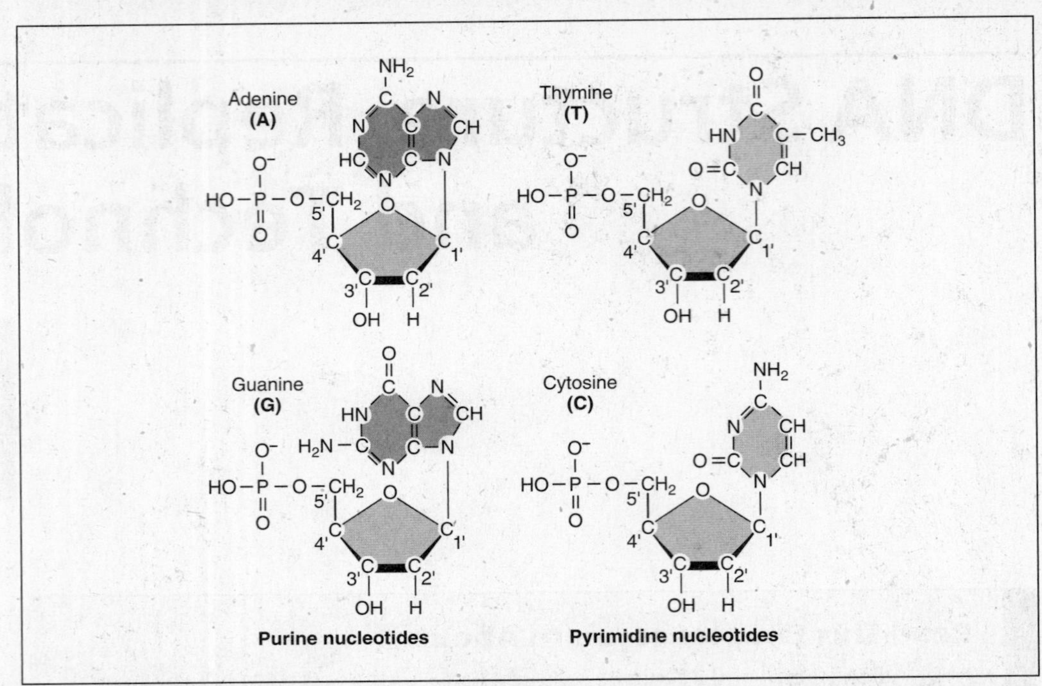

FIGURE 3-1: Nucleotide structure. All nucleotides contain a sugar, a PO_4, and a nitrogenous base. SOURCE: From Sylvia S. Mader, *Biology*, 8th ed., McGraw-Hill, 2004; reproduced with permission of The McGraw-Hill Companies.

James Watson and Francis Crick proposed a model for the structure of DNA in 1953. By analyzing information from the studies of others, they knew that DNA existed in a double-stranded conformation and that the amount of A and T in a DNA molecule was always the same, as was the amount of C and G. They then developed a model of DNA structure seen in Figure 3-2.

A single stand of DNA has a sugar–phosphate backbone (where the nucleotides bond together using phosphodiester bonds). Two strands of DNA are hydrogen bonded together via their nitrogenous bases. The idea of complementary base pairing is essential to this model. Complementary base pairing means that a purine must pair with a pyrimidine. An A on one strand of DNA always bonds to a T on another strand of DNA using two hydrogen bonds. A C on one strand always bonds to a G on another strand using three hydrogen bonds. This base pairing holds together the two strands of DNA, which then twist to take on a double-helix conformation. Knowing the sequence of bases in one DNA strand makes it possible to determine the sequence of bases in the complementary strand.

Each strand of DNA has a specific polarity or direction in which it runs. This polarity is referred to as $5'$ and $3'$. The complementary strand of DNA always runs antiparallel, in the opposite direction of the original strand. So if one DNA strand runs $5'$ to $3'$, the other strand of the double helix runs $3'$ to $5'$.

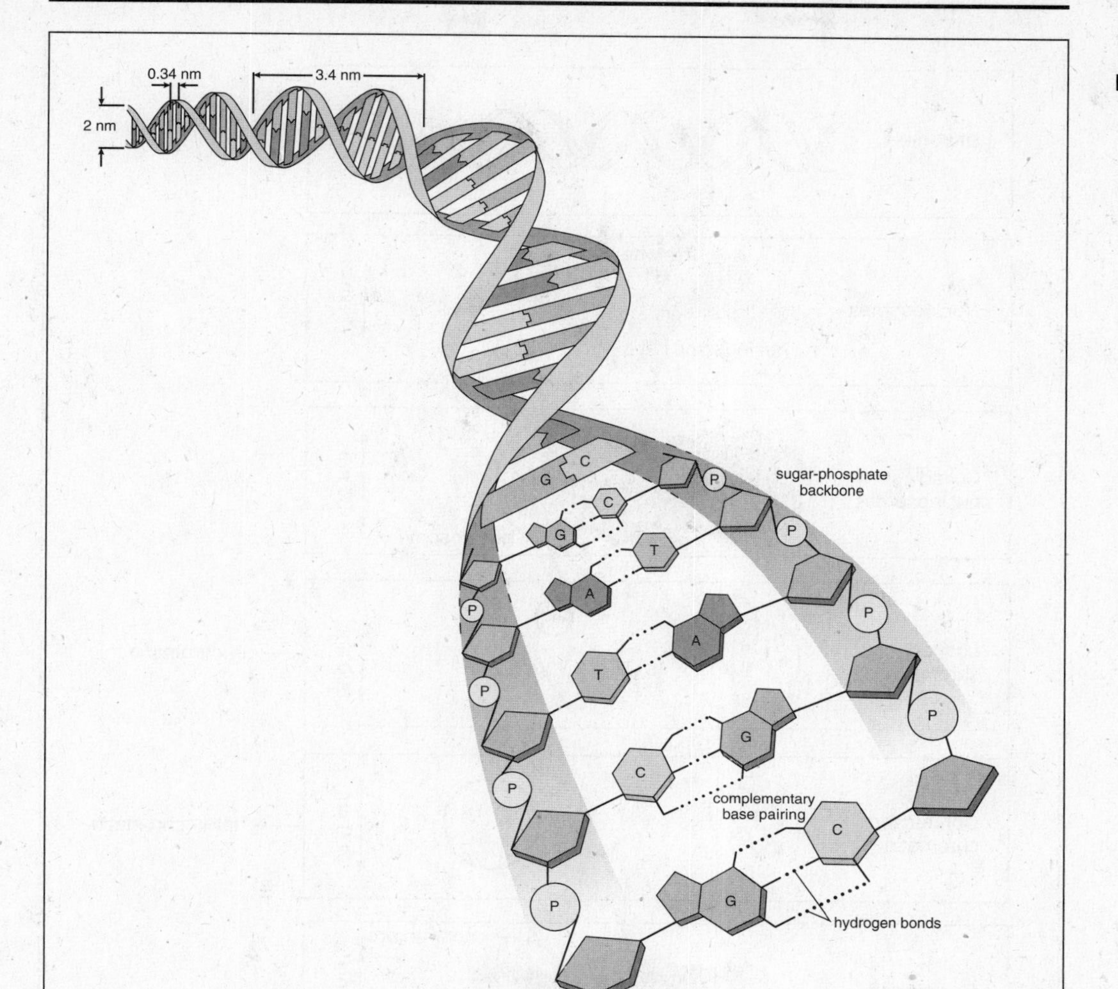

FIGURE 3-2: DNA double helix. The Watson and Crick model of DNA structure shows two strands of DNA that run antiparallel to each other. SOURCE: From Sylvia S. Mader, *Biology*, 8th ed., McGraw-Hill, 2004; reproduced with permission of The McGraw-Hill Companies.

CHROMOSOME STRUCTURE

In eukaryotic cells, DNA is organized in linear **chromosomes**. Humans have 23 pairs, or a total of 46 chromosomes, per **somatic** (nonreproductive) cell. A single chromosome consists of one DNA double helix wrapped around specialized **histone** proteins that form **chromatin**. Each chromosome contains an enormous amount of DNA, all of which must fit into the nucleus of the cell. Organizing the DNA around histones and other specialized proteins helps compact the DNA, as shown in Figure 3-3. During cell division, the chromatin coils even more to form a compact chromosome. When chromosomes replicate in preparation for cell division, the new copies stay attached to the original copy at a location called the **centromere**.

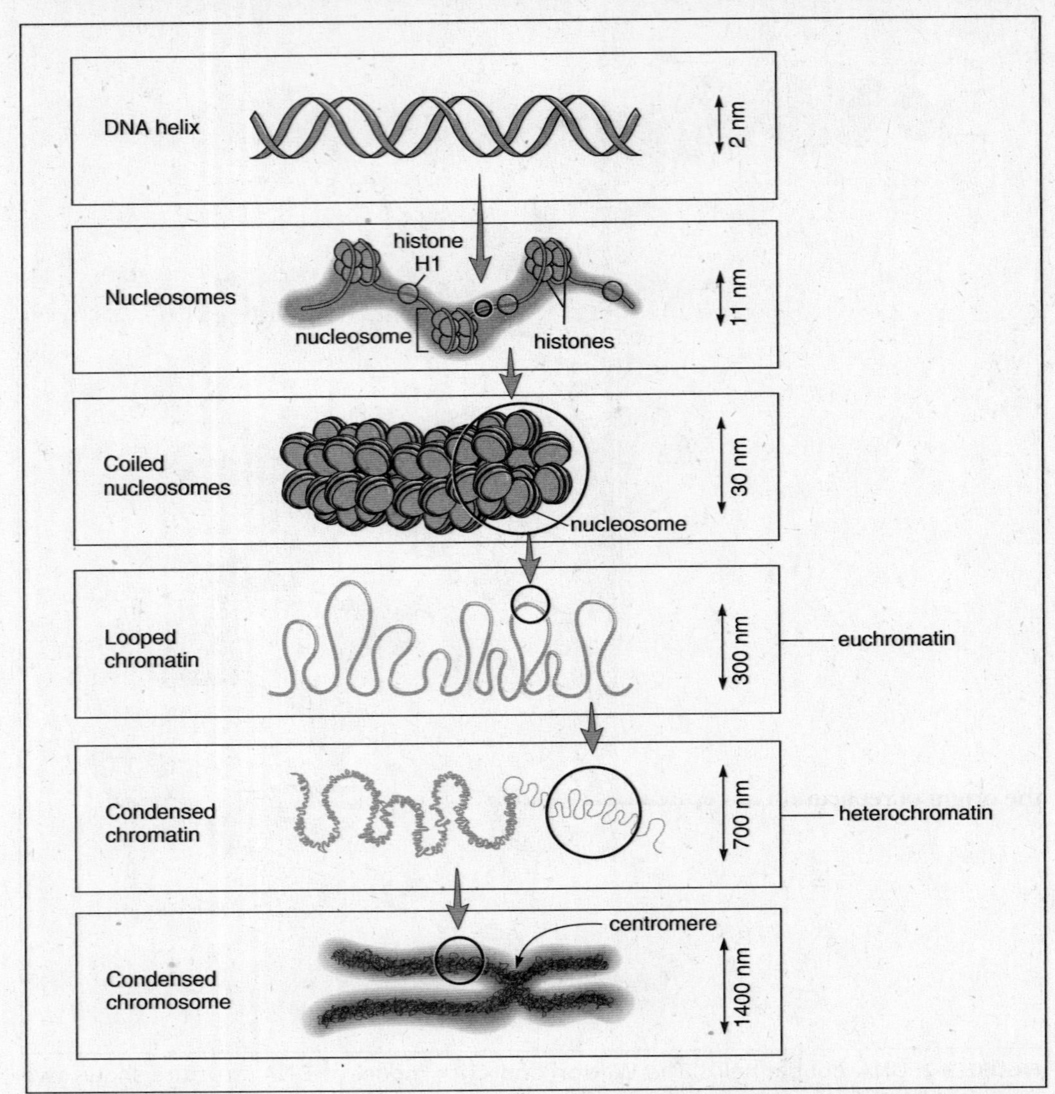

FIGURE 3-3: Chromosome structure. Chromosomes exhibit multiple levels of organization.
SOURCE: From Sylvia S. Mader, *Biology*, 8th ed., McGraw-Hill, 2004; reproduced with permission of The McGraw-Hill Companies.

At the end of each chromosome, there are repetitive sequences of DNA called **telomeres** that act as a sort of protective cap for the ends of the chromosomes. During each round of cell division, the telomeres become shorter. If all is working properly, the cell dies when the telomeres shorten or erode too much, leaving the critical nucleotides at the end of the chromosomes open to damage. This mechanism is necessary so that older cells are destroyed before they incur too much damage. In cancerous cells, this mechanism fails and cells achieve immortality.

DNA REPLICATION

During normal cell division, it is essential that all components of the cell, including the chromosomes, replicate so that each progeny cell receives a copy of the chromosomes from the parent cell. The process of replicating DNA must happen accurately to ensure that no changes to the DNA are passed on to the progeny cells.

The process of DNA replication is termed **semiconservative replication**. One double helix must be replicated so that two double helices result—one for each progeny cell. Because the DNA double helix has two strands, each strand can serve as a template to produce a new strand, as shown in Figure 3-4.

The process of semiconservative replication has three basic steps. First, the original DNA double helix must unwind. This process is achieved using the enzyme **helicase**. Next, the hydrogen bonds that hold the nitrogenous bases together must be broken. This "unzips" the double helix in a localized area of the chromosome called the **origin of replication**. Finally, each template strand produces a complementary strand of DNA using the normal rules of complementary base pairing. **DNA polymerase** is the key enzyme in this process. This enzyme binds to the DNA template and chemically reads the nucleotide sequence while assembling the complementary nucleotides to produce the new strand. The synthesis of DNA occurs in both directions moving outward from the origin of replication in **replication forks**.

As DNA is synthesized during replication, the DNA polymerase reads the template DNA strand from the $3'$ to $5'$ direction, which means that the new DNA being synthesized runs in the $5'$ to $3'$ direction. Because one DNA template runs in the $3'$ to $5'$ direction, DNA polymerase is able to read it and produce a continuous complementary strand called the **leading strand**. However, the other DNA template runs in the $5'$ to $3'$ direction, so the complementary strand (the **lagging strand**) is synthesized in a discontinuous manner because the replication fork is moving against the direction of DNA synthesis, as seen in Figure 3-5. In order to synthesize the discontinuous strand of DNA, a primer (a short sequence of nucleotides) must bind to the DNA and DNA polymerase begins to synthesize the new DNA strand until it runs into the next primer. This results in small pieces of DNA, termed **Okazaki fragments**, that must eventually be linked together. The primers are eventually degraded and the Okazaki fragments are linked using the enzyme **DNA ligase**.

MUTATIONS AND DNA REPAIR

During the process of DNA replication, it is possible for DNA polymerase to make mistakes by adding a nucleotide that is not complementary to the DNA template or

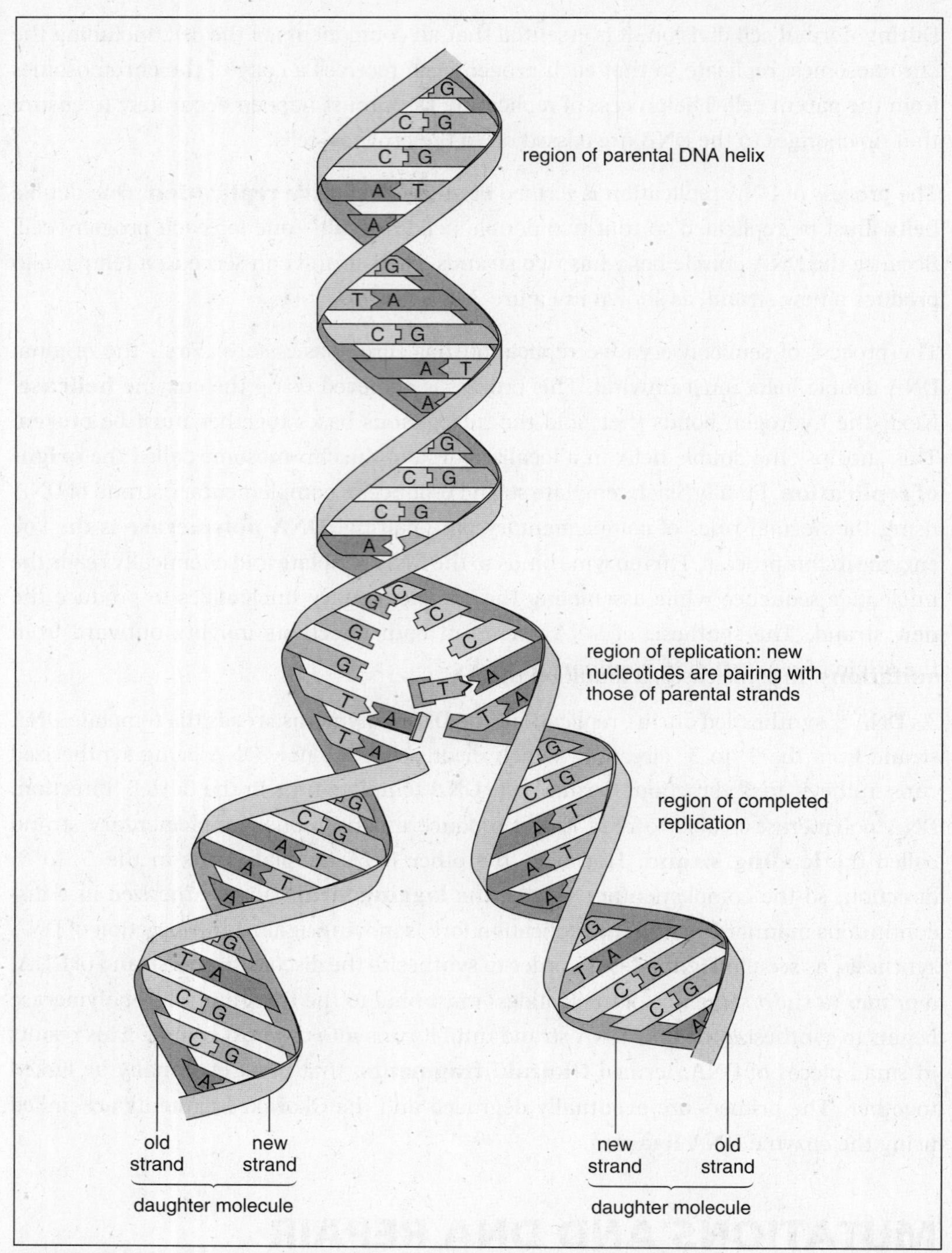

FIGURE 3-4: Semiconservative replication of DNA. During DNA replication, the DNA double helix unwinds and each strand serves as a template for the formation of a new strand. SOURCE: From Sylvia S. Mader, *Biology*, 8th ed., McGraw-Hill, 2004; reproduced with permission of The McGraw-Hill Companies.

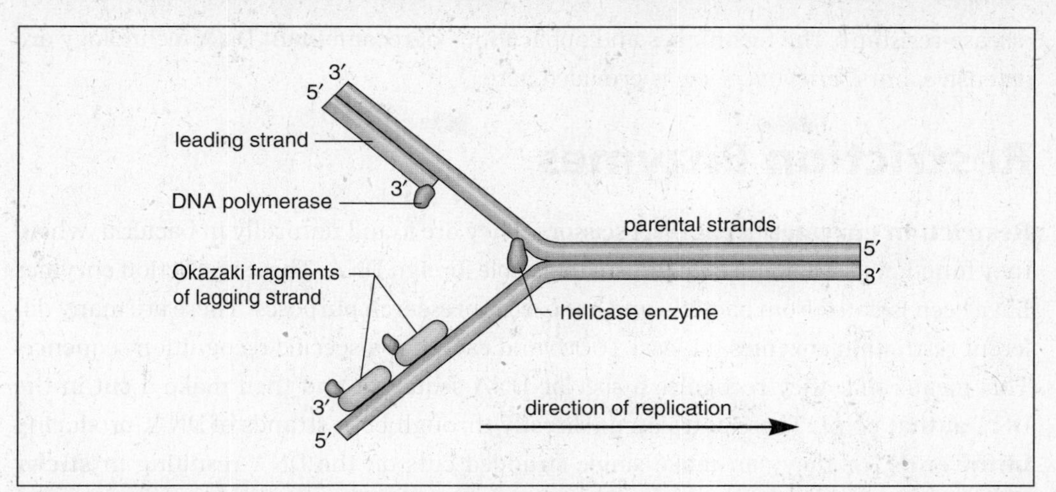

FIGURE 3-5: Okazaki fragments. DNA replication is continuous on the leading strand of DNA and discontinuous on the lagging strand, resulting in the formation of Okazaki fragments on the lagging strand. SOURCE: From Sylvia S. Mader, *Biology*, 8th ed., McGraw-Hill, 2004; reproduced with permission of The McGraw-Hill Companies.

by adding or deleting nucleotides on the new DNA strand. Luckily, DNA polymerase has a proofreading ability that usually detects these errors and repairs them. However, if these errors are not corrected, there are permanent changes to the DNA known as **mutations**. In some cases, a mutation that cannot be repaired successfully triggers the process of cellular suicide (apoptosis) to destroy the damaged cell. When this mechanism does not engage, the mutations remain and can be passed on to progeny cells. Mutations occur spontaneously at an estimated rate of one per billion nucleotides. Although these odds sound quite good, keep in mind that the DNA in a single cell has about 3 billion nucleotides. This means mutations inevitably occur each time the DNA is replicated. Mutations are sometimes harmless, but in some cases they can code for faulty proteins that can drastically alter the functioning of cells. This can lead to consequences such as a genetic disease or cancer.

Mutations occur naturally through the process of DNA replication, but there are certain factors that can greatly increase the mutation rate. These factors are referred to as **mutagens**. The mutagens that cause cancer are referred to as **carcinogens**. Should any of these mutations occur in **gametes** (egg or sperm cells), they are passed on to the next generation.

DNA TECHNOLOGY

A DNA molecule that contains DNA from two or more sources is termed **recombinant DNA**. An organism that contains recombinant DNA is termed **transgenic**. Recombinant DNA technology is a field that is advancing exponentially. It has wide-ranging applications, including the production of new drugs and the engineering of crops to be disease-resistant. The techniques and applications of recombinant DNA technology are extensive, but a brief overview is provided here.

Restriction Enzymes

Restriction enzymes act as DNA scissors. They are found naturally in bacteria, where they function as a defense mechanism to disable foreign DNA. These restriction enzymes have been isolated from bacteria and are used for research purposes. There are many different restriction enzymes (at least 1000) and each has a specific recognition sequence. This means that they recognize a specific DNA sequence and then make a cut in the DNA at that point. These cuts can go directly through both strands of DNA, producing **blunt ends**, or they can make single-stranded cuts on the DNA resulting in **sticky ends**, as seen in Figure 3-6. Researchers can use particular restriction enzymes to cut specific sequences of DNA in a predictable manner. DNA from more than one source can be cut with the same restriction enzyme. The resulting fragments can then be pasted together using the enzyme **DNA ligase**. The result is recombinant DNA.

Recombinant Plasmids

Plasmids are small, self-replicating loops of DNA found naturally in bacterial cells. These plasmids can be isolated and manipulated in the lab to produce recombinant DNA. The recombinant plasmid can then be reintroduced into bacteria in a natural process called **transformation** that allows bacteria to take in plasmids from their environment. Any foreign DNA that has been inserted into the plasmid is then **expressed** (meaning the proteins that it encodes for are made) by the bacterial cell.

If a researcher wants to take a **gene** (a segment of DNA with information to code for one protein) from a specific species and insert it into a plasmid, the first step would be to use the same restriction enzyme to cut the circular plasmid and to excise the gene of interest. The DNA ligase enzyme could be used to "glue" the gene into the plasmid, which then serves as a **vector** to carry the gene to another cell. Then, the plasmid could be introduced into a bacterial cell by transformation, and the bacterial cell begins to express the gene and produce the protein associated with that gene.

Gene Cloning

Gene cloning is the production of multiple copies of a particular gene. Introducing a recombinant plasmid into a bacterial cell and allowing for bacterial replication qualifies as gene cloning. Viruses can also be engineered to serve as vectors to carry foreign genes. These viruses can be introduced into host cells that replicate the recombinant viral DNA, thus cloning the gene.

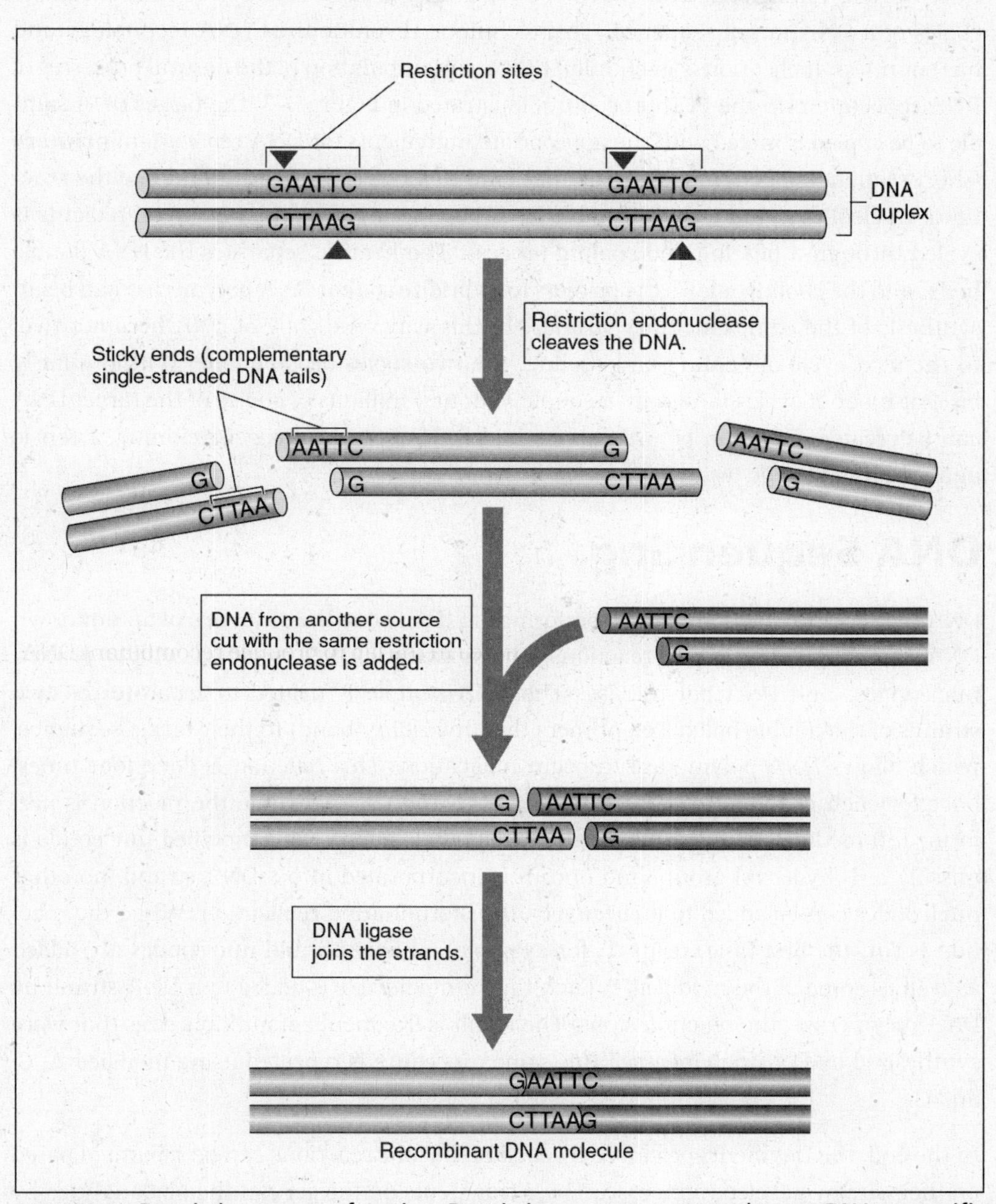

FIGURE 3-6: Restriction enzyme function. Restriction enzymes are used to cut DNA at specific sequences, which often results in the production of sticky ends. SOURCE: From George B. Johnson, *The Living World*, 3d ed., McGraw-Hill, 2003; reproduced with permission of The McGraw-Hill Companies.

DNA Amplification

Polymerase chain reaction (PCR) is a technique used in research to make multiple copies of a DNA target sequence. This technique revolutionized DNA technology and has countless applications. Essentially, PCR is a manipulation of the natural processes of DNA replication. In the PCR procedure, illustrated in Figure 3-7, the target DNA sample to be copied is mixed with the appropriate ingredients for DNA replication: **primers** (short sequences of engineered DNA that bind to a target sequence to initiate the reaction), nucleotides, and a thermostable DNA polymerase. The mixture of ingredients is cycled through a heating and cooling process. The heating separates the DNA double helix, and the cooling allows the primers to hybridize so that DNA polymerase can begin synthesis of the complementary strands. In this way, one copy of DNA becomes two. In the next cycle of heating and cooling, the two copies become four. Within a fairly brief number of cycles (and only a couple of hours) millions of copies of the target DNA can be created. PCR can be used for multiple purposes such as gene cloning, forensic analysis, and genetic testing.

DNA Sequencing

DNA sequencing reactions can be performed to determine the sequence of an unknown DNA sample. During these reactions, the unknown DNA is mixed with a primer, nucleotides, and DNA polymerase. The DNA sample is heated to separate the two strands of the double helix. The primers then hybridize (bond) to their target sequence, which allows DNA polymerase to begin replication. This reaction is done four times, once for each of the four nucleotides (A, T, C, and G). Each time the reaction is performed, a modified version of a single nucleotide is added. This modified nucleotide is missing a $3'$ hydroxyl group, and once it is incorporated into a DNA strand, no other nucleotides can be added to it effectively, thus terminating replication. When the reaction is run the first time, using T, for example, all four normal nucleotides are added as well as some of the modified T. Each time a modified T is added to a DNA strand by DNA polymerase, the reaction stops. The result is fragments of multiple sizes that were synthesized by DNA polymerase. This same procedure is repeated using modified A, C, and G.

In the end, the fragments produced by each of the four reactions can be separated based on their size by gel electrophoresis. These fragments on the gel can then be analyzed by a computer to indicate the original sequence of the DNA sample.

DNA Hybridization

To assess the similarities between two samples of DNA, it is necessary to perform a DNA hybridization reaction. In this procedure, two samples of DNA are heated to

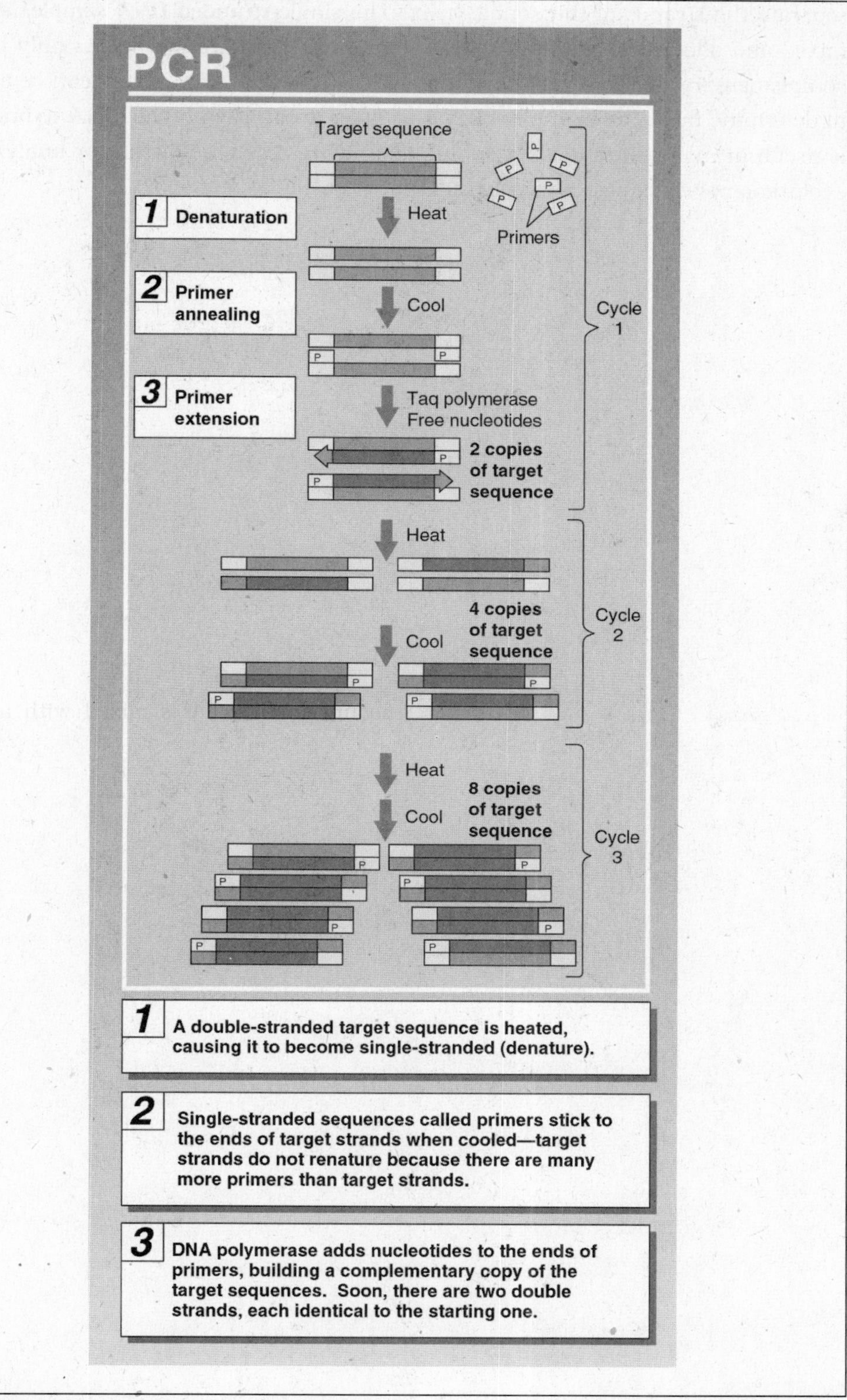

FIGURE 3-7: PCR. During the cycles of PCR, the strands of target DNA sequences are separated, primers anneal to the single strands, and DNA polymerase is used to extend the strands. Each cycle of PCR doubles the amount of the target DNA sequence. SOURCE: From George B. Johnson, *The Living World*, 3d ed., McGraw-Hill, 2003; reproduced with permission of The McGraw-Hill Companies.

separate the strands of the double helix. The single-stranded DNA samples are then mixed and allowed to hybridize with each other. Hybridization occurs only between complementary nucleotides. The amount of hybridization that occurs can be analyzed to determine how similar the two DNA samples are to each other. DNA hybridization is useful in many situations, including identifying a genetic disease or analyzing the evolutionary relatedness of two species.

CRAM SESSION

DNA Structure, Replication, and Technology

1. **DNA** exists as a **double helix**. Semiconservative replication is used to synthesize complementary strands of DNA. In this process, each strand of the double helix serves as a template to be copied by DNA polymerase. The result is two identical double helices, each having one strand from the original molecule and one newly synthesized strand.
2. **Chromatin** is a DNA double helix wrapped around histone proteins. When chromatin coils and condenses during cell division, the result is a **chromosome**.
3. Mistakes can happen during DNA replication leading to **mutations**. Carcinogens and other mutagens increase the natural rate of mutation.
4. **Recombinant DNA** consists of DNA from more than one source.
5. Many technologies and a variety of tools have been developed to create recombinant DNA and to analyze and compare DNA. Among them are the following:
   a. **Restriction enzymes** have recognition sequences and cut DNA at a specific target. They can be used to excise particular segments of DNA.
   b. **Plasmids** are small circular pieces of DNA naturally found in bacteria. They can be cut open with restriction enzymes and augmented with DNA segments from other species. These recombinant plasmids can be introduced into bacteria, which then express the new genes added to the plasmid.
   c. **Gene cloning** is used to produce multiple, identical copies of a particular gene. Bacteria carrying recombinant plasmids can be coerced to reproduce, creating multiple copies of a specific gene sequence. Viruses can be used in similar ways.
   d. **Polymerase chain reaction (PCR)** is used to amplify specific DNA sequences. A target DNA sequence, primers, nucleotides, and DNA polymerase are mixed together and go through heating and cooling cycles. Heat separates DNA strands and the cooling allows for primer annealing and extension by DNA polymerase. Each cycle doubles the amount of target DNA.
   e. In order to determine the sequence of DNA in a specific sample, **DNA sequencing** reactions are used. These reactions rely on modified forms of nucleotides that can terminate DNA replication producing DNA segments of variable length. Gel electrophoresis allows for the visualization and analysis of sequencing reactions.
   f. **DNA hybridization** reactions can be used to analyze the relatedness or similarity of two DNA samples.

CHAPTER 4

Protein Synthesis

Read This Chapter to Learn About:

- DNA and RNA
- The Genetic Code
- Transcription and Translation
- Mutations
- Regulation of Gene Expression

In Chapter 3, you learned that DNA is the basic hereditary material of the cell and that DNA is located on chromosomes in the nucleus of a cell. A segment of DNA located on a chromosome that has information to encode for a single protein is known as a **gene**. Proteins are made in the cytoplasm of the cell with assistance from ribosomes, but unfortunately, the genes carrying the instructions cannot leave the nucleus, and the ribosomes where the proteins are to be synthesized cannot enter the nucleus.

To get around this problem, the DNA message in the nucleus is converted to an intermediate **ribonucleic acid** (RNA) message that can travel out of the nucleus to the cytoplasm and be read by the ribosomes to produce a protein.

The process of protein synthesis is a two-step process: the conversion of DNA to RNA is **transcription**, and the conversion of the RNA message to a protein is **translation**. This process describes the flow of genetic information in the cell and is the **central dogma** of molecular biology.

DNA → RNA → protein

RNA

RNA is a form of nucleic acid that is a critical player in the process of protein synthesis. RNA molecules are very similar to DNA with a few exceptions, as shown in the following table.

Differences between DNA and RNA (Nucleic Acids)		
	DNA	**RNA**
Number of strands	2, double helix	Single
Sugar used in the nucleotide	Deoxyribose	Ribose
Nitrogenous bases used	Adenine, thymine, guanine, cytosine	Adenine, uracil, guanine, cytosine

Within the cell, there are three types of RNA: **ribosomal RNA** (rRNA), **transfer RNA** (tRNA), and **messenger RNA** (mRNA). Each type has a specific role in the process of protein synthesis. The functions of each type of RNA are shown in the following table.

Types of RNA	
Type	**Function**
Ribosomal (rRNA)	rRNA is made in the nucleolus of the nucleus. It is a structural component of ribosomes.
Transfer (tRNA)	tRNA is located in the cytoplasm of the cells. It is used to shuttle amino acids to the ribosome during the process of translation.
Messenger (mRNA)	mRNA is copied from DNA and serves as the messenger molecule to carry the DNA message to the ribosomes in the cytoplasm.

Production of mRNA—Transcription

The first step of protein synthesis is the production of mRNA from DNA. This process of transcription initially resembles the process of semiconservative DNA replication. At the point where transcription is to begin, the DNA double helix unwinds. In this local area, the hydrogen bonds holding together the base pairs break. Because only one strand of mRNA needs to be produced, only one strand of the DNA serves as a template.

The enzyme **RNA polymerase** recognizes sequences of DNA called **promoters** and binds to them. The RNA polymerase chemically reads the sequence of DNA and assembles the complementary RNA nucleotides in the $5'$ to $3'$ direction. The rules of complementary base pairing during transcription are similar to those of DNA replication, with one major change: RNA has the base uracil (U) instead of thymine (T). If the DNA contains the base A, then the complementary RNA contains the base U (not T), whereas C and G pair together. RNA polymerase continues to synthesize the complementary RNA strand until it reaches a termination sequence on the DNA. At this point, the RNA molecule is released and the DNA double helix reforms. The process of transcription can be seen in Figure 4-1.

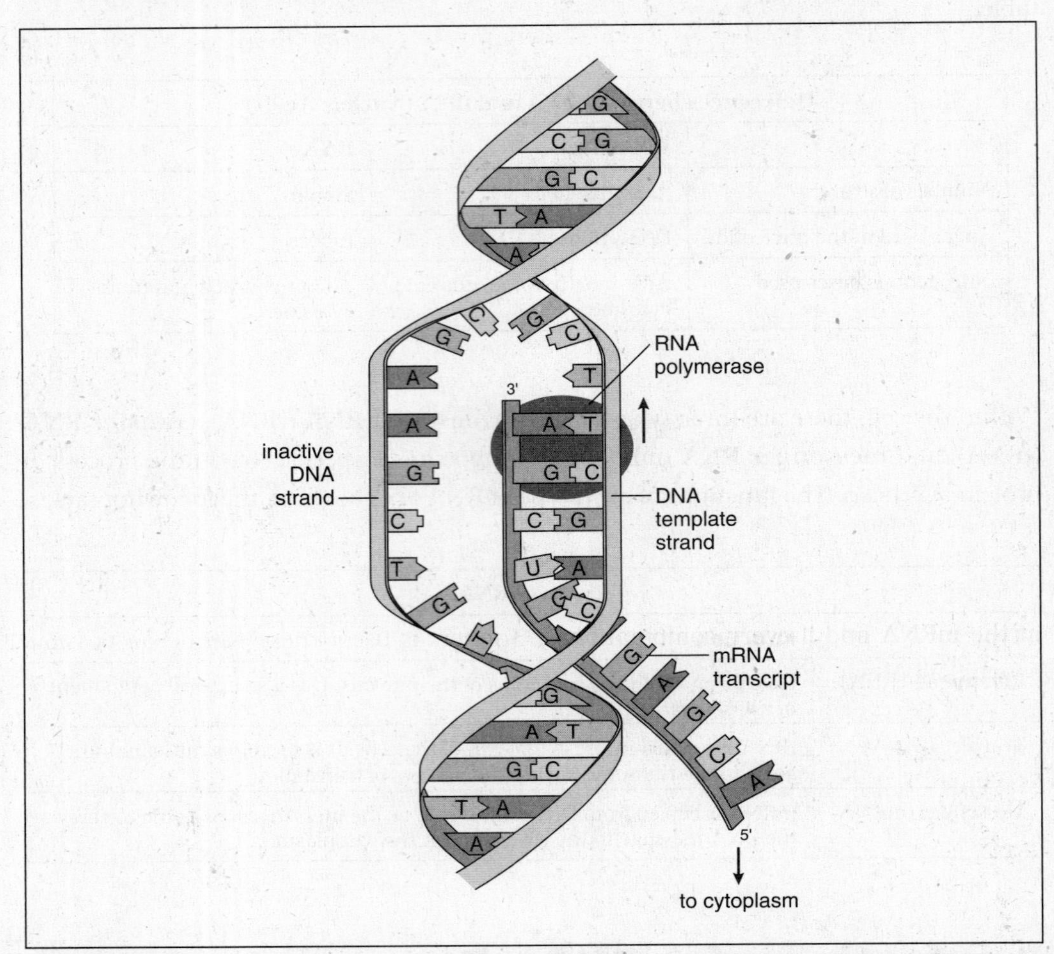

FIGURE 4-1: During transcription, complementary RNA is made from a DNA template using the enzyme RNA polymerase. SOURCE: From Sylvia S. Mader, *Biology*, 8th ed., McGraw-Hill, 2004; reproduced with permission of The McGraw-Hill Companies.

Modification of mRNA

Once RNA has been produced from the DNA template, it must be modified before it can be translated into a protein. First, **a 5′ cap** is added to the 5′ end of the RNA. This cap is a chemically modified nucleotide that helps regulate translation. Next, a poly-A tail is added to the 3′ end of the RNA. This tail consists of many A nucleotides placed on the end of the RNA. The purpose of the tail is to prevent degradation of the RNA molecule.

Although some of the chromosomal DNA has information that is needed to code for proteins, the majority of the DNA does not have information to code for proteins.

The coding DNA is termed **exons**, and the noncoding DNA is termed **introns**, or "junk DNA." Unfortunately, the introns are located within the exons, disrupting their sequence.

During transcription, the RNA that is copied from the DNA contains the sequences of both the introns and the exons. Prior to translation, the introns must be removed and the exons must be spliced together to form functional mRNA. Several unique RNAs can be produced by splicing the same exons in different sequences. Figure 4-2 demonstrates the **RNA splicing** process. This process occurs in the nucleus. Once the splicing is complete, the mRNA molecule moves through the nuclear pores to the cytoplasm where translation occurs.

THE GENETIC CODE

Now that the mRNA has been produced, it must be translated into a protein. In this case, there is a "language" barrier. The mRNA is written using a 4-letter code (A, U, C, and G), whereas proteins are made using a 20-letter code (there are 20 different amino acids used to make proteins). So how does a 4-letter language get converted to a 20-letter language? The mRNA is read as **codons**, 3 nucleotides at a time. Each codon has the information to specify one amino acid. Mathematically, there are 4 nucleotides in the mRNA and if every combination of 3 letters is used, there will be 64 possible codons, all of which are listed in the **genetic code** seen in Figure 4-3. Because there are only 20 amino acids used to make proteins, there is an overlap, or redundancy, in the code where more than one codon can code for the same amino acid. The significance of the redundancy of the genetic code will become apparent later when mutations are discussed.

Knowing the sequence of codons on the mRNA makes it possible to use the genetic code to decipher the sequence of amino acids that is used to build the protein in translation. Any change to the DNA, which, in turn, changes the mRNA codons, can potentially change the order of amino acids and thus the shape and function of the intended protein.

PROTEIN SYNTHESIS—TRANSLATION

The process of translation occurs in the cytoplasm. The codons on the mRNA are read and the appropriate amino acids needed to produce the protein are assembled. This process requires assistance from various enzymes, ribosomes, and tRNA.

RIBOSOMES

In eukaryotic cells, ribosomes are composed of two subunits, one large and one small, of rRNA and various proteins. Once the ribosome assembles on the mRNA,

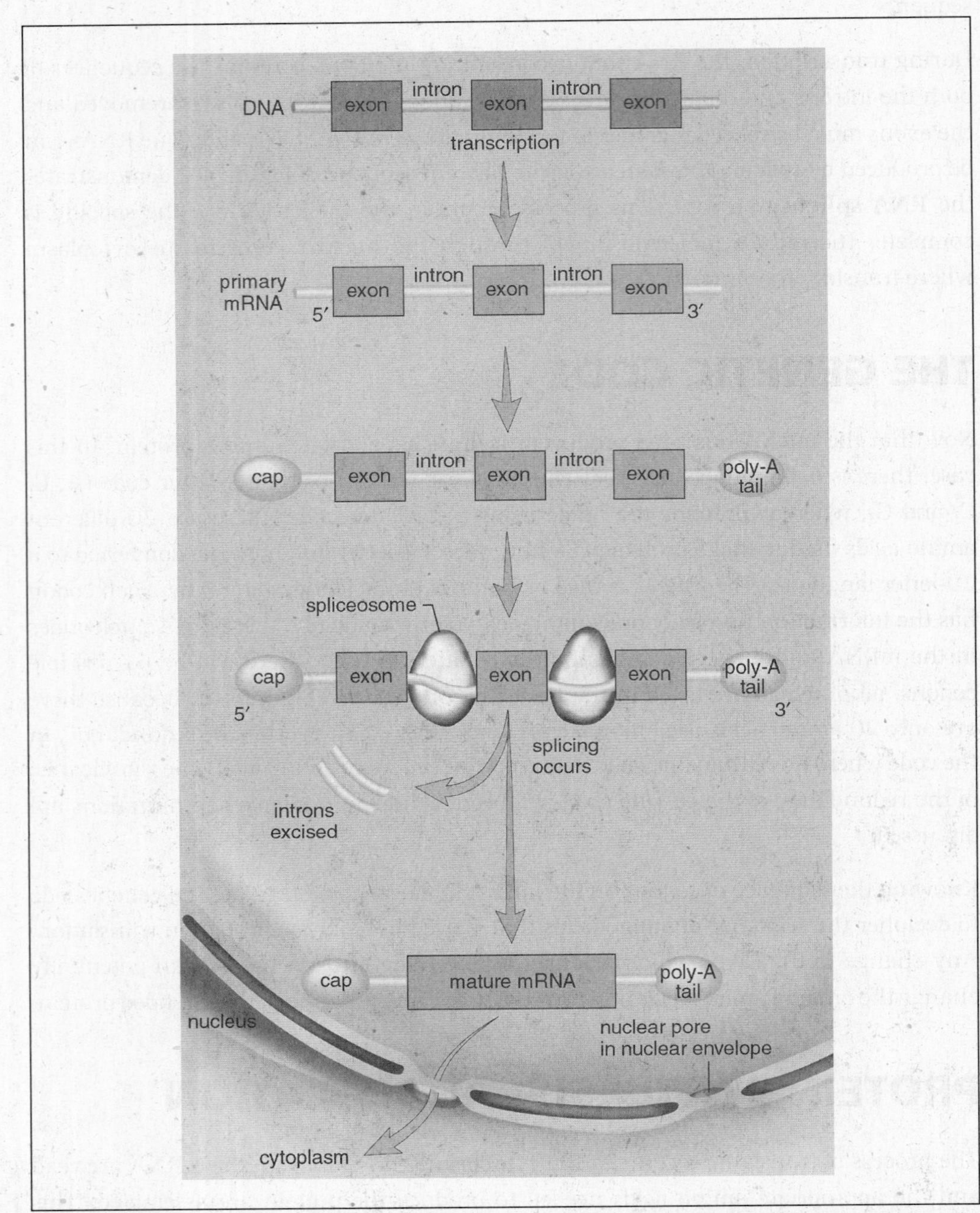

FIGURE 4-2: During mRNA processing, introns are removed from the mRNA and the exons are spliced together. Capping and the addition of a poly-A tail are also part of mRNA processing.

SOURCE: From Sylvia S. Mader, *Biology*, 8th ed., McGraw-Hill, 2004; reproduced with permission of The McGraw-Hill Companies.

there are two RNA binding sites inside the ribosome: the peptidyl (P) site and the aminoacyl (A) site.

First letter	Second letter: U	Second letter: C	Second letter: A	Second letter: G	Third letter
U	UUU, UUC: Phe UUA, UUG: Leu	UCU, UCC, UCA, UCG: Ser	UAU, UAC: Tyr UAA: Stop UAG: Stop	UGU, UGC: Cys UGA: Stop UGG: Try	U C A G
C	CUU, CUC, CUA, CUG: Leu	CCU, CCC, CCA, CCG: Pro	CAU, CAC: His CAA, CAG: Gln	CGU, CGC, CGA, CGG: Arg	U C A G
A	AUU, AUC, AUA: Ile AUG: Met or start	ACU, ACC, ACA, ACG: Thr	AAU, AAC: Asn AAA, AAG: Lys	AGU, AGC: Ser AGA, AGG: Arg	U C A G
G	GUU, GUC, GUA, GUG: Val	GCU, GCC, GCA, GCG: Ala	GAU, GAC: Asp GAA, GAG: Glu	GGU, GGC, GGA, GGG: Gly	U C A G

FIGURE 4-3: The genetic code. The codons on mRNA can be read on the genetic code to predict the sequence of amino acids produced by a particular mRNA. SOURCE: From Eldon D. Enger, Frederick C. Ross, and David B. Bailey, *Concepts in Biology*, 11th ed., McGraw-Hill, 2005; reproduced with permission of The McGraw-Hill Companies.

tRNA

The tRNA molecules shuttle the appropriate amino acids to the ribosomes as dictated by the codons on the mRNA. The tRNA itself is a piece of RNA folded into a specific configuration. On one end, the tRNA contains an **anticodon** that is complementary to the codon on the mRNA. On the other end of the tRNA, a specific amino acid is attached.

The Steps of Translation

Translation occurs as a three-step process. First, the ribosome must assemble on the mRNA. Next, the amino acids dictated by the codons must be brought to the ribosome

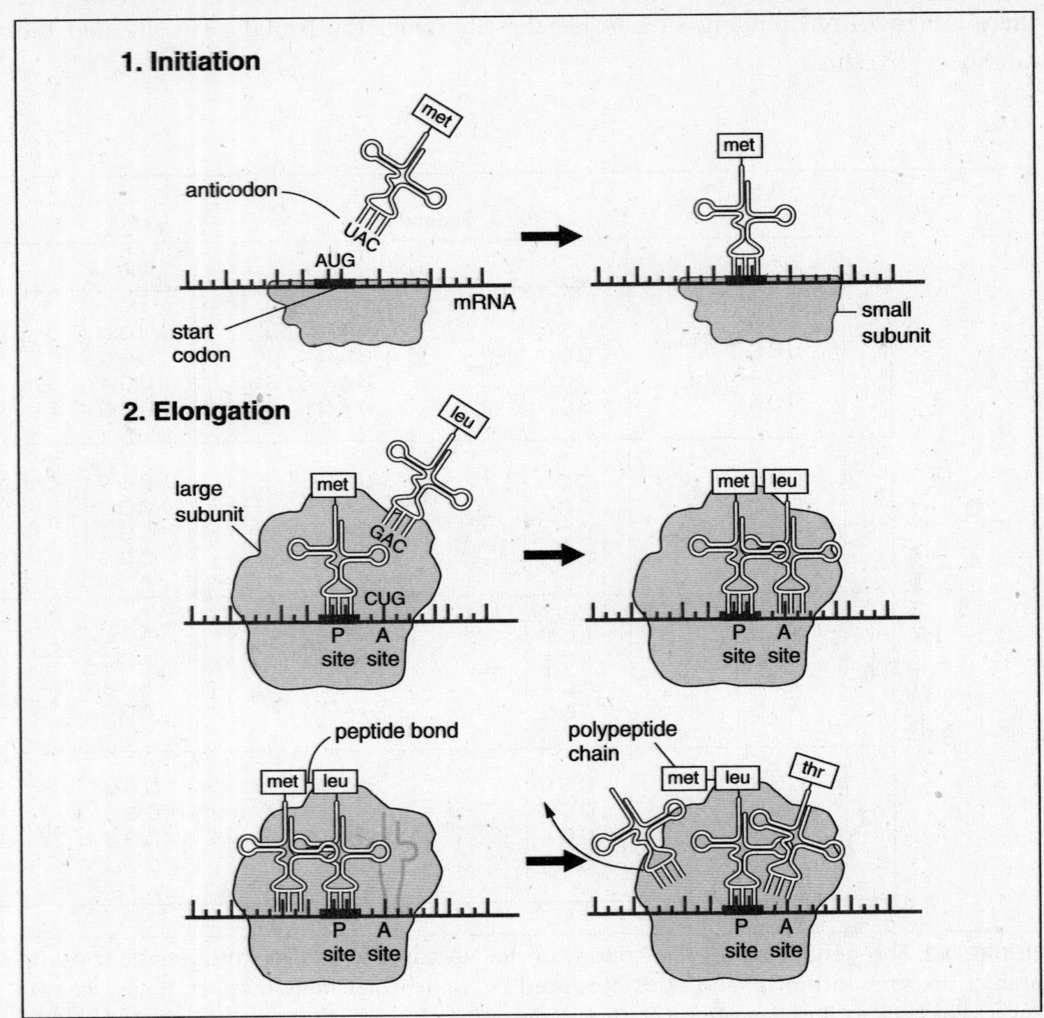

FIGURE 4-4: Translation. Translation initiates when the first tRNA interacts with the start codon. The tRNAs continue to deliver their amino acids to the ribosome during elongation.

and bonded together. Finally, the resulting protein must be released from the ribosome. The entire process of translation can be seen in Figure 4-4.

INITIATION

The process of translation begins when the ribosome assembles on the mRNA. The location for ribosomal assembly is signaled by the **start codon** (AUG) found on the mRNA. The small ribosomal subunit then binds to the mRNA. The first tRNA enters the P site of the ribosome. This tRNA must have the appropriate anticodon (UAC) to hydrogen bond with the start codon (AUG). As seen in the genetic code, the amino acid specified by the start codon is methionine. Thus the first amino acid of every protein will be methionine. Now, the large subunit of the ribosome can assemble on the mRNA.

ELONGATION

At this point, the P site of the ribosome is occupied but the A site is not. A tRNA bearing the appropriate anticodon to bind with the next codon of the mRNA will enter the ribosome and hydrogen bond to the codon. A key enzyme will be used at this point to form a peptide bond between the two amino acids in the P and A sites. This enzyme is **peptidyl transferase**. The two amino acids are now attached to the tRNA in the A site. The tRNA in the P site breaks off (leaving behind its amino acid) and leaves the ribosome. The ribosome now moves over one codon to the right, putting the remaining tRNA in the P site and leaving an empty A site. This process of a new tRNA entering, a peptide bond forming between amino acids, the tRNA in the P site leaving, and the ribosome shifting over by one codon occurs over and over again.

TERMINATION

There are three mRNA codons (UAA, UAG, and UGA) that act as **stop codons** and do not code for amino acids. When one of these codons reaches the A site of the ribosome, no more tRNAs enter and the protein is released from the ribosome. The ribosomal subunits dissociate. This signals the end of translation. In some cases, it is necessary for the released protein to be modified before it can be functional. This often occurs in the endoplasmic reticulum or the Golgi complex.

MUTATIONS

Mutations change the coding sequence of DNA. When the DNA changes, the mRNA codons change, and the amino acid sequence of the protein made may change. In some cases, this may produce a protein that functions better than the one intended by the DNA (thus providing an advantage); in other cases, it may produce a protein that functions equivalent to the intended protein, or, in the worst case, the change may lead to a protein that functions worse than the intended protein or does not function at all. Recall that mutations happen spontaneously and that the rate of mutation is increased by exposure to mutagens.

Mutations can occur in several different ways—by the change in a single nucleotide, by the addition or deletion of a nucleotide, or by the movement of nucleotides.

Point Mutations

When a single nucleotide is swapped for another, the resulting mutation is termed a **point** (substitution) **mutation**. This ultimately changes a single codon on the mRNA. In some cases, this mutation is silent, meaning that if the changed codon still codes for the intended amino acid, there will be no detectable consequence. However, sometimes even a single point mutation can have major consequences. If the new codon codes for a different amino acid than what was intended (a **missense mutation**), the new protein may not function properly. This can lead to a genetic disease such as sickle cell. It is also possible for a change in a single nucleotide to produce a stop codon in a new location causing a **nonsense mutation**. In this case, the protein produced would be too short and most likely nonfunctional.

Frameshift Mutations

A **frameshift mutation** is the result of the addition or deletion of nucleotides. Unlike the point mutation where the overall number of nucleotides does not change, adding or deleting nucleotides changes the total number of nucleotides. Because mRNA is read in codons, an addition or deletion alters all of the codons from the point of the mutation onward. This disrupts the reading frame of the mRNA. Because many codons are changed, the frameshift mutation generally produces a damaged or nonfunctional protein.

Transposable Elements

Certain segments of DNA are known to be **transposable elements (transposons)**. These pieces of DNA have the ability to become mobile and insert themselves into specific genes, which is why they are sometimes called **jumping genes**. The insertion of a transposon into a specific gene causes a mutation that disrupts the coding sequence in a way similar to that of a large frameshift (insertion) mutation.

REGULATION OF GENE EXPRESSION

Gene expression refers to the control of which genes are transcribed and translated. Each cell has many genes and it is not necessary for every cell to express every gene it has. In order to be efficient, cells are selective about which genes they express, making only the proteins that are necessary at a given time.

Gene expression can be regulated on a permanent level due to a process called **differentiation**. Because nearly all of the cells within the body are specialized, it makes sense that these cells really need only to express the genes related to that cell's particular function. Even though all cells have the same genes, each specialized cell is only capable of expressing a small subset of those genes. So although brain cells have the gene to produce the protein insulin, they are unable to express this gene, as it is not needed for the functioning of a brain cell and it would be inefficient to make an unnecessary protein. The process of differentiation happens early in development and is thought to be an irreversible process. Cells that have yet to differentiate are referred to as **stem cells.**

Once a cell has differentiated and "decided" on a set of genes that it must express, genes within this set can be regulated on a minute-to-minute basis. There are a number of ways to regulate the process of transcription and translation. These methods can completely prevent gene expression or control the rate of the process, in turn influencing the amount of protein that is produced. The table titled "Mechanisms for Regulating Gene Expression" summarizes the major mechanisms for the control of gene expression.

Transcriptional Regulation

One way to control the expression of a particular gene is by physically regulating access to the gene. In order for transcription to occur, DNA binding proteins (also called **transcription factors**) must be able to bind to the promoter sites on DNA. The binding of these transcription factors attracts RNA polymerase to the area so that transcription can begin. Recall that each chromosome is coiled. If coiling is particularly tight in a specific region, the transcription factors and RNA polymerase are unable to access the gene, thus preventing transcription and the expression of the gene. If the chromosome was to loosen its coiling, transcription factors and RNA polymerase would be able to access the promoter region of the gene and transcription would begin.

An extreme example of how the coiling of chromosomes affects gene expression is a process called **X chromosome inactivation** that occurs in females. Gender is determined by the **sex chromosomes**, X and Y. Females have a pair of X chromosomes inherited from each parent, whereas males have one X inherited from their mother and a Y inherited from their father. In females, one of the two X chromosomes in each cell is randomly selected for inactivation. The X chromosome to be inactivated is coiled so tightly that RNA polymerase will never be able to access the genes, and thus no information on the inactive X chromosome is ever expressed.

Post-Transcriptional Regulation

When transcription does occur, there are a variety of other regulatory methods that can be used to determine if translation will ultimately occur. The RNA that is produced during transcription must be spliced and modified before it is translated. If this does not occur, no protein is made and the gene is not expressed. Furthermore, even if the mRNA is properly modified, the rate at which it leaves the nucleus (via nuclear pores) influences if and how quickly it is translated. The faster the mRNA enters the cytoplasm, the more it is translated, leading to an increased amount of protein produced.

Translational Control

Once in the cytoplasm, the mRNA begins to degrade quickly, usually within minutes. One reason for this quick degradation is that enzymes begin destroying the poly-A tail that was added during RNA modification in the nucleus. Without the poly-A tail, translation cannot occur. The longer the tail is, the longer the mRNA exists for translation and the more protein that is made. The shorter the tail, the less protein that is made.

Post-Translational Regulation

Once transcription and translation are complete, a protein exists. However, there is one last way to regulate gene expression at this point. In some cases, the protein is inactivated immediately following synthesis. In other cases, the protein may not go through its normal modification procedure thus rendering it useless.

Mechanisms for Regulating Gene Expression	
Stage at Which Regulation Occurs	**Mechanisms Used**
Transcriptional regulation	Coiling of chromosomes to physically prevent or allow the access of transcription factors and RNA polymerase to the promoter regions of DNA
Post-transcriptional regulation	Whether or not mRNA is properly spliced Control over the rate at which mRNA leaves the nucleus via nuclear pores
Translational regulation	Life span of mRNA, which is influenced by the length of the poly-A tail added during RNA modification in the nucleus
Post-translational regulation	Degradation of protein immediately following synthesis Failure to properly modify the protein rendering it useless

CRAM SESSION

Protein Synthesis

1. The central dogma of molecular biology describes the flow of genetic material in the cell: DNA → RNA → protein.
2. The two types of nucleic acids are DNA and RNA. RNA is different from DNA in that it is single stranded, uses the sugar ribose instead of deoxyribose in its nucleotides, and uses the nitrogenous base uracil instead of thymine.
3. RNA comes in three varieties. rRNA is used to produce the ribosomes, tRNA shuttles amino acids, and mRNA, made during the process of transcription, carries the message for the synthesis of a protein from the nucleus to the cytoplasm.
4. Protein synthesis is a two-step process. Transcription converts DNA to mRNA in the nucleus. Translation converts mRNA to a protein in the cytoplasm.
   a. Transcription produces a single mRNA molecule from a DNA template. The enzyme RNA polymerase facilitates this process. The mRNA contains the codons needed to dictate the order of amino acids in the protein. The genetic code is used to provide information on the amino acid dictated by each codon. Once transcription is complete, the RNA must be modified by splicing, adding a $5'$ cap, and a poly-A tail to produce functional mRNA.
   b. Translation requires assistance from ribosomes and tRNA. Initiation begins with ribosome assembly at the start codon (AUG). As each tRNA enters the ribosome, its anticodon must bind to the codon on the mRNA to ensure that the right amino acid is being delivered. The amino acids delivered by tRNAs are bound by peptide bonds during elongation. Termination occurs when a stop codon is reached.
5. Mutations that occur in the DNA are copied into mRNA, which may affect the amino acid sequence of a protein. Point mutations change a single nucleotide, whereas frameshift mutations add or delete nucleotides to shift the reading frame of the ribosome.
6. Transcription and translation are regulated processes. Differentiation dictates which genes can be expressed within a cell, leading to specialization. Within a specialized cell, the expression of specific genes is regulated during transcription, after transcription, during translation, and after translation. See the table titled "Mechanisms for Regulating Gene Expression" for a summary of regulatory methods for gene expression.

CHAPTER 5

Genetics

Read This Chapter to Learn About:

- Basic Mendelian Concepts
- Exceptions to Mendel's Laws
- Linked Genes and Multiple Alleles
- Incomplete Dominance and Codominance
- Sex Linkage
- Pedigree Analysis
- Environmental Influences on Genes

The basic principles of genetics were proposed by Gregor Mendel in the 1860s. His work with several traits in pea plants led him to propose several theories of inheritance. Mendel did all his work and postulated his theories at a time when the genetic material had not even been discovered, so the fact that his theories hold true today could be considered quite a stroke of luck.

An understanding of several basic terms is essential to discuss genetics. The exact genetic makeup of an individual for a specific trait is referred to as the **genotype**, whereas the physical manifestation of the genetic makeup is referred to as a **phenotype** for a specific trait. A **gene** has information to produce a single protein or enzyme. However, genes can exist in different forms termed **alleles**. In some cases, mutations can cause the production of alleles that produce faulty enzymes needed for metabolism. This leads to a class of genetic disorders known as **inborn errors of metabolism**.

Through his studies of pea plants, Mendel formulated several laws to explain how particular traits were inherited. These laws addressed issues concerning how specific traits were sorted and passed on to progeny and how some traits dominated over others.

MENDEL'S LAW OF SEGREGATION

A basic law developed by Mendel was the Law of Segregation. There are several important ideas found in this law. They can be summarized as follows:

- For every given trait, an individual inherits two alleles for the trait, one from each parent.
- As an individual produces egg and sperm cells (**gametes**), the two alleles segregate so that each gamete contains only a single allele per trait. During fertilization, each gamete contributes one allele per trait, providing the offspring with two alleles per trait.

There are exceptions to the law of segregation. These include the alleles carried on sex chromosomes in males. Because males contain one X chromosome and one Y chromosome, the male does not have two alleles per trait for genes on the sex chromosomes. Another exception is that mitochondria contain their own DNA (in single copy) that is inherited separately from chromosomal DNA. All mitochondrial DNA is inherited maternally. Occasionally, alleles from the mitochondrial DNA may incorporate into the chromosomal DNA in a process termed **genetic leakage**.

COMPLETE DOMINANCE

Mendel also proposed the concept of dominance to explain how some traits are expressed, whereas others are hidden. Individuals can inherit two of the same allele (**homozygous**) or two different alleles (**heterozygous**) for any given trait. In the heterozygous individual, only one allele is normally expressed yet the other allele is hidden. The **dominant** allele is the one expressed, whereas the **recessive** allele is hidden in the presence of a dominant allele. When an individual is heterozygous for a particular trait, the dominant trait appears in the individual's phenotype, but the individual still carries—and can pass on via their gametes—the recessive allele. A recessive phenotype is only observed when the individual is homozygous for the recessive allele. Keep in mind that dominant traits are not necessarily more common or more advantageous than recessive traits. Those labels refer only to the pattern of inheritance that the allele follows. The most common allele in the population is usually referred to as the **wild type**.

By convention, a single letter is selected to represent a particular trait. The dominant allele is always notated with a capital letter, and the recessive allele is notated with a lowercase letter. An example of possible allelic combinations can be seen in the following table.

Possible Allelic Combinations		
Alleles Inherited	**Genotype**	**Phenotype**
AA	Homozygous dominant	Dominant
Aa	Heterozygous	Dominant
aa	Homozygous recessive	Recessive

PREDICTING THE GENOTYPE OF OFFSPRING AND PARENTS

When the genotypes of both parents are known for a specific trait, the genotypes of the offspring can be determined. The tool used for this is known as the Punnett square.

A **monohybrid cross** is a breeding between two parents (the P generation) in which a single trait is studied. The offspring of this cross are called the **F_1 generation** (first filial generation). A breeding between two F_1 offspring produces the next generation, F_2, and so on. The potential gametes of each parent are determined and every possible combination of gametes is matched up in a matrix (the Punnett square) in order to determine every possible genotype of potential offspring. A ratio of the phenotypes of the offspring is expressed as dominant:recessive.

Mendel worked with many traits in the pea plant. He found that when he crossed a true breeding (homozygous) plant of a dominant phenotype to a true breeding plant of a recessive phenotype, 100% of the F_1 offspring had the dominant phenotype. However, when he bred two of the F_1 offspring, he found that 75% of the F_2 offspring had the dominant phenotype, yet 25% had the recessive phenotype. Although the recessive phenotype disappeared in the F_1 generation, it reappeared in the F_2 generation. The F_1 offspring were all heterozygous. When two heterozygotes are bred, the offspring always show Mendel's observed 3:1 **phenotypic ratio**. A cross between two heterozygotes that results in a 3:1 phenotypic ratio can be seen in Figure 5-1.

The genotype of a parent with a dominant phenotype can also be determined using a method known as a **test cross** (or **backcross**). An organism with the dominant phenotype may be either homozygous or heterozygous. In the testcross, the parent with the dominant phenotype is always crossed to a homozygous recessive mate. The outcome of the phenotypic ratio of the offspring reveals the genotype of the unknown parent. If 100% of the offspring have the dominant phenotype, then the unknown parent is homozygous dominant. If the offspring display a 1:1 ratio, the genotype of the unknown parent is heterozygous. The possible outcomes of a testcross can be seen in Figure 5-2.

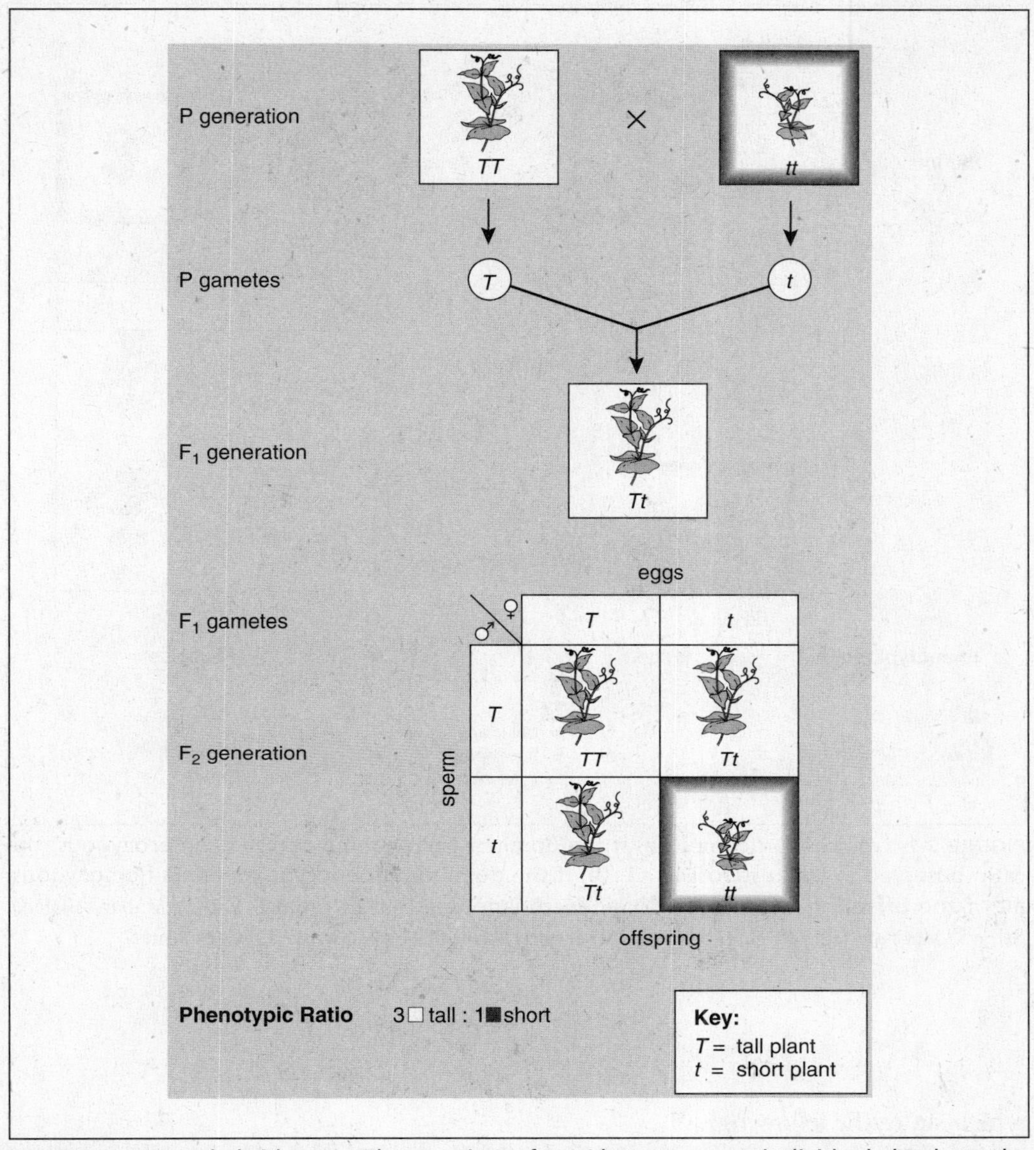

FIGURE 5-1: Monohybrid cross. The crossing of two heterozygous individuals leads to the typical 3:1 phenotypic ratio observed by Mendel. SOURCE: From Sylvia S. Mader, *Biology*, 8th ed., McGraw-Hill, 2004; reproduced with permission of The McGraw-Hill Companies.

MENDEL'S LAW OF INDEPENDENT ASSORTMENT

A **dihybrid cross** considers the inheritance of two different traits at the same time. The same rules of the monohybrid cross apply as long as the traits involved meet certain criteria. Those criteria are developed from Mendel's **law of independent assortment**,

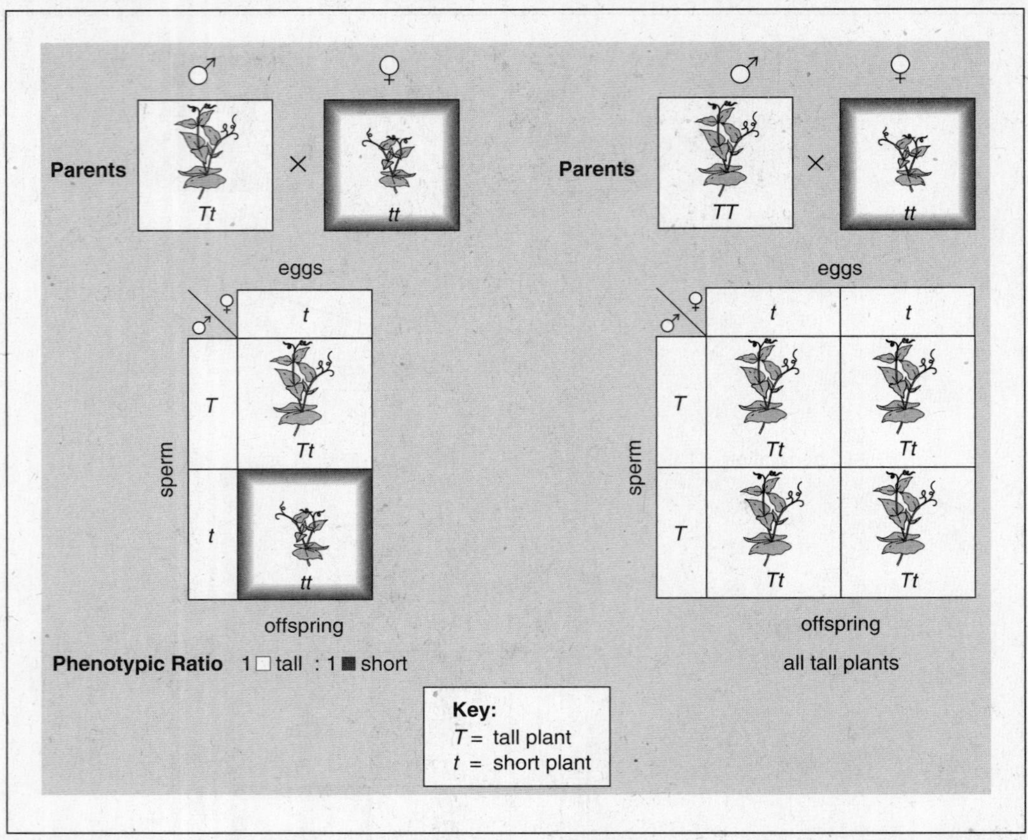

FIGURE 5-2: Testcross outcomes. (a) If the dominant phenotype parent is heterozygous, the ratio observed in the testcross is 1:1. (b) If the dominant phenotype parent is homozygous, all of the offspring exhibit the dominant phenotype. SOURCE: From Sylvia S. Mader, *Biology*, 8th ed., McGraw-Hill, 2004; reproduced with permission of The McGraw-Hill Companies.

which states the following:

- The alleles must assort independently during gamete formation, meaning that the distribution of alleles for one trait has no influence on the distribution of alleles for the other trait.
- If two genes are linked, meaning they occur on the same chromosome, they do not assort independently, and thus are inherited together, changing the expected outcomes in the offspring.

Two unlinked traits can be considered together in a Punnett square. When two traits are involved in a dihybrid cross, each trait is assigned a different letter. In order to predict the possible offspring, all possible gamete combinations of each trait for the parents must be considered. Suppose two parents have the genotypes AABB and aabb. All F_1 offspring will be AaBb. If two F_1 offspring are bred, a 9:3:3:1 ratio is seen in the F_2 generation. See Figure 5-3 for an example of a dihybrid cross.

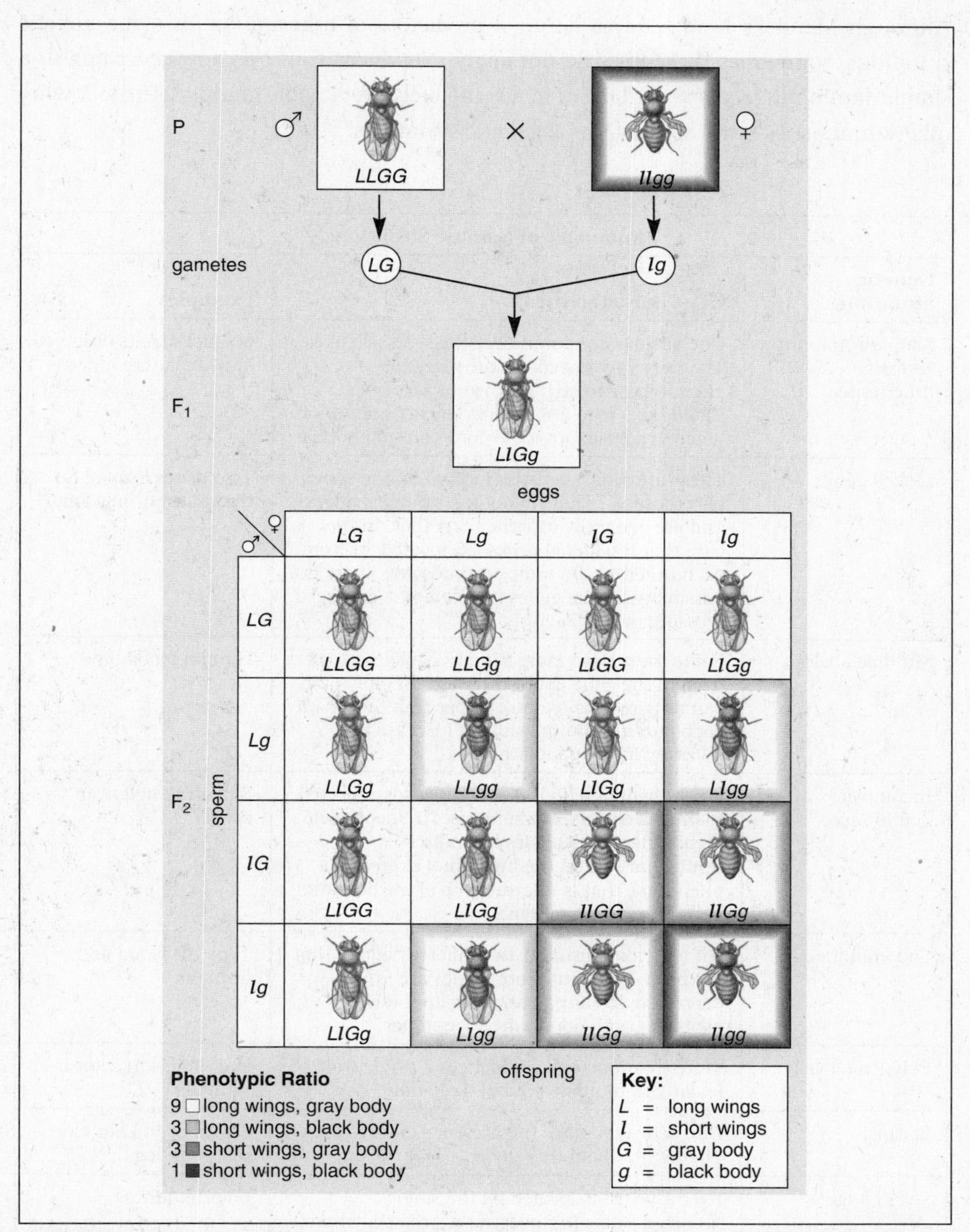

FIGURE 5-3: In a dihybrid cross, the inheritance of two unlinked traits are considered simultaneously. In this cross, Mendel's 9:3:3:1 phenotypic ratio is observed. SOURCE: From Sylvia S. Mader, *Biology*, 8th ed., McGraw-Hill, 2004; reproduced with permission of The McGraw-Hill Companies.

EXCEPTIONS TO MENDEL'S LAWS

Although Mendel's laws tend to be good predictors of inheritance for some genetic situations, sometimes these laws do not apply. Not every trait operates according to a simple dominant/recessive pattern or in a completely predictable manner. The following table summarizes some of the various genetic situations.

Summary of Genetic Situations

Genetic Situation	Key Characteristics	Examples
Simple dominant/ recessive inheritance	One allele is dominant over the recessive allele. The only way to express the recessive phenotype is to be homozygous recessive. Individuals who are homozygous dominant or heterozygous express the dominant phenotype.	Mendel's traits observed in pea plants
Linked genes	These are separate genes located on the same chromosome. They do not assort independently and are generally inherited together. In the case that the linked genes are located far from each other on the same chromosome, there is a possibility for the genes to recombine during crossing over in meiosis.	Two genes located on the same chromosome
Multiple alleles	Some traits have more than two alleles to select from in the gene pool. Although an individual can only receive two alleles per trait (one from each parent), multiple alleles increase the diversity in the population.	Human blood type
Incomplete dominance	An individual who is heterozygous is expected to have a dominant phenotype. In incomplete dominance, both alleles are expressed somewhat so that the individual expresses a phenotype that is intermediate of the dominant and recessive phenotypes.	Snapdragon flower color
Codominance	An individual inherits two different alleles that are both dominant. Both alleles are fully expressed, leading to an individual who expresses both dominant phenotypes.	Type AB blood in humans
Polygenic traits	More than one gene influences a single trait, leading to multiple potential phenotypes.	Hair and skin color in humans
Epistasis	One gene can mask the presence of an expected phenotype of another gene.	Fur color in Labrador retriever dogs
Sex linkage	Recessive traits located on the single X chromosome in males are expressed, whereas they must be inherited on both X chromosomes to be expressed in women.	Colorblindness

Linked Genes

The location of a gene on a chromosome is referred to as the **locus** of the gene. Genes that are linked occur on the same chromosome, which means that if one allele is found in a gamete, the other is as well because they are on the same chromosome. In the case of linkage, the combination of gametes produced is not as diverse as would be the case with nonlinked alleles. In some cases, the loci of the alleles are so close together that they are always inherited together. However, if the loci of the alleles are far away from each other on the chromosome, then there is a possibility for crossing over and genetic recombination to occur. This process is discussed in more detail with meiosis in Chapter 6.

Multiple Alleles

For the traits Mendel observed with pea plants, there were always two alleles. One was dominant and one was recessive. Although an individual can inherit only two alleles (one from each parent) for any given trait, there is the possibility that there may be more than two alleles to select from in the **gene pool**, which consists of all genotypes in the population. These new alleles arise due to mutation and increase diversity in the population.

Human blood type is an example of **multiple alleles**. The ABO system has three alleles: I^A, I^B, and i. The alleles I^A and I^B are dominant, whereas the allele i is recessive. Each allele codes for either the presence or absence of particular **antigens** on the surface of red blood cells. With simple dominant/recessive traits, two phenotypes are expected—a dominant phenotype and a recessive phenotype. Any time multiple alleles are involved with a trait, more that two potential phenotypes are expected. This is the case in blood types, where four phenotypes can be observed: Type A, Type B, Type AB, and Type O.

Incomplete Dominance

According to Mendelian rules, a heterozygous individual always expresses the dominant phenotype. If alleles behave by **incomplete dominance**, this is not the case. Flower color in snapdragons is a classic example. If the allele R codes for red flowers and the allele r codes for white flowers, Mendelian rules would suggest that the heterozygote (Rr) would have red flowers. However, because this trait behaves according to incomplete dominance, both alleles are expressed to some degree, leading to a pink (intermediate) phenotype in the heterozygous offspring. In the case of incomplete dominance, only two alleles are involved, yet there are three potential phenotypes that can arise.

Codominance

Codominance is similar to incomplete dominance. For this to occur, the trait involved must first have multiple alleles, and more than one of them must be dominant. If a heterozygous individual inherits two different dominant alleles, both alleles are expressed, leading to an individual who has both phenotypes (as opposed to a blended phenotype seen with incomplete dominance).

Human blood type is an example of codominance as well as multiple alleles. Should an individual inherit the genotype of I^AI^B, they express the A phenotype as well as the B phenotype. In this case, the result is Type AB blood. See the following table for more details on human blood type.

The Genetic Basis of Human Blood Types		
Blood Type	**Potential Genotypes**	**Antigens Found on the Red Blood Cell Surface**
Type A	I^AI^A or I^Ai	A
Type B	I^BI^B or I^Bi	B
Type AB	I^AI^B	A and B
Type O	ii	none

Polygenic Traits

Generally, a single gene influences one trait. **Polygenic traits** involve gene interaction. This means that more than one gene acts to influence a single trait. Skin color and hair color are both examples of polygenic traits. Because more than one gene is involved, the number of potential phenotypes is increased, resulting in continuous variation.

Epistasis

Epistasis is a unique genetic situation where one gene interferes with the expression of another gene. In many cases, epistasis can lead to the masking of an expected trait. An example is coat color in Labrador retrievers. These dogs have black, chocolate, or yellow fur. In addition to the gene that controls fur color, the B gene, there is another allele that controls how pigment is distributed in the fur, the E gene. The B gene produces an enzyme that processes brown pigment to black pigment. Dogs that have the genotype BB or Bb produce black pigment, whereas those with the genotype bb produce brown pigment. The E gene allows the pigment to be deposited into the hair follicle. If the Labrador is EE or Ee, it is able to deposit the pigment. However, dogs with the genotype ee do not. Therefore, the gene B determines if a dog produces black or brown pigment, but these phenotypes can only be expressed if the dog is homozygous dominant or heterozygous for the E gene. Any dog that is homozygous recessive for the E gene, ee, is yellow.

Pleiotropy

Pleiotropy occurs when a single gene influences two or more other traits. Most frequently, the effects of pleiotropy are seen in genetic diseases. In sickle cell disease, the mutation in the hemoglobin gene results in the production of hemoglobin protein with a reduced oxygen-carrying ability. This, in turn, affects multiple organ systems in the body, explaining the multiple symptoms of the disease.

Sex Linkage

When alleles are found on the X and Y sex chromosomes, the normal rules of genetics may not apply. Although the sex chromosomes do contain genes to influence gender, there are other traits found on these chromosomes that have nothing to do with gender. Women inherit an XX genotype, whereas men inherit an XY genotype. In men, traits that occur on the sex chromosomes are the exception to the normal rule of always having two alleles per trait. Because the sex chromosomes in men are not a true pair, they do not have two alleles per trait on their sex chromosomes. The Y chromosome contains relatively few genes as compared to the X chromosome.

When a recessive trait is located on the X chromosome, women must receive two copies of the recessive allele (one from each parent) to express the recessive trait. However, men who inherit a recessive allele on their only X chromosome express the recessive phenotype. Color blindness and hemophilia are examples of sex-linked traits. Although women can express these traits, they must receive the recessive alleles on both X chromosomes (meaning they must receive it from both of their parents). Therefore, these traits are more commonly observed in men, as they must only receive the recessive trait on their single X chromosome.

Women who are heterozygous for a trait on the X chromosome do not express the trait, yet they are carriers for this and can pass the traits to their sons. Since women are genotypically XX, every egg cell they make contains the X chromosome. Men are XY, and thus half their sperm contain the X chromosome, and half contain the Y chromosome. In males, the Y chromosome must come from the father and the X comes from the mother.

PEDIGREE ANALYSIS

A **pedigree** is a diagram that is used to help determine a pattern of inheritance over multiple generations. In pedigrees, males are indicated by a square and females by circles. Shading in the square or circle indicates individuals in the pedigree that are affected by a certain trait. No shading in the square or circle indicates those that are not affected. In some cases, a square or circle may be half shaded to indicate a carrier or a heterozygote. Horizontal lines indicate matings and vertical lines show offspring. An example of a pedigree indicating recessive inheritance is seen in Figure 5-4.

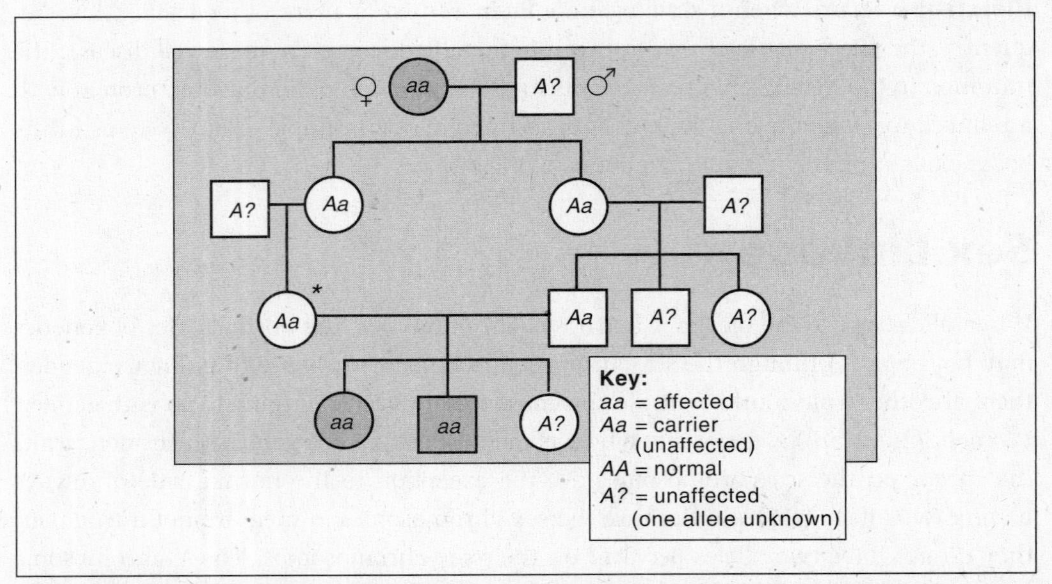

FIGURE 5-4: This pedigree shows a recessive pattern of inheritance. SOURCE: From Sylvia S. Mader, *Biology*, 8th ed., McGraw-Hill, 2004; reproduced with permission of The McGraw-Hill Companies.

If a pedigree shows many more males than females being affected, sex-linked inheritance should automatically be suspected. If males and females both seem affected equally, look for skipping of generations. Dominant traits usually appear in each generation, whereas recessive traits often skip generations.

ENVIRONMENTAL INFLUENCES ON GENES

Although some genes behave according to very predictable rules, there are many cases where some external or internal environmental factor can interfere with the expression of a particular genotype. **Penetrance** of a genotype is a measure of the frequency at which a trait is actually expressed in the population. If a trait were described at 80% penetrance, it would mean that 80% of the people with the genotype for the particular trait would have the phenotype associated with the genotype. Although some traits always show 100% penetrance, others do not. Within an individual, **expressivity** is a measure of the extent of expression of a phenotype. This means that in some cases, expression of a phenotype is more extreme than others.

There are many examples of how the environment affects the expression of a particular phenotype. Hydrangea plants may have the genotype to produce blue flowers, but depending on the acidity of the soil that they are grown in (an environmental factor), they may express a different phenotype than expected (such as pink flowers). Women who have the BRCA 1 and 2 alleles are at a high, but not guaranteed, risk for developing breast cancer, meaning that something other than the allele determines the expression of the allele.

There are many traits that cannot be predicted by genotype alone (like intelligence, emotional behavior, and susceptibility to cancer). In many cases, the interaction of genes and the environment is a complicated relationship that is impossible to predict. Factors in humans such as age, gender, diet, and so forth are all factors known to influence the expression of certain genotypes.

CRAM SESSION

Genetics

1. Mendelian laws govern the inheritance of many genetic traits. These laws assume that one allele is dominant, asserting itself over a recessive allele.
   a. The law of segregation states that an individual inherits two alleles per trait—one from each parent—and that the gametes made by an individual contain one allele per trait.
   b. The law of independent assortment states that in order to accurately predict the offspring of a particular breeding for several traits, that all of the traits must assort independently. Linked genes violate this law.
2. Punnett squares are used to predict the offspring if the genotypes of the parents are known. Testcrosses can be used to determine the genotype of a parent with the dominant phenotype. Monohybrid crosses consider one trait, whereas dihybrid crosses consider two traits simultaneously.
   a. The breeding of two heterozygotes always gives a 3:1 ratio for monohybrid crosses and a 9:3:3:1 ratio for dihybrid crosses.
3. Mendel's theories do not consider certain types of genetic situations, such as those discussed in the table entitled "The Genetic Basis of Human Blood Types."
   a. Some traits work by incomplete dominance or codominance.
   b. Many traits have multiple alleles.
   c. In some cases, the normal situation of one allele influencing one trait is violated, such as in polygenic traits and pleiotropy. Epistasis involves one gene interfering with the expression of another.
   d. Sex linked traits are inherited uniquely. Men have only one X and one Y chromosome, so any recessive genes found on these chromosomes are expressed.
4. Pedigrees can be used to attempt to determine patterns of inheritance for particular traits. Recessive traits tend to skip generations, whereas dominant traits do not. Situations in which men are affected much more frequently than women suggest sex-linked inheritance.
5. The penetrance of a genotype is not always 100%. External and internal environmental factors other than the inheritance of a particular allele can influence expression of that allele.

CHAPTER 6

Cell Division

Read This Chapter to Learn About:

- Cell Division and Chromosomes
- Mitosis
- Cytoplasmic Inheritance
- Meiosis
- Mistakes in Meiosis

As stated in previous chapters, the cell is the basic unit of structure and function in an organism. For life to continue, cells must divide and reproduce. Cell division in eukaryotes happens through two processes: mitosis and meiosis. **Mitosis** is normal cell division used for growth and the replacement of cells. In mitosis, a parent cell is copied in order to produce two identical daughter cells. However, there are times when producing genetically identical offspring cells is not appropriate, such as during sexual reproduction. During the process of sexual reproduction, genetically diverse gametes must be created. These gametes are produced by the process of **meiosis.** Both mitosis and meiosis have many features in common. During either division process, it is critical that the chromosomes be properly replicated and allotted to each of the daughter cells.

CHROMOSOMES

Chromosomes occur in **homologous pairs** as can be seen in the karyotype in Figure 6-1. For each pair of chromosomes found in an individual, one member of the pair came from the maternal parent and the other member of the pair came from the paternal parent. Recall the genetic inheritance of two alleles per trait—one allele per trait from each parent.

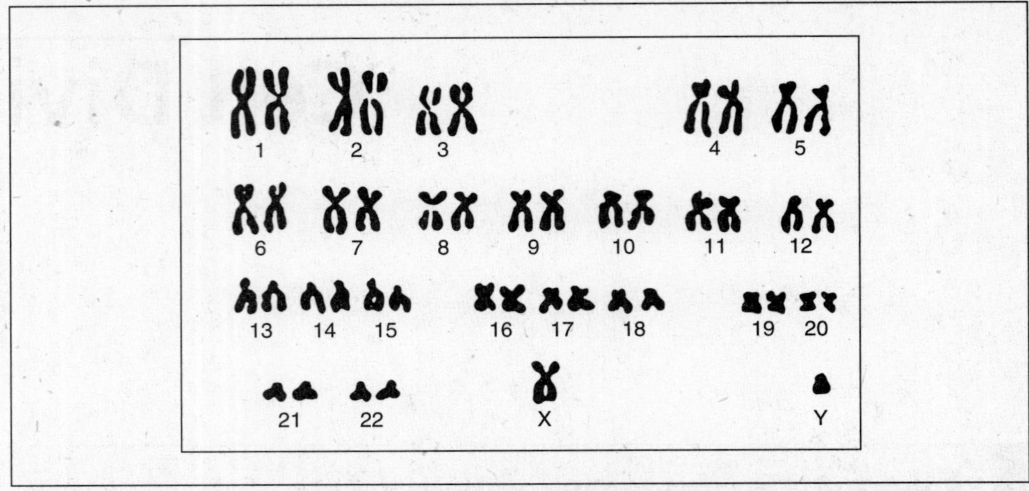

FIGURE 6-1: A karyotype. Chromosomes from a single cell are arranged in pairs to construct a karyotype. This karyotype is from a male. SOURCE: From Eldon D. Enger, Frederick C. Ross, and David B. Bailey, *Concepts in Biology*, 11th ed., McGraw-Hill, 2005; reproduced with permission of The McGraw-Hill Companies.

The total number of chromosomes found in an individual is called the **diploid** (2N) number. When individuals reproduce, this number must be cut in half to produce **haploid** (N) egg and sperm cells. The human diploid number is 46, and the human haploid number is 23. The process of mitosis begins with a diploid cell and ends with two identical diploid cells. In the process of meiosis, a diploid cell begins the process and produces four haploid gametes.

When a cell is not dividing, each chromosome exists in a single copy called a **chromatid**. However, when the cell is preparing to divide, each chromosome must be replicated so that it contains two chromatids, sometimes called **sister chromatids.** Each chromosome has a compressed region called the **centromere,** and when the chromosomes replicate, the sister chromatids stay attached to each other at the centromere. Figure 6-2 shows the structure of a replicated chromosome.

MITOSIS

Mitosis is the process of normal cell division in eukaryotic cells. It occurs in most cells with the exception of gametes and mature nerve and muscle cells in animals. It begins with a single parent cell that replicates all components within the cell, divides the components into two piles, and then splits to form two genetically identical daughter cells. The most critical components for replication and division are the chromosomes, so particular care is taken to ensure an equal distribution of chromosomes to each daughter cell.

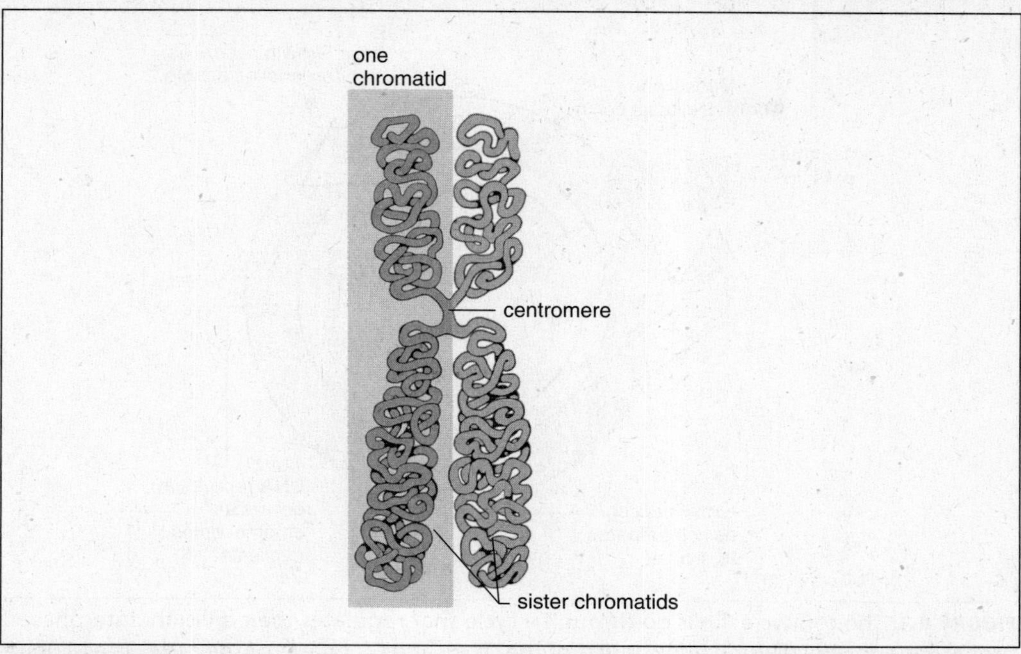

FIGURE 6-2: Chromosome structure. A replicated chromosome consists of two sister chromatids attached to each other at the centromere. SOURCE: From Sylvia S. Mader, *Biology*, 8th ed., McGraw-Hill, 2004; reproduced with permission of The McGraw-Hill Companies.

The Cell Cycle

Mitosis is used for the growth of organisms because it takes an increased number of cells for an organism to get bigger. When an individual has stopped growing, mitosis is needed only to replace cells that have died or been injured. For this reason, mitosis must be a regulated process that occurs only when new cells are needed. The **cell cycle** is used to regulate the process of cell division in each individual cell. A normal cell cycle has the following stages that can be seen in Figure 6-3.

- G_1: This is the first gap phase of the cell cycle. In this stage, the parent cell is growing larger, adding additional cytoplasm, and replicating organelles.
- S: During this phase of DNA synthesis, the chromosomes are all being replicated. Once this stage is complete, each chromosome consists of two sister chromatids connected at the centromere.
- G_2: This is the second gap phase. The cell continues to grow in size and make final preparations for cell division.
- M: During the M phase, mitosis actually occurs. The replicated chromosomes and other cellular components are divided to ensure that each daughter cell receives equal distributions. The division of the cytoplasm at the end of the M phase is referred to as **cytokinesis.**

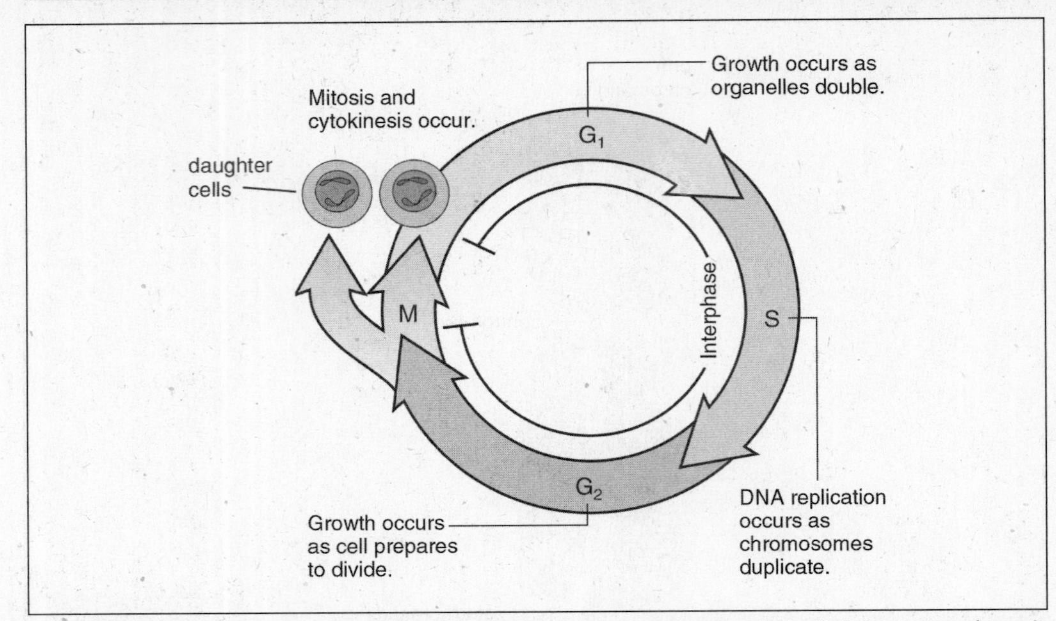

FIGURE 6-3: The cell cycle. Cells go through a cycle that regulates their division. Interphase is preparation for cell division and consists of the G_1, S, and G_2 phases of the cycle. The M phase is where the cells actually divide. SOURCE: From Sylvia S. Mader, *Biology*, 8th ed., McGraw-Hill, 2004; reproduced with permission of The McGraw-Hill Companies.

The first three phases of the cell cycle, G_1, S, and G_2, are collectively called **interphase,** which simply means preparation for cell division. The actual cell division occurs during the M phase of the cycle.

Some cells, such as mature human nerve and muscle cells, lose the ability to progress through the cell cycle and are thus unable to divide. Such cells are considered to be in the G_0 phase of the cell cycle, where division never resumes.

M PHASE

The M phase of the cell cycle is subdivided into four stages: **prophase, metaphase, anaphase,** and **telophase.** The primary concern in these stages is alignment and splitting of sister chromatids to ensure that each daughter cell receives an equal contribution of chromosomes from the parent cell. A visual summary of the events of the M phase can be seen in Figure 6-4.

PROPHASE: Chromosomes are located in the nucleus. Prior to division, the chromosomes are not condensed and thus are not visible. Leaving the chromosomes in an uncondensed state makes it easier to copy the DNA but makes the chromosomes very stringy and fragile. Once the DNA is replicated, the chromosomes must condense so that they are not broken as they are divided up into the two daughter cells.

Another major event of prophase is a breakdown of the nuclear membrane releasing the chromosomes into the cytoplasm of the cell. The **centrioles** present in the cell replicate and move to opposite ends of the cell. Once they have migrated to the poles of the cell, they begin to produce a spindle apparatus consisting of spindle fibers that radiate outward, forming **asters.** The spindle fibers are made of microtubules that ultimately attach to each chromosome at the **kinetochore.** The kinetochore appears at the centromere of each chromosome.

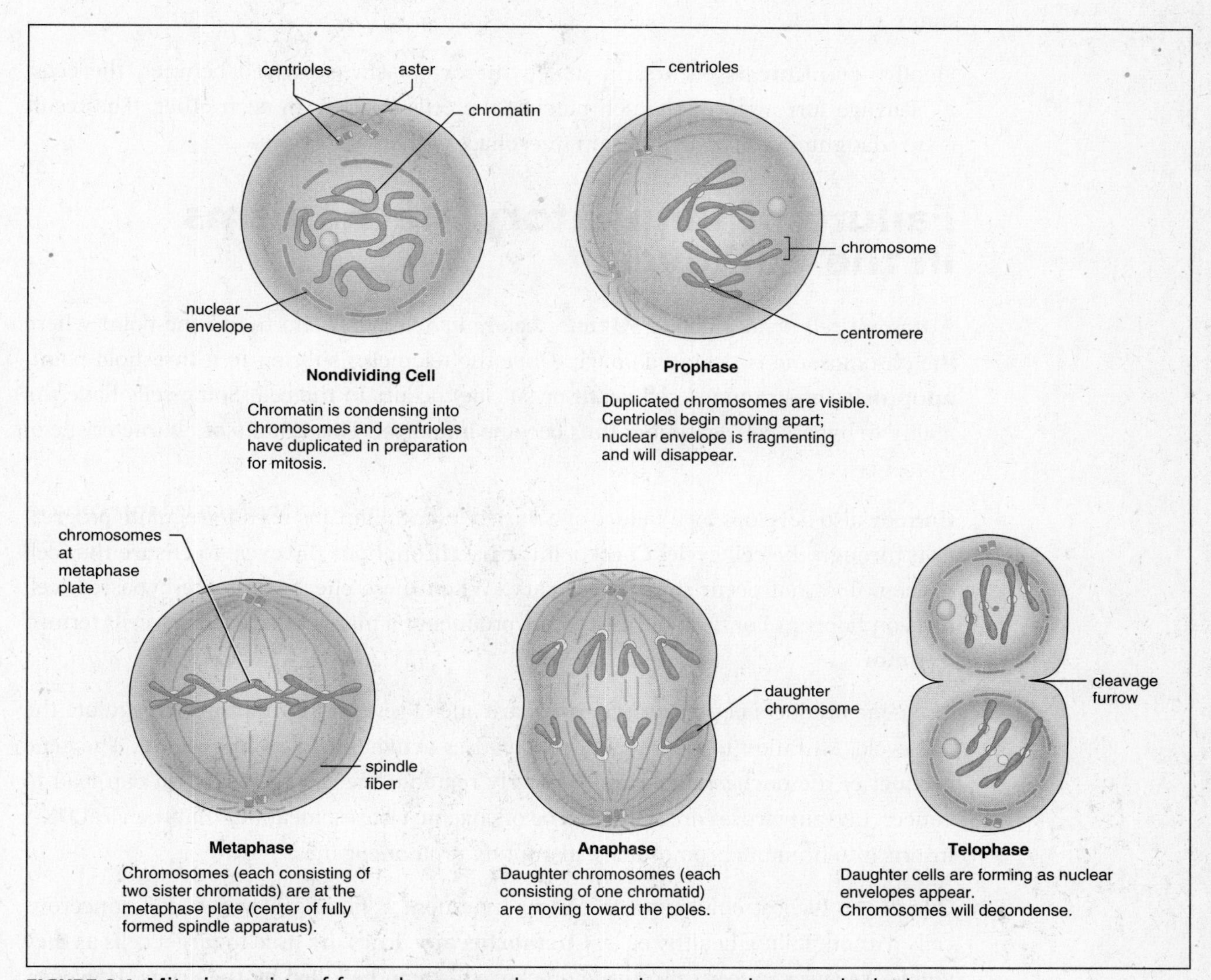

FIGURE 6-4: Mitosis consists of four phases: prophase, metaphase, anaphase, and telophase.
SOURCE: From Sylvia S. Mader, *Biology*, 8th ed., McGraw-Hill, 2004; reproduced with permission of The McGraw-Hill Companies.

METAPHASE: In metaphase, each chromosome is attached to a spindle fiber at the kinetochore. The chromosomes are aligned along the center of the cell at the metaphase plate.

ANAPHASE: During anaphase, the centromere splits, allowing each chromatid to have its own centromere. At this point, the chromatids can be separated from each other and are pulled toward opposite poles of the cell, separating the chromosomes into two distinct piles, one for each daughter cell.

TELOPHASE: Now that the chromosomes have been divided into two groups, the spindle apparatus is no longer needed and disappears. A new nuclear membrane forms around each set of chromosomes and the chromosomes uncoil back to their original state.

Finally, **cytokinesis** occurs, in which the cytoplasm is divided between the cells. A cleavage furrow forms, which pinches the cells apart from each other. The result is two daughter cells ready to begin interphase of their cell cycles.

Failure of Regulatory Mechanisms in the Cell Cycle

A normal cell divides about 50 times before its telomeres shorten to the point where the chromosome is risking damage. Once the telomeres shorten to a threshold point, **apoptosis** (programmed cell death or suicide) occurs in the cell. Some cells have the ability to bypass cell death and thus become immortal. This is a major characteristic of cancer cells.

Cancer also develops by a failure of a variety of mechanisms used to regulate progression through the cell cycle. Checkpoints exist throughout the cycle to ensure that cell division does not occur unless necessary. When these checkpoints are bypassed, cell division happens continually, ultimately producing a mass of unnecessary cells termed a **tumor.**

The gene products of **protooncogenes** are one of several mechanisms to regulate the cell cycle. Mutation to a protooncogene causes activation to an **oncogene.** The gene product of the oncogene does not properly regulate the cell cycle, which can lead to cancer. Certain viruses are known to be oncogenic viruses, meaning that the viral DNA inserts into human chromosomes, disrupting protooncogenes.

One of the biggest challenges to cancer treatment is finding a way to kill cancerous cells without killing healthy cells. **Chemotherapy** drugs are used to target cells as they divide. Because cancer cells divide quickly, they can be killed by the drugs. However, other cells in the body that are dividing are also damaged. This is the cause of many of the side effects related to cancer treatments.

MEIOSIS

Because mitosis produces genetically identical diploid daughter cells, it is not appropriate for sexual reproduction. If diploid cells were used for reproduction in humans, each egg would contain 46 chromosomes, as would each sperm. This would result in embryos having 96 chromosomes. This number would double each generation if mitosis was used to produce gametes.

The process of meiosis begins with a diploid parent cell in the reproductive system that has completed interphase and then follows stages similar to mitosis, twice. The result is four haploid gametes that are genetically diverse. A summary of the events of meiosis can be seen in Figure 6-5.

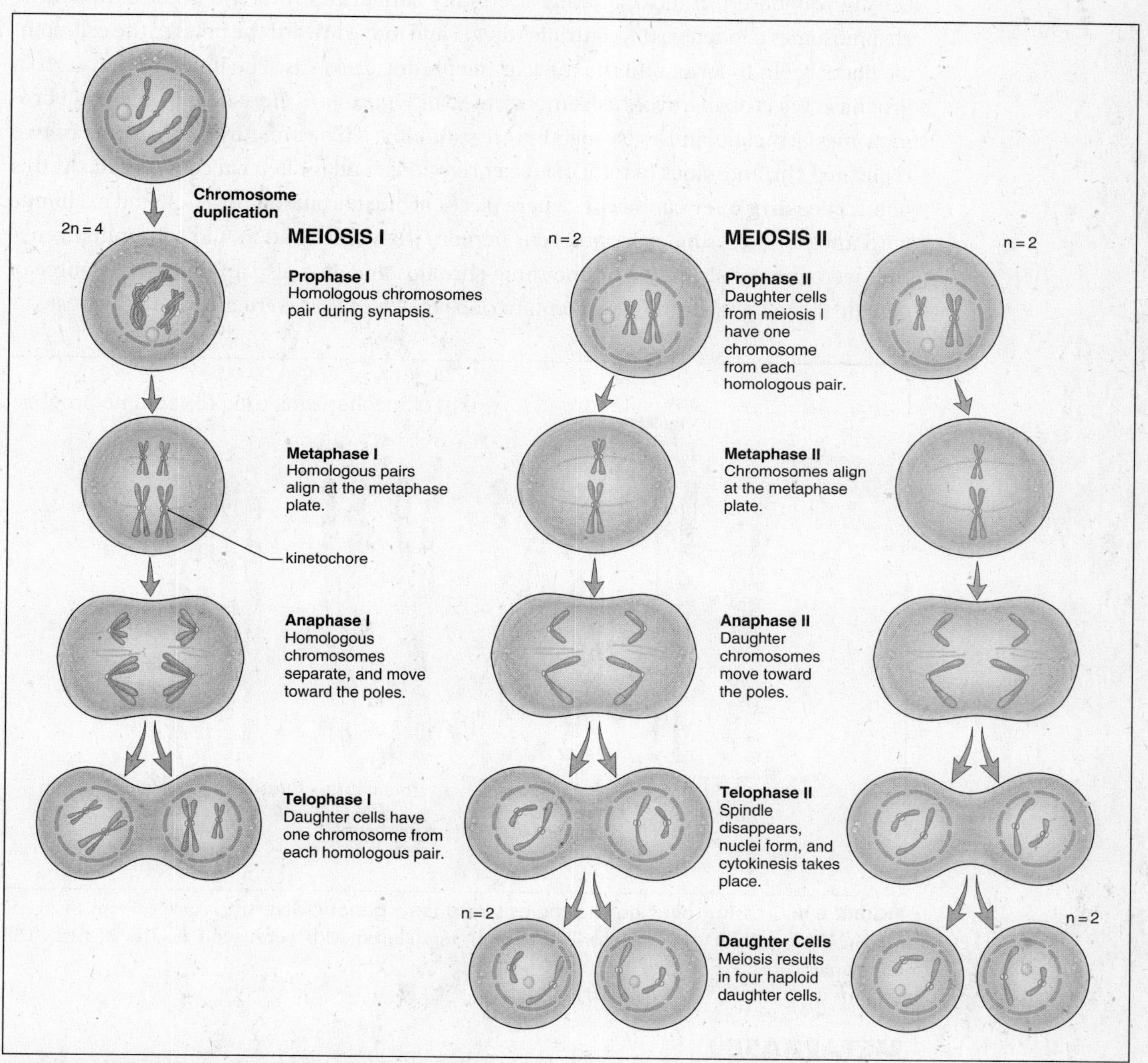

FIGURE 6-5: Meiosis consists of two rounds of cell division. SOURCE: From Sylvia S. Mader, *Biology*, 8th ed., McGraw-Hill, 2004; reproduced with permission of The McGraw-Hill Companies.

Meiosis I

Meiosis I encompasses stages similar to mitosis with two major changes. The first involves genetic recombination between homologous pairs, and the second involves the alignment of chromosome pairs during metaphase of meiosis I.

PROPHASE I

During prophase I of meiosis, there are many similarities to prophase of mitosis. The chromosomes condense, the centrioles divide and move toward the poles of the cell, spindle fibers begin to form, and the nuclear membrane dissolves. The unique event seen in prophase I is crossing over, as demonstrated in Figure 6-6. Homologous pairs of chromosomes associate and twist together in **synapsis.** This configuration consists of two replicated chromosomes (a total of four chromatids) and is often called a **tetrad.** At this point, **crossing over** can occur, where pieces of one chromatid break off and exchange with another. Crossing over can occur in more than one location and can unlink genes that were previously linked on the same chromosome. It is also an important source of genetic diversity, creating new combinations of alleles that were not seen previously.

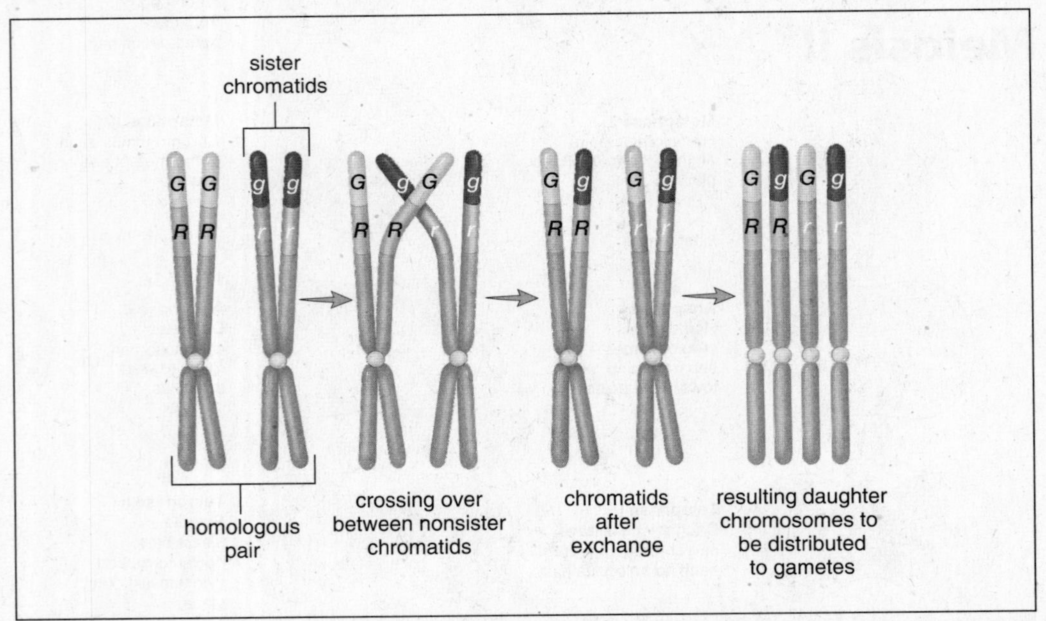

FIGURE 6-6: Crossing over during meiosis results in genetic diversity. SOURCE: From Sylvia S. Mader, *Biology*, 8th ed., McGraw-Hill, 2004; reproduced with permission of The McGraw-Hill Companies.

METAPHASE I

In metaphase of mitosis, chromosomes aligned single file along the center of the cell. In metaphase I of meiosis, the chromosomes align as pairs along the center of the cell.

This alignment of pairs is the critical factor in creating haploid daughter cells. Recall from genetics the law of independent assortment. The alignment of each member of the homologous pair during metaphase I is random so that each daughter cell has a unique combination of maternal and paternal alleles.

ANAPHASE I

The homologous pairs separate from each other during anaphase I and are pulled to the poles of the cells. This separation is referred to as **disjunction.**

TELOPHASE I

The events of telophase I are similar to those of the telophase of mitosis. The spindle apparatus dissolves, nuclear membranes form around each set of chromosomes, and cytokinesis occurs to form the two daughter cells. At this point, each daughter cell is genetically unique and contains half the number of chromosomes of the parent cell. However, these chromosomes are still in their replicated form, consisting of two chromatids.

Meiosis II

Meiosis II is only necessary to split the chromatids present in the daughter cells produced during meiosis I. There is no interphase between meiosis I and II because the chromosomes are already replicated. The events of meiosis II are as follows:

- Prophase II: Centrioles replicate and move toward the poles of the cell, chromosomes condense, and the nuclear membrane dissolves.
- Metaphase II: Chromosomes align along the center of the cell.
- Anaphase II: Sister chromatids are separated and move toward the poles of the cell.
- Telophase II: Nuclear membranes re-form and cytokinesis occurs to produce daughter cells.

At the end of meiosis II, there are four daughter cells. Each is haploid with a single copy of each chromosome. Each cell is genetically diverse as a result of crossing over and independent assortment.

Mistakes in Meiosis

Mistakes that happen during meiosis can have drastic consequences. Because the gametes are used for reproduction, any chromosomal damage to the gametes is passed on to the next generation. There are several ways in which mistakes can occur, changing the number of chromosomes or damaging them.

If chromosomes fail to separate properly during meiosis, a **nondisjunction** has occurred. This leads to gametes that have the wrong number of chromosomes. If those gametes are fertilized, the resulting embryo will have the wrong diploid number. An example of this is Down's syndrome, which often is the result of a nondisjunction in the female gamete. If a female egg contains 24 chromosomes instead of the expected 23 and is fertilized by a normal sperm, the resulting embryo will have 47 chromosomes, which is one more than expected. This condition is referred to as a **trisomy.** In the case where a gamete is missing a chromosome as the result of a nondisjunction and is fertilized by a normal gamete, the result is an embryo with 45 chromosomes. This is termed a **monosomy.** With the exception of Down's syndrome, which is a trisomy of human chromosome 21 (which is very small) and certain trisomies and monosomies of the sex chromosomes (X and Y), most embryos with trisomies and monosomies do not survive development.

There are other forms of chromosomal damage that can occur in meiosis. They typically have serious, if not fatal, consequences. They are as follows:

- **Deletion:** A deletion occurs when a portion of a chromosome is broken off and lost during meiosis. Although the total chromosome number is normal, some alleles have been lost.
- **Duplication:** A duplication occurs when a chromosome contains all of the expected alleles and then receives a duplication of some alleles.
- **Inversion:** An inversion occurs when a portion of a chromosome breaks off and reattaches to the same chromosome in the opposite direction.
- **Translocation:** Translocation occurs when a piece of a chromosome breaks off and reattaches to another chromosome.

GAMETOGENESIS

Meiosis results in four gametes. In men, all four of these gametes becomes sperm. In women, only one of these gametes becomes a functional egg cell, which is released once every 28 days during **ovulation.** If all four gametes became functional eggs and were released each cycle, there would be the potential for four embryos. The three gametes that do not become functional eggs in women are termed **polar bodies.** Some of the major differences between meiosis in men (**spermatogenesis**) and women (**oogenesis**) are described in the following table.

Differences between Spermatogenesis and Oogenesis		
Characteristic	**Spermatogenesis**	**Oogenesis**
Time at which the process begins	At puberty	Before a female is born (during development)
Time at which the process ends	Theoretically, never	At menopause
Time needed to complete meiosis	65–75 days	Many years
Number of gametes made	Unlimited numbers are possible	The number of follicles is set at birth in females
Fates of the daughter cells	All four are sperm	One is the egg and the other three are polar bodies
Age of the gametes	Not applicable—old sperm are degraded	Women are born with a set number of follicles which are the same age as the woman
Presence of arresting stages in meiosis	None	Meiosis I starts before birth and then arrests. Meiosis I resumes only after puberty. Only one cell is selected to complete meiosis I per month. Meiosis II happens only if fertilization occurs.

CYTOPLASMIC INHERITANCE

Recall that organelles such as mitochondria contain their own DNA. This DNA is circular and is derived from a single source. Therefore there are no pairs of alleles, just a single copy of each allele. Any genes present on mitochondrial DNA are inherited by the daughter cells during cytokinesis of mitosis or meiosis. Unlike chromosomal inheritance, any genes passed through mitochondrial DNA do not follow the normal Mendelian laws of genetics. All mitochondrial DNA in an individual is derived from mitochondrial DNA in an egg cell from the mother. This is sometimes referred to as maternal inheritance.

CRAM SESSION

Cell Division

1. Cell division in eukaryotes occurs in two forms: mitosis and meiosis.
2. Prior to cell division, all chromosomes must be replicated resulting in two sister chromatids attached at the centromere. During cell division, each daughter cell receives a single chromatid.
3. Mitosis is used for growth, repair, and replacement of cells in all cells excluding gametes and mature nerve and muscle cells.
   a. The process is regulated through the cell cycle. When unregulated, cancer can develop.
   b. Mitosis begins with a diploid (2N) parent cell and ends with two identical diploid daughter cells.
   c. The phases of mitosis include prophase, metaphase, anaphase, and telophase.
4. Meiosis is used for sexual reproduction to produce gametes (egg and sperm).
   a. It begins with a diploid cell and goes through two rounds of cell division to produce four genetically diverse haploid gametes.
5. The key differences between mitosis and meiosis occur in meiosis I.
   a. Crossing over that occurs during prophase I generates new combinations of alleles and diversity.
   b. The alignment of chromosomes in homologous pairs in metaphase I of meiosis is key to generating haploid daughter cells.
   c. Meiosis II is needed only to separate the chromatids of the daughter cells generated in meiosis I. This process resembles mitosis.
6. Meiosis in men is spermatogenesis and in women is oogenesis. There are some major differences between the two processes pointed out in the table entitled "Differences Between Spermatogenesis and Oogenesis."
7. Mistakes that occur during meiosis lead to damaged gametes. Nondisjunction results in gametes with an incorrect chromosome number, whereas translocation, deletion, inversion, and duplication lead to damaged chromosomes. If a gamete with one of these chromosomal mutations is fertilized, the results can be fatal early in development.

CHAPTER 7

Evolution

Read This Chapter to Learn About:

- ➤ Mechanisms for Evolution
- ➤ Genetic Basis for Evolution and the Hardy–Weinberg Equation
- ➤ Types of Natural Selection
- ➤ Speciation
- ➤ Types of Evolution
- ➤ Evidences for Evolution
- ➤ Chordate Evolution
- ➤ The Origin of Life

Evolution simply means change. The changes referred to are genetic ones; this puts the concept of mutation at the center of the process of evolution. Evolution is something that occurs over time, so a single individual does not evolve, but populations of individuals do evolve. **Microevolution** deals with genetic changes within a population, whereas **macroevolution** is concerned with changes that occur to a species on a larger scale over a longer period of time.

MECHANISMS FOR EVOLUTION

There are a variety of factors responsible for the microevolution of a particular population. Natural selection, based on mutation, tends to be the major driving force for evolution, whereas genetic drift and gene flow can also influence the process.

Mutation

New alleles are created by mutation. These new alleles code for proteins that may be beneficial, neutral, or detrimental as compared to the original protein intended by the allele. New alleles that code for beneficial proteins can provide advantages that are ultimately selected for by natural selection and are passed to the next generation, whereas detrimental alleles are selected against.

Natural Selection

A central concept to the process of how evolution occurs is that of Darwin's natural selection. **Natural selection** explains the increase in frequency of favorable alleles from one generation to the next. This results from differential reproductive success in which some individuals reproduce more often than others and thus are selected for; this increases the frequency of their alleles in the next generation. Those that reproduce less, decrease the frequency of their alleles in the next generation.

FITNESS CONCEPT

The concept of evolutionary **fitness** is key to natural selection. In this context, *fit* refers to the reproductive success of an individual and his/her contribution to the next generation. Those individuals that are more fit are more evolutionarily successful because their genetic traits are passed to the next generation, thus increasing the frequency of specific alleles in the gene pool.

Over generations, selective pressures that are exerted on a population can lead to adaptation. When selective pressures change, some organisms that may have been considered marginally fit before may now be extremely fit under the new conditions. Furthermore, those individuals who may have been very fit previously may no longer be fit. Their genetic adaptations are selected against. Although an individual cannot change their genetics, over time, the population changes genetically, which is termed **adaptation**.

DIFFERENTIAL REPRODUCTIVE SUCCESS AND COMPETITION

Competitive interactions within a population are another critical factor for natural selection. The ability to out-compete other individuals for resources, including mates, is a key feature of fitness. In any given population, some individuals are better able to compete for resources and are considered more fit than others. This leads to differential reproductive success. This concept assumes that mating in the population is random. In some cases, such as with humans, mating is nonrandom, which leads to another form of selection to be discussed shortly.

Competition between species can also influence the evolutionary progression of all species involved. In some cases, **symbiotic** relationships exist where two species exist together for extended periods of time. In **mutualistic** relationships, both species benefit from the association. In **parasitic** relationships, one species benefits at the expense of the other species. In **commensalism**, one species benefits whereas the other species is relatively unaffected.

When two species are competing for the same ecological requirements or **niche**, the reproductive success and fitness, as well as the growth of one or both populations, may be inhibited by their ability to compete for resources. This changes the microevolutionary course of the population.

Gene Flow

When individuals leave a population, they take their alleles with them, resulting in **gene flow**. This can decrease genetic variation within the gene pool of the population and ultimately affect the evolution of the population. **Outbreeding** occurs with the individuals that leave the population. They can add diversity to the gene pool of their new population by adding new alleles to it.

Genetic Drift

Genetic drift involves changes to the allelic frequencies within a population due to chance. Although this is generally negligible in large populations, it can have major consequences in smaller populations. The **bottleneck effect** is a form of genetic drift where catastrophic events may wipe out a large percentage of a population. When the population is small, the few remaining alleles in the gene pool may not be characteristic of the larger population. Generally, genetic diversity is lost due to inbreeding by the remaining members of the population. The **founder effect** is a form of genetic drift that occurs when a small number of individuals leave a larger population and form their own small population where inbreeding is necessary. The new population only has the diversity brought to it by the founding members.

GENETIC BASIS FOR EVOLUTION AND THE HARDY–WEINBERG EQUATION

The **Hardy–Weinberg equation** can be used to calculate allelic frequencies within a population if the population is large and microevolution is not occurring—not necessarily a realistic situation. The Hardy–Weinberg equation is expressed as

$$p^2 + 2pq + q^2 = 1$$

where p represents the frequency of a dominant allele and q represents the frequency of a recessive allele such that $p + q = 1$.

The equation can be used to show the frequency of homozygous dominant individuals (p^2), the frequency of heterozygotes ($2pq$), and the frequency of homozygous-recessive individuals (q^2). Given information on the frequency of a single allele, all other pieces within the equation can be determined. Although these frequencies are temporarily accurate, any evolution occurring within the population would shift these predicted values.

In order for Hardy–Weinberg allelic frequencies to hold true, certain criteria must be met. If any of these criteria are violated, the allelic frequencies change over time. Any of the following will negate Hardy–Weinberg equilibrium:

- Nonrandom mating
- Gene flow
- Populations with a small number of individuals
- Mutations
- Bottleneck effect
- Founder effect

TYPES OF NATURAL SELECTION

For any given trait, there can be several different phenotypes. If two phenotypes are present for a particular trait, **dimorphism** is the case. If three or more phenotypes are seen for a particular trait, **polymorphism** is at work. For example, flower color in snapdragons exhibits polymorphism with red, white, and pink phenotypes. Some phenotypes can be considered intermediates (like pink flowers) or can be extremes from either end of the intermediate phenotype (like red and white flowers). When natural selection occurs, it may select for intermediate phenotypes, either extreme phenotype, or both extreme phenotypes as seen in Figure 7-1.

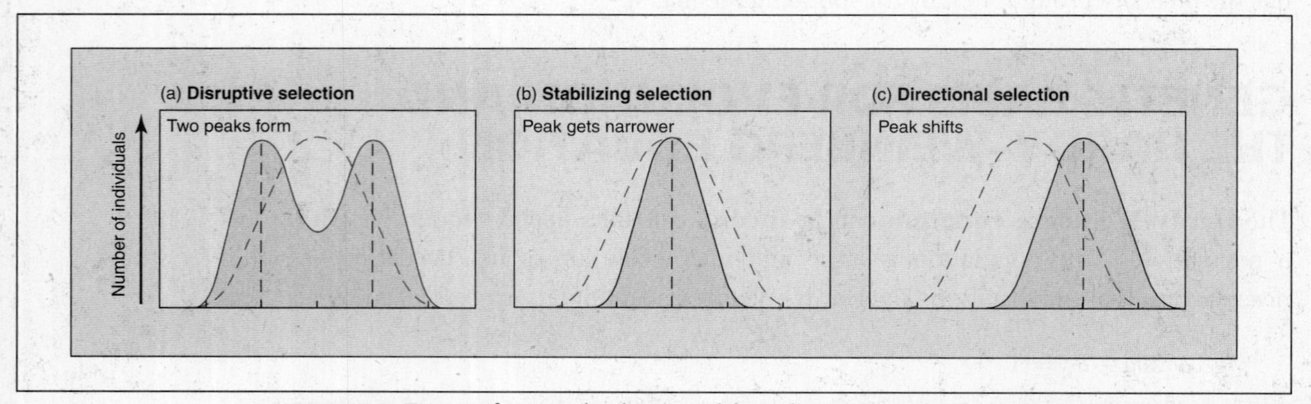

FIGURE 7-1: Types of natural selection. (*a*) In disruptive selection, the extreme phenotypes are selected for whereas the intermediate phenotype is selected against. (*b*) In stabilizing selection, the intermediate phenotype is favored. (*c*) In directional selection, one extreme phenotype is favored. SOURCE: From George B. Johnson, *The Living World,* 3d ed., McGraw-Hill, 2003; reproduced with permission of The McGraw-Hill Companies.

Directional Selection

In **directional selection**, an allele that is considered advantageous is selected for. The allelic frequency continues to shift in the same direction generation after generation. In this case, one allele that produces an extreme phenotype is selected for. The selection of antibiotic resistance alleles in bacteria is an example of directional selection. Over time, selective pressures can result in an entire population possessing the same allele for a particular trait.

Stabilizing Selection

Stabilizing selection leads to favoring the alleles that produce an intermediate phenotype. Human birth weight is an example of stabilizing selection. Babies of an intermediate weight are favored over those that are too small to survive or too large to be easily delivered.

Disruptive Selection

In some cases, the environment favors two extreme phenotypes at the same time. In this case, **disruptive selection** occurs in which individuals with either extreme phenotype are favored, whereas those with the intermediate phenotype are selected against. Over time, the continued selection of both extremes may eventually lead to the evolution of two distinct species.

Artificial Selection

When particular alleles are purposely selected for based on nonrandom mating, **artificial selection** occurs. The breeding of domesticated dogs is an excellent example of the results of artificial selection. All breeds of dogs are members of the same species, all of which have been selectively breed from wolves for specific traits that are appealing to the breeder. Both toy poodles and Great Danes are examples of the extreme phenotypes that can be selected for when artificial selection is used. Traits that are artificially selected for are not necessarily the result of the most fit alleles. Many breeds of dog have medical conditions or predispositions as a result of artificial selection.

SPECIATION

Natural selection can ultimately result in the formation of new species. By definition, a **species** is a group of individuals who can breed with each other and not with members of other species. When populations become geographically isolated from each other, members of the same species may evolve differently in different locations. This type of speciation is referred to as **allopatric**. Even when geographic barriers do not exist to divide a population, there are factors that can ultimately prevent some members of the population from breeding with others. This is **sympatric** speciation. Over time, they may evolve into two different species that can no longer breed with each other.

There are a variety of mechanisms that occur in order to prevent interbreeding between species. When these mechanisms do not work, hybrid species may develop.

Prezygotic Isolation

Prezygotic isolation mechanisms occur to prevent fertilization between the gametes of members of two different species. They are as follows:

- Temporal isolation: Two different species may live in the same environment but have different breeding seasons, or may have overlapping breeding seasons but breed during different times of the day.
- Ecological isolation: The two species live in different habitats and thus rarely encounter each other to breed.
- Behavioral isolation: The mating behaviors of the two species are not compatible.
- Reproductive isolation: The reproductive structures of the members of different species are not compatible, so attempts to mate are not successful.
- Gamete isolation: The gametes of one species cannot fertilize the gametes of the other species, so reproduction is unsuccessful even if successful copulation occurs.

Postzygotic Isolation

If prezygotic isolation mechanisms fail, there are a variety of mechanisms that occur after fertilization in order to prevent successful reproduction between members of different species. The postzygotic isolation mechanisms are as follows:

- Hybrid inviability: If fertilization occurs between the gametes of two different species, the zygote is unable to continue in development.
- Hybrid sterility: If fertilization and subsequent development successfully occurs, the hybrid offspring is sterile and unable to reproduce.
- Hybrid breakdown: Some hybrid offspring are fertile and can reproduce. However, the second-generation offspring are infertile.

TYPES OF EVOLUTION

The evolutionary process can proceed in a variety of directions or patterns such as convergent, divergent, and parallel evolution.

Convergent Evolution

When two populations exist in the same type of environment, which provides the same selective pressures, the two populations evolve in a similar manner via **convergent evolution**. Although the populations may not be closely related, they may develop similar structures or mechanisms, termed **analogous structures**, to allow them to function in similar environments. Fish in Antarctica have evolved the ability to produce specialized glycoproteins that serve as a sort of antifreeze to prevent their tissues from freezing in the low temperature water. Fish on the opposite side of the world, in the Arctic, have evolved the same kind of antifreeze protection mechanism. Genetic studies show that the two species of fish produce antifreeze proteins that are very different from each other, strongly suggesting that two independent events led to the evolution of these mechanisms.

Divergent Evolution

In a single population, it is possible that individuals within the population evolve differently. Over time, this may lead to the development of new species via **divergent evolution**. In many cases, changes to the population or geographic isolation may cause different adaptations within the population. This sort of evolution can lead to **homologous structures**. Vertebrate limbs are an excellent example of divergent evolution. The forearms of different vertebrate species have different structures and functions; however, they all diverged from a common origin.

Parallel Evolution

When two species share the same environment, the evolution of one species can affect the evolution of the other species. This is called **parallel evolution** or **coevolution**. Any changes to one species requires adaptations to the other species in order for them to continue to exist in the same environment. An example might be how the predation patterns of birds influence the evolution of butterfly species sharing the same space. Some butterflies have evolved the ability to store poisonous chemicals that deter birds from eating them, whereas other butterflies simply mimic the poisonous ones to avoid being preyed on.

EVIDENCES FOR EVOLUTION

There is a large body of evidence to support evolutionary theory. Most evidence comes from studies of paleontology, biogeography, molecular biology, comparative anatomy, and comparative embryology.

- **Paleontology:** The field of paleontology provides evidence for evolution in the form of the fossil record. **Fossilization** occurs when parts of or a whole organism become embedded in sediments. These fossils serve as physical evidence of organisms that lived in the past.
- **Biogeography:** The study of the natural distribution of organisms, **biogeography**, can also lend evidence to evolutionary theory. Looking at similarities and differences between different organisms living in different locations can help decipher evolutionary relationships between organisms as well as common ancestors that may have been shared.
- **Biochemical analysis:** The ability to compare DNA and other molecules such as proteins made by various organisms also helps support evolutionary theory. Organisms that are more closely related share more DNA similarities than organisms that are distantly related. Evolutionary time can be measured by genetic changes over time. Comparisons of mutation rates in conserved gene sequences can be used to construct **molecular clocks** that can estimate when specific lineages emerged.
- **Comparative anatomy:** An analysis of anatomic features in organisms can be used to make comparisons between different species. **Homologous structures** such as the forelimbs of mammals are similar in structure between various organisms and they share a similar function thus indicating that these structures came from a common ancestral species. **Analogous structures** have similar functions between various organisms, but these structures have different lines of descent. **Vestigial structures** seen in organisms have no functional value, yet are evolutionary remnants from ancestral species. Examples of vestigial structures are the human appendix, bone structure for hind limbs in whales, and the coccyx (tailbone) in humans. Because these structures remain even without apparent functions, they are evidence of prior evolutionary forces at work.
- **Comparative embryology:** Although many organisms look distinctly different in their adult forms, they may share striking similarities during their developmental periods. Comparative embryology is used to study these similarities during the developmental process. A modification of Haeckel's theory of ontogeny recapitulate phylogeny (or **recapitulation**) suggests that species that have an evolutionary relationship generally share characteristics during embryonic development.

CHORDATE EVOLUTION

Chordates are a branch of animals that include humans. In their adult form, they are a very diverse group including tunicates, lampreys, fish, amphibians, reptiles, birds, and mammals. Although a comparison of an adult tunicate and a human shows dramatic differences, a comparison of their developmental stages shows striking similarities and helps to categorize these organisms. Chordates share the following characteristics during development:

- The presence of a **notochord**, which is a piece of tissue that helps support the body
- The presence of a **dorsal nerve cord** that runs parallel to the notochord, one end becoming the brain
- The presence of **gill slits**
- The presence of a **post-anal tail**

Vertebrates are categorized within the chordate group. They differ from other chordates in that they have bone tissue that is used to form vertebrae that eventually replace the notochord in development. Other unique adaptations seen in vertebrates are the presence of jaws, fins, gills, and lungs. The major groups of vertebrates are described in the following table.

Major Groups of Vertebrates		
Group	**Major Characteristics**	**Examples**
Cartilaginous fish	Skeleton made of cartilage Gill slits Teeth used to collect prey	Sharks, skates, and rays
Bony fish	Skeleton made of bone Most are classified as ray-finned fish with flexible fin supports Some are lobe-finned fish with skeletal support in their ventral fins Lungfishes have gills and small pouches on the gut wall that take in oxygen; they must surface for air	Sturgeons, eels, herrings, grouper, bass, tuna, seahorses, salmon, and barracudas
Amphibians	Tetrapods—have four limbs Lungs and a three-chambered heart allow for survival out of water Skin must be kept moist to allow for gas exchange across the skin Most lay eggs in water	Frogs, toads, salamanders, and newts
Reptiles	Cold-blooded—the environment regulates body temperature Four-chambered heart Internal fertilization Females lay amniotic eggs, usually on land	Turtles, lizards, snakes, crocodiles, alligators
Birds	Warm blooded—body temperature is regulated internally Have feathers Internal fertilization Amniotic eggs Potential for flight	Hummingbird, ostrich, sparrow, eagle
Mammals	Hair or fur Young are born after a period of gestation Females provide milk for the young Parental care for young Some species are monotremes (lay eggs), marsupials (young complete development in a maternal pouch), and placental	Three groups: monotremes—spiny anteaters, duck-billed platypus; marsupials—kangaroos, koalas, Tasmanian devils; placental—rabbits, horses, seals, elephants, manatee, goats, whales, dolphins, bats, and primates (including humans)

THE ORIGIN OF LIFE

The origin of life encompasses an enormous number of detailed events. This section briefly summarizes the major events.

It is assumed that life began approximately 4 billion years ago when the solar system formed. The **Big Bang theory** is used to explain how the solar system might have come about. This theory supposes that a large, hot mass of material in the galaxy broke apart to evenly distribute matter and energy throughout the universe. This distribution of material caused a major cooling of temperature. Nuclear fusion created many of the major elements that over time collided and condensed into stars that provided light and heat. As stars died, heavier elements were released, eventually producing the sun.

The primitive atmosphere consisted of a variety of organic and inorganic substances such as ammonia, methane, water, carbon monoxide, carbon dioxide, nitrogen gas, and hydrogen gas. What was lacking was oxygen gas. Water vapors formed rain, which collected and contained mineral runoff from rocks. A source of energy within these bodies of water, perhaps from lightening or sunlight, fueled the reactions needed to create organic molecules such as sugars, amino acids, nucleotides, and fatty acids. As these monomers collected in clay, they may have polymerized to form molecules such as proteins, carbohydrates, and nucleic acids.

Some proteins produced within clay may have had enzymatic properties allowing them to interact with other molecules. The formation of a membrane around these primitive enzymes produced the first **protocell** that had the ability to self-replicate. An association of the protocell with RNA may have produced the first actual cell that resembled modern prokaryotes. These cells lacked organelles and lived in an anaerobic environment. Mutations that occurred in some populations of the first cells provided **photoautotrophic** abilities. The development of photosynthesis had an important impact in that it produced oxygen gas, thus creating an aerobic environment where aerobic cellular respiration evolved as the dominant energy-producing process.

Eukaryotic cells evolved from the prokaryotic-like cells. The theory of **endosymbiosis** explains the presence of certain organelles within eukaryotic cells. The engulfment of bacterial cells led to mitochondria and the engulfment of photosynthesizing prokaryotes led to the development of chloroplasts seen in modern day plants. **Membrane infolding** explains the presence of organelles such as the nucleus and endoplasmic reticulum in eukaryotic cells. Internalizing some of the cell membrane provided greater surface area for reactions to occur and thus this adaptation was selected for. These first eukaryotic cells resembled **protists** in structure.

Over time, symbiotic relationships occurred between eukaryotic cells that lead to specialization and the division of labor needed to eventually produce multicellular organisms. All modern plants, animals, and fungi evolved from these primitive eukaryotic cells.

CRAM SESSION

Evolution

1. Evolution means change. Individuals do not evolve; populations do.
2. Microevolution deals with changes within a population, whereas macroevolution is concerned with changes in species over longer periods of time.
3. Natural selection is the primary mechanism for evolution. The following are key features of natural selection:
   a. There is always genetic variation within a population.
   b. The inheritance of certain alleles enables some individuals to better compete for resources within the population.
   c. This leads to differential reproductive success where some individuals reproduce more than others.
   d. Those that reproduce more pass their alleles to the next generation, thus changing the allelic frequencies in the population. Those that reproduce more and pass on their alleles are considered the most fit in that particular environment. Fitness is always relative.
4. Natural selection can favor or select for certain phenotypes. Directional selection selects for an extreme phenotype, stabilizing selection favors an intermediate phenotype, and disruptive selection favors both extreme phenotypes simultaneously. Artificial selection occurs when mating is nonrandom, thus altering the normal process of natural selection.
5. The Hardy–Weinberg equation is used to calculate allelic frequencies in a population. The two equations are: $p + q = 1$ and $p^2 + 2pq + q^2 = 1$. The equation is violated by the following conditions: small populations, nonrandom mating, gene flow, mutations, and genetic drift (bottleneck and founder effects). Most populations violate the assumptions of the Hardy–Weinberg equation.
6. Speciation can be sympatric or allopatric. There are prezygotic and postzygotic methods in place to prevent interbreeding between species.
7. Chordates are a group of animals that includes the vertebrates. All members of this group share common embryonic features.
8. Evidence for evolution occurs in the form of the fossil record, biogeography, comparative anatomy, and comparative embryology.
9. The origin of life is explained by several key theories including the Big Bang theory, endosymbiotic theory, and membrane infolding theory.

CHAPTER 8

Bacteria and Fungi

Read This Chapter to Learn About:

- ➤ Classification of Bacteria
- ➤ Bacterial Structure and Shapes
- ➤ Bacterial Reproduction
- ➤ Basic Fungal Structure
- ➤ Life Cycles and Reproduction of Fungi
- ➤ Classification of Fungi

Living organisms can be classified into three domains: **Eukarya**, **Bacteria**, and **Archaea**. All eukaryotic organisms classify in the domain Eukarya, leaving all prokaryotic cells to be classified as either Bacteria or Archaea. Although both of these domains share the characteristics of being single-celled, absorbing their nutrients, having a single loop of DNA, and lacking organelles, there are some differences between the two groups. Archaea used to be mistakenly classified as bacteria, but their molecular and cellular structures were found to be quite different. They have a unique cell wall, ribosomes, and membrane lipids. Archaea live in diverse environments, and some species are termed **extremophiles** due to their extreme habitats such as polar ice caps, thermal vents, jet fuel, and others. Most Archaea species are **anaerobes**; none are known to be **pathogens**.

This chapter focuses on bacteria, single-celled organisms with medical and environmental significance, and on a group of eukaryotes, the fungi, which may be unicellular or multicellular.

Bacteria

Bacteria are extremely diverse. They may be classified according to the way they obtain nutrients from the environment or by their oxygen requirements. The following are the basic bacterial classifications:

- **Photoautotrophs:** These are bacterial species that produce their own nutrients through the process of photosynthesis using carbon dioxide from the environment.
- **Photoheterotrophs:** These are bacteria that perform photosynthesis but cannot use carbon dioxide from the environment. They extract carbon from a variety of other sources.
- **Chemoautotrophs:** These species get their energy from inorganic compounds and their carbon needs are obtained from carbon dioxide.
- **Chemoheterotrophs:** These bacteria obtain energy from inorganic substances and carbon from a variety of sources excluding carbon dioxide. These species are further subdivided based on the source of carbon they use. Some species can extract carbon through parasitic or symbiotic interactions with a host or through the decomposition of other organisms.

Bacteria can also be classified based on their oxygen requirements, or lack thereof, for cellular respiration.

- **Obligate aerobes** always require oxygen for aerobic cellular respiration.
- **Obligate anaerobes** never need oxygen and generally do not divide and, in some cases, are killed by exposure to oxygen.
- **Facultative anaerobes** sometimes use oxygen and sometimes do not require oxygen for cellular respiration.

STRUCTURE OF BACTERIA

The structure of bacterial cells is less complex than that of eukaryotic cells due to a lack of membrane-bound organelles, but bacterial and eukaryotic cells do share some characteristics in their cytoplasm and ribosomes. Figure 8-1 shows the basic structure of a bacterium, and the following table shows the major structures present in bacterial cells as well as their functions.

Major Bacterial Structures	
Structure	**Function**
Cell membrane	Outer boundary of the cell
Cytoplasm	Liquid portion of the cell where chemical reactions occur
Ribosomes	Site of protein synthesis
Chromosome	Contains the genes needed to produce proteins required for the cell. The bacterial chromosome consists of a single loop of DNA located in the nucleoid region of the cell.

Major Bacterial Structures (*Cont.*)	
Structure	**Function**
Plasmids	Small additional loops of DNA called plasmids found in some bacterial cells. The plasmids often contain genes to code for resistance to an antibiotic.
Cell wall	Most bacteria have a cell wall that contains peptidoglycan. The cell wall is found on the outer surface of the cell membrane and typically occurs in one of two conformations, which can be determined using the Gram-staining method. Some bacteria are Gram-positive, with their cell wall consisting of a thick layer of peptidoglycan, whereas other bacteria are Gram-negative, having one layer of peptidoglycan and another layer of lipids.
Capsule and slime layers	A layer of sugars and proteins on the outer surface of some bacterial cells. It forms a sticky layer that can help the cell attach to surfaces.
Flagella	Bacteria may have a single flagellum, multiple flagella, or no flagella. Those with one or more flagella are motile, as the flagella rotate to propel the cell. The bacterial flagella are different from eukaryotic flagella in structure. Bacteria flagella consist of the protein flagellin in a hollow, helical conformation that anchors into the cell membrane. A proton pump in the membrane provides power to rotate each flagellum.
Pili	Tiny proteins that cover the surface of some types of bacterial cells. They assist the cell in attaching to surfaces.
Spores	A few species of bacteria are capable of creating spores when environmental conditions are not favorable. When bacteria exist in spore form, they are capable of surviving adverse conditions for many years. When conditions become favorable again, the spores germinate into the vegetative cell form again.

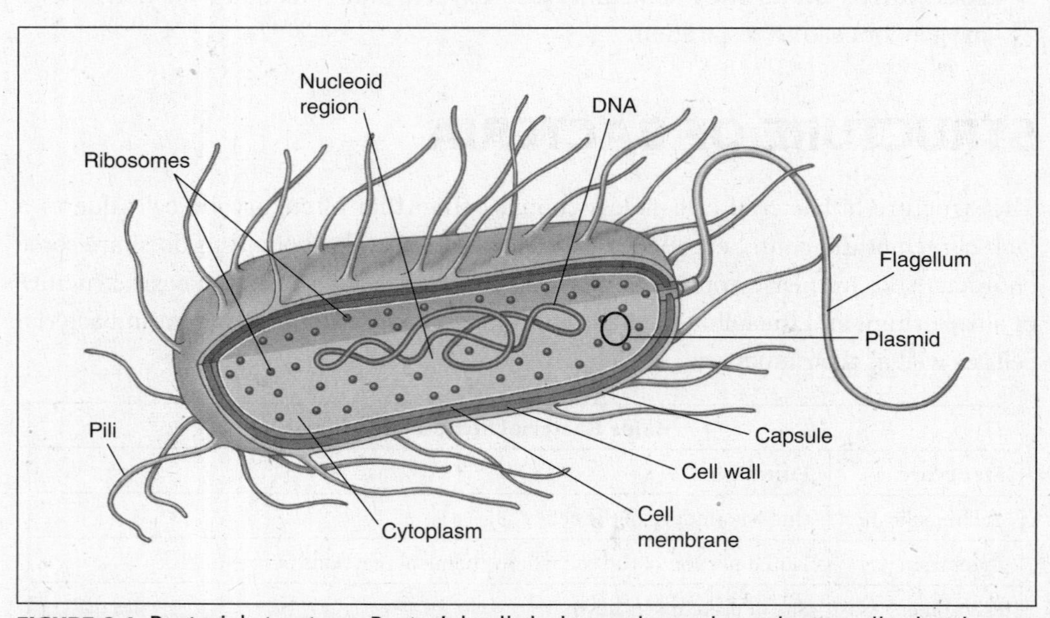

FIGURE 8-1: Bacterial structure. Bacterial cells lack membrane-bound organelles but have a variety of cell structures. SOURCE: From George B. Johnson, *The Living World,* 3d ed., McGraw-Hill, 2003; reproduced with permission of The McGraw-Hill Companies.

SHAPES OF BACTERIA

Most bacteria have shapes that correspond to one of three typical conformations. These shapes and organization among cells can be used as diagnostic features. The shapes exhibited by most bacteria are as follows and can also be seen in Figure 8-2.

- **Cocci** are circular in shape. They may exist singly, in pairs (diplococci), in clusters (staphylococci), or in chains (streptococci).
- **Bacilli** are rod- or oblong-shaped; they may occur in chains
- **Spirilli** have a spiral shape

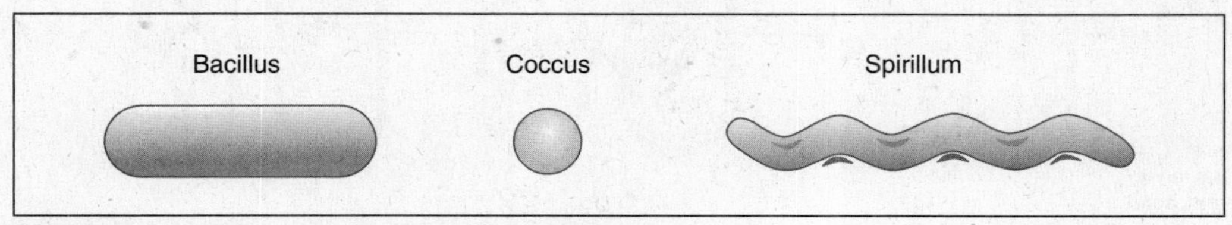

FIGURE 8-2: Bacteria come in a variety of shapes.

REPRODUCTION IN BACTERIA

Bacteria lack the structures needed to perform mitosis and because they have only a single chromosome, they really do not have a need for cell division as complex as mitosis. Bacteria do, however, have several ways of passing genetic material to other bacteria.

Binary Fission

Bacteria divide by the process of **binary fission**. It involves the replication of the single loop of DNA and the passing of a copy of the DNA to each of two daughter cells. This process can occur fairly quickly, in some cases as often as once every 20 minutes. Because bacteria are unicellular, creating a new cell means creating a new organism. This process is a type of asexual reproduction as each division produces genetically identical offspring. The only way to introduce variation into the population is by mutation, conjugation, or transformation.

Conjugation

Some bacteria have another means of passing genetic material to other bacteria. During the process of **conjugation**, illustrated in Figure 8-3, a single bacterial cell may copy its **plasmid** and pass it to another cell. The most commonly studied type of plasmid to be passed is called the F plasmid, or F factor (the F standing for **fertility**). In order to pass the plasmid to another cell, a physical connection must be established. This connection is referred to as the **sex pilus**, and is made by the cell that contains the plasmid (termed the male, or F^+). The sex pilus connects to a cell lacking the plasmid (the female, or F^-) and serves as a bridge to pass a copy of the plasmid to the female. Once complete, both cells are male and contain the plasmid.

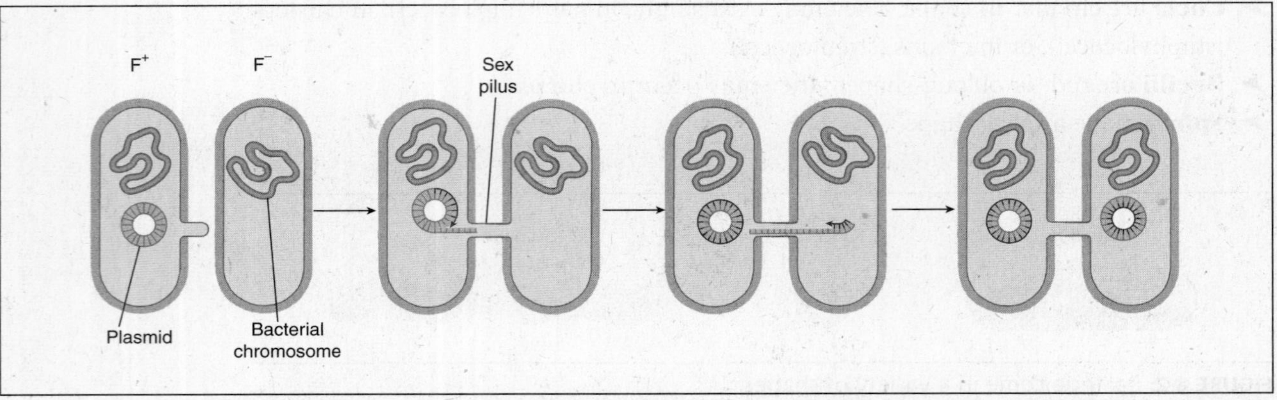

FIGURE 8-3: During conjugation, a bacterial cell containing a plasmid forms a sex pilus, which allows for the transfer of a copy of the plasmid to another bacterial cell that is lacking the plasmid. SOURCE: From George B. Johnson, *The Living World*, 3d ed., McGraw-Hill, 2003; reproduced with permission of The McGraw-Hill Companies.

Conjugation provides a rapid mechanism to pass plasmids within a population. Occasionally, plasmids become integrated into the chromosome, and when the plasmid is transferred via conjugation, some of the bacterial chromosome may be transferred as well.

Because many plasmids encode for resistance to things such as antibiotics, rapid conjugation can quickly render an entire bacterial population resistant to a particular antibiotic under the right selective pressures. This has important medical significance in that an antibiotic used to kill infection-causing bacteria becomes useless if the bacteria are resistant to that antibiotic. Some bacteria are resistant to multiple antibiotics as a result of picking up several plasmids via conjugation.

Transformation

Another way that bacteria can pick up genetic variations is through **transformation**. Some bacteria are able to pick up DNA from their environment and incorporate it into their own chromosomal DNA. Bacteria that are able to pick up foreign DNA are termed **competent**. Although some bacteria are naturally competent, others can be coerced to develop competence by artificial means.

GENE EXPRESSION AND REGULATION

Just like other organisms, bacteria go through the processes of transcription and translation. Gene expression in bacteria is regulated primarily by **operons**, seen in Figure 8-4, that control the access of RNA polymerase to the genes to be transcribed. They do this primarily via **repressor proteins**. There are many different operons that have been described, but they all have some basic features:

- A **promoter** sequence on the DNA where RNA polymerase must bind. If the promoter is inaccessible, the gene will not be transcribed.
- An **operator** sequence on the DNA where a **repressor protein**, if present, can bind. When a repressor is bound to the operator, the promoter sequence is blocked, and RNA polymerase cannot access the site.
- A **regulator gene** that produces a repressor protein when expressed.
- **Structural genes** that are the actual genes being regulated by the operon.

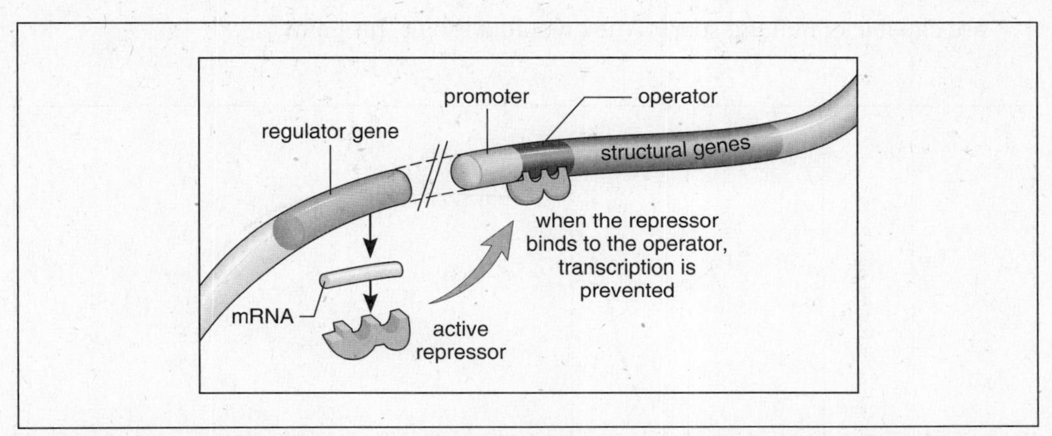

FIGURE 8-4: The operon system is used to regulate gene expression in prokaryotic cells. SOURCE: From Sylvia S. Mader, *Biology*, 8th ed., McGraw-Hill, 2004; reproduced with permission of The McGraw-Hill Companies.

Operons come in two basic categories: inducible and repressible. **Inducible operons** are normally "off," whereas **repressible operons** are normally "on." In an inducible operon, the repressor always binds to the operator so that transcription is always prevented unless an inducer molecule is present. When the inducer is present, it binds to the repressor, preventing the repressor from binding to the operator. This allows transcription to occur. In repressible operon systems, the repressor is always inactive so that transcription always occurs. Only when a co-repressor is present to interact with the repressor can transcription be inhibited. When the repressor and co-repressor are bound, they can then interact with the operator site and prevent access by RNA polymerase, thus turning off transcription.

GROWTH CYCLE

Bacteria follow a typical growth cycle shown in Figure 8-5 that is limited by environmental factors as well as the amount of nutrients available. The following stages of the growth cycle occur:

- **Lag:** There is an initial lag in growth that occurs when a new population of bacteria begins to reproduce. This lag time is normally brief.
- **Logarithmic growth:** As bacteria begin to perform binary fission at a very rapid rate, logarithmic growth occurs. This can last only for a limited amount of time.
- **Stationary:** As the number of bacteria increases, resources decrease; and while some bacteria are still dividing, some are dying. This evens out the population count.
- **Decline (death):** As the population hits its maximum, the lack of nutrients along with a variety of wastes means that the population will begin to decline and more cells die than are being replaced by cell division. For the few species of bacteria that are capable of making spores, they would do so at this point.

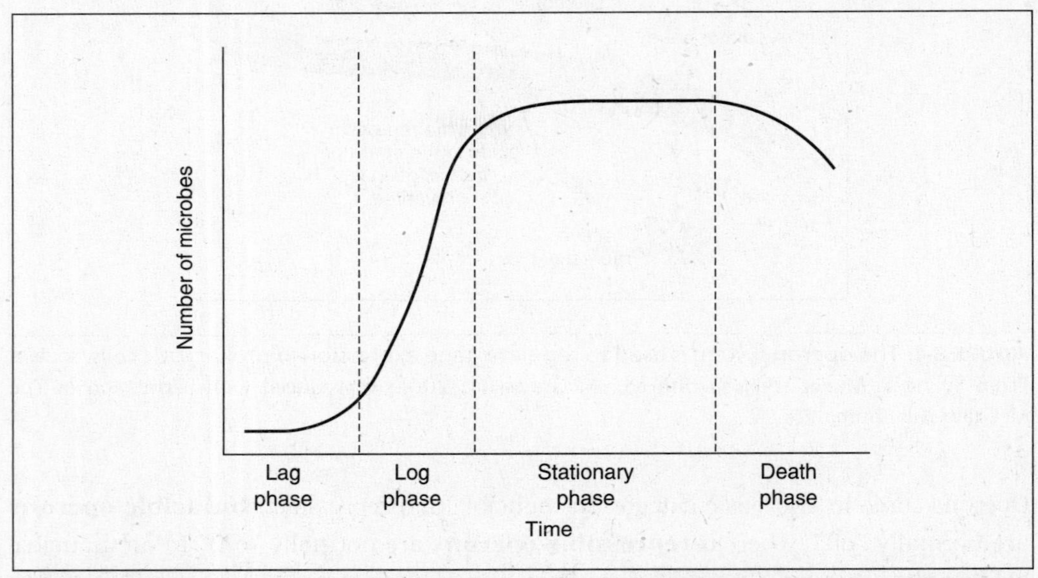

FIGURE 8-5: A growth curve for bacteria. Bacteria exhibit logarithmic growth until conditions are no longer ideal. At this point, a stationary phase occurs followed by a decline phase in which bacterial numbers decrease.

FUNGI

Fungi constitute a diverse number of species within the Eukarya domain. Some fungi, such as yeasts, are unicellular, whereas others, such as mushrooms and molds, are multicellular. Many are harmless, but some are pathogenic. All are heterotrophs, gaining their nutrients from other organisms. They secrete enzymes that break down organic molecules to a size small enough to be absorbed through the cell membrane. They may do this by feeding on dead and decaying organisms or by parasitic relationships with living organisms.

BASIC STRUCTURE OF A FUNGUS

Although some fungi such as yeasts are unicellular, most are multicellular and fairly complex, as seen in Figure 8-6. Within a multicellular fungus, the **mycelium** is the structure that grows near food sources in order to obtain nutrients for the fungus. Within the mycelium, there are **hyphae** filaments where the nucleus of each cell is located. Other structures such as those needed for reproduction are present. Their structure differs depending on the species of fungus, and in fact, fungi are classified according to their reproductive structures and mechanisms of reproduction.

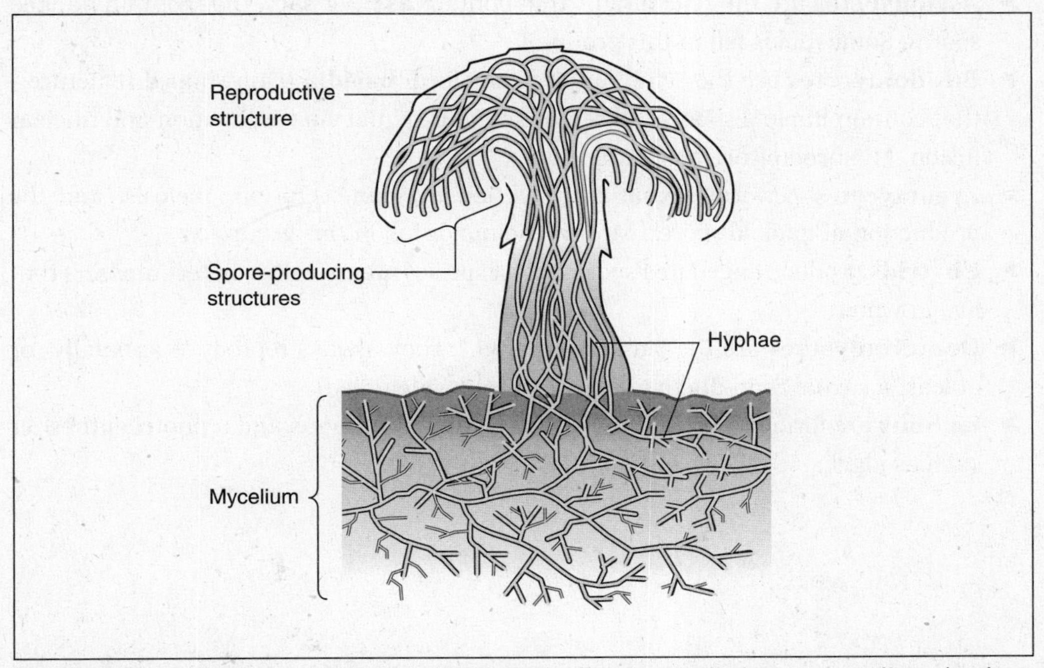

FIGURE 8-6: Typical fungal structure. Multicellular fungi typically have a mycelium, hyphae, and reproductive structures.

LIFE CYCLES AND REPRODUCTION OF FUNGI

Depending on the species, fungi are able to reproduce either sexually or asexually, or sometimes by both methods. During asexual reproduction, **spores** are formed in specialized structures in the fungus and perform mitosis to generate offspring. The structure of spores used for fungal reproduction is very different from the spores used as survival structures in certain species of bacteria. In some cases, spores are not used at all and the cells fragment to form new cells in the process of budding. This is common in unicellular fungi.

Sexual reproduction is a less common means of reproduction in fungi and often occurs only when environmental conditions are poor. During sexual reproduction, gametes are made by specialized structures in the fungus. The two gametes fuse, leaving a diploid cell that performs meiosis and produces haploid spores.

CLASSIFICATION OF FUNGI

Fungi are classified according to their method of reproduction. The following are the major groups of fungi:

- **Yeasts** are single-celled organisms that reproduce by budding.
- **Ascomycetes** are the "sac fungi," that contain **asci**, or sacs, that contain haploid spores. Some molds fall in this group.
- **Basidomycetes** are the "club fungi." They form **basidia** (club-shaped structures) that contain haploid spores. All reproduction is sexual via conjugation and nuclear fusion. Mushrooms are an example.
- **Zygomycetes** perform sexual reproduction by gamete fusion, meiosis, and the production of haploid spores. Most bread molds fall in this group.
- **Chytrids** produce flagellated spores. Most species are parasites or decomposers that live in water.
- **Deuteromycetes** are the "imperfect fungi." They always reproduce asexually (or at least a sexual reproductive phase cannot be identified).
- **Lichens** are formed from an interaction between a fungus and a photosynthesizer such as algae.

CRAM SESSION

Bacteria and Fungi

1. Organisms in the domains Bacteria and Archaea consist of prokaryotic cells. Archaea have unique habitats and unusual cell structures that differentiate them from bacteria.
2. Bacteria are classified based on how they obtain nutrients and energy. Phototrophs perform photosynthesis, whereas chemotrophs do not. Autotrophs use carbon dioxide as a carbon source, whereas heterotrophs receive carbon from a source other than carbon dioxide.
3. Bacteria contain a variety of structures that assist in their ability to attach to surfaces and survive adverse condition. See the table entitled "Major Bacterial Structures" to review bacterial structures.
4. Bacteria come in the following typical shapes: cocci, bacilli, and sprilli.
5. Reproduction in bacteria is by binary fission. Mutation, conjugation, and transformation serve as mechanisms to introduce variation into bacterial populations.
   a. Conjugation involves the transmission of a copy of a plasmid from one cell to another.
   b. Transformation occurs when bacteria take up DNA from their environment and incorporate it into their genome.
6. Bacterial gene expression is controlled by operon systems.
   a. The genes of repressible operons are always expressed unless a repressor and co-repressor block the operator sequence.
   b. The genes of inducible operons are never expressed unless an inducer removes the repressor from the operator sequence.
7. Bacterial population growth is limited by environmental factors.
8. Fungi can be unicellular or multicellular and always consist of eukaryotic cells. Classification is based on methods of reproduction and reproductive structures such as spores.
9. Fungal reproduction can involve asexual and sexual reproductive cycles. Certain fungi have haploid and diploid phases of the life cycle.

CHAPTER 9

Viruses

Read This Chapter to Learn About:

- ➤ Viral Life Cycles
- ➤ Animal Viruses
- ➤ Retroviruses
- ➤ Bacteriophages
- ➤ Transduction

Viruses are a unique biological entity in that they do not resemble typical cells. There is debate over whether viruses are living organisms at all because they are unable to reproduce without a host cell; nor can they perform many of the activities associated with living organisms without the help of a host. Because viruses lack typical cell structures, they are much smaller than any form of prokaryotic or eukaryotic cells.

Although some viruses contain more sophisticated structures, the only items required for a virus are a piece of genetic material, either DNA or RNA, and a protective protein coat for the genetic material. The viral **genome** (a collection of all the genes present) can consist of only a few genes upward to a few hundred genes. Viruses are categorized as animal viruses, plant viruses, or bacteriophages. They can also be classified according to which nucleic acid they contain.

LIFE CYCLE OF A VIRUS

Viruses are specific to the type of host cell that they infect. In order for a cell to be infected, it must have a receptor for the virus. If the receptor is absent, the cell cannot be infected by the virus. Although it seems odd that cells would evolve receptors for viruses, it is usually a case of mistaken identity. The viruses actually mimic another substance for which the cell has a legitimate need and thus has a receptor present. The process of viral infection is seen in Figure 9-1.

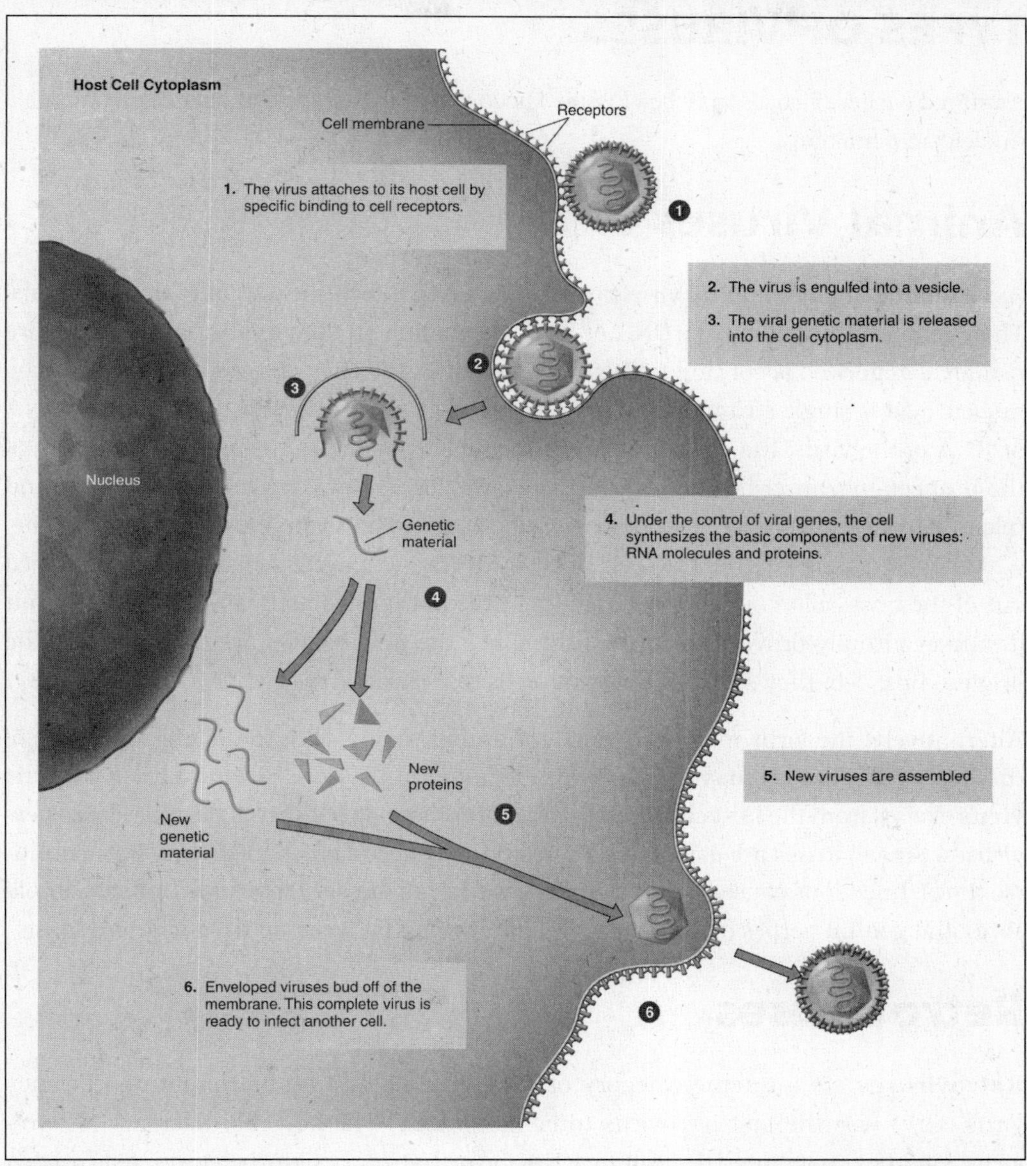

FIGURE 9-1: Viral infection. SOURCE: From Marjorie Kelly Cowan and Kathleen Park Talaro, *Microbiology: A Systems Approach*, McGraw-Hill, 2004; reproduced with permission of the The McGraw-Hill Companies.

Once a virus binds to a receptor on the membrane of the host cell, the viral genetic material enters the host either by injecting itself across the cell membrane or by being taken in via endocytosis. At some point, the viral genes are transcribed and translated by the host cell. The nucleic acid of the virus is also replicated. Eventually new viruses are produced and released from the host cell. Because each virus contains a copy of the original genetic material, they should all be genetically identical. Mutations are the primary way to induce variation into the viral population.

TYPES OF VIRUSES

As stated earlier, viruses may be classified according to their typical host cell or by their nucleic acid makeup.

Animal Viruses

As the name implies, animal viruses are designed to infect the cells of various animals. Their genetic material may be DNA or RNA, depending on the virus. Animal viruses are usually categorized according to the type of nucleic acid they possess, and whether the nucleic acid is single stranded or double stranded. Once the host cell takes in the DNA or RNA of the virus, the virus may immediately become active using the machinery of the host cell to transcribe and translate the viral genes. New viruses are assembled and released from the host cell by one of two methods: **lysis** of the host's cell membrane, which immediately kills the host cell, or by **budding**, where the new viruses are shipped out of the host cell via exocytosis. Budding does not immediately kill the host cell, but it may eventually prove fatal to the host. Once the new viruses are released from the original host cell, they seek out new host cells to infect.

Alternatively, the virus may become latent, integrating itself into the chromosomes of the host cell, where it may stay for variable amounts of time. Eventually, the **latent virus** excises from the host chromosome and becomes active to produce and release new viruses. Some viruses are capable of alternating between active and latent forms multiple times. Infections caused by specific viruses such as the herpes viruses that cause cold sores and genital herpes are notorious for alternating between active and latent forms.

Retroviruses

Retroviruses are a unique category of RNA viruses. The human immunodeficiency virus (HIV) was the first retrovirus to be discovered. The key characteristic of retroviruses is that they enter the cell in RNA form that must then be converted to DNA form. This is the opposite of the normal flow of information in the cell, which dictates that DNA produces RNA during transcription. The process of converting viral RNA backward into DNA is called **reverse transcription**, and is achieved by an enzyme called **reverse transcriptase**.

The viral genome codes for reverse transcriptase. When the retrovirus enters the host cell, its RNA is immediately transcribed, and one result of this is the production of reverse transcriptase. The reverse transcriptase then produces a DNA copy of the viral genome. In the case of HIV, the DNA then integrates into the host cell's chromosomes (the host cell is a specific cell type in the immune system) and enters a latent phase that may last more than 10 years. When the viral DNA excises from the host chromosome, it becomes active and begins producing new viruses. When this happens on a mass scale, the death of host cells signals the beginning of deterioration in the immune system, which causes acquired immunodeficiency syndrome (AIDS).

Bacteriophages

Bacteriophages are DNA viruses that infect bacteria exclusively. They always inject their DNA into the host bacterial cell and then enter either a lytic cycle or a lysogenic cycle, as seen in Figure 9-2. Viruses can also transport parts of the bacteria's genetic makeup to other cells.

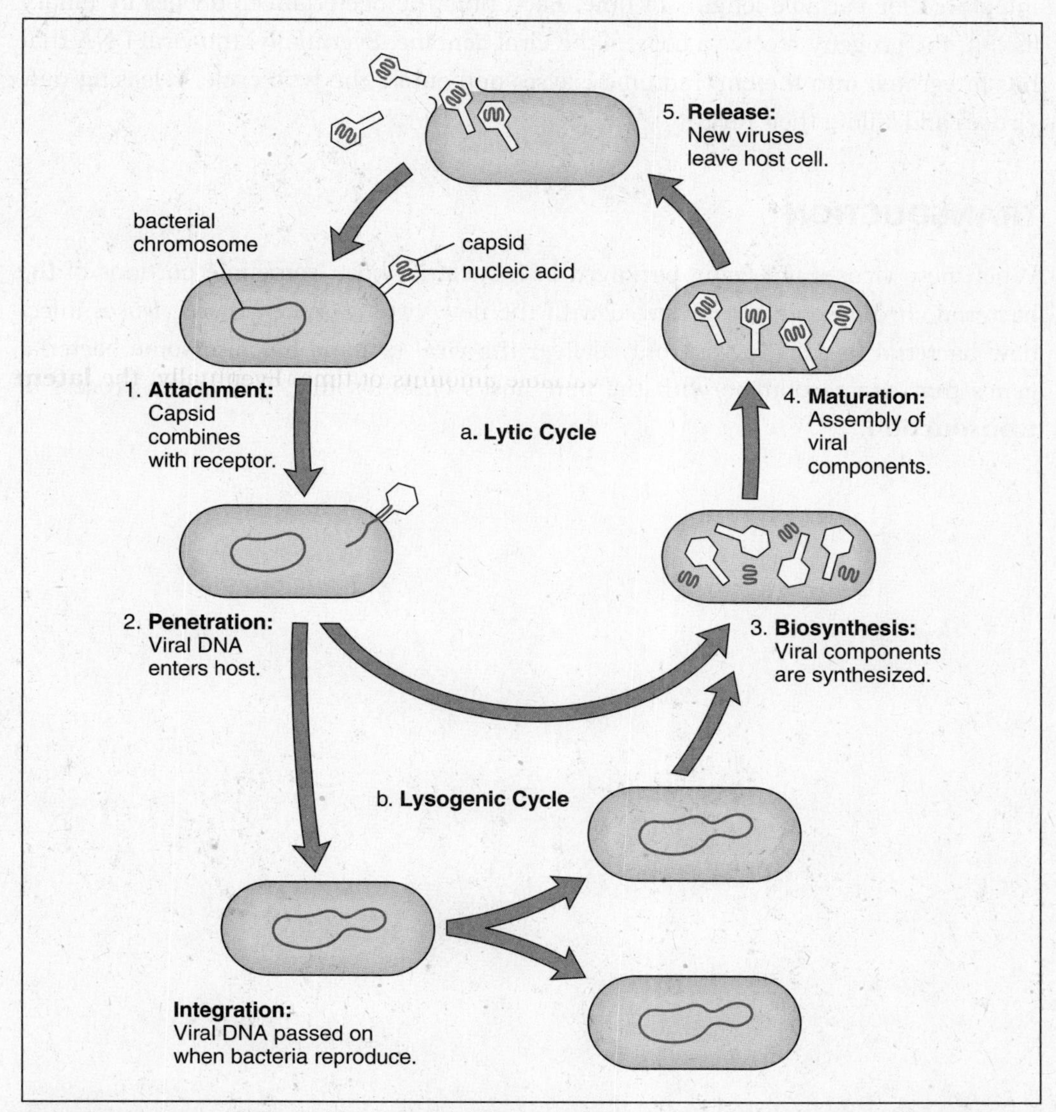

FIGURE 9-2: Bacteriophage life cycles. (a) In the lytic cycle, viral particles are released when the cell is lysed. (b) In the lysogenic cycle, viral DNA integrates into the host cell chromosome. The lysogenic cycle can be followed by the lytic cycle. SOURCE: From Sylvia S. Mader, *Biology*, 8th ed., McGraw-Hill, 2004; reproduced with permission of The McGraw-Hill Companies.

THE LYTIC CYCLE

In the **lytic cycle**, a bacteriophage immediately activates once it is inside its host. New viruses are synthesized and leave the host cell via lysis, always killing their bacterial host. The new viruses then go out and infect new host cells.

THE LYSOGENIC CYCLE

The **lysogenic cycle** is a variation that some viruses use. After injecting DNA into their host, the viral DNA integrates into the bacterial chromosome. The viral DNA may stay integrated for variable lengths of time. Each time the bacterial cell divides by binary fission, the progeny receive a copy of the viral genome. Eventually, the viral DNA that has integrated into the chromosome excises and enters the lytic cycle, releasing new viruses and killing their host.

TRANSDUCTION

When new viruses are being packaged in a bacterial host, sometime portions of the bacterial chromosome get packaged with the new viruses. When these viruses infect new bacterial hosts, they not only deliver the viral genome but also some bacterial genes that can recombine with the new host's chromosome. This is the process of **transduction**.

CRAM SESSION

Viruses

1. Viruses can exhibit complex structures, but many are simple. There are only two required materials for a virus: a piece of genetic material (either DNA or RNA) and a protective protein coat.
2. Viruses are specific in that they can infect only host cells that have a receptor for that virus.
3. Animal viruses may take over the host cell immediately after they infect the cell, or they may become latent, integrating into the chromosomes of the host.
   a. Active viruses coerce the host cell to produce new copies of the virus that are released by host cell lysis or budding.
   b. Latent viruses eventually excise from the chromosomes and enter the active cycle.
4. Retroviruses are a unique class of animal viruses—they are latent viruses that use RNA as their genetic material.
   a. The viral genome codes for the production of reverse transcriptase, which allows the virus to convert itself to DNA form, enabling it to insert itself into the host cell chromosomes.
   b. HIV is an example of a retrovirus.
5. Bacteriophages are viruses that infect bacteria. They can enter the lytic or lysogenic cycles.
   a. Viruses in the lytic cycle immediately take over their host, produce new viruses, and kill the host as the cell is lysed to release new viruses.
   b. Viruses that enter the lysogenic cycle insert themselves into the bacterial chromosome and are transmitted to all progeny of the bacteria during binary fission. The bacteriophage eventually excises and enters the lytic cycle.
6. Transduction occurs when a virus in the lysogenic cycle excises from the bacterial chromosome, taking a portion with it. The new viruses that are packaged contain viral DNA and bacterial DNA that can be transferred to new host cells.

CHAPTER 10

Tissues and Skin

Read This Chapter to Learn About:

- ➤ Epithelial Tissue
- ➤ Connective Tissue
- ➤ Muscular Tissue
- ➤ Nervous Tissue
- ➤ Membranes
- ➤ The Integumentary System

TISSUES AND MEMBRANES

A **tissue** is a group of similar type cells that perform specialized functions. Multiple tissue types associate with each other to form **organs**. Organs are organized into body systems that have specific functions in the body—for example, digestion or reproduction.

TYPES OF TISSUES

There are four major types of tissue found in animals: epithelial, connective, muscular, and nervous.

Epithelial Tissue

Epithelial tissues are generally found on surfaces of any part of the body in contact with the environment. The cells that form epithelial tissue occur in sheets or layers where one cell is directly connected to the next. A **basement membrane** of sticky polysaccharides and proteins made by the cells attaches the tissue to other underlying tissues.

Epithelial tissues are named according to the shape of their cells as well as the number of layers that compose the tissue. The cells that compose epithelial tissues come in the following shapes, seen in Figure 10-1:

- **Squamous**: flat cells
- **Cuboidal**: cube-shaped cells
- **Columnar**: oblong-shaped cells

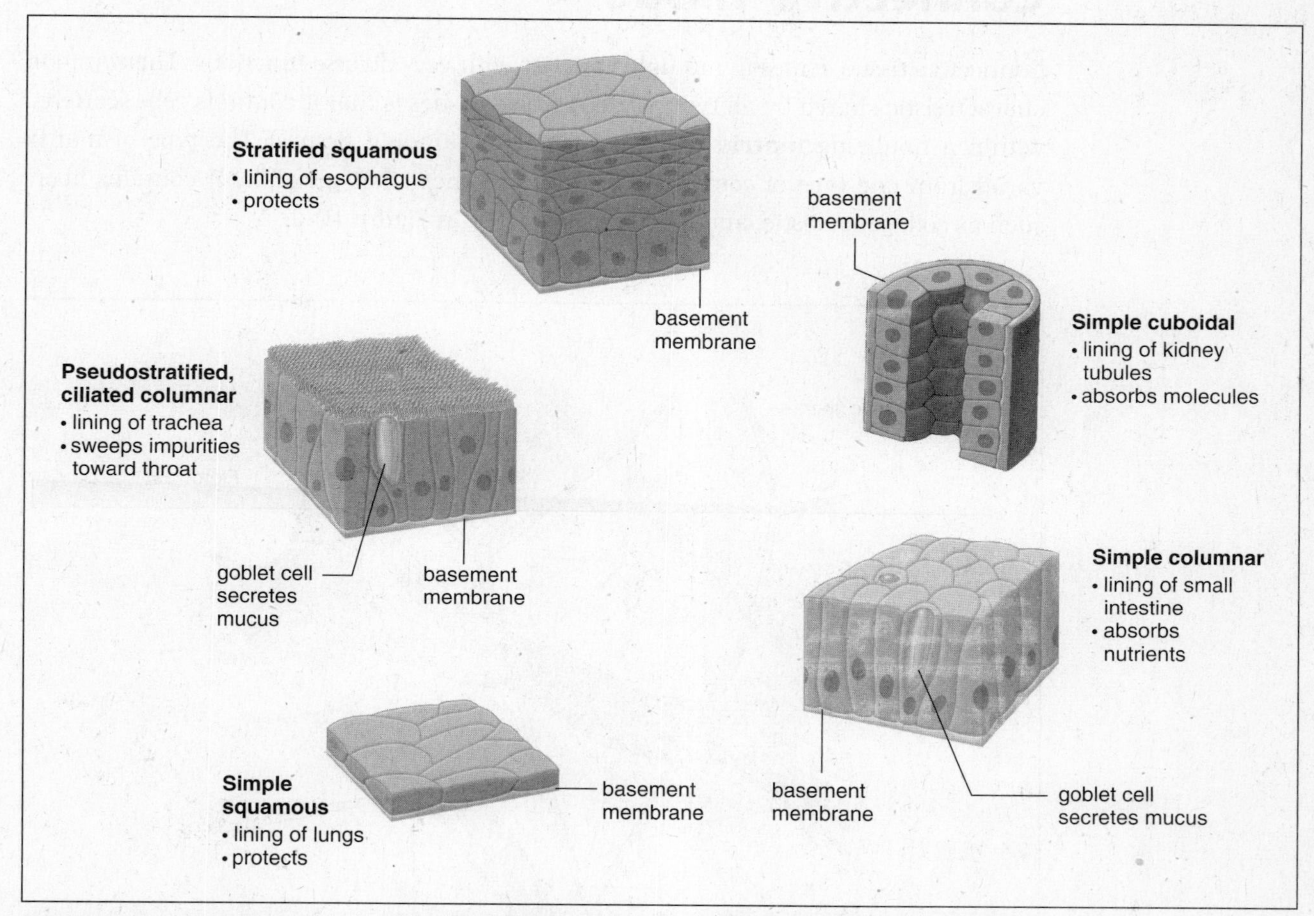

FIGURE 10-1: Epithelial tissues are classified based on the shape of the cells that compose the tissue and the number of layers found within the tissue. SOURCE: From Sylvia S. Mader, *Biology*, 8th ed., McGraw-Hill, 2004; reproduced with permission of The McGraw-Hill Companies.

If there is a single layer of cells, the tissue is termed **simple**. If multiple layers of cells are present, the tissue is termed **stratified**. Epithelial tissue always has two names—one to indicate the cell shape and one to indicate the number of layers present.

There are multiple types of epithelial tissues that all have specific functions. Generally, cuboidal and columnar tissues are well suited to secreting products such as mucus or digestive enzymes and are usually found in simple form. They are also used for absorption in areas such as the digestive tract. Simple squamous epithelium is well suited to diffusion in places such as the alveoli (air sacs) of the lungs because it is so thin. Stratified squamous is used for structures such as skin, where entire layers of the tissue might be lost on a regular basis. Simple forms of epithelial cells that line closed spaces in the body, including body cavities, blood vessels, and lymphatic vessels are termed **endothelium**. Most epithelial cells are replaced as often as they are shed.

Connective Tissue

Connective tissue comes in multiple varieties with very diverse functions. The common characteristic shared by all types of connective tissues is that it contains cells scattered within a nonliving **matrix** separated from surrounding tissues. The type of matrix varies from one type of connective tissue to the next, but generally it contains fibers such as collagen, elastic, and reticular fibers seen in Figure 10-2.

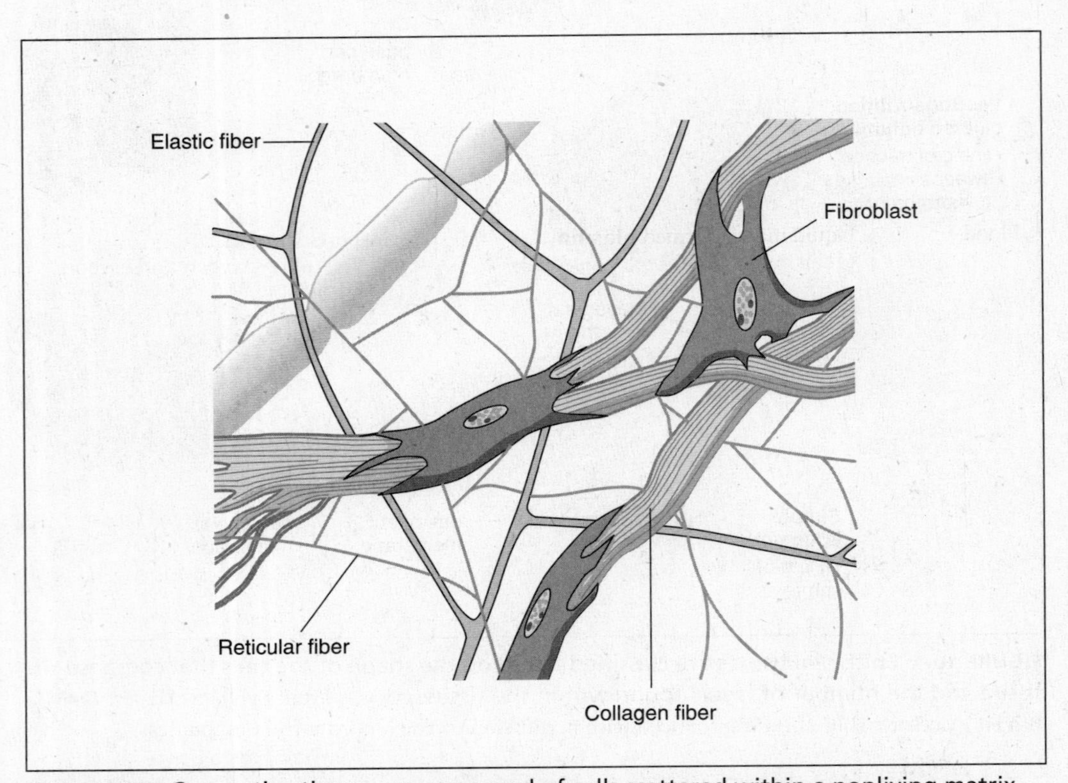

FIGURE 10-2: Connective tissues are composed of cells scattered within a nonliving matrix.

Fibroblasts are the primary cell type that produces matrix fibers. **Collagen fibers** have great strength, whereas **elastic fibers** have stretchability. **Reticular fibers** help attach one type of connective tissue to another type.

There are several types of connective tissue, including loose tissue, dense tissue, cartilage, bone, blood, and lymph. Each type of connective tissue is characterized by its specific cell type as well as by the properties of the matrix. The characteristics of the major types of connective tissues and their functions are given in the following table.

Major Connective Tissue Types		
Tissue Type	**Characteristics**	**Functions**
Loose	Collagen and elastic fibers are abundant in the matrix to form a fairly loose consistency Fibroblasts are the main cell types Adipose tissue is a specialized form that stores fat for energy reserves and insulating properties	Fills space in body cavities, attaches skin to underlying tissues, stores fat
Dense	Collagen fibers are abundant and tightly packed to provide tensile strength Fibroblasts are the main cell type	Tendons (attach muscles to bones) and ligaments (attach bone to bone)
Cartilage	Collagen fibers are embedded in a gel-like matrix Chondrocytes are the major cell type	Support of body structures such as ears, nose, trachea, and vertebrae
Bone	Collagen fibers are abundant within a rigid calcium phosphate matrix Osteocytes, osteoblasts, and osteoclasts are the major cell types	Structural support for the body, protection of internal organs, storage of calcium
Blood	Liquid matrix termed **plasma** Major cell types are red blood cells, white blood cells, and platelets	Transports substances within the body, including oxygen and carbon dioxide; fights infection; blood clotting
Lymph	Liquid matrix Major cell type is white blood cells	Fights infection, transports substances within the body, regulates fluid levels in other tissues

Muscular Tissue

There are three types of muscle found within the body: smooth, cardiac, and skeletal. All are composed of bundles of muscle cells that have the ability to contract. Each contains many mitochondria that produce the ATP required for contraction. The appearance of the muscle as well as its relationship with the nervous system determines the tissue type. The three muscle types seen in Figure 10-3 are as follows:

- **Smooth muscle** is under involuntary control by the nervous system. This means that it contracts with no conscious effort. Its appearance is smooth, hence the name. Smooth muscle is found throughout the digestive tract, urinary tract, and reproductive system. It is also found in veins, where it helps blood move toward the heart.
- **Cardiac muscle** is found only in the heart, and is also under involuntary control. It has a **striated** (striped) appearance when viewed under a microscope. Any damage to the cardiac muscle, such as a heart attack, can have major, if not fatal, consequences.

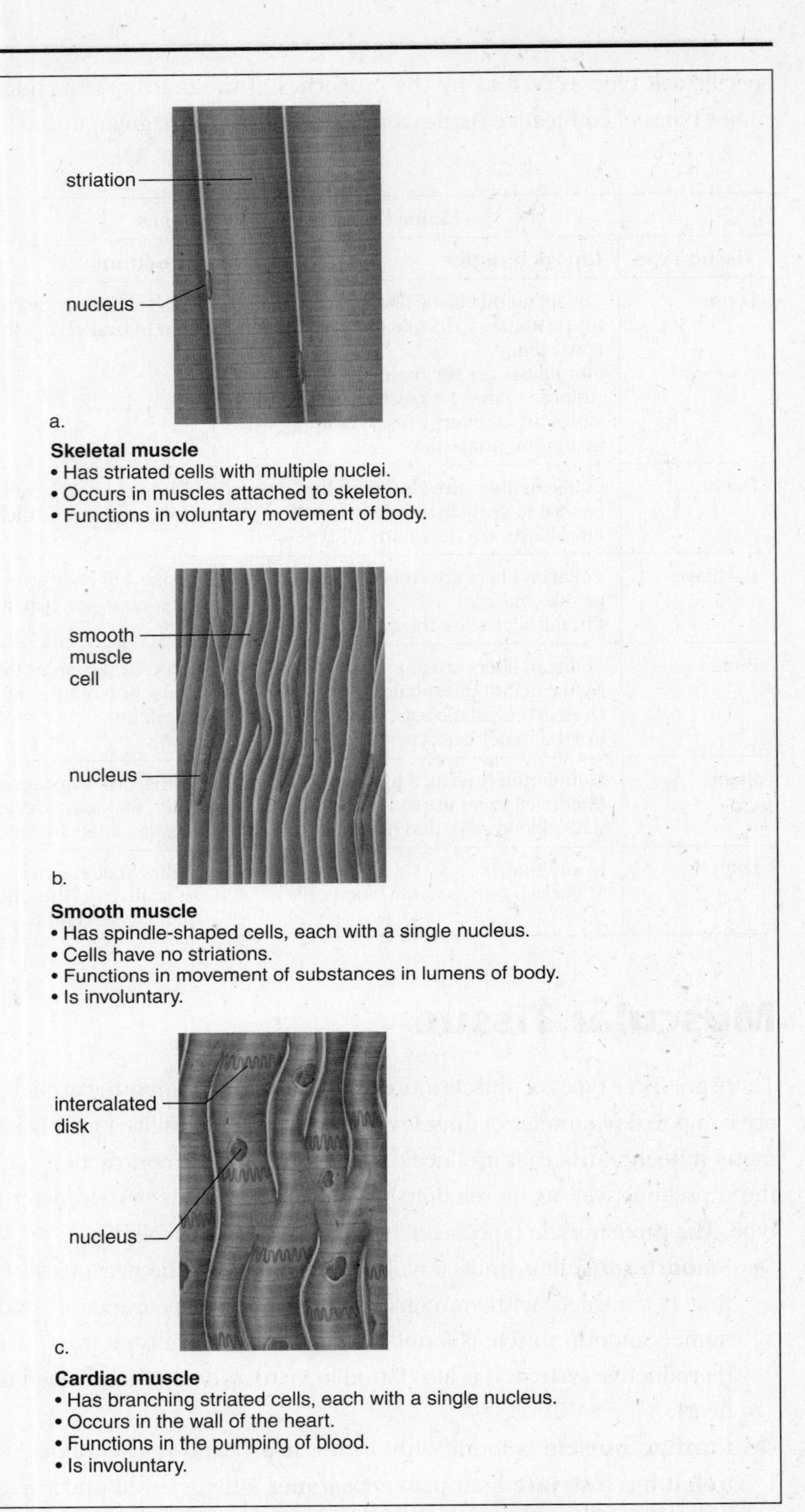

FIGURE 10-3: Types of muscular tissue. (*a*) Skeletal muscle is striated and is under voluntary control. (*b*) Smooth muscle is nonstriated and is under involuntary control. (*c*) Cardiac muscle is striated and is under involuntary control. SOURCE: From Sylvia S. Mader, *Biology,* 8th ed., McGraw-Hill, 2004; reproduced with permission of The McGraw-Hill Companies.

- **Skeletal muscle** is under voluntary or conscious control. It has a striated appearance. Skeletal muscles are usually connected to bones and are used for movement.

Nervous Tissue

Nervous tissue is composed of two primary cell types: neurons and glial cells. **Neurons** have the ability to communicate with each other as well as with other cells in the body, whereas **glial cells** provide supporting functions to neurons. Mature neurons are unable to perform mitosis to replace themselves, which is one reason that neurologic injuries and illnesses are so serious.

All neurons have a cell body that contains the nucleus and most of the cell's organelles. Projections reaching out from the **cell body** are **dendrites** and **axons**. Dendrites pick up messages and send them to the cell body. The cell body then processes the message and sends electric impulses out through a long projection called the **axon**. In some cases, axons may be more than a meter in length. The axon terminates in **synaptic knobs**, which are extensions of the axon. The synaptic knobs can send messages to the dendrites of other neurons via chemicals known as **neurotrans-mitters**. The space between a synaptic knob of one neuron and the dendrite of another neuron is the **synapse**. The structure of a generalized neuron can be seen in Figure 10-4.

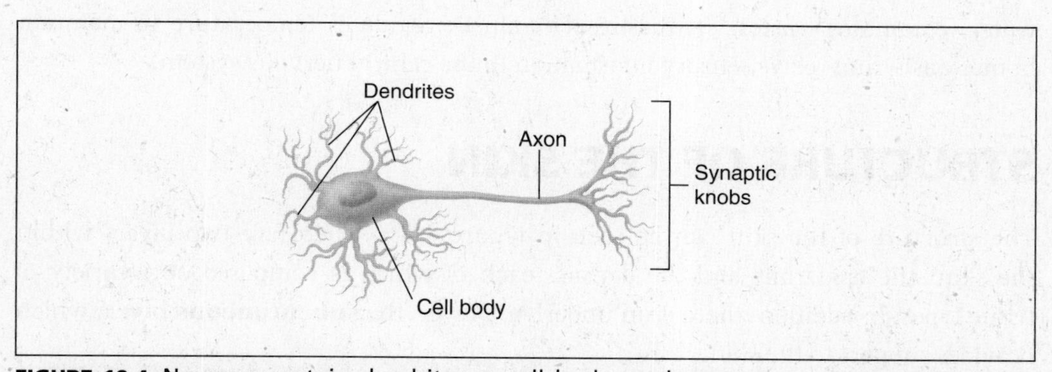

FIGURE 10-4: Neurons contain dendrites, a cell body, and an axon. SOURCE: From George B. Johnson, *The Living World*, 3d ed., McGraw-Hill, 2003; reproduced with permission of The McGraw-Hill Companies.

Many neurons have **Schwann cells** wrapped around their axons. These cells produce a hydrophobic lipoprotein called **myelin**, which forms a sheath to insulate the axon and help impulses propagate along the axon at the fastest possible rate. The myelin sheath contains **nodes of Ranvier**, which are gaps in the sheath. As impulses are sent down the length of the axon, they jump from node to node.

MEMBRANES

Membranes within the body are composed of an association of two tissue types. There are three major types of membranes: serous, mucous, and cutaneous.

- **Serous membranes** are made of epithelial and connective tissues and line internal body cavities and cover organs.
- **Mucous membranes** are also composed of epithelial and connective tissues. Their job is to secrete mucus onto surfaces of the body that are in contact with the outside environment, such as the mouth, nose, trachea, and digestive tract.
- The **cutaneous membrane** covers the outer surface of the body and is also known as the **skin**. The upper layers of the skin are composed of epithelial and connective tissues.

THE INTEGUMENTARY SYSTEM

The **integumentary system**, or skin, is composed of a variety of specialized tissue types to meet its diverse functions. The skin has also been described as a membrane, but because it is made of two or more tissue types, it also qualifies as an organ.

FUNCTIONS OF THE SKIN

The skin serves as a protective barricade against abrasion and infection. It also conserves water, eliminates wastes, synthesizes vitamin D, regulates temperature to maintain homeostasis, and relays sensory information to the central nervous system.

STRUCTURE OF THE SKIN

The structure of the skin can be seen in Figure 10-5. There are two layers within the skin: the epidermis and the dermis, each of which is composed of a variety of tissue types. In addition, there is an underlying layer, the **subcutaneous layer**, which provides support to the skin.

The Epidermis

The **epidermis** is a thin layer found on the surface of the body. It is composed of five layers of stratified squamous epithelial tissue, which is connected to the dermis via a basement membrane. Cells in the bottom (basal) layer of the epidermis constantly divide by mitosis to provide enough skin cells to replace those lost due to shedding. Because the epidermis contains only epithelial tissues, there are no blood vessels present to provide oxygen and nutrients. These items must be provided from the dermis via diffusion.

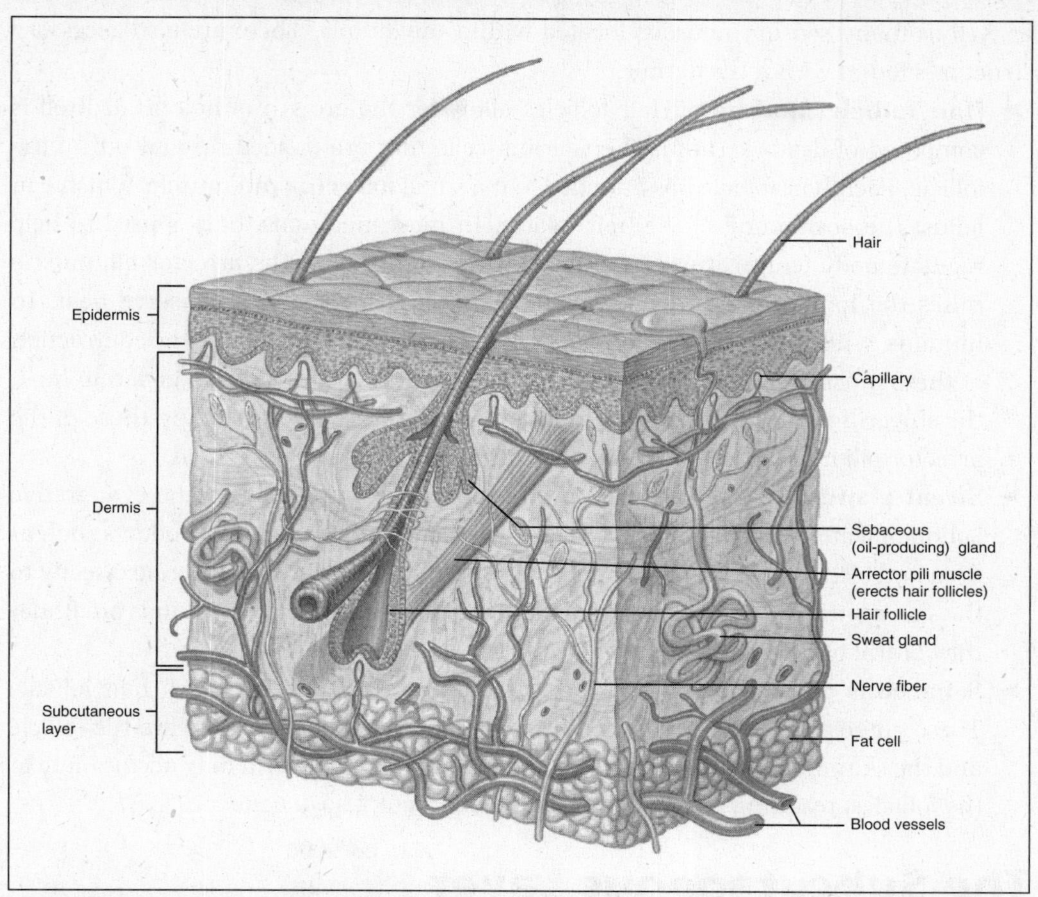

FIGURE 10-5: The skin is composed of a variety of tissue types and contains accessory structures in the form of glands, follicles, and hair. SOURCE: From George B. Johnson, *The Living World,* 3d ed., McGraw-Hill, 2003; reproduced with permission of The McGraw-Hill Companies.

As cells move upward in the epidermis, they die due to lack of oxygen and nutrients. They are also filled with the waterproofing protein called **keratin**. The cells near the top of the epidermis are dead, keratinized cells that are relatively impenetrable to water. **Nails** are a specialized structure containing highly keratinized cells growing from a **nail bed**. They are used for protection at the ends of the extremities.

In the bottom layers of the epidermis, **melanocytes** are present. These cells make the protein **melanin**, which protects the cells from UV damage and provides the skin with pigmentation. Those with more melanin, and thus darker skin, have better protection against UV damage to the skin.

The Dermis

The **dermis** is much thicker than the epidermis. Its primary function is to support the epidermis and anchor it to deeper tissues. The main tissue type is relatively dense connective tissue containing collagen and elastic fibers. There is a good blood supply as well as many sensory neurons located within the dermis. There are also accessory structures found within the dermis.

- **Hair follicles and hair:** Hair follicles allow for the growth of hair. Hair itself is composed of dead, keratinized epidermal cells that are pushed up and out of the follicle. Each hair follicle has attached to it a small **arrector pili** muscle, which can adjust the positioning of the hair follicle. In most mammals, hair is used to help regulate body temperature. In animals with a lot of hair, the arrector pili muscle raises the hair to help provide insulation around the body to conserve heat. In humans, who lack significant amounts of hair on most of their body, the contraction of these muscles causes goosebumps. Because muscle contractions generate heat, the shivering that occurs due to rapid muscle contractions, including those of the arrector pili muscles, is a way to generate heat when the body is cold.
- **Sweat glands:** Sweat glands are used for cooling of the body by the evaporative action of water on the skin's surface and for the excretion of waste products such as urea and electrolytes in small concentrations. Some sweat glands secrete directly to the skin's surface, whereas others secrete into hair follicles. Sweat glands are under the control of the nervous system.
- **Sebaceous glands:** A sebaceous (oil) gland is associated with each hair follicle. These glands secrete a fluid called **sebum** into the follicles to lubricate the follicle and the skin. In individuals who excrete excess sebum, bacteria may accumulate in the follicles, resulting in the inflammation characteristic of acne.

The Subcutaneous Layer

The **subcutaneous layer** is also known as the **hypodermis**. It supports the skin and contains a large proportion of loose connective tissue as well as an excellent blood supply. The subcutaneous layer anchors the skin to tissues and muscles deeper within the body. It also serves as an insulator that helps with **thermoregulation** in the body due to the presence of **adipose tissue**, also known as **subcutaneous fat**.

Because blood is warmer than body temperature, the patterns of circulation in the surface capillaries of the dermis and subcutaneous layer can be used to conserve or release heat as needed. During **vasoconstriction** blood vessels constrict, keeping blood and heat near the body's core. During **vasodilation** the blood vessels dilate, allowing some of the heat from the blood to escape through the surface of the skin.

CRAM SESSION

Tissues and Skin

1. A **tissue** is a group of specialized cells performing similar functions in close proximity to one another.
2. Tissues come in four varieties: epithelial, connective, muscular, and nervous.
   a. Epithelial tissue is usually found on surfaces and is characterized by layers of closely packed cells in various shapes (squamous, cuboidal, and columnar). Simple epithelium has one layer, stratified has more than one layer.
   b. Connective tissue is characterized by cells scattered in a nonliving matrix. The type of cells and the qualities of the matrix differentiate the connective tissues. Typical fibers found in the matrix are collagen, elastic, and reticular. Some connective tissues have a typical matrix (loose), some have a tightly packed matrix (dense), some have a liquid matrix (blood and lymph), some have a calcified matrix (bone), and some have a gel-like matrix (cartilage).
   c. Muscle tissue comes in three varieties. Skeletal muscle is striated and under voluntary control. Cardiac muscle is striated and under involuntary control. Smooth muscle lacks striations and is under involuntary control.
   d. Nervous tissue consists of neurons and glial cells. Neurons are used for communication and cannot reproduce once mature. Glial cells support neurons.
3. The membranes of the body contain a combination of two types of tissue and are serous (lining internal cavities and covering organs), mucous (secreting mucus to surfaces), and cutaneous (the skin).
4. The primary functions of the skin are to serve as a barricade for deeper tissues, conserve water, eliminate wastes, regulate body temperature, relay sensory information to the nervous system, and synthesize vitamin D.
   a. The skin consists of the epidermis (stratified squamous epithelium) and the dermis (connective tissues) as well as the supportive tissues of the subcutaneous layer (connective tissues). Accessory structures include sweat glands, sebaceous glands, hair follicles, hair, and nails.

CHAPTER 11

The Nervous System and Senses

Read This Chapter to Learn About:

- ➤ Neuron Structure
- ➤ Communication Between Neurons
- ➤ The Central Nervous System
- ➤ Structure and Function of the Brain
- ➤ The Spinal Cord and Reflex Actions
- ➤ The Peripheral Nervous System
- ➤ The Senses

The nervous system has the daunting task of coordinating all the body's activities. The **central nervous system (CNS)** is composed of the brain and spinal cord, and the **peripheral nervous system (PNS)** is composed of all nervous tissue located outside of the brain and spinal cord. **Nerves** are the primary structures within the PNS. In order to understand the functioning of the CNS and PNS, it is necessary to look at the detailed function of **neurons**, the basic unit of function of the nervous system.

THE NEURON

Neurons perform the critical function of transmitting messages throughout the body. There are several types of neurons: sensory neurons, motor neurons, and interneurons.

- ➤ **Sensory (afferent) neurons** exist in the PNS, pick up sensory impulses, and direct their messages toward the CNS.
- ➤ **Motor (efferent) neurons** exist in the PNS and direct their messages from the CNS to the peripheral parts of the body.
- ➤ **Interneurons**, which transfer messages, are found only in the CNS.

There are also a large number of **glial cells** present in the nervous system. Glial cells provide support to neurons and, unlike mature neurons, are capable of mitosis.

Basic Structure of a Neuron

The basic structure of neurons was described in Chapter 10 and can be seen in Figure 11-1. To review and summarize, the major structures within the neuron are

- **Dendrites:** projections that pick up incoming messages
- **Cell body:** processes messages and contains the nucleus and other typical cell organelles
- **Axons:** projections that carry electrical messages down their length
- **Synaptic knobs:** extensions occurring at the ends of an axon that send electrical impulses converted to chemical messages in the form of neurotransmitters to other neurons

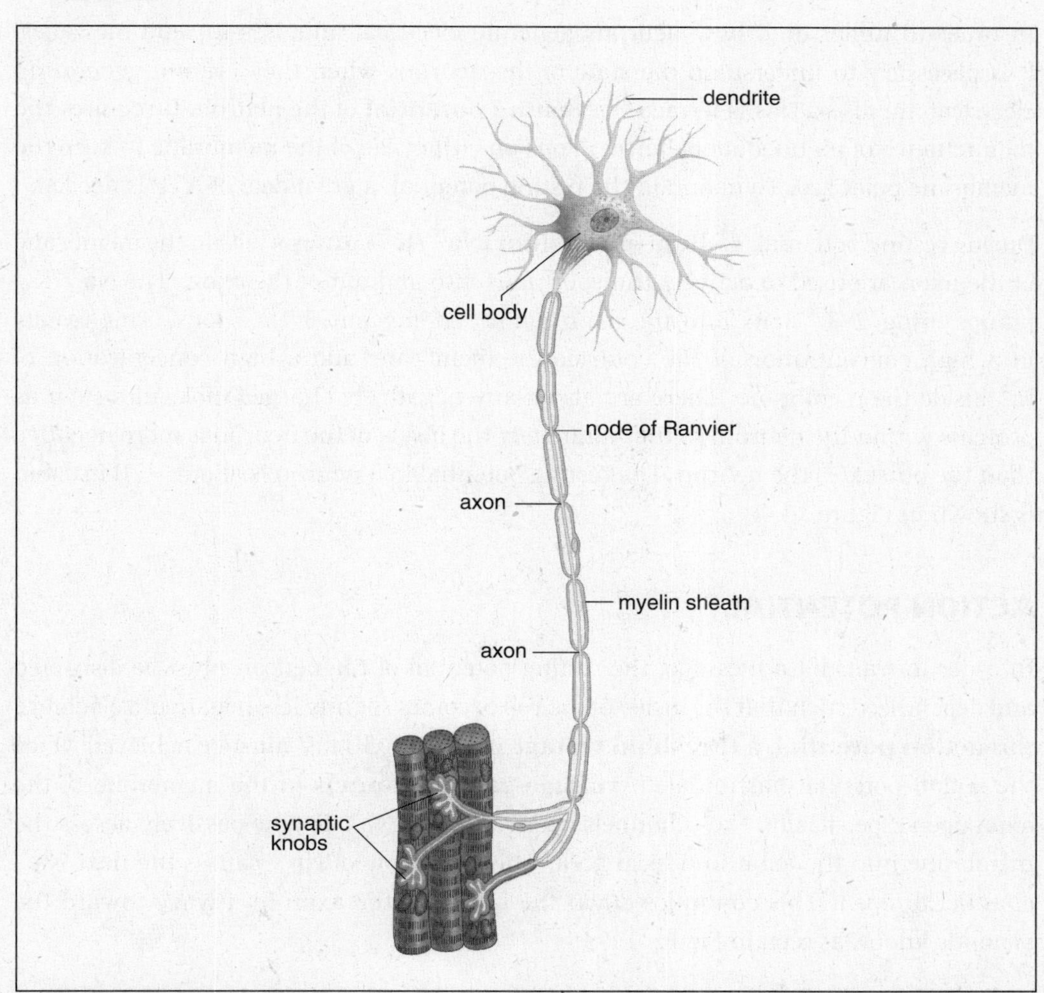

FIGURE 11-1: Neuron structure. A typical neuron consists of dendrites, a cell body, and an axon. Some neurons have myelinated axons. SOURCE: From Sylvia S. Mader, *Biology*, 8th ed., McGraw-Hill, 2004; reproduced with permission of The McGraw-Hill Companies.

- **Myelin sheath:** covering, produced by **Schwann cells** (specialized glial cells), that surrounds the axon of some neurons; gaps between the myelin are called **nodes of Ranvier**
- **Synapse:** the space between the synaptic knobs of one neuron and the dendrites of another neuron

Basic Function of a Neuron

Neurons send messages in the form of electrical impulses throughout the body. They do this through a complex series of processes involving changes from a resting potential to an action potential and communication via neurotransmitters.

RESTING POTENTIAL

In order to understand how neurons generate electrical impulses to send messages, it is necessary to understand the state of the neurons when they are not generating electrical impulses. This is termed the **resting potential** of the neuron. It requires the maintenance of an unequal balance of ions on either side of the membrane to keep the membrane polarized. To maintain the resting potential, a great deal of ATP is needed.

During resting potential, **sodium–potassium (Na^+/K^+) pumps** within the membrane of the axon are used to actively transport ions into and out of the axon. The Na^+/K^+ pumps bring 2 K^+ ions into the axon while sending out 3 Na^+ ions. This results in a high concentration of Na^+ outside the membrane and a high concentration of K^+ inside the membrane. There are also many negatively charged molecules such as proteins within the neuron, so that ultimately the inside of the neuron is more negative than the outside of the neuron. The resting potential in a neuron is about -70 mV and is shown in Figure 11-2.

ACTION POTENTIAL

In order to transmit a message, the resting potential of the neuron must be disrupted and depolarized such that the inside of the cell becomes slightly less negative. To achieve this **action potential**, a **threshold voltage** of about -50 mV must be achieved. Once the action potential has initiated, **voltage-gated channels** in the membrane of the axon open. Specifically, Na^+ channels open, allowing Na^+ to flow passively across the membrane into the axon in a local area. This local flow of Na^+ causes the next Na^+ channel to open. This continues down the length of the axon in a wave toward the synaptic knobs, as seen in Figure 11-3.

Although the speed of the action potential varies depending on the axon diameter and whether the axon is myelinated, the strength of the action potential cannot. Action potentials are an all-or-nothing event. If the threshold voltage is not hit, the action

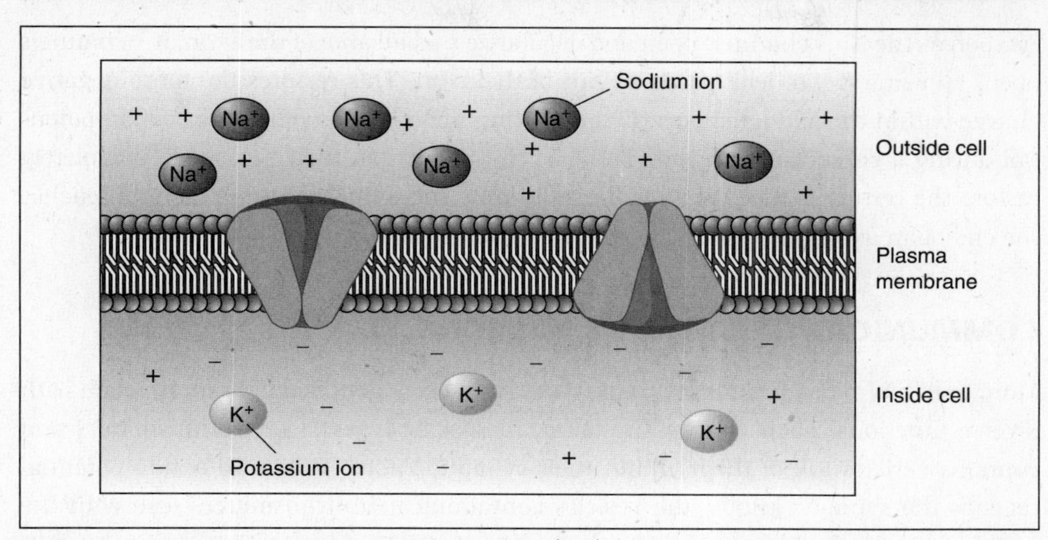

FIGURE 11-2: During a resting potential, the Na^+/K^+ pump is used to maintain an unequal balance of ions inside and outside the cell.

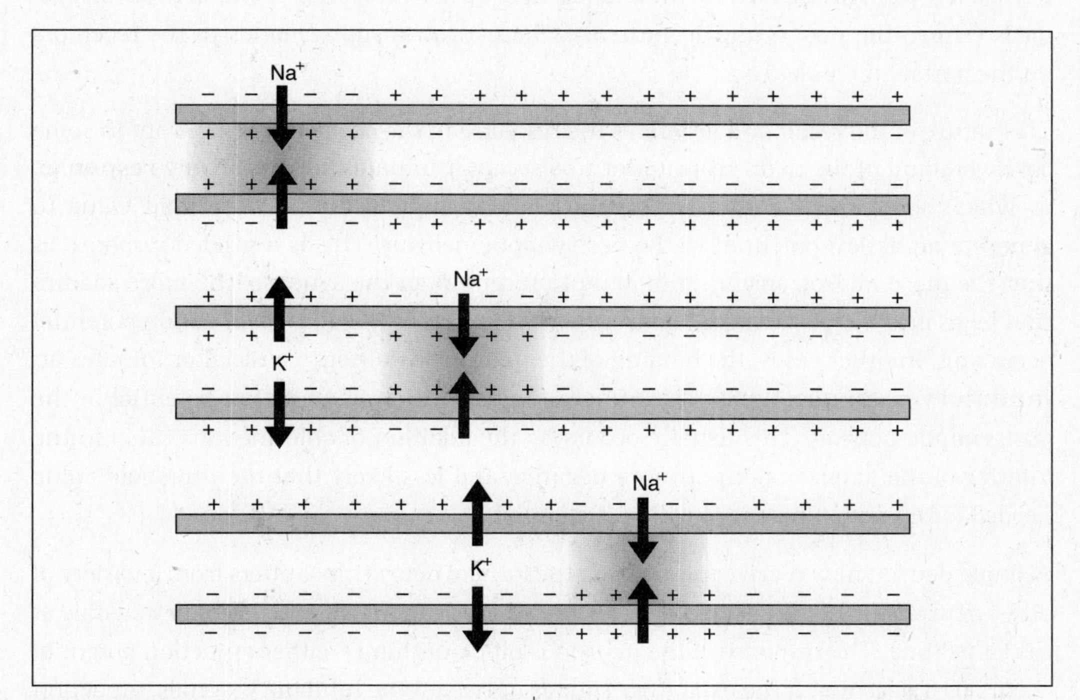

FIGURE 11-3: During the action potential, Na^+ enters the neuron, causing depolarization.

potential does not happen. If the threshold value is achieved or exceeded, the action potential occurs with the same electrical charge of about +35 mV each time.

In myelinated neurons, the voltage-gated ion channels are permeable only to ions at the nodes of Ranvier. This allows for the action potential to jump from one node to the next in the process of **salutatory conduction**.

As soon as the Na^+ channels open and depolarize a small area of the axon, K^+ channels open, allowing K^+ to leak passively out of the axon. This restores the more negative charge within the axon, temporarily preventing the initiation of another action potential during a **refractory period**. The Na^+/K^+ pump can then be used to completely restore the resting potential by repolarization. By the time an action potential reaches the end of an axon, the rest of the axon is already repolarized.

COMMUNICATION BETWEEN NEURONS

More than fifty types of **neurotransmitters** have been identified in humans, each with diverse functions. Each neuron specializes in specific types of neurotransmitters and contains vesicles full of them within their synaptic knobs. When an action potential reaches the synaptic knobs, the vesicles containing neurotransmitters fuse with the membrane by exocytosis and release their contents to the synapse. This neurotransmitter release requires calcium to occur. The neuron that releases the neurotransmitter is termed the **presynaptic neuron**, and the neuron that responds to the neurotransmitter is termed the **postsynaptic neuron**. The neurotransmitter binds to the receptors on the postsynaptic neurons.

The nature of the receptor determines the response in the postsynaptic neuron. In some cases, binding of the neurotransmitter to a receptor initiates an **excitatory response**, in which some sodium channels open in an attempt to hit the threshold value to generate an action potential in the postsynaptic neuron. This is a graded response in that the more neurotransmitter binding to receptors in the synapse, the more sodium that leaks into the postsynaptic neuron, which increases the odds of an action potential occurring. In other cases, the binding of the neurotransmitter to a receptor initiates an **inhibitory response**, which discourages the generation of an action potential in the postsynaptic neuron. This usually occurs by the addition of chlorine ions (Cl^-) to the interior of the axon, making it more negative and less likely that the threshold value needed for an action potential will be generated.

A single neuron may receive messages in the form of neurotransmitters from a variety of other neurons. In some cases, a neuron can receive excitatory and inhibitory signals at the same time. The response of the neuron is all-or-nothing—either an action potential occurs or it does not. If the excitatory signals outweigh the inhibitory signals, an action potential occurs. If the inhibitory signals outweigh the excitatory signals, an action potential does not occur.

Once a neurotransmitter has been released into the synapse and has interacted with a receptor, it must be cleared from the synapse to avoid sending repeated messages. Depending on the type of neurotransmitter, removal may be via **reuptake** where the neurotransmitters are taken back into the presynaptic neuron or by enzymatic degradation of the neurotransmitter in the synapse.

THE CENTRAL NERVOUS SYSTEM

The central nervous system (CNS) is composed of the brain and spinal cord. The **brain** and spinal cord both consist of many neurons and supporting glial cells. **White matter** within the brain and spinal cord consists of myelinated axons. **Gray matter** consists of clusters of cell bodies of neurons.

Cranial bones and vertebrae protect the CNS, as do protective membranes called the **meninges**. There are three meninges (i.e., dura mater, arachnoid, and pia mater). Between two of the meninges and within cavities of the brain, there is **cerebrospinal fluid**. This fluid has several critical functions: It provides nutrients and removes wastes as well as cushioning and supporting the brain. Cerebrospinal fluid is made by the brain and is eventually reabsorbed by the blood.

The **blood–brain barrier** is another mechanism of protection for the brain. Although the name implies that the blood–brain barrier is a structure, it is not. It is a mechanism that selects the components in the blood that are allowed to circulate into the brain via brain capillaries. This selection is based on a unique membrane permeability that allows for the easy passage of most lipid soluble molecules while preventing other molecules from entering the brain tissue. The benefits of the blood–brain barrier are that it protects the brain from substances in the blood that might cause damage and maintains a constant environment for the brain, which is not very tolerant of fluctuations.

Structure and Function of the Brain

The brain is the central command center of the nervous system. It processes conscious thought and sensory information, it coordinates motor activities of skeletal muscle and other organ systems within the body, and it maintains vital functions such as heart rate and ventilation.

The brain can be divided into the **cerebrum**, **cerebellum**, **brain stem, and diencephalon** (see Figure 11.4). The cerebrum in particular has extremely diverse functions. The right and left hemispheres process information in different ways. The right side of the brain tends to specialize in spatial and pattern perception, whereas the left side of the brain tends to specialize in analytical processing and language. The connection of the two hemispheres via the **corpus callosum** is essential to integrating the functions of both sides of the brain.

Integration of functions of the brain is also accomplished via the **limbic system**. Complex activities such as mood and emotions, as well as memory, cannot be achieved in a single area of the brain. Instead, several areas of the brain must interact. The limbic system, a tract of neurons that connects several areas of the brain, including the cerebrum, hypothalamus, medulla, and areas associated with the sense of smell,

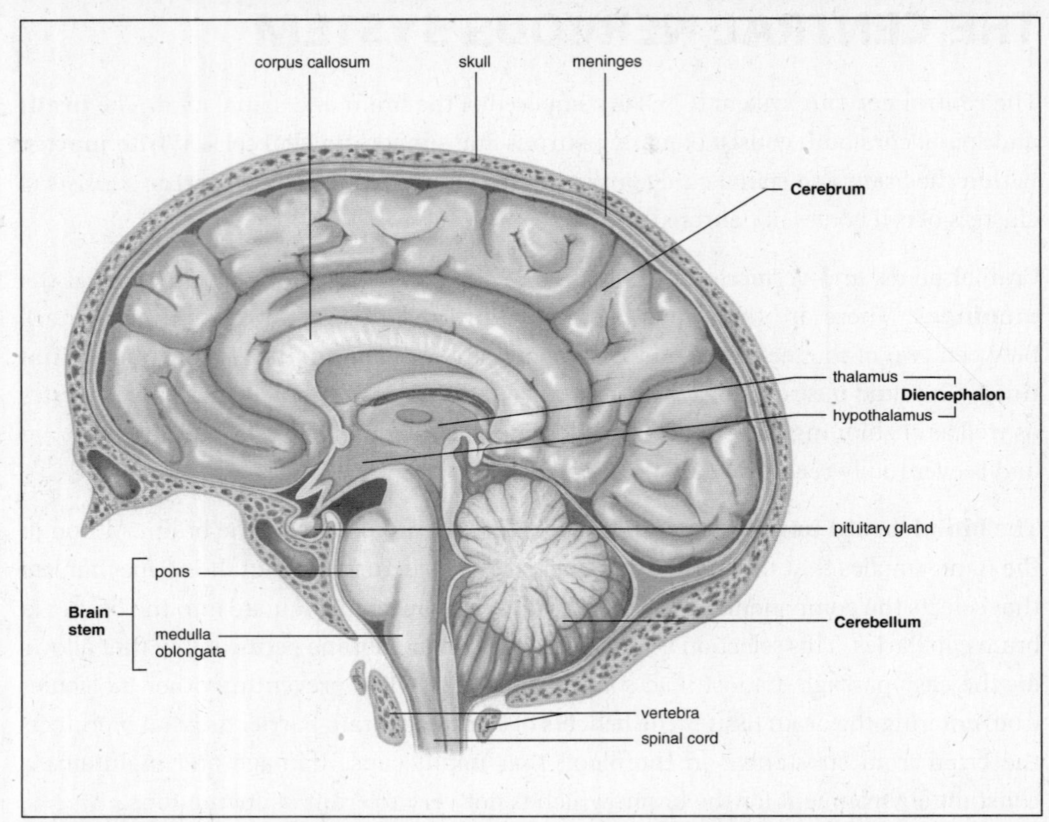

FIGURE 11-4: Brain structure. The cerebrum of the brain is divided into right and left hemispheres connected by the corpus callosum. SOURCE: From Sylvia S. Mader, *Biology*, 8th ed., McGraw-Hill, 2004; reproduced with permission of The McGraw-Hill Companies.

provides for this interaction. Within the limbic system, the **hippocampus** helps convert short-term memories to long-term memories.

The functions of each of the parts of the brain, including the cerebrum, are shown in the following table.

Major Structures of the Brain and Their Functions	
Structure	**Function**
Cerebrum	The cerebrum is the largest portion of the brain and is divided into right and left hemispheres as well as into four lobes (frontal, parietal, occipital, and temporal). Within the cerebrum there are specific areas for each of the senses, motor coordination, and association areas. All thought processes, memory, learning, and intelligence are regulated via the cerebrum. The cerebral cortex is the outer tissue of the cerebrum.
Cerebellum	The cerebellum is located at the base of the brain. It is responsible for sensory–motor coordination for complex muscle movement patterns and balance.

Major Structures of the Brain and Their Functions (*Cont.*)	
Structure	**Function**
Brain stem	The brain stem is composed of several structures and ultimately connects the brain to the spinal cord. ➤ The pons connects the spinal cord and cerebellum to the cerebrum and diencephalon. ➤ The medulla oblongata (or medulla) has reflex centers for vital functions such as the regulation of breathing, heart rate, and blood pressure. Damage to the medulla is usually fatal. Messages entering the brain from the spinal cord must pass through the medulla. ➤ The reticular activating system (RAS) is a tract of neurons that runs through the medulla into the cerebrum. It acts as a filter to prevent the processing of repetitive stimuli. The RAS is also an activating center for the cerebrum. When the RAS is not activated, sleep occurs.
Diencephelon	The diencephalon is composed of two different structures. ➤ The hypothalamus is used to regulate the activity of the pituitary gland in the endocrine system. In addition, the hypothalamus regulates conditions such as thirst, hunger, sex drive, and temperature. ➤ The thalamus is located adjacent to the hypothalamus and serves as a relay center for sensory information entering the cerebrum. It routes incoming information to the appropriate parts of the cerebrum.

The Spinal Cord and Reflex Actions

The spinal cord serves as a shuttle for messages going toward and away from the brain. It also acts as a reflex center, having the ability to process certain incoming messages and provide an autonomic response without processing by the brain. Spinal reflexes are important because they are faster than sending a message to the brain for processing.

The **reflex arc** seen in Figure 11-5 is a set of neurons that consists of a receptor, a sensory neuron, an interneuron, a motor neuron, and an effector. It involves both the CNS and the PNS. The receptor transmits a message to a sensory neuron, which routes the message to an interneuron located in the spinal cord. The interneuron processes the message in the cord and sends a response out through the motor neuron. The motor neuron passes the message to an effector, which can carry out the appropriate response.

THE PERIPHERAL NERVOUS SYSTEM

The peripheral nervous system (PNS) is composed of pairs of nerves that are bundles of axons. There are 12 pairs of **cranial nerves** branching off the brain stem and 31 pairs of **spinal nerves** branching off the spinal cord. Some nerves are composed of only sensory neurons, others of only motor neurons, and others of a combination of sensory and motor neurons. The nerves that exist in the PNS are categorized into one of two divisions: the somatic nervous system or the autonomic nervous system.

The **somatic nervous system** controls conscious functions within the body, such as sensory perception and voluntary movement due to innervation of skeletal muscle.

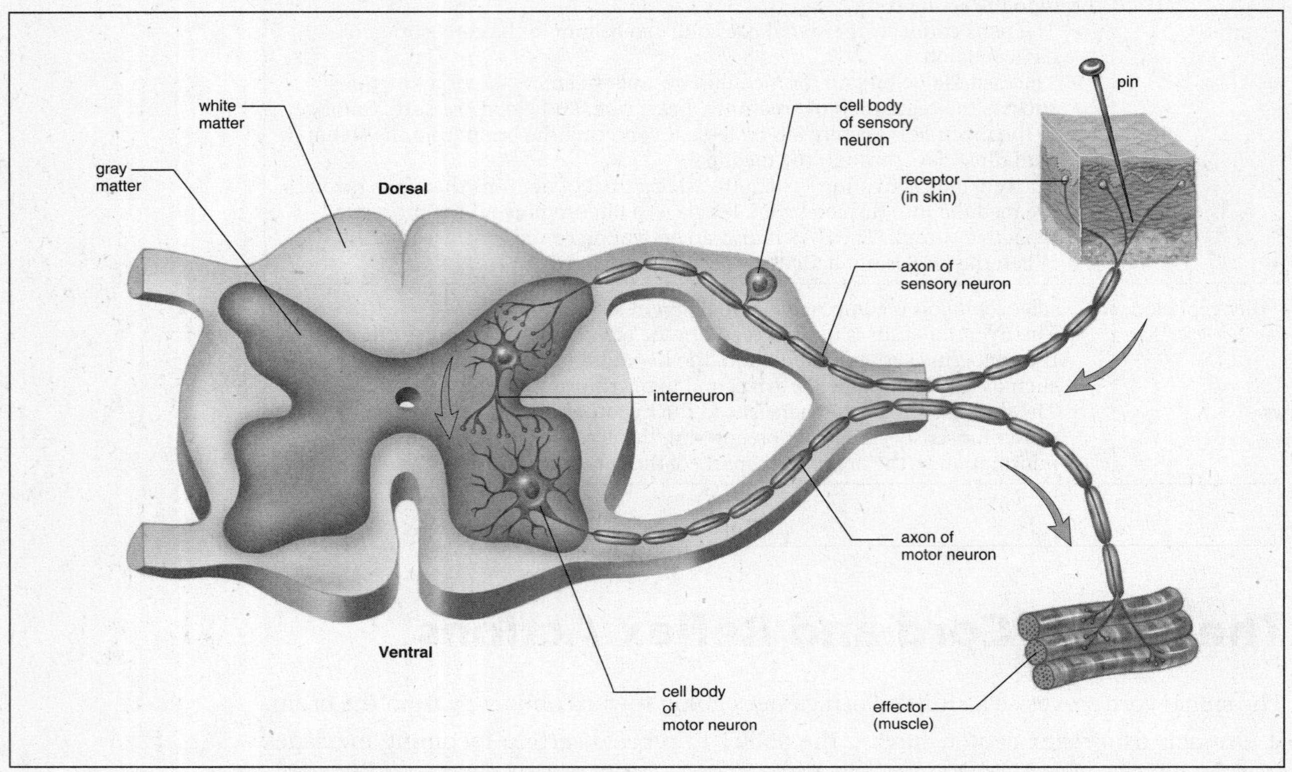

FIGURE 11-5: A reflex arc. A stimulus is carried by a sensory neuron to interneurons located in the spinal cord, which relays messages to motor neurons. SOURCE: From Sylvia S. Mader, *Biology*, 8th ed., McGraw-Hill, 2004; reproduced with permission of The McGraw-Hill Companies.

The **autonomic nervous system** controls the activity of involuntary functions within the body in order to maintain homeostasis. The autonomic nervous system is further subdivided into the **sympathetic** and **parasympathetic** branches. Both branches innervate most internal organs. The sympathetic branch is regulated by the neurotransmitters epinephrine and norepinephrine. When activated, the sympathetic branch produces the fight-or-flight response, in which heart rate increases, ventilation increases, blood pressure increases, and digestion decreases. These responses prepare the body for immediate action.

The parasympathetic branch is antagonistic to the sympathetic branch and is the default system used for relaxation. Generally, it decreases heart rate, decreases ventilation rate, decreases blood pressure, and increases digestion. The neurotransmitter acetylcholine is the primary regulator of this system.

SENSORY RECEPTORS

Sensory receptors located throughout the body are able to communicate with the nervous system, ultimately allowing for the perception of sensory information via the senses. There are various types of **sensory receptors** that differ based on the stimulus to which they are sensitive. The typical sensory receptors in the body as well as the stimulus they are sensitive to are shown in the following table.

Sensory Receptors and Their Stimuli		
Receptor	**Stimulus Detected**	**Functions or Senses**
Chemoreceptors	Chemicals	Gustation (taste) and olfaction (smell)
Thermoreceptors	Temperature	Monitoring body temperature
Photoreceptors	Light	Vision
Mechanoreceptors	Pressure	Tactile perception in the skin, propioception (sense of body awareness), hearing, and equilibrium
Pain receptors	Pressure and chemicals	Conveys messages to the CNS concerning tissue damage

Receptor Potential

Each sensory receptor is sensitive to a particular stimulus. The intensity of the stimulus is conveyed by a graded **receptor potential**. The greater the stimulus, the larger the receptor potential. When the receptor potential hits the threshold level, an action potential is generated. Over time, most sensory receptors stop responding to repeated stimuli in the process of **sensory adaptation** so that action potentials are no longer generated and we are no longer aware of the stimulus.

The Special Senses

The special senses of the body include taste, smell, hearing, balance (equilibrium) and vision. The receptors for each of the special senses are located within the head in specialized structures.

TASTE

The sense of taste relies upon chemoreceptors located within taste buds on the tongue and, to a lesser degree, in other parts of the mouth. There are four primary sensations that can be perceived by the receptors of the tongue, including sweet, sour, salty, and bitter. These can be stimulated independently or in combinations to produce the perception of a variety of tastes. Different regions of the tongue have different sensitivities to particular tastes. In addition to information provided by chemoreceptors located within the taste buds, a large portion of the perception of taste is actually dependent on the sense of smell.

SMELL

The chemoreceptors for olfaction (smell) are located in small patches in the top of each nasal cavity and are covered in mucus. When chemicals from the air dissolve in mucus, they stimulate the receptors. The message from the receptor eventually makes its way to the cerebrum as well as to the limbic system. Unlike the chemoreceptors in the mouth, those in the nose are sensitive to about 1000 different chemicals. Combinations of signals from several receptors allow us to perceive more than 10,000 different scents.

HEARING AND EQUILIBRIUM

The structures of the ear, seen in Figure 11-6, are responsible for the sense of hearing as well as the sense of balance or equilibrium. Mechanoreceptors in the ear are sensitive

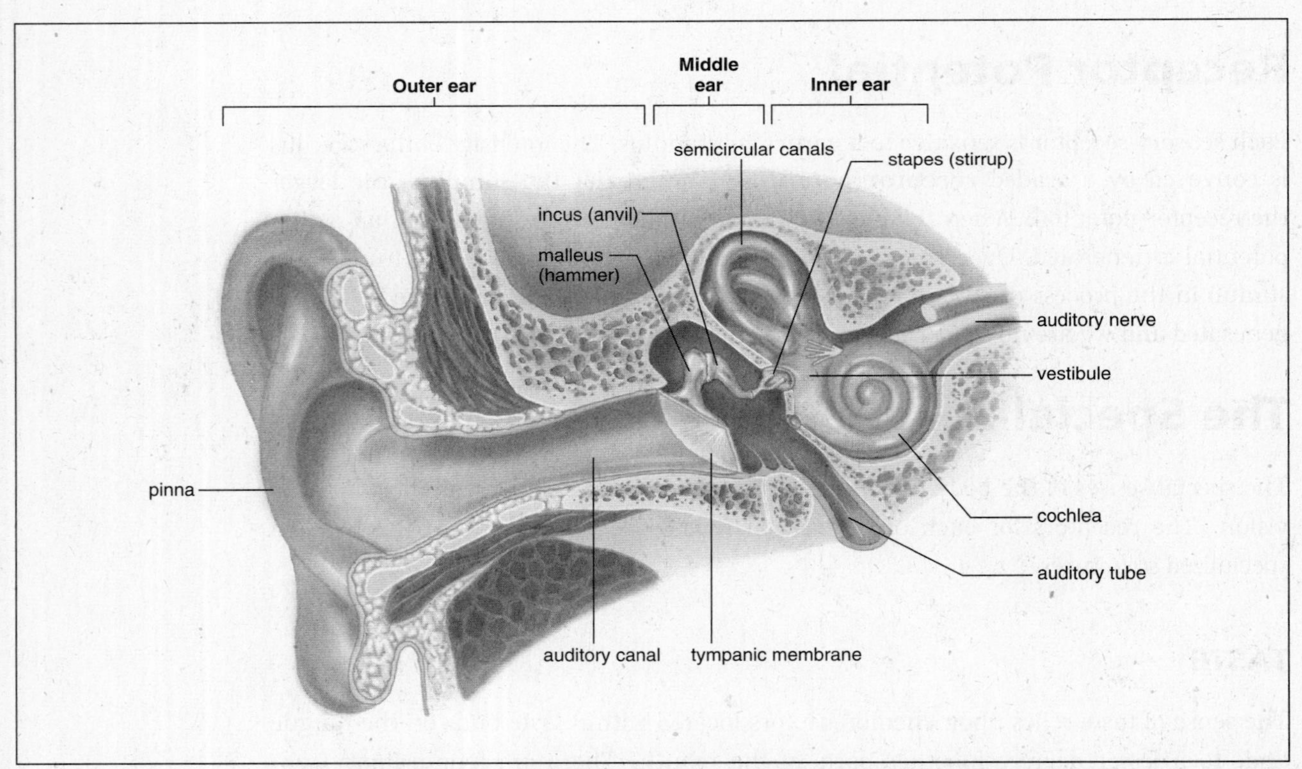

FIGURE 11-6: Ear structure. The outer ear collects sound waves, which then pass to the middle ear, where the sound waves are amplified. The inner ear is responsible for both hearing and equilibrium. SOURCE: From Sylvia S. Mader, *Biology*, 8th ed., McGraw-Hill, 2004; reproduced with permission of The McGraw-Hill Companies.

to pressure and sound waves that enter the outer ear. The structure of the **outer ear** consists of the **pinna**, which funnels sound waves toward the auditory canal. Once in the auditory canal, sound waves travel to the middle ear.

The **middle ear** contains the **tympanic membrane** (eardrum), which produces vibrations in response to sound waves. Three bones (**ossicles**) in the middle ear (the **malleus**, **incus**, and **stapes**) amplify the signal as it moves toward the inner ear.

The **inner ear** consists of a variety of structures. The amplified signals from the ossicles reach the **cochlea**. The cochlea contains the **organ of Corti**, which has specialized mechanoreceptors called **hair cells** that contain fluid with small "hairs" on the surface. When the fluid in the hair cells vibrates, the hairs send a message that is transduced into action potentials. These action potentials travel via the **auditory nerve** to the cerebrum for processing.

The **vestibular apparatus** within the inner ear is used to maintain a sense of equilibrium. Within the vestibular apparatus, **semicircular canals** of the inner ear contain hair cells filled with fluid. This fluid moves during motion of the head, changing the positioning of the hairs within the cell. The hair cells then send a message regarding the positioning of the head to the cerebrum for processing. The vestibular apparatus also contains the **vestibule**, which helps in the perception of balance when the head and body are not moving.

VISION

The eyes are responsible for vision and contain two types of photoreceptors that are sensitive to different forms of light. The **rods** are used for night vision and black and white vision, whereas the **cones** are used for color vision. Cones come in three varieties and are sensitive to red, blue, or green light.

The eye has three layers. The outer layer is the **sclera**, or the white of the eye. The middle layer of the eye is the **choroid**, which is used to supply oxygen and nutrients to other tissues of the eye. The inner layer of the eye is the **retina**, which contains the photoreceptors and is connected to the optic nerve.

Light waves enter the eye through the transparent **cornea**, which protects the underlying **lens** of the eye, as seen in Figure 11-7. As light enters the cornea, it moves through the **pupil**. The diameter of the pupil can be adjusted by the **iris** that surrounds it. The lens focuses the light into an image on the retina located at the back of the eye. The rods and cones convert the image into patterns, which are sent to the cerebrum for processing via the **optic nerves**. The area where the optic nerve leaves the retina is lacking in photoreceptors and is commonly referred to as the **blind spot**.

The shape of pigments located in the rods and cones is modified when a stimulus is received. These modified pigments change the membrane permeability, resulting in a

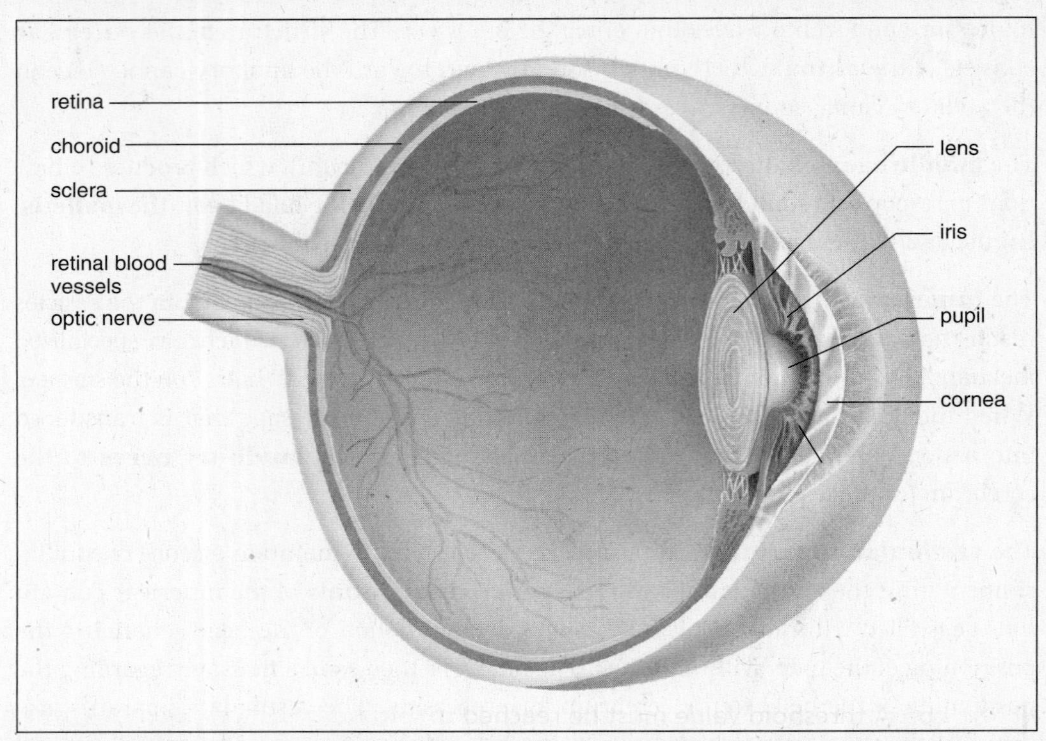

FIGURE 11-7: Eye structure. The photoreceptors for vision are located in the retina on the inner layer of the eye. SOURCE: From Sylvia S. Mader, *Biology*, 8th ed., McGraw-Hill, 2004; reproduced with permission of The McGraw-Hill Companies.

receptor potential that ultimately produces an action potential that can be transported by the optic nerves. Rods and cones contain one pigment called **retinal** that binds to the protein **opsin**. Rods and each kind of cone all have a different form of opsin that can be distinguished from each other. The specific opsin found in rods is **rhodopsin**. Modification of any type of opsin results in stimulation of the receptor and a message being sent to the cerebrum for processing.

CRAM SESSION

The Nervous System and Senses

1. The central nervous system consists of the brain and spinal cord, whereas the peripheral nervous system consists of nerves outside of the brain and spinal cord.
   a. Neurons are the functional units of both systems.
2. Neurons contain dendrites that receive messages from other cells, a cell body that processes the message, an axon that conveys an electrical impulse, and synaptic knobs that release chemical signals in the form of neurotransmitters to other cells. The axons of some neurons are myelinated to allow for faster conduction.
   a. The resting potential is maintained by the Na^+/K^+ pumps, which require a lot of ATP to maintain. Na^+ is pumped out of the neuron and K^+ is pumped in, resulting in the interior of the neuron having a more negative charge than the outside. Resting potential is maintained, at about -70 mV.
   b. A threshold value must be reached to initiate an action potential, which opens Na^+ channels to create a more positive charge in the neuron. A typical action potential reaches $+35$ mV.
   c. Action potentials propagate down the length of an axon.
   d. Neurotransmitters are used to communicate at the synapse. They can either encourage the initiation of an action potential in the postsynaptic neuron (excitatory) or discourage the initiation of an action potential (inhibitory).
3. The brain is divided into several distinct regions, as shown in the table in the chapter.
4. Reflex actions can be processes through the spinal cord in a reflex arc, which consists of a sensory neuron, interneuron, and motor neuron.
5. The peripheral nervous system innervates the entire body.
   a. The somatic division is used for voluntary motor functions.
   b. The autonomic system is used for involuntary motor functions. The autonomic system is divided into the sympathetic and parasympathetic branches.
      (1) The sympathetic branch responds to epinephrine and norepinephrine for the fight-or-flight response.
      (2) The parasympathetic branch responds to acetylcholine and causes relaxation in the body.
6. Sensory information is detected by receptors in the body. They can be sensitive to light, pressure, chemicals, and temperature.

7. The special senses include taste, smell, vision, hearing, and equilibrium.
   a. Tastes and smells are detected by chemoreceptors.
   b. Hearing involves specialized hair cells, a form of mechanoreceptors. The complex structures of the ear amplify sound waves and create vibrations, which are transduced into action potentials.
   c. Equilibrium is achieved through communication of the vestibular apparatus in the inner ear with the brain.
   d. Vision relies on rods and cones (photoreceptors) located on the retina of the eye.

CHAPTER 12

Muscular and Skeletal Systems

Read This Chapter to Learn About:

- ➤ Skeletal, Smooth, and Cardiac Muscle
- ➤ Cartilage Structure
- ➤ Bone Structure
- ➤ Bone Cells
- ➤ Joints

THE MUSCULAR SYSTEM

Muscles provide structural support, help maintain body posture, regulate openings into the body, assist in thermoregulation via contractions (shivering) that generate heat, and contracts to help move blood in veins toward the heart, thus assisting in peripheral circulation. In Chapter 10, three types of muscle tissue were introduced. Skeletal muscle and cardiac muscle are striated, whereas smooth muscle is not. Cardiac and smooth muscle are involuntary, whereas skeletal muscle is under voluntary control.

SKELETAL MUSCLE

Skeletal muscles are responsible for voluntary movement. The cells in skeletal muscle have multiple nuclei as the result of the fusing of multiple cells. The muscle cells also contain high levels of mitochondria to provide ATP needed for contraction and the protein **myoglobin** that acts as an oxygen reserve for muscles.

The fibers of skeletal muscle can be classified as fast-twitch or slow-twitch fibers. **Fast-twitch fibers** are designed for a fast rate of contraction, but they lack stamina and fatigue easily because their primary energy source is anaerobic cellular respiration. They have less myoglobin and fewer mitochondria than slow-twitch cells. The **slow-twitch fibers** contain more mitochondria and more myoglobin, which gives them longer endurance as they obtain most of their ATP from aerobic cellular respiration. Creatine-PO_4 can serve as a short-term energy reserve for either type of muscle fiber when ATP levels are low.

Structural Organization of Skeletal Muscle

Muscles are a bundle of muscle cells held together by connective tissue as seen in Figure 12-1. The muscle cells have **sarcoplasm** (cytoplasm), a modified endoplasmic reticulum called the **sarcoplasmic reticulum**, and a cell membrane called the **sarcolemma**, which interacts with the nervous system via the **transverse tubule**

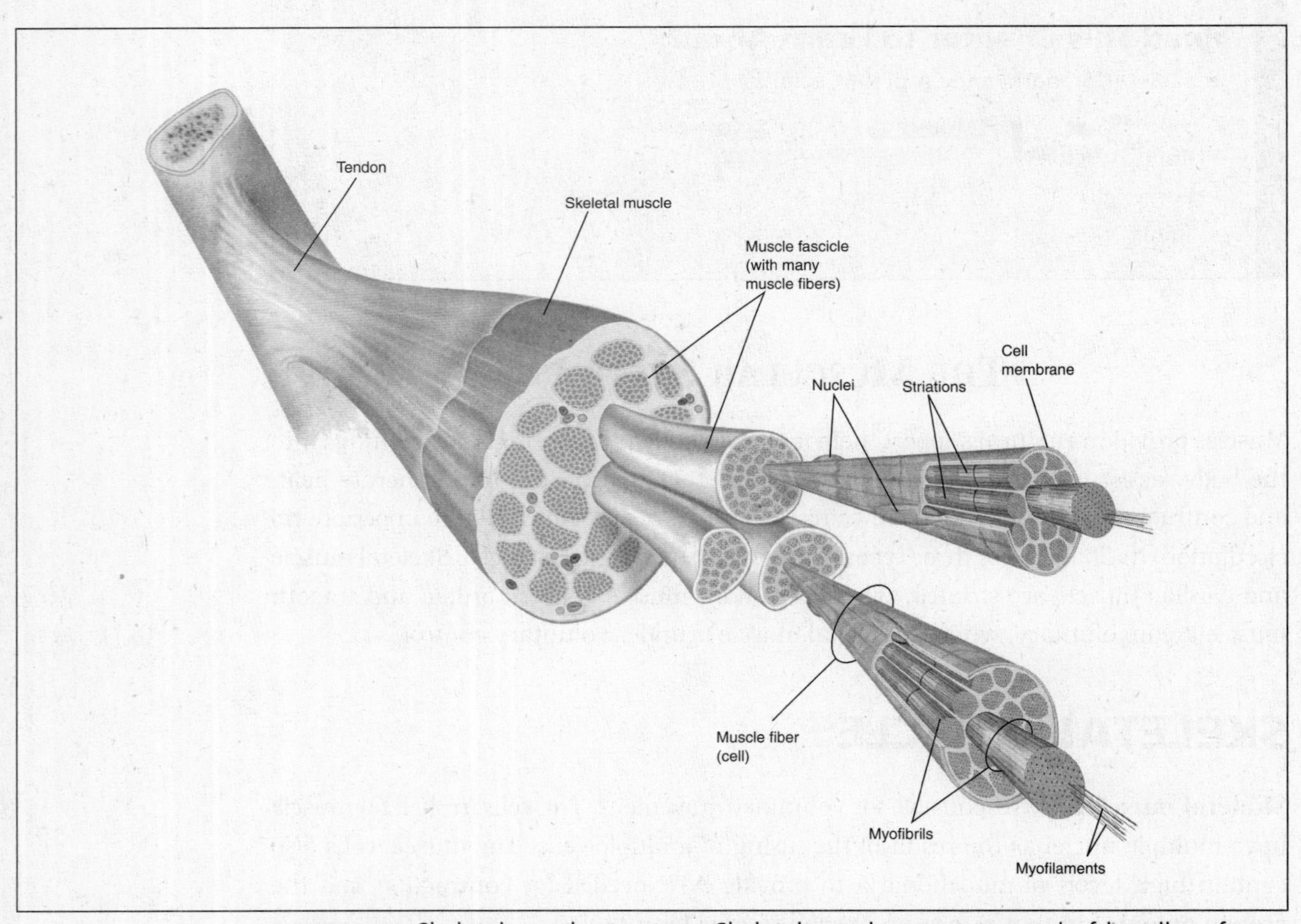

FIGURE 12-1: Skeletal muscle structure. Skeletal muscles are composed of bundles of muscle cells or fibers. Each fiber is composed of myofibrils, which are in turn composed of myofilaments. SOURCE: From George B. Johnson, *The Living World,* 3d ed., McGraw-Hill, 2003; reproduced with permission of The McGraw-Hill Companies.

system (T tubule). This system provides channels for ion flow through the muscle and has anchor points for sarcomeres.

Within the muscle cells are bundles of muscle fibers called **myofibrils**, which are made of the proteins actin, troponin, tropomyosin, and myosin. **Actin** fibers have a thin diameter and associate with the proteins **troponin** and **tropomyosin** to produce thin filaments. **Myosin** fibers have a thick diameter with protruding heads and are called **thick filaments**.

In skeletal muscles, the actin and myosin fibers overlap each other in highly organized, repeating units called **sarcomeres**. The overlapping of the fibers is what causes striation of the muscle. The shortening of sarcomeres is what causes muscle contraction. The structure of a sarcomere can be seen in Figure 12-2.

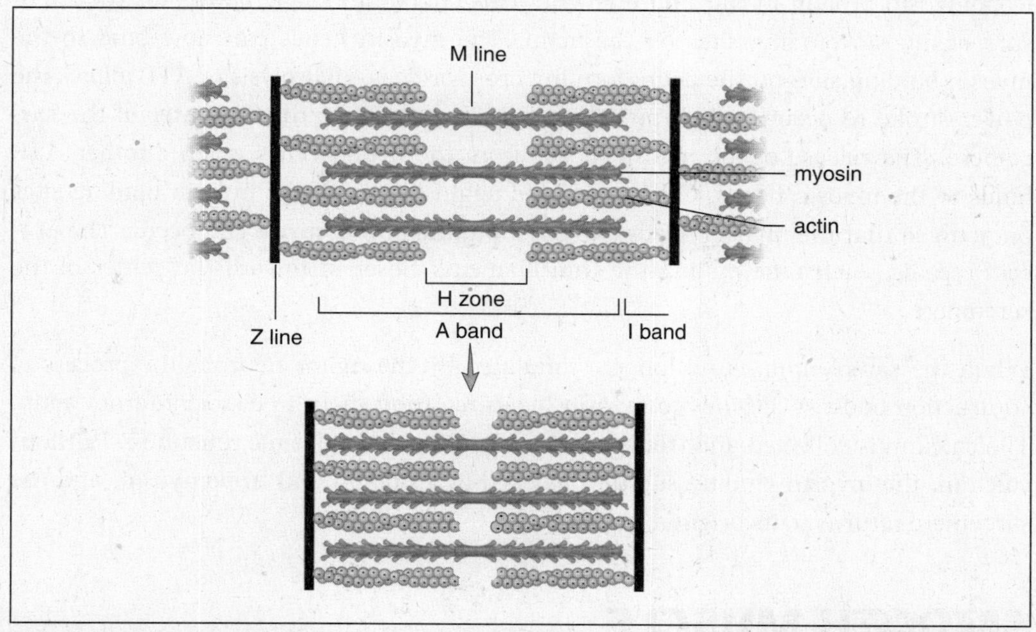

FIGURE 12-2: Sarcomere structure. Shortening of the sarcomere occurs when actin filaments move toward the center of the sarcomere. This shortening of the sarcomere is responsible for skeletal muscle contraction. SOURCE: From Sylvia S. Mader, *Biology*, 8th ed., McGraw-Hill, 2004; reproduced with permission of The McGraw-Hill Companies.

The major regions of the sarcomere are as follows:

- **M line:** line that marks the center of the sarcomere
- **Z line:** line that separates one sarcomere from the next
- **H zone:** the area where only thick filaments are present; it shortens during contraction
- **I band:** the area where only thin filaments are present; it shortens during contraction
- **A band:** the area where thick and thin filaments overlap

The Sliding Filament Model of Skeletal Muscle Contraction

Muscle tissues have regions where the sarcolemma is in contact (via a synapse) with the synaptic knobs of a motor neuron from the somatic branch of the peripheral nervous system. This area is the **neuromuscular junction**. A neurotransmitter called **acetylcholine** is released from the motor neuron and binds to receptors on the sarcolemma. This causes an action potential that initiates shortening of the sarcomeres. The muscle fibers influenced by a single neuromuscular junction are termed a **motor unit**.

The action potential that occurs based on stimulation from the motor neuron causes the release of calcium from the sarcoplasmic reticulum into the sarcoplasm. The calcium binds to the troponin in the thin filaments. This causes a conformational shift in the tropomyosin protein in the thin filament. This change in shape allows for the exposure of myosin binding sites on the actin. The myosin heads can now bind to the myosin binding sites on the actin, forming cross-bridges. Hydrolysis of ATP allows the power stroke to occur, which pulls the thin filaments toward the center of the sarcomere. The release of the myosin heads from the actin occurs when another ATP binds to the myosin heads. Calcium is used again to expose the myosin binding sites on actin so that the myosin heads can bind and the power stroke can occur. The process repeats, each time pulling the thin filaments closer in toward the center of the sarcomere.

When the sarcolemma is no longer stimulated by the motor neuron, the process of contraction ends. ATP binds to myosin heads, causing them to dissociate from actin. The calcium is collected and transported back to the sarcoplasmic reticulum. Without calcium, the myosin binding sites are blocked by troponin and tropomyosin, and the sarcomere returns to its original length.

SMOOTH MUSCLE

Smooth muscle can be found in multiple parts of the body including the bladder, digestive tract, reproductive tracts, and surrounding blood vessels. Each cell in smooth muscle contains a single nucleus, as opposed to the multiple nuclei found in skeletal muscle. Smooth muscle contains actin and myosin, but it is not organized as sarcomeres, which is why smooth muscle lacks striations. The actin and myosin slide over each other. This sliding is regulated by calcium and requires energy provided by ATP.

The autonomic branch of the peripheral nervous system innervates smooth muscles via sympathetic and parasympathetic stimulation to produce involuntary contractions. The **sympathetic response** generally uses the neurotransmitters **epinephrine** (adrenaline) and **norepinephrine** (noradrenaline) to prepare the body for physical activity. The **parasympathetic response**, on the other hand, responds to acetylcholine and uses it to return the body and muscles to a relaxation state. Smooth muscle can perform **myogenic activity**, meaning it can also contract without stimulation from the nervous system.

CARDIAC MUSCLE

Cardiac muscle is found only in the myocardium of the heart. It is striated due to the presence of sarcomeres (which require calcium and ATP for contraction, just as in skeletal muscle) but is not multinucleated like skeletal muscle. A typical cardiac muscle cell would have one or possibly two nuclei. Cardiac muscle is innervated by the autonomic branch of the peripheral nervous system. Like smooth muscle, it can also perform myogenic activity, contracting without stimulation from the nervous system.

THE SKELETAL SYSTEM

Skeletons can exist as **exoskeletons** found on the exterior of the body or **endoskeletons** found on the interior of the body. The disadvantage of an exoskeleton (found in many arthropods such as lobsters and crabs) is that it does not grow with the organism, making it necessary to shed the skeleton and produce a new one to accommodate growth. An endoskeleton is found in vertebrates, including fish, birds, and mammals.

The human endoskeleton, seen in Figure 12-3, is made of bone and associated cartilage. It is divided into two major parts: the axial skeleton and the appendicular skeleton. The **axial skeleton** is composed of the skull, the vertebral column, the sternum, and the rib cage. The pelvic and shoulder girdles and limbs in the body are part of the **appendicular skeleton**.

The skeleton is used for protection of internal organs, support, storage of calcium and phosphates, production of blood cells, and movement.

STRUCTURE OF CARTILAGE

Cartilage is a connective tissue. The matrix is termed **chondrin** and the primary cell type is the **chondrocyte**. During embryonic development, the skeleton begins as cartilage. During the developmental period, much of the cartilage is subject to **ossification**, through which it is turned into bone by a calcification process. Only a small amount of cartilage remains in the adult skeleton, because most of it has been turned to bone. The primary areas where cartilage is found in the adult human skeleton are the nose, the ears, the discs between vertebrae, the rib cage, the joints, and the trachea. Cartilage is unique in that it contains no blood vessels, nor is it innervated.

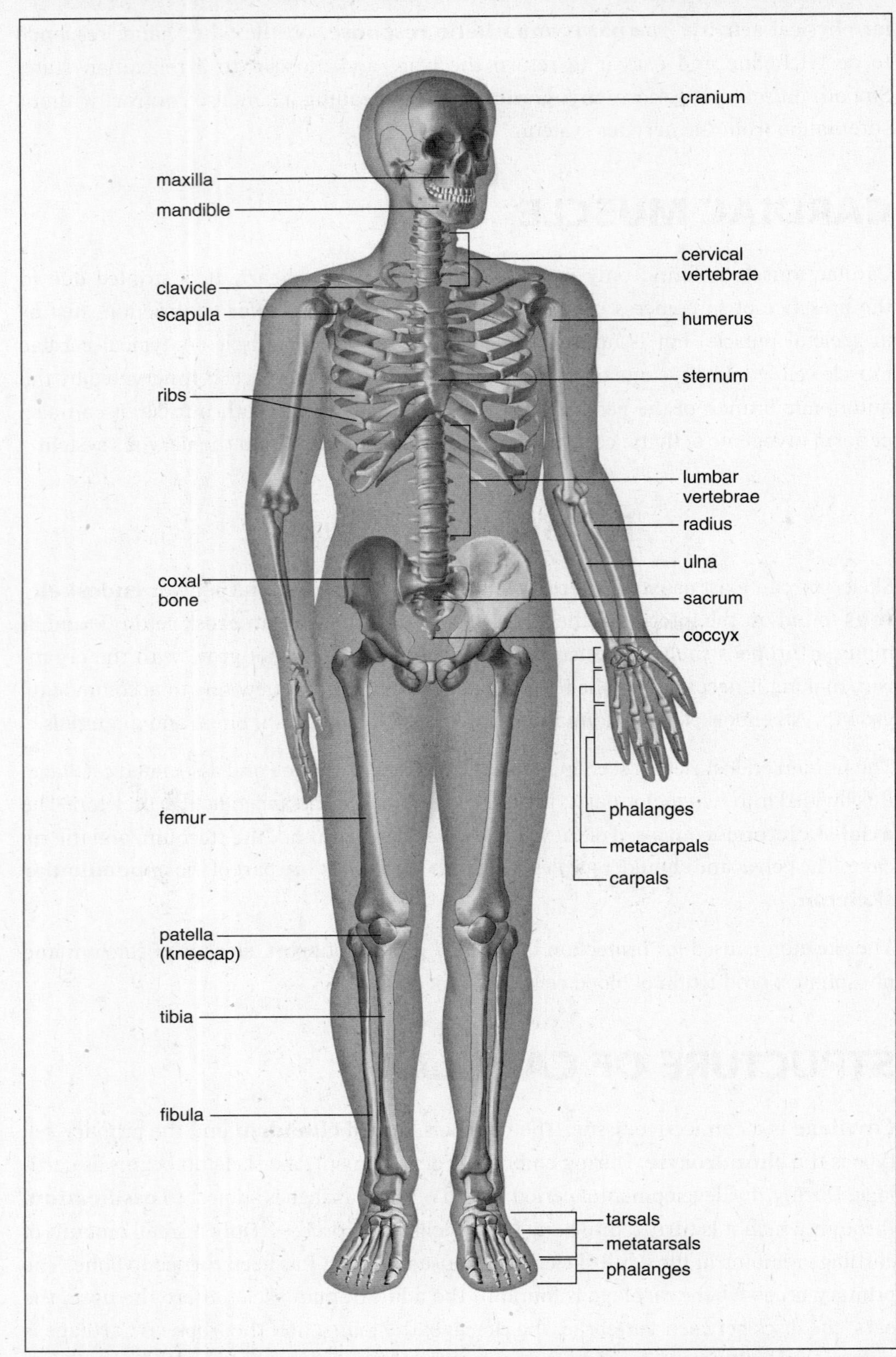

FIGURE 12-3: The human skeleton. The axial skeleton is composed of the skull, vertebral column, sternum, and ribs. The remaining bones in the body are part of the appendicular skeleton. SOURCE: From Sylvia S. Mader, *Biology*, 8th ed., McGraw-Hill, 2004; reproduced with permission of The McGraw-Hill Companies.

STRUCTURE OF BONE

Bone tissue is found as compact bone and spongy bone. **Compact bone** is very dense, whereas **spongy bone** is less dense and contains marrow cavities. Within marrow cavities, there is yellow and red bone marrow. **Red bone marrow** contains the stem cells that differentiate into red blood cells, white blood cells, and platelets. **Yellow bone marrow** is primarily a reserve for adipose (fat) tissue.

Long bones within the body have a characteristic structure, as seen in Figure 12-4. The ends of the bone are typically covered in cartilage and are termed the **epiphyses**.

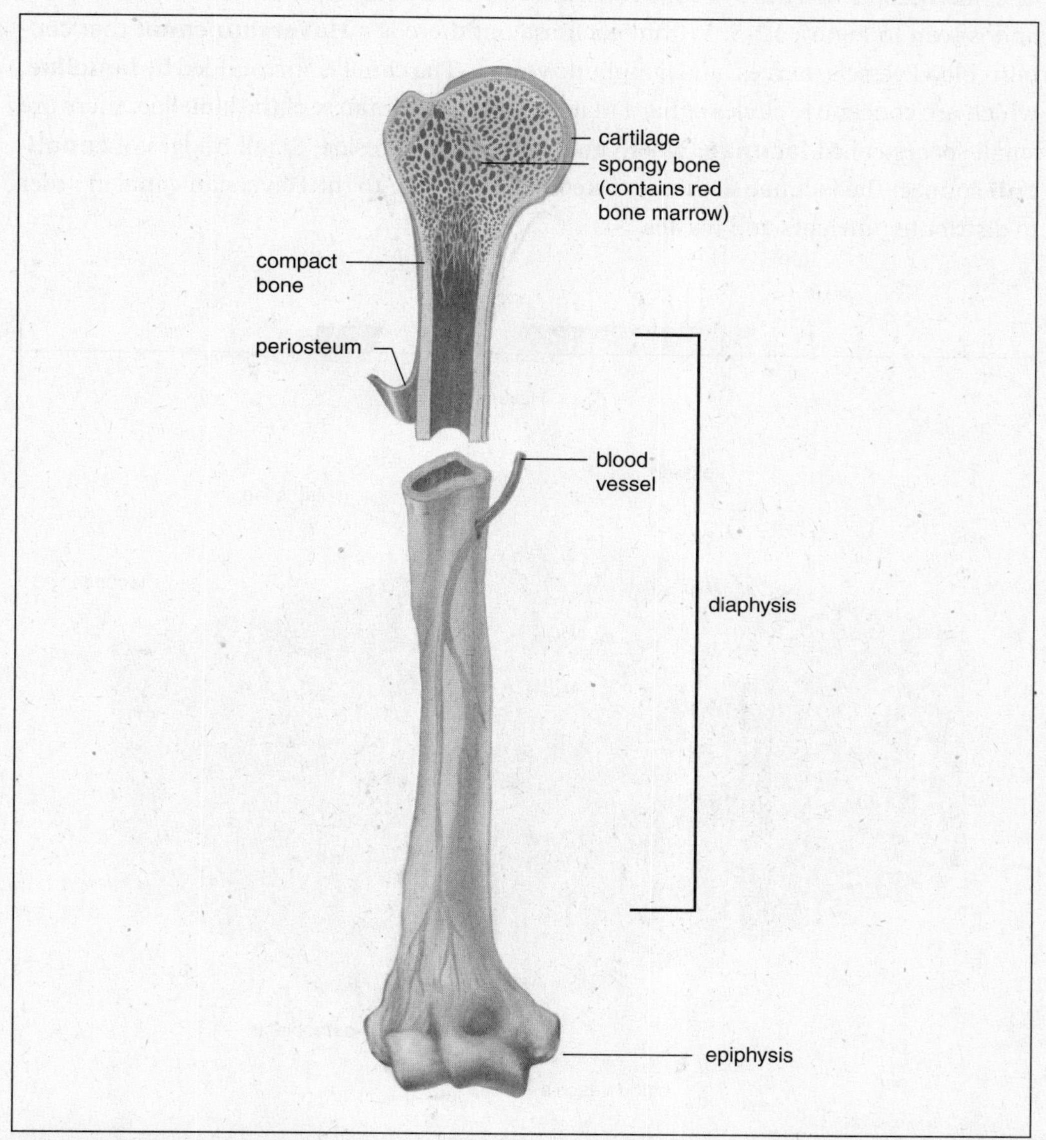

FIGURE 12-4: Long-bone structure. Spongy bone tissue contains red bone marrow and is located at the ends of long bones. The hollow marrow cavity located in the shaft of the bone contains yellow bone marrow. SOURCE: From Sylvia S. Mader, *Biology*, 8th ed., McGraw-Hill, 2004; reproduced with permission of The McGraw-Hill Companies.

The ends are made primarily of spongy bone covered in a thin layer of compact bone. The shaft of the bone (the **diaphysis**) is made of compact bone surrounding a marrow cavity. The **epiphyseal plate** is a disc of cartilage that separates the diaphysis from each epiphysis, and this is where bone lengthening and growth occurs. The **periosteum** surrounds the bone in a fibrous sheath and acts as a site for the attachment of muscles via **tendons**.

Microscopic Structure

The microscopic structure of bone consists of the matrix, which is found within **osteons** and is seen in Figure 12-5. Within each osteon, there is a **Haversian canal** that contains blood vessels, nerves, and lymphatic vessels. The canal is surrounded by **lamellae**, which are concentric circles of hard matrix. Within the matrix of the lamellae, there are small spaces called **lacunae**, where mature bone cells reside. Small bridges of **canaliculi** connect the lacunae within an osteon and merge into the Haversian canal in order to distribute nutrients and wastes.

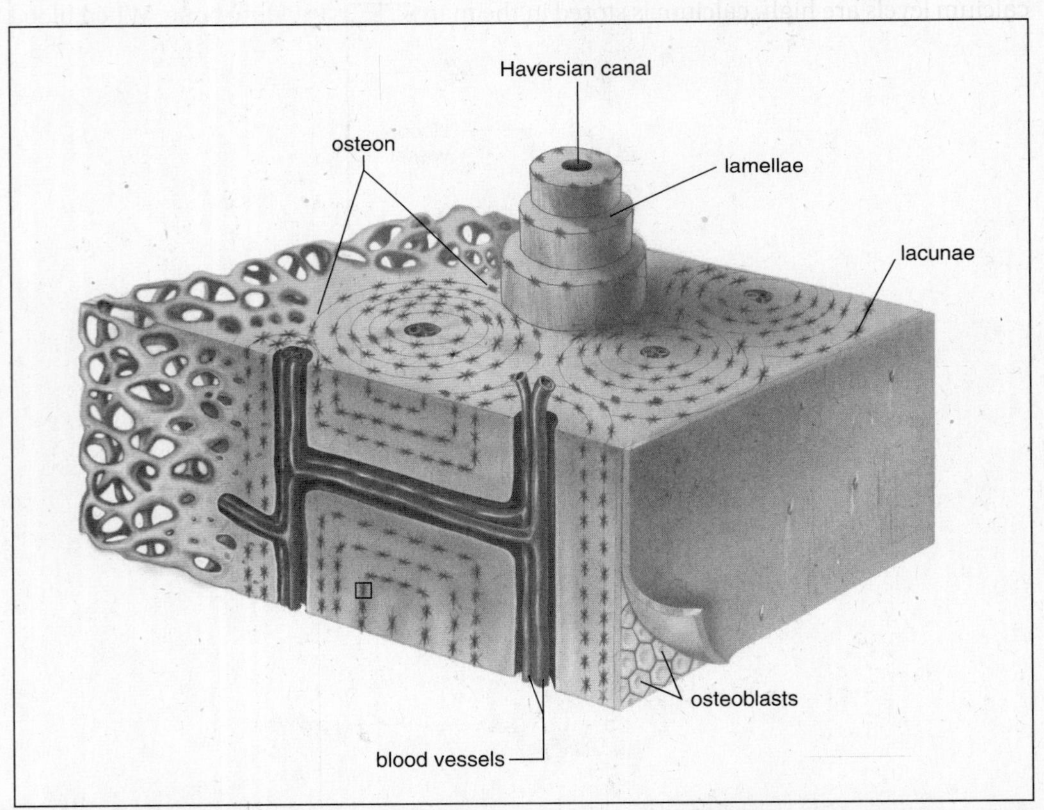

FIGURE 12-5: Bone tissue structure. SOURCE: From Sylvia S. Mader, *Biology*, 8th ed., McGraw-Hill, 2004; reproduced with permission of The McGraw-Hill Companies.

Bone Cells

Within the bone, there are three major cell types: osteocytes, osteoblasts, and osteoclasts.

- **Osteocytes** are found within the lacunae of osteons. They are mature bone cells involved in the maintenance of bone tissue.
- **Osteoblasts** and **osteoclasts** are found within bone tissue as well, and are immature cells.

Both osteoblasts and osteoclasts are involved in the constant process of breaking down and building bone known as **bone remodeling**. Osteoblasts build bone by producing components of the matrix, whereas osteoclasts break down bone in the process of bone reabsorption. Eventually, osteoblasts and osteoclasts become trapped within matrix of bone tissue and become osteocytes. Osteoblasts are also responsible for bone growth and ossification during development. Ideally, the levels of matrix break down and building will be in equilibrium once growth is complete.

The hormones calcitonin from the thyroid gland and parathyroid hormone from the parathyroid glands are responsible for the process of bone remodeling. When blood calcium levels are high, calcium is stored in the matrix, thus building bone. When blood calcium levels are low, calcium is released from the matrix by the breakdown of bone tissues. The levels of calcium in the blood must be carefully regulated, as calcium is also needed for muscle contraction, nervous system communication, and other functions.

JOINTS

Joints are areas where two bones meet and are composed of connective tissues. Immovable joints such as those involved with skull bones do not move at all. Partially movable joints have some degree of flexibility, such as the vertebrae in the spinal column. **Synovial joints**, such as the hip or knee, have a much larger range of motion and have a fluid-filled joint cavity. **Ligaments** are made of dense connective tissue and attach one bone to another within a synovial joint.

CRAM SESSION

Muscular and Skeletal Systems

1. Muscles are used to support the body, regulate openings into the body, generate heat through contractions, and help blood move through veins.
2. The fibers of skeletal muscles are organized in sarcomeres, which are the units of contraction.
   a. The sarcomere has thick filaments made of myosin and thin filaments made of actin and associated troponin and tropomyosin. This configuration is responsible for the striation of skeleton muscle.
   b. The sliding filament model explains how myosin heads can attach to actin binding sites, pulling the thin filaments toward the center of the sarcomere, causing it to shorten.
   c. ATP and calcium are both needed for this process.
3. Smooth muscle contains sliding actin and myosin filaments, but they are not organized as sarcomeres, which explains why smooth muscle lacks striations.
   a. Smooth muscle is involuntary and can contract on its own without signals from the nervous system.
4. Cardiac muscle is also striated and contracts in a manner similar to skeletal muscle.
   a. Cardiac muscle is also capable of involuntarily contracting on its own without stimulation from the nervous system.
5. The skeletal system begins as cartilage in embryos, most of which is replaced by bone in the process of ossification.
6. Bone matrix is a major storage reserve for calcium and phosphates.
7. Compact bone tissue is dense, whereas spongy bone is less dense and contains red and yellow marrow.
   a. Red bone marrow contains stem cells that can differentiate into any type of blood cell.
   b. Yellow marrow stores fat.
8. Osteons are the structural unit of bone tissue. Osteocytes are mature bone cells that perform maintenance within the osteons.
   a. During bone remodeling, osteoblasts build new bone when blood calcium levels are high, and osteoclasts break down bone when blood calcium levels are low.
9. Joints form where two bones meet. Joints can be immovable, partially movable, or free-moving (synovial).

CHAPTER 13

The Endocrine System

Read This Chapter to Learn About:

- Hormone Secretion
- Hormone Specificity
- Hormone Functions
- Major Endocrine Glands and Their Products

In order to maintain homeostasis in the body, it is necessary to regulate the functioning of specific targets within the body. The endocrine system functions in this regulation. It is made up of **endocrine glands** located throughout the body. These glands secrete **hormones** that function to achieve this regulation.

ENDOCRINE GLANDS

As stated previously, endocrine glands secrete hormones into the bloodstream. Some endocrine structures also have the ability to serve exocrine functions as well. An **exocrine gland** secretes its products onto a surface, into a body cavity, or within organs. The pancreas is an example of a gland with both endocrine and exocrine functions. Its ability to secrete the hormones insulin and glucagon into the blood qualifies it as an endocrine gland, whereas its ability to secrete digestive enzymes into the small intestine qualifies it as an exocrine gland.

The major endocrine glands can be seen in Figure 13-1.

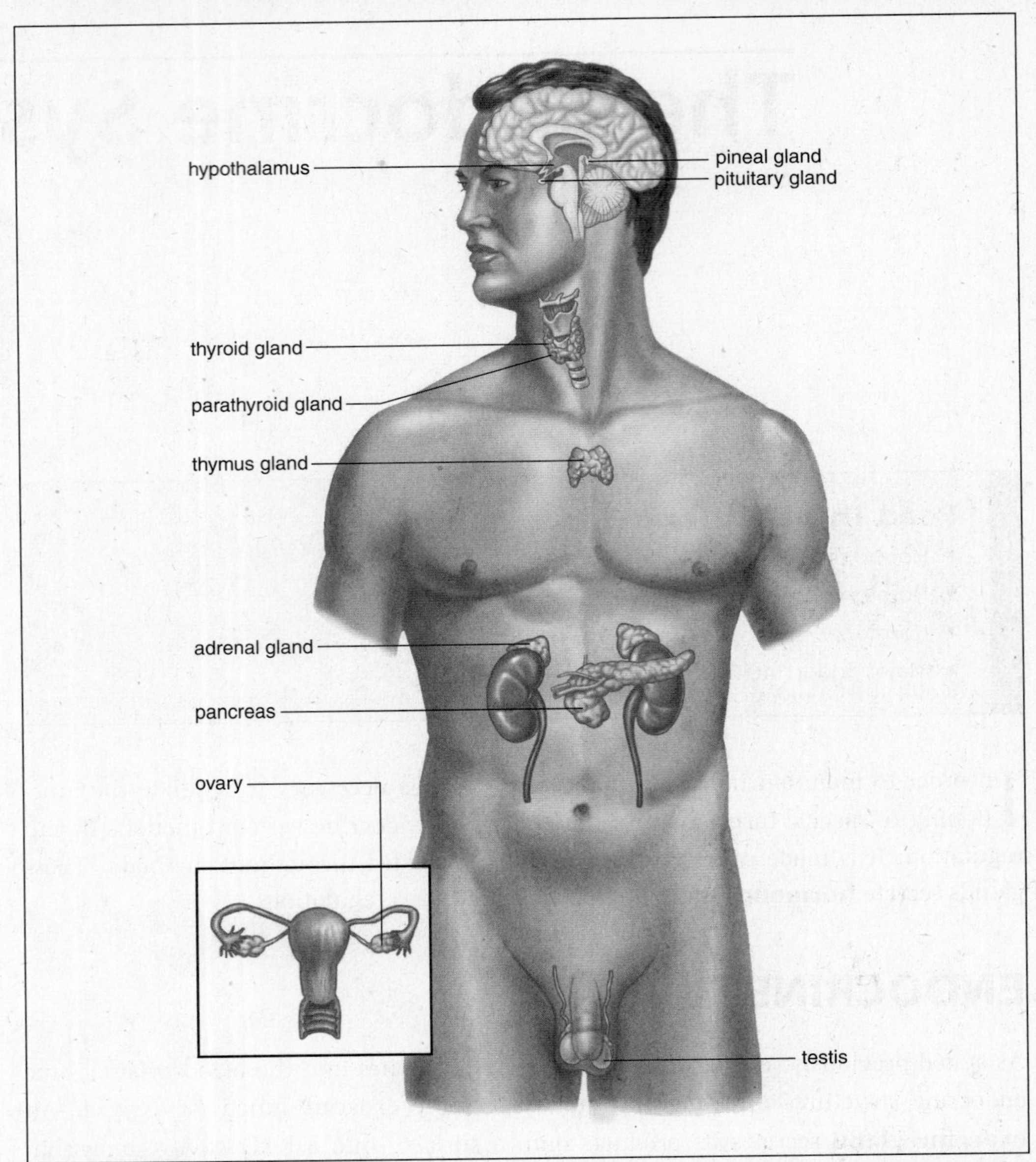

FIGURE 13-1: The endocrine system. Anatomic location of major endocrine structures of the body. SOURCE: From Sylvia S. Mader, *Biology*, 8th ed., McGraw-Hill, 2004; reproduced with permission of The McGraw-Hill Companies.

HORMONES

A hormone is a chemical messenger secreted from an endocrine gland into the bloodstream that travels to a specific target in the body and changes the functioning of that target. The target can be individual cells, tissues, or entire organs.

More than 50 different hormones have been identified. They are grouped in two major categories: steroid (lipid-soluble) hormones and nonsteroid (water-soluble, peptide) hormones. **Steroid hormones** are derivatives of the lipid **cholesterol**. Nonsteroid hormones are made of either modified amino acids or small proteins. The target cell receptors for steroid hormones exist in the cytoplasm of the cell, whereas the receptors for nonsteroid hormones exist on the cell membrane of the cell.

Hormone secretion is regulated by several mechanisms, and the different types of hormones function in different ways in the cell.

Regulation of Hormone Secretion

The secretion of hormones is usually regulated via negative feedback mechanisms. In **negative feedback,** the response of the endocrine system or a target is the opposite of a stimulus. For example, if the level of a specific hormone gets particularly high (the stimulus), then the secretion of that hormone is reduced (opposite of the stimulus). In addition, some conditions, such as low blood calcium (the stimulus), may trigger the release of hormones to cause an opposite response (the breakdown of bone tissue to increase blood calcium levels). It is not uncommon to see **antagonistic hormones**—two hormones with opposing functions such as a hormone to raise blood sugar and another to lower blood sugar. Being able to adjust a particular situation in the body from high and low ends is necessary to maintain homeostasis. Failure of the endocrine system to maintain homeostasis can lead to conditions such as diabetes, hyper- or hypothyroidism, growth abnormalities, and many others.

Although not nearly as common as negative feedback mechanisms, **positive feedback** mechanisms do exist. In this case, the stimulus causes actions in the body, regulated by hormones, that further amplify that stimulus, moving the body away from homeostasis. Although this may sound like a bad thing, it is necessary in some cases, such as childbirth, where one hormone amplifies another. Positive feedback mechanisms are short lived, and eventually homeostasis is returned.

The nervous system can override endocrine feedback mechanisms in some cases. When the body is extremely stressed or experiencing trauma, the nervous system can exert control over the endocrine system in order to make adjustments to help the body cope with the situation. The **hypothalamus** in the brain is the main link between the endocrine and nervous systems. The hypothalamus monitors body conditions and makes changes when it deems appropriate. It produces regulatory hormones that influence glands such as the pituitary, which, in turn, regulates other glands in the endocrine system.

Hormone Specificity

Because hormones travel through the bloodstream, they encounter countless potential targets as they circulate. The specificity of hormones is based on their interaction with a receptor on the target cells. The hormone affects only those cells with a receptor for that specific hormone. Once a hormone binds to the receptor, a series of events that change the functioning of a cell in some way begins. These changes can involve gene expression, chemical reactions, membrane changes, metabolism, and so forth. Because the hormones travel through the blood, their actions and effects involve a relatively slow process.

Mechanisms of Action of Hormones

As stated earlier, the two types of hormones differ in their chemical composition, receptor sites, and in their mechanisms of action. The target cell receptors for steroid hormones exist in the cytoplasm of the cell, while the receptors for nonsteroid hormones exist on the cell membrane of the cell. These differences affect the mechanism of action of the two types of hormones.

MECHANISM OF ACTION IN STEROID HORMONES

Steroids are derivatives of cholesterol, which are lipid soluble and can easily cross the cell membrane. Once inside a cell, the steroid locates and binds to a cytoplasmic receptor. The steroid–receptor complex moves into the nucleus and interacts with DNA to cause activation of certain genes. This serves as the signal to initiate transcription and translation so that the cell expresses a new protein. This new protein changes how the cell is functioning in some way. A summary of steroid hormone action can be seen in Figure 13-2.

MECHANISM OF ACTION IN NONSTEROID HORMONES

Peptide hormones are composed of amino acid derivatives or small proteins and do not cross the cell membrane. They recognize a receptor on the cell membrane surface. The hormone itself is termed a **first messenger**, because it never enters the cell and only triggers a series of events within the cell, many of which are moderated by **G proteins** found in the cell membrane. The binding of the hormone to the receptor initiates a series of reactions in the cell that ultimately lead to the production of a **second messenger** molecule within the cell. A common second messenger of peptide hormones is **cyclic adenosine monophosphate** (cAMP), a derivative of ATP. The second messenger changes the function of the target cell by altering enzymatic activities and cellular reactions. A summary of nonsteroid hormone action can be seen in Figure 13-3.

OTHER CHEMICAL MESSENGERS

In addition to steroidal and nonsteroidal hormones, there is another category of chemical messenger—**prostaglandins**. These are lipid-based molecules released from cell

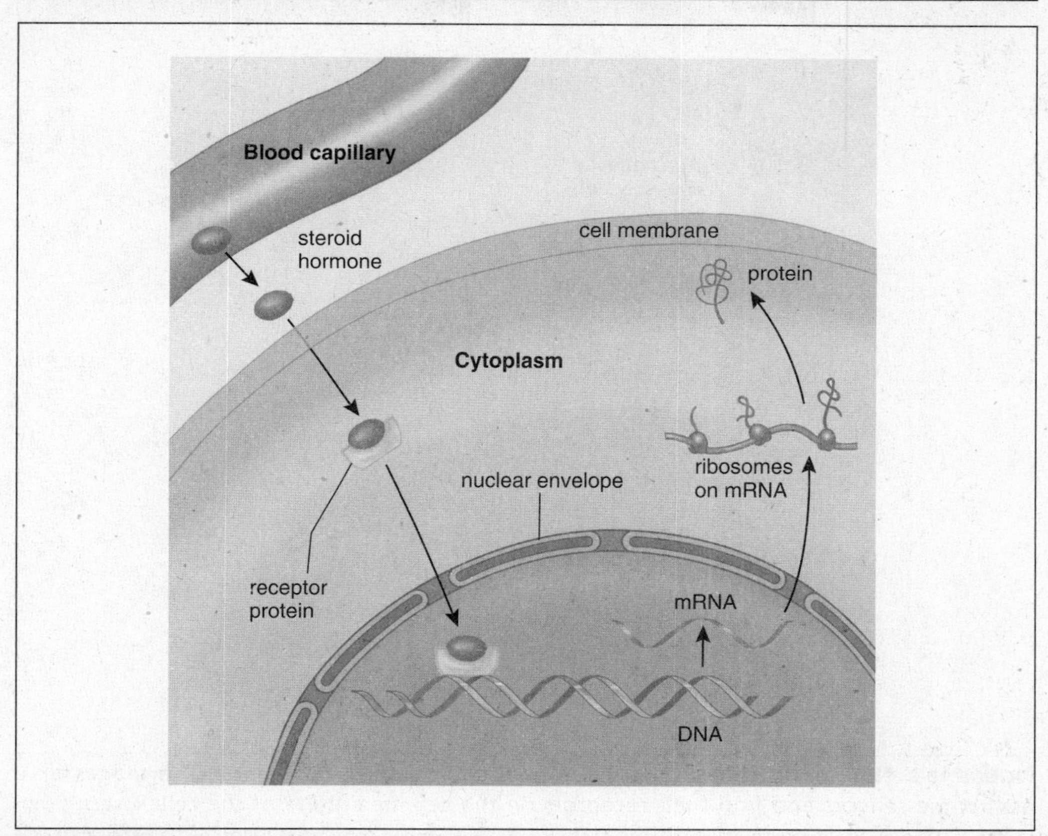

FIGURE 13-2: Steroid hormone mechanism of action. Steroid hormones act only on the cells in which they find their receptors in the cytoplasm of the cell. SOURCE: From Sylvia S. Mader, *Biology*, 8th ed., McGraw-Hill, 2004; reproduced with permission of The McGraw-Hill Companies.

membranes, not from endocrine glands. Prostaglandins function as a sort of local hormone involved in functions as diverse as regulation of body temperature, blood clotting, the inflammatory response, and menstrual cramping caused by uterine contractions.

THE ENDOCRINE SYSTEM

The major endocrine glands of the body include the hypothalamus, pituitary gland (separated into the anterior lobe and posterior lobe), pineal gland, thyroid gland, parathyroid glands, and adrenal glands. Some organs within the body also have endocrine functions, including the thymus gland, ovaries, testes, pancreas, heart, placenta, kidneys, stomach, and small intestine.

The hypothalamus and pituitary gland have a unique relationship based on their proximity to each other in the brain, as seen in Figure 13-4. The pituitary gland secretes many hormones, some of which influence the secretion of hormones from other endocrine glands. The regulatory hormones made by the hypothalamus control the secretion of hormones from the anterior pituitary. The hypothalamus produces

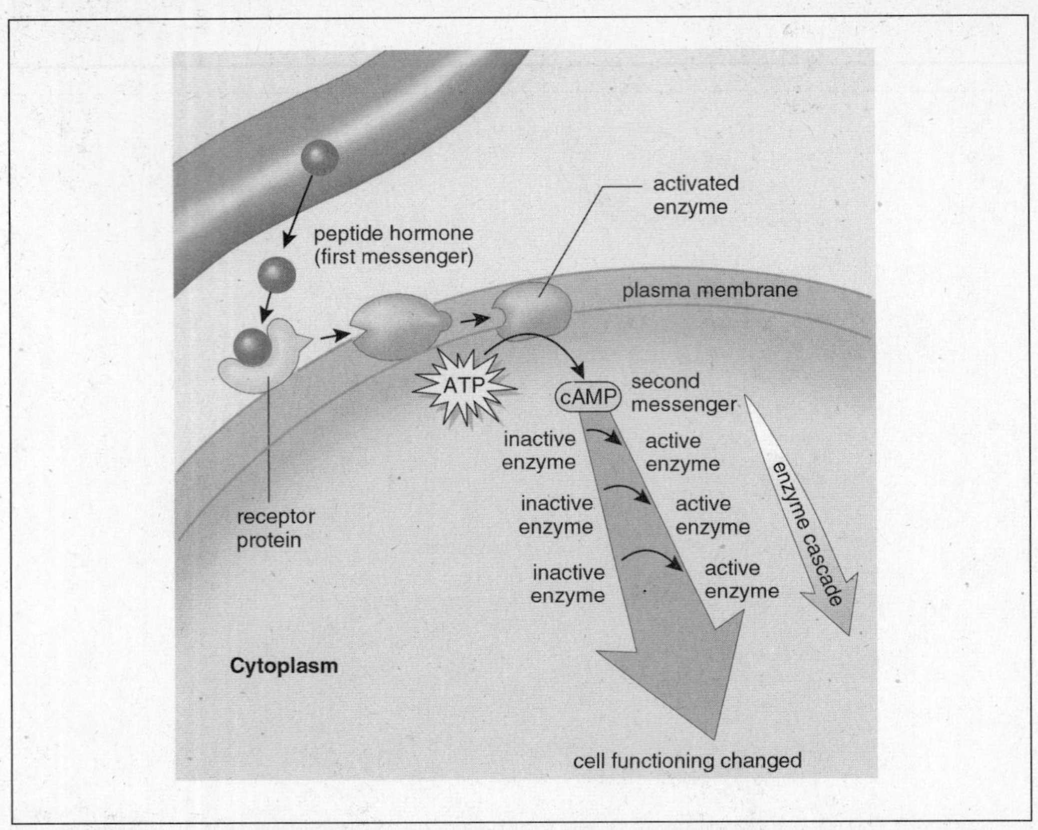

FIGURE 13-3: Nonsteroid hormone mechanism of action. Nonsteroid (peptide) hormones serve as first messengers and find their receptors on the cell membrane of the cell. SOURCE: From Sylvia S. Mader, *Biology*, 8th ed., McGraw-Hill, 2004; reproduced with permission of The McGraw-Hill Companies.

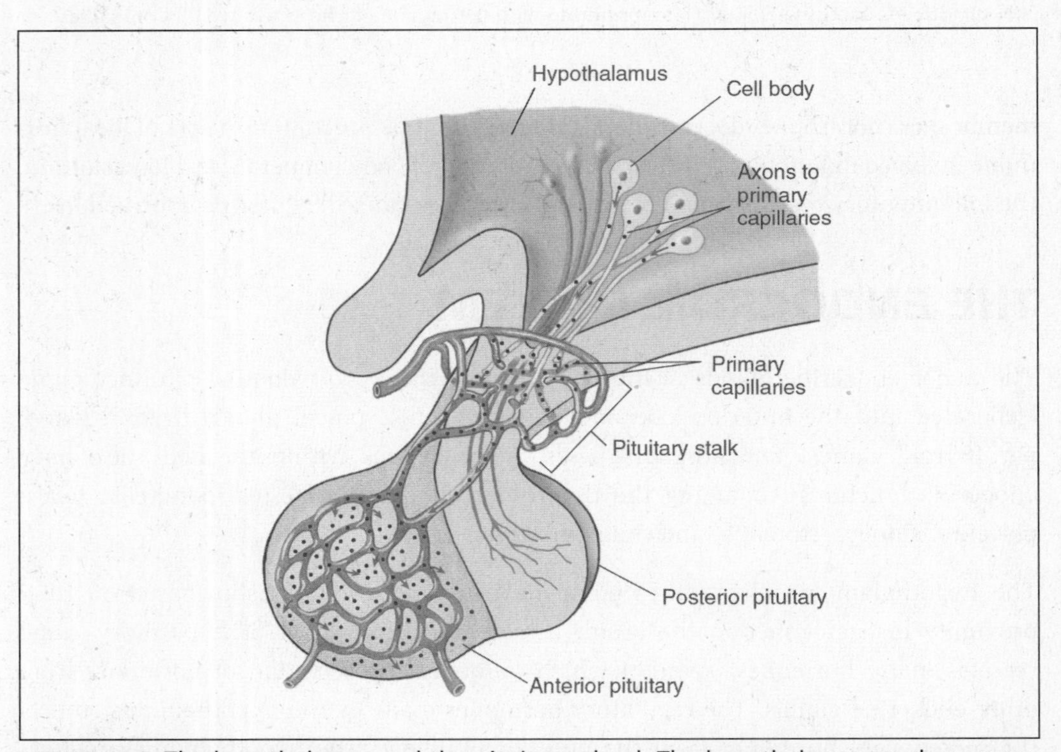

FIGURE 13-4: The hypothalamus and the pituitary gland. The hypothalamus produces regulatory hormones that travel to the pituitary gland. SOURCE: From George B. Johnson, *The Living World*, 3d ed., McGraw-Hill, 2003; reproduced with permission of The McGraw-Hill Companies.

releasing hormones, which stimulate the release of anterior pituitary hormones, or **inhibiting hormones**, which inhibit the release of hormones from the anterior pituitary. The hypothalamus also makes antidiuretic hormone and oxytocin, but both are stored and released from the posterior pituitary.

The following table lists the major endocrine glands, the hormones they produce, and the function of those hormones. Unless marked with an asterisk, each of these hormones is a peptide.

Endocrine Structures and the Hormones They Make

Endocrine Structure	Hormones Made and Their Function
Anterior pituitary	➤ Follicle stimulating hormone (FSH): in women, FSH stimulates the secretion of estrogen in the ovaries and assists in egg production via meiosis; in men, FSH has a role in sperm production ➤ Luteinizing hormone (LH): in women, LH causes ovulation; in men, LH is involved in testosterone secretion from the testes ➤ Thyroid stimulating hormone (TSH): stimulates the thyroid gland ➤ Growth hormone (GH): stimulates growth of muscle, bone, and cartilage ➤ Prolactin (PRL): stimulates the production of milk ➤ Adrenocorticotropic hormone (ACTH): stimulates the cortex of the adrenal glands ➤ Endorphins: act on the nervous system to reduce the perception of pain
Posterior pituitary (These hormones are made by the hypothalamus but are released by the posterior pituitary)	➤ Antidiuretic hormone (ADH): allows for water retention by the kidneys and decreases urine volume; also known as vasopressin ➤ Oxytocin (OT): causes uterine contractions during childbirth; also stimulates milk ejection
Pineal gland	➤ Melatonin: influences patterned behaviors such as sleep, fertility, and aging
Thyroid gland	➤ Thyroid hormones (TH): regulate metabolism throughout the body; also acts on the reproductive, nervous, muscular, and skeletal systems to promote normal functioning; T3 and T4 require iodine to function properly ➤ Calcitonin (CT): influences osteoblasts, which build bone in response to high blood calcium levels; ultimately lowers blood calcium levels
Parathyroid glands	➤ Parathyroid hormone (PTH): influences osteoclasts, which break down bone in response to low blood calcium levels, this ultimately increases blood calcium levels; PTH is antagonist to CT
Adrenal medulla (inner portion of the adrenals)	➤ Epinephrine: released in response to stress; causes fight-or-flight response; also known as adrenaline ➤ Norepinephrine: released in response to stress; causes fight-or-flight response; also known as noradrenaline
Adrenal cortex (outer portion of the adrenals)	➤ Glucocorticoids*: help cells convert fats and proteins into molecules that can be used in cellular respiration to make ATP; high levels inhibit the inflammatory response of the immune system; examples are cortisol and cortisone ➤ Mineralocorticoids*: increase sodium retention by the kidneys and potassium excretion; an example is aldosterone ➤ Gonadocorticoids*: secreted in small amounts; examples are androgens and estrogens

Endocrine Structures and the Hormones They Make (*Cont.*)	
Endocrine Structure	**Hormones Made and Their Function**
Thymus	➤ Thymopoietin: stimulates the maturation of certain white blood cells involved with the immune system (T cells); decreases with age as the thymus gland atrophies
Ovaries	➤ Estrogen*: involved in the development of female secondary sex characteristics as well as follicle development and pregnancy ➤ Progesterone*: involved in uterine preparation and pregnancy
Testes	➤ Testosterone*: a type of androgen needed for the production of sperm as well as for the development and maintenance of male secondary sex characteristics.
Pancreas	➤ Insulin: decreases blood sugar after meals by allowing glucose to enter cells to be used for cellular respiration; a lack of insulin or lack of response by cell receptors to insulin is the cause of diabetes mellitus; made by the beta islet cells ➤ Glucagon: increases blood sugar levels between meals by allowing for the breakdown of glycogen; antagonistic to insulin; made by the alpha islet cells
Heart	➤ Atrial natriuretic peptide (ANP): made by the heart to lower blood pressure
Kidneys	➤ Renin/angiotensin: Used to regulate blood pressure by altering the amount of water retained by the kidneys ➤ Erythropoietin (EPO): stimulates the production of red blood cells from stem cells in the red bone marrow
Stomach	➤ Gastrin: released when food enters the stomach; causes the secretion of gastric juice needed to begin the digestion of proteins
Small intestine	➤ Cholecystokinin (CCK): stimulates the release of pancreatic digestive enzymes to the small intestine; also stimulates the release of bile from the gallbladder to the small intestine ➤ Secretin: stimulates the release of fluids from the pancreas and bile that are high in bicarbonate to neutralize the acids from the stomach
Placenta (temporary organ during pregnancy)	➤ Human chorionic gonadotropin (HCG): signals the retention of the lining of the uterus (endometrium) during pregnancy ➤ Relaxin: used to release ligaments attaching the pubic bones to allow for more space during childbirth ➤ Estrogen: needed to maintain pregnancy ➤ Progesterone: needed to maintain pregnancy

*Steroid hormones

CRAM SESSION

The Endocrine System

1. Hormones are chemical messengers that travel via the blood to change the functioning of their target. Targets can be cells, tissues, or organs that contain a receptor for a particular hormone.
2. Endocrine tissues or glands secrete hormones into the bloodstream.
   a. Some endocrine structures have other functions unrelated to their endocrine role.
3. Hormones are used to help the body maintain homeostasis, usually through negative feedback loops, in which hormones are used to produce the opposite affect of a specific stimulus.
4. Hormones come in two major classes: steroid (lipid soluble) or nonsteroid (peptide, water soluble).
   a. Steroid hormones are derived from cholesterol and easily pass through the cell membrane. They find their receptors in the cytoplasm of their targets and bind to the receptor.
      (1) Interactions with the DNA result in the expression of new genes and new proteins that can change the functioning of the cells.
   b. Peptide hormones find their receptors on cell membranes and act as first messengers.
      (1) The binding of the hormone to the receptor initiates reactions in the cell that ultimately lead to the activation of second messengers, such as cAMP, which can alter the functioning of target cells.
5. The major endocrine structures and hormones of the body can be reviewed in the table given in the chapter.

CHAPTER 14

The Cardiovascular System

Read This Chapter to Learn About:

- Blood and Blood Types
- Blood Vessels
- The Structure of the Heart

The cardiovascular system in humans consists of a four-chambered **heart** to pump blood and a series of vessels that transport blood in the body. **Blood** is a connective tissue used to deliver oxygen, nutrients, water, hormones, and ions to all the cells of the body. It is also used to pick up the carbon dioxide and wastes produced by cells and to move these to the appropriate areas for elimination. Furthermore, it assists in thermoregulation in the body as well as in fighting infectious agents.

The cardiovascular system is closely linked to the following organ systems in the body:

- The **respiratory system** for the elimination of carbon dioxide, the pick-up of oxygen, and the regulation of blood pH
- The **urinary system** for the filtration of blood, removal of nitrogenous wastes, regulation of blood volume and pressure, and regulation of blood pH
- The **digestive system** for the pick-up of nutrients to be distributed to the body

The critical functions of the cardiovascular system are achieved by blood, which is transported through the system. For this reason, the composition of blood must be examined closely.

BLOOD

Blood consists of a liquid matrix, plasma, and formed elements, or cells. Humans contain 4 to 6 liters of blood, and this entire volume can be circulated through the body in less than 1 minute. The pH of blood is 7.4 (slightly basic), and the temperature is slightly warmer than body temperature. Because the temperature of blood is warmer than the body, changing patterns of circulation can help distribute heat to where it is needed in the body. **Vasoconstriction** decreases the diameter of vessels, keeping blood closer to the core to warm the body, whereas **vasodilation** increases the diameter of the vessels, allowing them to release heat toward the surface of the skin to cool the body.

Plasma

Plasma is the liquid portion of the blood, occupying approximately 55% of the total volume of blood. The primary component of plasma is water. In order to adjust the volume of blood in the body, the water levels of plasma can be altered. This is one role for the kidneys, which can retain or release water via urine to adjust the blood volume. An increase in blood volume increases blood pressure, whereas a decrease in blood volume decreases blood pressure.

In addition to water, plasma also contains nutrients, cellular waste products, respiratory gases, ions, hormones, and proteins. There are three classes of plasma proteins produced by the liver: immunoglobulins, albumins, and fibrinogens. **Immunoglobulins** are primarily used in the immune response, **albumins** are used to transport certain molecules within the blood, and **fibrinogen** is an inactive form of one protein that is needed to clot blood.

Formed Elements

The **formed elements**, or cells, of the blood are all derived from stem cells in the bone marrow. The three types of cells that are found within the blood are **erythrocytes** (red blood cells), **leukocytes** (white blood cells), and **thrombocytes** (platelets), all seen in Figure 14-1.

The **hematocrit** value of blood is a relative comparison of cell volume to plasma volume. The percentage of blood occupied by cells is considered the hematocrit value and is generally about 45%. Because red blood cells are by far the most abundant blood cell, hematocrit values are primarily influenced by red blood cells.

ERYTHROCYTES

Erythrocytes, or red blood cells, are the most abundant type of blood cell. As they mature from stem cells in the bone marrow, they do something odd in that they loose their organelles. Without organelles, these cells are unable to perform aerobic cellular respiration, and they cannot perform mitosis to replace themselves. These cells live only about 120 days, at which point they are destroyed by the liver and spleen. In order to make new red blood cells, more stem cells in the bone marrow must be coerced to differentiate into red blood cells by the hormone **erythropoietin (EPO)**. Red blood cells also have an unusual biconcave disc shape that provides them with increased surface area and the ability to be flexible as they move through small vessels.

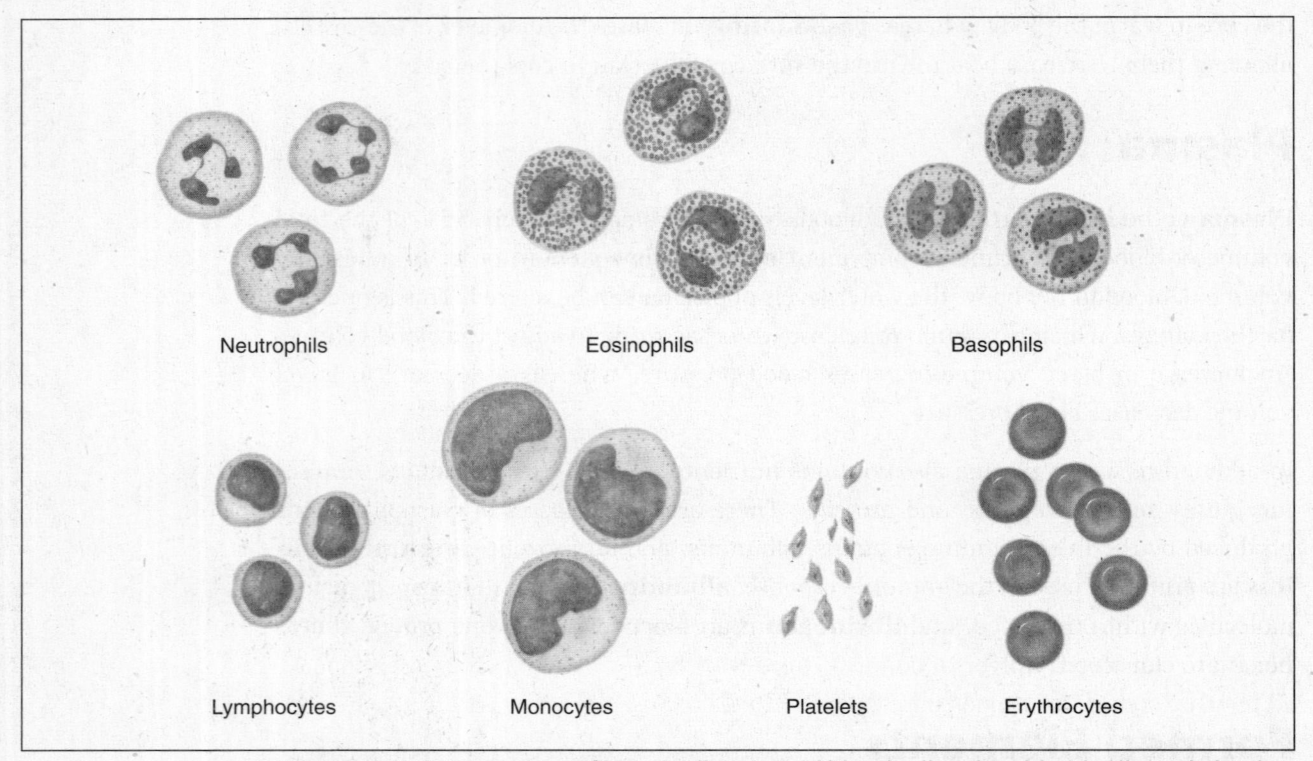

FIGURE 14-1: A comparison of blood cell types: red blood cells, white blood cells, and platelets all have different structures and functions. SOURCE: From Eldon D. Enger, Frederick C. Ross, and David B. Bailey, *Concepts in Biology*, 11th ed., McGraw-Hill, 2005; reproduced with permission of The McGraw-Hill Companies.

Erythrocytes are essential to the transport of oxygen throughout the body. They also determine an individual's blood type.

TRANSPORT OF GASES: The critical component of red blood cells is the protein **hemoglobin**. Each red blood cell contains about 250 million hemoglobin molecules. Functional hemoglobin consists of four protein chains each wrapped around an iron (heme) core. A molecule of hemoglobin is capable of carrying four molecules of oxygen (O_2). In total, a single red blood cell can carry about a billion O_2 molecules. The affinity of hemoglobin for O_2 is good; fetal hemoglobin has a higher affinity for O_2 than does adult hemoglobin. However, the respiratory poison carbon monoxide (CO) has a much greater affinity for hemoglobin than does O_2. CO binds to hemoglobin at the expense of O_2, ultimately starving the cells of O_2.

As hemoglobin binds to one oxygen molecule, a conformational change in the shape of hemoglobin occurs to allow for the loading of the next three oxygen molecules.

The same process occurs during the unloading of O_2. Once O_2 is unloaded in the capillary beds of the body, some of the CO_2 produced by the cells is carried by hemoglobin. Carbon dioxide also combines with water to produce carbonic acid, which dissociates into **hydrogen ions** and **bicarbonate ions**. The hemoglobin carries the hydrogen ions whereas the bicarbonate ions are carried by plasma. The **Bohr effect** states that increasing concentrations of hydrogen ions (which decrease blood pH) and increasing concentrations of CO_2 decrease hemoglobin's affinity for O_2. This allows for O_2 to unload from hemoglobin into tissues of the body, such as muscle, when CO_2 levels are high in tissues. In the lungs, a high level of O_2 encourages the dissociation of hydrogen ions from hemoglobin; these ions join with bicarbonate ions in the plasma to form CO_2 and water. The CO_2 is exhaled. The enzyme **carbonic anhydrase** catalyzes the formation and disassociation of carbonic acid. The end product of hemoglobin breakdown is **bilirubin**, which is ultimately excreted into the small intestine via bile from the liver.

BLOOD TYPE: Blood type is genetically determined by the presence or absence of specific antigens on the red blood cells, as discussed in Chapter 5. The immune system does not normally produce antibodies against any antigens that are present in the body, but it does produce antibodies against any antigens that are not in the body. If blood with foreign antigens is transfused—an incompatible blood transfusion—antibodies attack the foreign antigens of the incompatible blood type, causing **agglutination**, or clumping, of the blood. Type O blood has no surface antigens, and people with that type blood are considered universal donors. Those with type AB blood make no antibodies, so they are considered universal recipients. The following table displays some important characteristics of blood types and compatibilities.

Blood Types and Compatibilities

Blood Type	Antigens Present	Antibodies Produced	Can Donate to Types	Can Receive from Types
Type A	A	Anti-B	A, AB	A, O
Type B	B	Anti-A	B, AB	B, O
Type AB	A and B	None	AB	A, B, AB, O
Type O	None	Anti-A and anti-B	A, B, AB, O	O

There is another red blood cell antigen, the **Rh factor**, that is controlled by a separate gene. If the Rh factor is present, this is considered Rh^+ and no antibodies are made against the Rh factor. If the Rh factor is absent, this is considered Rh^- and anti-Rh antibodies can be made. Those who are Rh^+ can receive blood matched for type that is Rh^+ or Rh^-. Those who are Rh^- can receive only blood matched by type and that is also Rh^-.

LEUKOCYTES

Leukocytes, or white blood cells, are a diverse collection of cells, all of which are derived from stem cells in the red bone marrow and function throughout the body. They are found in much lower levels than red blood cells; however, the white blood cell level can fluctuate greatly, particularly when a person is fighting infection. Details of the specific types of white blood cells are discussed in Chapter 18.

White blood cells can be categorized in the following manner and are distinguished based on their microscopic appearance:

- **Granulocytes** have cytoplasm with a granular appearance. These cells include **neutrophils**, **basophils**, and **eosinophils**. Neutrophils are used to perform phagocytosis. Basophils are involved in inflammation and allergies, whereas eosinophils are involved in dealing with parasitic infections.
- **Agranulocytes** have cytoplasm that does not have a grainy appearance. They include **monocytes**, which mature into **macrophages**, and **lymphocytes**, which are further subdivided into T cells and B cells. Monocytes and macrophages perform phagocytosis, whereas lymphocytes function as the specific defenses of the immune system.

THROMBOCYTES

Thrombocytes, or platelets, are fragments of bone marrow cells called **megakaryocytes**. Platelets live only 10–12 days once mature, so they are replaced often. An injury to blood vessels initiates a complex series of reactions that ultimately converts the inactive plasma protein fibrinogen to fibrin. The platelets release **thromboplastin**, which converts the inactive plasma protein **prothrombin** to the active form, **thrombin**. Thrombin then converts **fibrinogen** to **fibrin**. Fibrin forms a meshwork around the injury; this network serves to trap other cells to form a clot. The process of blood clotting requires multiple plasma proteins as well as calcium and vitamin K.

BLOOD VESSELS

Blood flow progresses in unidirectional loops as seen in Figure 14-2. One loop is the **systemic circuit**, which moves blood from the heart throughout the body and back to the heart. The other loop is the **pulmonary circuit**, which moves blood from the heart to the lungs and then back to the heart. The blood flows in these loops through a series of vessels, the major ones being arteries and veins.

Arteries and Veins

Arteries are large blood vessels leaving the heart. The arteries have thick walls and are very elastic to accommodate blood pressure. As arteries leave the heart, they branch

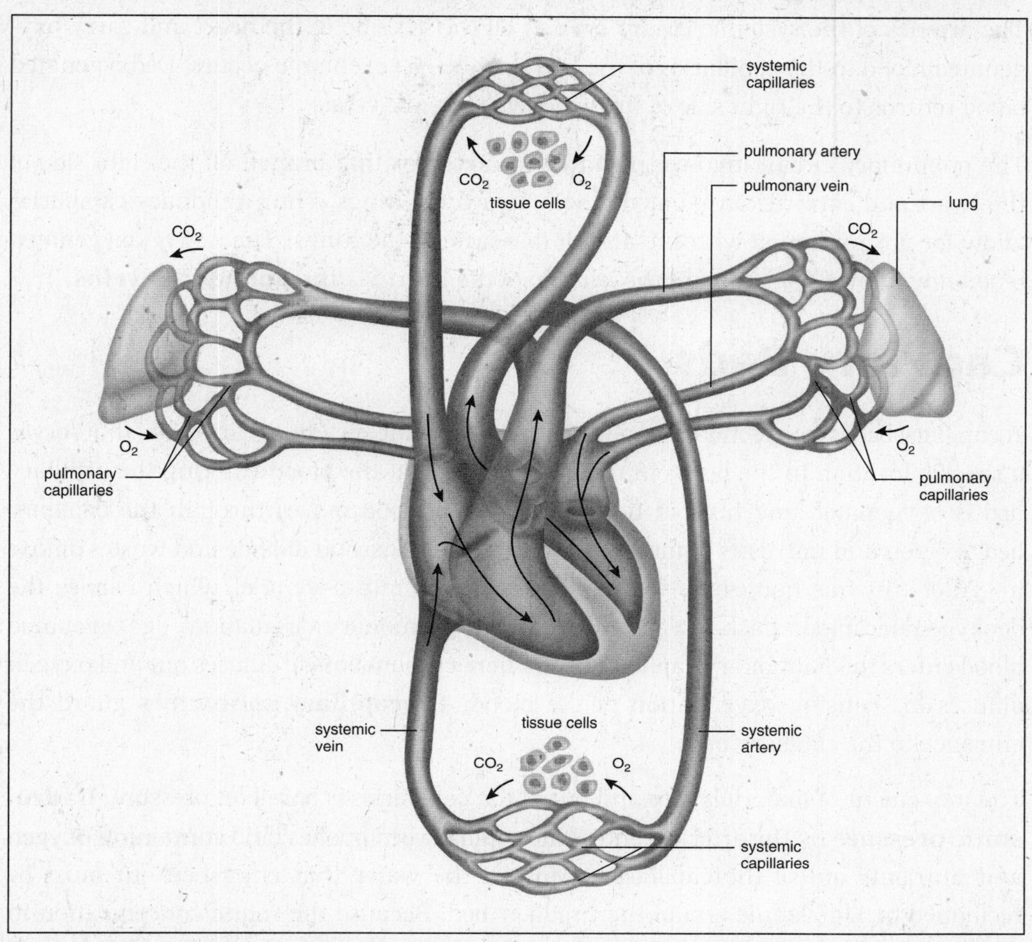

FIGURE 14-2: Blood flow through the body. The right side of the heart pumps blood to the lungs, where external respiration occurs, whereas the left side of the heart pumps blood to the body, where internal respiration occurs. SOURCE: From Sylvia S. Mader, *Biology*, 8th ed., McGraw-Hill, 2004; reproduced with permission of The McGraw-Hill Companies.

into smaller vessels called **arterioles**, which become more and more narrow, eventually forming the **capillaries**, which are the smallest vessels. **Capillary beds** are the site of gas exchange within tissues and are so small that red blood cells have to line up single file to pass through them. In the capillary beds the blood gives up its oxygen and picks up the waste product carbon dioxide. Once the gases have been exchanged, the capillaries become wider in diameter and become **venules**, which head back toward the heart. The venules become larger **veins** that ultimately merge into the heart. Veins are not as thick walled as arteries because they do not have to deal with the forces exerted by blood pressure. Although blood pressure pushes blood through arteries and arterioles, the movement of blood in venules and veins is facilitated by smooth muscles that contract to push the blood along and by valves that close to prevent the backflow of blood. **Vasoconstriction** and **vasodilation** of arteries serve as a means of regulating blood flow, pressure, and temperature.

The arteries of the systemic circuit branch off the left side of the heart and carry oxygenated blood to the capillaries of the body where gas exchange occurs. Deoxygenated blood returns to the right side of the heart via systemic veins.

The pulmonary circuit involves **pulmonary arteries** that branch off the right side of the heart and carry deoxygenated blood toward the lungs. The pulmonary capillaries allow for gas exchange with the alveoli (air sacs) of the lungs. The newly oxygenated blood now moves back toward the left side of the heart via the **pulmonary veins**.

Capillary Beds

A capillary bed is a collection of capillaries, all branching off a single arteriole, that serves a specific location in the body. In the systemic circuit, the blood entering the capillary bed is oxygenated and high in nutrients. As the blood moves through the capillary bed, oxygen and nutrients diffuse out into tissues and carbon dioxide and wastes diffuse in. After this has happened, the capillaries merge into a venule, which carries the deoxygenated blood back toward the heart. In pulmonary circulation, deoxygenated blood enters the pulmonary capillary bed, where carbon dioxide diffuses out and oxygen diffuses in, causing oxygenation of the blood. **Precapillary sphincters** guard the entrance to the capillary beds.

The movement of materials into and out of the capillaries is based on pressure. **Hydrostatic pressure** on the arteriole end of the capillary bed pushes fluid containing oxygen and nutrients out of the capillaries. Most of the water that is pushed out must be reclaimed on the venule end of the capillary bed. Because the solute concentration in the capillaries is higher than the fluids surrounding them, osmosis draws the water back into the capillaries at the venule end of the capillary bed as carbon dioxide and wastes diffuse in. Any excess water that is not reclaimed is returned to circulation by the lymphatic system. Some materials may enter or exit the capillaries via endocytosis or exocytosis.

STRUCTURE AND FUNCTION OF THE HEART

The structure of the heart can be seen in Figure 14-3. The **myometrium** is the cardiac muscle of the heart. Other tissues for supporting structures such as valves and chamber linings are also present. The right and left sides of the heart have very distinct functions and are kept separate from each other by the **septum**, which is a thick barricade between the two sides of the heart. Each side of the heart has two chambers. The upper chamber is the **atrium** and the lower chamber is the **ventricle**. The atria and ventricles are separated by **atrioventricular (AV) valves**. **Semilunar valves** regulate the flow of blood out of the ventricles. A fluid-filled sac called the **pericardium** surrounds the entire heart.

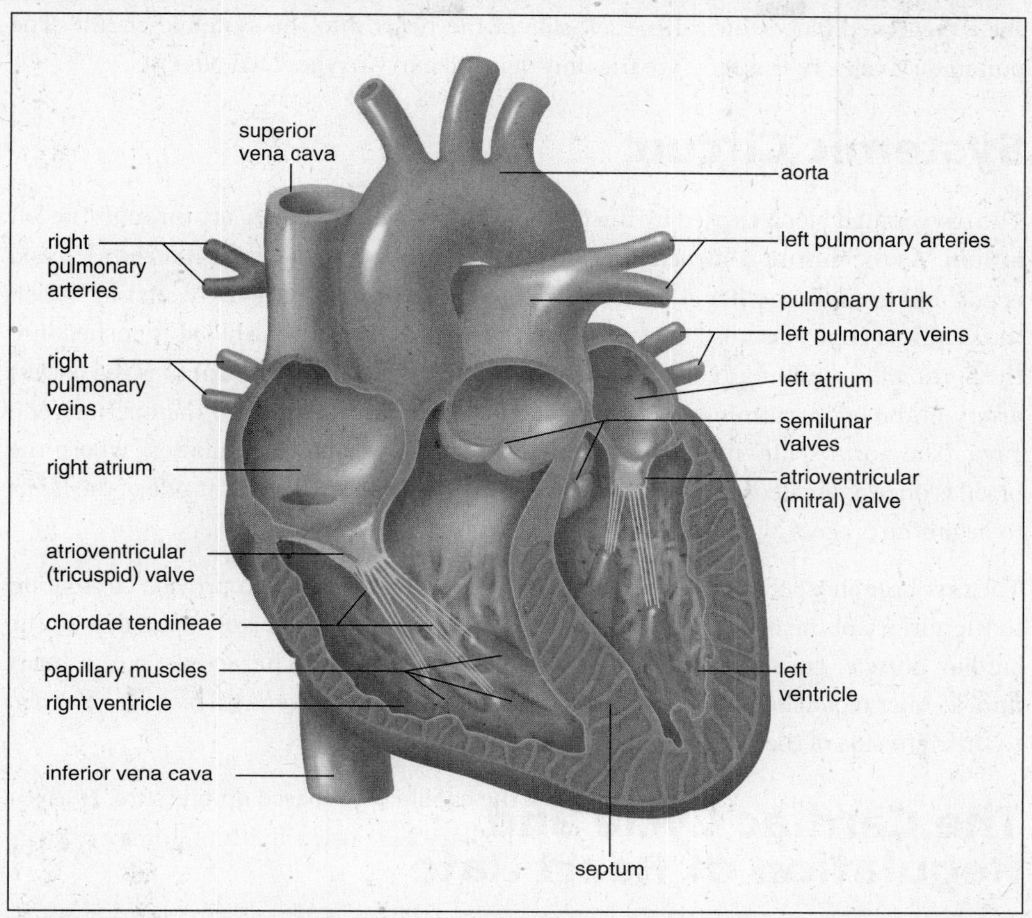

FIGURE 14-3: Heart structure. The vena cava carries deoxygenated blood into the right side of the heart. The aorta carries oxygenated blood out of the left side of the heart. SOURCE: From Sylvia S. Mader, *Biology*, 8th ed., McGraw-Hill, 2004; reproduced with permission of The McGraw-Hill Companies.

Pulmonary Circuit

The pulmonary circuit begins when veins within the body eventually merge into the **venae cavae**, which lie on the dorsal wall of the thoracic and abdominal cavities. The **superior vena cava** comes from the head and neck, whereas the **inferior vena cava** comes from the lower extremities. These vessels carry deoxygenated blood and merge into the right atrium of the heart. As the atrium contracts, blood passes through an AV valve (the **tricuspid valve**) into the right ventricle. As the ventricle contracts, blood passes through a semilunar valve (the **pulmonary semilunar valve**) into the pulmonary arteries. These **pulmonary arteries** carry blood to the lungs and are the only arteries in the body that do not carry oxygenated blood. They branch into capillaries that surround the **alveoli** (air sacs) in the lungs. There, gas exchange occurs to oxygenate the blood. Once gas exchange has occurred, **pulmonary veins** carry the oxygenated blood toward the left side of the heart into the systemic circuit. The pulmonary veins in the body are the only veins to carry oxygenated blood.

Systemic Circuit

The oxygenated blood carried by the pulmonary veins enters the heart through the left atrium. As the atrium contracts (in synch with contraction of the right atrium), blood is pushed through another AV valve (the **bicuspid valve**) to the left ventricle. When the ventricle contracts (also in synch with the right ventricle), the blood is pushed into the aorta via a semilunar valve (the **aortic semilunar valve**). The **aorta** is the largest artery in the body, running along the dorsal wall of the body next to the inferior vena cava. The aorta splits into arteries, arterioles, and eventually capillaries, where the blood is once again deoxygenated and must be pushed back to the right side of the heart to begin the process all over again.

The first branches off the aorta are the **coronary arteries**, which provide circulation to the surface of the heart. Blockage of the coronary arteries can stop blood flow to the cardiac muscle, causing death of the cardiac muscle that is characteristic of a heart attack. After blood flows through the coronary arteries, deoxygenated blood is returned to the right side of the heart by **coronary veins**.

The Cardiac Cycle and Regulation of Heart Rate

During the **cardiac cycle**—the events of a single heartbeat—two contractions occur. First, the two atria contract simultaneously, pushing blood into the ventricles. Next, the two ventricles contract (**systole**), pushing blood out of the heart. A brief resting period (**diastole**) occurs to allow the two atria to refill with blood and then the cycle begins again. An **electrocardiogram** (ECG or EKG) visualizes the electrical currents that are generated by the heart during the cardiac cycle.

Cardiac muscle is involuntary and has the ability to contract on its own without stimulation from the nervous system. The impulses that generate heart contraction are spread through the conducting system of the heart as seen in Figure 14-4. The **sinoatrial** (SA) **node**, also known as the **pacemaker**, is a bundle of conducting cells in the top of the right atrium that initiates contractions. The SA node sends electrical impulses through the two atria, causing them to contract. The impulse arrives at the **AV node** and then is spread through the **bundle of His** and through **Purkinje fibers** in the walls of the ventricles, causing ventricular contraction.

Although the SA node generates its own rate of contraction at an average of 70 contractions per minute, the heart is innervated by the autonomic nervous system, which can adjust the rate of contraction. The sympathetic branch of the autonomic nervous

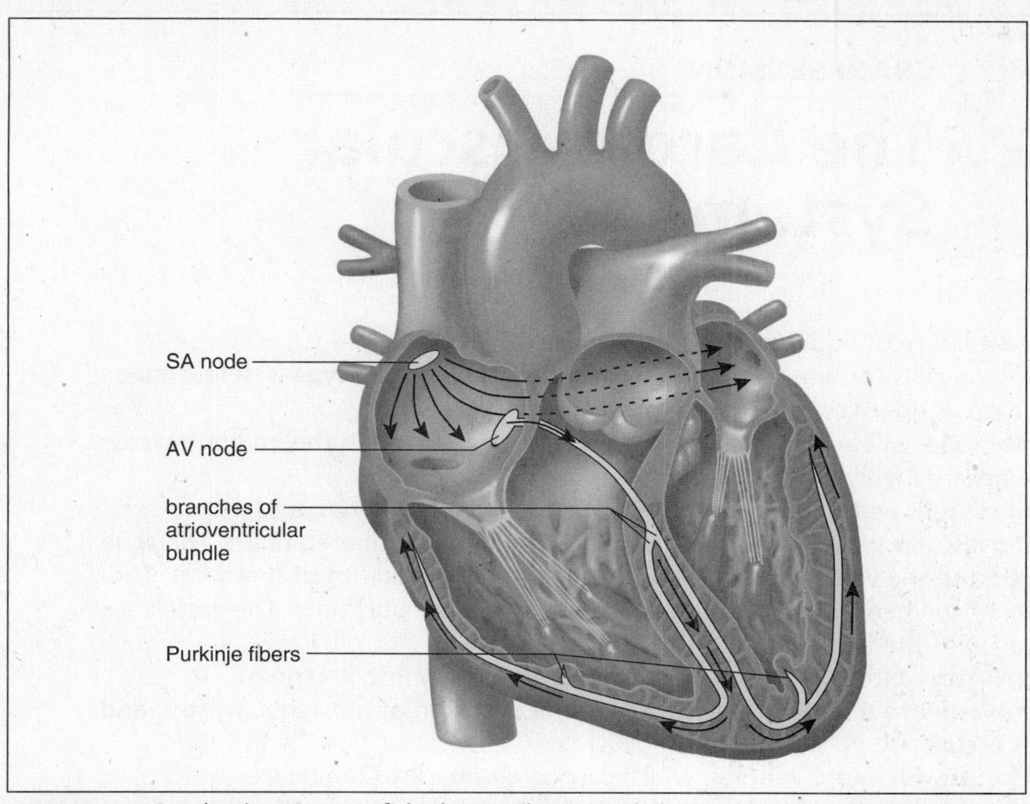

FIGURE 14-4: Conducting system of the heart. The SA node located in the right atrium serves as the pacemaker of the heart. SOURCE: From Sylvia S. Mader, *Biology*, 8th ed., McGraw-Hill, 2004; reproduced with permission of The McGraw-Hill Companies.

system can increase the rate of contraction, whereas the parasympathetic branch can decrease the rate of contraction. The medulla oblongata in the brain monitors conditions such as blood pH level (which is an indicator of CO_2 and O_2 levels) and signals adjustments in the heart rate as appropriate for the situation.

Blood Pressure

Blood pressure is a measurement of the force that blood exerts on the walls of a blood vessel. Typically it is measured within arteries. The pressure has to be enough to overcome the **peripheral resistance** of the arteries and arterioles. It is expressed with two values: a systolic pressure and a diastolic pressure. The **systolic pressure** is the higher value and is the pressure exerted on arteries as the ventricles contract. The **diastolic pressure** is the lower value and is a measurement of pressure on the arteries during ventricular relaxation. Blood pressure is regulated primarily by regulation of blood volume through the kidneys. The higher the blood volume, the higher the blood pressure is. A **sphygmomanometer** is used to measure blood pressure.

CRAM SESSION

The Cardiovascular System

1. Blood is a mixture of liquid plasma and formed elements.
   a. Red blood cells contain hemoglobin to transport O_2. Blood type is determined based on antigens on the surface of the red blood cells.
   b. White blood cells are involved in fighting infection through the immune system.
   c. Platelets are involved with blood clotting.
   d. Plasma carries water, gases, nutrients, wastes, ions, hormones, and more.
2. The cardiovascular system of humans consists of a four-chambered heart. The atria pump blood to the ventricles and the ventricles pump blood out of the heart. The chambers of the heart are separated by valves to prevent backflow. The vessels associated with the heart are as follows:
   a. Arteries carry blood away from the heart and narrow into arterioles.
   b. Arterioles narrow into capillaries, where the diffusion of nutrients, wastes, and gases occur.
   c. Capillaries widen into venules, which carry blood back to the heart.
   d. Venules widen into veins, which ultimately merge into the heart.
3. The right side of the heart is associated with the pulmonary circuit (pumping to the lungs) and the left side of the heart is associated with the systemic circuit (pumping to the body).
4. The contractions of the heart are initiated by the SA node (pacemaker) that sends signals through the conducting system of the heart.
   a. When needed, the autonomic branch of the peripheral nervous system can intervene and change the rate of contraction.
   b. CO_2 and pH levels are monitored closely to determine if the heart rate needs to increase or decrease.
5. Blood pressure is the force blood exerts on arteries. It can be adjusted by altering the blood volume. The kidneys can alter blood volume by adjusting the volume of water in plasma.
   a. Increased blood volume means increased pressure.
   b. Decreased blood volume means decreased pressure.

CHAPTER 15

The Respiratory System

Read This Chapter to Learn About:

- The Structures of the Respiratory System
- The Act of Ventilation
- Gas Exchange in the Alveoli

The respiratory system has the primary job of providing the body with oxygen and eliminating carbon dioxide. **Pulmonary arteries**, which are low in O_2 and high in CO_2, come off the right side of the heart and carry deoxygenated blood to the lungs. These arteries branch off into capillaries surrounding the alveoli (air sacs) in the lungs. CO_2 diffuses from the pulmonary capillaries into the alveoli to be exhaled. O_2 that enters the lungs is distributed to hemoglobin in the erythrocytes within the capillaries. The newly oxygenated blood travels via **pulmonary veins** back to the left side of the heart, from which it is distributed throughout the body.

In addition to oxygenating blood, the respiratory structures are responsible for pH regulation, vocal communication, the sense of smell, and protection from infectious agents and particles.

STRUCTURE OF THE RESPIRATORY SYSTEM

The respiratory system is essentially a series of tubes that conducts air into the alveoli located in the lung tissues. The major structures of the system can be seen in Figure 15-1.

Air is inhaled through the nose or mouth. Because the respiratory system is an open system, it is particularly vulnerable to infection. In the nose and **pharynx** (back of the throat), air is warmed up to body temperature, moisturized so that gas exchange can occur, and filtered. Both areas are covered with a mucous membrane that helps

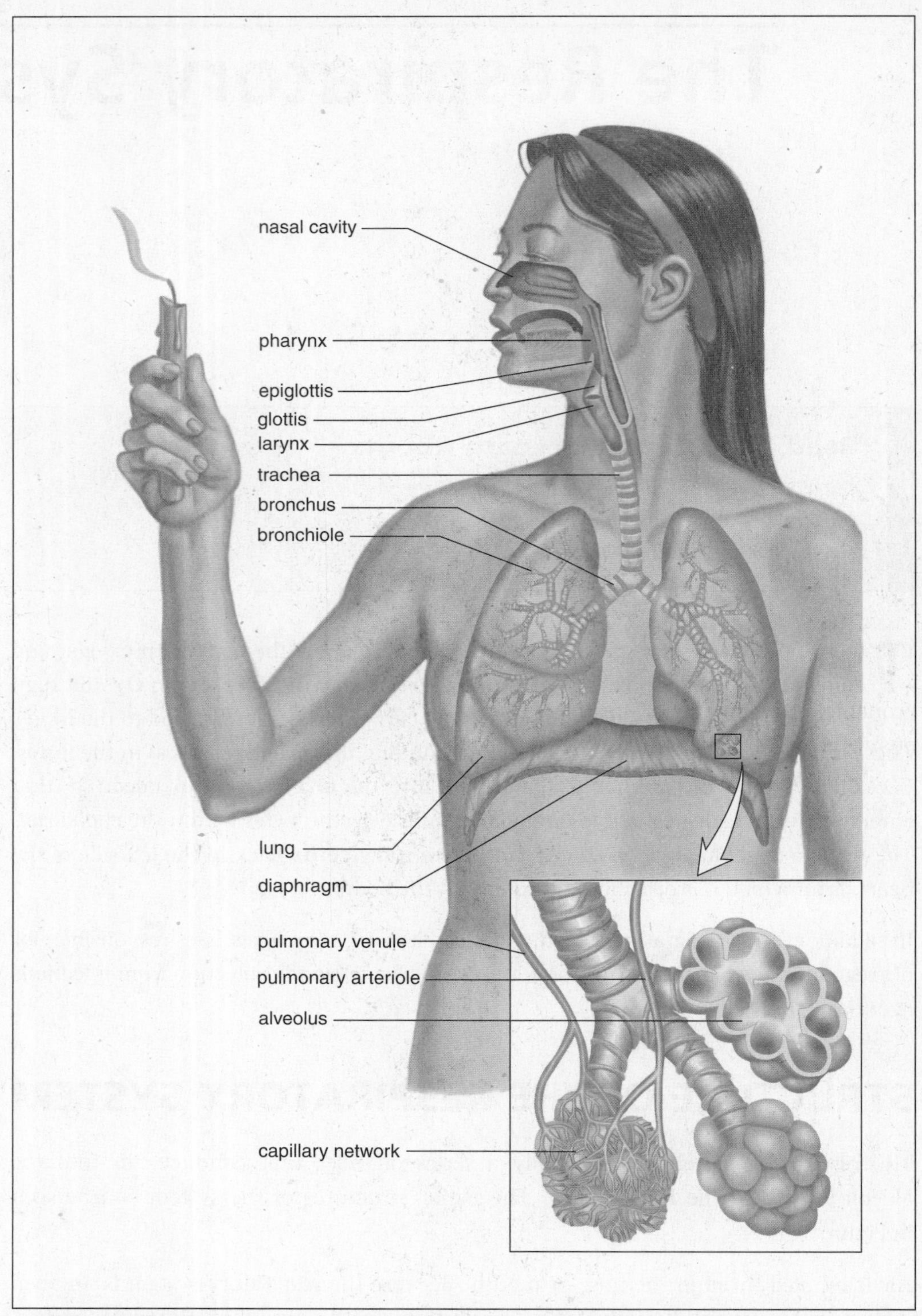

FIGURE 15-1: The respiratory system. Gas exchange occurs between the alveoli of the lungs and surrounding capillaries. SOURCE: From Sylvia S. Mader, *Biology,* 8th ed., McGraw-Hill, 2004; reproduced with permission of The McGraw-Hill Companies.

prevent desiccation of the tissues and collects particles and microbes that may enter the system. The nose is particularly well suited to filtration because it has cilia and hair to help trap substances that enter the respiratory system. Although filtration in the nose and pharynx does not catch all particles, it catches many of them. The nose has the additional function of olfaction.

Air that passes through the nose or mouth moves into the pharynx, where there are two passageways: the **esophagus** and the **larynx**. During breathing, air flows through the **glottis**, which is the opening of the larynx. The larynx is the voice box; it is made of cartilage and has vocal cords that vibrate, producing sound. Unless a person is swallowing, the esophagus is closed off and the glottis is open. However, if a person is swallowing, a small piece of cartilage called the **epiglottis** covers the glottis and stops food from entering the larynx.

As air flows through the larynx, it eventually makes its way into the trachea and the lower respiratory tract. The trachea is supported by C-shaped rings of cartilage. The interior surface of the trachea is covered with mucus and cilia to further trap any materials that may not have been caught in the nose or pharynx.

The trachea branches off toward the left and right into **bronchi**. The two bronchi branch into smaller and smaller tubes called the **bronchioles**. Smooth muscles surrounding the bronchioles can adjust their diameter to meet oxygen demands. The bronchioles terminate in tiny air sacs called the **alveoli**. The alveoli are numerous to provide a large amount of surface area for gas exchange. They are made of simple squamous epithelium that allows for easy gas exchange with the capillaries that surround them.

The **lungs** are a collection of resilient tissue, encompassing the bronchioles and alveoli. In humans, the right lung has three lobes of tissue, whereas the left lung has only two lobes. A fluid-filled **pleural membrane** surrounds each lung. A **surfactant** fluid produced by the tissues decreases surface tension in the lungs, which keeps the alveoli inflated and functioning and helps prevent alveolar collapse. Without surfactant to relieve surface tension, the lungs are unable to function.

VENTILATION

Gas exchange within the lungs results from pressure gradients. To get air into the lungs, the volume of the chest cavity must increase to decrease pressure in the chest cavity. This allows air to flow from an area of more pressure (outside the body) to an area with less pressure (the chest cavity). This is the process of **inhalation** (or **inspiration**). It occurs when the **diaphragm**, the thin muscle that separates the thoracic cavity from the abdominal cavity, contracts and pushes down, as seen in Figure 15-2. The **intercostal muscles** of the rib cage also assist in inhalation by contracting to help move the ribcage up and out. When the diaphragm and intercostal muscles relax, the volume of the chest cavity decreases, which results in a higher level of pressure inside the chest cavity as compared to outside. This forces air to leave the lungs by the process of **exhalation (expiration)**.

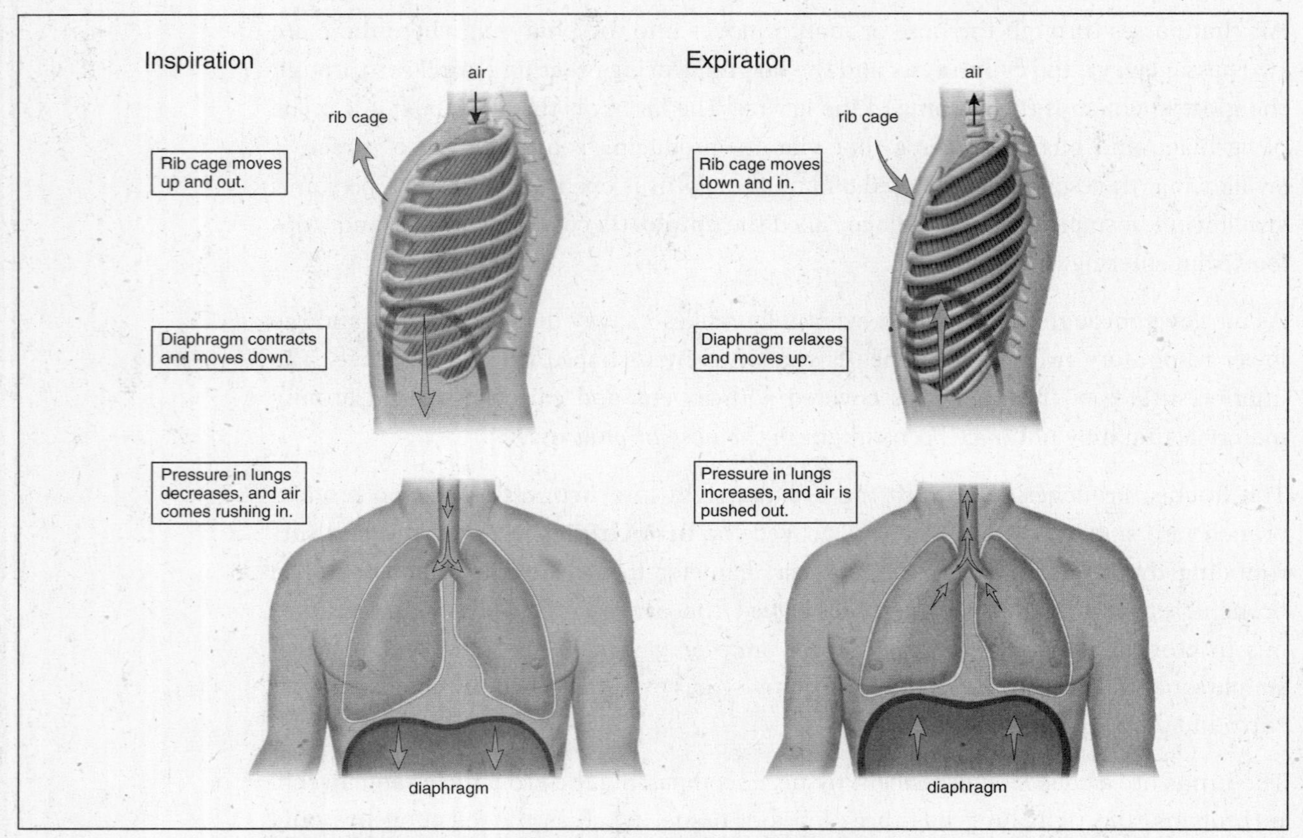

FIGURE 15-2: Ventilation. Contraction of the diaphragm allows for inspiration, whereas relaxation of the diaphragm allows for exhalation. SOURCE: From Sylvia S. Mader, *Biology*, 8th ed., McGraw-Hill, 2004; reproduced with permission of The McGraw-Hill Companies.

The medulla oblongata of the brain controls the ventilation rate. The diaphragm is innervated and connected to the area of the medulla that controls breathing. The inspiratory neurons are active, causing contraction of the diaphragm, followed by a period of inactivity that allows for relaxation of the diaphragm and exhalation. In a relaxed situation, the diaphragm is stimulated between 12 and 15 times per minute. During times of increased O_2 demand and excessive CO_2 production, this rate can increase significantly. Although it might be expected that oxygen levels are the primary influence on breathing rate, it turns out that the primary trigger is the CO_2 level, which is monitored by chemoreceptors located in the brain and in certain large blood vessels. As CO_2 levels increase, the pH decreases, and the breathing rate increases to eliminate the excess CO_2. This, in turn, increases the O_2 levels.

GAS EXCHANGE

The concentration of gases can be measured as partial pressures. After an inhalation, the amount of O_2 or **partial pressure** of O_2 in the alveoli is greater than the amount of O_2 or partial pressure in the capillaries surrounding the alveoli. Gases flow from an area of high concentration (partial pressure) to an area of low concentration (partial pressure), so O_2 moves from the alveoli into the capillaries and there binds to hemoglobin. Furthermore, immediately following inhalation, CO_2 levels are low in the alveoli and high in the capillaries. Diffusion moves the CO_2 into the alveoli where it can be exhaled. At this point, the blood in the pulmonary capillaries is oxygenated and ready to move back to the left side of the heart via the pulmonary veins.

The role of CO_2 exchange is important in the maintenance of acid–base balance within the body. When CO_2 interacts with water, it forms carbonic acid. The carbonic acid is converted to bicarbonate ions and hydrogen ions, as described in Chapter 14. The bicarbonate ions help buffer pH in the body. When the pH of the body becomes too acidic, the reaction can be reversed. The bicarbonate and hydrogen ions join together to produce carbonic acid, which is then converted to water and CO_2. The CO_2 is exhaled in order to adjust pH.

CRAM SESSION

The Respiratory System

1. The respiratory system is intimately connected with the circulatory system. The primary role of the respiratory system is to perform gas exchange to oxygenate blood coming from the right side (pulmonary circuit) of the heart.
2. The pathway of airflow moves from the nasal cavity to the pharynx through the glottis to the larynx, down the trachea, into the bronchi and bronchioles, ultimately terminating in the alveoli, where gas exchange occurs.
3. Surfactant fluid is necessary to reduce surface tension and the capacity of alveoli to collapse on themselves.
4. Ventilation occurs as the result of differing pressures inside and outside the thoracic cavity. The rate of ventilation is controlled by the medulla in the brain, which monitors blood pH.
   a. As the diaphragm contracts during inspiration, it pushes down into the abdominal cavity, making more room in the chest and decreasing pressure. This lets air flow into the lungs.
   b. During exhalation, the diaphragm relaxes, pushes up, and increases the pressure in the chest cavity. This pushes air out of the lungs.
5. Gas exchange in the alveoli is the result of differences in partial pressures.
   a. Immediately following inhalation, the partial pressure of O_2 is higher in the alveoli than the pulmonary capillaries, thus O_2 flows into the pulmonary capillaries.
   b. CO_2 partial pressure is higher in the pulmonary capillaries than in the alveoli, so CO_2 diffuses into the alveoli.
6. The exchange of CO_2 is critical to pH maintenance. The conversion of CO_2 to carbonic acid to bicarbonate and hydrogen ions is a reversible reaction that is used to regulate pH.

CHAPTER 16

The Digestive System

Read This Chapter to Learn About:

- ➤ The Structures of the Digestive System
- ➤ The Pathway of Food Through the Digestive Tract
- ➤ The Oral Cavity, the Stomach, the Small Intestine, and the Large Intestine
- ➤ The Liver, Pancreas, and Gallbladder
- ➤ The Absorption of Nutrients

The digestive system, also known as the **gastrointestinal (GI) tract**, is designed to extract nutrients (i.e., glucose, amino acids, and fatty acids) from food and eliminate wastes. Specifically, the digestive system has the following functions: the mechanical digestion of food achieved by chewing, the chemical digestion of food achieved by assorted digestive enzymes, the absorption of nutrients into the bloodstream, and the elimination of waste products. The system is set up as a series of modified tubes to keep food and digestive enzymes sequestered from the body, as shown in Figure 16-1.

TISSUES OF THE DIGESTIVE SYSTEM

Any contact of the digestive enzymes with the rest of the body could actually result in the digestion of self tissues. Furthermore, the digestive system is an open system where infectious organisms can enter. The contents of the system must be kept away from the rest of the body. In order to make sure that the components of the digestive system are kept separate from the rest of the body, the digestive tubes are composed of four tissue layers, as follows:

- ➤ The **muscosa** layer is a mucous membrane that actually comes in contact with food. It serves as a lubricant and provides protection from desiccation, abrasion, and digestive enzymes. The mucosa lacks blood vessels and nerve endings.

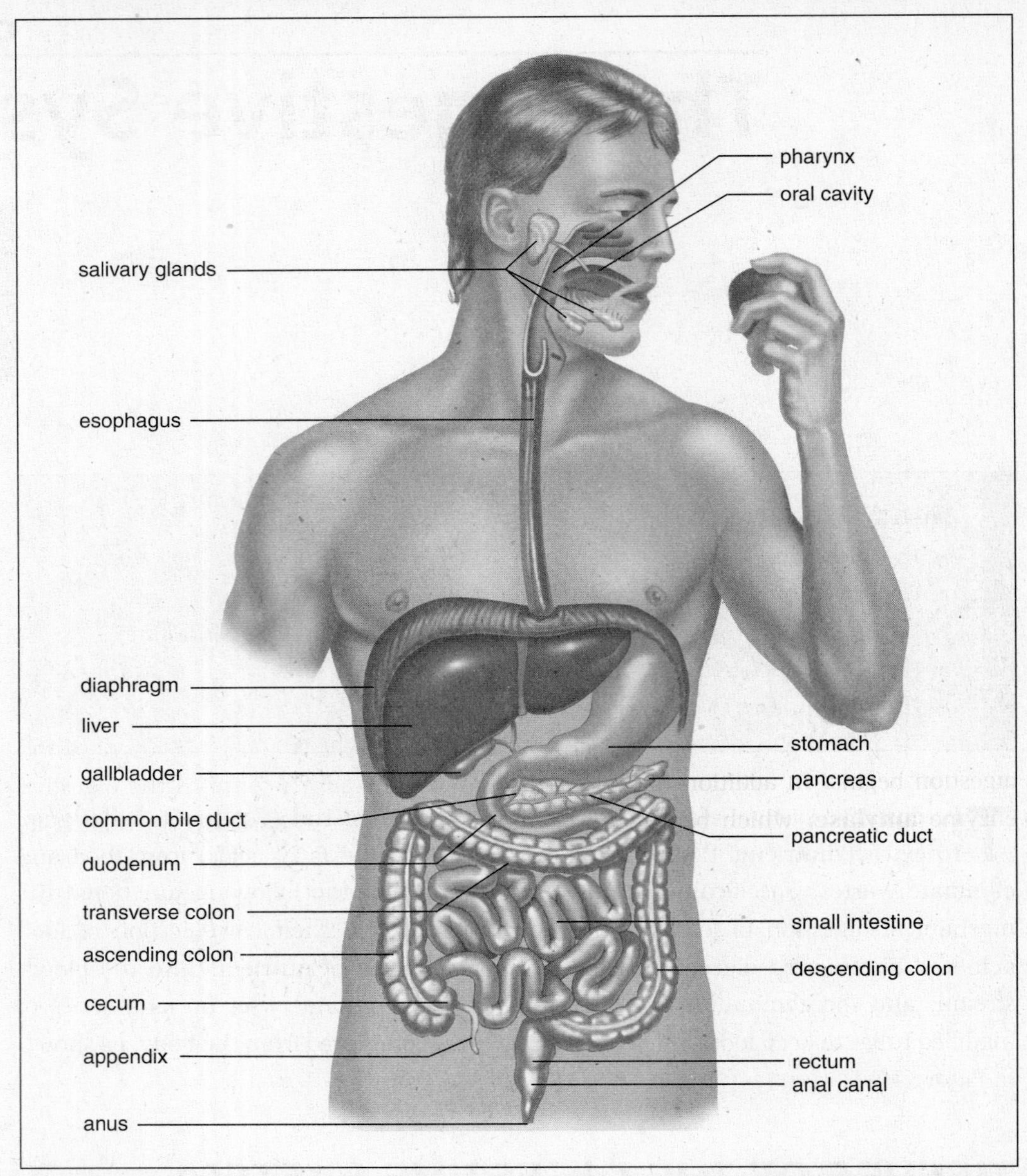

FIGURE 16-1: The digestive system. The digestive system includes the gastrointestinal tract as well as the accessory structures of the liver and pancreas. SOURCE: From Sylvia S. Mader, *Biology*, 8th ed., McGraw-Hill, 2004; reproduced with permission of The McGraw-Hill Companies.

- The **submucosa** layer is below the mucosa. It contains blood vessels, lymphatic vessels, and nerve endings. Its primary function is to support the muscosa and to transport materials to the bloodstream.
- The **muscularis** layer is composed of two layers of smooth muscle that run in opposing directions. The nerve endings in the submucosa serve to stimulate the muscularis layer to produce contractions that propel food through the system. The muscular contractions are termed **peristalsis**.

- The **serosa** is a thin connective tissue layer that is found on the surface of the digestive tubing. Its purpose is to reduce friction with other surfaces in contact with the GI tract.

THE PATHWAY OF FOOD THROUGH THE DIGESTIVE TRACT

Food enters the digestive tract at the oral cavity. From there, it moves to the esophagus and then the stomach. Small bits of the stomach contents are released to the small intestine. The small intestine completes digestion with some help from secretions from the liver and pancreas, and is the site where nutrients are absorbed into the bloodstream. Finally, the waste products of digestion are solidified in the large intestine and released.

The Oral Cavity

As food is ingested and enters the mouth, three sets of salivary glands begin to secrete **saliva**. The teeth are responsible for mechanical digestion of food by breaking it into smaller pieces by chewing. As saliva mixes with the chewed food, chemical or enzymatic digestion begins. In addition to its lubricating function, saliva contains the digestive enzyme **amylase**, which begins the chemical breakdown of carbohydrates such as starch. Because food does not stay in the mouth for long, amylase rarely gets to complete its job in the oral cavity. During chewing, the food is rolled into a **bolus** (small ball) by the tongue and is ready to be swallowed.

The Esophagus

As food is ready to be swallowed, it must pass by the **pharynx**. Recall that the pharynx has two openings: one to the larynx and one to the esophagus. Normally, the esophagus is closed during breathing so that air passes through the larynx. When food touches the pharynx, a reflex action occurs that pushes the **epiglottis** over to cover the **glottis** of the **larynx**. This allows for the bolus to proceed down the esophagus. Once in the esophagus, muscular contractions force the food toward the stomach by **peristalsis**.

The Stomach

The **stomach** is a relatively small, curved organ when empty, but due to the presence of many folds in its interior lining, it can expand greatly when full of food. The stomach is unique in that it has a very acidic environment, and its secretions must be retained in the stomach. Tightly closing muscular **sphincters** guard the top and bottom of the stomach and make sure this happens.

The top sphincter opens to allow the bolus to enter. Once food is inside the stomach, it is mixed with **gastric juice** for the purpose of liquefying the food as well as initiating the chemical digestion of proteins. The hormone **gastrin** signals the gastric glands of the stomach to begin producing gastric juice as well as to start churning.

Gastric juice is composed of a mixture of mucus to protect the stomach lining itself from being digested; pepsinogen, which is an inactive form of the enzyme that digests protein; and **hydrochloric acid**, which is needed to activate the pepsinogen to its active form called **pepsin**. The hydrochloric acid secreted in the stomach provides an overall pH of 1–2, which is highly acidic. Normally a pH this low would denature enzymes, but pepsin is unusual in that it is inactive except at a low pH. The low pH of the stomach is also helpful in killing most infectious agents that may have entered the digestive tract with food.

After the food mixes with gastric juice, the resulting liquid is called **chyme**. The chyme leaves the stomach in small bursts as the bottom stomach sphincter opens. Depending on the size and nutritional content of the meal, it takes on average about 4 hours for the stomach to empty its contents into the small intestine. Strangely, the only items that are absorbed into the bloodstream from the stomach are alcohol and aspirin.

The Small Intestine

The **small intestine** is a tube of approximately 6 meters in length. Its primary job is to complete the chemical digestion of food and to absorb the nutrients into the bloodstream. The small intestine relies on secretions from the liver and pancreas to complete chemical digestion. As the bottom sphincter of the stomach opens, small amounts of chyme enter the top region of the small intestine, which is termed the **duodenum**. The acidity from gastric juice must be neutralized. This is done by the secretion of sodium bicarbonate from the pancreas into the small intestine. In addition to receiving secretions made from the pancreas, the duodenum also receives secretions from the liver. These secretions help with chemical digestion, which occurs in the middle region of the small intestine, termed the **jejunum**, and in the lower end of the small intestine, termed the **ileum**.

THE LIVER AND GALLBLADDER

The **liver** is composed of several lobes of tissue and is one of the larger organs in the body. The liver has countless functions. In the case of the digestive system, the liver produces **bile**, which is a fat emulsifier. Although bile is not an enzyme, it helps break fats into smaller pieces so that they are more susceptible to digestion by enzymes secreted from the pancreas. Bile contains water, cholesterol, bile pigments, bile salts, and some ions. Bile from the liver is stored in the **gallbladder**, a small structure on the underside of the liver. As food enters the small intestine, the hormones **secretin** and **CCK** signal for the release of bile to the small intestine via the **common bile duct**.

The liver has some other functions within the digestive system. After nutrients are absorbed in the small intestine, blood from the capillaries in the small intestine travel directly to the liver via the **hepatic portal vein**. Once in the liver, the glucose levels of the blood are regulated. When blood glucose levels get high, the liver stores the excess as **glycogen** under the influence of insulin. When blood sugar levels are low, the liver breaks down glycogen to release glucose under the influence of **glucagon**. The liver also packages lipids in lipoproteins to allow them to travel throughout the body. The smooth endoplasmic reticulum within the liver cells produces enzymes to detoxify certain harmful substances. The liver also stores vitamins A, E, D, and K (the fat-soluble vitamins). After these functions occur in the liver, blood re-enters general circulation.

THE PANCREAS

The **pancreas** secretes pancreatic juice into the small intestine via the **pancreatic duct**. Although the pancreas has cells involved in endocrine functions (producing insulin and glucagon), it also has exocrine cells that produce **pancreatic juice**. Signaled by the hormones secretin and CCK, the pancreas secretes pancreatic juice when food enters the small intestine.

Pancreatic juice contains the following substances:

- **Bicarbonate ions**, which act as a neutralizer of stomach acid
- **Amylase**, which completes carbohydrate (starch) digestion that began in the oral cavity to release glucose
- **Proteinases**, which complete protein digestion that was started in the stomach to release amino acids. There are three specific proteinases found in pancreatic juice—trypsin, chymotrypsin, and carboxypeptidase.
- **Lipase**, which breaks down fats to fatty acids and glycerol
- **Nucleases**, which break down DNA and RNA to nucleotides

ABSORPTION OF NUTRIENTS

Once the bile and pancreatic juice are mixed with the contents of the small intestine, chemical digestion is nearly complete. Although the pancreatic enzymes are most essential to chemical digestion, there are a few additional enzymes that are needed. They are made by the small intestine and include

- **Maltase**, which breaks down the disaccharide maltose to glucose
- **Sucrase**, which breaks down the disaccharide sucrose to glucose and fructose (another simple sugar)

- **Lactase**, which breaks down the disaccharide lactose to glucose and galactose (another simple sugar)
- **Aminopeptidase**, which breaks down small pieces of proteins to amino acids

Once the food has been exposed to the secretions of the pancreas, liver, and small intestine, chemical digestion is complete. Then, the nutrients must be absorbed and the wastes eliminated. It can take anywhere from 3 to 10 hours for nutrients to be absorbed from the small intestine.

The small intestine has an internal anatomy with a tremendous surface area that makes it well suited for absorption, as seen in Figure 16-2. The mucosa in the small intestine is folded into **villi**, which form the brush border. The villi are then further folded into microscopic **microvilli**. Within each **villus**, there are capillaries and a **lacteal** (a lymphatic capillary). Nutrients such as glucose and other simple sugars, amino acids, vitamins, and minerals diffuse into the capillaries within each villus. From there they are carried into the blood circulation. The products of fat digestion take another route.

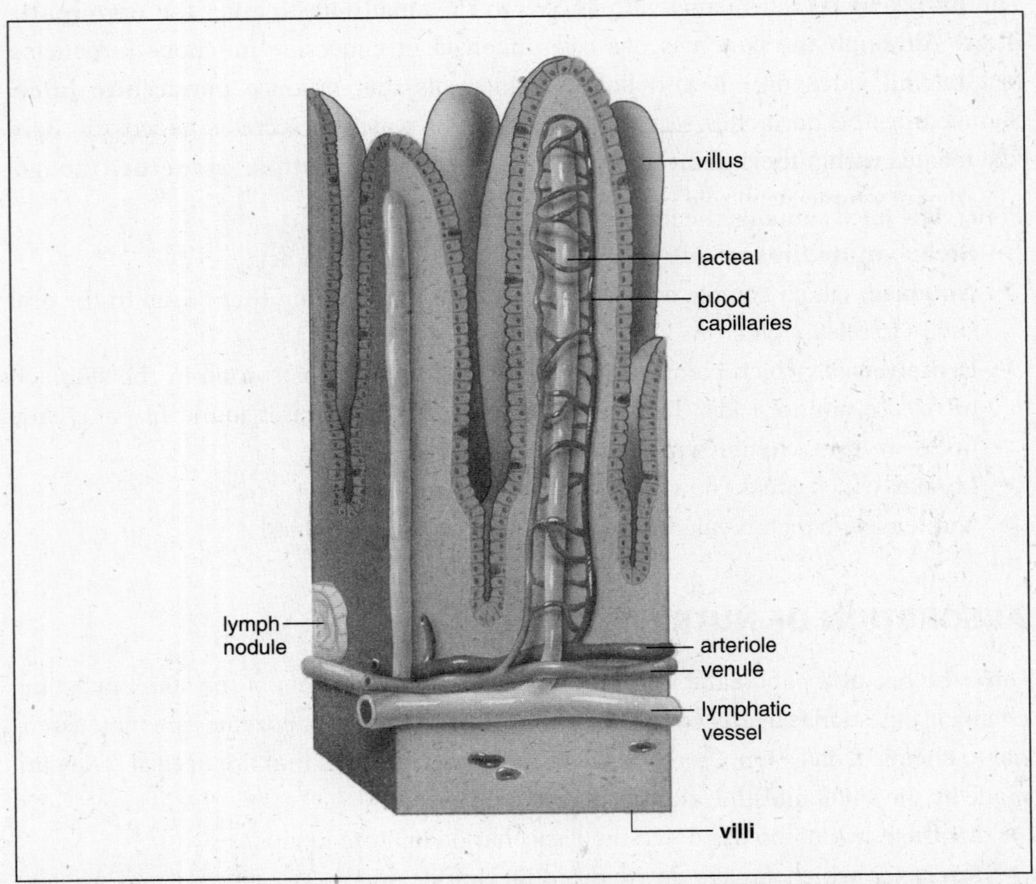

FIGURE 16-2: Absorption of nutrients. Nutrients are absorbed by villi, which are located in the small intestine. SOURCE: From Sylvia S. Mader, *Biology*, 8th ed., McGraw-Hill, 2004; reproduced with permission of The McGraw-Hill Companies.

The fat products are assembled into a triglyceride and packaged in a special coating, including cholesterol, which creates a **chylomicron**. These structures cannot diffuse into capillaries, so they enter the lacteals. The lymphatic fluids deliver the chylomicrons to the bloodstream at the **thoracic duct**, which is a merger between the circulatory and lymphatic systems.

The Large Intestine

Now that the nutrients have been absorbed into the bloodstream, the remnants of digestion have made their way to the **large intestine**. Now water must be reclaimed by the body to solidify the waste products. These waste products are stored by the large intestine and released at the appropriate time. In addition, the large intestine contains a large population of **normal flora**, or harmless resident bacteria. These bacteria are responsible for the synthesis of certain vitamins that the body needs.

The large intestine has a much larger girth than the small intestine, but its length is reduced. The large intestine is about 1.5 meters long. There are four regions within the large intestine:

- The **cecum** is a small area where the large intestine connects with the small intestine on the right side of the body. There is an outgrowth of this area that constitutes the appendix. The **appendix** has nothing to do with digestion, but it happens to be located within the digestive system. The appendix is a **vestigial structure** thought to play a noncrucial role in the lymphatic system.
- The **colon** constitutes the majority of the large intestine. The **ascending colon** moves up the right side, the **transverse colon** moves horizontally across the abdomen, and the **descending colon** runs down the left side of the body. The primary role of the colon is water absorption in order to solidify the feces. Vitamin absorption can also occur in the colon. It can take up to 24 hours for materials to pass through the colon.
- The **rectum** is the ultimate destination for feces in the large intestine. Stretching of this area stimulates nerves and initiates the defecation reflex.
- The **anal canal** receives the contents of the rectum for elimination. There are two sphincters regulating exit from the anal canal. The first internal sphincter operates involuntarily and the second external sphincter is under voluntary control.

CRAM SESSION

The Digestive System

1. The GI tract is structured to make sure that the contents of the digestive system do not leak into other areas of the body.
   a. Four layers of tissue are found in the digestive organs: the muscosa, submucosa, muscularis, and serosa.
2. The role of the digestive system is to chemically break down nutrients to their monomer subunits. These nutrients then must be transported to the circulatory system for distribution to the cells of the body.
3. In the oral cavity, mechanical digestion occurs and chemical digestion begins. The enzyme amylase is used to break down carbohydrates like starch. Food then travels through the esophagus to the stomach.
4. In the stomach, food is churned with gastric juice. The secretions from the stomach include mucus (for protection), hydrochloric acid, and pepsinogen. HCl converts pepsinogen to its active form—pepsin, which is capable of protein digestion.
5. As food enters the small intestine, secretions from the pancreas neutralize stomach acid.
   a. Bile from the liver (and stored in the gallbladder) begins to emulsify fats.
   b. Amylase, proteinases, lipase, and nucleases from the pancreatic juice continue chemical digestion of carbohydrates and proteins and begin the digestion of fats and nucleases.
   c. The small intestine produces other enzymes to further degrade disaccharides and small protein fragments.
   d. Glucose and amino acids are absorbed through the villi into the blood circulation, whereas the products of fat digestion enter lacteals and eventually merge into the blood circulation.
6. The large intestine contains bacteria that synthesize certain vitamins. It also reclaims water from the remaining components and solidifies, stores, and eliminates wastes.

CHAPTER 17

The Urinary System

Read This Chapter to Learn About:

- Organs of the Urinary System
- Kidney Structure
- Regulation of Blood Volume and Pressure
- Maintenance of Acid–Base Balance
- Properties of Urine
- Other Functions of the Kidneys

The urinary system consists of the kidneys that produce urine and supporting structures—ureters, bladder, and urethra—that store, transport, and eliminate urine from the body. The kidneys are the main excretory organs of the body; however, the skin can also act as an excretory organ. In addition to producing urine as a means of eliminating nitrogenous cellular waste products, the urinary system also regulates blood pressure by adjusting blood volume, adjusts blood pH, and regulates the osmotic concentrations of the blood.

The two **kidneys** of the urinary system filter blood to produce urine. The urine then moves toward the bladder via two **ureters**, which are tubes that connect each kidney to the **bladder**. Once urine moves into the bladder, it is stored until it eventually leaves the body through the **urethra**. The anatomy of the urethra is different in males and females. In males, the urethra is relatively long and is shared with the reproductive system so that sperm can move through it when appropriate. In females, the urethra is shorter, and it is only used for urine passage. The structures of the urinary system are shown in Figure 17-1.

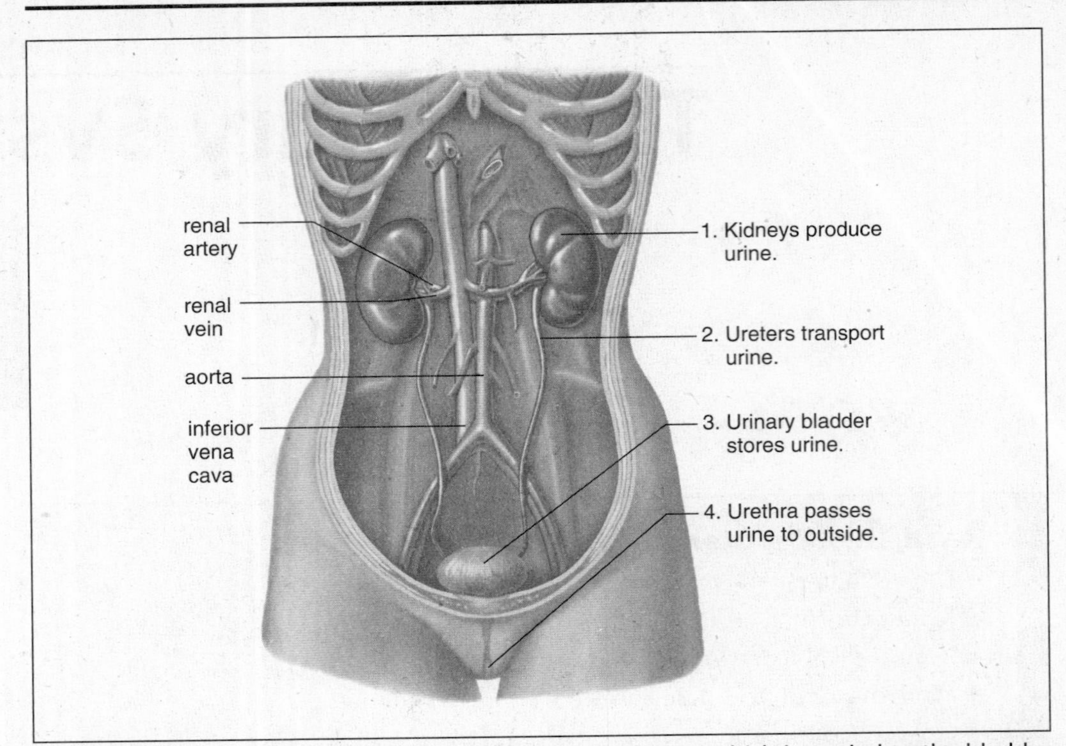

FIGURE 17-1: The urinary system. The kidneys produce urine, which is carried to the bladder by the ureters, where it is eliminated from the system. SOURCE: From Sylvia S. Mader, *Biology*, 8th ed., McGraw-Hill, 2004; reproduced with permission of The McGraw-Hill Companies.

THE KIDNEY

The kidneys are the workhorses of the urinary system. They are located along the dorsal surface of the abdominal wall, above the waist, and are secured by several layers of connective tissue, including a layer of fat. Each kidney has an **adrenal gland** located on top of it.

The outer region of the kidney is the **renal cortex**, the middle portion is the **renal medulla**, and the inner portion is the **renal pelvis**, as shown in Figure 17-2.

The kidneys are responsible for filtering blood, so they have an excellent blood supply. **Renal arteries**, which branch off the aorta, carry blood into the kidneys, whereas the **renal veins** carry blood away from the kidneys toward the inferior vena cava. The indentation where the ureter, renal artery, and renal vein attach to each kidney is the **renal hilus**.

Within the renal medulla of each kidney, there are triangular chunks of tissue called **renal pyramids**. Within these renal pyramids and extending into the renal cortex are about 1 million **nephrons** per kidney. The nephrons, seen in Figure 17-3, are microscopic tubules that are the basic functional units of the kidney and actually produce urine.

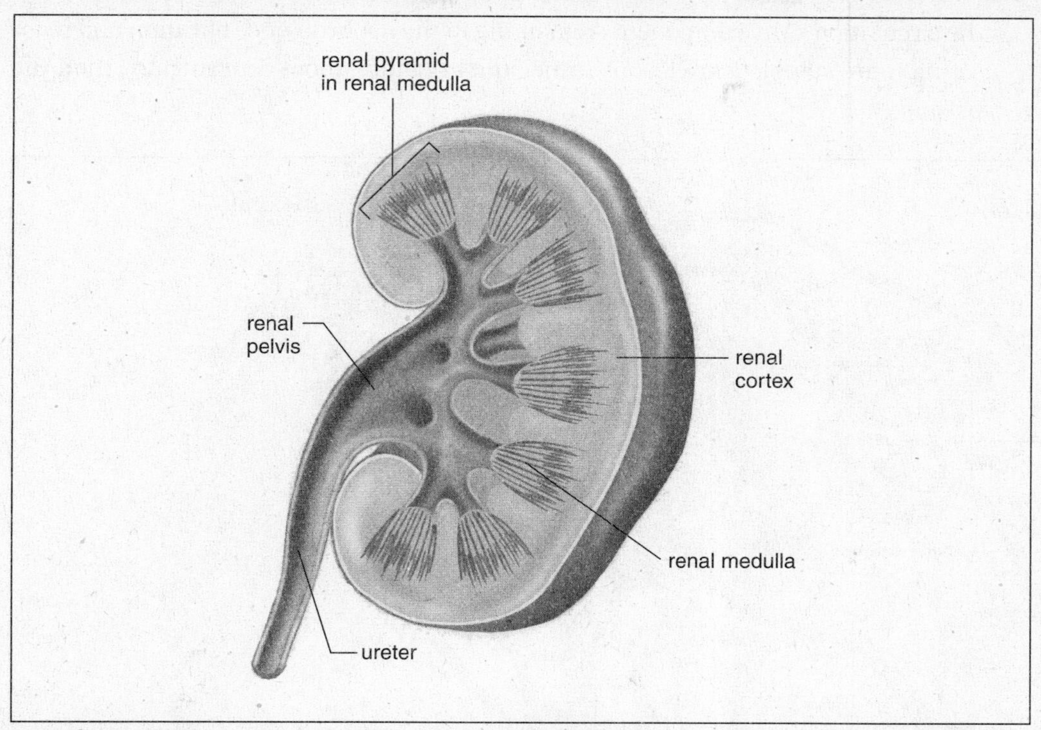

FIGURE 17-2: Kidney structure. Urine is produced in the nephrons of the kidney, which are found within the renal cortex and renal medulla. SOURCE: From Sylvia S. Mader, *Biology*, 8th ed., McGraw-Hill, 2004; reproduced with permission of The McGraw-Hill Companies.

THE NEPHRON

A network of capillaries surrounds each nephron. Any items leaving the nephron are picked up by the capillaries and returned to the bloodstream.

The important parts of the nephron and their roles in the filtration of blood and production of urine are discussed next.

➤ The **renal corpuscle** has two parts. The first is the **glomerulus**, which is a network of capillaries, and the second is **Bowman's capsule**, which surrounds the glomerulus. There is no direct connection between the glomerulus and Bowman's capsule; rather, there is a space between the two. Afferent arterioles carry blood into the glomerulus, where the pressure of the blood pushes certain components of the blood into Bowman's capsule. Efferent arterioles carry blood out of the glomerulus. Only blood components that are small in size enter Bowman's capsule. This means that blood cells and plasma proteins should not enter Bowman's capsule, whereas plasma components such as water, ions, small nutrients, nitrogenous wastes, gases, and others should enter Bowman's capsule. The materials that enter Bowman's capsule are referred to as **filtrate**, and have approximately the same osmotic concentration as the plasma. A large percentage (approximately 99%) of filtrate that enters the nephron should be reabsorbed back into the circulation. Any components remaining in the nephron once filtration and reabsorption are complete are lost as urine, and are much more concentrated than the plasma.

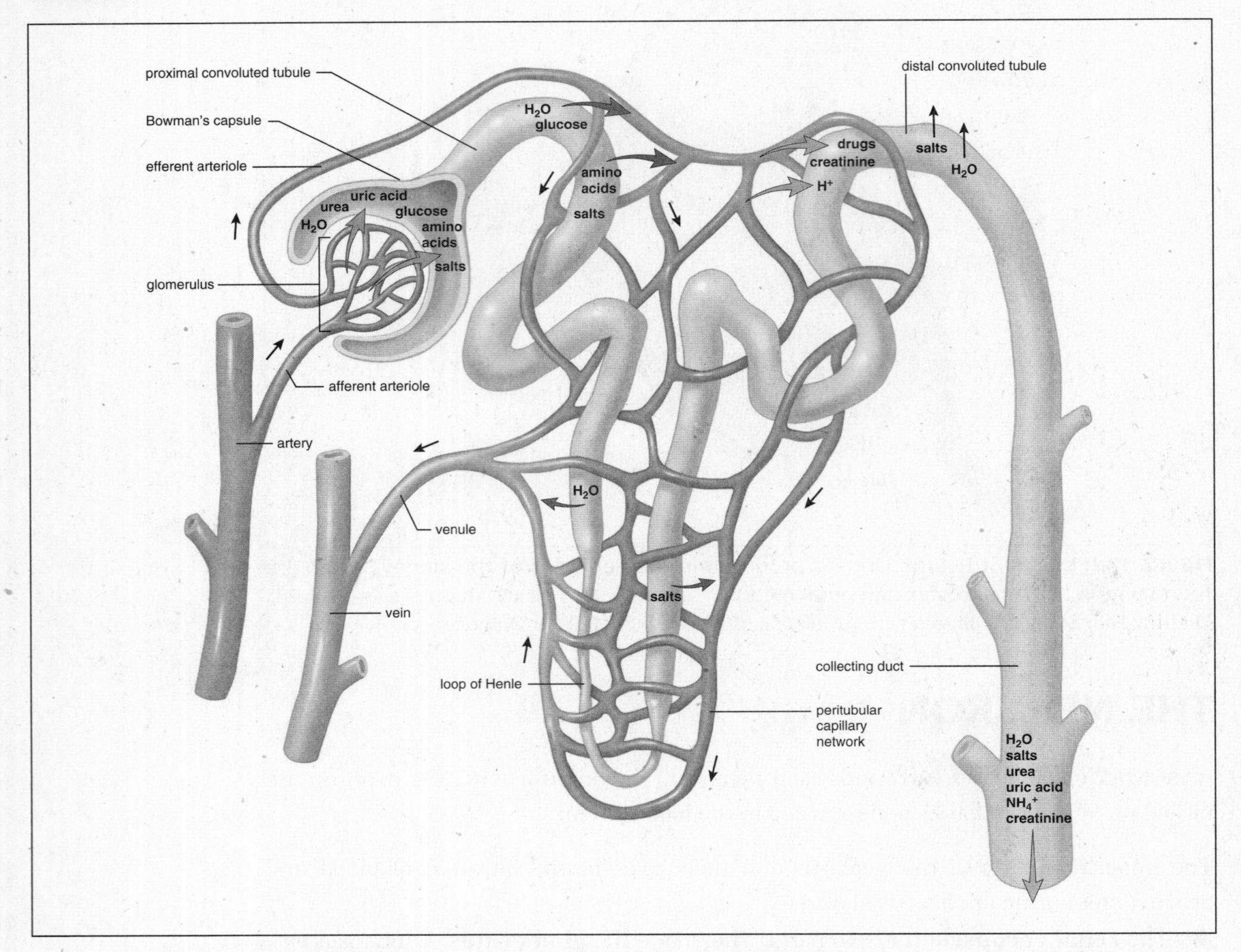

FIGURE 17-3: Nephron structure. Urine is produced as filtrate moves through the nephron. In reality, a nephron is twisted along itself, but for ease of viewing, the nephron shown here has been untwisted. SOURCE: From Sylvia S. Mader, *Biology*, 8th ed., McGraw-Hill, 2004; reproduced with permission of The McGraw-Hill Companies.

- The **proximal convoluted tubule** allows for the reabsorption of nutrients such as glucose and amino acids, water, salt, and ions. The majority of reabsoprtion occurs here.
- The **loop of Henle** also allows for reabsorption, primarily of salt (NaCl) and water by osmosis. A fairly complex **countercurrent multiplier system** is in effect in the loop of Henle. This area is a loop with a descending side and an ascending side that are located in close proximity to each other. Each limb of the loop has a different osmotic concentration. As salt is actively pumped out of the ascending

limb, it creates a high osmotic pressure that draws water out of the descending limb via osmosis. Fresh filtrate then enters the loop of Henle, pushing the existing filtrate from the descending limb into the ascending limb. The process of pumping salt out of the ascending limb and the osmotic movement of water out of the descending limb is repeated several times.

- The **distal convoluted tubule** is where the fine-tuning of filtrate concentration begins; its activities are regulated by specific hormones described later. The more water reabsorbed in this section of the nephron, the more concentrated the urine becomes; the lower the urine volume, and the higher the blood volume.
- The **collecting duct** can be shared by several nephrons. The remaining urine empties into the collecting duct, where it moves toward the renal pelvis and ultimately into the ureters to be carried to the bladder. Hormones can also be used in the collecting duct to allow for the reabsoprtion of more water, making the urine even more concentrated.

FACTORS AFFECTING AND RESULTING FROM NEPHRON FUNCTION

Specific hormones regulate the fine-tuning of filtrate concentration and reabsorption of water, which, in turn, affect blood volume and pressure and the acid–base balance of the blood.

Regulation of Blood Volume and Pressure

If **antidiuretic hormone (ADH)** is present, more water is reabsorbed in the distal convoluted tubule and the collecting duct. This increases the concentration of urine and decreases urine volume. If **aldosterone** is present, more salt is reabsorbed from the distal convoluted tubule and collecting ducts. Water follows the movement of salt by osmosis. This results in an increased concentration of filtrate and a decrease in urine volume. Both ADH and aldosterone have the same effects on filtrate concentration. By increasing water reabsorption, blood volume has been increased. The increase in blood volume is one way to increase blood pressure.

The secretion of ADH and aldosterone is regulated by renin produced by the kidneys. Renin secretion is triggered by low blood pressure in the afferent arterioles. **Renin** converts a protein made by the liver called **angiotensin I** into angiotensin II. **Angiotensin II** then triggers the release of ADH by the posterior pituitary gland and aldosterone by the adrenal cortex.

Diuretics are substances that increase urine volume. Alcohol qualifies as a diuretic because it interferes with the activity of ADH. Caffeine is also a diuretic because it interferes with salt reabsorption and aldosterone's function. Large amounts of alcohol or caffeine have a noticeable effect by increasing urine volume. This in turn decreases blood volume and pressure.

One last hormone that alters nephron function is **atrial natriuretic peptide (ANP)**, which is secreted by the heart. When the heart stretches due to elevated blood pressure, ANP is released. ANP decreases water and salt reabsorption by the nephrons. This results in less concentrated urine, a higher urine volume, and a lower blood volume. The reduced blood volume means a decrease in blood pressure.

Maintenance of Acid–Base Balance

As the kidneys filter blood, they also balance the pH of blood; this is one of their most important functions. Even a relatively minor change to blood pH can have drastic consequences, which is one of the reasons that kidney failure can be deadly. Luckily, **dialysis** methods are available to mimic normal kidney functions for patients whose kidneys do not work properly.

Recall that CO_2 interacts with water to produce carbonic acid. Carbonic acid can then dissociate into hydrogen ions and bicarbonate ions, both of which influence pH. The pH of blood can be adjusted by changing the amount of bicarbonate ions being reabsorbed and by altering the amount of hydrogen ions being retained in the nephron. When the blood pH drops and becomes acidic, more bicarbonate ions return to circulation and hydrogen ions are released in urine, which gives urine an acidic pH.

PROPERTIES OF URINE

The substances remaining at the end of the collecting duct constitute urine. Urine always contains water, ions (such as Ca^{2+}, Cl^-, Na^+, and K^+), and nitrogenous wastes. Depending on the diet and function of other organs in the body, other components might be present in the urine. Because it was filtered directly from blood, the urine should be sterile. The presence of proteins, blood cells, or nutrients within the urine is considered abnormal.

The three primary nitrogenous wastes, all produced by cells, are as follows:

- **Urea:** As cells deaminate amino acids during protein metabolism, the resulting product is **ammonia**, which is highly toxic. The liver converts ammonia to a less toxic waste called urea. The kidneys concentrate urea and release it via urine. Urea is the most abundant of nitrogenous waste products.
- **Uric acid:** During nucleic acid metabolism in the cell, uric acid is produced as the waste product.
- **Creatinine:** As muscle cells use creatine phosphate to produce ATP needed to fuel muscular contraction, creatinine is produced as a waste product.

ADDITIONAL FUNCTIONS OF THE KIDNEYS

In addition to their role in blood filtration and urine production, the kidneys have two additional jobs. First, the kidneys act to convert vitamin D from the diet into its active form that can be used by the cells. The kidneys convert vitamin D to **calcitriol**, which helps the body absorb calcium and phosphorous. Second, the kidneys secrete the hormone erythropoietin (EPO), which is used to stimulate red blood cell production in the red bone marrow.

CRAM SESSION

The Urinary System

1. The primary organs of excretion are the kidneys. The urinary tract is composed of two kidneys and two ureters that merge into the bladder and the urethra, which excretes urine. The urethra in males is shared with the reproductive system for sperm passage and is longer than the urethra of females.
2. The renal medulla and cortex of each kidney contain nephrons that filter blood to produce urine. Filtrate from the blood is pushed into the nephron via blood pressure at the glomerulus of the nephron. The majority of items that enter the nephron as filtrate must be reabsorbed into the blood. The parts of the nephron are as follows:
   a. The glomerulus and Bowman's capsule: Blood pressure pushes filtrate from the glomerulus into Bowman's capsule.
   b. The proximal convoluted tubule: nutrients and water are reabsorbed.
   c. The loop of Henle: A countercurrent multiplier system is used to draw salt and water out of the nephron and back into circulation.
   d. The distal convoluted tubule: More salt and water are reabsorbed. The hormones ADH and aldosterone adjust water reabsorption.
   e. The collecting duct: More salt and water are reabsorbed. The hormones ADH and aldosterone adjust water reabsorption. The remains are moved to the bladder to be stored as urine.
3. The three nitrogenous waste products found in urine are urea, uric acid, and creatinine.
4. The pH of the blood can be adjusted by the selective reabsorption or excretion of bicarbonate and hydrogen ions.
5. The amount of water excreted has a direct influence on blood volume and pressure.
   a. The higher the urine volume, the lower the blood volume and blood pressure.
   b. The lower the urine volume, the higher the blood volume and blood pressure.

CHAPTER 18

The Lymphatic and Immune Systems

Read This Chapter to Learn About:

- ➤ Structures of the Lymphatic System
- ➤ Function of the Immune System
- ➤ Nonspecific and Specific Immune Defenses
- ➤ Cellular Immunity
- ➤ Immunodeficiency

THE LYMPHATIC SYSTEM

The **lymphatic system** consists of a series of vessels running throughout the body, lymph, and lymphoid tissue as shown in Figure 18-1. The system serves to return fluids that were unclaimed at the capillary beds to the circulatory system, picks up chylomicrons from the digestive tract and returns them to circulation, and fights infection via leukocytes.

The vessels of the lymphatic system carry the fluid **lymph**, which is the same composition as plasma and interstitial fluid. Lymph moves through the vessels primarily due to the influence of muscular contractions that push lymph and valves that prevent the backflow of lymph.

The lymphoid tissues within the system are as follows:

- ➤ **Lymph nodes** are swellings along lymphatic vessels that contain macrophages for phagocytosis of pathogens and cancer cells and lymphocytes for immune defenses. The lymph is filtered through the nodes before moving on in the system.

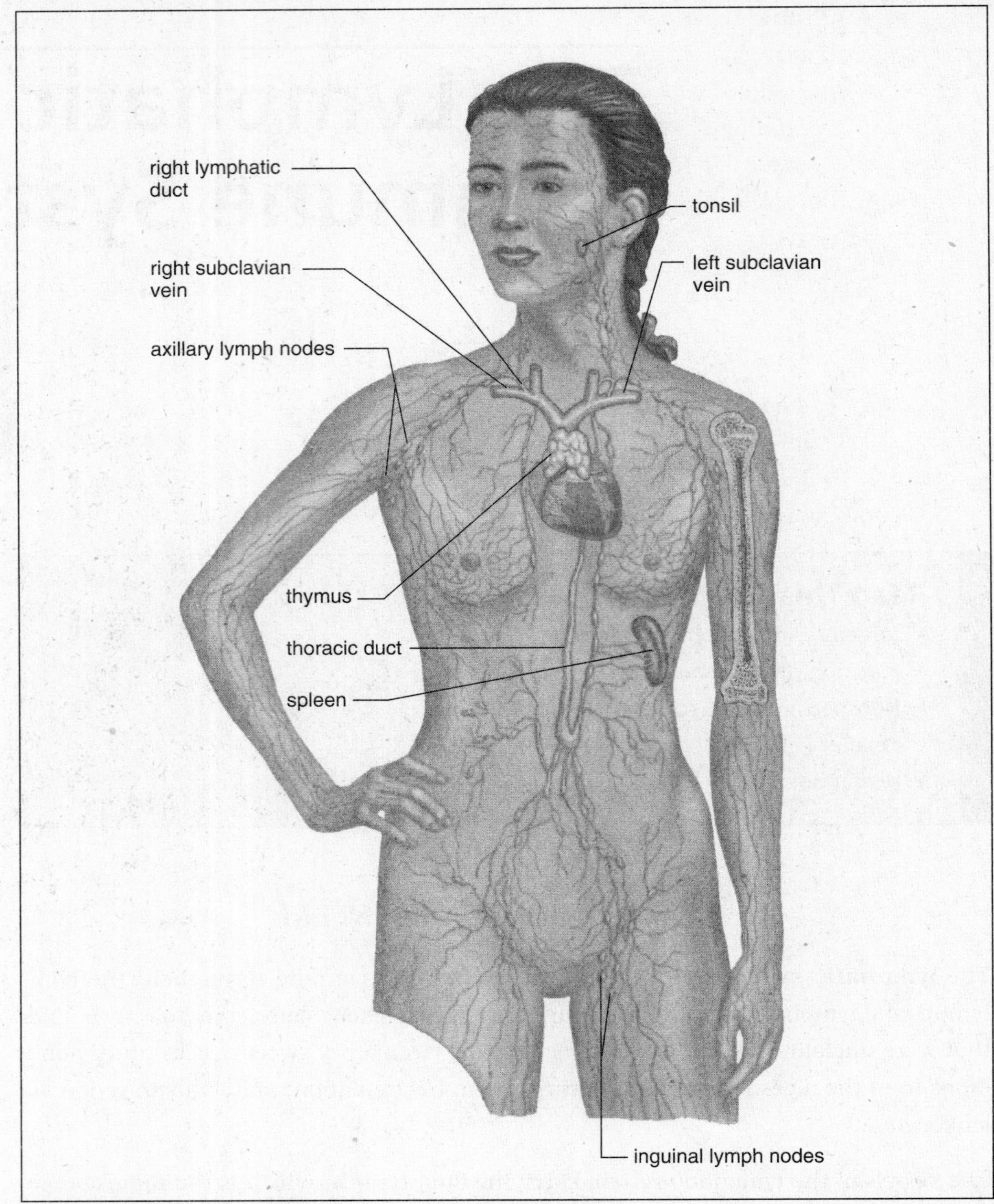

FIGURE 18-1: The lymphatic system is composed of vessels, lymph nodes, the tonsils, spleen, and thymus gland. SOURCE: From Sylvia S. Mader, *Biology*, 8th ed., McGraw-Hill, 2004; reproduced with permission of The McGraw-Hill Companies.

Clusters of lymph nodes exist in the neck, under the arms, and in the groin. Swelling of the lymph nodes is a sign of infection.

- **Tonsils** resemble a lymph node in their ability to prevent infection by pathogenic organisms in the throat.

- **Peyer's patches** in the small intestine are clusters of lymphatic tissues that serve to prevent infectious organisms from crossing the intestinal wall into the abdomen.
- The **thymus gland** allows for the maturation of T cells, which are a form of lymphocytes that are needed for specific immune defenses.
- The **spleen** is located on the left side of the body and also acts as a blood filter. In addition, the spleen has an excellent blood supply and acts to destroy old red blood cells and platelets.

The Immune System

The immune system exists anywhere white blood cells are found. This includes the blood, lymph, and within tissues of the body. The job of the immune system is to differentiate "self-cells" from "non-self" (foreign cells) and to eliminate both foreign cells and abnormal self-cells.

Immune defenses begin with nonspecific responses, which try to prevent foreign cells from entering the body and attack them if they do enter, and later, if needed, move to specific responses. A **nonspecific immune defense** always works the same way, no matter what the offending invader, whereas **specific immune defenses** are activated and tailor-made to a specific invader.

NONSPECIFIC DEFENSES

Nonspecific defenses come in several varieties. They include

- **Physical and chemical barricades** that prevent foreign cells from entering the body. The skin is an example of a barricade that generally prevents infection. Mucous membranes provide another good barricade. Chemicals such as sweat, stomach acid, and lysozyme also generally prevent infection.
- **Defensive leukocytes** that include neutrophils, monocytes, and macrophages (mature monocytes), all of which are capable of phagocytosis to destroy pathogens that may have entered the body. Eosinophils are another type of defensive leukocyte. They enzymatically destroy large pathogens, such as parasitic worms, that cannot be phagocytized. Finally, natural killer (NK) cells find self-cells that seem to have odd membrane properties and destroy them. Cancerous cells are notorious for having altered cell membranes and are usually destroyed by NK cells.
- **Defensive proteins** function against both viral invaders and bacterial invaders. A virally infected cell can secrete proteins called **interferons** and release these proteins as messengers to other cells that are yet to be infected. This limits the spread of the virus within the body. Interferons are not specific: they work against all types of viruses. The **complement system** is a series of multiple plasma proteins

that are effective at killing bacteria by causing lysis of their cell membrane. The complement system also enhances phagocytosis within the area of invasion.

- **Inflammation** occurs when there is damage to tissues. It is characterized by redness (due to increased blood flow), heat (due to increased blood flow), swelling, and pain. Increased blood flow to the area is caused by the chemical **histamine**, which is secreted by basophils. This increased blood flow brings in other white blood cells, proteins, and other components needed to fight infection. Histamine makes capillaries more permeable than normal, and this, in turn, results in increased fluid in the area, which causes swelling. This swelling can put pressure on pain receptors, causing the sensation of pain.
- **Fever** occurs when the body temperature is reset to a higher level by chemicals called **pyrogens**. Controlled fevers are beneficial, because they increase metabolism and stimulate other immune defenses. When fevers get too high, they are dangerous and can cause the denaturing of critical enzymes needed to sustain life.

SPECIFIC DEFENSES

When nonspecific defenses fail to control infection, specific defenses must be used. Because these are customized to the specific invader, they take at least a week to respond strongly to a new **antigen**. An antigen is a substance, including microbes, that elicits an immune response.

There are two types of specific defenses: humoral immunity, involving B cells, and cellular immunity, involving T cells. Both types of specific defense use **lymphocytes**, which are derived from stem cells in the red bone marrow. B cells complete their maturation in the bone marrow, whereas T cells mature in the thymus gland. Both types of cells are designed to destroy antigens.

In **humoral immunity**, the B cells ultimately secrete antibodies to destroy foreign antigens. In **cellular** (or **cell-mediated**) **immunity**, T cells are used to directly destroy infected or cancerous cells. A specific variety of T cell known as the **helper T cell** is the key coordinator of both humoral and cellular responses, which happen simultaneously.

Humoral Immunity

Humoral immunity involves the production of specific **antibody** proteins from B cells that have been activated. Each B cell displays an antibody on its cell membrane, and each of the million or more B cells in the body has a different antibody on its membrane. The activation of a particular B cell by a specific antigen is based on shape recognition between the antibody on the B cell membrane and the antigen. The activation process is also dependent on chemicals from a helper T cell, which is discussed later in the chapter.

The activation of a particular B cell causes proliferation of that B cell, leading to a population of **plasma cells** that actively secrete antibodies and **memory B cells** that produce the same type of antibody. This is referred to as **clonal selection**. It is the key event of the **primary immune response**, which leads to **active immunity**. It takes at least a week for this response to reach peak levels. Once antibodies are produced in large quantities by plasma cells, they circulate through blood, lymph, and tissues where they seek out their antigen and bind to it, forming a complex. Once an antibody binds to an antigen, the complex is either phagocytized or agglutinates and is later removed by other phagocytic cells.

The primary immune response and active immunity can be achieved by natural exposure to an antigen or by vaccination. On secondary and subsequent exposures to the same antigen, the memory B cells that were created during the primary exposure proliferate into plasma cells that produce the needed antibodies, providing a much faster response to the antigen. Although antibodies do not circulate for long once an antigen has been destroyed, memory B cells can last for years, if not forever.

Sometimes antibodies are passed from one person to another, which leads to **passive immunity**. This occurs during pregnancy, when maternal antibodies cross the placenta, and during breast-feeding. Breast milk contains antibodies that are transmitted to the newborn. Passive immunity can be induced by the injection of antibodies from one individual to another. Passive immunity is short-lived and declines within a few months.

ANTIBODY STRUCTURE

An antibody consists of four protein chains bonded together. Two chains are identical heavy chains and two chains are identical light chains. The structure of an antibody can be seen in Figure 18-2. **Variable regions** on the chains bind to the antigen, while **constant regions** on the chains assist in the destruction of the antigen.

There are five types of organizations of constant regions, which result in five different classes of antibodies (also called **immunoglobulins**): IgG, IgM, IgA, IgD, and IgE. The incredible diversity of antibodies produced by the immune system is based on a genetic recombination system that produces countless variable regions. The genes that produce the variable regions of antibodies are broken into segments that can be arranged in variable orders during RNA splicing to produce many variations of the variable region.

Cellular Immunity

Cellular immunity is based on the actions of T cells, which come in several varieties. T cells have a cell membrane receptor that, like antibodies on the surface of B cells,

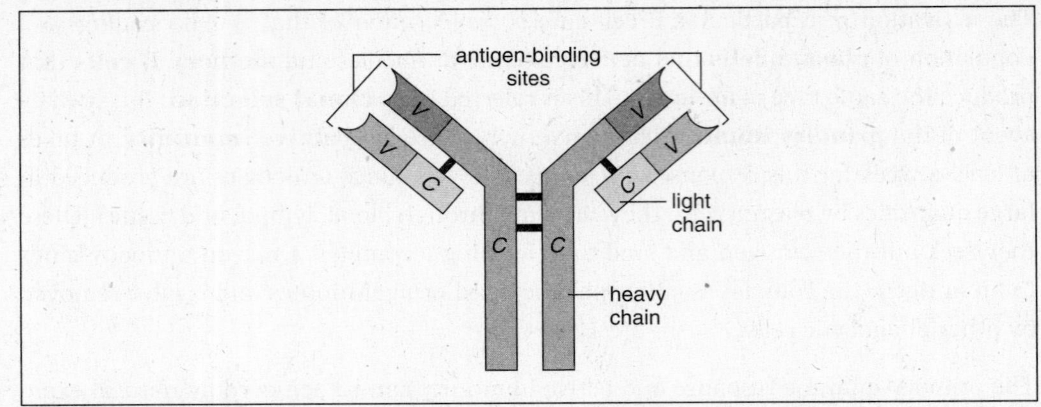

FIGURE 18-2: Antibody structure. Antibodies contain heavy chains and light chains. The variable regions of the antibody allow for antigen binding. SOURCE: From Sylvia S. Mader, *Biology*, 8th ed., McGraw-Hill, 2004; reproduced with permission of The McGraw-Hill Companies.

recognizes the shape of one particular antigen. However, T cells cannot be directly activated by contact with the antigen. The antigen must be presented to a **helper T cell** by a self-cell that is infected with the antigen or a macrophage that has engulfed the antigen. The cell presenting the antigen secretes the chemical **interleukin-1** as it binds to the helper T cell. The helper T cell then secretes **interleukin-2**. This step is crucial for both humoral and cellular responses. The activation of the helper T cell and secretion of interleukin-2 allows for activation of B cells as well as the activation of **cytotoxic** (killer) **T cells**, which can now bind to the antigen. The cytotoxic T cell proliferates and produces effector cells by clonal selection; all the cells produced this way have the ability to seek and destroy the foreign antigen. As with plasma cell activation, this primary response takes at least a week to occur. **Memory T cells** are also produced. Once the antigen has been completely destroyed, **suppressor T cells** are used to stop the response of cytotoxic T cells. Only memory T cells remain. They can be activated quickly to cytotoxic T cells on secondary and subsequent exposures to the same antigen.

IMMUNODEFICIENCY DISORDERS

There are a variety of immune system problems that are characterized as **immunodeficiencies**. These disorders occur when the immune system is lacking one or more component of its specific defenses. For example, severe combined immunodeficiency syndrome (SCID) occurs when an individual is born without lymphocytes. This renders the person with no specific immune defenses and quickly causes death unless the individual is maintained in a sterile, isolated environment.

The human immunodeficiency virus (HIV) induces immunodeficiency in an individual. It selectively infects helper T cells. When the virus activates after a latent period, the helper T cells begin to die. Without them, the interleukin-2, which is needed to stimulate B cell activation and cytotoxic T cell activation, is not present. This leads to an inability to mount humoral and cellular responses to antigens. This leads to acquired immunodeficiency syndrome (AIDS), which causes death due to an overwhelmed immune system.

CRAM SESSION

The Lymphatic and Immune Systems

1. The lymphatic system is composed of vessels carrying lymph, associated lymph nodes, the tonsils, the thymus gland, the spleen, and a few other lymphoid tissues. This system reclaims tissue fluid and returns it to the blood, delivers the products of fat digestion to the blood, and fights infection via white blood cells.
2. The immune system exists anywhere white blood cells are, including the lymphatic system, the blood, and within tissues. The primary duty of the immune system is to destroy foreign cells as well as self-cells that are infected or cancerous.
3. The nonspecific immune defenses are used to prevent foreign cells from entering the body or to quickly disable foreign cells that have entered. These defenses include physical and chemical barricades, defensive white blood cells, interferons, the complement system, inflammation, and fever.
4. Specific defenses are used when the nonspecific defenses fail to destroy foreign invaders.
   a. B cells can be cloned to form plasma cells that secrete antibodies. Antibodies circulate, bind to a foreign antigen, and then are marked for destruction. This is the humoral response.
   b. T cells can be cloned into cytotoxic T cells that seek out the antigen, bind to it, and kill it. This is the cellular response.
   c. In order to activate plasma cells or cytotoxic T cells, helper T cells must first be activated by antigen presentation. This allows for the secretion of interleukin-2 by the helper T cells, which, in turn, permit the cloning of plasma cells and cytotoxic T cells.
   d. The process of performing clonal selection occurs when a new antigen is encountered and is a slow process.
      (1) Secondary exposure to the same antigen results in a rapid immune response due to memory cells produced during the primary response.
      (2) Although plasma cells and cytotoxic T cells do not stick around forever, memory cells can last for many years.
5. HIV selectively targets and eventually kills helper T cells, thereby disabling the humoral and cellular immune responses.

CHAPTER 19

Reproduction and Development

Read This Chapter to Learn About:

- The Female Reproductive System
- The Male Reproductive System
- Fertilization
- Embryonic Development
- Fetal Development
- Birth

THE REPRODUCTIVE SYSTEM

The female and male reproductive systems have a common function of producing gametes for sexual reproduction in the form of egg cells in females and sperm cells in males, both of which are produced by the **gonads**. The process of gamete production via meiosis has already been discussed in Chapter 6. In addition, the female reproductive system must be structured to accept sperm from the male system and allow for embryonic and fetal development.

THE FEMALE REPRODUCTIVE SYSTEM

The functions of the female reproductive system are carried out in special structures under the influence of a complex system of hormones that regulate the development of egg cells (oogenesis), their release from the ovary (ovarian cycle), and the preparation of the woman's body for embryo implantation and development (menstrual cycle).

Structure

The female reproductive system is enclosed within the abdominal cavity and is open to the external environment. Although having an opening to the outside environment is necessary for reproduction and childbirth, it presents some unique problems in terms of the ability of pathogens to enter the system.

The structures of the female reproductive system consist of two ovaries, where egg production occurs, and supporting structures, as shown in Figure 19-1. After an egg is released from an ovary, it is swept into the **fallopian tube (oviduct)** that is associated with that ovary. If sperm are present and they meet with the egg in the fallopian tube, **fertilization** occurs. The fallopian tubes merge into the **uterus**, which is composed of a muscular **myometrium** and a vascularized lining called the **endometrium**. If the egg has been fertilized, the embryo implants in the endometrium, where development continues. If fertilization has not occurred, the egg is lost with the shedding of the endometrium, which occurs about every 28 days during **menstruation**. The **vagina** serves as an entry point for sperm to enter the system, as an exit point for menstrual fluids, and as the birth canal during childbirth. The pH of the vagina is acidic, which can discourage the growth of certain pathogens. The **cervix** regulates the opening of the uterus into the vagina and is normally very narrow.

The external genitalia of the female system is known as the **vulva**. The **labia** are folds of skin that surround the opening to the vagina and occur as an outer and inner pair. The **clitoris** is located under the anterior portion of the inner labia. It has multiple nerve endings and is associated with female sexual arousal.

The breast tissue contains **mammary glands**, which serve the primary purpose of producing milk for a newborn. The tissue that surrounds the glands is fibrous connective tissue. The mammary glands are under the control of several hormones including estrogen, progesterone, oxytocin, and prolactin.

The Menstrual Cycle

The female reproductive cycle lasts about 28 days (or 4 weeks) on average. Characteristic changes occur within the uterus during this time. These changes are referred to as the **menstrual** (or **uterine**) **cycle**.

The menstrual cycle, which has three phases, is as follows:

- **Menses:** During the first 5 or so days of the cycle, the existing endometrium is lost via menstrual fluid as arteries serving the endometrium constrict, cutting off the cells from oxygen and nutrients, resulting in tissue death.

- **The proliferative phase:** During the second week of the cycle, the primary event in the uterus is the proliferation of cells to replace the endometrium that was lost during menstruation
- **The secretory phase:** During the last 2 weeks of the cycle, the endometrium is prepared by the action of specific hormones for the implantation of an embryo. If an embryo is not present, as the 28th day of the cycle approaches, the endometrium deteriorates and menses soon begins as the cycle restarts.

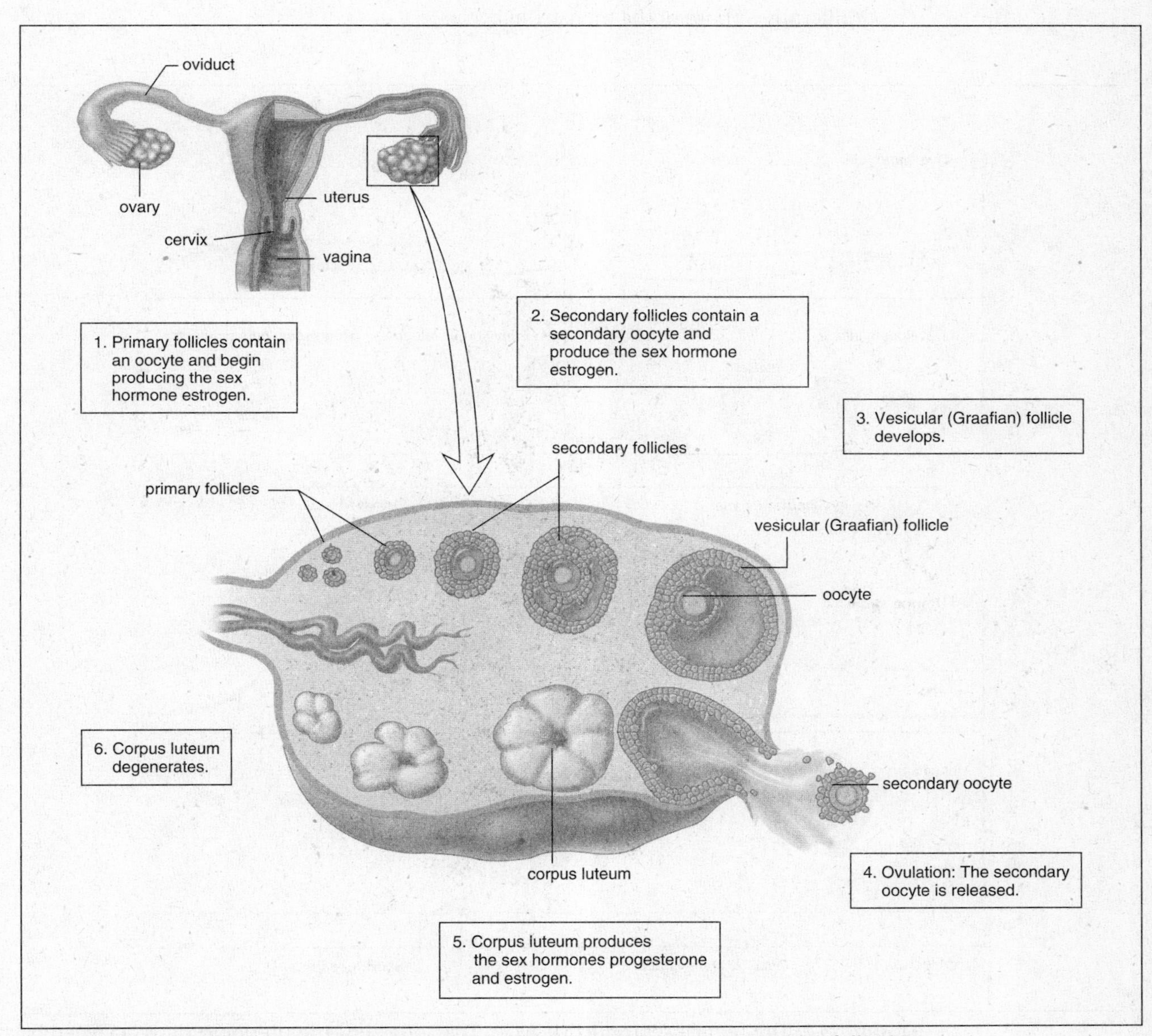

FIGURE 19-1: The female reproductive system. The ovaries release an egg into the fallopian tube during each reproductive cycle. Should the egg be fertilized, the resulting embryo implants in the uterus. SOURCE: From Sylvia S. Mader, *Biology*, 8th ed., McGraw-Hill, 2004; reproduced with permission of The McGraw-Hill Companies.

The Ovarian Cycle

Eggs are produced through the ovarian cycle, whose timing must be carefully choreographed to the menstrual cycle, both of which are shown in Figure 19-2. Like the menstrual cycle, the ovarian cycle also typically lasts 28 days.

The ovarian cycle consists of the following phases:

➤ **The preovulatory phase:** This phase consists of the events prior to ovulation and lasts from days 1 to 13 of the cycle. This timing corresponds to menses and the proliferative phase of the menstrual cycle.

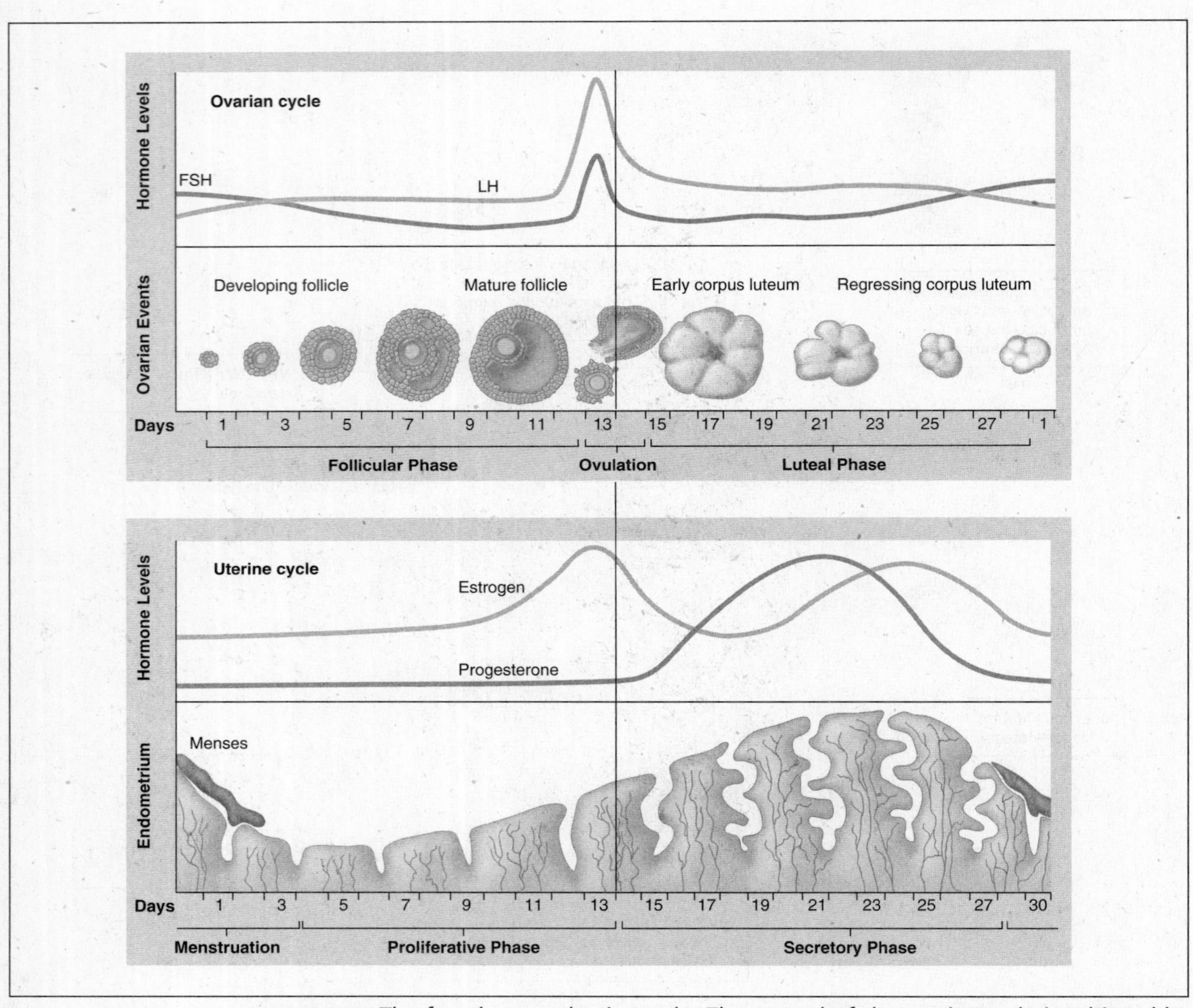

FIGURE 19-2: The female reproductive cycle. The control of the uterine cycle is achieved by estrogen and progesterone. The ovarian cycle is regulated primarily by FSH and LH. SOURCE: From Sylvia S. Mader, *Biology*, 8th ed., McGraw-Hill, 2004; reproduced with permission of The McGraw-Hill Companies.

- **Ovulation:** The rupture of a follicle in the ovary and subsequent release of an egg to a fallopian tube constitutes **ovulation**. It occurs on day 14 of the cycle.
- **The postovulatory phase:** During this phase, the egg may be fertilized. This phase lasts from days 15 to 28 and corresponds with the secretory phase of the menstrual cycle. Should fertilization occur, the embryo implants in the endometrium. If fertilization has not occurred, the menstrual cycle restarts, causing the egg to be lost.

Oogenesis

The events that lead to the development of an egg are termed **oogenesis** and are regulated through the ovarian cycle. Within the ovaries of a female, the process of meiosis has begun before she was even born. This process results in the creation of **primary follicles** within the ovaries. A follicle consists of a potential egg cell surrounded by a shell of follicular cells to support the egg. The number of primary follicles is set at birth and is usually around 700,000. As the female is born and ages, many of these follicles die. By the time a female reaches **puberty** at age 12–14, as few as 200,000 follicles remain. Although the number has drastically declined and continues to decline with age, there are still more than enough follicles to support the reproductive needs of a female, because only one egg is released every 28 days. Until puberty begins, these follicles stay in an arrested phase of meiosis.

After puberty, the ovarian cycle begins. During the preovulatory phase of the ovarian cycle, a few primary follicles resume meiosis and begin growing as primary oocytes. Starting at day 1 of the cycle, the anterior pituitary secretes **follicle stimulating hormone (FSH)** and **luteinizing hormone (LH)**. Recall that the anterior pituitary is under the control of the hypothalamus. The hypothalamus produces **gonadotropin-releasing hormone (GnRH)**, which stimulates the release of FSH and LH. FSH causes the growth of several follicles, which begin to produce estrogen. The more that the follicles grow, the more estrogen is produced. Thus estrogen is at its highest level during the second week of the cycle, which corresponds to the rebuilding of the endometrium after menstruation.

During the second week of the cycle, the high levels of estrogen actually inhibit FSH such that no more follicles begin to grow on this cycle. As estrogen from the growing follicles continues to rise, there is a massive surge in LH. This surge causes the completion of the first round of meiosis, leading to the formation of a **secondary oocyte** and the rupture of the follicle within the ovary. This is ovulation, and it happens on day 14 of the cycle. The oocyte is released to the fallopian tubes and the remnants of the follicle remain in the ovary.

If the oocyte is fertilized, meiosis II occurs, resulting in a mature egg (**ovum**). Only one mature egg is needed during ovulation, so the remaining three cells produced during meiosis are termed **polar bodies**. They are much smaller than the egg and are degraded. If more than one egg is released on a given cycle, the potential exists for multiple fertilizations and multiple embryos.

The remains of the follicle in the ovary continue secreting estrogen and become the **corpus luteum**, which also secretes progesterone. The estrogen and progesterone suppress FSH and LH so that no more eggs are released in this cycle. These hormones also keep the endometrium prepared to receive an embryo during the secretory phase (third and fourth weeks) of the menstrual cycle.

At the end of the fourth week of the cycle, if an embryo has not implanted in the endometrium, the cycle restarts. At this point, the corpus luteum degrades. Without the corpus luteum, the levels of estrogen and progesterone decline. The lack of these hormones, particularly the lack of progesterone, is the cause of menstruation, which can occur only when these hormone levels are low. Furthermore, the lack of estrogen and progesterone allows the pituitary gland to begin secreting FSH and LH again to begin the process of a new cycle.

The ability to perform the ovarian cycle ends at **menopause**, when the ovaries are no longer sensitive to FSH and LH. The levels of estrogen and progesterone in the body decrease and the ovaries atrophy. This typically occurs between the ages of 45–55.

Because estrogen and progesterone have the ability to suppress the actions of FSH and LH, they are the hormones of choice for use in birth control methods such as pills, patches, injections, rings, or implants. The use of synthetic estrogen and/or synthetic progesterone can be used to manipulate the ovaries into not ovulating because FSH and LH are suppressed.

THE MALE REPRODUCTIVE SYSTEM

Structure

In contrast to the female system, the male reproductive system is not housed completely within the abdominal cavity and is a closed system, which can be seen in Figure 19-3. The male gonads or **testes** produce sperm. The remaining reproductive structures serve as a means to transport sperm out of the body and into the female system.

Sperm begin their development in the **seminiferous tubules** of the testes, where they are nourished by **Sertoli cells**. The testes are housed in the scrotum outside of the abdominal cavity, where the temperature is a few degrees cooler than body temperature. Oddly enough, sperm require a temperature cooler than normal body temperature to become functional. As sperm develop in the testes, they move into the **epididymis** associated with each testis and also located in the scrotum. Once in the epididymis, the sperm acquire motility and are stored.

When ejaculation occurs, the sperm must be moved toward the male urethra. The sperm enter the **vas deferens**, which are tubes that move up into the abdominal cavity. From there, the two vas deferens merge into the **ejaculatory duct** and into the urethra. Recall that the urethra is also used for the passage of urine. When sperm are moving through, the urethra is unavailable to the bladder.

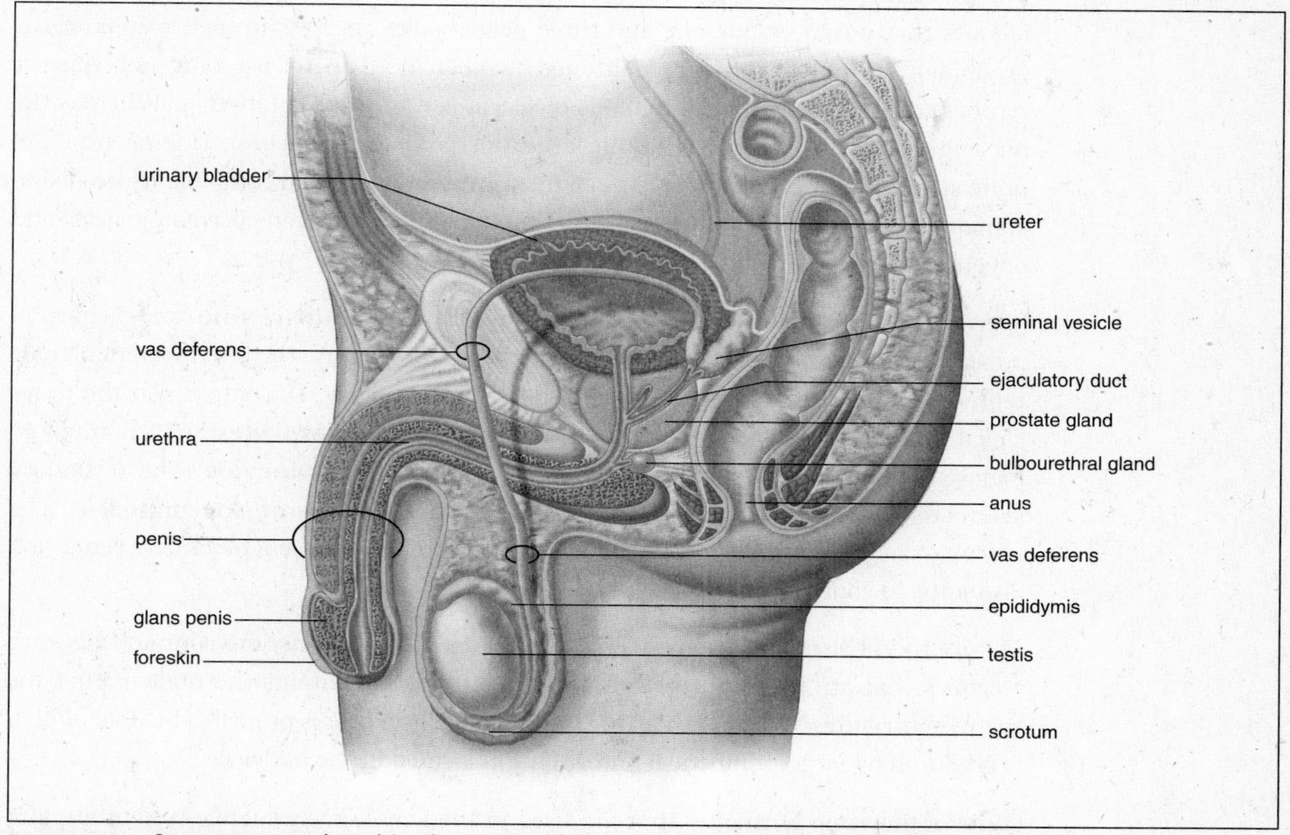

FIGURE 19-3: Sperm are produced in the testes of the male reproductive system. As sperm are released from the male system, secretions from a variety of glands are added to produce semen. SOURCE: From Sylvia S. Mader, *Biology*, 8th ed., McGraw-Hill, 2004; reproduced with permission of The McGraw-Hill Companies.

Once in the urethra, three types of glands add their secretions to the sperm as they pass by. This creates **semen**, which is a mixture of sperm and secretions. The urethra progresses through the length of the penis where it can be introduced into the female vagina.

The glands of the male reproductive system that provide secretions to semen are as follows:

- The **seminal vesicles** provide a fluid rich in nutrients to serve as an energy source for the sperm.
- The **prostate gland** wraps around the urethra and deposits a secretion that is alkaline to balance the acidic environment of the vagina.

- The **bulbourethral glands** secrete a fluid prior to ejaculation, which may serve to lubricate the urethra for sperm passage.

Spermatogenesis

The process of **spermatogenesis** produces sperm through meiosis. Unlike meiosis in females that produces one egg and three polar bodies, meiosis in men results in the production of four sperm cells. Although women need to release only one egg per reproductive cycle, men need millions of sperm per fertilization attempt. Whereas the one egg produced in oogenesis is quite large, the sperm produced in spermatogenesis are quite small. This is because the egg cell must contain additional components needed to support embryonic development. Additional differences between spermatogenesis and oogenesis are presented in Chapter 6.

Spermatogenesis requires the hormonal influence of **testosterone** and begins at puberty. Testosterone is secreted during development to cause the development of male reproductive structures, but it is then halted until puberty. Diploid cells in the testes called **spermatogonia** differentiate into **primary spermatocytes**, which undergo meiosis I, which produces two haploid **secondary spermatocytes**. The secondary spermatocytes undergo meiosis II to produce four mature sperm (**spermatozoa**). The spermatozoa then move to the epididymis to mature. The process takes between 2 and 3 months to complete.

Mature sperm structure can be seen in Figure 19-4. The **acrosome** contains digestive enzymes that are used to penetrate the egg, the head contains the nucleus that the sperm contributes to the egg, and the tail is a flagellum that is propelled by ATP, which is produced by large numbers of mitochondria located in the midpiece.

Some of the same hormones that are used in the female reproductive system are also used to regulate spermatogenesis. GnRH from the hypothalamus allows for the secretion of LH from the anterior pituitary. LH causes the production of testosterone by cells in the testes. The secretion of GnRH from the hypothalamus also results in the release of FSH from the anterior pituitary. Although testosterone is needed to stimulate spermatogenesis, FSH is also needed to make the potential sperm cells sensitive to testosterone. The levels of testosterone regulate sperm production in a manner that resembles a thermostat. Elevated testosterone levels temporarily inhibit GnRH, which in turn inhibits FSH and LH.

DEVELOPMENT

As the haploid nucleus of a sperm cell is contributed to an egg cell (also containing a haploid nucleus) during fertilization, the resulting cell is termed a **zygote**. The zygote begins cell division by mitosis. This reduces a ball of identical cells that is the **embryo**.

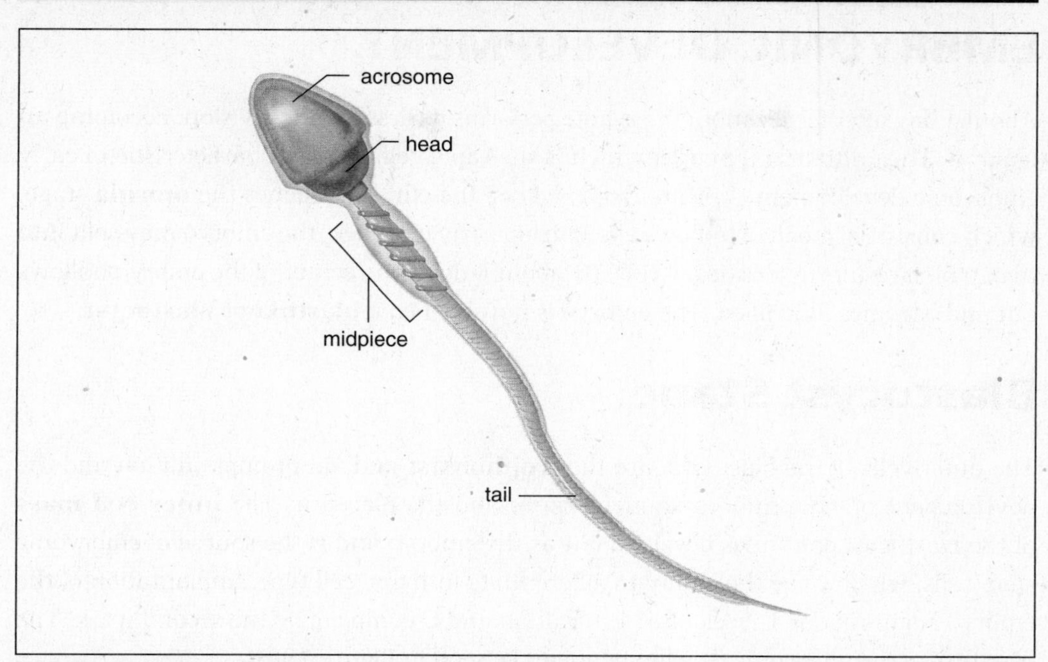

FIGURE 19-4: Sperm structure. The head of the sperm contains the nucleus, which is needed to fertilize an egg. SOURCE: From Sylvia S. Mader, *Biology*, 8th ed., McGraw-Hill, 2004; reproduced with permission of The McGraw-Hill Companies.

In humans, the first 8 weeks of development constitute **embryonic development**, and all development after 8 weeks constitutes **fetal development**. The human gestation (development) period is 266 days, or about 9 months. These 9 months are divided into trimesters. Embryonic development is complete within the first trimester.

FERTILIZATION

Sperm have the ability to survive about 48 hours in the female reproductive system, whereas an egg cell survives only about 24 hours. Sperm deposited prior to or right after ovulation are capable of fertilizing the egg, which usually happens in the upper third of a fallopian tube. Whereas 200–500 million sperm are typically released during ejaculation, only about 200 sperm make it to the egg cell.

Secretions from the female system change the membrane composition of the sperm near its acrosome. This membrane instability causes the release of the acrosomal contents. This allows the sperm to penetrate the **corona radiate** (outer layer) of the egg. Then, the sperm must pass through the next layer of the egg, the **zona pellucida**. The first sperm to pass through the zona pellucida passes its nucleus into the egg. This causes a depolarization in the membrane of the egg, which makes it impenetrable to fertilization by other sperm. The nuclei of the egg and sperm fuse, creating the zygote.

EMBRYONIC DEVELOPMENT

About 1 day after fertilization, the zygote performs its first mitotic division, becoming an embryo. This initiates **cleavage**, which is the rapid cell division characteristic of early embryonic development. Within about 4 days, the embryo reaches the **morula** stage, which consists of a ball of hollow cells. During early cleavage, the embryo may split into two, which results in identical twins. By about 6 days, the center of the embryo hollows out and becomes fluid filled. The embryo is now termed a **blastula** or **blastocyst**.

Blastocyst Stage

The outer cells of the blastocyst are the **trophoblast** and aid in implantation and the development of extraembryonic membranes and the placenta. The **inner cell mass** of the blastocyst continues development as the embryo and is the source of embryonic stem cells, which have the ability to differentiate into any cell type. Implantation of the embryo begins about 1 week after fertilization and is complete by the second week. The events of early embryonic development can be seen in Figure 19-5.

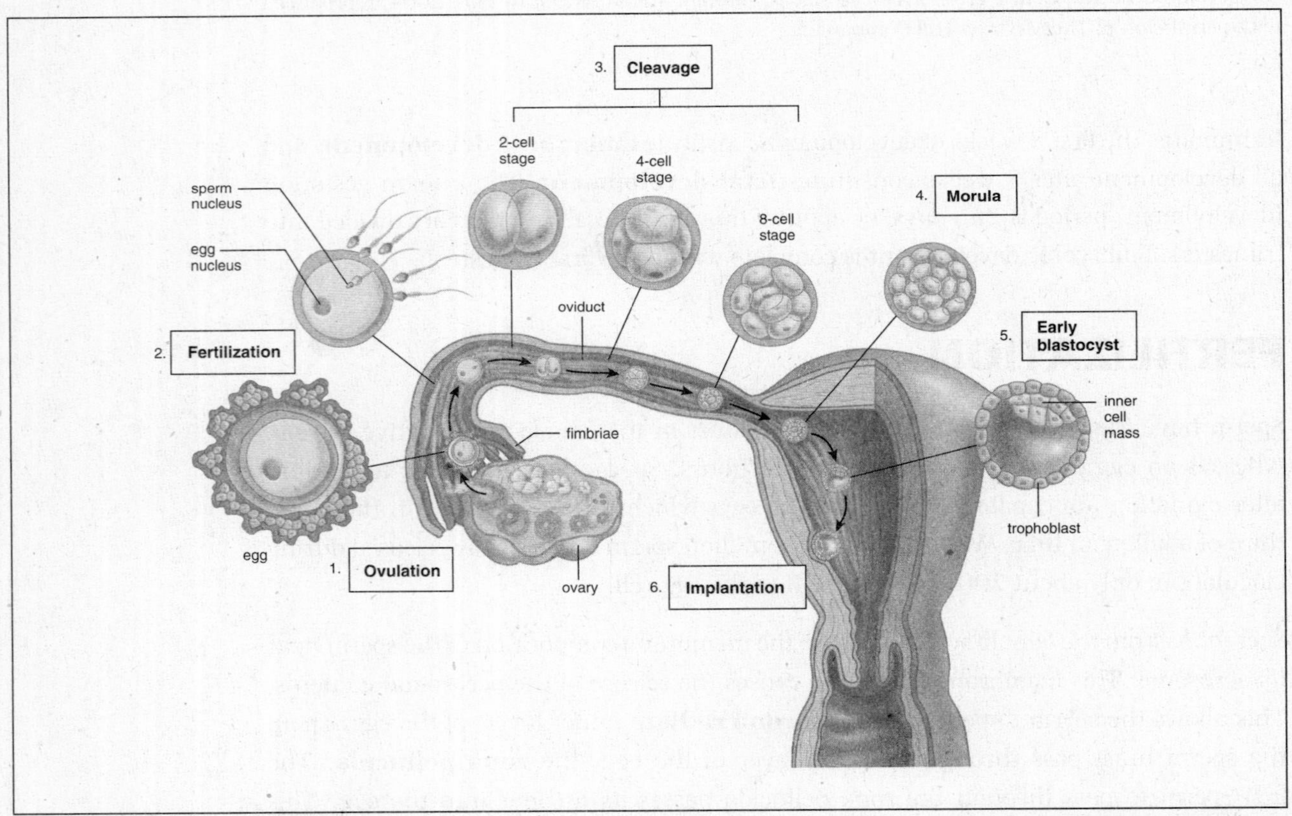

FIGURE 19-5: Early embryonic development. Fertilization occurs in the fallopian tube. The developing embryo moves down the tube to eventually implant in the endometrium of the uterus at the blastocyst stage of development. SOURCE: From Sylvia S. Mader, *Biology*, 8th ed., McGraw-Hill, 2004; reproduced with permission of The McGraw-Hill Companies.

The blastocyst produces a critical hormone that is important in the maintenance of pregnancy. **Human chorionic gonadotropin (HGC)** is the signal to the corpus luteum to not degrade. Normally, the degradation of the corpus luteum causes a decline of estrogen and progesterone and triggers menstruation. At this point in development, menstruation would mean a loss of the embryo in a **spontaneous abortion**. HCG ensures that the corpus luteum continues to secrete estrogen and progesterone so that menstruation is delayed.

Gastrulation and Neuralization

The next event of embryonic development is the **gastrula** stage. During **gastrulation**, three primary germ layers are formed as the cells in the embryo shift into layers, as seen in Figure 19-6. Once a cell enters a **germ layer**, its ability to differentiate into specific cell types is limited.

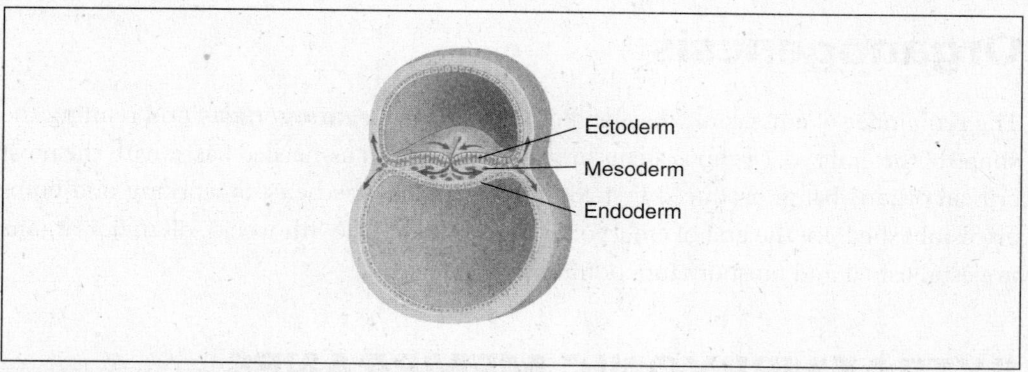

FIGURE 19-6: During gastrulation, embryonic cells shift into the primary germ layers. SOURCE: From George B. Johnson, *The Living World*, 3d ed., McGraw-Hill, 2003; reproduced with permission of The McGraw-Hill Companies.

The three germ layers and the fates of cells in these layers are as follows:

- **Ectoderm:** cells in this layer express the genes needed to become skin cells and cells of the nervous system
- **Mesoderm:** cells in this layer express the genes needed to become muscles, bones, and most internal organs
- **Endoderm:** cells in this layer express the genes needed to become the lining of internal body cavities as well as the linings of the respiratory, digestive, urinary, and reproductive tracts

Once the germ layers are complete, **neuralization** occurs to begin the development of the nervous system. Mesoderm cells form the **notochord**. Ectoderm above the notochord starts to thicken and folds inward, forming neural folds that continue to deepen and fuse to produce a **neural tube**, which eventually develops into the central nervous system. At this point, a head and tail region have been established in the embryo.

Induction and Gene Expression

As differentiation continues, certain cells can influence the gene expression of other cells in the process of **induction** via chemical messengers. Communication between cells is also used to establish positional information in the embryo, which is critical to the formation of internal organs as well as the limbs. **Homeobox genes** produce proteins that are essential for guiding the development of the shape of the embryo. The proteins produced by the homeoboxes are transcription factors that serve to turn on specific genes within cells at specific times.

Induction helps ensure that the right structures occur in the right places. An additional process that is necessary during embryonic development is **apoptosis** of certain cells. Although it seems odd to talk about cell death during development, it is necessary. For example, the separation of fingers and toes is the result of apoptosis of the cells that at one time joined the structures.

Organogenesis

The remainder of embryonic development deals with **organogenesis** and refining the shape of the embryo. Organ systems are developed on an as-needed basis, with the most critical organs being produced first. By the 4th week, the heart is working and limbs are established. By the end of embryonic development (the 8th week), all major organs are established and most are functioning.

EXTRAEMBRYONIC MEMBRANES

While the embryo is in the process of implanting into the endometrium, four membranes are formed outside of the embryo. They are as follows:

- The **amnion** surrounds the embryo in a fluid-filled sac that serves a protective function and provides cushioning for the embryo and fetus.
- The **allantois** is a membrane that ultimately forms the umbilical cord, which is the connection between the embryo and the **placenta** (the organ that delivers nutrients and oxygen and removes carbon dioxide and wastes).
- The **yolk sac** is where the first blood cells develop. In other species, it serves as a source of nutrients.
- The **chorion** eventually becomes the embryo's side of the placenta.

THE PLACENTA

The **placenta** develops from the chorion and grows in size during development. It provides nutrients and oxygen to the embryo and removes wastes. Recall that fetal hemoglobin has a greater affinity for oxygen than adult hemoglobin. The placenta produces HCG, estrogen, and progesterone to maintain the pregnancy. It also produces the hormone relaxin to release the ligaments that attach the pubic bones to provide more space in the birth canal. It takes about 3 months for the placenta to fully develop.

FETAL DEVELOPMENT

Fetal development is primarily a refinement of the organ systems that are already established during embryonic development. The fetus enlarges in size and the organ systems are refined, so that they are all functioning, or are capable of functioning at the end of gestation.

BIRTH

Labor is triggered by the hormone **oxytocin**, which is released by the posterior pituitary gland. Oxytocin causes contractions of the uterus, which intensify with time. Initially, the cervix must dilate, which can take hours. The amnion usually ruptures during the dilation stage. Once the cervix is dilated, contractions continue, which lead to expulsion of the baby. After the baby is delivered, the umbilical cord is clamped and cut, which severs the connection to the placenta. Finally, the placenta is delivered at the end of labor.

CRAM SESSION

Reproduction and Development

1. Both the male and female reproductive systems are under hormonal control to produce gametes.
   a. The female system has the additional responsibility of housing embryonic and fetal development if fertilization occurs.
2. The female system is composed of ovaries that release eggs to the fallopian tubes. Fertilization should occur in the fallopian tubes. As the embryo moves down the fallopian tube, it reaches the uterus, where it implants in the endometrium. The cervix is the narrow opening at the bottom of the uterus. The vagina is the connection to the outside environment.
3. The uterus performs the menstrual cycle on a 28-day cycle. The first week involves loss of the endometrium when estrogen and progesterone levels are low. The second week involves replacement of the endometrium. The third and fourth weeks involve the endometrium preparing to receive an embryo. If no embryo is present at the end of the 28 days, the cycle restarts.
4. The ovarian cycle regulates the release of eggs. Review Figure 19-2. Follicles in the ovaries grow in response to FSH. As follicles grow, they secrete estrogen, which influences the production of greater amounts of LH. On day 14, an LH surge causes the follicle to rupture and the egg to be released to the fallopian tube during ovulation. The remnants of the follicle secrete estrogen and progesterone needed to prepare the uterus for an embryo.
5. The male reproductive system produces sperm in the testes; the sperm then move to the epididymis for maturation. During ejaculation, sperm move into the vas deferens and the urethra, where they pick up secretions from the seminal vesicles, prostate gland, and bulbourethral glands.
6. Spermatogenesis is controlled by testosterone. However, testosterone is produced only in the presence of LH. Once testosterone is made, FSH is necessary to coerce sperm cells to respond to testosterone.
7. During fertilization, one sperm penetrates the egg, which results in a diploid zygote. Cleavage occurs to form the embryo. The embryo goes through the blastocyst stage, gastrulation, neuralization, and organogenesis during the first month of development.
8. Embryonic development concludes after 8 weeks and fetal development commences for the remainder of the gestation period. Fetal development consists of the growth and refinement of structures that were developed in embryonic development.

ON YOUR OWN

MCAT Biology Practice

This section contains 19 short quizzes, one for each of the 19 chapters of the biology review section that you have just completed. Use these quizzes to test your mastery of the concepts and principles covered in the biology review. Explanations for every question are provided following the last quiz. If you need more help, go back and reread the corresponding review chapter.

CHAPTER 1 THE CELL

Quiz

1. A cell is exposed to a substance that prevents it from dividing. The cell becomes larger and larger. This situation

   A. should present no problem to the cell because it can continue to perform all other necessary functions

   B. should present no problem to the cell because the surface area of the cell increases as the volume of the cell increases

   C. will eventually be problematic to the cell because the surface area to volume ratio is decreasing

   D. will eventually become problematic to the cell because the surface area to volume is increasing

2. Human sperm cells require energy in order to move toward an egg cell. What cell structure would enable the sperm to move?

   A. Mitochondria

   B. Lysosomes

   C. Flagellum

   D. Smooth endoplasmic reticulum

3. Osmosis always moves water from

   A. outside the cell to inside the cell

   B. inside the cell to outside the cell

   C. an area of low water concentration to an area of high water concentration

   D. an area of low solute concentration to an area of high solute concentration

4. The "rough" in rough endoplasmic reticulum refers to the presence of ________ which ________.

   A. cytoskeleton fibers; help maintain the structure of the cell

   B. ribosomes; produce proteins

   C. Golgi complex; sorts endoplasmic reticulum contents

   D. chloroplasts; produce energy

5. Pancreatitis is a disorder that causes human pancreas cells to destroy all components in the cell and die. This disorder would involve a problem with which of these cell structures?

   A. Lysosomes

   B. Ribosomes

   C. Cell membrane

   D. Mitochondria

CHAPTER 2 ENZYMES, ENERGY, AND CELLULAR METABOLISM

Quiz

6. A child is taken to the hospital because she has an extremely high fever. The physician immediately orders an ice bath to lower her body temperature. The most logical explanation for this treatment is that an increased body temperature

   A. could cause denaturation to her enzymes, causing critical cellular reactions to be unable to occur

   B. would prevent substances from being transported properly across the cell membrane

   C. would increase the rate of cellular respiration to a dangerous level

   D. could prevent cell division

7. The final result of ATP production after anaerobic respiration is

   A. the same as the amount of ATP made in aerobic respiration

   B. equal to the amount of ATP made in glycolysis

   C. low relative to the amount of ATP made in aerobic respiration

   D. high relative to the amount of ATP made in aerobic respiration

8. The drug DNP destroys the H^+ gradient that forms in the electron transport chain. The most likely consequence would be

   A. the cells are forced to perform fermentation

   B. ATP production increases

   C. glycolysis stops

   D. oxygen consumption increases

9. A child is born with a rare disease in which mitochondria are missing from his skeletal muscles (the mitochondria are present in all other body cells). Physicians find that this child's muscle can function. They also find

   A. the muscles contain large quantities of lactic acid even after very mild activity

   B. the muscles require extremely large amounts of oxygen to function

   C. the muscle cells cannot convert glucose to pyruvate

   D. the muscle cells require large amounts of CO_2 to function

10. The carbon dioxide exhaled by animals is produced in

A. glycolysis

B. lactate fermentation

C. Kreb's cycle

D. electron transport chain

CHAPTER 3 DNA STRUCTURE, REPLICATION, AND TECHNOLOGY Quiz

11. A DNA analysis shows that an organism has 21% cytosine. The amount of thymine this organism would have is

A. 21%

B. 29%

C. 50%

D. 23%

12. A lab technician allows a cell to perform semiconservative replication in the presence of radioactive nucleotides. Which of the following would occur?

A. The DNA would be radioactive in one of the double helices, but not in the other double helix.

B. The DNA in each of the double helices would be radioactive.

C. The DNA would not be radioactive in either of the double helices.

D. The mRNA made from this DNA would be radioactive.

13. A DNA template reads: T A C G A C. Which of the following is the complementary DNA sequence?

A. U A C G A C

B. T A C G A C

C. A T G C T G

D. A U G C U G

14. Which of the following is a type of enzyme that cuts double-stranded DNA at a specific site within a specific base sequence often in such a way to create "sticky ends"?

A. Reverse transcriptase

B. Restriction enzyme

C. DNA polymerase

D. Ligase

15. A small amount of DNA was extracted from a blood sample left at a crime scene. How could technicians obtain a greater quantity of this DNA to use for further testing?

A. By performing electrophoresis

B. By cutting the DNA sample with restriction enzymes

C. By performing PCR

D. By using DNA hybridization

CHAPTER 4 PROTEIN SYNTHESIS Quiz

16. Protein X is 40 amino acids long. The number of nucleotides that were present in the mRNA used to make this protein was

A. 120

B. 60

C. 40

D. 20

17. RNA

A. is usually found as a double helix

B. is made during translation

C. is distinguished from DNA by the use of ribose sugar in the nucleotides

D. is produced in the cytoplasm

18. The DNA of a cell contains instructions for making ______________; the instructions are carried away from the DNA by ________ and are taken to the cell's ___________, where translation occurs.

A. genes; proteins; nucleus

B. ribosomes; tRNA; nucleus

C. chromosomes; tRNA; rough endoplasmic reticulum

D. proteins; mRNA; ribosomes

19. Why aren't mutations that cause changes to amino acid sequences more common?

A. Because the DNA polymerase enzyme proofreads and usually corrects its mistakes.

B. Because tRNAs can correct mistakes and bring the proper amino acid to the ribosome.

C. Because RNA polymerase can correct DNA mistakes during transcription.

D. Any of the above acceptable explanations.

CHAPTER 5 GENETICS Quiz

20. A particular genetic disease is caused by a recessive allele. If two parents have the dominant phenotype but a heterozygous genotype, what are the chances of their child having the disease?

A. 0%

B. 25%

C. 50%

D. 100%

21. An allele known as BRCA1 is often found in women with breast and ovarian cancer. However, some women develop these cancers and do not have the BRCA1 allele. In addition, some women who have BRCA1 never develop breast or ovarian cancer. How can this be explained?

A. This must be an example of pleiotropy.

B. These types of cancers are in no way associated with specific alleles.

C. Something other than the allele must factor into whether cancer develops.

D. The BRCA1 allele must follow a dominant pattern of inheritance.

22. Phenylketonuria (PKU) is caused by a recessive allele. A man is the carrier of the disorder; his wife does not have PKU and is not a carrier of the disorder. What is the probability that some of their offspring will have the disorder and that others will be carriers?

A. 0% of the offspring will have the disorder; 0% of the males would be carriers; 100% of the females would be carriers

B. 0% of the offspring will have the disorder; 50% will be carriers

C. 50% of the offspring will have the disorder; 50% will be carriers

D. 25% of the offspring will have the disorder; 75% will be carriers

23. Duchene muscular dystrophy is caused by a sex-linked allele. Its victims are almost always boys who die before the age of 20. Why is this disorder rarely seen in girls?

A. The allele is carried on the Y chromosome.

B. Girls must receive the allele from both their mother and their father to have the disease.

C. Males carrying the allele don't usually live long enough to reproduce.

D. Sex-linked traits are never seen in girls.

24. The phenotypic ratio obtained in a Punnett square is 1:2:1. Based on this ratio, you can conclude that which genetic "rules" where involved?

A. Multiple alleles

B. Pleiotropy

C. Normal dominance

D. Incomplete dominance

CHAPTER 6 **CELL DIVISION**

Quiz

25. Which stage of mitosis is not matched correctly with a key characteristic of that stage?

A. Telophase: cytokinesis occurs

B. Metaphase: chromatids separate

C. Prophase: nuclear membrane disappears temporarily

D. Anaphase: the spindle fibers pull toward the poles of the cell

26. If a daughter cell contains 14 chromosomes at the end of meiosis I, how many chromosomes does each daughter cell contain at the end of meiosis II?

A. 7

B. 14

C. 23

D. 46

27. How do mitosis and meiosis differ?

A. The goal of mitosis is to produce cells that are genetically identical to the original parent cell; the goal of meiosis is to produce cells that contain twice the number of chromosomes as the original parent cell.

B. The cells formed by mitosis are diploid; the cells formed by meiosis are haploid.

C. Mitosis results in the production of gametes; meiosis results in the production of cells that are used for the organism's growth and replacement of damaged cells.

D. Synapsis and crossing over occur in mitosis to assure new combinations of genetic material; there is no synapsis and crossing over in meiosis.

28. If a sperm cell contains 12 chromosomes, it must have come from an original parent cell that contained ________ chromosomes.

A. 4

B. 6

C. 12

D. 24

CHAPTER 7 EVOLUTION Quiz

29. When a catastrophe wipes out a large percentage of a population and reduces the gene pool of the population, this is an example of

    A. gene flow

    B. directional selection

    C. artificial selection

    D. genetic drift

30. Darwin's theory of natural selection is often referred to as "survival of the fittest." The "fittest" organisms in a population are those that

    A. are the biggest and strongest

    B. mutate the fastest

    C. live the longest

    D. reproduce the most

31. In a certain population, 81% of individuals have a homozygous recessive genotype. Using the Hardy–Weinberg equations, what percentage of individuals in the next generation will be heterozygous?

    A. 10%

    B. 19%

    C. 18%

    D. 9%

32. Two amphibian species live within the same range. Both species reproduce in large, permanent bodies of water; however, each species mates at different times of the day. This is an example of

    A. ecological isolation

    B. temporal isolation

    C. behavioral isolation

    D. reproductive isolation

33. "Endosymbiosis" and "membrane infolding" refer to

A. how the first cells evolved

B. how multicellular organisms evolved

C. how eukaryotic cells evolved

D. how terrestrial plants and animals evolved

CHAPTER 8 BACTERIA AND FUNGI

Quiz

34. When a certain bacterium encounters the antibiotic tetracycline, the antibiotic molecule enters the cell and attaches to a repressor protein. This keeps the repressor from binding to the bacterial DNA, which allows certain genes to be transcribed. These genes code for enzymes that break down tetracycline. This set of genes is best described as

A. an operator

B. an inducible operon

C. a repressible operon

D. a regulator

35. The genes for antibiotic resistance are often carried on bacterial plasmids. These plasmids can be transferred from one bacterium to another. The most likely method for plasmid transfer would be

A. transduction

B. transformation

C. conjugation

D. binary fission

36. An unknown cell type has been isolated. It is suspected that this cell is prokaryotic, but tests are performed in order to confirm this suspicion. The presence of which of these cell structures would confirm that the cells are prokaryotic?

A. Cytoplasm

B. A flagellum

C. A peptidoglycan cell wall

D. Spores

37. Fungi are typically characterized based on reproductive methods and structures. A fungus that always reproduces asexually is most likely classified as

A. Zygomycetes

B. Ascomycetes

C. Deuteromycetes

D. Basidomycetes

38. An organism that does not perform photosynthesis and receives carbon from carbon dioxide would be classified as a

A. photoautotroph

B. chemoautotroph

C. chemoheterotroph

D. photoheterotroph

CHAPTER 9 VIRUSES

Quiz

39. A bacteriophage that is in a latent phase

A. is in the lytic cycle

B. is in the lysogenic cycle

C. is actively making new viruses

D. is not inserted into the host chromosome

40. A bacteriophage that is in the lysogenic cycle will do all of the following except:

A. lyse the host

B. be passed on as the bacterial host cell divides

C. eventually enter the lytic cycle

D. insert into the host's DNA

41. Which of the following characteristics distinguish retroviruses from all other types of viruses?

A. They are viruses that use RNA as their genetic material.

B. They must interact with a receptor on the host to infect it.

C. They enter latent phases.

D. They use the enzyme reverse transcriptase.

42. Host cells are often killed by a virus as the direct result of

A. integrating into the host cell chromosomes

B. lysis of the host to release new viruses

C. replication of the viral genetic material within the host

D. toxic enzymes that damage the host cell that are coded for in the viral genome

CHAPTER 10 TISSUES AND SKIN
Quiz

43. You are looking at a piece of tissue under the microscope, trying to identify what type of tissue it is. You know that it came from the lining of the trachea of the respiratory tract. Based on this information, the tissue is most likely

A. epithelial

B. muscular

C. nervous

D. connective

44. A prescription medication lists a rare side effect of impairing the movement of smooth muscle. This medication could potentially affect

A. the heart

B. the walls of the small intestine

C. the muscles in the arms and legs

D. the diaphragm

45. A basement membrane is characteristic of

A. epithelial tissue

B. connective tissue

C. nervous tissue

D. muscular tissue

46. A physician is examining a patient who has some extreme swelling in their hands and feet. This chronic swelling and fluid buildup is most likely related to what type of tissue?

A. Blood

B. Lymph

C. Adipose

D. Bone

CHAPTER 11 THE NERVOUS SYSTEM AND SENSES Quiz

47. A person is having some strange vision problems. They can see well in dim light or when viewing black and white objects, but their color vision is impaired. The problem must be with their

A. rods

B. cones

C. optic nerve

D. rhodopsin

48. A drug that causes sodium to leak into the neuron would

A. cause the outside of the neuron to become more positive

B. make it easier to trigger action potentials in the neuron

C. cause the inside of the neuron to become more negative

D. make it more likely for chloride ions to enter the neuron

49. You touch a hot stove and your hand jerks back. Which of the following describes the pathway of nerve impulses responsible for this reflex?

A. Motor neuron → sensory neuron → interneuron

B. Sensory neuron → interneuron → motor neuron

C. Interneuron → sensory neuron → motor neuron

D. Motor neuron → interneuron → sensory neuron

50. Damage to which of the following brain structures would be most likely to cause fatal consequences?

A. The cerebral cortex

B. The corpus callosum

C. The medulla oblongata

D. The cerebellum

51. Overall, the ______________ nervous system slows down the body and diverts energy to digestion, whereas the ______________ nervous system slows down digestion and increases overall body activity.

A. autonomic; somatic

B. sympathetic; autonomic

C. sensory; motor

D. parasympathetic; sympathetic

CHAPTER 12 MUSCULAR AND SKELETAL SYSTEMS Quiz

52. An x-ray reveals that a patient has a fracture in her clavicle. The bone cells that would be most active at the site of the injury are

A. osteocytes

B. osteoblasts

C. osteoclasts

D. cartilage

53. Concerning bone formation, if blood calcium levels are high, bone will be __________________ due to the effects of the hormone ______________________.

A. built; calcitonin

B. built; parathyroid hormone

C. broken down; calcitonin

D. broken down; parathyroid hormone

54. During a quick sprint, the most likely source of energy is ________ and the most active muscles rely on _____________ muscle fibers.

A. anaerobic respiration; fast-twitch

B. creatine–PO_4; fast-twitch

C. aerobic respiration; slow-twitch

D. aerobic respiration; fast-twitch

55. Acetylcholine is a neurotransmitter that is responsible for initiating skeletal muscle contractions. Which of the following drugs might cause skeletal muscle paralysis?

A. A drug that interferes with acetylcholine breakdown in the synapse.

B. A drug that causes sodium channels to open in the neurons serving the skeletal muscles.

C. A drug that prevents release of acetylcholine to the synapse.

D. Any of the above could cause paralysis.

CHAPTER 13 THE ENDOCRINE SYSTEM Quiz

56. The hormones secreted from the ovaries and testes are produced from

A. cholesterol

B. carbohydrates

C. a long protein

D. a single amino acid

57. If you were stranded on a desert island with no fresh water to drink, the level of which of the following would rise in your bloodstream in an effort to conserve water?

A. Insulin

B. Oxytocin

C. Glucagon

D. Antidiuretic hormone

58. A patient is diagnosed with a tumor in an endocrine gland. If the patient has major problems in maintaining a constant level of blood glucose, what endocrine gland was affected?

A. Pancreas

B. Parathyroid

C. Adrenal

D. Anterior pituitary

59. Steroid hormones

A. bind to cell membrane receptors

B. enter cells and bind to cytoplasmic receptors

C. rely on cAMP to produce their effects

D. are derived from proteins

CHAPTER 14 THE CARDIOVASCULAR SYSTEM Quiz

60. Many poor, starving college students get paid to donate plasma. Which of the following substances is not donated when the plasma is removed?

A. Antibodies

B. Water

C. Red blood cells

D. Calcium

61. Which of the following would have the same oxygen content?

A. Blood entering the lungs – blood leaving the lungs

B. Blood entering the right side of the heart – blood entering the left side of the heart

C. Blood entering the right side of the heart – blood leaving the right side of the heart

D. Blood entering the tissue capillaries – blood leaving the tissue capillaries

62. A person has an irregular heartbeat and must have an artificial pacemaker implanted to regulate their heart rate. What normal structure is being replaced by the artificial pacemaker?

A. The AV node

B. The SA node

C. A nerve

D. The tricuspid valve

63. The highest blood pressure in the body should be found in

A. veins

B. pulmonary capillaries

C. systemic capillaries

D. arteries

64. In response to dehydration, which of the following would be expected?

A. Plasma volume would increase

B. Blood pressure would increase

C. Hematocrit levels would increase

D. Red blood cell counts would increase

CHAPTER 15 THE RESPIRATORY SYSTEM Quiz

65. Immediately after inhalation, in the alveoli,

A. O_2 concentration in the pulmonary capillaries is high, CO_2 concentration in alveoli is high

B. O_2 concentration in the pulmonary capillaries is low, CO_2 concentration in alveoli is low

C. O_2 concentration in the pulmonary capillaries is low, CO_2 concentration in alveoli is high

D. O_2 concentration in the pulmonary capillaries is high, CO_2 concentration in alveoli is low

66. Which of the following is the primary trigger that would increase your ventilation rate?

A. High oxygen levels

B. Low oxygen levels

C. High carbon dioxide levels

D. High pH levels

67. When the diaphragm contracts, the thoracic cavity volume ________, and you ________.

A. gets smaller; exhale

B. gets larger; inhale

C. gets smaller; inhale

D. gets larger; exhale

68. Which of the following choices indicates the correct pathway of CO_2 leaving the body?

A. Alveoli, bronchioles, bronchi, trachea, larynx, pharynx

B. Pharynx, larynx, trachea, bronchi, bronchioles, alveoli

C. Alveoli, bronchi, bronchioles, larynx, trachea, pharynx

D. Alveoli, bronchioles, bronchi, trachea, pharynx, larynx

CHAPTER 16 THE DIGESTIVE SYSTEM Quiz

69. A person is on medication to increase the pH of their stomach. A side effect is that the increased pH causes the normal stomach enzymes to be unable to function. Which component of this person's diet may not be completely digested?

A. Proteins

B. Carbohydrates

C. Fats

D. Nucleic acids

70. If you chew on a piece of bread long enough, it will begin to taste sweet, because

A. pepsin is breaking down the proteins to amino acids

B. lipases are breaking down the fats

C. amylase is breaking down the starch into glucose

D. saliva tastes sweet

71. The large intestine

A. absorbs amino acids and glucose

B. completes protein digestion

C. contains bacteria that can synthesize certain vitamins

D. allows for lipid absorption

72. Most nutrients are absorbed in the

A. oral cavity

B. small intestine

C. stomach

D. large intestine

CHAPTER 17 **THE URINARY SYSTEM**

Quiz

73. How does the filtrate that enters the nephron differ from urine?

A. It contains very little water

B. It has a higher concentration of glucose

C. It has a lower concentration of cells

D. It has a lower concentration of proteins

74. Which of these does not enter the nephron from the blood?

A. Water

B. Amino acids

C. Proteins

D. Glucose

75. Initial water reabsorption (leaving the nephron) along with reabsorption of glucose and amino acids occurs in this part of the nephron:

A. proximal convoluted tubule

B. loop of Henle

C. distal convoluted tubule

D. collecting duct

76. Drinking alcohol _____ antidiuretic hormone. This results in a(n)_____ of urine volume.

A. enhances; increase

B. enhances; decrease

C. inhibits; increase

D. inhibits; decrease

CHAPTER 18 THE LYMPHATIC AND IMMUNE SYSTEMS

Quiz

77. After getting scratched by a cat, you notice that your skin becomes swollen, red, and painful. The chemical responsible for these symptoms is ____________, which is released from ____________.

A. interferon; virally infected cells

B. histamine; basophils

C. interleukin; helper T cells

D. complement; blood

78. A person suffering from influenza (flu) is unlikely to get another viral infection at the same time. This is because of the secretion of ________ by the immune system.

A. histamine

B. interferon

C. interleukins

D. antibodies

79. Development of a secondary immune response is based on

A. antibodies

B. memory cells

C. interleukins

D. T cells

80. A patient has been diagnosed with an immunodeficiency disease. The doctor suspected this because

A. she rejected an organ transplant

B. she seems to attack self-antigens

C. she has many allergies

D. she has prolonged, repeated infections

CHAPTER 19 REPRODUCTION AND DEVELOPMENT

Quiz

81. A fertility doctor is trying to determine why a couple is infertile. It has been determined that the woman is fertile, so the problem must be with the man. In analyzing his semen, the doctor notices that many of the sperm are not viable. Further analysis shows a low level of nutrients in the semen. The structure malfunctioning must be the

A. bulbourethral glands

B. seminal vesicles

C. prostate

D. urethra

82. A female is not pregnant, but is not menstruating as expected. Testing shows that the reason she is not menstruating is because her _______ levels are high during the time menstruation should be occurring.

A. estrogen

B. follicle-stimulating hormone

C. luteinizing hormone

D. progesterone

83. The process of organogenesis in development begins ________ after fertilization.

A. 7 weeks

B. 12 weeks

C. 2 weeks

D. 4 weeks

84. Muscular tissue in the embryo is derived from which germ layer?

A. Mesoderm

B. Endoderm

C. Ectoderm

D. Trophoblast

85. Which part of the sperm is needed to actually penetrate the egg?

A. The head

B. The tail

C. The midpiece

D. The acrosome

CHAPTER 1 THE CELL
Answers and Explanations

1. **C** This question is essentially asking why cells must be small in size. As a cell gets larger, its surface area and volume increase. Cells typically need a large surface area relative to a small volume, or a large surface area to volume ratio. This allows the cell to get the substances it needs into and out of the cell quickly. The larger a cell gets, the smaller the surface area to volume ratio becomes.

2. **C** This question relies on knowledge of cell structures. Mitochondria are used for the process of cellular respiration, which produces ATP. Lysosomes perform cellular digestion. The smooth endoplasmic reticulum is responsible for lipid synthesis. The only structure that enables a cell to move is the flagellum.

3. **D** Osmosis is a specialized form of passive transport that moves water from the side of the membrane that has more water to the side of the membrane that has less water. The higher the water concentrations, the lower the solute concentrations, and vice versa. Choices A and B suggest that osmosis can occur in only one direction (either into or out of the cell), which is incorrect. Choice C directly contradicts the definition of osmosis. This leaves choice D as the correct answer. Water moves from an area of low solute concentration (high water) to an area of high solute concentration (low water).

4. **B** This question relies on an understanding of the rough endoplasmic reticulum and its function. The endoplasmic reticulum is divided into smooth and rough sections. The difference between the two sections is that the rough section has ribosomes on the surface. Ribosomes produce proteins during the process of translation.

5. **A** The question asks about a cell structure that would be able to break down (digest) cellular components. Of the structures listed, the lysosomes would be the best option. Ribosomes produce proteins, the cell membrane serves as the outer boundary of the cell, and the mitochondria are used for cellular respiration.

CHAPTER 2 ENZYMES, ENERGY, AND CELLULAR METABOLISM Answers and Explanations

6. **A** This question presents a situation in which a very high fever is increasing temperature in the body. Of the choices listed, some make more sense than others. It does not seem plausible that increased temperature would increase the cellular respiration rates to dangerous levels, that increased temperature would prevent substances from crossing the membrane, or that increased temperature would alter cell division. However, increased temperatures are one factor that can cause denaturation of enzymes, thus possibly preventing critical cellular reactions from occurring. The high temperature must be quickly reduced, and an ice bath would accomplish this.

7. **C** Aerobic respiration produces far more ATP than anaerobic respiration. The process of anaerobic respiration produces 2 ATP (both from glycolysis), whereas aerobic respiration produces about 36 ATP from glycolysis, Kreb's cycle, and the electron transport chain. The most accurate choice would be C, anaerobic respiration produces far less ATP than aerobic respiration.

8. **A** If the H^+ (proton) gradient were to be destroyed, the electron transport chain would be affected, as it is the only step of cellular respiration that relies on a concentration gradient. You must look for the choice that relates to how a cell performs cellular respiration without an electron transport chain. The only option is to move to anaerobic respiration, which requires fermentation, as indicated by choice A.

9. **A** Because the Kreb's cycle and electron transport chain of aerobic respiration occur in the mitochondria, a situation in which mitochondria are absent would prevent aerobic respiration. The only option would be anaerobic respiration. In animals, the form of anaerobic respiration used would be lactic acid fermentation, which converts pyruvate to lactic acid.

10. **C** This question relies on direct recollection of the steps of aerobic respiration. During the Kreb's cycle, CO_2 is released as citric acid is broken down and rearranged. There is no release of CO_2 in any other steps of aerobic respiration.

CHAPTER 3 DNA STRUCTURE, REPLICATION, AND TECHNOLOGY
Answers and Explanations

11. B The amount of cytosine and guanine are always equal. The same holds true for adenine and thymine. If the percentage of cytosine is 21%, the percentage of guanine must also be 21%. This totals 42%, leaving 58% to be shared equally between adenine and thymine. This means that the percentage of thymine would be 29%.

12. B During semiconservative replication, the DNA double helix unwinds and unzips, and each strand serves as a template. DNA polymerase copies each strand, producing a complementary strand. The resulting double helices contain one original template strand and one newly synthesized strand.

13. C This question relies on knowing the rules of complementary base pairing. Cytosine always pairs with guanine, and adenine always pairs with thymine. The only time uracil is present would be in RNA.

14. B The question asks for an enzyme that can cut DNA and produce "sticky ends," which are single-stranded pieces of DNA. Of the enzymes listed, only one is capable of producing sticky ends. Reverse transcriptase converts RNA to DNA, DNA polymerase synthesizes complementary strands of DNA, and ligase is used to seal fragments of DNA. Restriction enzymes are capable of cleaving DNA, and some types have recognition sites that cause uneven cutting, resulting in sticky ends.

15. C Polymerase chain reaction (PCR) can be used to amplify small amounts of DNA. The PCR process goes through cycles and heats DNA to separate the strands, cools the single-stranded pieces to allow for primer hybridization, and uses DNA polymerase and nucleotides to extend the DNA fragments. Over many cycles, millions of copies of DNA can be produced. Gel electrophoresis is used to separate different-size pieces of DNA or proteins, restriction enzymes cut pieces of DNA, and DNA hybridization joins two strands of DNA.

CHAPTER 4 PROTEIN SYNTHESIS Answers and Explanations

16. A If a protein is 40 amino acids in length, the instructions for those 40 amino acids came from the mRNA. A codon of the mRNA consists of 3 nucleotides. Each codon provides information for a single amino acid. A protein with 40 amino acids was produced from an mRNA with 40 codons, or 120 nucleotides.

17. C RNA and DNA differ in several ways. RNA exists as a single strand, which is copied from DNA in the nucleus during the process of transcription. The nucleotides that are used to make RNA are cytosine, guanine, adenine, and uracil. The sugar that is used in RNA nucleotides is ribose (as compared with deoxyribose in DNA).

18. D This question can easily be answered based on elimination. The first blank of the question asks about the function of DNA. The instructions provided by DNA are transcribed into RNA and eventually translated into a protein by the ribosomes. This points us to choice D.

19. A Mutations are changes to the DNA. Although it does not catch all mistakes, DNA polymerase does catch many errors and allows for their correction during a proofreading process. RNA polymerase does not have proofreading or correcting abilities, nor do tRNAs correct errors in the process of translation.

CHAPTER 5 GENETICS
Answers and Explanations

20. **B** A Punnett square can be used to solve this problem. If both parents are heterozygous, the cross is Aa × Aa. This leads to a 3:1 phenotypic ratio, with 75% of the offspring showing the dominant phenotype and 25% of the offspring showing the recessive phenotype.

21. **C** The question indicates that breast cancer can be associated with the BRCA1 allele. However, not all cases of breast cancer are associated with this allele, and some individuals who have the allele do not develop cancer. The only reasonable explanation in this case is that something other than a genetic factor must be involved. Environmental influences, in addition to certain alleles, are known to influence the development of certain phenotypes in some cases.

22. **B** The parents in this case are a carrier father and a normal (noncarrier) mother. This is not a sex-linked trait, so we can indicate the genotype of the father as Aa and the mother as AA. If you work out the Punnett square, 50% of the offspring will be carriers (Aa) and 50% of the offspring will be normal (AA). None of the children will have the disease.

23. **B** The question indicates that Duchenne muscular dystrophy is a sex-linked allele that affects males more so than females. However, the question indicates that the disorder does occur in females, although it is more rare than in males. This indicates that the allele is on the X chromosome. Because a male only has one X chromosome that he inherited from his mother, he will express whatever allele is on that X. Therefore, males need to inherit the allele from only their mother to be affected. For a female to have this disease, she must receive it on both X chromosomes (from both parents). It would be more likely for a male to inherit this from one parent (his mother) than for a female to inherit the allele from both parents.

24. **D** The phenotypic ratio contains three numbers, which indicates that three phenotypes are present. Of the choices listed, only one can lead to the presence of three different phenotypes. In the heterozygous condition, incomplete dominance can lead to the expression of an intermediate phenotype.

CHAPTER 6 CELL DIVISION Answers and Explanations

25. B The major events of each stage of mitosis are as follows: During prophase, the spindle fibers begin to form, the chromosomes coil and condense, and the nuclear membrane dissolves. In metaphase, the chromosomes are aligned down the center of the cell by the spindle fibers. In anaphase, the spindle fibers retract toward each pole of the cell, splitting the chromatids. In telophase, cytokinesis occurs. Of the choices listed, choice B is incorrectly paired.

26. B This question requires an understanding of the phases of meiosis I and II. At the end of meiosis I, each daughter cell has half the chromosomes of the original parent cell. However, those chromosomes will still be replicated. The purpose of meiosis II is to split the chromatids so that each gamete contains unreplicated chromosomes. If the cell at the end of meiosis I has 14 chromosomes (replicated), the same number are present in each cell at the end of meiosis II, but they will be unreplicated.

27. B This question requires knowledge of the purposes of mitosis and meiosis. Mitosis is used for growth, repair, and replacement of cells. Mitosis begins with a diploid cell and ends with identical diploid daughter cells. The purpose of meiosis is to produce genetically diverse daughter cells. Meiosis begins with a diploid parent cell and ends with four genetically unique haploid daughter cells. The only choice that reflects correct statements concerning mitosis and meiosis is choice B.

28. D A sperm cell would be the result of cells that had completed meiosis II. The question indicates that the number of chromosomes in the haploid sperm cell would be 12. The original parent cell that started meiosis would be diploid, or twice that of the haploid number. In this question, the diploid number of the starting cell would be 24.

CHAPTER 7 EVOLUTION Answers and Explanations

29. D Gene flow occurs when individuals leave the population, taking their genes with them. Directional and artificial selection do not deal with changes to the gene pool based on allelic flow. However, genetic drift can be used to explain a major change in the gene pool as the result of the loss of a large number of individuals in the population.

30. D This question relies on an understanding of the fitness concept. Fitness is relative to the specific population. Fit individuals carry alleles that make them more likely to reproduce and pass those alleles to the next generation. Therefore, those that reproduce the most in a certain population would be considered the most fit under the conditions of that population.

31. C The two equations of Hardy–Weinberg are $p + q = 1$ and $p^2 + 2pq + q^2 = 1$, where $p =$ the frequency of the dominant allele and $q =$ the frequency of the recessive allele. The question says that the homozygous recessive genotype is 81% of the population. This means that $q^2 = 0.81$. Now solve for q and p. If $q^2 = 0.81$, then $q = 0.9$ and $p = 0.1$. The question asks for the frequency of heterozygous individuals, which is $2pq$, or $2(0.9)(0.1) = 0.18$.

32. B This question relies on knowledge of prezygotic isolation mechanisms. The question indicates that the species live in the same range. Ecological isolation occurs when species rarely interact with each other. Behavioral isolation occurs when the mating behaviors of two species are incompatible, and reproductive isolation deals with incompatibilities between reproductive structures. Of the choices given, only temporal isolation explains the fact that the two species exist in the same range yet reproduce at different times.

33. C Endosymbiosis refers to the engulfment of a prokaryotic cell by another cell and explains the presence of organelles such as mitochondria and chloroplasts. Membrane infolding refers to the internalization of the cell membrane, which explains the presence of organelles such as the endoplasmic reticulum. Both of these theories collectively explain the presence of organelles, which are characteristic to the evolution of eukaryotic cells.

CHAPTER 8 BACTERIA AND FUNGI
Answers and Explanations

34. B The question describes a system of gene regulation using repressors. Because the organisms described are bacterial, and the question is asking about gene regulation, this should narrow the choices to operons. Because the enzyme produced in this question is only used when the antibiotic tetracycline is present, this would be an example of an inducible operon.

35. C To answer this question, you must be familiar with methods of bacterial gene transfer. Because the question says that a plasmid is being transferred, this would indicate that conjugation is the most likely method. Transformation involves taking up excess DNA from the environment, and transduction involves bacteriophage transfer. Binary fission is normal bacterial cell division.

36. C This question is asking you to identify a cell structure that is only found in bacterial cells. Cytoplasm can be eliminated because it is found in all types of cells. Flagella can be found in many types of cells that must move. Spores can be found in bacteria, but also in fungi. The only choice that is unique to bacteria is a cell wall that contains peptidoglycan.

37. C Fungi tend to be classified based on their means of reproduction and reproductive structures. Of the groups listed, only Deuteromycetes lacks a sexual reproduction phase and only reproduces asexually. Although the other classes of fungi can also reproduce asexually, they have sexual phases of the life cycle as well.

38. B Phototrophs perform photosynthesis, whereas chemotrophs do not. Autotrophs obtain carbon from carbon dioxide, whereas heterotrophs obtain carbon from other sources. Because the organism described in this question does not perform photosynthesis, it must be a chemotroph. It obtains carbon from carbon dioxide. This organism would be classified as a chemoautotroph.

CHAPTER 9 VIRUSES
Answers and Explanations

39. B Latent phases are characteristic of the lysogenic cycle. During this cycle, the bacteriophage inserts into the chromosome of the host, initiating the latent period. During latency, new viruses are not made.

40. A During the lysogenic cycle, the virus inserts into the host chromosome and stays latent. Each time the cell divides, the bacteriophage DNA is passed on to the progeny. Eventually, lysogeny ends and the lytic cycle begins. During the lysogenic cycle, the host cell is not damaged or lysed.

41. D Each of the choices listed is characteristic of a retrovirus; however, some of these characteristics are common to other viruses. For example, viruses other than retroviruses use RNA as their genetic material. The unique feature of retroviruses is that they code for the enzyme reverse transcriptase. This enzyme allows the RNA virus to convert itself into DNA form.

42. B Host cells tend to be killed as newly formed viruses try to exit the cell. Lysis is one means a virus can use to leave the cell. This ruptures the cell membrane, which ultimately kills the host cell.

CHAPTER 10 TISSUES AND SKIN Answers and Explanations

43. A The question indicates that the unknown tissue came from the lining of the trachea. Tissues that line surfaces are typically some form of epithelial tissue. Connective, muscular, and nervous tissues tend to be found somewhere other than the surface of a structure.

44. B Muscle comes in three varieties: skeletal, cardiac, and smooth. Smooth muscle is characterized as lacking striation and being involuntary. Of the choices listed, only the muscles of small intestine would qualify as smooth muscle.

45. A A basement membrane is a substance that attaches one type of tissue to another. This substance is unique to epithelial tissues, which are often on surfaces and must be attached to other tissues deeper in the body.

46. B The question introduces the idea of swelling due to fluid buildup. The type of tissue that typically reclaims excess fluids in the body is lymph. Swelling that occurs in the body is the result of the lymphatic system not working properly to reclaim the excess fluids.

CHAPTER 11 THE NERVOUS SYSTEM AND SENSES Answers and Explanations

47. B The eyes contain two types of photoreceptors that are sensitive to light. Rods are used for black and white vision, whereas cones are used for color vision. Cones come in three varieties that work together to process color images.

48. B During resting potential, sodium is restricted to the outside of the neuron as the result of the action of the Na^+/K^+ pump. During an action potential, Na^+ enters the neuron causing depolarization. If a drug where to allow Na^+ to enter the neuron at will, then action potentials would be more likely to occur.

49. B Messages are picked up by sensory neurons in the peripheral nervous system, passed to interneurons in the central nervous system, and then responses are passed on to motor neurons in the peripheral nervous system.

50. C The cerebral cortex is responsible for conscious thought and the processing of sensory information. Although damage to this structure can have serious consequences, it typically is not fatal. The corpus callosum connects the two halves of the cerebral cortex and allows for the transfer of information between the two halves; it is not vital. The cerebellum controls functions such as the coordination of complex motor patterns, which is also not vital. Of the choices listed, only the medulla oblongata has functions that are vital, meaning that if they were disrupted, death would occur.

51. D This question is asking about the branches of the autonomic nervous system. The sympathetic innervation relies on adrenaline to exert its effects and it typically functions to increase heart rate, ventilation rate, etc., while slowing down digestive function. During relaxation, the parasympathetic system takes over, reducing heart rate and ventilation rate and increasing the digestive rate.

CHAPTER 12 MUSCULAR AND SKELETAL SYSTEMS Answers and Explanations

52. B This question relies on your knowledge of bone cells. Osteoblasts build new matrix in response to increased blood calcium levels. Osteoclasts break down bone tissue in response to low blood calcium levels. Osteocytes are mature bone cells involved in tissue maintenance. If an individual had a bone injury, the osteoblasts would be most active during the repair process.

53. A When blood calcium levels are high, they must be lowered. This occurs by the building of new bone matrix. The cells responsible for building new matrix are osteoblasts, which are influenced by the hormone calcitonin produced by the thyroid gland.

54. A Energy can be provided via aerobic or anaerobic respiration. For a burst of quick energy, anaerobic respiration is typically used to provide ATP. For long-term energy needs, aerobic respiration is used. Muscle fibers can be classified as slow twitch or fast twitch. Fast-twitch muscle fibers lack stamina, but are ideal for quick bursts of contractions, whereas slow-twitch fibers are used for longer contractions and endurance. In the case of a quick sprint, the energy source would be provided by anaerobic respiration, which would fuel fast-twitch fibers.

55. C If acetylcholine is the neurotransmitter used to initiate skeletal muscle contraction, it would need to be released to the synapse to exert its effects. If it were not present in the synapse, no muscle contraction should occur. If a drug were to prevent the release of acetylcholine, paralysis should occur in the muscle.

CHAPTER 13 THE ENDOCRINE SYSTEM Answers and Explanations

56. A The ovarian hormones are estrogen and progesterone, whereas the testes produce testosterone. Each of these is a steroid hormone that is produced from cholesterol in the body.

57. D This question asks about which hormone would be involved with water retention in the body. Of the choices listed, insulin and glucagon are used to regulate blood sugar, and oxytocin deals with uterine contractions and labor. Antidiuretic hormone, secreted from posterior pituitary, acts on the kidneys to increase water retention in the body.

58. A To answer this question, you must be familiar with blood sugar regulation. The antagonistic hormones insulin and glucagon are used to regulate blood sugar. Insulin decreases blood glucose levels, whereas glucagon increases blood glucose levels. Both of these hormones are secreted from the pancreas.

59. B Steroid and nonsteroid (peptide) hormones have different mechanisms of action. Nonsteroid hormones are derived from amino acids or proteins and bind to cell membrane receptors. They act as a first messenger, causing signal transduction that ultimately activates cAMP (a second messenger) in the cell. Steroid hormones are derived from cholesterol and enter cells and find their receptors in the cytoplasm. The steroid–receptor complex moves to the nucleus to interact with the DNA so that gene expression is altered.

CHAPTER 14 THE CARDIOVASCULAR SYSTEM Answers and Explanations

60. C Plasma consists of the liquid portion of the blood, which excludes cells. Of the choices listed, the only item that is not part of the plasma would be red blood cells.

61. C This question relies on your knowledge of heart structure and function. The right side of the heart is the pulmonary circuit. Deoxygenated blood enters the right atrium, moves into the right ventricle, and then on to the lungs to be oxygenated. The left side of the heart is the systemic circuit. Oxygenated blood from the lungs enters the left atrium, the left ventricle, and leaves through the aorta to be carried through the body. Gas exchange occurs again in the systemic capillaries, deoxygenating the blood. Of the choices listed, only blood entering the right side of the heart and leaving the right side of the heart (heading for the lungs) would have the same oxygen content.

62. B You must understand the conducting system of the heart to answer this question. The natural pacemaker of the heart is the SA node located in the right atrium. Electrical impulses are generated by the SA node and pass through the atria to initiate atrial contraction. This electrical impulse is picked up by the AV node, which eventually causes ventricular contraction. An implanted artificial pacemaker replaces the function of the SA node.

63. D Blood pressure is force exerted on the walls of the arteries. By far, arteries have the greatest pressure of any vessel in the body in order to move blood through the arteries and arterioles.

64. C During dehydration, the water levels of plasma decrease. As plasma volume decreases, blood volume and pressure decrease as well. Dehydration would not have any direct influence on cell counts. However, as plasma volume decreases, hematocrit (the relative comparison of plasma volume to cell volume) would increase.

CHAPTER 15 THE RESPIRATORY SYSTEM Answers and Explanations

65. C Immediately following inhalation (before gas exchange occurs), the O_2 levels in the alveoli should be high and the CO_2 levels should be low. Within the pulmonary capillaries, the blood has come from the right side of the heart. This would mean that the blood is low in O_2 and high in CO_2. Of the choices listed, only choice C is accurate.

66. C An increase in ventilation rate is triggered primarily by high CO_2 levels, which are detected by chemoreceptors.

67. B When the diaphragm contracts, it pushes down toward the abdomen, increasing space in the thoracic cavity. As the volume of the chest cavity increases, air flows in to cause inhalation.

68. A The alveoli are the sites of gas exchange in the lungs. This is where CO_2 is picked up. Moving up from the alveoli leads to the bronchioles, the bronchi, the trachea, the pharynx, and the larynx.

CHAPTER 16 THE DIGESTIVE SYSTEM Answers and Explanations

69. A The question is essentially asking about what substances are normally digested in the stomach. The stomach produces the enzyme pepsin, which is activated by the low pH (high acidity) of the stomach. Pepsin is used to digest proteins. If stomach pH was increased, pepsin would not be activated and protein digestion in the stomach may not happen properly.

70. C Bread is high in starch. The only digestive enzyme in saliva is amylase, which digests starch (carbohydrates). If you leave a high-starch item in the mouth long enough, the amylase begins chemical digestion, releasing glucose, which tastes sweet.

71. C All chemical digestion is completed in the small intestine. Absorption of all nutrients occurs in the small intestine. The large intestine serves as a means to reabsorb water and solidify waste products. The large intestine also has a large number of normal flora (resident bacteria) that synthesize certain vitamins.

72. B Chemical digestion occurs in the oral cavity, stomach, and small intestine. Once chemical digestion is complete, all nutrients are absorbed into the blood via the small intestine.

CHAPTER 17 THE URINARY SYSTEM
Answers and Explanations

73. B The substances that enter the nephron from the blood are termed filtrate. As filtrate moves through the nephron, many substances are routed back into the bloodstream. Whatever remains at the end of the nephron is excreted as urine. Choices C and D can immediately be eliminated because neither cells nor proteins should enter the nephron as part of the filtrate. The filtrate is much higher in water content then urine is, which eliminates choice A. The primary difference is that filtrate contains glucose, which is sent back to the bloodstream at the proximal convoluted tubule. This means that although filtrate contains glucose, urine should not.

74. C Only substances small in size are pushed into the nephron. Of the choices listed, proteins are large polymers and should not typically enter the nephron.

75. A Nutrients such as amino acids and glucose enter the nephron, but must be reclaimed back into circulation. This occurs at the proximal convoluted tubule, where water is also sent back to circulation.

76. C Alcohol is a known inhibitor of antidiuretic hormone. Antidiuretic hormone acts to pull more water out of the nephron to increase blood volume. By doing this, urine volume decreases. If antidiuretic hormone is inhibited by alcohol, more water stays in the nephron, which increases urine volume.

CHAPTER 18 THE LYMPHATIC AND IMMUNE SYSTEMS
Answers and Explanations

77. B This question is describing the inflammatory response. Local inflammation is the result of the chemical histamine. Histamine is secreted by basophils (mast cells).

78. B During viral infection, infected cells produce interferons. These interferons help prevent other cells from viral infection. Interferons are a nonspecific defense of the immune system. If the person was infected by a virus, the interferons produced in response to that virus may be protective against other viruses.

79. B Secondary immune responses occur as the result of a primary exposure to a new antigen. During the primary response, cytotoxic (killer) T cells are activated, as are plasma cells. Memory cells are also produced. The memory cells are capable of living for extended periods of time. Should the antigen return, memory cells activate quickly to proliferate into killer T cells and plasma cells.

80. D An immunodeficiency implies that something is missing from the immune system so that it does not function properly. Choices A and B both indicate a strong immune system that is attacking a foreign antigen. Choice C mentions allergies, which are immune system hypersensitivities. Only choice D suggests a situation that might be consistent with an immunodeficiency. Repeated and prolonged infections indicate that the immune system is not functioning properly.

CHAPTER 19 REPRODUCTION AND DEVELOPMENT Answers and Explanations

81. B The male reproductive system has three glands that add secretions to sperm to produce semen. Of those glands, the seminal vesicles specialize in adding a nutrient-rich fluid to the sperm. If the levels of nutrients in semen are low, this is most likely a malfunction with the seminal vesicles.

82. D This question relies on knowledge of female sex hormones and the reproductive cycle. Menstruation occurs during the first week of the cycle. During the first week of the cycle, the only hormone that is absent is progesterone. If a female had consistently high levels of progesterone, menstruation would not occur.

83. D Organogenesis deals with the development of organs and organ systems during embryonic development. This process begins during the fourth week of embryonic development.

84. A There are three primary germ layers in the embryo—the endoderm, mesoderm, and ectoderm. All forms of muscle are derived from the mesoderm.

85. D The head of the sperm contains the DNA that the sperm must provide to the egg. The midpiece is high in mitochondria, which provide the ATP needed to move the flagellum (tail) that propels the sperm toward the egg. The acrosome contains enzymes that are released when the sperm encounters the egg, helping the sperm penetrate through layers of the egg.

PART VI

REVIEWING MCAT ORGANIC CHEMISTRY

SECTION I

STRUCTURE

CHAPTER 1

Bonding

Read This Chapter to Learn About:

- ➤ Types of Bonds
- ➤ Lewis Structures and Resonance Forms
- ➤ Condensed Formulas
- ➤ Line Notation
- ➤ Keeping Track of Lone Pairs

TYPES OF BONDS

Nobel laureate Roald Hoffmann described molecules as "persistent groupings of atoms." Such assemblies are held together by chemical bonds, which conventionally fall into one of two categories: ionic and covalent. Both bonds involve electrostatic forces—ionic bonds are present between a positively charged cation and a negatively charged anion (as in sodium chloride), whereas covalent bonds arise from the mutual attraction of two positively charged nuclei to negatively charged electron density shared between them (as in diatomic nitrogen). However, covalent and ionic bonds are just two ends of a bonding continuum.

When the electronegativity difference between the atoms is quite small, the electron density is equitably distributed along the internuclear axis, and it reaches a maximum at the midpoint of the bond. Such a bond is said to be purely covalent. However, when the two nuclei have widely divergent electronegativities, the electron density is not "shared" at all: Two ionic species are formed and the electron density approaches zero along the internuclear axis at the edge of the ionic radius.

However, organic chemistry rarely operates at the boundaries of purely covalent or purely ionic bonding. Instead, most examples lie along a continuum between these two

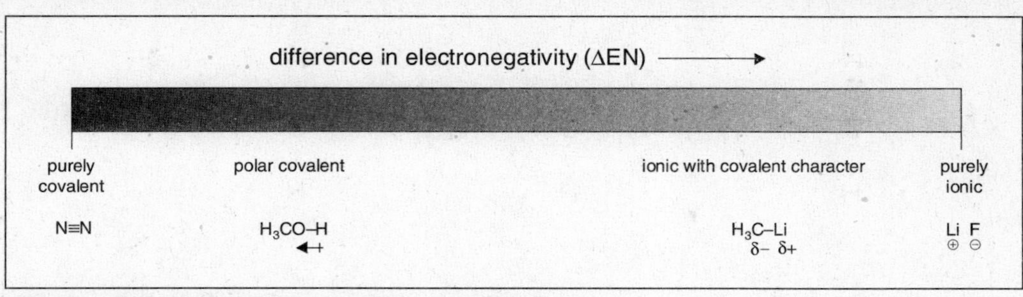

FIGURE 1-1: Types of bonding in organic molecules.

extremes (Figure 1-1). For example, polar covalent bonds result from an uneven sharing of electron density, a situation that sets up a permanent dipole along the bond axis. The O—H and C—F bonds are examples of polar covalent bonds. Such bonds are stronger than one would expect, because the covalent attraction is augmented by the Coulombic forces set up by the dipole.

On the other end of the spectrum, there are many examples of essentially ionic compounds that exhibit covalent character. In other words, even though there are practically two ionic species bound together, electron density is still shared between them. Almost all carbon–metal bonds fall into this category. For example, methyllithium (H_3CLi) can be thought of as a methyl anion (H_3C^-) with a lithium counterion (Li^+). Even though this is not a strictly accurate representation (i.e., there is indeed shared electron density), it still allows us to make sound predictions about its chemical behavior, as we see later in Section III.

LEWIS STRUCTURES AND RESONANCE FORMS

Examining electronegativity trends allows us to make predictions about bond polarity, but we must also understand the larger bonding picture: How many bonds are formed with each atom? The Lewis dot diagram represents a surprisingly simple device for representing global molecular bonding on a primary level. These diagrams are built up by considering the valence electrons brought to the table by each atom (conveniently remembered by counting from the left on the periodic table) and then forming bonds by intuitively combining unpaired electrons. For example, methane (CH_4) and formaldehyde (H_2CO) are constructed in Figure 1-2.

Notice that there are two types of electron pairs in the molecules shown in Figure 1-2: (1) shared pairs (or bonds), which are represented by lines (each line representing two shared electrons); and (2) lone pairs, which are depicted using two dots. When calculating formal charges—which, incidentally, should always be done—assign to a given atom all of its lone pair electrons and half of each shared pair; then compare the sum to the number of valence electrons normally carried. For example, consider the

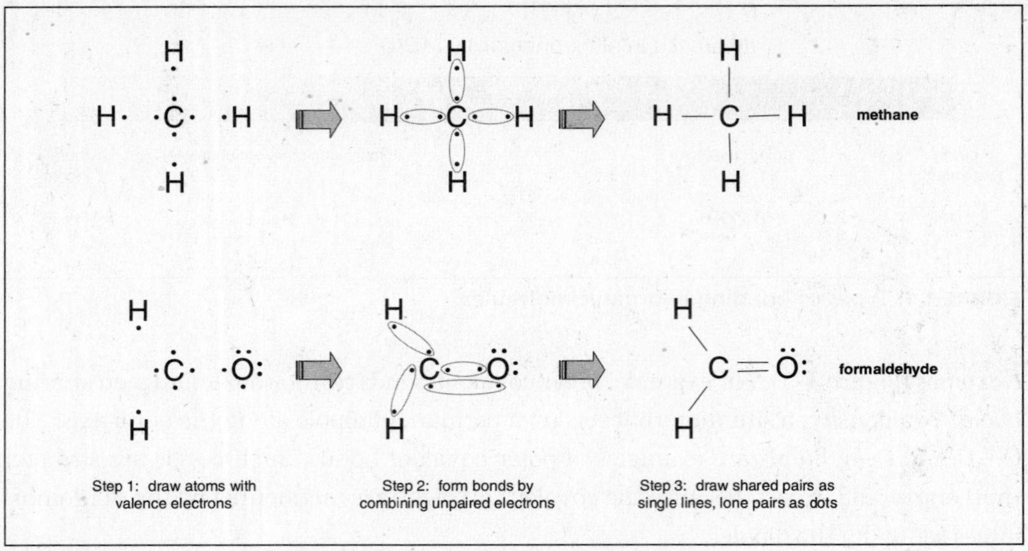

FIGURE 1-2: Lewis structures for methane and formaldehyde.

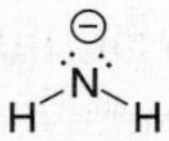

FIGURE 1-3: Lewis structure for the amide species.

amide species (Figure 1-3). The nitrogen atom is surrounded by two lone pairs (nitrogen "owns" all four) and two shared pairs (nitrogen "owns" only one in each pair), giving a total of six electrons assigned to nitrogen. Compared to the five valence electrons normally carried by nitrogen, this represents an excess of one electron; therefore, a formal charge of –1 is given to the nitrogen atom.

Occasionally, the Lewis structures do not coalesce right away into such tidy packages. Therefore, we must often select the most reasonable Lewis representation from a collection of candidates. These are known as resonance forms, and although all reasonable candidates tell us something about the nature of the molecule they represent, some resonance structures are more significant contributors than others. In making such an assessment, the following guidelines are helpful.

Rules of Charge Separation

1. All things being equal, structures should have minimal charge separation.

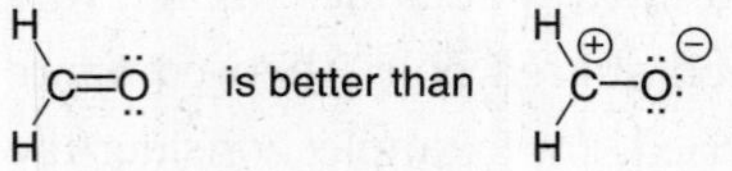

is better than

2. Any charge separation should be guided by electronegativity trends.

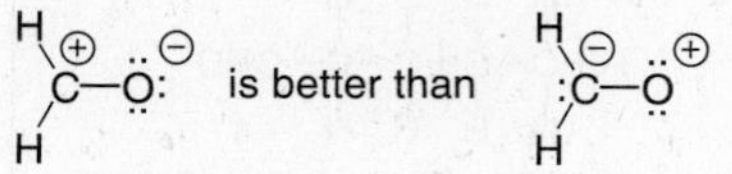

is better than

Octet Rules

1. Little octet rule: All things being equal, each atom should have an octet.

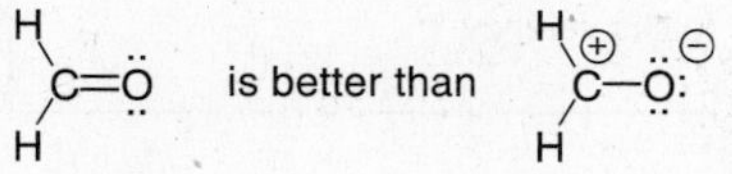

is better than

2. Big octet rule: No row 2 element can accommodate more than 8 electrons.

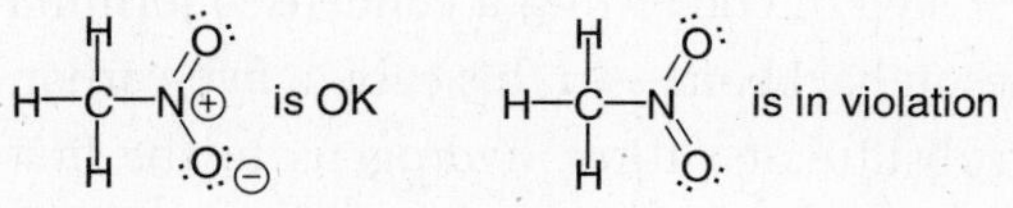

is OK

is in violation

Of all these, only the big octet rule is inviolable. Structures that break the first three rules are less desirable—those that break the last one are unreasonable and unsupportable. Keep in mind that Lewis structures are gross simplifications of a more complex reality, and sometimes no one representation is adequate to describe the total bonding within a molecule. Even when a structure satisfies all the rules—as with the first depiction of formaldehyde—other structures (resonance forms) may need to be considered to predict the properties of a molecule. The concept of resonance is considered in Chapter 3.

CONDENSED FORMULAS

Another useful principle for constructing structural representations from molecular formulas involves the idea of valency, or the number of bonds typically formed by a given element. For example, carbon normally has a valency of four; nitrogen, three; oxygen, two; hydrogen and the halogens, one. It is valency that underlies the hidden code of so-called condensed formulas, as illustrated in Figure 1-4 for the compound 3-hydroxypentanal. Condensed formulas can appear to be ambiguous about structure, but in fact, they are rich in structural information as long as you are aware of a few simple rules:

1. Formulas are read from left to right.
2. Each carbon atom in the formula is connected to the next carbon in the formula.
3. Everything to the right of a carbon (but before the next) is connected to that carbon.
4. Parentheses are used to indicate whole groups attached to a carbon.
5. Normal valencies must be satisfied (C = 4; N = 3; O = 2; H = 1; etc.).

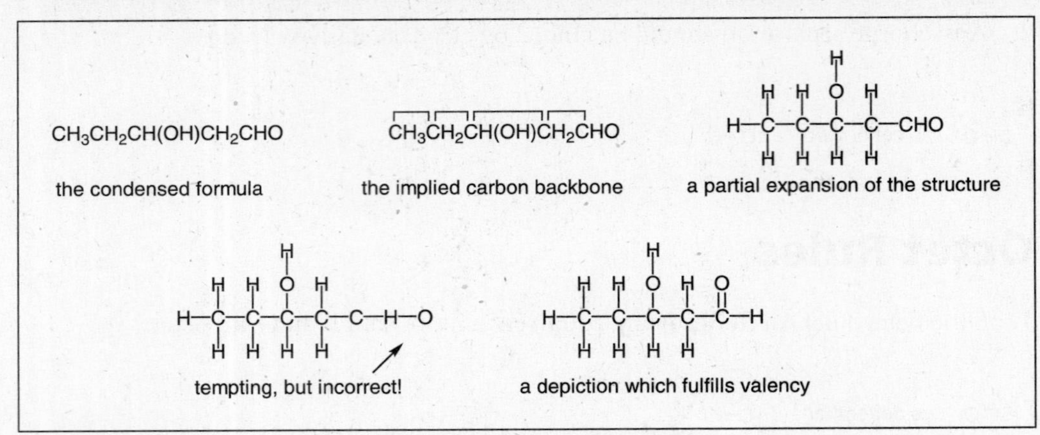

FIGURE 1-4: Converting condensed formulas to full formulas.

As the example in Figure 1-4 shows, the first step in converting a condensed formula into a structural formula is to lay out the carbon backbone—in this case, a five-carbon chain. We then connect all the indicated substituents: three hydrogens to the first carbon, two hydrogens to the second, and so on, recognizing that the hydroxy (OH) moiety is treated as an entire group attached to the third carbon. The tricky part is the expansion of the CHO group. It is tempting to imagine a structure in which the carbon is attached to the hydrogen, which in turn is attached to the oxygen (lower left depiction). However, this would violate three valency guidelines: oxygen has only one bond, carbon only two, and hydrogen one too many. A proper reading of the condensed formula would be, "Hydrogen belongs to the fifth carbon, and so does oxygen." With a bit of thought, the only reasonable arrangement would be the one shown in the lower right, which includes a carbon–oxygen double bond.

LINE NOTATION

Condensed formulas are one type of shortcut for depicting structure—their claim to fame is that they can be produced by word processors even when graphical software is unavailable. However, an even more important and widespread shortcut is line notation, which is universally used by organic chemists when they sketch compounds, its advantage being that the salient features of complex structures can be depicted quickly and conveniently. Figure 1-5 shows 3-hydroxypentanal in line notation. The assumptions underlying the simplification are as follows:

1. Every unlabeled vertex or terminus is a carbon atom.
2. Any unused valency of carbon is filled by hydrogen.

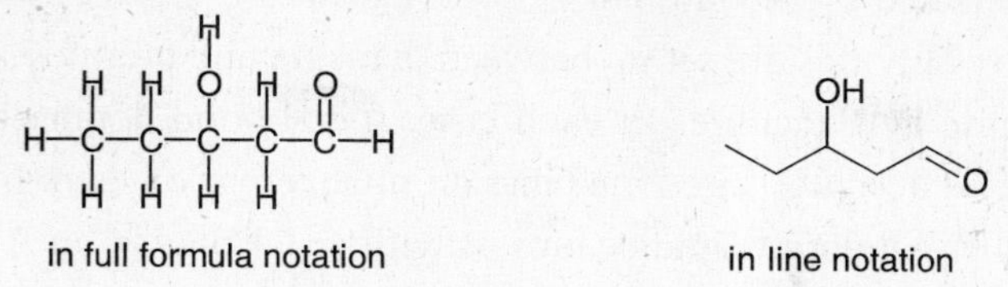

in full formula notation

in line notation

FIGURE 1-5: Two valid depictions of 3-hydroxypentanal.

KEEPING TRACK OF LONE PAIRS

You may have noticed that the depictions shown in Figure 1-5 do not include lone pairs—in fact, the depiction of lone pairs is an arbitrary matter, and they are often neglected in drawings. However, it is absolutely crucial to know where they are, as the presence or absence of lone pairs can define the chemistry of a species. For example, consider the two seemingly similar compounds, diethylborane and diethylamine (Figure 1-6)—both have row 2 central atoms that are trivalent, but their behaviors could not be more different. There is no lone pair on diethylborane—the central boron has only six electrons around it, thereby breaking the little octet rule. The molecule is starved for electron density, and it therefore reacts with electron-rich substrates. However, diethylamine does have a lone pair on the nitrogen, even though some depictions may not include this detail. Thus, nitrogen fulfills the octet rule, and the lone pair can even function as a source of electron density in many of its characteristic reactions.

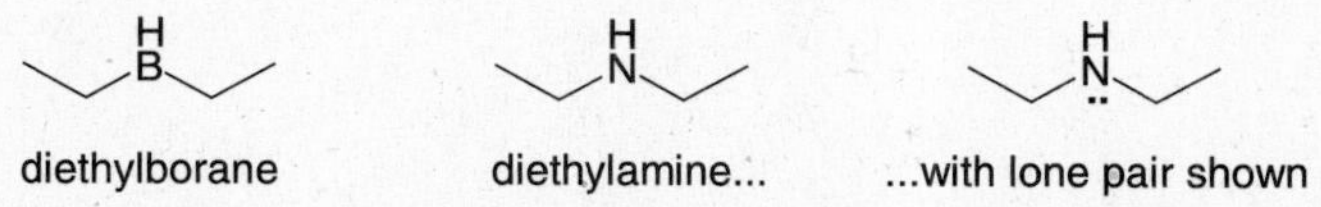

diethylborane

diethylamine...

...with lone pair shown

FIGURE 1-6: Two very different trivalent compounds.

The lone pair is a pivotal participant in an important intermolecular phenomenon known as the hydrogen bond (H-bond). Two components are necessary for hydrogen bonding: an H-bond donor and an H-bond acceptor. An H-bond donor typically comes from a hydrogen bound to a heteroatom, which for organic chemists almost always means nitrogen, oxygen, or sulfur (and sometimes fluorine). These N—H, O—H, and S—H bonds are quite polar, with most of the electron density being hoarded by the heteroatom. As a result, the hydrogen starts to look rather like a proton (H^+) in need of extra electron density. Thus, an H-bond acceptor is anything that can provide this electron density—and more often than not it comes in the form of a lone pair.

This is nicely illustrated by the familiar H-bonding motif in DNA base pair recognition (Figure 1-7). Three H-bonds are set up between cytosine and guanine, whereas two are enjoyed by thymine and adenine. In each case, the H-bond donors are N—H bonds, whereas the H-bond acceptors are lone pairs on nitrogen or oxygen. Inspection reveals that the degree of H-bonding would be far less with a G-T pair or an A-C pair.

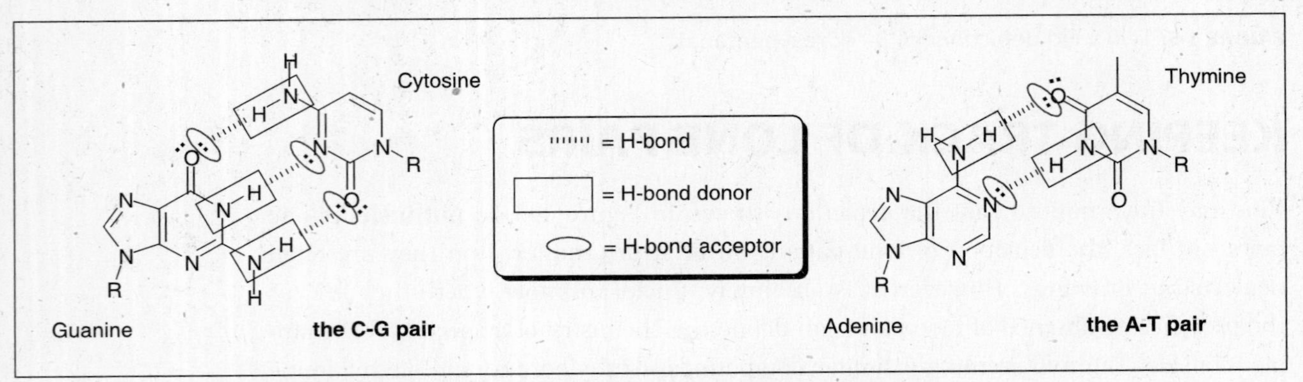

FIGURE 1-7: Hydrogen bonding in DNA base-pair recognition.

CRAM SESSION

Bonding

Chemical bonds can take many forms. The most common bonds in organic chemistry are

- **Covalent bonds** between atoms of very similar electronegativity
- **Polar covalent bonds** between atoms of moderately differing electronegativity
- **Ionic bonds** between atoms of vastly differing electronegativity in their ionic form

The forces important in bonding are

- Mutual attraction of two positive nuclei to the shared negative electron density between them
- Coulombic attraction of a positively charged cation to a negatively charged anion

The former is dominant in purely covalent bonds; the latter in ionic bonds; and both are significant in polar covalent bonds.

The connectivity of bonds in molecules can be shown by **Lewis structures,** which sometimes are depicted using multiple **resonance forms.** These forms are evaluated using the following criteria:

- Each atom should have an octet of electrons (although fewer are possible).
- No row 2 atom can have more than an octet.
- Structures should minimize charge separation.
- Charge separation should follow electronegativity.

One important aspect of full Lewis structures is that they show the location of lone pairs, which are an important component of **hydrogen bonding.**

Organic chemists use many different modes of molecular depiction, including **condensed formulas,** which are purely textual, and **line notation,** which are spartan graphical drawings. Both of these often neglect lone pairs, so it is very important to be able to recognize centers bearing lone pairs even when they are not explicitly shown.

CHAPTER 2

Molecular Shape

Read This Chapter to Learn About:

- Geometry of Atoms within Molecules
- Asymmetric Centers and Enantiomerism
- Molecular Conformations
- Conformations of Cycloalkanes

GEOMETRY OF ATOMS WITHIN MOLECULES

Now that you have an understanding of connectivity among atoms within molecules, it is appropriate to consider the three-dimensional arrangement of these atoms. There are three frequently encountered geometries in organic chemistry: digonal (or linear), trigonal, and tetrahedral (Figure 2-1). Each atom within a molecule is almost always characterized by one of these geometries, the chief hallmark of which is the associated bond angle: 180° for linear arrays, 120° for trigonal planar centers, and 109.5° for tetrahedral arrangements. Deviation from these ideal bond angles does occur, but significant deformation usually has a destabilizing effect known as bond angle strain.

There are at least three ways to conceptualize molecular geometry. One classical approach is through the hybridization of atomic orbitals. If orbitals are derived from wave functions, then these functions can be combined mathematically to obtain hybrid descriptions. Thus, if we combine an *s* orbital, which is spherically symmetric, with a single *p* orbital, which has directionality along a single axis, it stands to reason that the outcome (two equivalent *sp* orbitals) should also have directionality along one dimension. Likewise, the combination of an *s* orbital with two *p* orbitals gives a result (three equivalent sp^2 orbitals) that defines a plane (Figure 2-2).

geometry	arrangement	bond angle	hybridization of x
digonal	a—x—b	180°	sp
trigonal	a, c—x—b	120°	sp^2
tetrahedral	a, b, c, d—x	109.5°	sp^3

FIGURE 2-1: Common central atom geometries in organic chemistry.

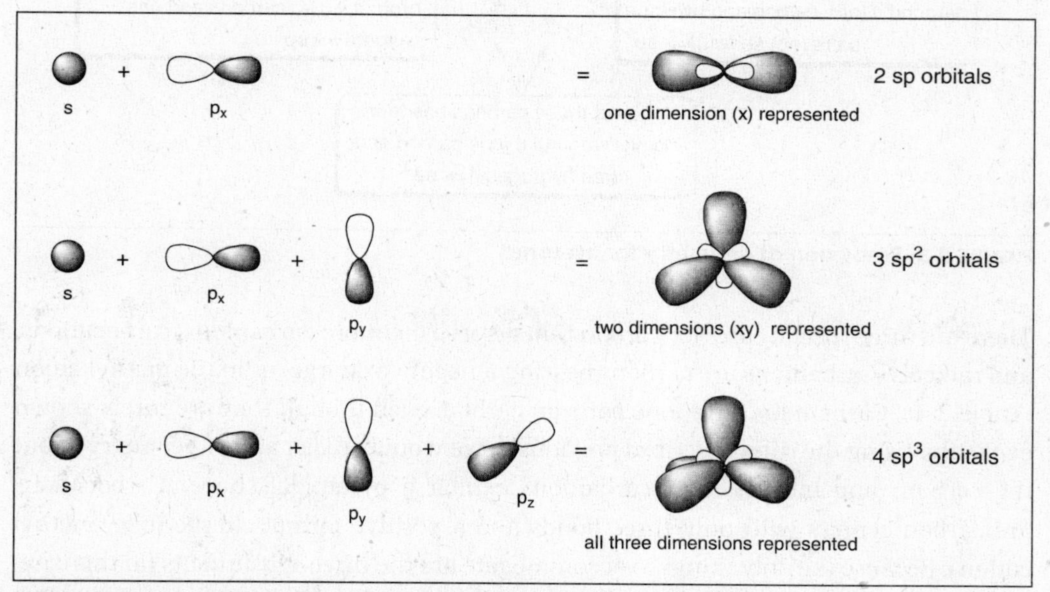

FIGURE 2-2: Central atom geometry as described by orbital hybridization.

Another framework is conceptually more straightforward. Known as valence shell electron pair repulsion (VSEPR) theory, this approach asks the question of how negatively charged electron clouds can most effectively stay out of each other's way. If there are only two electron clouds, then the happiest arrangement is to be diametrically opposed to each other; for three clouds, a trigonal planar array; and for four clouds, a tetrahedral arrangement provides the maximum distance among them. Interestingly, these considerations predict exactly the same outcomes as hybrid orbital theory, although they are fundamentally different approaches. Strictly speaking, neither are theoretically accurate as compared to a strict quantum mechanical analysis—a third method not addressed in this work—however, they are useful predictive models nonetheless.

Thus, to predict the geometry about a central atom, one must ask only how many things surround it—where "things" are understood to be either atoms or lone pairs. Figure 2-3 offers an illustration of this method, and experimental evidence backs up the predictions (e.g., the C—C—C bond angle is practically 120°). A generalization can also be derived from this example: Because carbon is almost always tetravalent, then if only single bonds are attached to a carbon, it must have four other atoms surrounding it; if a double bond is attached to that carbon, then only three atoms can surround it; and so on. As a consequence, carbons with only single bonds tend to be tetrahedral (sp^3), carbons that are part of a double bond tend to be trigonal (sp^2), and triply bonded carbons are always digonal (*sp*).

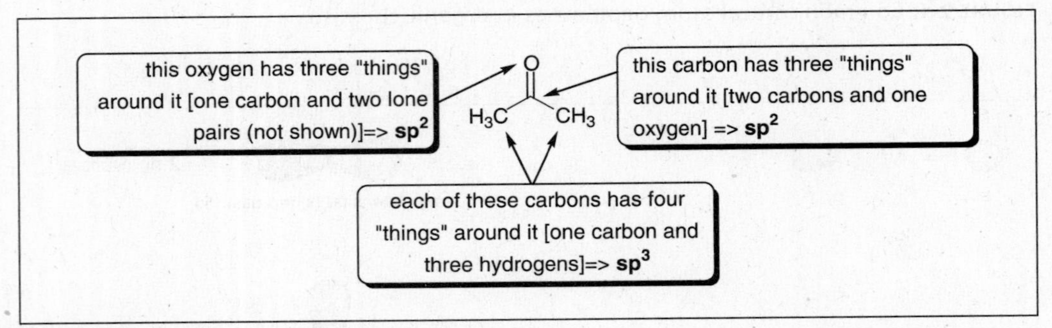

FIGURE 2-3: Prediction of geometry for acetone.

There are three special cases for carbon that deserve mention: carbanions, carbocations, and radicals. Carbanions are carbons bearing a negative charge, as in the methyl anion (Table 2-1). Carbanions have lone pairs on carbon, even though they are rarely shown explicitly. Using the rules described previously, we would predict an sp^3 geometry about the carbon, and indeed, most carbanions exhibit a pyramidal shape. Carbocations are carbon centers with only three bonds and a positive formal charge (e.g., methyl cation). Because the only things to accommodate are the three substituents (in this case,

TABLE 2-1 Geometries of Some Special Carbon Centers

Species	Formula	Structure
Methyl anion	$^{\ominus}CH_3$	
Methyl cation	$^{\oplus}CH_3$	empty
Methyl radical	$\cdot CH_3$	

hydrogen atoms), the predicted geometry of carbon is sp^2—and experimental evidence supports the idea that carbocations are planar. Radicals lie somewhere between these two extremes. Because the *p* orbital is only half filled, it does not demand the same space that a doubly filled orbital would, but still requires more than an empty one. The most accurate way to think about such centers is as a shallow pyramid that very rapidly inverts. A time-averaged representation would approximate a planar species; therefore, frequently radicals are depicted as being planar.

ASYMMETRIC CENTERS AND ENANTIOMERISM

Because they are inherently three-dimensional, tetrahedral carbons can impart a special characteristic to molecules, a property known as asymmetry. However, before we can understand asymmetry, we first have to examine the idea of symmetry. The uninitiated think of symmetry as a binary state—that an object is either symmetric or not symmetric. However, there are degrees of symmetry, and many so-called elements of symmetry. This is a topic best treated in the domain of mathematics, but we shall very briefly scratch the surface here to gain some underpinnings for practical application.

First, let us consider methane (CH_4)—an unassuming molecule, yet full of symmetry. Figure 2-4 shows three types of symmetry elements (there are others not shown here) belonging to methane. The sigma (σ) plane is an imaginary mirror that slices through

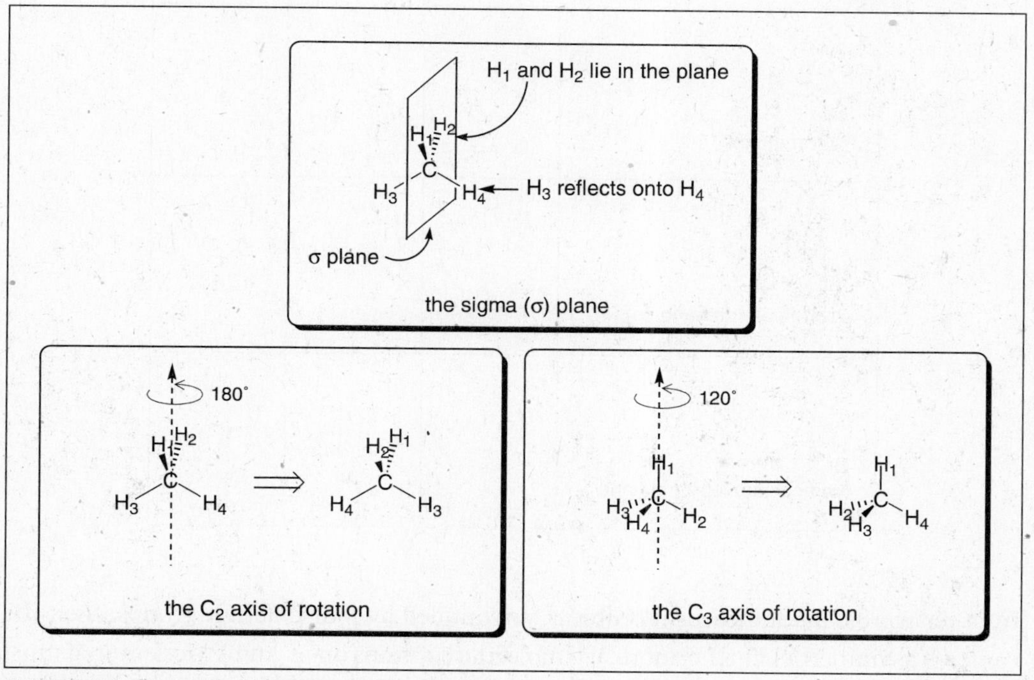

FIGURE 2-4: Symmetry elements of methane.

the molecule—reflection through the plane results in an image identical to the starting depiction. Methane actually has six such planes: one with H_1 and H_2 in the plane; one with H_1 and H_3; one with H_1 and H_4; one with H_2 and H_3; one with H_2 and H_4; and one with H_3 and H_4. There are also two types of rotational axes. A C_2 axis has two "clicks" in a 360° rotation—180° each turn; methane has six of these, much like the sigma planes. A C_3 axis has three clicks in a full rotation—120° each turn; methane has four, one coinciding with each of the C—H bonds.

As we add substituents to methane, however, symmetry starts to drop off rapidly. For example, the addition of a single substituent, such as fluorine (Table 2-2) reduces the number of sigma planes by half (each sigma plane must include the fluorine), slashes the number of C_3 axes to one (the axis that coincides with the C—F bond), and eliminates the C_2 axes altogether. A second substituent results in an even more impoverished symmetry environment, such that chlorofluoromethane has only a single sigma plane. After a third substituent is introduced (i.e., bromochlorofluoromethane), there are no symmetry elements left. Therefore, such a molecule is said to be asymmetric.

TABLE 2-2 Symmetry Elements of Methane Derivatives

Compound	σ Planes	C_2 Axes	C_3 Axes
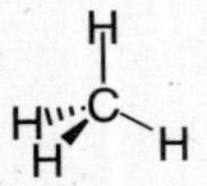 methane	6	6	4
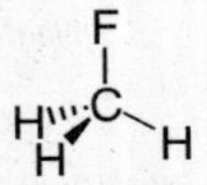 fluoromethane	3	0	1
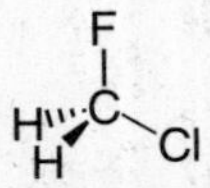 chlorofluoromethane	1	0	0
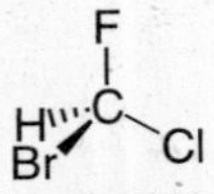 bromochlorofluoromethane	0	0	0

In other words, if a tetrahedral center is surrounded by four different groups, then the center is asymmetric. Furthermore, if a molecule contains an asymmetric center, it must exist as one of two enantiomers, or antipodes (literally, opposite feet). Enantiomers are

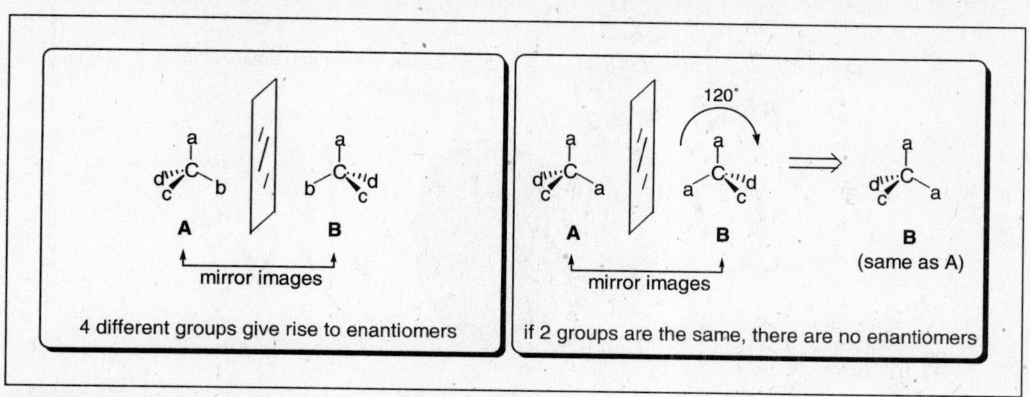

FIGURE 2-5: Tetrahedral centers and enantiomerism.

non-superimposible mirror images, just like hands and feet (Figure 2-5), and they represent two distinct compounds. Consequently, asymmetric centers are also called stereogenic centers and chiral (handed) centers.

If any two or more groups are the same, then the center possesses a sigma plane, and the mirror image becomes a replica of the same molecule, so no enantiomerism exists. In other words, the mirror image is the compound itself, and there is only one chemical entity. Thus, the presence of a sigma plane in a molecule destroys chirality (asymmetry). With more complex molecules, multiple chiral centers may be present; however, even in these cases a global sigma plane may be present. This is a topic addressed in more detail in Chapter 5.

Nevertheless, chiral compounds may be found in either of two enantiomeric forms. The enantiomers are identical in almost all physical properties: melting point, boiling point, dielectric constant, and so on, with one notable exception. A solution of a chiral compound interacts with light in such a way that the plane of polarized light is rotated on passing through a sample of the compound, a phenomenon known as optical rotation (Figure 2-6, top). The other enantiomer rotates light to the same degree, but in the opposite direction (Figure 2-6, bottom). Enantiomers that rotate light clockwise (i.e., to the right) are known as dextrorotatory isomers, and those that rotate light in a counterclockwise direction (i.e., to the left) are known as levorotatory isomers. It is important to understand that there is no straightforward way to correlate structure to behavior. In other words, we cannot tell just by looking at a molecule whether it would be dextrorotatory or levorotatory. By the same token, optical rotation tells us nothing about a molecule except that it is chiral.

A sample that rotates plane-polarized light is said to be optically active. The degree to which the sample rotates light is called the optical rotation (in degrees). If a sample is optically inactive (that is, it does not rotate plane-polarized light), it could mean one of two things: (1) the sample does not contain chiral molecules, or (2) the sample contains

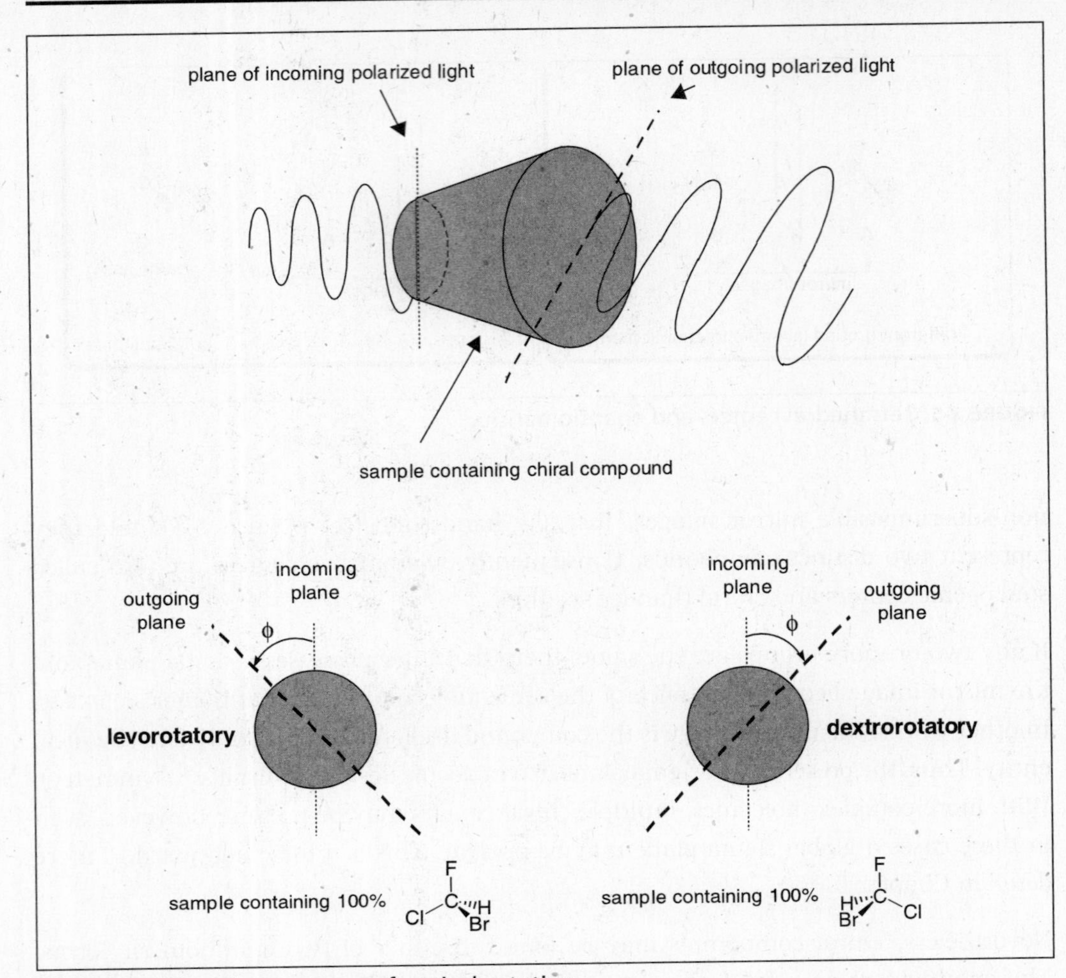

FIGURE 2-6: The phenomenon of optical rotation.

exactly equal quantities (50/50) of two enantiomers. The latter situation, known as a racemic mixture, produces no optical activity, because for each levorotatory molecule there is a dextrorotatory counterpart—equal numbers pulling in opposite directions maintain the status quo.

The term **racemic** refers to the composition of an enantiomeric mixture, in which the components are exactly equal. Another term related to composition is optically pure, which means that only one enantiomer is present with no contamination from the other enantiomer. Of course, there are other compositions possible (a 60/40 mixture, for example) that are neither racemic nor optically pure. Such lopsided mixtures are called **scalemic;** another term is **enantiomerically enriched,** because there is more of one enantiomer than another. Such scalemic mixtures can be defined by **enantiomeric excess** (or % ee), which is a quantitative term derived by subtracting the percentage of the lesser component from that of the greater. Therefore, a 55/45 mixture would have a 10% ee, a 95/5 mixture would have a 90% ee, and so forth.

All the vocabulary that surrounds this topic can easily get muddled. Always remember that chirality is a property of a molecule (it is a geometric reality derived from the molecular shape), whereas optical activity is a property of a sample (that is, a macroscopic collection of molecules). In other words, molecules can be chiral or achiral, and samples can be optically active or inactive. Furthermore, the ideas of optical purity and optical activity are often confused. Here it is useful to keep in mind that optical purity is a **state** of a sample (i.e., a reflection of its composition), whereas optical activity is a **behavior** of a sample. Optically pure samples are always optically active, but not all optically active samples are optically pure, since scalemic mixtures are also optically active. Table 2-3 correlates these various properties and behaviors of enantiomeric mixtures.

TABLE 2-3 Properties and Behaviors of Enantiomeric Mixtures

State of Purity	Optically Pure	Scalemic	Racemic	Scalemic	Optically Pure
Optical behavior					
Dextrorotatory enantiomer (%)	0	25	50	75	100
Levorotatory enantiomer (%)	100	75	50	25	0
Enantiomeric excess (% ee)	100	50	0	50	100
Optical rotation	$-2x$	$-x$	0	x	$2x$
Optically active?	Yes		No	Yes	

MOLECULAR CONFORMATIONS

From this cursory inspection, it is clear that a fair amount of complexity surrounds the environment of a single carbon atom. But what about larger arrays of atoms? Because most molecules have considerable flexibility, it is important to understand what kinds of shapes are most stable and the energetics involved in their interconversion. Before we do this, however, it is necessary to come to terms with two additional methods of depiction for molecular structure: the sawhorse (dash-wedge) projection and the Newman projection.

Consider a two-carbon array with three substituents on each carbon. Figure 2-7 shows such an array in two different depictions. The sawhorse projection views the molecule from the side. Substituents that come out of the plane toward us are depicted with wedges; those that go away from us are shown with dashes. If neither dash nor wedge, a plain line (and attached substituent) is assumed to lie in the plane of the paper. "Plain bonds lie in the plane" is a good mnemonic device in this regard.

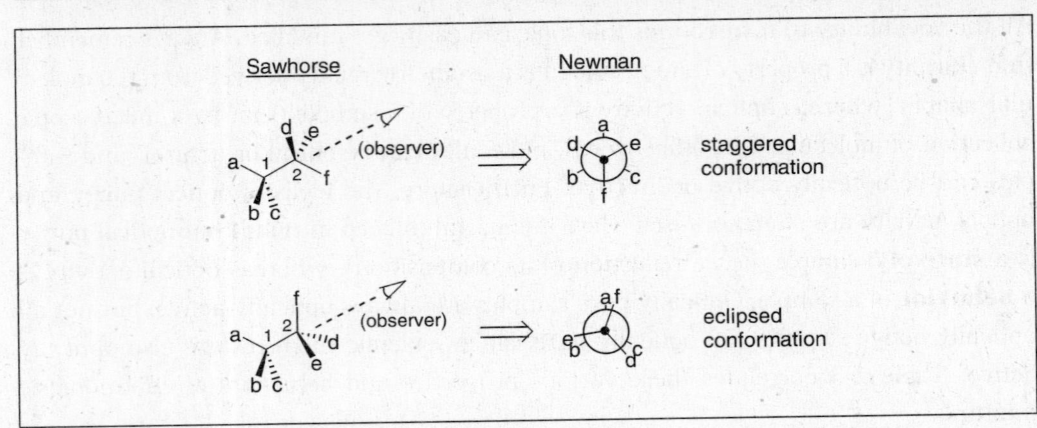

FIGURE 2-7: Sawhorse and Newman projections.

A Newman projection simply looks at the molecule from a different angle, namely along a carbon–carbon bond (indicated by the observer's eye). This is a much more straightforward way of showing the spatial relationship of substituents. Just remember that the small dot represents the front carbon, whereas the large circle represents the back carbon. Thus, in the projections shown, the small dot is carbon-2 and the large circle is carbon-1; in other words, we have drawn a C2 → C1 Newman projection. However, we could have just as easily drawn a C1 → C2 variant, in which the small dot would be carbon-1.

Because there can be rotation about carbon-carbon bonds, this molecule can adopt a variety of conformations. In general, these conformations fall into one of two categories: staggered and eclipsed. Note that in the sawhorse depictions, the plain bonds (i.e., neither dash nor wedge) in the staggered conformation describe a "Z," or zigzag pattern, whereas in the eclipsed conformation they form a "U." This is a quick and easy way to distinguish one from another.

In staggered conformations, any two substituents are characterized by one of two relationships. Substituents are said to be **gauche** with respect to each other if they are side-by-side. In the illustration, there are six gauche relationships: a—e, e—c, c—f, f—b, b—d, and d—a. The other relationship is **antiperiplanar,** in which case the two substituents are as far away as possible from each other. In the same depiction, there are three antiperiplanar relationships: a—f, e—b, and d—c. In eclipsed conformations, there is only one type of relationship between substituents we care about, namely eclipsed. In the eclipsed Newman projection, there are three pairs of eclipsed substituents: a—f, c—d, and b—e. Do not be confused by the double duty of "eclipsed"—there are staggered and eclipsed **conformations,** which describe the global molecular attitude, and there are gauche, antiperiplanar, and eclipsed **relationships** between substituents. There are only gauche and antiperiplanar relationships in staggered conformations, and there are only eclipsed relationships in eclipsed conformations.

It should come as no surprise that eclipsed conformations are of higher energy than staggered conformations. This is due to the steric interactions that result from the very close proximity of substituents in the eclipsed relationships. By the same token, staggered conformations that place large groups in a gauche arrangement are of higher energy than those that have those groups antiperiplanar with respect to each other. Using these basic principles, we can construct a diagram showing the relative energies of a conformational ensemble, as illustrated in Figure 2-8 for butane ($CH_3CH_2CH_2CH_3$). This is a process known as **conformational analysis.**

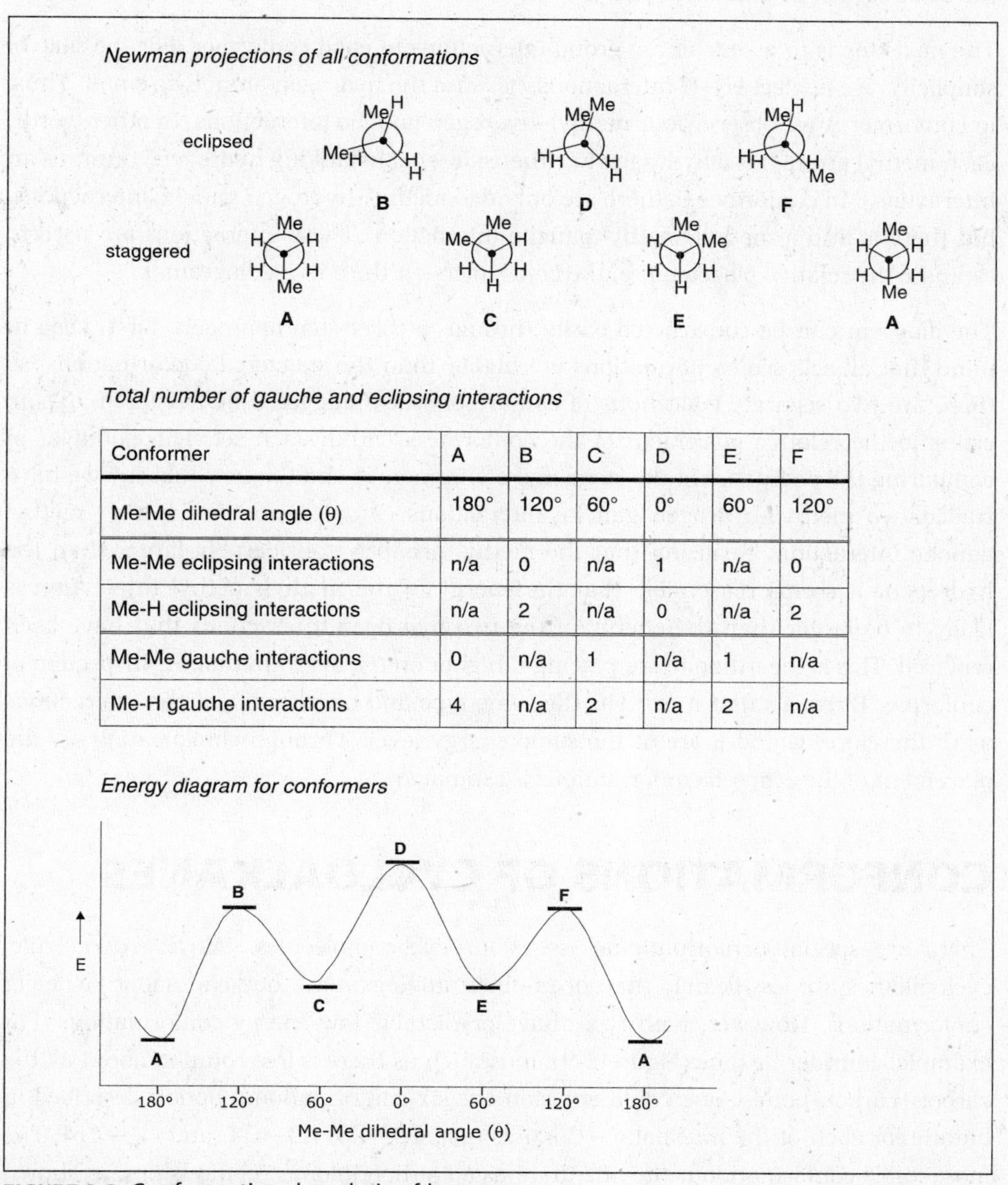

Conformer	A	B	C	D	E	F
Me-Me dihedral angle (θ)	180°	120°	60°	0°	60°	120°
Me-Me eclipsing interactions	n/a	0	n/a	1	n/a	0
Me-H eclipsing interactions	n/a	2	n/a	0	n/a	2
Me-Me gauche interactions	0	n/a	1	n/a	1	n/a
Me-H gauche interactions	4	n/a	2	n/a	2	n/a

FIGURE 2-8: Conformational analysis of butane.

To begin the conformational analysis, all the possible conformations for the molecule, both eclipsed and staggered, must be drawn. This might seem more difficult than it really is—first, just jot down any conformation and then methodically convert it to the remaining possibilities. For example, in Figure 2-8 the first conformation (A) is chosen arbitrarily. Conformation A is converted to C by rotating the front carbon (small dot) clockwise—notice that the back carbon does not change. This maneuver has the net effect of bringing the front methyl (CH_3) group from the 6 o'clock position to the 10 o'clock position. To get from A to C, the molecule must pass through the high-energy eclipsed conformer B. The remaining conformations are derived by continuing to rotate the front carbon in a clockwise fashion until we arrive back at A.

The next step is to assess all the group interactions in each conformer. For the sake of simplicity, we neglect H—H interactions, because the hydrogen atom is so small. Thus, in conformer A we observe four methyl–hydrogen gauche interactions (in other words, each methyl group has a hydrogen to either side—each flanking hydrogen counts as an interaction). In conformer C, there are only two methyl–hydrogen gauche interactions, but there is also a methyl–methyl gauche interaction. These interactions are used to estimate the relative placement of the conformers on the energy diagram.

The diagram can be constructed easily through a three-step approach. First, keep in mind that all eclipsed conformations are higher than the staggered conformations, so there are two separate collections of conformers (i.e., staggered and eclipsed). Then, estimate the relative placement of the conformers within each set. For example, in comparing the energetics of the staggered conformers A and C, one could say we have traded two methyl–hydrogen gauche interactions (2 vs. 4) for one methyl–methyl gauche interaction. Realizing that the methyl group is considerably larger than the hydrogen, it seems reasonable that the energy of the methyl–methyl interaction is likely to be larger than the energy of the two hydrogen interactions that have been replaced. This is the rationale for placing C higher on the diagram than A. Inspection of conformer E reveals that it exhibits the same type and number of gauche interactions as C; therefore, C and E are at the same energy level. Through similar analysis, the placement of the eclipsed conformations is estimated.

CONFORMATIONS OF CYCLOALKANES

There are special conformational issues for cyclic molecules. As a general rule, cycloalkanes are less flexible than open-chain analogs, and they can adopt far fewer conformations. However, both types have predictable low-energy conformations. For example, consider hexane (Figure 2-9): inasmuch as there is free rotation about all the carbon–carbon bonds, one could envision the six conformational option described in butane for each of the internal C—C bonds (i.e., C2—C3, C3—C4, and C4—C5). The most stable conformation is the one that has all antiperiplanar relationships, as shown

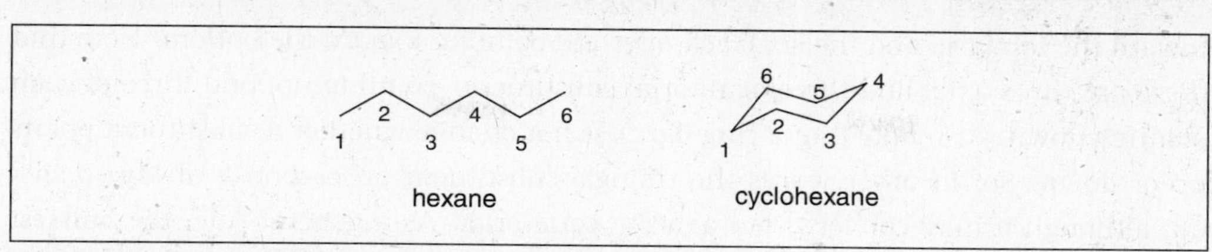

FIGURE 2-9: Stable conformations of hexane and cyclohexane.

in the illustration. Similarly, cyclohexane can and does adopt many conformations, but the most stable arrangement is the so-called chair conformation.

Although it may not be evident, there are two alternate chair forms that are related by the chair flip. It is important to understand that the flip is not a rotation of the molecule, but a reconformation. As the top depiction in Figure 2-10 shows, carbon-1 flips from pointing downward to pointing upward, carbon-4 does the reverse, and the entire molecule reconforms in place. As a side note, always keep in mind that the chair projection is a side view of the ring, and the lower bond (i.e., C2—C3) is assumed to be in front.

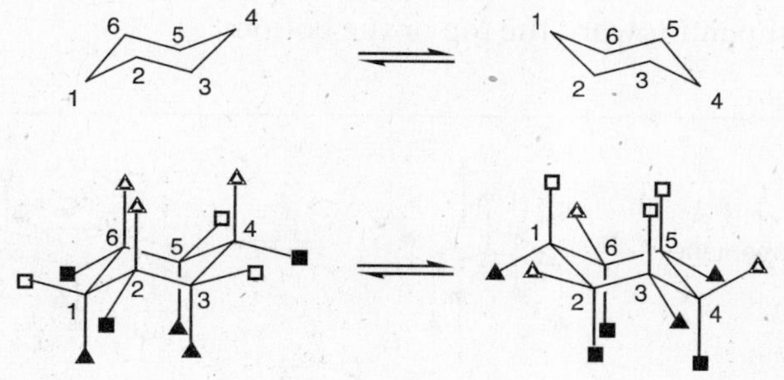

FIGURE 2-10: The chair flip of cyclohexane.

When substituents are added to the cyclohexane, the ring flip takes on special significance. To understand this, we must first come to terms with some characteristics of the substituents on a cyclohexane ring. Examination of Figure 2-10 (bottom) reveals that substituents can adopt one of two attitudes: axial (shown as triangles in the left chair) or equatorial (shown as squares). Every chair conformation has six of each type. Note that a chair flip interchanges axial and equatorial substituents, so that all of the triangles become equatorial on the right side. It is a good idea to construct a model and physically induce a ring flip to see how this works.

It is equally important to understand what does *not* change during a ring flip. Any ring has two faces—as these depictions are drawn, we can call them the top and bottom faces. Any substituent points toward either the top or the bottom face. Again, each chair cyclohexane has six of each type—the six white substituents are all pointing toward the top face, and the six black ones are pointing toward the bottom. Note that there are three axial and three equatorial substituents pointing up, and three of each pointing down. Also note that a ring flip does not change whether a substituent points up or down. So, in other words the triangle substituent on carbon-4 always points up, although it may convert from axial to equatorial. As a general rule, the bulkiest substituent prefers the equatorial position, a topic examined in quantitative detail in Chapter 6.

Other ring sizes have different lowest energy conformations, and a brief survey is worthwhile. For example, the most stable conformation of cyclopentane (Figure 2-11) is the so-called envelope, which (like the chair) has two forms that equilibrate through the flipping of the envelope flap. Cyclobutane adopts a so-called puckered conformation, which again has two forms that equilibrate through a ring flip. Notice that the same rules for cyclohexanes hold true for these cycloalkanes—namely, that there are two types of substituent attitudes (here, called pseudoaxial and pseudoequatorial) that interconvert on ring flips; also, the substituents point toward a certain face, and that face remains constant throughout conformational changes. Finally, cyclopropanes have practically no conformational flexibility—there is really only one possibility, and all substituents have equivalent attitudes. However, of course, there are still two faces to the molecule, so substituents can point toward the top or the bottom.

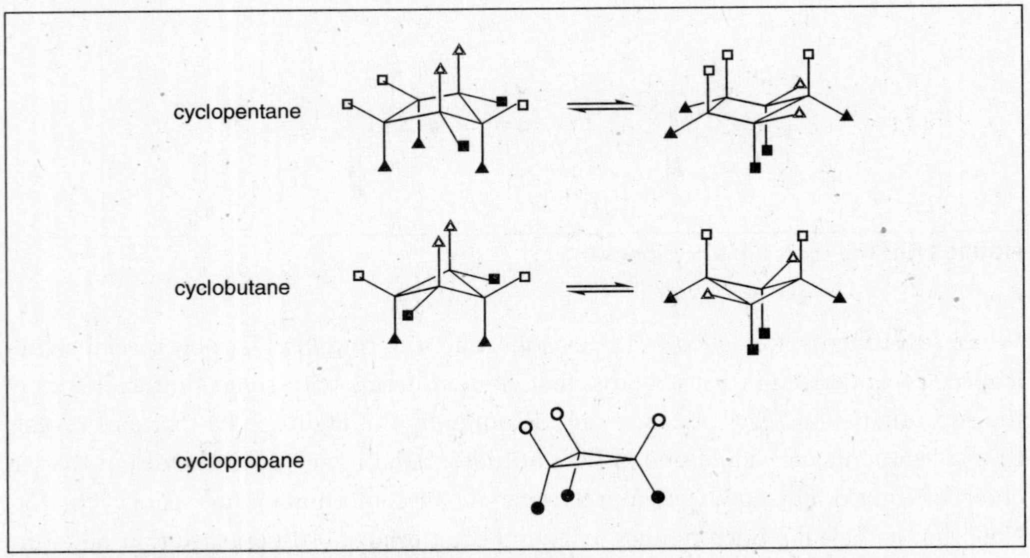

cyclopentane

cyclobutane

cyclopropane

FIGURE 2-11: Stable conformations of other cycloalkanes.

So what factors govern the adoption of a given stable conformation? The main determinant is the minimization of ring strain. In cyclic molecules, there are essentially two sources of strain: bond angle strain and torsional strain. The former derives from a compression of the ideal sp^3 bond angle (109.5°) to accommodate a cyclic array. Not surprisingly, this is worst for the three-membered ring, and is almost nil for the five- and six-membered rings. The other source of strain is less obvious. Torsional strain derives from the torque on individual bonds from eclipsed substituents trying to get out of each other's way. Again, this is most pronounced in cyclopropane (see Figure 2-11), and all the other cycloalkanes twist in ways to minimize or eliminate this strain. For example, inspection of a molecular model of chair cyclohexane reveals that all carbon–carbon bonds have a perfectly staggered conformation. Interestingly, the five-membered ring is not so lucky. Even in the envelope conformation, substituents tend toward eclipsing each other. Table 2-4 summarizes the individual components, but a good overall take-home message is that ring strain decreases according to the trend: $3 > 4 \gg 5 > 6$. This has implications in reactivity of cyclic molecules, as we see in Section II.

TABLE 2-4 Ring Strain in Cycloalkanes

Cycloalkane	Torsional Strain	Bond Angle Strain
Cyclopropane	A lot	A lot
Cyclobutane	Not much	A lot
Cyclopentane	A little	Practically none
Cyclohexane	None	None

CRAM SESSION

Molecular Shape

The shape of a molecule is ultimately driven by the **central atom geometry** at each atomic center, which in turn is determined by **atomic orbital hybridization.** For carbon, these are

- ***sp* hybridization,** which results in linear geometry and 180° bond angles
- ***sp*^{2} hybridization,** which results in trigonal planar geometry and 120° bond angles
- ***sp*^{3} hybridization,** which results in tetrahedral geometry and 109.5° bond angles

When a tetrahedral center is surrounded by four unique groups, the center is said to be an **asymmetric center** (or stereogenic). Compounds with asymmetric centers are almost always **chiral** (or handed), and they exist as two **enantiomers** (or nonsuperimposible mirror images). The way that the four unique groups are arranged around a stereocenter is referred to as **configuration.** Chiral compounds exhibit the property of **optical rotation,** and enantiomers are identical in every physical property except for their optical behavior. Compounds that rotate light to the right are called **dextrorotatory** and those that rotate to the left are **levorotatory.**

Molecules can adopt many **conformations** (relative arrangements of atoms or groups), which are usually rapidly interconverting. For open-chain compounds (such as butane), the most important conformations are **staggered** (most stable) and **eclipsed** (least stable). In staggered conformations, substituents on adjacent atoms that are farthest from each other are said to be **antiperiplanar;** those that are next to each other are **gauche.** In eclipsed conformations, all substituents are **eclipsed.**

Cycloalkanes adopt particular conformations depending upon the ring size. Cyclopropanes are planar; cyclobutanes are **puckered;** cyclopentanes are **envelopes;** and cyclohexanes are **boats** (less stable) and **chairs** (more stable). Substituents in rings can assume **axial** positions (more crowded) or **equatorial** positions (less crowded). The strain in a ring system arises from **torsional strain** (eclipsing interactions) and **bond angle strain.** Cyclohexane, however, is almost free from either type of strain. Cyclopropane is the most strained ring.

CHAPTER 3

Electronic Structure

Read This Chapter to Learn About:

- ➤ Molecular Orbitals
- ➤ Extended π Systems
- ➤ Aromatic Systems
- ➤ Molecular Orbital Description of Resonance

MOLECULAR ORBITALS

The concept of hybridization is discussed in Chapter 2, and this is a useful way to predict molecular shape. To understand the electronic behavior of molecules, however, it is necessary to introduce molecular orbital (MO) theory. Although in fact there are many manifestations of MO theory, the basic premise is that when atoms are close enough to each other to form bonds, their orbitals combine in ways that produce a new MO outcome. This is very similar to the concept of hybridization, in which *s* and *p* orbitals combine to form new hybrid orbitals. For example, the unhybridized carbon atom has a 2*s* orbital and three identical 2*p* orbitals in its valence shell. To accommodate three things (atoms or lone pairs) around carbon, two of the 2*p* orbitals and the 2*s* orbital combine to form three identical sp^2 orbitals, leaving one 2*p* orbital untouched (Figure 3-1).

Although useful, the idea of hybridization is a simplification. It is tempting to imagine (and, in fact, some textbooks propose) that these hybrid orbitals interact in predictable ways to make new molecules. In reality, MOs are governed by complex quantum chemical considerations that can be predicted fully accurately only by sophisticated computer calculations. However, that is not to say that molecular orbitals are devoid of conceptual power. To explore some of these principles, let us examine the molecular orbital scenario for ethene (Figure 3-2).

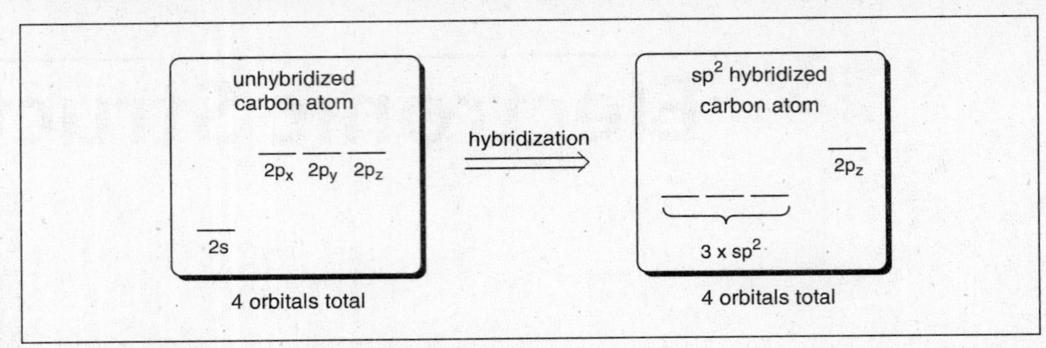

FIGURE 3-1: Atomic orbital hybridization.

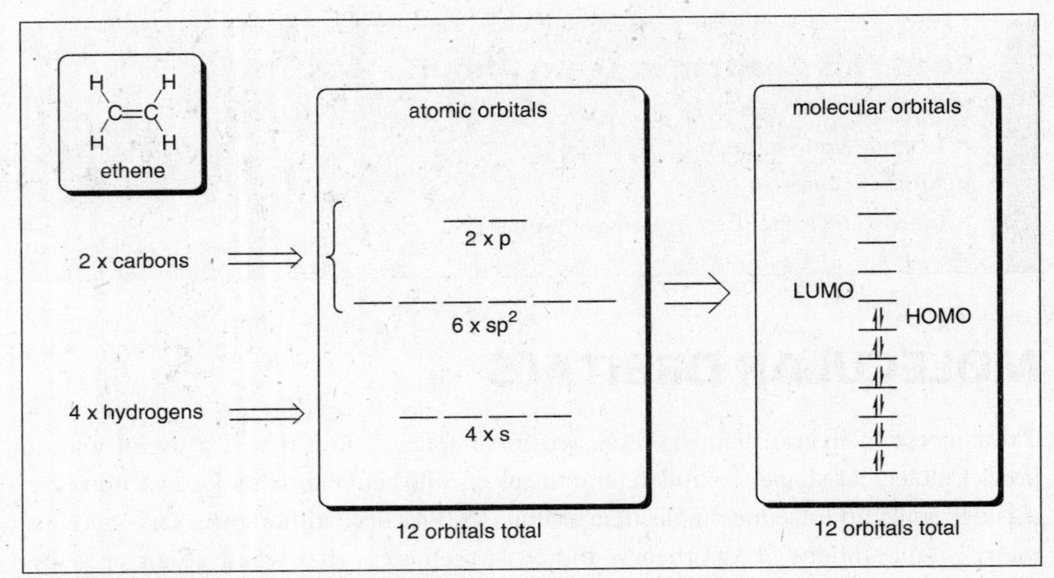

FIGURE 3-2: Molecular orbitals of ethene.

Ethene is constituted from two carbon atoms and four hydrogen atoms. Each carbon atom has four atomic orbitals (we can think of them as being sp^2 hybridized), and each of the four hydrogen atoms has a single 1*s* orbital, for a total of 12 atomic orbitals. These atomic orbitals combine to form 12 new and unique MOs, all of different energies. Hund's rule and the Pauli exclusion principle apply to molecules as well as atoms, so once these molecular orbitals are formed, they fill up from bottom to top. Ethene has 12 electrons total (4 valence electrons from each carbon and 1 electron from each hydrogen), so the bottommost 6 MOs are filled, and the topmost 6 are empty.

In general, the shapes of these orbitals (as calculated by computer algorithms) are complex and nonintuitive. However, there are special cases that can be predicted very accurately using frontier molecular orbital (FMO) theory. Here, the frontier can be thought of as separating the filled from the unfilled orbitals—the frontier orbitals are thus the highest occupied molecular orbital (HOMO) and the lowest unoccupied molecular orbital (LUMO). It turns out that these orbitals govern much of the reactivity of molecules; moreover, their shape can be predicted by manipulating the unhybridized *p* orbitals using a method called the linear combination of atomic orbitals (LCAO).

The principle of LCAO is that adjacent *p* orbitals mix in predictable ways. If each orbital is a mathematical wave function, then they can be combined by addition and subtraction to result in new MOs. This principle is illustrated in Table 3-1. If the wave function of the first *p* orbital (φ_1) is added to that of the second *p* orbital (φ_2), then a new molecular orbital is formed in which the two *p* orbitals are in-phase (lower row). A constructive relationship results, in which the electron density is shared between the two carbon atoms, providing a bonding interaction. This type of bond is designated a π bond. Conversely, if one *p* orbital is subtracted from the other, then a destructive combination results, and the electron density drops to zero between the carbon atoms—this point of no electron density is called a node (designated by the dashed line in the schematic). In general, a node occurs every time the sign of the phase (+/−) occurs.

TABLE 3-1 Frontier Molecular Orbitals for Ethane

LCAO	Schematic	Model	Nodes	Symm	Population	Frontier
$+\varphi_1 - \varphi_2$	+ - / - +		1	A	—	LUMO
$+\varphi_1 + \varphi_2$	+ + / - -		0	S	⥮	HOMO

EXTENDED π SYSTEMS

These π molecular orbitals arise any time there are contiguous arrays of unhybridized *p* orbitals. Extended π systems can also develop when more than two adjacent *p* orbitals are present, as is the case with butadiene (Table 3-2), a molecule that has two adjacent double bonds. Historically, such double bonds have been called **conjugated,** a term that stems from their tendency to undergo chemistry together, rather than as isolated double bonds. This behavior is explained by FMO theory: Each carbon bears an unhybridized *p* orbital, and this contiguous array interacts to give four new π molecular orbitals. Because we constructed them using only the *p* orbitals, each of which housed a single electron, we have only four electrons to accommodate. Therefore, only the first two molecular orbitals are filled.

TABLE 3-2 π Molecular Orbitals for Butadiene

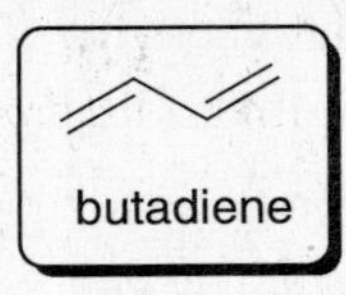

LCAO	Schematic	Model	Nodes	Symm	Population	Frontier
$+\varphi_1 - \varphi_2 + \varphi_3 - \varphi_4$			3	A	—	
$+\varphi_1 - \varphi_2 - \varphi_3 + \varphi_4$			2	S	—	LUMO
$+\varphi_1 + \varphi_2 - \varphi_3 - \varphi_4$			1	A	⇅	HOMO
$+\varphi_1 + \varphi_2 + \varphi_3 + \varphi_4$			0	S	⇅	

As it turns out, the two lowest energy molecular orbitals in butadiene have a net constructive effect, so they are called bonding orbitals and designated π. The two highest energy molecular orbitals, however, are overall destabilizing; Therefore, they are called antibonding orbitals and designated π*. Notice that increasing energy correlates with increasing number of nodes; this is a general trend. There is another guideline that derives from symmetry. An outcome of the mathematics behind LCAO dictates that all π molecular orbitals must be either symmetric (S) or antisymmetric (A). The former (and more familiar) term means that the orbital reflects on itself across a vertical center line, with phases matching exactly. Antisymmetric means that reflection across a center line results in the exact **opposite** phase for each and every point. When the π molecular orbitals are arranged properly, the progression usually alternates between symmetric and antisymmetric orbitals.

Generally speaking, the larger the extended π system, the lower the energy. Therefore, molecules with multiple double bonds are more stable if those bonds are arranged in alternating arrays. For example, 1,3-pentadiene is more stable than 1,4-pentadiene, and 2-cyclohexanone is more stable than 3-cyclohexanone (Figure 3-3). This is a

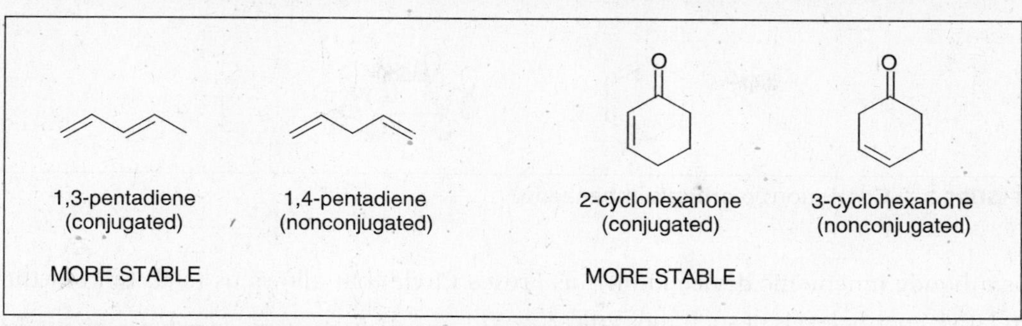

FIGURE 3-3: Conjugated versus nonconjugated double bonds.

phenomenon known as conjugation. Because we know that the origin of the effect can be described by MO theory, it comes as no surprise that conjugated double bonds often behave as a collective, whereas nonconjugated double bonds behave as two isolated species.

True conjugation arises from a mixing of adjacent *p* orbitals. A similar (but weaker) effect can arise from the interaction of *p* orbitals with adjacent bonds. For example, the methyl cation is not a very stable species, largely owing to the empty *p* orbital on carbon and the resulting violation of the octet rule. However, the ethyl cation is a bit more stable, because the C—H sigma bond on the methyl group can spill a bit of electron density into the empty *p* orbital, thereby stabilizing the cationic center (Figure 3-4). This is an effect known as hyperconjugation, and although the sharing of electron density is not nearly as effective here as in true conjugation, it is still responsible for the following stability trend in carbocations and radicals: methyl < primary $\ll$ secondary < tertiary.

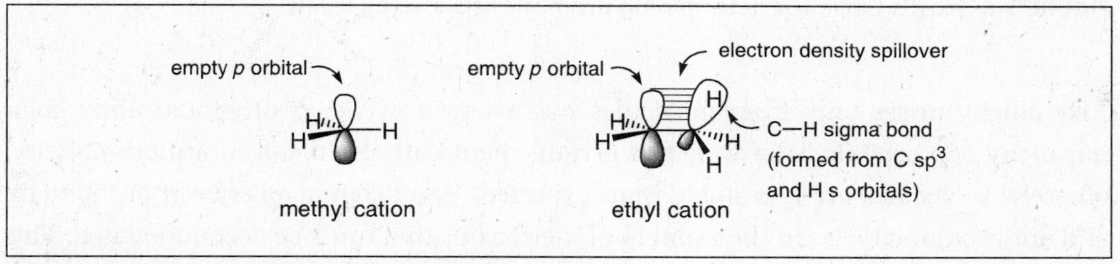

FIGURE 3-4: Hyperconjugation in the ethyl cation.

AROMATIC SYSTEMS

In comparison with linear arrays, when *p* orbitals are arranged in a circle without interruption, as in benzene (Figure 3-5), unexpected stabilization can emerge. As an outcome of the mathematics of LCAO (which we do not discuss here), there are usually multiple sets of degenerate orbitals (i.e., orbitals of the same energy level). Although it is not intuitively straightforward to construct molecular orbital diagrams themselves, there

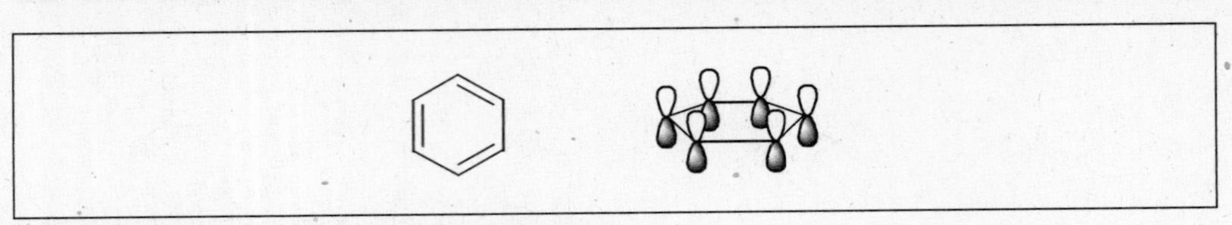

FIGURE 3-5: Contiguous *p* orbitals in benzene.

is a handy mnemonic device known as Frost's Circle that allows us to sketch out the relative energy levels of such molecules.

Frost's Circle is really quite simple—it starts by drawing a regular polygon of the appropriate size with one point down. A circle is then drawn around the polygon, and each point of contact represents a molecular orbital energy level. For example, benzene is a six-membered ring, so its array of molecular orbitals starts with a single lowest-energy MO, followed by two sets of two MOs of the same energy, then a single highest-energy MO (Figure 3-6). Because benzene has 6 π electrons, only the three lowest-energy orbitals are filled. This electronic arrangement accounts for the fact that benzene is particularly stable, a phenomenon known as aromaticity.

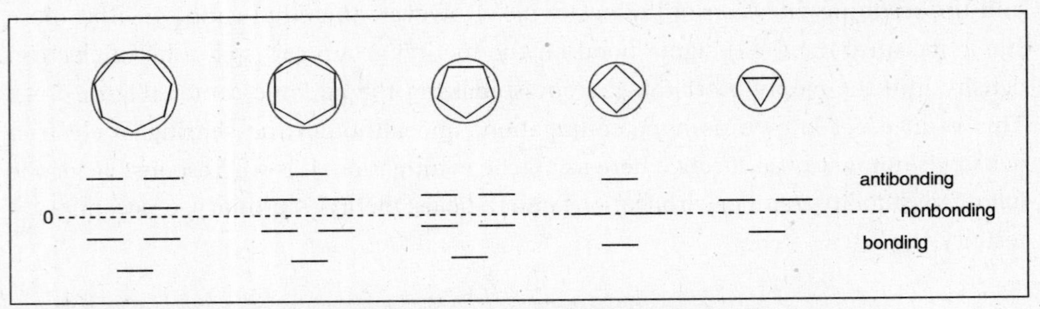

FIGURE 3-6: Frost's Circle for determining aromatic MO energy levels.

Aromaticity arises only from molecules possessing a cyclic, contiguous, and coplanar array of *p* orbitals. However, this arrangement can also result in antiaromaticity, whereby molecules are less stable than expected. The difference between aromaticity and antiaromaticity lies in the number of π electrons that must be accommodated. This can be predicted using Hückel's rule, which states that systems having $4n$ π electrons, where n is simply an integer value (i.e., 4, 8, 12 electrons, etc.) tend to be antiaromatic, and those having $(4n + 2)$ π electrons (i.e., 2, 6, 10 electrons, etc.) tend to be aromatic.

Figure 3-7 demonstrates the molecular orbital basis of Hückel's rule using cyclopentadienyl ions. The molecular orbital energy levels are drawn using Frost's Circle, and then populated with the appropriate number of *p* electrons. The cyclopentadienyl cation has only 4 electrons in the π system, whereas the corresponding anion has 6 (the negative charge represents a lone pair of electrons on carbon). Using Hückel's rule, we would

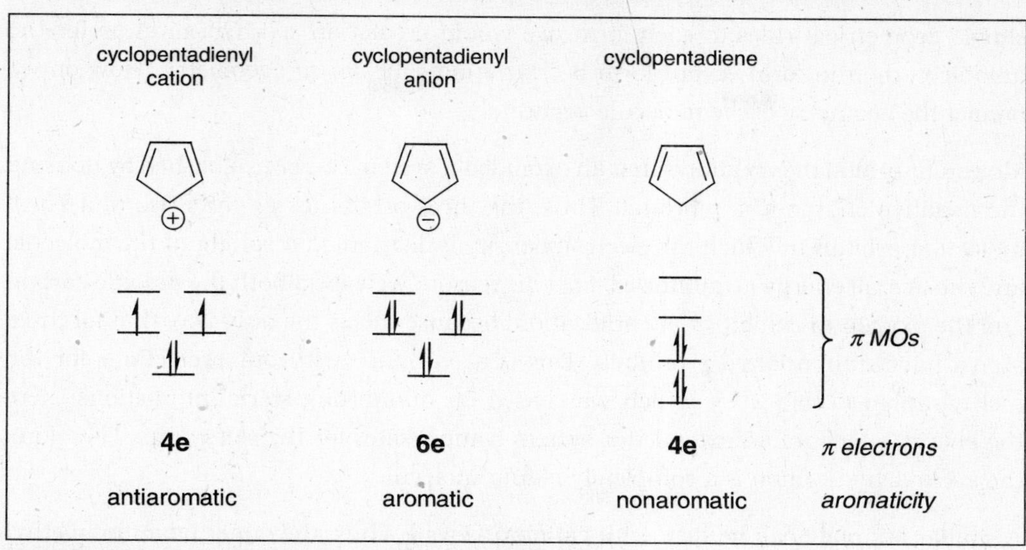

FIGURE 3-7: The molecular orbital origin of aromaticity.

predict the anion to be stable (aromatic) and the cation to be unstable (antiaromatic), which is indeed the experimentally observed result. However, the MO picture provides more insight into why the cation is so unstable—the molecule has two unpaired electrons. Finally, cyclopentadiene itself (far right) is considered nonaromatic. To be aromatic or antiaromatic, there must be a cyclic, contiguous, coplanar array of *p* orbitals, cyclopentadiene is out of the running because there is no *p* orbital at the methylene (CH_2) center, which interrupts the π system. Therefore, its MO diagram is analogous to that of butadiene (Table 3-1).

MOLECULAR ORBITAL DESCRIPTION OF RESONANCE

Molecular orbital theory also provides a more thorough understanding of resonance. For example, the acetaldehyde anion (Figure 3-8) is stabilized by resonance delocalization, whereby the negative charge is distributed between the carbon and the oxygen. Resonance forms do not "equilibrate"—that is to say, the molecule does not oscillate between forms A and B—rather, the two resonance structures are an attempt to represent a more complete truth than any single Lewis structure can. However, if we apply

A B

FIGURE 3-8: Resonance stabilization of the acetaldehyde anion.

simple geometrical rules to each form, we would predict an sp^3 hybridization for the anionic carbon in form A, but form B clearly indicates an sp^2 geometry. How do we predict the geometry of the molecule accurately?

Molecular orbital theory shows that an extended π system can be established by housing the negative charge in a *p* orbital. Thus, the three-orbital array gives rise to a set of molecular orbitals in which the electron density is distributed throughout the molecule and the overall energy is minimized. For this reason, we expect both the anionic carbon and the oxygen to exhibit sp^2 hybridization, because this is the only way that all three atoms have unhybridized *p* orbitals. This is at variance with our predictions for the methyl anion (Table 2-1), which was based on minimizing steric interactions; here the energy benefit of an extended π system compensates for the steric cost. Therefore, the acetaldehyde anion is a completely planar molecule.

A similar anomaly can be seen with cations, as well. Thus, the dimethyl aminomethyl cation (Figure 3-9) can be shown as two resonance forms (A and B), in which the nitrogen lone pair participates in stabilizing the adjacent positive charge on carbon. However, we are faced with a dilemma similar to the acetaldehyde anion: Form A would predict an sp^3 hybridization for nitrogen, whereas form B appears to have sp^2 geometry. The only way for form B to have meaning is if a π bond is established between the two *p* orbitals on carbon and nitrogen (Figure 3-9, inset). Therefore, the species is planar.

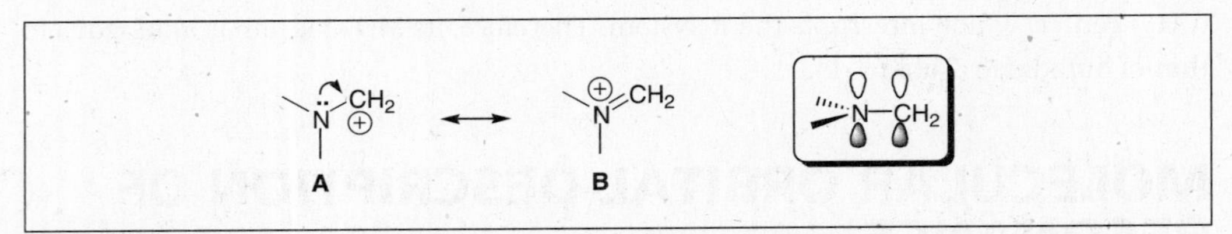

FIGURE 3-9: Resonance stabilization of the dimethylaminomethyl cation.

Aside from the geometric considerations, these two examples show the action of two specific types of functional groups. The carbonyl group (C=O) is an electron-withdrawing group (designated EW), whereas the amino group (NR_2) is an electron-donating group (ED). This idea is important for a variety of chemical reactivities, but the easiest way to keep the categories straight is to examine whether the group would best stabilize a cation or an anion. Generally speaking, electron density can be pushed **into** EW groups, thus stabilizing a negative charge, whereas electrons can be pushed **from** ED groups, which stabilizes adjacent positive charges. For this reason, ED groups generally have lone pairs.

There are two components of EW and ED behavior: resonance and induction. As the term suggests, the resonance effect can be represented by electron-pushing arrows to give different resonance forms. This effect may be weak or strong, but it can exert influence over very large distances through extended π systems. The inductive effect stems primarily from electronegativity; it can be quite strong, but its influence is local—the magnitude drops off sharply as distance from the functional group increases. Table 3-3 summarizes these effects for some common functional groups. In most cases the inductive and resonance effects work in concert, whereas in some instances they are at odds. The amino and methoxy substituents are almost always classified as ED groups, even though the central atoms are electronegative—the activity of the lone pairs dominates their behavior. The halogens are more subtle (e.g., chlorine). Generally speaking, they are grouped with the EW substituents, although their behavior can be more akin to ED groups in aromatic chemistry (Chapter 8).

TABLE 3-3 Some Common ED and EW Groups

Functional Group	Formula	Resonance Effect	Inductive Effect
Electron-withdrawing (EW) groups			
Acetyl	$-COCH_3$		
Carbomethoxy	$-CO_2CH_3$		
Chloro	$-Cl$		
Cyano	$-CN$		
Nitro	$-NO_2$		
Phenylsulfonyl	$-SO_2Ph$		
Electron-donating (ED) groups			
Amino	$-NH_2$		
Methoxy	$-OCH_3$		
Methyl	$-CH_3$		
		← more EW | more ED →	← more EW | more ED →

CRAM SESSION

Electronic Structure

Atomic orbitals are combined by the method of linear combination of atomic orbitals (LCAO) to form new **molecular orbitals** (MOs). The number of molecular orbitals obtained is equal to the number of atomic orbitals used to make them. Usually, there are an equal number of **bonding orbitals** (of lower energy than disconnected atoms) and **antibonding orbitals** (of higher energy than disconnected atoms). Electrons fill up molecular orbitals from lowest to highest according to **Hund's rule** and the **Pauli exclusion principle;** electrons in bonding orbitals stabilize a molecule, whereas those in antibonding orbitals are destabilizing. Much of the chemistry of molecules is determined by their **frontier orbitals:**

- The highest occupied molecular orbital (HOMO)
- The lowest unoccupied molecular orbital (LUMO)

In double and triple bonds, one bond is a σ **bond** formed by the end-to-end interaction of hybridized orbitals; the remaining bonds are π **bonds** arising from side-to-side interaction of unhybridized *p* orbitals. When π bonds are next to each other, they form **extended** π **systems,** in which the π bonds are said to be **conjugated.** These conjugated systems form molecular orbitals that are easily predictable from the combination of adjacent *p* orbitals. In general, conjugated π bonds are more stable than nonconjugated ones. The MO description of conjugated systems offers a fuller understanding of Lewis resonance forms.

Certain cyclic conjugated π bonds can form **aromatic systems.** To be aromatic, a molecule must have an array of *p* orbitals that is

- Cyclic
- Contiguous
- Coplanar

Systems that do not fulfill all of these criteria are considered **nonaromatic.** For those that do satisfy all three conditions, **Hückel's rule** states that aromatic systems must contain $(4n + 2)$ π electrons, whereas those that contain $4n$ π electrons are considered **antiaromatic.** Aromatic systems are more stable than otherwise predicted, and antiaromatic systems are less stable. The position of MO levels for aromatic and antiaromatic molecules can be predicted using **Frost's Circle.** From an MO standpoint, antiaromatic systems usually have unpaired electrons in degenerate orbitals, and thus behave as diradicals.

Aromatic and conjugated systems can be affected by attached substituents. Generally speaking, substituents are classified as either **electron-donating** (ED) groups, such as amino and methoxy, or **electron-withdrawing** (EW) groups, such as nitro and cyano.

The impact can be broken down into **inductive effects**, which arise from electronegativity differences and are communicated through σ bonds, and **resonance effects,** in which electron density is moved across extended π systems. Sometimes the two effects are in opposition, in which case the classification of the substituent is governed by its resonance behavior (i.e., methoxy is considered ED even though it is inductively electron-withdrawing).

CHAPTER 4

Nomenclature

Read This Chapter to Learn About:

- The Syntax of Nomenclature
- Naming Cyclic Species
- Naming Aromatic Species
- Colloquial Nomenclature
- Naming Stereoisomers

THE SYNTAX OF NOMENCLATURE

The entrée into organic nomenclature is typically the class of compounds known as the simple alkanes, straight-chain hydrocarbons that are named according to the number of carbons in the chain (Table 4-1). The first four alkanes bear nonintuitive, historically derived names; however, from pentane onward they are constructed from the Latin roots. If used as a substituent, the names are modified by changing the suffix "-ane" to "-yl"; thus, a two-carbon alkane is ethane, but a two-carbon substituent is ethyl.

In the event of branched hydrocarbons, IUPAC rules dictate that the longest possible chain be used as the main chain, then all remaining carbons constitute substituents. If two different chains of equal length can be identified, then the one giving the fewest number of substituents should be chosen. For example, in Figure 4-1, two seven-membered chains are possible. However, the left-hand choice is preferred, because it gives two substituents—as opposed to the right-hand option, which only gives one substituent. This rule may seem arbitrary, but more substituents means less complex substituents, so naming is easier.

The key to understanding IUPAC nomenclature is realizing that it follows syntactic rules just like any other language. A basic premise of this syntax is that there can only be one main chain in a compound, which is grammatically analogous to a noun. Once identified, this main chain may be appended with as many substituents as necessary, but they are always auxiliaries, much like adjectives and adverbs. Table 4-2 illustrates

TABLE 4-1 Simple Alkanes

Name	Carbons (No.)	Formula
Methane	1	CH_4
Ethane	2	CH_3CH_3
Propane	3	$CH_3CH_2CH_3$
Butane	4	$CH_3(CH_2)_2CH_3$
Pentane	5	$CH_3(CH_2)_3CH_3$
Hexane	6	$CH_3(CH_2)_4CH_3$
Heptane	7	$CH_3(CH_2)_5CH_3$
Octane	8	$CH_3(CH_2)_6CH_3$
Nonane	9	$CH_3(CH_2)_7CH_3$
Decane	10	$CH_3(CH_2)_8CH_3$
Undecane	11	$CH_3(CH_2)_9CH_3$
Dodecane	12	$CH_3(CH_2)_{10}CH_3$

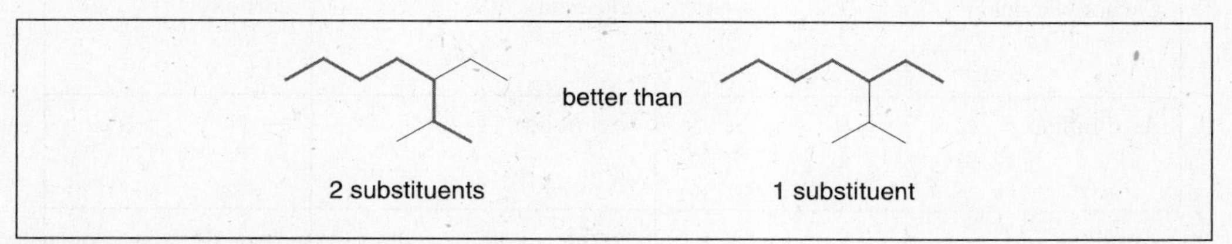

FIGURE 4-1: Maximizing substituents.

TABLE 4-2 Syntax of Nomenclature

Expression	Diagram
embellishment	embellishment
embellished **woodwork**	woodwork / embellished
3 2 1 propane	propane
3 2 1 / 10 9 8 7 6 5 4 3 2 1 5-propyldecane	decane / propyl

this concept with the use of sentence diagrams. Thus, in the name, "5-propyldecane," "propyl" is an adjective and "decane" is the noun it modifies.

Generally, compounds contain functional groups, which are atomic assemblies associated with certain properties or reactivities, such as the hydroxy group (—OH) or the carbonyl group ($R_2C{=}O$). For the purposes of nomenclature, IUPAC has assigned priorities to these functional groups, as illustrated in Table 4-3. When molecules contain functional groups, then the parent chain is defined as the longest possible carbon chain **that contains the highest order functional group.** Once that parent chain is identified, then all other functionalities become substituents (modifiers).

For example, consider the two compounds in Table 4-4. Both contain the hydroxy (—OH) functional group, but in the top it defines the parent chain. Therefore, the

TABLE 4-3 IUPAC Priorities of the Functional Groups

Functional Group	Formula	As a Parent Compound	As a Modifier
Carboxylate ester	O, O, R	alkyl-oate	—
Carboxylic acid	O, OH	-oic acid	carboxy-
Acid halide	O, X (X = Br, Cl, F)	-oyl halide	—
Amide	O, NH_2	-amide	—
Nitrile	—CN	-nitrile	cyano-
Aldehyde	O, H	-al	oxo-
Ketone	O	-one	oxo
Thioketone	S	-thione	thioxo-
Alcohol	—OH	-ol	hydroxy
Thiol	—SH	-thiol	mercapto-
Amine	$—NH_2$	-amine	amino-
Ether	—OR	(used only as modifier)	alkoxy-
Nitro	$—NO_2$	(used only as modifier)	nitro-
Halide	—X X = F, Cl, Br, I	(used only as modifier)	halo- (e.g., chloro-)

TABLE 4-4 Functional Groups in the Main Chain and as Substituent

Name	Diagram
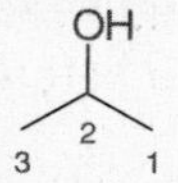 OH 3 2 1 propan-2-ol	propan-2-ol
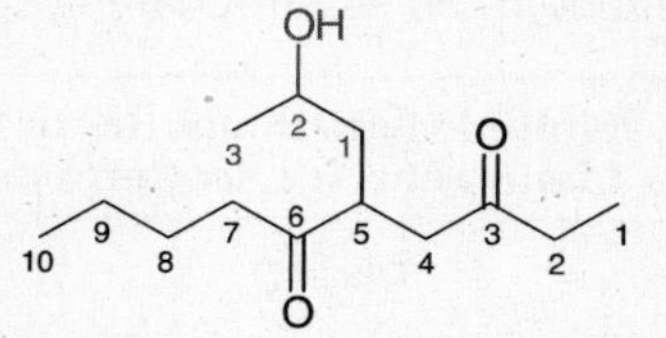 OH 3 2 1 O 10 9 8 7 6 5 4 3 2 1 O 5-(2-hydroxypropyl)decan-3,6-dione	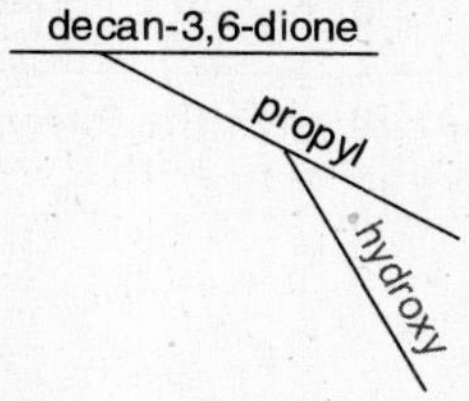 decan-3,6-dione propyl hydroxy

"-ol" form is used, and the compound is named "propan-2-ol." However, the lower compound has both the hydroxy and the carbonyl functional groups. Because the carbonyl has the higher IUPAC priority, it defines the parent chain; the hydroxy group thus serves only as a modifier, and the "-ol" suffix is not used.

Double bonds add an interesting wrinkle in the process. Although they do not carry much priority alone, the parent chain should contain the highest order functional group and as many double bonds as possible, even if this means that a shorter parent chain is obtained or if other functionality is excluded. For example, the compound in Figure 4-2 contains four functional groups: the carboxylic acid, carbonyl, hydroxy, and alkenyl functionalities. Of these, the carboxylic acid has highest priority, so the parent chain must include it. It is tempting to also include the carbonyl group to make a 13-carbon chain; however, IUPAC dictates that both double bonds be included (as highlighted), even though this results in a shorter 9-carbon chain that does not encompass the carbonyl functionality.

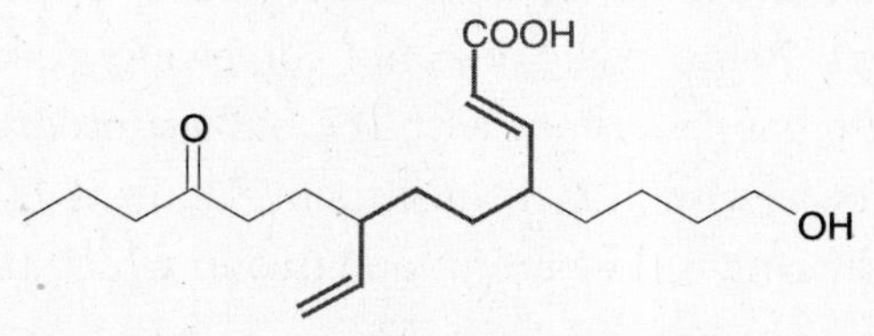

COOH
O
OH

4-(4-hydroxybutyl)-7-(3-oxohexyl)-nona-2,8-dienoic acid

FIGURE 4-2: Naming a multifunctional acyclic compound.

As the previous examples illustrate, a parent chain may contain multiple functional groups of the same type, or a functional group with one or more double bonds. Using our grammatic analogy, this is like having a compound noun (e.g., "boxcar"). In the former case, appropriate prefixes are used (i.e., di-, tri-, etc.); in the latter, the "en" prefix is used, as shown in Table 4-5. However, aside from the alkene, two different functional groups may not be included in the same parent chain. For example, if an alcohol and a ketone are present in the same molecule, it must be named as a ketone with a hydroxy substituent.

TABLE 4-5 Examples of Multiple Functionalities in the Parent Chain

Multiple Functionalities in the Parent Chain That Are Consistent with IUPAC		Multiple Functionalities in the Parent Chain That Are *Not* Consistent with IUPAC
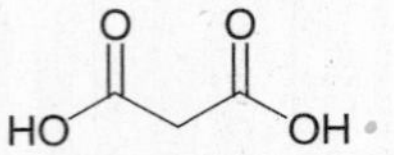 propanedioic acid	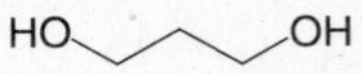 propan-1,3-diol	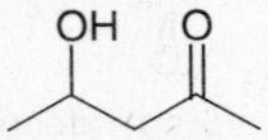 IUPAC: 4-hydroxypentan-2-one (NOT pent-2-on-4-ol)
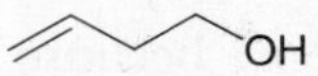 but-3-en-1-ol	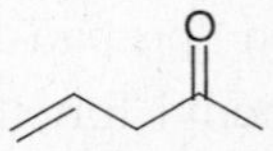 pent-4-en-2-one	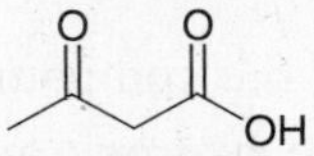 IUPAC: 3-oxobutanoic acid (NOT butan-4-on-1-oic acid)

In summary, the parent chain should be established using the following guidelines (in order of priority):

1. Include the highest order functional group
2. Include as many double bonds as possible
3. Select the longest carbon chain (with the maximum number of substituents)

NAMING CYCLIC SPECIES

Compounds containing cyclic arrays of carbon atoms are known as carbocycles. The presence of a ring does not change any of the considerations discussed so far; however, it does force us to make a choice. Either the ring becomes the parent chain or it is part of some substituent. Being cyclic does not confer any special consideration—if it happens to represent the longest chain with the highest order functional group, then it is the parent chain; otherwise, it is a substituent. Figure 4-3 shows two examples of compounds in which the ring is the parent, and two in which the ring is a substituent.

Compounds can also exhibit bicyclic ring structures, which fall into three categories: fused, bridged, and spiro. In fused and bridged bicyclic compounds, there are always two carbons which are part of both rings. For the sake of simplicity, we first consider the simple (unfunctionalized) bicyclic structures. The parent alkane is considered to contain all the carbon atoms in both rings. The parent name is then preceded by bicyclo[l.m.n]-, where l, m, and n represent the number of carbons between the two

1,2,3-trimethylcyclopropane 2-cyclopropylpentane 2-(2-hydroxypropyl)-cyclohexanone 1-(2-hydroxycyclohexyl)-propan-2-one

FIGURE 4-3: Examples of carbocyclic compounds.

shared carbons. Figure 4-4 shows some examples of these structures (arrows indicate the shared carbons). In bridged systems, the shared carbons are called **bridgehead carbons;** otherwise, they are known as **ring fusion carbons.**

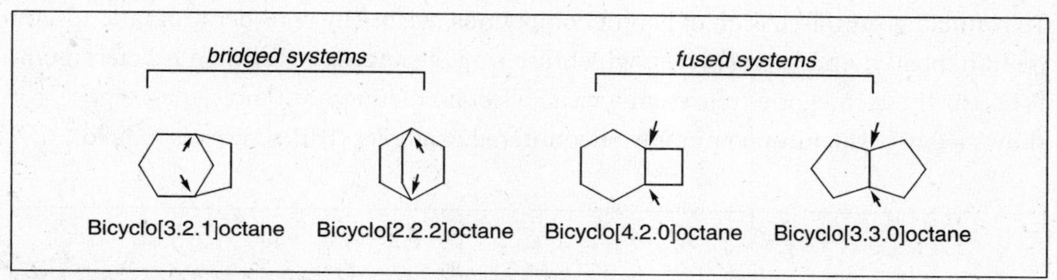

FIGURE 4-4: Bridged and fused bicyclic compounds.

Numbering on bridged and fused systems begins at one shared carbon atom, then goes around the largest ring first, in such a way that gives the lowest number to any functionality on the ring. For example, in the bicyclo[4.2.0]octanone derivative in Figure 4-5, either the top or bottom ring fusion carbon must be carbon-1; the bottom was chosen because it gives the keto functionality a lower number than would be obtained had the top carbon been carbon-1.

7-methyl-bicyclo[4.2.0]octan-3-one

FIGURE 4-5: Numbering a bicyclic compound.

In spiro compounds there is only one shared carbon between the two rings. The nomenclature rules are very similar to the bridged and fused systems, with a couple of exceptions. First, the parent name is preceded by spiro[l.m]-, where l and m represent the number of carbons between the spiro carbon (here only two strands are possible). Second, the numbering starts at the spiro carbon and then proceeds around

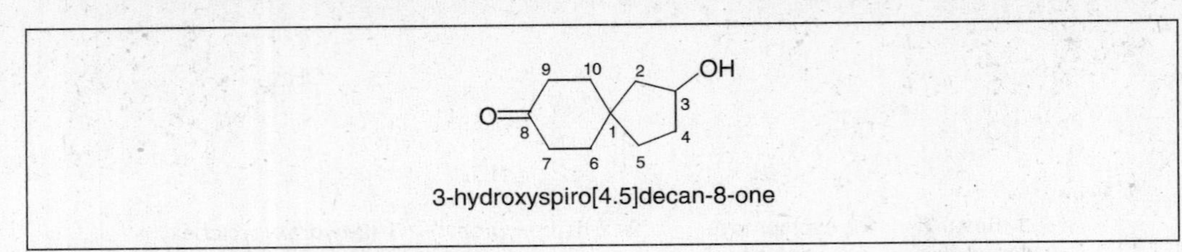

FIGURE 4-6: Naming a spiro compound.

the **smaller** ring first. Thus, in Figure 4-6, the numbering proceeds first around the five-membered ring, even though the higher order functional group is on the larger ring.

NAMING AROMATIC SPECIES

To round out our discussion of parent compounds, we briefly consider aromatic species (see Chapter 3) and heterocycles, which are ring structures containing a heteroatom. Whereas these categories represent a vast collection of important molecules, Figure 4-7 shows a few of the most commonly encountered examples (IUPAC names in bold).

benzene **pyridine** **furan** **oxacyclopropane** oxirane epoxide **azacyclopropane** aziridine **thiacyclopropane** thiirane episulfide

FIGURE 4-7: Common aromatic rings and heterocycles.

These structures are treated as parent compounds when connected to simple substituents (e.g., methylbenzene, 2-ethyl-3-methylfuran), but are named as substituents when high-priority functionality is contained on another chain, as illustrated in Figure 4-8. Note that no number is required for phenylethanoic acid, because there is only one possible position for substitution (namely, the 2-position). Also, the substituent's point of attachment is given in some cases: the expression, "2-furanyl," indicates that the furan is attached to the parent chain at the furan's 2-position.

Phenylethanoic acid 4-(2-furanyl)but-3-en-2-one 2-methyl-3-(2-oxacyclopropyl)propanal

FIGURE 4-8: Aromatics and heterocycles as modifiers.

Sometimes when functionality is attached directly to an aromatic ring, special nomenclature results (Figure 4-9). These names are a jumble of common and IUPAC names. For example, phenol (hydroxybenzene) is a common name grandfathered by IUPAC, whereas aminobenzene was called aniline historically (and still colloquially), but IUPAC has established the new name of benzeneamine. When used as modifiers, these species follow the standard rules; thus, the phenol residue would become a hydroxyphenyl group as a substituent.

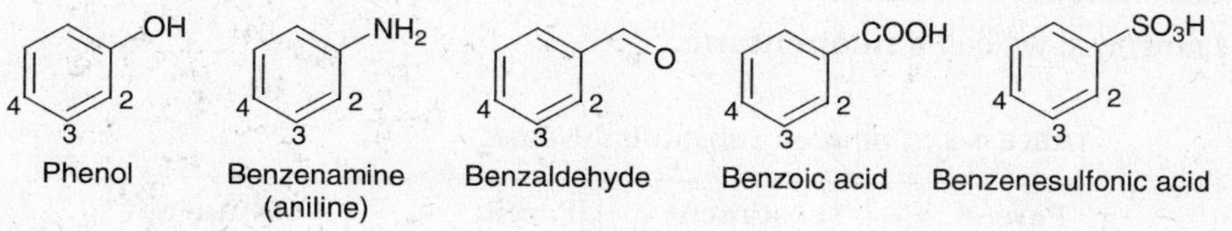

Phenol | Benzenamine (aniline) | Benzaldehyde | Benzoic acid | Benzenesulfonic acid

FIGURE 4-9: Special cases for aromatics.

COLLOQUIAL NOMENCLATURE

Now that we have a pretty good understanding of nomenclature rules in hand, we should pause to consider some miscellaneous vocabulary used by organic chemists to describe carbon centers (Figure 4-10). One set of terms arises from considering how many other carbons are attached to a given center. If a carbon has only one other carbon attached to it, then it is designated a **primary (1°) center;** if two other carbons are attached, it is a **secondary center;** and so on. This terminology is also used for substituents, so if a chloro substituent is attached to a carbon that is attached to two other carbons, it is said to be a **secondary chloride.**

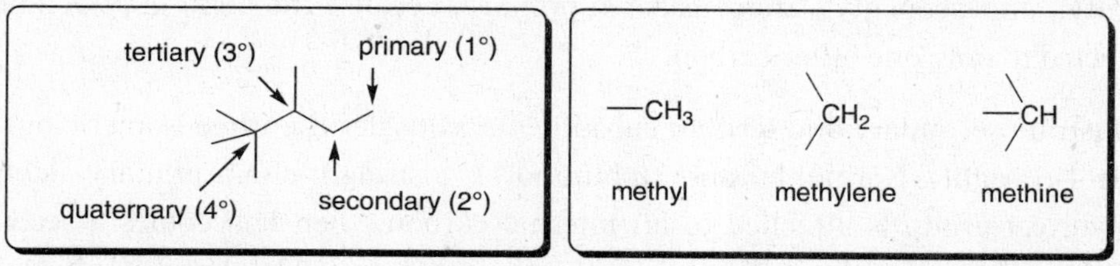

tertiary (3°), primary (1°), quaternary (4°), secondary (2°) | $-CH_3$ methyl, CH_2 methylene, CH methine

FIGURE 4-10: Two ways to describe carbon centers.

The other way to describe carbon centers is by examining how many hydrogens are attached. Carbons with three hydrogens are called **methyl groups;** those with two are **methylene groups;** and those with only one are known as **methines.** Obviously, these are closely related terminologies, but they have subtle differences and are used for slightly different purposes.

Because we have touched on common nomenclature, it would also be useful to come to terms with the subject so that we can be conversant with names still found in everyday usage. First, consider the pentane molecule (Table 4-6, left). When all five carbon atoms are lined up in a row, it is called **normal pentane,** or *n*-pentane. However, we can imagine an isomer of pentane in which a methyl group is shifted to the next carbon. Because this isomer has five carbons, old nomenclature still considers it a pentane, but it is clearly a different molecule. Therefore, it is designated **isopentane.** Finally, a third isomer is possible by connecting all four methyl groups to a common atom. Again, a different molecule is formed, but the name **isopentane** is already taken; thus, focusing on its newness, we call it **neopentane.**

TABLE 4-6 Common Substituent Names

Parent	Substituent	Parent	Substituent
n-pentane	*n*-pentyl	OH *n*-butanol	*n*-butyl
isopentane	isopentyl	OH *sec*-butanol	*sec*-butyl
neopentane	neopentyl	OH *tert*-butanol	*tert*-butyl

These three pentane structures can also be used as substituents, as shown in Table 4-6 (squiggly line showing point of attachment). Notice that the isopentyl group ends with a bifurcated carbon chain, or a fork. Likewise, the neopentyl group terminates with a three-pronged carbon array. However, all three of these substituents (*n*-pentyl, isopentyl, and neopentyl) are considered primary, because the point of attachment is connected to only one other carbon.

To illustrate secondary and tertiary substituents, consider the three isomeric butanols (Table 4-6, right). Normal butanol (*n*-butanol) is a straight-chain primary alcohol. If the hydroxy group is attached to an internal carbon, then it becomes a secondary alcohol, so old nomenclature designates this *sec*-butanol (IUPAC: butan-2-ol). One can also imagine four-carbon alcohol in which the hydroxy group is attached to a tertiary site—this is known as *tert*-butanol, or *t*-butanol (IUPAC: 1,1-dimethylethanol). There is one notable exception to these patterns, namely isopropanol—strictly speaking, it should be called *sec*-propanol, but convention has settled on the former.

In addition to substituent oddities, old nomenclature also has different ways of naming parent compounds, some of which are illustrated in Figure 4-11. Sometimes the roots are different, as in formic acid and formaldehyde (vs. methanoic acid and methanal).

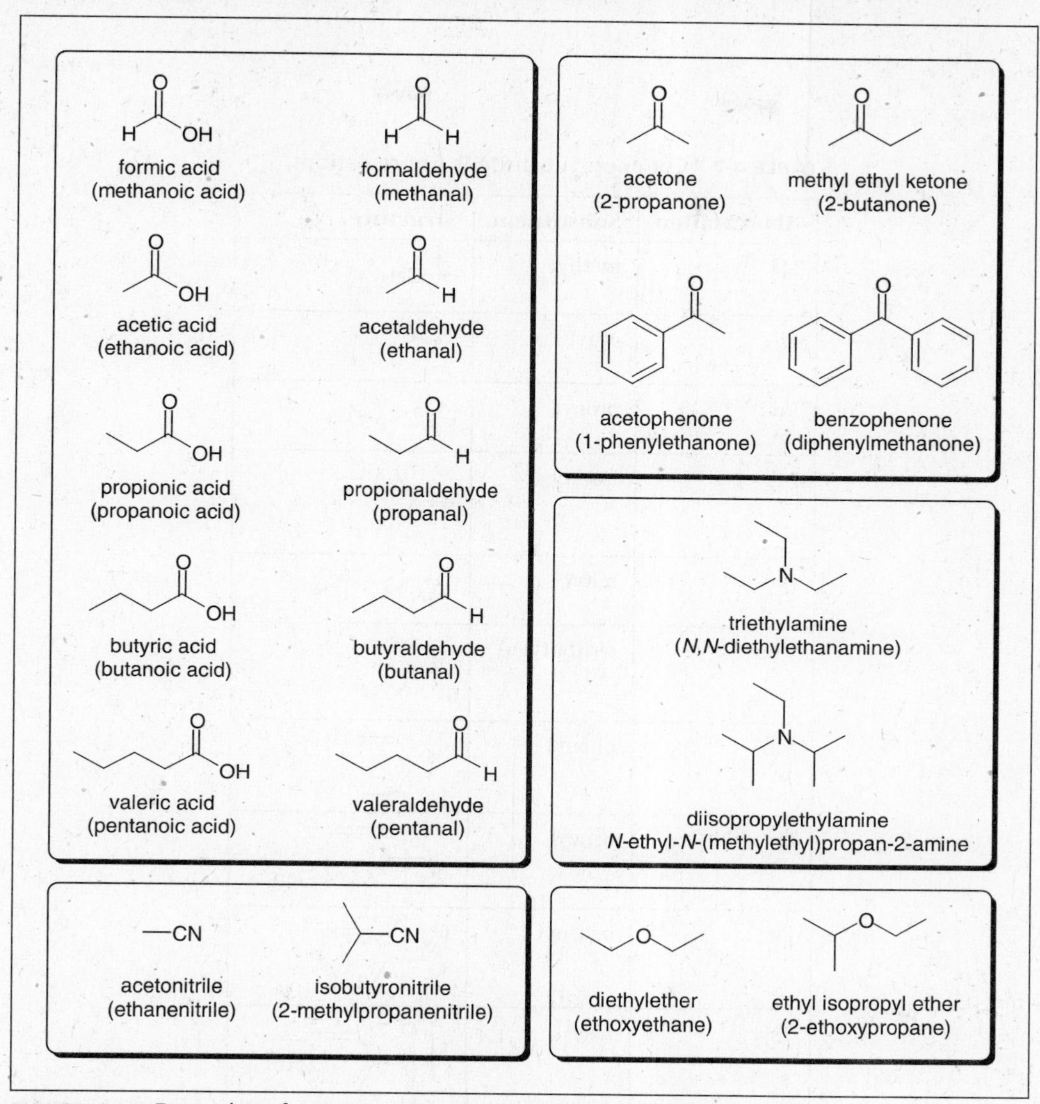

FIGURE 4-11: Examples of common nomenclature (with IUPAC equivalents).

Other cases seem to be fairly arbitrary, as in acetone and benzophenone. However, there is a logic that underlies some of the common nomenclature. For example, ketones and ethers were named as if the carbonyl or oxygen were a bridge—so any ketone would be named by the two substituents plugged into the carbonyl (e.g., methyl ethyl ketone; ethyl isopropyl ether). Similarly, amines were named by the three substituents plugged into the nitrogen.

From a depictional standpoint, structures in this text and other sources are often simplified using abbreviations. Table 4-7 provides a sampling of some of the more common abbreviations, along with their names and corresponding structures. The last three entries are used for generic substituents, so that one structure can represent many possibilities.

TABLE 4-7 Common Substituent Abbreviations

Abbreviation	Substituent	Structure
Me	methyl	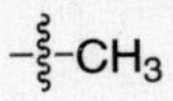 $-CH_3$
Et	ethyl	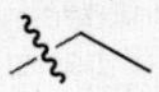
Pr	propyl	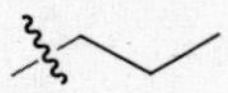
i-Pr	isopropyl	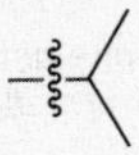
Bu	butyl	
t-Bu	*tert*-butyl	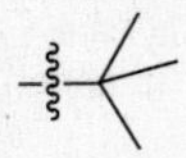
Ph	phenyl	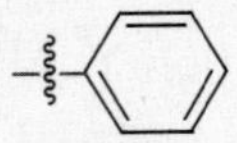
Bn	benzyl	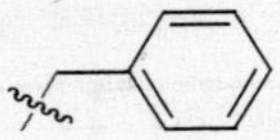
Bz	benzoyl	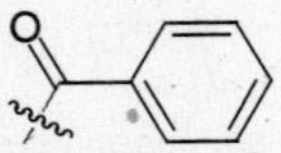 O
Ts	tosyl	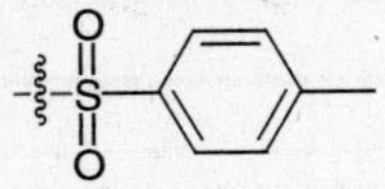 O, S, O
Ms	mesyl	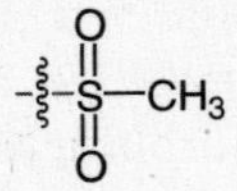 O, S, O, CH_3
R	alkyl	Any aliphatic group
Ar	aryl	Any aromatic group
Ac	acyl	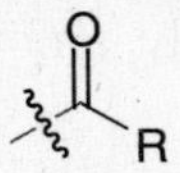 O, R

NAMING STEREOISOMERS

So far we have discussed nomenclature as it relates to connectivity. We must also be able to specify unequivocally how substituents are oriented in space, which enters into the realm of stereochemistry. As far as nomenclature is concerned, there are two types of stereochemistry: absolute and relative.

Absolute stereochemistry deals with the geometry about a chiral center. In Chapter 2 the concept of enantiomerism was introduced; here we discuss how to name each enantiomer. Since any chiral carbon is surrounded by four different things, the first order of business is to prioritize the substituents according to the IUPAC convention known as the Cahn-Ingold-Prelog (CIP) rules. The algorithm for CIP prioritization is as follows:

1. Examine each atom connected directly to the chiral center (we call these the "field atoms"); rank according to atomic number (high atomic number has priority).
2. In the event of a field atom tie, examine the substituents connected to the field atoms.
   a. Field atoms with higher-atomic-number substituents win.
   b. If substituent atomic number is tied, then field atoms with more substituents have priority.

For the purposes of prioritization, double and triple bonds are expanded—that is, a double bond to oxygen is assumed to be two oxygen substituents. If two or more field atoms remain in a tie, then the contest continues on the next atom out for each center, following the path of highest priority.

To illustrate the application of the CIP rules, consider the assigment of an enantiomer of 2-ethyl-3-methylbutan-1,2-diol shown in Figure 4-12. The sawhorse depiction in the center shows us the three-dimensional arrangement of the atoms. To assign CIP priorities, it is useful to think of the structure with all atoms written out explicitly, as shown on the left. Here, the field atoms are an oxygen and three carbons, labeled C_α, C_β, and C_γ. Of the four atoms, the oxygen wins and therefore bears highest priority (a). The next step is to examine the substituents on the remaining field carbons: C_α is

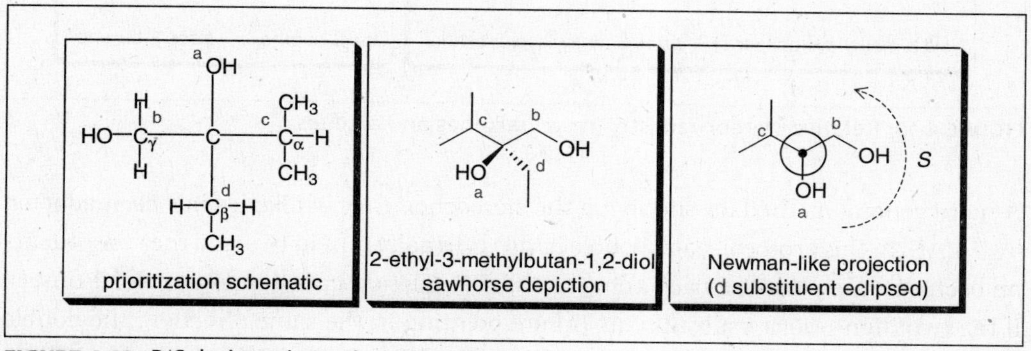

prioritization schematic

2-ethyl-3-methylbutan-1,2-diol sawhorse depiction

Newman-like projection (d substituent eclipsed)

FIGURE 4-12: *R*/*S* designation using CIP rules.

connected to 2 carbons and 1 hydrogen; C_β is connected to 1 carbon and 2 hydrogens, and C_γ is connected to 1 oxygen and 2 hydrogens. We thus assign C_γ as second priority (b), because it is connected to the highest atomic number substituent (oxygen); C_α then assumes third priority (c), because it is connected to 2 carbons rather than 1.

The next step is to orient the molecule so that the lowest-priority field atom (d) is going directly away from us, as is shown on the right side of Figure 4-12. Viewing this projection, we observe whether the progression of a → b → c occurs in a clockwise or counterclockwise sense. If clockwise, the enantiomer is labeled *R*; if counterclockwise, *S*. Please note that this is simply a nomenclature convention, and it has no relationship to physical reality. Recall from Chapter 2 that enantiomers are called dextrorotatory (*d*) if they rotate plane-polarized light to the right, and levorotatory (*l*) if they rotate light to the left. There is **no correlation** between *R*/*S* nomenclature and *d*/*l* behavior.

To add the enantiomeric designation to a chemical name, simply append the information to the far left side and include it in parentheses. The position of the chiral center is also included. Thus, 2-ethyl-3-methylbutan-1,2-diol becomes (2*S*)-2-ethyl-3-methylbutan-1,2-diol. If the molecule contains multiple chiral centers, then the designations are simply arranged by number in the same set of parentheses.

As the descriptor suggests, relative stereochemistry describes the relationship of two substituents with respect to each other. For example, in cycloalkanes two groups may be situated on the same face (*cis*) or on opposite faces (*trans*), as illustrated in Figure 4-13. The same nomenclature applies to disubstituted alkenes, where the two substituents can be directed toward the same side (*cis*) or opposite sides (*trans*) of the double bond. Unlike enantiomers, these stereoisomers often have markedly different physical properties (boiling points, etc.).

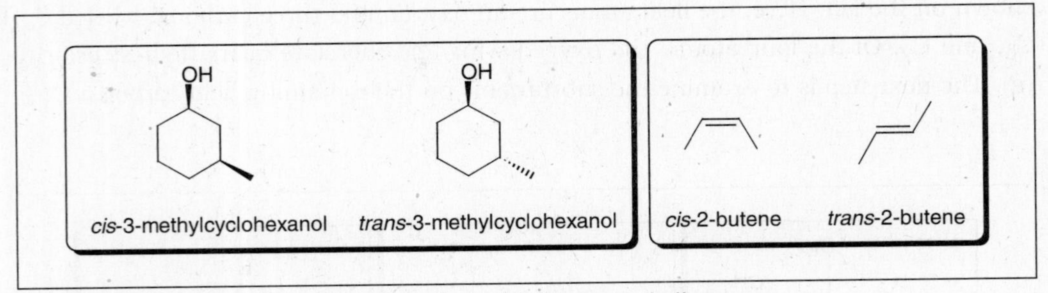

FIGURE 4-13: Relative stereochemistry in cycloalkanes and alkenes.

A more general method for specifying the stereochemistry of alkenes has been adopted by IUPAC. In this protocol, each double bond is virtually rent in two and the substituents on each side are prioritized according to the CIP rules described in Figure 4-14 (inset). If the two high-priority substituents (a) are pointing in the same direction, the double bond is specified as Z (from the German *zusammen*, or together); if they point in opposite

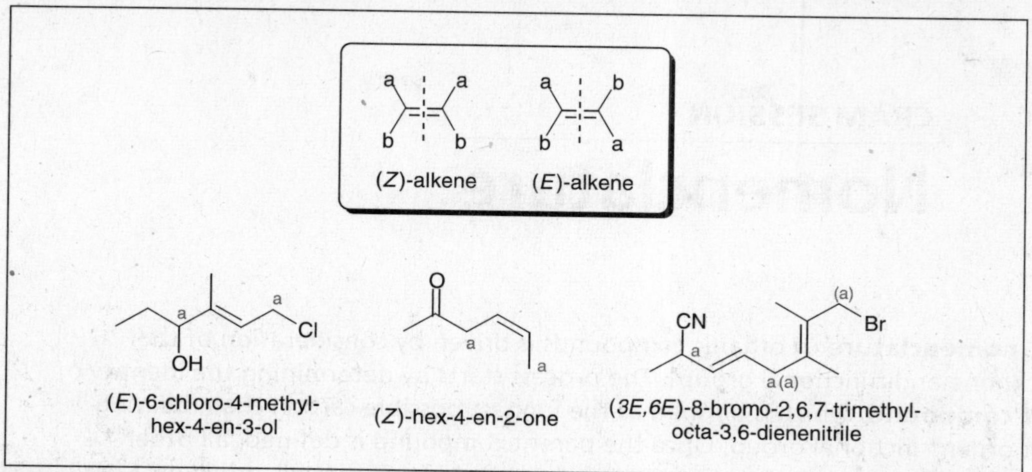

FIGURE 4-14: Stereochemistry of alkenes using CIP (*E*/*Z*).

directions, the double bond is designated *E* (*entgegen*, or across from). Adding this stereochemical information to the chemical name proceeds exactly as described for *R*/*S* designation.

CRAM SESSION

Nomenclature

The **IUPAC nomenclature** of organic compounds is driven by consideration of the carbon backbone and functional groups. The process starts by determining the identity of the **parent compound,** which is defined as the longest possible carbon chain bearing the highest order functional group. Once the parent compound is defined, all other groups become **substituents.** Substituents can be **simple** (e.g., methyl, ethyl, etc.) or **compound** (e.g., 2,2-dimethylpropyl). When numbering a compound substituent, the point of attachment is always considered carbon 1. The conversion of alkane chains into substituents is simple: methane becomes methyl. However, functional groups are more subtle: an alkanol becomes a hydroxalkyll substituent, and an alkanone becomes an oxoalkyl substituent.

A **cyclic species** is no different from its acyclic analogs. If the cycloalkane bears the highest order functional group, then it becomes the parent compound; otherwise, it is named as a substituent. Bicyclic species come in three varieties:

- **Fused bicyclic** systems, in which the two rings share a common side
- **Bridged bicyclic** systems, which share two atoms (the bridgeheads)
- **Spirobicyclic** systems, which have only one atom (the spiro center) in common

These systems are defined by the total number of carbon atoms in the ring system, and by counting the number of atoms between the ring fusion atoms, bridgehead atoms, or spiro center, respectively.

At its most basic level, nomenclature specifies the connectivity (or **constitution**) of a molecule. However, a given constitution can give rise to many **stereoisomers,** which differ in the spatial arrangement of substituents. **Relative stereochemistry** refers to the relative position of groups with respect to each other (e.g., *cis*- vs. *trans*-1,2-dimethylcyclopropane), whereas **absolute stereochemistry** deals with the **configuration** about a chiral center, which is specified using the **Cahn-Ingold-Prelog** (CIP) rules of priority. A brief recap of this procedure is as follows:

- Prioritize atoms around an asymmetric center according to atomic number.
- Orient the center so that the lowest-priority atom is farthest away.
- If the priority of the remaining atoms proceeds clockwise from highest to lowest, then it is an ***R* center.**
- If the priority of the remaining atoms proceeds counterclockwise from highest to lowest, then it is an ***S* center.**

It is important to remember that there is no obvious connection between *R* or *S* configuration and optical rotation!

The CIP rules are also used to specify the geometry around double bonds. The methodology is as follows:

- Orient the double bond so that it is horizontal.
- On each side, label the two substituents according to their CIP priority.
- If the two top-priority substituents are pointing in the same direction, it is a ***Z* alkene.**
- If the two top-priority substituents are oriented opposite each other, it is an ***E* alkene.**

All stereochemical information is included in the IUPAC name as a prefix, thus:

(2*R*,3*S*,4*R*)-2-bromo-3-chloro-4-iodopentane

CHAPTER 5

Reconciling Visual Meaning

Read This Chapter to Learn About:

- ➤ Depiction and Meaning
- ➤ Manipulating Structures in Three-Dimensional Space
- ➤ Rapid Comparison of Depictions

DEPICTION AND MEANING

Coming to terms with the three-dimensionality of organic chemistry is often a challenge for students, yet this is possibly the most important transferable cognitive skill developed by the study of the subject. The novice is confronted with a jumble of alternative depictional devices that appear to be interchangeable. However, a structural drawing is a way of communicating, and specific information is carried in these depictions. For example, if we are presented with a simple line drawing (Table 5-1), we can see the landscape of the molecule quickly—what the regiochemistry is and which functional groups are present—but we are told nothing about the stereochemistry. However, the Fischer projection was developed specifically for quickly conveying the absolute stereochemistry (configuration) of a molecule. We must choose the right depiction carefully for the information we wish to convey, and we must also be able to fully interpret the messages given by specific structures.

One particularly thorny depictional issue centers around relative and absolute stereochemistry. For example, we can draw a structure for *cis*-5-methylcyclohex-2-enol, which is unequivocal and easily distinguishable from the corresponding *trans*-isomer (Table 5-2). However, we have arbitrarily chosen to draw the substituents with two

TABLE 5-1 Summary of Structural Depictions

Type of Depiction	Example	Best for Depicting
Line structure	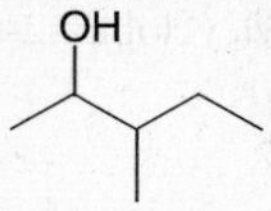 OH	Constitution and connectivity
Sawhorse	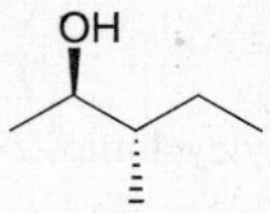 OH	Relative and absolute stereochemistry
Newman	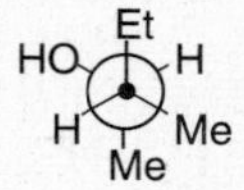 Et, HO, H, H, Me, Me	Conformation
Chair	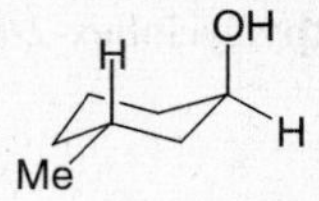 H, OH, H, Me	
Fischer	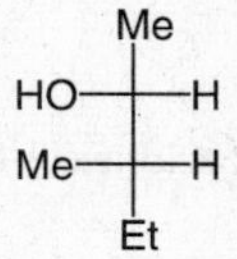 Me, HO—H, Me—H, Et	Configuration (absolute stereochemistry)

TABLE 5-2 Relative Stereoisomers

Structural depiction	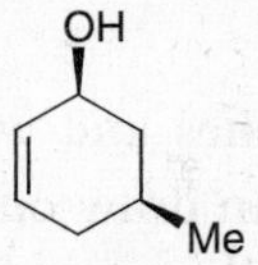 OH, Me	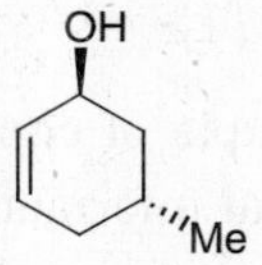 OH, Me
Corresponding name	*cis*-5-methylcyclohex-2-enol	*trans*-5-methylcyclohex-2-enol

wedges—a structure with two dashes would have been equally valid. In this example, our only intent was to show that the two substituents are on the same side of the molecule, that is, their relative stereochemistry. Whether carbon-1 is an *R* or *S* center is unknown.

Unfortunately, we use the same kind of depiction to show absolute stereochemistry. For example, if we were to draw specifically the 1*S*, 5*S* enantiomer of *cis*-5-methylcyclohex-2-enol, both substituents would be attached with wedges (Table 5-3)—but here we are depicting their *absolute* placement in space, not just their positions relative to each other. Conversely, the *R*, *R* enantiomer would be drawn with dashes. Note that when the configuration is included in the name, *cis*/*trans* designation is unnecessary. Converting a name to a structure is relatively easy—if *R*/*S* information is given, we simply represent it accurately in the structure; if only relative (*cis*/*trans*) stereochemical information is

TABLE 5-3 Relative versus Absolute Stereochemistry

Structural Depiction	Corresponding Name	*cis/trans*
OH Me (optically pure)	(1*S*, 5*S*)-5-Methylcyclohex-2-enol	*cis*
OH Me (optically pure)	(1*R*, 5*R*)-5-Methylcyclohex-2-enol	
OH Me (optically pure)	(1*S*, 5*R*)-5-Methylcyclohex-2-enol	*trans*
OH Me (optically pure)	(1*R*, 5*S*)-5-Methylcyclohex-2-enol	

given, we have a couple of choices for the dashes and wedges; if we are provided with no stereochemical information, then we can draw only a simple line structure. However, the reverse operation is trickier: Properly interpreting a sawhorse structure requires context. A good rule of thumb is that any sawhorse structure containing chiral centers is **assumed to be a racemic mixture** unless somehow identified as a single enantiomer. This context can be in the form of optical rotation data or explicit statements, such as "single enantiomer" or "optically pure."

Occasionally, we must be deliberately ambiguous about the stereochemistry of a compound—for example, if the stereochemical arrangement has not been determined, or if we know that there is a mixture of stereoisomers. For this, there is a device known colloquially as the "squiggly line," which indicates the orientation of the substituent can be up, down, or both. In cyclic structures, the use of the squiggly line has the same effect as using plain line structure (Figure 5-1, left). However, for alkenes, there is really no other alternative to show a mixture of *cis* and *trans* isomers than to use this handy depictional device (Figure 5-1, right).

In summary, our choice of depiction must be chosen carefully to reflect what we know about a particular molecule or collection of molecules, and it must be suited to the task of

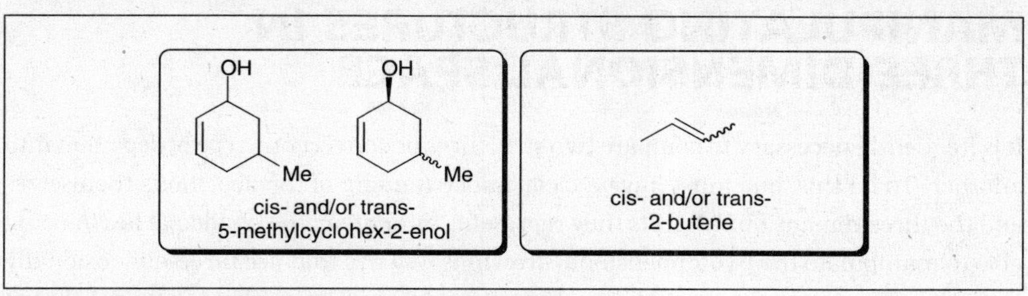

FIGURE 5-1: Deliberately ambiguous stereochemistry.

representing this information. In addition, a given structural representation has specific meaning that we must interpret properly. As an organizing principle, it is very useful to think of structure (and representation) as layers of detail (Figure 5-2), the lowest level of detail being constitutional (how the atoms are connected), then stereochemical, and ultimately conformational (the particular shape a molecule adopts)—not unlike the primary, secondary, and tertiary structure of proteins.

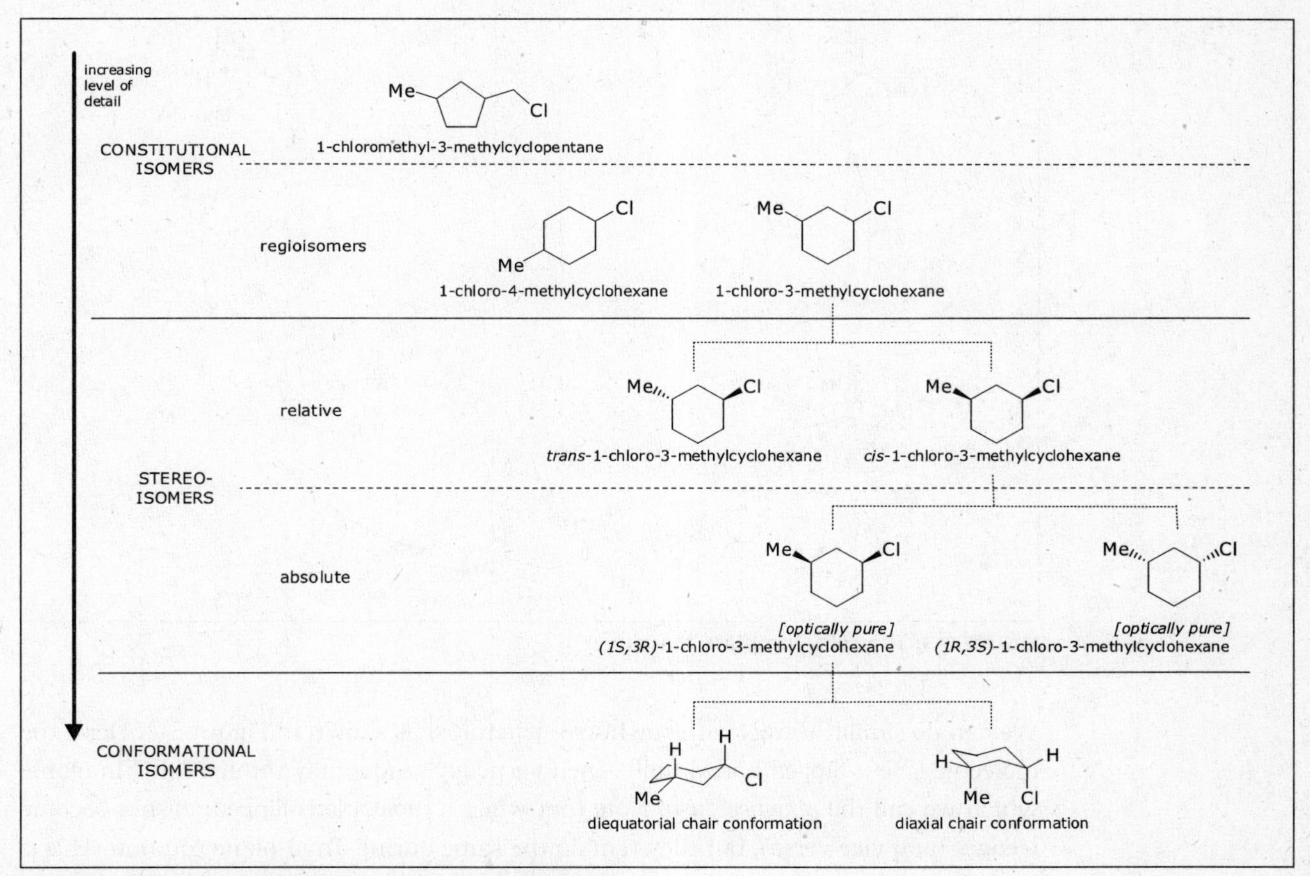

FIGURE 5-2: Levels of detail in structural representation.

MANIPULATING STRUCTURES IN THREE-DIMENSIONAL SPACE

It is frequently necessary to compare two structures or convert one type of depiction into another. To do this, one must have a clear understanding of the depictions themselves and the three-dimensional objects they represent. Ultimately, the challenge lies in being able to manipulate the two-dimensional drawings in ways that are three-dimensionally competent. For example, consider the chair form of *trans*-3-methylcyclohexanol shown in Figure 5-3. The six depictions shown are simply reorientations of the molecule by rotating in a plane perpendicular to the page, a process we call the *spinning plate.* Note that we have not reconfigured the molecule—these are not ring flips, but a static structure viewed from different angles, as if perusing it in our hands.

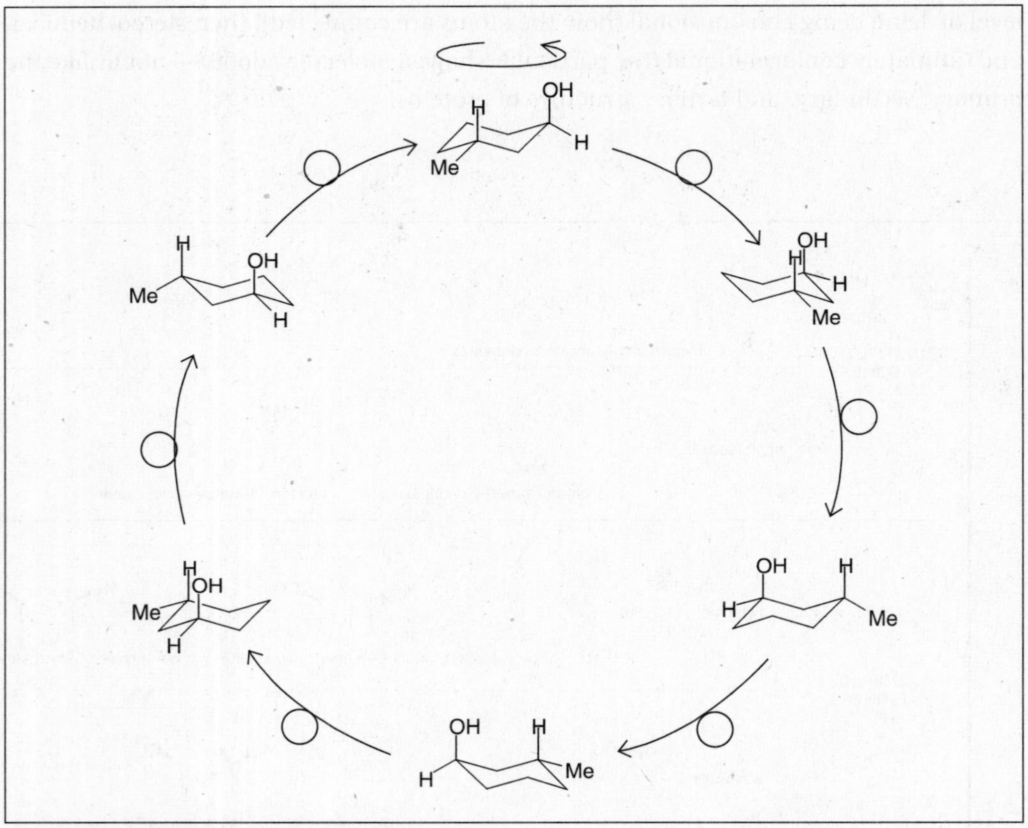

FIGURE 5-3: The spinning plate maneuver.

We can do similar things with sawhorse structures, as shown in Figure 5-4. Here, the molecule is first flipped horizontally, then vertically, and finally rotated 180° in plane, which we call the *pinwheel turn.* Note that when a molecule is flipped, dashes become wedges (and vice versa), but they remain the same during an in-plane rotation. This is an obvious statement, but it is critical to be conscious of the fact as we manipulate

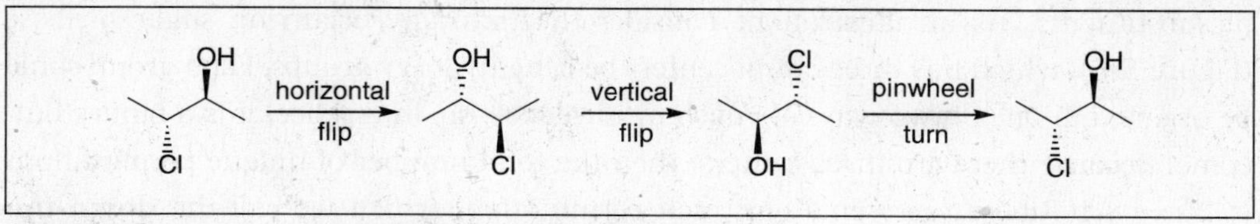

FIGURE 5-4: Some manipulations of sawhorse structures.

structures later on. We can also perform these maneuvers on chair cyclohexanes as well.

When converting specialized depictions—like Newman or Fischer projections—to a more conventional drawings (e.g., sawhorse structures), bear in mind that this is ultimately a change in our point of view. So for a Newman projection, we sight the molecule down a carbon–carbon bond, but in a sawhorse, we view the same bond side-on. An easy way to think about the conversion is to imagine that the Newman projection is hinged at the back carbon and we push the structure into the page, much like we would shut an open door (Figure 5-5); thus, the bond that points directly toward us in structure A (the C2—C3 bond) lies in the plane in structure B. We can use the same technique with Fischer projections, as long as we remember that the stylized representation (C) really implies the dashes and wedges shown in structure D. Note that two-centered Fischer projections are always eclipsed.

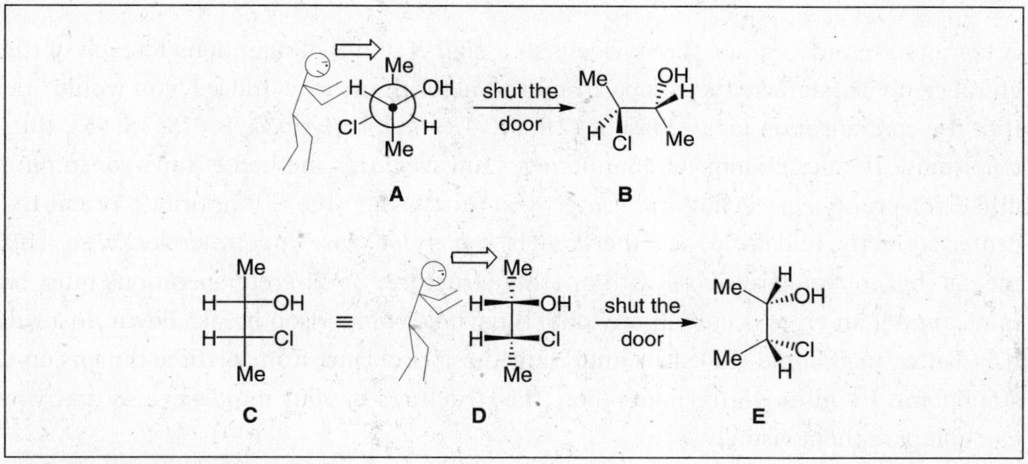

FIGURE 5-5: Shutting the door.

RAPID COMPARISON OF DEPICTIONS

Now imagine that we were asked to evaluate the two representations A and C, and decide whether they are enantiomers, diastereomers, or the same thing. First, let us review the difference between enantiomers and diastereomers. Both fall under the broader umbrella term of **stereoisomers,** which refers to different spatial orientation of substituents. As an illustration, consider the naturally occurring sugar D-ribose (Figure 5-6), which has three chiral centers bearing hydroxy groups. Each group could be oriented in one of two ways—a dash or a wedge—so, in essence, it is a binary outcome. Because there are three centers, then the total number of unique permutations is $2^3 = 8$. D-Ribose represents one such permutation, which we call the **down-up-down isomer** (referring to the orientation of the hydroxy groups). The enantiomer of D-ribose is L-ribose (the D and L designations are peculiar to carbohydrate nomenclature), and notice that **each and every chiral center is inverted**—this is true for any set of enantiomers. So, L-ribose accounts for one stereoisomer of D-ribose, but there are six others—all of them diastereomers of D-ribose. Figure 5-6 shows only one of the diastereomers, namely D-xylose, which has a down-down-down arrangement of hydroxy substituents. Therefore, we can define a diastereomer as a stereoisomer in which one or more, **but not all,** chiral centers have been inverted.

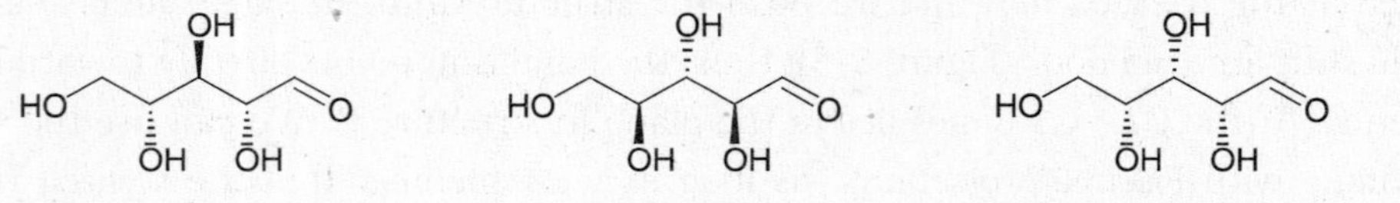

D-ribose

L-ribose (an enantiomer of D-ribose)

D-xylose (a diastereomer of D-ribose)

FIGURE 5-6: Enantiomers versus diastereomers.

A common temptation for the novice is to assign *R* and *S* designations to each of the chiral centers and make the comparison in an algorithmic way. Indeed, you would find that the configuration for D-ribose is (2*R*, 3*R*, 4*R*) and for L-ribose is (2*S*, 3*S*, 4*S*), thus confirming the designation of enantiomer. However, this method is time consuming and error-prone. For each chiral center, you must assign the CIP priorities, orient the center correctly, and decide whether a $\rightarrow$ b $\rightarrow$ c is clockwise or counterclockwise. This means that to compare ribose with another structure, 36 discrete operations must be made, and if an error occurs in any one, the whole comparison breaks down. Instead, it is better to become skilled in analyzing these problems from a three-dimensional standpoint. In other words, manipulate the structures in your mind's eye so that you can compare them visually.

With this backdrop, let us return to the question of comparing Newman projection A to Fischer projection C in Figure 5-5. The first order of business is to convert them to sawhorse representations (Figure 5-7B and E, respectively). Then the question is how to compare the two very different sawhorses. To make a reliable comparison, we must first reorient the structures so that they share elements of commonality. It is generally easier to simply take one as a reference and reorient the other—arbitrarily take structure B as a reference and manipulate E to adopt the elements of commonality. So what are these elements? If we examine structure B, we find that (1) it is a staggered conformation;

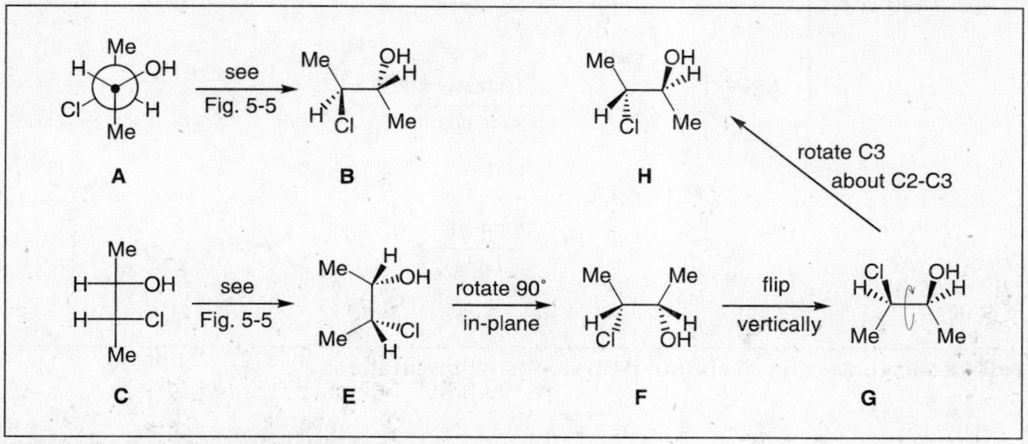

FIGURE 5-7: Comparing two representations.

(2) C1, C2, C3, and C4 are all in the plane; (3) the hydroxy group is on the right and pointed toward the top of the page; and (4) the chloro substituent is on the left and pointed toward the bottom of the page (never mind for the moment whether they are dashes or wedges).

We now set about the task of reorienting E to adopt these four elements, starting with an in-plane rotation to get structure F, followed by a vertical flip to obtain structure G. Notice that the dashes and wedges remain the same during the rotation, but change during the vertical flip. Our motivation for this maneuver was to place the hydroxy group on the right side and pointing toward the top of the page. Notice, however, that the conformation is eclipsed (as, indeed, all Fischer projections must be). To make a comparison, we must then rotate about the C2—C3 bond 180° to arrive at structure H. Unlike the previous operations, this is a reconformation of the molecule—but the identity of the compound remains the same. Now we are in a position to evaluate the two structures (B and H). Notice that every chiral center in H is the opposite of that in B; therefore, the two must be enantiomers.

You can carry out similar operations on cyclic molecules. Again, it is generally easier to convert all representations into dash-wedge drawings before making manipulations and comparisons. Figure 5-8 demonstrates two mnemonics for re-envisioning chair cyclohexane depictions. Imagine that the back carbon–carbon bond is a hinge, and that the cyclohexane ring is a flat surface, like a table or an ironing board. You can push the molecule up on the hinge, much like a foldaway ironing board, in which case the hydroxy group becomes a dash at the 3 o'clock position, and the methyl group is a wedge at the 10 o'clock position. Alternatively, you can let the surface fall suspended by the hinge, as if letting down a drop-leaf table. Here, the hydroxy group would still be at 3 o'clock, but as a wedge, and the methyl group would be a dash at 7 o'clock.

You can apply these techniques to the comparison of a chair representation to a two-dimensional dash-wedge cyclohexane drawing, for example, structures A and B

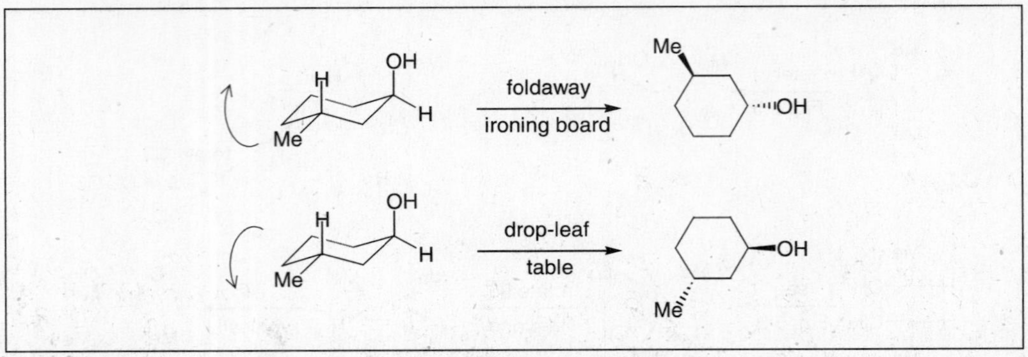

FIGURE 5-8: Converting chairs into dash-wedge representations.

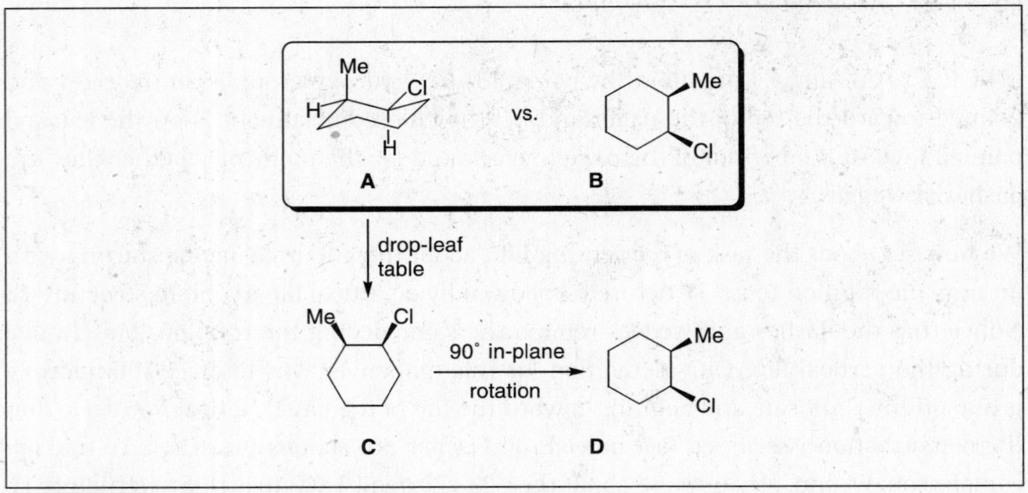

FIGURE 5-9: Comparing chair cyclohexane derivatives.

in Figure 5-9. Use B as a reference structure and manipulate A to adopt the required orientation. Here, the elements of commonality are quite straightforward: methyl is at 2 o'clock and chloro is at 4 o'clock. First, use the drop-leaf table maneuver on A to obtain the dash-wedge structure C, which has methyl at 11 o'clock and chloro at 1 o'clock. Then it is easy to see that a simple in-plane rotation of the molecule puts the substituents in the proper attitude for comparison. Inspection reveals that A and B are, indeed, the same thing.

Again, your goal should be the ability to carry out these manipulations in your mind's eye. As you practice this skill, sketching out intermediate steps with pencil and paper can be helpful, as shown in the figures in this chapter. Also, molecular models are great support devices to help you see the three-dimensional reality of two-dimensional drawings. Once you develop this facility, comparing different representations becomes extremely rapid and reliable. Moreover, you hone the valuable (and increasingly rare) skill of three-dimensional visualization.

CRAM SESSION

Visual Meaning

Molecules can differ from each other on many levels. If two molecules have the same molecular formula but are not identical, they are said to be **isomers** of each other. Following are the main categories of isomerism:

- **Constitutional isomers** differ in the way they are put together. If the isomers differ only in the placement of substituents, they are called **regioisomers.**
- **Stereoisomers** have the same connectivity but differ in the spatial positioning of substituents. Like constitutional isomers, these cannot be interconverted.
- **Conformational isomers** arise from flexibility in molecules that allow them to adopt a variety of shapes. These isomers are usually easily interconvertible.

Chemists use many different ways to depict molecules, and each method of depiction is used for a particular purpose:

- **Line notation** quickly reveals constitution and connectivity.
- **Sawhorse** (or dash-wedge) notation shows stereochemical relationships.
- **Newman projections** are used to examine conformations of freely rotating systems.
- **Chair** drawings are helpful for depicting conformational elements of cyclohexanes.
- **Fischer projections** are used to quickly convey absolute stereochemistry.

It is extremely important to develop the cognitive skill to imagine the three-dimensional structure of a two-dimensional depiction, especially when trying to compare two isomers. Keep in mind that **enantiomers** have the opposite configuration at each and every stereocenter, whereas **diastereomers** differ in the configuration of at least one, but not all, stereocenters. A useful technique for making these comparisons involves first converting the two structures to the same type of depiction, and then manipulating them so that they are oriented in the same fashion. It is strongly discouraged to attempt such a comparison by assigning *R* and *S* configurations.

Conceptual Map of Section I

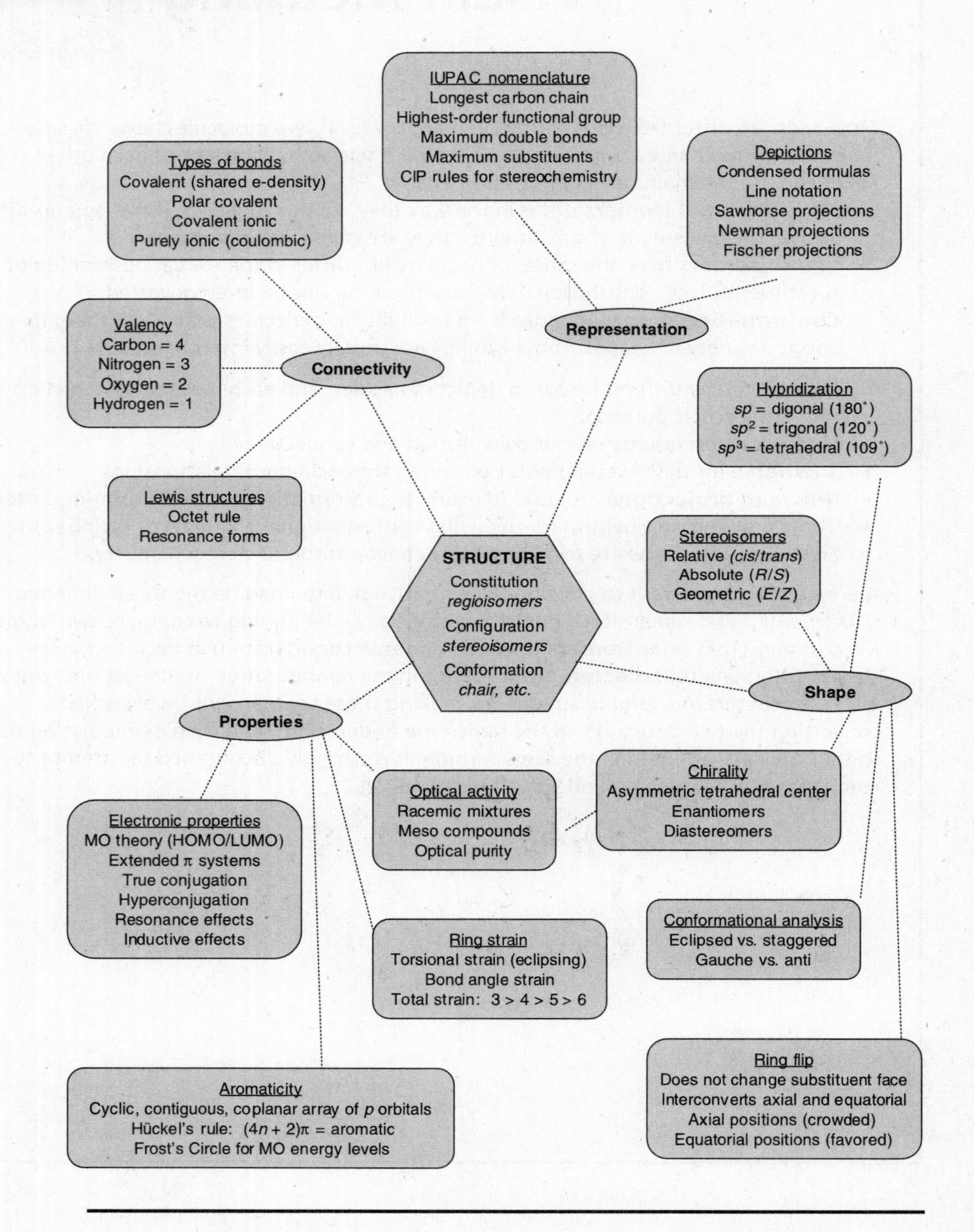

SECTION II

REACTIVITY

CHAPTER 6

Energy Changes in Molecules

Read This Chapter to Learn About:

- ➤ Equilibrium and Free Energy
- ➤ The Kinetic Basis of Equilibrium
- ➤ Equilibrium versus Irreversible Reaction
- ➤ The Nature of the Transition State

EQUILIBRIUM AND FREE ENERGY

Chemistry is driven by energetics; consequently, it is necessary to understand the basic principles of energy changes in chemical systems to comprehend and predict chemical reactions. As a beginning framework, consider the prototypical chemical reaction shown in Figure 6-1, in which a system is transformed from state A to state B. In principle, we should consider all such chemical systems to be in equilibrium, as described in the chemical equation

$$A \underset{}{\overset{K_{eq}}{\rightleftharpoons}} B \qquad (1)^*$$

whereby the ratio of A to B is governed by the equilibrium constant (K_{eq}), which has the form

$$K_{eq} = [B]/[A] \qquad (2)$$

*Equations and reactions in this part of the book are numbered for easy reference. Note that equations and reactions referred to again in later discussion are referenced by the same number.

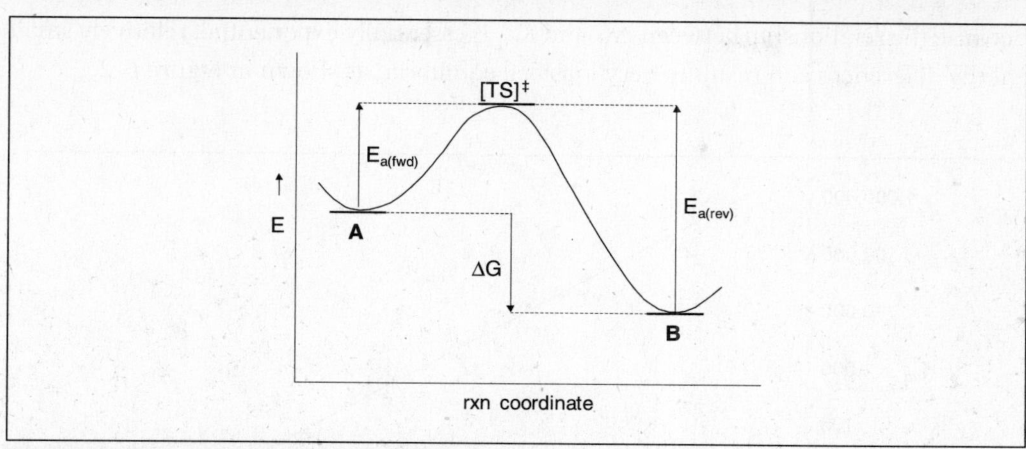

FIGURE 6-1: Energy diagram for chemical reaction.

The equilibrium constant is also related to the change in free energy (ΔG) in going from state A to state B (Figure 6-1), as given in the equation (solved for ΔG and then solved for K_{eq})

$$\Delta G = -RT \ln K_{eq} \tag{3}$$

$$K_{eq} = e^{-\Delta G/RT} \tag{4}$$

Thus, if you know the change in free energy, you can predict the relative concentrations of species at equilibrium, and vice versa.

We say the process shown here is **spontaneous**, that is, it tends to go forward.

Despite popular use, spontaneity has nothing to do with the **speed** of a reaction, but rather the extent to which it occurs. It is easiest to think about this in terms of equilibrium processes: A negative ΔG leads to an equilibrium constant that is greater than unity; therefore, the equilibrium tends to lie to the right, and the process is spontaneous. The opposite is true for a nonspontaneous process. However, if a process is nonspontaneous, it does not mean that **no** product is formed (although that can be the case)—it simply means that the product (or right side of the equation) does not **predominate**. Table 6-1 correlates these concepts for a general equilibrium process.

TABLE 6-1 Thermodynamic Trends for Equilibrium Processes

ΔG	K_{eq}	Spontaneity	Equilibrium Lies to the
Negative	>1	Spontaneous	Right
0	1	N/A	Dead center
Positive	<1	Nonspontaneous	Left

Because the relationship between ΔG and K_{eq} is essentially exponential, relatively small energy differences can result in very lopsided equilibria, as shown in Figure 6-2.

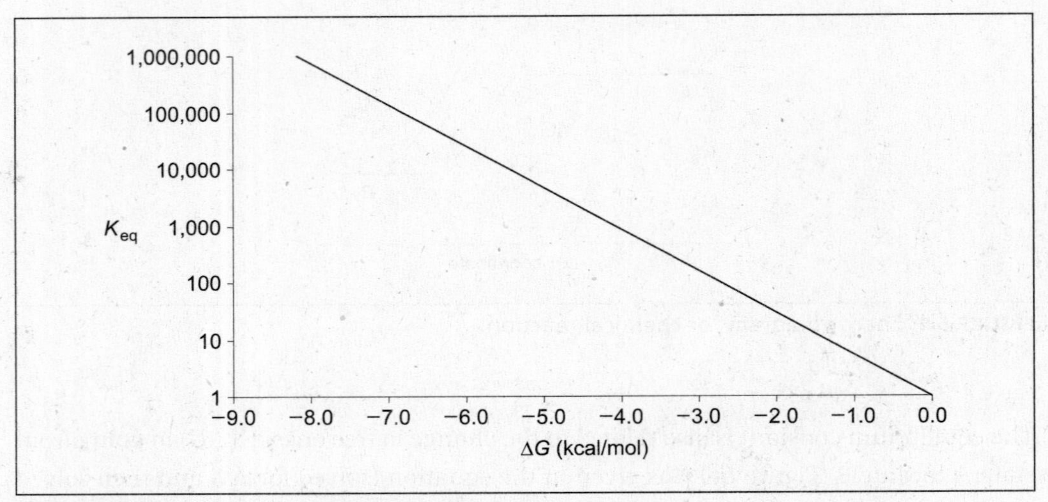

FIGURE 6-2: Dependence of equilibrium constant on ΔG.

An excellent example of this trend can be seen in the conformational equilibria of substituted cyclohexanes. For example, fluorocyclohexane exists as two chair conformers in equilibrium (Figure 6-3, top). The chair form with equatorial fluorine is 0.25 kcal/mol more stable than the corresponding chair with an axial fluorine; the equilibrium lies to the right, although there is still about 40% of the axial fluoro at room temperature. However, when the substituent is methyl rather than fluorine, the steric demand of the substituent is greater, and the preference for an equatorial attitude is more pronounced. Thus, the magnitude of ΔG is 1.7 kcal/mol, which leads to a more spontaneous process with an equilibrium constant of 18; in this case, only about 5% of the axial methyl is present at room temperature.

F
H
F
H
$\Delta G = -0.25$ kcal/mol
$K_{eq} = 1.5$
1.0 : 1.5

Me
H
Me
H
$\Delta G = -1.70$ kcal/mol
$K_{eq} = 18$
1.0 : 18

FIGURE 6-3: Equilibria of substituted cyclohexanes.

In short, free energy (G) is a state function that predicts the extent of spontaneity for a chemical process. However, free energy is composed of two other thermodynamic functions, according to the equation

$$\Delta G = \Delta H - T\Delta S \tag{5}$$

where enthalpy (H) is driven largely by bond formation and cleavage, nonbonded (steric) interactions, and strain; whereas entropy (S) is impacted most significantly by a change in the number of molecules in the system. Note that the latter term includes a temperature factor, so it can be leveraged by experimental conditions.

THE KINETIC BASIS OF EQUILIBRIUM

There is another way to think about equilibrium in a more dynamic way, namely, as the competition between two simultaneous reactions; thus the buildup of product is given by

$$\mathrm{A} \xrightarrow{k_{\mathrm{fwd}}} \mathrm{B} \tag{6}$$

whereas at the same time product is being depleted by the process

$$\mathrm{B} \xrightarrow{k_{\mathrm{rev}}} \mathrm{A} \tag{7}$$

where k_{fwd} and k_{rev} refer to the rate constants for the reactions A → B and B → A, respectively. The rate constant for a reaction is related to its activation energy by the Arrhenius equation

$$k = Ae^{-E_a/RT} \tag{8}$$

where R is the universal gas constant (1.986 cal/K·mol) and A is sometimes called the **frequency factor**, because it relates to the fraction of sufficiently energetic collisions that have the proper orientation for productive reaction—this constant is peculiar to a given system.

Thus, you can express the rate constants for the forward and reverse reactions by the following set of equations:

$$k_{\mathrm{fwd}} = Ae^{-E_{a(\mathrm{fwd})}/RT} \tag{9}$$

$$k_{\mathrm{rev}} = Ae^{-E_{a(\mathrm{rev})}/RT} \tag{10}$$

where $E_{a(\mathrm{fwd})}$ and $E_{a(\mathrm{rev})}$ correspond to the activation energies for the forward and reverse reactions, respectively, as identified in Figure 6-1. If we express the two competing rate constants as a ratio, we get

$$\frac{k_{\mathrm{fwd}}}{k_{\mathrm{rev}}} = \frac{Ae^{-E_{a(\mathrm{fwd})}/RT}}{Ae^{-E_{a(\mathrm{rev})}/RT}} \tag{11}$$

assuming that the frequency factors cancel, then the equation reduces to

$$\frac{k_{fwd}}{k_{rev}} = e^{(-E_{a(fwd)}/RT)-(-E_{a(rev)}/RT)} = e^{(E_{a(rev)}-E_{a(fwd)})/RT} \tag{12}$$

Inspection of Figure 6-1 reveals the following graphical relationship:

$$E_{a(rev)} - E_{a(fwd)} = -\Delta G \tag{13}$$

Substituting this value into Equation 12 gives

$$\frac{k_{fwd}}{k_{rev}} = e^{-\Delta G/RT} \tag{14}$$

Finally, comparing Equations 4 and 14 allows us to establish the following fundamental relationship between kinetics and thermodynamics:

$$K_{eq} = k_{fwd}/k_{rev} \tag{15}$$

Let us explore the implication of Equation 15 even further. If we assume that both the forward and reverse reactions are unimolecular, then we can express the rate of each reaction as follows:

$$\text{rate}_{fwd} = k_{fwd}[A] \tag{16}$$

$$\text{rate}_{rev} = k_{rev}[B] \tag{17}$$

or, solving for the rate constants,

$$k_{fwd} = \frac{\text{rate}_{fwd}}{[A]} \tag{18}$$

$$k_{rev} = \frac{\text{rate}_{rev}}{[B]} \tag{19}$$

Substituting Equations 18 and 19 into Equation 15, we get

$$K_{eq} = \frac{k_{fwd}}{k_{rev}} = \frac{\left(\frac{\text{rate}_{fwd}}{[A]}\right)}{\left(\frac{\text{rate}_{rev}}{[B]}\right)} \tag{20}$$

Because at equilibrium the rate of the forward reaction (rate_{fwd}) equals that of the reverse reaction (rate_{rev}), the two nominator terms cancel, and the equation reduces to

$$K_{eq} = \frac{[B]}{[A]} \tag{21}$$

which is, of course, an expression for the Law of Mass Action.

EQUILIBRIUM VERSUS IRREVERSIBLE REACTION

With this background, let us turn our attention to the question of why we sometimes write some chemical processes as equilibria (two opposing arrows) and other processes simply as "reactions" (i.e., A → B). In part, this depends on what we are intending to emphasize in the representation. For example, in a mechanism, if it is important to understand that two species are in equilibrium, then we present it as such; however, if it is necessary only to know that species A leads to species B, which in turn leads to species C, then single arrows are generally used.

However, sometimes there is a more subtle issue at play. The Principle of Microscopic Reversibility states that if a pathway exists to convert a reactant to a product, then that same pathway exists for the product to be converted back to the reactant. Thus, from a first approximation we should assume that all reactions are equilibrium processes. Yet there are some cases that for all practical purposes can be called **irreversible reactions.** For example, Figure 6-4 shows a reaction profile for what we might consider an irreversible process, where the activation energy for the reverse process (E_{ar}) is much greater than that for the forward process (E_{af}). If, at room temperature, there is a large fraction of molecules with energy greater than E_{af}, but a vanishingly small fraction of molecules with energy greater than E_{ar}, then it is reasonable to say the process is irreversible.

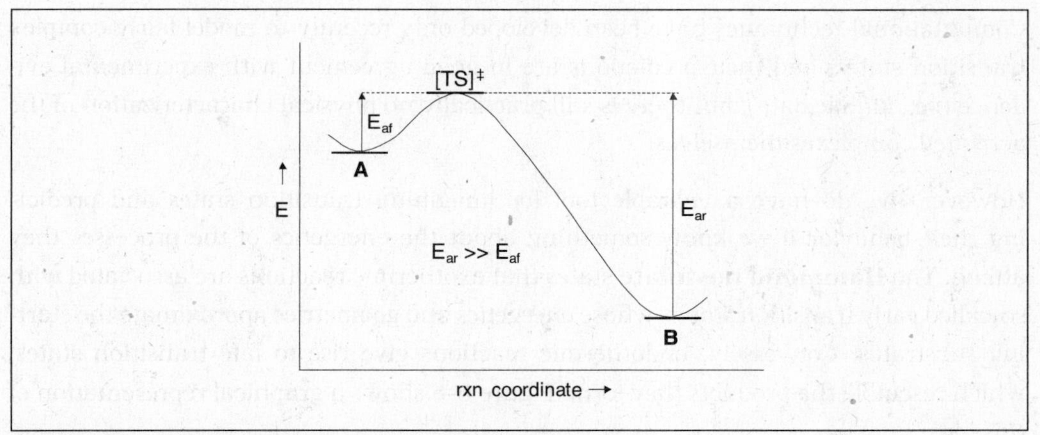

FIGURE 6-4: Reaction profile of an "irreversible" reaction.

Indeed, relatively small changes in activation energies can result in dramatic impacts on the reaction rates. As Figure 6-5 demonstrates (note the logarithmic y scale), raising an activation barrier from 0.5 kcal/mol to 2.5 kcal/mol results in a 30-fold slower reaction—in other words, a reaction that took an hour to complete would now take more than a day. If the barrier were raised to 10 kcal/mol, that same reaction would take more than 1000 years. This is exactly the basis of catalysis—accelerating reactions by lowering their activation energies.

THE NATURE OF THE TRANSITION STATE

So far we have not addressed the transition state complex (TS), which arguably defines the activation energy. In fact, transition states are notoriously difficult to

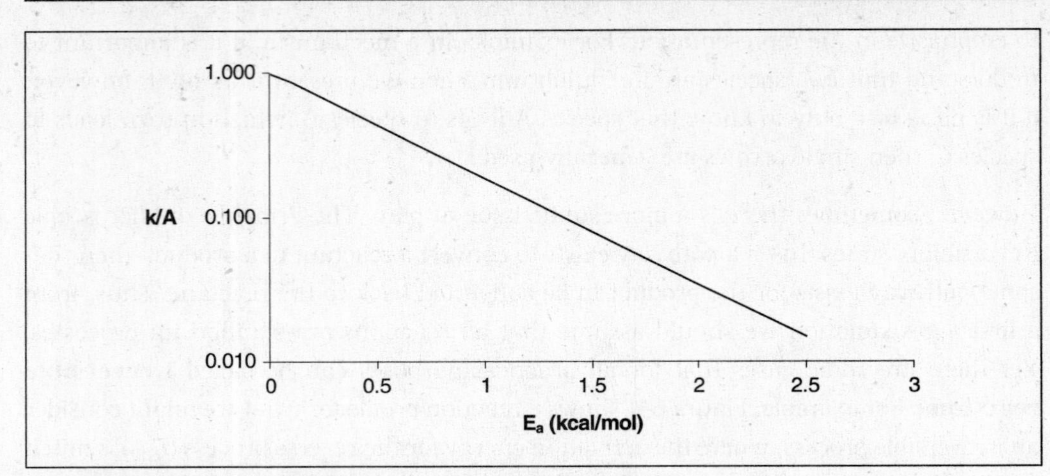

FIGURE 6-5: Impact of activation energy on reaction rate.

study—because they sit atop the crest of a hill, their lifetimes are fleeting to nonexistent; thus, traditional spectroscopic methods fall short of being able to observe them. Computational techniques have been developed only recently to model fairly complex transition states, and their predictions are in good agreement with experimental evidence (i.e., kinetic data), but there is still practically no physical characterization of the activated complexes themselves.

However, we do have a valuable tool for imagining transition states and predicting their behavior if we know something about the energetics of the processes they attend. The **Hammond postulate** states that exothermic reactions are associated with so-called early transition states, whose energetics and geometries approximate the starting substrates. Conversely, endothermic reactions give rise to late transition states, which resemble the products they form. Figure 6-6 shows a graphical representation of this idea.

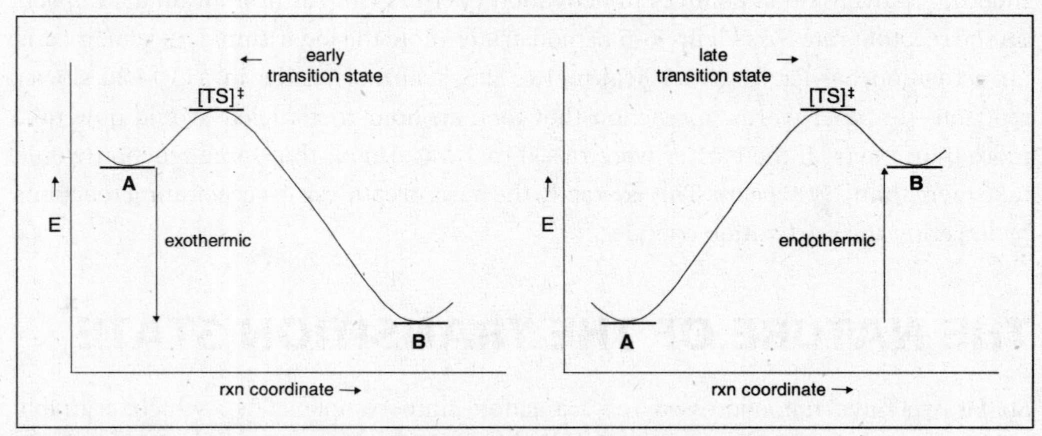

FIGURE 6-6: Two transition states predicted by the Hammond postulate.

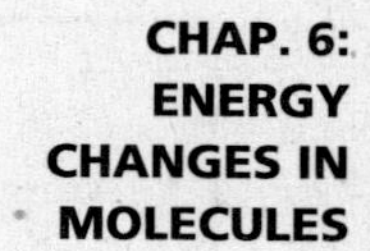

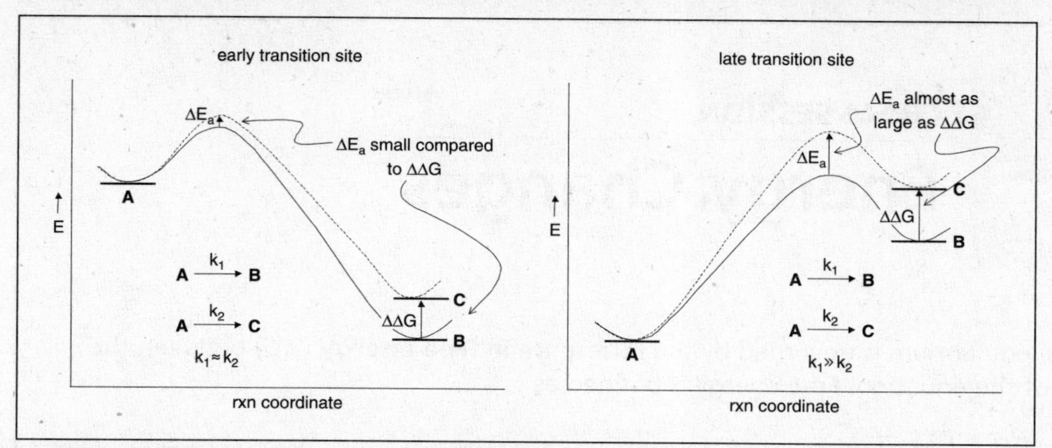

FIGURE 6-7: Effect of differing product stabilities on early and late transition states.

This rather straightforward assumption can be used to rationalize and predict the kinetic behavior of competing processes. For example, Figure 6-7 shows a hypothetical energy profile for a substrate that can provide two products (B or C) of differing thermodynamic stabilities. Because the ΔG values of the two reactions are different, we use the term $\Delta\Delta G$ to discuss the different product stabilities. Likewise, the activation energies leading to the products are also different (ΔE_a). If the reaction is exothermic (left reaction profile), then the transition state in both cases resembles the starting material, so the ΔE_a is small compared to the $\Delta\Delta G$. However, in a late transition state scenario, the energy differences of the products are fairly efficiently translated back to the transition states, so the ΔE_a approximates the $\Delta\Delta G$. Therefore, an increase in product stability affects the rate of an exothermic reaction only slightly, whereas the rate of an endothermic process can be accelerated significantly. This idea is the conceptual cornerstone for rationalizing many product distribution outcomes.

CRAM SESSION

Energy Changes

A chemical equilibrium is governed by the difference in **free energy** (ΔG) between the two sides of the equation. Free energy is defined as

$$\Delta G = \Delta H - T\Delta S$$

where ΔH is the change in **enthalpy** (or heat) and ΔS is the change in **entropy** (or disorder) in the system. It is ΔG that determines the **spontaneity** of a reaction (that is, whether or not it will go forward). Therefore, equilibria with negative ΔG values lie to the right, while those with positive ΔG values lie to the left. The greater the difference in ΔG, the more lopsided the equilibrium. In other words, ΔG impacts the **equilibrium constant** (K_{eq}) according to the equation

$$\Delta G = -RT \ln K_{eq}$$

where R is the universal gas constant.

In a chemical equilibrium, the **forward reaction** converts species from the left side to species on the right side, and the **reverse reaction** does the opposite. The rate of each reaction depends upon the **activation energy** (E_a) which must be overcome in going from one side to the other. This barrier is related to the rate constant (k) by the **Arrhenius equation**:

$$k = Ae^{-E_a/RT}$$

where A is the **frequency factor** unique to a given reaction.

At the energy crest in the path from left to right lies the **transition state**, which has an infinitesimally short lifetime. According to the **Hammond postulate**, endothermic processes have a **late transition state**, which is product-like in geometry, whereas exothermic processes proceed through an **early transition state**, which resembles reactant.

An **irreversible reaction** occurs when the ΔG is an extremely large negative number. Under these circumstances, the reverse reaction is slow enough to be negligible compared to the forward reaction.

CHAPTER 7

Radical Chemistry

Read This Chapter to Learn About:

- Basic Radical Processes
- Propagation Sequences

BASIC RADICAL PROCESSES

We begin here a treatment of the fundamental reaction steps in organic chemistry, which are remarkably few. Just as almost all the compositions of Western music are constructed from the twelve notes of the chromatic scale, so too are the myriad mechanisms in organic synthesis composed of a dozen or so fundamental steps. In understanding these principle components, we thus amass a repertoire of building blocks for assembling more involved reaction mechanisms.

Generally speaking, mechanistic steps fall into three broad categories: (1) radical reactions, which involve species with unpaired electrons and the motion of single electrons; (2) polar reactions, which engage the activity of electron pairs and usually involve coulombic charges in the mechanistic pathway; and (3) pericyclic reactions, in which four or more electrons move in concert without the generation of charge. Each category contains an interesting array of useful mechanistic steps.

Within the radical paradigm, there are three electronic maneuvers that capture the majority of chemistry encountered in radical processes (Figure 7-1). In the process of **atom abstraction**, a radical center ($R\cdot$) induces the homolytic cleavage of an existing σ bond, forming a new bond in the process and leaving behind a new radical center. Radicals can also engage in **addition to a π bond**, which exhibits electronic motion identical to the first example, except that a single, more complex molecule results from the transaction. In fact, this is the process responsible for many

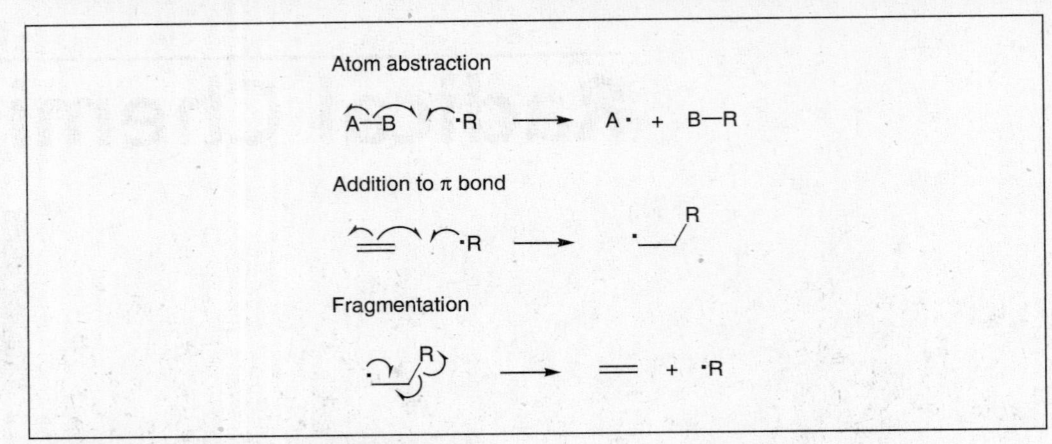

FIGURE 7-1: Three basic radical processes.

polymerization reactions. Finally, a radical species can undergo **fragmentation**, which is essentially the reverse of addition to a π bond.

Note that in each case, the number of radical centers is constant in going from left to right. That is, the radical is neither created nor destroyed, but rather transformed into a new species. But what is the origin of the first radical center? One of the most common radical-producing processes is the homolytic cleavage of σ bonds, in which the two electrons of the shared pair go in separate directions to produce two new radicals (Figure 7-2, top). As part of a mechanism, this would fall under the heading of an initiation step. By the same token, the combination of any two radicals to form a new molecule would constitute a termination sequence.

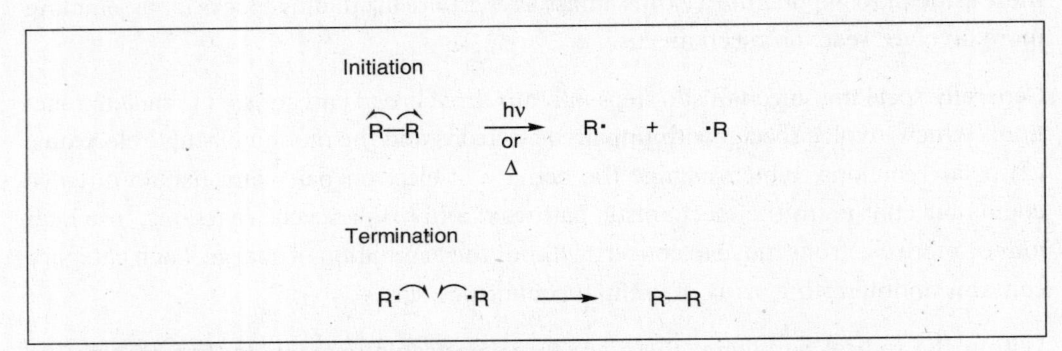

FIGURE 7-2: Radical initiation and termination.

PROPAGATION SEQUENCES

Between initiation and termination, there is propagation. Because homolytic cleavage requires quite a bit of energy, if we relied on the stoichiometric generation of radicals through this process, then the rate would be so slow as to be synthetically useless. Instead, once a single radical is formed, it can engage in a series of self-sustaining

$$CH_4 + Br_2 \longrightarrow CH_3Br + HBr$$

Initiation

$$Br\text{—}Br \xrightarrow[\text{or } \Delta]{h\nu} 2\ Br\cdot$$

Propagation

Step 1: $H_3C\text{—}H \quad \cdot Br \longrightarrow H_3C\cdot + H\text{—}Br$

Step 2: $H_3C\cdot \quad Br\text{—}Br \longrightarrow H_3C\text{—}Br + \cdot Br$

FIGURE 7-3: Radical bromination of methane.

propagation steps to generate new compounds. This is nicely illustrated with the radical bromination of methane (Figure 7-3). The initiation step involves the homolytic cleavage of molecular bromine into two bromine radicals. In principle, this step (at a costly 46 kcal/mol) must occur only once. Afterward, the bromine radical attacks methane, engaging in hydrogen atom abstraction to form hydrobromic acid and methyl radical (step 1). The methyl radical, in turn, attacks molecular bromine (present in stoichiometric quantity), which suffers bromine atom abstraction and produces methyl bromide and regenerates bromine radical (step 2). Notice that each of the propagation steps consumes stoichiometric reactant and generates stoichiometric product.

Moreover, the combination of propagation steps comprises a self-sustaining catalytic system. That is, the methyl radical of step 1 feeds step 2, and the bromine radical of step 2 feeds step 1. Another way of representing the reaction in a way that focuses on the catalytic cycle is shown in Figure 7-4. The two radical species anchor the radical chain,

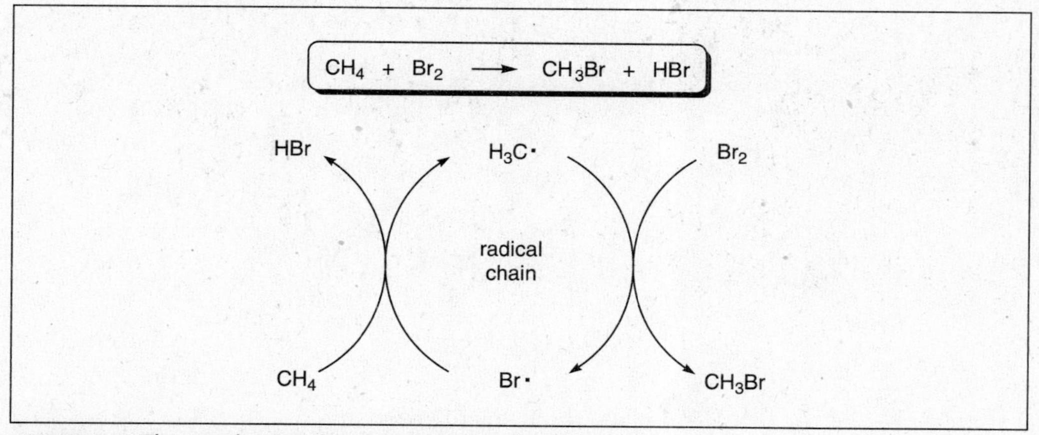

FIGURE 7-4: The catalytic cycle of radical bromination.

TABLE 7-1 Energetics of the Bromination of Methane

Propagation Step	Bonds Broken		Bonds Formed		Net ΔH (kcal/mol)
	Bond	ΔH (kcal/mol)	Bond	ΔH (kcal/mol)	
Step 1	C—H	+105	H—Br	−87	+18
Step 2	Br—Br	+46	C—Br	−71	−25
Overall process					−7

and the process continues indefinitely as long as there is reactant left. Such a system is a bonus for synthetic protocols, but sometimes a liability in the environment—for example, in the catalytic depletion of ozone.

You can also analyze the energetics of the propagation sequence simply by examining the bonds formed and broken in each step. Because bond enthalpies are representative of homolytic bond cleavage, we can simply consult values tabulated in textbooks or reference materials. Thus, the first step in bromination sacrifices a carbon–hydrogen bond (+105 kcal/mol) and forms a hydrogen–bromine bond (−87 kcal/mol); because the bond formed is weaker than the bond broken, the first step is endothermic (Table 7-1). However, the subsequent step more than compensates for the initial investment, so that the overall stoichiometric process is slightly exothermic.

CRAM SESSION

Radical Chemistry

Radicals are species with unpaired electrons, and they are generally very reactive. They are most commonly formed by the thermal or light-induced homolytic cleavage of a weak σ bond, a process sometimes termed **initiation**. Once formed, these radicals engage in three common modes of reaction:

- Atom abstraction
- Addition to a π bond
- Fragmentation

In all of the above cases, a new radical center is formed by the reaction; therefore, these are often called **propagation** steps.

In many synthetic processes (such as the chlorination of methane), these reactions form a **catalytic cycle**. Each propagation step for a catalytic cycle should usually either (a) consume a stoichiometric reactant or (b) generate a stoichiometric product. The **reaction enthalpy** (ΔH) of each step can be easily calculated by summing the bonds formed and subtracting this value from the sum of the bonds broken (in other words, enthalpy released vs. enthalpy invested).

Two radicals can also combine to form a bond, thus ending the radical chain. This is a process known as **termination**.

CHAPTER 8

Polar Chemistry

Read This Chapter to Learn About:

- ➤ Proton Transfer
- ➤ Trends in Acidity and Basicity
- ➤ Ionization
- ➤ Nucleophilic Capture
- ➤ Nucleophilic Displacement (S_N2)
- ➤ The S_N1/S_N2 Continuum
- ➤ E1 Elimination
- ➤ E2 Elimination
- ➤ Global Trends in sp^3 Reactivity
- ➤ β-Elimination
- ➤ Nucleophilic Attack at an sp^2 Carbon
- ➤ Electrophilic Capture of a Double Bond

PROTON TRANSFER

Perhaps the most fundamental (but also sorely misapprehended) polar reaction step is the proton transfer. Often obscured by the rubric of "acid–base" chemistry, confusion about the proton transfer begins with neglect of the physical reality behind the process. A proton is simply that: a single atomic particle adorned with only its positive charge, like a lacrosse ball being shuttled back and forth between so many tiny electronic nets. A proton transfer is like a hydrogen bond gone to the extreme, and transferable protons are typically those that would be part of hydrogen bond donors (Chapter 1); in fact, the transition state for a proton transfer is best described as a hydrogen bond. Having said this, let us come to terms with the terminology of acids and bases.

TABLE 8-1 Three Acid–Base Paradigms

Paradigm	Acid		Base	
	Definition	Example of Species Embraced	Definition	Example of Species Embraced
Arrhenius	Proton donor	HCl	Hydroxide donor	NaOH
Brønsted	Proton donor	HCl	Proton acceptor	NaOH, **NH_3**
Lewis	Electron density acceptor	HCl, **Fe^{3+}**, **BF_3**	Electron density donor	NaOH, NH_3

There are three general paradigms of acid–base chemistry, as shown in Table 8-1. Arrhenius developed one of the earliest theories, which led to a broad understanding of acids as proton donors (e.g., HCl) and bases as hydroxide donors (e.g., NaOH). However, in time there were species recognized to exhibit basic behavior that harbored no apparent hydroxide moiety (ammonia, for example). The Brønsted theory reconciled this difference by redefining bases as proton acceptors: Ammonia accepts a proton from water and, in so doing, generates hydroxide. Thus, the Brønsted theory did not really affect the roster of recognized acids, but it expanded our understanding of bases considerably.

Lewis did for acids what Brønsted did for bases. Brønsted's focus was on the proton and how it is liberated and sequestered by chemical entities. Lewis looked at this situation from the other angle—that is, in terms of what is accepting the proton. If a proton transfer is indeed the extreme form of hydrogen bonding, and if a hydrogen bond donor (Chapter 1) is potentially anything with a lone pair, then the act of accepting a proton could also be viewed as a donation of electron density. Thus, Lewis defined bases as electron density donors and acids as electron density acceptors. This definition really did not expand the realm of bases—in other words, ammonia was still a base whether it was seen as accepting the proton or donating a lone pair for the proton's capture. However, it revolutionized our view of acids. Arrhenius acids were still Brønsted acids, and Brønsted acids were still Lewis acids, but their acidity was based on the activity of a proton—and although the proton is undoubtedly an acceptor of electron density, it is only one of a wide array of electron density acceptors. For example, the iron(III) cation is known to lower the pH of aqueous solutions, but it clearly has no source of protons. However, under the Lewis acid definition, we can see how it functions as an acid in terms of accepting electron density from the six water molecules that form an octahedral arrangement in solution. Ultimately, this complex also provides Brønsted acidity by liberation of a proton (Figure 8-1).

However, the Lewis definition also extends both acidic and basic behavior into nonaqueous environments, which has obvious significance for organic chemistry. For example, boron trifluoride is a stereotypical Lewis acid, because it has only six electrons about boron and therefore is eager to accept an extra electron pair. So Lewis acidic is BF_3

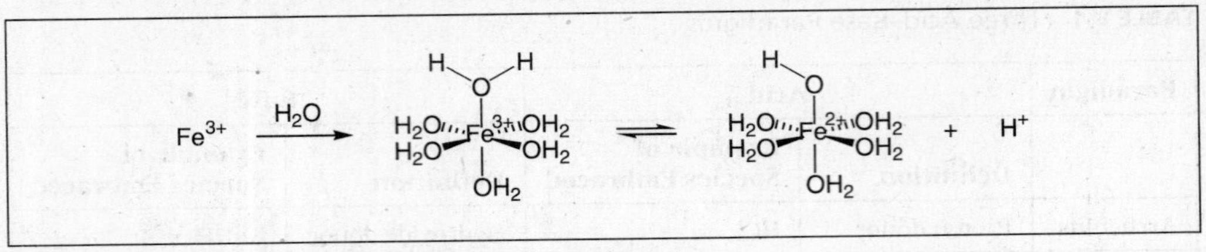

FIGURE 8-1: The aqueous acidity of iron(III) cation.

that when dissolved in diethyl ether, it forms a very strong complex with the solvent (Figure 8-2) in which the ether oxygen bears a formal positive charge and the boron bears a formal negative charge—a species known as a Lewis acid–Lewis base complex.

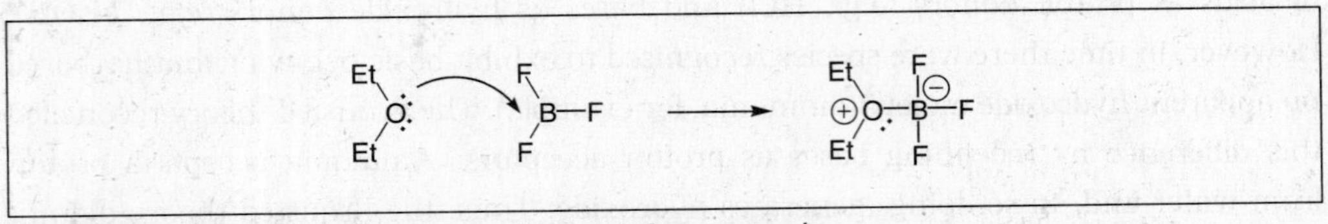

FIGURE 8-2: Formation of a Lewis acid–Lewis base complex.

We use the Lewis acid–Lewis base concept as a workhorse for understanding much of the mechanistic underpinnings of polar chemistry. However, proton transfer can be treated using the Brønsted notion of acids and bases. Here we must pause and remind ourselves of some basic principles related to Brønsted acidity. Recall from general chemistry that water provides very low background levels of hydroxide and protons through an equilibrium process known as **autoionization** (Figure 8-3).

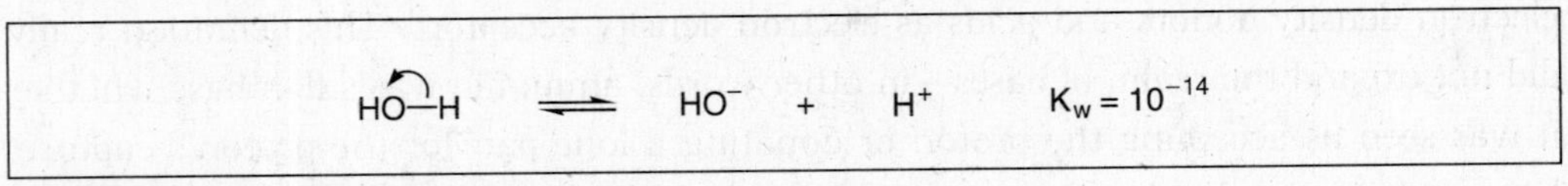

$$HO{-}H \rightleftharpoons HO^- + H^+ \qquad K_w = 10^{-14}$$

FIGURE 8-3: The autoionization of water.

This process is governed by an equilibrium constant (given the special designation of K_W) on the order of 10^{-14}. If the equilibrium constant is written in terms of chemical species, we have

$$K_w = \frac{[H^+][HO^-]}{[H_2O]} = [H^+][HO^-] \qquad (22)$$

where the denominator term drops out of the expression because the concentration of water does not change in the system. If we assume that the only source of protons (or more accurately, hydronium) and hydroxide in pure water is the autoionization process, then the two quantities must be equal; therefore,

$$[H^+] = [HO^-] = \sqrt{10^{-14}} = 10^{-7} \qquad (23)$$

To avoid dealing with very small numbers, the convention of pH has been established to express these concentration values, such that

$$pH = -\log_{10}[H^+] \tag{24}$$

Therefore, if the proton concentration is 10^{-7}, then the pH is 7. An analogous scale, pOH, is used for the hydroxide concentration. In pure water, then, the pH and pOH are both 7. When acids or bases are added to the water, the proton or hydroxide levels are increased, respectively. However, in aqueous systems, Equation 22 must still be satisfied. Thus, as hydroxide levels increase, proton levels must decrease, and vice versa. We can express this constraint in terms of pH and pOH as follows:

$$pH + pOH = 14 \tag{25}$$

Similar conventions have been constructed around the expression of acid and base strength. For example, we can represent the experimentally derived equilibrium constant for the ionization of acetic acid as shown in Figure 8-4, where the equilibrium constant bears the special designation of **acidity constant** (K_a).

$$\text{AcO–H} \rightleftharpoons \text{AcO}^- + \text{H}^+ \qquad K_a = 2.0 \times 10^{-5}$$

FIGURE 8-4: The ionization of acetic acid.

You can also express the acidity constant in terms of chemical species, which has the generic form

$$K_a = \frac{[H^+][A^-]}{[HA]} = \frac{[H^+][AcO^-]}{[HOAc]} \tag{26}$$

In addition, an analogous scale of pK_a can be used to express the numeric values of the acidity constant according to the relationship

$$pK_a = -\log_{10}(K_a) \tag{27}$$

Therefore, acetic acid is associated with a pK_a value of 4.7.

There are a few general trends that can be gleaned from the inspection of pK_a values. First, if you understand a strong acid to be one that ionizes to a large extent (i.e., the ionization is a right-handed equilibrium), then you would expect the acidity constant to be greater than unity and, consequently, the pK_a to be less than zero. By the same token, a weak acid does not completely ionize, so the equilibrium would lie to the left, the acidity constant would be less than unity, and the pK_a value would be greater than zero. However, the more important general trend for the organic chemist is: **The lower the pK_a, the stronger the acid.**

So far, so good—but organic chemists do something that is considered incorrect by the rest of the chemical community, namely, we use pK_a values to characterize bases as well. That we might come to terms with this useful (if unorthodox) convention, let us first examine the formally correct method of assessing base strength. Toward this end, consider the activity of ammonia in water (Figure 8-5).

$$H_3N: \quad H{-}OH \rightleftharpoons H_3\overset{\oplus}{N}{-}H + \overset{\ominus}{O}H \qquad K_b = 1.8 \times 10^{-5}$$

FIGURE 8-5: The activity of ammonia in water.

Because this equilibrium generates hydroxide, the equilibrium constant bears the special designation of the **basicity constant**. As we did earlier, we can also express the basicity constant in terms of chemical species, which has the generic form

$$K_b = \frac{[BH^+][HO^-]}{[B]} = \frac{[NH_4^+][HO^-]}{[NH_3]} \tag{28}$$

Similarly, the scale of pK_b can be used to express the numeric values of the basicity constant according to the relationship

$$pK_b = -\log_{10}(K_b) \tag{29}$$

Therefore, ammonia is associated with a pK_b value of 4.7, which is only coincidentally the same as the pK_a value for acetic acid.

However, keep in mind that every base has a conjugate acid—for ammonia, the conjugate acid is ammonium ion (NH_4^+). Therefore, in principle you could examine the behavior of ammonium ion in water as a way of understanding the basicity of ammonia. The equilibrium involved is represented in Figure 8-6.

$$H_3\overset{\oplus}{N}{-}H \quad :OH_2 \rightleftharpoons H_3N: + H{-}\overset{\oplus}{O}H_2$$

FIGURE 8-6: Behavior of ammonium ion in water.

This is an equilibrium that generates hydronium ion (or more simply, H^+) on the right side. Therefore, you can write an expression for the acidity constant in terms of the chemical species, as follows:

$$K_a = \frac{[H_3O^+][NH_3]}{[NH_4^+][H_2O]} = \frac{[H^+][NH_3]}{[NH_4^+]} \tag{30}$$

Converting to pK_a, we obtain

$$\begin{aligned} pK_a &= -\log\left(\frac{[H^+][NH_3]}{[NH_4^+]}\right) \\ &= -\log[H^+] + (-\log[NH_3]) - (-\log[NH_4^+]) \end{aligned} \tag{31}$$

If you use the convention that $pX = -\log[X]$, then we can rewrite Equation 31 as follows:

$$pK_a = pH + pNH_3 - pNH_4^+ \qquad (32)$$

Using the same symbology, the pK_b derived from Equation 28 becomes

$$pK_b = -\log\left(\frac{[NH_4^+][HO^-]}{[NH_3]}\right) \qquad (33)$$
$$= pOH + pNH_4^+ - pNH_3$$

Leveraging the relationship in Equation 25, we can express the pK_b in terms of pH as follows:

$$pK_b = (14 - pH) + pNH_4^+ - pNH_3 \qquad (34)$$

Recognizing the commonality with Equation 32, we can rearrange Equation 34 to obtain

$$14 - pK_b = pH - pNH_4^+ + pNH_3 \qquad (35)$$

Thus, we can show with Equations 32 and 35 that for a conjugate acid/conjugate base pair

$$pK_a = 14 - pK_b \qquad (36)$$

The outcome expressed in Equation 36 is extremely important, because it establishes the mathematic linkage between pK_a and pK_b. In essence, the information for one is bound up in the other. Therefore, organic chemists take the shortcut of dealing only with pK_a values, whether we are talking about acids or bases. In other words, ammonia is associated with a pK_b of 4.7, but also with a pK_a of 9.3. Not surprisingly, when we associate values to species in this backward way, the trend is also reversed—that is, the stronger the base, the higher the corresponding pK_a.

It is best to think of the pK_a as belonging not to a particular species, but to an equilibrium between a conjugate pair. For example, the ammonia/ammonium equilibrium can be expressed as shown in Figure 8-7. Thus, ammonium **acting as an acid** (i.e., giving up a proton) has a pK_a of 4.7, and ammonia **acting as a base** (i.e., accepting a proton) is associated with the same pK_a of 4.7. This way, we can use the same data to predict behavior for two different species. In other words, we would expect ammonium to behave as a weak acid and ammonia to behave as a weak base.

$$NH_4^{\oplus} \underset{}{\overset{pK_a = 4.7}{\rightleftharpoons}} NH_3 + H^{\oplus}$$

FIGURE 8-7: The ammonia/ammonium conjugate pair.

To keep you on your toes, however, ammonia is amphoteric—that is, it can also function as an acid, giving up a proton to form the amide anion (Figure 8-8). This equilibrium is associated with a much higher pK_a of 36. Using the acidity and basicity trends discussed previously, we would thus be forced to conclude that ammonia is an extremely weak acid and amide is an extremely strong base.

$$NH_3 \overset{pK_a = 36}{\rightleftharpoons} {}^{\ominus}NH_2 + H^{\oplus}$$

FIGURE 8-8: The ammonia/amide conjugate pair.

Water is also an amphoteric species—it can act as an acid and lose a proton to form hydroxide, a process with a pK_a of 15.7, or it can function as a base, accepting a proton to become hydronium, which is associated with a pK_a of −1.7. With these two data points, we can draw the following conclusions: (1) water is a very weak acid; (2) water is a very weak base; (3) hydronium is a strong acid; and (4) hydroxide is a strong base. Figure 8-9 tabulates benchmark pK_a ranges for some commonly encountered species. Memorizing these benchmarks will help you make sense of other pK_a values discussed in subsequent sections.

Acid (increasing strength ↓)		Base (increasing strength ↑)	pK_a	
NH_3		NH_2^-	38	amines acting as acids
$(i\text{-}Pr)_2NH$		$(i\text{-}Pr)_2N^-$	36	
MeOH		MeO^-	15.7	water/alcohols acting as acids
H_2O		HO^-	15.7	
RNH_3^+		RNH_2	10-11	amines acting as bases
NH_4^+		NH_3	9.2	
HCN		CN^-	9.2	weak acids/ weak bases
RCO_2H		RCO_2^-	4-5	
$RCH_2OH_2^+$		RCH_2OH	-2	water/alcohols acting as bases
H_3O^+		H_2O	-1.7	
HCl		C I-	-7	strong acids
HBr		B r-	-9	

FIGURE 8-9: Some pK_a benchmarks for acids and bases.

TRENDS IN ACIDITY AND BASICITY

Now that you are conversant with how the strengths of acids and bases are conveyed in organic chemistry, turn your attention to understanding the trends in acid and base strength. Two principles are useful in this discussion. First, the strength of an acid and its conjugate base are inversely proportional. In other words, the stronger the acid, the weaker its conjugate base—for example, hydrochloric acid is a very strong acid, whereas chloride is all but nonbasic. Second, it is easiest to understand acid strength in terms of the stability of the anion produced—that is, as the consequent anion becomes more stable, the more likely an acid is to donate a proton to generate that anion. Thus, if we can come to terms with trends in anion stability, by extension we can predict acidity of compounds.

One trend in anion stability can be related to electronegativity. For example, if you examine the simple anions in the latter region of the second row on the periodic table, a left-to-right periodic trend is revealed. Thus, from the methyl anion (pK_a 49) to the fluoride anion (pK_a 3.2) there is a steady increase in anion stability going from left to right, which corresponds to an increasing electronegativity of the central atom (C $\rightarrow$ N $\rightarrow$ O $\rightarrow$ F). If you consider that electronegativity relates to the tendency of an atom to hold onto electron density, then the tighter the electron density is held, the less likely it is to engage in reactivity (such as capture of a proton). Therefore, the more electronegative species tend to be weaker bases (Figure 8-10).

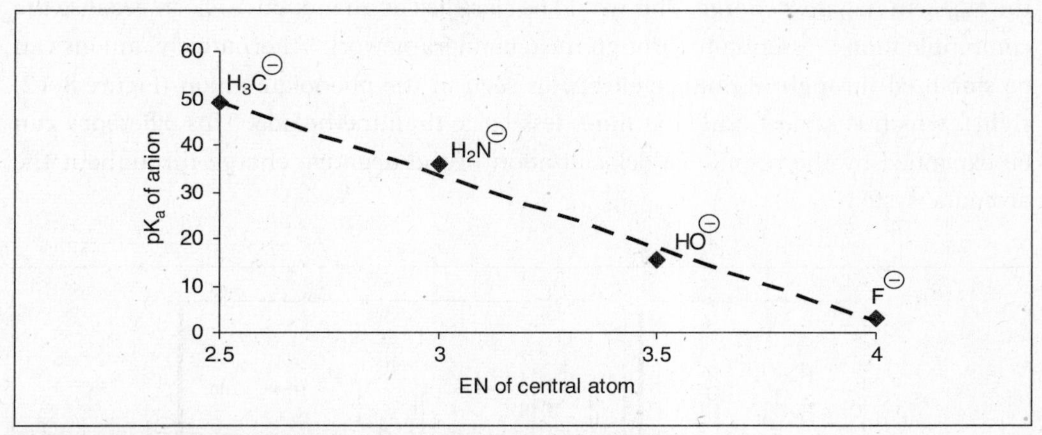

FIGURE 8-10: Anion stability as a function of electronegativity.

A similar trend is observed in a vertical cross section of the periodic table. For example, comparing the simple anions of Group 16, you observe steadily decreasing pK_a values as you go from top to bottom (Figure 8-11). This trend can be rationalized on the basis of atomic radius—that is, the larger the volume that the negative charge must occupy, the more stable the species. Another way to understand this is that larger molecules can minimize electron–electron repulsions more effectively. In summary, anion stability (and thus conjugate acid strength) increases from left to right and from top to bottom on the periodic table. Making diagonal comparisons becomes more difficult and nonintuitive. For example, silane (SiH_4) is just as acidic as ammonia (NH_3),

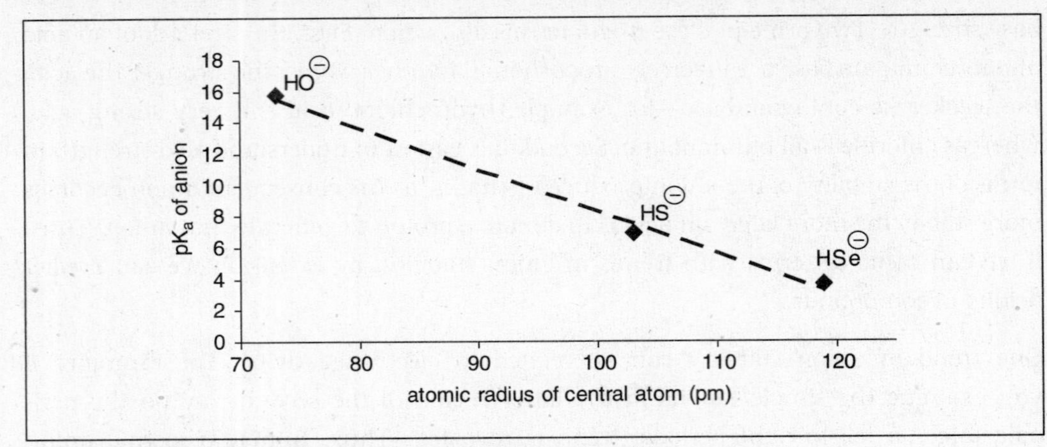

FIGURE 8-11: Anion stability as a function of atomic radius.

but arsine (AsH_3)—all the way down in the fourth period—is still 7 pK_a units (i.e., 10 million times) less acidic than water (H_2O).

In more complex molecules, acidity and basicity can be modulated through inductive and resonance effects as well. For example, compared to acetate's pK_a of 4.7, trifluoroacetate (Figure 8-12, left) has a pK_a of −0.3, an observation that can be rationalized based on the electron-withdrawing nature of the three fluorine atoms, which stabilize the adjacent negative charge. This would be classified as an inductive effect, because the communication is essentially through the σ bond framework. Alternatively, anions can be stabilized through resonance effects, as seen in the phenolate anion (Figure 8-12, right), which is almost a million times less basic than methoxide. This difference can be explained by the resonance delocalization of the negative charge throughout the aromatic system.

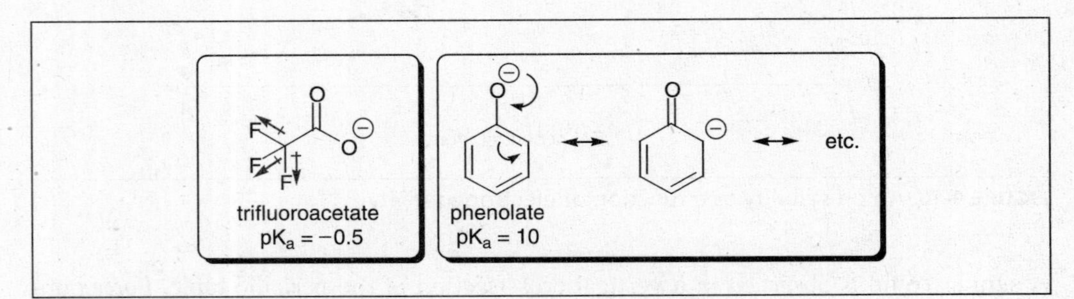

FIGURE 8-12: Inductive and resonance effects on anion stability.

With this thorough grounding in acid and base strength, you are now prepared to deal with the question of proton transfer *per se*. In other words, how do you use these concepts to predict whether a proton transfer is likely or not? First, you should consider any proton transfer to be an equilibrium at the first approximation. Next, the equilibrium should be cast as a competition between two bases for the same proton—only one base can have the proton at a given time. Consider the behavior of amide (H_2N^-)

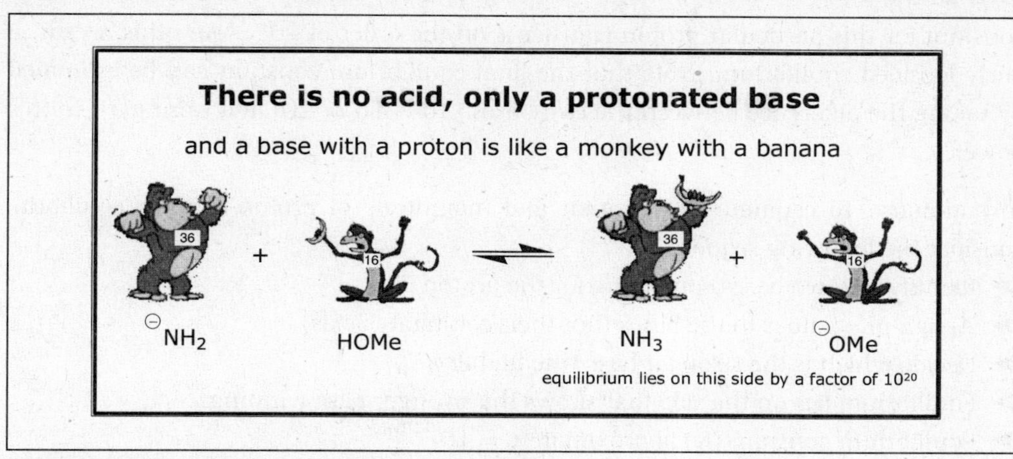

FIGURE 8-13: The semiamorphic proton transfer.

in methanol (HOMe). Figure 8-13 paints a somewhat unconventional picture of this battle in the form of a chimp and a gorilla fighting for a single banana—assuming pure brawn wins the day, then the gorilla will walk away with the prize. To put this in more chemical terms, the two bases vying for the proton are amide (pK_a 36) and methoxide (pK_a 16). Clearly, the amide is the stronger base; therefore the equilibrium lies to the right, because this is the side that shows the stronger base claiming the proton.

Not only can we predict the direction of the equilibrium, but we can also estimate the magnitude of the equilibrium constant. Figure 8-14 demonstrates the principle behind the process. The first line shows the equation related to the acidity of ammonia, with the corresponding equilibrium constant. The second line is simply the reverse of the first; keep in mind that when we reverse the direction of an equilibrium, we take the inverse of the equilibrium constant—in other words, if the equilibrium constant is right over left, then the numerator and denominator have been exchanged in reversing the equation. The third line shows the equation related to the acidity of methanol, along with the appropriate equilibrium constant. Finally, if the second and third equations are added together, we obtain the proton transfer reaction in question. When chemical equations are added, so are their equilibrium constants. Therefore, the equilibrium

$$NH_3 \rightleftharpoons H^{\oplus} + {}^{\ominus}NH_2 \qquad K_1 = K_a = 10^{-36}$$

$${}^{\ominus}NH_2 + H^{\oplus} \rightleftharpoons NH_3 \qquad K_2 = 1/K_a = 10^{36}$$

$$HOMe \rightleftharpoons H^{\oplus} + {}^{\ominus}OMe \qquad K_3 = K_a = 10^{-16}$$

$${}^{\ominus}NH_2 + HOMe \rightleftharpoons NH_3 + {}^{\ominus}OMe \qquad K_4 = K_2 + K_3 = 10^{(36-16)} = 10^{20}$$

FIGURE 8-14: Estimation of equilibrium constant for proton transfer.

constant for this particular proton transfer is on the order of 10^{20}—in other words, a fairly lopsided equilibrium. Note that the final equilibrium constant can be estimated by taking the difference between the two bases ($36 - 16 = 20$) and raising 10 to this power.

In summary, to estimate the position and magnitude of proton-transfer equilibria, consider the following sequence:

- Identify the two bases competing for the proton
- Assign pK_a values to the bases (not their conjugate acids)
- Decide which is the stronger base (the higher pK_a)
- Equilibrium lies on the side that shows the stronger base winning
- Equilibrium constant (K) approximately $= 10^{\Delta pK_a}$

From a mechanistic standpoint, the proton transfer is quite simple: a lone pair from the basic species attacks the proton and forces it to deposit the shared electron density onto the substrate, as shown in Figure 8-6. The proton transfer is also among the fastest reactions known—almost every molecular collision results in effective reaction; thus, the rate tends to be controlled by diffusion of the molecules.

IONIZATION

Acidity can be described as an ionization event, as shown in Figure 8-8, whereby a proton is liberated from a neutral molecule. Another form of ionization is also frequently encountered in organic chemistry—in a sense, it can be thought of as the reverse of a Lewis acid–Lewis base interaction shown in Figure 8-2. This process, known simply as **ionization**, is depicted in Figure 8-15. Here, a σ bond is cleaved heterolytically to form an anion and a carbocation (carbon-centered cation). For this reaction to be kinetically relevant, the leaving group (LG) must provide a very stable anion, and the carbocation must be secondary, tertiary, or otherwise stabilized (e.g., allylic or benzylic).

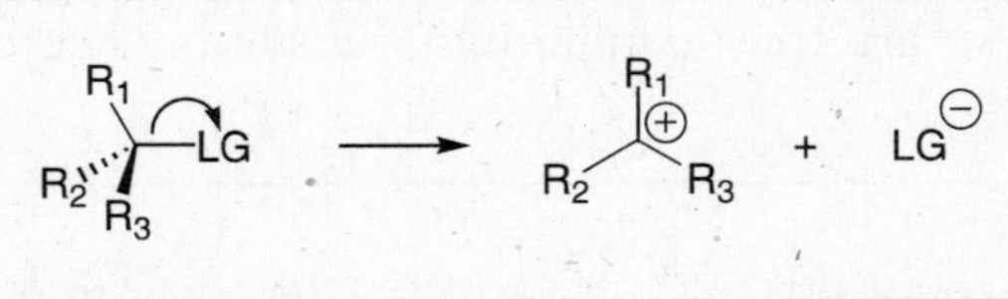

FIGURE 8-15: Unimolecular ionization.

It is easy enough to determine whether a substrate provides a secondary or tertiary carbocation, but the LG provides a somewhat more subtle challenge. Fortunately, we have a convenient way to evaluate so-called LG ability. In the previous discussion, we came to the conclusion that strong acids ionize because they form particularly stable anions. We also argued that the ionization in Figure 8-15 requires a stable ion to be formed. Thus, it follows that the conjugate bases of strong acids make good leaving groups. To state this in the form of a trend, **the stronger the conjugate acid, the better the leaving group**. For example, hydroxide is not a very active leaving group (Figure 8-16), and if we consider the conjugate acid of hydroxide (namely, water), we would agree that it is a very weak acid ($pK_a = 15.7$). However, chloride is a great leaving group; by the same token, its conjugate acid (HCl) is without question a very strong acid ($pK_a \ll 0$). Thus, simply by examining a table of pK_a values, we can fairly accurately assess the likelihood of a species to function as a leaving group for ionization reactions.

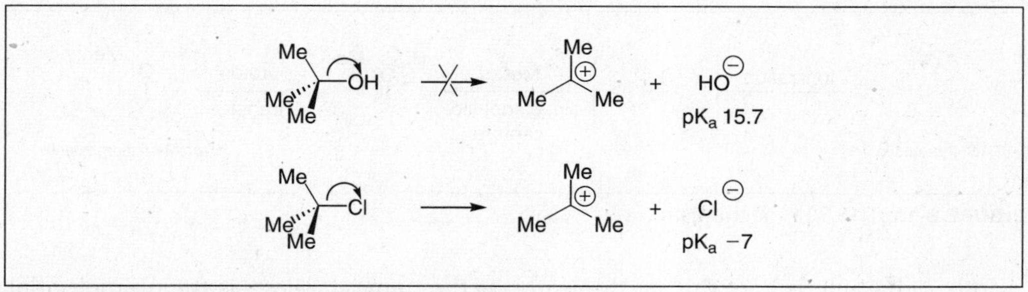

FIGURE 8-16: Impact of leaving group on ionization.

NUCLEOPHILIC CAPTURE

Once a carbocation has been formed, it can suffer subsequent capture by a nucleophile to form a new neutral species, as shown in Figure 8-17 (inset). In principle, any species that would be classified as a Lewis base is a potential nucleophile, because a nucleophile is also a source of electron density. Even though generic depictions often represent the nucleophile with a negative charge, this is largely stylistic: There are also neutral nucleophilic species. For example, the *t*-butyl carbocation can be captured by methanol (Figure 8-17, bottom) to form an initial adduct that undergoes subsequent proton loss to provide methyl *t*-butyl ether.

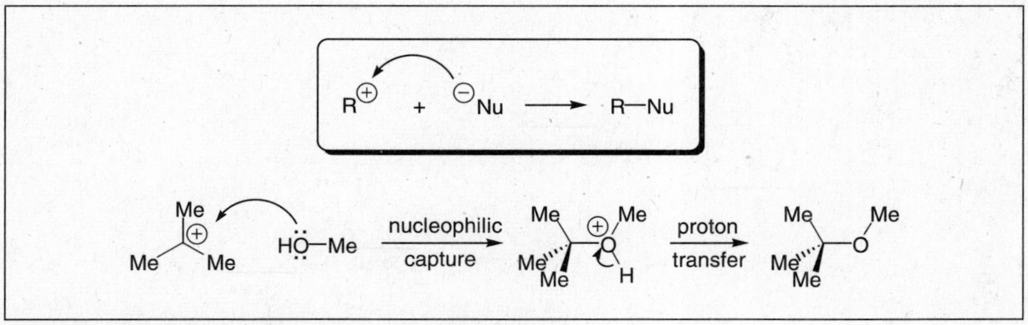

FIGURE 8-17: Nucleophilic capture of a carbocation.

When ionization is followed by nucleophilic capture, an overall process known as **unimolecular substitution** (S_N1) occurs. An illustrative example is the **solvolysis**

reaction, in which solvent molecules act as the nucleophilic species. Thus, when 2-bromobutane is dissolved in methanol, ionization ensues to form a secondary carbocation that is captured by solvent to give (after proton loss) 2-methoxybutane (Figure 8-18). Note that even though optically pure (*S*)-2-bromobutane is used, the product is obtained as a racemic mixture. The chirality is destroyed in the first step of the mechanism, and the planar carbocation has an equal probability of being attacked from either side by the nucleophile. Indeed, the generation of racemic products from optically pure starting materials is a hallmark of the S_N1 reaction.

Br
ionization
MeOH
nucleophilic capture
H ⊕ Me
proton transfer
Me
(*S*)-2-bromobutane
(*rac*)-2-methoxybutane

FIGURE 8-18: The S_N1 mechanism.

Another characteristic of this mechanism—as the name implies—is the unimolecular nature of its kinetics. In other words, the rate of the reaction is dependent only on the concentration of the substrate, according to the rate law

$$\text{Rate} = k[\text{2-bromobutane}] \qquad (37)$$

Inspection of the reaction profile (Figure 8-19) reveals why this is so. Even though there are three discrete steps in the overall mechanism, ionization exhibits the highest activation energy. According to the Arrhenius equation (Equation 8), this step is consequently the slowest, or the rate-determining step. Generally speaking, the kinetics of the rate-determining step determine the kinetics of the entire reaction—and because only the 2-bromobutane is involved in the first step, it is the only species that appears in the overall rate law.

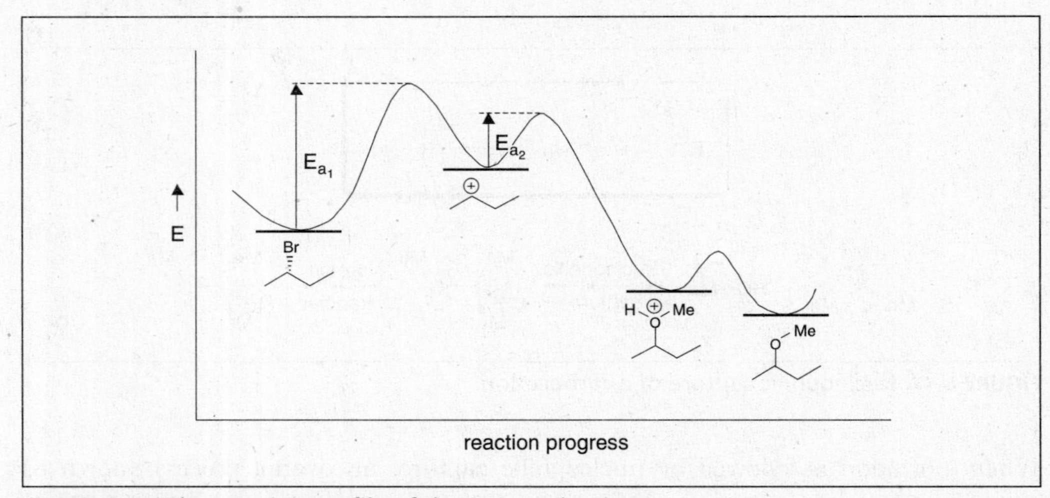

FIGURE 8-19: The reaction profile of the S_N1 mechanism.

NUCLEOPHILIC DISPLACEMENT (S_N2)

When the nucleophile is strong enough, substitution can take place directly, without the need for prior ionization. For example, treatment of 2-bromobutane with sodium cyanide leads to direct displacement in one step (Figure 8-20).

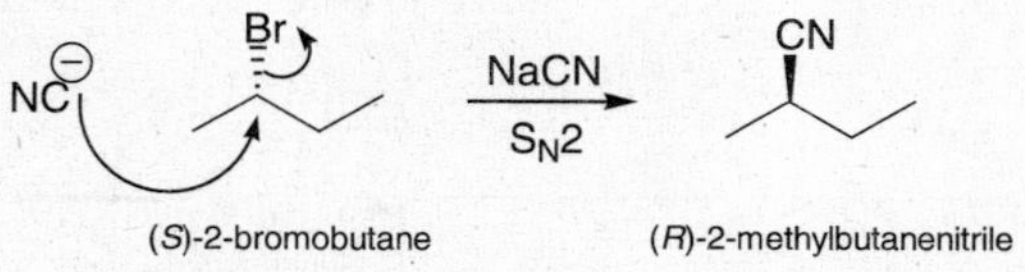

(*S*)-2-bromobutane — NaCN, S_N2 → (*R*)-2-methylbutanenitrile

FIGURE 8-20: Nucleophilic displacement by cyanide anion.

One significant outcome of the S_N2 mechanism is that stereochemical information is preserved. In other words, if the electrophilic center is chiral and the substrate is optically pure, then the product will be optically active, not a racemic mixture. Furthermore, the stereochemistry is inverted about the electrophilic carbon center. This is due to the so-called backside attack of the nucleophile, which approaches the electrophile from the side opposite the leaving group, leading to a trigonal bipyramidal transition state, which is itself a chiral entity (Figure 8-21). As the reaction proceeds, the three substituents fold back to the other side, much like an umbrella inverting in a gust of wind.

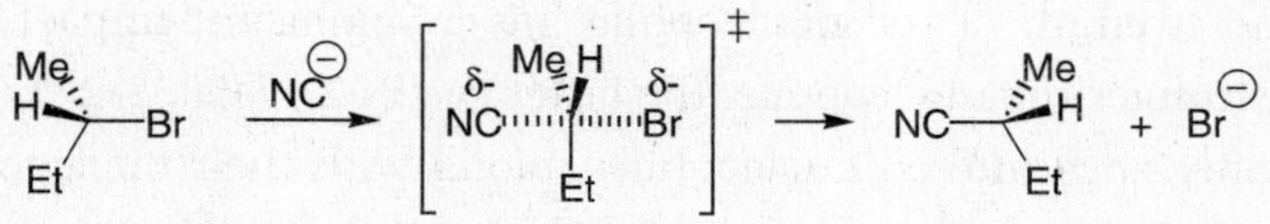

FIGURE 8-21: Inversion of stereochemistry in S_N2.

The S_N2 reaction proceeds by a concerted mechanism, unlike the stepwise S_N1 sequence. Thus, there is a single activation energy that governs the kinetics of the process. Because the transition state involves both the substrate and the nucleophile, the rate law must also include terms for both species, as follows:

$$\text{Rate} = k[\text{2-bromobutane}][\text{NaCN}] \quad (38)$$

For this reason, S_N2 reactions are particularly sensitive to concentration effects—a twofold dilution results in a fourfold decrease in rate (Figure 8-22).

Aside from concentration, the rate of the S_N2 reaction also depends on the nature of the two species involved. Because the backside attack is sterically demanding, S_N2 reactivity drops off as substitution about the electrophilic center increases. Figure 8-23 summarizes the effect of substrate structure on the rate of nucleophilic displacement.

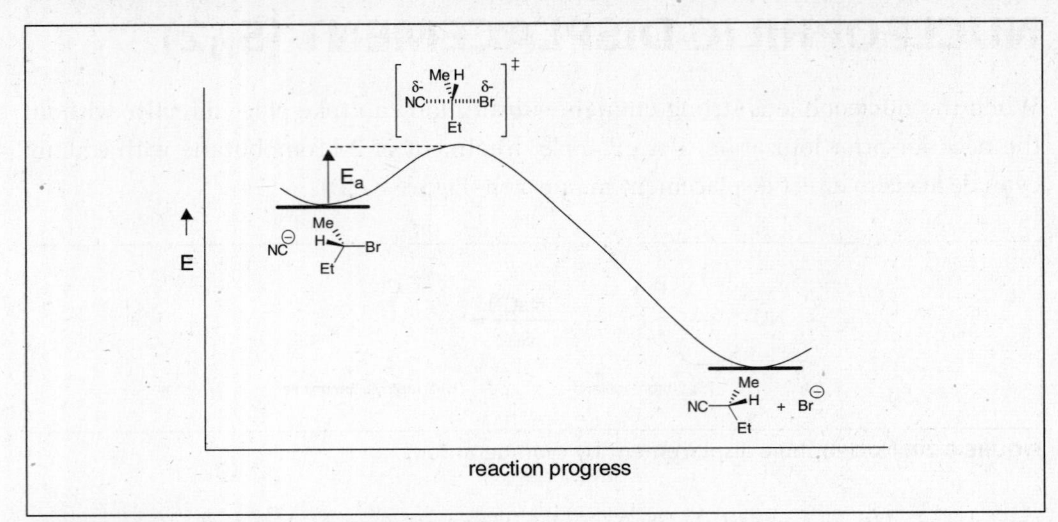

FIGURE 8-22: The reaction profile of the S_N2 mechanism.

Me—LG | —LG | —LG | —LG

methyl > primary > secondary >> tertiary (practically inactive)

FIGURE 8-23: Substrate reactivity series for S_N2.

In addition, the strength of the nucleophile has a significant impact on the rate of reaction—the stronger the nucleophile, the faster the S_N2. Table 8-2 shows a sampling of some frequently encountered nucleophiles, along with their nucleophilic constants (n_{MeI}). These values are derived experimentally from observing the rate at which the nucleophile engages methyl iodide (an extremely effective electrophile) in S_N2 displacement. As the value of n_{MeI} increases, so does the activity of the nucleophile. Note that these values are logarithmic in nature, much like pH, so that iodide ($n_{MeI} = 7.4$) is roughly ten times more nucleophilic than hydroxide ($n_{MeI} = 6.5$), which is already quite a competent S_N2 nucleophile.

The tabular data is useful for comparing the strengths of various nucleophiles, but perhaps more important is understanding the periodic trends behind the data. Unlike leaving groups, for which we could apply a simple rule of thumb, predicting nucleophilic behavior is a bit more subtle. However, two general trends are helpful guides. First, there is a left-to-right trend, which we can explore by comparing ammonia ($n_{MeI} = 5.5$) to methanol ($n_{MeI} = 0.0$) and hydroxide ($n_{MeI} = 6.5$) to fluoride ($n_{MeI} = 2.7$). In going from left to right in a row, nucleophilicity decreases, a phenomenon that can best be understood in terms of electronegativity: The more tightly an atom holds onto its electron density, the less likely that electron density is to engage in other activity, like nucleophilic displacement.

TABLE 8-2 Nucleophilic Constants

Species	n_{MeI}	pK_a
MeOH	0.0	−1.7
F^-	2.7	3.5
AcO^-	4.3	4.8
Cl^-	4.4	−5.7
Me_2S	5.3	
NH_3	5.5	9.3
N_3^-	5.8	4.8
PhO^-	5.8	9.9
HO^-	6.5	15.7
Et_3N	6.7	10.7
CN^-	6.7	9.3
Et_3As	7.1	
I^-	7.4	−11
Et_3P	8.7	8.7
PhS^-	9.9	6.5
$PhSe^-$	10.7	

Data taken from Pearson, R.G., Songstad, J., *J. Am. Chem. Soc.* **1967**, *89*, 1827; Pearson, R.G., Sobel, H., Songstad, J., *J. Am. Chem. Soc.* **1968**, *90*, 319; Bock, P.L., Whitesides, G.M., *J. Am. Chem. Soc.* **1974**, *96*, 2826. As tabulated in Carey & Sundberg, *Advanced Organic Chemistry*.

Next, there is an up-to-down trend, which is evident in the series of fluoride (n_{MeI} = 2.7), chloride (n_{MeI} = 4.4), and iodide (n_{MeI} = 7.4). Here, nucleophilicity increases as we go down a column. Unlike the first trend, this cannot be explained adequately on the basis of electronegativity; for example, sulfur and selenium have the same electronegativity (EN = 2.6), yet phenylselenide (n = 10.7) is almost ten times more effective than phenylsulfide (n = 9.9). The conventional rationale for this effect is that the attacking centers become larger and more polarizable, which allows the electron density to interact with the electrophilic center at larger intermolecular distances.

The third and most tenuous trend is that for nucleophiles **with the same attacking atom**, nucleophilicity tends to track basicity. Thus, the series acetate (AcO^-), phenoxide (PhO^-), and hydroxide (HO^-) describes successively stronger bases (pK_a = 4.8, 9.9, and 15.7, respectively), which are also increasingly more active nucleophiles (n_{MeI} = 4.3, 5.8, and 6.5, respectively). As Figure 8-24 demonstrates, this is a nonlinear relationship. Keep in mind that when the attacking atoms are different, all bets

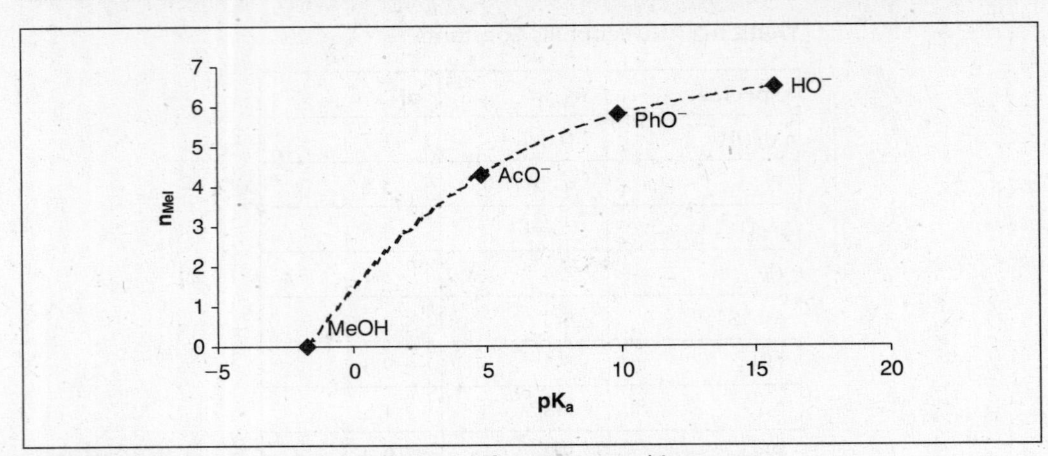

FIGURE 8-24: Correlation of pK_a and n_{MeI} for same attacking atom.

are off. For example, iodide (pK_a = −11) is a far weaker base than fluoride (pK_a = 3.5), but a far more effective nucleophile.

Moreover, structural features can impact the nucleophilicity of certain species. For example, potassium *t*-butoxide (*t*-BuOK) is a relatively strong base (pK_a = 19), but a poor nucleophile because of steric hindrance about the anionic center. Similarly, lithium diisopropylamide (LDA) is frequently used as a very strong, nonnucleophilic base, because the anionic nitrogen is imbedded within an almost inaccessible steric pocket. Hexamethyldisilazide (HMDS) is even more sterically encumbered—to function as a nucleophile is a hopeless aspiration (Figure 8-25).

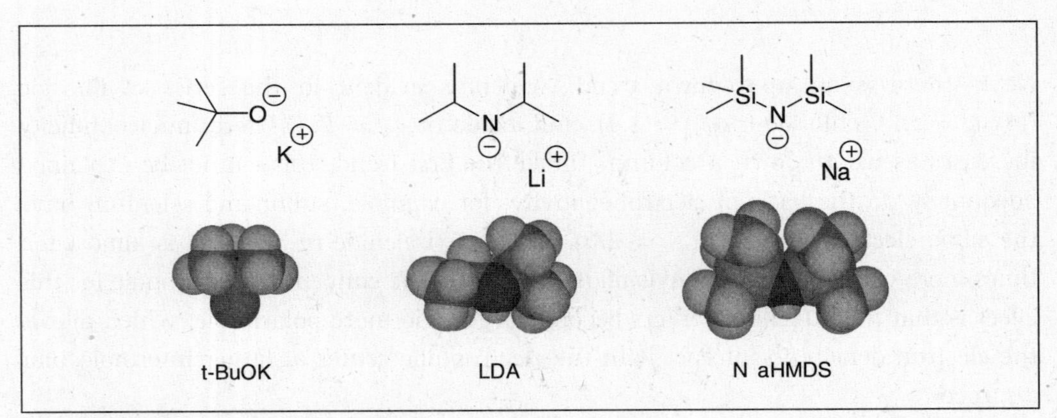

FIGURE 8-25: Sterically hindered anions.

THE S_N1/S_N2 CONTINUUM

Frequently, the two substitution mechanisms (S_N1 and S_N2) are presented as alternate possibilities for a binary outcome—pathways in competition for primacy: one must predominate, and sometimes we can exert influence over the decision. Indeed, the rate of S_N1 and S_N2 reactions can be impacted significantly by experimental conditions, particularly in the choice of solvent. If we consider the transition states for both mechanisms, we find very different behaviors as the system approaches the transition state. For example, in a typical S_N2 scenario, a negatively charged nucleophile attacks an electrophilic center; at the transition state, this negative charge is in the act of being transferred to the leaving group—thus, the charge is distributed throughout the transition state structure. However, in the rate-determining ionization step of the S_N1 reaction, a neutral molecule is converted into two charged species; at the transition state, this charge separation is already beginning to occur (Figure 8-26). Therefore, the stabilization of charges affects the two mechanisms in very different ways.

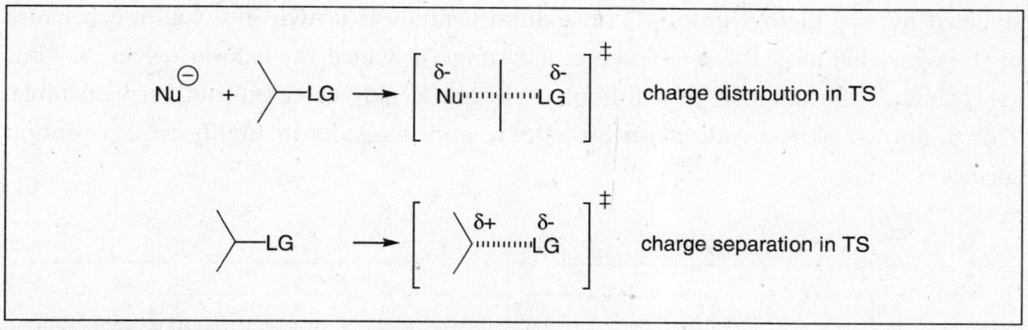

Nu⊖ + LG → [δ- Nu·····LG δ-]‡ charge distribution in TS

LG → [δ+ ·····LG δ-]‡ charge separation in TS

FIGURE 8-26: Differences in S_N1 and S_N2 transition states.

One of the simplest avenues available to modulate charge stabilization is by changing the reaction medium, namely solvent. A wide array of solvents are used in organic chemistry; a small sampling is shown in Figure 8-27. The choice of solvent is driven by many practical factors, such as expense, ease of removal, toxicity, flammability, and environmental impact—but there are also mechanistic drivers for running a reaction

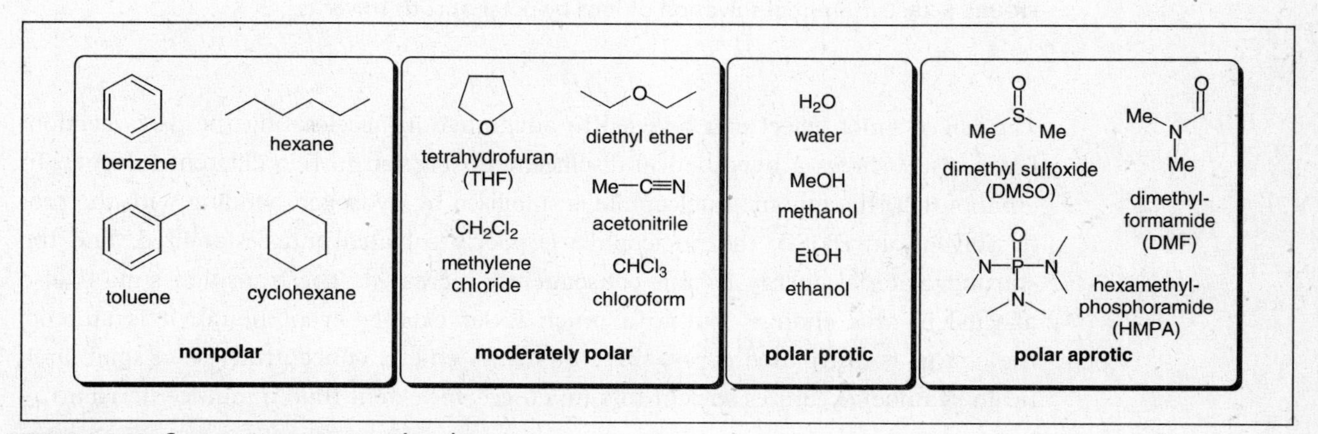

FIGURE 8-27: Some common organic solvents.

in a given solvent, particularly in light of the transition state differences discussed previously.

For example, polar aprotic solvents (DMSO, DMF, HMPA, etc.) are not particularly pleasant or convenient solvents to work with (due to high boiling points and toxicity)—however, they have some peculiar and useful properties, one of which is the differential solvation of ions. The term **aprotic** refers to the fact that these molecules have no protons on heteroatoms (as do water and the alcohols); therefore, they cannot serve as hydrogen bond donors. However, they can stabilize ions by virtue of their polarity. As an illustration, consider a solution of sodium cyanide in DMSO (Figure 8-28). The sodium cation is very well stabilized by an organized solvent shell, in which the electronegative oxygen atoms are aligned toward it. However, the cyanide ion is poorly solvated, because the electropositive region of the solvent polar bond is sterically shielded by two methyl groups. The sodium cyanide is drawn into solution because of the overwhelming thermodynamic advantage provided by sodium solvation—but even though the anion is in solution, it is only weakly solvated and very unstable. This is known as the **naked anion effect**, and it results in highly reactive anion species.

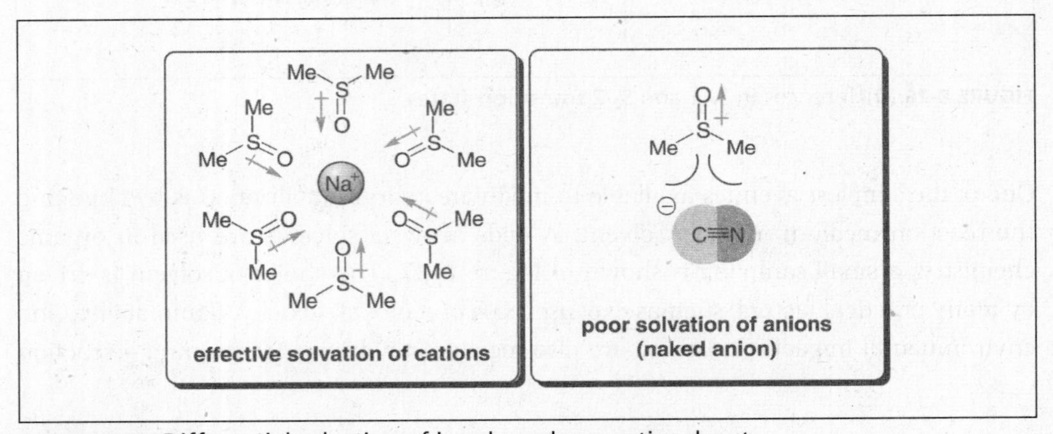

FIGURE 8-28: Differential solvation of ions by polar aprotic solvents.

The naked anion effect can be used to advantage in accelerating the S_N2 reaction. Figure 8-29 shows a hypothetical displacement reaction in two different solvents. In ethanol (left) the anionic nucleophile is stabilized by hydrogen bonding with the protic solvent. In DMSO, the nucleophile is poorly solvated and destabilized, and the starting material energy level is consequently increased. The transition state is also affected by this change, but to a much lesser extent—even though it is anionic, the charge is distributed across three atoms. A charge concentrated in a small area (as in cyanide) organizes solvent to a much greater extent than a multicentered array with diffuse charge, so it stands to reason that removal of this stabilization has an impact on the starting materials more so than the transition state. The result is that the activation energy is lowered, because the effective energy difference between reactants and transition state is decreased.

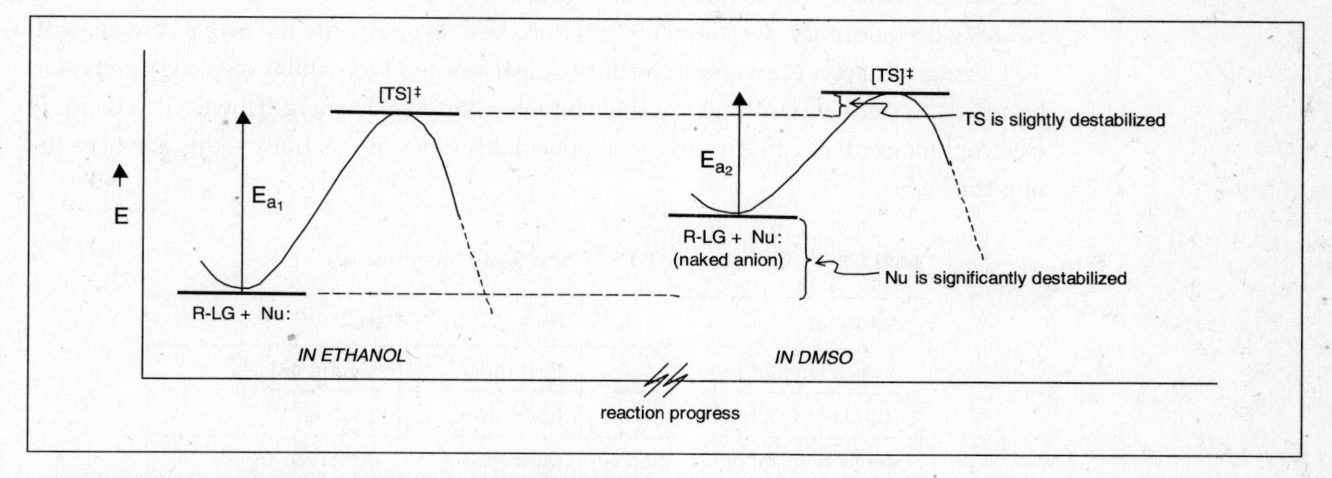

FIGURE 8-29: Acceleration of S_N2 using polar aprotic solvent.

The same principle can be used to rationalize the rate enhancement of S_N1 substitution in protic solvents (Figure 8-30). Compared to an aprotic solvent like THF, ethanol has the ability to stabilize the nascent charges in the transition state, thereby lowering the activation energy. Generally speaking, the starting material is also slightly stabilized by the more effective solvent, but this is small in comparison to the benefit enjoyed by the transition state. Therefore, the net result is a decrease in the activation energy and an acceleration of the reaction.

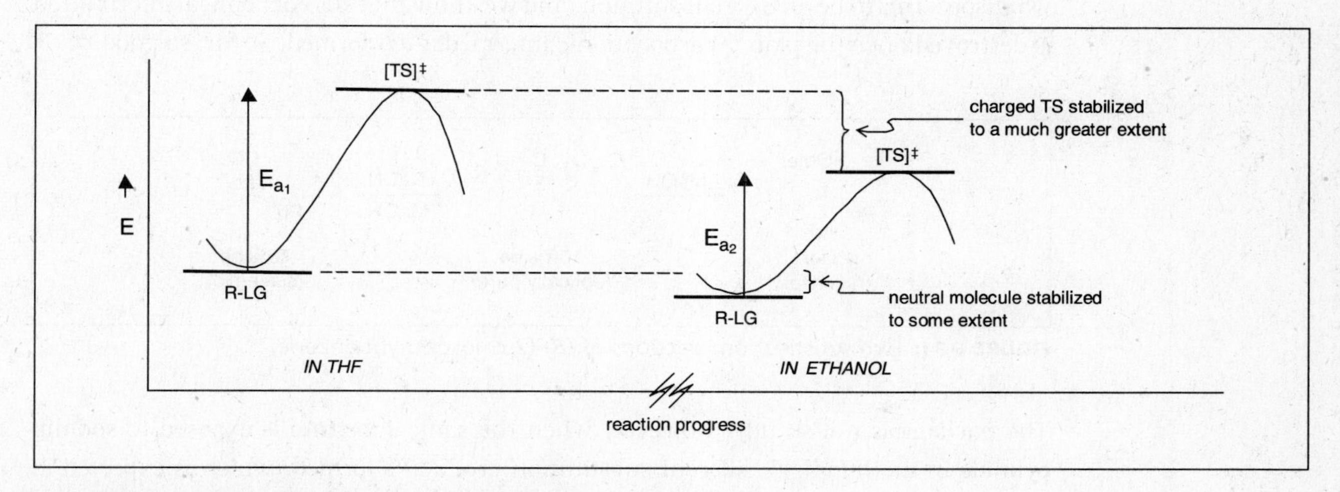

FIGURE 8-30: Acceleration of S_N1 reactions by protic solvents.

In short, the two substitution mechanisms do appear to construct a dichotomy. Table 8-3 compares several of the hallmarks for the two mechanisms. For example, the S_N2 reaction results in clean inversion of the electrophilic center, so that any stereochemical information is preserved, whereas the S_N1 pathway leads to scrambling of the stereochemistry at the reactive site. As another example of divergent outcomes, substrates with a primary electrophilic center proceed through only the S_N2 pathway, with S_N1 being unlikely. Conversely, tertiary substrates tend to exhibit only S_N1 behavior, because the steric environment forbids bimolecular displacement. However, secondary electrophilic centers can, in theory, support both types of reactivity—this is where the plot thickens.

TABLE 8-3 Comparison of S_N1 and S_N2 Mechanisms

Behavior	S_N1	S_N2
Overall kinetics	First order	Second order
Chiral electrophilic centers are	Racemized	Inverted
Stereochemical information is	Destroyed	Preserved
Typical nucleophiles	Weak	Strong
Typical electrophiles	2° or 3°	1° or 2°
Accelerated by	Protic solvents	Polar aprotic solvents

Consider the conundrum presented by the substitution reactions of (*R*)- (2-chloroethyl)-benzene (Figure 8-31). If the substrate is simply stored in methanol, a substitution reaction ensues, leading to the corresponding methyl ether. Even though the starting material is optically pure (100% ee), the product shows no optical activity—in other words, it is a racemic mixture of both enantiomers. This comes as no surprise, because we suspect this to be an S_N1 substitution, and we know that stereochemical information is destroyed once the planar carbocationic intermediate is formed. So far, so good.

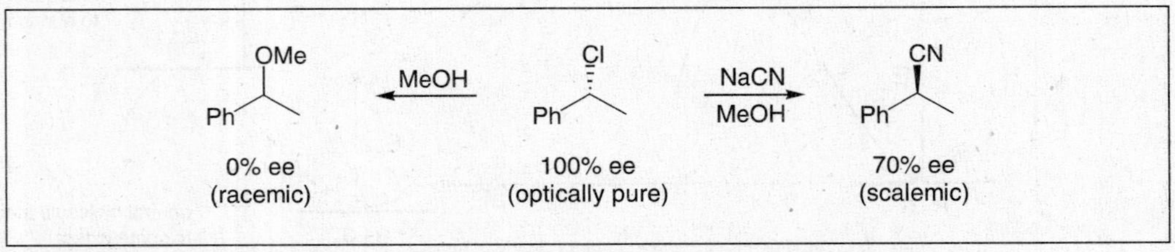

OMe
Ph
0% ee
(racemic)
MeOH
Cl
Ph
100% ee
(optically pure)
NaCN
MeOH
CN
Ph
70% ee
(scalemic)

FIGURE 8-31: Two substitution reactions of (*R*)-(2-chloroethyl)benzene.

The package is not so tidy, however, when the same substrate is exposed to sodium cyanide in methanol. A different substitution product is formed, which is expected in as much as cyanide is a far better nucleophile than methanol (Table 8-2). Cyanide is also powerful enough to engage in S_N2 substitution, and indeed, the majority of the substrate does appear to have undergone inversion of stereochemistry. However, this inversion is not universal—the result of 70% ee means that 85% of the product is the *S*-isomer and 15% is the corresponding *R*-form. It is neither optically pure nor racemic, a state called **enantiomerically enriched or scalemic.**

Yet how could such a result have come about? The conventional textbook explanation is that we are observing a competition between the two reaction mechanisms—that a portion of the substrate has undergone clean S_N2 inversion, whereas the rest has been converted by an S_N1 mechanism, which accounts for the partial scrambling of stereochemistry. However, a more complete conceptual framework emerges if we consider the S_N1 and S_N2 to be two bounds of a continuum. That is, S_N1 is a stepwise process in which the leaving group completely leaves before the nucleophile approaches, and the S_N2 is a concerted process in which bond formation between the nucleophile and electrophilic center is exactly matched with the bond cleavage between the electrophilic center and the leaving group. We can describe such a mechanism as being concerted and synchronous.

Between the stepwise and perfectly synchronous concerted pathway, however, lies a continuum of asynchronous reactions (Figure 8-32). In other words, the bond between

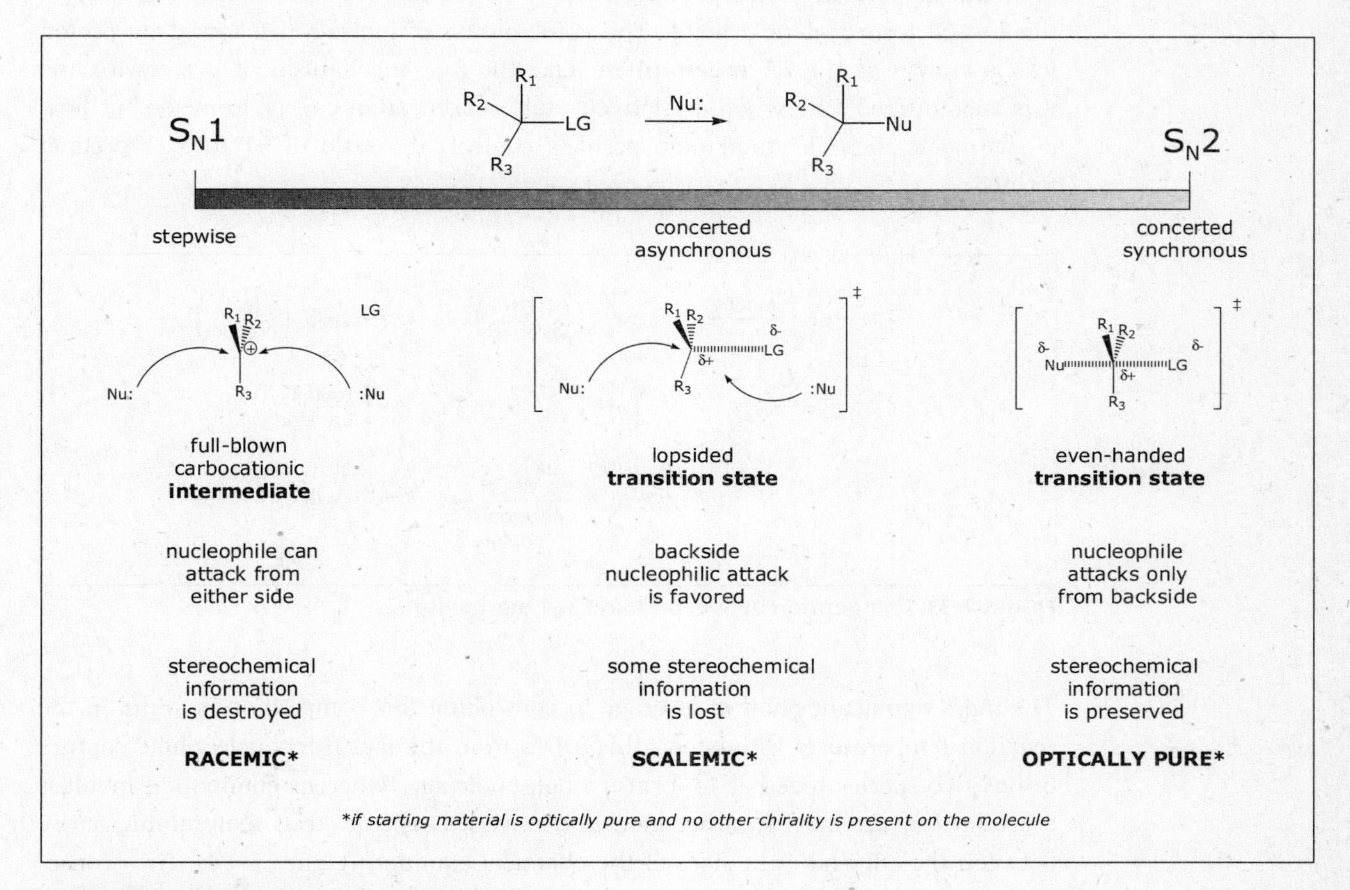

FIGURE 8-32: The spectrum of substitution.

the electrophilic center and the leaving group begins to weaken and lengthen, and the nucleophilic species begin to swarm around the developing electropositive center before the leaving group actually leaves. Depending on the progress of the bond cleavage, the leaving group may be far enough away (and the positive charge well enough developed) that a statistical fraction of the nucleophilic species actually engages in frontside capture of the electrophilic center. Another way to think about this situation is to imagine a carbocationic intermediate still shielded on one side by the leaving group that has not quite left, so that most (but not all) of the nucleophilic species choose to attack from the other side.

E1 ELIMINATION

In contrast to the S_N1/S_N2 dichotomy, there is a pair of reaction mechanisms that are truly in competition with each other most of the time. This idea is exemplified by the methanolysis of *t*-butyl bromide (Figure 8-33), a process that is often accompanied by the generation of 2-methylpropene. Both products arise from a common intermediate—the initially formed *t*-butyl carbocation can be captured by methanol to give the corresponding ether (S_N1 product), or it can suffer direct proton loss to form a double bond. The combination of ionization followed by proton loss is known as the **E1 mechanism**. Like the S_N1 mechanism, it is stepwise and it is encountered only when a relatively stable carbocation can be formed—but how do we come to understand (and perhaps control) the ratio of E1 to S_N1 product formed?

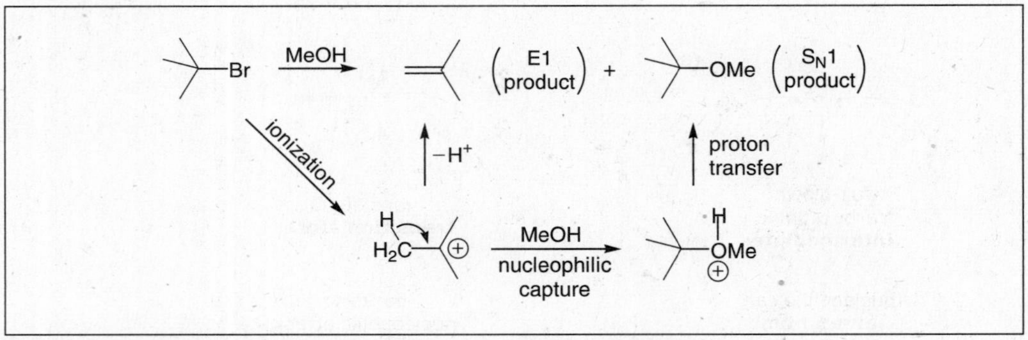

FIGURE 8-33: Competition between E1 and S_N1 mechanisms.

The most significant point of leverage in controlling this competition is found in the reaction temperature. Ultimately, this arises from the fact that nucleophilic capture brings two species together to form a single molecule, whereas elimination involves a proton transfer from a carbocation to a solvent molecule, thus maintaining parity between the left and right sides of the chemical equation (Figure 8-34). As a consequence, the entropic term for nucleophilic capture is large and negative (i.e., produces

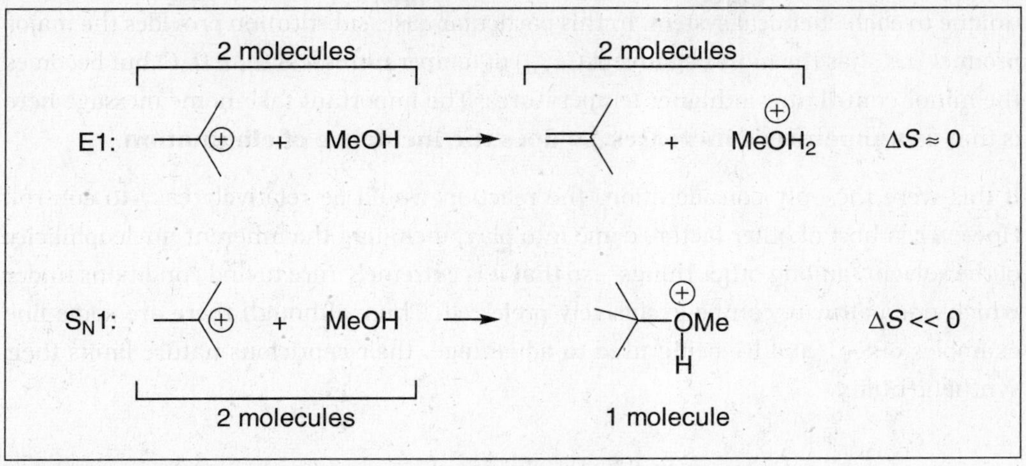

2 molecules → 2 molecules

E1: (carbocation) + MeOH → (alkene) + $MeOH_2^+$ $\Delta S \approx 0$

S_N1: (carbocation) + MeOH → (product)–$\overset{+}{O}$Me(H) $\Delta S << 0$

2 molecules → 1 molecule

FIGURE 8-34: Entropic differences between E1 and S_N1 mechanisms.

a more ordered system) as opposed to elimination, which is associated with near-zero and sometimes positive entropies of reaction.

The enthalpy term is also different: in the example above, elimination provides a C—C π bond worth about 65 kcal/mol, whereas nucleophilic capture results in a much stronger C—O σ bond worth about 90 kcal/mol. Thus, were it not for the entropic term, nucleophilic capture would always be the thermodynamically most favorable route. Remember, however, that free energy (and thus spontaneity) is governed by the relationship

$$\Delta G = \Delta H - T\Delta S \tag{5}$$

Thus, as temperature increases, so does the impact of the entropic term. If temperature is kept low, then the enthalpic term dominates and S_N1 is preferred. At higher reaction temperatures, entropic biases begin to emerge until eliminative pathways start to take precedence. Figure 8-35 graphically represents this temperature dependence, which is

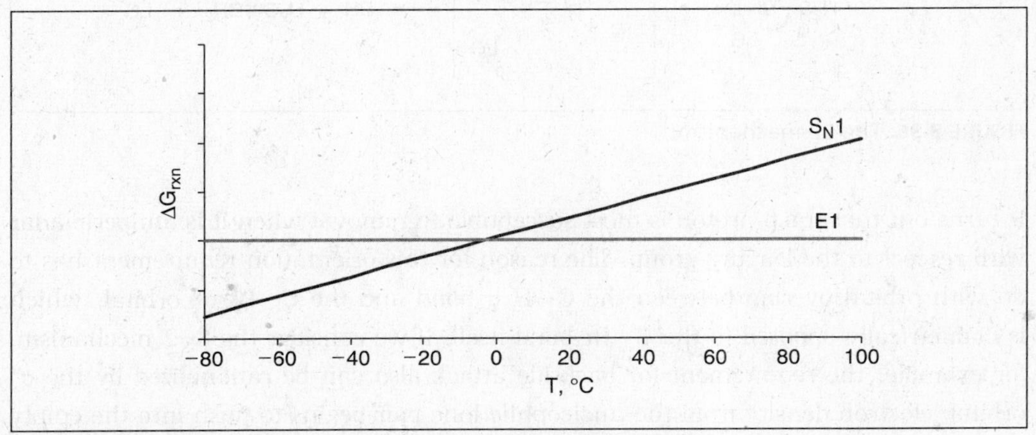

FIGURE 8-35: Influence of temperature on spontaneity of S_N1 versus E1.

unique to each chemical system. In this particular case, substitution provides the major product (i.e., has the most negative ΔG_{rxn}) at temperatures less than 0°C, but becomes the minor contributor at higher temperatures. The important take-home message here is that **as temperature increases, so does the incidence of elimination**.

If this were the only consideration, the reaction would be relatively easy to control. However, a host of other factors come into play, including the inherent nucleophilicity of the solvent, among other things—so that it is extremely rare to find conditions under which one pathway can be exclusively preferred. Thus, although there are some fine examples of S_N1 and E1 being used to advantage, their capricious nature limits their synthetic utility.

E2 ELIMINATION

With stronger bases another eliminative pathway becomes feasible, namely bimolecular elimination (E2). This is a concerted process in which the base removes a β-proton as the leaving group leaves. Like the S_N2 reaction, the E1 has a single transition state involving both the base and the substrate; therefore, the observed kinetics are second order and the rate is dependent on the concentration of both species (Figure 8-36). Note that as the reaction progresses, two molecules give rise to three molecules, an entropically very favorable result. Therefore, like the E1 mechanism, bimolecular elimination is made much more spontaneous with increasing temperature. A further implication is that, although the Principle of Microscopic Reversibility states that the reverse pathway exists, it becomes improbable that the three requisite species will encounter each other in exactly the right geometry to take advantage of that pathway. Thus, E2 eliminations tend to be practically irreversible.

$$B:^{\ominus} + H{-}H_2C{-}CH_2{-}LG \longrightarrow \left[B^{\delta-}{\cdots}H{\cdots}H_2C{=}CH_2{\cdots}LG^{\delta-} \right]^{\ddagger} \longrightarrow BH + H_2C{=}CH_2 + LG^{\ominus}$$

FIGURE 8-36: The E2 mechanism.

It turns out that the β-proton is most susceptible to removal when it is antiperiplanar with respect to the leaving group. The reason for this orientation requirement has to do with orbital overlap between the C—H σ bond and the C—Br σ* orbital, which is diametrically opposed to the C—Br bond itself. If we consider the S_N2 mechanism, for example, the requirement for backside attack also can be rationalized by the σ* orbital: electron density from the nucleophile lone pair begins to push into the empty antibonding orbital, weakening the corresponding leaving group σ bond and leading to

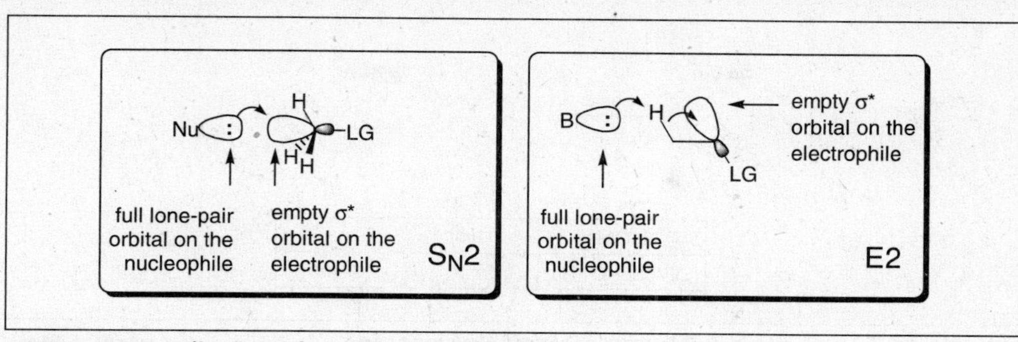

FIGURE 8-37: Implication of σ* orbitals in bimolecular mechanisms.

substitution (Figure 8-37). Similarly, if we think of a base not so much as "pulling off a proton" as "pushing off electron density," then the leaving group σ* would seem to be a likely receptacle for that electron density. However, because the electron density in this case is not coming directly from the base, but rather from the C—H bond being attacked, the C—H bond and the σ* orbital must be able to communicate—and this is at a maximum when the leaving group is antiperiplanar to the α-proton.

The net result of this constraint is that stereochemical information often transmits from reactant to product. For example, (1*S*, 2*S*)-(1-bromo-1-phenyl-2-methylethyl)benzene (Figure 8-38, top left) undergoes E2 elimination to provide *Z*-methylstilbene as the sole stereoisomer. To understand this selectivity, it is easiest to depict the substrate as a Newman projection (Figure 8-38, bottom). Rotating the entire molecule counterclockwise to bring the bromine into the 6 o'clock position, then rotating the back carbon about the C1—C2 bond to bring the β-proton into the 12 o'clock position, we now have the proper antiperiplanar alignment. Subsequent E2 elimination collapses the molecule to planarity, and we see that the two phenyl groups are indeed on the same side.

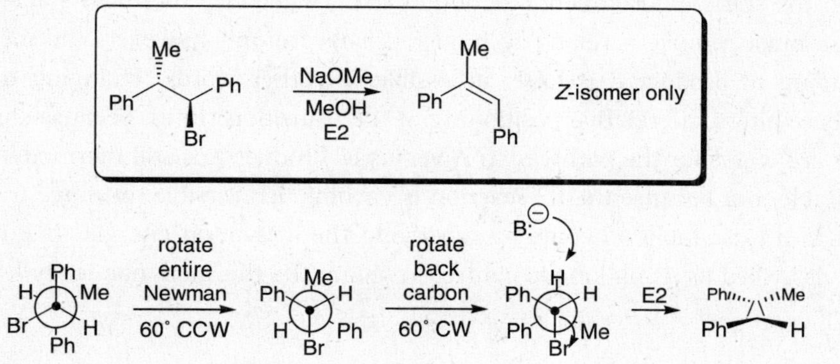

FIGURE 8-38: Stereochemical implications of the E2 elimination.

There are also regiochemical considerations in this reaction, some of which are under our control. For example, consider the E2 elimination of 1-bromo-1-methylcyclohexane

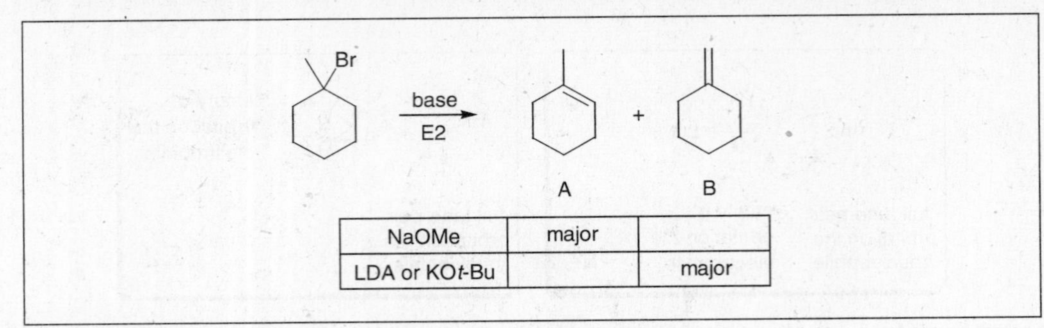

	A	B
NaOMe	major	
LDA or KO*t*-Bu		major

FIGURE 8-39: E2 elimination of 1-bromo-1-methylcyclohexane.

under two different sets of conditions (Figure 8-39). If sodium methoxide is used, the major elimination product is the internal alkene A; however, lithium diisopropylamide (LDA) as well as potassium *t*-butoxide deliver the product bearing the exo double bond (B) as the predominant isomer.

How do we rationalize these results? First, we must understand the outcome with sodium methoxide. Generally speaking, double bonds become thermodynamically more stable as the number of substituents increases. The conventional rationale for this observation is that substituents communicate with *p* orbitals in the π bond through hyperconjugation, in much the same way as we have seen them communicate with (and stabilize) *p* orbitals on radical and carbocationic centers. Using this argument, we would expect product A to be thermodynamically more stable by virtue of its more substituted double bond (trisubstituted vs. disubstituted for B). If this stability translates back to the transition state level, then we would expect product A to form faster and thus accumulate to a greater extent.

Apparently something different is happening with LDA and *t*-butoxide. Keep in mind that these are very bulky bases—as they approach the β-proton, they are very sensitive to the steric environment surrounding it. To produce the more stable product, the base must remove a relatively hindered ring proton—however, the methyl proton leading to product B is fairly accessible. In other words, changing to a more bulky base alters the relative positioning of the transition states, because steric interactions are worse for the pathway to A versus B. Product A is still thermodynamically more stable, but because the E2 reaction is virtually irreversible, we have trapped the system in a less stable outcome by modifying the activation energies (Figure 8-40). This is described as using kinetic control to obtain the thermodynamically least stable product.

GLOBAL TRENDS IN sp^3 REACTIVITY

The organic student is often confronted with having to predict whether a given sp^3 center will react via the S_N1, S_N2, E1, or E2 mechanism. Unfortunately, there is no

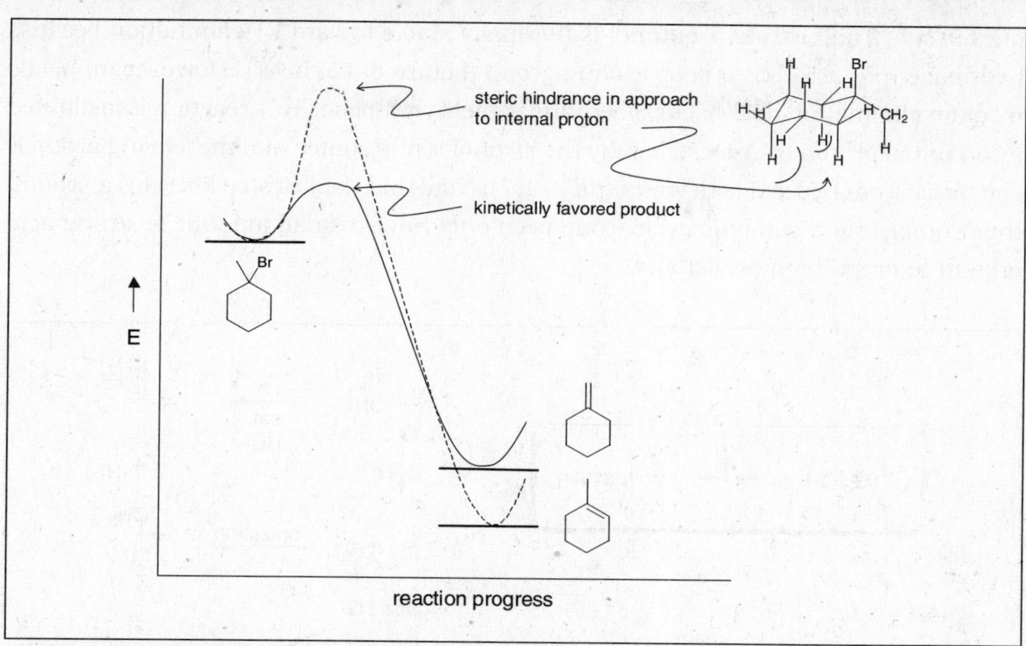

FIGURE 8-40: Kinetic control in the E2 elimation using LDA.

universal algorithm that generates pat answers; we can, however, establish a systematic method for assessing each situation and making supportable predictions about reactivity. Two useful categories of evaluation are the substrate itself and the experimental conditions in which the substrate is found.

First, all four of the reaction mechanisms require a good **leaving group**—without that prerequisite, no system progresses very far. Moreover, the better the leaving group, the faster the reaction. Next, we should characterize the **electrophilic center**—is it primary, secondary, or tertiary? If primary, we can rule out S_N1 and E1; if tertiary, we can rule out S_N2. Figure 8-41 summarizes these opposing reactivity trends.

S_N2 fast → slower → **stops**
Me—LG
LG
LG
LG
stops ← slower — fast S_N1/E1

FIGURE 8-41: Substrate trends for S_N2 and S_N1/E1.

Turning our attention to the conditions, one important piece of data is whether the medium is *acidic, neutral,* or *basic*. For example, very basic conditions can favor elimination pathways, because the key step in those mechanisms is proton abstraction. Alternatively, acidic conditions can actually convert poor leaving groups into good ones *in situ*. To illustrate, *t*-butanol is thermally stable toward E1 elimination because hydroxide (pK_a 15.7) is a poor leaving group (Figure 8-42, inset). However, in acidic medium elimination does occur, liberating 2-methylpropene. This reaction is facilitated by an initial proton transfer, whereby the alcohol is protonated and the leaving group is converted from hydroxide to water (pK_a −1.7). Note that the last step liberates a proton, thus completing a catalytic cycle—one need only have a small amount of strong acid present to promote this reactivity.

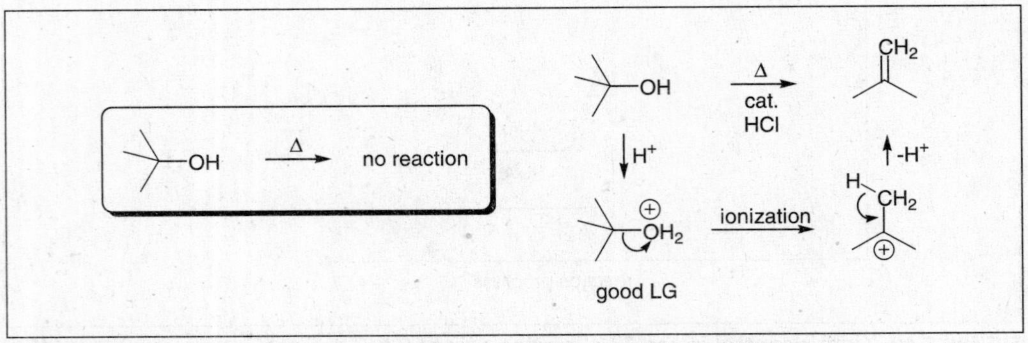

OH Δ no reaction

OH Δ cat. HCl CH₂

H⁺ -H⁺

OH₂ ionization H CH₂

good LG

FIGURE 8-42: Reactivity of *t*-butanol under neutral and acidic conditions.

Next, we consider the nature of the nucleophile/base, or more simply put, that which is interacting with the electrophilic substrate. Sometimes, the behavior is straightforward—for example, iodide is an excellent nucleophile, but not at all basic; however, LDA is extremely basic but nonnucleophilic because of its steric baggage. Very often, however, we must anticipate both basic (i.e., proclivity toward proton abstraction) and nucleophilic (i.e., tendency toward attacking electrophilic carbon) behavior from any given species. Methoxide is a handy example—it is a competent base as well as a nucleophile. Figure 8-43 provides a useful framework for thinking about these properties, that is, in terms of behavior and activity. Because many species can act as either a base or a nucleophile, the behavior axis is a way to visualize the idea of a nucleophilicity-to-basicity (Nu/B) ratio. For example, phenylsulfide (PhS^-) would have a high Nu/B ratio—because it is more nucleophilic than basic—and *t*-butoxide would have a low Nu/B ratio, in this case because its nucleophilicity is reduced by sterics. The activity axis is simply a way to think about the magnitude of the nucleophilic or basic property. To illustrate, sodium methoxide would represent a fairly aggressive reagent that would fall within the upper half of the reactivity space where bimolecular reactions are favored, whereas methanol would be a relatively passive medium, conducive to unimolecular processes in which the leaving group simply leaves without outside influence.

Combined with our knowledge of the substrate, we can use this framework to begin to understand and predict the subtle interplay among the reaction mechanisms. Figure 8-44 outlines the various outcomes possible from the interaction of ethyl and *t*-butyl bromide with methanol and methoxide. In the first example (ethyl bromide

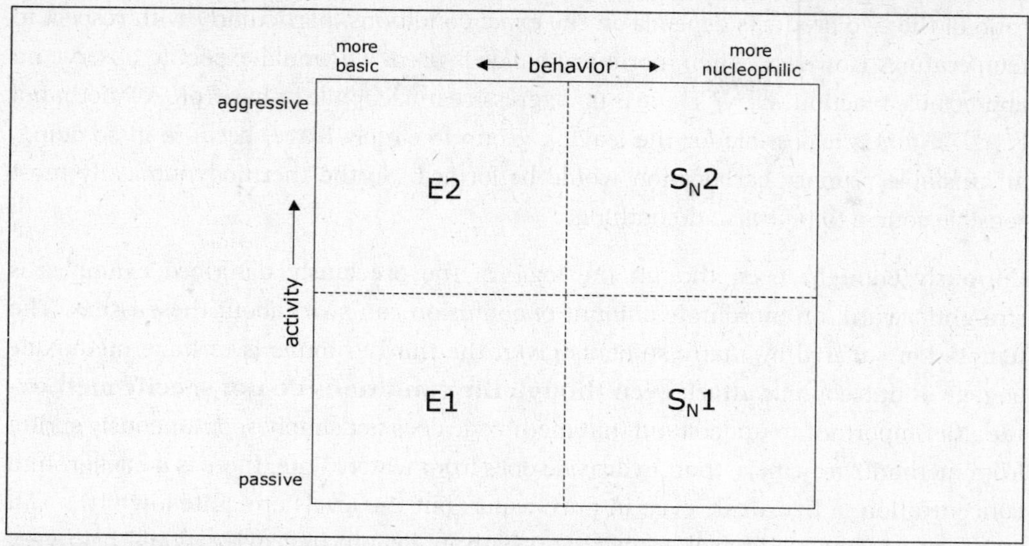

FIGURE 8-43: Cartesian space of the nucleophile/base (Nu/B).

Br — NaOMe / MeOH → OMe S_N2

Br — NaOMe / MeOH → E2

Br — MeOH → no reaction

Br — MeOH → E1 + OMe S_N1

FIGURE 8-44: Behavior of 1° and 3° bromides toward methanol and methoxide.

with methoxide), we would characterize the conditions as aggressive (i.e., in the upper two quadrants), so the top mechanistic suspects would be S_N2 and E2. Because the electrophile is sterically unencumbered, it is reasonable to expect S_N2 to predominate, although E2 is certainly a contender at high temperatures. In the second example (*t*-butyl bromide with methoxide), the conditions are still aggressive, but we can rule out S_N2 altogether because of the steric inhibition of the tertiary center. Thus, E2 is the only reasonable choice.

Things get more subtle when only methoxide is absent (second and third examples). Now the conditions are passive, so that the lower two quadrants (S_N1 and E1) become most plausible. With the tertiary bromide, we must consider both unimolecular mechanisms, because they are almost always in competition with each other. The exact ratio of the two products depends on the exact conditions, particularly with respect to temperature. However, when a primary halide is used, we would expect to observe no appreciable reaction. Why? There is no aggressive nucleophile or base (pK_a of methanol is –1.7), nor is it possible for the leaving group to simply leave, because in so doing, an unstable primary carbocation would be formed. So the thermodynamically most sensible course to take is to do nothing.

Curiously enough, even though the logic of the previously described examples is straightforward, an inordinate amount of confusion can swirl about these issues. The temptation suffered by many students given the third example is to have methoxide engage in nucleophilic attack **even though the conditions do not specify methoxide**. It is important to understand that methoxide does not simply spontaneously spring from methanol any more than hydroxide does from water. True, there is a background concentration of hydroxide even in pure water, but the levels are quite low (10^{-7} M, to be exact). One could well argue that a bath in sodium hydroxide would be a very different experience than a bath in water—by the same token, running a reaction in the presence of methoxide generally has a different outcome than the same reaction in only methanol. Even though methoxide is derived from methanol, it is methanol's conjugate base, and therefore a much more aggressive species—a different beast, as it were.

β-ELIMINATION

Whenever a leaving group is situated next to an anionic center, elimination is likely to occur. This process is known as **β-elimination** and is shown graphically in Figure 8-45. Even though a σ bond is sacrificed for a much weaker π bond, the thermodynamic drivers are the stability of the resulting anion and a favorable entropy term.

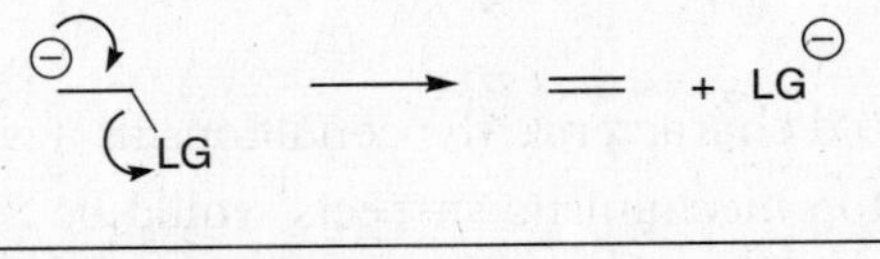

FIGURE 8-45: β-elimination of a carbanion.

The β-elimination gives rise to a more general phenomenon known as the $E1_{CB}$ elimination, which arises when a leaving group is situated next door to an electron-withdrawing (EW) moiety (Figure 8-46). The protons adjacent to the EW group are somewhat acidic, because the resulting anion is stabilized. Moreover, this negatively charged species can suffer β-elimination to give an alkene. In a sense, the $E1_{CB}$ is a stepwise analog of the E2 mechanism—unlike the E2, however, the $E1_{CB}$ tolerates even marginal leaving groups (such as alkoxide).

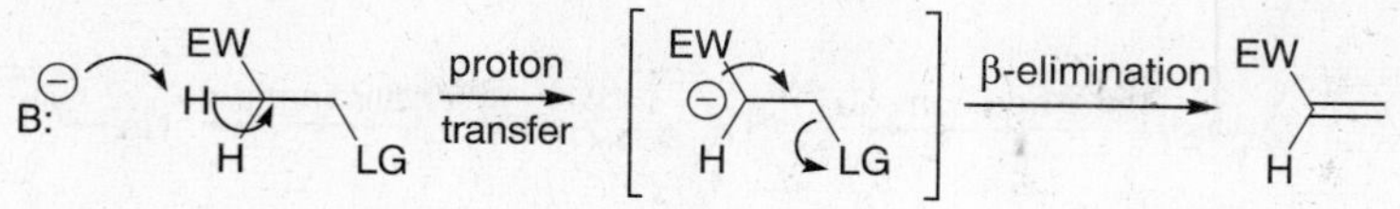

FIGURE 8-46: The $E1_{CB}$ elimination.

NUCLEOPHILIC ATTACK AT AN *sp*² CARBON

So far, we have discussed nucleophilic attack at an sp^3-hybridized electrophilic center bearing a leaving group. If we think of polar mechanisms being a flow of electron density from source to sink, then the nucleophile is the electron source and the leaving group is the electron sink. However, we can imagine another form of electron sink—which, after all, is just another way to describe a stabilized anionic charge. Figure 8-47 shows two such possibilities. In the first, a nucleophile adds to an electron-deficient double bond in such a way as to place the anionic center adjacent to the stabilizing electron-withdrawing group. This type of reaction is often called **conjugate addition,** because the stabilizing group is usually a π system in conjugation with the double bond.

The other common mode of addition onto an sp^2 center is by nucleophilic attack onto a carbonyl carbon (Figure 8-47, bottom). This provides, after aqueous workup, an alcohol—and various modes of carbonyl addition are frequently used for the synthesis of complex alcohols.

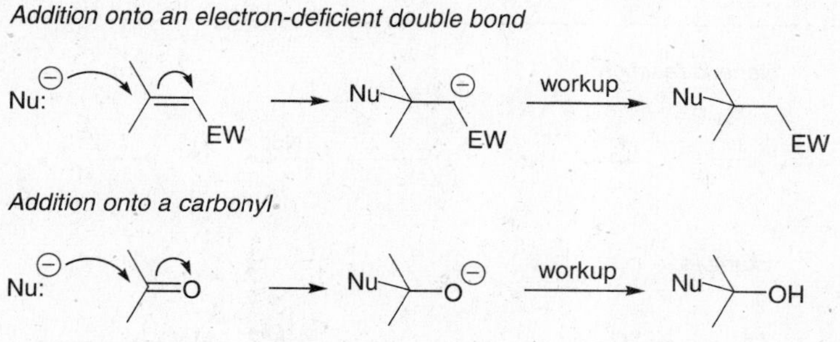

FIGURE 8-47: Nucleophilic attack at an sp^2 carbon.

In aromatic chemistry, certain substitution patterns give rise to a specific reactivity involving nucleophilic addition. When a leaving group is positioned in a 1,4-arrangement with respect to an electron-withdrawing group on a benzene ring, the aromatic system can undergo what is known as **ipso** (*lit.*, at this site) attack of the nucleophile (Figure 8-48). Formally, it looks as if the leaving group has been displaced in an S_N2-type reaction, but this would be improbable from a molecular orbital point

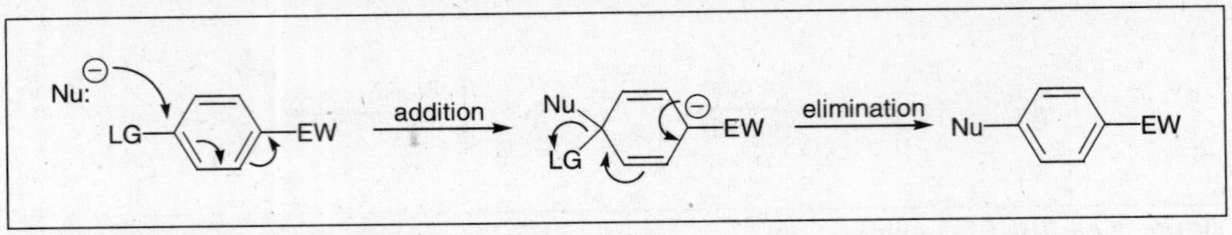

FIGURE 8-48: Nucleophilic aromatic substitution.

of view: the σ* orbital that the nucleophile starts to fill in an S_N2 reaction is positioned directly inside the benzene ring, making it quite inaccessible. Instead, the mechanism involves initial conjugate addition to provide an anionic intermediate, which collapses through eliminative ejection of the leaving group to restore aromaticity with the new substituent on board. This sequence also works with a 1,2-substitution pattern.

ELECTROPHILIC CAPTURE OF A DOUBLE BOND

In the preceding examples, we have looked at electron-deficient π systems that are susceptible to attack from nucleophiles. However, the large majority of alkenes are actually somewhat nucleophilic. This makes sense if we consider the highest occupied π molecular orbital (HOMO) of ethene, for example (Table 3-1). The π electron density is riding rather high atop the molecule, far removed from the coulombic tether of the nuclei. It seems intuitively supportable, then, to posit that the π bond might be polarized toward electropositive species, and in some cases undergo electrophilic capture, as illustrated in Figure 8-49.

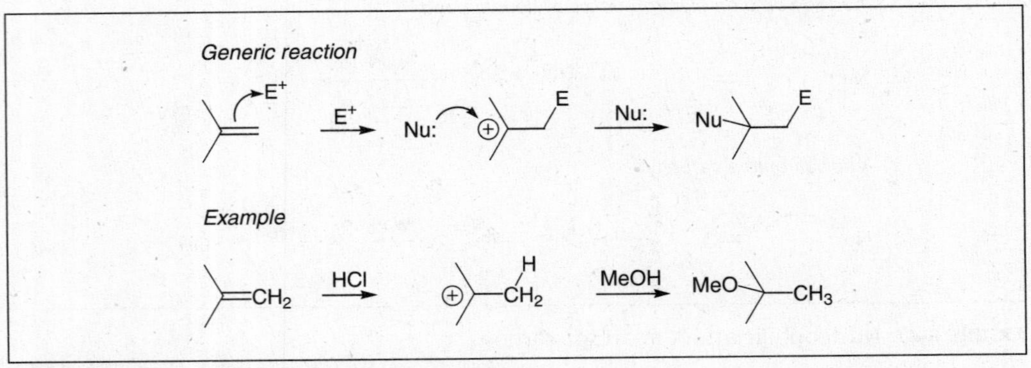

FIGURE 8-49: Electrophilic attack of an alkene.

The carbocation generated by the electrophilic attack is usually trapped by some weak nucleophile (e.g., solvent), much like the second step of an S_N1 reaction. If the alkene is unsymmetric (i.e., differently substituted on either side of the double bond), then the electrophilic attack will occur in such a way as to give the more substituted (hence, more stable) carbocation. Thus, the pendant nucleophile is generally found at the more

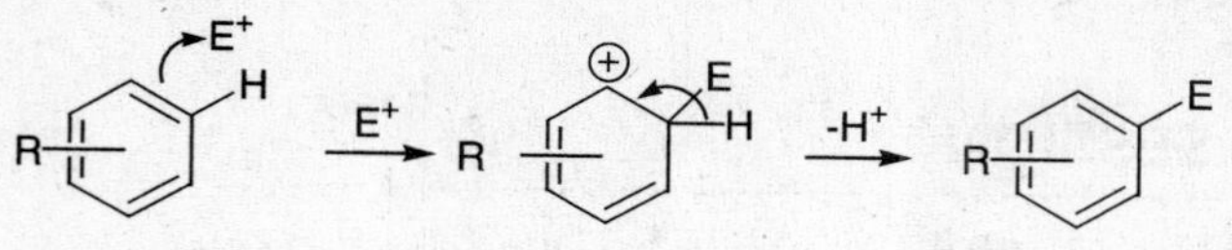

FIGURE 8-50: Electrophilic aromatic substitution.

substituted position. This regiochemical outcome is known as the **Markovnikov effect,** and the more substituted product is dubbed the **Markovnikov product.**

This mode of reactivity is also enormously important in aromatic chemistry. As shown in Figure 8-50, an aromatic ring can capture an electrophile to form a doubly allylic carbocationic intermediate, which quenches the positive charge by loss of a proton to form a new aromatic species. The first step is energetically challenging, because it destroys aromaticity; consequently, only very strong electrophiles engage in aromatic substitution. By the same token, proton loss to regain aromaticity is usually far more preferred than the nucleophilic capture observed in nonaromatic π systems (Figure 8-49).

CRAM SESSION

Polar Chemistry

The lion's share of reactions encountered in organic chemistry fall into the category of polar chemistry, or processes that involve the generation of charged species through **heterolytic** processes. The most common polar reactions are

- Proton transfer
- Ionization
- Nucleophilic capture
- Nucleophilic displacement (S_N2)
- Bimolecular elimination (E2)
- Loss of a proton to form a π bond
- β-elimination of a carbanion

Each of these is a **concerted** reaction involving a single transition state. There are also some common combinations of these basic steps:

- S_N1, ionization + nucleophilic capture
- E1, ionization + loss of a proton to form a π bond
- $E1_{CB}$, proton transfer + β-elimination of a carbanion

These are **stepwise** mechanisms involving multiple transition states and proceeding through an **intermediate** (a carbocation in the case of S_N1 and E1; a carbanion in the case of $E1_{CB}$).

Most polar reactions can be described as the interaction of a Brønsted base with a proton source, or as a **nucleophile** (donor of electron density) with an **electrophile** (acceptor of electron density). Generally speaking, a nucleophile is characterized by an active lone pair, whereas an electrophile can be described as an electron-deficient center. Common forms of the latter are carbocations, tetrahedral carbons attached to a **leaving group**, and electron-deficient double bonds (olefins and carbonyls).

In predicting which particular mechanistic pathway will be dominant, the following factors should be taken into account:

- The nucleophile (How nucleophilic is it? How basic is it?)
- The electrophile (Is it primary, secondary or tertiary?)
- The conditions (temperature, solvent, and concentration)

For example, high temperatures and basic environments favor elimination reactions over substitution, and aggressive conditions (strong nucleophiles and/or bases) favor bimolecular processes over unimolecular ones. Polar protic solvents such as ethanol favor ionization processes (i.e., S_N1 and E1), whereas polar aprotic solvents such as DMSO and DMF favor S_N2 reactions due to the "naked anion" effect.

In terms of **trends**, both basicity and nucleophilicity tend to decrease from left to right on the periodic table (e.g., NH_3 vs. H_2O). However, basicity decreases from top to bottom, whereas nucleophilicity increases (e.g., F^- vs. I^-). When evaluating leaving groups, remember that the stronger the conjugate acid, the better the leaving group (i.e., HCl is a strong acid; Cl^- is a good leaving group). The stability of carbocations (and thus the likelihood of ionization) increases from primary to tertiary centers, and the rate of S_N2 decreases along the same series due to steric constraints.

Steric considerations can also be used for control and selectivity in E2 reactions. Thus, an elimination promoted by a small base like methoxide tends to give the more substituted (and more stable) alkene, while the use of a **hindered base** such as LDA often yields the least substituted alkene through a phenomenon known as **kinetic control**.

Sometimes stereochemical outcomes can provide clues to reaction mechanism. For example, S_N2 displacement leads to **inversion of configuration** as a result of **backside attack**, and stereochemical information is preserved. Thus, an optically pure starting material will yield an optically pure product. In contrast, an S_N1 reaction results in loss of stereochemical information, so that an optically pure starting material will yield a **racemic** product (50/50 mixture of both enantiomers).

CHAPTER 9

Pericyclic Chemistry

Read This Chapter to Learn About:

- Sigmatropic Rearrangements
- Cycloadditions
- Electrocyclizations

The final class of reactions we consider fall under the umbrella of pericyclic processes. These are reactions that involve the concerted rearrangement of electrons (usually, as the name suggests, in some cyclic array) without the generation of charge or radical character. Pericyclic chemistry encompasses a wide variety of reactions, but here we discuss three broad categories that are illustrative, frequently encountered, and synthetically relevant.

SIGMATROPIC REARRANGEMENTS

The etymologic origin of the sigmatropic rearrangement relates to the apparent relocation of a σ bond in concert with π-bond migration. Figure 9-1 illustrates one such event, known as the **Cope rearrangement.** At face value, a σ bond has moved from the right side to the left—however, there are actually six electrons in play. The classical depiction of electron motion uses three double-headed arrows (left inset); however, another way of imagining this process is that two allyl radicals form by σ-bond cleavage and the resonance forms of each reconnect at the other end (suggested by the right inset)—this allows for a more detailed molecular orbital treatment (which we do not pursue here), even though radical species are never formed *per se.*

Formally, these processes are known as **[3,3] rearrangements,** because there is a three-atom array walking across a three-atom array. If you imagine the hand movements to the kindergarten song "The Itsy Bitsy Spider," then your hands are the

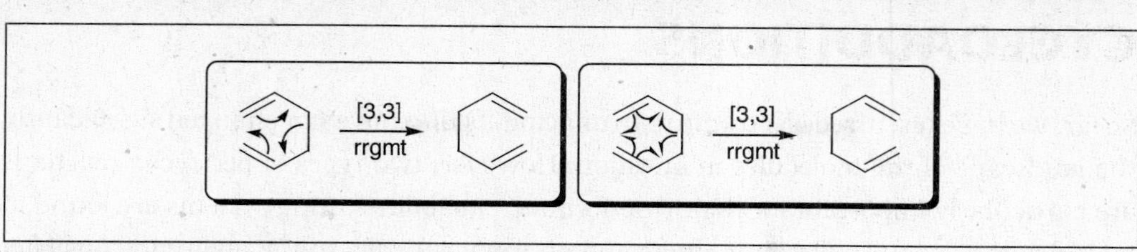

FIGURE 9-1: Two electronic interpretations of a [3,3]-rearrangement.

three-carbon fragments, and your thumbs and pinkies are the ends of those fragments. Other fragment sizes are certainly possible (the [2,3] is common), but they are outside the scope of the present discussion.

Atoms other than carbon are also fair game. For example, oxygen can replace one of the sp^3 carbons in what is known as the **Claisen rearrangement** (Figure 9-2). Unlike the Cope, which converts a diene to another diene, the Claisen actually provides different functionality. Thus, we start with an allyl vinyl ether, and the rearrangement results in a γ,δ-unsaturated carbonyl compound. In fact, this is a good way to scout the Claisen in a reaction sequence: allyl vinyl ethers and/or γ,δ-unsaturated carbonyls are often smoking guns that a Claisen is in play.

FIGURE 9-2: The Claisen rearrangement.

There is another manifestation of the Claisen that can be tricky to apprehend initially, as it does not provide said γ,δ-unsaturated ketone as a final product. In fact, this was the original version of the Claisen rearrangement (Figure 9-2, inset). Here, an allyl phenyl ether—obtained by the allylation of the phenoxide anion—undergoes thermal [3,3] rearrangement indeed to provide a cyclic γ,δ-unsaturated ketone, but it is thermally unstable. Through the process of tautomerization, whereby a π bond and a hydrogen atom swap places (which can be catalyzed by either catalytic acid or base), aromaticity is restored. Thus, the overall outcome is a migration of the allyl moiety from the oxygen to the carbon atom two doors down.

CYCLOADDITIONS

So far, we have discussed electronic rearrangements that have not changed significantly the landscape of the molecules in question. However, two types of pericyclic reactions are particularly important for their ring-forming capabilities. Ring systems are found in biologically relevant natural products and pharmaceuticals, so the ability to construct complex cyclic architecture is a concern for the synthetic chemist.

One workhorse protocol is the **Diels–Alder reaction**, named after the reaction's discoverers (Otto Diels and Kurt Alder), but also known more generically as the **[4 + 2] cycloaddition.** Here the formal nomenclature is not based on the number of atoms involved, but the number of electrons. Thus, a 4π component (called the **diene**) and a 2π component (called the **dienophile**) come together to form a cyclohexene, as shown in Figure 9-3. Again, we can see that the reaction involves the concerted motion of six electrons—this is important, because the transition state has considerable aromatic character, an outcome that provides for a particularly low activation barrier.

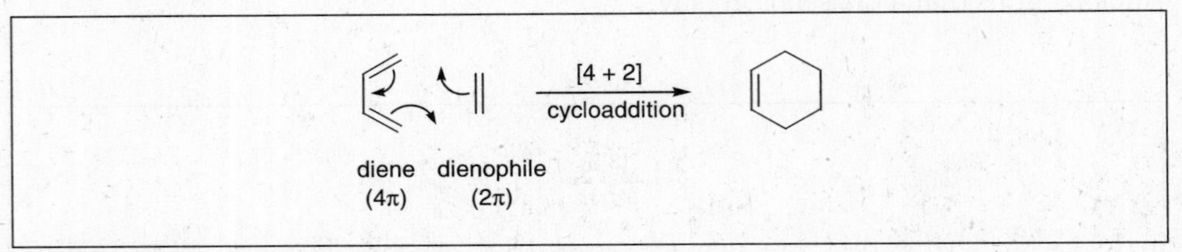

FIGURE 9-3: The prototypical Diels–Alder reaction.

From a molecular orbital point of view (Figure 9-4), the Diels–Alder reaction involves the interaction between the highest occupied molecular orbital (HOMO) of the diene and the lowest unoccupied molecular orbital (LUMO) of the dienophile. Notice that the phases between those two frontier orbitals align constructively during initial approach (inset). Another practical consideration emerges from the MO treatment—as the energy gap between the two interacting orbitals diminishes, so the interaction itself is magnified. Therefore, typical Diels–Alder dienes are electron rich (raising the energy of the HOMO) and typical dienophiles are electron poor (lowering the energy of the LUMO).

The very defined nature of the transition state is also responsible for the remarkable degree of stereochemical control during the reaction. Notice that the Diels–Alder can produce as many as four contiguous chiral centers (as shown in Figure 9-5), and we can predict the stereochemistry of these centers with a certain amount of accuracy. First, notice that the ***trans***-geometry of the dienophile is preserved in the product—that is, the methyl and carbonyl were *trans* before and they remain *trans*. The somewhat more subtle consideration revolves around the relative stereochemistry between the

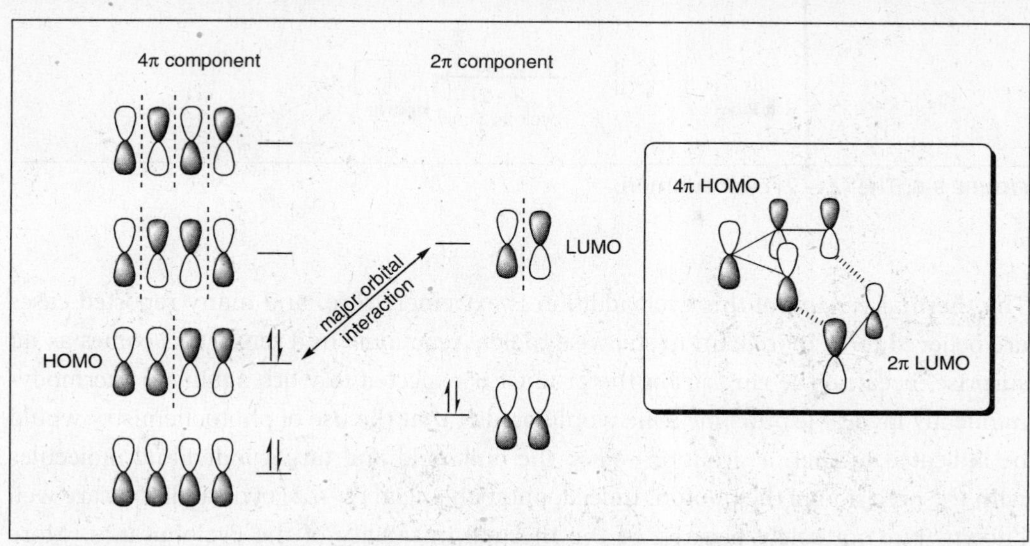

FIGURE 9-4: The molecular orbital underpinnings of the Diels–Alder reaction.

FIGURE 9-5: Stereochemical implications for the Diels–Alder reactions.

methylene bridge and the carbonyl group. It turns out that in the initial approach, the carbonyl prefers to tuck into the underbelly of the diene. Why this is so is the matter of some debate—the conventional wisdom rationalizes it on the basis of favorable secondary orbital overlap between the carbonyl π system and the diene, a phenomenon known as π **stacking.** Whatever the reason, the observable outcome is that the carbonyl group ends up in the endo position of the product (inset). This is surprising, because the sterically more favorable real estate lies toward the shallower top face, or **exo** position. Therefore, this anomaly is known as the **endo rule.**

Cycloadditions are not limited to the [4 + 2] variety. In fact, there are many types, but we treat only one other here, namely the [2 + 2] cycloaddition (Figure 9-6). This is a very useful protocol for constructing cyclobutanes, which can be a synthetic challenge. In this reaction, two "enes" come together to build the cyclic species.

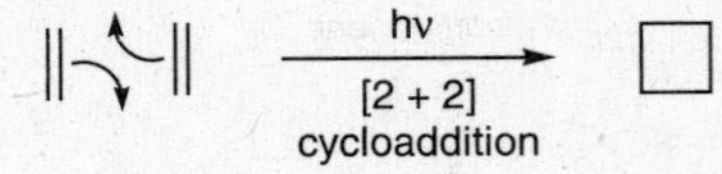

hv
[2 + 2]
cycloaddition

FIGURE 9-6: The [2 + 2] cycloaddition.

The thermal version of this cycloaddition is extremely rare, and many reported cases are believed to be more akin to stepwise radical reactions. In a sense, this comes as no surprise, because the ring strain the product is expected to work against a thermodynamically favorable outcome. One would predict that the use of photochemistry would be indicated in such a situation—keep the bulk cold and target individual molecules with the precision of the photon. Indeed, photochemical [2 + 2] cycloadditions are well known, but not solely because of the thermal instability of the cyclobutanes. More importantly, MO theory (Figure 9-7) makes clear that the interaction of the ground state HOMO with the ground state LUMO would not provide constructive interaction of molecular orbitals. However, if one molecule is promoted to the first excited state, then what once was a LUMO is now the highest occupied orbital (designated the **SOMO,** or **singly occupied molecular orbital**). Thus, the SOMO of the excited molecule interacts with the LUMO of another ground state molecule to provide constructive orbital overlap (inset).

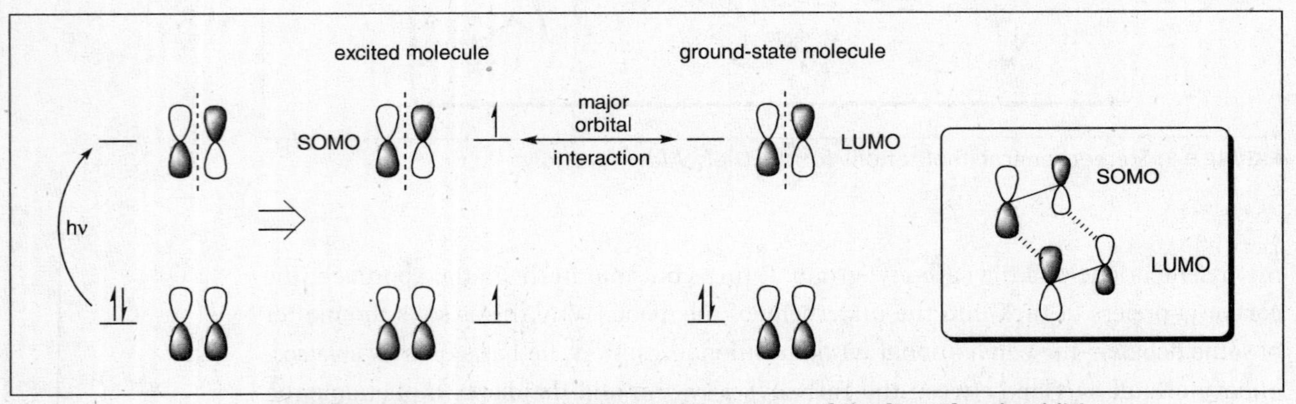

FIGURE 9-7: The molecular orbital underpinnings of the [2 + 2] cycloaddition.

ELECTROCYCLIZATIONS

Another way to produce cyclic molecules from acyclic precursors is through electrocyclization. This is a process involving an extended π system, and could encompass a rather wide array of *p* orbitals. We consider here two types: 4π and 6π. The overall process is the conversion of a conjugated diene to a cyclobutene (4π) or a conjugated triene to a cyclohexatriene (6π), as shown in Figure 9-8. Both are best considered equilibrium reactions, and the reverse process is known as a cycloreversion.

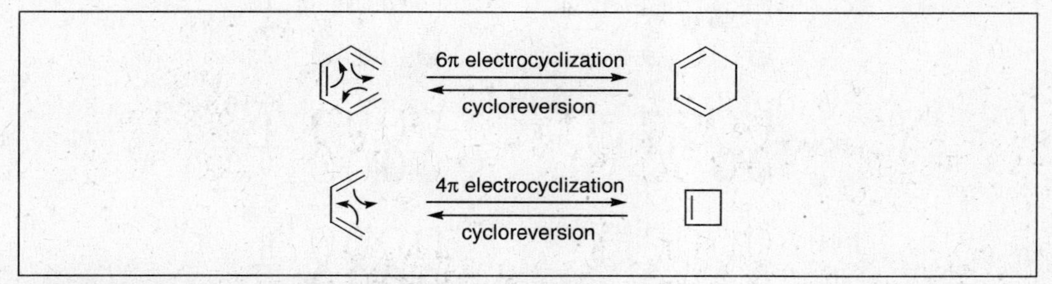

FIGURE 9-8: Two types of reversible electrocyclization.

The stereochemistry of this reaction is remarkably fixed and predictable. In fact, the observation that the stereochemical outcome always fell into certain patterns opened up many of the applications of MO theory that we use today (and also earned a Nobel prize). For example, if *trans,trans*-2,4-hexadiene (Figure 9-9) is subjected to photochemical conditions, only the *cis*-3,4-butadiene is formed. The question is, why?

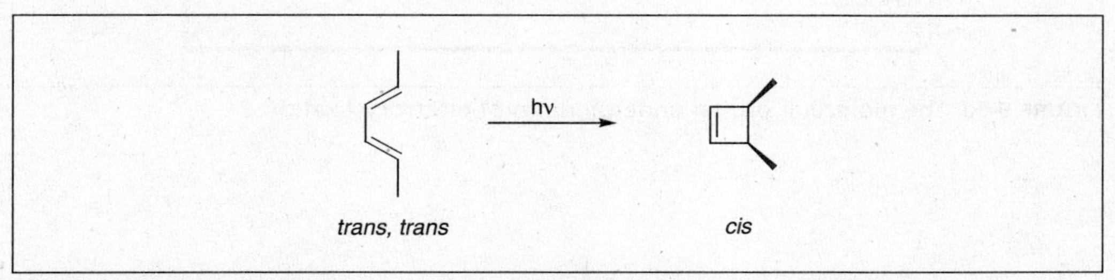

FIGURE 9-9: The stereochemical outcome of photolytic 4π electrocyclization.

An examination of molecular orbital theory provides an explanation (Figure 9-10). Exposure to light of the appropriate wavelength promotes an electron into the LUMO, thus converting it to the SOMO, which now controls the course of the reaction. If we superimpose the SOMO onto a three-dimensionally competent depiction of the diene (inset), we see that the two end lobes must turn in toward each other to enjoy constructive orbital overlap. This is known as a **disrotatory ring closure,** because the right orbital is turning counterclockwise and the left orbital is turning clockwise, thus in opposite senses. If the ring closure is attempted thermally, then the HOMO dictates reactivity and the electrocyclization proceeds through a conrotatory process (both orbitals turning clockwise), and the cyclobutene exhibits *trans* stereochemistry.

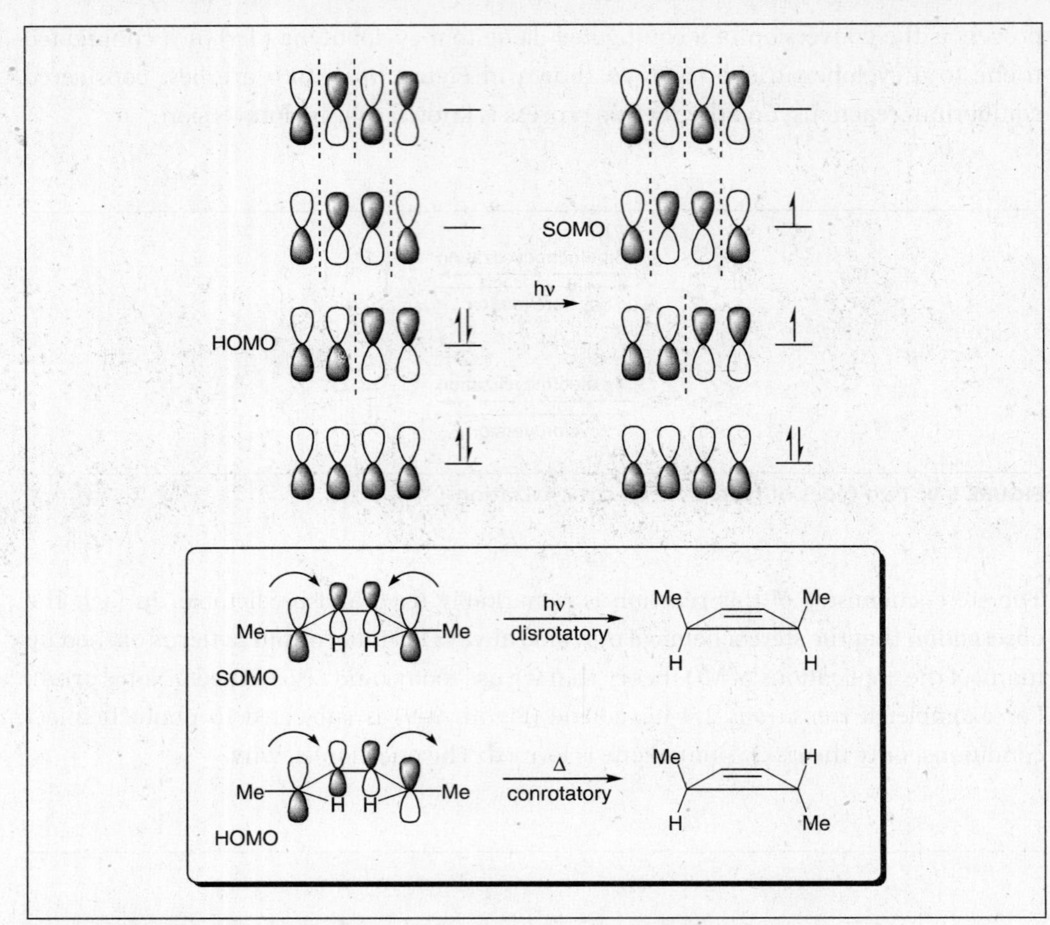

FIGURE 9-10: The molecular orbital underpinnings of electrocyclization.

CRAM SESSION

Pericyclic Chemistry

Another broad category of chemical processes, pericyclic reactions involve the concerted motion of electrons in such a way that no charges are generated. As the name implies, they also proceed through a cyclic transition state. Pericyclic reactions fall into three broad categories:

- Sigmatropic rearrangements
- Cycloadditions
- Electrocyclizations (and cycloreversions)

The common sigmatropic rearrangments include the **Cope rearrangement** and the **Claisen rearrangement.** Both of these reactions occur with 1,5-diene systems whereby a σ bond appears to migrate from one location to another. No rings are formed in the process.

In contrast, cycloadditions create ring systems from acyclic precursors. The most frequently encountered examples are the thermal **Diels–Alder reaction,** in which a **diene** (or 4π component) reacts with a **dienophile** (or 2π component); and the photochemical **[2 + 2] cycloaddition,** in which two simple olefins combine to form a cyclobutane. The stereochemistry of both processes is governed by orbital symmetry rules, which can be easily understood (and predicted) by considering **HOMO–LUMO interactions.** When cyclopentadiene reacts with acrolein, the carbonyl group ends up in the sterically unfavorable endo position of the product, a phenomenon known as the **endo rule.**

In electrocyclizations, an extended π system converts to a closed ring form, whereby one of the π bonds is transformed into a σ bond. This process is also governed by orbital symmetry rules: by examining the HOMO of the substrate, one can predict whether the electrocyclization will be **conrotatory** (that is, both ends rotate in the same sense) or **disrotatory** (the ends rotate in opposite senses). Photochemical electrocyclizations proceed with a different stereochemical outcome than their thermal counterparts because an electron is promoted to the LUMO (making it a SOMO), which then governs the orbital symmetry of the reaction.

Conceptual Map of Section II

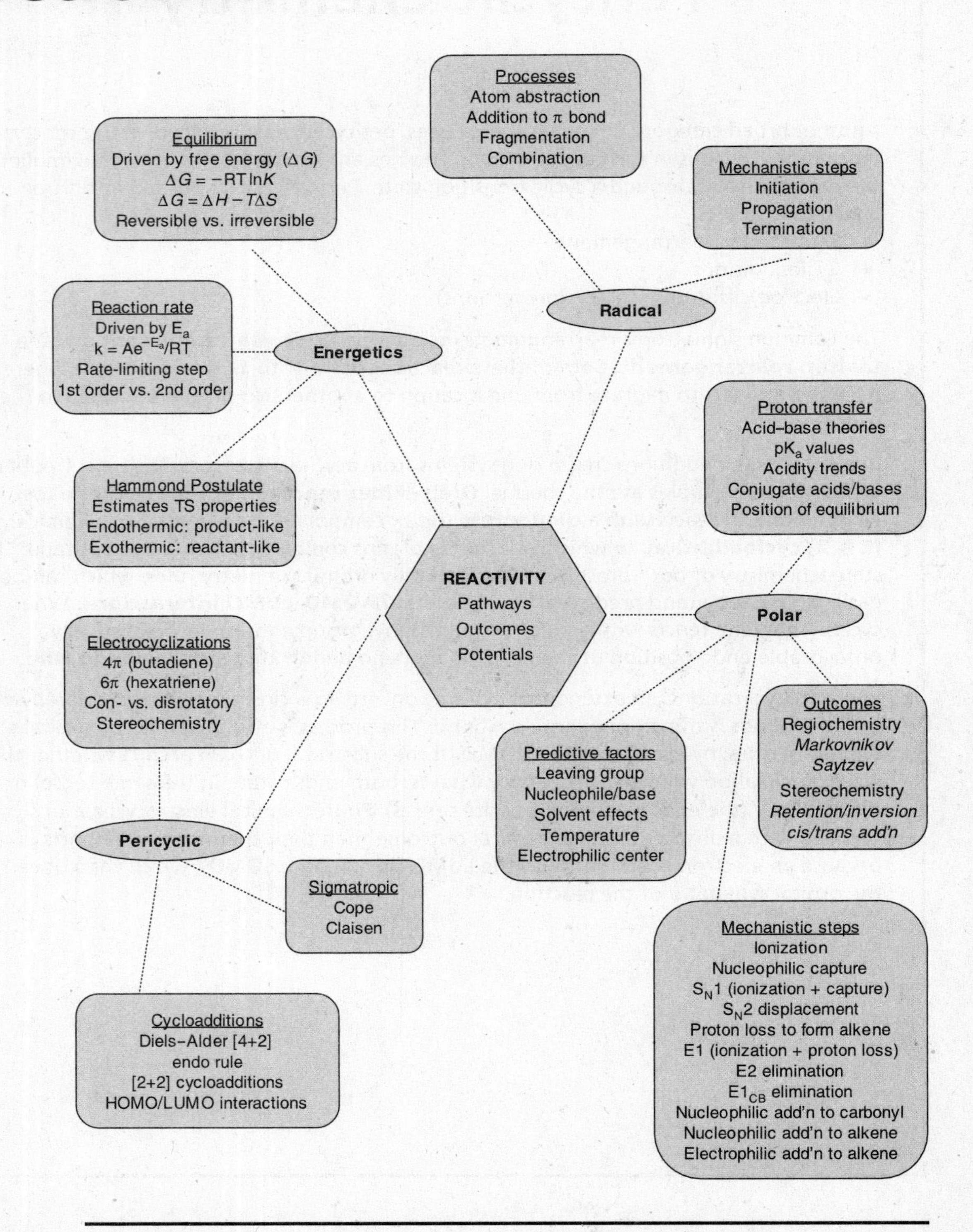

SECTION III

METHODOLOGY

CHAPTER 10

Alcohols and Ethers

Read This Chapter to Learn About:

- Preparation of Alcohols
- Reactions of Alcohols
- Preparation of Ethers
- Reactions of Ethers

PREPARATION OF ALCOHOLS

Alkenes are susceptible to acid-catalyzed hydration (Reaction 1) to form the more substituted alcohol—the so-called Markovnikov product. Under these conditions, *p*-toluenesulfonic acid (*p*TsOH) is used as a convenient, organic-soluble source of proton, similar in acidity to sulfuric acid. The regioselectivity is driven by the formation of the more stable carbocation, which is captured by water. A drawback of the methodology is that the carbocations can undergo rearrangement. One way to avoid this unwanted chemistry is to hydrate the double bond in the presence of mercury(II) cation (which acts as a surrogate proton), and then reduce the resulting organomercury intermediate with sodium borohydride, a protocol known as **oxymercuriation/demercuriation** (Reaction 2). The advantage here is that the initially formed mercurinium species never fully opens to a carbocation; therefore, rearrangement is suppressed.

cat. *p*TsOH; H_2O, dioxane → HO (R1)

$Hg(OAc)_2$; H_2O, dioxane → $NaBH_4$ → HO (R2)

Alkenes may also be hydrated in an anti-Markovnikov fashion by first being converted to an organoborane intermediate through hydroboration. This species is then subjected to basic oxidative conditions to provide the less substituted alcohol and sodium borate as a by-product. The regioselectivity here is driven not by carbocation stability, but by sterics—the boron (which attaches to what will be the alcohol center) is delivered to the less-substituted position.

BH_3, THF; NaOH, H_2O_2 → OH (R3)

Many synthetically important methods for alcohol synthesis involve ketones. For example, ketones and aldehydes can be reduced to the corresponding alcohols by hydride sources, such as lithium aluminum hydride (LAH) and sodium borohydride (Reaction 4). Carbon-based nucleophiles also add to carbonyls. For example, alkylmagnesium halides serve as sources of carbanions, which add to carbonyls in a protocol known as the **Grignard reaction** (Reaction 5). The nice aspect of this strategy from the synthetic standpoint is that a new carbon–carbon bond is formed in the process, and more complex molecules can be accessed. [Note: As a general convention, we assume aqueous workup (H_3O^+/H_2O) for all reactions.]

O → $NaBH_4$ or LAH → OH (R4)

O → RMgBr, THF → OH, R (R5)

Reductive strategies can also launch from other carbonyl compounds, such as esters. For example, the use of a strong hydride source (such as LAH) leads to the complete reduction of esters to the corresponding alcohols (Reaction 6). Notice that the alcohols produced are necessarily primary.

O, OR → LAH, THF → OH (R6)

Alcohols can be synthesized through the S_N2 displacement reaction using hydroxide as a nucleophile (Reaction 7). Suitable substrates incorporate unhindered electrophilic centers with good leaving groups, such as alkyl halides, tosylates, and mesylates (for structures, please see Appendix). Because competing elimination is sometimes problematic, primary substrates are preferred.

R X → NaOH, dioxane → R OH (R7)

X = Br, Cl, I, OTs, OMe

Finally, epoxides suffer ring opening in the presence of a variety of nucleophiles, including hydroxide, alkoxides, amines, and carbanions, to give the corresponding ethanol derivatives (Reaction 8). The finer details of epoxide chemistry are discussed in Chapter 14.

Epoxide —Nu→ Nu–CH2–CH2–OH (R8)

REACTIONS OF ALCOHOLS

Secondary alcohols are oxidized to the corresponding ketones (Reaction 9) in the presence of the Jones reagent (potassium dichromate in aqueous sulfuric acid). These conditions are simply a way to produce chromic acid *in situ* (that is, directly in the reaction vessel), which in turn provides equilibrium quantities of chromium trioxide, the active oxidizing agent. Primary alcohols are oxidized all the way to the carboxylic acid, because the initially formed aldehyde undergoes hydration to the **gem**-diol (in which two hydroxy groups are on the same carbon) under the conditions—an equally competent substrate for oxidation (Figure 10-1). The reaction can be stopped at the aldehyde stage, however, by using pyridinium chlorochromate (PCC). This reagent also provides chromium trioxide, but in a way that does not liberate water (Figure 10-2). Without water, the aldehyde cannot form the gem-diol and is thus trapped.

$RCH_2OH \xrightarrow{CrO_3} RCHO \underset{}{\overset{H_2O}{\rightleftharpoons}} RCH(OH)_2 \xrightarrow{CrO_3} RCOOH$

FIGURE 10-1: The role of water in the Jones oxidation of primary alcohols.

Jones conditions

$K_2Cr_2O_7 \xrightarrow[H_2O]{H^+} 2\ HO\text{–}CrO_2\text{–}OH \rightleftharpoons CrO_3 + H_2O$

Pyridinium chlorochromate (PCC)

$C_5H_5NH^+\ {}^-O\text{–}CrO_2\text{–}Cl \longrightarrow CrO_3 + Cl^-$ *(anhydrous)*

FIGURE 10-2: Two sources for chromium trioxide for oxidation of alcohols.

$(CH_3)_2CHOH \xrightarrow[H_2SO_4,\ H_2O]{K_2Cr_2O_7} (CH_3)_2C{=}O$ (R9)

R–CH₂–OH $\xrightarrow[H_2SO_4,\ H_2O]{K_2Cr_2O_7}$ R–C(=O)–OH (R10)

$CH_3CH_2OH \xrightarrow[CH_2Cl_2]{PCC} CH_3CHO$ (R11)

It is often useful to convert the hydroxy functionality of alcohols into some other moiety that serve as a good leaving group. For example, alcohols can be treated with phosphorous tribromide (PBr_3) to give the corresponding alkyl bromide (Reaction 12), or thionyl chloride ($SOCl_2$) to obtain the alkyl halide (Reaction 13). Other good leaving groups include the tosylate (Reaction 14) and the mesylate (Reaction 15) groups. The resulting substrate can then be subjected to nucleophilic substitution of base-catalyzed elimination.

$CH_3CH_2OH \xrightarrow{PBr_3} CH_3CH_2Br$ (R12)

$CH_3CH_2OH \xrightarrow{SOCl_2} CH_3CH_2Cl$ (R13)

$CH_3CH_2OH \xrightarrow{TsCl} CH_3CH_2OTs$ (R14)

$CH_3CH_2OH \xrightarrow{MsCl} CH_3CH_2OMs$ (R15)

Alcohols are dehydrated in the presence of acid and heat through an S_N1 process. Two things can be problematic with this protocol. First, regiochemistry can be very hard to control in unsymmetric alcohols. Second, rearrangements of the intermediate carbocations can lead to product mixtures. A synthetically more relevant method might be to convert the alcohol to a tosylate and then subject it to a base-promoted E2 elimination.

$CH_3CH(OH)CH_2CH_3 \xrightarrow[\Delta]{p\text{TsOH}} CH_3CH{=}CHCH_3$ (R16)

Esters are produced when alcohols are treated with acyl chlorides (Reaction 17). This reaction proceeds even without added base. However, because a full equivalent of HCl is liberated in the course of the alcoholysis, a mild base (such as pyridine or triethylamine) is often included.

R–OH $\xrightarrow{R'C(=O)Cl}$ R'–C(=O)–O–R (R17)

PREPARATION OF ETHERS

In principle, ethers can be formed from alcohols in either basic or acidic conditions. In practice, however, the base-catalyzed variant (Reaction 18) is the more general. This reaction is known as the **Williamson ether synthesis,** and it involves the deprotonation of an alcohol to form the corresponding alkoxide, which then engages in S_N2 displacement to provide an ether. There is, however, at least one instance where the acid-catalyzed variant provides us with an avenue not otherwise available, namely in the formation of tertiary ethers (Reaction 19). The mechanism is akin to the acid-catalyzed hydration of alkenes (Reaction 1), whereby an intermediate carbocation is intercepted by the alcohol.

R—OH base [R—O⊖] R′ LG R O R′ (R18)

R—OH pTsOH R—O (R19)

REACTIONS OF ETHERS

From a synthetic standpoint, ethers do not undergo a great many reactions under normal conditions—which is why they are well suited to be organic solvents (for example, THF and diethyl ether). In severely acidic environments, ethers suffer proton-catalyzed elimination, but under conditions in which almost any organic compound decompose. As one notable exception to this stability, however, tertiary ethers are relatively sensitive to acid. For example, when warmed in the presence of *p*-toluenesulfonic acid, *t*-butyl ethers undergo E1 elimination to liberate 2-methylpropene and an alcohol (Reaction 20). This reaction is so facile that the *t*-butyl group is often used for the temporary protection of alcohol functionality.

R—O cat. pTsOH Δ R—OH + (R20)

CRAM SESSION

Alcohols and Ethers

Alcohols can be prepared from

- The hydration of **alkenes** (Markovnikov and anti-Markovnikov)
- The reduction of **ketones, aldehydes,** and **esters**
- The addition of organometallics to ketones and aldehydes (the **Grignard reaction**)
- The reaction of hydroxide with **alkyl halides** and **tosylates**
- The nucleophilic ring opening of **epoxides**

Common reactions of alcohols are

- Oxidation of primary alcohols to **carboxylic acids** (Jones reagent)
- Oxidation of primary alcohols to **aldehydes** (PCC)
- Oxidation of secondary alcohols to **ketones** (Jones reagent or PCC)
- Conversion of the hydroxyl to a **good leaving group** (Cl, Br, OTs, OMs)
- Dehydration to an **alkene**
- Reaction with an acyl halide to form an **ester**

Ethers can be prepared from

- The reaction of an **alkoxide with an alkyl halide** (the Williamson ether synthesis)
- The acid-catalyzed reaction of an **alkene with an alcohol**

Ethers tend to be chemically inert under most conditions, but in a strongly acidic environment they can fragment to their **alcohol and alkene components.**

CHAPTER 11

Ketones and Aldehydes

Read This Chapter to Learn About:

- ➤ Preparation of Ketones and Aldehydes
- ➤ Reactions of Ketones and Aldehydes

PREPARATION OF KETONES AND ALDEHYDES

Carbonyl compounds are available from the corresponding alcohols through oxidation. Many reagents can be used for this protocol, but two common examples are the oxidation of secondary alcohols to ketones using Jones conditions (Reaction 9) and the selective oxidation of primary alcohols to aldehydes using PCC (Reaction 11). Of course, PCC is effective with secondary substrates, as well; in fact, it is the reagent of choice for acid-sensitive alcohols.

$$\text{(CH}_3)_2\text{CHOH} \xrightarrow[\text{H}_2\text{SO}_4,\ \text{H}_2\text{O}]{\text{K}_2\text{Cr}_2\text{O}_7} (\text{CH}_3)_2\text{C=O} \qquad \text{(R9)}$$

$$\text{CH}_3\text{CH}_2\text{OH} \xrightarrow[\text{CH}_2\text{Cl}_2]{\text{PCC}} \text{CH}_3\text{CHO} \qquad \text{(R11)}$$

When the alcohol is allylic or benzylic, a particularly mild method is available, namely treatment with manganese dioxide in a solvent like methylene chloride (Reaction 21). There are very few other functionalities that are affected by MnO_2, including other alcohols. Thus, allylic and benzylic alcohols can be selectively oxidized in the presence of other alcohol functionality.

$$\text{C}_6\text{H}_5\text{CH(OH)CH}_3 \xrightarrow[\text{CH}_2\text{Cl}_2]{\text{MnO}_2} \text{C}_6\text{H}_5\text{COCH}_3 \qquad \text{(R21)}$$

Ketones and aldehydes are also available from the ozonolysis of olefins or alkenes (Reaction 22). Here two carbonyl compounds are liberated from a single alkene—usually, only one of the products is the desired target. A similar strategy launches from 1,2-diols (also known as *vic*-diols), which oxidatively cleave when treated with periodic acid (*not* pronounced like Mendeleev's table!). Because 1,2-diols can be derived from alkenes by dihydroxylation (Reaction 48), this represents an alternative to ozonolysis, which requires very specific (and not universally available) equipment.

$$R_1R_2C{=}CR_3R_4 \xrightarrow[CH_2Cl_2]{O_3} \xrightarrow{Me_2S} R_1R_2C{=}O + O{=}CR_3R_4 \qquad (R22)$$

$$R_1R_2C(OH){-}C(OH)R_3R_4 \xrightarrow{HIO_4} R_1C(O)R_2 + R_3C(O)R_4 \qquad (R23)$$

Aldehydes are also available through a few reductive protocols, including the hydride reduction of acyl chlorides using lithium tri-*t*-butoxyaluminum hydride (Reaction 24), and the selective reduction of esters and nitriles using DIBAL (Reactions 25 and 26, respectively). In the latter case, the initial product is an imine, which is liberated during workup and undergoes acid-catalyzed hydrolysis to the aldehyde under those conditions (Figure 11-1). In the reduction of both esters and nitriles, the selectivity derives from the nature of the initial adduct, in which the isobutyl groups on the aluminum sterically hinder the addition of another equivalent of hydride.

$$R{-}C{\equiv}N \xrightarrow{DIBAL} \left[(i\text{-}Bu)_2Al{-}N{=}CH{-}R \right] \xrightarrow[workup]{aqueous} RCH{=}NH \xrightarrow[H_2O]{H^+} RCHO$$

FIGURE 11-1: DIBAL reduction of nitriles.

$$RC(O)Cl \xrightarrow[THF]{LiAl(O\text{-}t\text{-}Bu)_3H} RC(O)H \qquad (R24)$$

$$RC(O)OR' \xrightarrow[THF]{DIBAL} RC(O)H \qquad (R25)$$

$$R{-}C{\equiv}N \xrightarrow[THF]{DIBAL} RC(O)H \qquad (R26)$$

Terminal alkynes are hydrated in predictable ways to give ketones and aldehydes. For example, the mercury(II)-catalyzed Markovnikov hydration of a terminal alkyne provides the corresponding methyl ketone (Reaction 27). The initially formed product is a so-called enol, which is rapidly isomerized to the carbonyl through a process known as the keto-enol tautomerization (Figure 11-2). This is really an equilibrium process mediated by either acid or base—in almost all cases, the position of equilibrium lies strongly in favor of the keto form. Anti-Markovnikov hydration of alkynes is also possible using bulky dialkylboranes, such as dicyclohexylborane or disiamylborane (Sia_2BH), followed by oxidative workup (Reaction 28). The role of the bulky alkyl groups is to provide greater steric bias in the addition.

R—C≡C—H $\xrightarrow[H_2O]{Hg^{2+}}$ (enol: OH, H, H, R) $\underset{\text{cat. } H^+ \text{ or } HO^-}{\overset{\text{tautomerization}}{\rightleftharpoons}}$ R–C(=O)–CH_3

FIGURE 11-2: The keto-enol tautomerization.

R—≡ $\xrightarrow[H_2O,\ \text{dioxane}]{HgSO_4}$ R–C(=O)–Me (R27)

R—≡ $\xrightarrow{\text{dicyclohexylborane (B–H)}}$ $\xrightarrow[H_2O_2]{NaOH}$ R–CH_2–CHO (R28)

REACTIONS OF KETONES AND ALDEHYDES

Almost all reactions involving the carbonyl group can be understood in terms of three zones of reactivity (Figure 11-3). First and the most familiar of these is the electrophilic carbonyl carbon, made electron-deficient by polarization of both the σ and the π bonds. As a consequence, this center is susceptible to nucleophilic attack. Second, the oxygen's lone pairs can serve as a locus of weak Lewis basicity, which can coordinate to Lewis acids (including protonation). Finally, anions α to the carbonyl are stabilized by a resonance form that places the negative charge on oxygen; therefore, carbonyl compounds can be deprotonated at the α position in strong base.

the carbonyl oxygen is weakly Lewis basic

α-protons are weakly Brønsted acidic

the carbonyl carbon is electrophilic

FIGURE 11-3: Three zones of reactivity for the carbonyl group.

Starting with reactions that involve the carbonyl carbon, ketones and aldehydes can be reduced to the corresponding alcohols using sodium borohydride or LAH—ketones give secondary alcohols, whereas aldehydes provide primary alcohols (Reaction 4). Carbonyls are also attacked by carbon-based nucleophiles, such as organolithiums and organomagnesium compounds (Grignard reagents). The products of these reactions are alcohols in which a carbon–carbon bond has been formed (Reaction 5).

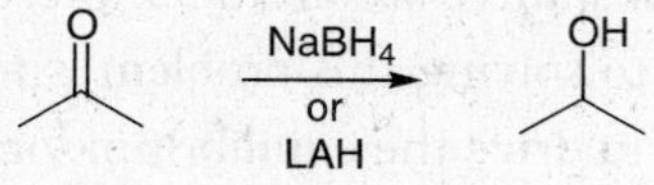

$NaBH_4$ or LAH (R4)

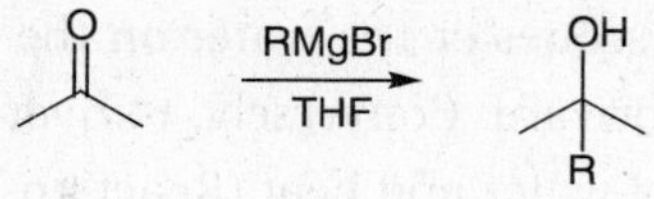

RMgBr, THF (R5)

Ketones and aldehydes are converted to alkenes by the action of phosphonium ylides in a protocol known as the **Wittig reaction** (Reaction 29), which proceeds through initial nucleophilic attack on the carbonyl carbon. The key intermediate in this reaction is the four-membered heterocyclic ring, the oxaphosphetane, which thermally fragments (retro [2+2] cycloaddition) to provide the alkene and triphenylphosphine oxide as a by-product.

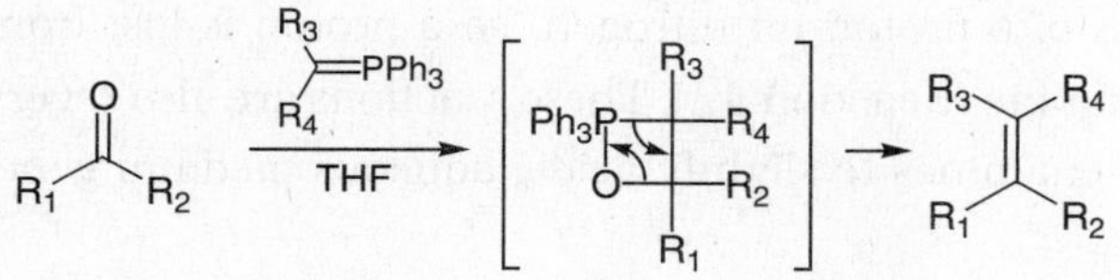

$R_3R_4C{=}PPh_3$, THF (R29)

As a marginal note, although the Wittig reagents look like arcane species, in fact they are quite straightforward to synthesize using triphenylphosphine, a powerful nucleophile that reacts with a broad array of alkyl halides to give alkylated phosphonium salts (Figure 11-4). These phosphonium salts are acidic by virtue of the strong electron-withdrawing character of the positively charged phosphorus center—deprotonation with *n*-butyllithium provides the corresponding ylide.

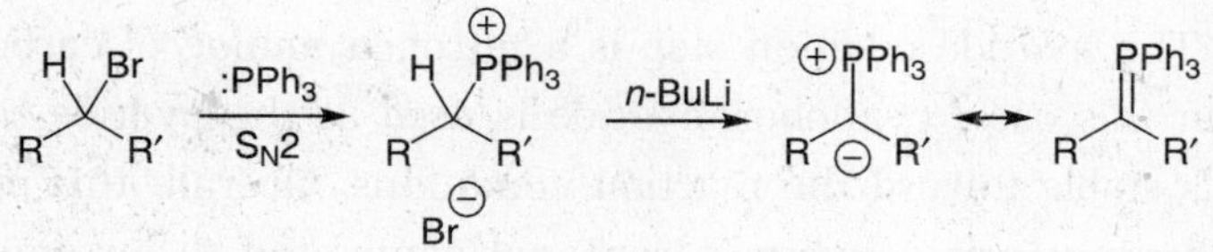

:PPh_3, S_N2; *n*-BuLi

FIGURE 11-4: Preparation of a Wittig reagent.

Carbonyl compounds are converted to acetals by treating with an excess of alcohol in the presence of catalytic quantities of a strong acid, such as *p*-toluenesulfonic acid. This is an equilibrium system, which has a favorable enthalpic term (trading a C—O π bond for a C—O σ bond) but an unfavorable entropic term (three molecules going to two molecules). The latter issue is often addressed by using ethylene glycol (1,2-ethanediol) as the alcohol component, so that both hydroxy groups are carried by a single molecule. However, even in the best of circumstances we are often presented with an equilibrium mixture. The classical approach to solving this problem is to use a large excess of alcohol (usually used as the solvent) to drive the equilibrium forward through a Le Châtelier's principle consideration. If the alcohol happens to be precious and/or the limiting reagent, then an alternative is to sequester the water on the right side with a desiccant, thereby "pulling" the reaction forward. Conversely, to hydrolyze the acetal, the strategy is to add plenteous quantities of water and heat (Reaction 30).

$R_1C(=O)R_2$ + 2 MeOH ⇌ (cat. *p*TsOH) $R_1C(OMe)_2R_2$ + H_2O (R30)

Primary amines add to carbonyls to give imines through a sequence of nucleophilic addition, followed by loss of water. The initially formed iminium ion loses a proton to give the observed product (Reaction 31). In the case of a secondary amine, the iminium ion cannot be quenched by loss of a proton on nitrogen, so a proton is lost from the adjacent carbon, yielding an enamine (Reaction 32). These reactions are also reversible, so that exposure of imines and enamines to slightly acidic aqueous medium generally yields the starting ketones.

$R_1C(=O)CH_2R_2$ → (R_3—NH_2, MeOH) $R_1C(=NR_3)CH_2R_2$ (R31)

$R_1C(=O)CH_2R_2$ → (pyrrolidine, MeOH) $R_1C(NC_4H_8)=CHR_2$ (R32)

Imines can also be intercepted with hydride *in situ* (i.e., in the same pot) to provide a new amine. The hydride addition step is a nitrogen analog of carbonyl reduction (Reaction 4); in this case, cyanoborohydride is used as the hydride source, primarily because it is stable toward the reaction conditions. Overall, this process, known as **reductive amination,** converts a carbonyl compound to an amine in one pot (Reaction 33).

$R_1C(=O)CH_2R_2$ → (R_3NHR_4, $NaBH_3CN$, MeOH) $R_1CH(NR_3R_4)CH_2R_2$ (R33)

Next we turn our attention to reactivity at the α carbon. Due to the electron-withdrawing character of the carbonyl group, aldehydes and ketones can be deprotonated to form α carbanions. We have encountered carbanion chemistry before with organomagnesium (Grignard-type) compounds, but those are relatively basic species. The so-called enolate anions derived from carbonyls, however, are less basic and—to some extent—more versatile in their chemistry. One frequently enountered reactivity involves subsequent alkylation with typical electrophiles (Reaction 34).

$$R_1C(=O)CH_2R_2 \xrightarrow[\text{THF}]{\text{Base}} \xrightarrow{R_3CH_2LG} R_1C(=O)CH(R_2)CH_2R_3 \quad \text{(R34)}$$

The ease of deprotonation varies widely, depending on the substitution pattern of the carbonyl. For example, aldehydes (pK_a = ca. 16) are acidic enough to be completely deprotonated by potassium *t*-butoxide (pK_a = ca. 19); however, ketones (pK_a = ca. 20) are only partially deprotonated under those conditions. Esters are even less acidic (pK_a = ca. 26), and amides (pK_a = ca. 35) require LDA for full deprotonation. Bearing in mind that the pK_a scale is logarithmic, this knowledge has significant predictive value—furthermore, we can leverage this differential for selectivity and control (Figure 11-5). For example, a molecule containing both an aldehyde and an ester functionality could, in principle, be selectively deprotonated at the aldehyde.

Aside from the issue of chemoselectivity (targeting one functional group over another), there is also the issue of regioselectivity. Most ketones have α-protons on either side of the carbonyl group. In some cases we can modify conditions to prefer one particular mode of deprotonation. For example, when 2-methylcyclohexanone (Figure 11-6) is deprotonated with *t*-butoxide under protic conditions at relatively high temperature, the thermodynamically more stable enolate (i.e., more substituted double bond) is formed. This is because the *t*-butoxide is not strong enough to completely deprotonate the substrate; therefore, a proton-transfer equilibrium will be set up in which the product distribution is governed by ΔG considerations.

However, when an excess of very strong, hindered base is used (e.g., LDA), the less substituted enolate is formed. To understand this, we must remember that removal of the methine α-proton is sterically hindered; thus, it is kinetically favorable to remove one of the methylene α-protons, even though the product is not the most stable. An excess of base is used to prevent the enolate from equilibrating by way of residual excess ketone, which would serve as a proton source. Low temperature is advantageous, because it allows the reaction to better discriminate between the two activation barriers.

Once formed, these carbanions can interact with many different electrophiles, including another carbonyl compound (Reaction 35). This is analogous to a Grignard addition in

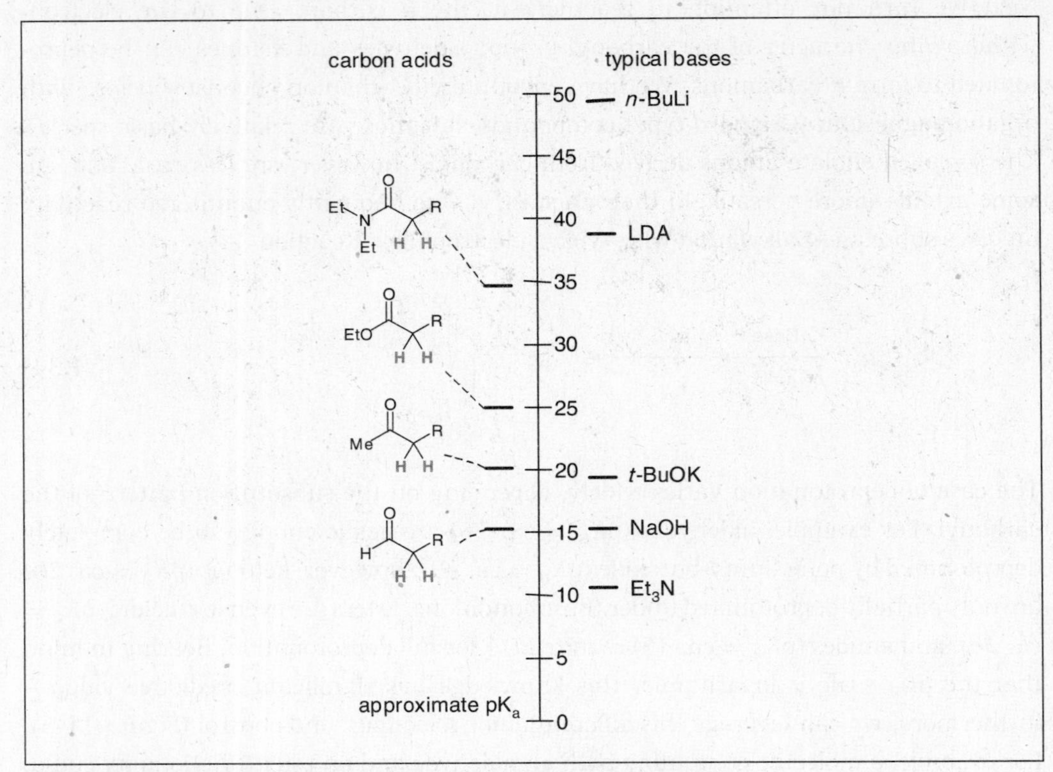

FIGURE 11-5: pK_a values of carbon acids relative to typical bases.

FIGURE 11-6: Kinetic and thermodynamic enolate formation.

some ways, except that the nucleophile is not an organomagnesium species, but an enolate. This process, known as the **aldol reaction,** can take place between two carbonyl compounds of the same type (the so-called homo-aldol condensation) or between two different species (the "mixed," or "crossed," aldol). The latter version works best if one substrate is a ketone and the other an aldehyde, because the aldehyde tends to function exclusively as the electrophilic species. Alternatively, one could use a carbonyl with no α-protons, for which it would be impossible to generate an enolate anion; therefore, it is relegated to electrophile status.

Experimental conditions for the aldol condensation can be controlled such that the initially formed adduct is isolated. However, with extended reaction times or higher temperature, an elimination ensues to provide that α,β-unsaturated carbonyl compound, which is also very synthetically useful.

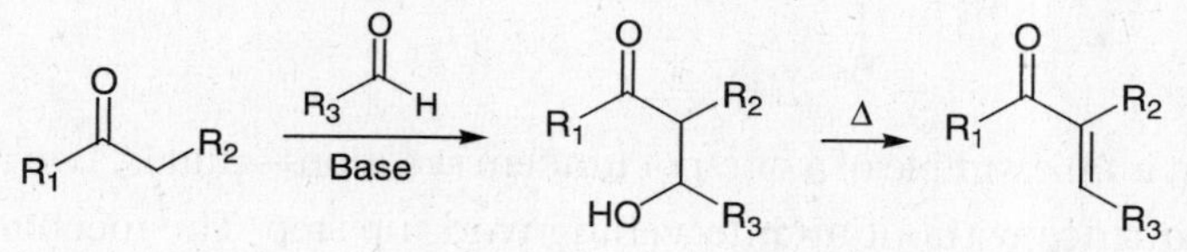

(R35)

Electron-deficient double bonds can also serve as competent electrophiles toward enolates (Reaction 36) through a process known as **conjugate addition** (also referred to as the **Michael addition** in some cases). Here only catalytic amounts of base are needed.

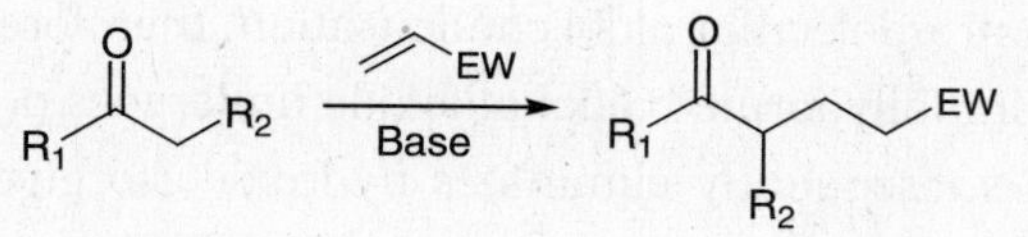

(R36)

A clever combination of the Michael addition and the aldol condensation are revealed in the base-catalyzed reaction of a ketone with the reagent methyl vinyl ketone (MVK), a protocol known as the **Robinson annulation** (Reaction 37). This methodology has

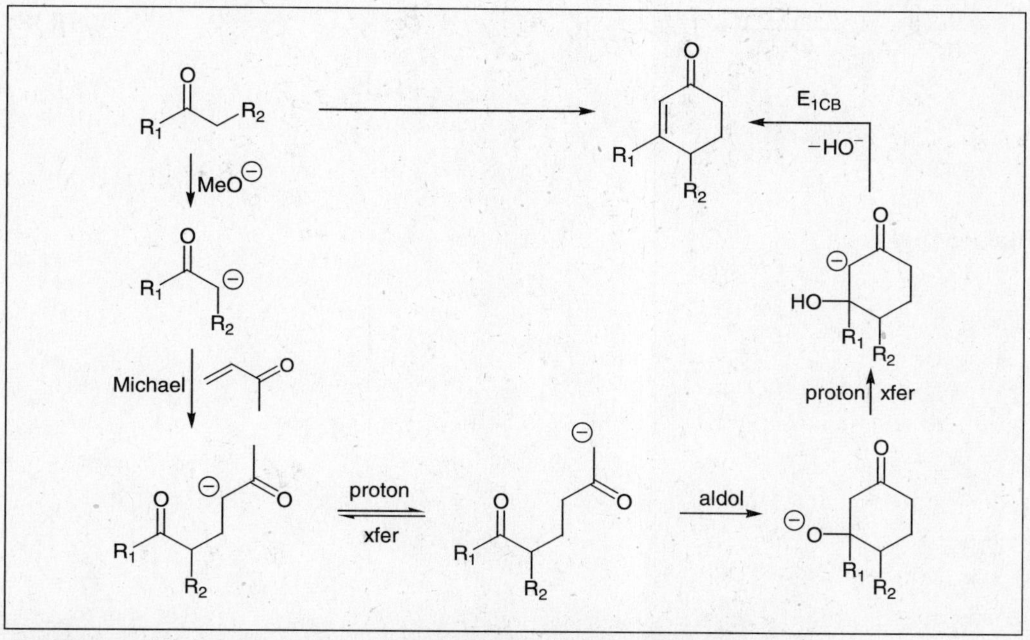

FIGURE 11-7: Mechanistic intermediates in the Robinson annulation.

the distinction of forming a ring from two acyclic species.

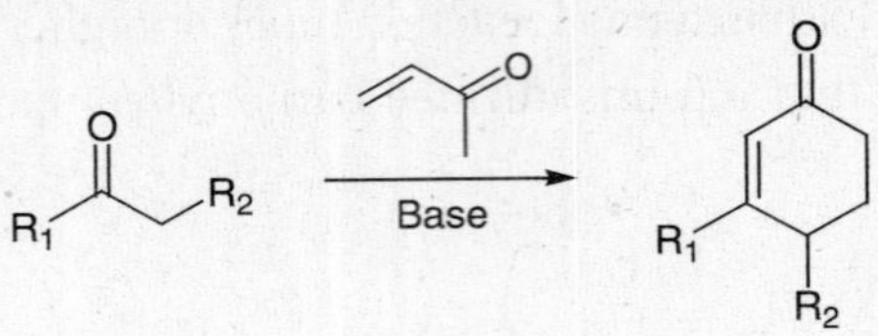

(R37)

The Robinson annulation is an example of a one-pot tandem reaction—that is, two reactions are occurring in sequence without an intervening work-up step. The mechanism (Figure 11-7) starts with deprotonation of the substrate to form an α carbanion. Michael addition onto MVK provides an initial anionic adduct in which the enolate anion is internal (i.e., more substituted). This species is in equilibrium with the external (less substituted) enolate through a proton transfer reaction. Even though this is an unfavorable equilibrium (i.e., lies to the left), the small equilibrium concentration of the external enolate is much more reactive toward intramolecular aldol condensation; therefore, the equilibrium is driven to the right. The initially formed aldol alkoxide undergoes proton transfer to provide an enolate, which subsequently eliminates hydroxide to give the cyclic enone product observed.

Finally, under the rubric of functional group transformation, ketones are converted to esters by the action of *m*-chloroperbenzoic acid (*m*CPBA), which can be viewed as an oxygen-transfer reagent, in a protocol known as the **Baeyer-Villiger oxidation.** In principle, either alkyl group could migrate to the oxygen center—as a general rule, the center that can best support a positive charge has the higher migratory aptitude (Reaction 38).

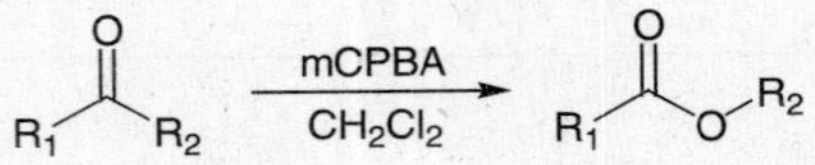

(R38)

CRAM SESSION

Ketones and Aldehydes

Ketones can be prepared from

- The oxidation of **secondary alcohols** (Jones, PCC, MnO_2 for benzylic substrates)
- The ozonolysis of **alkenes**
- The hydration of **alkynes** ($HgSO_4$)

Aldehydes can be prepared from

- The oxidation of **primary alcohols** (PCC)
- The reduction of **nitriles** (DIBAL)
- The reduction of **esters** (DIBAL)
- The reduction of **acyl halides** ($LiAl(O\text{-}t\text{-}Bu)_3H$)
- The hydration anti-Markovnikov hydration of **terminal alkynes** (hydroboration)

The reactions of ketones and aldehydes arise from three properties of the carbonyl group:

- Electrophilicity of the carbonyl carbon
- Brønsted acidity of the α-protons
- Lewis basicity of the carbonyl oxygen

Some common transformations of ketones and aldehydes are

- Conversion to **alcohols** (reduction and Grignard reaction)
- Conversion to **alkenes** (Wittig condensation)
- Formation of **acetals**
- Formation of **imines** and **enamines**
- Conversion to **amines** (reductive amination)
- Base-mediated **alkylation** (alkyl halides and Michael addition)
- Formation of **enones** (aldol condensation)
- Formation of **cyclohexenones** (Robinson annulation)
- Conversion to **esters** (Baeyer-Villiger oxidation)

CHAPTER 12

Alkenes and Alkynes

Read This Chapter to Learn About:

- Preparation of Alkenes
- Reactions of Alkenes
- Preparation of Alkynes
- Reactions of Alkynes

PREPARATION OF ALKENES

There are several known eliminative techniques aimed at the synthesis of olefins. One classical method is the acid-catalyzed dehydration of alcohols (Reaction 16). As these would be considered thermodynamic conditions, the expected product would be the most stable alkene. In practice, however, this method suffers from the specter of hydride shifts and carbocationic rearrangements, as well as uncontrolled migration of the double bond. A generally better behaved protocol is the base-promoted E2 elimination of alkyl halides or tosylates (Reaction 39), where the regiochemistry can often be controlled by the steric bulk of the base.

$$R_1CH(OH)CH_2R_2 \xrightarrow[\Delta]{pTsOH} R_1CH{=}CHR_2 \qquad \text{(R16)}$$

$$R_1CH(LG)CH_2R_2 \xrightarrow[THF]{base} R_1CH{=}CHR_2 \qquad \text{(R39)}$$

There are a few functional group transformations that provide alkenes, including the Wittig reaction (Reaction 29) and reductive protocols. Alkynes undergo catalytic hydrogenation in the presence of Lindlar's catalyst gives the *cis*-alkene (Reaction 40). The key here is the use of Lindlar's catalyst, which is highly attenuated to prevent over-reduction to the fully saturated species. The *trans*-alkene is available from the dissolving metal reduction of the alkyne (Reaction 41).

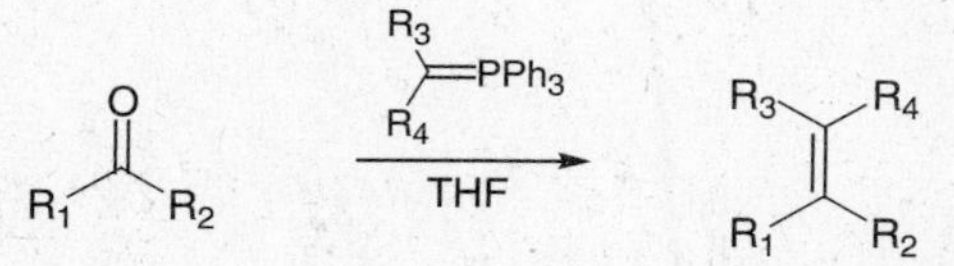

(R29)

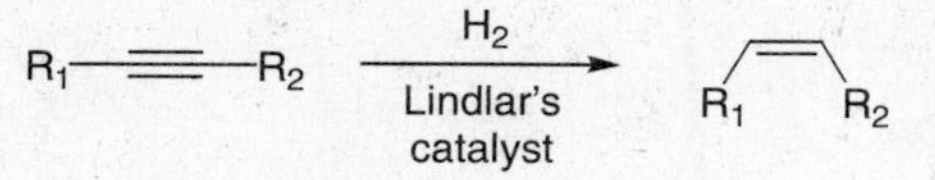

(R40)

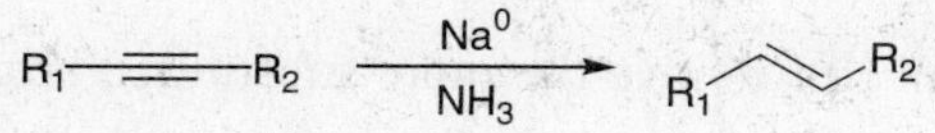

(R41)

Primary amines can provide alkenes through a two-step sequence, starting with the conversion to an ammonium salt—the so-called exhaustive methylation reaction. In the presence of silver(I) oxide under thermal conditions, these ammonium salts suffer elimination to give generally the less substituted double bond, a process known as the Hofmann elimination (Reaction 42). One of the functions of silver here is to sequester the potentially nucleophilic iodide. From a cost standpoint, however, silver also happens to be a significant drawback of the methodology, particularly for large-scale reactions.

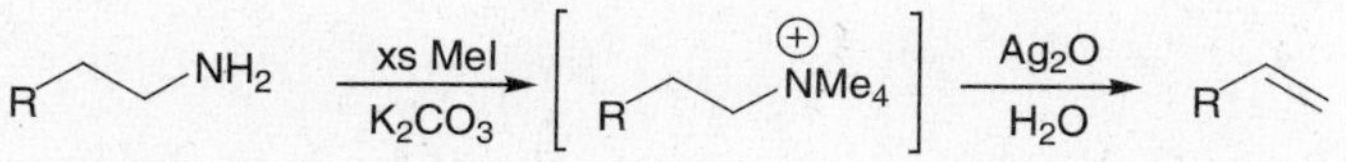

(R42)

Finally, the aldol condensation (Reaction 35) is unique among other techniques in this section for preparing alkenes, as it provides a route for electron-deficient olefins.

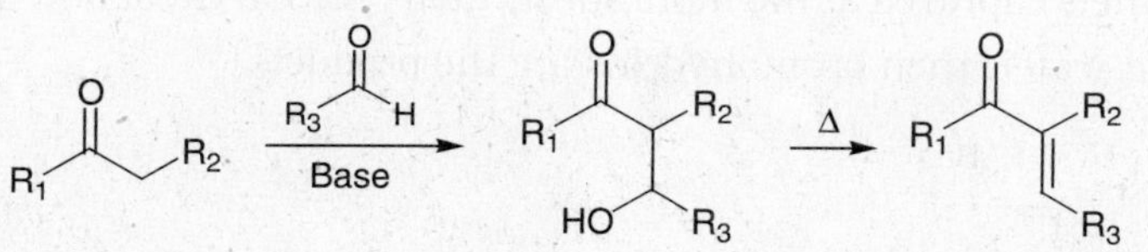

(R35)

REACTIONS OF ALKENES

Unsymmetric alkenes undergo Markovnikov hydration in the presence of strong acids (Reaction 1), a procedure limited to relatively robust substrates. A milder alternative is to use the oxymercuriation-demercuriation sequence (Reaction 2), which has the advantage of not forming a full-fledged carbocation, thus suppressing potential alkyl shifts. Anti-Markovnikov hydration is available through hydroboration followed by oxidative workup (Reaction 3).

(R1)

(R2)

(R3)

Alkenes also suffer Markovnikov addtion of HBr to give the more substituted bromide (Reaction 43). In the presence of some radical initiator, such as di-*t*-butyl peroxide or AIBN, the anti-Markovnikov product is obtained (Reaction 44). The latter reaction proceeds via a radical mechanism in which the first step is the addition of bromine radical to the double bond in such a way as to give the most stable radical intermediate. This carbon-based radical then abstracts hydrogen from HBr to regenerate the bromine radical and sustain the chain mechanism.

(R43)

(R44)

Treatment of alkenes with molecular bromine in inert solvent leads to the formation of the *trans*-dibromoalkane (Reaction 45). The stereochemistry is driven by the initial formation of a cyclic bromonium ion, which is ring-opened in an S_N2 fashion by bromide. However, if the reaction is run in a mildly nucleophilic solvent, such as methanol, the bromonium is captured at the more substituted position (Reaction 46). If the bromination is run in water, then bromohydrins are the products.

(R45)

(R46)

Alkenes can be reduced to the corresponding alkanes by catalytic hydrogenation (Reaction 47), or they can be oxidized to two carbonyl compounds using ozone (Reaction 22). Treatment with catalytic osmium tetraoxide in hydrogen peroxide leads to *cis*-dihydroxylation (Reaction 48). The resultant 1,2-diols are sometimes interesting synthetic targets in their own right; however, since they can also be oxidatively cleaved to the carbonyl compounds under mild conditions, this is sometimes used as an alternative methodology for ozonolysis.

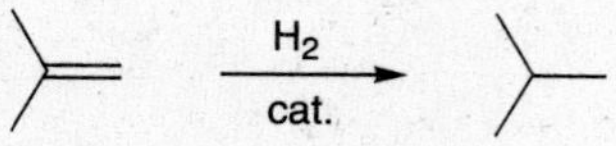

H_2, cat. (R47)

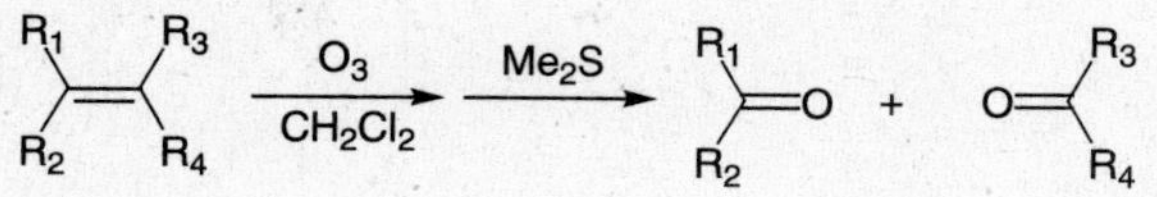

O_3, CH_2Cl_2; Me_2S (R22)

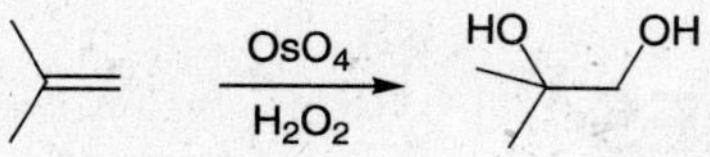

OsO_4, H_2O_2 (R48)

Another oxidative technique for olefins is epoxidation (Reaction 49). This is a very general protocol that can be applied to a great many types of alkenes. Two reagents are worthy of consideration here. The classical oxygen transfer agent is *m*-chloroperbenzoic acid (*m*CPBA), which has the advantage of being a relatively shelf-stable crystalline solid. The drawback is that the *m*-chlorobenzoic acid by-product must be removed after the reaction is complete. An alternative oxidant is dimethyldioxirane (DMD), which has the advantage of producing acetone as a by-product (Figure 12-1). Thus, the reaction mixture can simply be concentrated *in vacuo* to isolate the product. The downside of DMD is that it must be freshly prepared before each use.

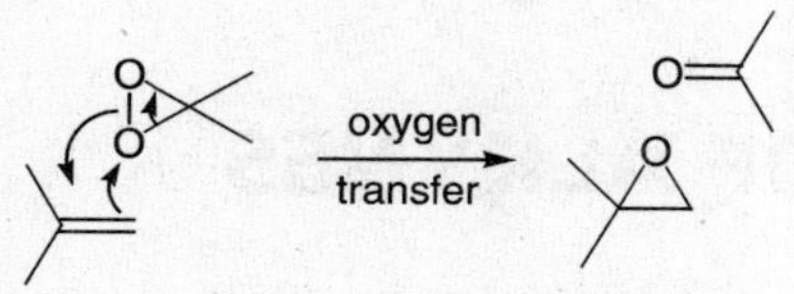

oxygen transfer

FIGURE 12-1: Mechanism of DMD epoxidation.

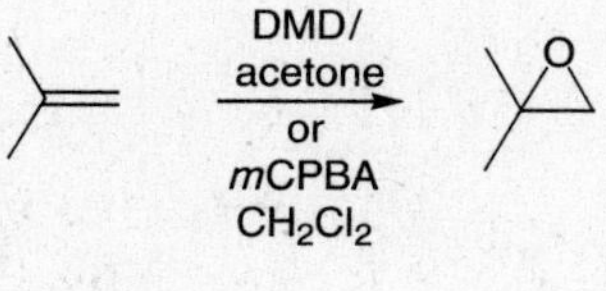

DMD/acetone or *m*CPBA CH_2Cl_2 (R49)

Three-membered carbocycles can be prepared from alkenes using one of two cyclopropanation protocols (Reaction 50)—either diiodomethane in the presence of a zinc–copper couple (the Simmons–Smith conditions) or diazomethane in the presence of some copper catalyst. In both cases the reactive intermediate is a carbene that adds to the alkene (Figure 12-2).

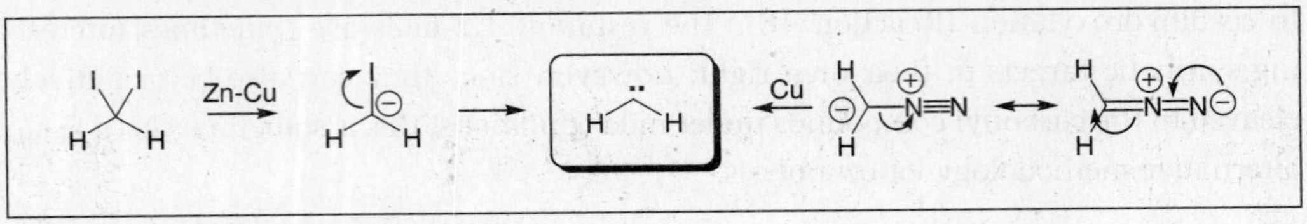

FIGURE 12-2: Two routes to carbene.

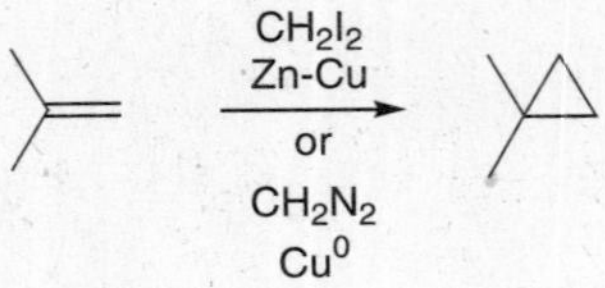

$$\text{alkene} \xrightarrow[\text{or } CH_2N_2,\ Cu^0]{CH_2I_2,\ \text{Zn-Cu}} \text{cyclopropane} \qquad \text{(R50)}$$

PREPARATION OF ALKYNES

Alkynes are available from alkenes through initial dibromination (i.e., oxidation) followed by double elimination (Reaction 51). Owing to their acidity, terminal alkynes can also undergo deprotonation followed by alkylation to give more complex substrates (Reaction 52).

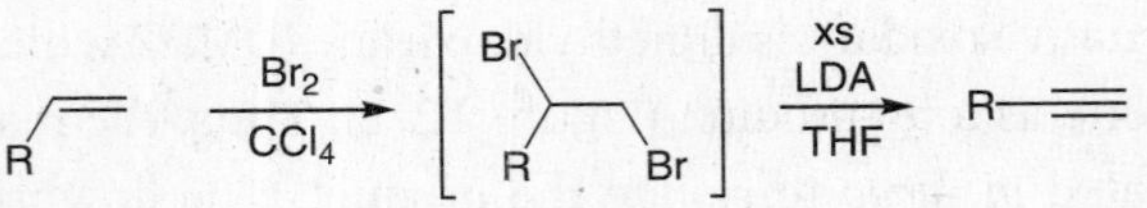

(R51)

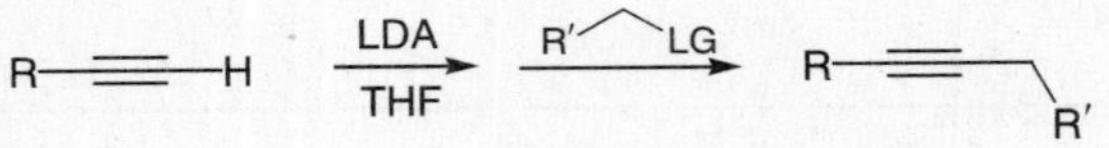

(R52)

REACTIONS OF ALKYNES

The π bond in an alkyne undergoes some of the same chemistry seen in the π chemistry of alkenes—the only difference is that a double bond is left over. Thus, mercuric ion mediates Markovnikov hydration leading to methyl ketones (Reaction 27), and boron reagents can be used for anti-Markovnikov hydration to give aldehydes (Reaction 28).

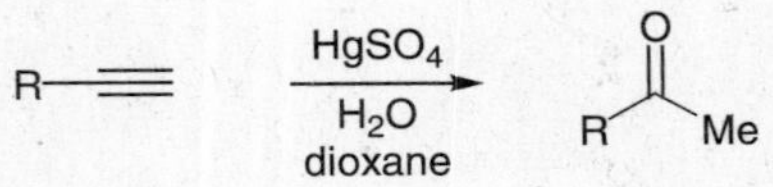

(R27)

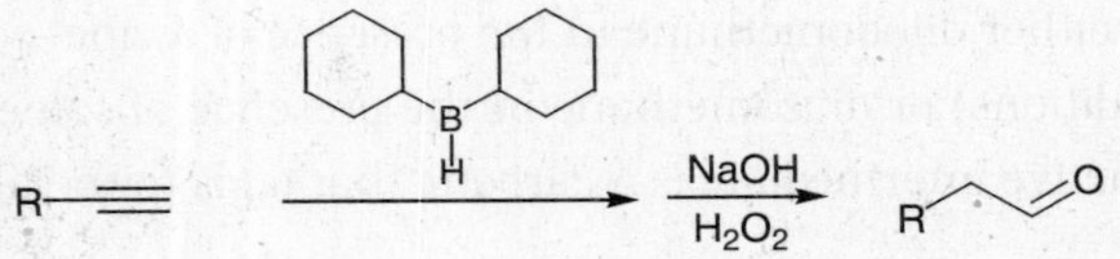

(R28)

Alkynes are reduced to *cis*-alkenes through catalytic hydrogenation (Reaction 40) in the presence of Lindlar's catalyst, an attenuated palladium catalyst (5% Pd on $CaCO_3$/Pb). Alternatively, *trans*-alkenes are available through the dissolving metal reduction using sodium in ammonia (Reaction 41).

$$R_1{-}C{\equiv}C{-}R_2 \xrightarrow[\text{Lindlar's catalyst}]{H_2} \textit{cis-}R_1CH{=}CHR_2 \qquad \text{(R40)}$$

$$R_1{-}C{\equiv}C{-}R_2 \xrightarrow[NH_3]{Na^0} \textit{trans-}R_1CH{=}CHR_2 \qquad \text{(R41)}$$

CRAM SESSION

Alkenes and Alkynes

Alkenes can be prepared from

- The acid-catalyzed dehydration of **alcohols**
- The base-mediated E2 elimination of **alkyl halides**
- The Wittig reaction of **ketones** and **aldehydes**
- The reduction of **alkynes** (Lindlar and dissolving metal)
- The exhaustive methylation and elimination of primary **amines** (Hoffmann)

Some common transformations of alkenes are

- Hydration to **alcohols** (Markovnikov and anti-Markovnikov)
- Hydrohalogenation **to alkyl halides** (Markovnikov and anti-Markovnikov)
- Reaction with halogens to form **dihalides** and **halohydrins**
- Reduction to **alkanes**
- Ozonolysis to form **ketones** and **aldehydes**
- Dihydroxylation to form **vicinal diols**
- Oxidation to **epoxides**
- Reaction with carbenes to form **cyclopropanes**

Alkynes can be prepared from

- The bromination and double-elimination of **alkenes**
- The deprotonation and alkylation of **terminal alkynes**

Some common transformations of alkynes are

- Hydration to **ketones**
- Anti-Markovnikov hydration of terminal alkynes to **aldehydes**
- Reduction to **alkenes** (Lindlar and dissolving metal)

CHAPTER 13

Carboxylic Acid Derivatives

Read This Chapter to Learn About:

- ➤ Preparation of Carboxylic Acids
- ➤ Reactions of Carboxylic Acids
- ➤ Preparation of Carboxylate Esters
- ➤ Reactions of Carboxylate Esters
- ➤ Preparation of Amides
- ➤ Reactions of Amides

The various carboxylic acid derivatives (acids, acid chlorides, esters, amides, and nitriles) exhibit vastly differing properties, yet they share many synthetic pathways. As we launch into a discussion that examines the interconversion of these derivatives, it will be helpful for us to understand the conceptual framework known as the electrophilicity series of carboxylic acid derivatives (Figure 13-1). If we bear in mind that the electrophilicity of the carbonyl carbon is rationalized on the basis of a charge-separated resonance structure (inset), then modifications to the structure that destabilize that form should increase reactivity. Thus, for acyl chlorides (X=Cl), the inductive effect of the chlorine makes the carbonyl carbon even more electron deficient; therefore, the center is expected to be much more electrophilic. Indeed, most acyl chlorides must be stored over a desiccant to prevent hydrolysis from adventitious water vapor. However, amides ($X=NR_2$) are expected to be less reactive by virtue of the fact that the nitrogen lone pair can provide resonance stabilization to the carbonyl carbon. This prediction is supported by the behavior of amides, which tend to be fairly nonreactive.

This series is also a useful framework for thinking about interconversion among functional groups. In other words, it is difficult to move uphill through the series

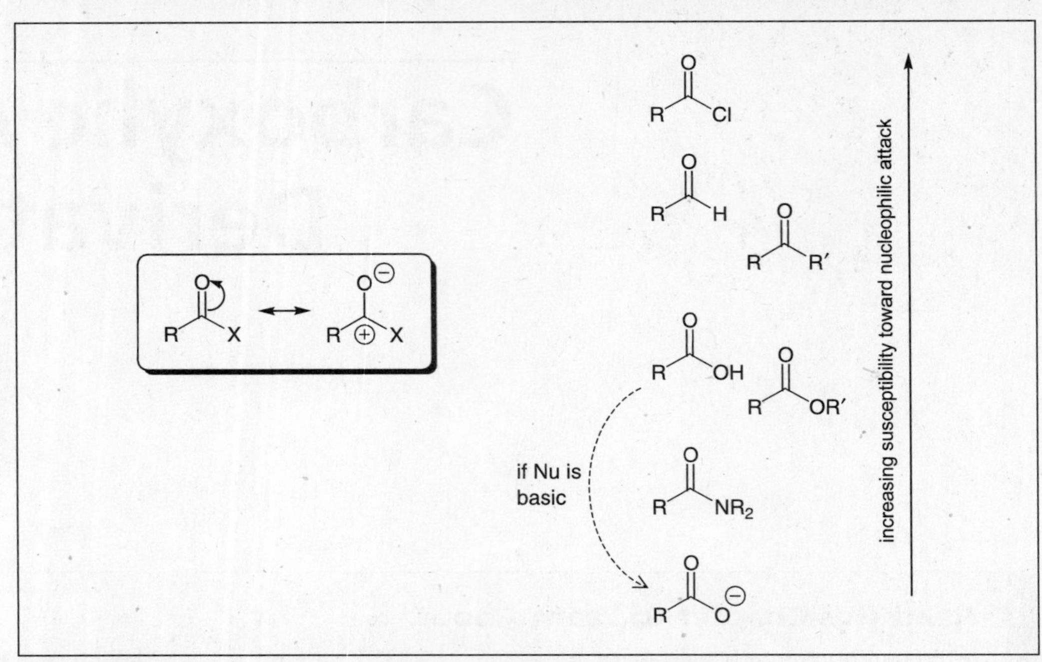

FIGURE 13-1: The electrophilicity series of carbonyl compounds.

(e.g., amide → ester) but relatively easy to cascade down from the top (e.g., chloride → acid). Thus, many methods for lateral or uphill functional group conversions take advantage of auxiliary high-energy species to drive the chemistry. For example, transforming an acid into an acid chloride (Reaction 57) is an inherently unfavorable uphill event, but we take advantage of the extremely high energy of thionyl chloride to drive the reaction.

PREPARATION OF CARBOXYLIC ACIDS

Carboxylic acids are prepared by the oxidation of primary alcohols under Jones conditions (Reaction 10); the method can also launch from the aldehyde stage (Reaction 53). Generally speaking, secondary alcohols and ketones make less suitable substrates.

$$RCH_2OH \xrightarrow[H_2SO_4,\ H_2O]{K_2Cr_2O_7} RCOOH \qquad \text{(R10)}$$

$$RCHO \xrightarrow[H_2SO_4,\ H_2O]{K_2Cr_2O_7} RCOOH \qquad \text{(R53)}$$

Carboxylic acids can also be obtained through hydrolytic approaches. For example, exposing an ester to high concentrations of potassium hydroxide at elevated temperatures leads to the production of the corresponding carboxylate salt, a process known as **saponification** (Reaction 54). The reaction proceeds via initial nucleophilic attack of hydroxide to give an anionic tetrahedral intermediate, which collapses to eject either hydroxide or alkoxide. In principle, these two species are similar leaving groups, but when alkoxide is ejected, the carboxylic acid just formed is immediately and irreversibly deprotonated, thus driving the equilibrium to the right. The carboxylate is protonated during workup to give the carboxylic acid. Nitriles are hydrolyzed to carboxylic acids as well (Reaction 55), although this is usually carried out under acidic conditions.

$$RCOOR' \xrightarrow[H_2O]{KOH} RCOOK \xrightarrow[\text{workup}]{\text{acidic}} RCOOH \tag{R54}$$

$$R{-}C{\equiv}N \xrightarrow[H_2O]{H_2SO_4} RCOOH \tag{R55}$$

A very clever and versatile Grignard methodology allows for the effective functional group transformation from alkyl bromide to alkanoic acid with one extra carbon (Reaction 56). Thus, the alkyl halide is treated with magnesium metal to give the corresponding alkylmagnesium halide, which is quenched with carbon dioxide to yield the acid through Grignard addition.

$$R{-}Br \xrightarrow[THF]{Mg^0} R{-}MgBr \xrightarrow{CO_2} RCOOH \tag{R56}$$

REACTIONS OF CARBOXYLIC ACIDS

Carboxylic acids are converted to the corresponding acyl chlorides by treatment with thionyl chloride or by using oxalyl chloride in the presence of catalytic dimethylformamide (DMF). The driving force in the former set of conditions is the liberation of SO_2 gas, whereas the oxalyl chloride methodology ends up releasing carbon dioxide and carbon monoxide. The latter conditions are advantageous from the practical standpoint of materials handling—thionyl chloride is very corrosive to human tissue and scientific equipment (Reaction 57).

$$RCOOH \xrightarrow[\text{ClCOCOCl, cat. DMF}]{SOCl_2 \text{ or}} RCOCl \tag{R57}$$

Carboxylic acids are usually sluggish toward reduction, because most hydride sources first remove the acid proton. The resulting carboxylate anion is then relatively resistant to nucleophilic attack. However, the use of very active hydride sources (such as LAH) results in reduction to the primary alcohol (Reaction 58).

$$RCOOH \xrightarrow[THF]{LAH} RCH_2OH \quad (R58)$$

PREPARATION OF CARBOXYLATE ESTERS

Esters are conveniently prepared by the treatment of an acyl chloride with an alcohol (Reaction 59). Inasmuch as a stoichiometric amount of HCl is liberated as the reaction proceeds, sometimes a buffering tertiary amine is added, such as triethylamine or pyridine.

$$RCOCl \xrightarrow{R'\text{—}OH} RCOOR' \quad (R59)$$

If the alcohol is in plentiful supply, then the classic approach of mixing acid and alcohol in the presence of strong acid would be appropriate (Reaction 60). This is essentially an equilibrium reaction, so taking measures to drive the equilibrium is sometimes necessary (e.g., adding desiccant to sequester water).

$$RCOOH \xrightarrow[\text{cat. } p\text{TsOH}]{R'\text{—}OH} RCOOR' \quad (R60)$$

An extremely useful methodology for preparing methyl esters involves treatment of the acid with diazomethane (Reaction 61). The mechanism involves an initial proton transfer from the acid to diazomethane; the subsequently formed carboxylate anion then displaces nitrogen gas from the protonated diazomethane. The reagent is very selective for carboxylic acids, the conditions are practically neutral, and the only by-product is an inert gas.

$$RCOOH \xrightarrow[Et_2O]{CH_2N_2} RCOOMe \quad (R61)$$

REACTIONS OF CARBOXYLATE ESTERS

Esters are saponified to the corresponding acids by treatment with potassium hydroxide at high temperatures (Reaction 54). Similarly, esters can engage in a process known as transesterification (Reaction 62), whereby the alcohol portion of the ester is exchanged for another species. This is another equilibrium reaction, which can be catalyzed by acid or base.

R–C(=O)–OR′ —(KOH, H_2O)→ R–C(=O)–OH (R54)

R–C(=O)–OR′ —(R″OH, cat. *p*TsOH or NaOR″)→ R–C(=O)–OR″ (R62)

The reduction of esters can provide alcohols on the treatment with strong hydride sources like LAH (Reaction 6), or aldehydes when the selective reagent DIBAL is employed (Reaction 25). Since acids can be difficult to manipulate, esters are sometimes used as an intermediate from acid to aldehyde via esterification.

CH_3–C(=O)–OR —(LAH, THF)→ CH_3CH_2–OH (R6)

R–C(=O)–OR′ —(DIBAL, THF)→ R–C(=O)–H (R25)

PREPARATION OF AMIDES

Analogous to esters, amides are conveniently prepared by the treatment of amines with acyl chlorides (Reaction 63). Primary and secondary amides can be deprotonated at nitrogen and treated with alkyl halides or tosylates to give the *N*-alkylated derivates (Reaction 64).

R–C(=O)–Cl —(R′–NH–R″)→ R–C(=O)–N(R′)(R″) (R63)

R–C(=O)–NH–R′ —(base)→ —(R″–CH_2–LG)→ R–C(=O)–N(R′)(R″) (R64)

Oximes (derived from ketones and hydroxylamine) are *N*-chlorinated by thionyl chloride (or PCl_5) to give an activated species that undergoes Beckmann rearrangement, ultimately providing an amide. In the rearrangement, generally speaking, the R-group opposite the oxime hydroxy group migrates (Reaction 65).

R–C(=O)–R′ —(HO–NH_2, cat. *p*TsOH)→ R–C(=N–OH)–R′ —($SOCl_2$)→ R–C(=O)–NH–R′ (R65)

REACTIONS OF AMIDES

For the most part, amides are notoriously sluggish in saponification reactions. However, there is a selective method for the rapid and selective hydrolysis of primary amides that relies on nitrous acid (HONO) generated *in situ* from sodium nitrite and HCl (Reaction 66). The mechanism proceeds via the diazotization of the amide nitrogen, creating a phenomenal leaving group easily displaced by water.

$$RC(=O)NH_2 \xrightarrow[HCl]{NaNO_2} RC(=O)OH \quad (R66)$$

Amides are reduced to amines when treated with strong hydride sources such as LAH (Reaction 67), but DIBAL selectively provides an imine, which is hydrolyzed during workup to yield ultimately an aldehyde (Reaction 68).

$$RC(=O)NR'R'' \xrightarrow[THF]{LAH} RCH_2NR'R'' \quad (R67)$$

$$RC(=O)NR'R'' \xrightarrow[THF]{DIBAL} RC(=O)H \quad (R68)$$

Finally, an interesting and very useful functional group transformation is represented by the Hofmann rearrangement (Reaction 69), in which an amide is converted to an amine with one less carbon. The mechanism proceeds via initial *N*-chlorination of the amide, which is subsequently deprotonated. The *N*-chloroamide anion suffers loss of chloride to provide an acyl nitrene intermediate, which rearranges to an isocyanate. Aqueous acidic workup releases the amine through hydrolysis.

$$RC(=O)NH_2 \xrightarrow[NaOH,\ H_2O]{Cl_2} R{-}NH_2 \quad (R69)$$

CRAM SESSION

Carboxylic Acid Derivatives

Carboxylic acids can be prepared from

- The oxidation of **primary alcohols** and **aldehydes** (Jones)
- The hydrolysis of **esters** and **nitriles**
- The reaction of **Grignards and carbon dioxide**

Some common transformations of carboxylic acids are

- Conversion to **acyl halides**
- Reduction to **primary alcohols** (LAH)

Carboxylic esters can be prepared from

- The reaction of **alcohols with acyl halides**
- The acid-catalyzed condensation of **alcohols with carboxylic acids**
- The **alkylation of carboxylic acids** (diazomethane)

Some common transformations of carboxylic esters are

- Base-catalyzed hydrolysis to **carboxylic acids** (saponification)
- Acid- or base-catalyzed conversion to **other esters** (transesterification)
- Reduction to **primary alcohols** (LAH)
- Reduction to **aldehydes** (DIBAL)

Amides can be prepared from

- The reaction of **amines with acyl halides**
- The base-catalyzed **alkylation of amides**
- Chlorination and **rearrangement of oximes** (Beckmann rearrangement)

Some common transformations of amides are

- Hydrolysis to **carboxylic acids** (nitrous acid)
- Reduction to **amines** (LAH)
- Reduction to **aldehydes** (DIBAL)
- Rearrangement to **amines with one less carbon** (Hoffmann rearrangement)

CHAPTER 14

Epoxide Chemistry

Read This Chapter to Learn About:

- Preparation of Epoxides
- Reactions of Epoxides

PREPARATION OF EPOXIDES

By far the most widely used method for synthesizing epoxides is the epoxidation of alkenes with oxygen transfer reagents such as dimethyldioxirane and *m*-chloroperbenzoic acid (Reaction 49). Alternatively, bromohydrins (themselves usually derived from alkenes) can be induced to undergo ring closure in the presence of base (Reaction 70). There are other less frequently encountered methods that are not discussed here.

DMD / acetone or *m*CPBA CH_2Cl_2 (R49)

HO ... X; X = Br, Cl; base (R70)

REACTIONS OF EPOXIDES

Epoxides undergo ring opening in the presence of a wide variety of nucleophiles. For example, Grignard reagents attack unsymmetric epoxides at the less-hindered position to give ring-opened products bearing the more substituted alcohol (Reaction 71).

Similar reactivity is observed for heteroatomic nucleophiles under basic conditions (Reaction 72). However, the regiochemical outcome changes when nucleophilic ring opening occurs in acidic conditions. The rationale for this crossover is that the epoxide is protonated in acidic media, allowing the heterocyclic ring to open up a bit in such a way that the bond between the oxygen and the more substituted position is lengthened. This places partial positive character on the carbon at the more substituted position; therefore, the nucleophile (which under acidic conditions is typically quite weak) is attracted to this electropositive center and ultimately promotes displacement (Reaction 73). However, this is unlike an S_N1 reaction, because the stereochemical information is preserved as a general rule.

O — RMgBr / THF → OH, R (R71)

O — NaOR / ROH → OH, OR (R72)

O — cat. *p*TsOH / ROH → OR, OH (R73)

CRAM SESSION

Epoxide Chemistry

Epoxides can be prepared from

- The oxidation of **alkenes** (DMD or mCPBA)
- The base-catalyzed ring closure of β–**halohydrins**

The most common transformation of epoxides is

- Nucleophilic ring opening to **alcohols**

It is important to remember that the regiochemistry of this reaction depends upon the conditions. As a general rule,

- Under acidic conditions, nucleophilic attack occurs at the more-substituted position.
- Under basic conditions, nucleophilic attack occurs at the less-substituted position.

CHAPTER 15

Amines

Read This Chapter to Learn About:

- Preparation of Amines
- Reactions of Amines

PREPARATION OF AMINES

The direct alkylation of amines is a difficult path to tread—the product of each alkylation is more nucleophilic than the previous species; therefore, the products compete with the starting material for the electrophile. Typically, unpleasant mixtures are obtained. Luckily, there are workarounds for this problem. One is the Gabriel synthesis (Reaction 74), in which phthalimide is deprotonated and monoalkylated in a smooth and controlled manner. The latent amine is then freed from the phthalate protecting group by treatment with hydrazine.

Phthalimide (NH) —base→ —R–CH₂–LG→ N-alkylphthalimide —$H_2N{-}NH_2$→ RCH_2NH_2 (R74)

Another way to address the problem is to use amide (H_2N^-) instead of amine (Reaction 75). Now, after alkylation occurs, the product is much less nucleophilic than the reactant (RNH_2 vs. H_2N^-). The downside to this approach is that amide is quite basic and aggressive, so if the substrate contains sensitive functionality, it is likely to undergo collateral decomposition. A considerably milder alternative is represented by the Staudinger reaction (Reaction 76), whereby azide anion (an excellent nucleophile) is used for the initial displacement, and then the resulting alkyl azide is reduced to the primary amine.

$$R\text{-}CH_2\text{-}LG \xrightarrow[\text{THF}]{NaNH_2} R\text{-}CH_2\text{-}NH_2 \qquad \text{(R75)}$$

$$R\text{-}CH_2\text{-}LG \xrightarrow[\text{THF}]{NaN_3} R\text{-}CH_2\text{-}N_3 \xrightarrow{\text{LAH}} R\text{-}CH_2\text{-}NH_2 \qquad \text{(R76)}$$

Other reductive techniques include the complete reduction of amides to amines (Reaction 67) and the reductive amination procedure (Reaction 33), which converts ketones into amines.

$$R\text{-}C(=O)\text{-}NR'R'' \xrightarrow[\text{THF}]{\text{LAH}} R\text{-}CH_2\text{-}NR'R'' \qquad \text{(R67)}$$

$$R_1\text{-}C(=O)\text{-}CH_2\text{-}R_2 \xrightarrow[\substack{NaBH_3CN \\ MeOH}]{R_3R_4NH} R_1\text{-}CH(NR_3R_4)\text{-}CH_2\text{-}R_2 \qquad \text{(R33)}$$

Two rearrangements yield amines—the Hofmann rearrangement (Reaction 69), which proceeds by the basic chlorination of a primary amide, and the Curtius rearrangement, which features an acyl azide (Reaction 77). Both mechanisms ultimately involve acyl nitrene intermediates, but each has a unique way of arriving at that road marker. Furthermore, since the acyl azide species is derived from an acyl chloride, which in turn is prepared from a carboxylic acid, Curtius chemistry is a way of converting the carboxylic acid functionality into an amine with the loss of a single carbon.

$$R\text{-}C(=O)\text{-}NH_2 \xrightarrow[\substack{NaOH \\ H_2O}]{Br_2 \text{ or } Cl_2} R\text{—}NH_2 \qquad \text{(R69)}$$

$$R\text{-}C(=O)\text{-}OH \xrightarrow[\text{2. } NaN_3]{\text{1. } SOCl_2} R\text{-}C(=O)\text{-}N_3 \xrightarrow[\substack{H_2O \\ H^+}]{\Delta} \left[R\text{-}N{=}C{=}O \right] \rightarrow R\text{—}NH_2 \qquad \text{(R77)}$$

REACTIONS OF AMINES

Amines can undergo alkylation in a couple of ways—one is through the reductive amination protocol (Reaction 33); another involves the conjugate addition of an amine onto an electron-deficient double bond (Reaction 78). Amines are also easily acylated in the presence of acyl chlorides to provide the corresponding amides (Reaction 63).

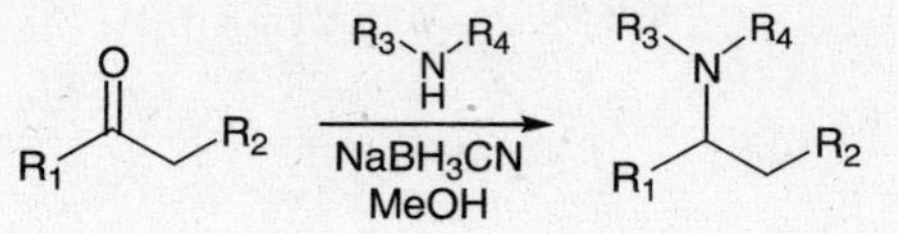

(R33)

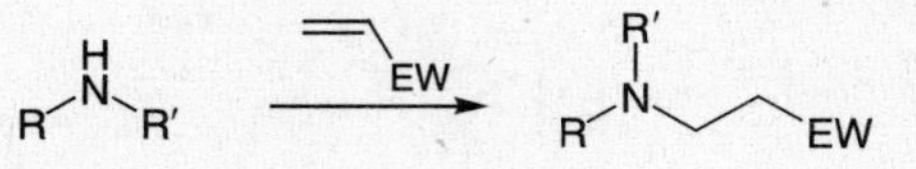

(R78)

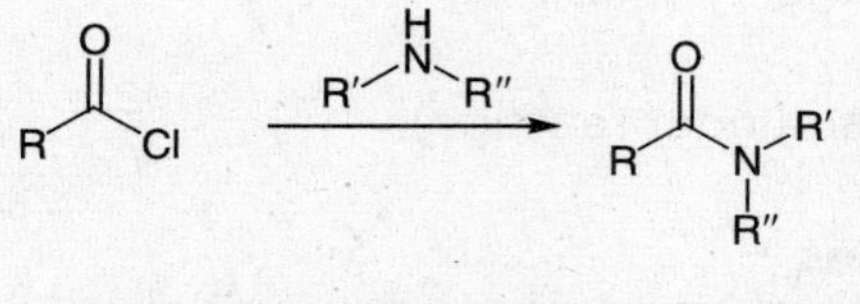

(R63)

Amines can be converted into unsaturated compounds in a few ways. For example, the combination of exhaustive methylation and subsequent Hofmann elimination (Reaction 42) delivers an alkene (usually the least stable). Primary amines condense onto ketones and aldehydes to provide imines (Reaction 31), while secondary amines give only enamines (Reaction 32).

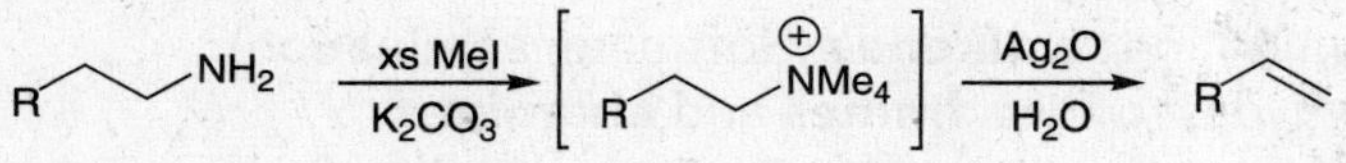

(R42)

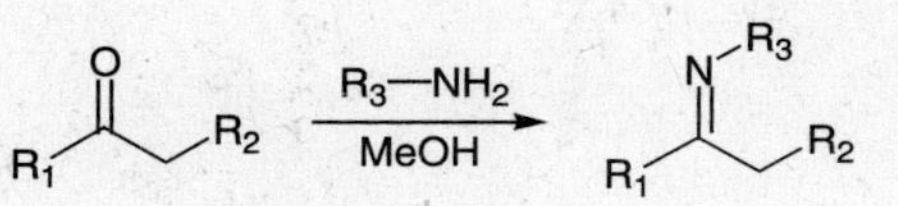

(R31)

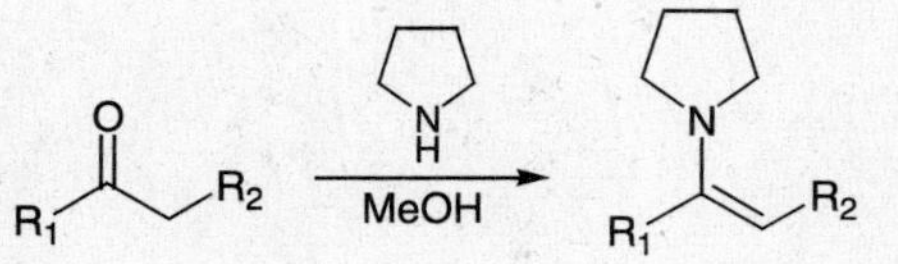

(R32)

Finally, an interesting and useful functional group transformation is represented by the double tosylation of a primary amine, followed by nucleophilic displacement using sodium iodide, thus converting an alkylamine into the corresponding alkyl iodide in high yield (Reaction 79).

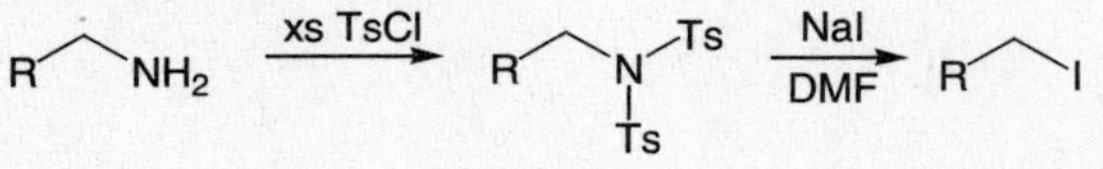

(R79)

CRAM SESSION

Amines

Amines can be prepared from

- The alkylation and hydrolysis of **phthalimide** (Gabriel synthesis)
- The alkylation of **sodamide**
- The alkylation and reduction of **sodium azide** (Staudinger reaction)
- The reduction of carboxylic **amides** (LAH)
- The reductive amination of **ketones** and **aldehydes**
- The halogen-mediated rearrangement of carboxylic **amides** (Hofmann)
- The rearrangement of **acyl azides** (Curtius)

Some common transformations of amines are

- Reaction with ketones and aldehydes under reductive conditions to form **amines**
- Conjugate addition onto alkenes to form other **amines**
- Reaction with acyl halides to form carboxylic **amides**
- Exhaustive methylation and elimination to **alkenes** (Hoffmann elimination)
- Reaction with ketones and aldehydes to form **imines** and **enamines**
- Reaction with excess tosyl chloride and sodium iodide to form **alkyl iodides**

CHAPTER 16

Aromatic Chemistry

Read This Chapter to Learn About:

- Functionalization of Benzene
- Functional Group Transformations

FUNCTIONALIZATION OF BENZENE

The lion's share of aromatic chemistry launches ultimately from benzene. Therefore, it would seem reasonable to examine first how we functionalize benzene, and then explore how to modify and embellish that functionality. Five basic reactions stand tall in the realm of benzene functionalization, all of which are classified as electrophilic aromatic substitution reactions (see Chapter 8). Mechanistically, they differ only with respect to the active electrophile that is initially captured by benzene.

The first three functionalizations introduce heteroatomic substituents. For example, nitration (Reaction 80) involves treatment with nitric acid in sulfuric acid, in which the nitronium cation (NO_2^+) is believed to be the relevant electrophilic species. Sulfonation (Reaction 81) is carried out in a variety of media, ranging from concentrated sulfuric acid to "monohydrate" (100% sulfuric acid) to oleum, which is sulfuric acid saturated with sulfur trioxide (SO_3), the species generally believed to function as the active electrophile. Chlorination (Reaction 82) proceeds by treatment with molecular chlorine in the presence of aluminum trichloride, which serves to activate the chlorine through Lewis acid coordination.

$$C_6H_6 \xrightarrow[H_2SO_4]{HNO_3} C_6H_5NO_2 \qquad (R80)$$

$$C_6H_6 \xrightarrow[H_2SO_4]{SO_3} C_6H_5SO_3H \qquad (R81)$$

Benzene $\xrightarrow[AlCl_3]{Cl_2}$ chlorobenzene (R82)

Two other methods introduce carbon-based substituents. The Friedel–Crafts alkylation (Reaction 83) involves an alkyl chloride activated by a Lewis acid catalyst. This works fine for fairly simple precursors; however, primary alkyl chlorides are not very well behaved toward this protocol, and cationic rearrangement is always an issue. Both of these concerns can be addressed somewhat by using the Friedel–Crafts acylation (Reaction 84). Here the pertinent electrophilic species is believed to be the acylium cation ($R{-}C{\equiv}O^+$), which is regiochemically stable.

Benzene $\xrightarrow[AlCl_3]{RCH(Cl)R'}$ $C_6H_5CH(R)R'$ (R83)

Benzene $\xrightarrow[AlCl_3]{RC(O)Cl}$ $C_6H_5C(O)R$ (R84)

Of course, the complexity increases when we begin to attach multiple substituents onto the same aromatic ring. It turns out that as a general rule, electron-donating (ED) substituents (e.g., methoxy, amino) have an activating influence on the aromatic ring, and they also direct subsequent reactivity to the *ortho* and *para* sites. Conversely, electron-withdrawing (EW) groups (e.g., nitro, carbonyl) have a deactivating influence on the ring, and they also direct subsequent electrophiles to the *meta* positions. Curiously, the halogens straddle the fence: they are deactivating, but *ortho*/*para* directing (Table 16-1).

TABLE 16-1 Substituent Effects on Electrophilic Aromatic Substitution

Electron-Donating (ED) Groups	Electron-Withdrawing (EW) Groups
Activating *ortho*/*para*-directing $-NR_2$ (amino, amido) $-OR$ (alkoxy, hydroxy) $-R$ (alkyl)	Deactivating *meta*-directing $-NO_2$ (nitro) $-CO_2R$ (esters, acids) $-C(O)R$ (ketones, aldehydes) $-SO_3H$ (sulfo) $-CN$ (cyano)
Halogens	
Deactivating *ortho*/*para*-directing $-F$, $-Cl$, $-Br$, $-I$	

At first these results appear to be confounding, until we carefully consider their origins. Two factors are playing a role here—inductive effects and resonance effects.

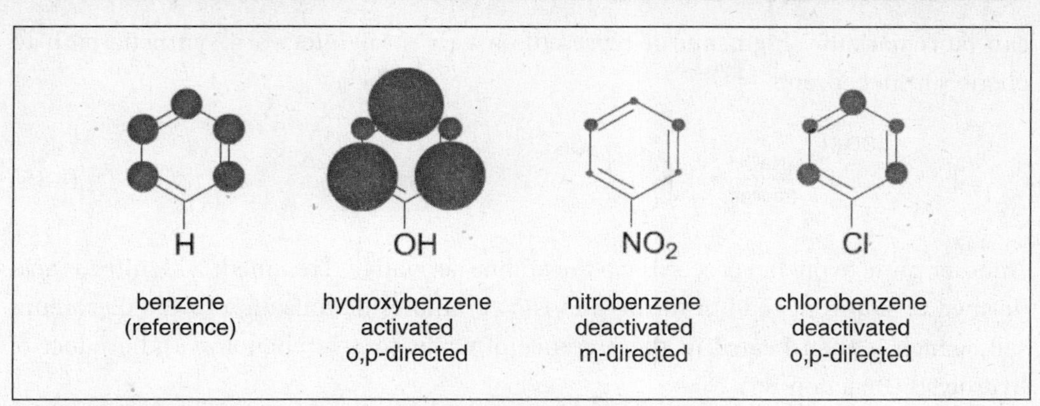

FIGURE 16-1: Electron density coefficients for substituted benzenes.

Figure 16-1 shows a graphical representation of some computational outcomes with respect to the coefficient of electron density throughout the benzene ring with various substituents. First, let us consider the hydroxy substituent. As for induction, the oxygen is electronegative, so we would expect electron density to be drawn from the ring (note the diminished *meta* coefficients compared to benzene itself); however, the lone pairs on the oxygen are able to contribute much more electron density through resonance. Because the resonance factor disproportionately affects the *ortho* and *para* sites, these are the areas of highest electron density and, thus, nucleophilicity.

Now we turn our attention to nitrobenzene. The nitro group is severely electron-withdrawing from an inductive and resonance standpoint. The inductive effect draws electron density out of the entire ring, whereas the resonance effect selectively removes even more electron density from the *ortho* and *para* sites. Therefore, the whole ring is deactivated, and the *ortho*/*para* positions are even more deactivated, so the most electron-rich positions are the *meta* sites.

Finally we examine chlorobenzene. The *chloro* substituent is inductively electronegative, but it also has lone-pair electron density that can be donated back. Thus, the electron density is skimmed off the top for the whole ring, but a little is added back at the *ortho* and *para* positions, thanks to resonance. Therefore, the ring is deactivated, but *ortho*/*para* is still more nucleophilic.

FUNCTIONAL GROUP TRANSFORMATIONS

The next set of reactions can be classified as functional group transformations on aromatic systems. In other words, an effective strategy is sometimes to simply get some functionality on board, and then convert it to the desired functional group in a subsequent step. One of the oldest (but still relevant) such conversions is the fusion reaction (Reaction 85), in which benzenesulfonic acid derivatives are cooked up in molten sodium hydroxide to provide the corresponding phenol. The yields for this reaction can be remarkably high, and it represents a very straightforward synthetic path to phenols from benzene.

$$C_6H_5SO_3H \xrightarrow[\text{(molten)}]{\text{NaOH}} C_6H_5OH \quad \text{(R85)}$$

Another route to phenol proceeds via the aniline derivative. Treatment with nitrous acid (formed *in situ* from sodium nitrite and HCl) promotes diazotization to the diazonium salt, which is then heated in the presence of water to give phenol as the product of hydrolysis (Reaction 86).

$$C_6H_5NH_2 \xrightarrow[\text{HCl}]{\text{NaNO}_2} C_6H_5N^{\oplus}{\equiv}N\ Cl^{\ominus} \xrightarrow[\Delta]{H_2O} C_6H_5OH \quad \text{(R86)}$$

When electron-withdrawing substituents (nitro, carbonyl, etc.) and potential leaving groups (OTs, Cl, etc.) are situated in a 1,2- or 1,4-relationship, the opportunity arises to engage in a reaction known as nucleophilic aromatic substitution, also known as the S_NAr mechanism (see Chapter 8). For example, *p*-chloronitrobenzene suffers attack by methoxide in an addition-elimination sequence to give the corresponding methoxy-nitrobenzene with the ejection of chloride (Reaction 87). Other nucleophiles can also be used in this protocol.

$$O_2N{-}C_6H_4{-}Cl \xrightarrow[\text{ROH}]{\text{NaOR}} O_2N{-}C_6H_4{-}OR \quad \text{(R87)}$$

It is useful to know how to modulate the oxidation state in the substitution patterns around the aromatic nucleus. Toward that end, nitrobenzenes can be reduced to anilines using either catalytic hydrogenation or iron in HCl (Reaction 88). Similarly, acylbenzenes are smoothly reduced to the corresponding alkylbenzenes using catalytic hydrogenation or zinc amalgam in HCl, otherwise known as the Clemmensen reduction (Reaction 89). It is through the Clemmensen that we can circumvent the problems caused by the Friedel–Crafts alkylation using primary halides. Instead, we can use acyl halides (i.e., Friedel–Crafts acylation) and then reduce the initial product.

$$C_6H_5NO_2 \xrightarrow[\text{or Fe / HCl}]{H_2\text{ / cat.}} C_6H_5NH_2 \quad \text{(R88)}$$

$$C_6H_5C(=O)R \xrightarrow[\text{or Zn(Hg) / HCl}]{H_2\text{ / cat.}} C_6H_5CH_2R \quad \text{(R89)}$$

Equally useful is the ability to oxidize an alkyl substituent to an acyl derivative through a protocol known as benzylic oxidation (Reaction 90). This reaction is surprisingly general and can use a variety of Cr^{VI} reagents, such as PCC or chromium trioxide in acetic acid.

PCC or CrO_3 / HOAc

(R90)

CRAM SESSION

Aromatic Chemistry

Because the energetics of aromatic systems are significantly different from that of their aliphatic counterparts, the chemistry of benzene involves methods not found in other areas. Generally speaking, the synthesis of benzene derivatives involves two components:

- The introduction of functional groups onto benzene and its derivatives
- The modification of these functional groups

The common methods for functionalizing benzene are

- **Nitration**
- **Sulfonation**
- **Halogenation**
- Friedel–Crafts **alkylation** and **acylation**

The regiochemistry of the introduction of these groups is governed by the **directing effects** of functionality already present on the aromatic ring. As a rule of thumb:

- Electron-donating (ED) groups are activating and *ortho/para* directing
- Electron-withdrawing (EW) groups are deactivating and *meta* directing
- Halogens are deactivating, but *ortho/para* directing

Some important **functional group transformations** of benzene derivatives are

- Conversion of **benzenesulfonic acids to phenols** (fusion)
- Conversion of **anilines to phenols** (diazotization)
- Nucleophilic aromatic substitution (S_NAr)
- Reduction of **nitrobenzenes to anilines**
- Reduction of **acyl benzenes to alkyl benzenes** (Clemmensen)
- Oxidation of **alkyl benzenes to acyl benzenes** (benzylic oxidation)

CHAPTER 17

Cycloadditions

Read This Chapter to Learn About:

- [4+2] Cycloaddition
- [2+2] Cycloadditions

[4+2] CYCLOADDITION

The highest profile [4+2] cycloaddition from a synthetic standpoint is the Diels–Alder reaction (Reaction 91), in which a diene and a dienophile react to form a cyclohexene derivative. The so-called normal Diels–Alder works best with electron-rich dienes (4π components) and electron-poor dienophiles (2π components).

EW + EW —Δ, toluene→ EW (R91)

Although there are plenty of examples of acyclic dienes giving reasonable yields, they suffer from the sad circumstance that their most stable conformation (*s-trans*) is the least reactive. Therefore, the Diels–Alder is often carried out using cyclic dienes, such as cyclopentadiene (Reaction 92). At this point, a stereochemical feature arises, namely the relative stereochemistry between the methylene bridge and the electron-withdrawing (EW) group. Because the EW group is tucked into the endo pocket of the molecule, it is said to obey the "endo rule." This is a purely empirical rule with little (if any) theoretical underpinning.

+ EW —Δ, toluene→ EW (R92)

Furans are also suitable $4p$ components for the Diels–Alder reaction (Reaction 93), giving bicyclic adducts not unlike those of cyclopentadiene. These initial adducts can be isolated, if desired—or they can be subjected to thermolysis, whereby the oxygen bridge is eliminated to gain aromaticity. Tetrahydronaphthalene derivatives can be accessed by trapping *bis*(methylene)-cyclohexadiene with a dienophile (Reaction 94). These are generally very facile reactions, owing to the instability of the 4π component and the aromaticity gained in undergoing cycloaddition.

EW, R, Δ, O, EW, R, Δ, EW, R (R93)

Δ, EW, EW (R94)

[2+2] CYCLOADDITIONS

There are several [2+2] cycloadditions that are of synthetic interest; we examine only two here. The first is a thermal [2+2] between an alkene and ketene (Reaction 95), which provides a cyclobutanone derivative. Thermal [2+2] cycloadditions are sluggish in general because of molecular orbital considerations, but the strain in ketene provides somewhat of a thermodynamic boost.

R_1, R_2, R_3, R_4 + O=C=CH_2 Δ → R_2, R_1, O, R_3, R_4 (R95)

The other type of [2+2] is the photochemical reaction of an alkene with a benzaldehyde derivative to give an oxetane derivative (Reaction 96). This is a protocol known as the Paterno–Büchi reaction, and it represents a synthetically valuable way to arrive at oxetane skeletons.

R_1, R_2, R_3, R_4 + O, Ph hv → R_2, R_1, O, R_3, R_4, Ph (R96)

CRAM SESSION

Cycloadditions

The synthetically most useful cycloaddition is the **Diels–Alder reaction,** which is a [4+2] cycloaddition between a diene (4π component) and dienophile (2π component), a process that results in a **cyclohexene derivative.** Common diene components are **cyclopentadiene** and **furans;** cycloadducts of the latter generally undergo thermal elimination of oxide to form **benzene derivatives.** The normal Diels–Alder reaction works best with electron-rich dienes and electron-poor dienophiles.

Two common [2+2] cycloadditions are thermal **alkene-ketene cycloadditions**, which produce **cyclobutanone derivatives,** and the photochemical cycloaddition of an alkene onto a carbonyl (the **Paterno–Büchi reaction**), which forms an **oxetane derivative** (oxacyclobutane).

CHAPTER 18

Carbohydrates and Peptides

Read This Chapter to Learn About:

- Carbohydrates
- Peptides

There are two classes of biologically relevant compounds that merit examination as a class. Both polysaccharides and polypeptides are highly chiral biopolymers derived from a relatively small subset of monomer units—yet their function in nature is incredibly diverse. Although each field is worthy of a course on its own, we review here only a few of the salient features of vocabulary and chemistry in each.

CARBOHYDRATES

Polysaccharides are composed of sugars, which can be thought of chemically as polyhydroxyketones and aldehydes. As a general rule, each sugar contains one carbonyl group and several hydroxyl groups. If the carbonyl group is terminal in the chain (i.e., an aldehyde), then the sugar is termed an **aldose;** if the carbonyl is internal, then the sugar is a **ketose.** Furthermore, sugars are characterized by the number of carbons in the backbone. Thus, three-carbon sugars are generically classified as **trioses,** with the next higher homologs being **tetroses** (four carbons), **pentoses** (five carbons), and **hexoses** (six carbons). Thus, for example, glucose is an aldohexose whereas fructose is a ketohexose (Figure 18-1).

In solution, sugars are in equilibrium between the open-chain form and a cyclic hemiacetal form, a transformation catalyzed by either acid or base. Any pentose and hexose can, in principle, form either a five-membered ring (a furanose) or a six-membered ring

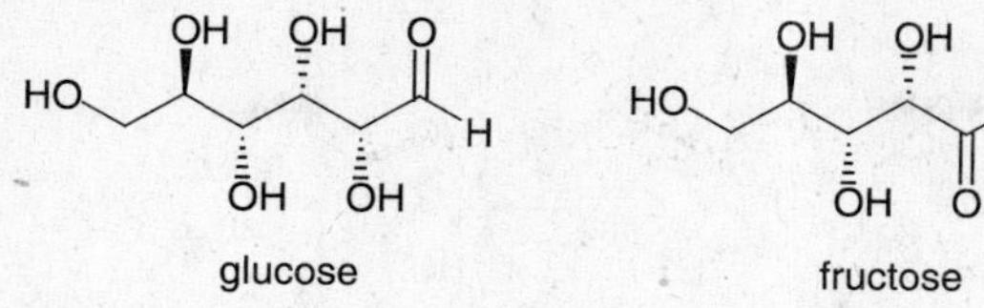

glucose fructose

FIGURE 18-1: Glucose and fructose.

(a pyranose); however, often one form predominates for a given sugar. The outcome is highly substrate-dependent. For example, glucose forms almost predominantly the pyranose ring (Figure 18-2), whereas fructose provides a 1:2 mixture of furanose and pyranose structures, respectively.

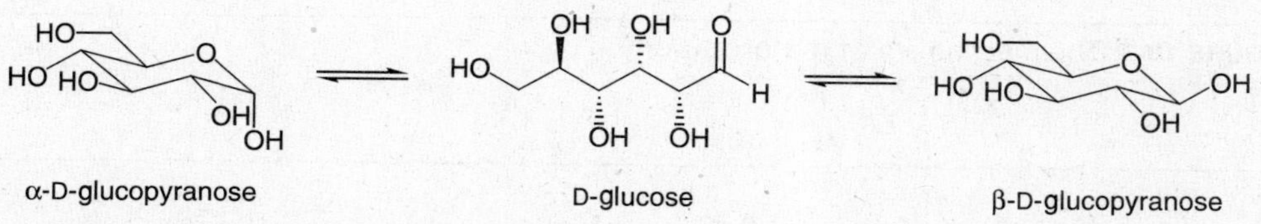

α-D-glucopyranose D-glucose β-D-glucopyranose

FIGURE 18-2: Equilibrium mixture of open-chain and pyranose forms of D-glucose.

Notice that the hydroxyl substituent at C1 (the anomeric carbon) can point toward the top or the bottom face of the ring—the former isomer is designated β and the latter α. Either isomer can be prepared in pure crystalline form, but in solution 9:16 equilibrium mixture of α:β isomers is eventually established. As the composition shifts from pure α or pure β to the equilibrium ratio, the bulk optical rotation changes as well—thus, this phenomenon has been called **mutarotation.** Chemically speaking, mutarotation is nothing more than an acid-catalyzed hemiacetal cleavage, followed by a reformation of the hemiacetal with attack on the opposite face of the carbonyl group (Figure 18-3).

If we examine the chair conformations of the two pyranose structures, an interesting question arises. The β form would appear to be much more thermodynamically favored, because all substituents are in the equatorial attitude—and although this is indeed the major isomer, the amount of α-pyranose is not insignificant. In fact, it is far greater than would be predicted by steric factors alone. Therefore, some other factor must be playing a role in privileging this axial hydroxyl substituent.

The answer lies in dipole–dipole interactions. If we examine Newman projections of both isomers, sighting down the bond between C1 and the ring oxygen (Figure 18-4), we find that the β-isomer places the C1 hydroxyl group in a position that bisects the lone pairs on the ring oxygen. This, in turn, brings the dipole of the hydroxyl group and ring oxygen into coalignment—a situation that is electrostatically disfavored. However, the α-isomer allows for the two dipoles to adopt a more favorable alignment with respect to

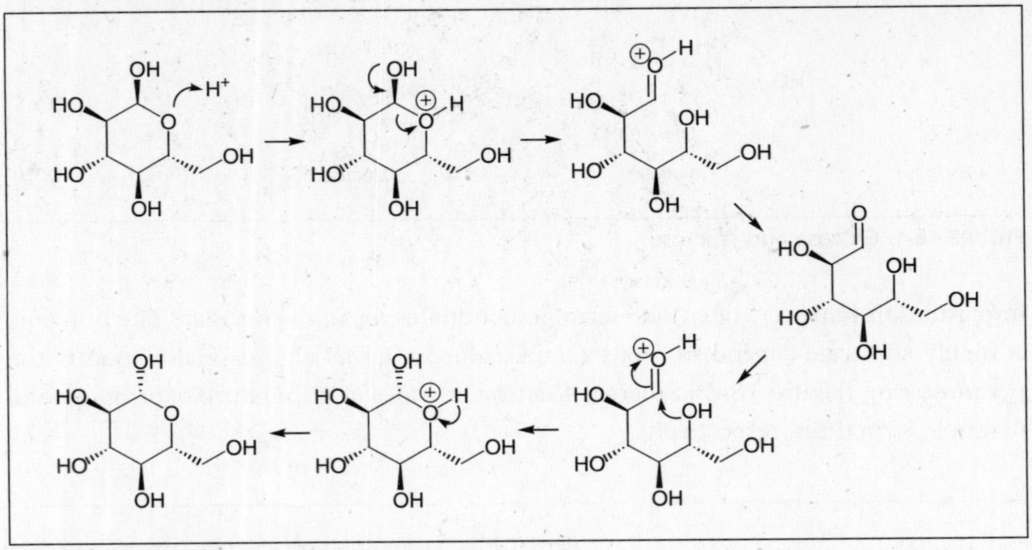

FIGURE 18-3: The mechanism of mutarotation.

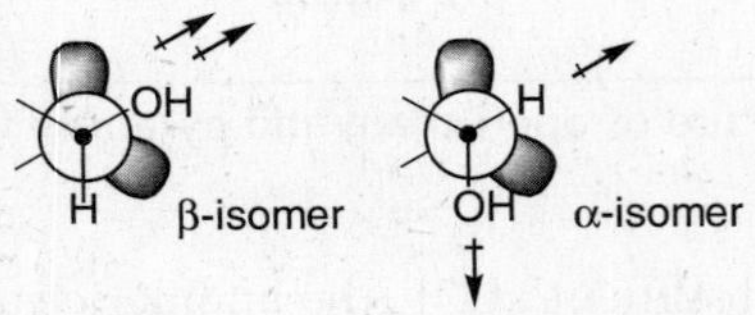

FIGURE 18-4: The anomeric effect.

each other. This phenomenon is known as the **anomeric effect**, and in many sugars it results in the predominant formation of the α-isomer.

The anomeric position is also where sugars link together to form polysaccharides. Chemically speaking, this linkage is produced when the hemiacetal of one sugar molecule is converted to an acetal by condensation with a hydroxyl group from another sugar molecule. In addition to the identity of the sugars involved, these polysaccharides are characterized by the configuration at the anomeric carbon, as well as the hydroxyl group used to form the acetal linkage. For example, cellulose is constructed from D-glucose linked at the 4-hydroxyl position with β-anomeric configuration (Figure 18-5).

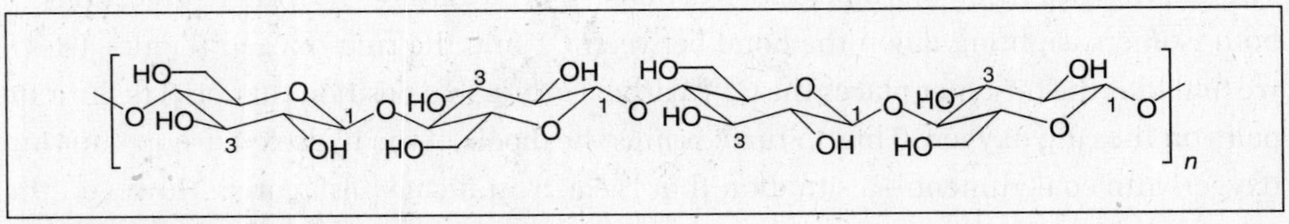

FIGURE 18-5: Cellulose, a polysaccharide.

There are many qualitative ("wet chemical") tests for the classification of saccharides, most of which are mainly of historical interest. The most familiar (Tollens, Benedict, Fehling) are those that demonstrate the presence of a reducing sugar. By definition, a reducing sugar possesses functionality that is easily oxidizable. Essentially, this includes any hemiacetal, aldehyde, or α-hydroxy ketone moiety. Thus, practically all monosaccharides classify as reducing sugars.

PEPTIDES

Proteins are constructed from an ensemble of 20 or so naturally occurring amino acids. In contrast with the saccharides, which possess only carbonyl and hydroxyl functionality, the amino acids boast a wide array of functional groups, as shown in Table 18-1. As monomers at physiological pH, all amino acids exist as ionized species,

TABLE 18-1 The Amino Acid Side Chains

Amino Acid	Abbr.	Side Chain Structure	Relevant Functionality	Side Chain pK_a
Glycine	gly	–H	None	
Alanine	ala	–Me	Alkane	
Valine	val	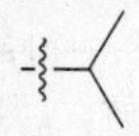	Branched alkane	
Leucine	leu	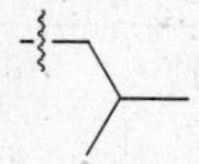	Branched alkane	
Isoleucine	ile	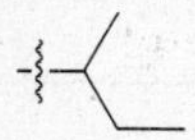	Branched alkane	
Phenylalanine	phe	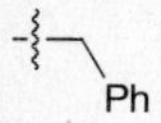 Ph	Phenyl ring	
Tryptophan	trp	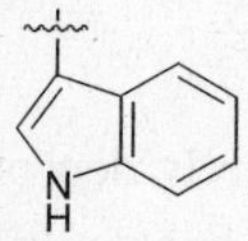 N H	Indole	
Histidine	his	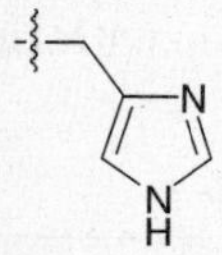 N N H	Imidazole	pK_a 6.1
Tyrosine	tyr	 OH	Phenol	pK_a 10.1
Serine	ser	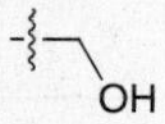 OH	Primary alcohol	

TABLE 18-1 The Amino Acid Side Chains (*Cont.*)

Amino Acid	Abbr.	Side Chain Structure	Relevant Functionality	Side Chain pK_a
Threonine	thr	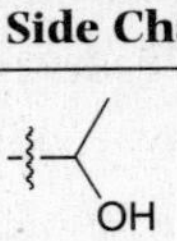 OH	Secondary alcohol	
Methionine	met	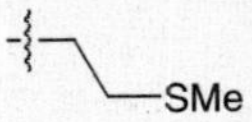 SMe	Dialkyl sulfide	
Cysteine	cys	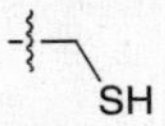 SH	Mercaptan	pK_a 8.2
Asparagine	asn	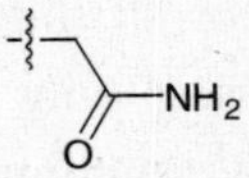 NH_2, O	Amide	
Glutamine	gln	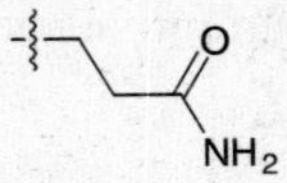 O, NH_2	Amide	
Aspartic acid	asp	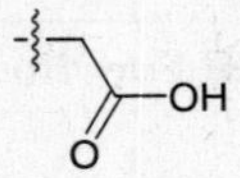 OH, O	Carboxylic acid	pK_a 3.7
Glutamic acid	glu	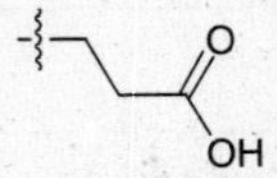 O, OH	Carboxylic acid	pK_a 4.3
Lysine	lys	 NH_2	Primary amine	pK_a 10.5
Arginine	arg	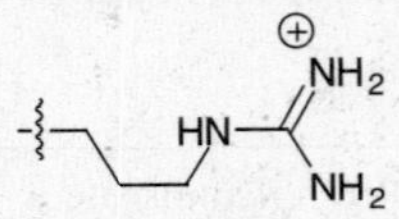 HN, $NH_2^{\oplus}$, NH_2	Guanidine	pK_a 12.5
Proline	pro	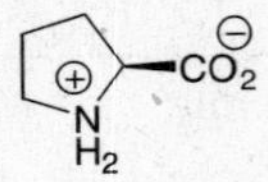 $\oplus$ N H_2, $CO_2^{\ominus}$	None	

as shown in Figure 18-6. The carboxylic acid is entirely deprotonated, and the amino group is completely protonated—so even though the molecule is ionized, there is a net coulombic charge of zero. Such a species is known as a zwitterion.

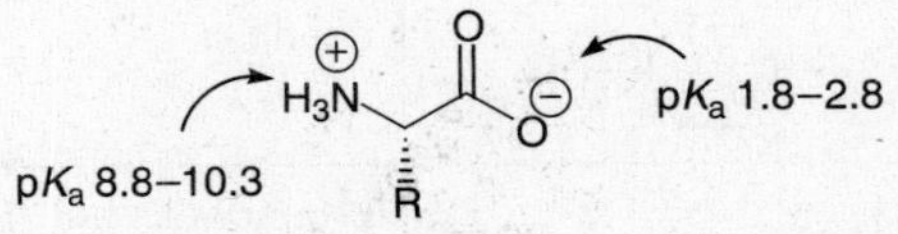

FIGURE 18-6: Zwitterionic nature of amino acids.

Nature takes advantage of these diverse amino acids, particularly in the realm of enzyme catalysis, by assembling them together into synergistic arrangements. In contrast to the polysaccharides, which are connected by acetal linkages, amino acids are bound together by a relatively robust amide linkage. Thus, the primary structure of proteins can be described as a polyamide backbone embellished with functionalized side chains at regular intervals (Figure 18-7). Although there is hindered rotation about the amide C—N bond, there is relatively free rotation about the C—C bonds in the backbone. Thus, the polymer can adopt a variety of conformations, the energetics of which are determined by many complex factors, including intermolecular hydrogen bonding, hydrophobic interactions, and solvation effects.

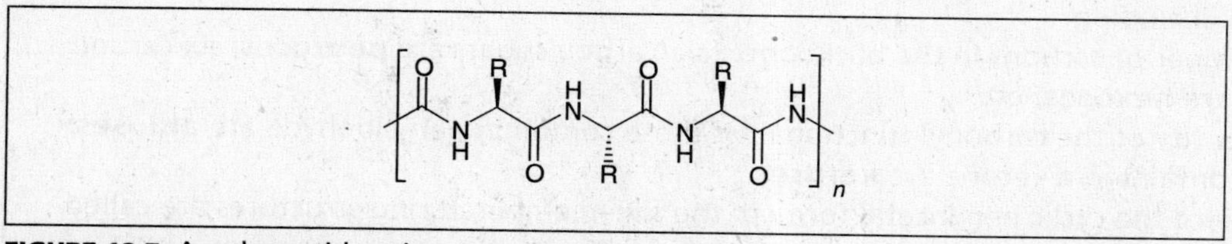

FIGURE 18-7: A polypeptide primary structure.

CRAM SESSION

Carbohydrates and Peptides

Sugars (or saccharides) are polyhydroxyketones and aldehydes, many of which occur in nature. In solution, these compounds tend to isomerize to cyclic hemiacetals, from which carbohydrates derive much of their structural rigidity. Sugars are characterized by some key criteria, including:

- The number of carbons in the backbone: five-carbon sugars are **pentoses;** six-carbon sugars are **hexoses;** etc.
- The identity of the carbonyl functionality: those containing an aldehyde are **aldoses;** those containing a ketone are **ketoses**
- The size of the cyclic hemiacetal formed: the five-membered ring structures are called **furanoses;** the six-membered rings are **pyranoses**

Sugars can polymerize into **polysaccharides,** which are linked at the acetal functionality, also known as the **anomeric** position. The substituent at the anomeric carbon can isomerize between axial and equatorial attitudes through a process known as **mutarotation.**

Proteins (or **polypeptides**) are polymers of **amino acids,** which are connected through an amide linkage (also called a **peptide bond**). In their monomeric form, amino acids in solution exist as **zwitterions,** in which the amine and carboxylic acid functionalities are both ionized.

Conceptual Maps of Section III

Aliphatic/Olefinic

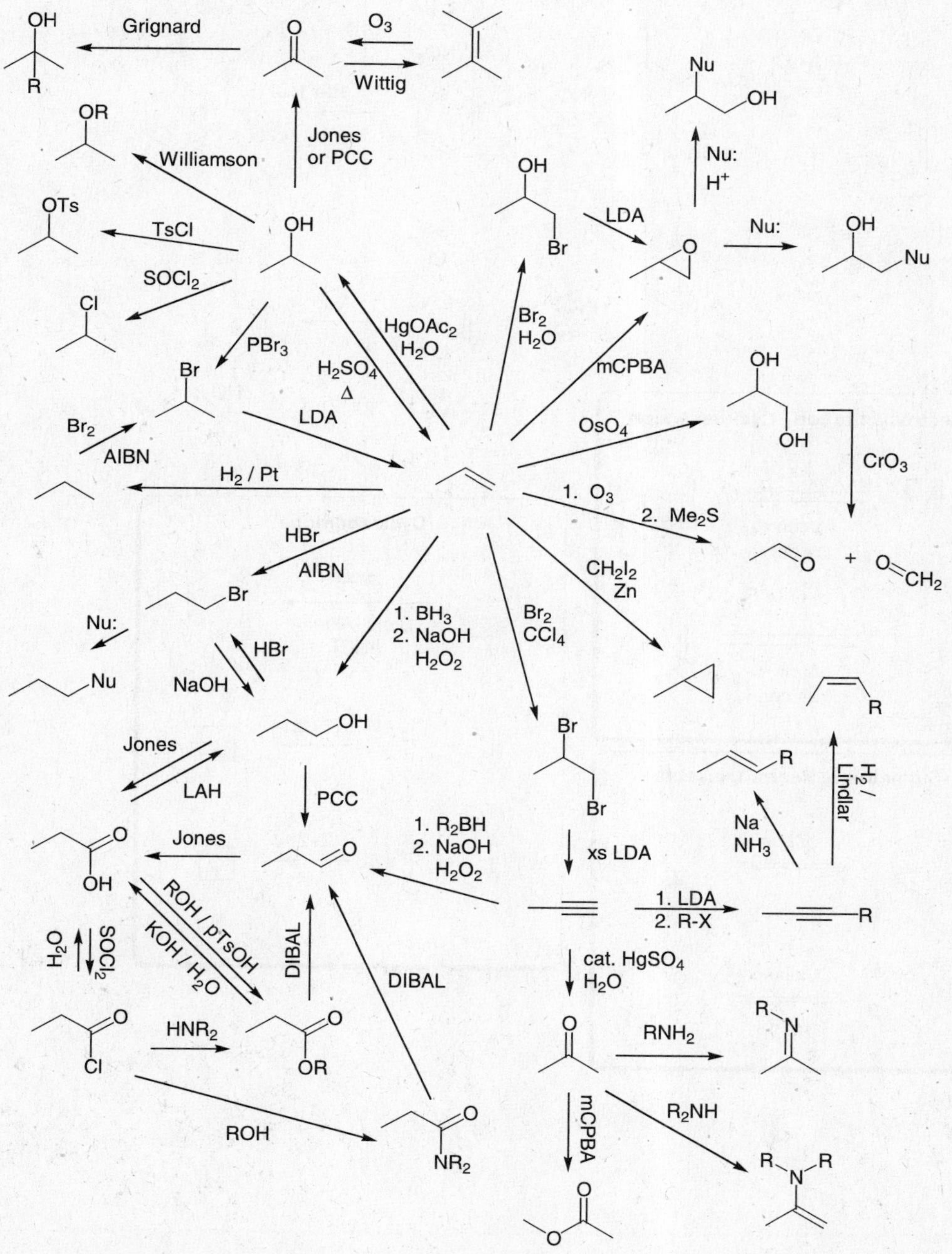

Aromatic/Pericyclic

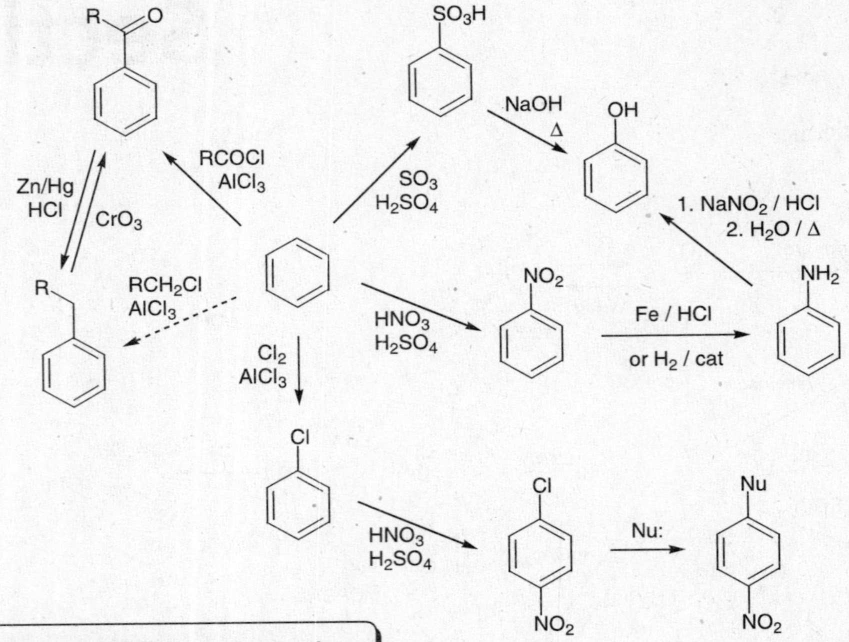

R
O
SO_3H
NaOH
Δ
OH
Zn/Hg
HCl
CrO_3
RCOCl
$AlCl_3$
SO_3
H_2SO_4
1. $NaNO_2$ / HCl
2. H_2O / Δ
R
RCH_2Cl
$AlCl_3$
NO_2
NH_2
HNO_3
H_2SO_4
Fe / HCl
or H_2 / cat
Cl_2
$AlCl_3$
Cl
Cl
Nu
HNO_3
H_2SO_4
Nu:
NO_2
NO_2

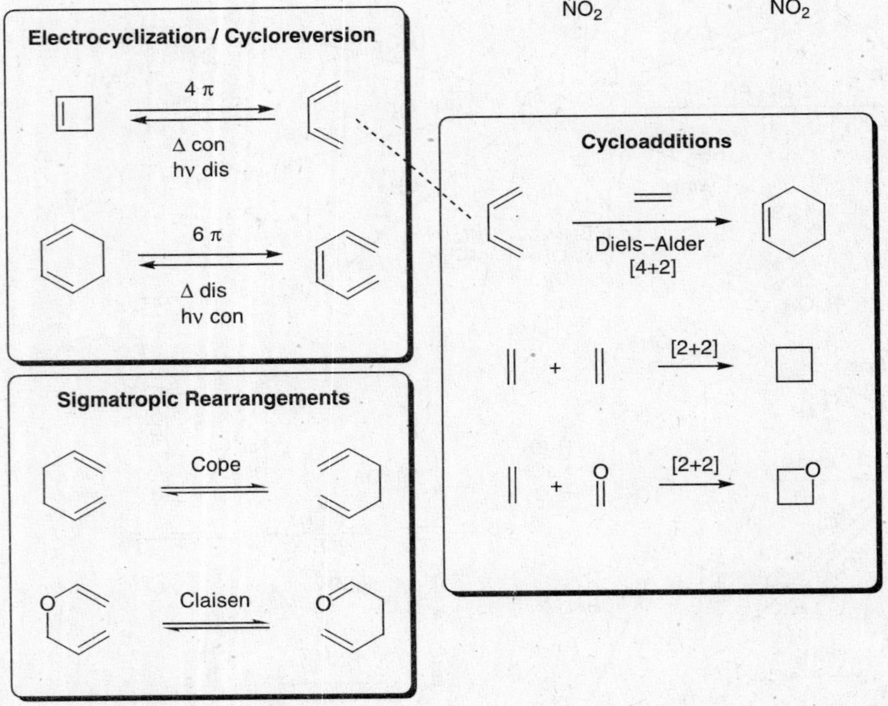

Electrocyclization / Cycloreversion

4 π
Δ con
hν dis

6 π
Δ dis
hν con

Cycloadditions

Diels–Alder
[4+2]

+
[2+2]

+
O
[2+2]
O

Sigmatropic Rearrangements

Cope

O
Claisen
O

SECTION IV

SEPARATION AND PURIFICATION

Having synthesized a given compound, it is incumbent on the experimentalist to isolate the desired product from what can be a complex reaction mixture. This is necessary for proper characterization and identification (Section V), as well as for the purposes of end use. For example, impurities are often detrimental to subsequent synthetic steps, and in the case of pharmaceuticals, by-products can be harmful to health or even deadly. Furthermore, pure compounds are generally more shelf-stable and storable than the crude reaction mixtures. Therefore, it is advantageous to be familiar with the conventional methods of purification in the synthetic organic laboratory.

The most common techniques fall into four broad categories (see table), each based on slightly different chemical principles—in some respects overlapping, in others complementary. From a practical standpoint, a significant point of differentiation is scalability. For example, distilling a kilogram of solvent poses no particular technical challenges; by the same token, recrystallizing 100 grams (g) of a solid is easily done with common laboratory equipment, although the technique itself can be challenging. However, purifying even 20 g of a reaction mixture by chromatography is an expensive and laborious proposition. Extending this to an industrial scale on the order of 100 kilograms (kg) or more is no less of a challenge, although it can be done.

General Methods of Separation and Purification

Method	Extraction	Chromatography	Distillation and Sublimation	Recrystallization
Principle of Separation	Differential solubility between two or more solvent systems (e.g., water and dichloromethane)	Differential affinity between a stationary phase (e.g., silica or cellulose) and a mobile phase	Differential vapor pressures	Differential solubility in a single solvent system
Common Applications	Initial aqueous workup of reactions; separation of inorganic salts from organic products	Final purification of reaction mixtures (e.g., flash columns); analysis of products (e.g., HPLC and GC)	Purification of solvents and volatile products; concentration of reaction mixtures	Purification of products and reagents
Scope and Limitations	Broadly applicable; can be used with any two immiscible solvents	Can be used on a wide variety of compounds and mixtures	Limited to products with appreciable vapor pressure at or below 300°C	Limited to crystalline products; often fails with complex reaction mixtures
Scalability	Excellent	Limited	Very good	Excellent
Relative Expense	Low	High	Moderate	Low

By the same token, the various methods also differ in their scope and range of application. Clearly, a substance must be a solid to be purified by recrystallization, and a compound must exhibit some degree of volatility under reasonable conditions to be distilled. However, column chromatography can be used to purify a wide array of compounds, whether solid or liquid, polar or nonpolar, volatile or nonvolatile. It is sometimes advantageous to use combinations of methods—for example, a "quick and dirty" chromatography column to remove baseline impurities from a complex reaction mixture, followed by a careful recrystallization to obtain an analytically pure sample.

The following chapters are meant to give a brief and general overview of each technique, with a focus on theoretical underpinnings and practical applications.

CHAPTER 19

Extraction

Read This Chapter to Learn About:

- Background
- Aqueous Workup
- pH-Controlled Extraction

BACKGROUND

The basis of extractive techniques is the "like dissolves like" rule. Water typically dissolves inorganic salts (such as lithium chloride) and other ionized species, whereas solvents (ethyl acetate, methylene chloride, diethyl ether, etc.) dissolve neutral organic molecules. However, some compounds (e.g., alcohols) exhibit solubility in both media. Therefore, it is important to remember that this method of separation relies on partitioning—that is, the preferential dissolution of a species into one solvent over another. For example, 2-pentanol is somewhat soluble in water [ca. 17 grams (g)/100 milliliters (mL) H_2O], but infinitely soluble in diethyl ether. Thus, 2-pentanol can be preferentially partitioned into ether.

AQUEOUS WORKUP

One of the most common uses of extraction is during aqueous workup as a way to remove inorganic materials from the desired organic product. On a practical note, workup is usually carried out using two immiscible solvents—that is, in a biphasic system. If a reaction has been carried out in THF, dioxane, or methanol, then it is generally desirable to remove those solvents by evaporation before workup, because they have high solubilities in both aqueous and organic phases, and can set up single-phase systems (i.e., nothing to separate) or emulsions. Typical extraction solvents include ethyl acetate, hexane, chloroform, methylene chloride, and diethyl ether. All of these form crisp delineations between phases.

The two layers are commonly referred to as the **aqueous phase** and the **organic phase.** It is important to keep track of the phases, as their positions are solvent-dependent. For example, diethyl ether is lighter than water, so the organic phase rests on top in the separatory funnel, whereas methylene chloride is heavier than water and so sinks to the bottom (Figure 19-1). Many students realized only too late that they have discarded the wrong layer from an extraction.

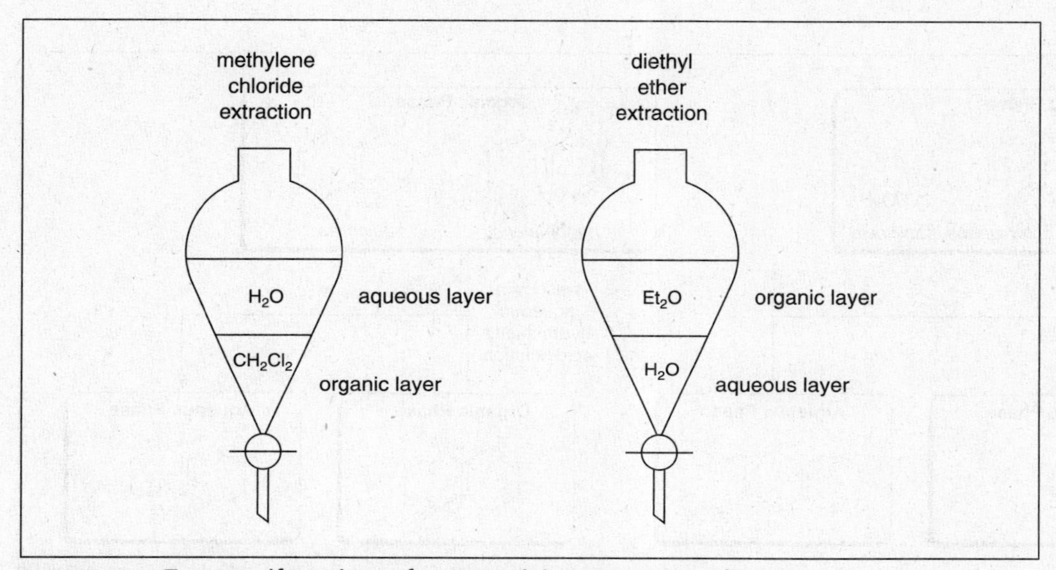

FIGURE 19-1: Two manifestations of an organic/aqueous extraction.

By way of vocabulary, two operations are actually encountered in the separatory funnel. When components are removed from an organic layer by shaking with an aqueous solution, the organic phase is said to be **washed** (e.g., "The combined ether extracts were washed with aqueous sodium bicarbonate solution"). However, when components are removed from water by treatment with an organic solvent, the aqueous phase is said to be **extracted** (e.g., "The aqueous layer was extracted with three portions of ethyl acetate"). Thus, aqueous layers are extracted, and organic layers are washed—although these two terms are sometimes (erroneously) used interchangeably.

Aqueous workup can involve more than just separation. For example, reactions that produce anions (e.g., Grignard reactions) are usually "quenched" with a mildly acidic aqueous solution (e.g., saturated ammonium chloride) at the end of the reaction to neutralize any residual base. The same holds true for very acidic reactions. Thus, aromatic nitration reactions (HNO_3/H_2SO_4) are usually quenched by pouring onto a large quantity of ice, which dilutes the acidic environment.

pH-CONTROLLED EXTRACTION

Although extractions are usually carried out with a neutral aqueous phase, sometimes pH modulation can be used to advantage. For example, a mixture of naphthalenesulfonic acid and naphthalene can be separated by washing with bicarbonate, in which case the sulfonic acid is deprotonated and partition into the aqueous phase. Similarly, a mixture of naphthalene and quinoline can be separated by an acid wash (Figure 19-2), taking advantage of the basic nature of the heterocyclic nitrogen (pK_a 9.5). If it is necessary to isolate the quinoline, it is neutralized with bicarbonate and extracted back into an organic solvent.

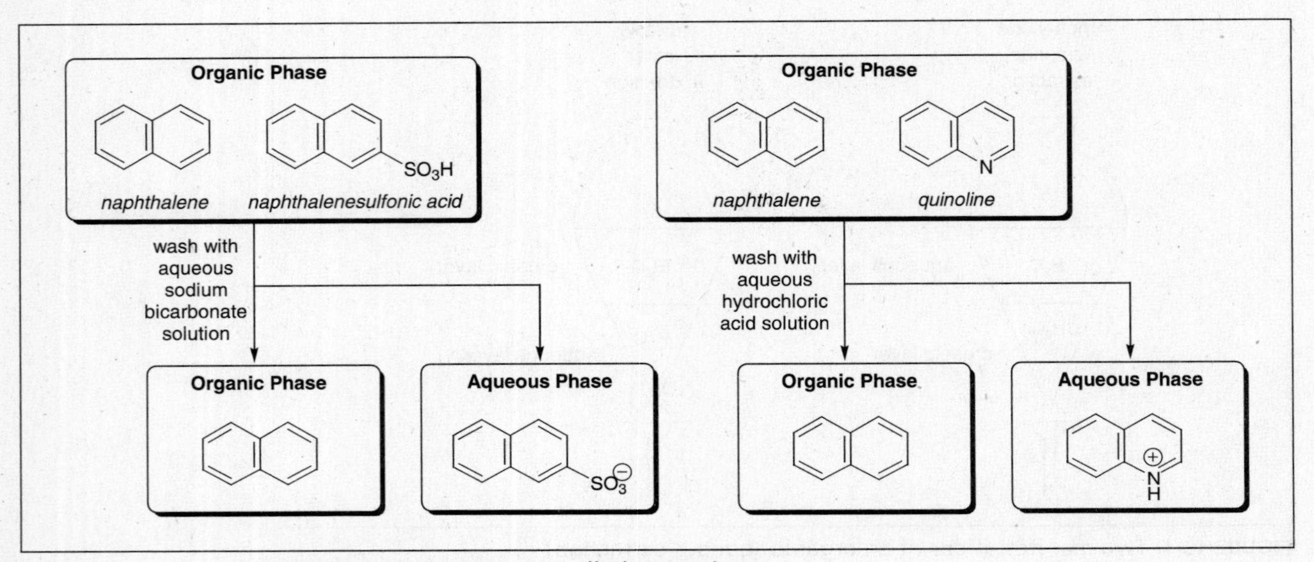

FIGURE 19-2: Two pH-controlled extractions.

Organic solvents are used for workup because they are easily removed by evaporation, leaving behind the organic compound of interest. A common problem encountered at this point is residual water from the aqueous washes. Ethyl acetate and diethyl ether both dissolve large quantities of water (3.3% and 1.2%, respectively). Therefore, it is advantageous to wash these organic layers with brine (saturated NaCl solution) at the end of the extraction sequence—the brine draws out the dissolved water through an osmotic-like effect. For methylene chloride and chloroform a brine wash is unnecessary, since the solubility of water in these solvents is quite low. Once freed from the bulk of residual water, the organic layer is dried over a desiccant, such as sodium sulfate, calcium chloride, or magnesium sulfate, and then decanted or filtered before evaporation.

CRAM SESSION

Extraction

Extraction is a method of purification based on differential solubility or **partitioning** of species between two immiscible solvents (one of which is usually water). A pair of such solvents is described as a **two-phase system:** water constitutes the **aqueous phase,** and the immiscible solvent is termed the **organic phase.** Generally speaking, charged species (and other extremely polar compounds) migrate to the aqueous phase, whereas neutral and moderately polar components remain in the organic phase.

It is common to employ extraction as a first step in working up an organic reaction, a process called **aqueous workup.** This step serves the dual purpose of quenching any remaining organic anions and cations created during the reaction and removing unwanted inorganic by-products (such as salts). The piece of glassware used for this process is known as a **separatory funnel.**

Organic compounds with ionizable groups (acids and bases) can be separated using **pH-controlled extraction** techniques, in which the aqueous layer is adjusted to a pH so that one species is neutral and the other is ionized, thus forcing the former into the organic phase and the latter into the aqueous phase.

CHAPTER 20

Chromatography

Read This Chapter to Learn About:

- Background
- Elution Trends
- Normal- and Reverse-Phase Systems

BACKGROUND

Chromatography represents the most versatile separation technique readily available to the organic chemist. Conceptually, the technique is very simple—there are only two components: a stationary phase (usually silica or cellulose) and a mobile phase (usually a solvent system). Any two compounds usually have different partitioning characteristics between the stationary and mobile phases. Because the mobile phase is moving (thus the name), then the more time a compound spends in that phase, the farther it travels.

Chromatographic techniques fall into one of two categories: analytical and preparative. Analytical techniques are used to follow the course of reactions and determine purity of products. These methods include gas chromatography (GC), high-performance liquid chromatography (HPLC), and thin-layer chromatography (TLC). Sample sizes for these procedures are usually quite small, from microgram to milligram quantities. In some cases, the chromatograph is coupled to another analytical instrument, such as a mass spectrometer or NMR spectrometer, so that the components that elute can be easily identified.

Preparative methods are used to purify and isolate compounds for characterization or further use. The most common techniques in this category are preparative HPLC, preparative TLC, and column chromatography. One popular modification of column chromatography is the flash column, which operates at medium pressures [about 10 pounds per square inch (psi)] and provides very rapid separation (ca. 5 min elution times). Column chromatography is suitable for sample sizes ranging from a few milligrams to several grams.

ELUTION TRENDS

For gas chromatography, the mobile phase is an inert carrier gas, such as helium or argon. Therefore, for the components to move, they must be volatile under the analytical conditions. Toward this end, GC columns are usually heated during analysis. The retention time (t_R) can be controlled by the oven temperature—the higher the temperature, the more quickly the sample elutes (Figure 20-1).

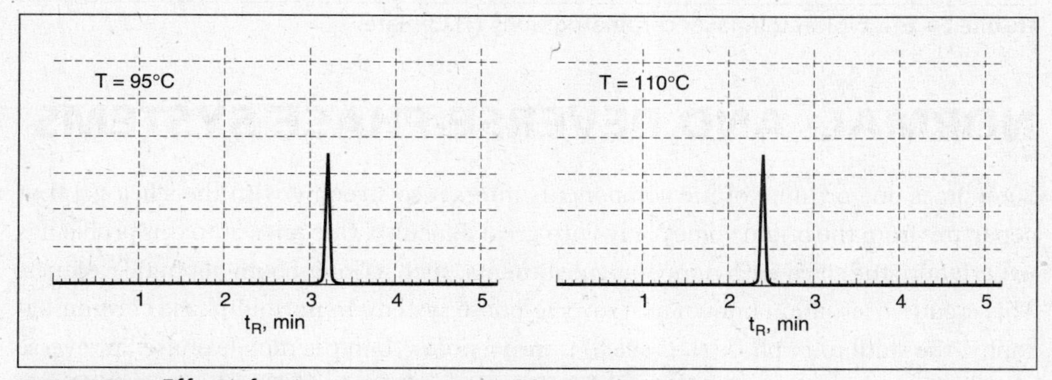

FIGURE 20-1: Effect of oven temperature on GC retention time.

The other methods fall under the category of LC (liquid chromatography), where the mobile phase is a solvent system, which can be used instead of temperature to leverage retention. Occasionally, this is a single solvent, but more often than not it is a binary mixture of solvents with different polarities. The advantage of the latter is that the bulk polarity can be modulated by varying the ratio of the two solvents.

For example, consider a typical TLC plate (Figure 20-2) developed in a 1:1 mixture of ethyl acetate and hexane, which exhibits two well-separated components. The spots can be characterized by their R_f value, which is defined as the distance traveled from the origin divided by the distance traveled by the mobile phase. Generally speaking, the slower moving component (R_f 0.29) is either larger, more polar, or both. If we wanted a larger R_f value, we could boost the solvent polarity by increasing the proportion of ethyl acetate in the mobile phase. Conversely, more hexane would result in lower-running spots.

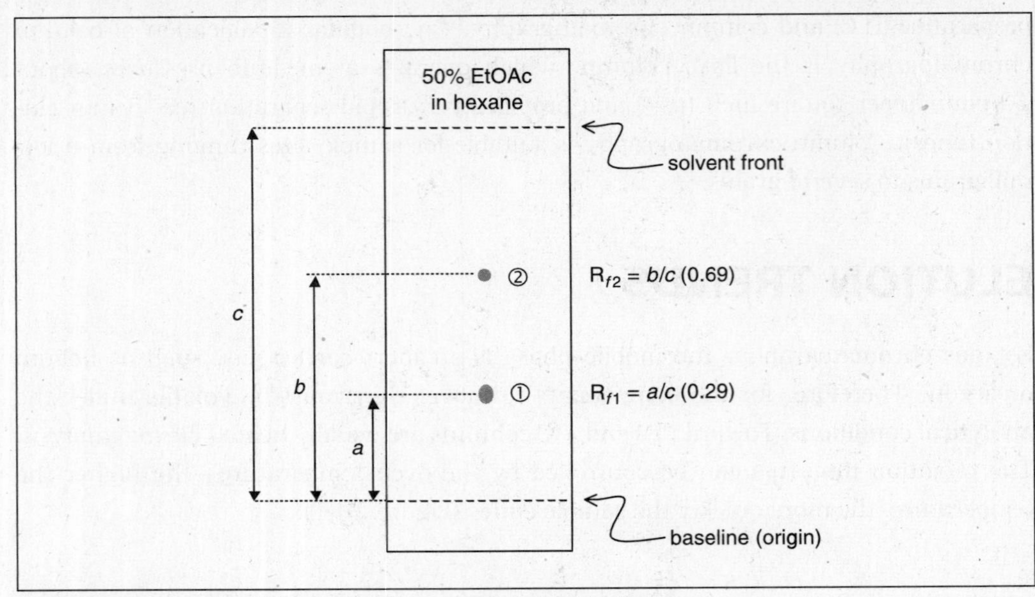

FIGURE 20-2: A typical thin-layer chromatography (TLC) plate.

NORMAL- AND REVERSE-PHASE SYSTEMS

Sometimes one or more of the components interact so strongly with the silica gel that departure from the origin comes only with great difficulty. One answer to this problem is to derivatize the silica with nonpolar substituents, such as long-chain aliphatic residues. This creates a situation known as a reverse-phase system. In normal-phase chromatography, the stationary phase (i.e., silica) is **more polar** than the mobile phase; in reverse phase, the stationary phase is **less polar** than the mobile phase. The derivatized silica is not only much less polar itself, but it also allows for the use of very polar solvents, such as water and methanol. The reversed phase results in some counterintuitive outcomes. For example, more polar components actually elute **faster,** and a more polar solvent system results in an **increased** retention time (i.e., lower R_f value). These parameters are summarized in Table 20-1.

TABLE 20-1 Comparison of Normal- and Reverse-Phase Chromatography

Parameter	Normal Phase	Reverse Phase
Typical stationary phase	Underivatized silica gel	C8-hydrocarbon derivatized silica
Representative mobile phase	Ethyl acetate/hexane mixture	Acetonitrile/water mixture
More polar components	Have lower R_f values	Have higher R_f values
Increasing solvent polarity	Increases R_f values	Decreases R_f values

Generally speaking, most analytical methods employ reverse phase systems, whereas the majority of preparative techniques are based on normal silica gel (an outcome largely driven by cost, as the derivatized silica is quite expensive). Among preparative techniques, flash columns are the most prevalent. Typical columns range from 0.5 to 5.0 centimeters (cm) in diameter, with a silica column height of about 6 inches. A sample is loaded to the top of the column and then rapidly eluted by forcing solvent through under medium pressure. A series of fractions are collected, the volume of which is determined by the size of the column—for a 2-centimeter column, one usually collects 10-milliliter (mL) fractions.

Once collected, the fractions are spotted on a long TLC plate and developed in the same solvent system used for eluting the flash column. Figure 20-3 shows an ideal scenario, in which three components have been successfully isolated in separate fractions. The next step is to combine like fractions into "cuts"—for example, the first cut might contain fractions 3–6; the second cut, fractions 8–13; and the third cut, fractions 15–17. The solvent is then removed by evaporation.

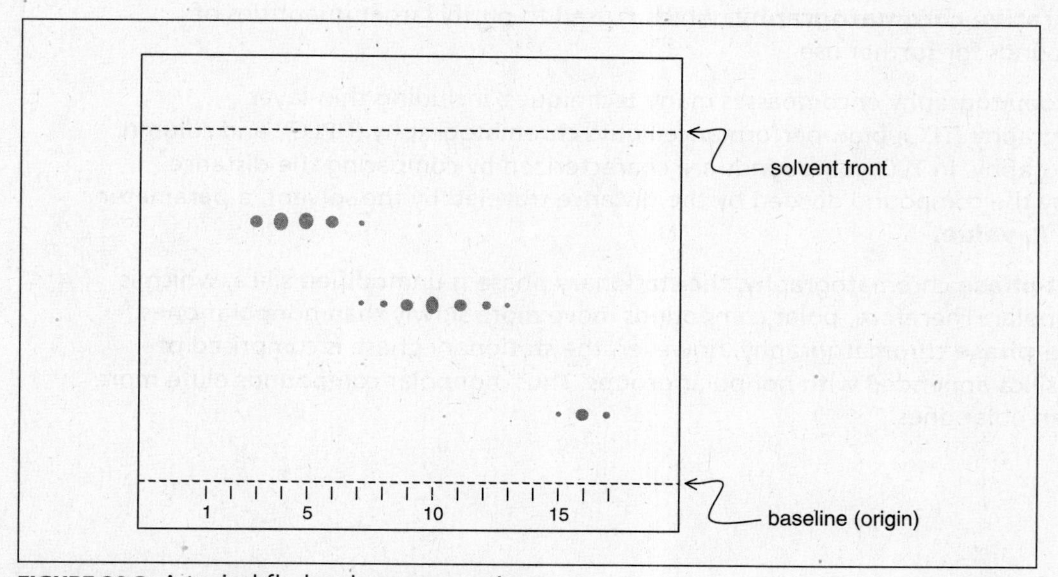

FIGURE 20-3: A typical flash column separation.

CRAM SESSION

Chromatography

Chromatography is a method of purification based on differential partitioning of compounds between a sorbent, or **stationary phase,** and a carrier, or **mobile phase.** The mobile phase can be an inert gas, which gives rise to **gas chromatography** (GC), or a solvent system, which is the basis of **liquid chromatography** (LC). Chromatographic techniques can be divided into two categories:

- **Analytical chromatography,** which separates small quantities of a mixture for the purpose of establishing the composition
- **Preparative chromatography,** which is used to purify larger quantities of compounds for further use

Liquid chromatography encompasses many techniques, including thin-layer chromatography (TLC), high-performance liquid chromatography (HPLC), and column chromatography. In TLC, compounds are characterized by comparing the distance traveled by the compound divided by the distance traveled by the solvent, a parameter called the **R_f value.**

In **normal-phase** chromatography, the stationary phase is unmodified silica, which is relatively polar. Therefore, polar compounds move more slowly than nonpolar ones. In **reverse-phase** chromatography, however, the stationary phase is comprised of modified silica appended with nonpolar groups. Thus, nonpolar compounds elute more slowly than polar ones.

CHAPTER 21

Distillation and Sublimation

Read This Chapter to Learn About:

- ➤ Simple and Fractional Distillation
- ➤ Distilling Compounds with Low Volatility
- ➤ Rotary Evaporation
- ➤ Sublimation

SIMPLE AND FRACTIONAL DISTILLATION

If chromatography is the most versatile separation method in the laboratory, it might be argued that distillation is the most common. This technique is used very frequently for purifying solvents and reagents before use. When volatile components are being removed from nonvolatile impurities, the method of **simple distillation** is used (Figure 21-1). In this familiar protocol, a liquid is heated to the boil, forcing vapor into a water-cooled condenser, where it is converted back to liquid and is conveyed by gravity to a receiving flask. When volumes are small [1–10 milliliters (mL)] a special apparatus known as a **short-path distillation head** is used. This piece of equipment is designed to minimize dead volume in the assembly and maximize recovery of the distillate. Otherwise, the principle is the same.

When separating two liquids with similar boiling points—or substances that tend to form azeotropes—**fractional distillation** is used. In this method, a connector with large surface area (such as a Vigreux column) is inserted between the still pot and the distillation head. The purpose of this intervening portion is to provide greater surface

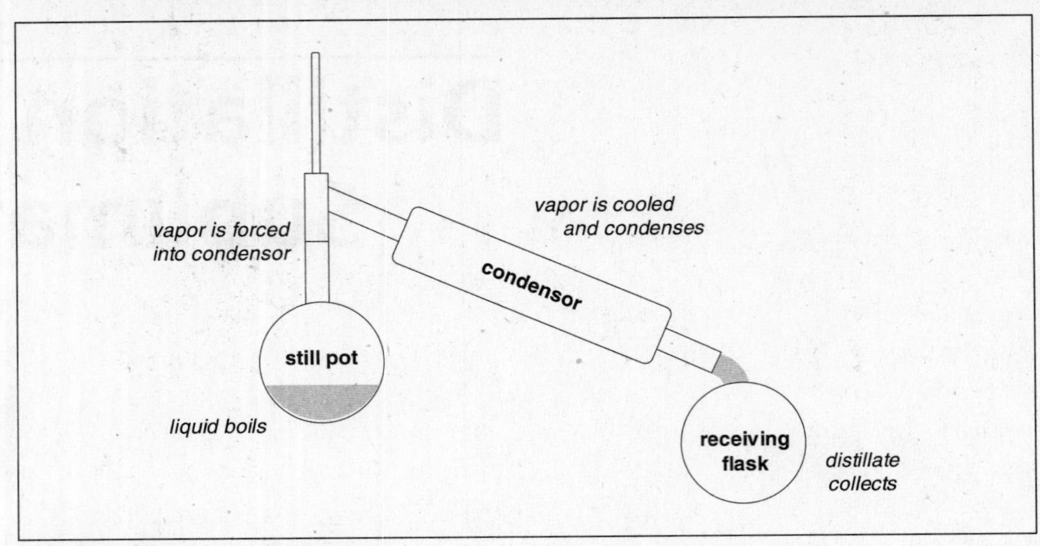

FIGURE 21-1: Simple distillation.

area upon which the vapor can condense and revolatilize, leading to greater efficiency of separation. In more sophisticated apparatus, the Vigreux column and condenser are separated by an automated valve that opens intermittently, thus precisely controlling the rate of distillation.

DISTILLING COMPOUNDS WITH LOW VOLATILITY

For compounds with limited volatility, the technique of **bulb-to-bulb distillation** (Figure 21-2) is sometimes successful. In this distillation, the liquid never truly boils—that is to say, the vapor pressure of the compound does not reach the local pressure of the environment. Instead, the sample is placed in a flask (or bulb) and subjected to high vacuum and heat, which sets up a vapor pressure adequate to equilibrate through a passage to another bulb, which is cooled with air, water, or dry ice. The temperature differential drives the vapor equilibrium into the cooler bulb, where the

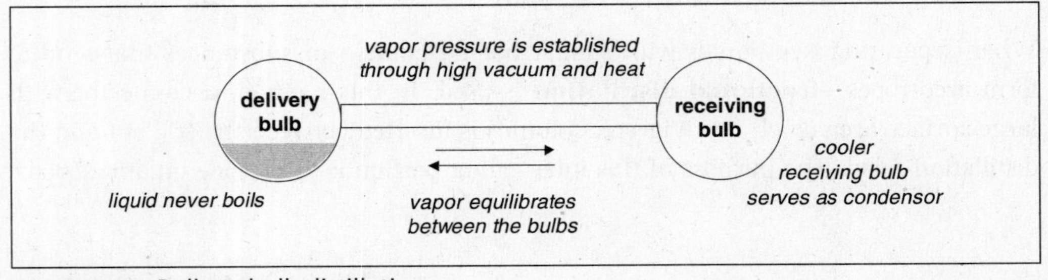

FIGURE 21-2: Bulb-to-bulb distillation.

compound condenses. This is the same principle behind the Kugelrohr (*German*, "bulb and tube") distillation.

The effect of temperature on the volatility of compounds is well known, but the impact of reduced pressure is much less appreciated. However, distillation at reduced pressure, or **vacuum distillation,** brings many advantages. For example, consider a liquid that has a boiling point of 180°C at atmospheric pressure (760 Torr). If we were to attempt that distillation, we should not be surprised to observe charring of the compound at such elevated temperature. Fortunately, reducing the pressure using a water aspirator (ca. 20 Torr total pressure) results in a significant lowering of the boiling point. Figure 21-3 shows this effect using a pressure–temperature nomograph. First, the normal boiling point at atmospheric pressure is located on the center scale; then, a straightedge is used to span that point and the distillation pressure on the rightmost scale; tracing the straightedge back to the leftmost scale provides an estimate of the boiling point at that pressure. Thus, a compound with a boiling point of 180°C at 760 Torr boils at about 80°C under a vacuum of 20 Torr. At 1 Torr (easily attainable with a vacuum pump), the boiling point is well below room temperature! For the purposes of purification, we

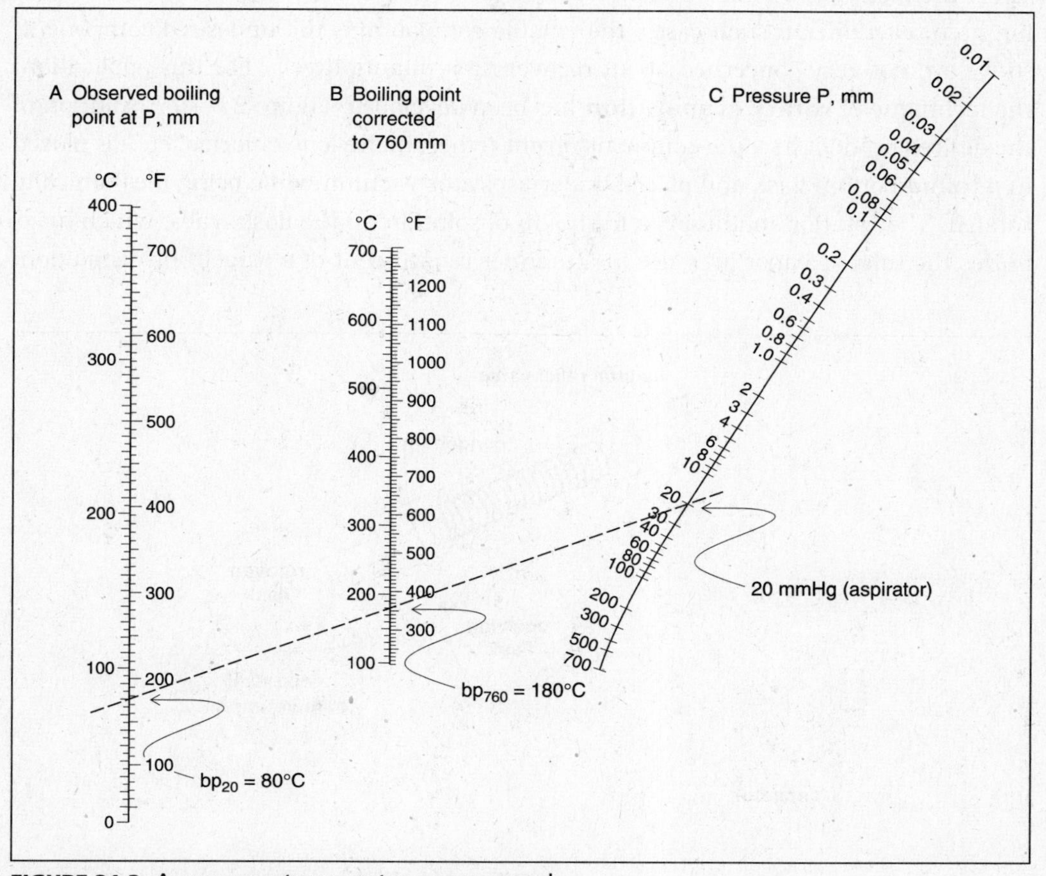

FIGURE 21-3: A pressure–temperature nomograph.

are better off with the aspirator, however, because at 1 Torr we could not recover the compound using a water-cooled (ca. 20°C) condenser.

Another way to distill sparingly volatile compounds is by **steam distillation.** In a true steam distillation, live steam is introduced into the still pot by means of a metal tube, or dipleg. The steam heats up the pot and carries any vaporized material from the headspace to the condenser, where it collects and drains into the receiving flask with the condensed water. In a common modification, water is simply added to the still pot and the mixture is heated in the same fashion as a simple distillation—the steam is thus produced *in situ*. The underlying principle of steam distillation is that an azeotrope forms with water, which has a lower boiling point than the pure compound itself. For example, naphthalene has a boiling point of 218°C, but in the presence of steam an azeotrope is formed, which contains 16% naphthalene and boils at 99°C. In addition, the continual physical displacement of the headspace by water vapor also aids in the collection of slightly volatile components.

ROTARY EVAPORATION

There are many occasions when all we need do is remove the solvent (e.g., after working up a reaction). In such cases, the volatile compound is the undesired component, so we are not very concerned about recovering it quantitatively. For this application, the technique of **rotary evaporation** has been developed (Figure 21-4). A solution of the desired product in some common solvent (ether, methylene chloride, etc.) is placed in a round-bottom flask and placed under aspirator vacuum while being mechanically rotated. The rotation maintains a fresh film of solution on the flask walls, which maximizes the solvent vapor pressure in a manner reminiscent of a Kugelrohr distillation.

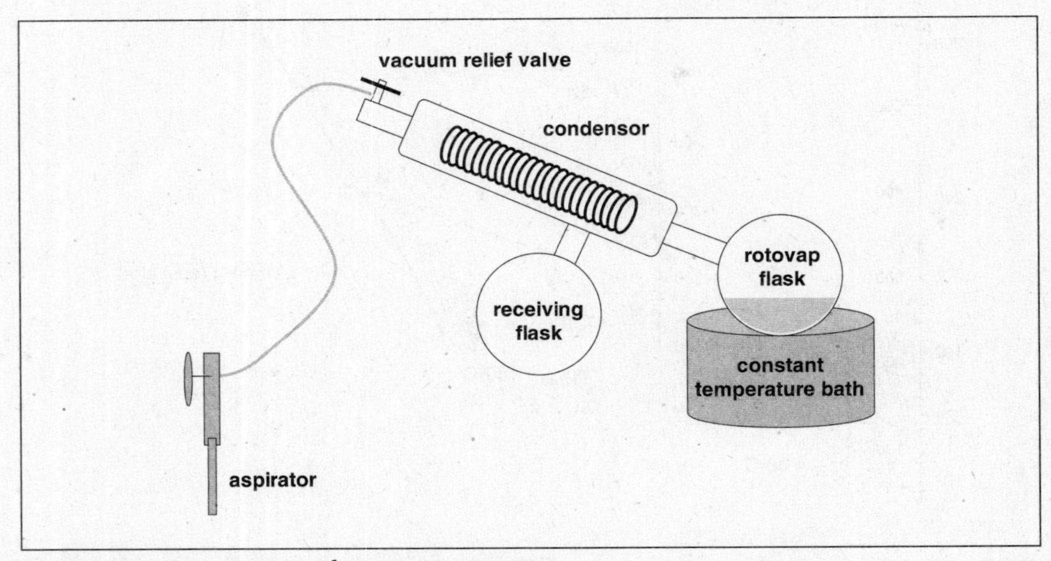

FIGURE 21-4: Components of a rotary evaporator.

The vapor rises to a water-cooled glass coil, where it condenses and collects into a receiving flask. The delivery flask is kept in an ambient temperature bath to counteract the evaporative cooling effect. In this way, 50 mL of methylene chloride can be removed in about 5 min. However, if the desired product is somewhat volatile (bp_{760} < 250°C), care must be taken to prevent inadvertent loss on the rotary evaporator.

SUBLIMATION

The method of **sublimation** is identical in principle to distillation, the only difference being that the distillate is a solid at the temperature of distillation. For example, potassium *tert*-butoxide (KO*t*-Bu) is a solid at room temperature (mp 257°C), but at 1 Torr it can be sublimed at 220°C. Because most sublimations are carried out under vacuum at high temperature, the necessary safety requirements limit the availability of large apparatus. Furthermore, because the cooling surface is quickly covered with solid—unlike a distillation, in which the liquid is removed from the condenser by gravity—the amount of material recoverable from a single sublimation tends to be small. Typical laboratory quantities range from the milligram to the 10-gram (g) scale.

CRAM SESSION

Distillation and Sublimation

Distillation and sublimation are methods of purification based on differential vapor pressures. There are several techniques used in common practice, including

- **Simple distillation,** for compounds with very different volatilities
- **Fractional distillation,** for compounds with similar volatility and azeotropes
- **Vacuum distillation,** for compounds with very high boiling points
- **Short-path distillation,** for small quantities of substances
- **Bulb-to-bulb distillation** (or Kugelrohr), for compounds with extremely low volatility
- **Steam distillation,** for high-boiling compounds that form an azeotrope with water
- **Rotary evaporation,** for the rapid removal of solvent from a nonvolatile compound
- **Sublimation,** for room-temperature solids with some volatility

CHAPTER 22

Recrystallization

Read This Chapter to Learn About:

- ➤ Background and Theory
- ➤ Methodology and Pitfalls
- ➤ Inadequate Separation

BACKGROUND AND THEORY

Chromatography is the most versatile method of purification, and distillation the most common, but recrystallization is the most elegant. Unfortunately, it is also the hardest to master and the most poorly comprehended by the beginning experimentalist. In essence, the principle is quite simple: Coax one compound out of solution while leaving any impurities or by-products dissolved, and then separate the precipitate from the solution. In practice, however, it can be the ultimate test of patience and endurance, especially if there is a poor understanding of the underlying concepts.

To explore this technique, let us say we have a 40-gram (g) sample containing a mixture of compounds (90% A, 10% B) with the solubility curves shown in Figure 22-1. If the sample were added to 100 milliliters (mL) of ethyl acetate (EtOAc) and heated to the boil (77°C), we should obtain a solution. Why? According to the solubility chart, 100 mL of ethyl acetate dissolves about 37 g of compound A at 77°C and 28 g of compound B. We have only 36 g and 4 g, respectively, so both compounds are dissolved. However, if we cool the solution in an ice bath (practical internal temperature = ca. 5°C), the solubility limits drop to 8 g of A and 6 g of B. If the solvent can accommodate only 8 g of A in solution, then 28 g of A precipitates out. However, 100 mL of ethyl acetate at 5°C can dissolve 6 g of B, and there is only 4 g present, so it remains in solution. If this heterogeneous mixture is filtered, then pure compound A can be recovered on a Büchner funnel, and a mixture of A and B passes through into the filtrate.

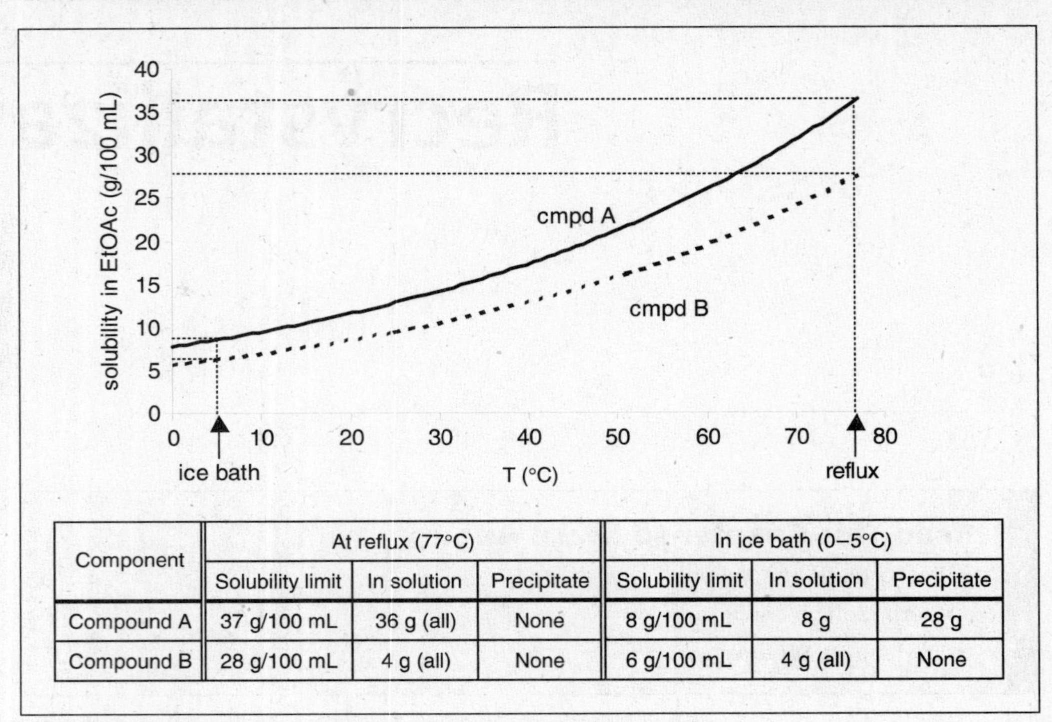

Component	At reflux (77°C)			In ice bath (0–5°C)		
	Solubility limit	In solution	Precipitate	Solubility limit	In solution	Precipitate
Compound A	37 g/100 mL	36 g (all)	None	8 g/100 mL	8 g	28 g
Compound B	28 g/100 mL	4 g (all)	None	6 g/100 mL	4 g (all)	None

FIGURE 22-1: Leveraging solubility curves in recrystallization.

With this example in mind, it is easy to see that recrystallization works best if the sample is already relatively pure. Because compound B is present in relatively small amounts, its relative concentration is significantly lower than that of compound A. In fact, paradoxically we are able to remove the **less soluble component** by this method. However, if we had been presented with a 1:1 mixture, the situation would not have been so straightforward.

Another circumstance that works well for recrystallization is when the undesired component is inherently more soluble—for example, if component A had a solubility limit of 37 g/100 mL and component B had solubility of 150 g/100 mL. It is rare to find such a pleasant constellation of factors. Furthermore, it is sometimes even necessary to work with a binary solvent system. If a single solvent cannot be found that dissolves the sample and allows crystallization at reasonable temperatures, then a mixture of miscible solvents (one more polar, one less polar) must be fine-tuned to the necessary solvating power. Diethyl ether and pentane constitute one such pair.

METHODOLOGY AND PITFALLS

Operationally, when using a binary system, the sample is first dissolved at the boil in the smallest possible quantity of the more polar solvent. Then, while reflux is maintained, the less polar solvent is slowly added until the solution just becomes cloudy (the so-called cloud point), which is an indication that the solubility limit has been reached. A few drops of the more polar solvent are added to remove the cloudiness, and the solution is allowed to stand undisturbed while cooling slowly.

Although the method appears simple, a seasoned practitioner is acquainted with the subtleties and challenges encountered in the laboratory. Following is a select list of common pitfalls in the execution of recrystallization:

- **Adding too much solvent.** In the example used here, if the 40-g sample had been tossed into 500 mL of solvent rather than 100 mL, the solvating capacity at 5°C would be 40 g for compound A (8 g/100 mL × 500 mL). Therefore, no product would precipitate. Although 500 mL sounds excessive, this is the same as dissolving 1 g in 12.5 mL.
- **Adding too little solvent.** If the solution is too concentrated, small amounts of the undesired component can precipitate out as well. Furthermore, if the filtrate/precipitate slurry is too thick, difficulties can be encountered at the filtration stage and filtrate can become trapped in the product.
- **Cooling too rapidly.** Ideally, the solution should constantly ride the edge of the solubility limit, whereby the desired product is slowly and continually precipitating out of solution. This gives rise to robust crystals that are easy to harvest. However, if the system is shocked (i.e., suddenly brought far below the limit of solubility), the insoluble component can present itself as an oil, a dreaded event known as "oiling out." This is undesirable, as it usually traps impurities and seldom provides satisfactory product.

INADEQUATE SEPARATION

It is important always to bear in mind that recrystallization is not a purification *per se*—the subsequent filtration is! No matter how much care is taken with the recrystallization itself, nothing can make up for a botched filtration step. Keep in mind that even the driest filtercakes are only about 85% solids content (i.e., 15% filtrate), and many are much less than that (some as low as 20% solids!). If we take the example used previously and collect the 28 g of compound A on a filter without washing, the residual filtrate would leave behind compound B at a level of almost 1%. Therefore, it is imperative to wash the filtercake with a small amount of fresh, cold solvent to displace any filtrate from the precipitate.

CRAM SESSION

Recrystallization

Recrystallization is a method of purification based on differential solubilities. The technique involves three discrete steps:

1. **Dissolving** a mixture into a minimal amount of solvent
2. Careful **crystallization** of the desired component by gradual cooling
3. Separation of precipitate from supernatant using **filtration**

It is important to bear in mind that the actual purification comes from the final step of separation. Recrystallization works best with samples that are already relatively pure. Common pitfalls in the technique are:

- Using too much solvent (too dilute to recrystallize)
- Using too little solvent (unwanted contaminants co-crystallize)
- Rapid cooling or other shocks (causes "oiling out")
- Inadequate separation of precipitate from supernatant

Conceptual Map of Section IV

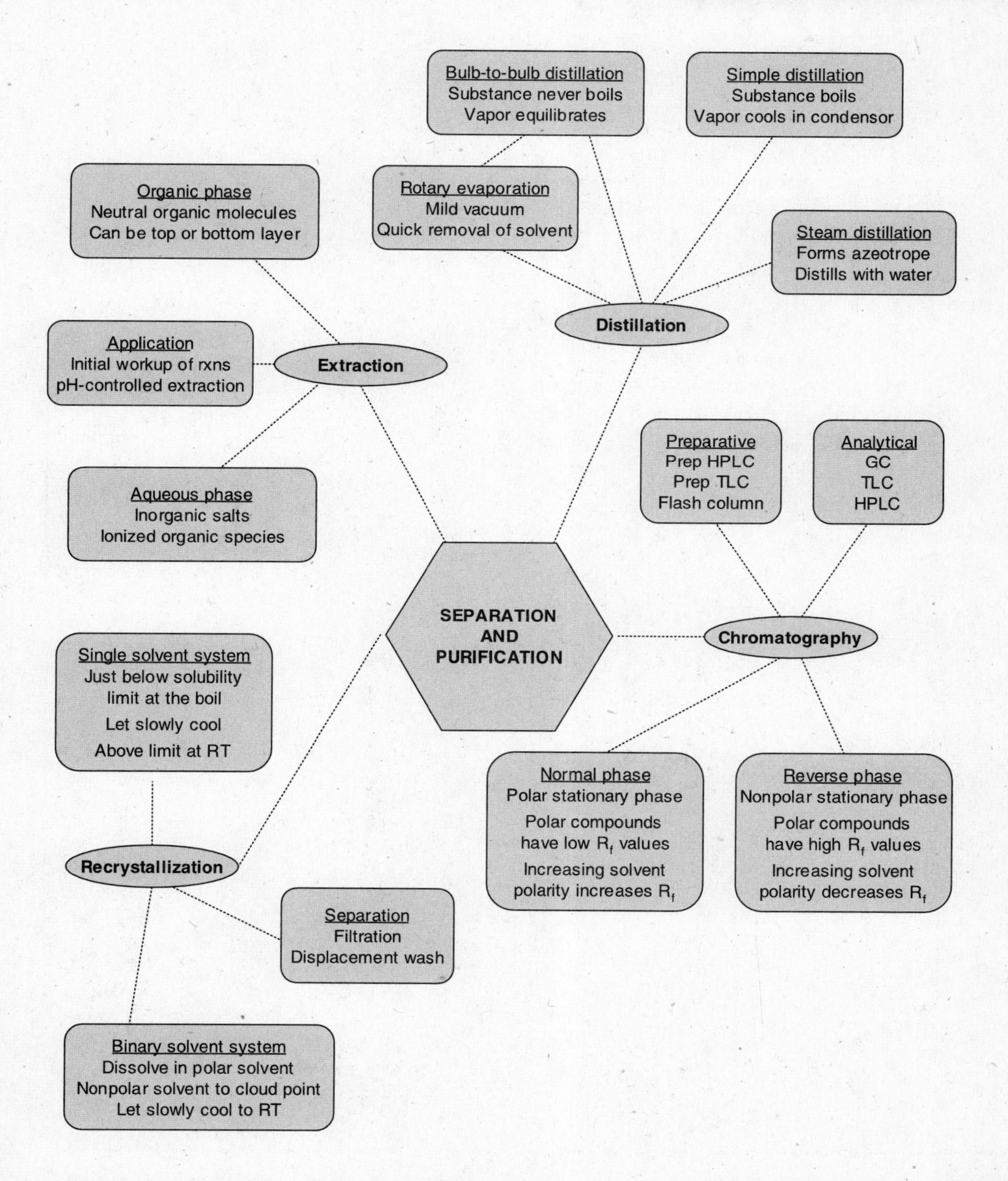

SECTION V

CHARACTERIZATION

Now that we know how to make complex organic compounds and obtain them in their pure form, we must be able to back up our synthetic claims—that is, prove that we have made what we say we have made. In an alternate scenario, our experiment may have resulted in an unexpected (and as yet unknown) compound that must be identified. This brings us into the realm of characterization. In the not too distant past (i.e., in the 1950s), proof of structure had to be conducted through the literal process of analysis—that is, the controlled deconstruction of the complex molecule into smaller parts that were well known. This was an arduous task that could easily span months for the more complex synthetic targets.

Luckily, the process of characterization is now a much more rapid proposition, thanks to the development of several powerful analytic protocols. We examine here five of the more frequently encountered techniques that can be divided into two broad categories. For example, the following table summarizes three nondestructive analytic methods based on absorption spectroscopy—that is, the interaction of light with matter. Please note that each technique brings to the table a different packet of information. Taken together, however, they provide an enormously useful ensemble of structural data.

Methods of Nondestructive Electromagnetic Spectroscopy

Method	Transition Observed	Absorption Range	Provides Information About
Ultraviolet–Visual Spectrum	Electronic	200–800 nm	Extended π systems
Infrared	Vibrational	600–4,000 cm^{-1}	Functional groups
Nuclear Magnetic Resonance	Nuclear Spin flip	25–1,000 MHz (instrument-dependent)	Skeletal framework and three-dimensional arrangement of atoms

There are also two destructive methods (see following table) that are still workhorses of structure determination—and although the sample is sacrificed for the analysis, the methods require only milligram quantities.

Two Important Destructive Analytical Techniques

Method	Principle	Provides Information About
Combustion analysis	Samples undergo perfect combustion in excess O_2; combustion products are quantified	Molecular composition
Mass spectrometry	Samples are ionized and accelerated through a magnetic field; fragments are separated based on mass/charge ratio	Molecular weight and structure

Another very powerful method that is becoming increasingly available to the organic chemist is x-ray crystallography, which uses a single crystal of an organic compound and bombards it with x-rays. The resulting diffraction pattern is analyzed to provide the three-dimensional arrangement of nuclei within the molecule. In fact, an x-ray crystal analysis is considered an ironclad proof of structure.

Although the theory of x-ray crystallography is beyond the scope of this work, the following chapters provide brief overviews of the remaining five techniques, which compose the canon of analytic methods most often used for structure proof of small organic molecules.

CHAPTER 23

Combustion Analysis

Read This Chapter to Learn About:

- Principle and Background
- Finding Empirical Formulas
- Degree of Unsaturation

PRINCIPLE AND BACKGROUND

One of the oldest analytic methodologies still in use is **combustion analysis** (also called elemental analysis and C, H, and N analysis). Simply put, a small amount of compound undergoes carefully controlled combustion in the presence of excess oxygen to provide the typical products of combustion—for compounds containing carbon, hydrogen, and nitrogen, these are CO_2, H_2O, and N_2. In classic methodologies, which measured only carbon and hydrogen content, water was sequestered by a desiccant module and carbon dioxide was trapped by a strongly basic module, and the differential weights of the sequestering modules were used for the analytic data. The current method involves passing the combustion gases through a chromatography column, which allows for the detection of CO_2, H_2O, and N_2 by thermal conductivity (much like a conventional gas chromatograph). Because the analysis is carried out in the presence of excess oxygen, determining the oxygen content of the compound is impossible by this method, although there are specialized (and expensive) techniques available for oxygen and many other elements.

FINDING EMPIRICAL FORMULAS

Combustion analysis data is presented as weight percent of each given element, as shown in the boxed area of Figure 23-1. To be useful, however, this data must be

workup of analytical data

provided by analysis

Element	wt%	at wt	χ	mol%	mol ratio
C	**66.10**	12.011	5.503	31.80	7.00
H	**10.31**	1.0079	10.229	59.11	13.02
N	**11.01**	14.007	0.786	4.54	1.00
O	12.58	15.999	0.786	4.54	1.00
Total	100.00		17.305	100.00	

FIGURE 23-1: Treatment of combustion analysis data.

converted to a mole ratio of elements. The first thing to recognize here is that the weight percent values for carbon, hydrogen, and nitrogen do not add to 100%. Given the practical considerations mentioned earlier, we assume the balance is made up by oxygen (in this case, 12.58%), unless we have evidence from alternative analyses that other elements are present. Next, because we must leverage the mass quantities into molar ratios, a third column is introduced containing the atomic weights of all elements involved. Dividing the weight percent by the atomic weight gives a modified value designated χ, or molar contribution (fourth column), which has no particular meaning on its own; however, if we divide each χ value by the sum of all the χ values, we can express the elemental abundance as a mole percent value (fifth column). The final step is to find the molar ratio—the best way to approach this is to divide all mole percent values by the smallest value. For example, carbon is present in 31.80 mol% and nitrogen is present in 4.54 mol%; dividing 31.80 by 4.54, we can say that the molar ratio of carbon to nitrogen is 7.00:1.00 (sixth column).

It is not uncommon for the ratios to be nonperfect (e.g., molar ratio of 13.02 for hydrogen)—after all, they are experimentally derived values. However, occasionally we encounter values far from round integers, and this can be indicative of a fractional empirical formula. For example, if our calculations yield an empirical formula of $C_{3.52}H_{5.99}O_{1.00}$, this probably means that a realistic empirical formula is twice those values, or $C_7H_{12}O_2$.

The empirical formula already can tell us much about the structure of the molecule, aside from the obvious benefit of knowing the identity and ratio of elements on board. One useful feature we can extract from the elemental data is the degree of unsaturation (DoU), also known as the index of hydrogen deficiency (IHD). The idea of unsaturation originally sprang from alkene chemistry—olefinic hydrocarbons consume hydrogen (i.e., undergo hydrogenation) until all the double bonds are reduced, at which point the hydrocarbon is said to be "saturated" with hydrogen. However, carbonyl groups also consume hydrogen to form the corresponding saturated molecule (Figure 23-2), as do cyclic structures (at least formally). Therefore, each double bond or ring in a

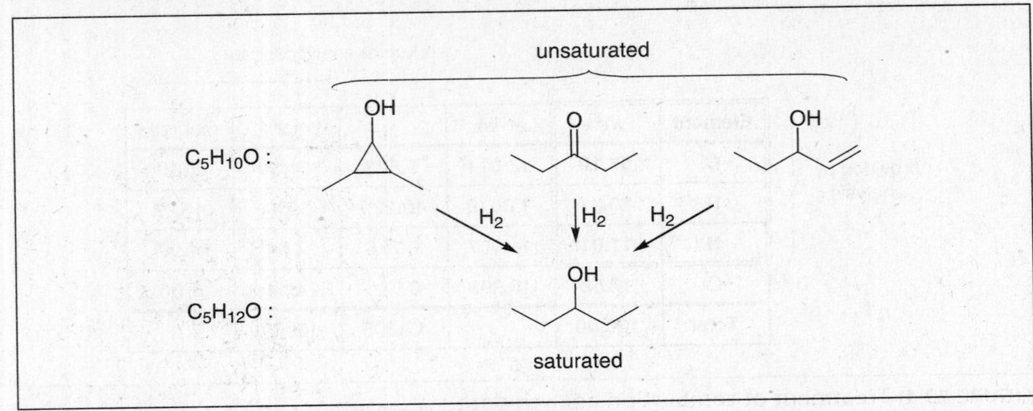

FIGURE 23-2: Saturated and unsaturated compounds.

compound confers one degree of unsaturation. For example, benzene has four degrees of unsaturation—one for the ring, and one for each double bond in the ring.

DEGREE OF UNSATURATION

It turns out that there is a fundamental pattern we can use to our advantage. For all hydrocarbons, the relationship of carbons to hydrogens in saturated molecules is given by the expression

$$\text{Saturated molecule} = C_nH_{(2n+2)} \tag{39}$$

As each degree of unsaturation is introduced, the number of hydrogens drops by two. Thus, ethane (saturated) has a formula of C_2H_6, but ethene (1 DoU) is C_2H_4, and ethyne (2 DoU) is C_2H_2. Thus, given a molecular formula, all we have to do is compare the actual number of hydrogens to the number required for saturation, remembering the factor of 2 hydrogens for every DoU.

The only subtlety here is what happens when molecules have elements besides carbon on board. Rather than introduce a set of meaningless equations, let us simply consider a collection of saturated molecules with various common elements (Figure 23-3). Propane follows the rule of $C_nH_{(2n+2)}$, and introducing an oxygen into the mix (i.e., propanol) appears to have no effect on the C/H ratio. This makes sense if we think of propane as a Tinkertoy or a molecular model—if we pull off a hydrogen atom and plug in an oxygen atom, the oxygen atom still has a raw edge that can be capped with the hydrogen we

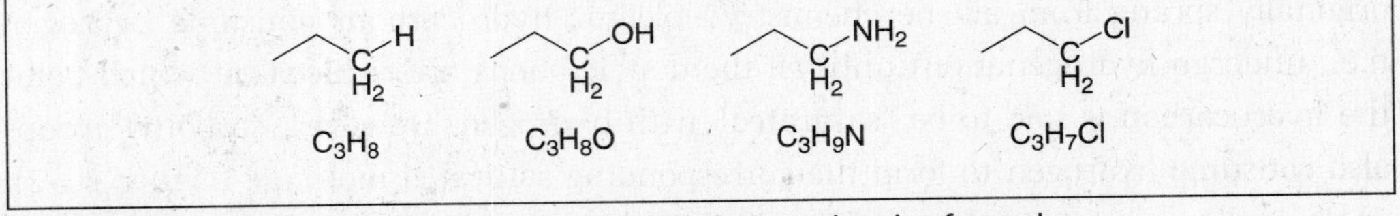

C_3H_8 C_3H_8O C_3H_9N C_3H_7Cl

FIGURE 23-3: Some saturated compounds and their molecular formulas.

just pulled off. In other words, because oxygen is divalent, it can be "inserted" in a C—H bond without changing the overall hydrogen count.

However, propylamine has one extra hydrogen atom than propane. Again, this is entirely reasonable if we consider that inserting a nitrogen into a C—H bond leaves one valency spot on the nitrogen unfilled—that is, we have to reach into our model kit for another hydrogen atom to plug into the third valency of nitrogen. Conversely, pulling off a hydrogen and replacing it with any monovalent species (such as chlorine) leaves us with a hydrogen atom in our hand. Consequently, the number of hydrogens in the molecular formula drops by one.

In summary, the following method is useful in determining the degree of unsaturation for a given molecular formula:

1. Calculate the hydrogen count for the corresponding saturated molecule
   a. Number of hydrogens = 2 (number of carbons) + 2
   b. Oxygen (or any divalent species) has no effect on this count
   c. For every nitrogen (or any trivalent species), add 1
   d. For every halogen (or any monovalent species), subtract 1
2. Compare the actual hydrogen count to the saturated hydrogen count
   a. Divide the difference by 2 to get the DoU
   b. If the difference is an odd number, recheck the math in step 1

Applying these guidelines to an example, let us consider the empirical formula $C_5H_7N_2OCl$ (Figure 23-4). Because the formula has 5 carbons, we would predict 12 ($2n + 2$) hydrogen atoms for a simple hydrocarbon. The presence of oxygen has no effect on this value, but we adjust the hydrogen count up by 2 because of the 2 nitrogen atoms, and then down by 1 because of the monovalent chlorine. The adjusted saturation count for hydrogen is thus 13. The empirical formula has only 7, for a difference of 6—dividing this value by 2 gives 3 degrees of unsaturation. With this in hand, we can limit our consideration to those molecules that exhibit the proper degree of unsaturation—one such example is shown in the inset.

DoU Calculation for $C_5H_7N_2OCl$

"normal" saturation count	12
modification for oxygen	0
modification for nitrogen	+ 2
modification for chlorine	− 1
adjusted saturation count	13
actual hydrogen count	7
Δ between actual and saturated	6
degrees of unsaturation (Δ ÷ 2)	3

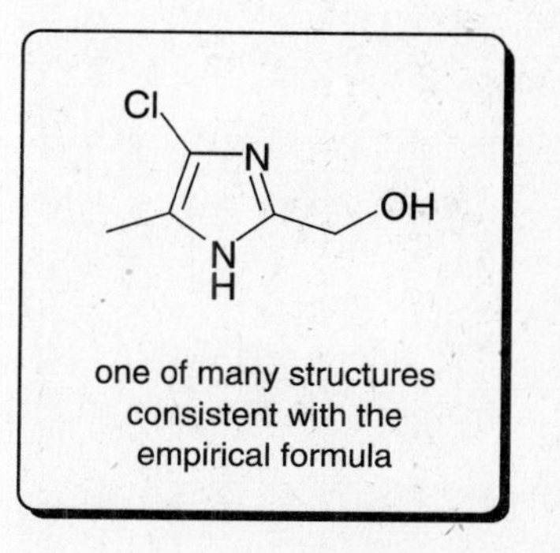

FIGURE 23-4: Application of the DoU calculation.

CRAM SESSION

Combustion Analysis

Combustion analysis is used to determine the exact atomic composition of a compound. This information can be used to establish an empirical formula, as follows:

1. Sum all weight percent values given; if less than 100%, assume the balance is oxygen.
2. Convert weight percent to mole percent using atomic weights.
3. Divide all mole percent values by the smallest mole percent to give a molar ratio.
4. If the molar ratio is not integral, explore multiples (e.g., $C_{2.0}H_{4.0}O_{0.5}$ becomes C_4H_8O).

It is important to remember that this is an empirical formula only. The molecular formula can only be determined if the molecular weight is also known (for example, using mass spectrometry).

The empirical formula provides useful information about a molecule by revealing degrees of unsaturation (DoUs), whereby a degree of unsaturation is a double bond or a ring system. The method of calculation is as follows:

1. Calculate the hydrogen count for a saturated molecule.
   a. Number of hydrogens = 2(number of carbons) + 2.
   b. Oxygen (or any divalent species) has no effect on this count.
   c. For every nitrogen (or any trivalent species), add 1.
   d. For every halogen (or any monovalent species), subtract 1.
2. Compare the actual hydrogen count to the saturated hydrogen count.
   a. Divide the difference by 2 to get the DoU.
   b. If the difference is an odd number, then recheck the math in Step 1.

CHAPTER 24

UV-Vis Spectroscopy

Read This Chapter to Learn About:

- ➤ Background and Theory
- ➤ Information from Wavelength
- ➤ Information from Intensity

BACKGROUND AND THEORY

If a molecule exhibits multiple degrees of unsaturation, the question naturally becomes whether they manifest themselves as rings or double bonds—and if the latter, whether the π bonds are isolated or in conjugation. To address this dual question, we turn to the first of three spectroscopic methods. Recall that π systems arise from the combination of adjacent *p* orbitals, and that the number of molecular orbitals is determined by extent of the *p* orbital array. Furthermore, as the π system encompasses more orbitals, the energy spacing becomes more compact and the distance between the HOMO and LUMO diminishes (Figure 24-1). With simple dienes, this distance is already close enough that absorption of light in the near-UV results in electronic excitation. As conjugation is extended, this absorption moves to longer wavelengths (lower energy) and eventually creeps into the visible region. Therefore, these extended π systems are often called **chromophores,** as they are responsible for the color in many organic compounds. In any event, the wavelength of light absorbed is dependent on the degree of conjugation in the chromophore, and so this absorption can tell us much about the nature of the π system in an unsaturated molecule.

INFORMATION FROM WAVELENGTH

It turns out that the π system is also remarkably sensitive to substituent patterns, a phenomenon that can be leveraged for extracting structural information. Through the

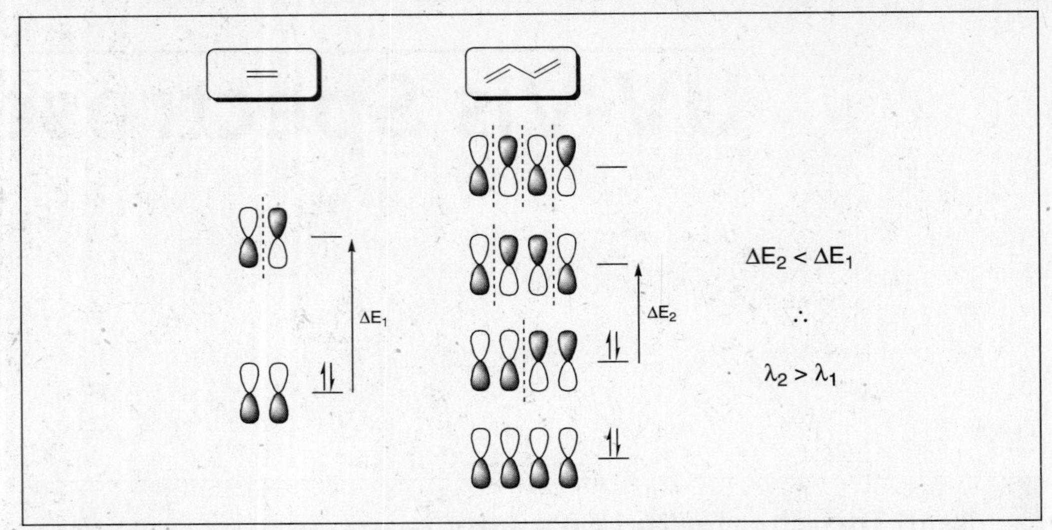

FIGURE 24-1: The impact of extended π systems on absorbance.

examination of many experimentally derived values, sets of rules have been developed for predicting the wavelength of maximum absorbance (λ_{max}) for various substrates as a function of structure, work carried out primarily by Woodward, Fieser, and Nielsen. A brief summary of these rules is presented in Figure 24-2. Although not immediately

	dienes	enones and enals			benzophenones, benzaldehydes, and benzoic acids	
base chromophore						
base absorbance λ_{max} (nm)	214	202	215	245 208 (R=H)	246 250 (R=H)	230

modifications to the base absorbance

		all	α	β	γ	δ	o	m	p
substituents	position	all	α	β	γ	δ	o	m	p
	-R	5	10	12	18	18	3	3	10
	-OR	6	35	30	17	31	7	7	25
	-Br	5	25	30	25	25	2	2	15
	-Cl	5	15	12	12	12	0	0	10
	$-NR_2$	60		95			20	20	85
π system architecture	double bond extending conjugation	+ 30 nm					not typically relevant		
	exocyclic character of a double bond	+ 5 nm							
	homoannular diene component	+ 39 nm							

FIGURE 24-2: Abbreviated rules for predicting λ_{max} for UV-Vis absorptions.

intuitive, they are straightforward to apply once the framework is understood, and they are very powerful predictive tools.

There are a couple of subtleties that tend to be pitfalls for the beginning spectroscopist. First, we must approach the calculation as an accountant, recognizing that we must be very mindful of what is already in the ledger, as well as those items that do not have an impact on the bottom line. Thus, when a double bond is added to the system to extend conjugation, the base value is increased by 30 nanometers (nm), and we have essentially a new chromophore. Also, for carbonyl-containing chromophores, the —R or —OR groups on the other side of the carbonyl group are already accounted for in the base absorbance value, so it is important not to double-count these substituents later on.

Two specialized terminologies also deserve mention. First, a **homoannular diene** refers to any two double bonds that are incorporated into the same ring. Therefore, 1,3-cyclohexadiene would be considered a homoannular diene, but cyclohex-2-enone would not, because only one double bond is incorporated into the ring itself. For every instance of a homoannular component, we add 39 nm to the base absorbance. Second, even though the etymology of an exocyclic double bond means "outside the ring," it is best thought of as a double bond that terminates in a ring. Each time this occurs, we add 5 nm to the base absorbance.

To get some practice in applying these rules, consider the tricyclic enone shown in Figure 24-3. We choose the base absorption of 215 (six-membered-ring enone) and tack on three double bonds to extend the conjugation (30 nm each for 90 nm total). We have thus established the domain of the chromophore (shown in bold). It is often helpful to highlight the chromophore for accounting purposes, even by darkening in with a pencil. Examination of the chromophore thus reveals that ring C houses one homoannular diene component, so we add 39 nm to the ledger. Careful scrutiny also reveals two occurrences of exocyclic double bonds: both the α,β and the ε,ζ olefins terminate in ring B. We add 5 nm for each occurrence.

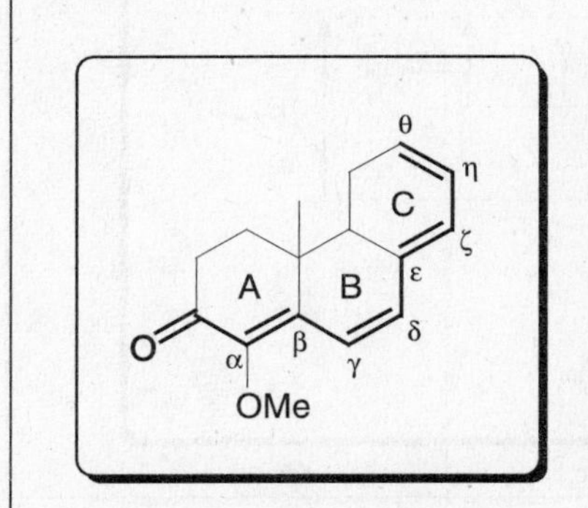

Base chromophore (6-ring enone)	215
3× double bond extensions	+ 90
Homoannular diene (in ring C)	+ 39
2× exocyclic double bonds	+ 10
α-alkoxy substituent	+ 35
β-alkyl substituent	+ 12
ε-alkyl substituent (same as γ)	+ 18
θ-alkyl substituent (same as γ)	+ 18
Predicted λ_{max}	437

FIGURE 24-3: Example in the application of predictive UV-Vis rules.

The accounting of substituents is sometimes tricky, particularly for cyclic molecules. First, we must recognize that the substituent connected to the carbonyl group is already accounted for, so we do not double-count it. We start instead with the methoxy substituent in the α position, which adds 35 nm to the chromophore. The rest of the substituents are dealt with in the same manner. To help sort out what are bona fide substituents, imagine the chromophore (bold portion) is a hallway we could walk through—how many doorways would we see, and in what positions? Thus, starting from the carbonyl, we would see a door to our right at the α position, one on our left at the β position, another left door at ε, and finally a left door at θ. For accounting purposes, it matters not that there are substituents attached to those substituents (it also is inconsequential whether the "doors" are on the right or left—it simply helps us visualize the virtual corridor).

INFORMATION FROM INTENSITY

The intensity of the absorbance can also be diagnostic. For example, in the tricyclic enone shown previously we would expect to see at least two bands in the UV-Vis spectrum. The absorbance calculated in Figure 24-4 corresponds to the $\pi \rightarrow \pi^*$ transition from the HOMO to the LUMO. However, the lone-pair electrons on oxygen can also undergo excitation. Because they are technically not connected to the extended π system, they are neither bonding nor antibonding; therefore, they are considered nonbonding (n) electrons. The absorption is thus designated an $n \rightarrow \pi^*$ transition (Figure 18-7). Because the energy gap is smaller, this absorption occurs at higher wavelengths than the $\pi \rightarrow \pi^*$ transitions.

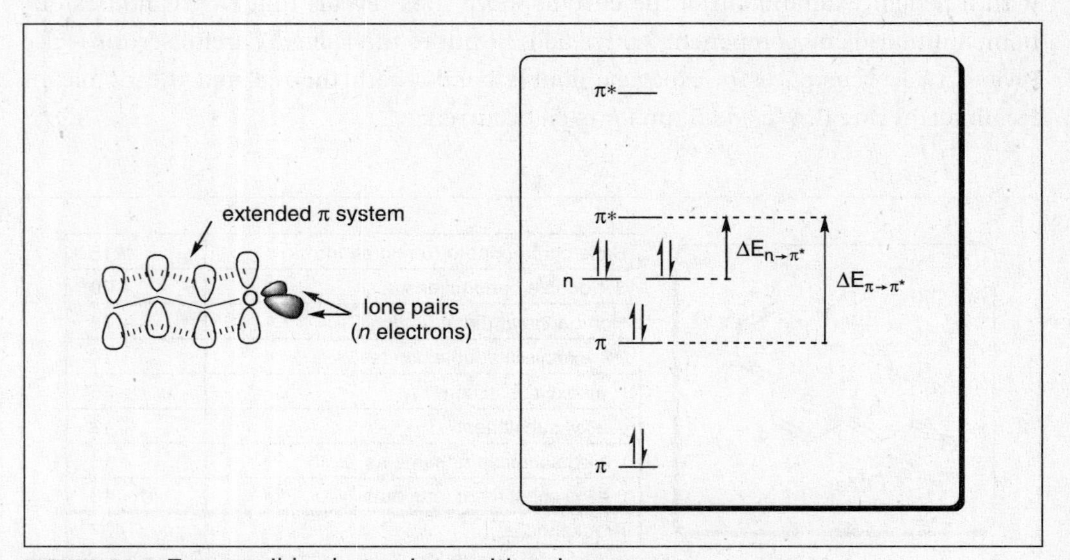

FIGURE 24-4: Two possible electronic transitions in enones.

The two types of excitation events differ not only in the maximum wavelength, but also in intensity. Because the lone pairs are essentially orthogonal to the extended π system, it would seem difficult to imagine how electrons could be promoted from one to the other. Indeed, these $n \rightarrow \pi^*$ events are known as symmetry forbidden electronic transitions. Like so many other forbidden things, they still happen, but the quantum efficiency is much lower; therefore, the intensity of the absorption is considerably weaker than the $\pi \rightarrow \pi^*$ absorption.

The intensity of a specific absorption is usually reported in terms of the molar extinction coefficient (ε), which is given by the relationship

$$\varepsilon = \frac{A}{cl} \tag{40}$$

where A is the absorbance at the λ_{max}, c is the molar sample concentration, and l is the path length in centimeters (in most instruments, this is 1 cm). For allowed transitions, the molar extinction coefficients are on the order of 10^3 to 10^6. Forbidden transitions are typically in the hundreds.

CRAM SESSION

UV-Vis Spectroscopy

The UV-Vis spectrum arises from electronic transitions, most notably from the HOMO to the LUMO. Thus, the position and intensity of absorption bands can reveal much about extended π systems in a molecule. The absence of significant UV-Vis absorptions above 240 nm is usually interpreted as a lack of conjugated π bonds within a molecule. When a conjugated system (such as a diene) is indeed present, it is called a **chromophore,** as the absorption often gives rise to color.

Although detailed empirical rules can be used to predict the frequency of absorption, it is not usually expected that these rules have been memorized. However, an understanding of the general trends is extremely useful. For example, the following structural modifications result in a shift to higher wavelength absorption:

- Extension of the π system by another double bond in conjugation
- A **homoannular diene** component (both double bonds are within the same ring)
- An **exocyclic double bond** in the chromophore

Furthermore, the intensity of the absorbance can give clues about its origin. Those with very large **molar extinction coefficients** ($>10^3$) are usually the result of allowed **π to π* transitions,** whereas those with low extinction coefficients ($<10^3$) arise from forbidden ***n* to π* transitions.**

CHAPTER 25

Mass Spectrometry

Read This Chapter to Learn About:

- ➤ Background and General Techniques
- ➤ Structural Information from Mass Spectrometry

BACKGROUND AND GENERAL TECHNIQUES

Mass spectrometry is based on the principle of differentiating molecules by accelerating charged species through a strong magnetic field or across a voltage potential, in which behavior is dictated by the charge-to-mass ratio of the ions. From a technical standpoint, two general areas have been the focus of dramatic innovation: (1) the method of ion generation and (2) the technique used for separation of ions.

The classic method for ion generation is known as EI, or electron impact. Here a sample is bombarded with very high-energy electrons, which transfer energy into the molecules, much like photons of visible light induce the formation of an excited state in UV-Vis spectroscopy. However, these excited species are so energetic that the only way to relax is by releasing an electron, thereby forming a radical cation, as shown in Figure 25-1. This species is known as the molecular ion.

Once formed, these ions are accelerated through some differentiating field. The classic approach for differentiation is to pass the beam of charged particles through a magnetic field, which refracts the ions based on their charge-to-mass ratios and velocities. In a very broad sense, this is analogous to the refraction of white light into a spectrum of colors based on the differential interaction of variously energetic photons with the medium of the prism. Meanwhile, this technique has been supplanted largely by the time-of-flight (TOF) approach. In this method, ions are imbued with constant kinetic energy and then directed across a fixed distance under a voltage potential. The time of

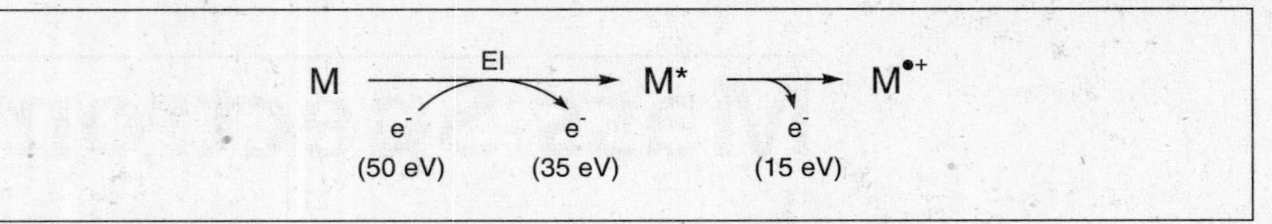

FIGURE 25-1: Formation of a molecular ion by electron impact (EI).

flight (t) is given by the equation

$$t = \sqrt{\left(\frac{m}{z}\right)\left(\frac{d^2}{2Ve}\right)} \qquad (41)$$

where m/z is the mass/charge ratio, d is the distance traveled, V is the voltage, and e is the electronic charge. Thus, the larger the mass/charge ratio, the longer the time of flight at constant kinetic energy.

If we assume that only one electron is lost (that is, the charge of all species is $+1$), then the time of flight can be directly correlated to mass—thus, the name "mass spectrometry." Indeed, some of the most useful information from this technique derives from the mass of the molecular ion. Recall from Chapter 18 that combustion analysis gives information about the atomic ratios within a molecule, but it does not provide a definitive value for the molecular weight. In other words, a compound with molecular formula $C_6H_{12}O$ would give the same elemental analysis results as a $C_{12}H_{24}O_2$ compound. However, mass spectrometry can provide this missing information: a molecular ion peak at 200 amu is very strong evidence for the latter molecular formula. Thus the combination of elemental analysis and low-resolution mass spectrometry can be used to establish the molecular formula of an unknown compound or help prove the identity of a synthesized molecule—elemental analysis provides the ratio of elements and mass spectrometry fixes the total weight.

However, the more sensitive protocol of high-resolution mass spectrometry (HRMS) can do both. In this technique, molecular ion masses can be obtained at four-decimal resolution. This allows us to take advantage of the fact that every element has a unique noninteger mass, so that any number of combinations should also add up to a unique noninteger quantity. Thus, the nominal MS molecular ion peak of 127 corresponds to several molecular formulas, including the two structures shown in Figure 25-2. Whereas nominal MS cannot distinguish between the two entities, HRMS data can specify a particular molecular formula with a high degree of accuracy. Thus, a molecular ion peak of 127.0995 would be much more consistent with the structure on the left.

Electron impact (EI) is a fairly harsh ionization method, and many times it is difficult to observe the molecular ion peak in a MS spectrum generated this way. Luckily, there are now milder methods to ionize compounds, including electrospray ionization (ESI) and chemical ionization (CI), which provide ions by protonation (or in some cases

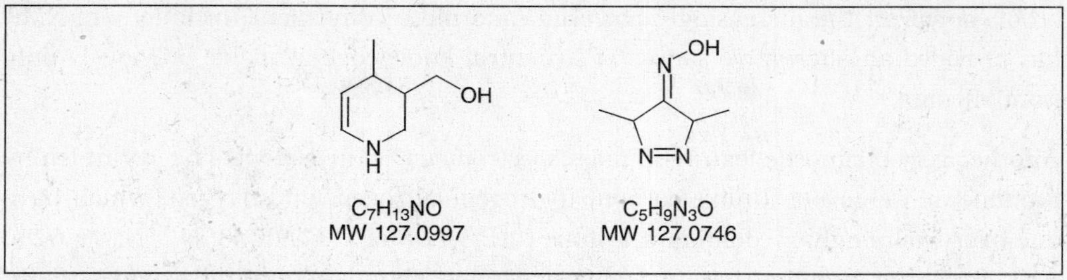

$C_7H_{13}NO$
MW 127.0997

$C_5H_9N_3O$
MW 127.0746

FIGURE 25-2: Two molecules with nominal molecular weight 127.

deprotonation), thereby avoiding the very unstable radical cation species produced by EI—not to mention blasting the substrate with 50+ electron volts (eV) of energy.

STRUCTURAL INFORMATION FROM MASS SPECTROMETRY

On the other side of the coin, the very decomposition processes that diminish the molecular ion peak can themselves provide important structural information about the substrate. As one example, radical cations derived from ketones undergo fragmentation in the vicinity of the carbonyl. Figure 25-3 shows two such processes for 2-methylhexan-3-one. The first involves σ bond cleavage, in which the isopropyl radical and an acyl cation are formed from the initial radical cation. The molecular ion can also undergo the McLafferty rearrangement, a six-electron process that liberates ethene and a lower molecular weight radical cation. Keep in mind that only charged species are detected, although the lost pieces can be inferred from the difference between the molecular ion peaks and the fragment peaks. Without a rigorous analysis, it is at least intuitively straightforward that knowledge about these fragmentation processes can be useful in piecing together the structure of the original compound. Indeed, the painstaking interpretation of fragmentation patterns was once *de rigeur* for structure

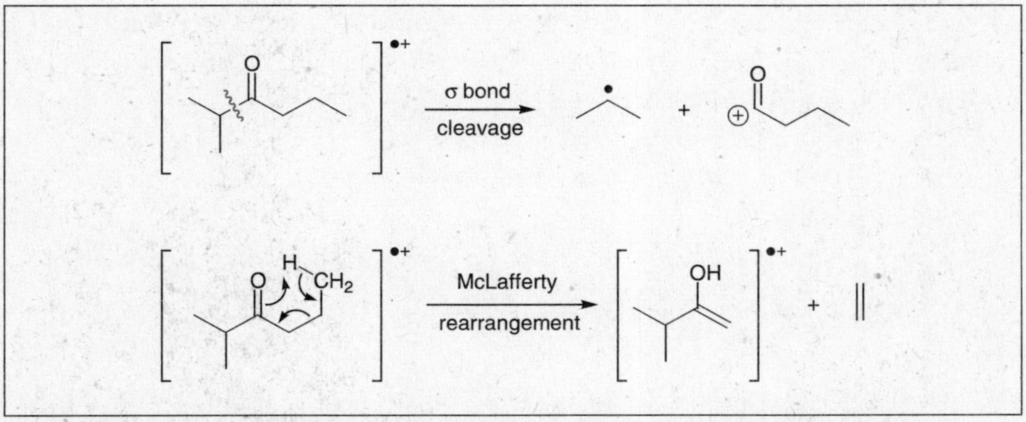

σ bond cleavage

McLafferty rearrangement

FIGURE 25-3: Two fragmentation processes for 2-methylhexan-3-one radical cation.

proof; however, the increasingly powerful (and more convenient) technique of NMR has provided an alternative source of structural knowledge available previously only from MS data.

Another very diagnostic feature in mass spectrometry is the isotopic fingerprint left by bromine and chlorine. Unlike carbon, hydrogen, nitrogen, and oxygen, which have one overwhelmingly predominant isotope (^{1}H, 99.98%; ^{12}C, 98.93%; ^{14}N, 99.62%; ^{16}O, 99.76%), bromine has two almost equally abundant natural isotopes (^{79}Br, 50.51%; ^{81}Br, 49.49%), and chlorine has a roughly 1:3 ratio of isomers in nature (^{35}Cl, 75.47%; ^{37}Cl, 24.53%). The practical result of this phenomenon is that molecules containing chlorine or bromine leave characteristic M+2 patterns in a ratio of 1:1 or 1:3, depending on which halogen is present.

CRAM SESSION

Mass Spectrometry

Mass spectrometry is a technique used to differentiate molecules based on molecular weight. Simply put, a mass spectrum is derived from a two-step sequence:

1. Ionization of the compound
2. Acceleration of the ionized compound across an electric field

Common methods of ionization include

- Electron impact (EI)
- Electrospray ionization (ESI)
- Chemical ionization (CI)

By examining the time-of-flight (TOF) through the field, a mass-to-charge ratio can be calculated, which is usually correlated directly to molecular weight. When molecules are subjected to **high-resolution mass spectrometry** (HRMS), the unique molecular formula can be determined as well as the molecular weight.

In many mass spectrometry experiments, especially those involving EI, the initially formed radical cations (which are very unstable) undergo decomposition (e.g., the **McLafferty rearrangement**) during analysis to provide well-defined fragments. These fragments can reveal structural details about the molecule (for example, whether an isopropyl or benzyl group is present). Furthermore, multiplicities in the parent ion peak can indicate the presence of atoms that have multiple naturally abundant isotopes, such as bromine and chlorine.

CHAPTER 26

Infrared Spectroscopy

Read This Chapter to Learn About:

- ➤ Background and Theory
- ➤ Extracting Information from IR Spectra

BACKGROUND AND THEORY

Molecules have many degrees of freedom, and each electronic state (ground, first excited, second excited, etc.) has an array of energy levels corresponding to various vibrational and rotational states. Just like the electronic transitions, movement among vibrational energy levels is quantized and can be studied through spectrophotometric methods. These energy changes are, of course, much smaller; therefore, the electromagnetic radiation required for excitation is of correspondingly lower energy. For most vibrational excitations, absorption occurs in the infrared region of the spectrum.

Although governed by quantum considerations, many vibrational modes can be modeled using classical physics. For example, a stretching vibration between two nuclei can be characterized using Hooke's law, which predicts that the frequency of an absorbance is given by the relationship

$$\bar{\nu} = 4.12\sqrt{\frac{f}{\left(\frac{m_1 m_2}{m_1 + m_2}\right)}} \tag{42}$$

where $\bar{\nu}$ is the frequency in wavenumbers (equal to $1/\lambda$), m_1 and m_2 are the masses of the two nuclei in amu units, and f is the force constant, which can be roughly correlated to bond strength (Table 26-1).

Hooke's law provides a useful conceptual framework for predicting the position of various stretches in the IR spectrum. Essentially, there are two deciding terms: the force

TABLE 26-1 Nominal Force Constant Values for Estimating IR Absorptions

Type of Bond	Nominal Force Constant (f) (dynes/cm)
Single	5×10^5
Double	10×10^5
Triple	15×10^5

constant and the reduced mass, which is given by $m_1m_2/(m_1 + m_2)$. The former is fairly straightforward—for example, the higher the force constant (i.e., the stronger the bond), the higher the wavenumber of the absorbance. Therefore, the C=C bond is predicted to absorb at higher wavenumber than the C—C bond, and the C≡C bond higher still. The reduced mass is a bit more subtle. Consider the C—C bond in relation to the C—H bond. The reduced mass for two carbon atoms would be $144/24 = 6$. In contrast, a carbon atom and hydrogen atom give a reduced mass of $12/13 = 0.923$. Since this term is in the denominator, the lower reduced mass translates into a higher wavenumber.

EXTRACTING INFORMATION FROM IR SPECTRA

These two theoretic factors help understand the overview of IR absorptions given in Figure 26-1. The IR spectrum can be broken into four very broad (and fuzzy) categories. At very high frequencies (2500–4000 cm^{-1}) we see what can be dubbed the X—H

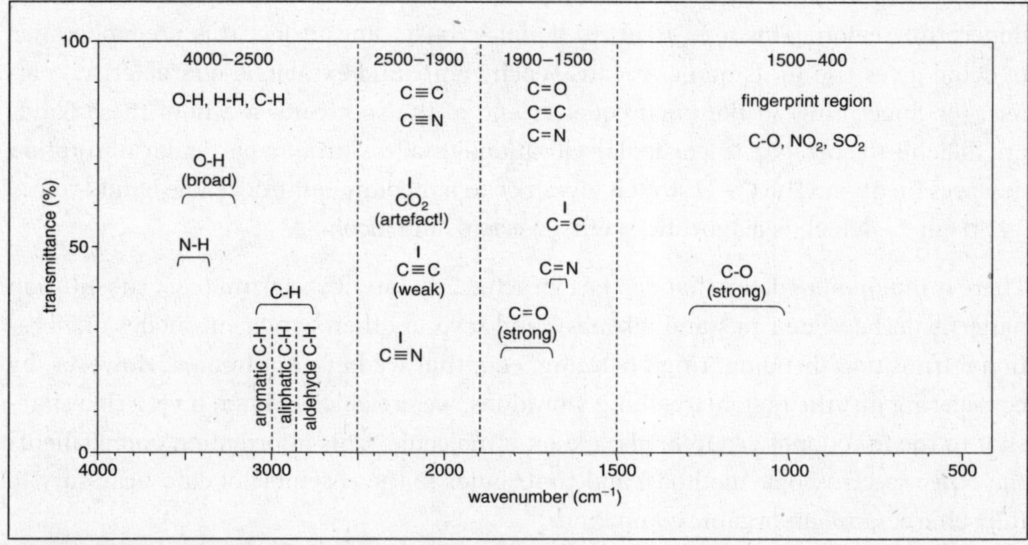

FIGURE 26-1: Overview of infrared absorbance ranges.

stretches, positioned at high energy by virtue of the very light hydrogen atom (reduced mass consideration). There are a few features here worth mentioning. First, the almost ubiquitous C—H stretches congregate around 3000 cm^{-1}, but the exact positioning is diagnostic: aromatic C—H stretches tend to be just above 3000 cm^{-1}, and aliphatic stretches tend to be just below that mark. The carbonyl C—H stretch also has a characteristic position, showing up at just less than 2800 cm^{-1}. Another telltale feature in this region belongs to the O—H stretch, which tends to be very broad and dominating. Presence of such a band is a sure sign of an O—H group in your compound (or for the sloppy experimentalist, adventitious water in your IR sample!).

Next we move to the region between 1900 and 2500 cm^{-1}, which is home to the triple bond. This area is usually fairly empty, so that absorptions are easy to spot. The alkyne stretch is typically sharp but very weak, and sometimes absent altogether, particular for symmetrically substituted triple bonds. Nitriles, however, tend to be more prominent, although they still tend toward the weak side. The only thing to watch out for here is that carbon dioxide also absorbs in this vicinity—because most instruments now take measurements in normal atmosphere and subtract out the background CO_2, artifacts can arise from quirky subtraction errors.

The real estate between 1500 and 1900 cm^{-1} houses many of the doubly bonded species, including alkene and carbonyl stretches. Alkenes give rise to a medium-intensity band just higher than 1600 cm^{-1}, whereas the garden-variety carbonyl stretch for a ketone shows up as an intense absorption at about 1715 cm^{-1}. Other carbonyl-containing functional groups wobble around this value: Esters absorb at slightly higher wavenumber (>1730 cm^{-1}), whereas amides have a somewhat lower-frequency absorbance (<1700 cm^{-1}). Often the carbonyl stretch, when present, is the predominant feature of the IR spectrum.

Finally, the portion of the spectrum bounded by 400 and 1500 cm^{-1} is known as the fingerprint region. This area tends to be fairly busy, and in fact this preponderance of detail gives rise to its name, because each compound exhibits a characteristic pattern (or fingerprint) in the low-frequency end of the spectrum. Many of these bands are difficult to correlate to particular vibrational modes, but one particular absorption deserves mention. The C—O stretch gives rise to a medium but noticeable band around 1100 cm^{-1}, which is diagnostic of ethers, esters, and alcohols.

There is much more detail that can be extracted from an IR spectrum (e.g., substitution patterns on benzene rings and alkenes), and several other significant modes of vibrational transition (bending, ring breathing, etc.) that we have not treated. However, by considering just the major stretching vibrations, we are able to obtain a very rich snapshot of the functional group landscape on a molecule. This information complements the other spectroscopic methods, and contributes to the ensemble of data necessary to fully characterize an organic compound.

CRAM SESSION

Infrared Spectroscopy

Infrared (IR) spectra are usually reported in **wavenumbers** (measured in cm^{-1}), which track frequency. Infrared absorptions arise from vibrational transitions within molecules. Consequently, the theoretical basis of IR can be modeled by **Hooke's law,** a relationship governing the macroscopic oscillation of two masses connected by a spring. The two major variables in the equation are

- The force constant (analogous to **bond strength**)
- The reduced mass term (dependent upon the **masses of the atoms** involved)

The trends revealed by considering these two variables are

- Weaker bonds absorb at lower wavenumbers
- Vibrations between a light and a heavy atom resonate at higher wavenumbers than those between two heavy atoms

Simply put, IR data provides information about functional groups within a molecule. The specific positions of absorbances can be very diagnostic (e.g., carbonyl stretch at 1740 cm^{-1}), but an overview of the IR spectrum can also help tie together these smaller details. Thus, the spectrum can be divided into the following very general zones:

- 4000–2500 cm^{-1}: vibrations between heavy and light atoms (OH, CH, NH)
- 2500–1900 cm^{-1}: vibrations of triply bonded heavy atoms (nitriles, alkynes, etc.)
- 1900–1500 cm^{-1}: vibrations of doubly bonded heavy atoms (carbonyls, olefins, etc.)
- 1500–400 cm^{-1}: vibrations of singly bonded heavy atoms

Because the last zone tends to be very busy (and because it is unique for a given compound), it is typically called the **fingerprint region.**

CHAPTER 27

Proton Nuclear Magnetic Resonance

Read This Chapter to Learn About:

- Background and Theory
- Principle of Chemical Shift
- Information from Chemical Shift
- Splitting Patterns and Integration
- Information from Coupling Constants

BACKGROUND AND THEORY

Nuclear magnetic resonance (NMR) represents the organic chemist's most powerful tool for structure elucidation. The principle of this spectroscopic method is based on the transition of nuclei between spin states, the so-called spin flip. Recall from general chemistry that a proton can have a quantum spin number of $+1/2$ or $-1/2$; these two states are also called "spin up" and "spin down." In the absence of a magnetic field, these spin states are of equivalent energy, or degenerate. However, when a strong magnetic field is applied, the energy states differentiate themselves (Figure 27-1). Nuclei aligned with the magnetic field find a lower energy state (designated α), whereas those opposed to the field are at higher energy (β).

Analogous to electronic transitions in UV-Vis spectroscopy, nuclei in the lower energy state can be promoted to the next level by the absorption of electromagnetic radiation of the appropriate energy. In the case of spin flips, this absorption occurs in the radio frequency region—in other words, they are relatively low-energy transitions. In principle, this resonance can be observed with any nucleus of odd mass or odd atomic number. We examine here two of most indispensable methods, ^{1}H-NMR and ^{13}C-NMR.

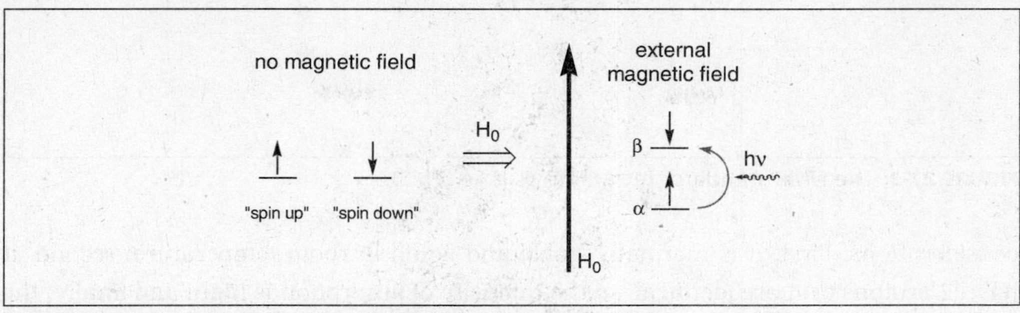

FIGURE 27-1: The splitting of degenerate nuclear spin states.

However, methods for observing other nuclei have become widely available, including ^{11}B, ^{19}F, and ^{31}P (of obvious importance for biochemists).

Very high magnetic fields must be used for useful data, and much of the innovation in NMR hardware centered around magnet technology (e.g., superconducting magnets). For all our efforts, however, the nucleus never experiences the full, unadulterated magnetic strength of the external field (H_0). This is because the electronic cloud surrounding the molecule shields the nuclei from the field. Instead, the α and β levels are split by a somewhat attenuated magnetic field, called the effective magnetic field (H_{eff}). Electron density, however, is a variable parameter. In a local environment in which the electron cloud has been impoverished (e.g., by the inductive effect of an electronegative atom), the nucleus is shielded less and thus experiences a stronger H_{eff} (Figure 27-2). Consequently, the energy splitting is enhanced and the resonance occurs at correspondingly higher frequency. This phenomenon of higher-frequency absorbances in electron-poor regions is known as deshielding.

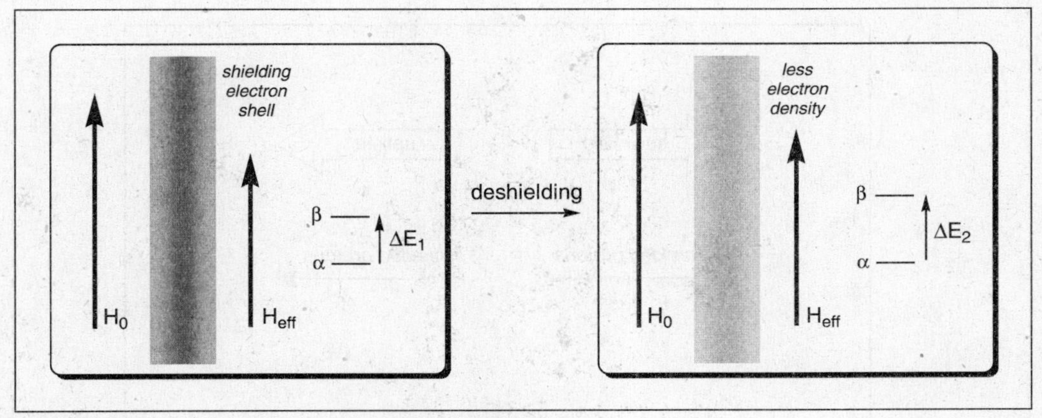

FIGURE 27-2: The shielding effect and deshielding.

Because the resonance frequency is field-dependent, no two spectrometers yield exactly the same resonance values. To alleviate this issue, spectroscopists historically have used an internal standard, against which all other signals are measured. The choice of tetramethylsilane (TMS, Figure 27-3) as an NMR standard was driven by a few practical

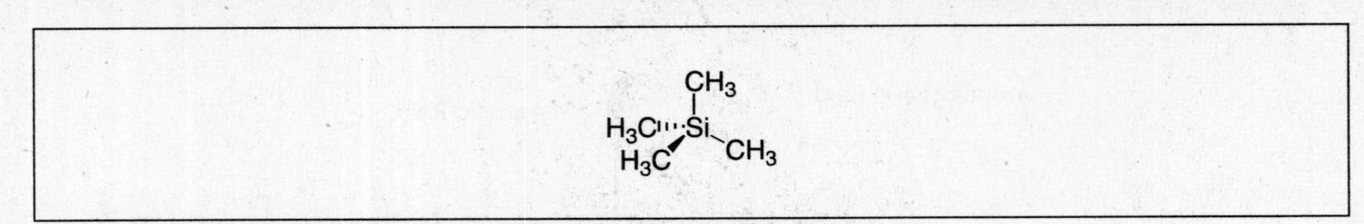

FIGURE 27-3: The NMR standard, tetramethylsilane (TMS).

considerations. First, it is thermally stable and liquid at room temperature; second, it has 12 protons that are identical, so the intensity of absorption is high; and finally, the protons in TMS are quite shielded (silicon is not particularly electronegative), so almost all other proton resonances we observe occur at higher frequencies relative to TMS.

PRINCIPLE OF CHEMICAL SHIFT

Instruments are characterized by the resonance frequency of TMS in the magnetic field of that instrument. For example, a 400-megahertz (MHz) NMR spectrometer incorporates a superconducting magnet in which TMS resonates at 400 MHz. Stronger magnets lead to higher values (e.g., 600 MHz, 900 MHz); in fact, magnets are so often specified by their TMS frequencies that we sometimes forget that the unit of MHz is a meaningless dimension for a magnetic field. That is, a 400-MHz NMR has a magnetic field strength of about 9.3 Tesla (T), yet for whatever reason, this parameter is almost never mentioned.

Nevertheless, we can use the TMS resonance to talk about the general landscape of the NMR spectrum. Figure 27-4 shows a typical measurement domain for a 400-MHz spectrometer, which is bounded on the right by the resonance frequency

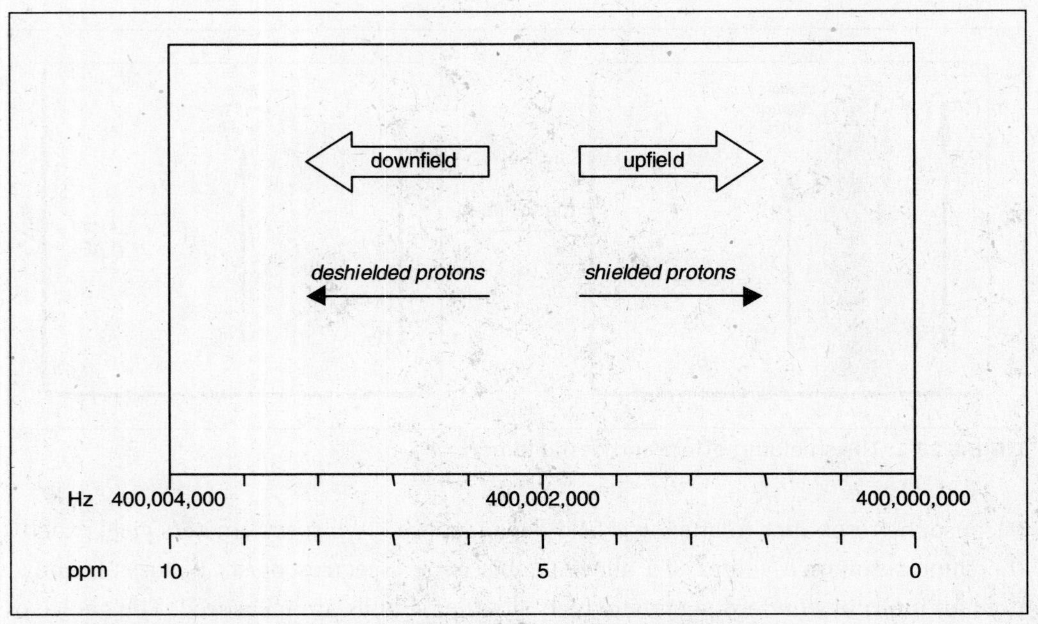

FIGURE 27-4: Important architectural features of the ^{1}H-NMR spectrum.

of TMS (400,000,000 Hz). The range through which most organic molecules absorb extends to about 400,004,000 Hz, or 4000 Hz relative to TMS. However, these values change as magnetic field strength is altered—an absorbance at 2000 Hz on a 400-MHz NMR would absorb at 1000 Hz on a 200-MHz NMR. To normalize these values, we instead report signals in terms of parts per million (ppm), which is defined as

$$\text{ppm} = \frac{\nu}{R} = \frac{\text{Hz}}{\text{MHz}} \tag{43}$$

where ν is the signal frequency (in Hz) relative to TMS and R is the base resonance frequency of TMS (in MHz), a parameter known as **chemical shift**. Thus, on a 400-MHz instrument a signal at 2000 Hz would be reported as having a chemical shift of 5 ppm (2000 Hz/400 MHz). This allows us to establish a universal scale for proton NMR ranging from 0 to about 10 ppm—although it is not terribly uncommon to find protons that absorb outside that range.

Working within this framework, it is important to understand certain features and terminology related to position in the NMR spectrum. First, the ppm scale obscures the fact that frequency increases to the left, so this must be borne in mind. Thus, as protons are deshielded, they are shifted to the left—spectroscopists have dubbed this a "downfield" shift (Figure 27-4). Similarly, migrating to the right of the spectrum is said to be moving upfield. Phenomenologically, a downfield shift is evidence of deshielding, and this understanding helps us interpret NMR data.

INFORMATION FROM CHEMICAL SHIFT

Using this idea, we can establish general regions in the NMR spectrum where various proton types tend to congregate (Figure 27-5). For example, most purely aliphatic compounds (hexane, etc.) absorb far upfield in the vicinity of 1 ppm. The attachment of electron-withdrawing groups (e.g., carbonyls) tends to pull resonances downfield. Thus, the protons on acetone show up at around 2.2 ppm. In general, the region between 0 and 2.5 ppm can be thought of as home to protons attached to carbons attached to carbon (H—C—C). This is also where terminal alkyne protons absorb.

As electronegative elements are attached, however, resonances are shifted even farther downfield. Thus, the methyl protons on methanol resonate at about 3.4 ppm, and the protons on chloromethane appear at about 3.1 ppm. Multiple electronegative elements have an additive effect: compared to chloromethane, dichloromethane resonates at 5.35 ppm, and chloroform (trichloromethane) absorbs at 7.25 ppm. As a rough guideline, the region between 2.5 and 5 ppm belongs to protons attached to carbons that are attached to an electronegative element (H—C—X).

Moving farther downfield, we encounter protons attached to sp^2 carbons, starting with the olefinic (alkenyl) variety, which absorb in the region between about 5 and

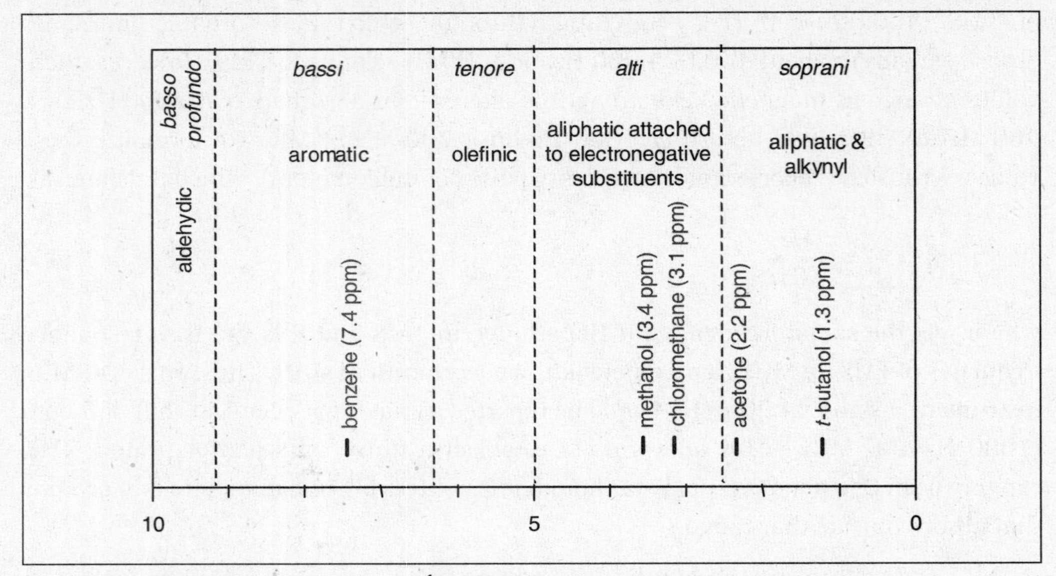

FIGURE 27-5: General regions of the ^{1}H-NMR spectrum, with benchmarks.

6.5 ppm. Next come the aromatic protons, which range from about 6.5 to 9 ppm. As a benchmark, benzene—the prototypical aromatic compound—absorbs at 7.4 ppm. Just like their aliphatic cousins, the olefinic and aromatic protons are shifted downfield as electron-withdrawing groups are attached to the systems. Finally, signals in the region between about 9 and 10 ppm tend to be very diagnostic for aldehydic protons—that is, the protons connected directly to the carbonyl carbon.

Now a word of caution. These rough guidelines are just that—they do not represent strict demarcations. Rather, they paint the NMR spectrum in broad brush strokes, so that we can take away an immediate impression from the data to start our thought processes. Think of these regions as voices in a choir, whose ranges often cross each other. Just as tenors often sing the alto line in Early Music, so too can olefins exhibit resonances above 5 ppm.

SPLITTING PATTERNS AND INTEGRATION

It would be gift enough if NMR told us only about a given proton's chemical environment—but we can gain much more information from the spectrum. One source of knowledge derives from a phenomenon known as scalar (or spin–spin) coupling, through which a proton is influenced by its nearest neighbors. Figure 27-6 explores this effect with the hypothetical situation of one proton (H_x) being next door to a pair of protons (H_a and H_b). The latter two protons are either spin up or spin down, and in fact we can imagine four permutations (2^2) of two nuclei with two states to choose from. If both H_a and H_b are aligned with the external magnetic field, they will serve to enhance H_{eff}, thereby increasing the energy gap between the $H_x\alpha$ and β states, which in turn

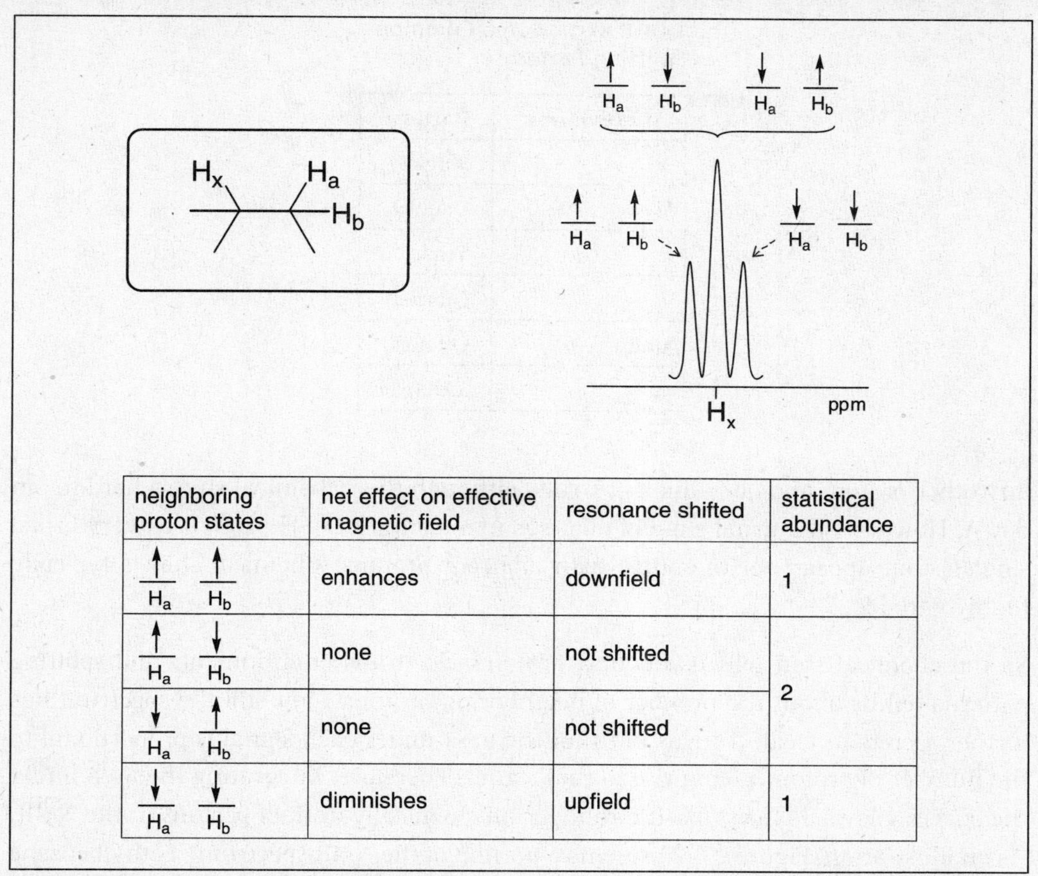

neighboring proton states	net effect on effective magnetic field	resonance shifted	statistical abundance
$\uparrow H_a$ $\uparrow H_b$	enhances	downfield	1
$\uparrow H_a$ $\downarrow H_b$	none	not shifted	2
$\downarrow H_a$ $\uparrow H_b$	none	not shifted	
$\downarrow H_a$ $\downarrow H_b$	diminishes	upfield	1

FIGURE 27-6: The origin of scalar splitting.

leads to a higher frequency absorbance (a downfield shift). Conversely, the situation in which both H_a and H_b oppose the external magnetic field diminishes the H_{eff}, leading to an upfield shift. The two remaining permutations have one spin up and the other spin down, thereby canceling each other out. The result is a pattern known as a triplet, in which the three prongs of the signal are present in a 1:2:1 ratio.

Had there been only one neighboring proton, there would have been only two possibilities: The neighboring proton is either spin up or spin down. The resulting pattern would have been a doublet. By the same token, three neighboring protons would have eight permutations (2^3), which could be all up, all down, two up and one down, or one up and two down. This would lead to a quartet pattern. If we summarize the trends in these examples, one neighboring proton gives a doublet, two give a triplet, and three give a quartet. In other words, the pattern exhibited (or "multiplicity") has one more prong than the number of neighboring protons. Patterns of this type are said to obey the "$n + 1$ rule," where $n =$ the number of neighboring protons, and $n + 1 =$ the multiplicity of the signal. The most common splitting patterns and their abbreviations are summarized in Table 27-1. It should be noted here that heteroatomic protons (N—H, O—H) are capricious beasts—in very dilute (and pure) solution, they behave much like

TABLE 27-1 Some Common Splitting Patterns

Abbreviation	**Pattern**
s	Singlet
d	Doublet
t	Triplet
q	Quartet
quint	Quintet
m	Multiplet

any other proton, and obey the $n + 1$ rule, although their chemical shift is hard to pin down. However, the usual state of affairs is to encounter O—H protons as very broad singlets that appear not to couple with adjacent protons. Chemical shift is typically highly variable.

So the chemical shift tells us about a proton's electronic environment, and splitting patterns tell us about the number of neighboring protons—but still the spectrum has further secrets to yield. It turns out that the area under each signal is proportional to the number of protons giving rise to that signal. Therefore, integrating the area under the curves allows us to establish a ratio for all chemically distinct protons in the NMR. As an illustration, Figure 27-7 presents a portion of the NMR spectrum of ethylbenzene corresponding to the ethyl moiety. There are two signals (at 1.15 ppm and 2.58 ppm), corresponding to the two sets of chemically distinct protons on the ethyl substituent

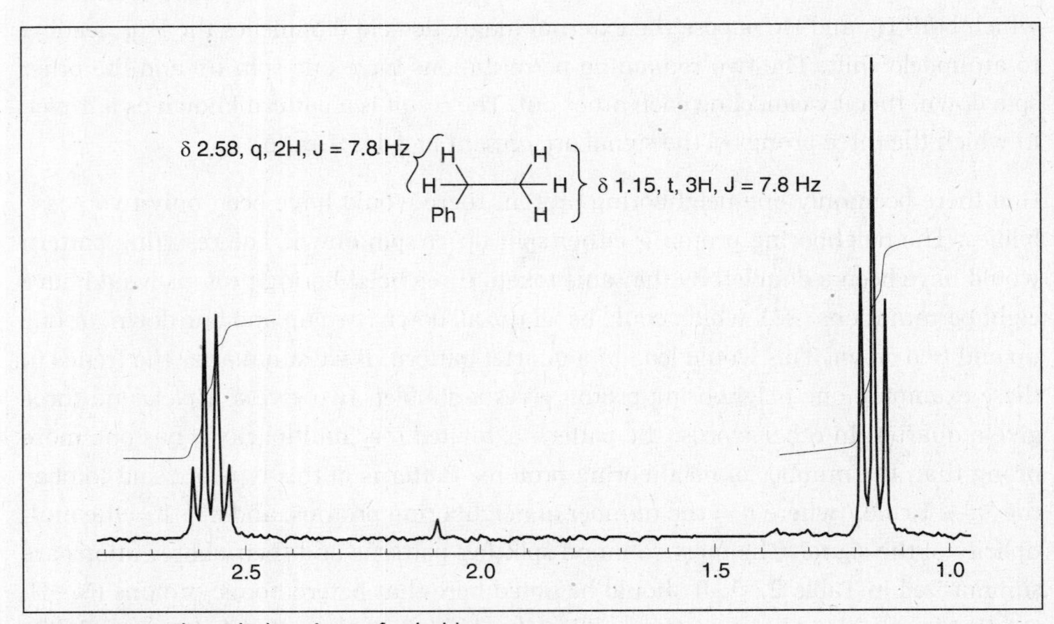

FIGURE 27-7: The ethyl moiety of ethyl benzene.

(a methyl and methylene group, respectively). The downfield shift of the methylene group is an indication of its being closer to the slightly electron-withdrawing benzene ring. The splitting patterns tell us that the protons resonating at 1.15 (the triplet) are next to two other protons, and the protons at 2.58 (the quartet) are adjacent to three other protons. Furthermore, the integral traces reveal that the two sets of protons are present in a 3:2 ratio. We extract the integration data simply by taking a ruler and measuring the heights of the integral traces. The absolute values mean nothing by themselves, but taken together they establish a ratio of the various protons.

Within a given pattern, the spacing between the peaks is known as the coupling constant, or *J*-value, usually reported in Hz. Notice in Figure 27-7, the proper way to report NMR data is as follows (explanatory comments in parentheses): δ (indicates the values are in ppm) 1.15 (the chemical shift), *t* [the multiplicity (here, a triplet)], 3H (the number of protons given by the integration), $J = 7.8$ Hz (the measured coupling constant). This is a very inflexible format, and should always be used when reporting NMR data. A natural question, then, is how to extract the *J*-values from an NMR spectrum. Again, the ruler is our friend. By carefully measuring against the bottom scale, we can assign ppm values to each prong of the signal (Figure 27-8). Applying Equation 43, if ppm = Hz/MHz, then Hz = (ppm)(MHz). In other words, to convert ppm to Hz, simply multiply by the "field strength," or more accurately, the native resonance of TMS. The spectrum in Figure 23-2 was obtained on a 300-MHz instrument, so the prong at 1.186 ppm corresponds to 355.8 Hz (1.186×300). Once the peaks are converted to units of Hz, the *J*-value is simply the distance between the peaks. It is worth noting at this point that chemical shifts are reported in ppm because their ppm values do not change with

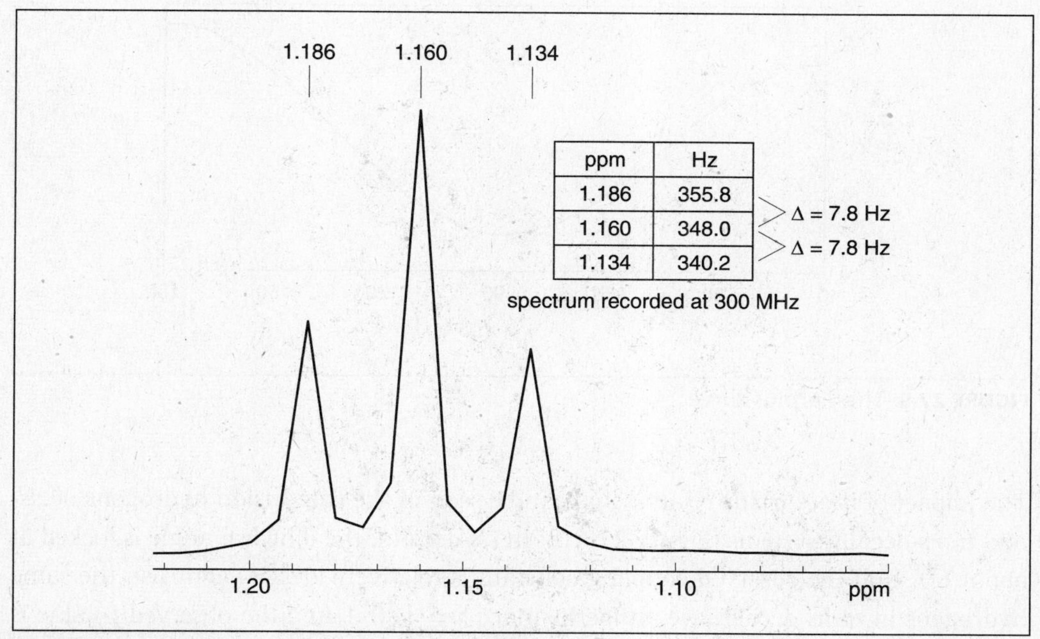

ppm	Hz
1.186	355.8
1.160	348.0
1.134	340.2

FIGURE 27-8: Measuring *J*-values from a spectrum.

field strength; by the same token, *J*-values in Hz remain constant regardless of magnetic field, so this is how they are reported.

INFORMATION FROM COUPLING CONSTANTS

As luck would have it, the magnitude of the coupling constant can also provide useful information about the structure of a molecule. It is important to understand here, that scalar coupling is not a through-space interaction, but a through-bond effect. In other words, the information about the neighboring nuclear spin states is actually communicated through the electronic network bonding the two sets of nuclei. Also bear in mind that this information must travel through three bonds (H—C, C—C, C—H). Consequently, the orientation relationship between the two C—H bonds is very important. In fact, an empirical relationship has been elucidated between the observed *J*-value and the dihedral angle (ϕ) set up between the two C—H bonds. Figure 27-9 shows this relationship in graphical format, which is known as the Karplus curve. The important outcomes are these: When the dihedral angle is at 0° (eclipsed) and 180° (antiperiplanar), the *J*-value is at a maximum (ca. 12 Hz), and at a 90° dihedral angle, the *J*-value is at a minimum (ca. 2 Hz). This makes sense, if one considers that when the two C—H bonds are orthogonal, communication between the σ bonds is practically nonexistent.

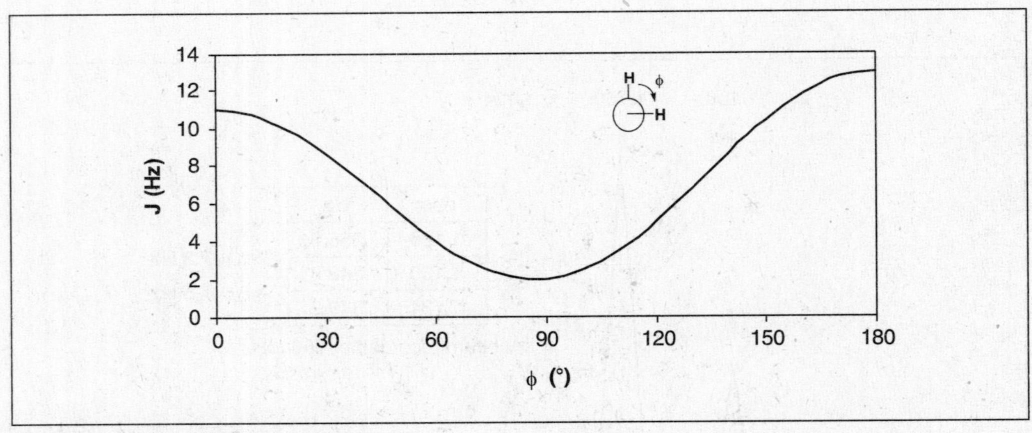

FIGURE 27-9: The Karplus curve.

The impact of the Karplus relationship can be seen in the ring-fusion hydrogens of *cis*- and *trans*-decalin systems (Figure 27-10). In *cis*-decalin, the dihedral angle is locked at about 60° and the observed coupling constant is relatively low. By contrast, the same hydrogens in *trans*-decalin are antiperiplanar (ϕ = 180°), and the observed *J*-value is quite large. The Karplus curve also explains the coupling constants in freely rotating

FIGURE 27-10: Impact of the Karplus relationship in constrained systems.

systems, which exhibit a time-averaged value of about 7.5 Hz (our garden-variety *J*-value for most acyclic aliphatic protons).

Generally speaking, protons connected to the same carbon (geminal protons) do not split each other—with one significant exception. Figure 27-11 shows three types of geminal protons: homeotopic, enantiotopic, and diastereotopic. To better understand the terminology, consider the imaginary products formed by replacing each of the two geminal protons with another atom, say a chlorine substituent. In the homeotopic example, the two "products" are identical (2-chloropropane); for the enantiotopic protons, two enantiomeric "products" are formed [*R*- and *S*-(1-chloroethyl)benzene]; and replacing H_a versus H_b in the cyclopropane derivative gives diastereomers (*cis*-dichloro vs. *trans*-dichloro).

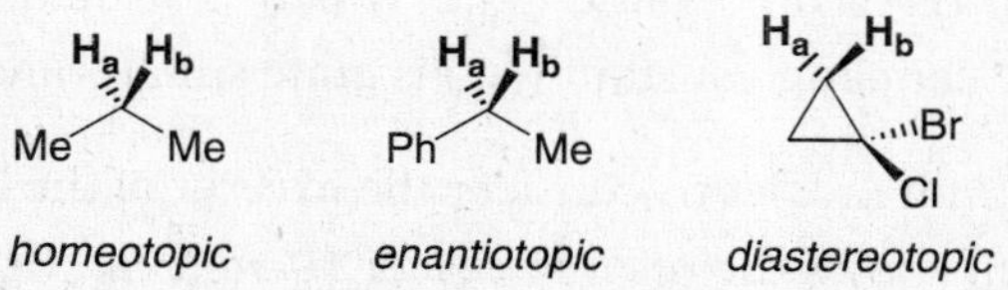

FIGURE 27-11: Three types of geminal protons.

Homeotopic and enantiotopic geminal protons do not split each other, because they are chemically equivalent. In other words, they are identical through a plane of symmetry (the plane of this page). However, diastereotopic protons are not chemically equivalent—in the example in Figure 27-11, H_a is in proximity to the chloro substituent, whereas H_b is close to the bromine. In short, they are in different worlds and therefore behave as individuals. Therefore, in these special cases, geminal protons do split each other, and the magnitude of the *J*-value is relatively large. Figure 27-12 lists this value, along with other representative coupling constants. This summary is useful for the interpretation of spectra.

Some interesting patterns arise when a proton is coupled to two different neighbors with divergent *J*-values. For example, terminal alkenes have three olefinic protons that are related with three different coupling constants (Figure 27-12). The *trans*-olefinic coupling constant (J_{ac}) is the largest, with typical values ranging from 12 to 16 Hz;

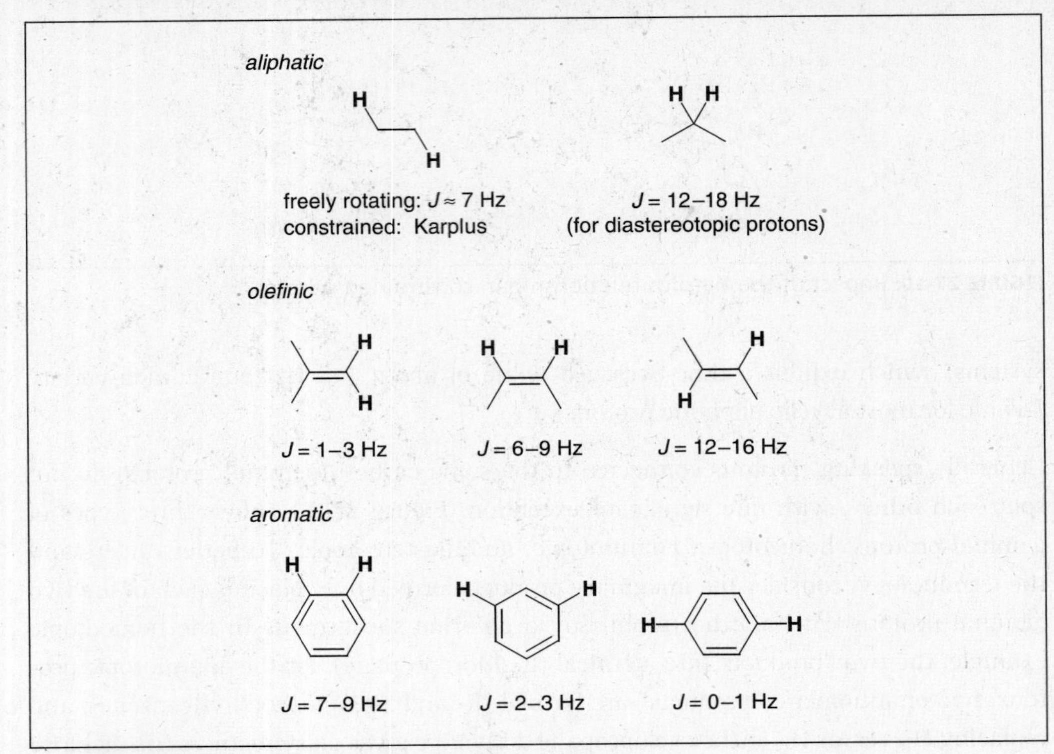

FIGURE 27-12: Some representative coupling constants.

the next largest is the *cis*-olefinic value (J_{ab}), which is usually around 6–9 Hz; and finally the *gem*-olefinic coupling constant (J_{bc}) is quite small—about 1–3 Hz.

With this information in hand, let us consider the excerpt of the NMR spectrum shown in Figure 27-13, which shows the resonances for H_c and H_b. H_a is split into a doublet

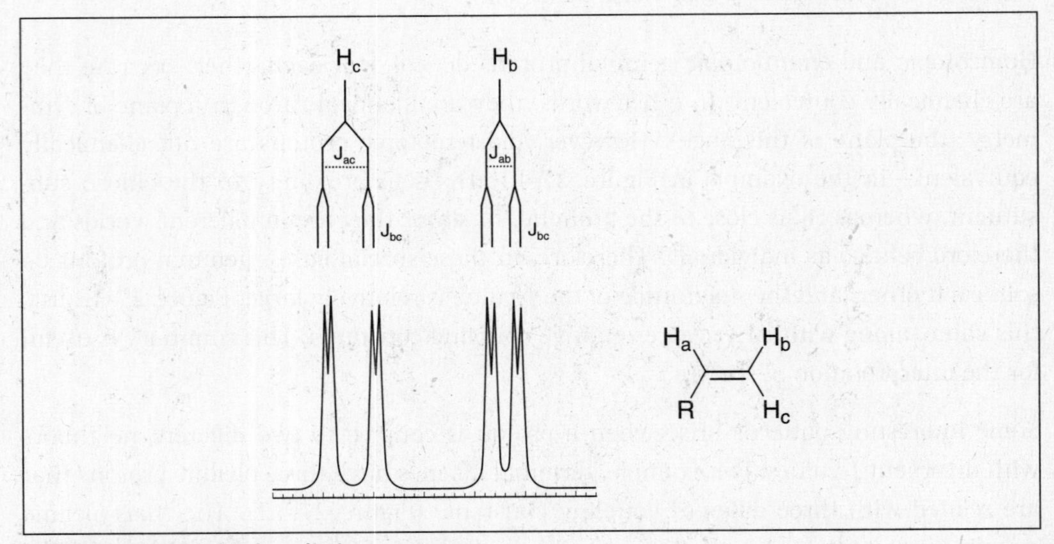

FIGURE 27-13: Multiple splitting patterns.

by H_c (with a large *J*-value), and again into a doublet by H_b (with a small *J*-value). This pattern is known as a doublet of doublets (or colloquially, as a double doublet) and is designated with the abbreviation "dd." Similarly, H_b is split into a doublet by H_a (with a medium *J*-value), and again into a doublet by H_c (with a small *J*-value), to give another double doublet. The two double doublets are qualitatively different, and even casual visual inspection reveals that the downfield pattern is a combination of the **large** and small coupling constants, whereas the upfield pattern incorporates the **medium** and small coupling constants. This alone would allow us to make a fairly confident assignment of these peaks to H_c and H_b, respectively.

CRAM SESSION

Proton NMR

Nuclear magnetic resonance (NMR) arises from the absorption of energy in a nuclear **spin flip,** a transition that is extremely sensitive to the electronic density around it. The position of an NMR signal relative to the internal standard TMS is called the **chemical shift,** and this data reveals the nature of a proton's chemical environment. Protons that resonate at much higher frequency than TMS are said to be **downfield,** and those that appear close to the standard are said to be **upfield.** A downfield shift is indicative of an electron-deficient (or **deshielded**) environment, usually caused by the proximity of an electron-withdrawing (EW) group. The **integration** of NMR signals is used to calculate the ratio of the respective protons. The typical range for a proton NMR spectrum is 0–10 ppm, and this can be divided into the following general categories:

- 0 to ca. 2.5 ppm: aliphatic and alkynyl C—H protons
- ca. 2.5 to ca. 5 ppm: aliphatic C—H protons where carbon is attached to an electronegative element
- ca. 5 to ca. 6.5 ppm: alkenyl (sp^2) C—H protons
- ca. 6.5 to ca. 9 ppm: aromatic (sp^2) C—H protons
- ca. 10 ppm: aldehydic C—H protons

In addition, signals often exhibit **splitting patterns,** many of which can be interpreted using the ***n* + 1 rule,** which states that the **multiplicity** corresponds to the total number of adjacent protons plus one. Thus, a proton with only one next-door proton would give rise to a doublet and so on. This phenomenon of **scalar splitting** is a **through-bond** effect communicated by the σ backbone. The magnitude of the splitting (otherwise known as the **coupling constant** or ***J*-value**) is also diagnostic. Freely rotating systems usually exhibit a *J*-value of about 7 Hz; however, this value changes in constrained systems such as cycloalkanes. In these situations, the *J*-value can be predicted by consulting the **Karplus curve,** which correlates the coupling constant to the **dihedral angle** between the two adjacent C—H bonds.

CHAPTER 28

Carbon Nuclear Magnetic Resonance

Read This Chapter to Learn About:

- ➤ Background
- ➤ Information from Chemical Shift
- ➤ Carbon Signals and Symmetry
- ➤ Solving Structure Elucidation Problems

BACKGROUND

The principle of carbon nuclear magnetic resonance (^{13}C-NMR) is identical to that of hydrogen NMR (^{1}H-NMR); however, there are a few practical differences that deserve mention. First, although the natural abundance of ^{1}H is almost 100%, the ^{13}C isotope makes up only slightly more than 1% of carbon found in nature. This means that the NMR-active carbon nucleus is quite dilute in the typical molecule. As a result, signal-to-noise ratios tend to be lower and data collection times are usually longer. Moreover, although scalar coupling can exist (in principle) between two ^{13}C nuclei, the statistical likelihood that a pair of this species are situated adjacent to each other is vanishingly small.

Carbon can also couple to ^{1}H nuclei, which are plentiful, but most experimental methods wipe out this interaction with a technique known as off-resonance decoupling in order to simplify the spectrum. The carbon nucleus resonates at one-quarter the frequency of a proton nucleus; for example, in a 400-megahertz (MHz) instrument, the carbons of TMS absorb at 100 MHz. Because the two spectral windows are so far apart, all the protons in a molecule can be promoted to the β state by intense irradiation in that region, whereas the carbon spectrum is being taken. Because all the protons are in the same spin state, they all exert the same influence on H_{eff} for carbon, so the ^{13}C signals are not split. In other words, every carbon shows up as a singlet. However, unlike proton NMR, integration is generally useless in carbon NMR.

INFORMATION FROM CHEMICAL SHIFT

Another happy circumstance regarding the ^{13}C spectrum is that the carbon nuclei are spread over a much wider territory, so instead of a range of 10 parts per million (ppm), the typical carbon sweep width is 200 ppm. Figure 28-1 provides a broad overview of ^{13}C resonances, combined with a conceptual framework for understanding these trends. In the 0–50 ppm range, we find mostly aliphatic carbons attached to nothing in particular. When π systems are introduced, a dramatic downfield shift occurs through an effect known as anisotropy—areas above and below the π bonds tend to be shielded, whereas the middle areas tend to be strongly deshielded. The upshot is that alkenes and aromatic carbons show up in the general region of 100–150 ppm. When we progress to alkynes an opposite anisotropic effect occurs: the barrel-like π structure of the alkyne sets up an electronic current whose field lines actually shield at the ends of the triple bond. Thus, acetylenic carbons resonate in the 50–100 range.

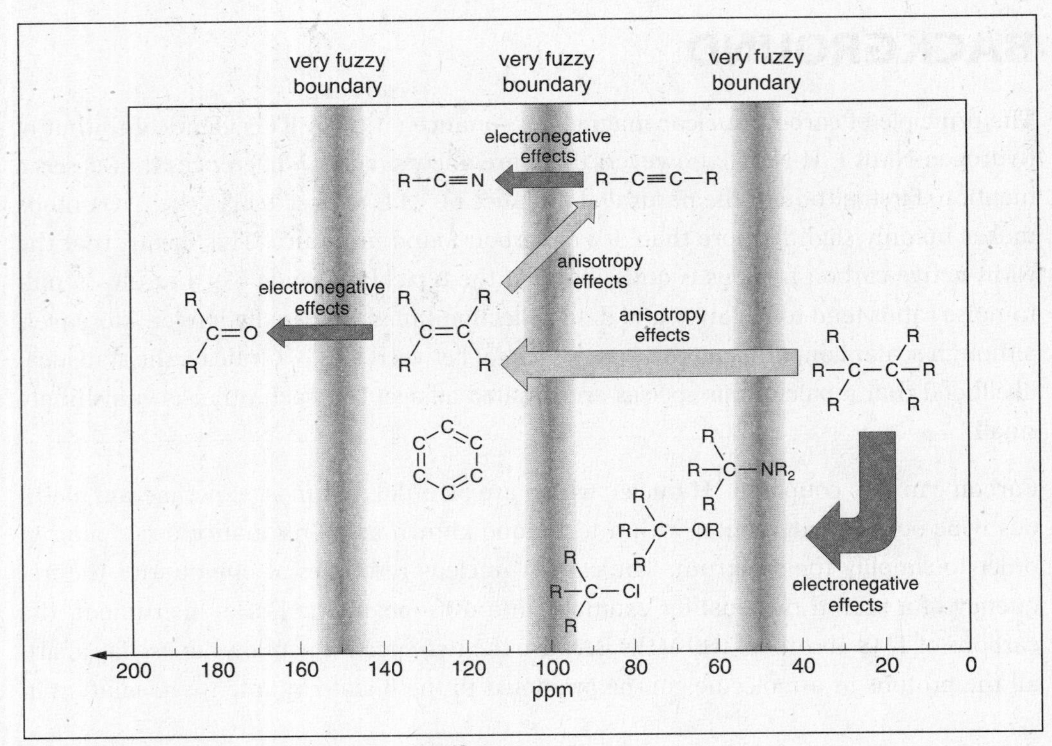

FIGURE 28-1: Conceptual landscape of ^{13}C-NMR.

The next general trend arises from an electronegativity effect. When carbons of any type (*sp*, sp^2, sp^3) are attached to electronegative elements, a downfield shift occurs. Thus, aliphatics are moved into the 50–100 ppm region, acetylenics (i.e., nitriles) are moved into the 100–150 range, and olefinics (i.e., carbonyls) migrate to the 150–200 ppm region.

CARBON SIGNALS AND SYMMETRY

The combination of this wide landscape with the sharp singlet signals means that every unique carbon in a molecule usually can be resolved in a ^{13}C spectrum. This is useful because it can reveal important information about symmetry within a molecule. For example, 2-nitroaniline and 3-nitroaniline (Figure 28-2, top) exhibit six carbon peaks, as we would expect. However, 4-nitroaniline only has four carbon signals. Why? Because the two carbons *ortho* to the amine are identical, as are two carbons *meta* with respect to the amine. We can recognize this by the σ mirror plane that exists in the last molecule, which relates these two sets of carbons by symmetry. Similarly, the three isomers of pentane (Figure 28-2, bottom) exhibit different numbers of ^{13}C absorbance, as they have differing degrees of symmetry.

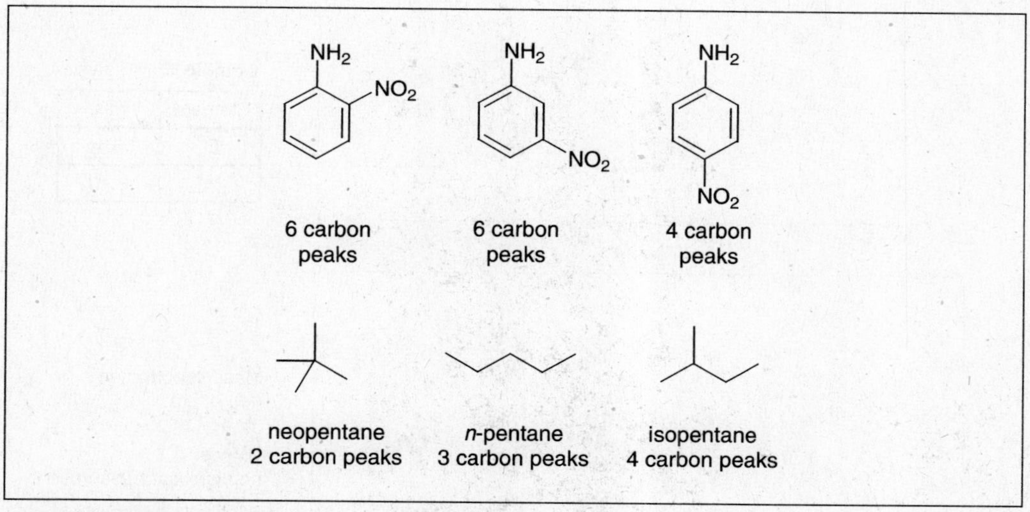

NH_2 NO_2 — 6 carbon peaks

NH_2 NO_2 — 6 carbon peaks

NH_2 NO_2 — 4 carbon peaks

neopentane
2 carbon peaks

n-pentane
3 carbon peaks

isopentane
4 carbon peaks

FIGURE 28-2: Impact of symmetry on ^{13}C-NMR spectrum.

SOLVING STRUCTURE ELUCIDATION PROBLEMS

It is the author's contention that collections of structure elucidation problems should be sold in supermarkets as alternatives to crosswords. Indeed, these are puzzles of the most exciting kind—piecing together clues from a variety of sources to converge on a supportable proposal for structure. As with all puzzles, they are best approached with a healthy combination of algorithm with trial and error. Consider, for example, the collection of data presented in Figure 28-3. How would we go about making sense of all that?

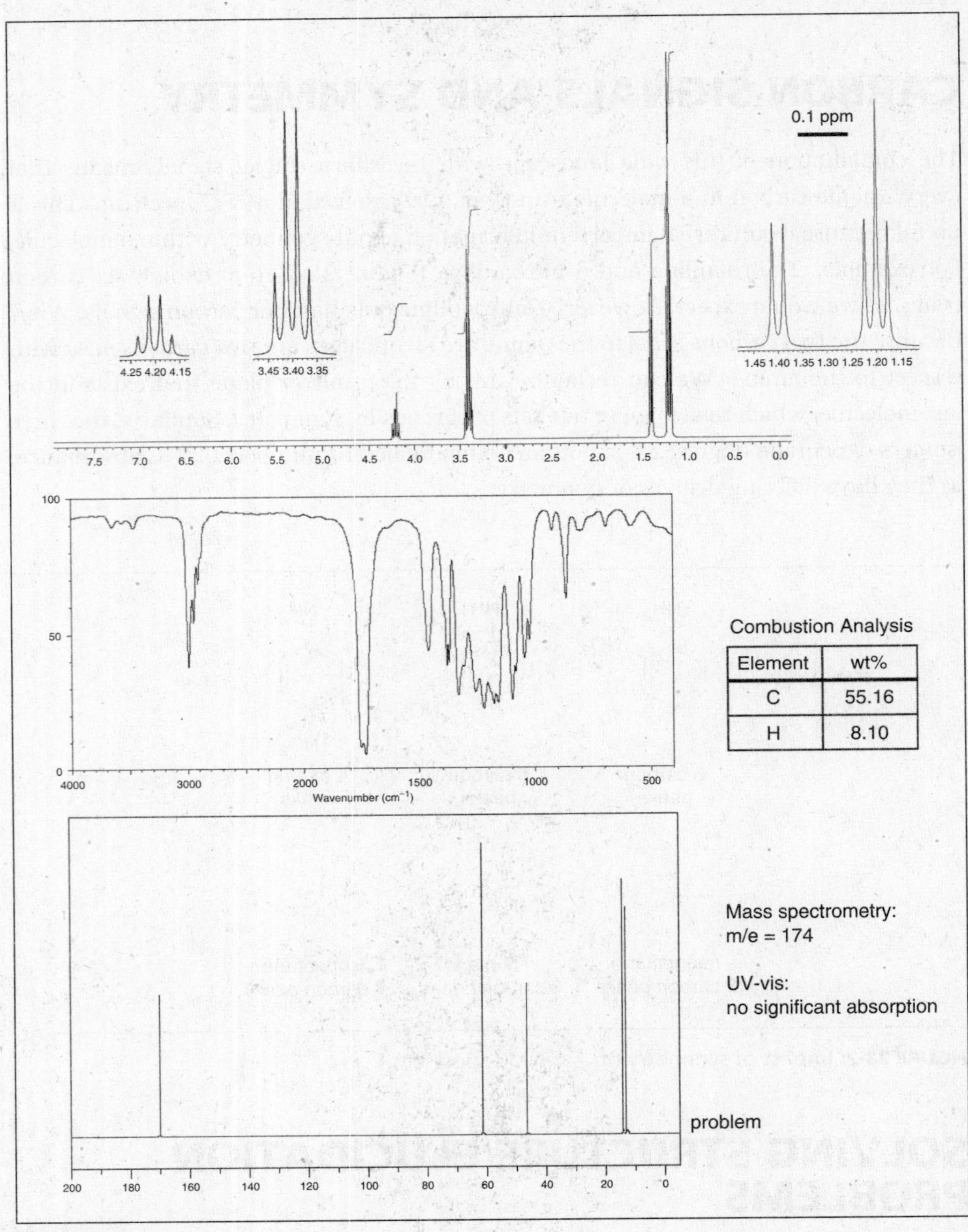

FIGURE 28-3

The first thing we do is work up the combustion analysis data, where we find an empirical formula of $C_4H_7O_2$. The mass spectrometry data gives us a molecular weight of 174, so our molecular formula must be twice the empirical formula, or $C_8H_{14}O_4$. A DoU calculation yields two degrees of unsaturation, but the absence of a significant UV-Vis absorption allows us to rule out any conjugated diene or enone systems. We can also rule out aromatic systems, because we have an insufficient degree of unsaturation (a benzene ring represents four degrees of unsaturation).

The next thing we do is to take a look at the IR, where we find the dominant feature to be a carbonyl stretch at about 1740 cm^{-1}. Aside from the aliphatic C—H stretches and possibly a C—O stretch buried in the fingerprint region, there is not really much more to take away. This is also important, as it allows us to rule out alcohols, aldehydes (no C—H stretch at 2800 cm^{-1}), and carboxylic acids right away. Furthermore, we know that at least one of our degrees of unsaturation is bound up in a carbonyl group.

Moving along to the ^{13}C data, we find two signals in the general aliphatic region, one signal in the downfield aliphatic region, one signal in the C—X category, and one carbonyl carbon signal (confirming our suspicions from the IR). The discrepant number of five ^{13}C signals versus eight carbons in the molecular formula points toward a high degree of symmetry in the molecule, a clue that is very valuable as we start to piece together structures.

Finally, we move to the proton NMR, which exhibits the following signals:

δ 4.2, 1H, q
δ 3.4, 4H, q
δ 1.4, 3H, d
δ 1.2, 6H, t

The signals at 3.4 and 1.2 are particularly telling—particularly in light of our previous assertion of symmetry from examining the carbon NMR. Specifically, there is a three-proton limit for any one signal (i.e., a methyl group), so the integration ratios of 4 and 6 must mean that we are dealing with two *identical* sets of signals, as follows:

2 × δ 3.4, 2H, q
2 × δ 1.2, 3H, t

With this information in hand, we can sketch a fragment that is consistent with this data, as follows:

```
       H  H
       |  |
2 ×  —C—C—H
       |  |
       H  H
```

If we look closely at the chemical shift, we conclude that the ethyl groups must be connected to an electronegative element (δ 3.4 ppm). Because we have oxygen in the molecular formula, we propose

```
          H  H
          |  |
2 ×  —O—C—C—H
          |  |
          H  H
```

Now let us turn our attention to the other two signals in the spectrum, namely, the quartet at 4.2 and the doublet at 1.4. Because we insulated our ethyl groups with oxygen atoms, the only way to get splitting is for these two signals to be coupled to each other. This would add to our palette as follows:

```
          H  H               H
          |  |          |    |
2 ×  —O—C—C—H  +  H—C—C—H
          |  |          |    |
          H  H               H
```

A quick atom count reveals that we have now accounted for all our hydrogens, and all but two of our carbons and oxygens. We know from IR that we must have a carbonyl, from ^{13}C-NMR that the molecule is highly symmetrical, and from the molecular formula that we have two degrees of unsaturation. Therefore, we propose two carbonyl fragments

```
          H  H               H              O
          |  |          |    |              ||
2 ×  —O—C—C—H  +  H—C—C—H  + 2 ×  —C—
          |  |          |    |
          H  H               H
```

Now the solution to the puzzle becomes the challenge of plugging these fragments into each other so that the required symmetry is met and that no raw edges (unfilled valencies) are left. With a bit of trial and error, we propose the following, which satisfies all spectral data:

```
    H  H     O
    |  |     ||
H—C—C—O—C
    |  |     |   H
    H  H     |   |
          H—C—C—H
    H  H     |   |
    |  |     |   H
H—C—C—O—C
    |  |     ||
    H  H     O
```

CRAM SESSION

Carbon NMR

Carbon NMR is based on the same principle as proton NMR. However, carbon signals are spread across a wider span than proton absorbances (ca. 200 ppm vs. 10 ppm). Furthermore, only carbon-13 is observed by NMR, and its natural abundance is about 1%, so practically no carbon–carbon splitting is observed. Carbon does couple with hydrogen, but this coupling is artificially suppressed by **off-resonance decoupling.** Consequently, each carbon signal shows up as a singlet, and each unique carbon can usually be resolved.

For this reason, carbon NMR is very useful in identifying **symmetry** within a molecule. In other words, if the molecular formula indicates the presence of 10 carbon atoms, but the NMR only reveals 8 signals, this is strong evidence that a pair of carbon atoms are related by symmetry.

Like proton NMR, chemical shift can also provide important information. In very general terms, the carbon NMR spectrum can be divided into four regions:

- 0 to ca. 50 ppm: aliphatic (sp^3) carbons
- ca. 50 to ca. 100 ppm: sp^3 carbons attached to heteroatoms; alkynyl (*sp*) carbons
- ca. 100 to ca. 150 ppm: aromatic and olefinic (sp^2) carbons; nitrile (*sp*) carbons
- ca. 150 to ca. 200 ppm: carbonyl carbons

Conceptual Map of Section V

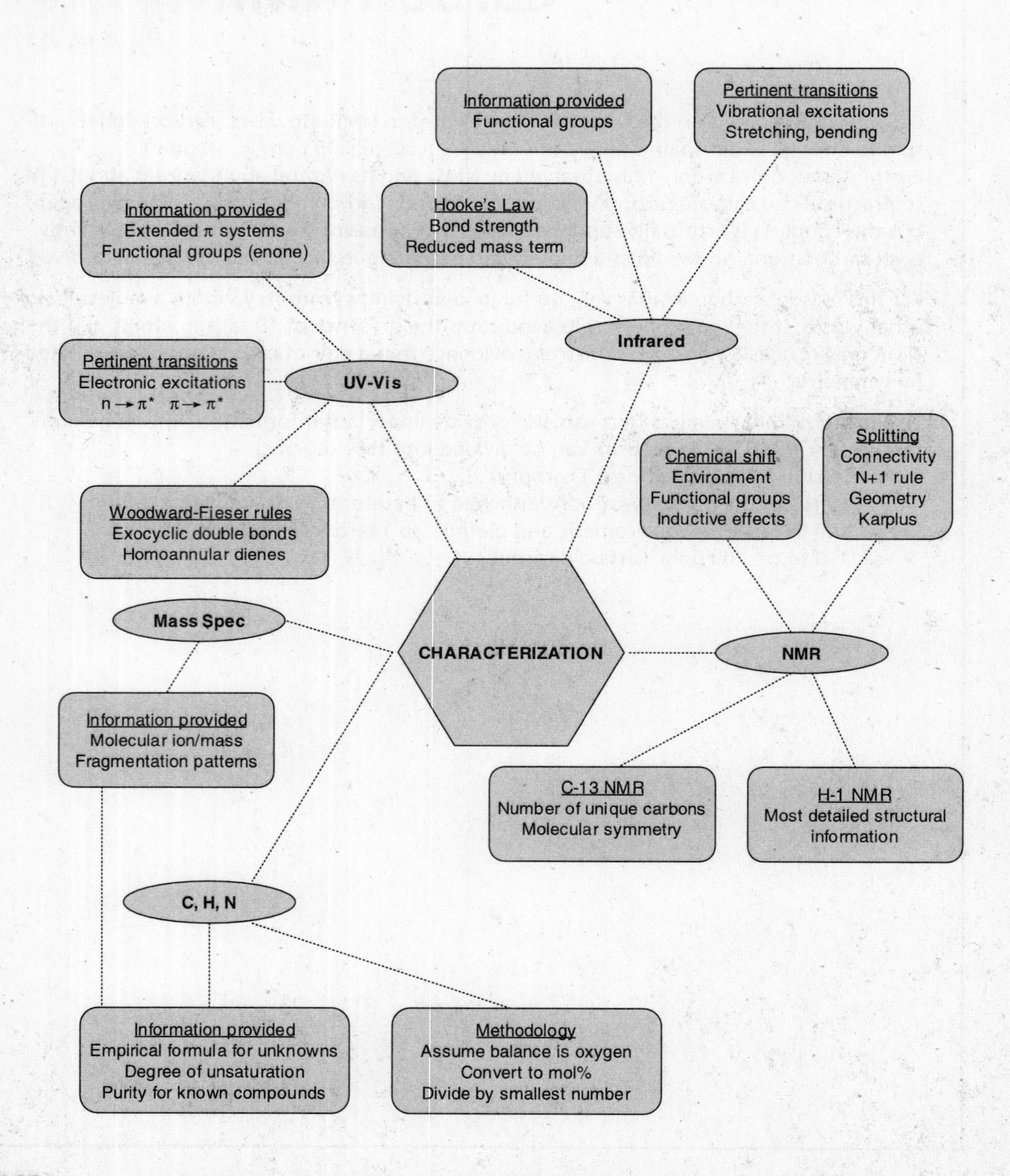

APPENDIX

Reagents

Formula or Abbreviation	Structure	Used for
AIBN [azobis (isobutyronitrile)]	NC–C–N=N–C–CN	Radical initiator
$AlCl_3$ (aluminum trichloride)	Cl, Cl, Al, Cl	Lewis acid catalyst
BF_3 (boron trifluoride)	F, F, B, F	Lewis acid catalyst
BH_3 (borane)	H, H, B, H	Hydroboration
Br_2 (bromine)	Br—Br	Radical bromination; dibromination
CCl_4 (carbon tetrachloride)	Cl, Cl, Cl, Cl	Nonpolar inert solvent
$CHCl_3$ (chloroform)	H, Cl, Cl, Cl	Polar, nonflammable solvent
CH_2Cl_2 (dichloromethane)	Cl, H, Cl, H	Polar, nonflammable solvent
CH_2I_2 (diiodomethane)	I, H, I, H	Simmons–Smith cyclopropanation
CH_2N_2 (diazomethane)	⊖ ⊕ N=N=C, H, H	Making methyl esters from acids Cyclopropanation (w/Cu catalyst)
DIBAL (diisobutylaluminum hydride)	Al, H	Selective reduction of esters, amides, and nitriles to aldehydes
Dicyclohexylborane	H, B	Hydroboration of alkyne derivatives; anti-Markovnikov hydration
Dioxane	O, O	Good solvent for dissolving water and organic substrates
DMD (dimethyldioxirane)	O, O	Epoxidation of alkenes
DMF (dimethylformamide)	O, N, H	Polar aprotic solvent

Formula or Abbreviation	Structure	Used for
DMSO (dimethylsulfoxide)	O=S(CH₃)₂	Polar aprotic solvent
Et_2O (diethyl ether)	⁄⁀O⁀⁄	Medium polarity solvent
$FeBr_3$ (iron tribromide)	Br–Fe(–Br)–Br	Lewis acid catalyst
H_2 (hydrogen)	H—H	Hydrogenation; reduction of nitro
H_2O_2 (hydrogen peroxide)	H–O–O–H	Oxidative workup of hydroboration
$Hg(OAc)_2$ (mercuric acetate)	n/a	Oxymercuriation
$HgSO_4$ (mercuric sulfate)	n/a	Markovnikov hydration of alkynes
HIO_4 (metaperiodic acid)	O=I(=O)(=O)–OH	Oxidative cleavage of 1,2-diols
HMPA (hexamethylphosphoramide)	O=P(NMe₂)₃	Preventing aggregation (polar aprotic solvent)
$K_2Cr_2O_7/H_2SO_4$ [potassium dichromate (Jones reagent)]	K⊕ ⊖O–Cr(=O)₂–O–Cr(=O)₂–O⊖ K⊕	Oxidation of alcohols
LAH (lithium aluminum hydride)	Li⊕ H–Al⊖(H)(H)–H	Very strong hydride source Reduces esters to alcohols
$LiAl(Ot\text{-}Bu)_3H$ [lithium tri(*t*-butoxy)aluminum hydride]	Li⊕ H–Al⊖(OtBu)₃	Modified hydride source Reduces acid chlorides to aldehydes
LDA (lithium diisopropylamide)	(iPr)₂N⊖ Li⊕	Strong, hindered base
Lindlar's catalyst	$Pd/CaCO_3/Pb(OAc)_2$/quinoline	Reducing alkynes to *cis*-alkenes
*m*CPBA (*m*-chloroperbenzoic acid)	Cl–C₆H₄–C(=O)–O–OH	Epoxidation of alkenes
MnO_2 (manganese dioxide)	O=Mn=O	Selective oxidation of allylic alcohols

Formula or Abbreviation	Structure	Used for
MsCl [methanesulfonyl chloride (mesyl chloride)]	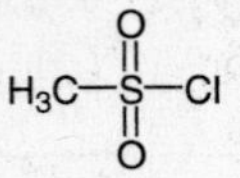 H_3C-SO_2-Cl	Converting hydroxyl to a good LG
$NaBH_4$ (sodium borohydride)	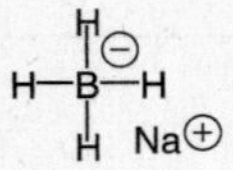 BH_4^{-} Na^{+}	Mild source of hydride
$NaBH_3CN$ (sodium cyanoborohydride)	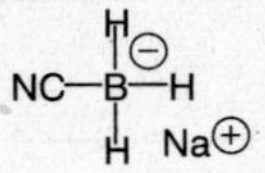 $NC-BH_3^{-}$ Na^{+}	Reductive amination Hydride source stable to mild acid
$NaNO_2$ (sodium nitrite)	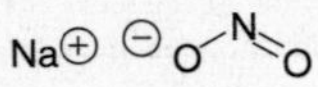 Na^{+} $^{-}O-N=O$	Diazotization of amines (with HCl)
NBS (*N*-bromosuccinimide)	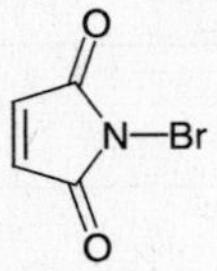 N–Br	Bromine surrogate
n-BuLi	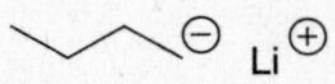 Li^{+}	Strong base
NCS (*N*-chlorosuccinimide)	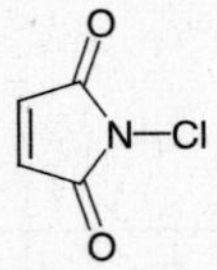 N–Cl	Chlorine surrogate
O_3 (ozone)	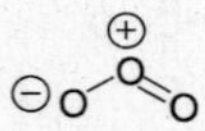 $^{-}O-O^{+}=O$	Oxidative cleavage of alkenes
OsO_4 (osmium tetroxide)	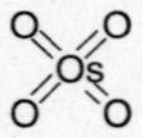 OsO_4	Dihydroxylation of alkenes
PCC (pyridinium chlorochromate)	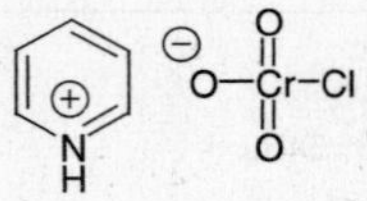 $C_5H_5NH^{+}$ $^{-}O-CrO_2-Cl$	Selective oxidation of primary alcohols to aldehydes
PPh_3 (triphenylphosphine)	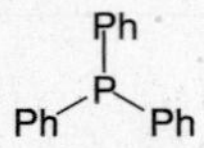 $P(Ph)_3$	Making Wittig reagents
$SOCl_2$ (thionyl chloride)	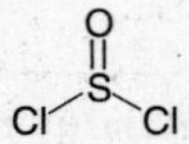 $Cl-S(=O)-Cl$	Converting alcohols to alkyl chlorides
THF (tetrahydrofuran)		Medium polarity solvent
*p*TsCl [*p*-toluenesulfonyl chloride (tosyl chloride)]	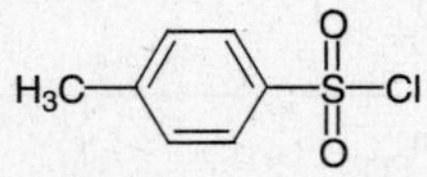 $H_3C-C_6H_4-SO_2-Cl$	Converting hydroxyl to a good LG

Formula or Abbreviation	Structure	Used for
*p*TsOH [*p*-toluenesulfonic acid (tosic acid)]	$H_3C-C_6H_4-SO_2-OH$	Organic-soluble source of strong acid
Zn(Hg) (zinc amalgam)	n/a	Clemmensen reduction (with HCl)

ON YOUR OWN

MCAT Organic Chemistry Practice

This section contains five quizzes, one for each of the five main sections of the organic chemistry review section that you have just completed. Use these quizzes to test your mastery of the concepts and principles covered in the organic chemistry review. Explanations for every question are provided following the last quiz. Each explanation includes the number of the corresponding chapter in the organic chemistry review. If you need more help, go back and reread the corresponding review chapter.

SECTION I STRUCTURE (Chapters 1–5) Quiz

1. Which one of the following molecules has a polarity that is **different** from that of the other three choices?

   A. H_2O

   B. CH_3Cl

   C. CH_3OH

   D. CO_2

2. Which equation below correctly calculates the formal charge of nitrogen in the nitrate ion?

   A. $6 - 4 - \frac{1}{2}(4) = 0$

   B. $6 - 6 - \frac{1}{2}(2) = -1$

   C. $5 - \frac{1}{2}(8) = +1$

   D. $6 - \frac{1}{2}(8) = +2$

3. Which of the following statements about intermolecular forces is **incorrect**?

   A. Van der Waals forces exist between nonpolar molecules such as $C_{20}H_{42}$.

   B. Dipole–dipole interactions exist between molecules of methanol.

   C. A chloride ion will be attracted to an oxygen atom when NaCl is dissolved in water.

   D. Hydrogen bonds will be found between molecules of a carboxylic acid.

4. Hydrogen bonding is found between the molecules of all of the following **except**

   A. NH_3.

   B. DNA.

   C. proteins.

   D. HCl.

5. Which of the following statements is most accurate?

   A. Ethene has two *sp* hybridized carbon atoms and a liner molecular geometry.

   B. Acetylene has a linear geometry and the hydrogen atoms are 109.5 degrees apart.

C. Ethyne has four atoms that lie in a straight line.

D. Ethane has two carbon atoms that lie 120 degrees apart.

6. Which of the following formation of a dimethyl cyclohexane is expected to be most stable?

A. *cis*-1,3-Dimethyl cyclohexane where both methyl groups are in the equatorial position

B. *trans*-1,4-Dimethyl cyclohexane where both methyl groups are in the axial position

C. *cis*-1,2-Dimethyl cyclohexane

D. A boat conformation of the compound where the methyl groups are in the flagpole positions

7. What do the following two compounds have in common?

CH=CH–C(CH_3)=CH–CH=CH–C(CH_3)=CH–CH_2OH

A. They both contain carbon atoms that are *sp* hybridized.

B. They both contain isoprene units.

C. They are both conjugated hydrocarbons.

D. They are both steroids.

8. How many σ and π bonds are found in a molecule of 1,4-octadiyne?

A. Ten σ and two π.

B. Seventeen σ and four π.

C. Eleven σ and six π.

D. Twenty-one σ and zero π.

9. Which explains the stability of benzene?

A. Benzene is resonance stabilized.

B. Benzene is stabilized by nonpolar covalent bonds.

C. Benzene has a planar shape to its molecule.

D. Benzene can complex a positive ion much like a crown ether can.

10. When 1,3-pentadiene is reacted with Cl_2, two products are formed:

$ClCH_2{-}CH(Cl){-}CH{=}CH{-}CH_3$ and $ClCH_2{-}CH{=}CH{-}CH(Cl){-}CH_3$.

Which choice below does **not** explain why this occurs?

A. The system is conjugated.

B. The chlorine is a more reactive and less selective element when it reacts.

C. There are multiple stable intermediates.

D. Resonance structures are formed during the reaction.

11. The order that shows an increasing priority of substituents is

A. H, Cl, $CH_3CH_2{-}$, Br

B. $-CH_3$, F, $-NH_2$, Br

C. $-CH{=}CH_2$, $-CH_2NH_2$, $-CHO$, $-C({=}O)NH_2$

D. $-CHBr_2$, $-C_6H_5$, $-CH_2F$, Cl

12. Which substance is **not** paired up with its correct name?

A. 3-Ketopentanoic acid

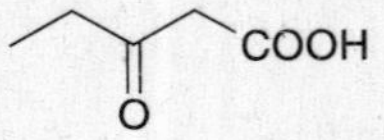

B. Cyclopentanoic acid

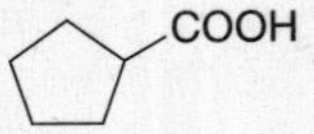

C. 6-Formylhexanoic acid

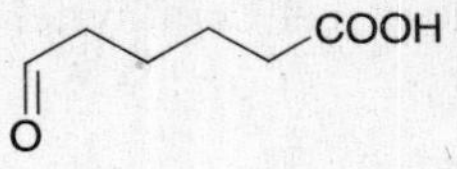

D. 1,4-Octandiol

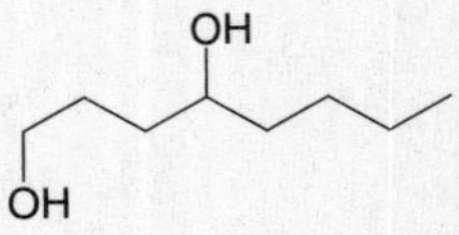

13. The compound shown here is called

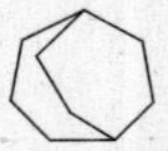

A. Bicyclo[3.2.2]nonane.

B. Bicyclo[3.2.2]heptane.

C. Bicyclo[4.4.4]nonane.

D. Bicyclo[4.4.4]heptane.

14. Which compound is correctly paired up with the groups found on its benzene molecule?

A. A xylene has three methyl groups present.

B. A cresol has a hydroxyl group and a methyl group present.

C. A toluic acid has a —COOH group and a hydroxyl group present.

D. An aniline has a $—NH_2$ group and a —CHO group present.

15. The molecule below has two stereocenters:

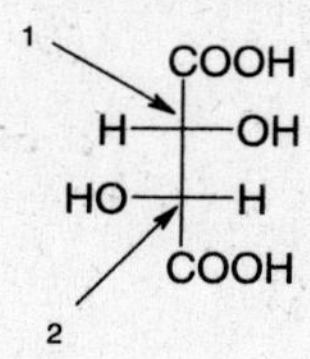

What is the expected R/S configuration of the carbon atoms labeled 1 and 2?

A. Both have an R configuration.

B. Both have an S configuration.

C. Carbon 1 has an R configuration, whereas 2 has an S configuration.

D. Carbon 1 has an S configuration, whereas 2 has an R configuration.

16. What is the relationship between the two molecules below?

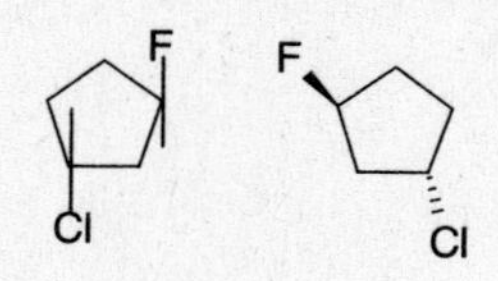

A. Diasteriomers

B. Enantiomers

C. Meso compounds

D. Structural isomers

17. Which statement below is incorrect about the set of structures below?

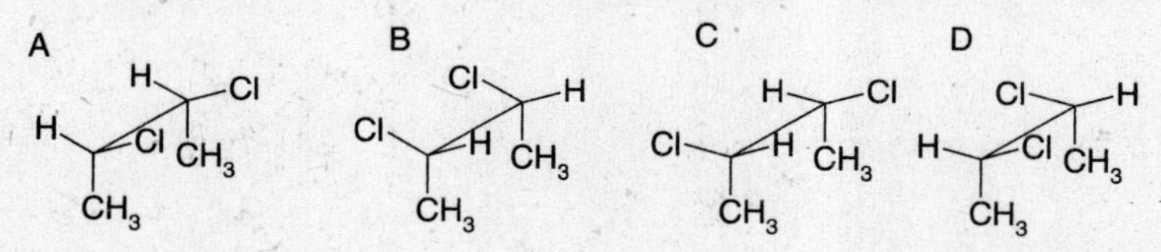

A. A and B are meso compounds.

B. C and D are enantiomers.

C. B and C are diasteriomers.

D. A and D are enantiomers.

SECTION II REACTIVITY (Chapters 6–9) Quiz

18. Consider the termination step shown here.

$$H_3C\cdot + H_3C\cdot \rightarrow H_3C{-}CH_3$$

Select the most accurate description of this reaction's thermodynamic properties:

A. ΔH is positive; ΔS is positive

B. ΔH is positive; ΔS is negative

C. ΔH is negative; ΔS is positive

D. ΔH is negative; ΔS is negative

19. Which intermediate shown here is considered to be the most stable?

A. $CH_3\cdot$

B. $R\text{-}{}^{+}C{=}CH_2$

C. $R_2C{=}C^{+}{-}R$

D. $(Ar)_3C\cdot$

20. Which set of conditions shown here will **not** favor an S_N2 reaction?

A. A strong nucleophile

B. A leaving group on a primary carbon atom

C. A lower temperature

D. A solvent of a higher polarity

21. The approximate pK_a value of an organic compound is 50. This organic compound is most likely

A. an alcohol.

B. a carboxylic acid.

C. a terminal alkyne.

D. an alkane.

22. 6-Bromohexanoic acid is reacted with NaOH. The resulting product is

A. classified as a lactone.

B. 6-hydroxyhexanoic acid.

C. 5-hexenoic acid.

D. classified as a lactam.

23. Of the four acids that present themselves here, which one is expected to be the least acidic?

A. H_2SO_4

B. CH_3COOH

C. CF_3COOH

D. CCl_3COOH

24. Which of the following statements does not describe an S_N2 reaction?

A. The reaction occurs via a backside attack.

B. The products are completely inverted in configuration.

C. The intermediate is pentacoordinate.

D. The reaction requires just one molecule to start the reaction.

25. 1,3-Butadiene reacts with ethylene to form cyclohexene. This occurred via

A. polymerization of a diene.

B. Grignard cycloaddition.

C. a Diels–Alder cycloaddition.

D. a Friedel–Crafts cycloaddition.

SECTION III METHODOLOGY (Chapters 10–18) Quiz

26. Of the methods shown here, which is **not** an appropriate method for making isopropyl alcohol?

A. Acetone reacts with $NaBH_4$.

B. Acetone reacts with LAH.

C. Acetaldehyde reacts with CH_3MgBr in THF.

D. Propene is reacted with BH_3/THF and then $NaOH/H_2O_2$.

27. A primary alcohol is reacted with PCC/CH_2Cl_2. The resulting functional group is going to be

A. an aldehyde.

B. a carboxylic acid.

C. a ketone.

D. an ester.

28. A Williamson synthesis is demonstrated here by which of the following reactions?

A. $Me{-}OH + CH_2{=}C(CH_3)_2/p\text{TsOH} \rightarrow MeOC(CH_3)_3$

B. $Me{-}OH + NaOH \rightarrow Me\text{-}O^-$ and then $Me{-}O^- + CH_3CH_2CH_2Cl \rightarrow MeO{-}CH_2{-}CH_2CH_2CH_3$

C. $CH_3CH(OH)CH_3 + p\text{TsOH/heat} \rightarrow CH_3CH{=}CH_2$

D. 1-butanol + $SOCl_2$ → 1-chlorobutane

29. $n{-}C_4H_9MgBr$ is reacted with ethylene oxide to form compound A. Compound A is then reacted with $KMnO_4$, and then acidified to make compound B. Compound B is then reacted with $SOCl_2$ to form compound C. Which of the following is true?

A. Compound A is a six-carbon alcohol.

B. Compound B is an aldehyde.

C. Compound C has four carbon atoms.

D. Compound B will test positive when carried out in the iodoform reaction.

30. Consider the following reactions:

I. Ethanol + acetic acid + H_2SO_4 catalyst →

II. Carboxylate ion + primary halide →

III. Butanoyl chloride + 1-propanol →

Which of these reactions forms an ester?

A. I only

B. II only

C. I and III only

D. I, II, and III

31. After the ozonolysis of one mole of a compound, two moles of CHO—CH_2CH_2—CHO were formed. What was the original compound?

A. 1,5-Cyclooctadiene

B. 4-Octene

C. CHO—$CH_2C(=O)$—$C(=O)CH_2$—CHO

D. *cis*-$CH_3CH_2CH_2CH=CHCH_2CH_2CH_3$

32. Butanal reacts with two moles of ethanol and dry HCl is bubbled through the reaction. Which of the following statements regarding this reaction is **false**?

A. A hemiactal will form as an intermediate.

B. Two —OR groups will be on the carbon atom that was once the aldehyde functional group.

C. This reaction can also be carried out using a base.

D. This reaction will have a product with the same functional group formed when glucose forms a ring.

33. When $(CH_3)_2C=PPh_3$ reacts with CHO—CH_2CH_3, the final product is

A. $(CH_3)_2CH$—O—CH_2CH_3

B. $(CH_3)_2C=CHCH_2CH_3$

C. $(CH_3)_2$—CH—$CH(OH)CH_2CH_3$

D. $(CH_3)_2C=P\equiv C$—CH_2CH_3

34. Two methods for producing 2-chlorobutane are

A. obtaining the major product of a substitution reaction with butane and Cl_2 in light; reacting 1-butene with HCl.

B. obtaining the minor product of a substitution reaction with butane and Cl_2 in light after a substitution reaction; reacting 2-butene with Cl_2.

C. reacting 1-butene with Cl_2; shaking butane with NaCl(aq).

D. reacting butane with chlorobenzene; reacting 2-butene with CH_3CH_2MgCl.

35. In order to convert an aldehyde to a carboxylic acid, you would use

A. H_2SO_4/H_2O

B. $K_2Cr_2O_7/H_2SO_4/H_2O$

C. R-MgBr + CO_2

D. $SOCl_2$

36. A carboxylic acid is treated with LAH in THF. The expected functional group is

A. an ester.

B. a ketone.

C. an aldehyde.

D. a primary alcohol.

37. A compound is treated with $NaNO_2$/HCl and a carboxylic acid is formed. The initial reactant was

A. an aldehyde.

B. an amine.

C. an amide.

D. an ester.

38. Aniline is reacted with $CH_3C(=O)Cl$. The expected major product is

A. $NHC(=O)CH_3$

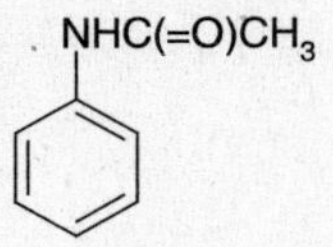

B. NH_2 $C(=O)CH_3$

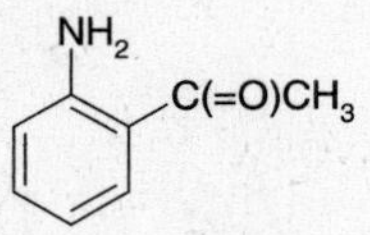

C. NH_2 C(=O)CH_3

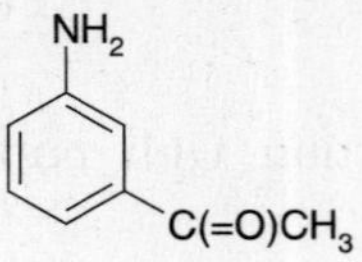

D. NH_2 C(=O)CH_3

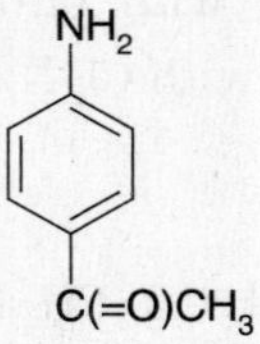

39. What is the major product of the reaction:

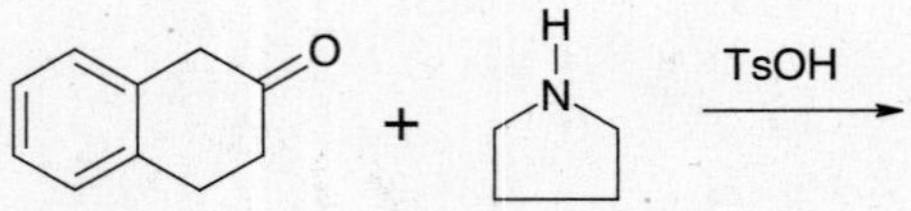

TsOH

A.

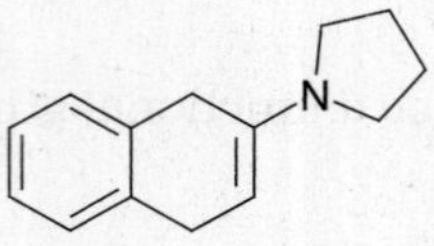

B. OH N

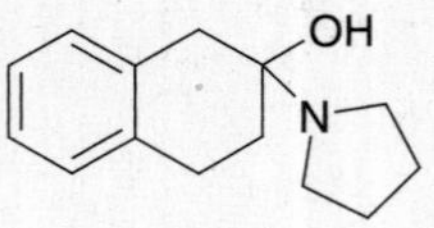

C. N

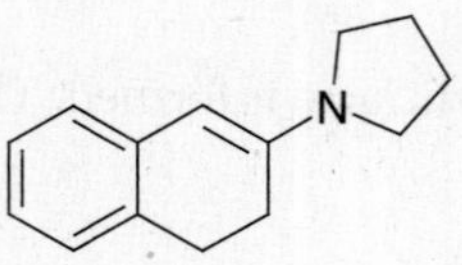

D. OH N H

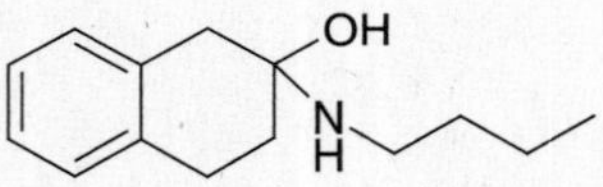

40. Consider the proposed synthesis of methylamine shown here:

$$NH_3 \xrightarrow{MeI} MeNH_2$$

This is **not** an efficient synthesis of methylamine because

A. ammonia is an inadequate nucleophile.

B. iodomethane is an inadequate electrophile.

C. methylamine is more nucleophilic than ammonia.

D. methylamine is more volatile than ammonia.

41. Which reaction shown here is a feasible method for the production of α-tetralone?

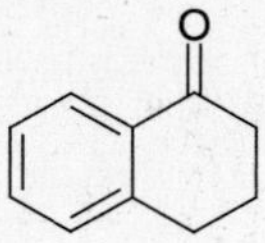

A. 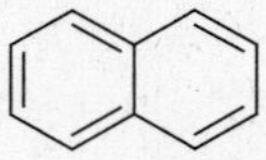 [naphthalene] is reacted to form a Grignard reagent and then CO_2 is used to complete the process.

B. 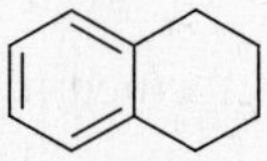 [tetralin] is reacted with HNO_3 and H_2SO_4.

C. 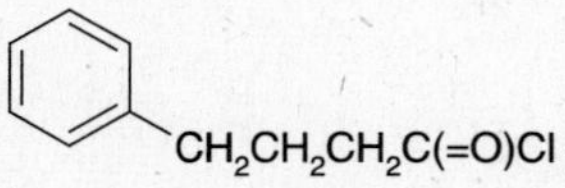 [benzene]$CH_2CH_2CH_2C(=O)Cl$ is reacted with $AlCl_3$.

D. 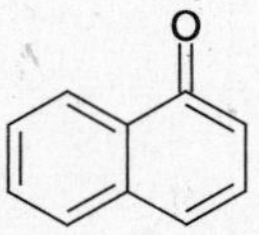 [1-naphthalenone] is reacted with H_2 at high pressures.

42. Which reagent is best for carrying out the reaction that forms the following ketone?

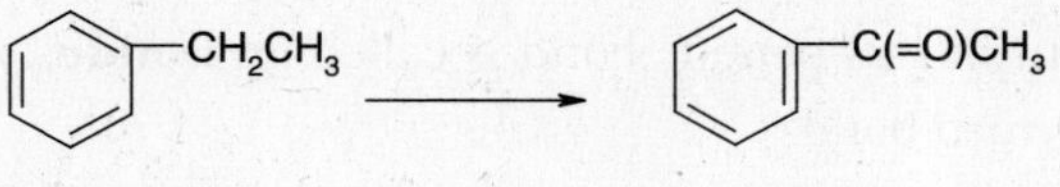

[benzene]$-CH_2CH_3$ ⟶ [benzene]$-C(=O)CH_3$

A. Zn(Hg)/HCl

B. DIBAL/THF

C. OsO_4/H_2O_2

D. PCC

43. Toluene is reacted with HNO_3/H_2SO_4 and the *para* product is then reacted with Sn/HCl and then with NaOH. The final product is

A. 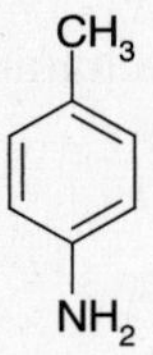 CH_3 [benzene ring] NH_2

B. 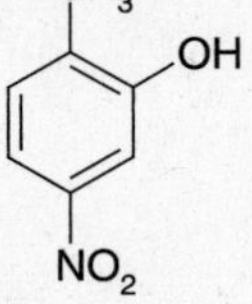 CH_3, OH [benzene ring] NO_2

C.

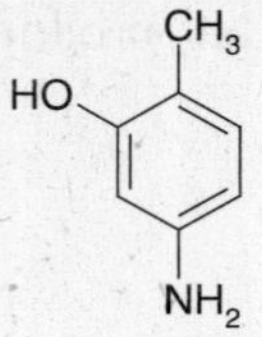

D.

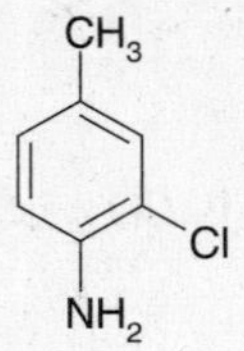

44. An aldehyde is reacted with Tollen's reagent in base. What is formed in the test tube during this reaction?

A. A solution that turns from orange to green

B. A silver mirror

C. A yellow precipitate

D. A cloudy white solution

45. The amine group of one amino acid reacts with the carboxyl group of another amino acid. Water is removed from the reaction and a peptide bond is formed. The functional group present in this peptide bond is called an **amide.** Which is **false** regarding this newly formed bond?

A. The bond gives rise to a planar portion of the dipeptide.

B. The bond between the carbonyl carbon and nitrogen has some double bond character.

C. The α carbon atoms that have a hydrogen atom and R group bonded to them must be in a *trans* position to maintain stability.

D. The nitrogen of the amide bond must be sp^3 hybridized to help the protein maintain its structure.

46. A compound is treated with Benedict's reagent and a brick red precipitate forms. Which of the following was tested for?

A. L-Arginine

B. Glucose

C. Adenine

D. Cholesterol

47. There are two anomeric forms of glucose, the α and β. Which of the following processes can occur between these two forms?

A. Keto-enol tautomerization

B. Inversion of configuration

C. Mutarotation

D. Fries rearrangement

48. At physiologic pH values an amino acid can have its amino group as NH_3^+ and its carboxyl group as COO^-. The result is an electrically neutral species that has both negative and positive charges. This is called

A. an acetal.

B. a hemiacetal.

C. a zwitterion.

D. an anomer.

49. Which of the following statements is **false** regarding the structure of proteins?

A. The secondary structure consists of only an α helix.

B. The primary structure is the sequence of amino acids in the chain.

C. The quaternary structure is many tertiary subunits put together.

D. The tertiary structure can be held together by disulfide bridges.

SECTION IV SEPARATION AND PURIFICATION (Chapters 19–22) Quiz

50. When using a separatory funnel, you would **not**

A. test the polarity of the layers separated by mixing a small sample of the layer with water.

B. turn the separatory funnel upside down while holding the stopper top as to open the stopcock and release the pressure inside.

C. shake the contents of the separatory funnel while the liquids inside are hot.

D. give ample time for the layers inside to separate.

51. Thin-layer chromatography involves all of the following **except**

A. making sure that the spots are below the level of the solvent in the development unit.

B. closing the development system to saturate the air inside with solvent vapor.

C. the use of an immobile phase such as alumina or silica gel.

D. using substances like iodine or sulfuric acid to help visualize colorless compounds.

52. Which of the following is proper in setting up and performing a distillation?

A. Making sure the water flow through the condenser is from a higher point to a lower point

B. Using boiling chips or a stir bar during the distillation

C. Distilling to dryness

D. Making sure that the thermometer's bulb is above the sidearm of the distillation flask

53. Which of the following is **not** a good technique when performing a recrystallization?

A. Using suction filtration to collect and dry the crop of purified crystals on a Büchner funnel

B. Dissolving the impure compound in a cold solvent

C. Allowing the filtrate and flask to cool at room temperature after filtering

D. Testing a number of solvents to use for the recrystallization

SECTION V CHARACTERIZATION (Chapters 23–28) Quiz

54. Which compound shown here is expected to have the highest degree of unsaturation?

A. Cyclopropane

B. Cyclopentene

C. 1-Hexyne

D. Benzene

55. A compound is analyzed via mass spectroscopy. An M^+ peak with a relative abundance of 100% appears at $m/z = 108$. An M^{+2} peak appears with an abundance of 98% at an $m/z = 110$. This compound most likely contains the element

A. chlorine.

B. sulfur.

C. bromine.

D. oxygen.

56. The following information has been obtained about a compound: The compound turned a yellow solution of $FeCl_3$ to a purple color; the IR spectrum of this compound shows a strong, broad peak at 3300 cm^{-1} and at 1230 cm^{-1}; the H NMR spectrum shows the following:

- Many peaks, δ 7.2 (4H)
- Singlet, δ 4.7 (1H)
- Doublet, δ 1.2 (6H)
- Multiplet, δ 3.3 (1H)

This compound probably

A. lacks a benzene molecule.

B. contains a secondary alcohol.

C. is a phenol with an isopropyl group.

D. contains an iodine atom.

57. Consider the following infrared spectrum for an oxygen-containing compound.

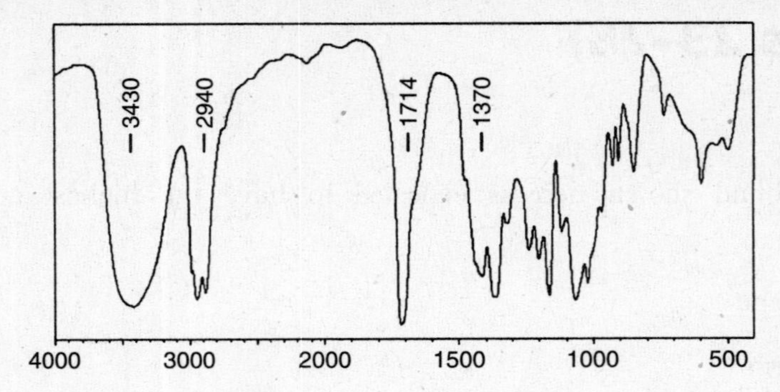

The compound most likely to have generated this spectrum is

A. 5-Oxopentanal

B. 5-Hydroxypentan-2-one

C. 4-Penten-2-ol

D. 4-Penten-2-one

58. Toluene is treated with hot potassium permanganate. Afterward, HCl is used to complete this particular reaction and the product is isolated and purified. The infrared spectra of this product features

A. a broad peak at 3000 cm^{-1} and a strong sharp peak at 3300 cm^{-1}.

B. a broad peak ranging from 3600 to 3200 cm^{-1} along with a strong sharp peak at 1710 cm^{-1}.

C. a strong sharp peak at 1700 cm^{-1} and a sharp peak at 2250 cm^{-1}.

D. only strong, sharp peaks between 3000 and 2850 cm^{-1}.

59. A compound with the molecular formula of C_9H_{12} shows the following H NMR spectrum: δ 7.2 singlet 5H, δ 2.9 multiplet, δ 1.2 doublet. The most probable structure of this compound is

A. Isopropyl benzene.

B. *p*-Methyl ethyl benzene.

C. Propyl benzene.

D. 1,3-Nonadiyne.

60. Butanamide is treated with NaOBr and the product is separated and purified. The product

A. is butanamine.

B. indicates three signals in the C-13 NMR.

C. shows amide I and II bands overlapping in its IR spectra.

D. shows a strong sharp peak at 1700 cm^{-1} in its IR spectra.

SECTION I STRUCTURE (Chapters 1–5) Answers and Explanations

1. **D** Water is famous for being a polar molecule and because of their structures, choices B and C also have an overall dipole moment. Because of its linear shape, carbon dioxide, despite having polar bonds, is a nonpolar molecule. The symmetric shape of the carbon dioxide molecule, O=C=O, makes it a nonpolar molecule. **(Chapter 1. Bonding)**

2. **C** The equation to use to calculate formal charge is

   (Valence electrons of the atom) – (number of electrons unshared)

   – 1/2(number of shared electrons)

   In the nitrate ion, the nitrogen atom makes one double bond with one oxygen atom and two single bonds with the other two oxygen atoms. Also, there are no unshared electrons about the nitrogen atom. Because nitrogen has five valence electrons and shares eight electrons, choice C is correct. Choice A shows the formal charge on the one oxygen atom that (temporarily) has a double bond and choice B shows the formal charge of either oxygen that has a single bond. **(Chapter 1. Bonding)**

3. **C** The four intermolecular forces are presented in this question. van der Waals exists between nonpolar molecules and explains why these compounds can still exist as solids; as the molecular weight between the nonpolar molecules increases, the van der Waals force increases as well. Methanol is a polar molecule, which explains its solubility in water, and it experiences dipole–dipole interactions. Hydrogen bonds form between molecules that have hydrogen atoms bonded to oxygen atoms, as is the case with a carboxylic acid. Choice C presents the molecule–ion attraction, where a positive ion attracts the negative pole of a molecule and vice versa. Choice C is incorrect, because it has the negative chloride ion attracted to the negative oxygen end of a water molecule. **(Chapter 1. Bonding)**

4. **D** Hydrogen bonding is a weak force found between molecules that have a hydrogen atom bonded to F, O, or N. HCl does not meet these requirements. The complimentary base pairs in DNA are held together by hydrogen bonds, the perfect force to hold the two strands together yet still be weak enough to allow for the unzipping of DNA in order to undergo replication. The α helixes and β pleated sheets of proteins also exhibit hydrogen bonds, as does ammonia. Hydrogen bonding is also responsible for the unusually high boiling point of water. **(Chapter 1. Bonding)**

5. **C** Ethyne has a structure of H—C≡C—H and all four atoms lie in a straight line. This molecular geometry is called **linear** and the hydrogen atoms are

180 degrees apart. Ethene is sp^2 hybridized and the atoms are 120 degrees apart, making the molecular geometry around each carbon atom trigonal planar. Acetylene is the same compound as ethyne, and its atoms are not 109.5 degrees apart, as this is characteristic of molecules with a tetrahedral geometry (such as methane). **(Chapter 2. Molecular Shape)**

6. **A** The most stable place for groups to be on a molecule of cyclohexane is in the equatorial position. In this position there is little chance of the groups crowding each other. In the axial position, groups crowd other atoms and groups above and below the ring. Choices B and C demonstrate this instability. Finally, the boat conformation of cyclohexane is unstable because of the interaction of the substituents in the flagpole positions. **(Chapter 2. Molecular Shape)**

7. **B** Isoprene units contain the following structure:

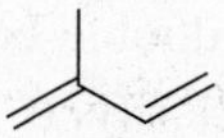

These units help to form the compounds called terpenes, which are found in many natural substances. These units have the general formula of C_5H_8 and they can be found in the compounds limonene and vitamin A, which are shown in the problem. There are no *sp* hybridized carbon atoms in the molecules, as they lack any $-C\equiv$ or $=C=$ type of bonds. Limonene lacks any conjugated system of carbon atoms. Finally, if they were steroids they would have the common four-ring structure of

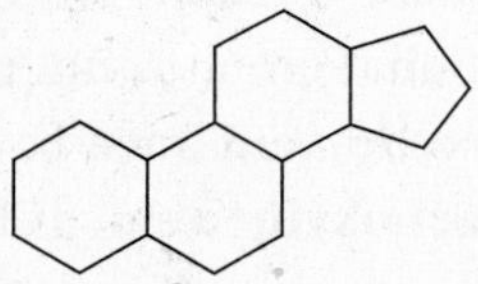

(Chapter 2. Molecular Shape)

8. **B** First let's take a look at the molecule

$$H-C\equiv C-CH_2-C\equiv C-CH_2-CH_2-CH_3$$

In total, this molecule has fifteen single bonds and two triple bonds. Of the triple bonds, two of each of the three bonds are π and one of each is σ. All of the hydrogen to carbon bonds are σ bonds as they are all single bonds. The carbon-to-carbon single bonds are also σ bonds. This gives a total of seventeen σ and four π bonds. **(Chapter 3. Electronic Structure)**

9. **A** The bonds between the carbon atoms in benzene are conjugated and are stabilized by resonance of six delocalized π electrons. The bonding within benzene is nonpolar, thus making it a nonpolar solvent. The sp^2 hybridization of the carbon atoms makes the entire molecule planar in shape as well. Benzene, unlike a crown ether, cannot complex a positive ion. **(Chapter 3. Electronic Structure)**

10. B The diene reacting in this problem is conjugated. When it forms, resonance stabilizes the carbocation intermediates. Questioning the selectivity and reactivity of chlorine is reserved for when determining the major product of a monochlorination or monohalogenation of an alkane. **(Chapter 3. Electronic Structure)**

11. C The priority of groups is determined by the atomic number of the element. The lower the atomic number, the lower the priority of the atom. If there are two of the same element on the stereocenter, then the next set of atoms are examined. At the first point of difference is where a priority is assigned. In the case of double bonds, the priorities are assigned as if the atoms were duplicated. If there is a triple bond present then the priorities are assigned as if the atoms were triplicated. In choice A, the chlorine and ethyl group are out of order. In choice B, the amine and fluorine are out of order. Choice D, the group containing bromine as the second atom to examine in a group, is of the highest priority. Choice C shows the correct order of increasing priority. After expanding the double bonds to be single bonds and examination, this choice has an increasing priority. **(Chapter 4. Nomenclature)**

12. B This question tests nomenclature of groups with higher/lower priorities than others. Choice A has a ketone and a carboxylic acid. In this case, the acid keeps its name and the ketone is identified by the term "keto." Choice C has a similar situation to choice A, where there is an aldehyde group present. Because the aldehyde has a lower priority, it is termed **formyl.** Choice D is correct as it has two OH groups on carbon atoms 1 and 4. Choice B should not be named using "-oic acid," because the carboxylic acid functional group is not part of the longest chain. This is a case where the term **carboxylic acid** must be used. The correct name for this compound is **cyclopentyl carboxylic acid. (Chapter 4. Nomenclature)**

13. A There are nine carbon atoms total in this compound, so the compound is named as a nonane. Between the fused carbon atoms are 3 carbon atoms, then 2 carbon atoms, then 2 carbon atoms. The correct name is Bicyclo[3.2.2]nonane. **(Chapter 4. Nomenclature)**

14. B A xylene is a dimethyl benzene, whereas toluic acid has a methyl group on a benzoic acid. Aniline has just one NH_2 group. A cresol is a phenol with a methyl group on the benzene molecule. **(Chapter 4. Nomenclature)**

15. A Converting the structure given into a wedge-dash diagram you get

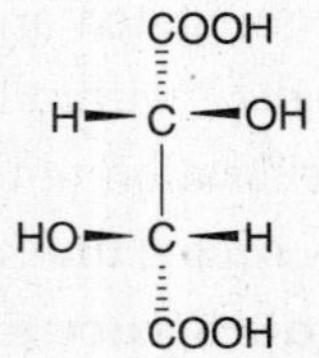

```
      COOH
       :
  H—  C —OH
       |
 HO—  C —H
       :
      COOH
```

The priority of the groups is OH>COOH>$C_2H_3O_3$>H. In both chiral carbon atoms the movement of the eye from highest priority to lowest priority is in a clockwise direction. Therefore, both chiral centers have an R configuration. **(Chapter 4. Nomenclature)**

16. B This diagram shows that the fluorine atom is above the ring and the chlorine is below the ring. The wedged fluorine indicates that the fluorine is above the ring and the dashed chlorine indicates that the chlorine is below the ring. Because the two diagrams are mirror images of each other that are not superimposable, these two compounds are enantiomers. **(Chapter 5. Reconciling Visual Meaning)**

17. D Compounds A and B are meso compounds as they are superimposable and have stereocenters. C and D are non-superimposable mirror images of each other and are enantiomers. B and C differ at one chiral center and are diasteriomers. A and D are diasteriomers of each other. **(Chapter 5. Reconciling Visual Meaning)**

SECTION II REACTIVITY (Chapters 6–9) Answers and Explanations

18. D This step represents bond formation, which is always exothermic; therefore, ΔH is negative. Because two species are combining to form one, translational degrees of freedom are lost, and ΔS is negative, as the disorder has decreased. **(Chapter 6. Energy Changes in Molecules)**

19. D A tertiary intermediate is more stable than a secondary one, as dictated by the following order of stability: $3^\circ > 2^\circ > 1^\circ >$ Methyl. In the case where alkyl and aryl groups are present, the aryl groups stabilize the carbon intermediate far better than any alkyl group can. Choice D, besides being a tertiary intermediate, also has aryl groups, making this the clear choice. **(Chapter 7. Radical Chemistry)**

20. D S_N2 reactions take place with stronger nucleophiles, leaving groups on primary carbon atoms and at lower temperatures. A polar solvent favors the formation of ions, a characteristic **not** found in S_N2 reactions. **(Chapter 8. Polar Chemistry)**

21. D Ethanol has a pK_a of about 16 and, as expected, acetic acid has a pK_a of about 4.8 as the lower pK_a value, the stronger the acid. Terminal alkynes can give up their terminal hydrogen atoms in base as to react and extend the carbon chain when reacted with the proper alkyl halide. A terminal alkyne has a pK_a value of about 25 for their terminal hydrogen atoms. Alkanes are not very acidic at all and have a very high pK_a value of about 50. **(Chapter 8. Polar Chemistry)**

22. A The NaOH reacts with the most acidic hydrogen atom from the carboxylic acid group and form a conjugate base. This then reacts in an S_N2 fashion with the carbon atom that holds the bromine atom. The result is a cyclic ester, a lactone, as shown here. This is not to be confused with a cyclic amide, a lactam.

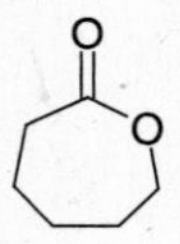

(Chapter 8. Polar Chemistry)

23. B By far, sulfuric acid is the strongest acid in the group. The other three choices present the inductive effect. The high electronegativities of the fluorine and chlorine make them inductive electron withdrawing groups. In turn, they make the acetic acid even more acidic. Because fluorine has the highest electronegativity, it is expected to have the greatest impact on the acidity of the acetic acid. **(Chapter 8. Polar Chemistry)**

24. D The S_N2 reaction showcases the nucleophile reacting from the backside of the leaving group forming a pentacoordinate intermediate and completely inverts the configuration. The S_N2 reaction also requires two molecules to start the reaction. An S_N1 reaction needs just one molecule to start the reaction by forming an ion. **(Chapter 8. Polar Chemistry)**

25. C The Diels–Alder reaction is a [4 + 2] cycloaddition. In this reaction, a diene will react with an alkene to form a cyclohexene. This reaction does not include any polymerization or use of Grignard reagents. Also, Friedel–Crafts syntheses are used to place various groups on benzene compounds. **(Chapter 9. Pericyclic Chemistry)**

SECTION III METHODOLOGY (Chapters 10–17) Answers and Explanations

26. **D** Choices A and B are reductions of a ketone to produce a secondary alcohol. Choice C outlines a Grignard reaction to make a secondary alcohol. In order to do this, begin with an aldehyde and add on an alkyl group. Choice D demonstrates an anti-Markovnikov reaction, and the OH group is added to the double-bonded carbon atom on the end of the molecule. **(Chapter 10. Alcohols and Ethers)**

27. **A** This reaction starts with a primary alcohol, a group that is at the end of a molecule. The resulting product is most likely to also have its functional group at the end of the molecule as well, pointing to either choice A or B as the correct answer. This particular reaction leads to the formation of an aldehyde. **(Chapter 10. Alcohols and Ethers)**

28. **B** The Williamson synthesis is used to form an ether from an alcohol. In this reaction a base is used to remove the most acidic hydrogen atom from the alcohol and the resulting alkoxide ion is reacted with a primary halide to complete the reaction. **(Chapter 10. Alcohols and Ethers)**

29. **A** Compound A is 1-hexanol as ethylene oxide is used to the carbon chain of a Grignard reaction by two carbon atoms. Compound B is hexanoic acid and compound C is $CH_3(CH_2)_4COCl$. Choice D questions the iodoform reaction, a test for methyl ketones. This reaction produces a yellow precipitate that has a melting point range of 120° to 123°C. **(Chapter 10. Alcohols and Ethers)**

30. **D** Reaction I forms an ester using sulfuric acid as a catalyst. Reaction II occurs via an S_N2 reaction and forms an ester. Finally, HCl is produced along with an ester as reaction III removes chlorine from the acyl chloride and the hydrogen from the alcohol. **(Chapter 10. Alcohols and Ethers)**

31. **A** Examining the products of the reaction, there could be two possible reactants, cyclobutene and 1,5-cyclooctadiene. Because two moles of the product are formed, only the 1,5-cyclooctadiene could have been the reactant, as cyclobutene would have been in a 1:1 ratio between reactant and product. **(Chapter 11. Ketones and Aldehydes)**

32. **D** This reaction is an acid-catalyzed acetal formation and it can also be carried out using a base, RO^- in ROH. A hemiacetal forms as an intermediate and the acetal is the final product. When an open chain of glucose forms a ring, the resulting functional group at the anomeric carbon atom is a hemiacetal, and not an acetal. **(Chapter 11. Ketones and Aldehydes)**

33. B The final product of the Wittig reaction is an alkene. Choice B shows the carbonyl carbon reacting with the carbon doubly bonded to phosphorus to produce the alkene 2-methyl-2-pentene. **(Chapter 11. Ketones and Aldehydes)**

34. A Reacting butane with Cl_2 gives a major product of 2-chlorobutane and a minor product of 1-chlorobutane. Reacting 1-butene with HCl gives a Markovnikov addition and the hydrogen atom will go to the carbon of the double bond which already has more hydrogen atoms present. Keep in mind, "He who has more hydrogen gets the hydrogen." This puts the Cl from HCl on the double-bonded carbon with fewer hydrogen atoms. Reacting 1-butene with Cl_2 gives an addition reaction and produces 1,2-dichlorobutane. Shaking butane with NaCl(aq) does nothing as the nonpolar butane does not mix with the polar solution. Neither reaction in choice D produces 2-chlorobutane. **(Chapter 12. Alkenes and Alkynes)**

35. B In order to convert the aldehyde to an acid, you need an oxidizing agent. The use of $K_2Cr_2O_7$, H_2SO_4, and H_2O do the trick nicely. Choice A is useful in converting a nitrile to a carboxylic acid, whereas the Grignard reagents in choice C convert the Grignard reagent into a carboxylic acid. Finally, choice D is useful in producing an acyl chloride from an acid. **(Chapter 13. Carboxylic Acid Derivatives)**

36. D This reaction calls on reducing agents to convert the carboxylic acid functional group into a primary alcohol. Also, take note that the functional group is on an end carbon in both the reactants and products. To form an ester, the carboxylic acid must react with an alcohol, whereas it requires a different number of steps and/or reagents to form an aldehyde or ketone from a carboxylic acid. **(Chapter 13. Carboxylic Acid Derivatives)**

37. C This reaction is an outline of the hydrolysis of a primary amide to form a carboxylic acid. **(Chapter 13. Carboxylic Acid Derivatives)**

38. A This reaction looks tempting to first say that the amine group is an *ortho para* director. But this reaction does not take place on the benzene ring at all. Instead, the chlorine is removed from the reaction along with a hydrogen atom from the aniline and the reaction takes place at the nitrogen to form an amide. **(Chapter 15. Amines)**

39. C This reaction forms an enamine, C=C—N. Choices B and D are not the correct products for the reaction. Further inspection of the remaining choices shows that the double bond has been positioned in two different places. Choice C is more likely to be the enamine formed because it is a conjugated enamine, and its intermediates formed during the reaction are more likely to be stable. **(Chapter 15. Amines)**

40. C As alkyl groups are attached to nitrogen centers, the nitrogen becomes more electron-rich, just as more highly substituted carbocations are stabilized by the donation of electron density from alkyl substituents. This means that methylamine is a better nucleophile than ammonia, so it competes more effectively for the iodomethane. Thus, instead of getting clean monoalkylation of ammonia, a complex reaction mixture is obtained. This is why indirect methods are required, such as the Gabriel synthesis. **(Chapter 15. Amines)**

41. C This reaction is a Friedel–Crafts acylation. In this particular case, the reaction takes place on the very same molecule instead of a benzene ring on another molecule. Close examination of choice D shows a molecule that violates the octet rule, where the double bond is attached to the naphthalene. Choice B has the reagents to carry out a nitration, and choice A makes no sense at all. **(Chapter 16. Aromatic Chemistry)**

42. D PCC is one reagent that can be used to carry out this reaction, as $CrO_3/HOAc$ is also a good choice. Choice A is used to carry out a Clemmenson reduction, which is exactly the reverse of the reaction presented in the problem. Choice B is used to make an aldehyde from an amide. Finally, choice C is used to perform a *syn* addition of two OH groups to an alkene. **(Chapter 16. Aromatic Chemistry)**

43. A This reaction starts by nitrating a toluene molecule. The reaction calls for the *para* product of the reaction to be the focus. This gives the molecule

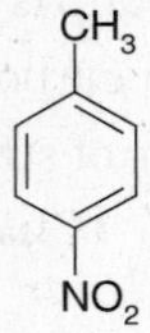

CH_3

NO_2

Reacting this molecule with tin in HCl and then with a base causes the nitrate group to become an amine group. **(Chapter 16. Aromatic Chemistry)**

44. B Tollens reagent is $Ag(NH_3)_2^+$ and it reacts with aldehydes. The silver coats the sides of the test tube, and you could actually see their reflection in this "mirror." **(Chapter 18. Carbohydrates and Peptides)**

45. D The formation of a partial double bond between the nitrogen and carbon in an amide does not allow the nitrogen to be sp^3 hybridized. The other three choices hold true as the double-bond character rotates between the carbon and oxygen and the carbon and nitrogen bonds. **(Chapter 18. Carbohydrates and Peptides)**

46. B Benedict's reagent has Cu^{2+} ion in solution, which is reduced to Cu^{1+} by a reducing sugar. The resulting brick red compound is Cu_2O. **(Chapter 18. Carbohydrates and Peptides)**

47. C Mutarotation is the formation of the two forms of glucose. The β formation dominates over the α form. Choice A applies to ketones and enols, whereas choice B applies to S_N2 reactions. Finally, choice D applies phenyl acetate and its ability to form two isomers of hydroxyacetophenone. **(Chapter 18. Carbohydrates and Peptides)**

48. C A **zwitterion** is an electrically neutral substance that has both positive and negative charges. Acetals and hemiacetals are formed from the reaction between alcohols and aldehydes. An anomer is one of the forms that glucose can take on once its long-chain form has closed into a ring. **(Chapter 18. Carbohydrates and Peptides)**

49. A Choices B, C, and D correctly describe the structure of proteins. The missing piece to choice A is that the secondary structure of a protein can also include a β pleated sheet. **(Chapter 18. Carbohydrates and Peptides)**

SECTION IV SEPARATION AND PURIFICATION (Chapters 19–22) Answers and Explanations

50. C Although using the separatory funnel it is proper to give time for the layers to separate, vent the funnel while being held upside down (be sure to hold the stopper top in place) and to test the polarity of the layers after draining them through the stopcock's bore. You would not place hot liquids in the funnel and shake them. This increases the surface area of the volatile substances inside, causing them to evaporate even faster and build up pressure inside the funnel. **(Chapter 19. Extraction)**

51. A The spots that have been created via the capillary tubes must remain above the solvent in the unit. Placing under the level of the solvent will wash them away. The other three choices are all necessary or good techniques to use as needed during TLC. **(Chapter 20. Chromatography)**

52. B Boiling chips or use of a stir bar is effective in refluxing or distilling because they help to distribute the heat evenly inside the flask's bulb and prevent bumping from occurring. As for the other choices, you should always make sure that

- The water enters the lower part of the condenser and exits through the top.
- There are liquids remaining in the distilling flask so as not to heat to dryness.
- The bulb of the thermometer is below the sidearm of the distilling flask.

(Chapter 21. Distillation and Sublimation)

53. B Choosing a proper solvent and dissolving the compound in a minimum of warm solvent are good techniques for a recrystallization. The resulting solution is passed through warm filter paper that has been moistened with the solvent. The filtrate is allowed to cool and undisturbed, preferably overnight. The flask is then cooled in ice and the crystals are collected and dried on a Büchner funnel. Tests for purity are then performed. **(Chapter 22. Recrystallization)**

SECTION V CHARACTERIZATION (Chapters 23–28) Answers and Explanations

54. D The degrees of unsaturation can be determined by the formula [(2C + 2) – H]/2. Cyclopropane, C_3H_6, has [(2 × 3 + 2) – 6]/2, giving a DoU of 1. Cyclopentene, C_5H_8, has [(2 × 5 + 2) – 8]/2, giving a DoU of 2. 1-Hexyne, C_6H_{10}, has [(2 × 6 + 2) – 10]/2, giving a DoU of 2. Benzene, C_6H_6, has [(2 × 6 + 2) – 6]/2, giving a DoU of 4. **(Chapter 23. Combustion Analysis)**

55. C The compound being analyzed is bromoethane. Because nearly equal amounts of Br^{79} and Br^{81} are found in nature, they have peaks of almost equal height in the mass spectrum. These peaks have an *m/z* difference of two. Chlorine has relative heights of Cl^{35} 100% to Cl^{37} 33%. Sulfur would have relative heights of S^{32} 100% to S^{34} 4.4%. Finally, oxygen would have relative heights of O^{16} 100% to O^{18} 0.2%. **(Chapter 25. Mass Spectrometry)**

56. C The iron (III) chloride test turns purple on reacting with a phenol. The peak at 3300 cm^{-1} is that of an alcohol, as its broadness is indicative of hydrogen bonding. The peak at 1230 cm^{-1} lies in the C—O fingerprint region and is indicative of an Ar—O bond. The H NMR spectrum shows an aromatic compound containing four hydrogen atoms, meaning that this compound is disubstituted. The singlet indicates the hydrogen atom of the phenol, and the isopropyl group explains the doublet and multiplet more upfield. **(Chapter 26. Infrared Spectroscopy)**

57. B The three functional groups in question are the carbonyl, the hydroxyl, and the alkene. There is no absorption band in the alkenyl region, but there are strong bands in the carbonyl (1714 cm^{-1}) and hydroxyl (3430 cm^{-1}) regions. You can also take note that there are various strong, broad bands in the 1100–1000 cm^{-1} region, indicating single C to O bonds of a primary alcohol. This molecule is the only choice that exhibits both of these functionalities.

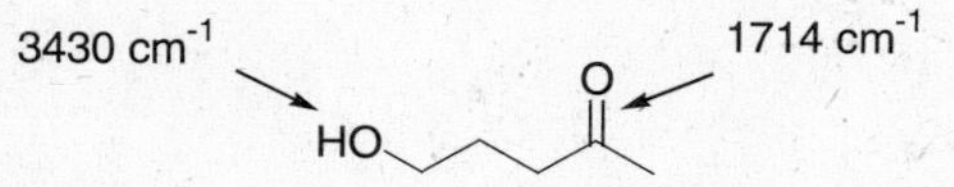

3430 cm^{-1} 1714 cm^{-1}

HO O

(Chapter 26. Infrared Spectroscopy)

58. B This reaction has turned toluene into benzoic acid. The characteristic peaks of this functional group are a broad peak ranging from 3600 to 3200 cm^{-1} along with a strong sharp peak at 1710 cm^{-1}. Other peaks described in this problem are H—C≡ terminal alkyne stretch at 3300 cm^{-1}, C≡N nitrile stretch at 2250 cm^{-1}

and H—C alkyl stretch between 3000 and 2850 cm^{-1}. **(Chapter 27. Proton Nuclear Magnetic Resonance)**

59. **A** This problem features a peak with a chemical shift of 7.2 and an area under the curve that indicates five hydrogen atoms. This peak correlates to a monosubstituted benzene ring, C_6H_5-. Choices B and C are eliminated immediately, because you know that there is a benzene present and that it is not disubstituted. Subtracting this from the molecular formula of the compound leaves C_3H_7-. Because the H NMR displays a doublet and a multiplet, an isopropyl group is appropriate to choose. If there were a propyl group, then there would be three H NMR signals in addition to the one from the benzene. The fact that only two additional signals are viewed means that the isopropyl group is present. **(Chapter 27. Proton Nuclear Magnetic Resonance)**

60. **B** The reaction that has occurred here is called the **Hofmann degradation.** The reactant is butanamide, $CH_3CH_2CH_2C(=O)NH_2$, and it has become $CH_3CH_2CH_2NH_2$. This indicates three carbon atoms in the C^{13} NMR. Remember that this is a degradation reaction and the amine product has one fewer carbon atoms than the amide reactant. Because the carbonyl group is removed from the reactant, there are no peaks at 1700 cm^{-1} in the IR spectrum, and there are no amide I and II bands in the 1700–1600 cm^{-1} region of the IR spectrum either. **(Chapter 28. Carbon Nuclear Magnetic Resonance)**

Two Sample MCATs on CD

Your Goals for This Part:

- Take two complete sample MCATs for practice
- Practice pacing yourself to meet time limits
- Use your scores to evaluate your test readiness
- Review the explanations to assess strengths and weaknesses

The CD-ROM component of *McGraw-Hill's New MCAT* contains two sample tests designed to be just like the real MCAT in terms of content coverage and level of difficulty. The tests on CD-ROM simulate the actual computerized MCAT and also offer automatic timing and scoring functions.

Each test is in four sections, as shown in the following table:

Sample MCAT Format		
Section	**Number of Questions**	**Time Allowed (minutes)**
Physical Sciences	52 multiple-choice questions (8 passages with 5–6 questions each, plus 10 stand-alone questions)	70
Verbal Reasoning	40 multiple-choice questions (6 passages with 6–8 questions each)	60
Writing Sample	2 essay questions (30 minutes for each)	60
Biological Sciences	52 multiple-choice questions (8 passages with 4–6 questions each, plus 10 stand-alone questions)	70
Totals	**146**	**260 (= 4 hours, 20 minutes)**

Mark your answers to this test by clicking in the corresponding ovals. When you are finished with each test, determine your score and carefully read the answer explanations, especially for any questions you answered incorrectly. Identify any weak areas by determining the areas in which you made the most errors. Then go back and review the corresponding chapters in this book. If time permits, you may also want to review your stronger areas.

These practice tests will be an accurate reflection of how you'll do on test day if you treat each one as the real examination. Here are some hints on how to take each test under conditions similar to those of the actual exam:

- Find a time when you will not be interrupted.
- Complete each test in one session, following the specified time limits for each section. Allow yourself a break of 10 minutes between sections, for a total of three, just as on the real exam. The MCAT with its breaks lasts nearly 5 hours, so it is a test of your endurance as well as a test of your knowledge. Use these practice tests to assess your test-taking stamina.
- If you run out of time on any section, take note of where you ended when time ran out.
- Become familiar with the directions to the test. You'll save time on the actual test day by already being familiar with this information.
- Remember that there is no penalty for guessing on the MCAT, so if time is running out before you can finish a section, feel free to guess on the remaining questions.

After you finish a practice test, the program will check your answers against the Answer Key and calculate your score. For the Writing Sample, use the given scoring rubric to score your essay (or have a friend score it for you).